D1529114

HANDBOOK OF INDUSTRIAL ENGINEERING AND MANAGEMENT

WILLIAM GRANT IRESON *is currently professor and chairman of the industrial engineering department at Stanford University. From 1948 to 1951, before coming to Stanford, he was professor of industrial engineering at the Illinois Institute of Technology.*

Mr. Ireson is a native of Virginia and holds B.S. and M.S. degrees in industrial engineering from the Virginia Polytechnic Institute. Upon graduation from this institution, he joined the Wayne Manufacturing Corporation as an engineering trainee in production. Although serving the company in a number of capacities, he worked primarily in production control, where he was in charge of production schedules for the entire plant.

In 1941, he returned to V. P. I. as an instructor in industrial engineering. He advanced steadily through the ranks of assistant and associate professorships, becoming an acting professor and the acting head of the department in 1947.

Mr. Ireson has specialized in production, quality control, reliability, engineering economy, and industrial development. He is author of Factory Planning and Plant Layout *and coauthor, with Eugene L. Grant, of the fifth edition of* Principles of Engineering Economy. *He is editor of both the* Reliability Handbook *and the Prentice-Hall series of monographs in industrial engineering and management science. He has written many articles and technical reports.*

Mr. Ireson is a registered professional industrial engineer in California and a registered professional engineer in Illinois, and has served as a consultant to many government agencies, international organizations, and industrial firms. He is a member of the American Society of Mechanical Engineers, the Society for International Development, the American Society for Engineering Education, Sigma Xi, and the Institute of Management Science. He is also a Fellow of the American Institute of Industrial Engineers and the American Society for Quality Control.

EUGENE L. GRANT *is professor of economics of engineering, emeritus, at Stanford University. He came to Stanford in 1930 from Montana State College where he was professor of industrial engineering.*

He has written extensively in the area where engineering impinges upon management. His Principles of Engineering Economy (*fifth edition, with W. G. Ireson, 1970*) *has been a standard text and reference book in its field for more than 40 years. His* Statistical Quality Control (*third edition, 1964*) *also is a leading work in its field. In addition, he is co-author with P. T. Norton, Jr. of* Depreciation (*1949*) *and with L. F. Bell of* Basic Accounting and Cost Accounting (*second edition, 1964*).

Professor Grant holds the degrees of B.S. and C.E. from the University of Wisconsin and the degree of A.M. (in economics) from Columbia University. He is an honorary member of the American Society for Quality Control, a fellow of the American Society of Civil Engineers, the American Statistical Association, and the American Association for the Advancement of Science, and is a member of the American Society for Engineering Education and the International Association for Statistics in Physical Sciences. He received the 1944 Thomas Fitch Rowland Prize of the ASCE, the 1952 Shewhart medal of the ASQC, a Distinguished Service Citation from the College of Engineering of the University of Wisconsin in 1964, and a Founders' Award from the American Institute of Industrial Engineers in 1965. The ASQC has honored him by establishing an annual E. L. Grant award for distinguished contributions to quality control education. The Engineering Economy Division of ASEE has honored him by establishing an annual Eugene L. Grant award for the best paper in each volume of The Engineering Economist.

HANDBOOK OF INDUSTRIAL ENGINEERING AND MANAGEMENT

SECOND EDITION

EDITED BY

W. GRANT IRESON

AND

EUGENE L. GRANT

PRENTICE-HALL INC.

ENGLEWOOD CLIFFS, NEW JERSEY

Current Printing (last digit)
3 2 1

13–378463–0

Library of Congress Catalog Number 71–139954
Printed in the United States of America

Prentice-Hall International, Inc., *London*
Prentice-Hall of Australia, Pty. Ltd., *Sydney*
Prentice-Hall of Canada, Ltd., *Toronto*
Prentice-Hall of India Private Limited, *New Delhi*
Prentice-Hall of Japan, Inc., *Tokyo*

PREFACE

This handbook will be useful to other people besides practicing industrial engineers and students of industrial engineering and management. For example, it will be helpful to persons who have been trained in other engineering fields but who are now engaged in management activities. Moreover, it will help all persons—engineers and others—who require concise reference material on various matters related to industrial engineering and management. People already engaged in industrial management and others aiming at management positions, for instance, will find this book a valuable guide.

In contrast to the situation in some of the older engineering fields, the topics to be included in an industrial engineering handbook are not clearly defined by custom. Certain traditional industrial engineering subjects, such as motion and time study, plant layout and materials handling, and tool engineering, for example, obviously call for coverage. Other closely related subjects that have been included are engineering economy, capital budgeting, standardization, safety engineering, and inspection and quality control. The trend of the times is toward more and better use of the techniques of statistical inference in solving managerial problems; a great deal of useful reference material is included in the section on industrial statistics.

The trend of the times is also toward a more analytical approach in making many different types of managerial decisions. The introductory section on industrial systems and organizations provides explicit guidance in the design of systems to accomplish the numerous activities of every organization and to make the organization function efficiently through integration of the operational controls. The section on managerial economics summarizes many of the contributions that economic thinking can make to the formulation of business policies.

Sections on computers and data processing, critical path methods, reliability analysis, and linear programming and its applications constitute an important segment of the new material included in this second edition. Computers have become a daily tool of industrial engineering. Without them, the more analytical techniques employed in critical path methods, inspection and quality control, reliability analysis, and linear programming would be severely restricted. More and more industrial engineers are using mathematical models and simulation as means of solving complex problems that would not have been considered solvable only a few years ago. The combination of analytical techniques and the availability of high-speed computers have opened up a great many new opportunities for the professional industrial engineer.

The section discussing the attitudes of organized labor toward industrial engineering methods gives an extremely helpful picture of certain considerations that enter into decision-making today. The short section on industrial climatology presents useful information, not readily available elsewhere, regarding important factors in certain types of decisions concerned with plant location, construction, and the like.

In the preparation of the handbook sections, the authors have tried to stress general principles and to illustrate their applications to industry rather than merely to describe details of industrial practice. In presenting the various topics, the effort has been to stress the best current practice, avoiding, on the one hand, mere description of average practice, and, on the other hand, the presentation of untried theories.

Most duties of industrial engineers deal in one way or another with efforts to control costs. In a sense, this entire handbook is a treatise on the many-sided subject of cost control.

In general, the handbook sections are presented in readable style so that each section may be viewed as a concise text in its particular field, a text written by an authority on the subject. Thus the handbook is useful for home study and is also suitable as text material for courses in industrial management.

We wish to express our sincere thanks to the authors of the handbook sections for their dedication to this project throughout the many stages of its development. We believe that their contributions will have important effects on the future of the industrial engineering profession because they have presented modern approaches to complex material in easily understood form. We are indebted to the Literary Executor of the late Sir Ronald A. Fisher, F.R.S., to Dr. Frank Yates, F.R.S., and to Oliver & Boyd Ltd., Edinburgh, for permission to reprint our Tables 13.25 and 13.27 from the books *Statistical Methods for Research Workers and Statistical Tables for Biological, Agricultural, and Medical Research.*

Stanford University

W. Grant Ireson
Eugene L. Grant

CONTENTS

HANDBOOK OF INDUSTRIAL ENGINEERING AND MANAGEMENT

PAUL LEBENBAUM, JR., *after graduating from Stanford University and obtaining his master's degree in electrical engineering from the Massachusetts Institute of Technology, joined the General Electric Company on their test and advanced engineering programs. In his more than 20 years with General Electric, he held development, design, and managerial engineering positions. He was in charge of product planning and engineering and manufacturing planning for a redesign of a major line of direct-current motors and generators and was responsible for the development and introduction of a new line of packaged, adjustable-speed, direct-current motor drives. He was a member of the first multifunctional team established to develop automated operational and managerial information systems for internal General Electric operating departments.*

Since 1960 Mr. Lebenbaum has been associated with the electronics and aerospace industries. He has been general manager of the Palo Alto tantalum capacitor plant of International Telephone and Telegraph Company; assistant to the director of engineering of the Space Systems Division of Lockheed Aircraft Corporation; and director of management systems of the Autonetics Division and executive staff adviser in management technology and financial planning and analysis in the Executive Offices of the Aerospace and Systems Group of the North American Rockwell Corporation. He has been a lecturer in business systems design in the Industrial Engineering Department of Stanford University.

Mr. Lebenbaum is the author of technical papers on the altitude rating of electric apparatus and the design of direct-current motors for use in automatic control systems. He is a member of Phi Beta Kappa, Tau Beta Pi, and Sigma Xi.

SECTION 1

INDUSTRIAL SYSTEMS AND ORGANIZATION

PAUL LEBENBAUM, JR.

1.1 INTRODUCTION

Planning, organizing, operating, and measuring are the functional elements of the work of the professional manager. This section discusses his organizational work, including both the organizational structuring of the business and the designing of systems and procedures which tie the organization together and establish the methods by which the business attains its objectives.

Since the elements of planning, operating, and measuring have an impact on organizing a business, we shall define thoroughly the meanings of all terms before proceeding with the discussion of organization.

A. *Planning:* Setting the objectives, goals, and standards of performance for the business

B. *Organizing:* Classifying the work required to accomplish the objectives of the business, dividing the work into manageable components and jobs, grouping the components into an orderly organizational structure, staffing the organization, and formulating the systems and procedures for performing the work to be done

C. *Operating:* Performing the tasks of decision-making, directing, delegating, communicating, motivating, integrating, interpreting, and working to achieve the overall objectives of the business

D. *Measuring:* Establishing control systems, recording and reporting performance against preestablished standards, continuously evaluating actual performance against standards, informing employees and managers responsible for the variances between actual and standard, and continuously following up on corrective actions taken to reduce the variances.

Systems and organizations are the means by which a manager carries out his work. Industrial systems must be *planned* so that their outputs meet the requirements for which they were designed; they must be simple so that their purposes and methods of *operation* can be easily communicated to those using them; they must provide the necessary motivation to those involved in their applications; and they must have *measurable* outputs so that the task or tasks for which they have been designed can be controlled.

Organizations must be *planned* so that the objectives of the business can be met in the most economic and timely manner; they must be designed to ease the *operating* task by assigning clear lines of responsibility and reducing organizational interfaces in order to shorten communication channels; and they must be established in such a manner that their outputs can be *measured* in order to establish effective means of managerial control.

Because they have a common objective—to provide the means by which a manager carries out his work—systems and organizations will be discussed together. The approaches used in the design of a system or of an organization are similar. In either case, the objectives of the system or the organization must be established and the output requirements specified; the work activities that must be performed to generate the output must be defined; and the actual output must be measured and compared with the required output to inform the manager when corrective action is necessary.

1.2 SYSTEMS ENGINEERING APPLIED TO THE DESIGN OF INDUSTRIAL BUSINESS SYSTEMS AND ORGANIZATIONS

This section applies systems engineering, which has proven effective in the design of military- and industrial-equipment systems, to the design of industrial-business systems and organizations. The approach, called *industrial-business systems engineering*, gives the manager and systems designer a unifying methodology which simplifies the design task. Industrial-business systems engineering selects the primary subsystems of the total business system and divides the design effort into a series of more easily manageable parts. The categorization emphasizes the similarities of work activities of various businesses and thus helps to standardize design methods so that the manager and designer can apply the methods used to solve problems in one business to the solution of problems in another.

1.2.1 Methodology of Systems Engineering

Systems engineering is the discipline used in the engineering design of large-scale military- and industrial-equipment systems to assure proper functioning of the total system. It employs a top-down approach by first determining the overall input-output requirements of the system in terms of the functions the system is to perform. It then divides the total system into successively smaller subsystems ("black boxes"), specifies the input-output requirements of each subsystem and each smaller system component, and determines the method of interconnection of the various subsystems and components in order to accomplish the overall objective of the total system. With this approach, a large system can be synthesized from a number of smaller subsystems with the assurance that total system performance will meet specifications when the subsystems are assembled and interconnected.

The design of an industrial-business system or organization parallels that of a missile system. It would be almost impossible to design a missile system without dividing it into more manageable design tasks. Major missile subsystems include guidance, propulsion, control, and structure. Each subsystem can be designed essentially independently—only the knowledge of its input-output requirements, its environmental constraints, and its interfaces with other subsystems are needed for design work to proceed with the assurance that the total system will perform properly when all the components are connected.

In general, the following steps are part of the systems-engineering approach to the design of complex, highly engineered equipment systems.

1. Define system input-output requirements
2. Select subsystems
3. Define subsystems
4. Develop system and subsystem functional flow diagrams
5. Write system and subsystem specifications
6. Write subsystem interface specifications
7. Determine overall system performance and compare with operational requirements.

Step 1 is primarily devoted to defining the problems, Steps 2 to 6 to solving it, and Step 7 to checking the solution to determine whether the designed system meets the original requirements. These three generalized steps are required in the design of any system (equipment, industrial-business, or organizational). The objective of the system must be defined, the system elements must be designed and interconnected or grouped, and finally the resulting grouping of elements must be checked out (either theoretically or actually) to determine whether the original system requirements have been met.

Systems-engineering methodology offers the advantages of a standardized approach, a savings in design time and effort, and a design method that will be readily understood and accepted by the increasing numbers of technically trained personnel now entering the management field.

1.2.2 Definition of System Input-Output Requirements

The first step in the systems-engineering approach to the design of industrial-business systems is to define the inputs and outputs of the enterprise being considered. The input to a manufacturing business is an order—either a purchase order from a customer or a manufacturing order from the marketing organization of the firm—and the output is the finished product. Although the capture of the order through the marketing organization and the collection of the customer's payment through the financial organization should be taken into account in closing the entire customer-enterprise loop,

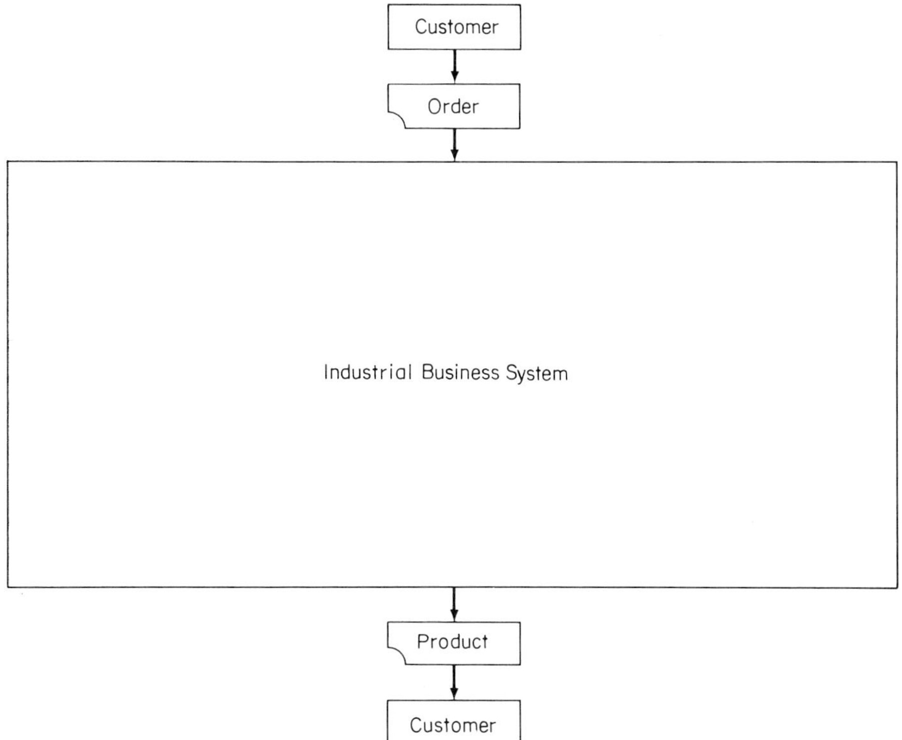

Fig. 1.1 *Overall order-product conversion cycle.*

it is assumed for the purposes of this section that the industrial-business system starts with the receipt of the order and ends with the shipment of the product.

Since input is the customer's order and output is the finished product, the industrial-business system can be looked upon as the transfer function which converts the order into the product. The system must collect and store the data; disseminate the information; generate the documents; order and receive the material; and fabricate, assemble, and test the parts and assemblies necessary to make the conversion from order to product—all in a manner which will best satisfy both the customer's requirements and the objectives of the business.

Figure 1.1 is a block diagram of the relationship between the customer, his order, the industrial system, and the product. Since this type of diagram will be used throughout this section, its conventions will be discussed in detail at this point. The square rectangles represent systems or organizations, and the notched rectangles represent the documents or hardware which are the outputs of the systems or organizations. Timephasing of actions is represented by moving from the top of the diagram to the bottom. Thus Fig. 1.1 can be interpreted as follows:

A. Customer issues order.

B. Industrial system receives customer order.

C. Industrial system converts customer order to finished product.

D. Customer receives finished product.

This conversion is common to all industrial systems regardless of the type of business involved. By using this conversion, we may develop a generalized industrial-system model which can be tailored as required in designing the information systems and organizational structures for a wide variety of businesses.

1.2.3 Industrial Subsystems

The second step in the industrial systems-engineering approach is to divide the total industrial system into a logical series of subsystems. This is the single most important step in systems design. Past divisions were horizontal in nature, and involved such subsystems as production control, purchasing, accounting, engineering, warehousing, and shipping. But the problem of integrating such obviously diverse activities was difficult. However, if one develops proper vertical subsystems, the design task becomes simpler and more capable of being successfully completed.

Since the primary purpose of the industrial system is to convert the customer's order into a finished product, the subsystems selected must help to make this conversion in a simple, easily understood, and practical manner. Preferably, each subsystem should be relatively independent of the others so that any changes or perturbations will have a minimal impact on the total system; the activities carried out within each subsystem should be of the same general type; and the number of interfaces between the subsystems should be kept to a minimum. When an interface is required, however, it should be capable of being clearly defined. These requirements also apply to the subsystems of a military- or industrial-equipment system.

The subdivision which most nearly meets these requirements is based on the elements of the manager's job—what he must do and what information he needs to do it. Thus if one divides the total business system into subsystems of planning, operations, and control, the framework for the categorization of the industrial system begins to develop. If one then adds a financial and an information subsystem, the breakdown is complete and every activity in any business enterprise can be dropped into one of the five boxes. Figure 1.2 divides the industrial system into these five subsystems.

Each of the five subsystems has a definite purpose and output. The purpose of the planning subsystem is the generation of future plans for the enterprise; operations accomplishes the physical job of receiving the order and designing and manufacturing the product; control measures actual performance against plans in order to assure that the customer's requirements and the objectives of the enterprise are both met; financial takes care of the fiscal aspects of the business and feeds cost data into other subsystems for planning, control, and information purposes; and finally information collects the relatively fixed data needed as inputs for other subsystems.

Proper grouping of business activities is extremely important in industrial systems and organizational design. Difficulties arise when diverse activities are grouped into a single subsystem or organizational entity. The ultimate operation of the system is more efficient when common or interrelated activities are tied together.

Each subsystem selected has two character-

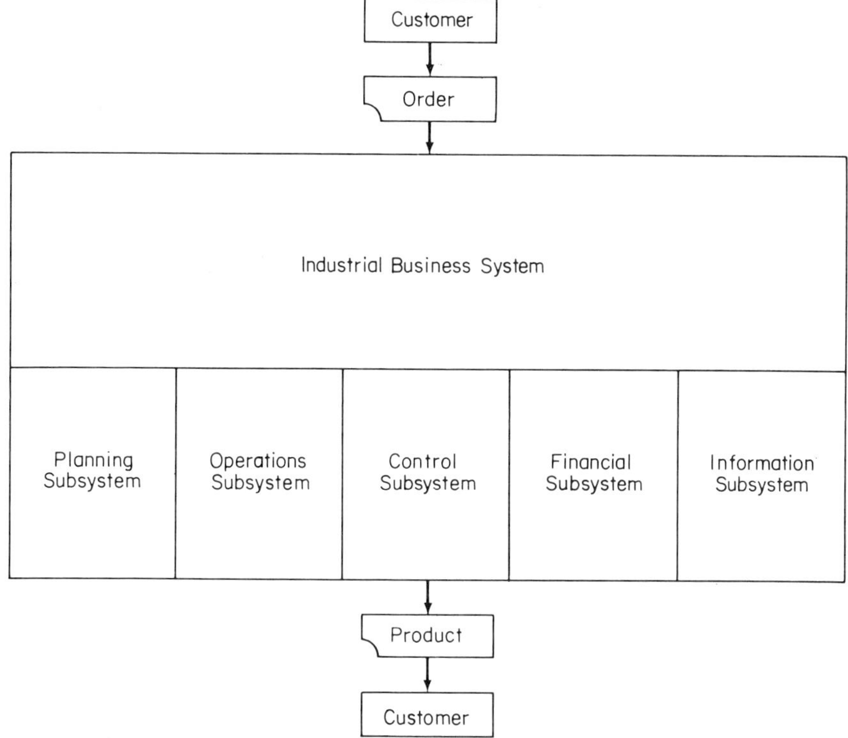

Fig. 1.2 *Overall order-product conversion cycle with major industrial subsystems.*

istics which simplify the systems-design task—the time relationships of the activities assigned to it and the type of data required to perform these activities. Planning concerns itself primarily with the future; operations, control, and financial with the present; and information with the past. The data required by the subsystems can be classified as fixed, variable, or derived. Planning, operations, control, and financial subsystems use primarily variable data or data derived from other variable or fixed data; the information subsystem uses relatively fixed, historical data.

These two characteristics—time and variability—are also important in the establishment of data or information files used in conjunction with industrial systems. Although the *master file* concept—a single file of information for an entire business—has attracted many designers because of its apparent simplicity, implementation is almost impossible because of the magnitude of the task of determining what data should go into the file and how they can be retrieved for use. The systems-engineering approach divides the single master file into a series of subfiles grouped into the categories of past, present, future, fixed, variable, and derived. The resulting files are smaller, better integrated, more easily maintained, and more accessible.

1.2.4 Work Activities of Industrial Businesses

In the preceding section, a framework has been developed for categorizing the various work activities of a business. Now, as the activities are defined, they can be assigned to the proper subsystem for future systemization or organizational grouping. Also, the primary subsystems enable the manager or systems designer to draw on his own experience to determine those work activities which must ultimately be grouped by system or organization to fulfill the objectives of the business most effectively. If the work activities are not specified and defined, the systems and organizations developed for accomplishing the work become amorphous, lack direction, are difficult to understand, assign dual responsibilities for the same tasks, and ultimately are not accepted by the people affected. A clear definition of the work activities required, however, provides a firm basis for logical system and organizational design.

Another feature of this approach is that the order-product conversion cycles of most manufacturing businesses are similar when viewed from the standpoint of the *work activities* required to make the conversion. When viewed from an *organizational* standpoint, each business seems to be an entirely separate entity. For example, one instinctively feels that the business of manufacturing steam turbines is entirely different from that of manufacturing paint. It is obvious that investment, facilities, manufacturing time cycles, marketing and distribution, engineering, and scheduling are all of different magnitudes and importance in the two businesses. It is not so apparent at first that both businesses must perform the same work activities in the order-product conversion cycle because each general manager will insist that his business is the more complex and is entirely different from that of the other.

When the work-activity approach to industrial system and organizational design is used, the differences among businesses seem smaller and the similarities become more apparent. The design problem becomes simpler, and understanding of both the principles and the method of operation increases. The manager and designer can draw on the best methods used for the performance of similar work activities in a wide variety of businesses and thus broaden the base of their information and experience.

1.2.5 Types of Industrial Businesses

Since the ultimate objective of this section is to develop a generalized model of an industrial system and organization, it is necessary at this point to discuss the various types of industrial businesses to which the model will apply. Industrial businesses are generally classified into four major categories:

A. Continuous-process (paper, steel, cement, chemicals, etc.)

B. Standard (household appliances, radio and television, electronic components, toys, etc.)

C. Semistandard (automobiles, machine tools, power transformers, etc.)

D. Job-shop (missiles, space vehicles, steam turbines, large motors and generators, etc.)

Each industry is characterized by its method of product manufacture (continuous or batch), its method of assembly (production line or individual work station), or its method of meeting customers' requirements (standard models, semistandard models assembled from standardized subassemblies, or custom-designed models). Regardless of facilities, assembly techniques, or degree of special design, however, all industries require similar functions in order to sell, design, and manufacture the finished product. All must sell the product, accept the customer's order, convert the customer's order into some form of internal-order document, plan and schedule the design and production of the product, purchase raw and/or finished materials or parts, manufacture the product, test the product, and finally ship or distribute the product or both. The magnitude and relative importance of each function vary considerably from industry to industry, but all functions must be present. For example, the engineering or design function is normally larger in relation to other functions in a job-shop industry than in a continuous-process industry. The purchasing function in the continuous-process, standard, or semistandard industries buys materials and parts primarily on a rate basis, whereas in the job shop purchasing normally buys in small-quantity lots. The warehousing and distribution functions are almost negligible in a job-shop industry but are of major magnitude and importance in the others.

1.2.6 Definition of Industrial Subsystems

The third step in the industrial systems-engineering approach requires the definition of the selected subsystems. Each subsystem is best defined by its input-output requirements and by the work activities required to generate the outputs from the specified inputs. Each of the subsystems established in Par. 1.2.3 and shown in Fig. 1.2 will now be discussed in depth, and the work-activity distribution will be established. In the discussion, a job-shop business will be used because it usually contains all the activities necessary to manufacture a product and thus presents the most generalized case. Other types of businesses—those with standard product lines, with products and markets requiring large finished-goods inventories both at the factory and in regional warehouses, with products assembled to customers' orders from standard parts, and with products manufactured by a continuous process—are variations of the job shop. Since a generalized model of a business is being developed here, we shall start with the most generalized case.

1.2.6.1 Planning Subsystem. The planning subsystem contains those work activities required both to generate the plans for meeting long-range objectives and to determine the products and resources necessary to fulfill the plans. The input to the planning subsystem is a statement of the long-range objectives of the business, and its outputs are the future product, facilities, financial, and manpower plans required to meet those objectives. Without objectives and plans, a business can drift like a ship without a rudder. The history of business is replete with failures caused by lack of proper planning. In the short range, it is possible to operate a successful business with no future plans. There are many examples of single or one-shot successes. But the companies which continue to be successful over a period of years are those with firm long-range objectives and continuously updated plans to meet them.

The most important reason for the establishment of objectives and plans is to ensure progress and growth. One of the best examples of the usefulness of objectives in furthering progress is found in the field of sports. Without the continual publicity given to local, national, and world records in all areas of athletic endeavor and the precise documentation of such records, the challenge to better them would not be present and the number of new records set in a given period of time would decrease greatly. Another example of the effectiveness of goal-setting in motivating people is the classic reply of the English mountain climber G.H.L. Mallory, when asked why he wanted to climb Mt. Everest. His answer: "Because it is there."

Setting objectives is equally important in assuring the continued success of an enterprise. Objectives must be set well ahead of the date of their accomplishment since the acquisition of manpower, facilities, and financial resources requires time. Plans represent the step-by-step process by which the objectives of the business are attained. Activities within the planning subsystem include:

A. Manpower planning
B. Facilities planning
C. Financial planning
D. Research-and-development planning
E. Product planning
F. Market planning
G. Management planning

1.2.6.2 Operations Subsystem. The most important of the five industrial subsystems is the operations subsystem for it includes those activities directly related to the physical conversion of the customer's order to a finished product. Without a properly designed operations subsystem, well-designed planning, control, financial, and information systems are useless. This obvious fact can be overlooked, however, and the emphasis placed instead on adding controls to a poor operating system to improve its performance. It is often true that when controls are added some improvement results. However, major improvements can be made only when the operations subsystem is optimized *prior* to the design and the application of the control subsystem. Separation of operations from control is extremely important in applying the systems-engineering approach.

We shall at this point discuss the relationship of the operations to the control subsystem before proceeding farther. Basically, any industrial-equipment system consists of a means (the operations subsystem) for converting its input to the required output and a means (the control subsystem) for measuring and comparing the output with the input. The comparison continuously corrects the input so that the desired output is maintained. If the output is too high, the control subsystem feeds back a signal which subtracts from the original input and thus reduces the output to its desired value. Similarly, if the output is low, the feedback signal will add to the input and thus increase the output. Figure 1.3 is a pictorial representation of such a system.

There is a basic difference, however, between an industrial-equipment system and an industrial-business system. In the former the feedback is compared with the input and the correction is automatic; in an industrial-business system, however, any variance between the feedback information and the plan normally must be reviewed by a manager, and corrective measures must then be taken by management action. Most industrial-business systems therefore are primarily reporting systems and not control systems in the true sense of the terms. Control is exerted after variances from plan are reported by the feedback system.

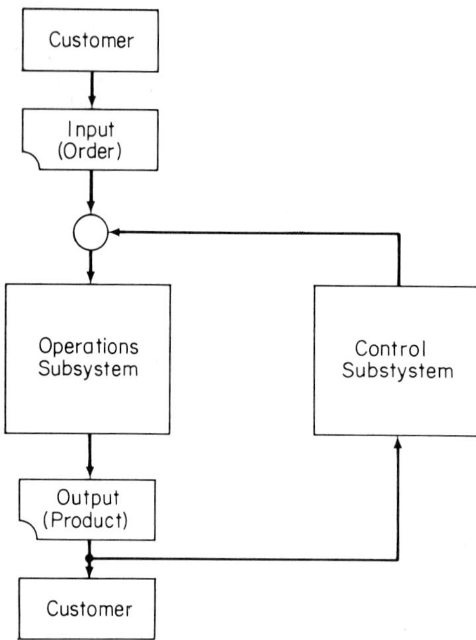

Fig. 1.3 *Operations and control subsystem relationships.*

The basic activities of the generalized operations subsystem are:

A. Receipt of customer's order

B. Conversion of customer's order to internal order (order editing)

C. Establishment of schedules (scheduling)

D. Design of product (engineering)

E. Establishment of manufacturing plan for product (planning)

F. Issuance of tool, fabrication, and assembly orders, and of material requisitions (ordering)

G. Issuance of purchase orders for outside material and parts (purchasing)

H. Receipt of outside material and parts (receiving)

I. Fabrication and assembly of parts

J. Testing of parts and assemblies

K. Stocking and warehousing of parts and assemblies

L. Shipping of finished products

Figure 1.4 is a generalized flow diagram of the operations subsystem. A brief description of the diagram follows:

1. A customer's order is received and converted to an internal order by an "order-editing" function.

2. The internal order is received and a schedule developed by a "scheduling" function.

3. The internal order is received by an "engineering" function and drawings, specifications, and parts lists for the product are developed and issued.

4. The drawings, specifications, and parts lists are received by a "manufacturing scheduling-planning-and-ordering" function, which plans the shop operations for the product, places fabrication orders for tools and parts fabrication, orders outside materials and parts from vendors through the purchasing function, issues assembly orders, and prepares detail schedules for all manufacturing operations.

5. The orders are received by the tool, fabricating, and assembly shops, and by purchasing; outside materials and parts are ordered, received, and stocked; and the work as specified by the orders is completed.

6. The assembled products are tested.

7. The tested products are warehoused or shipped to the customer or both.

It should be noted that in the development of the diagram, a functional approach has been used. No attempt has been made to associate the work to be done with an organizational entity, although in many cases the activity title agrees with classical organizational nomenclature. This distinction is important since it is in the *functional* activities that one finds the similarities among various businesses. Businesses may differ in their organizational structure, but they all must carry out similar work in their order-to-product cycle. They all perform the functions of "order editing"; "engineering"; "manufacturing scheduling, planning, and ordering"; "purchasing"; "stocking and receiving"; "assembling and testing"; and "warehousing and shipping." The magnitude and importance of each operational element may differ, but nevertheless each is required. In a job-shop business, all these elements have approximately equal magnitude. In a continuous-process industry, such as paper manufacturing, the activities of (1) "manufacturing scheduling, planning, and ordering," (2) "assembling and testing," and (3) "warehousing and shipping" and the primary ones.

Automobile manufacture gives an excellent example of both the generality of the flow diagram and the modifications which can be made to adapt it to various businesses. During the design, tooling, and inventory buildup of a new model, the business is essentially of the job-shop

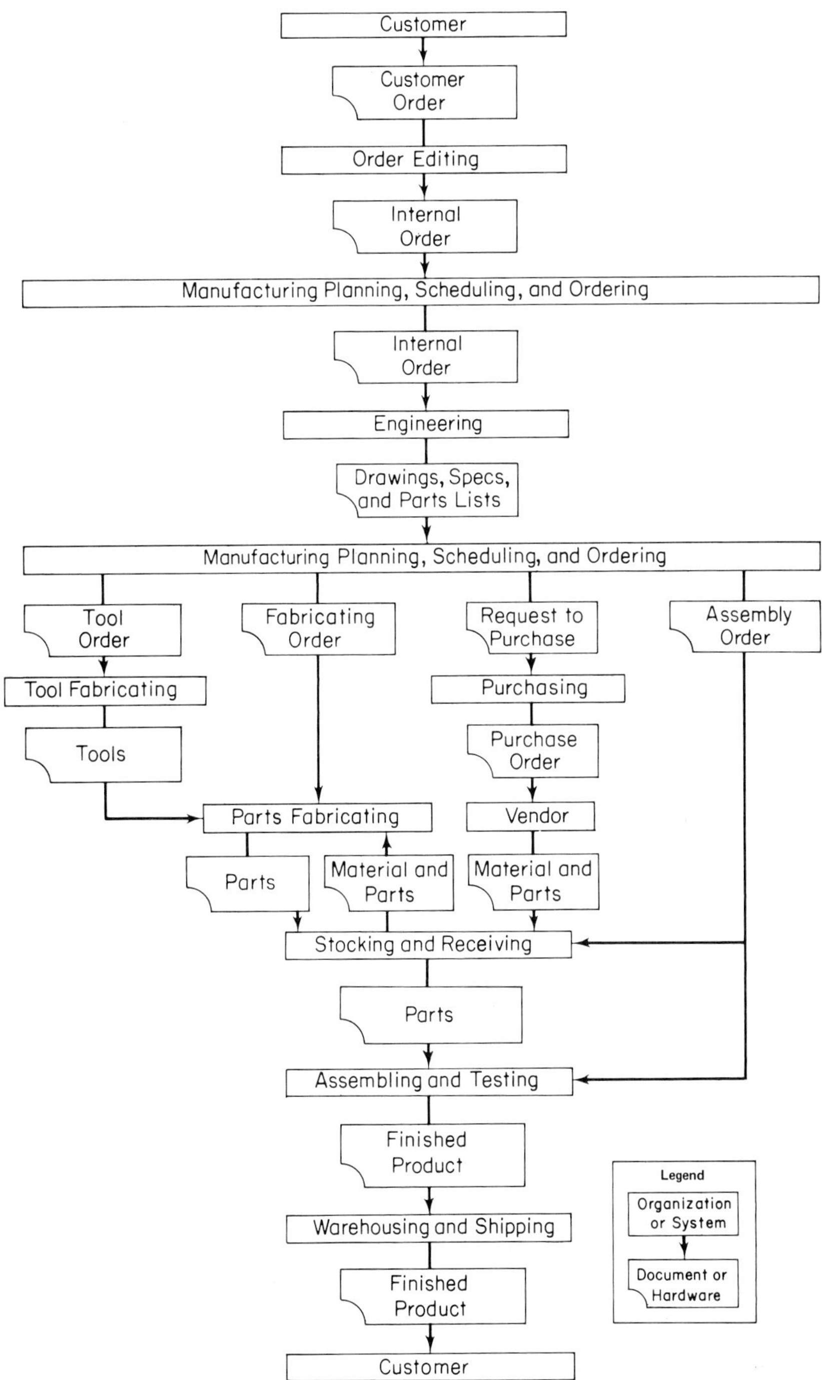

Fig. 1.4 *Generalized flow diagram of the operations subsystem.*

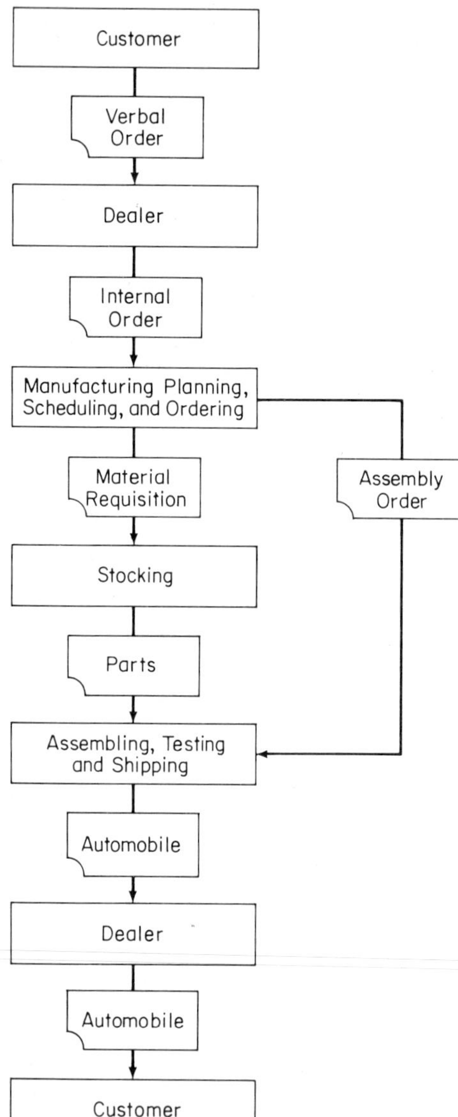

Fig. 1.5 *Flow diagram of the operations subsystem for the automotive industry.*

type—the customer is the marketing organization of the manufacturer. All activities are required in order to fill dealer stocks with finished cars as well as to build up stocks of standard parts from which finished cars can be assembled later to meet special customer requirements. Once sales of the new models start, the business converts to the consumer-goods type. The marketing function estimates sales rates of finished products by model, and the manufacturing function produces sufficient parts to build the cars to fill dealer stocks and keeps sufficient finished parts on hand to assemble cars to special customer requirements. At this point, engineering and tool fabrication are no longer active, and the diagram is much simplified. In the case of a special customer order, the dealer acts as the order-editing function and, by using standardized nomenclature for the desired optionals, he can transmit an internal order directly to the factory. The function of manufacturing planning, ordering, and scheduling issues the proper orders to stocking and assembling, and the car is built and shipped. Figure 1.5 is a modified flow diagram for the automotive industry after the initial filling of dealer inventories.

A second example of the application of the flow diagram is in the heavy electrical-equipment industry. Here the marketing organiza-

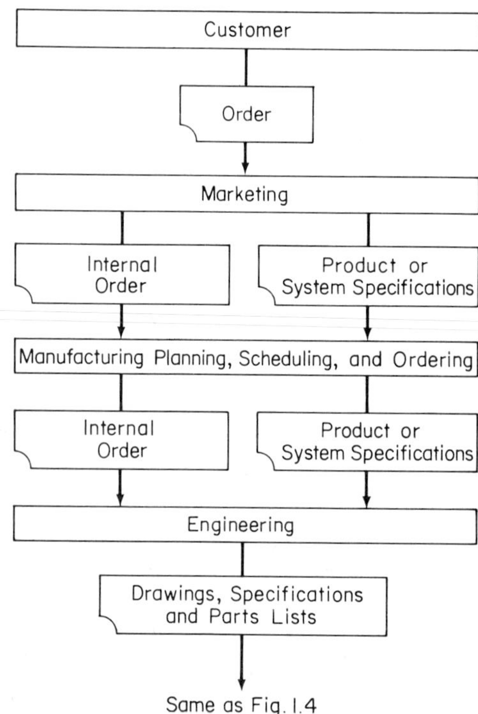

Fig. 1.6 *Flow diagram of the operations subsystem for the heavy electrical-equipment industry.*

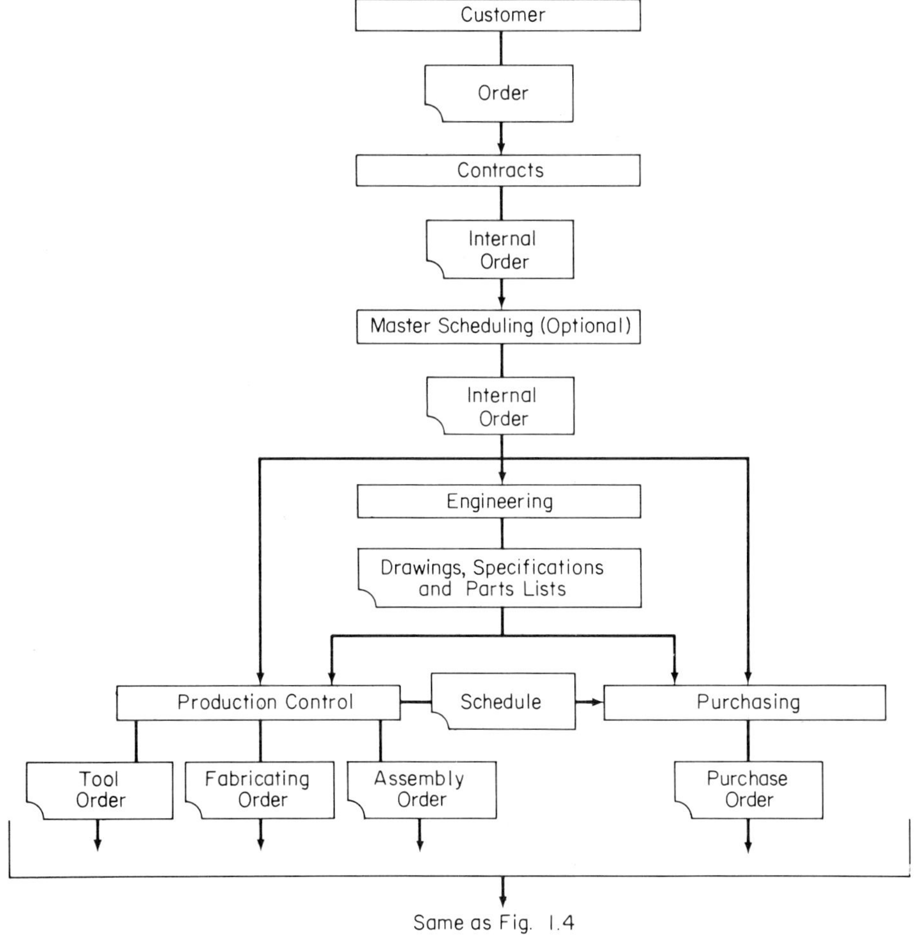

Fig. 1.7 *Flow diagram of the operations subsystem for the aerospace industry*

tion obtains the customer's order and normally generates an internal order. If the equipment is special or a part of a large system, further definition of customer requirements is made by means of a systems specification by a systems- or application-engineering group which is also normally part of the marketing organization. The rest of the information and hardware flow is similar to that in the generalized diagram. Figure 1.6 is the modified diagram for this industry.

A third example is in the defense and aerospace industry, where normally two organizational changes are made, as Fig. 1.7 shows. First, the customer's order is usually received by a contracts organization, which issues the internal order in the form of a contract notice, general order, or other document authorizing the internal organization to start work on the contract and assigning schedule dates and budget dollars for the work to be performed. Second, the purchasing or material function is normally a separate organization because of the magnitude and importance of the subcontracting and purchasing effort. When the customer's contract calls for specific subcontract effort, the material function operates in parallel with the production-control function and handles the subcontracted effort as well as the purchasing of materials and parts required for in-house use.

A fourth and final example is found in industries employing distributors for marketing their products. Such industries include standard, small electric motors; electronic components; and hardware and plumbing supplies. If the product is not in the distributor's stock, it is ordered from a local or regional warehouse by an internal order which describes the product with a standard catalog number. In this case the distributor acts as the order-editing function. Figure 1.8 is a simplified flow diagram for this type of operation.

The point here is that the specific information and hardware flow diagram of any industry can be developed from the generalized diagram. An industrial system designed to fit the general case can be more easily applied to special cases since the functions performed and the documents generated have a commonality among various industries. The systems-engineering approach gives the functional manager a better overview of the total industrial process and also gives the systems designer a better understanding of the interrelationship among the various work activities of industrial businesses, so that he can more easily integrate the total business system. As will be shown later, the information needed in the performance of each of the work activities is common to many of them. If the total information requirements can be established, much duplication of effort can be eliminated and information-storage requirements can be greatly reduced. Article 1.4 discusses more details of the operations subsystem.

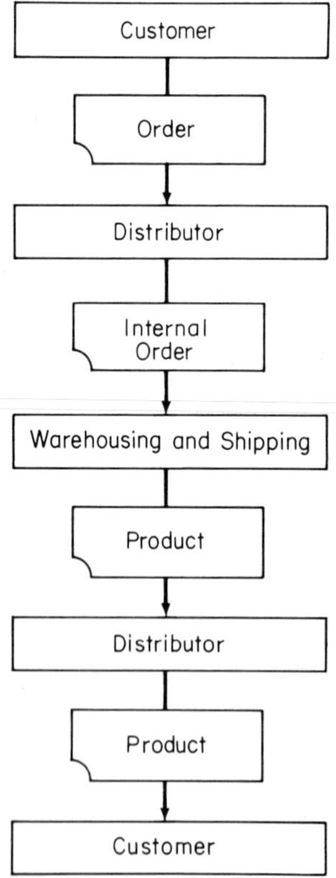

Fig. 1.8 *Flow diagram of the operations subsystem for light industries using distributors.*

1.2.6.3 Control Subsystem. In recent years, much emphasis has been placed on business control systems. However, controls can only improve the performance of the industrial-business system up to a certain point. If the operations subsystem is not properly designed, no amount of additional control can improve performance. Although this point appears obvious, it is often overlooked in the haste to correct a specific operating problem.

Control systems have become extremely complex, and the data required to operate them have required more and more file storage because controls have been added indiscriminately. Control systems must be planned and should not be added on a hit-or-miss basis. If they are not planned, both those controlling the business and those controlled become confused.

The systems-engineering approach separates the design of the operating subsystem from that of the control subsystem. These two subsystems are treated separately because once one begins to assemble the vast amounts of control data required, one tends to lose sight of the basic operational data needed to manufacture the product. Of the total information requirements of a business, it is estimated that 75 per cent are in the control area and only 25 per cent in the operational area. However, because the operations subsystem has the most important effect in reducing the overall cost of doing business, it is the subsystem which should be designed alone and first.

A second reason for the separation is that methods of control are much more personal and controversial than the activities required to design and manufacture a product. Control techniques are tied closely to the management philosophy of the individual manager and thus are more difficult to generalize. If, however, agreement can be reached on the operational work and information requirements, the control subsystem can be custom designed for the business concerned.

It is generally recognized that proper control requires the measurement of performance against predetermined, quantitative standards. Such measurements in the major areas of cost, schedule, technical performance, and quality are usually the basis for any business control system. A standard is set for each element to be controlled in the form of a budget, promised schedule date, test limit, or quality specification. Actual performance is then measured against these standards. The establishment of the standards and of the points in the operational cycle at which performance should be measured is the ultimate responsibility of the head of the business. His agreement and approval should be obtained *before* any design work is started on the control subsystem. He should also specify the form in which management control information is to be presented to him so that lower-level information can be designed and collected in a consistent manner. If the control information system is not planned in advance at all levels, reports and information will be duplicated in the files of all managers and supervisors in a constant effort to be ready for any telephone call from a higher-level manager. Preplanning the control subsystem is extremely important since control information constitutes such a high percentage of the total information requirements of a business.

Although the general objective of any control subsystem is to improve the cost, schedule, and quality performance of the business, primary emphasis is usually placed on cost. The cost elements to be controlled include:

A. Direct labor cost
B. Material cost
C. Indirect salaries and wages
D. Indirect expenses
E. Engineering salaries and wages
F. Engineering expenses
G. Development expenses
H. General and administrative salaries and wages
I. General and administrative expenses

Depending on the industry, all or combinations of these cost elements are used in establishing budgets and then in measuring actuals against such budgets.

One effective way to control direct, indirect, and administrative salaries and wages is to set head-count budgets rather than dollar budgets in these areas. This method is simple, and head-count data are usually easily available. Care must be taken, however, to assure that both low- and high-salaried employees are removed in order to meet head-count targets. This can be done by establishing a managerial and supervisory ratio—the ratio of total employees to the total number of managers and supervisors. If this ratio is maintained as the total number of employees is decreased, higher-paid managerial and supervisory classifications (normally classified as indirect) will also be reduced. Since, on the average, one clerical employee is assigned to a manager or supervisor, additional, indirect, head-count reduction is also obtained when the ratio is maintained during periods of declining total employment or when the ratio is increased in an effort to reduce the ratio of indirect-to-direct expense under any condition.

Direct labor costs are controlled by measuring actual costs against predetermined standards for accomplishing a given job. If calculated standards cannot be developed, estimated standards should be on the basis of previous experience, judgment, or previous actuals for performing the same work. Later sections of this handbook describe methods for establishing standards in the manufacturing area. Establishment of standards in other functional areas is not so precise but, nevertheless, is just as important.

Areas of cost other than salaries and wages are equally effectively controlled by the establishment of standards and budgets for each cost element and by periodic measurement of actuals made against such standards. The degree of detailed cost breakdown and the periodicity of the measurement depend on the depth of control necessary. Also, the cost-accounting system (standard or job), the industry (defense or commercial), and the competitive influences affect the cost control system. Cost control systems can be expensive and their cost of operation should be continually measured against the improvements attributable to them.

The second major area of control is the measurement of schedule performance. This subsystem is easier to establish than is the cost control subsystem since schedule performance is more easily measured—either the scheduled work was completed on the promised date or not.

Of prime importance, of course, especially if good customer relations are to be maintained, is the date on which the product has been promised to the customer. In order to meet this final date the total task of converting the customer's order into a finished product must be broken down into a series of subtasks, and schedule dates must be established for each of these subtasks. Various methods of establishing such in-work dates are available—Gantt charting, line-of-balance method, Pert-time method, and Critical-Path Method (CPM) (see Sec. 12)—but all rely on the development of schedule dates for the completion of small elements of the total job and on the proper interrelationship of such dates. Again, the complexity of the product, the total time available for the order-product conversion process, and the priority placed on meeting the final shipment date determine the degree of detail required by the schedule subsystem.

The simplest measurement of schedule performance is a report of "per cent promises kept." If, for example, in any given time period a certain organizational entity promises to complete fifty specific tasks but only completes forty, its schedule performance as measured by per cent promises kept is 80 per cent. Tasks completed ahead of schedule should not be included in this calculation. Only those tasks which were scheduled to be completed during the time period should be measured. Schedule performance below 90 per cent as measured by this standard should be critically reviewed. This standard is sometimes separated into two parts—"per cent original promises kept" and "per cent all promises kept." In many cases rescheduling is necessary, so that measurement by only the per cent original promises kept is sometimes unfair. If rescheduled dates are met, credit should be given for such performance, even though the original promise date was not met.

The control of quality is the third major control subsystem. Section 14 of this handbook describes in detail the various systems available for the control of quality. Every effort should be made to relate the checkpoints in the quality control system with those in the cost, schedule, and configuration control systems. The better the integration of these systems, the more logical the overall system design and the less data required to operate the various subsystems. Government specifications for contractor planning and control systems are calling for increasing relationships among the cost, schedule, quality, and configuration control systems.

Control subsystems in addition to these include manpower control, inventory control, and configuration control.

A. *Manpower control.* As mentioned above, cost and manpower controls are intimately tied together. Cost can best be controlled by control of manpower since wages and salaries, employee benefits, travel expenses, occupancy costs, etc., all depend on the number of people on the payroll.

If historical records are available and the gross number of employees can be established for a given order or sales volume, an effective way to control manpower is to vary it as the orders received or the sales billed by the business increase or decrease. A good rule is to vary the number of employees in proportion to the square root of the volume. Thus, if the orders or sales volume doubles, the employees should be increased by only 41 per cent; if the volume is halved, the number should be dropped to 71 per cent. This square-root ratio takes into account the fact that certain expenses are fixed and are somewhat independent of volume.

Paragraph 1.3.3.1 discusses in detail the question of whether to adjust employees on the basis of orders received or sales billed or an average of the two. In general, sales billed should be used in industries having long manufacturing cycles and orders received used in those having short cycles.

B. *Inventory control.* Control of inventory is an important factor in controlling costs as well as schedules. It has been estimated that for an industrial concern the cost of carrying inventories averages 25 per cent of the average annual inventory. The charge includes the cost of capital to purchase the inventory; handling, storage, and insurance charges; costs of ordering, expediting, reviewing, and inspecting; and costs incurred through obsolescence and physical deterioration of the inventory.

Since a large percentage of company investment is represented by inventory, this tie-up of capital has a direct impact on cash flow and profit. Thus, on one hand, inventories should be kept low to increase cash flow and improve profits; but on the other hand, inventories must be large enough and must contain the proper quantities of raw material and parts to ensure continuous production flow and meeting of customer schedules.

Inventory control, therefore, must balance these two factors—cost and protection. The accuracy with which inventory can be controlled is based primarily on the accuracy of forecasted usage of material and parts and the inventory dollar investment to which the management of the business is willing to commit itself.

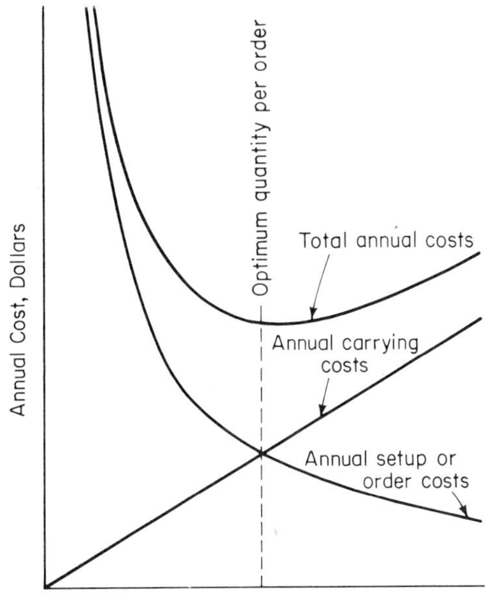

Fig. 1.9 *Economic order quantity.*

Once these two factors are determined, the theoretical, economical manufacturing- and purchase-order quantities can be calculated by a variety of formulas. And by controlling order quantities, inventory is also controlled.

The economical order-quantity principle is based on the fact that for every manufactured or purchased part, there is an optimum order quantity that results in the lowest possible total cost per part. This total cost is the sum of the cost of carrying the part in inventory and the fixed cost of placing the order, including the setup costs for producing the parts. The larger the quantity per order, the greater the annual carrying cost, but the lower the annual fixed cost of placing the order. Figure 1.9 is a graphical representation of the problem.

Optimum quantity per order theoretically occurs when the annual carrying costs equal the annual order costs. Mathematically, the optimum quantity per order can be shown to be:

Optimum quantity per order

$$= \sqrt{\frac{2\,(\text{Annual unit usage})\,(\text{Cost per setup or order})}{(\text{Cost per unit})\,(\text{Per cent annual carrying cost})}}$$

Although the formula is simple for internally manufactured parts, the calculation is complicated for purchased parts and material, which normally have a price break for varying quantities per order.

Some of the factors which must be considered in making inventory decisions include:

1. Annual cost of storage per unit
2. Desired annual return on capital
3. Number of pieces used per day
4. Number of square feet of net storage space per unit
5. Storage limitations on in-process or stockroom storage space
6. Extreme fluctuations in raw material or parts
7. Limited working capital
8. Multiple price-quantity combinations
9. Variable setup charges and tool costs
10. Multiple vendors
11. Lead times

Thus, the inventory problem is a complex one. Large computers have been used to test inventory rules ahead of time and to determine the effect of such rules on operating performance. Because of the variety of factors involved, the problem of arriving at a single optimum solution is a difficult one. One very effective way to control inventory is to withdraw all approval authority for placing purchase or manufacturing orders and to place such authority in the hands of the head of the business only. Although the action is drastic and may appear at first glance to shut down completely the production process, it has been generally found that a more careful review of essential inventory requirements is made prior to submittal to the chief executive and that the situation quickly comes under control.

C. *Configuration control.* The task of accounting for and recording the planned and actual design of each product of a business throughout its production and service life is configuration control. Whether the product is built for the durable-goods, consumer-goods, or defense market, configuration control is necessary in order to identify properly the parts and subassemblies which make up the final product. Without such identification, servicing, supplying of spare parts, retrofitting, and the other aspects of proper support of the product after delivery are impossible.

The importance of configuration control has long been recognized in the durable- and consumer-goods industries. Without proper customer service and adequate supplies of spare parts, the position of a company in the marketplace can rapidly deteriorate. In addition, the renewal-parts business can be made extremely profitable if proper attention is given to it and adequate records of product configuration are initiated and continually updated.

In the aerospace industry the importance of configuration control lies in the necessity of keeping an accurate record of product design in the face of many design changes. These changes are caused by operating in areas well in advance of existing state-of-the-art knowledge and by rapid expansion of technological knowledge in these same areas. Since many of the products of the aerospace industry are developed to obtain data, a knowledge of the exact configuration of the product is necessary if the data are to be properly interpreted.

Proper maintenance of the drawings, specifications, and parts lists defining the product is the first step in exercising configuration control. Once the decision has been made to change the design of a part, it must be determined whether the change affects the form, fit, or function of the part being changed or of any higher level assembly with which the part is associated. If there is no effect on form, fit, or function, the part is completely interchangeable and on reidentification of any drawing or specification is required. However, if the part is not interchangeable, it must be reidentified by a change in its part number and those of all higher-level subassemblies and assemblies, including the top assembly. Without such a procedure, there is no way of knowing exactly what configuration of parts is contained in the finished product.

The Department of Defense, the prime customer of the aerospace industry, has formalized and documented the procedures to be followed in *configuration management*. They have defined configuration management as the task of controlling, accounting for, and reporting the planned and actual design of each article throughout its production and service life. The main difference between the procedures specified by the government and those required in the durable- and consumer-goods manufacturing field is that customer approval is normally required in the defense industry, whereas changes are made at the manufacturer's discretion in other industries. The Department of Defense requires that all changes that affect interchangeability of components subsequent to a designated point in time must have its approval. The configuration at this point is often called the *baseline configuration*. It has been defined as the technical description, comprising detailed, uniform specifications and engineering drawings, against which the first article configuration inspection is performed for the particular end item being purchased.

Configuration control is the systematic evaluation, coordination, and approval or disapproval of all changes to the vehicle or end item, ground-support equipment, and facilities subsequent to the establishment of baseline configuration. Configuration accounting is the documentation of proposed and implemented changes made to systems and equipment in order to maintain knowledge of configuration status. It provides records of proposed changes to end items (such as missiles, aircraft, and space vehicles) and their supporting ground and surface equipment and facilities and records the implementation of these changes. Configuration reporting includes the preparation of management control reports establishing planned configuration; engineering and manufacturing documentation determining the engineering configuration and assuring that authorized changes have been incorporated in the actual manufactured article; and reliability assurance reports recording the location of critical or time-sensitive components in each article in order to relate failure data to the component manufacturing, test, and operating history.

The system designed to perform the configuration control function will vary from filing the sketch (sometimes made on the back of an envelope) to the elaborate procedures specified by the government. In all cases, the system should be designed to give the best balance between the needs of the business and the requirements of the customer.

In summary, the control subsystem should be designed only after the operating subsystem has been developed, inputs and outputs of the various functional activities have been determined, and quantitative standards have been established for the measurement of individual and organizational performance against such standards. The number of standards should be kept to the minimum necessary to ensure that the business will meet its goals and objectives. An excessive number of measurements makes it difficult for both higher and lower levels of management to operate effectively and adds to the cost of operating the business. The measurements selected should be well understood, should be used on a regularly scheduled basis, and should allow for continual follow-up action when variances between standards and actuals occur.

1.2.6.4 Financial Subsystem. In this presentation, the financial subsystem contains those functions relating to the fiscal and monetary activities of the business. As such, it has responsibility for the following functions:

A. Payroll
B. Accounts receivable
C. Accounts payable
D. Collection and reporting of
   1. Direct labor costs
   2. Indirect costs
   3. Material costs
   4. Engineering costs
   5. General and administrative costs
E. Company financial statements
F. Tax accounting
G. Inventory accounting
H. Financial estimating

The financial subsystem is considered separately since it requires highly specialized knowledge and discipline. It receives its inputs from all the internal organizations of the business as well as from external customer organizations. Its outputs are used throughout the business, and its operations are influenced to a great extent by legal, contractual, and governmental regulations. Its records must be auditable by a wide variety of both internal and external organizations and agencies, and the accuracy and timeliness of such records determine to a great extent the future success of the business since the majority of management decisions are made on the basis of financial data. (See also Secs. 2,3, and 4.)

Perhaps the most important output of the financial subsystem is the series of cost reports which it submits to the cost control activity within the control subsystem. The comparison of the actual costs with the cost standard (or budgets) allows all management levels to effectively control the cost performance of the various elements of the business and also allows the chief executive to measure the overall performance of the business against its stated profit objectives.

Because of the close relationship between the financial subsystem and the cost control portion of the control subsystem, the responsibility for both establishing budgets and reporting actuals against them is assigned to the financial subsystem in most industrial businesses. However, in the aerospace industry the responsibility for cost control is normally assigned on a project or contract basis to a separate project- or program-oriented management organization and the financial subsystem acts as a cost-collection system, feeding contract cost information to the project office. Overall control of costs is exercised by the project organization. Great emphasis is placed at the project level on forecasting costs and anticipating any contract overruns or schedule slippages. Contract overrrun is the term used in government contracting to describe the incurring of costs in excess of those negotiated. The contractor must report anticipated overruns to the contracting office.

In industrial businesses, however, the financial subsystem usually develops for management the various financial budgets and then reports costs against them. It also establishes product costs from time standards furnished by the manufacturing organization and from material costs furnished by the purchasing group. This single source of responsibility for budgeting, costing, and collecting actual costs allows the manager to obtain easily the proper information for making decisions which will affect the overall profitability of the business.

Certain other activities of the financial subsystem have very close relationships with portions of the operations, control, and information subsystems. In designing any of the subsystems, such interfaces must be kept in mind so that subsystem integration can be easily achieved without major redesign. The interfaces occur among the following subsystems:

A. *Purchasing, receiving, accounts payable, and inventory accounting*. In order to establish proper vendor relationships as well as to control internal inventory, means must be provided for purchasing and receiving to notify the financial organization both at the time vendor material is ordered and when it is received. This notification is necessary in order to forecast cash requirements for the payment of vendor invoices, to pay vendor invoices promptly after receipt of material in order to take advantage of discounts, to forecast inventory balances, and to properly account for raw-material and finished-parts inventory.

B. *Scheduling, warehousing, and inventory accounting*. These three areas must have close interfaces in order to ensure the maintenance of proper inventory records. Work-in-process, raw-material, and finished-parts inventories are affected by work completions at various internal work stations, by withdrawals of raw materials from warehouse stock, and by the use of parts in the internal fabrication and assembly processes. Warehouse inventory must be debited when withdrawals are made and work-in-process inventory credited with both the raw material used and the labor added to the material and parts. The inventory-accounting subsystem must, therefore, be tied closely to the material - requirements, stock - disbursement, and scheduling subsystems in order to account correctly for all inventory balances.

C. *Shipping, accounts receivable, and inventory accounting.* These three subsystems must be interrelated if customers are to be invoiced for products shipped and finished-goods inventory credited when the material is delivered.

D. *Payroll and employee relations.* Much of the information in payroll records is required in employee-relations activities and the reverse is also true. Information on rates of pay, salary history, payroll deductions, vacation payments, absences, etc., are all required by both functions. An integration of the payroll and employee-relations files can aid greatly in reducing processing time and in acquiring statistics relating to employees. Proper design of a master employee-payroll file can result in improved information for salary-planning, employee-benefit, and managerial-development programs.

1.2.6.5 Information Subsystem. Of all the five subsystems described here—planning, operations, control, financial, and information—the information subsystem is the most comprehensive. Much has been written on master files and most of the literature has been devoted to the concept of a single master file of information for an entire business. As described in this section, however, the information subsystem and resulting file essentially contain only the relatively fixed information required by the four other subsystems of the business as well as the historical information on which future management decisions may be based. The separation of fixed and historical information from variable, present, and future information greatly simplifies the total business systems-design problem.

The most significant savings in the operation of business systems can come from the proper collection and dissemination of the common information used throughout the business. The distribution of data can be performed either manually or by computers. In either case it is necessary to determine the information requirements of various using functions, categorize the information, eliminate duplication, file the information, and finally determine the most useful format in which the information can be distributed to the using functions.

A specific example of common information is the data contained in the parts list, bill of material, or list of material which is normally generated by the engineering or designing function. This list describes the parts or subassemblies required for the next higher-level assembly, which may be either another subassembly or the final assembly. The list contains as a minimum the following information:

A. Part or identifying number
B. Part nomenclature or description
C. Material
D. Specifications or notes
E. Item number on parts list
F. Quantity required for next higher assembly

Figure 1.10 is a typical parts list. The format is that prescribed by the Department of Defense in MIL–STD–100, dated March 1, 1965, but the information contained in it is generally standard throughout industry. The information on the parts list is used by the engineering, planning, scheduling, cost, and purchasing functions to delineate the parts necessary to manufacture the finished product. Unless the business system is properly designed, each function will recopy the parts list according to its own particular format and then add the additional information necessary to carry out its work activity. For example, planning needs the parts list to add the operations required for the fabrication or assembly of each part, the work station at which each operation is to be performed, the standard setup and run time for each operation, the normal flow time, the amount of raw material from which the part is to be fabricated, the identifying number of the tool to be used, etc. Scheduling must "explode" the parts lists for a given finished product in order to determine total quantities of material, parts, and assemblies required; schedule dates for receipt of vendor material; schedule dates for start and completion of work to be performed on the parts by various internal departments; etc. Similarly, costs for material and labor are calculated for each part and then added to determine the standard material and labor cost of the finished product.

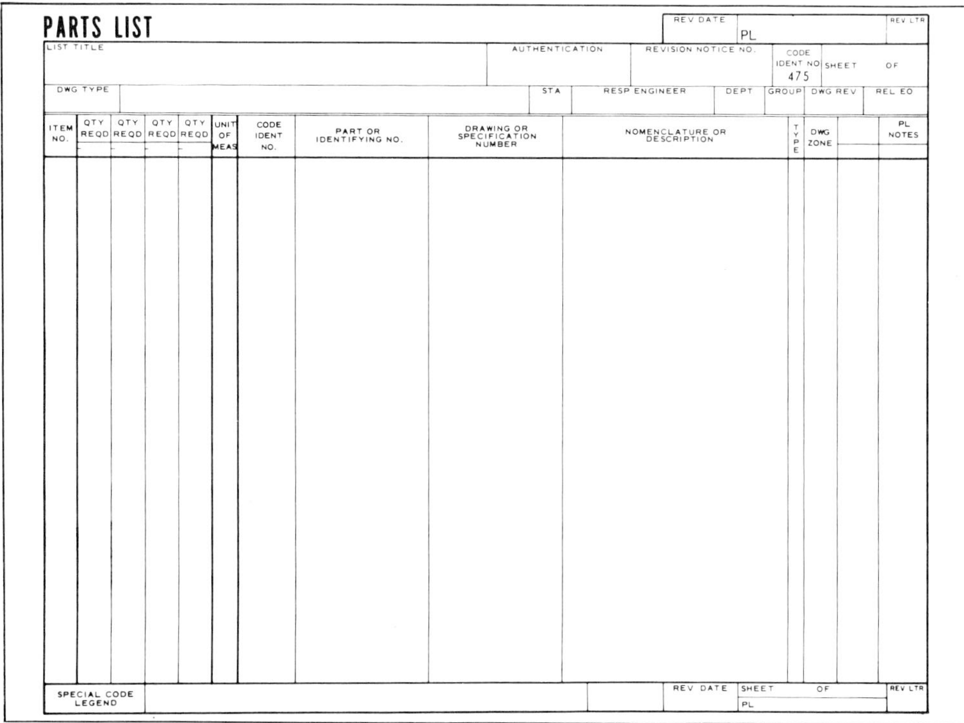

PARTS LIST

REV DATE | PL | REV LTR

LIST TITLE | AUTHENTICATION | REVISION NOTICE NO. | CODE IDENT NO 475 | SHEET | OF

DWG TYPE | STA | RESP ENGINEER | DEPT | GROUP | DWG REV | REL EO

ITEM NO.	QTY REQD	QTY REQD	QTY REQD	QTY REQD	UNIT OF MEAS	CODE IDENT NO.	PART OR IDENTIFYING NO.	DRAWING OR SPECIFICATION NUMBER	NOMENCLATURE OR DESCRIPTION	TYPE	DWG ZONE		PL NOTES

SPECIAL CODE LEGEND | REV DATE | SHEET | OF | REV LTR | PL

Fig. 1.10 *Typical engineering parts list.*

Avoiding duplication of information and effort in designing the information subsystem, therefore, requires a thorough knowledge of the information needed by each work activity within the business, the method by which the information is generated or collected, and the interrelationship of the information needed by one work activity and that needed by other activities.

In almost all manufacturing industries, the engineering-drawing number assigned to a drawing or parts list is the primary means of information identification. Over the years most work activities within various businesses have filed their information by this drawing number. It therefore affords a natural heading under which information can be collected and tabulated.

The information subsystem can be viewed as a data bank from which information can be drawn as required. Data which can be stored in it include:

A. Engineering parts lists
B. Parts-usage ("where-used") file
C. Reliability information
D. Specifications
E. Manufacturing-planning information
F. Labor-standards information
G. Quality information
H. Personnel information
I. Capital-equipment inventory
J. Standard-costs information
K. Capital-equipment-usage information

One method used to collect the data to be stored by the information subsystem employs an information matrix. The matrix enables one to list the information requirements of the various activities of the business and then to determine the common elements of data to be stored in the master file so that duplication can be eliminated. It consists of a tabulation of the work activities in the vertical columns and a listing of the items of information needed by each activity in the horizontal rows. If an element of information is needed by a given activity, the fact is indicated by a mark at the intersection of the vertical column and horizontal row. If the information element is required

Information Element \ File	Engineering Parts List	Parts Usage	Manufacturing Planning	Labor Standards	Standard Cost
Part Number	X	X	X	X	X
Nomenclature	X	X	X	X	X
Item Number	X	X	X	X	X
Material Specifications	X	X	X		
Next Higher Assembly	X	X	X		
Quantity per Next Higher Assembly	X	X	X		
Make Department			X	X	X
Operation Number			X	X	
Operation Name			X	X	
Tool Type			X		
Tool Number			X		
Normal Flow Time			X	X	
Minimum Flow Time			X	X	
Setup Time				X	X
Run Time				X	X
Direct Labor Cost					X
Material Cost					X

Fig. 1.11 *Information matrix.*

by several activities, several marks appear in the same horizontal row, thus indicating that the information is common. This display of information can then be used in the design of either a manual or a computerized master file, as well as in the design of forms to be used by various functional elements of the business.

Figure 1.11 is an information matrix for the engineering-parts-lists, parts-usage, manufacturing-planning, labor-standards, and standard-costs files of a manufacturing business. If the information were stored in separate files, almost twice the manual or computerized storage capacity would be needed because of duplication. Storing in matrix form reduces the size of the file and also eases file maintenance since only a single file has to be updated compared with the five which would normally be changed.

The matrix has been constructed using the engineering-drawing number of the part, subassembly, or finished product as the basic element of data. The easiest way to develop the matrix is to collect existing output documents from each of the present functional organizations of the business. Normally each functional group has tailored and refined its output documents to meet its own and the recipient's needs in the best way. This fact is true within a given industry as well as throughout many different industries. The manufacturing-planning documents from a wide variety of manufacturing departments all contain almost the same information. It can be assumed, therefore, that the

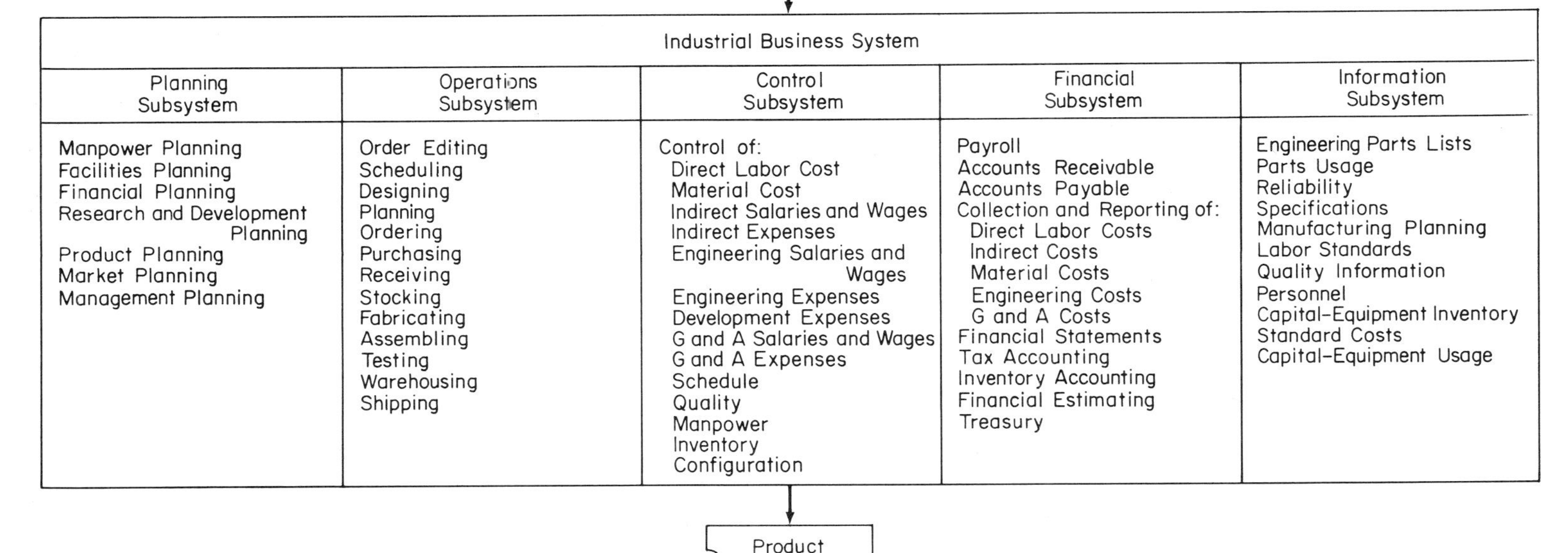

Fig. 1.12 *Overall order-product conversion cycle with major industrial subsystems and their work activities.*

information included on existing planning documents is the information which has proved over a period of years to meet best the needs of the majority of manufacturing organizations. The same is true of such documents as engineering parts lists, shop orders, requests to purchase, and purchase orders. The title and format may vary, but the information requirements are relatively constant from company to company and from industry to industry.

The matrix in Fig. 1.11 indicates that the engineering, planning, and cost activities require thirty-one pieces of information to generate the documents for which they are responsible. Since only seventeen different items of data are in the matrix, there would be almost a two-to-one duplication of data elements if files were set up for each activity.

Good planning and design of the information subsystem by the use of an information matrix can result in substantial cost savings. Elimination of duplication alone decreases the manpower required to prepare, copy, recopy, disseminate, and maintain the necessary information. Efficient handling of data also shortens manufacturing-cycle time and further improves the overall operation of the business.

1.2.6.6 Summary of Industrial Subsystems. The industrial system was first viewed as a transfer function which converts the customer's order into a finished product (Fig. 1.1). The total system was then divided into subsystems based on the three major elements of the work of a manager—planning, operating, and controlling. Financial and information subsystems were added in order to supply the data needed by the three major subsystems. Figure 1.2 depicted the order-product conversion cycle and the five subsystems of manufacturing businesses.

The work activities of the business were then discussed, developed, and assigned to one of the five subsystems. Figure 1.12 summarizes the order-product conversion cycle, the major subsystems, and the work activities of the subsystems.

The data used in the activities of each subsystem are characterized by their relative time relationships—future, present, or past—and by their relative constancy—fixed, variable, or derived. This grouping of data greatly simplifies the tasks of systems design and file organization.

In addition, the concept of the information matrix (Fig. 1.11) was introduced. This matrix is the basis of industrial systems design and affords a convenient way to collect and depict the data required by the business system.

1.3 INDUSTRIAL ORGANIZATION

In preceding paragraphs of this section, the functions and activities of industrial businesses have been delineated. Before discussing in more detail the design of operating systems, it is necessary to group like types of work and to assign the responsibility for the performance of such work to elements in an organizational structure.

Organizing is the process of defining and grouping the activities of a business so that they may be most logically assigned and effectively executed. The steps in the organizational process are the following:

A. The determination of the various kinds of work which must be performed in order to meet the objectives of the business

B. The assignment of like kinds of work into a minimum number of organizational components

C. The placement of each component into the simplest possible organizational structure

D. The staffing of each component with competent personnel

E. The development of written policies and procedures through which the objectives of the business and the responsibilities of each component are made known

1.3.1 Purpose of Organizational Structuring

The prime purpose of organizational structuring is to divide the total business task into a number of smaller, more manageable tasks so that the work of managing may be accomplished in a more efficient manner and the objectives of the business met more easily and effectively. Jethro, Moses' father-in-law, was the first management consultant to recommend an organizational change in order to improve the effectiveness of a leader in meeting his objectives. After watching Moses teaching, advising, and judging his people from dawn to dusk, Jethro pointed out (Exodus 18: 13–23) that what Moses was doing was good for neither himself nor the people; that both would "wilt" away; and that the burden was too great to be performed alone. He then advised: "Thou shalt provide out of all the people able men such as fear God, men of truth, hating covetousness; and place such over them to be rulers of thousands, and rulers of hundreds, rulers of fifties, and rulers of tens; and let *them* judge the people at all seasons; and it shall be that every *great* matter they shall bring unto thee, but every *small* matter they shall judge; so shall it be easier for thyself, and they shall bear the burden with thee. If thou shalt do this thing, and God command thee so, then thou shalt be able to endure and all this people shall also go to their place in peace."

Adam Smith wrote in *Wealth of Nations:* "The division of labor, however, so far as it can be introduced, occasions in every art, a proportional increase in the productive powers of labor. The separation of different trades and employments from one another seems to have taken place in consequence of this advantage." His comments also apply to the division of management. As industrial enterprises grew in size, this division became a necessity. No longer was it possible for a single man to conceive of a product and then sell, design, build, test, ship, and install it. He soon required a helper, and then he was forced to organize in the sense that certain assignments of responsibility had to be made. As more and more people were added, more and more tasks were delegated. The organizational structure became more complex; policies had to be written to inform people of the objectives of the business and to establish guidelines for the operation of the business. Finally, systems had to be designed and procedures written in order to instruct people in the performance of the day-to-day activities of the business.

1.3.2 Organizational Design

Organizational design is the process by which one builds the optimum organizational structure and assigns the necessary responsibilities and authorities to the manager or supervisor of each component within the structure in order to best carry out the objectives of the enterprise. The design process consists of implementing the first steps outlined in Art. 1.3.

1.3.2.1 Determination of Objectives. Before a business is organized, its objectives must be established. Although much has been written about various theoretical goals which might be set, generation of profit dollars is normally considered the prime reason for the existence of any privately owned business. Subgoals include furnishing needed services or products or both; obtaining favorable customer acceptance; providing good jobs, wages, and working conditions; attracting investor capital; maintaining good supplier and subcontractor relationships; and meeting social, civic, and governmental responsibilities. The organization should be designed to assure the attainment of each of the subgoals. The organizational structure depends on the nature of the product or service, the type of customer, the number of suppliers or subcontractors, and the geographical locations of the business and the market.

1.3.2.2 Major Functional Activities of Industrial Businesses. The three major operating functions of almost all industrial businesses are marketing, engineering, and manufacturing. Marketing includes market research, product planning, sales promotion, advertising, and direct sales contact with the customer. Engineering includes research, development, new-product design, and old-product improvement to meet the needs of the customer. And manufacturing includes planning, ordering, scheduling, purchasing, fabricating, assembling, and controlling the quality of the product to be delivered to the customer.

Basically, marketing should be customer-oriented, engineering should be product-oriented, and manufacturing should be process-oriented. These three different orientations sometimes make the organizing process a difficult one. Application of this concept can best be described by using the automotive, heavy-electrical-equipment, and aerospace industries as examples.

Each automotive division within a multidivision company serves a certain price segment of the total market for automobiles. Thus, it is primarily market- or customer-oriented. Within each division, however, engineering is subdivided on a product basis—engines, bodies, chassis, etc.—and the manufacturing function is organized by process—assembly, sheet-metal (bodies), machining (engines), etc. Assembly plants are good examples of the process orientation of manufacturing since all models of several different makes of automobiles are interchangeably assembled on a single production line.

In the large electrical-equipment industry, major organizational groupings are customer-oriented. Thus, one finds industrial products, consumer products, aerospace and defense products, and electric utility products as the primary organizational groups immediately below the chief executive. Each group services a certain segment of the total company market. Within the major groupings, divisions are usually product-oriented (steam turbines, motors and generators, refrigerators, electronic equipment, transformers, missiles, atomic-power equipment, etc.). Within each product line, engineering is organized around the characteristics of the product, and manufacturing is divided by process—sheet-metal, punch-press, screw-machine, assembly, test, etc.

With essentially a single customer—the Department of Defense (DOD) and the National Aeronautics and Space Administration (NASA)—the aerospace industry is organized primarily by product. Thus, aerospace and defense companies have aircraft, marine, communications, navigation, propulsion, missile, space-systems, ordnance, and radar divisions as their top level organizations. Because of rapidly expanding technologies, the engineering function in this industry is highly specialized. One finds propulsion, structures, guidance-and-control, radar, and computer units within the engineering organization. Manufacturing still maintains its classic process orientation. One peculiarity in the aerospace and other defense industries will be discussed in detail later but should be mentioned here. Although there is nominally only one customer for the output of the defense industry, both DOD and NASA are highly project-oriented—each major project or procurement is organized on a project or program basis. Since each major effort is managed by a separate office, any given enterprise finds itself doing business with many program offices. Thus, although the products and technologies are similar, the multiprogram nature of the business creates special interface problems, which must be considered in any organizational structuring, especially since customer influence is extremely strong in this industry.

In addition to the marketing, engineering, and manufacturing activities, three other supporting activities are needed to carry out completely the work of the enterprise. These additional functions are financial, employee relations, and legal.

To summarize, then, the six basic work activities of an industrial business and their functions are:

1. *Marketing:* to determine the markets and customers potentially available; to plan and request the products necessary to meet market needs; to sell and distribute the products; and to develop proper relations among the company and its shareholders, the public, and governmental agencies

2. *Engineering:* to innovate, develop, and design salable products and services at the required price and reliability

3. *Manufacturing:* to procure, fabricate, and process the materials necessary to furnish the salable products of the business with the required quality and cost on the proper schedule

4. *Financial:* to provide cost information and analysis; to raise and account for the money required to carry on the operation of the business; and to account for all the assets of the business

5. *Employee relations:* to plan for and develop the human resources of the business and to develop proper relations between the company and its employees

6. *Legal:* to ensure that the business is conducted in accordance with the basic laws and regulations of the community and nation.

1.3.2.3 Work Activities of Industrial Businesses. Paragraph 1.2.3 divided the total industrial system into the subsystems of planning, operations, control, financial, and information without regard to organizational structure. The division was based on the elements of the work of the manager, who must plan, operate, and control the business, basing his decisions on financial and other supporting information. However, the chief executive officer cannot carry out the work alone and so must delegate work to others through a formal organizational structure.

Paragraph 1.3.2.2 identified the major functional activities of industrial businesses as marketing, engineering, manufacturing, financial, employee and public relations, and legal. Over the years, it has been found that each of the detailed work activities involved in converting a customer's order into a finished product or service falls into, or can be assigned to, one of these functional groupings. Thus, it is convenient to use them as the basis for developing an organizational structure.

It should be noted that each functional organization requires subsystems of planning, operations, control, financial, and information. Thus, the marketing function must develop plans, methods of operating and controlling, and means of collecting the information necessary to accomplish its work. The internal functional subsystems must be compatible with those of the enterprise. This interrelationship between the business system and the organizational structure is extremely important in both system and organizational design. The business system ties together the organizational structure.

1.3.3 Organizational Structuring

The following principles of good organizational design should be reviewed prior to the start of any organizational structuring:

1. A given work activity should be assigned to a single organizational component in such a manner that the manager of the component can be held solely responsible for its successful completion.
2. There should be a minimum number of interfaces between organizational components in order to reduce coordination problems and to eliminate divisions of responsibility.
3. The number of organizational levels between the lowest-level individual worker and the chief executive officer should be kept to a minimum in order to improve channels of communication.
4. Staff positions and organizations should be kept to a minimum and, if used at all, should act only in an advisory or consulting capacity.
5. The use of assistants should be discouraged in order to keep communication channels as direct and as short as possible.
6. The outputs of organizational components should be measurable in terms of cost, schedule, and quality so that the manager of the component can be held accountable for the work assigned to him and his performance measured against a predetermined standard.
7. Similar types of work requiring generally similar employee capabilities, training, backgrounds, and salary and wage requirements should be grouped together into single organizational components.
8. In general, organizational components whose inputs or outputs or both are closely related should be located together geographically in order to reduce interface and communication problems.

1.3.3.1 Organizational Size. Before starting to design an organizational structure for an enterprise, the required number of employees must be estimated. This procedure appears obvious on the surface but is many times overlooked in organizational design—both in the design of new organizations and in organizational modifications necessary to meet changing business conditions. The number of organizational components, the number of organizational levels, and the number of managers and supervisors required are all based on the total number of employees of the enterprise. Unless this fact is recognized, the business can be over- or undermanaged, with consequent inefficient operation and loss of profit.

Whether the organization is being established or has been in existence for some time and requires change, the orders and sales forecasts primarily determine the total number of employees. On a long-term basis, employees can be paid only from income received from customers in return for goods and services. The amount of such income available for employee wages, salaries, and benefits over a given period of time can be obtained by subtracting from the orders or sales (or the average of orders and sales) forecasts for the period of the amount paid for purchased materials and services, fixed expenses (rent, heat, light, power, taxes, depreciation, etc.), variable expenses (travel, supplies, advertising, etc.), and profit before taxes. This difference, when divided by the average period earnings and benefits per employee, determines the allowable number of employees which the business can support during the period being considered. If the forecasted number of employees derived in this manner is insufficient to accomplish the work required to meet the objectives of the business, the assumptions on which the number is based should be reviewed and corrective action taken. In a new business, more employees will probably be needed than the orders or sales dollars can support and additional capital will be required to maintain the enterprise during its formative period. In an established business, great care must be taken in establishing employee head-count forecasts if the business is to remain profitable.

In using orders and sales forecasts for the establishment of head count, the question often arises as to which forecast (orders or sales) should be used or whether an average of the two should be taken as the best means of estimating work load. This decision has to be based on the maturity of the business (new or established), the overall market trend (rising or falling), the relative distribution of effort

TABLE 1.1. *1968 AVERAGE SALES DOLLARS PER EMPLOYEE**

The Ten Highest	
Needham Packing	$272,610
Iowa Beef Packers	242,664
American Beef Packers	210,293
Coastal States Gas Producing	161,579
Land O'Lakes Creameries	158,453
Hess Oil & Chemical	155,827
Spencer Packing	144,450
Archer Daniels Midland	143,544
Farmers Union Central Exchange	122,643
Amerada Petroleum	121,721
The Five Lowest	
Otis Elevator	11,811
Blue Bell	11,739
General Instrument	11,568
Mattel	11,012
Fairchild Camera & Instrument	9,511
The Industry Medians	
Mining	$75,832
Petroleum refining	75,753
Tobacco	51,311
Food and beverage	45,804
Soaps, cosmetics	33,505
Chemicals	30,844
Metal manufacturing	29,859
Publishing and printing	29,514
Paper and wood products	29,057
Motor vehicles and parts	27,522
Pharmaceuticals	26,662
Shipbuilding and railroad equipment	25,145
Glass, cement, gypsum, concrete	25,051
Farm and industrial machinery	23,738
Aircraft and parts	22,991
Rubber	22,806
Metal products	20,782
Measuring, scientific, and photographic equipment	20,757
Appliances, electronics	20,354
Textiles	18,252
Office machinery (includes computers)	18,045
Apparel	15,228
All industry	26,235

* October 28, 1969. Reprinted from the *Fortune Directory*; © 1969, Time, Inc.

between engineering and manufacturing, and the overall length of the manufacturing cycle from receipt of order to delivery of product. Orders forecasts tend to be used by new businesses, by businesses having a high engineering content, by businesses having a short manufacturing cycle, or in a rapidly declining market. Sales forecasts are used by older businesses, by businesses having a high manufacturing content, by businesses having a long manufacturing cycle, or in a gradually declining market. The average of orders and sales dollars is a good first approximation which can be refined later.

A simpler but much more approximate method of determining the number of employees required to generate a given number of sales dollars uses statistical data on the average sales dollars per employee for a variety of companies and industries. Each year, *Fortune* (published monthly by Time, Inc.) supplies such data for the five hundred largest industrial companies in the United States. Table 1.1 shows data derived from the 1968 operating results of the companies in the survey. It should be noted that, although the range by individual company is broad ($272,610 to $9511 sales per employee), the range by industry is much narrower ($75,832 to $15,228). Fifty per cent of the industries are grouped within plus or minus 20 per cent of the median of $26,235 sales per employee. Thus, if the forecasted sales output of a company for future periods is available, the number of employees it requires can immediately be estimated.

This method of estimating assumes a relatively stable level of inventory and a constant relationship of material and purchased-services costs to total cost of sales. If sizable increases in inventories or contributed value are expected, the number of employees required for a given sales output will be higher; and conversely, a decrease in inventories or contributed value will reduce the number.

Although the methods presented here for determining the total number of employees required by a company to produce its goods and services are admittedly approximate, they do indicate the total number of employees which the business can afford to support. After detailed manpower requirements have been developed, these approximate indicators can be used as guides and checkpoints.

1.3.3.2 Span-of-Control Theory. Once the total number of employees required has been determined, the span-of-control theory can be used to estimate the number of organizational units and levels needed to properly delegate authority and responsibility. It is based on the assumption that a manager can effectively manage and supervise a certain number of either subordinate managers or, in the case of a first-line supervisor, individual workers. The number of subordinates will vary with work activity and the level of the manager in the

TABLE 1.2. *TOTAL EMPLOYEES SUPERVISED FOR VARIOUS MANAGERIAL SPANS OF CONTROL*

Level Number	Total Number of Levels 2	3	4	5	6
Managerial Span: 4	First-Line Span: 15				
1	1	1	1	1	1
2	15	4	4	4	4
3		60	16	16	16
4			240	64	64
5				960	254
6					3810
Total Empl.	16	65	261	1045	4149
Managerial Span: 5	First-Line Span: 15				
1	1	1	1	1	1
2	15	5	5	5	5
3		75	25	25	25
4			375	125	125
5				1875	625
6					9375
Total Empl.	16	81	406	2032	10,156
Managerial Span: 6	First-Line Span: 15				
1	1	1	1	1	1
2	15	6	6	6	6
3		90	36	36	36
4			540	216	216
5				3240	1296
6					19,440
Total Empl.	16	97	583	3499	19,995
Managerial Span: 7	First-Line Span: 15				
1	1	1	1	1	1
2	15	7	7	7	7
3		105	49	49	49
4			735	343	343
5				5145	2401
6					36,015
Total Empl.	16	113	792	5545	38,816

total organization. The more uniform and routine the work activity and the lower the organizational level, the greater the span of control can be. Thus, the number of people in a filing activity reporting to a first-line supervisor will be much greater than the number reporting to a supervisor in a research-and-development activity. Also, the number of first-line supervisors reporting to a manager can be normally much larger than the number of high-level managers reporting to a chief executive.

Even though the spans of control for different activities, companies, and industries vary, average values can be used in developing a preliminary organizational structure. One of the purposes of the span-of-control theory is to reduce the number of levels in an organization in order to improve communication channels between the chief executive and the individual worker. The greater the number of levels, the slower the reaction time of the organization and the greater the dilution of responsibility.

It is generally assumed that four to seven managers or supervisors can report effectively to a manager and fifteen individual workers can report to a first-line supervisor. Table 1.2 shows the total number of employees who may be supervised with managerial spans of four, five, six, and seven; with a first-line supervisory span of fifteen; and with varying numbers of organizational levels (two to six). The number of employees supervised can range from sixteen, with two organizational levels and a managerial span of four, to 271,715, with seven levels and a managerial span of seven. In general, a maximum span and minimum number of levels should be selected.

Although the spans shown are overall averages, actual spans vary from function to function. Thus, manufacturing generally supports higher overall spans and engineering lower ones. Typical managerial ratios are in Table 1.3. They are determined by dividing the total number of employees less managers and supervisors by the number of managers and supervisors.

Again, it is emphasized that the span-of-control theory should be used with caution and the ratios should be applied as measurements only after experience has been gained in the operation of the business. Indiscriminate use of the ratios in designing the organization can result in poor overall performance. Indirect costs, for example, can be reduced by increasing spans of control (two people—the displaced manager and his secretary—are normally removed when spans of control are increased), but the ability of managers to perform their work properly may be seriously jeopardized if the spans are too large.

TABLE 1.3. *TYPICAL MANAGERIAL RATIOS*

Functional Area	Managerial Ratios
Marketing	8: 1 to 10: 1
Research	8: 1 to 10: 1
Engineering	10: 1 to 12: 1
Manufacturing (total)	15: 1 to 20: 1
Quality control	12: 1 to 14: 1
Purchasing	10: 1 to 12: 1
Financial	10: 1 to 12: 1
Employee relations	10: 1 to 12: 1

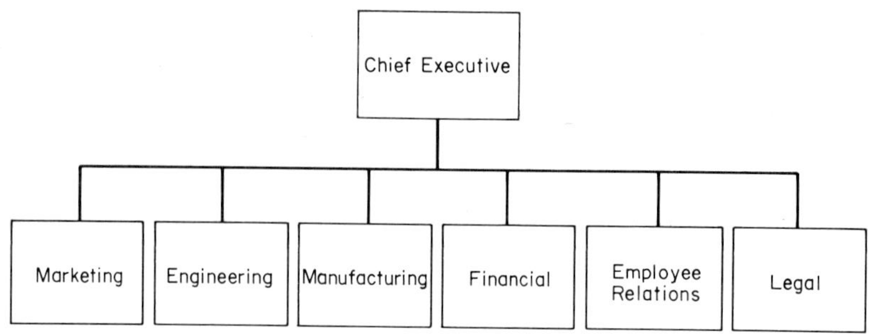

Fig. 1.13 *Basic functional organizational structure.*

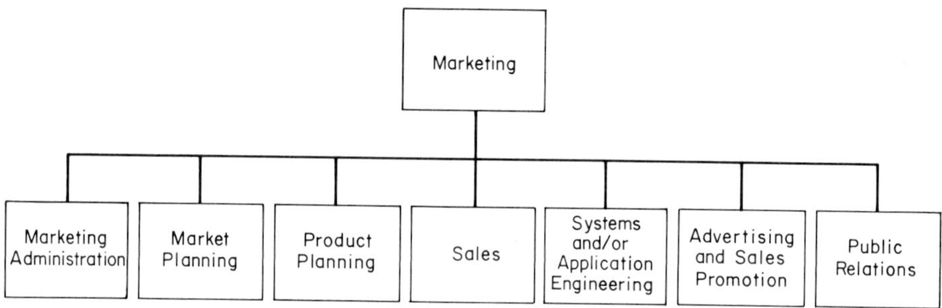

Fig. 1.14 *Basic marketing organization.*

1.3.4 Functional Organization—Single-Product-Line Businesses

Once the total manpower requirements and the approximate number of organizational units and levels for the business have been determined, it is possible to begin detailing the organizational structure. A function-oriented organization is usually the building block of this structure in most manufacturing businesses. It contains, as a minimum, organizational units devoted to the three major activities of the business—marketing, engineering, and manufacturing; and depending on the size of the business, may have, in addition, units representing the financial, employee and public relations, and legal functions. If the latter three activities are not represented by fully staffed organizations, individuals either inside or outside the enterprise must be assigned responsibility for them. Figure 1.13 is the basic organizational structure for a business having a single product line.

Within each functional organization, subfunctional units are established to perform the subactivities. Figures 1.14 through 1.18 are organizational charts for each of the major functions of a business having a single product line. The legal organization is not detailed since in most cases it is rather small and consultative in nature. Court cases are normally handled by outside legal firms on a fee or retainer basis.

There are obviously many organizational variations throughout industry because each organizational structure must be adapted to the markets, the products, or the processes of the business. In selecting subfunctional organizational units here, every effort has been made to make the selection as general as possible. Paragraph 1.3.4.1 discusses some of the major variations in functional organizations in different businesses and industries.

A. *Marketing* (Fig. 1.14). In order to fulfill the marketing charter of determining the markets and customers potentially available; planning and requesting the products necessary to meet market needs; selling and distributing the products; servicing the products after sale; and developing proper relations between the company and its shareholders, the public, and governmental agencies, the marketing function should be structured so that an assignment of responsibility for each of its activities can be made. If the enterprise is of sufficient size to require the full scope of marketing effort, organizational units can be established to be responsible for each activity.

The two main operating units of the marketing organization are sales and systems and application engineering (if not assigned to engineering); the planning units are market and product planning; and the support units are marketing administration, advertising and sales promotion, and public relations. Marketing administration also serves as the overall control unit.

1. *Marketing administration.* Responsible for marketing administrative matters (including overall marketing manpower and indirect-expense forecasts and budgets, internal systems and procedures, personnel development, wage and salary administration, office facilities, and organizational planning); coordination with those concerned of standardized contractual language, clauses, and terms and conditions; order service (including receipt of customers' orders, order edit, conversion of customers' orders to internal orders, and maintenance of records of order schedule status); and coordination with those concerned with company pricing policy.

2. *Market planning.* Responsible for market research, market surveys, economic planning, determination of the total available market for products of the enterprise and expected share of the total market, and plans for market development of products.

3. *Product planning.* Responsible for establishment of customer requirements and functional specifications for future company products (including physical parameters, reliability and quality requirements, selling price, and performance requirements).

4. *Sales.* Responsible for customer relations, preparation of proposals, field sales and service, spare-parts and logistics sales and service, installation and testing services, sales forecasts, and interfaces between customers and all company personnel.

5. *Systems and/or application engineering.* Responsible for direct support to the sales organization in establishing overall system specifications to meet customer systems requirements or in applying company products to existing customer systems or products. (May be assigned to engineering.)

6. *Advertising and sales promotion.* Responsible for presenting products to the public and the customers through the use of existing advertising media and sales promotional material (handbooks, descriptive literature on products, catalogs, etc.) in direct support of the sales organization.

7. *Public relations.* Responsible for direction of media relations and preparation, clearance, and distribution of news releases and related informational material to external media, including general press, trade and technical publications, and radio and television; coordination of public-information activities related to sales promotion; direction and control of activities necessary to assure good relations with community officials and residents; administration of in-plant charitable drives; and preparation of company annual report.

B. *Engineering* (Fig. 1.15). In order to fulfill the engineering charter of innovating, developing, and designing salable products and services at the required price and reliability, the engineering function should be structured so that an assignment of responsibility for each of its activities can be made. As in the case of marketing, if the enterprise is of sufficient size to re-

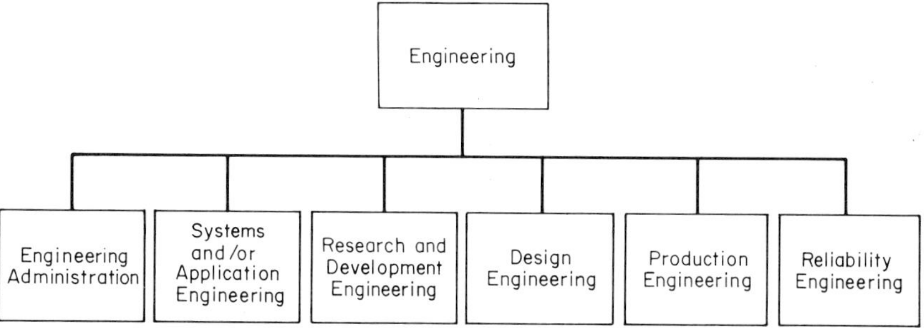

Fig. 1.15 *Basic engineering organization.*

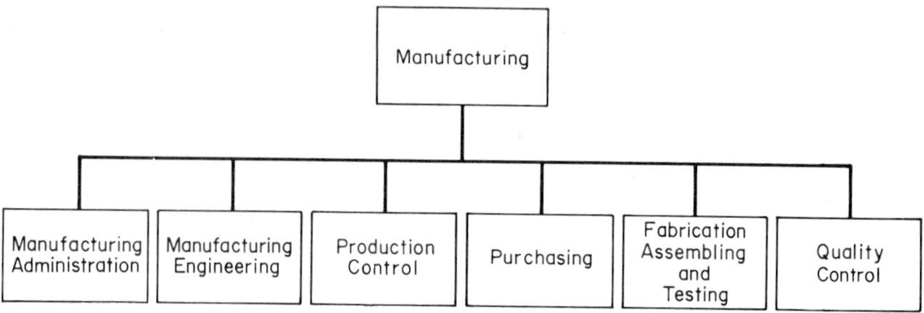

Fig. 1.16 *Basic manufacturing organization.*

quire the full scope of engineering effort, organizational units can be established to be responsible for each activity.

The main operating units of the engineering organization are systems and/or application engineering (if not assigned to marketing), design engineering, and production engineering; the planning unit is research-and-development engineering; and the support units are engineering administration and reliability engineering. Engineering administration also serves as the overall control unit.

1. *Engineering administration.* Responsible for engineering administrative matters (including overall engineering manpower and indirect-expense forecasts and budgets, internal systems and procedures, personnel development, wage and salary administration, office facilities, and organizational planning), engineering standards, engineering reproduction services, blueprint and configuration control, and engineering-drawing release and status.

2. *Systems and/or application engineering.* Responsible for establishment of overall systems specification to meet customer systems requirements or for application of company products to existing customer systems or products. (May be assigned to marketing.)

3. *Research-and-development engineering.* Responsible for research into basic scientific principles and concepts which may be applied to company products and the determination of the technical feasibility of such concepts and principles through laboratory production and development of breadboard samples.

4. *Design engineering.* Responsible for reduction to practice of developmental samples (including standardized process and testing specifications, material specifications, design methods and specifications, parts specifications, and standardized module and assembly designs and drawings) and design and drafting of specialized products outside the standardized line.

5. *Production engineering.* Responsible for production and issuance of drawings, parts lists, and instructions to the manufacturing organization for the fabrication and assembly of products ordered by the customer; liaison engineering with the manufacturing fabrication, assembly, and testing organizations; and value and producibility engineering and cost reduction.

6. *Reliability engineering.* Responsible for development of reliability standards; determination of product reliability through in-house tests, quality-control reports, and field experience; determination of component-part and subassembly reliabilities; and review of product design to assure maintenance of established reliability requirements.

C. *Manufacturing* (Fig. 1.16). To fulfill the manufacturing charter of procuring, fabricating, and processing the materials necessary to furnish salable products with the required quality and cost on the proper schedule, the manufacturing function should be structured so that an organizational assignment of responsibility for each activity can be made. As in marketing and engineering, if the enterprise is of sufficient size to require the full scope of manufacturing effort, organizational units can be established to be responsible for each activity.

The main operating units of the manufacturing organization are the fabrication, assembly, and testing shops; the planning unit (long-range plans for facilities, new methods, etc.) is manufacturing engineering; and the support units are manufacturing administration, manufacturing engineering, production control, purchasing, and quality control. Manufacturing administration and production control also act as control units.

1. *Manufacturing administration.* Responsible for manufacturing administrative matters (including overall manufacturing manpower and indirect-expense forecasts and budgets, internal systems and procedures, personnel development, wage and salary administration, office facilities, and organizational planning) and manufacturing cost analysis.

2. *Manufacturing engineering.* Responsible for industrial engineering (including operations analysis and methods improvements, work measurement studies, engineering economic analyses, plant layouts, material handling systems, value engineering, equipment development, and advanced manufacturing techniques and methods); company facilities (including facilities planning, acquisition, administration, maintenance, and disposition); production planning (including parts and assembly operations and routings, time standards, tool ordering, and make-or-buy decisions); and tool and test-equipment design and manufacture.

3. *Production control.* Responsible for parts and assembly scheduling; establishment of material requirements and ordering of such material from purchasing; issuance of shop orders specifying quantities, operations, routing, time standards, and schedules of parts and assemblies to be manufactured; maintenance of order and schedule status records; and inventory control.

4. *Purchasing.* Responsible for external purchases of all productive and nonproductive materials and services required by the business; execution and administration of all purchase orders for materials and subcontracts; and warehousing, traffic, transportation, and shipping.

5. *Fabrication, assembly, and testing.* Responsible for physical production of company products in accordance with requirements received from production control and with drawings and specifications developed by engineering, quality control, and manufacturing engineering.

6. *Quality control.* Responsible for establishment of procedures and methods governing the control of quality of workmanship, materials, processes, tooling, and testing of products as necessary to ensure compliance with both enterprise and customer requirements; development of improved quality-control methods, techniques, and standards; inspection of all parts, assemblies, and finished products to assure that finished products conform to engineering instructions, drawings, and specifications, and that the quality of workmanship and materials meets established standards; approval of sources, products, and processes of qualfied suppliers and assurance of conformance with specifications of purchased materials, parts, assemblies, and services; calibration and certification of all mechanical and electrical inspection and test equipment used for deliverable products; maintenance and analysis of proper data in order to define problem areas and initiate corrective action; and review of discrepant material in accordance with established material-review policy.

D. *Financial* (Fig. 1.17). To fulfill the financial charter of providing cost information and analysis, raising and accounting for the money required to carry on the operation of the business, and accounting for all the assets of the business, the financial function should be structured so that an assignment of responsibility for each of its activitics can be made. If the enterprise is of sufficient size to require the full scope of financial effort, organizational units can be established to be responsible for each activity.

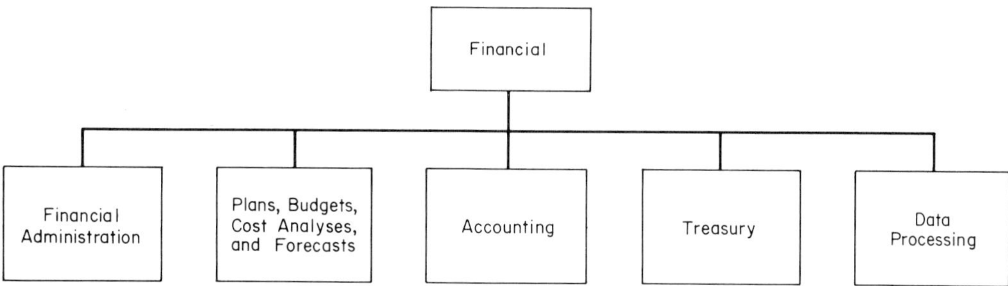

Fig. 1.17 *Basic financial organization.*

The main operating units of the financial organization are general accounting, payroll accounting, treasury and data-processing; the planning unit is the budget, cost-analysis, and forecast organization; and the support units are financial administration and cost accounting. Financial administration also serves as the internal control unit for the financial organization, and the budget, cost-analysis, and forecast organization serves as the financial control unit for the enterprise as a whole.

1. *Financial administration.* Responsible for financial administrative matters (including overall financial manpower and indirect-expense forecasts and budgets, internal systems and procedures, personnel development, wage and salary administration, office facilities, and organizational planning); property administration for the overall enterprise; documentation of overall financial systems and procedures required to establish uniform practices in financial matters throughout the entire enterprise (payroll routines, chart of accounts, travel-expense statements, cash disbursements, etc.); company insurance coverage and financial administration of the company retirement, group, and health-insurance plans; and audits to determine conformance with financial procedures.

2. *Plans, budgets, cost analyses, and forecasts.* Responsible for company and functional plans, budgets and forecasts for all cost elements; statistical reports, including reports of budgets vs. actuals and resulting variances; cost analyses to explain variances from budgets; and preparation and review prior to release of all financial data and reports for use external to the company.

3. *Accounting.* Responsible for custody and control of all accounts and records which may be maintained by the company, and activities of general, cost, payroll, and tax accounting.

4. *Treasury.* Responsible for management and custody of all funds and securities of the company; determination and administration of all company financing; transfer of money to and from appropriate banks as required by the needs of the company; management of all company cash disbursements and receipts; and approval of credit aspects of all transactions with customers and suppliers.

5. *Data-processing.* Responsible for operation of all company data-processing equipment; collection and preparation of input data for data-processing (including establish-

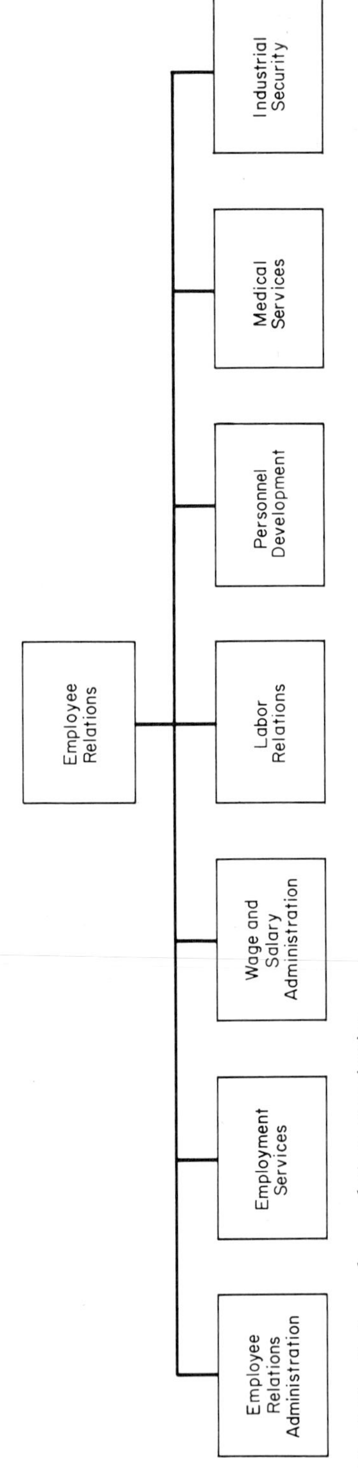

Fig. 1.18 *Basic employee-relations organization.*

ment of format, procedures, and due dates); procurement and utilization control of data-processing equipment; programing for business data-processing; and data-processing services in support of all functional activities.

E. *Employee relations* (Fig. 1.18). To fulfill the employee-relations charter of planning and developing the human resources of the business and developing proper relations between the company and its employees, the employee-relations function should be structured so that an assignment of responsibility for each of its activities can be made. Again, as in the case of other functions, if the enterprise is of sufficient size to require the full scope of employee-relations effort, organizational units can be established to be responsible for each activity. The whole employee-relations organization can be considered a support group for the entire business. No one unit of the organization is devoted to planning or control in the sense that units of other functional activities are. All efforts of the group are devoted to improving relationships between the company and its employees in order to better meet overall company objectives.

1. *Employee-relations administration.* Responsible for employee-relations administrative matters (including overall employee-relations manpower and indirect-expense forecasts and budgets, internal systems and procedures, personnel development, wage and salary administration, office facilities, and organizational planning); documentation of overall employee-relations systems and procedures required to establish uniform practices in employee-relations matters throughout the entire enterprise (hiring routines, employee transfers, keeping of employee records, vacation practices, union grievance procedures, etc.); interpretation of employee policies in all areas; development and administration of employee recreation and welfare activities; and direction of company unemployment-insurance activities.

2. *Employment services.* Responsible for all employment programs to recruit, interview, test, and select company employees; employee hiring, transfer, and termination; maintenance of all employee records covering experience, background, and employment history both within and without the company; and processing all paper work relating to employee rate changes, transfers, leaves of absence, terminations, changes of status, and vacation and sick leaves.

3. *Wage and salary administration.* Responsible for establishment and maintenance of company wage and salary program for all classifications of employees (including job descriptions, evaluation plans, and determination of rates of pay); and establishment and maintenance of external and internal wage and salary relationships through appropriate surveys and participation in the surveys of other organizations.

4. *Labor relations.* Responsible for coordination of company activities in all areas relating to labor and union relations; negotiation of all labor agreements for the company; development of the necessary policies and procedures to ensure company compliance with all applicable labor laws; and investigation, negotiation, and resolution of all grievances with union representatives.

5. *Personnel development.* Responsible for direction of all company activities in areas of personnel development (including establishment and direction of company education and training programs and salaried-employee performance-evaluation and rating systems); maintenance of employee-qualification records; and maintenance of file of employees qualified for promotion to higher-level supervisory and managerial positions.

6. *Medical and safety services.* Responsible for maintenance of medical facilities and programs to assure that the health and the physical conditions of new employees meet company standards; emergency health and accident treatment; maintenance of employee medical records; and direction of the company industrial-hygiene and -safety programs.

7. *Industrial security.* Responsible for plantwide police activities (including direction of on-site traffic, prevention of theft, identification of employees and visitors, and inspection of employee-carried incoming and outgoing property); plantwide fire department activities; investigation of backgrounds of applicants and employees; special investigations of alleged irregularities such as major theft or fraud, profiting, influence, and restrictive practices; and continuous surveillance over company and government classified material and property.

1.3.4.1 Variations in the Basic Functional Organization. Paragraph 1.3.4 grouped the various activities of a single-product-line business into a basic functional organization. The marketing, engineering, manufacturing, financial, and employee-relations functions were selected as the major organizational units, and

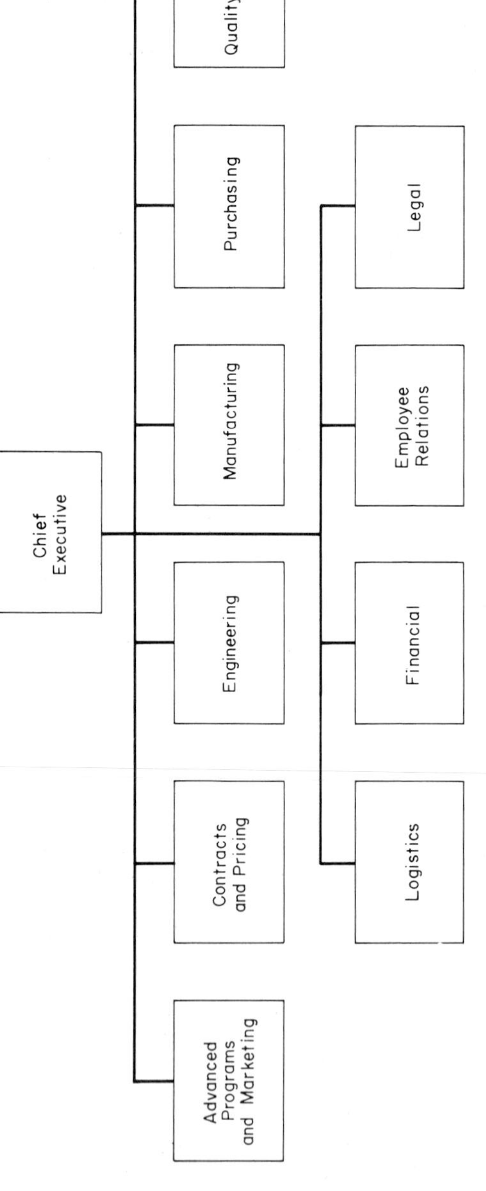

Fig. 1.19 *Typical aerospace and defense-industry functional organization.*

activities within these functions were grouped into smaller subunits. In actual practice, the grouping of functional and subfunctional organizational units may differ from company to company and industry to industry. This difference is due both to the size of the business and to the market in which the business operates. The purely functional organization is most common in single-product-line businesses in consumer- and durable-goods industries where the number of employees is less than 1000 to 2000 and annual sales volumes are less than $20,000,000 to $40,000,000.

The major variation in the functional organization described here occurs in the aerospace and defense industry. In this industry, the five basic, first-level units discussed above are normally subdivided into additional functional units. Figure 1.19 shows a typical aerospace and defense-industry functional organization. The main differences between this structure and that discussed previously lie in the marketing and manufacturing organizations. The activities of marketing and manufacturing are normally assigned to six units instead of two. The marketing function is divided into advanced programs and marketing, contracts and pricing, and logistics, and manufacturing is split into manufacturing (production operations), purchasing, and quality control.

The original marketing and manufacturing activities are reassigned as follows:

Advanced programs and marketing. Responsible for marketing administration (except for development of standard contractual language, pricing policy, and order service), market planning, product planning, sales (except field service, spare parts, installation, and testing), advertising and sales promotion, public relations, and systems engineering for advanced programs.

Contracts and pricing. Responsible for development of standardized contractual language and pricing policy, contract and proposal negotiations, contract administration, order service (including receipt of customer orders, order edit, conversion of customer orders to internal orders, and maintenance of records of contract cost and schedule status), and establishment of contract and proposal prices.

Logistics. Responsible for field service (including development of maintenance plans, operation of company repair facilities in the field, and field installation of company products); publication of technical material regarding the operation, maintenance, and repair of company products; provision of specialized training to customer and company personnel in the operation and maintenance of company products; determination of spare parts, technical data, and tools required to support the company products in the field; initiation of their procurement and scheduling of their shipment; and administration of in-house modification and repair programs for products returned by the customer.

Manufacturing. Responsible for manufacturing administration; manufacturing engineering; production control; and fabrication, assembling, and testing of company products.

Purchasing (or material or procurement). Responsible for external purchases of all productive and nonproductive materials and services; execution and administration of all purchase orders for material and subcontracts; and warehousing, traffic, transportation, and shipping.

Quality control. Responsible for control of quality of company products; inspection of company products to check adherence to established company and customer specifications; approval of qualified sources; calibration and certification of mechanical and electrical inspection and test equipment; maintenance of proper quality-control records and data; and review of discrepant material with customer representatives.

The main reasons for this further subdivision of functions lie in the nature of the aerospace and defense business. First, the size of most aerospace and defense companies makes increased spans of control necessary in order to keep the number of organizational levels at a workable maximum. Second, the magnitude and importance of certain subfunctions are greater in the defense industry. For example, the size and complexity of the logistics, field-installation, field-service, and spare-parts activities are much greater in the defense business than in the consumer- or durable-goods business and warrant the establishment of a separate organizational unit to handle them. Similarly, the problems of developing contract work statements and of pricing and administering contracts make a separate organizational unit desirable. Again, purchasing, in the nondefense industries, is concerned mainly with ordering raw materials and relatively standard parts and assemblies. However, in the defense industries, purchasing is greatly involved in subcontracts requiring large research-and-development and production-hardware expenditures. And, finally, in the area of quality control, governmental specifications and other requirements make it essential that a separate organization be established to assure the quality of company products.

Some of the variations described here also apply in the consumer- and durable-goods industries. In consumer industries, for example, where both good field service and an adequate supply of spare parts are necessary to maintain good customer relations, separate organizations provide this post-sales service. In any case, before any organizational changes are made, the activities to be performed by the new unit should be reviewed in order to determine whether the additional organizational interfaces created by the change are warranted and whether the number of people needed to support the activity is sufficient to require a separate organization.

1.3.5 Functional-Product Organization—Multiple-Product-Line Businesses

As the enterprise grows in size, the number of people required to support the activities of the business also grows, although normally at a somewhat slower rate than the increase in sales volume. At some point, the purely functional organization becomes unwieldy and some other organizational structure is needed. This condition becomes especially obvious as different products are added to the company product line. Variations in customers, markets, technologies, and manufacturing processes begin to cause complications within each functional activity. Also, the ability to measure profit performance, an extremely important motivational tool, becomes difficult in a functionally organized, multiproduct business. The allocation of functional costs to different product lines in order to arrive at product costs becomes expensive and requires an inordinate amount of clerical effort. A solution to the problem, in the case of a multi-product-line business whose products are somewhat similar in nature, is the establishment of subproduct organizations under the chief executive with each subproduct-line manager responsible for the sales, engineering, and manufacture of the product assigned to him. Those functions or activities which cut across all product lines can be handled by functional managers also reporting to the chief executive. However, each product manager must

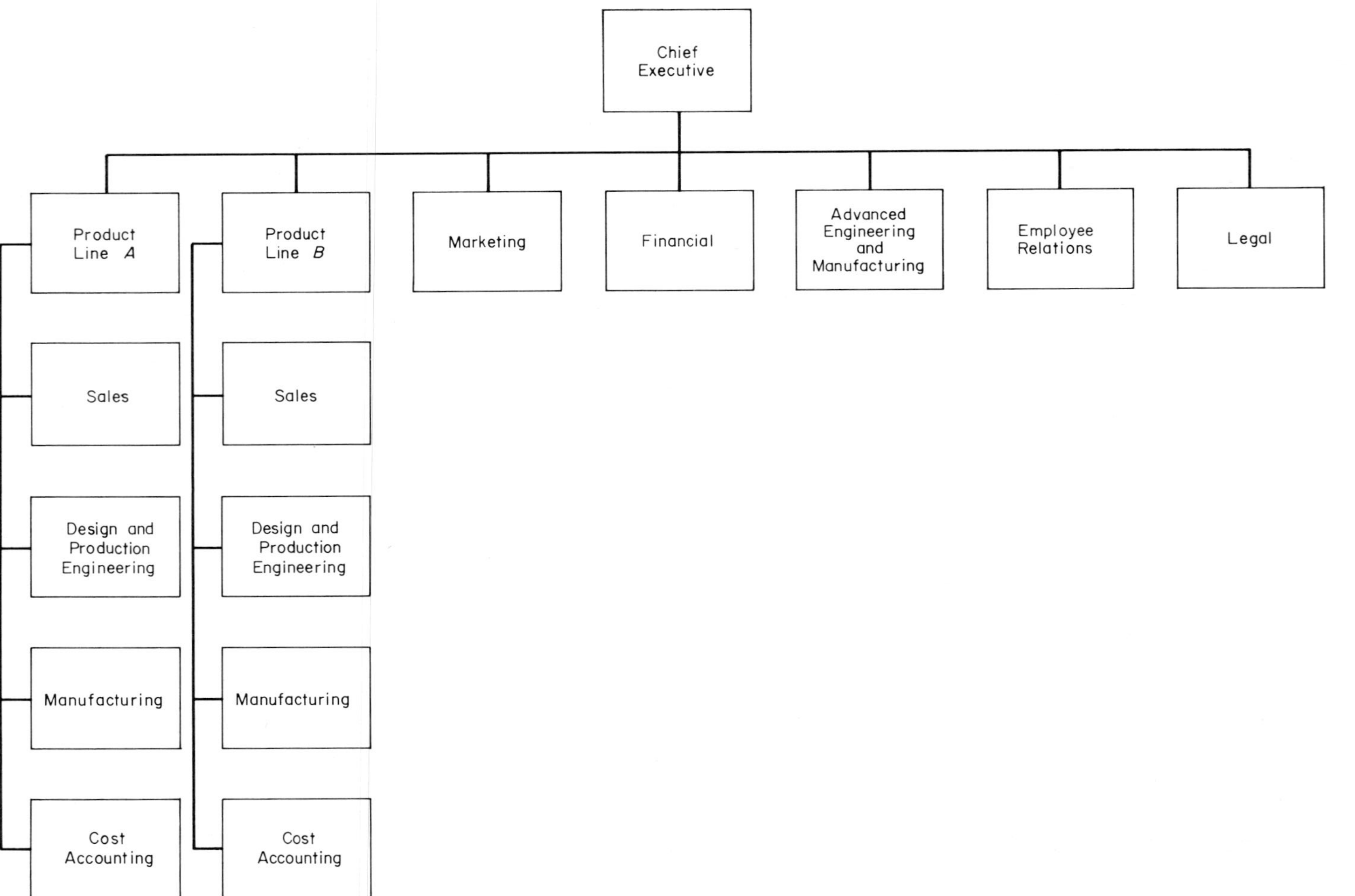

Fig. 1.20 *Typical functional-product organization for a medium-size business.*

have sufficient authority and resources to carry out his full profit-and-loss responsibility.

Figure 1.20 shows a typical functional-product organizational structure for a medium-size business. Here each product line has its own sales, design- and production-engineering, manufacturing, and cost-accounting units. The central marketing function still has responsibility for marketing administration, market and product planning, field sales (product-line sales-personnel work through field sales personnel in customer contacts in order to present a unified front to the customer), advertising and sales promotion, and public relations. The central financial function has all the activities described in Par. 1.3.4, but its cost-accounting personnel are decentralized to the product lines. Advanced engineering and manufacturing are on a centralized basis and work closely with the market and product-planning activities to develop new products and processes to assure the future growth of the business. Finally, employee relations and legal continue as separate centralized activities since they support all functions of the business.

In many cases there will be certain activities (maintenance, for example) which are required throughout the entire company. If no central function can handle the work, the activity can be assigned to one of the product-line functions with companywide responsibility for its execution.

The organizational structure shown here is applicable to businesses having from 2000 to 5000 employees, having annual sales ranging from \$40,000,000 to \$100,000,000, and having all facilities geographically centralized. For enterprises having a greater number of employees and greater annual sales and requiring geographical decentralization, each subproduct-line organization must take complete responsibility for all the basic functions. To establish overall corporate policy, it is then normal to set up at the level of the chief executive a series of functional services organizations to develop long-range plans, establish corporate policy, and advise the chief executive in the areas of their functional competence. They also keep the operating product-line functional managers informed of advanced developments in their respective areas.

Figure 1.21 shows the structure of a typical functional-product organization for a large-size business. The distinction between the operating product lines and the services organizations should remain clear cut, and the latter groups should act only as advisers and consultants to the chief executive and the operating divisions. As the number of product lines extends beyond six, group executives can be in charge of each of three or more product lines in order to keep the span of control of the chief executive down to a workable level.

1.3.6 Functional-Product-Program Organization

In many cases, a specific in-house customer program which cuts across all functional and product lines must be handled on a unified basis, with a single individual responsible for its completion. This method of operation is especially prevalent in the aerospace and defense industry, where contracts are placed on a program basis. The Air Force, for example, establishes a Systems Program Office for each major weapons-system program, with technical and fiscal surveillance over all contracts placed in support of the program. Most businesses, however, must be organized to handle more than a single contract or program. Since it would be uneconomical to organize a complete functional organization for each contract, some form of organizational interface must be established between the Air Force Systems Program Office and the normal functional organization of the contractor. This interface should provide a single point of contact between the customer and contractor organizations in all matters relating to the given program. It should allow the contractor to modify a relatively small interface organization in order to reflect possible customer organizational changes without changing the entire organizational structure of the business. Also, as programs move from the conceptual phase to the operational phase, organizational requirements change, and emphasis must be placed in different areas. In the early phases of a program, systems definition and systems engineering are important. As the product begins to be defined through specifications and drawings, configuration management must be established. In the production phase, procurement and production activities are of primary interest. And finally, in the operational phase, test and development are the primary activities. Across the whole life of the program, cost and schedule control of all the interrelated activities at any point in time must be exercised and the plans for future activities developed. A

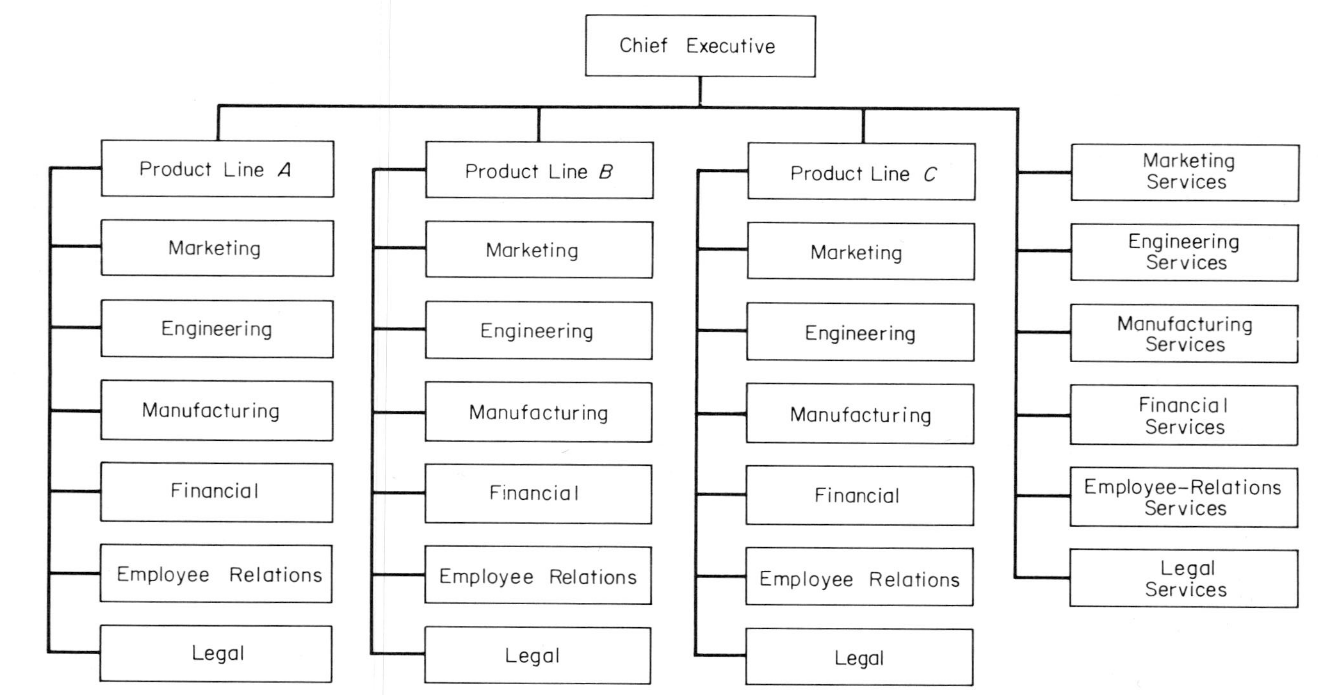

Fig. 1.21 *Typical functional-product organization for a large business.*

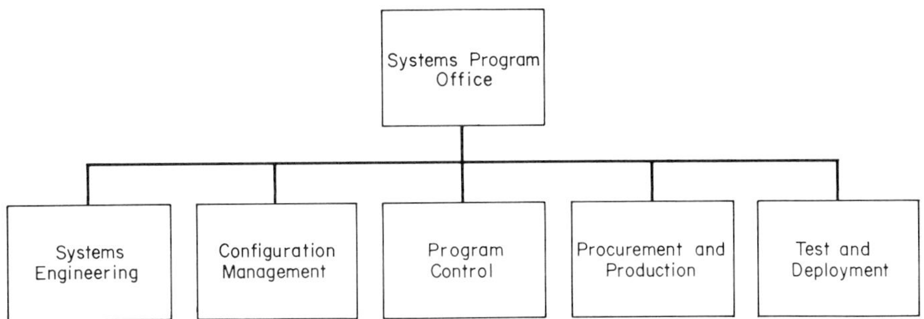

Fig. 1.22 *Typical Air Force systems program office organization.*

typical Air Force Systems Program Office is in Fig. 1.22. Figure 1.23 shows a typical medium-size aerospace-business functional-program organization, corresponding to that in Fig. 1.19 but including a program office reporting to the chief executive and program-administration units in the engineering, manufacturing, purchasing, quality-control, and logistics organizations. The program office is the lead organization in customer contacts and draws from the functional organizations for support. It is relatively small and acts as a management organization rather than as an operating or doing organization. In each of the functional organizations, a program-administration office, again small, is a contact point for the program office and the various organizational units within the function for program-oriented work. The program office acts as the internal contractor for the programs and subcontracts the program work to the functional organizations, in somewhat the same manner as a purchasing organization deals with vendors and suppliers. The program office should only define the work to be done, authorize the work, assign a cost and schedule for performance of the work, and monitor progress to assure successful completion. Great care must be exercised to prevent duplication of effort between the program office and functional organizations. Otherwise, costs become excessive, and responsibilities and accountabilities for the performance of work become difficult to assign and measure.

As indicated earlier, a small program management organization of this type gives the business organizational flexibility to meet changing customer or program requirements without disturbing the major functional units. With the number of new contracts, terminations, and changes in direction characteristic of this industry, any means which reduces the impact of such external influences on the major portion of the company organizational structure greatly improves overall operating efficiency.

1.3.7 Line-Staff Relationships

Much has been written about the relationships between the line organization and staff personnel. The original concept of the line-staff organization evolved from the military, where the line was the combat element and the planning and supporting functions were delegated to the staff. A famous general was reputed to have selected his line officers from those having the highest degrees of intelligence and initiative; his staff officers from those having high intelligence but less initiative; and his combat troops from those possessing the lowest intelligence and initiative. (He also pointed out that his major problems arose with those people having high initiative but low intelligence.)

In industry the line organizations have generally been those directly associated with securing the customer's order and converting it to a finished product, and the staff group has had responsibility for policy development and formulation, long-range planning, and coordination. This division of responsibility between line and staff is difficult to define. For example, the operating manager must be closely involved with the policy and planning decisions which affect his organization if he is to be held accountable for the successful performance of his work. The problem is not critical in a single-product-line organization (Fig. 1.13), where the six functional managers and the chief executive can act as a group to develop the required poli-

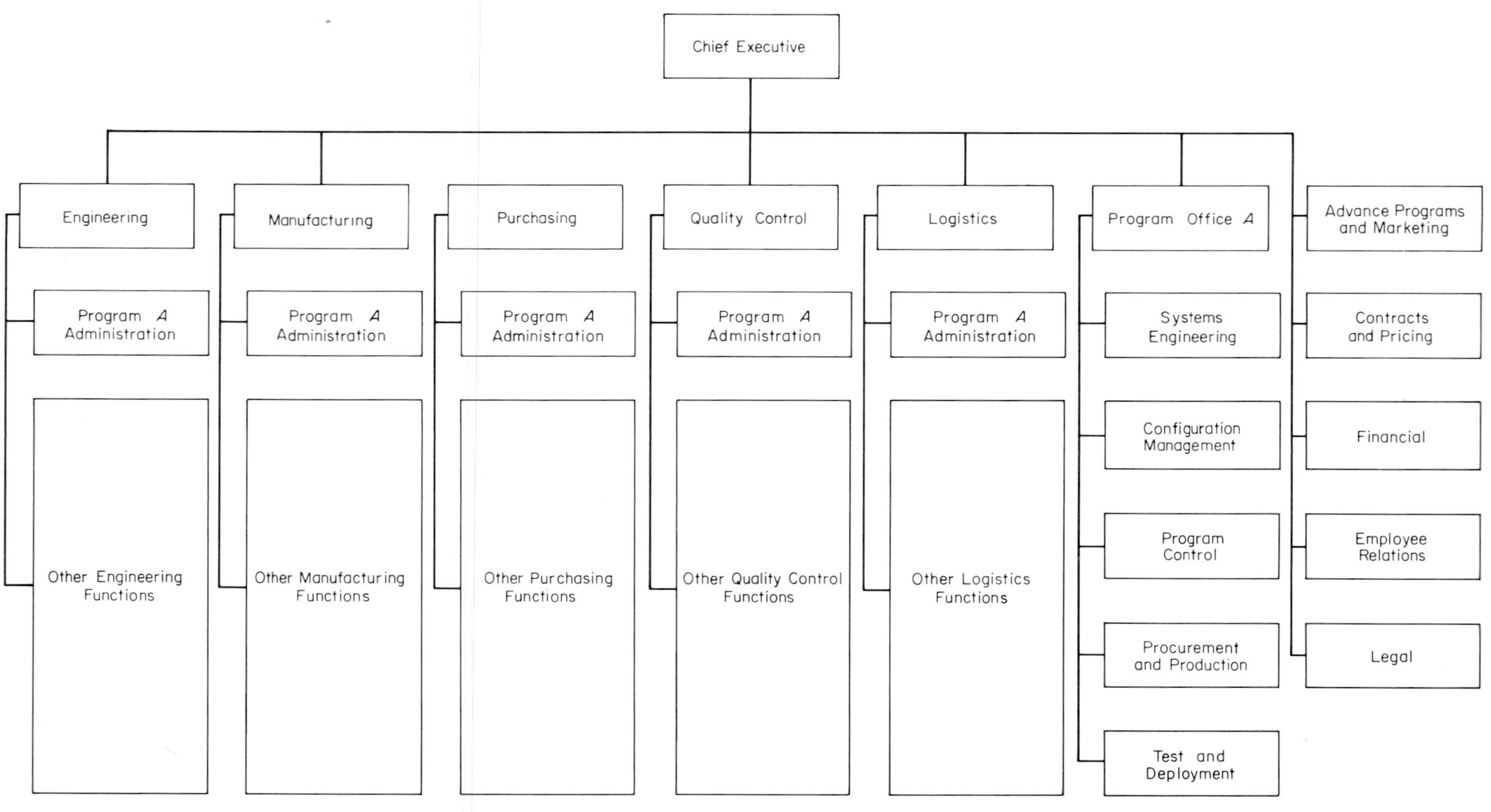

Fig. 1.23 *Typical medium-size aerospace business functional-program organization.*

cies and plans. It begins to appear as the organization grows larger and includes several product lines and ultimately several product groups.

One of the easiest ways to solve the problem, of course, is to eliminate all staff positions and work through the line organizations. Using this concept, the chief executive assigns specific planning or support responsibilities to each of the line operating managers reporting to him. If the organization is purely function-oriented, manpower planning is assigned to employee relations as the lead function with support from the other functional elements of the business in carrying out the assignment; responsibility for the development and operation of a management control system is assigned to the financial function since such a system is heavily cost-oriented; and development of company policies and procedures is assigned to the function responsible for the major activity of the business—manufacturing in a production-oriented company and engineering in a research-and-development organization. In any case, the effort can be assigned to a line organization without the establishment of additional policy-making or coordination groups. There is always the danger with this concept, however, that the planning function, so important to the future of any business, will be put aside because of the day-to-day operating problems of the function. Proper direction, motivation, and follow-up by the chief executive and the furnishing of adequate resources for the planning work to be done can overcome this problem.

As the organization grows and begins to add divergent product lines, the problem of developing corporatewide policies and plans solely through the product-line managers becomes more difficult. At this point, consideration should be given to the establishment of services organizations reporting to the chief executive. The word *services* has been chosen by many companies to describe the functions of those organizations in the enterprise and to easily distinguish their activities from those of the line organizations. Although this selection may appear to be a semantic trick, it nevertheless clearly defines the purpose of these organizations.

The services functions can be established on a one-to-one basis with the major functions already selected—marketing, engineering, manufacturing, financial, employee relations, and legal. Such an organization is in Fig. 1.21. As a group the services can act as consultants to the chief executive on corporatewide matters, and as individual organizations they can act as consultants and advisers to their functional counterparts in the product-line organizations. In this latter capacity, the service organization has responsibility for maintaining company leadership in the function for which it is responsible. This leadership is gained by doing functional research at a higher level than would be feasible for a function in a product line to carry out, by aiding in the recruiting and training of technical and professional personnel in all functional fields, and by planning and formulating suggestions and recommendations for company policies and plans affecting the various functional specialties.

Through the use of line managers to perform companywide tasks in smaller businesses and through the use of services organizations in larger companies, the line-staff relationship problem can be greatly simplified. Where work is assigned to other than line operating managers, great care must be exercised in defining the work, responsibility, and authority of the delegated individual or organization. In no case should the nonline organization have any authority over line managers, and the manager of the nonline organization should be responsible only to the individual assigning the work.

1.3.8 Centralization vs. Decentralization

Another favorite subject of numerous articles, books, and discussions is whether to centralize all the major functions of a multi-product-line business or to decentralize the business into a number of individual product lines, each having its own group of functional organizations. Figure 1.24 shows a centralized organization for a multi-product-line business, and Fig. 1.25 shows a decentralized organization. Figure 1.20, which was discussed in Par. 1.3.5, depicted a modified organization with decentralization of most of the business to product lines but with assignment of certain companywide activities to centralized groups.

Before decentralizing a centralized organization or centralizing a decentralized one, several points should be considered. The most important, of course, is whether the resulting organization can improve overall performance in meeting the goals and subgoals of the business. As discussed in Par. 1.3.2.1, the generation of profit dollars is normally the prime goal of a business. Hence, the question of centralization or decentralization is usually answered on the

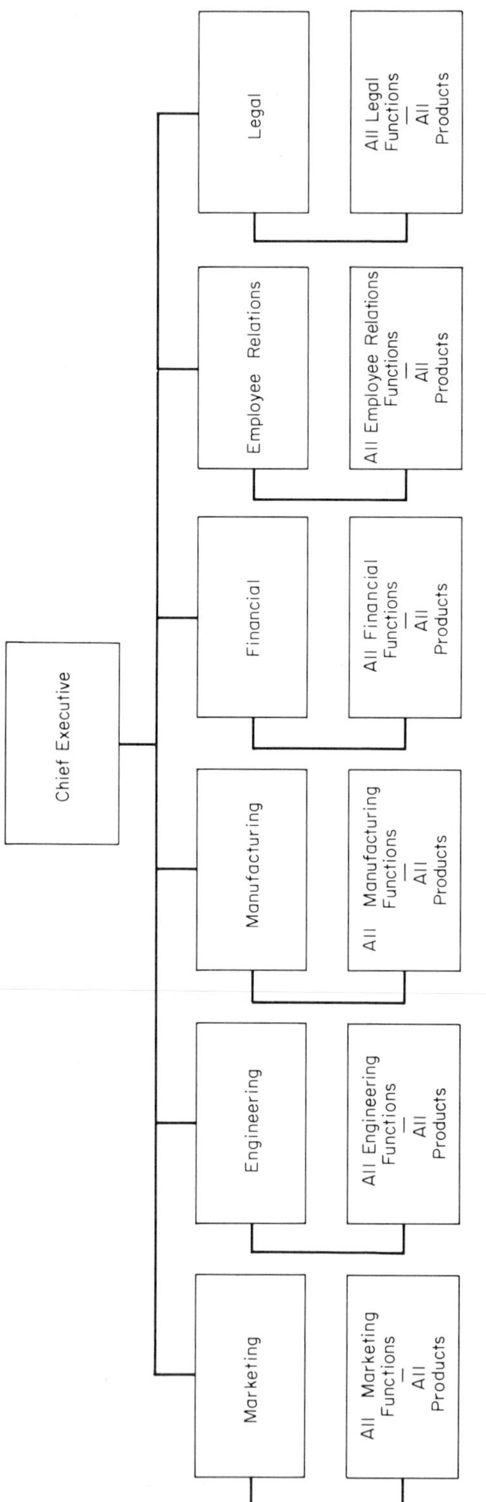

Fig. 1.24 *Centralized organization for a multi-product-line business.*

basis of which will result in the greatest improvement in company profits. From a purely economic standpoint, a centralized organization normally operates with fewer people and at lower cost. There is less duplication of effort, more uniformity, lower occupancy costs, better utilization of equipment, lower material costs because of greater volume of purchases, etc. On the other hand, the advantages of decentralization are normally noneconomic factors such as fewer organizational levels, improved communications, smaller organizational groupings, better employee (both managerial and individual) motivation, greater opportunities for manager development, more flexibility in decision-making, faster response time, improved community relations, and better dispersal of employment opportunities when geographical decentralization takes place.

In the final analysis, the organizational decision must be based on the tangible, economic improvements which can be gained by centralization compared with the intangible, noneconomic improvements gained by decentralization. Although it is difficult to establish fixed rules, every effort should be made to decentralize unless the purely economic costs of operating a centralized organization are at least 25 per cent less than those of operating a decentralized one.

Immediately after World War II, an upsurge in the trend toward decentralization occurred as rapid increases in the size and output of industrial businesses occurred. In many cases, the results of decentralization were disastrous, and a definite movement back to centralization began. An analysis of the failures can be used to develop the requirements for successful decentralization. Successful decentralization requires written and understood, uniform company objectives and policies, well-established and understood means of measuring the performance of decentralized organizational components against established standards, periodic reporting and review of such performance, and availability of a pool of managerial talent pretrained in the methodology of decentralized operations. The lack of attention to the first and last requirements—written, uniform company policies and availability of managerial talent—has caused many of the failures charged to decentralization. If these requirements are not met, anarchy results, the components of the company begin to head in a variety of directions, and the pendulum swings back to centralization with strong, centralized authority. A

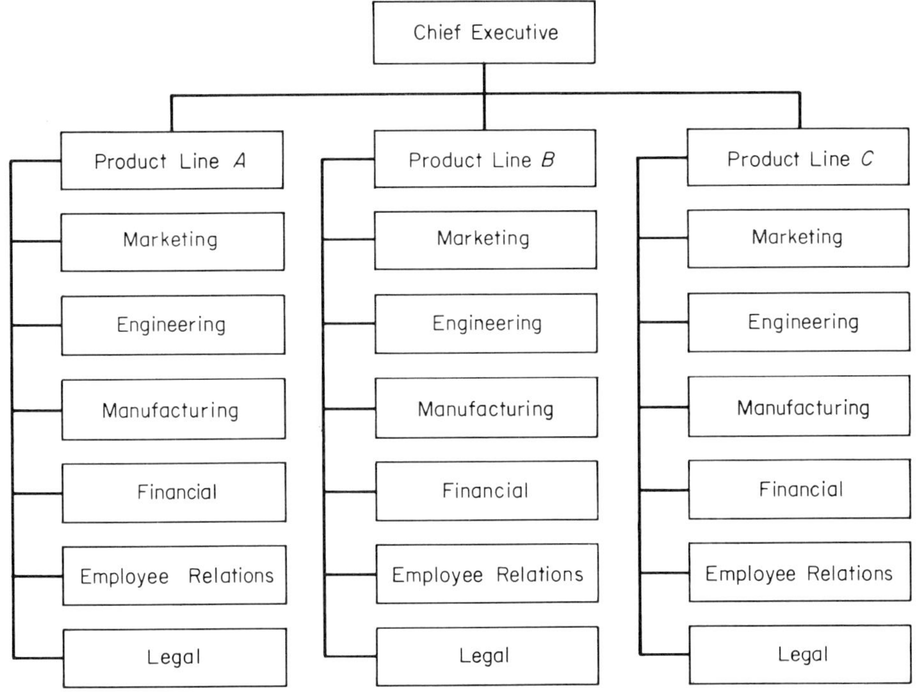

Fig. 1.25 *Decentralized organization for a multi-product-line business.*

review of the industrial histories of the largest corporations in the nation—General Motors, Ford, Du Pont, General Electric, Sears Roebuck, Montgomery Ward, etc.—give the reader excellent examples of the successes and failures of both centralized and decentralized organizational philosophies.

1.3.9 Organizational Nomenclature

One extremely important element of organizational design is the establishment of a uniform means of titling organizational components and corresponding managerial and supervisory positions. Unless organizational nomenclature is fully developed prior to establishing a new structure or to changing an old one, confusion and nonuniformity result and the chances of wholehearted acceptance of the new structure by the employees are greatly reduced. Poorly planned organizational nomenclature complicates performance measurement, wage and salary administration, indoctrination of new employees, generation of uniform policies and procedures, assignment of parking privileges and office space and facilities, and the wide variety of activities that are organization-oriented.

There are as many types of organizational nomenclatures as there are companies—and perhaps a few more. Table 1.4 illustrates the organizational and typical managerial titles of a large electrical-equipment manufacturer, a large aerospace corporation, and the U.S. Air Force. Uniformity in nomenclature is especially important in companies having more than one product-line organization and having geographical decentralization. Good preplanning in this area will prevent future changes in organizational and managerial titling. Such changes always have bad side effects and introduce a feeling of instability in company operation.

1.4 OPERATIONS SUBSYSTEM DESIGN

1.4.1 General

The industrial-business system has been divided into five subsystems—planning, operations, control, financial, and information.

TABLE 1.4. *ORGANIZATIONAL AND MANAGERIAL NOMENCLATURE*

Electrical Equipment		Aerospace		U.S. Air Force	
Organizational	Managerial	Organizational	Managerial	Organizational	Managerial
Corporation	President	Corporation	President	Headquarters	—
Group	Group executive	Group	Group executive	Command	—
Division	Division general manager	Operating division	President	Division	—
Department	Department general manager	Product division	Vice-president and general manager	Wing or office	—
Section	Manager–manufacturing	Department	Factory manager	Group or division	—
Subsection	Superintendent	Branch	General superintendent	Squadron or branch	—
Unit	General foreman	Section	Superintendent	Flight or section	—
Subunit	Foreman	Group	General supervisor		
		Unit	Supervisor		
		Subunit	Assistant supervisor		

Activities have been assigned to each subsystem based on commonality of purpose, time relationships, and the type of data (fixed, variable, or derived) required in order to generate the subsystem outputs. Also, each activity has been defined.

Although the industrial engineer is interested in all aspects of the business, he deals primarily with the physical conversion of the customer's order to a finished product and with the facilities required to accomplish this conversion. This paragraph, therefore, will discuss the design of the operations subsystem in detail since this subsystem encompasses the conversion process. The other subsystems are of equal importance in the overall functioning of a business but are essentially supporting to the operations subsystem. They should also be designed in accordance with the methodology developed here.

Paragraph 1.2.6.2 listed the activities of the operations subsystem and described the generalized flow diagram (Fig. 1.4) from receipt of the customer's order to shipment of the finished product. The discussion here will expand on the design of both the engineering and the manufacturing planning, scheduling, and ordering systems and of the documents used to transmit information from one function to another during the information-flow process prior to hardware fabrication. The manufacturing planning, scheduling, and ordering function has been divided into two subfunctions—production control, which schedules, plans, and issues tool, shop, and assembly orders and material requisitions to purchasing for outside material and parts; and purchasing, which orders and procures the raw material and parts from vendors and suppliers.

The approach here is to consider the operations subsystem as a continuous flow of information and hardware. The output from one function or organization is the input to the following one, and thus the information contained in the output documents from one group must be compatible with the input requirements of the receiving group. If the system is not designed with this thought in mind, duplication of effort, recopying of information, delays while awaiting proper information, system inefficiency, and increased cost of operation result. Because of this compatibility requirement, the operations subsystem should be designed by a cross-functional team of operating experts—both the generators of the information and the users of it. Too often functional (engineering, manufacturing, purchasing, etc.) systems are designed with no thought of the user. The result is a series of isolated systems, tied together by only a thin thread. Those using the total system then neither understand nor accept it, and the duplication of effort and inefficiency mentioned above are automatically built into the system.

In the descriptions of the engineering, production-control, and purchasing systems following, a functional description of the system will first be given; the input-output requirements (including the input information required and the output documents and formats used) will be established; and finally the internal processing of the input information in order to generate the output documents will be discussed (including a description of the steps taken in the conversion process, the files generated and used, and the manual and mechanized systems employed).

1.4.2 Engineering

1.4.2.1 Functional Description. In the operations subsystem, engineering is the process by which the requirements of the customer are converted into drawings; parts lists; and material, process, and test specifications for use by the manufacturing organization in the fabrication, assembling, and testing of the finished product. The customer's requirements may be defined by his own specifications or by internally generated product or systems specifications. The engineering process may include the development, fabrication, and testing of samples of parts and assemblies as well as the final-product definition by means of production drawings and specifications. The engineering organization is also responsible for the development of new products, whose specifications are established by the market and product-planning activities of the marketing organization.

1.4.2.2 Input-Output Requirements. The input to engineering is an internal order describing the product to be designed and manufactured or the work to be done, the engineering output required (drawings, parts lists, systems or part specifications, prototypes, etc.), and the date on which the engineering tasks are to be completed. In industries where engineering is job-costed, the internal order should also specify the budgeted manpower (in hours, man-days, man-weeks, etc.) or the budgeted dollar expenditures for engineering. The product or work definition can be part of the internal document, or reference can be made to a customer or internal systems or product specification.

The definition of the product or of the engineering work to be done is extremely important. Engineering converts customer requirements into detailed drawings and specifications so that the manufacturing organization can fabricate, assemble, and test the product. An improper or loose input definition results in inefficiencies in the whole operating system because of misunderstandings, changes, delays, added costs, etc. In many cases, firm product or work definition is not possible. However, a short delay in issuing the internal order in order to obtain better definition may shorten the overall engineering-manufacturing cycle substantially. If the order-issuing organization cannot properly define the work, it should be its responsibility to obtain the information and to issue to engineering only that portion of the work that can be defined. If assumptions are to be made concerning work definition, they should be made by the work requester, not by the work receiver. No time is gained by issuing an internal order to engineering stating, "Design one missile."

In establishing requirements, the systems or product specifications should be written from a functional standpoint. They should only define what the product is supposed to be, not tell how to design it. If the product is a device to fasten two pieces of steel together, only the shear and bending loads which will be transmitted by the pieces being fastened, the thickness of the plates being fastened, the standard or nonstandard tools required for its insertion or removal, and the environmental conditions which it must withstand should be specified. In no case should the material or diameter of the fastener be specified (unless clearance requirements are involved). Specifications other than those necessary to define the function or appearance of the product and the environment in which it will operate hamper the engineering and manufacturing groups in furnishing the best product at the lowest cost.

The output documents of engineering are the drawings, parts lists, and specifications required by manufacturing to purchase material and fabricate a product which will meet customer requirements. The engineering "paper" should be sufficiently complete so that reference to other external specifications is normally not necessary. In many cases industry or government standard specifications can be referenced if well understood by the manufacturing organization. In general, however, engineering should act as the interpreter of such specifications and record its interpretation in in-house documentation. The responsibility of quality control is then to assure that the product meets the requirements of this documentation.

Since engineering starts in-house effort in the design and fabrication of the product, the de-

TABLE 1.5. *INFORMATION SUBSYSTEM FILE ORGANIZATION*

A. Part Number
1. Nomenclature
2. Item number
3. Material specifications
4. Next higher assembly
5. Quantity per next higher assembly
6. Make department or vendor
7. Operation number
8. Operation name
9. Tool type
10. Tool number
11. Normal flow time
12. Minimum flow time
13. Setup time
14. Run time
15. Direct labor cost
16. Material

sign information it supplies should be compatible with the needs of the users—costing, scheduling, planning, ordering, purchasing, quality control, fabrication, testing, and shipping. The engineering parts list is not merely a reference document for engineering but also an action document for the rest of the business. Too often this fact has been overlooked with the result that the using functions recopy, add to, and reidentify the parts list to adapt it to their needs.

The reason for emphasizing the importance of the parts-list format and information at this point is to show that engineering can and should play a key role in the establishment of the basic information subsystem used throughout the business. If a manual system is employed, the parts-list format can help using functions. In mechanized systems, engineering can start the input of fixed data to the information subsystem. As discussed earlier, most activities in a business file a majority of their data by engineering drawing number. This fact is true for such files as engineering parts lists and parts-usage, planning, labor-standards, and cost information. Figure 1.11 depicted the information matrix for this series of files. Table 1.5 regroups the data elements of Fig. 1.11 and collects the data under the engineering drawing number. Each element can be added by the function generating it and all the information can be contained in a single file. Engineering can add data elements 1 through 5; planning, 6 through 14; and cost, 15 and 16. (If more than one department makes the part, more than one operation is performed, or more than one tool used, such information can be added in elements 6 through 14.)

To return to the drawings and parts lists issued by engineering as part of their formal output documentation, Fig. 1.26 is a sample engineering-drawing format and Fig. 1.27 is a sample model-breakdown list (sometimes called a model list). The drawing format includes provision for a list of material or for a parts list. However, every effort should be made to use a separate parts list of the type in Fig. 1.10, which has the same number as the drawing but has the prefix *PL* assigned to it.

The advantages of a separate parts list are many. First, it has been found that approximately 80 per cent of all engineering changes apply only to the parts list and approximately 20 per cent are made on the delineated drawing. Thus, when changing or revising drawings, time and cost savings can be obtained by using the smaller parts lists. Second, physical handling by the users is also made simpler since the majority of those requiring parts-list information do not need drawing and parts-list information at the same time. Third, when parts-list generation is mechanized, the smaller size of the separate parts list is adaptable to mechanized printout. And finally, using activities can add their own information to the separate parts list and thus eliminate much of the recopying necessary when the list is part of the drawing. For example, the last column on the right of Fig. 1.10 gives the using activities the opportunity to add information without recopying the entire list. The parts list can be furnished to the using activity as a transparency; after adding the necessary information, the user can reproduce the list and use it for his own purposes. Thus, the scheduling unit can explode the parts-list quantities, which are specified for a single, next higher assembly, into the total quantities required for the manufacturing order being produced and can indicate start and finish dates for each subpart or assembly. The planning unit can indicate departmental routings and labor standards for each drawing number; and the cost unit can apply labor and material costs.

The model-breakdown list (Fig. 1.27) is the top-assembly list for the finished product; it contains the first-level assemblies and parts which go into making the final assembly. In addition, reference assembly drawings and specifications can be called out on this list. With mechanized systems, it is fairly easy to

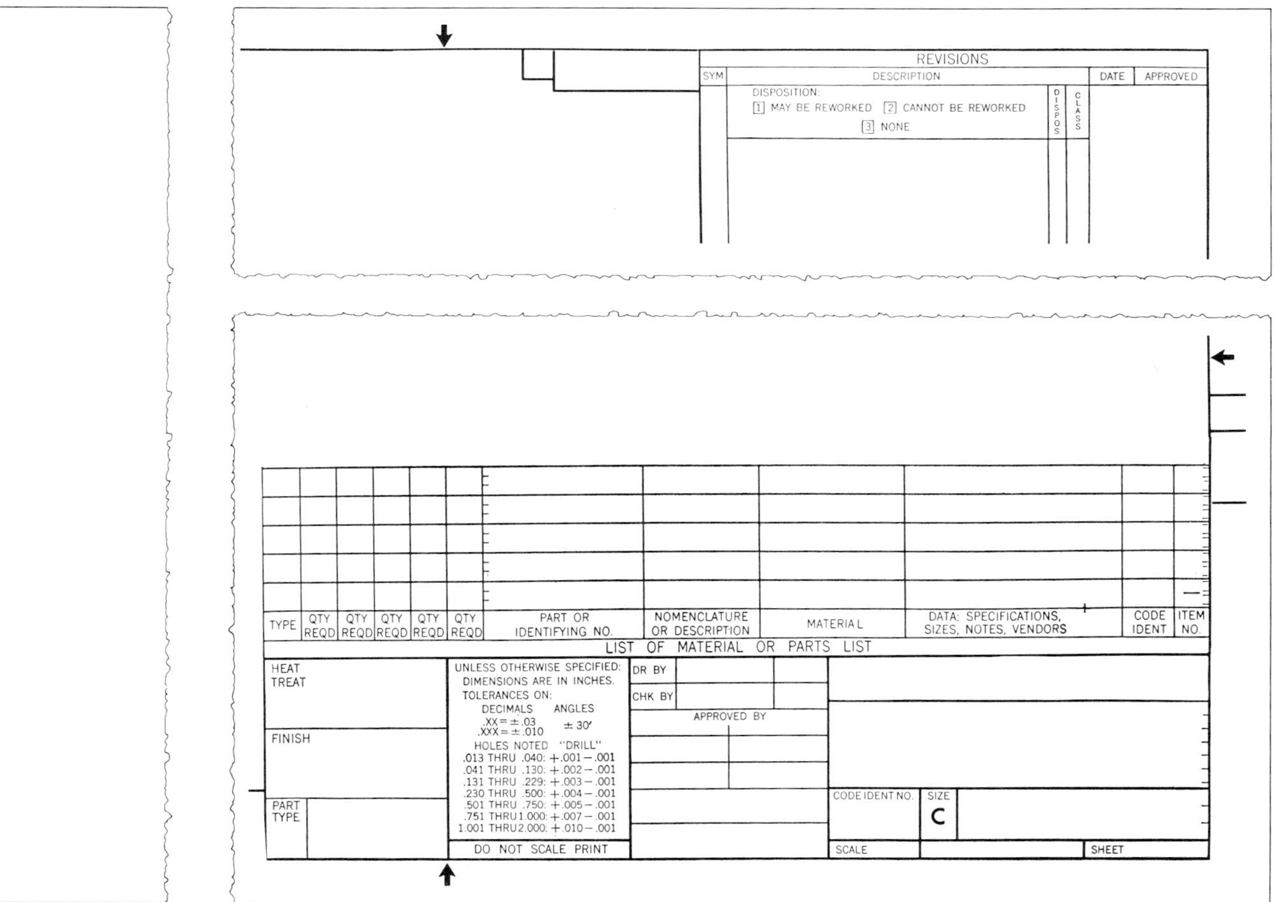

Fig. 1.26 *Sample engineering drawing format.*

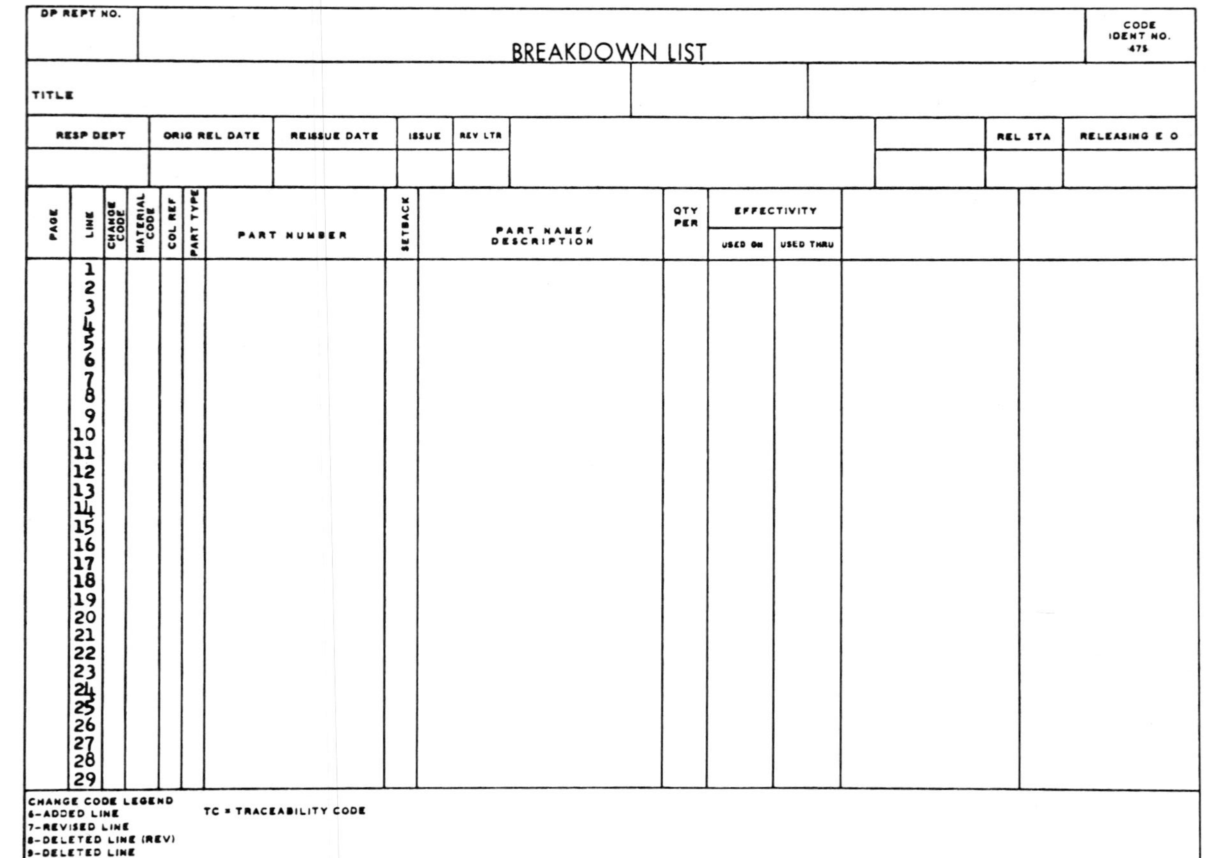

DP REPT NO.

BREAKDOWN LIST

CODE IDENT NO. 475

TITLE

RESP DEPT	ORIG REL DATE	REISSUE DATE	ISSUE	REV LTR			REL STA	RELEASING E O

PAGE	LINE	CHANGE CODE	MATERIAL CODE	COL REF	PART TYPE	PART NUMBER	SETBACK	PART NAME/ DESCRIPTION	QTY PER	EFFECTIVITY USED ON	EFFECTIVITY USED THRU		
	1												
	2												
	3												
	4												
	5												
	6												
	7												
	8												
	9												
	10												
	11												
	12												
	13												
	14												
	15												
	16												
	17												
	18												
	19												
	20												
	21												
	22												
	23												
	24												
	25												
	26												
	27												
	28												
	29												

CHANGE CODE LEGEND
6-ADDED LINE
7-REVISED LINE
8-DELETED LINE (REV)
9-DELETED LINE

TC = TRACEABILITY CODE

Fig. 1.27 *Sample model-breakdown list.*

generate a complete list of all drawings required for a given top assembly by starting with the model-breakdown list and then successively exploding each lower-level drawing until the last detail drawing on the "tree" has been called out. Such a complete list of drawings is called a *drawing-breakdown list*.

1.4.2.3 Internal Processing of Information. Within the engineering organization, input information (internal order and accompanying specifications) is processed in different ways depending on the complexity of the design task to be performed. These tasks range all the way from the slight modification of an existing design to the complete development of a new product. In the first case, the internal order is logged in, scheduled, and sent to the production-engineering group, which issues instructions to the design or detail drafting groups for the preparation and issuance of the necessary drawings and parts lists for the modified product. In the second case (complete development), the order is handled similarly, but it proceeds through the development, design, production-engineering, and drafting units prior to issuance. In both cases, after all the drawings and parts lists are prepared, they are released to manufacturing and other interested functional organizations through an engineering release activity.

The engineering documents involved in the operations subsystem include (1) production drawings and their subordinate documents (specification and procurement control documents), (2) model-breakdown lists, and (3) engineering orders or instructions. Production drawings provide all the engineering information necessary for the procurement, fabrication, assembly, and inspection of the product. Model-breakdown lists define the usage of all items relative to a finished product. Engineering orders or instructions are the official tools for releasing all new and changed production documents and for notifying all functional organizations of the release. The engineering release system reproduces the production documents and distributes copies to affected organizations (purchasing, manufacturing, quality control, etc.). The basic purpose of the release system is to transfer clear, usable information from the originator to the user as quickly as possible and at the same time to keep the necessary clerical records to ensure adequate control over the documentation.

Within the engineering organization, variations exist among companies in the steps taken in the preparation of the drawings and of the parts lists. In some cases, the drafting unit is an organization separate from the engineering unit and works from instructions issued by the engineers. The drafting unit prepares the drawings and all parts and model lists in final form and turns them over to the release unit. In other cases, the release unit makes up parts lists from rough drafts prepared by the drafting unit, and in addition, the release unit prepares final model lists from subordinate parts lists. This latter method relieves the drafting unit of the clerical tasks involved and is becoming more popular as parts lists are removed from drawings and are computer generated. When computers are used, the drafting unit can transmit information directly to the data-processing unit on special transmittal forms. As more sophisticated computer input devices are developed, a typewriter-keyboard device can directly enter information into the computer from a remote location and intermediate keypunching can be eliminated. The addition of cathode-ray tubes to the input device allows the operator to see what is being typed and to make corrections before entering the data into the computer file. It also allows the operator to recall a given parts list, display it on the tube, and make revisions to it directly, without using additional transmittal documents.

The main data files used in engineering include drawing and parts-list files; parts-usage files; engineering-order, or instruction, files; change- and release-record files; and standard-parts files. Of all these, the parts-usage file can be of most value in day-to-day operations. This file, sometimes also called the where-used file, shows all the next higher assemblies and in some cases all top assemblies on which a given part is used. The effect of changing a given part can be determined from the file for all the assemblies that the part has been used on and the decision of whether to make the change at all can be more intelligently made by the use of this information. The file is also useful in engineering standardization work since, if the part number includes a coding system to identify the family of parts (bolts, screws, electronic components, etc.) to which the specific part belongs, all similar parts can be reviewed at the same time. This parts-usage file is difficult to maintain on a manual basis if the number of parts is large. However, the establishment of such a file should be kept in mind during the design phases of an engineering information system since future mechanization may be practical.

Until the advent of the high-speed digital computer in the early 1950s, very little use was made of computers in engineering. The first applications of computers were in the area of scientific calculations, and until the late 1950s this was the primary use of all computers. In the engineering area, the use of computers for scientific calculations is still extremely high compared to use in the engineering information system.

The use of computers in establishing and maintaining parts-usage information is growing. Perhaps their most important use will be as the master file for the business information subsystem. Engineering will play an increasingly important role as the initiator of many of the data elements of this file and will make use of the file as well in its own operations.

1.4.3 Production Planning and Control*

1.4.3.1 Functional Description. Production planning and control has as many definitions as there are types and sizes of business. The variety stems from size and organizational structure rather than from basic differences in functions performed. It is most useful to describe production planning and control functionally. Then, a common understanding and approach can be achieved regardless of the kind of business, its organizational chart, or its size since the functions must still be performed. The activities, information, and documents described here apply primarily to large businesses of the job-shop type, in which products have high dollar value and are manufactured to customer specifications in relatively small quantities. The flow of information can be streamlined and the documents simplified as the product becomes less complex and is produced in larger quantities.

For our purposes, production planning and control includes those manufacturing planning, scheduling, and ordering activities which convert engineering and contractual paper into working manufacturing documents; and it also includes issuing the material, fabrication, assembly, and test orders necessary to ship to the customer a quality product on schedule and at a profit to the enterprise.

The conversion of a customer's order into a finished product requires the interaction of many functions within the organization. The design of the product, communicated to manufacturing in the form of engineering drawings and specifications, must be converted into a language more readily usable in the manufacturing process. Planning is the term most commonly used for the determination of the sequence and method of manufacturing the product. Once the design specifications and the sequence of manufacture have been developed, schedules for each level of production to support the end product ordered by the customer must be developed. Scheduling is the term most commonly used for this function. When the engineering design, the method of manufacture (planning), and the schedules have been completed, production orders, tooling orders, and material requests must be written and issued for all detail parts, assemblies, and tooling necessary to manufacture the end product. This function is known as *ordering*.

The following are functional descriptions of the planning, scheduling, and ordering activities which can be applied to most industries.

A. *Planning*. Production planning is the function of planning and describing the manufacturing sequences, processes, and tooling for the complete fabrication and assembly of company products. As engineering drawings and changes are released, production planning interprets these documents and determines the most efficient and economical method of production. Flow times for each part are established and are subsequently used to determine in-work and completion dates for details and assemblies in a setback pattern to support the end-item delivery schedule.

B. *Scheduling*. The scheduling function develops master and support schedules, coordinates design change-point effectivity, analyzes schedule progress, and coordinates the action required to correct problems occurring during the manufacturing process. Scheduling directs and coordinates manufacturing schedule revisions, such as those resulting from changes in engineering drawings, purchased parts, or material production rates.

Early in the order-product conversion cycle, the scheduling function develops a master pro-

---

* This paragraph was written by Kenneth J. Pascoe, manager of production engineering, Electro Sensor Systems Division of Autonetics, a division of North American Rockwell Corporation.

duction plan, which reflects agreements with engineering and purchasing and establishes schedules for engineering drawings, tooling, test equipment, fabrication, and receipt of purchased parts and material. In systems-type government contracts where make-or-buy decisions of major components are contractually specified, scheduling coordinates subcontractors' schedules and through purchasing maintains contact with subcontractors in order to prevent and alleviate problems which may affect schedules. These efforts keep management informed of progress and possible major problem areas so that corrective action can be taken quickly. Scheduling is a major manufacturing responsibility and requires personnel of the highest caliber with a broad background and experience in manufacturing operations and with a knowledge of the interrelated functions of the overall organization.

C. *Ordering:* When planning is completed and schedule information is available, the ordering function establishes raw-material and purchased-parts requirements to support inhouse fabrication, and issues production orders, tooling orders, and material requisitions on a schedule which ensures an optimum rate of production, with parts and materials available to each manufacturing department to support the delivery schedule.

Constant surveillance of the manufacturing process is maintained to guard against line stops due to lost parts, scrap, misrouted parts, etc. Rework and replacement orders are written on an expedited basis to maintain a smooth flow of the manufacturing process.

1.4.3.2 Input-Output Requirements. The flow of information through manufacturing organizations is similar, but the documents and terminology differ from company to company. The titles and formats of the documents described and depicted here are representative of those used in many industries. Other nomenclatures for the same documents are included in the text headings. This paragraph describes briefly the documents used and generated. Paragraph 1.4.3.3 discusses the preparation of the documents themselves in more detail.

A. *Internal order* (*general order*, *requisition*, *contract notice*, *sales order*). With the receipt of a contract or order, an internal authorizing document is issued. The internal order, used by properly designated management, authorizes effort to be performed and describes the products and services to be delivered to the customer. The internal order specifies the basis for budget constraints, if any, as established by contract provisions and by management. It provides a base against which costs can be controlled and accumulated. It interprets the contract or order and specifies the work to be performed, i.e., what work is to be accomplished, by whom, and when. Other documents and directives are incorporated into the internal order by reference to ensure clarity and understanding of work to be performed and to define contractual compliance requirements.

B. *Release setup* (*aerospace industry*). With the release of the internal order, the scheduling function establishes the manufacturing order release structure, subdividing total order quantities into economical lot sizes. The release setup becomes part of the internal order. Release numbers can be used to subdivide manufacturing costs for accounting purposes. Manufacturing also can use release numbers to establish change points and to identify lot quantities of parts.

C. *Master schedule*. Engineering, prior to the release of engineering drawings, develops a drawing-breakdown list, which displays all major assemblies and supporting parts in a setback sequence to support the end item. With this document, scheduling establishes a master schedule and acceleration curve. Customer delivery requirements are used as the basis for a master schedule, the purpose of which is to portray the relationship of concurrent and sequential activities or both, and to arrive at an overall manufacturing time spread to assure meeting the schedule commitment. The master schedule is developed after determining the flow times for the engineering, material, subcontract items, fabrication, assembly, and functional and systems tests. This information is then displayed in a form which makes it easy to see the time relationship for each level of effort building up to the end item. The master schedule establishes the schedule requirements for each major assembly to support the end-item delivery date.

D. *Acceleration curve*. In industries (such as aerospace and defense) where the length of the manufacturing cycle is affected by learning factors or extensive design changes, an acceleration curve can be used. It displays the compression of manufacturing time spreads as a function of the quantity of end items produced. It normally includes an end-item-by-end-item projection of assembly and test times in relation to schedule dates.

E. *Fabrication schedule*. Once the time relationships for the various activities have been established, schedules are developed for the major assemblies required to support the committed schedules. By using the drawing-breakdown list, the master schedule, the acceleration curve, and release setup, schedules for each major assembly are developed for the total order and displayed on a document called the fabrication schedule.

The fabrication schedule is the basic schedule control for all assemblies and is used by purchasing to establish reference points for material procurement and by production planning or control to schedule release of production orders to the shop. To more clearly understand the fabrication schedule, visualize a bicycle as the end item. The major assemblies are the frame, wheels, handle bars, and gears. To produce 200 bicycles a week, 400 wheels per week must be built. If the final assembly time is one week and the wheel assembly two weeks, the wheels should be scheduled with an in-work date three weeks prior to the completion date of the bicycle. This method of scheduling would be the most optimistic and the ideal one to keep inventories at a minimum, but in real situations there are scrap parts, lost parts, lost paper work, schedule slippages, etc. To compensate for these problems, most companies oversupport the final-assembly department by a certain number of weeks based on the product and business. This allows for any problems which may arise and assures delivery of the product on schedule.

F. *Planning worksheet*. The planning sequence of operations to manufacture a part cannot start until engineering drawings are released. As engineering completes the design and drawings, it releases them to manufacturing by means of an engineering order, or instruction. The engineering order is the authorizing document to release or change engineering information. It is also the authorizing document to stop manufacturing effort.

With the possession of engineering information, the planning function is ready to prepare planning worksheets, which describe the manufacturing operations, machines, flow times, tools, and materials required to fabricate each part or assemble each part or both. This description results from studying the engineering drawing of the part, determining the type of part or assembly or both, and, with the knowledge of manufacturing capabilities, developing the most efficient and economical method of manufacture. This can be compared to assembling a child's toy for which a picture (engineering drawing) and instructions (planning worksheet) are supplied. The worksheet is the set of instructions to fabricate or assemble a given part or both.

G. *Request for tool order*. During the development of planning, a determination of tooling requirements is made and a tool request issued to the ordering section. Tooling consists of mechanical devices, other than machines and equipment, which contribute to the fabrication or assembly of parts in the most feasible manner. A predetermined tooling policy, based on consideration of quantities in relation to cost and available time, establishes guidelines.

H. *Production, tooling, material orders*. After the master schedules, fabrication schedules, engineering drawings, and planning worksheets have been completed, the production orders, tooling orders, purchase requests, and material requisitions must be written to authorize the procurement of material and the manufacture of the parts and assemblies required to support a specific order. The ordering section accomplishes this function. Utilizing the fabrication schedule for each major assembly and the planning worksheet for each part, the ordering section establishes the schedule dates for tooling, material receipt, and fabrication.

1.4.3.3 Internal Processing of Information. Scheduling is the first step in processing information within the production-control function.

A. *Release setup*. Upon receipt of the internal order (Fig. 1.28) and using master-schedule data, the scheduling function establishes the release structure for the order. There is no firm ground rule for establishing a release setup since many variables must be considered. The customer requirement for delivery of a set quantity of items per month, which may be written into the order, overrides all other considerations. The scheduler, armed with the knowledge of the product and the type of order (i.e., research and development, production, or spares), determines the following details:

1. Similarity with other products produced
2. Number of assemblies, details, machine parts, etc.
3. Material requirements
4. Shop capacity

At this point, the scheduler, using judgment,

(5) ☐ MASTER ☐ SUBSIDIARY ☐ SINGLE

GENERAL ORDER

			(10) G.O. DATE / /	(1) G.O. NUMBER
CONTRACT DATE / /	(8) ASSIGNED DIVISION	(9) LEAD DIVISION	SUPPLEMENT (3) (4) P.C. (6)	(7) C.C.
(11) CONTRACT DPC / /	(12) CONTRACT NUMBER	(13) CUSTOMER	(14) SUPPLEMENT DATE / /	(15) M.G.O. NO.
DMS RATING	PRIME CONTRACT NUMBER	FINAL USER	SECURITY ACCESS REQ'D	DEAL NUMBER
(16) SUBJECT		(17) PROGRAM		CONTRACT CODE

CONTRACT TYPE (18)	☐ (F) FIRM FIXED PRICE ☐ (I) FIXED PRICE INCENTIVE ☐ (R) REDETERMINABLE TYPE	☐ (C) COST PLUS FIXED FEE ☐ (D) COST PLUS INCENTIVE FEE ☐ (T) TIME & MATERIAL	☐ (A) COST PLUS AWARD FEE ☐ (O) OTHER (SPECIFY) ☐ TYPE OF IDWA
RELEASE SYSTEM	☐ PRODUCTION	☐ ENGINEERING ☐ LOGISTICS	☐ OTHER (SPECIFY)
ACCOUNTING INFORMATION	☐ 2000 SERIES ACCOUNTS ☐ 3000 SERIES ACCOUNTS	☐ OTHER: ☐ TOTAL COST CENTER	☐ NO ☐ YES } RELEASES AND SUBACCOUNTS REQUIRED

IF THIS IS A SUBSIDIARY GENERAL ORDER, BLANKS BELOW NEED NOT BE COMPLETED			(19) TARGET COST $
ISSUANCE AUTH	☐ DEFINITIVE CONTRACT ☐ LETTER CONTRACT	☐ LETTER OF INTENT ☐ NOTICE OF AWARD	☐ NR AUTHORIZED
FINANCIAL	☐ PROGRESS PAYMENTS ☐ STANDARD NEG OVERHEAD RATES OR → ☐ IDWA PROFIT NOT APPLICABLE OR → ☐ RENEGOTIATIATION ACT—1951 OR →	☐ FOR OTHER BILLING INSTRUCTIONS (SEE T's & C's) ☐ SPECIAL NEG OVERHEAD RATES OR → ☐ INCURRED COST/TOTAL COST OR → ☐ VINSON—TRAMMEL ACT OR →	 ☐ NOT A COST-TYPE CONTRACT ☐ OTHER—SEE TERMS AND CONDITIONS ☐ NOT SUBJECT TO RENEGOTIATION
PAYMENT OFFICE			
ADMINISTRATIVE OFFICE			
FUNDING LIMITATIONS	CUSTOMER $	(NR $ EXPIRING:	) TOTAL $
CONFIG CONT	ANA BULLETIN ☐ 390a ☐ 391a ☐ 455	☐ OTHER DATED:	AS REVISED:
QUALITY ASSURANCE	☐ SOURCE INSPECTION	☒ SEE APPLICABLE SPECIFICATIONS LISTED BELOW	
MISCELLANEOUS	☐ SMALL BUS. REPORT	☐ BASIC AGREEMENT NO.	FOR CLAUSES SEE T's & C's

CONTRACT DOCUMENT(S) AND SUPERSEDURE INFORMATION:

(20) PREPARED BY NAME EXT	(21) PROGRAM MANAGER APPROVAL	(22) PP&C	☐ YES ☐ NO EXECUTIVE APPROVAL
LEAD DIVISION APPROVAL	ASSIGNED DIVISION APPROVAL		FINANCIAL APPROVAL ☐ TACIT

Fig. 1.28 *Sample general order.*

establishes lot sizes which will be the most economical and yet will meet committed delivery schedules.

The following are some of the things which must be considered in determining a release setup.

1. Detail parts should be produced in large quantities to take advantage of material price breaks; to reduce paper-work handling through ordering, purchasing, and production control; and to reduce machine setup time and inspection time.

2. Large detail releases may not be the most economical or most desirable because on new research-and-development (R&D) programs, the engineering design may not be firm and many design changes may be expected. If a large quantity of detail parts is produced, the quantity scrapped because of a design change could cost a considerable amount. On new R&D programs, therefore, the release size should be limited for change control purposes.

3. Assembly release size is subject to the same considerations. However, since changes at the assembly level may involve scrapping complete assemblies or expensive equipment items, the assembly release size on R&D programs is usually small enough to provide efficient change control during the initial phase. One detail release may support several assembly releases.

4. Both detail and assembly releases should be made larger as the program progresses. It is assumed that debugging of the product design, tooling, and test specifications will be, for the most part, completed on the first few units.

5. When a follow-on order is received for items already in production and if no changes are anticipated, release sizes should be established on the basis of optimum economical production quantities. However, if the technology of the product is progressively changing, it is advisable to establish only those detail releases that will support units scheduled for delivery in the relatively near future.

After performing the above analysis, the scheduler establishes the release setup (Fig. 1.29), obtains approval signatures, and issues it to other functional organizations.

B. *Master schedule and acceleration curve* (Fig. 1.30). Utilizing the drawing-breakdown list and the customer's schedule requirements, the scheduling function develops a firm master schedule. The scheduler requires much information before he can reach decisions. For example, the following are some of the things which a scheduler in the electromechanical business must know before he can establish a firm schedule.

1. Product complexity and description; comparison with other known products

2. New or exotic materials required, if any; castings or forgings or both required; long-lead purchased-equipment items required

3. Date major subcontract items will be released, based on weeks from go-ahead

4. Procurement time required for known long-lead items

5. Schedule for release of engineering drawings and specifications, based on go-ahead; date preliminary test specifications and schematics will be released

6. Total number of new assemblies

7. Type of functional test and systems test required; burn-in requirements; special power requirements

8. Need for dust-free area for precision assemblies

9. Need for complex dimensional tooling and time required for design and fabrication

10. Test-equipment requirements

a. Type and quantity of new equipment needed and possibility of using existing equipment on a noninterference basis

b. Design and fabrication time required

c. Certification time required and availability of assemblies or instruments required for certification

11. Detail fabrication time, based on long-lead items

12. Normal shop assembly and test time, based on shop experience with similar items

13. Type of inspection required

14. Subcontract experience on similar items and subcontract flow-time and schedule-support history

15. Shop capacity and facilities

16. Time required to increase capacity, augment facilities, etc.

17. Effect of new load and proposed business on firm in terms of shop capacity, facilities, and equipment

When the required data have been gathered, the scheduler prepares a master schedule to correlate all commitments and time spreads, based on weeks or months from go-ahead. On the basis of this schedule and of other facts, the

Subject: RELEASE SET-UP

Page 1 of 1

Part: C
Original Date:
Revised:
Replaces:
Page: 1 of 1

GENERAL ORDER BULLETIN NUMBER CCCC
RELEASE SET-UP

To: Standard Distribution List.
From: Mfg. Project Administration & Cost Analysis
Subject: Release Set-Up for G.O. CCCC Models Noted
Ledger Account 02XXX Series

DETAIL			ASSEMBLY			Program
Rel. No.	Qty/Rel.	Cum. Units	Rel. No.	Qty/Rel.	Cum. Units	Model Code
			SYSTEM ANALYZER			
			Model C71 Model Index			B0001
1001	1	1	1101	1	1	
2001	4	2-5	2101	4	2-5	
3001	10	6-15	3101	4	6-9	
			3201	4	10-13	
			3301	2	14-15	
			MODULE TESTER			
			Model C72 Model Index			
1002	1	1	1102	1	1	B0002
2002	4	2-5	2102	4	2-5	
3002	10	6-15	3102	4	6-9	
			3202	4	10-13	
			3302	2	14-15	
			TEST UNIT RECEIVER			
			Model C73 Model Index			
1003	10	21-30	1103	5	21-25	B0003
			1203	5	26-30	
2003	20	31-50	2103	5	31-35	
			2203	5	36-40	
			2303	5	41-45	
			2403	5	46-50	

SMALL PARTS

Rel. No.	Qty./Rel.	Detail Releases Covered
1903	30	1003 thru 2003

PROPERTY DISPOSITION

Ledger Account	Sub-Account
02320	9800

NOTE: Prior Production

Units	G.O.
1-10	1234

Factory Manager

General Supervisor
Mfg. Project Administration

Contracts and Pricing

General Supervisor
Mfg. Cost Analysis

Financial

Fig. 1.29 *Sample order release setup.*

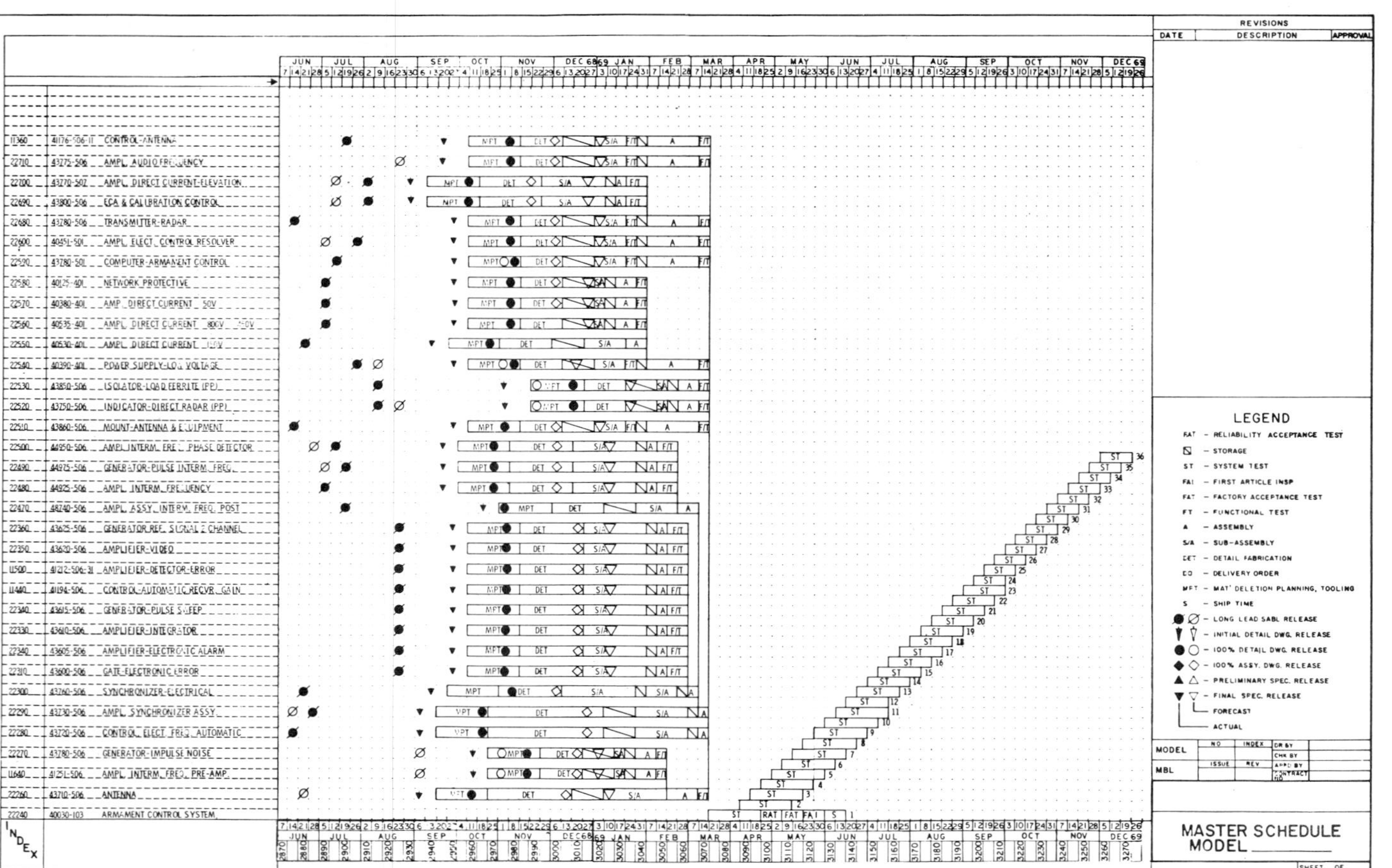

Fig. 1.30 *Sample master schedule and acceleration curve.*

capability for meeting the requested first-unit delivery and for attaining and maintaining the required delivery rate is determined.

Developed concurrent with the master schedule, an acceleration curve represents the scheduled completion rate for the order quantity of systems charted. It displays the time spread in weeks for each level of effort, starting with the first unit and going through to the completion of the order. The earlier units usually have a longer time spread because learning and design changes are major factors. As experience is gained and the design matures, the time spread is compressed.

Figure 1.30 is a combination model-breakdown list, master schedule, and acceleration curve for thirty-six systems. The tabulation on the left is the model-breakdown list; it identifies the items required for assembly of the final system and their part numbers. To the right of the model-breakdown list is the master schedule, depicting the time relationships of parts and activities needed to support the schedule date of the systems test (ST) of the first unit. It schedules release of long-lead items (ϕ), material procurement, planning, and tooling (MPT), detailed parts fabrication (Det), release of subassemblies (S/A), and functional test (F/T). The master schedule is developed for the first unit but is also used for determining release dates of subsequent units to support the systems-test dates merely by moving it to the right. The acceleration curve shows the time allowed for systems tests. Note that the total time for systems and associated tests decreases from eleven weeks for the first prototype system (1) to three weeks for the last unit (36).

C. *Fabrication schedule*. The master schedule described earlier gives the overall picture of what the end item consists of and the time and flow patterns of the details, subassemblies, and major assemblies required to assemble and test the deliverable end item. The detail-scheduling group receives the master-scheduling documents and, in turn, prepares detail schedules as working documents for the planning, purchasing, and manufacturing shops.

These schedules are called *fabrication schedules*. The information for preparing a fabrication schedule is taken from the master schedule, acceleration curve, and release setup. The assembly and functional-test flow spans for the major assemblies are taken from the master schedule and acceleration curve. With this information, a fabrication schedule is developed. It displays in weekly increments the quantity of major assemblies required for each release, and schedules for and quantities of details. Preparing such a schedule involves taking each major assembly, determining its level of support to the end item, and establishing a completion schedule for a week's worth of parts.

From the fabrication schedule and other authorizing requirement documents, a composite completion schedule is developed for each major assembly, showing all known firm requirements in weekly and cumulative quantity spreads. Since it sums up all requirements for each major assembly regardless of specific customer order, it is the basis for accumulating and maintaining cognizance of shop loads and for preparing internal shop completion documents.

The initial action taken by production planning, the second phase in processing internal production-control information, is the conversion of engineering documentation into detailed instructions for fabricating or assembling parts. After the release of drawings by engineering, planning immediately receives one or more copies. Drawings are accompanied by copies of the release authority, the engineering order (E.O.) (Fig. 1.31). Each E.O. is scheduled for completion and distributed to the responsible planner or planning subgroup. When drawings are initially released, the planner prepares a detailed, step-by-step plan for the fabrication of the part or parts shown on the drawing. The nature of the part may require use of a specific tool. In other cases, the use of a specific tool is included in the planning in the interest of reducing cost or maintaining a rate of production which will support requirements. It is essential that tooling be requested at the earliest possible point to allow time for design or fabrication or both.

The planning action consists of listing the individual operations required to convert the raw material into a completed part in their proper sequence on a planning worksheet (Fig. 1.32). Consideration must be given to the size and condition of the material at the beginning of the manufacturing cycle. Processing, such as cleaning, plating, or painting, must be included, usually at the end of the cycle. These requirements are normally defined by process specifications, which are referenced on the affected drawings.

The individuals assigned to production planning should be thoroughly familiar with manufacturing techniques and practices. A back-

ENGINEERING ORDER

A 449501

SUBJECT	JUSTIFICATION/REASONS/REMARKS	RELEASE: DEVEL. ☐ PRODUC. ☐	DISPOSITIONS: 1 MAY BE REWORKED 2 CANNOT BE REWORKED 3 NONE
INITIATOR — DATE / DESIGN APPROVAL — DATE		DOCUMENT (NOT) TO BE CHANGED MADE	

DOCUMENT or P/N	REV	EFFECTIVITY	MODELS	PROJECT OFFICE

1 2 3 4 5 6 7 8 9 10 11 12 13 14 15 16 17 18 19 20 21 22 23 24 25 26 27 28 29 30 31 32 33 34 35 36

ATTACH EO COPIES TO:

SENSITIVITY CHECK LIST-MARK EFFECT ON EACH ITEM:

CONTRACT	YES	NO	ENGINEERING	YES	NO		YES	NO	LOGISTICS	YES	NO	OTHER	YES	NO
SCOPE	☐	☐	CRITERIA	☐	☐	DESIGN PROOF	☐	☐	MAINT. ANAL.	☐	☐	______	☐	☐
PRICE	☐	☐	WEAPON SYS. SPEC.	☐	☐	SAFETY	☐	☐	RETROFIT	☐	☐	______	☐	☐
SCHEDULE	☐	☐	CEI SPEC.	☐	☐	RFI	☐	☐	PUBLICATIONS	☐	☐	______	☐	☐
GUARANTEES	☐	☐	INTERFACE	☐	☐	NOMENCLATURE	☐	☐	SUPP. EQUIP.	☐	☐	______	☐	☐
INCENTIVES	☐	☐	INTERCHANGE	☐	☐	COMP. PROG.	☐	☐	SPARES	☐	☐	______	☐	☐
DEFICIENCY	☐	☐	MAINTAINABILITY	☐	☐	______	☐	☐	KIT PROOFING	☐	☐	______	☐	☐

MFG. PROJ. ADMIN.	E.O.C.	RELEASE STATION
MATERIAL		
STRESS	CHECK	CONTROL NO.
CUSTOMER		
CONFIG. MGMT.		

A 449501

Fig. 1.31 *Sample engineering order.*

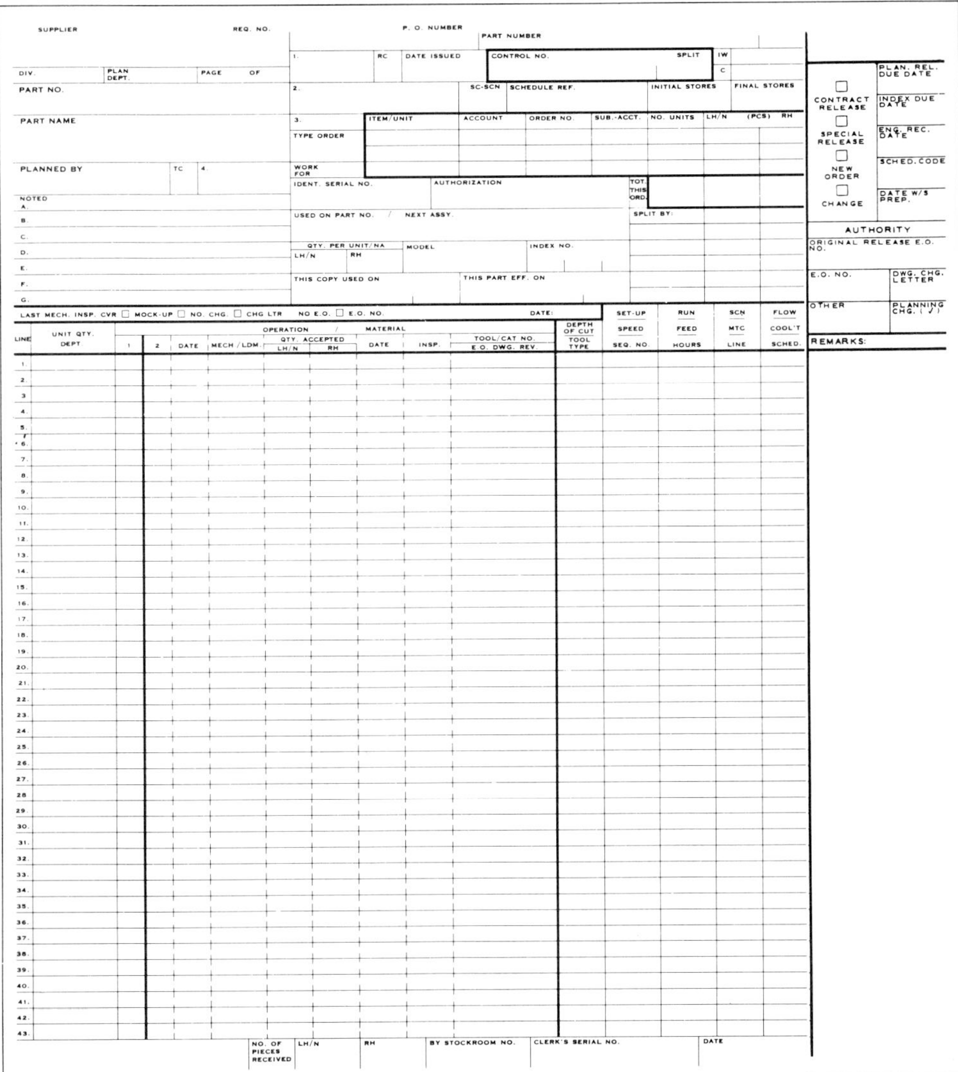

SUPPLIER
REQ. NO.
P. O. NUMBER
PART NUMBER
1. | RC | DATE ISSUED | CONTROL NO. | SPLIT | IW | C
DIV. | PLAN DEPT. | PAGE OF
PART NO.
2. | SC-SCN | SCHEDULE REF. | INITIAL STORES | FINAL STORES
PART NAME
3. | ITEM/UNIT | ACCOUNT | ORDER NO. | SUB.-ACCT. | NO. UNITS | LH/N | (PCS) | RH
TYPE ORDER
PLANNED BY | TC | 4.
WORK FOR
IDENT. SERIAL NO. | AUTHORIZATION | TOT. THIS ORD.
NOTED
A.
B.
C.
D.
E.
F.
G.
USED ON PART NO. / NEXT ASSY.
SPLIT BY:
QTY. PER UNIT/NA: LH/N | RH
MODEL
INDEX NO.
THIS COPY USED ON
THIS PART EFF. ON

☐ CONTRACT RELEASE
☐ SPECIAL RELEASE
☐ NEW ORDER
☐ CHANGE
PLAN. REL. DUE DATE
INDEX DUE DATE
ENG. REC. DATE
SCHED. CODE
DATE W/S PREP.

AUTHORITY
ORIGINAL RELEASE E.O. NO.
E.O. NO. | DWG. CHG. LETTER
OTHER | PLANNING CHG. (✓)

LAST MECH. INSP. CVR ☐ MOCK-UP ☐ NO. CHG. ☐ CHG LTR NO E.O. ☐ E.O. NO. DATE:

LINE	UNIT QTY. DEPT	OPERATION / MATERIAL 1	2	DATE	MECH / LDM.	QTY. ACCEPTED LH/N	RH	DATE	INSP.	TOOL/CAT NO. E.O. DWG. REV.	DEPTH OF CUT / TOOL TYPE	SET-UP / SPEED / SEQ. NO.	RUN / FEED / HOURS	SCN / MTC / LINE	FLOW / COOL'T / SCHED.
1.															
2.															
3															
4.															
5.															
6.															
7.															
8.															
9.															
10.															
11.															
12.															
13.															
14.															
15.															
16.															
17.															
18.															
19.															
20.															
21.															
22.															
23.															
24.															
25.															
26.															
27.															
28.															
29.															
30.															
31.															
32.															
33.															
34.															
35.															
36.															
37.															
38.															
39.															
40.															
41.															
42.															
43.															

REMARKS:

NO. OF PIECES RECEIVED | LH/N | RH | BY STOCKROOM NO. | CLERK'S SERIAL NO. | DATE

Fig. 1.32 *Sample planning worksheet.*

ground of shop experience is often helpful, particularly in the areas of machine parts, complex welded assemblies, electrical and electronic parts and assemblies, etc. Since the routing of any part is determined by indicating the department or group responsible for each operation listed, a planner must be familiar with the capabilities, machines, equipment, and facilities of each manufacturing department.

Production planning has the responsibility of reflecting the impact of the latest engineering release on the planning of any part. Within the limits prescribed by change control procedures, engineering is authorized to make such changes as are required to assure that the part will serve its intended purpose. Such changes and the authority for their accomplishment are conveyed to planning by released E.O.s or changed drawings or both. On receipt of authority to change a part, the planner must document the change in such a manner that all parts fabricated in the future will conform. He must also initiate the action to bring previously fabricated parts up to the current configuration when required. The particular circumstances dictate the exact action to be taken. In many cases, the planner must decide the most feasible action by an analysis of conditions, exercising good judgment.

The output of planning is the formal planning master, which is typed from the planning worksheet. The planning master for any given part number is filed and maintained on a continuing basis; it is used by the ordering group when the specific part number is to be ordered from the fabrication or assembly shops. Designing planning-worksheet, planning-master, and production-order forms so that duplication of one from another is possible reduces the cost of preparation and shortens the paper-work time cycle.

As stated previously, the ordering function, the last phase in processing internal production-control information, is responsible for the issuance of production orders, tooling orders, purchase requests, and material requisitions required to support a specific order. Other functional areas provide the basic elements of data required for the ordering operation. The internal order is the official authorizing document, the scheduling function supplies release-setup and scheduling data, and production planning supplies the routing or method of fabrication. When the internal-order, schedules, release-setup, and planning data are available, the ordering group is in a position to proceed.

To establish a level-by-level support pattern of all in-house fabricated items and material to support the end item, the ordering section must develop a setback list, starting with a major assembly and listing all supporting parts of the next lower level. It repeats this step at each level until the lowest level of detail or material items has been determined. Utilizing the setback list, the ordering section derives the schedules for the major components from the fabrication schedule. In-work (IW) and completion (C) dates are established for each lower-level supporting component, based on the flow time and the oversupport time previously established. The in-work and completion dates are determined mathematically. The flow time is subtracted from the completion date of the major assembly to determine the in-work date. Subtracting the oversupport time from the in-work date establishes the completion date for the next-lower-level supporting component. Its flow time subtracted from its completion date establishes its in-work date and so on. As these dates are calculated, production orders (Fig. 1.33), requisitions to purchasing (Fig. 1.34), and material requisitions (Fig. 1.35) are prepared for each part.

Pertinent data, such as internal-order number, customer-order number, and release infor-

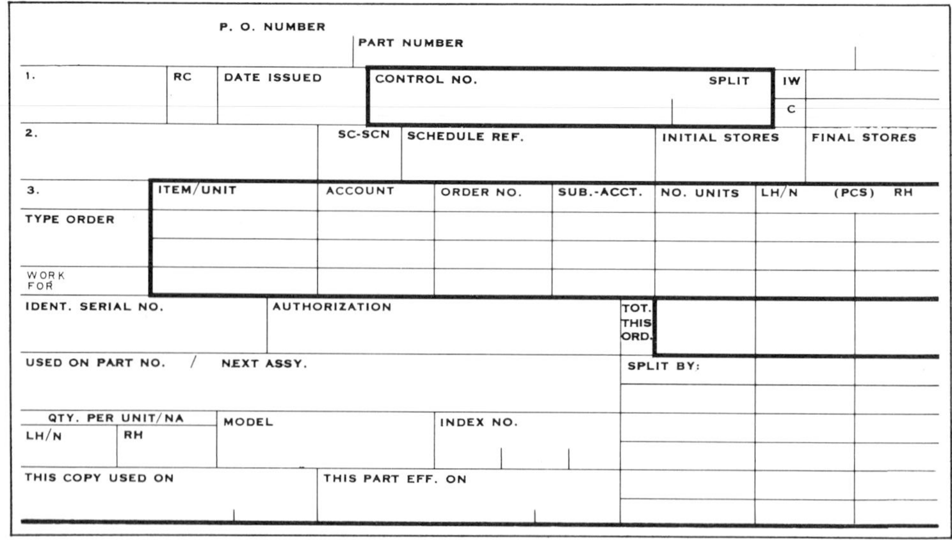
P. O. NUMBER
PART NUMBER
1. | RC | DATE ISSUED | CONTROL NO. | SPLIT | IW | C
2. | SC-SCN | SCHEDULE REF. | INITIAL STORES | FINAL STORES
3. | ITEM/UNIT | ACCOUNT | ORDER NO. | SUB.-ACCT. | NO. UNITS | LH/N | (PCS) | RH
TYPE ORDER
WORK FOR
IDENT. SERIAL NO. | AUTHORIZATION | TOT. THIS ORD.
USED ON PART NO. / NEXT ASSY. | SPLIT BY:
QTY. PER UNIT/NA | LH/N | RH | MODEL | INDEX NO.
THIS COPY USED ON | THIS PART EFF. ON

Fig. 1.33 *Sample production order.*

SEE INSTRUCTIONS ON REVERSE SIDE OF WORK COPY.
KEEP WORK COPY FOR YOUR FILE AND FORWARD OTHER COPIES TO PURCHASING.

REQUISITION TO PURCHASING

·Req. No. 580576

SEQUENCE NO | TCN | P.O. NUMBER | DATE

SB | SUPPLIER CODE | PA | BUYER NO | REFERENCE | CONTRACT ("C" - COMMERCIAL, "X" - GOV'T) | PRIORITY RATING CERT. UNDER DMS REG. 1

IN-PLANT ROUTING | PROVISIONS ATTACHED HERETO AND MADE A PART HEREOF

CALIF. PERMIT NO.
RESALE YES ☐ NO ☐ | SHIP VIA

TO: | TERMS | SS
SHIP TO:

F.O.B. | BILL TO

SHIPPING SCHEDULES - QUANTITIES REPRESENT THE NUMBER OF PARTS OR MATERIAL ON DATE INDICATED

ITEM	QUANTITY & DATE	QUANTITY & DATE	QUANTITY & DATE	QUANTITY & DATE	QUANTITY & DATE

ITEM	QUANTITY	SUPPLIER'S PART NO	ITEM IDENTIFICATION NO	U/M	G.O.	UNIT PRICE

PQA | CONTROL NUMBER

USED FOR | CHARGE TO DEPT./PROJECT NO

ACCOUNTING USE ONLY

ITEM	DEPT	LEDGER	ORDER NO	S/A-REL

SUGGESTED SUPPLIERS | EST. UNIT PRICE

PREFERRED SCHEDULE

ITEM	QTY. & DATE REQ'D	QTY. & DATE REQ'D	QTY. & DATE REQ'D	QTY. & DATE REQ'D	QTY. & DATE REQ'D

REQUISITIONED BY DEPT EXT DATE | DEPT. SCREEN DEPT EXT DATE | AUTHORIZED SIGNATURE DEPT TITLE DATE

MATERIAL CONT. APPROVALS NAME TITLE DATE | P.O. COPY TO SUPPLIER YES ☐ NO ☐ | BUYER'S SIGNATURE DATE

P.O. ACKNOW. REQ'D YES ☐ NO ☐

BIDDERS	PRICE	F.O.B.	TERMS	DELIVERY	QUOTE NO

PURCHASING CASE FILE

BUYER CHECK NUMBER INDICATED FOR REASON SHOWN ON REVERSE SIDE: 1A ☐ 1B ☐ 1C ☐ 2 ☐ 3 ☐ 4 ☐ 5 ☐ 6 ☐ 7 ☐ 8 ☐ 9 ☐ 10 ☐ 11 ☐ 12 ☐ 13 ☐ 14 ☐ 15 ☐ 16 ☐ 17 ☐ 18 ☐ 19 ☐ 20 ☐

Fig. 1.34 *Sample requisition to purchasing.*

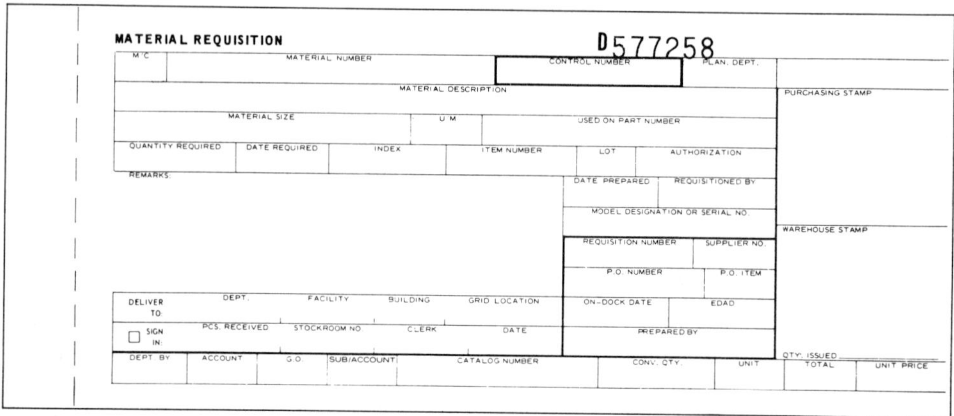

MATERIAL REQUISITION D577258

M/C | MATERIAL NUMBER | CONTROL NUMBER | PLAN. DEPT.

MATERIAL DESCRIPTION | PURCHASING STAMP

MATERIAL SIZE | U/M | USED ON PART NUMBER

QUANTITY REQUIRED | DATE REQUIRED | INDEX | ITEM NUMBER | LOT | AUTHORIZATION

REMARKS: | DATE PREPARED | REQUISITIONED BY

MODEL DESIGNATION OR SERIAL NO. | WAREHOUSE STAMP

REQUISITION NUMBER | SUPPLIER NO.

P.O. NUMBER | P.O. ITEM

DELIVER TO: | DEPT. | FACILITY | BUILDING | GRID LOCATION | ON-DOCK DATE | EDAD

☐ SIGN IN: | PCS. RECEIVED | STOCKROOM NO. | CLERK | DATE | PREPARED BY

DEPT. BY | ACCOUNT | G.O. | SUB/ACCOUNT | CATALOG NUMBER | CONV. QTY. | UNIT | QTY. ISSUED TOTAL | UNIT PRICE

Fig. 1.35 *Sample material requisition.*

mation, are also included on the various documents generated. Orders are subject to various checks and controls. It is normal to record that an order has been issued and completed. The number of copies produced and distributed will vary based on the business and internal systems within the company. For example, in one case, a copy may be filed as a record of issuance, but in another case the issuance may be recorded manually on a posting record. Internal controls in the shop departments may also require a copy or copies for specific purposes.

After the orders have been prepared, they are issued to the responsible manufacturing and material functions for further processing. Requisitions to purchasing are normally released far in advance of need date to allow for procurement, vendor fabrication, and delivery to the warehouse. Production orders and material requisitions (authority to withdraw warehouse material) are normally released to the shop departments a predetermined period of time in advance of the in-work date to allow for removal of material from the warehouse, kitting of parts, and line loading.

Material-ordering techniques vary from company to company. There are two main methods of ordering raw material and purchased parts. In the first case, production control has the responsibility for establishing quantity and schedule requirements for purchased material and parts and for ordering such parts from purchasing or the warehouse by means of requisitions to purchasing (Fig. 1.34) or material requisitions (Fig. 1.35). In this case, purchasing operates primarily as a services organization. In the second case, purchasing establishes material and parts requirements on receipt of engineering drawing breakdown lists and schedule requirements from production control. Paragraph 1.4.4 discusses both these methods of operation in more detail.

Two basic methods are commonly used to remove material from the warehouse and deliver it to the using department. One is the forwarding of suitable copies of the production order, including a listing of the material required, to the material department or purchasing department, which are normally responsible for warehousing. One copy usually routes the material, and the other copy is sent to the financial organization for accounting purposes. Although this approach is convenient for use by planning, it is generally unsatisfactory for purchasing if delivery of multiple items is requested. A more satisfactory approach is the establishment of single-item material requisitions. Processed individually by the purchasing organization, these documents are not subject to the delay or complex handling common to multiple-item documents.

The ordering of tools (Fig. 1.36) is usually a routine operation initiated by a request from the production planner. Appropriate accounting information is provided. Tooling requirements are described and a schedule is established. The in-work date of the first part to be produced with the tool determines the schedule. A tooling policy for a particular program may establish a point or points at which tooling is to be available. Parts may be fabricated prior to this point by a process which does not require the use of tooling. Duplicate tools may be ordered to support increased rates, as indicated by the tooling policy.

After authorizing documents have been released, maintaining released orders to reflect changes in configuration, quantity, effectiveness, etc., resulting from engineering changes is normally the responsibility of the planning function. Engineering changes affecting future release of orders must be communicated to the ordering section by means of new or changed planning and schedule data.

1.4.3.4 Manual Systems. The preceding paragraphs have described the steps necessary to convert engineering information into planning, scheduling, and ordering information. The documents described have been manually prepared and generated. Some type of manual system must be in operation prior to any attempt to develop a mechanized system in order to determine the information required and the documents to be used to transmit the information.

It is obvious that communication of information is one of the most critical problems confronting the manufacturing organization. The complexity and technical nature of products produced today bring with them a staggering problem in information-handling. For example, in the aerospace industry it may require 80,000 drawings and thousands of man-years to develop the basic design specifications for a ballistic missile. The control of men, materials, and machines and the assimilation of thousands of engineering changes which affect all phases of the organization are never-ending problems confronting management.

From the start of the engineering development to the shipment of the final product, tons of paper work and millions of elements of data pass through the entire organization. People, being human, make mistakes which cause lost, misrouted, and late data, misinterpretation of information, etc. To avoid these errors, many functional groups establish redundant checks to verify information fed to them from other functions. This operation could be called crutching and is inefficient and costly. That is not to say that adequate checks and controls are not required. To produce a quality product

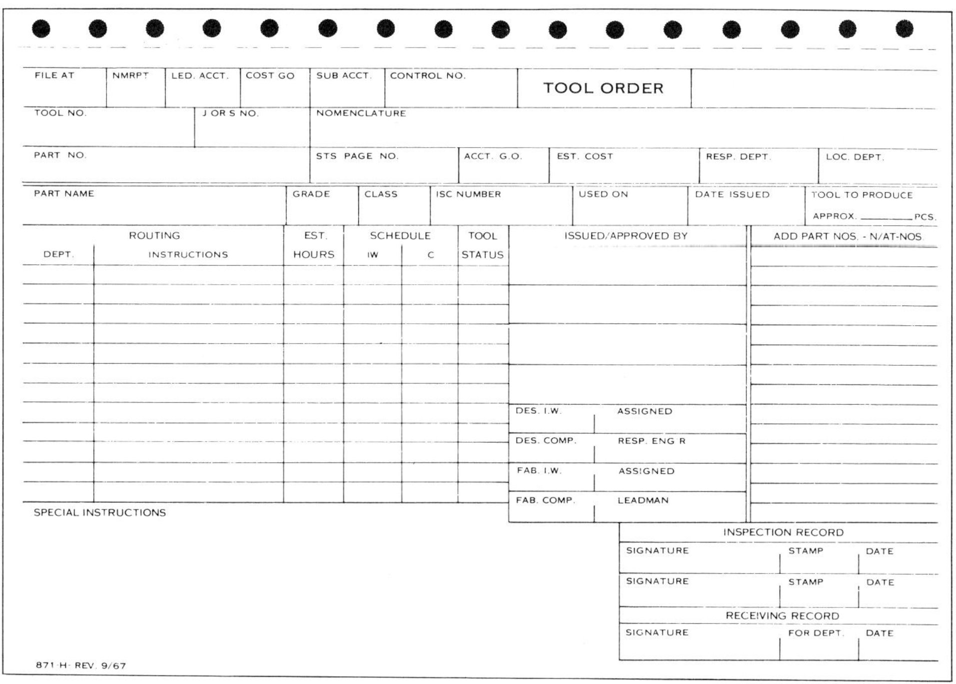

FILE AT	NMRPT	LED. ACCT.	COST GO	SUB ACCT.	CONTROL NO.	TOOL ORDER

TOOL NO.	J OR S NO.	NOMENCLATURE

PART NO.	STS PAGE NO.	ACCT. G.O.	EST. COST	RESP. DEPT.	LOC. DEPT.

PART NAME	GRADE	CLASS	ISC NUMBER	USED ON	DATE ISSUED	TOOL TO PRODUCE APPROX. _____ PCS.

ROUTING		EST.	SCHEDULE		TOOL	ISSUED/APPROVED BY	ADD PART NOS. - N/AT-NOS
DEPT.	INSTRUCTIONS	HOURS	IW	C	STATUS		
						DES. I.W. / ASSIGNED	
						DES. COMP. / RESP. ENG R	
						FAB. I.W. / ASSIGNED	
						FAB. COMP. / LEADMAN	

SPECIAL INSTRUCTIONS

INSPECTION RECORD		
SIGNATURE	STAMP	DATE
SIGNATURE	STAMP	DATE
RECEIVING RECORD		
SIGNATURE	FOR DEPT.	DATE

871-H- REV. 9/67

Fig. 1.36 *Sample tool order.*

on schedule at the lowest possible cost requires accurate and timely information, but redundancy must be scrutinized closely to keep cost at a minimum. Companies whose prices are continually high as a result of high costs jeopardize their competitive position and eventually become extinct.

To combat this and other problems, any system whether manual or mechanized must establish adequate procedures and controls. There are normally two types of procedures: policy and operating. Policy procedures establish organizational responsibilities and authority and operating parameters, and delineate the management objectives of the business. Operating procedures describe at a functional level the steps and controls necessary to accomplish a given task. A policy procedure is one that states that make-or-buy decisions are the responsibility of the manufacturing organization and that buy decisions in excess of a specified number of dollars require the approval of certain other members of functional and general management. In contrast, an operating procedure is one that describes the step-by-step method of processing and communicating a make-or-buy decision to the affected functional groups.

Within the planning, scheduling, and ordering functions, operating procedures should describe in detail the steps necessary to achieve the output objectives of the respective groups. A policy procedure should make clear the responsibilities and authority of each function so there is no question as to the role and mission of each group or department within the organization.

The key to a successful manual system lies in the ability of management to give sound and clear direction and to maintain the respect of all lower-level employees within the organization. The morale, interest, and pride of the worker, who must carry out certain daily tasks to achieve management objectives, are important factors in the success or failure of any system. The organizational structure, which will vary from company to company, also has a bearing on efficiency. It would be ideal if all functions could be performed within one department, but as organizations grow, responsibilities must be split into workable areas. This division of work creates new problems in information communication. Each function should recognize its own responsibilities and problems as well as those of other functions. Each function must perform its contribution to the overall business objective accurately, on time, and in harmony with other functions.

In the design of a manual system for production planning and control, the approach of systems engineering should be employed. The input-output documents required by the production-control function and the information they should contain have been described in earlier paragraphs. Information flow diagrams should be prepared, similar to that in Fig. 1.4, which depicted the generalized information flow diagram of the total operations subsystem. Responsible organizations and the documents they receive or issue should be delineated on the diagram prior to the development of any written procedures. The interface or information specifications which define the information contained in each document should be developed after an information matrix for the production planning and control function has been prepared. Preferably the information matrix for the entire business enterprise should be available prior to developing a business system for any given function. However, in many cases this total matrix is not available and so the functional organization must proceed to prepare one for its own internal operations. Knowledge of the internal operations of the functions furnishing the input documents and those receiving the output documents will eliminate duplication of information and effort. Once the information elements of each document have been specified, forms can be designed and procedures written to cover the day-to-day operations of the production planning and control function.

Manual systems require close scrutiny to ensure that functions are being performed accurately and on time. The establishment of clear and concise procedures, combined with daily direction and motivating techniques, is mandatory to attain a high degree of efficiency. Auditing on a periodic basis to ensure compliance with procedure eliminates many costly problems which can affect the image of the company. Most important is that all employees, from those in management to workers in the shop, understand the objectives of the business and their role and mission in meeting these objectives.

1.4.3.5 Mechanized Systems. The problem of paper-work handling and communication of information in industry is staggering. The administrative problems created by the greater technology of products and the never-ending increase of clerical and paper-work costs can be solved only by developing a more efficient method of communicating, disseminating, and retrieving information. Mechanization of information-handling is the best solution to the problem.

Some forms of mechanization have been available for many years. However, the application of computers to manufacturing information systems began basically after 1960. Industry used punched-card equipment in World War II; it caught hold as a business tool in the late 1940s and early 1950s. Originally it was primarily used for accounting applications such as payroll and labor distribution. In the mid 1950s industry began to use computers but mainly for engineering and scientific applications. The computers had high internal-speed processing but were still limited in their input and output speeds. Because of this limitation and their high cost, computers were not widely used in production planning and control systems. In 1960 the announcement of a second generation of computers appeared to open the way for development of production-control applications since computer manufacturers were able to produce a less costly, smaller, and more efficient computer. High internal speeds, the development of random-access files for storage of data, and short input-output processing times seemed to be the answer for handling the massive amount of data required in such systems.

Lack of experience in the use of computers and reluctance on the part of functional personnel created problems in the development of mechanized systems. The natural tendency of functional personnel to fear the loss or degradation of employment was a psychological problem that had to be overcome. Lack of functional experience by data-processing personnel and lack of computer experience by functional personnel caused conflicts in approaches used for mechanizing various functional information systems. The usual approach was to view each function independently with no regard for where the source data came from. With this method many functional systems were designed, and the duplication of information in the different systems resulted in redundant and costly design effort, excessively large computer files, and difficult updating problems as the basic information was changed. The systems-engineering approach developed in Par. 1.2, whereby a total plan is developed and subsystems implemented on a modular basis with each piece independent but fitting into the overall plan, offers the best solution to the systems-design problem. Through this technique various modules can be developed and implemented simultaneously with a minimum of redundancy. In addition to using the modular approach, it is important that in the design phases of any system functional and data-processing personnel work as a team to ensure the most efficient method of design. Mechanization does not cause functional organizations to relinquish their responsibilities. Rather it enhances their abilities to plan and to improve their decision-making roles by performing clerical tasks for them. Hence, functional personnel should accept the prime responsibility for the development, design, and implementation of mechanized systems.

The configuration of computer hardware has a definite bearing on the success of the system. The fact that computer-equipment design was originally oriented toward engineering and accounting functions has hampered the progress and development of mechanized business systems. In these other areas, particularly in the engineering and scientific fields, the problem is usually one of performing a large number of calculations with a relatively small amount of input data. Engineering data-processing operations can easily be planned in an orderly, sequential manner. In engineering and scientific applications, speed is very important in the computing operation but is of secondary importance in information storage and retrieval. For this reason, magnetic-tape or punched-card storage of primary data provides a suitable answer to the need. Production planning and control systems, in contrast, involve only a comparatively small number of simple arithmetic operations but involve very large volumes of constantly changing primary data. Here a major requirement is rapid information collection, storage, and retrieval on a random or variable basis. Speed of inquiry and answer is important in production control since the need to know is *now*. Ideally, the mechanized business system should record and reflect all information and changes on a random basis. Neither punched-card nor magnetic-tape storage systems, including those with high-speed and multiple-tape drive units, can meet these requirements. For this reason, information-storage equipment, known as random-access, disc-storage units, was designed to meet the need of business systems for both rapid and random input, storage, and output of large volumes of information. Coupled with a computer and certain types of data-collection equipment, it is ideal for manufacturing applications.

As described earlier, one of the first groups in the manufacturing organization to make use of the engineering data is the production-planning function. Development of the production plan starts with the order and continues through to its completion. While the order is in process, the plan may be continuously corrected and updated. Fast response to change is a key requirement of a production-control system. Use of mechanized manufacturing assembly parts lists, which can be derived from mechanized engineering parts lists, provides an effective means for formulation, communication, and updating of the plan. Random-access storage devices are most suitable for this purpose. They allow rapid processing and updating of data and, in addition, provide readily available information to operating departments. Planning-data availability is important to a company because this data is used in such vital activities as parts replacements, shop-order preparation and scheduling, shop load forecasting, raw-material requirements, tool requirements, personnel skill requirements, planning status, configuration management, cost estimating, and cost and schedule reporting.

The planning portion of a mechanized system offers these benefits:

1. Accurate and readily accessible data on methods of manufacture
2. Shorter data flow time between engineering and production
3. Up-to-date manufacturing parts lists for parts and raw-material requirements planning, configuration management, and cost estimating
4. Machine-prepared documents for use in executing the manufacturing plan (detail and assembly shop orders), which decrease the chance of misinterpretation of handwritten documents
5. Timely planning status data, used to highlight potential schedule problems

Certain companies have become disillusioned with mechanization because data-processing equipment has been sold on the basis of its internal technical excellence with little regard for the interfaces between it and existing or proposed business systems. Many companies bought the equipment even before improving manual systems and developing a master plan for future development of a mechanized system. Many costly mistakes were made, and as a result, only qualified acceptance has been gained from the functional operating people. There is no question that mechanization of information-handling will become an absolute requirement during the next decade. Sound planning, trained personnel, and functional leadership and participation are all necessary if the mechanization is to be successful.

1.4.4 Purchasing (or Material or Procurement)

1.4.4.1 Functional Description. The charter of the purchasing function (called material or procurement in many companies) is to make all external purchases of all productive and nonproductive material used by the enterprise; to execute and administer all purchase orders for such material and for subcontracting efforts; and to provide stocking, receiving, warehousing, traffic, transportation, and shipping services. Purchasing can be looked on as the production-control function for "buy" materials, just as production control plans, schedules, and orders "make" parts and assemblies. The vendor can be viewed as an external fabrication shop, with the purchase order similar to the internal tool or fabrication order. Figure 1.4 showed the parallelism of purchasing for external parts and materials and production control for internal parts.

1.4.4.2 Input-Output Requirements. There are two means of inputting information to purchasing. Both are in general use, and the final selection should be made on the basis of the needs and nature of the business and the responsibilities which the chief executive wishes to assign to either purchasing or production control. In the first method of operation, all requirements and requests for productive and nonproductive material, parts, and services are generated outside the purchasing organization. The requesting organization transmits information concerning quantities required and need dates to purchasing on requisitions to purchasing, requests for material, requests to purchase,

ACCOUNTING COPY

AUTONETICS
A DIVISION OF NORTH AMERICAN ROCKWELL CORPORATION
3370 MIRALOMA AVENUE, ANAHEIM, CALIFORNIA 92803

SEQUENCE NO	TCN	P.O. NUMBER	DATE

SB	SUPPLIER CODE	PA	BUYER NO	REFERENCE	CONTRACT ("C"-COMMERCIAL "X"-GOV'T)	PRIORITY RATING CERT. UNDER DMS REG. 1

IN-PLANT ROUTING (AUTONETICS USE) — PROVISIONS ATTACHED HERETO AND MADE A PART HEREOF

CALIF. PERMIT NO. X6938 — SHIP VIA:
RESALE: YES ☐ NO ☐

TO: — TERMS — SS
SHIP TO:

F.O.B. — BILL TO

SHIPPING SCHEDULES - QUANTITIES REPRESENT THE NUMBER OF PARTS OR MATERIAL ON DATE INDICATED

ITEM	QUANTITY & DATE	QUANTITY & DATE	QUANTITY & DATE	QUANTITY & DATE	QUANTITY & DATE

ITEM	QUANTITY	SUPPLIER'S PART NO	ITEM IDENTIFICATION NO	U/M	G.O.	UNIT PRICE

FORM A 70 C REV 4-68

NR INVOICE	ITEM RR	ITEM INV	ITEM RR	ITEM INV	ITEM RR	ITEM INV	INVOICE SUPPLIER	INVOICE DATE	INVOICE AMOUNT	INVOICE F/T	ADV PAY DATE	VOUCHER DATE	AUDITOR

PURCHASE ORDER TOTALS

(a) On date of shipment, send original bill of lading, air bill, or express receipt of Buyer's Traffic Department and one copy of Notice of Shipment to Buyer's Purchasing Department. State shipping point on all invoices.

(b) Do not deliver ahead of schedule unless authorized by Buyer. Do not insure or declare value on shipments beyond F.O.B. point. If the value of any one shipment exceeds $200,000, notify Buyer's Traffic Department by collect wire in advance of shipment.

(c) Unless otherwise specified by Buyer, an increase in the quantities called for by this order will be accepted and payment adjusted accordingly when caused by conditions of loading, shipping, packing or allowances in manufacturing processes, provided the excess does not exceed 5% of the specified quantity for any item, or $25.00 in value, whichever is less. Overshipments may be returned by Buyer at Seller's expense.

(d) Mail quadruplicate invoices to Buyer's Accounting Department when goods are shipped. Each case or parcel and accompanying packing list of contents must show order number. If no packing list accompanies the shipment, Buyer's count will be conclusive on Seller.

V To the extent applicable hereto, Seller shall comply with the provisions of DMS Regulation No. 1, as the same may be amended from time to time, and all other pertinent regulations and orders of B.D.S.A. in obtaining controlled material and other products and materials to fill this order.

VI ACCEPTANCE OF THE OFFER REPRESENTED BY THIS ORDER IS EXPRESSLY LIMITED TO THE TERMS HEREOF. SIGNING AND RETURNING THE ACKNOWLEDGMENT COPY OF THIS ORDER (IF ATTACHED HERETO) OR, IN ANY EVENT, DELIVERY IN WHOLE OR IN PART OF THE ARTICLES TO BE FURNISHED HEREUNDER SHALL CONSTITUTE ACCEPTANCE OF THIS ORDER. IT IS THE ENTIRE CONTRACT AND NO CHANGES ARE BINDING ON BUYER UNLESS THEY ARE IN WRITING AND SIGNED BY AN AUTHORIZED REPRESENTATIVE OF BUYER'S PURCHASING DEPARTMENT.

NORTH AMERICAN ROCKWELL CORPORATION
AUTONETICS DIVISION

BY ______________________

FORM A 70 C REV 4-68

Fig. 1.37 *Sample purchase order set.*

etc. In this case the requesting group is held responsible for inventory since purchasing acts primarily as a services function, ordering and scheduling only on the basis of the requests of others. In the second method, purchasing internally generates requirements from engineering parts lists and drawings and schedule information from production control. Quantities are determined directly from the internal order and engineering documentation. Requesting organizations still order special and nonproductive materials and services through purchase requisitions. Figure 1.4 showed the information flow when all material requirements are generated external to purchasing, and Fig. 1.7 represented the flow when purchasing generates the requirements. In Fig. 1.4 requisitions were transmitted by means of a request to purchase; in Fig. 1.7 the input documents were the engineering drawings and parts lists, the internal order, and the schedule requirements.

The primary output document of purchasing is the purchase order, which orders the material, parts, or services from the selected vendor or supplier and specifies what is to be furnished, when it is to be delivered, and the negotiated price.

1.4.4.3 Internal Processing of Information. A. *Method 1—external generation of material requirements.* Purchasing receives the requisition to purchasing (Fig. 1.34) partially filled out by the requester. The buyer can add information to the form without recopying the basic information. After the buyer selects the source, establishes the price, selects the most economical method of shipment, schedules delivery on an optimum basis, etc., he fills in the remainder of the form and has the purchase order typed. The purchase-order set (Fig. 1.37) has a duplicating master with which the actual purchase order and its copies can be reproduced and the copies can be furnished to the receiving department, buyer, expediter, requisitioner, and purchasing-department files. At the time the purchase order is duplicated, vendor-acknowledgment, accounting, and commitment copies are also produced. Thus, starting with the original requisition, succeeding activities generate and use common information through the flow of the documents.

In this method of operation, production control establishes the requirements for production material, as described in Par 1.4.3, by exploding the engineering parts lists. Schedule dates for all raw material and parts are established by using the final shipment date and the setback pattern for each assembly and part. In manual systems, this process is a laborious one. However, with mechanized systems, the generation of material requirements can be a by-product of the explosion of the engineering parts list, also required for planning, scheduling, and internal-ordering purposes. Much of the fixed information needed by purchasing can be added to either the engineering or manufacturing information files. Ideally, as already discussed, engineering should generate engineering requirements, and manufacturing and purchasing should add their fixed information to a single master file.

Increasing use of mechanization will undoubtedly cause material requirements to be generated by functions external to the purchasing organization. A mechanized form can be designed to combine buying action, open-order status, and supplier-source data. A mechanized system can automatically compare new material requirements with existing open orders and stock condition and can generate action notices for the buyer if necessary. With a mechanized system, purchasing personnel are able to devote a greater portion of their time to the very important tasks of reducing the cost of purchased materials and services and improving vendor and supplier schedule and quality performance.

B. *Method 2—internal generation of material requirements.* When purchasing is assigned the responsibility for generating production-material requirements, a material release activity within the purchasing organization receives the internal order, engineering documentation, and scheduling requirements. This activity explodes the parts list and establishes need dates for incoming material in support of internal fabrication and assembly schedules. A material control group then processes the requirements and checks them against existing stocks. If material is on hand to support both new and old requirements, no action is necessary. If this is not the case, a requisition to purchasing (Fig. 1.34) is written and delivered to the buyer, who then proceeds to issue the purchase order as described previously.

It is obvious that this system can also be mechanized. However, in the long run, it is more economical to have material requirements externally generated by production control at the same time that organization is establishing in-house planning, scheduling, and ordering.

BIBLIOGRAPHY

Aljain, G. W., *Purchasing Handbook*. New York: McGraw-Hill Book Company, 1958.

Allen, Louis A., *Management and Organization*. New York: McGraw-Hill Book Company, 1958.

Anthony, Robert N., *Planning and Control System—A Framework for Analysis*. Boston, Mass.: Division of Research, Graduate School of Business Administration, 1965.

———, *et al.*, *Management Control Systems: Cases and Readings*. Homewood, Ill.: Richard D. Irwin, Inc., 1965.

Barnard, Chester I., *Organization and Management*. Cambridge, Mass.: Harvard University Press, 1948.

Bethel, L. L., *et al.*, *Industrial Organization and Management*. New York: McGraw-Hill Book Company, 1962.

Dale, Ernest, *Planning and Developing the Company Organization Structure*. New York: American Management Association, 1952.

———, *The Great Organizers*. New York: McGraw-Hill Book Company, 1962.

Drucker, Peter F., *The Practice of Management*. New York: Harper & Row, Publishers, 1954.

Fayol, Henri, *General and Industrial Management*. London: Sir Isaac Pitman & Sons, Ltd., 1949.

Forrester, J. W., *Industrial Dynamics*. Cambridge, Mass.: The M.I.T. Press and John Wiley & Sons, Inc., 1961.

Heinritz, S. F., *Purchasing Principles and Application*. Englewood Cliffs, N. J.: Prentice-Hall, Inc., 1959.

IBM Corporation, *Aerospace Information and Control Systems*, Manuals E20–8111 to E20–1827, inclusive. White Plains, N.Y.: IBM Corporation, 1962–64.

Johnson, R. A., *et al.*, *The Theory and Management of Systems*. New York: McGraw-Hill Book Company, 1963.

Lazzaro, V., ed., *Systems and Procedures: A Handbook for Business and Industry*. Englewood Cliffs, N.J.: Prentice-Hall, Inc., 1959.

Moore, Franklin G., *Production Control* (2nd ed.). New York: McGraw-Hill Book Company, 1959.

NICB Report, *Improving Staff and Line Relationships*. New York: National Industrial Conference Board, Inc., 1956.

Radamaker, T., ed., *Business Systems*, Vols. I and II, Cleveland, Ohio: Systems and Procedures Association, 1963.

Scheele, E. D., *et al.*, *Principles and Design of Production Control Systems*. Englewood Cliffs, N. J.: Prentice-Hall, Inc., 1960.

Starr, Martin K., *Production Management: Systems and Synthesis*. Englewood Cliffs, N.J.: Prentice-Hall, Inc., 1964.

Villers, Raymond, *Dynamic Management in Industry*. Englewood Cliffs, N.J.: Prentice-Hall, Inc., 1960.

JOEL DEAN, *Joel Dean Associates, economic and management consultants, Hastings-on-Hudson, New York, is the author of a number of pioneering books on management problems, including* Managerial Economics (*Prentice-Hall, 1951*) *and* Capital Budgeting (*Columbia University Press, 1951*). Managerial Economics *received the National Award of the American Marketing Association and is widely used, both as an executive handbook and as a university text. Dr. Dean is a graduate of Pomona College; he holds an M.B.A. from the Harvard Graduate School of Business Administration and a Ph.D. from the University of Chicago. He has honorary doctoral degrees from the University of Stockholm, Pace College, and the University of Torino. In addition to his work as an economic and management consultant, he is emeritus professor of business economics in the graduate faculties of political science and in the graduate school of business, Columbia University. He is also associate editor of the* Journal of Industrial Economics.

During World War II, Dr. Dean headed the machinery pricing section of the Office of Price Administration and later had charge of developing and operating gasoline and fuel-oil rationing. Before the war he was executive secretary of the Conference on Price Research of the National Bureau of Economic Research. He was director of the Institute of Statistics and a member of the faculty of the University of Chicago for five years. His consulting firm was established in 1939.

A recognized authority on marketing, Dr. Dean has written many articles on pricing and marketing policies. For several years, he was an editor of the Journal of Marketing. *His pioneering work in the statistical determination of cost is widely recognized as setting a standard for later researchers. Among his publications are "Measuring the Productivity of Capital"* (Harvard Business Review, *Jan.–Feb. 1954*); *"Pricing Policies for New Products"* (Harvard Business Review, *Nov. 1950*); *"Measuring the Real Economic Earnings of a Machinery Manufacturer"* (Accounting Review, *April 1954*); *"Decentralization and Intracompany Pricing"* (Harvard Business Review, *July–Aug. 1955*); *"A Positive Program for Anti-Trust"* (Business Horizons, *Winter 1965*); *"Does Advertising Belong in the Capital Budget?"* (Journal of Marketing, *Oct. 1966*); *"Measuring the Productivity of Investment in Persuasion"* (Journal of Industrial Economics, *March 1967*); *"Pricing Pioneering Products"* (Journal of Industrial Economics, *July 1969*).

SECTION 2

MANAGERIAL ECONOMICS

JOEL DEAN

2.1 PROFITS

2.1.1 Nature of Profits

Profits are the primary measure of the success of a business firm. But the word "profit" has different meanings to investors, tax collectors, workers, politicians, and economists. Consequently, before we can decide what is and is not a profit, and before we can improve a company's profit position, we must first understand how profits arise.

2.1.1.1 Types of Profit Theory. Economic theories on profits can be classified into three broad groups: The first looks upon profits as the reward for bearing risks and uncertainties; the second views profits as the consequence of frictions and imperfections in the competitive adjustment of the economy to dynamic changes; the third sees profits as the reward for successful innovation.

According to the first type of theory, profits are the factor payment for taking a risk—for agreeing to take what is left over after contractual outlays have been made. This profit theory

*This section of the Handbook is drawn largely from two books: Joel Dean, *Managerial Economics* (Englewood Cliffs, N. J.: Prentice-Hall, Inc., 1951); and Joel Dean, *Capital Budgeting* (New York: Columbia University Press, 1951).

is based upon the notion that most people prefer stable, moderate incomes to high incomes that have high risk. Hence those that are willing to assume the risk get higher earnings, not only when they are lucky but (as a group) over the long run as well.

The major risk for a businessman comes from holding special-purpose assets as opposed to holding general purchasing power. To be sure, price-level risks arise from holding cash or its equivalent, but an even greater risk exists in holding a resort hotel that is subject to the hazards of changes in tastes and technology as well as of business cycles.

According to the second type of theory, profits are either just a frictional effect of changes in the economy, or they result from imperfections in the market system that allow individuals to establish monopoly positions. Although frictional profits are transient and add to a zero sum over the long run, monopoly profits can supposedly be maintained and accumulated over the years if the monopoly is protected.

In the third type of profit theory, profits are viewed as a wage for the service of innovation. They are thus a functional income, as in the risk theory. But risk plays no essential part in the innovation theory. Innovation refers broadly to any purposeful change in production methods or consumer tastes that increases national output more than it increases costs. The increase in net output is the profit that comes from innovation.

Risk and uncertainty are not necessary to this theory. The innovator may be able to forecast just what his gain will be, whereas other firms are blissfully ignorant of the impending obsolescence hanging over them. If there is any risk, it is carried by the investor, not by the entrepreneur.

The concept of innovation becomes very broad in this theory, since it includes not only new products, such as synthetic fibers, but also new organizations, new markets, new promotion, and new raw materials. To an important degree, innovation has been built into the competitive system, complete with research laboratories and advertising staffs. In many industries everyone has to run fast to stay in the same place. True innovation should be distinguished from style rivalry. Credible evidence of true innovation is a high rate of company growth from plowed-back earnings.

Theories that ascribe profits to risk-bearing or chance when uncertainty exists can be fitted into advanced concepts of the stationary economy, but they have little relevance to an economy that is growing in size and developing new products and technology. For the dynamic manufacturing industries, which are the principal concern of this book, corporate profits as reported by conventional accounts come from sources that reflect each of these theories of profits. Book profits are usually a combination of (1) the market price of equity capital (which conventional accounting views as a free good, but which must in reality be paid for by taking risks), (2) the ephemeral market power which results from lags in the competitive adjustment to innovation, and (3) rewards for technological, merchandising and managerial innovation.

2.1.2 Profit Measurement

Economists are unhappy about conventional accounting methods for measuring business income. Many think them inadequate and sometimes misleading for decision-making analysis, which often requires a complete reshaping of the conventional income statement. In this section we shall sketch the broad outlines of this classic controversy and suggest the kinds of modifications of conventional income statements that are important to economists and appropriate for various managerial purposes.

The most important differences between the economist's approach and the accountant's approach center on: (1) the inclusiveness of costs, i.e., what should be subtracted from revenue to get profit; (2) the meaning of depreciation; (3) the treatment of capital gains and losses; and, perhaps most important, (4) the price-level basis for valuation of assets, i.e., current vs. historical costs.

In measuring profits, the role of futurity in economic values and in business decisions underlies all four of these issues. Economists look to the future as the basic source of value of today's assets, and the businessman recognizes that for his decisions the past is irrelevant, except as a forecaster of the future. But the accountant has a problem here. In an effort to maintain sound, conservative "standards of factuality," accountants want to report historical facts and eschew "speculation" about the future.

To an accountant, net income is essentially a historical record of the past. To an economist, net income is essentially a speculation about the future. For corporations, economists would measure net income as the maximum amount that can be distributed in dividends (theoretically from now into the indefinite future) without impairing the company's earning power. Hence, the concept aims at preservation of stockholders' real capital. To estimate income, then, requires a forecast of all future changes in demand, changes in production processes, cash outlays to operate the business, cash revenues, and price changes (to state cash flows in terms of constant purchasing power). That is, we need a cash budget (adjusted for purchasing power) that forecasts farther into the future than anyone can see. If this budget were available, a program could be planned for borrowing and investing cash so as to allow for an annual cash dividend payment equal to the uniform consumption of real goods that we here conceive of as the index of the firm's income.

The accountant's concept of income, like the economist's, requires for its measurement a consolidation of dated transactions of cash outlays and cash receipts. But there is a basic difference: The accountant uses *past* transactions instead of *future*.

2.1.2.1 What to Include in Costs. The first specific issue in profit measurement is determining what costs are to be deducted from revenues. Wages, materials, and interest on borrowed capital are indisputable costs, but earnings of management and return on owners' capital are less clearly so.

An executive's earnings are a combination of salary, bonus, and dividends, in proportions aimed partly at tax savings. They contain elements of both cost and profits. But how you distinguish the two depends on the kind of profit theory to which you subscribe. Profits from assuming risk are measured by the executive's differential income from the firm over the earnings that could be gained by putting his time and money into perfectly stable and safe activities. His profits from innovation are more difficult to measure, since they are the differential income over earnings that he could make in established activities of comparable risk. These bases of comparison are not easy to set up. The cost to the firm of owner's capital also requires supplemental analysis to separate from profits this element of true cost and thus measure the economic profitability of a company. Accounting treats equity capital as though it were costless, which is manifestly incorrect. The size of resulting overstatement of economic profits by accounted profits is indicated by the opportunity cost of this equity capital. The return on capital invested in marketable stocks of comparable risk is a rough measure of this overlooked cost.

2.1.2.2 Depreciation. Accountants make periodic depreciation charges to income in order to "recover" the cost of equipment before its usefulness is exhausted. The procedure is to estimate the equipment's useful life in years and to make the annual charge just large enough to equal the original cost within that estimated period.

The objective of this procedure is to allocate the total original cost of equipment to production during the period in which it will be used. The effect is to shelter against dividend payout (over the depreciable life of the asset) cash earnings in the amount of its original cost. The resulting corporate cash savings are the major source of capital for internal corporate investment. The time-shape of accounted depreciation governs the annual amounts of this dividend-shield, created by book depreciation.

The allocation of original cost is not, to the economist, an important question, since original cost should not, per se, affect current decisions. The correct objective of management in this problem is to earn a rate of return on stockholders' capital at least as high, all risks considered, as the stockholders could have made in other types of investment. If the company cannot yield that high a return, capital should properly be turned back to the stockholders.

In principle, management should take three steps: (1) It should measure the current company rate of return as the discount rate that, applied to future earnings (before depreciation charges), will make the sum of these earnings equal to the present disposal value of the company's assets. (2) It should plow back earnings into new investments that promise rates of return at least as high as stockholders could get in comparable companies (its cost of capital), and should make enough of these internal investments to maintain the dollar amount of the company's future earnings. (3) It should distribute as dividends those earnings that cannot, within a reasonable time, be plowed back in projects that better the cost of capital.

This version of management's job is not easy to live with and is not found in accounting or administration literature. This economic view of rate of return on corporate investment from current operations and of the criterion for investment of cash-earnings plowback supplies the backdrop for the Managerial Economist's view of depreciation. The capital wastage that counts is the decline in the power to generate cash earnings. This decline is moderately well measured by the drop in disposal value of equipment which has an active market. Prices of used equipment reflect deterioration of earning power relative to new equipment. To be comparable across time when price levels change, market value depreciation should be measured in dollars of constant purchasing power. This concept of depreciation is indispensable for correct measurement of the productivity of a capitali nvestment and for correct decisions as to obsolescence replacement. For decisions of this kind, book depreciation misleads. Tax depreciation matters only because it affects after-tax cash flow. The capital wastage that is crucial is loss in earnings-value, which is roughly reflected in the market value. There is an unbridgeable gulf between this economic view and the accounting notion that a depreciation charge somehow "recovers" the orginal investment in the equipment.

To summarize, depreciation accounting is too crude a device to be managerially useful in our dynamic economy of changing technology, geographic mobility, rapid growth and rising price levels. As a neatly computed but fictitious cost, book depreciation obscures managerial analysis more than it enlightens. For Managerial Economics, accounted depreciation has just two functions: (1) Tax depreciation, as an allowable deduction, steps up after-tax cash earnings. (2) Book depreciation, as a deduction from reported net earnings, steps up corporate savings by restraining distribution of cash income to stockholders.

2.1.2.3 Treatment of Capital Gains and Losses. Capital gains and losses, or "windfalls," as they are often called, may be defined loosely as unanticipated changes in the value of property relative to other real goods. That is, a windfall reflects a change in someone's anticipation of the property's earning power. Fluctuations in stock-market prices are almost all of this nature.

A sound accounting policy to follow concerning windfalls is never to record them until they have been turned into cash by a purchase or sale of assets, since it is never clear until then *exactly* how large they are in dollar terms. However, a fact-minded management ought to have some sort of balance sheet, if only an estimated one, that forecasts and reacts to capital gains long before they have become exact enough to be acceptable to accountants. For example, if prices are to be determined with the objective of producing a "reasonable" rate of return on the valuation of investment, they should reflect projectable windfalls even though not yet cashed. Otherwise, a target rate of return based on a historically "factual," but nevertheless fictitious, capital value may lead to later unpleasant surprises from the resulting price policies.

2.1.2.4 Current vs. Historical Costs. As measured by accountants, profits are a residual in a calculation that uses dollars of many different dates—today's cash dollars, last year's inventory dollars, and equipment dollars of many different earlier years and diverse price levels. To measure real profits, all these assets must be stated in dollars of the same purchasing power. This is an elaborate operation, and the desirable data on prices, products, and dates are usually hard to estimate. With some expediting assumptions, however, usable approximations can be made. In respect to price-level impacts, three kinds of earnings estimates may be distinguished: (1) jumbled-dollar profits, (2) contemporary-dollar profits, and (3) constant-dollar profits. The earnings reported by conventional accounting are, as we have seen, a jumble of dollars of different dates and usually of different purchasing power.

Contemporary-dollar profits can be estimated by making price-level adjustments that would make the revenues and costs of a particular year reflect dollars of that year's purchasing power. For many enterprises, contemporary-dollar profits can be approximated by a combination of replacement-value depreciation and LIFO (last-in-first-out) costing of materials.

By the LIFO method, the last materials purchased are the first charged to the cost of goods sold. When the inventory turnover is slow, business income is measured on the basis of more recent prices of materials than under the first-in-first-out (FIFO) method. When prices are rising, LIFO produces a lower income than FIFO (since stated material costs are higher), and when prices are falling, it shows a higher income. LIFO thus tends to wash out the paper profits that result from comparing a closing inventory with an equal opening inventory stated at different prices. However, to attain the ideal of economic realism, a full restatement of inventory in constant prices is required.

An interesting estimate of the effect of contemporary-dollar profit deflation, made by the Machinery and Allied Products Institute in May 1949, was that the figure $50 billion for total corporate profits for 1946–1948 was 38 per cent fictitious. Nineteen billion was actually inventory profits (i.e., increase in the dollar value of a constant physical inventory) plus under-depreciation (failure to charge enough depreciation for replacement purposes).

A thoroughgoing deflation of financial statements requires a restatement of all accounts, year by year, in terms of a dollar of constant purchasing power and then a recomputation of profits. Such a deflation is a lengthy operation, but during inflation the results can be striking, as shown by the disparity between book and real earnings (Fig. 2.1). The middle column shows how inadequate deflation is achieved by dividing reported profits by a single index number.

When a year-by-year restatement of assets in constant dollars is combined with economic analysis, executive judgment, and imagination, income estimates can be tailored to fit such diverse kinds of management decisions as cost minimization, pricing policy, capital rationing, and dividend policy.

To expect the accounts, without additional analysis, to supply information of universal relevance is unrealistic. The primary purpose of an accounting system is to supply standardized, comparable pecuniary history. The accounts are a source of basic historical data that should be analyzed, supplemented, and projected in different ways to fit different kinds of managerial decisions.

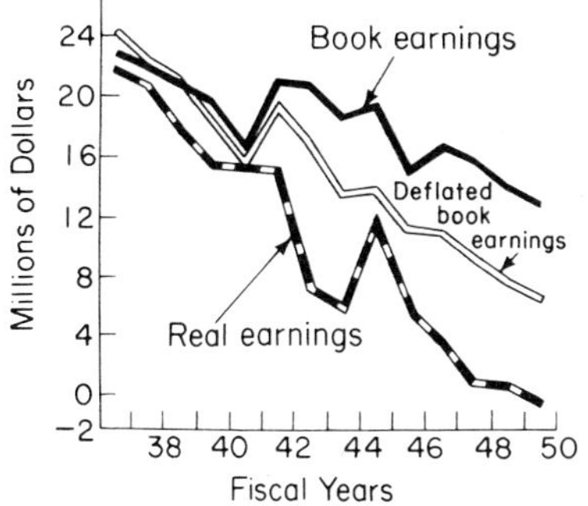

Fig. 2.1 *Reported earnings compared with real economic earnings of a machinery manufacturer. From* Accounting Review, *Vol. XXIX, p. 261.*

2.1.3 Profits for Control

In many large corporations profits are imperiled by bureaucratic inertia. Middle and lower management people tend to maximize security or comfort (e.g., routine) or personal influence in ways that conflict with the profit-making objective of the corporation. Three common deviationist tendencies appear when the profit motive is thus attenuated: (1) Energy is spent in expanding sales volume, maintaining "market position," and proliferating product varieties rather than in maximizing the present worth of long-run profits, which should be the main corporate objective. (2) Staff specialists strive for professional recognition and perfectionist standards of quality that cost more than they are worth. (3) Lower echelons of line management tend to be cautious and oversafe since there are no rewards for imaginative ventures commensurate with the perils of making mistakes.

For large, multiproduct companies, a way to prevent this kind of hardening of the profit-seeking arteries is to decentralize managerial responsibility in profit centers. The essence of this new kind of decentralization is to create within the corporation the counterpart of an enterprise economic system by breaking up a big firm into many quasi-autonomous business enterprises. The economic mechanics require five realignments:

1. Profit-center boundaries need to be marked off for a realignment of operating responsibility, usually shifting from a functional or territorial basis to a primarily product basis, with functional specialization and geographic reach subordinated.

2. Executives' responsibilities need to be restructured so as to give the profit-center manager authority over all functions in his division and thus make him able to control profits.

3. Transfer pricing needs to be market-oriented; economic valuations of products and services that are transferred inside the corporation should be substituted for accounting valuations.

4. Measurements of management performance need to be profit-oriented. Accounting reports should conform to the changed areas of executive responsibility; they should incorporate changed concepts and structures of profit performance; and they should be tuned to economic realities and be diagnostically useful by revealing the principal determinants of profit performance through subordinate budgetary controls.

5. Goals of profit center performance need to be competitively oriented and in line with economic realities. Such performance standards, which can be the basis for incentive compensation, should be predominantly based on short-run profit performance but should also reflect determinants of long-run profitability, such as market share, product modernity, and executive development.

Managerial decentralization achieved by profit responsibility is significantly superior to alternative devices. The establishment of profit centers with economically realistic intracompany prices and costs will permit measurement of executive performance in terms of achievement of profit goals that are adequately adjusted for circumstances. This sort of profit control increases the likelihood that independent decisions of component units will be in harmony with the objectives of the corporation as a whole. Measuring performances of component divisions in terms of economic profit forces integration of conflicting objectives (e.g., volume aspirations in the face of rising selling costs) at lower executive levels where greater familiarity with details is found.

2.1.4 Profit Forecasting and Break-even Charts

Three approaches to profit forecasting may be distinguished. (1) Spot projections: prediction of the entire income statement for a specified future period by forecasting each important revenue and cost separately. (2) Break-even analysis: identifying functional relations of both revenue and costs to output rate, with profits related to output as a residual, or, alternatively, relating profits to output directly by the usual data used in break-even analysis. (3) Environmental analysis: relating the company's profits to key variables in its economic environment, such as general business activity and general price level.

In recent years, break-even analysis has come into wide use by company executives, investment analysts, labor unions, and government agencies.

2.1.4.1 Break-even Analysis. A break-even chart (the central tool of break-even analysis) is a diagram of the short-run relation of total cost and of total revenue to rate of output. Figure 2.2 is a typical break-even chart, in which the vertical spread between the revenue line and the cost line defines the profit function.

Most breakeven analyses are aimed at the relation of short-run cost and profit behavior. But decidedly different kinds of data are used by different people in constructing break-even charts. At one extreme is the inductive method of asking accountants and engineers to estimate what costs should be at two or three hypothetical levels of output. At the other extreme is the simple empirical method of plotting costs against revenues over a long period of years, giving what is called a *migration-path* chart. Between the extremes is the statistical analysis of historical data that have been carefully selected for their relevance to the present product mix and plant, and have been corrected for irrelevant price changes and cost lead times.

The migration path probably has very little validity for current profit forecasting, since it is primarily a history of the growth of a firm and reflects all the changes that have occurred in plant, technology, product-mix, and labor unionization. What is needed is a chart for the conditions that management expects to meet during the coming forecast period.

The value of the inductive and statistical

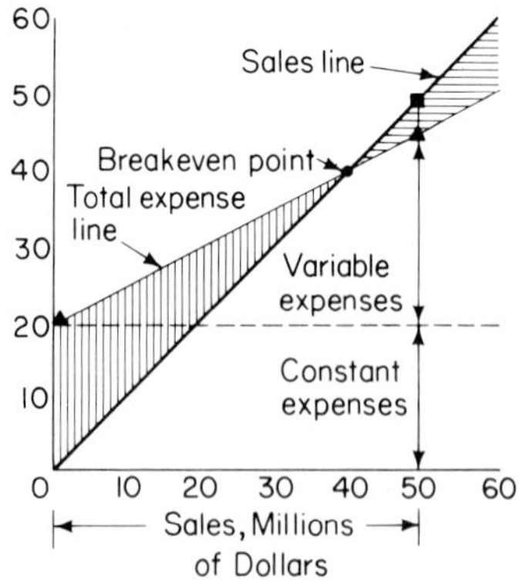

Fig. 2.2 *Basic static relation (breakdown chart). From Dean,* Managerial Economics, *p. 328.*

break-even charts depends partly on the care and skill with which they have been constructed, and few break-even charts seem to have met the important problems adequately. Probably the most serious pitfall has been the use of sales as a measure of output (the horizontal axis) in a company-wide break-even chart, even when the analysis covers a decade or more of fluctuating dollar values, changing product-mix, and changing patterns of plant utilization. To be meaningful for any extended period, a break-even chart should be limited to an individual plant, with an appropriate regrouping of company cost and sales records. And output should be measured in some kind of physical units for a product mix that is similar to the current and future mix. Perhaps the best practical measure of output for a multiproduct plant is production valued at some set of constant sales prices for the various products. For instance, an output series from 1945 to 1970 could be measured by valuing all products through the entire period at their 1970 prices. Of course, if output is in constant dollars, the total expense line must also be in constant dollars. That is, it must be adjusted to eliminate variations caused only by changes in wages and materials prices, since without this adjustment the cost of a given output in 1968 would be much above its cost in 1948—a spurious discrepancy for forecasting purposes.

With output thus measured physically, the slope of the sales line on the chart is no longer always 45 degrees. Its slope will instead depend on net selling prices and can therefore be manipulated to explore the profit consequences of changes in the firm's general level of prices.

Even after all these intricate and costly adjustments, the resulting break-even charts must be used with reservations. First, a single chart is generally good for only one product mix; hence, ideally, a whole set of charts is needed for the possible alternative combinations of products. Neither the data nor the money is usually available for such exhaustive analysis. Second, the variability of most costs with output is more subtle and sensitive than the crude dichotomy of perfectly fixed vs. proportionately variable, which is the underlying presumption of break-even analysis. To measure and forecast cost behavior with precision is expensive. Third, some costs are more manipulatable and postponable than is portrayed by break-even analysis. Substitutable materials, cost-cutting innovations and investments, avoidable overhead costs, postponable intangible investments (e.g., in promotion and R&D), and many other conditions are within management's discretion to tighten up when the compulsive forces of adversity make themselves felt.

2.2 DEMAND ANALYSIS

Demand analysis has two main managerial purposes: (1) forecasting sales, and (2) manipulating demand.

The sales forecast is the foundation for planning all phases of the company's operations, since purchasing commitments, production schedules, inventory plans, cash budgets, and capital expenditure programs all hinge on the short- and long-run view into the future. In an industry that is subject to wide seasonal swings in sales, there is an optimum production program that minimizes the total costs by balancing the costs (and risks) of inventory accumulation in slack seasons against the savings of larger lots and level production. In shooting for this operating program, forecasts of sales obviously play a basic role.

A second, more aggressive and creative use of demand analysis is that of formulating and comparing alternative business-getting plans and policies. Sales forecasting per se is largely passive. Starting with projections of external economic factors, it predicts industry sales and derives a forecast of the sales volume which will result if the firm continues on its present course.

To use demand studies in an active rather than a passive way, management must recognize the degree to which sales are a result of price policy, product improvements, advertising, and marketing effort. An estimate of what will happen with no changes in policy is useful, but this is only one of many alternative programs. Passive forecasts should be primarily a benchmark for estimating the consequences of other plans for adjusting prices, promotion, and/or products.

2.2.1 Demand Theory

To shape an empirical analysis of demand that will answer managerial questions, we should be familiar with various concepts of demand and know which concept is appropriate for each sort of management problem.

2.2.1.1 Demand Functions. Economic theory of demand has concentrated on the relation between the price of a product and the volume sold, which is frequently called the demand schedule. However, price is just one of many factors affecting sales, and the price-sales demand schedule is a single part of the complete functional relation between sales and its determinants. Promotional outlays, product styling, national income, and competition may affect sales even more strongly than does price, and these variables all interact with one another in such a way that the simple price-sales relation or income-sales relation depends on the level of all other variables. We have, in mathematical terms, demand functions in which several independent variables act jointly on the dependent variable—sales.

In empirical analysis, we never have sufficient data or money to establish complete demand functions. Hence the general practice is to go after important parts of the function—e.g., the income-sales relation—when other variables are reasonably steady. Which parts of the function are most important for a particular product depends on the type of product, type of buyer, and type of competition to be met. In general, the important determinants that are beyond management's control are incomes of buyers, lures of rivals, and prices of substitutes. The most important determinants that are controllable are price, promotion, and product.

2.2.1.2 Producers' Goods vs. Consumers' Goods. There are three general reasons for expecting distinctive demand behavior for producers' goods: (1) Buyers are professionals, and hence are more expert, price-wise, and sensitive to substitutes. (2) Their motives are more purely economic; products are bought, not for themselves alone but for profit prospects. (3) Their demand, being derived from demand for end-product consumers' goods, fluctuates differently and generally more violently.

2.2.1.3 Durable Goods vs. Perishable Goods. A durable good (e.g. a truck) is a stockpile of service-units (ton miles) which it gives off over its life. Durability makes demand analysis more complicated than for a perishable good, which gives a one-shot service. Demand is for the service units (ton miles) basically and determines the population and hence the annual demand for the durable good (trucks).

It is handy to study separately two components of this current demand for a durable good because their determinants are different: (1) replacement: replenishing the stockpile of service-units by maintaining the population of trucks and (2) expansion: increasing the stockpile of ton-miles by adding to the truck population.

The most important determinant of replacement demand is obsolescence, which governs the profitability of displacement. Physical deterioration should not be a deciding factor in the replacement of durable producers' goods. The precise timing of obsolescence replacements of producers' durables is influenced by cost of capital, by corporate cash generation, by tax stimuli (e.g., removal and reinstatement of the investment tax credit), and by expectations as to future utilization rates, future costs, and future life expectancy and salvage value. Most investments designed primarily to modernize also expand capacity, often relegating superseded equipment which is obsolete for first-line service to standby capacity (for which it is not obsolete). Thus obsolescence is individual, and rate of return is the test. But status and style are additional determinants of replacement demand even for some producer goods. As to modernization demand for consumer durables, all is vanity.

For the expansion component of demand for producers' durables the determinants are different. Growth in demand for the output of the equipment, economies of scale, aspirations as to share of productive capacity, and forecasts of intensity of competition, sales share, and profit margins are ingredients in the rate-of-return estimate that underlies expansion demand for producers' durables. For consumer durables, the expansion demand calculus is more simple and lyrical; e.g., for refrigerators: affluence, marriage, and mother-in-law.

2.2.1.4 Derived Demand vs. Autonomous Demand. When product demand is tied to purchase of some parent product, its demand is called "derived." Sometimes the dependent product is a component part (e.g., demand for doors derived from demand for houses). Sometimes dependence comes from complementary consumption (e.g., pretzels from beer).

The use of some products (e.g., steering wheels) is so closely tied to the use of others that they have no distinctive demand determinants of their own. But fixed proportions are rare; more often there is substitution leeway in the proportions as well as more than one parent use. Nylon-tire-cord demand, when related to the population of motor vehicles reflects both substitution and multiple uses.

As the number of uses and substitutability in each use increase, it becomes harder to tie demand down to parent products. For instance, small electric motors have no primary uses, but to analyze their demand in terms of their thousands of parent uses is impossibly tedious. The derived-demand approach facilitates forecasting when proportions of the two products are fairly stable and when there is a rigid time-lead in the parent product's demand.

2.2.1.5 Industry Demand vs. Company Demand. Many management problems require analyses that distinguish industry demand from company demand and explore the relationship between them. A projection of airline industry sales is usually an intermediate step in forecasting company sales; intelligent price leadership (and followership too) is based on an understanding of the relation of the company demand to that of competing firms.

An industry-demand schedule represents the relation of the price of the product to the quantity that will be bought from all firms. It has a clear meaning when the products of the various firms are close substitutes, when they differ markedly from those of bordering industries, and when they have a well-defined price level.

The conceptual distinction between industry demand and company demand is most useful when the boundaries of the industry are clearly definable in terms of a gap in the chain of substitutes (i.e., products that differ sharply in terms of substitutability from those of other industries), and when rival firms are large enough and similar enough to plan in terms of market share. Empirical research can then be framed in terms of price spreads or ratios over rivals, and of the effect of these price differentials upon market share.

As an illustration, estimates have been made of the sensitivity of gasoline market shares to changes in price differentials among brands. Narrowing or widening the differentials between the branded product of the price leader and the local private brand of the cut-rate distributor has a prompt effect on retail patronage. Similar market-share analyses can be carried out for other demand determinants, such as advertising.

"Market-share" concepts of demand are most usable in mature, well-defined oligopolistic industries whose products are relatively homogeneous—e.g., steel or cement. In this setting, market-share objectives are often dominantly defensive.

Planning marketing strategy to affect market share is, for established firms, largely a matter of competitive developments, e.g., overtaking product innovators, meeting price cutters, and countering advertising aggressors.

Market-share maintenance is sometimes a paramount corporate goal. It should not be. At most, market share is, first, a handy and indolent way to adjust corporate sales for fluctuations in external influences, and second, a rough and unlagged index of competitive competence.

2.2.1.6 Short-run Demand vs. Long-run Demand. A distinction between long-run and short-run demand functions is useful for some problems. Short-run demand refers to existing demand with its immediate reaction to price changes, income fluctuation, and so forth. Long-run demand is that which will ultimately exist as a result of the changes in pricing, promotion, or product improvement, after enough time has been allowed to let the market adjust itself to the new situation.

Thus, long-run demand does not refer to the time distance but instead to the completeness of the adjustment of sales to the altered determinants of demand. Because dissolving the obstacles to complete response takes time as well as education and investment, long-run demand is usually more sensitive to each of the determinants than short-run demand and more comprehensive in reaction to side effects.

Causes of differences between short-run and long-run demand are of two main kinds: (1) cultural lags in information and experience and (2) capital investments required of buyers to alter their consumption patterns. Cultural lags in long-run response of demand can be closed by education, experience, and research. Customer investments required to optimize response may be tangible as well as intangible. Cheaper electricity for heating houses increases demand: in the short run by warmer houses; in the long run by more electric homes.

2.2.1.7 Demand Fluctuations vs. Long-run Trends. The difference between forecasting year-to-year changes and forecasting underlying trends is analogous to that between business cycles and secular economic development. The external factors that are important for cyclical forecasts are different from trend projections. In year-to-year changes, much of the setting stays constant—competitive structure, market position, quality, and sometimes even prices (relative to substitutes and competitors, if not absolutely). The problem can then be narrowed down to the relation between sales pulsations and a few strategic variables, such as income, business activity, and competitive price differentials. For the long trend, in contrast, everything is fluid, and the effects of year-to-year determinants are buried by basic changes in the framework—e.g., shifts in taste, technology, and way of life in a laboristic, urban, welfare state.

2.2.1.8 Total Market vs. Market Segments. The comprehensiveness of the demand problem sets the framework for analysis. Some decisions, such as capacity expansion, call for an analysis that includes the total market. Other problems—notably in pricing, promotion, and distribution—call for analyses of separate market segments that have homogeneous demand characteristics. Examples are the Medicare market for pharmaceuticals or the young-married's market for Mustangs.

2.2.2 Methods for Forecasting Demand

Management uses of short-run sales forecasts differ from those of long-run forecasts. The near-future forecast supplies an economic foundation for operations planning: scheduling production, programing inventories and logistics of physical distribution, and projecting cash generation and operating profits. Long-run sales forecasts supply the framework for corporate investment planning: sourcing capital and rationing it among tangible investment proposals such as modernization and expansion of capacity, and among intangible investments such as research, promotion, and executive development.

Techniques usually are different for short-run forecasting than for long-run forecasting. The focus of most short-run forecasts is on the shifts of the demand curve caused by seasonal forces and by fluctuations in business conditions and in anticipations, e.g., buyers' incomes, expected income, job outlook, cash balances, and inventories. For durable goods, much of this short-run volatility of sales is made possible by the postponability of demand through short-run drawdowns of the stockpile of services (e.g., car miles embodied in the car population). In contrast, long-run forecasts focus on the trend of demand, abstracting from the seasonal pattern and averaging away cyclical fluctuations. Underlying relationships of sales to basic determinants of demand assert themselves in the long run because the determinants themselves range more widely and because response of sales to their changes will, in the long run, be more complete and sensitive. Long- and short-run forecasts overlap both in uses and in techniques.

When products of rivals are close substitutes, the forecast of company sales is usually made in two steps. The first is to predict total industry demand. The second is to forecast the company's sales share, usually separately for major sectors of the market, e.g., categories of users. The economic justification for a two-step approach is that the determinants of industry demand are different from those of sales share and that rival brands are close enough substitutes to make their aggregate sales more predictable than those of any one brand. The analysis that underlies the forecast of industry sales is usually more explicit and statistical than that for the market-share forecast.

2.2.2.1 Forecasting Sales of Established Products. This analysis centers on forecasting sales of consumer durable goods, which is believed to be particularly difficult, first because durability creates a carry-over of service units and second because consumers are more capricious in their investment outlays than are producers. Durability makes forecasting harder because part of the demand for new goods is for replacement purposes, a need that can be postponed and one that is affected by the believed adequacy of existing stock. Whether or not a

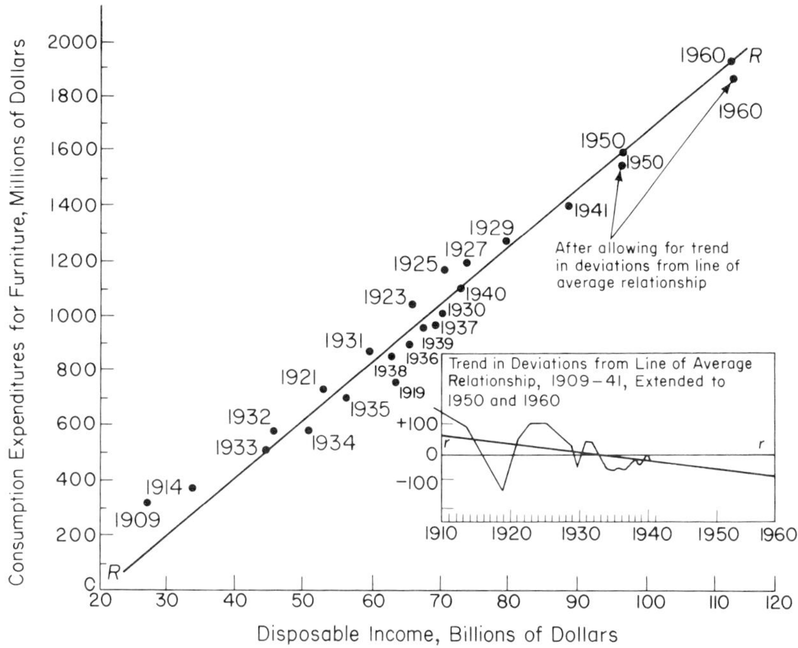

Fig. 2.3 *Relation between disposable income and consumption expenditures for furniture—1909–41, extended to 1950 and 1960. From Frederic Dewhurst and Associates,* America's Needs and Resources *(New York: Twentieth Century Fund, 1947), p. 728. Reprinted in Dean,* Managerial Economics, *p. 198.*

stock is adequate is determined more by attitudes of consumers than by objective physical or financial facts. Evidence of this is the dramatically low level of scrappage rate of automobiles during World War II. Unlike producers' goods, the adequacy of the existing stock of consumer durable goods apparently cannot be determined by obsolescence analysis, i.e., by precise calculations of the comparative costliness of keeping up an old product vs. buying a new one. One reason for this is that the quality of service supplied by the repaired durable good differs from the new, in the eyes of the consumer; a second reason is vanity, i.e., that his willingness to buy high-quality service is volatile, depending on the condition of his ego and his financial status and expectations.

Segregation of two components—replacement and expansion—which is useful for forecasting producers' durable demand is less clean cut and useful for consumer durables, partly because obsolescence does not come through a sharp, financial rate of return which supplies a rational basis for replacement demand. Compared with producers' durables the benefit from expansion of capacity of durable consumer goods (e.g., the family car-fleet) is also blurred by judgments about the quality of the services given off by the stockpile.

Techniques for forecasting sales of already marketed products are more formal and elegant than those for new products. For established products, four kinds can be distinguished: (1) time-series analysis, (2) regression analysis, (3) inertial analysis, and (4) intentions surveys.

Time-series analysis is designed to first separate the past behavior of sales into (1) a long-term trend and (2) a seasonal pattern. When these two are subtracted the residual fluctuations, when smoothed to eliminate random factors, are attributed to (3) general business conditions (cycle). For forecasting, these three separated patterns of the past can be reconstituted in a future projection whose precision is unfortunately quite dependent on the accuracy of the forecast of near-term future business conditions.

For some durable goods, predictions of long-term demand can be made by regression analysis. The average past relation of sales to the main independent determinants of demand —e.g., population, household formation, and income (or their growth rates)—is ascertained. Then the future value of each independent variable is predicted and then substituted in the equation. Figure 2.4 illustrates this technique. Regression forecasts are usually most successful for relatively mature products (e.g., automobiles) for which long experience is well documented and the rate of technological change is relatively slow.

High correlation overstates the forecasting ability of a regression equation. A high coefficient can be obtained by having enough variables and fiddling. But the forecasting power of the regression depends, not alone on past correlation but on the stability, causality, and projectability of the measured relationships. Even when regression relationships are close and remain stable, use for prediction requires forecasting the future magnitudes of the independent variables—a task which is sometimes as difficult as forecasting sales themselves. Regression analysis forecasts best when the independent variables have causal and stable relationships and when they can be independently forecasted with tolerable precision. Particularly successful are dynamic models which use the rate of growth of the independent and dependent variables rather than their absolute values. Regression analysis has greater promise for long-run forecasts of demand than for short-run forecasts and for established products than for new ones.

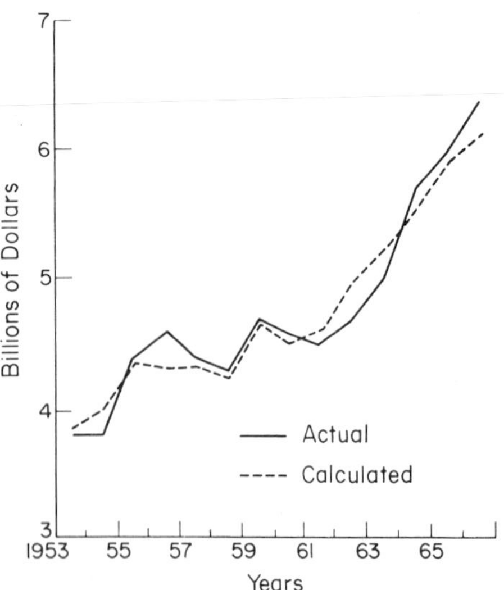

Fig. 2.4 *Personal consumption expenditures for furniture, 1953–66 (calculated).*

A third technique relies on inertia. The method is based upon a simple mechanical extrapolation of the current situation. The surprising success of this naive technique vs. more intricate economic analysis in some kinds of economic forecasting underscores the importance of momentum and habit, even in our dynamic and complex economy.

Surveys of consumer intentions supply a fourth basis for forecasting sales. This method has long been informally but successfully used for producers' durable goods, both for aggregates and for specific kinds of plant and equipment expenditures, e.g., capital appropriations surveys by NICB, McGraw-Hill, and SEC-FTC. Similar techniques are now used for consumer durables by the FRB "Survey of Consumer Finances" (using a sample of about 3000 households) and by the Survey Research Center of the University of Michigan in periodic surveys.

2.2.2.2 Forecasting Demand for New Products. Methods of forecasting demand for new products are quite different from those for established products. Forecasting methods have to be tailored to the particular product. Possible approaches can be classified as follows: (1) Project the demand for the new product as an outgrowth and evolution of an existing old product. (2) Analyze the new product as a substitute for some existing product or service. (3) Estimate the rate of growth and the ultimate level of demand for the new product on the basis of the pattern of growth of established products. (4) Estimate demand by making direct inquiries to the ultimate purchasers; then blow up the sample to full scale. (5) Offer the new product for sale in a sample market—e.g., by direct mail or through one chain store. (6) Survey consumers' reactions to a new product indirectly through the eyes of specialized dealers who are supposedly informed about consumers' needs and alternative opportunities.

2.2.3 Price Effects

Demand analysis that could isolate and measure the effect of price on sales would be useful for pricing decisions and product policy. To predict the sensitivity of volume to price, correlation analysis of the relation between short-run behavior of sales and short-run changes in prices has been widely and successfully used for agricultural commodities. For manufactured products, however, measurement of the effect of price on sales by correlation analysis is much harder. Usually prices have not been changed over a wide enough range to provide data. The product and the marketing mix keep changing; gestation lags and changes in competitive substitutes further obscure the effects of price on sales.

A promising method of measuring the effects of price on sales is the controlled experiment. Data can be created in various ways: by varying prices and price relations in a research design in the same markets or by simultaneous test marketing at different prices in different markets. The research problems are first to isolate price effects from the noise of other determinants, and second to appraise the reliability and projectability of the relationship.

Two kinds of effects of a price reduction must be distinguished: (1) effect upon the sales of the whole industry, assuming rivals match your price changes (market-demand elasticity) and (2) effect on sales share of the price cutter, assuming his price cut is not matched by rivals (share elasticity). The first kind of price elasticity is measured by statistical analyses of agricultural staples. For new manufactured products, such as frozen orange juice and instant coffee, market-demand elasticity was explored by test marketing the product in comparable sample cities at different prices.

The second kind (share elasticity) was measured for mattresses in a classic experiment. Identical mattresses, some bearing the Beauty Rest brand and others an unknown brand, were sold side-by-side, first at identical prices and later at various price spreads, to determine the effect of price disparity upon sales. At equal prices, Beauty Rest outsold the other brand fifteen to one; when the unknown was priced 12.5 per cent lower, it was outsold eight to one; and with a 25 per cent differential, sales were equal. Similar experiments to measure share elasticity of demand in respect to price are now common, particularly for products in supermarkets.

2.2.4 Income Effects

The relation of changes in demand to changes in buyers' income is basic for short-run sales forecasting. It has been the subject of much statistical study, particularly in terms of national disposable personal income. Similar analyses of the relation of corporate income (profits) to business investment show significant correlations.

The net relation of sales to national income is not simple, however, since it depends upon the underlying relations of the incomes of the particular group of buyers to the national aggregate. For example, if a function is fitted statistically to historical data for hay-baler sales and national disposable income, the curve is valid for the future only if hay-farmers' income fluctuations either conform rigidly to the national pattern or maintain the same relationship to national income as in the past. (This problem frequently arises when broad aggregates are used statistically, since such aggregates usually include much more than is relevant to the analysis of a single product's demand.

An illustration of a simple method for determining income relations is given in Fig. 2.3, which shows such a function for household appliances. The straight line in this chart is drawn freehand, using the data for 1909 to 1940, and extrapolated to full-employment income levels that were expected for 1950 and 1960 to make an estimate of potential appliance sales for those years. Deviations from the straight-line function show a distinct upward trend when plotted against time; if the trend is taken into account, the estimates are raised substantially, as shown.

2.2.5 Multifactor Effects

Analysis of demand in terms of a single controlling factor has only limited value, since the effect of price or of income on demand, though perhaps dominant, is measurably qualified by other variables such as the product improvement or promotion of competitors or substitutes. The natural route to a general demand function is therefore multiple-correlation analysis, where we can use several independent variables and get some idea of the relative influence of each on demand. Figure 2.4 illustrates this technique as applied to furniture demand.

Consumption expenditures for furniture in billions of dollars for 1953 through 1966 (actual) are compared with those predicted from a linear, least squares, multiple regression equation for the years 1953–1966 (calculated). The regression measures the past relationship between demand for furniture and three determinants of demand: disposable personal income, value of residential construction, and furniture/consumer goods price ratio.

The equation on which the "calculated" expenditures are based is:

$$F = -9.325 + 0.099Y + 0.076R + 0.088P$$

where F is personal compution expenditures for furniture, Y is disposable personal income, R is the value of private residential construction, and P is the ratio of the price index for furniture to the index of all consumer goods and services (with 1958 = 100). A high proportion of the fluctuations in furniture demand during this period was "accounted for" by these demand determinants: the multiple regression coefficient is $R^2 = 0.977$.

In using this statistical approach, the important problem is to find a few efficient causal factors, rather than to compile a catalog of influences on sales. Efficient indicators are found not only by insight into demand determinants, but by adaptation of statistical analysis to the problem and by experimentation with different functional relations. For instance, the durability of a demand function is often enhanced by converting dollar values into physical units and sometimes by using a time lag between variables.

2.3 COST

Business decisions are based upon plans for the future and require choices among rival plans. In making these choices, estimates are needed of the effect of each alternative plan upon future expenses and revenues; cost estimates must be tailored to the economic characteristics of the choices. The only costs that matter for business decisions are future costs; "actual" costs—i.e., current or historical costs—are useful solely as a benchmark for estimating the costs that will result from particular decisions.

What does it take to get good estimates of these decision-making costs? The first requisite is a precise picture of the alternative programs involved in the choice. An explicit definition of the differences among alternatives usually requires economic analysis of the situation. A second requirement is an understanding of different cost concepts, in order to select the one that is most relevant. In operating a modern multiple-product enterprise, the situations requiring decisions are so varied, not only among themselves but in the character of the alternatives available, that many different concepts of cost are needed for significant comparison of the alternative plans. The third requisite is flexible classification of the accounting records on several alternative bases. Since decision-making costs require classifications according to rival operating plans, multiple classifications are a desirable informational foundation for such cost estimates. A fourth requisite is ingenuity and skill in quantitative analysis, since typically the desired cost conjectures can be obtained only by economic and statistical analysis of accounting data.

Decision-making costs can be found from traditional accounting records only by reclassifications, deletions, additions, recombinations of elements, repricing of input factors, and analysis of projectable cost-behavior relationships. This is the process of shaping the cost conjectures to fit the concept of cost relevant for the management planning choice. Traditional accounting, because of the need for verifiable data to fulfill its financial reporting function, uses "original cost"—i.e., historical prices of input factors—so that costs do not (with a few exceptions, such as inventory valuation during inflation) reflect changes in price levels or managerial mistakes in investment. Traditional accounts are classified primarily on the basis of the object of expenditure, which is handy for verification. When a classification by function (process or department) is superimposed, it also is generally a formalized, unvarying classification, and the assumption is usually made that the original outlay can be spread among functions proportionately to one or several allocation bases, such as direct labor hours, square feet of space, and the like. Fair sharing of an overhead burden, which is the controlling concept for accounting allocation, does not measure the Managerial Economist's decision-oriented concept of cost causality.

Records of historical outlays, based upon these rigid classifications and formal proportionalities, need to be drastically reworked for decisions about the future. Classification should change from problem to problem.

Valuation as well as classification depends on the purpose of analysis and differs with the economic characteristics of alternative programs. The same piece of equipment will have different values in terms of its disposal price, replacement cost, value in its present job, and value in alternative jobs. The only values that are irrelevant for all decisions on what to do with a specific asset are its original cost and its book value.

Control and appraisal of executive performance require an approach to cost analysis that is essentially different from studies to aid decision-making. For managerial control, costs must be classified according to areas of executive responsibility, and according to the degree of authority over expense delegated to the executive. Once one of the alternative plans is chosen, responsibility for carrying it out in an acceptable manner is assigned, and expenses must be reclassified in a manner that will measure how the performance of each executive compares with some standard or budget. Thus the basic classification by objects of expenditure required for pecuniary history and financial reports must be modified to enable easy reclassification on a basis that parallels the structure of managerial organization.

2.3.1 Cost Concepts

The kind of cost concept to be used in a particular situation depends upon the business decision to be made. Hence an understanding of the meaning of various concepts is essential for clear business thinking. Several alternative bases of classifying costs and the relevance of each for different kinds of problems are discussed below.

Although it may be difficult, workable approximations of these concepts of cost can be developed, given, first, a clear understanding of the management problem and of the concept of cost that is relevant for it; second, familiarity with the business and its records; and, third, ingenuity and boldness. Most of the raw materials for making these special cost estimates are found in the accounting and statistical records of the company, though sometimes they need to be supplemented by special data.

Different combinations of cost ingredients are appropriate for various kinds of management problems. Disparities occur from deletions, from additions, from recombinations of elements, from price-level adjustments, and from the introduction of measurements that do not appear anywhere in the accounting records.

2.3.1.1 Opportunity vs. Outlay Costs. A distinction can be drawn between outlay costs and opportunity costs on the basis of the nature of the sacrifice. Since outlay costs involve financial expenditure at some time, they are recorded in the books of account. Opportunity costs take the form of profits from alternative ventures that are foregone when limited facilities are used for a particular purpose. Since they represent only sacrificed alternatives, they are never recorded as such in the financial accounts.

Opportunity cost is the cost concept to use when the supply of input factors is strictly limited. Such rigid supply may occur for technical reasons, e.g., the limited number of television channels available in a single locality; for social reasons, e.g., wartime rationing; for private reasons, such as lack of ready cash; or because the problem is too short-run to adapt facilities to their most profitable long-run relation to the job. In business problems the message of opportunity costs is that it is dangerous to confine cost knowledge to what the firm is doing. What the firm is not doing but could do is frequently the critical cost consideration that is perilous but easy to ignore. Under conditions of capital scarcity, for example, the cost of acquiring a $400,000 gasoline station in New York City is not usually the interest that would have to be paid on the borrowed money or the market cost of equity capital but, rather, the profits or cost savings that could have been achieved if the $400,000 had been invested in four suburban gasoline stations or in pipelines or refinery facilities. The rule should be opportunity cost of capital or market cost, whichever is higher.

2.3.1.2 Past vs. Future Cost. Most of the important managerial uses to which cost information is put actually require forecasts of future cost, rather than "actual costs," i.e., unadjusted records of past cost. An examination of some of the major managerial uses to which cost information is put reveals the universality of this principle. Among these uses are: expense control, projection of future income statements, appraisal of capital expenditures, decisions on new products and on expansion programs, and pricing. When historical costs are used, the assumption is made that unvarnished cost history is the best available estimate of probable future costs under the situations involved in the decision.

The fact that the future is always uncertain does not detract from the necessity for making explicit forecasts of future costs. We can usually make a more accurate projection than historical costs, which rarely represent the best guess that can be made concerning the future.

2.3.1.3 Short-run vs. Long-run Costs. The distinction between short-run and long-run cost behavior is basic in economic theory. Roughly, short-run costs are associated with variation in the utilization of fixed plant or other facilities, whereas long-run cost behavior encompasses changes in the size and kind of plant. The distinction is based upon the degree of adaptation of all input factors to rate and type of output. When there is perfect flexibility in the size of plant, labor force, executive talent, and so forth, long-run costs are identical with short-run costs. Anything short of complete adaptation of all production facilities to the change (e.g., in rate of output) results in costs that are higher than optimum. Costs could be made lower, given more time, a different size of plant, etc. Strictly, all such "incomplete adaptation" costs are short-run costs. The conventional dichotomy of long-run versus short-run cost curves needs to be expanded in economic theory to envision a whole family of cost curves that differ in degree of adaptation, so that the conventional long-run cost curve is the limiting case of perfect adaptation. In the real world, adjustments to higher output, new materials, or new product designs typically take a variety of forms that fall short of the perfect adaptation of the long-run cost curves.

Capacity is usually expanded and modernized *gradually* by widening a succession of bottlenecks rather than by adding an entire balanced unit. Completeness of adaptation to changes in facilities, methods and products can sometimes be measured and predicted by "learning curves." These progress cost functions relate (a) the reduction of cost toward complete adaptation to (b) some measure of the amount of adaptation experience; e.g., cumulative number of aircraft of new design produced in expanded, renovated plant.

2.3.1.4 Variable vs. Constant Cost. Variable costs are distinguished from constant costs on the basis of the degree to which they vary in total with changes in rate of output. Which cost items are fixed and which variable depends on the degree of adaptation of costs to output rate, i.e., the degree to which the adjustment is short run as opposed to long run. The distinction also depends on the size and the suddenness of the change in output, and on the amount of pressure put on management to increase efficiency and to defer postponable expenditures.

The distinction between variable and constant costs is important in forecasting the effect of short-run changes in volume upon costs and profits. The breakeven chart illustrates this application in predicting profits; the flexible budget illustrates an application to control of costs by setting standards that are adjusted for volume changes.

2.3.1.5 Traceable vs. Common Cost. A traceable cost is one that can be identified easily and indisputably with a unit of operation, e.g., a product, a department, or a process. In accounting terminology, direct costs are distinguished from indirect costs on the basis of traceability to products. Common costs are costs that are not traceable to individual final products, or to plant, department, and operation.

Traceability of costs becomes important when multiple products that incur common cost differ considerably in production or marketing processes, and when cost has significance in decisions on adding or subtracting from a product line, product pricing, or product merchandising. It is not necessary that costs be traceable all the way to the product for this distinction to be useful for management. The degree of traceability varies from cost to cost, some costs being traceable as far down as divisions, others down to departments, and others down to cost centers.

2.3.1.6 Out-of-Pocket vs. Book Cost. Out-of-pocket costs refer to costs that involve current payments to outsiders as opposed to book costs, such as depreciation, that do not require current cash expenditures. Not all out-of-pocket costs are variable—e.g., the night watchman's salary. Not all out-of-pocket costs are traceable—e.g., the electric power bill. Conversely, book costs are in some instances variable with volume—e.g., depletion of ore or oil—and in some instances are readily traceable. The distinction primarily affects the firm's cash position; but it is often used, mistakenly, to measure whether a cost varies in the short run with volume.

2.3.1.7 Incremental Costs vs. Sunk Cost. Incremental costs are the change in total costs resulting from a change in the level or nature of activity. They generally refer to short-run changes in output. But they can refer to any kind of change: adding a new product, changing distribution channels, adding new machinery. Sunk costs are the costs that are not altered by the change in question. Most business decisions require cost estimates that are essentially incremental, and costs that are not altered by the contemplated change are sunk and irrelevant.

Incremental costs are not necessarily variable, traceable, or cash costs. In many short-run problems, the most important incremental cost is the foregone opportunity of using limited facilities in their present work rather than shifting them to a new activity. Similarly, sunk costs can be cash costs (e.g., the president's salary); they can be variable (e.g., when the change in question is of customers rather than products); and they can be traceable.

2.3.1.8 Controllable vs. Noncontrollable Cost. The distinction between controllable and noncontrollable costs depends upon the level of management. Some costs are not controllable at the shop level since they depend on decisions upstairs, but at some level all costs come into the discretionary area of some executive. The controllability distinction is primarily useful for expense and efficiency control, since budgets can be set up that correspond to areas of managerial responsibility, as discussed in Art. 2.1, profits.

From the foregoing classification (see also Table 2.1), it is clear that cost is a relative matter. What is cost depends upon what sacrifices are really produced by a particular business decision. These different cost concepts do not necessarily correspond to any accounting category. It is better to use a rough approximation of the concept of cost that is correct for a particular decision than to have an accurate measurement of an irrelevant concept. The unsophisticated executive is in danger of taking the easier course of using conventional accounting costs as though they were appropriate for all purposes. Instead, it is better to modify reported costs as necessary to make the best possible guess at the concept theoretically relevant for each decision.

2.3.2 Cost and Output Rate

Generally, in the short run a functional relation exists between cost and a set of independent variables, which may include, for example, volume of production, size of production lot, prices of input services, and variety of output. The independent variables will be different for each type of manufacturing operation, although in general the most important variable is rate of output. The independent variables are considered to determine cost behavior.

The output relation is the most important to study because it is subject to faster and more frequent changes. The characteristics of the fixed equipment play a dominant role in the determination of a plant's pattern of short-run cost behavior—i.e., the shape of the relation between total cost and rate of output. The critical characteristic for this purpose is the degree of segmentation possible in the plant, that is, its potentiality of varying the rate of output flow without changing the proportions of variable inputs to fixed equipment in use. Segmentation in this sense refers to the technical nature of the fixed equipment that permits a wide range of choice in the machine-hours used per week.

TABLE 2.1. *CLASSIFICATION OF COST CONCEPTS**

Dichotomy			Basis of Distinction
Opportunity costs	vs.	outlay costs	Nature of the sacrifice
Past costs	vs.	future costs	Degree of anticipation
Short-run costs	vs.	long-run costs	Degree of adaptation to present output
Variable costs	vs.	constant costs	Degree of variation with output rate
Traceable costs	vs.	common costs	Traceability to unit of operations
Out-of-pocket costs	vs.	book costs	Immediacy of expenditure
Incremental costs	vs.	sunk costs	Relation to added activity
Escapable costs	vs.	unavoidable costs	Relation to retrenchment
Controllable costs	vs.	noncontrollable costs	Controllability
Replacement costs	vs.	historical costs	Timing of valuation

* From Dean, *Managerial Economics*, p. 271.

The point where a plant falls on this segmentation scale depends on the technical nature of the equipment, the success of managerial efforts to segmentize, and the nature of the labor contract. Three sources of segmentation can be distinguished. The first is physical divisibility, where fixed equipment consists of a large number of homogeneous units. An example of this is a hosiery mill where the knitting of stocking legs is done on many nearly identical knitting machines. It is possible, secondly, to introduce segmentation by varying the number and hours of the shifts per period that the fixed equipment is employed. Thirdly, if the technical nature of the fixed plant is such that it can be used at varying speeds, "speed" segmentation can be obtained by operating machines at faster or slower rates. In many industries the development of plants that can produce a variety of products in a wide range of proportions, and can operate efficiently at many rates of output, has brought about much time, speed, and unit segmentation.

2.3.2.1 Different Approaches to Determination of Cost Functions. For several kinds of management problems it is desirable to find an empirical approximation of short-run cost functions. Empirical functions are necessarily approximate, since they cannot include every kind of short-run cost factor, nor can they find the exact relation of cost to the factors used.

There are several approaches to an estimate of cost function: (1) Determination of the cost function and the degree of variation with output by statistical analysis. (2) Estimation of relationships of cost to output on the basis of engineering conjectures. (3) Classification of accounts into fixed, variable, semivariable, on the basis of judgment and inspection.

These three approaches are not always mutually exclusive. Often it is desirable to use two or more to supplement each other. Nevertheless, it is usually desirable to try to determine at the outset which of the three should receive greatest emphasis.

Statistical approach. When conditions are appropriate for its use, the statistical approach is likely to produce more reliable results than the other methods. In essence, it uses multiple-correlation analysis to find a functional relation between changes in costs and the important cost determinants, such as output rate, lot size, output fluctuations, and so forth. (See Par. 13.5.8.)

The power of statistical analysis lies in its ability to pick out the fixed-cost elements in each cost component, such as direct labor or fuel consumption, and to show whether costs vary with changes in particular cost determinants.

Figure 2.5 presents one finding of a statistical cost study. The technique has been used here to isolate the relation between output and cost when all other cost determinants are held constant at particular levels. One never finds such constancy of "other" cost determinants in empirical series of cost observations. The constancy shown here is rather one of the products of the statistical method, a highly useful product where a management problem relates to changes in a single cost determinant.

Accounting approach. The accounting approach is usually aimed at finding only the re-

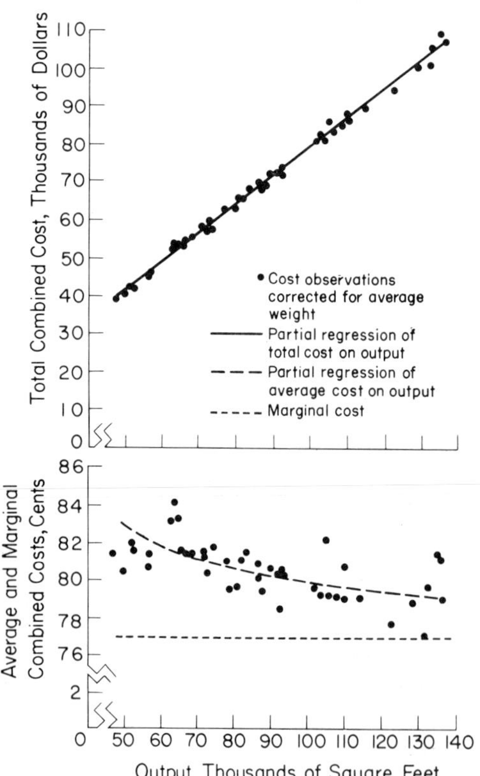

Fig. 2.5 *Partial regressions of total, average, and marginal combined cost on output: leather belt shop. From Dean,* Relation of Cost to Output for a Leather Belt Shop (*National Bureau of Economic Research, 1941*), *p. 27. Reprinted in Dean,* Managerial Economics, *p. 203.*

lation between output levels and costs. The method consists of classifying each cost item by inspection into one of three categories: (1) fixed, (2) variable, and (3) semivariable. Fixed costs are supposed to be the same (in monthly total) regardless of output rate; costs classed as variable are assumed to change proportionally with output; and semivariable costs are sometimes assigned roughly some proportion of variability and sometimes further analyzed into fixed and variable components.

This method is the simplest and least expensive of the three, since the analysis is based chiefly on inspection and experience. It tends to give an oversimplified model of cost behavior and should be supplemented by graphic statistical analysis as a test of cost variability.

Engineering approach. In essence, the engineering method consists of systematic conjectures about what cost behavior ought to be in the future on the basis of what is known about the rated capacity of equipment, modified by experience with manpower requisites and efficiency factors, and with past cost behavior. Hence, it depends upon knowledge of physical relationships supplemented by pooled judgments of practical operators. It should, and usually does, make use of whatever analyses of historical cost behavior are appropriate and available, as a means of improving the judgment. Typically, the engineering estimate is built up in terms of physical units—i.e., man-hours, pounds of material, and so forth—and is converted into dollars at current or prospective cost prices. The cost estimates are usually developed at a series of peg points that cover the contemplated or potential output range.

The engineering approach is the only feasible method when the inadequacy of experience and records provides little systematic historical basis for estimating cost behavior. The approach is also desirable when it is necessary to estimate the effect of major changes of technology or plant size upon cost behavior over a familiar or unfamiliar output range.

2.3.3 Cost and Size of Plant

In the studies of short-run cost behavior discussed in Par. 2.3.2, plant size was approximately constant. We now turn to the long-run problem of finding empirically what effect varying the size of plants has upon cost.

To clarify the measurement problem, it is desirable to distinguish plant size from two other phases of size that sometimes confuse the issue: size of firm, and depth of plant. Size of plant is different from size of firm, even when there is only one plant. The firm encompasses several economic functions in addition to production. Each of these operations usually has a different cost-size relationship and a different optimum size. Big companies find a practical solution for this disparity in having units that differ in size for each economic function—e.g., 60,000-barrel refineries and 1000-gallon gas stations. Despite oceans of speculation, little is really known about the relation of size of firm either to cost or to more subtle and relevant measures of overall economic efficiency, such as profitability. Our problem here is simply to find how cost varies with the size of plant for a particular function, such as retailing shoes or generating electricity.

The problem of size of plant must also be distinguished from depth of plant. Plant depth is a different dimension of "size" that is analogous to the degree of vertical integration for the firm. A pure assembly plant is not comparable to a plant that manufactures its own parts and subassemblies. Such differences in plant depth confuse the conception of the relation of plant size to cost or obscure its measurement.

Four approaches to the problem of cost and size of plant merit consideration: (1) Analysis of changes in actual cost that accompanied the growth of a single plant over a period of time. (2) Analysis of differences in actual cost of plants of different sizes operated by separate firms and observed at the same time. (3) Engineering estimates of the alternative cost where the same technology of manufacturing is used in plants of different sizes. (See Sec. 3.) (4) Analysis of differences in the actual costs of different-sized plants operated by one corporation.

The first approach encounters insuperable difficulties in correcting the data for changes in products, technology, and management, unless the firm displays very little technical advance during its growth. Successive observations of the firm as it adapts itself are likely to represent an expansion path that traces only growth in number of plants operated and seldom large-scale increases in plant size. Problems also arise in rectifying cost data for changes in prices, and for differences in technology and in accounting procedures.

The second approach—simultaneous observations of plants that differ in size and ownership—encounters comparable difficulties of appraising differences in products, techniques, accounting methods, valuation bases, price levels, and managerial effectiveness. Even for so homogeneous a product as electric power, there are variations in daily load patterns, age of plant, and many other factors that make cost comparisons all but meaningless for measuring the economies of large-scale plants.

The third method—engineering conjectures of costs of plants of different size—is the only one that has wide usage in the business community. However, the question always arises of how adequately such estimates take account of all operating circumstances. Recent experiments with pulverizing big plants and operating several smaller units have indicated that imponderables missed by engineering estimates are significant. Gains from the standpoint of employee morale and management efficiency seem to have offset economies of specialization.

The fourth method is promising for some types of problems. If a company has a large number of plants carrying on similar activities—for instance, a chain of grocery stores or warehouses—the question of the minimum-cost size of such plants inevitably arises. If the plants have uniform accounting systems and uniform management policies, this problem is susceptible to a statistical analysis similar to the methods discussed in the previous section. Figure 2.6 shows the results of such a study for a retail shoe chain (Thom McCann).

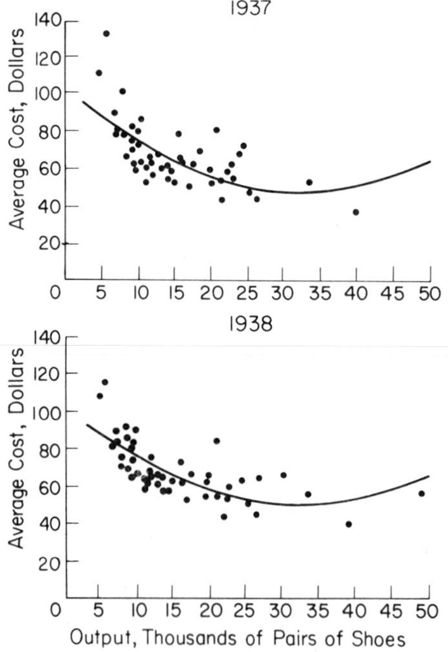

Fig. 2.6 *Average-cost function for shoe store chain. From Dean,* Long-Run Behavior of Costs in a Chain of Shoe Stores (*Chicago, University of Chicago Press, 1942*), *p. 25. Reprinted in Dean,* Managerial Economics, *p. 313.*

2.3.4 Costs of Multiple Products

Determination of the costs of individual articles produced and sold in multiple-product operations is of great practical importance for some problems. For example, it can be used to guide birth and death decisions that affect product line, to influence product modification and redesigning, to select the most appropriate system of price differentials among members of an existing product line, and to indicate opportunities and limits for nonprice competition by revealing the incremental profits of different products.

The individual product costs needed are the increments that will occur if and when the change in question occurs—i.e., the costs that are different if the decision goes one way rather than the other. These incremental and opportunity costs needed for decision-making have no necessary relation to the product costs obtained by conventional cost accounting. The conventional "full costs" are built from records of historical outlays and from necessarily arbitrary allocations of overheads, many of which are unaffected by (and therefore irrelevant to) the particular decision.

Some costs are traceable to individual products (e.g., purchased parts and components), while other costs are common to several products (i.e., they are not easily identified with a single product). For example, the cost of the factory building is common to all the types of products made there. The problem of product-costing arises in identifying parts of common costs with particular products.

Some common costs are unaffected by the kind of change that is up for decision. Common costs that are fixed do not need to be allocated, since they are irrelevant for any decision for which they are constant. It is the common costs that vary with the decision that must be allocated to individual products.

Some decisions involve such a major overhaul of the cost structure that special cost studies are needed to determine the incremental costs. But the occasions for such decisions are quite rare. More frequent are problems that are closely related to variations in rate of output of particular products. For these decisions short-run variable costs are most important. Hence special attention needs to be given the allocation of variable common costs.

2.3.4.1 Jointness of Products. For product-costing it is desirable to distinguish two broad categories of common products: joint products and alternative products. When an increase in the production of one product causes an increase in the output of another product, then the products and their costs are traditionally defined as joint. In contrast, when an increase in the output of a product is accompanied by a reduction in the output of other products, the products may be called alternative. Slag and steel are joint products, but steel rails and steel bars are alternative products.

In principle, two joint-cost situations can be distinguished: first, when the proportions of the end products can be varied; second, when these proportions are fixed. When proportions are fixed, separate product costs are indeterminate. There is not even much point in contemplating the separate costs of bringing hams and shoulders to the slaughterhouse, since each unavoidably accompanies the other. When one product is much less important than the other, it may be called a by-product—a gratuitous use of a waste material—but there is no real distinction between joint products and by-products. Where the march of technology is rapid, as in chemistry, by-products soon become joint products and may even take on senior status.

For joint products of variable proportions, cost problems relate commonly to the incremental effect of an increase in output rate to meet new demand for one of the joint products. Such an increase involves higher output for all the products and may therefore cause a reduction in prices of joint products in order to get rid of them—e.g., more slag from increased output of steel. Thus the added revenue from one joint product must cover not only the added cost of the whole product package, but also any loss of revenue from lowered prices of the other joint products as well.

In the case of common costs that are not joint between two products, increasing the output of one product is either unassociated with an increase in the output of the other, or requires a sacrifice or reduction of the output of the other. In the first instance, the separate incremental costs are, at least in principle, determinant; in the second case, where sacrificed production of other products is involved, the concept of opportunity costs is frequently the most important. For example, the principal cost of canning tomatoes may be the foregone opportunity to pack tomato juice.

2.3.4.2 Petroleum Refining—an Illustration. An illustration from home-heating fuel oil may help to point up the foregoing analysis.

Crude oil yields a mixture of joint and alternative products, in which, as a result of modern refinery processes—catalytic cracking and polymerization—the range of practical variation in proportions of gasoline and fuel oil is rather wide. Given this variability, the refiner's problem is to determine the real costs that will be incurred by a decision to increase the output of fuel oil without changing gasoline production.

The most satisfactory way to estimate the cost of heating oil for this management decision is its gasoline opportunity cost. The cost of heating oil is the foregone gains that could have been realized by converting the oil into gasoline, which is usually the most important product. Gasoline is the important product, because it is the one that justifies the refinery investment. It is the big-volume, high-margin product, the demand for which is inelastic, largely because there is little substitute competition. Hence there is also less risk of losing the market than there is of losing the market for heating oil. Thus costing heating oil on the basis of gasoline conversion value measures almost directly the minimum price at which it would be better to convert into gasoline rather than to sell as heating oil.

2.4 PRICE POLICY

2.4.1 Kinds of Policies and Environment

The overriding corporate objective should be to maximize profits in the long run—more specifically to maximize the present value of future earnings capitalized at the corporation's cost of capital.

Competitive action of the modern firm to

attain this objective has three main dimensions: (1) product—improvements in the product bundle, including packaging, service, etc., (2) promotion—advertising, personal selling, distribution, etc., and (3) pricing.

Pricing policies are of two main kinds: (1) price level: the basic or list price of a product or of a group of products and (2) price structure: the pattern of price discounts that tailors the net price to conditions of sale, e.g., trade discounts, quantity discounts, transport allowances, and payment discounts. Both kinds of pricing decisions are shaped by the competitive environment, but the first more profoundly than the second.

2.4.1.1 Kinds of Competitive Environment. Under perfect competition, sellers have no pricing problems because they have no price discretion; they sell at the market price or not at all. Price policy has practical significance only when there is enough imperfection of competition to permit the seller significant latitude in pricing. So our analysis of pricing is confined to those kinds of competitive structures that confer this required amount of price discretion.

Competitive situations can be classified in many ways. For pricing policies it is useful to classify on the basis of the degree and durability of product differentiation.

The kind of product distinctiveness that is important for competition is in the mind. It is not the physical and technical differences that matter, but instead the way they are seen, appraised, and acted upon by buyers. Thus it is the bundle of physical product, plus reputation and acceptance, plus services, which constitutes modern product differentiation. The differentiation of this product bundle from the product bundle of rivals in wanted ways is dynamic and is, in many industries, the major weapon of competitive warfare. Competitors' product bundles differ and buyers differ in their needs, in worldliness, knowledge, and in the weights they assign to the different dimensions of the product bundle. Consequently their degree of attachment to any particular seller varies over a wide spectrum, ranging from devotion through indifference to outright hostility. As a result it is common for each seller to have a sort of monopoly for his most ardent devotees and at the same time face vigorous competition for the patronage of other buyers and hopeless defeat for still others. Thus, every modern seller has some degree of monopoly power; but it is hedged and continually threatened by the efforts of the rivals to extend their own monopoly.

This competitive process in a technically advanced and affluent society causes an underlying tendency toward the erosion of the market power of a distinctive product innovation. This process of competitive erosion operates with different power and speed in different industries. The resulting diversity of degree and durability of the distinctiveness of the product bundle, twinned with its companion variable, the number of sellers, can serve as a basis for classifying competitive situations in ways that are instructive for pricing policies. Four types can be distinguished, depending on whether the product bundle has (1) lasting distinctiveness, which implies a single seller, (2) perishable distinctiveness, initially a single seller, (3) little distinctiveness and few sellers, or (4) little distinctiveness and many sellers. These four situations call for different kinds of price policy.

2.4.2 Pricing Products of Lasting Distinctiveness

Products whose wanted distinctiveness is pronounced and lasting have enduring monopoly power. In the short run monopoly pricing should be based on what the traffic will bear, i.e., the price which will maximize profits, taking into account the price sensitivity of demand and the incremental costs of production and of promotion. Thus, the controlling consideration is what the product is worth to the buyer, not what it costs the seller.

Long-run maximization of profits introduces modifications. One kind is designed to deter the entry and expansion of potential rivals by blunting incentives. A second kind of modification is based on the greater price sensitivity of long-run demand and on the effect of price on the growth of sales in a dynamic, high-substitutability economy.

The traditional sources of enduring monopoly—scarce raw materials, patents, and economies of large-scale production—have steadily weakened as a result of growth and advancing technology.

Raw-material monopoly is eroded (a) by exploration, which is stimulated and cheapened by technical advances, (b) by new man-made substitutes, and (c) by technologically intensified rivalry of existing substitutes. Patent monopoly is weakened by the fluidity of our rapidly advancing technology, which supplies alternative technical routes to the same end and makes it possible to "invent around" a patent barricade. Moreover, the ancillary economic power of a patent has been weakened by antitrust indictments and decisions. Hence cross-licensing and pooling of technical advances are becoming common practices. Scale economies as a source of monopoly power tail off with growth of the market and are threatened by technical breakthroughs. Hence, the probability is high that a new product which is distinctive at the outset will inevitably degenerate, under the erosion of competition, toward a modestly differentiated commodity; i.e., it will turn out to be the product of perishable distinctiveness.

2.4.3 Pricing Products of Perishable Distinctiveness

A high proportion of new products fail in the marketplace, which proclaims the difficulty of pricing pioneer products correctly. A pioneer product is here defined as one which incorporates a major innovation and hence possesses pronounced and wanted, though perishable, distinctiveness. Deterioration of the competitive situation of most new products makes the life cycle of the pioneer product's economic status a strategic consideration in practical pricing for two reasons. The first is that the appropriate pricing policy is likely to be different for each stage. Throughout the cycle, continual changes occur in promotional and price elasticity of demand and in costs of production and of distribution. These changes usually call for adjustments in price policy.

The second reason is that price itself influences the rate of competitive degeneration in the product's distinctiveness. The protected distinctiveness of pioneer products is doomed to progressive degeneration because of competitive inroads. As competitors' new products enter the field and their innovations narrow the gap of distinctiveness, the pioneer's zone of pricing discretion narrows. Pricing can alter this "competitiveness cycle" first by speeding growth and early advantages of market occupancy and economies of scale and second by profit margins that discourage entry.

Three different aspects of maturity usually move in parallel time paths: (1) technical maturity, indicated by declining rate of product development, increasing standardization among brands, and increasing stability of manufacturing processes and knowledge about them; (2) market maturity, indicated by consumer acceptance of the basic service idea, by widespread belief that the products of most manufacturers will perform satisfactorily, and by enough familiarity and sophistication to permit consumers to compare brands competently; and (3) competitive maturity, indicated by increasing stability of market shares and price structures.

Of course, interaction among these components tends to make them move together—that is, intrusion by new competitors helps to develop the market. But entrance is most tempting when the new product appears to be establishing market acceptance.

What are the factors that set the pace of degeneration? An overriding determinant is technical—the amount of capital investment needed to use the innovation effectively. But aside from technical factors, the rate of degeneration is controlled by economic forces that can be subsumed under (1) rate of market acceptance, and (2) ease of entry.

"Market acceptance" means the extent to which buyers consider the product a serious alternative to other ways of performing the same service. The speed of market acceptance varies widely, from the slow growth of garbage-disposal units to the spectacular acceptance of antihistamine cold tablets, ballpoint pens, and soil conditioners. Low unit cost (25¢ rather than $300) probably favors growth, and a past record of successful product innovations aids in giving consumers faith in the company's technical ability and honesty.

Ease of entry is even more difficult to analyze than market acceptance, but probably the most important factor to consider is competitors' capital resources for research and promotion. And of course the bigger the opportunity in a new product, the more capital there is available to invade your field.

2.4.3.1 Policies for Pioneer Pricing. The strategic decision in pricing a new product is the choice between: (1) a policy of high initial prices that skim the cream of demand; and (2) a policy of low prices from the outset serving as an active agent for market penetration. Although the actual range of choice is much wider than this, a sharp dichotomy clarifies the issues for consideration.

Skimming price. For products that represent a drastic departure from accepted ways of performing a service, a policy of relatively high prices coupled with heavy promotional expenditures in the early stages of market development (and lower prices at later stages) has proved successful for many products. There are several reasons for the success of this policy.

1. Demand is likely to be more inelastic with respect to price in the early stages than it is when the product is full-grown, particularly for consumers' goods. The public is still ignorant about the uses and limitations of the product, and there are frequently no readily apparent substitutes. Hence the people who are willing to buy tend to be adventuresome types who want to try out new ways of raising their living standards, and who are more susceptible to promotional effort than to price advantages.

2. Launching a new product with a high price is an efficient device for segmenting the market. This is actually a form of price discrimination. After selling to the market described in (1) above, the price is slowly lowered to reach successively less daring customers until market saturation is sufficient to rob the product of all novelty.

3. The skimming-price policy is safer, or at least appears so. That is, the company will not market the product at all unless initial prices cover the early high costs of production and selling—costs that, if success were certain, would be considered part of the investment outlay in the new product.

4. Many companies are not in a position to finance the product flotation out of distant future revenues, even when the effects on market expansion make a low initial price clearly more profitable than a high price. High initial prices thus finance the costs of raising a product family when uncertainties block the usual sources of capital.

Penetration price. The alternative policy is to use low prices as the principal instrument for penetrating mass markets early. This policy is the reverse of the skimming-price policy, in which price is lowered only as short-run competition forces it. The orthodox skimming policy has the virtue of safeguarding some profits at every stage of market penetration. But it prevents quick sales to the many buyers at the lower end of the income (or preference) scale who are unwilling to pay any substantial premium for novelty or reputation superiority. The active approach in probing possibilities for market expansion by early penetration pricing requires research, forecasting, and courage. The low-price pattern should be adopted with a view to long-run rather than to short-run profits, with the recognition that it usually takes time to attain the volume potentialities of the market.

What conditions warrant aggressive pricing for market penetration? First, there should be a high responsiveness of sales to reductions in price. Second, savings in production costs as the result of greater volume should be substantial. Third, the product must be of such a nature that it will not seem bizarre when it is first fitted into the consumers' expenditure pattern. Fluorescent lighting, which exemplifies these three traits, showed a dramatic growth of sales in response to early penetration pricing.

A fourth condition that is highly persuasive for penetration pricing is the threat of potential competition. One of the major objectives of most low-pricing policies in the pioneering stages of market development is to raise entry barriers to prospective competitors. But stay-out pricing is not always appropriate; its success depends on the costs of entry for competitors and on the expected size of market. When total demand is expected to be small, the most efficient size of plant may be big enough to supply over half the market. In this case, a low-price policy can capture the bulk of the market and successfully hold back "cheapie" competition, whereas high prices are an invitation for later comers to invade established markets by selling at discounts. In many industries, however, the important potential competitors are large multiple-product firms for whom the product in question is probably marginal. For such firms, present margins over costs are not the dominant consideration, because they are normally confident that they can get their costs down as low as competitors' costs if the volume of production is large. Thus, when the total market is expected to stay small, potential competitors may not consider the product worth trying, and a high-margin policy can be followed with impunity.

On the other hand, when potential sales appear to be great, there is much to be said for setting prices at their expected long-run level. A big market promises no monopoly in cost savings, and the prime objective of the first entrant is to entrench himself in a market share. Brand preference costs less at the outset than after the competitive promotional clamor has reached full pitch. An offsetting consideration is that, if the new product calls for capital recovery over a long period, there is a risk that later entrants will be able to exploit new production techniques which undercut the pioneer's original cost structure.

Profit calculation should recognize all the contributions that market-development pricing can make to the sale of other products and to the long-run future of the company. Often a decision to use development pricing will turn on these considerations of long-term impacts on the firm's total operation strategy rather than on the profits directly attributable to the individual product.

Some pricing precepts for new products. The important determinants in economic pricing of pioneering innovations are complex, interrelated, and hard to forecast. Experienced judgment is required in pricing and repricing the product to fit its changing competitive environment. Here are some pricing precepts suggested by the preceding analysis.

1. Be clear about corporate goals. Pricing a new product is an occasion for rethinking them. The overriding corporate goal should be long-run profit maximization, e.g., making the stock worth most.

2. Pricing a new product should begin long before its birth. Prospective prices, coupled with forecasted costs, should play the decisive role in product birth control.

3. Pricing a new product should be a continuing process of bracketing the truth by successive approximations. Rough estimates of the relevant concepts are preferable to precise knowledge of historical irrelevancies.

4. Costs can supply useful guidance in new product pricing, but not by the conventional wisdom (e.g., cost-plus pricing.). Costs of three persons are pertinent: the buyer, the seller, and the seller's rivals. The role of cost differs among the three, as does the concept of cost that is pertinent to that role: different costs for different purposes.

5. The seller's cost sets a reference base for picking the most profitable price. For this job the only costs that are pertinent to pricing a new product on the verge of commercialization (i.e., already developed and tested) are your incremental costs: the added costs of going ahead, at different plant scales. Costs of R&D and of market testing are at that time sunk and hence irrelevant.

6. The buyer's cost shapes his options. Look at your product through the eyes of your customer and price it just low enough to make it an irresistible investment, in view of his alternatives as he sees them. To estimate how much your product is worth to your prospect is never easy; but it can be rewarding.

7. Customers' rate of return should be the main consideration in pricing novel capital goods. A buyer's cost savings (and other earnings) expressed as a return on his investment in your new product are the key to predicting the price sensitivity of demand and to pricing profitably.

8. The costs of rivals shape their reaction to your prices. Recognize the pricing implications of the changing economic status and competitive environment of the product as it passes through its life cycle from birth to obsolescence. This cycle and the actions of the seller to influence it are of paramount importance for pricing policy.

9. The strategic choice between skimming and penetration pricing should be based on economics. The skimming policy—i.e., relatively high prices in the pioneering stage, cascading downward thereafter—is particularly appropriate for products whose sales initially are comparatively unresponsive to price but quite responsive to education. A policy of penetration pricing—i.e., relatively low prices in the pioneering stage in anticipation of the cost savings resulting from an expanding market—is best when scale economics are big, demand is price sensitive, and invasion is threatened. Low starting prices sacrifice short-run profits for long-run profits and discourage potential competitors.

2.4.4 Pricing Standard Products When Competitors Are Few

In this section we discuss oligopoly, the third competitive situation where price policy is an important management problem. Oligopoly is competition where three, four, or fifteen firms have not only similar products but roughly similar production costs. Usually, rivals' products are sufficiently different in buyers' minds to make brands an important feature in marketing and to allow differences in prices, yet sufficiently similar to make a seller watch rivals' prices closely.

In industries where a few competitors dominate the supply of relatively uniform products, periods of low demand and excess capacity create serious competitive problems. This is particularly so in industries with heavy plant investments and high barriers to entry. Each manufacturer is aware of the disastrous effects that an announced reduction of his own price would have on the prices charged by competitors. As a result, these companies have by experience developed a pronounced aversion for attempting to gain market share by open price cutting. Under these circumstances, market share is largely determined by secret price concessions and by nonprice competition.

When the dynamic changes in demand and cost conditions that prompt a given price change are viewed in much the same way by all rivals, they do not cause significant uncertainties concerning rivals' reactions. But these uncertainties do become serious: (a) when rivals are quite differently affected by the same general changes in conditions, (b) when rivals differ widely in their estimates concerning the future conditions for which they are pricing, and (c) when rivals have drastically different notions about the effectiveness of price changes.

Since these disruptive influences are continually at work to some degree in many industries, a critical problem of oligopoly is to devise industry practices that can reconcile the need for adjustments to changing demand and cost conditions with the desire to maintain the precarious competitive price structure (i.e., pattern of price spreads among rivals) and the uneasy market-share structure that avoids an uncontrolled price war.

Two important releases from this dilemma are (1) nonprice competition and (2) price leadership. When these are inadequate or slow to relax conflicting pressures in a new demand or cost situation, there are still several varieties of underground price competition that can be used to maintain a façade of industrial peace in the industry. We shall discuss these next.

2.4.4.1 Nonprice Competition. Nonprice competition means promoting and improving the product bundle, as contrasted with cutting the price of the same (or a shrunken) bundle—for example, pricing a better car or better appreciated car for the same money rather than the same car for less money. In the end it amounts to price competition, supplying more car per dollar. Nonprice competition is cold war. It is viewed with more equanimity than the hot war of overt price cutting for three reasons. First, precise matching is difficult, partly because the marketing mix of the product bundle is peculiar and is tailored to each firm's competitive situation, which deters speedy imitation and partly because a differential in competence produces great disparity in the efficiency of marketing activities, which underscores the importance of "know-how" barriers to precise matching. A second superiority is that retaliation is less likely to become vindictive and to get out of hand and degenerate into open warfare. Octane wars and advertising wars do occasionally occur in the gasoline industry but are much less common than price wars. Retaliation is diffused and less provocative since it often takes a different route. A campaign of free samples by one soap manufacturer is met by a contest or by couponing, rather than by matching or upping the sampling campaign. A third reason is that the effects upon sales and upon the market share of any particular promotion or product improvement are far less predictable than the effects of price cutting, and there is usually a less compelling necessity to retaliate.

2.4.4.2 Price Leadership. The institution of price leadership is another way for oligopoly competitors to achieve the delicate adjustment to changing cost and demand conditions without precipitating a price war. One firm (not necessarily always the same one) takes the initiative in raising the price when it thinks that increases in demand (relative to capacity) or increases in cost or both call for a price increase. Individual rivals then match, exceed, or shave under the leader's provisional price move. If shaved by a substantial rival, the leader usually has to back down to that lower figure. When concurrence of substantial and accepted rivals has, by this cut-and-dried process, been established, other sellers follow, matching the leader's price exactly or with established differentials which reflect disparity in market acceptance. The price leader usually leads only in price rises. Any competitor can lead de facto prices down; and usually the initiative in sub rosa price cutting is taken by smaller firms with inferior market acceptance that are normally followers in price increases. The action of a price leader in cutting price in a deteriorating market is usually a tardy official recognition of the lower level to which real prices have already sunk.

Price leadership in action may be seen most

clearly in a mature and stable industry with a standardized product, such as steel, oil, cement, or building materials. But it plays an important part in some industries that have considerable product differentiations.

Price leadership greatly reduces the number of possible reactions of a price change, and thus gives a modicum of certainty to the pricing aspects of market forecasting. All that is needed is one firm whose price policy is consistently acceptable to most of the industry. The form that leadership actually takes depends on the size structure of the industry's firms, on disparities in their cost functions, product differentiation, and geographical distribution, and on the pattern and stability of demand. In the more sophisticated industries, there is a feeling of mutual responsibility of the leader to set prices that other firms can live with, and of the followers to follow the leader in the best interests of the group.

Problems of the price leader. Broadly conceived, the problem of the price leader is a problem of industrial statesmanship, particularly when industry conditions are changing rapidly. If he fails to reconcile his own and the industry's interests with those of the followers, he may easily impair his own position in leadership and market share.

In many industries, price leadership has been fairly responsive to changes in cost and in demand, and at the same time has managed to dampen the amplitude of cyclical fluctuations. Under leadership, prices have not gone so high in boom periods or so low in depressions. Price leadership usually produces fewer but larger price changes over the cycle. It is important for the leader to know how long the lower price level will last. A temporary drop should be met only by informal concessions from the official price, since frequent changes in announced prices disrupt the followers' adjustments and weaken the leader's prestige. Only when market weakness indicates a fairly long-run shift in conditions should the major move of changing official prices be made. The price leader often merely formalizes what is generally recognized as inevitable in the trade, or merely forecasts sooner, and more accurately, what later becomes recognized. Forecasting is thus a critical part of leadership. It determines what attitude to take toward a deteriorating market and it signals the time for an advance. If the leader's forecasts of developing demand and cost conditions have generally been correct and are trusted by the industry, then the dissent that follows his upward price lead will be minimal and the price leader will have an unwarranted appearance of price-setting power. But if the price leader has forecasted badly and has a history of blunders, leadership may pass to a rival or be at most barometric, with only formalized changes that have occurred or that are generally recognized as inevitable.

2.4.4.3 Underground Price Competition. The discussion of price leadership has indicated that the most important device for short-run sales expansion is the secret price concession. The form of the concession is not important, so long as the buyer understands the real offer. It may take the form of high turn-in values, downgrading of high-quality lines, better payment terms, and so forth. Underground price competition in the tire industry takes such forms as concessions to mail-order houses, filling-station chains, and fleets, and special prices for "test tires" (allegedly slightly used).

The price leader's problem is to determine whether and how to reduce the level of official prices to meet undercover concessions. When are undercover concessions significant enough to warrant open reduction of prices? Here are a few strategic indicators:

1. When they spread over a wide geographic area.
2. When they continue for several months.
3. When first-line competitors indulge quite generally in undercover cutting.
4. When the price cutter is out to broaden the market. (If his motive seems to be to capture a bigger share of the same sized market by secret concessions, the leader may merely meet these informal concessions.)

One economic function of underground price competition is to give resilience and stability to an otherwise brittle oligopoly situation. Undercover price concessions make it possible to develop an informal hierarchy of prices which allows products of substandard acceptance a share of the market without setting off a price war.

Another function is to make the sensitivity of real prices to fluctuations in business activity much greater than the relative inflexibility of nominal prices as indicated. Thus responsiveness of the industrial economy to cyclical changes is more prompt and more curative as a result of unseen underground price competition.

2.4.5 Factors in Basic Pricing

Against this background of the dynamics of the structure of competition, we now examine some major determinants of the product's basic price (i.e., the product's price level as distinguished from structure of price differentials): (1) benchmarks of buyers' alternatives, (2) sensitivity of sales volume to price, (3) forecasts of cost behavior, (4) rate-of-return pricing, and (5) fully allocated cost pricing.

In terms of application, basic pricing is of two sorts: pricing new products and repricing old products. The discussion that follows is in terms of new products. In principle, much the same kind of analysis should monitor the repricing of established products.

2.4.5.1 Benchmarks of Buyers' Alternatives. In pricing, the buyers' viewpoint should be controlling. For every new product there are alternatives. Buyers' best alternatives are usually products already tested in the marketplace. The new product will, presumably, provide a superior solution to the problem of some buyers. The degree of superiority over substitutes usually also differs widely among buyers. And its average differs widely among new products.

Survey of alternatives. Because the prospective buyer of any new product has alternatives, these indirectly competitive products are the benchmark for the buyers' appraisal of the price-performance bundle of your new product. This comparison with benchmark products should ideally do each of the following:

1. Determine the major uses for your new product. For each application, determine your product's performance characteristics.
2. For each important usage area, specify the products that are the buyers' best alternatives to your new product. Determine the performance characteristics and requirements which buyers view as crucial for their product selection.
3. For each major use, determine how well your product's performance characteristics meet the requirements of customers compared with the performance of these buyers' alternative products.
4. Forecast the prices of alternative products in terms of transaction prices, adjusted for the impact of your new product and translated into units of use. From the prices of these benchmark substitutes estimate the alternative costs of the buyer per unit of the new product.
5. Estimate your superiority premium; i.e., price the performance differential in terms of what the superior solution supplied by the new product is worth to various buyers.
6. Figure a "parity price" for your product relative to the buyers' best alternative product in each use, for major categories of customers. Parity is a price which encompasses the premium a customer would be willing to pay for your comparative superiority in performance characteristics.

Pricing the superiority differential. Determining from this analysis that price premium of the new product's superiority over benchmark products which will be most profitable is an intricate and challenging problem.

The value to the customer of the innovational superiority of the new product is surrounded by uncertainties: whether the product will work, whether it will attain its designed superiorities, what its reliability and durability performance will be, and how soon it in turn will become obsolete. These uncertainties influence the price a customer would pay and the promotional outlay to persuade him to buy.

What matters is superiority as buyers value it, not superiority as calibrated by technicians' measurements or by the sellers' costs. Thus, more and better promotion can raise the premium-volume schedule and make a higher superiority premium achieve the same sales volume or rate of sales growth as would a lower premium without the promotion. This premium-volume schedule will be kicked about by retaliatory pricing of displaceable substitutes as well as by the imitative and innovative new-product competition of rivals.

The optimizing premium, in any specified time period, will depend upon your future costs as well as upon the hazy and dynamic demand schedule. It will be hard to find. Uncertainty about the future thus makes the appropriate pricing strategy for the long run a matter of sophisticated judgment.

2.4.5.2 Sensitivity of Volume to Price. Profitable pricing of a new product requires an estimate of how price will affect sales. This relationship can be explored in two steps: (1) find what range of price will make the product economically attractive to buyers; (2) estimate what sales volumes can be expected at various points in this price range.

Price Range. No product is really new; the most novel product merely plugs an abnormally large gap in the chain of substitutes. This gap marks out the potential range of its price.

For industrial products, quick and cheap ways to find this range usually tap the wisdom of people experienced in looking at comparative product performance in terms of buyers' costs: e.g., distributors and consulting engineers.

For consumers' goods, methods are different. In charting the price range of a radically novel product of small unit value, the concept of barter equivalent can be useful. A manufacturer of paper specialties tested a new product by spreading out on a big table a wide variety of consumer products totally unlike the new product. Representative housewives selected the products they would swap for the new product.

PRICE-VOLUME RELATIONSHIP. Within this range the effect of price upon its volume is the most difficult estimate in new-product pricing. We know in general that the lower the price, the greater the volume and the faster its rate of growth. The air-freight growth rate is about 18 per cent; priced higher, it will grow more slowly. But to know the precise position and shape of the price-quantity demand schedule or how much faster sales will grow if the price is 20 per cent lower is not possible. But guess it we must.

The best way to predict it is by controlled experiments: offering it at several different prices in comparable test markets under realistic sales conditions. For example, frozen orange juice was thus tested at three prices. When test marketing is not feasible, another method is to broaden the study of the cost of buyers' alternatives and to include forecasts of the sales volume of substitutes (and other indications of the volume of customers of different categories). This approach is most promising for industrial customers because performance comparisons are more explicit and measurable and economics more completely controls purchases.

2.4.5.3 Forecasts of Cost Behavior. To get maximum use from costs in new-product pricing, three questions must be answered: whose cost? which cost? and what role? As to whose cost, three are important: (1) the costs of prospective buyers, (2) the costs of existent and potential competitors, and (3) the costs of the producer of the new product. For each of the three, cost should play a different role and the concept of cost that is pertinent will differ accordingly.

BUYER'S COSTS. Costs of prospective customers should be used in setting the price of a new product by applying value analysis to prices and performance of alternative products to find the superiority premium that will make the new product attractive from an economic standpoint to buyers of specified categories. Rate-of-return pricing of capital goods illustrates this buyer's-cost approach, which is applicable in principle to all new products.

COMPETITORS' COSTS. Competitors' costs calibrate their capabilities. Costs of two kinds of competitive products can be helpful. For products already in the marketplace, the purposes are to estimate (1) their staying power and (2) the floor of retaliation pricing. For the first purpose, the pertinent cost concept is the competitors' long-run incremental cost; for the second, short-run incremental cost.

The second kind is the potential but as yet unborn competing product. Forecasts of competitors' costs for such products can help assess the effectiveness of a strategy of pricing the new product so as to discourage entry. For this purpose, the cost behavior to forecast is the relationship between unit production cost and plant size. The cost forecasts should take into account technological progress and should be spotted on a time scale.

SELLER'S COSTS. The costs of the producer play several roles in pricing a new product. The first is birth control. A new product must be prepriced provisionally early in the research and development stage and periodically as it progresses toward market. Forecasts of production and promotional costs at matching stages should play the role of forecasting its economic feasibility. The concept of cost that is relevant for birth-control role is the predicted full cost at a series of prospective volumes and corresponding technologies. It should include imputed cost of capital on intangible as well as tangible investment.

A second role of the seller's costs is toe stablish a price floor. This floor is also the threshold for selecting from candidate prices that which will maximize return on a new-product investment. For both jobs, the relevant concept is future, long-run incremental costs forecasted over a range of volume, production technologies, and promotional outlays in the marketing plan.

Two categories of forecasted cost have quite different impacts on new-product pricing:

a. Production costs (including physical distribution)

b. Persuasion costs, which are discretionary and rivalrous with price

The production costs that matter are the future costs over the long run that will be added by making (and peddling) this product on the predicted scale vs. not making it.

Initial promotion outlays are an investment in the product that cannot be recovered until some kind of market has been established. The innovator shoulders the burden of educating consumers.

The basic strategic problem is to find the right mixture of price and promotion to maximize long-run profits. You can choose a relatively high price in pioneering stages, together with large advertising and dealer discounts, and plan to get your promotion investment back early. Alternatively you can use low prices and lean margins from the very outset in order to discourage potential competition when the barriers of patents and investment in production capacity, distribution channels, or production techniques become inadequate.

Estimation of the costs of moving the new product through the channels of distribution to the final consumer must enter into the pricing procedure since these costs govern the factory price that will result in a specified final price. Distributive margins are partly pure promotional costs and partly physical distribution costs. Margins must at least cover the distributors' costs of warehousing, handling, and order-taking.

2.4.5.4 Rate-of-Return Pricing. Application of these principles of economic pricing is illustrated by rate-of-return pricing of new capital equipment. Industrial goods are sold to businessmen in their capacity as profit makers. Your product represents an investment by your businessman customer. The test of whether this investment is a desirable one should be its profitability to the customer. The pricing guide that this suggests is rate of return on the capital your customer will tie up by his investment in your product.

Rate-of-return pricing looks at your price through the investment eyes of the customer. It recognizes that your upper limit is the price which will produce the minimum acceptable rate of return on the customer's investment. The added profits obtainable from the use of your equipment differ among customers and among applications for the same customer. Cut-off criteria of required return also differ, so prospective customers differ in the rate of return which will induce them to invest in your product. Thus, the rate-of-return approach opens up a new kind of demand analysis for industrial goods. This analysis consists of inquiry into (1) the costs of buyers from displaceable alternative ways to do the job; (2) the cost-saving and profit-producing capability of your equipment in different applications and for different prospects; and (3) the capital-budgeting policies of your customers, with particular emphasis on their cost-of-capital and their minimum rate-of-return requirements.

2.4.5.5 Fully Allocated Cost Pricing. In sharp contrast to the customer orientation of rate-of-return pricing, fully allocated cost pricing is supplier-oriented. It is based on the seller's cost plus a uniform markup (which in practice gets grudgingly differentiated as the stubborn facts of the marketplace force retreat from this kind of formula pricing). The cost base is the fully allocated, accounted cost averaged over the output rates of a recent historical period.

This cost base has serious infirmities. Allocations of costs common to several products are necessarily arbitrary and allocations of fixed overheads are for many pricing decisions irrelevant. Incremental cost rather than full cost should set the floor in short-run pricing when capacity is underutilized. When it is not, opportunity cost, i.e., the passed-up profits per unit of the capacity bottleneck, should control pricing, and these costs are not reflected in accounted costs. Future costs alone are relevant, and they are not usually best forecasted by averaging past unit cost over happenstance series of output rates. The concept of cost relevant for long-run pricing is future long-run incremental cost. There are better ways to forecast it than to use fully allocated past cost. Another basic infirmity concerns the relevance of any cost of the seller. There is no necessary relation between (a) what it costs the manufacturer to make and sell a new product, and (b) what people will pay for it. Moreover, the historical costs of the innovator do not in themselves well predict the relevant concept, namely, future costs of potential competitors, and thus indicate a price which will discourage entry.

Full-cost pricing then adds to this full average unit-cost base a percentage markup designedly uniform among products, regardless of disparity in their superiority over rivals' products, differences in intensity of competition, and differences in production and promotional efficiency. It is this uniformity of percentage markup, added to the pseudouniformity built into accounting-cost allocations, that causes trouble, providing this cost-plus-pricing formula is actually adhered to.

The size of this markup does differ among industries and among firms. This difference occurs not only because of differences in competitive intensity and differences in turnover rate and risk, but also because of habit or custom and perhaps some notion of a "just price." In public-utility pricing, the size of the formula markup can be derived from an allowed reasonable rate of return on investment, which is based upon the composite cost of capital. To do this requires adjustment for differences among services (products) in capital turnover rate and calls for predictions of sales and costs of individual products.

For competitive industry, a similar analytical process can suggest a minimum companywide bogey rate of return on investment (at market value, not book value), based upon the corporation's composite cost of capital. Although this companywide minimum-earnings bogey could be apportioned by the public-utility process to individual products, it should not be used in competitive industry to set prices. Doing so will be to run the danger of pricing products out of the market at the bottom of the line and pricing the company out of profits at the top.

The widespread use of fully allocated cost-plus-pricing despite its intellectual frailties is puzzling. Perhaps it is a last resort in the absence of the forecasts needed for more defensible pricing methods. To predict at all precisely the way in which alternative prices will influence sales volume is difficult indeed. To foresee how today's price will affect tomorrow's demand and the entry of potential competitors is difficult. Faced with wide error margins and the high costs of making the predictions that ought to enter into scientific pricing, it is understandable that executives take refuge in the pseudocertainty of a price built up by piling an arbitrary and putatively uniform markup upon a cost base that is "known" and is often viewed as "the cost" even though not particularly relevant for pricing.

Safety is another attraction. A cost-plus formula can set a refusal price which prevents tying up facilities and manpower with work that would yield subnormal profits. Occasionally it can add assurance by helping to bring about pricing concurrence among rivals. When rivals are few and when their products and production processes are highly similar, full cost-plus pricing may supply a source of competitive stability. It can set a price more likely to be followed because it yields a rate of return on specialized investment acceptable to most other members of the industry.

A situation in which cost is an unusual limiter of price is in selling to a knowledgeable buyer who takes a high proportion of your output and can shift his patronage quickly. There are examples in the automobile-components industry, where buyers control the design of the product, know a lot about suppliers' costs, and have the resources and know-how to make the product themselves if unhappy about the sellers' prices.

Market-minus pricing of new products produces a relationship between costs and prices that may be close. However, the causality runs the other way. The new product must beat the prices and performance specifications of existing products, so the selling price and minimum performance traits are predetermined before birth. Working backward from this, the maximum cost which will produce a rate of return attractive enough to warrant commercialization is estimated. Tailoring the product to maximize competitive performance and still come within this cost ceiling sometimes produces a rigid relationship of cost to price, but this is the converse of the approach of cost-plus pricing.

2.4.6 Price Differentials

There are at least two parts to a complete price policy. The first is the method for determining the basic list price of a product, which has been the subject of this article so far. The second part, the subject of the rest of the article, is the system for determining the net price actually charged particular customers, that is, the system for price variation related to *conditions of sale:* (1) the trade status of the buyer, (2) the amount of his purchase, (3) the location of the purchaser, (4) the promptness of payment.

The average net realized price per unit obtained by the manufacturer depends not only upon the formal structure of price differentials for various classes of purchases, but also upon the proportion of sales made in each class, and the departures from formal price structures. Consequently, competitors usually differ in net realized price (yield) per unit, even when their formal price structures are quite similar. For these reasons, and because of differences in formal discount structures, there is often little relation between the average yield to the manufacturer per unit and the list prices of competitive products.

2.4.6.1 Goals of Differential Prices. From the seller's standpoint, the differential prices that result from the application of various discount structures and from product-line pricing may serve several purposes. It is, therefore, desirable to look first at the company's whole structure of price differentials in terms of these purposes, which may be grouped as follows.

1. *Implementation of marketing strategy.* The patterns of price differentials (product-price differentials and the various discount structures) should implement the company's overall marketing strategy. These price differentials should be efficiently geared with other elements in the marketing program (e.g., advertising and distribution channels) to reach the sectors of the market selected by strategy. In doing so, the job of a particular structure of discounts may be quite specific. For example, an oil company whose strategy was directed at large and few service stations served by giant transport trucks would grant large-quantity discounts for truck-load purchases.

2. *Market segmentation.* A major objective of differential prices is to achieve profitable market segmentation when legal and competitive considerations permit price discrimination.

The practical problem of putting price discrimination to work involves breaking the market into sectors that differ in price elasticity of demand. To the extent that it is feasible to seal off such segments of the market, charging different prices for different sectors can increase the total volume of sales. Price discounts of various sorts are a major means of achieving and profiting from market segmentation.

Having a line of products which covers a wide spectrum of quality, e.g., Zippo lighters, is a way to achieve profitable market segmentation. By spreading apart the prices of product-line members, based on differences among products in price sensitivity of demand, price discrimination can increase profits. One limitation is leakage between segments whose prices are disparate (e.g., substituting a cheap steel Zippo for a solid gold one). A study of the effect of widening the price differentials upon such sales switches is needed. Other examples of successful segmentation by means of the product line are limited-edition vs. hardcover vs. paperback books, and three grades of gasoline.

3. *Market expansion.* Differential pricing that is designed to encourage new uses or to woo new customers is a common goal of product-line pricing, but it also extends to various phases of the discount structure, depending upon the circumstances of a purchase by a new user, e.g., student subscriptions to the *Wall Street Journal.*

4. *Competitive adaptation.* Differential prices are a major device for selective adjustment to the competitive environment. Discounts are often designed to match what competitors charge under comparable conditions of purchase, in terms of net price to each customer class.

5. *Reduction of production costs.* Differential prices can sometimes help solve problems of production. Seasonal or other forms of time-period discounts may be partly for the purpose of regularizing output by changing the timing of sales. For example, since electricity cannot be stored, classifications of electric rates are designed to encourage off-season uses and to penalize uses that contribute to peaks.

6. *Cyclical adjustments.* The discount structure is the main area for adaptation of prices to changed economic conditions. A drop in demand hits the discount structure first, and its first impact there results in secret shading of quantities and classifications. Similarly, a competitor who is at a disadvantage in the quality, reputation, or distribution of his products uses the informal or formal structure of discounts to even up these weaknesses.

2.4.6.2 Distributor Discounts. Distributor (or trade-channel) discounts are deductions from list price that systematically make the net price vary according to the buyers' position in the chain of distribution. These differential prices distinguish among customers on the basis of their marketing functions (e.g., wholesaler vs. retailer), and are thus also called "functional discounts." Special prices given to manufacturers who incorporate the product in their own original equipment (e.g., tires and spark plugs sold to automobile manufacturers), special prices to other members of the same industry (e.g., gasoline "exchanges" among petroleum companies), and special prices to the federal government, to state governments, and to universities are examples of common forms of discounts that are close enough to trade-channel discounts to be grouped with them. Table 2.2 shows several typical trade-discount structures.

The economic function of distributor discounts is to induce independent distributors to perform marketing services. To build a discount structure on a sound economic basis, it is necessary to know: (1) the objectives of the discount structure, (2) distributors' operating costs, (3) discount structures of competitors, (4) opportunities for market segmentation.

Objectives. To find out exactly what services the manufacturer wants from each type of distributor requires a broad, carefully thought-out distribution plan that fits the product, the competitive position of the seller, and the folkways of the industry. The primary consideration in working out such a plan is the allocation of marketing functions between the manufacturer and the distributing chain and among the

TABLE 2.2. *TRADE-CHANNEL DISCOUNTS IN VARIOUS INDUSTRIES,* FROM LIST PRICE*

Type of Business	Manufacturer's Agent	Distributors	Wholesalers	Dealers
Air conditioning	50 and 10 per cent		50 per cent	40 per cent
Automotive accessories	50 and 10 per cent		50 per cent	40 per cent
Electrical appliances		40 and 20 per cent		35 or 40 per cent
Farm equipment			25 per cent	30 per cent
Heating controls	40 and 10 and 10 per cent individual units; 50 and 10 per cent package units		40 and 10 per cent individual units; 50 per cent package units	25 and 5 per cent individual units; 33⅓ per cent package units
Machinery	5 per cent			30 per cent stocking 10 per cent nonstocking
Motorcycles		30 per cent motorcycles; 50 per cent parts		25 per cent motorcycles; 45 per cent parts
Musical instruments—percussive			50 and 10 per cent	50 per cent
Office supplies	10 per cent commission		50 and 20 per cent	50 per cent
Radios	2 and 10 per cent commission		60 per cent, 60 and 5 per cent, 60 and 10 per cent, 50 and 10 per cent (varies with line and quality)	40 per cent; on certain line 33⅓ per cent
Scales	50 and 10 per cent		50 per cent	40 per cent
industrial division	50 per cent			
Stoves and heaters			33⅓ and 25 per cent or 40 and 25 per cent if bought in carload lots	33⅓ per cent; deluxe models 40 per cent
Toys	5 per cent		50 and 5 per cent	40 per cent
Water coolers			40 and 5 per cent maximum	30 and 5 per cent maximum

* Trade Discount Practices, Report No. 558, The Dartnell Corp., Chicago. Reproduced in Dean, *Managerial Economics*, p. 520.

links in that chain. The problem is to find which functionary can do each specific job most economically and effectively. For example, a large electrical manufacturer selling refrigerators through distributors and dealers decided that in one of its major markets the function of the retail dealer should be confined to displaying six basic models, taking orders for them, and arranging the terms of the individual transactions. In the plan that this manufacturer worked out, the wholesaler, in addition to his traditional function of selecting the dealers and helping them do a better selling job, receives the merchandise, inspects it, delivers it to the customer's premises, installs it, and takes complete charge of all subsequent mechanical service.

Distributors' operating costs. The most important function of trade-channel discounts is to cover the operating costs and normal profits of distributors. Discounts should be closely aligned to these costs if distributors are to play the part planned for them. Margins that are too rich produce excess selling effort or too many distributors, while margins that do not cover costs will not move the goods.

Should trade discounts be determined by the costs of the inefficient distributor or by the costs of the efficient distributor? One solution to this problem is to set trade discounts to cover the estimated operating costs (plus normal profits) of the most efficient two-thirds of the dealers. When cost estimates are uncertain, a practical test of excessive margins is the extent to which rehandlers pass margins on by knocking down realized prices.

Another check of distributors' costs is an estimate of the manufacturer's cost of performing the distributive function himself. Many companies periodically consider doing more of the marketing job themselves (e.g., bypassing the wholesaler), and such estimates are frequently available as by-products of these trade-channel policy studies. Moreover, some companies operate through different channels in different sections of the country, and thus have some cost experience in performing distributive functions.

Competitors' discount structures. In a sense, dealer discounts are a means of purchasing the dealer's sales assistance in a competitive market. In many industries the actual (as opposed to the nominal) discounts granted by rival sellers vary. The manufacturer must decide whether he is to be guided by the higher or by the lower discounts. Specifically, a manufacturer whose product is at some disadvantage in consumer acceptance may consider making an attempt to buy distribution by granting larger margins than do competitors. The success of such an effort usually depends on whether the margin incentive will actually induce the distributor to push the product and on whether competitors are likely to meet the wider margins.

The competition from substitute distribution channels should also be studied. Costs of alternative routes may place the selected channel at a disadvantage in terms of the ultimate price to the consumer for a comparable product-service combination. This factor may set ceilings on channel discounts for a chosen distribution route. Cheap substitute channels have been a salutory stimulant for the seller to seek more effective channels.

Opportunities for market segmentation. Trade-channel discounts can be one means of achieving profitable market segmentation.

In some industries, the market is broken down into several fairly distinct submarkets, each of which has its own peculiar competitive and demand characteristics. These submarkets provide a ready-made opportunity for market segmentation. In the tire market, for example, the following submarkets may be distinguished:

1. The original-equipment market, characterized by skill and bargaining strength of buyers and by big cyclical fluctuations in demand.

2. The individual-consumer replacement market, characterized by unskilled buying, brand preferences, and cyclical stability.
   a. The manufacturer's brand segment.
   b. The distributors' brand segment.

3. The commercial-operator replacement market, characterized by large buyers who are price wise and quality wise (e.g., bus companies).

4. The government-sales market, characterized by large orders, formal bids, and publication of successful bidder's price.

5. The export market, characterized by international competition.

Price discrimination among individual consumers in the retail market is a common form of market segmentation. The manufacturer's pricing problem here is whether to keep the initial margins high enough to permit dealers to make individual concessions to customers. Realized margins that are substantially lower than official margins do not necessarily mean that the official margins should be reduced.

This disparity may be justified in industries where competition at the dealer level is strong and where opportunities for personal differentiation are important. A dealer can then get the full price from some customers who are averse to shopping and bargaining and can give substantially lower prices, with the flavor of a bargain, to more careful shoppers. This kind of individual pricing can yield a higher dealer profit than can uniform pricing. A conspicuous example of such pricing is found in the operation of automobile dealers under normal competitive conditions. It is normally appropriate to permit the dealer considerable latitude when the unit cost of the article is high, when trade-ins and service concessions provide a convenient mechanism for veiled price reductions, and when the customer is not tied tightly to the dealer by strings of continuity of service or by customer relations.

2.4.6.3 Quantity Discounts. Quantity discounts are reductions in the net price that are systematically related to the amount purchased. Our analysis is confined to commercial discounts. It does not include package-size differentials at the consumer level. Illustrations of commercial quantity-discount structures are found in Fig. 2.7.

The essential problem for management in quantity discounts is to decide how big they shall be. What merchandising job do we want quantity discounts to do? One important job of quantity discounts is to reduce both the number of and the losses from small orders. It is common for a firm to find that 80 per cent of its orders account for only 20 per cent of its sales, and the cost of making these sales frequently causes an actual out-of-pocket drain of cash. Quantity discounts can help correct the size distribution of orders in three ways: (1) They may stimulate a given set of customers to order the same amount of business in bigger lots. (2)

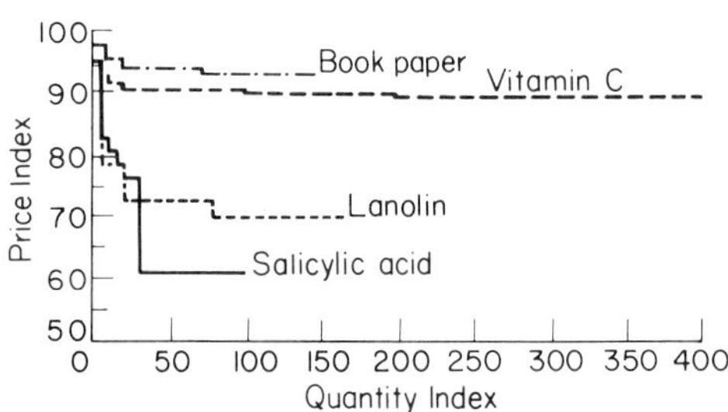

Fig. 2.7 *Quantity discount structures. From Dean,* Managerial Economics, *p. 530.*

They may induce the same customers to give the seller a larger share of their total business in order to get savings of quantity buying. (3) They may turn away small accounts and attract bigger accounts, thus altering the size distribution of the customers themselves.

To manipulate the size distribution of orders, the discount system must be framed in reference to competition. The seller must decide, by overall market strategy, which competitors he wants to better in what sectors of the market. The important factor competitively is the actual net prices charged to strategic customer classes, not the formal quantity-price structure per se. In some situations there is no room for such pinpointing of market targets, since discount structures of the industry are uniform, and deviations will be met by retaliation in some form. But frequently there are differences in the net quantity prices offered by various competitors to a given category of trade. The quantity-discount structure can then be integrated with the company's selling strategy and assigned a designated part of the total distribution job.

Legality. Quantity discounts have been a question of much litigation by the Federal Trade Commission, and any structure being considered must be scrutinized closely for its legality. This is a technical subject beyond the scope of a handbook. Legality hinges largely on proved cost savings resulting from large orders. Cost savings usually are provable only in the selling and distribution expenses of filling the order.

2.4.6.4 Cash Discounts. Cash discounts are reductions in the price which depend upon promptness of payment. Typical cash discount terms are 2 per cent off if paid in 10 days, full invoice price due in 30 days. The cash discount is a convenient way to identify bad credit risks. In some of the garment trades, where mortality is notoriously high, the cash discount is as high as 8 per cent, which makes the full wait extremely expensive. The higher price to the credit buyer thus reflects his weak bargaining position.

Unfortunately, there is no real information on the effects of cash discounts on bad-debt losses and on speed of collection.

2.4.6.5 Geographical Price Differentials. Geographical discount structures are important when transportation costs are high relative to selling price. They take a variety of forms:

- A. Uniform Delivered Pricing Methods
    1. Postage-stamp pricing
    2. Zone pricing
- B. Basing-point Pricing Methods
    1. Single-basing-point
    2. Multiple-basing-point
    3. Full-freight-equalization
- C. F.O.B. Pricing Methods
    1. Uniform F.O.B. price with no freight absorption
    2. Unrestricted F.O.B. price with unlimited freight absorption

2.5 MANAGEMENT OF CORPORATE CAPITAL

This article is concerned with the economics of capital budgeting, that is, with the kind of thinking necessary to design and carry through an optimum program for investing stockholders' money.

The criterion of optimization is stockholder wealth, specifically to maximize the present worth of the future stream of earnings capitalized at the cost of capital to the corporation. This should be the master goal of the corporation.

2.5.1 Meaning of Capital Expenditure Investment

A capital expenditure should be defined in terms of economics rather than in terms of accounting conventions or tax law. An investment is any outlay made today to get benefits tomorrow. The criterion is the speed of turnover into cash. Additions to plant capacity take several years to return their cash outlay. They tie up capital inflexibly for long periods. The same is largely true for intangible investments such as research on new products, advertising that has cumulative effects, and executive development.

This definition of investment does not correspond to the accounting distinction between capitalized and expensed outlays.

2.5.2 Nature of Capital Management Problem

The long-run profitability of the enterprise hinges on the solution of two problems of management of corporate capital: (1) sourcing (acquisition) of capital funds and (2) rationing (investment) of that capital. They should be quite separate. Investment proposals should compete for corporate funds on the basis of financial merit (the productivity of capital), independent of the source or cost of funds for that particular project. Investable funds of the corporation should be treated as a common pool, not compartmented puddles. Similarly, the problem of acquiring capital should be solved independent of its rationing and also on the basis of merit (the comparative costs and risks of alternative patterns of sourcing).

The investment of corporate capital is the ultimate responsibility of the company's high command. This responsibility is seldom delegated to any great degree, as it commits the company to a largely irrevocable course of development.

First, we shall study measurement of the economic worth of capital proposals, i.e., the productivity of capital. Second, we shall survey, for decision-making purposes, the various sources of supply of capital for the corporation. Third, we shall examine ways of estimating the cost of capital obtained from major sources. And fourth, we shall look at the mechanics of capital rationing.

2.5.3 Measurement of the Productivity of Capital

2.5.3.1 Demand for Capital. The underlying source of demand for capital expenditures is, or should be, prospective profitability. To develop an empirical approximation to the company's demand schedule for capital for internal investment during a specified time period, it is necessary: (1) to marshal all individual "needs" for capital expenditures that can be discovered and foreseen throughout the company; (2) to estimate for each proposal its prospective productivity in the form of rate of return on the added investment; (3) to array projects in a ladder of rate of return (as illustrated by Table 2.3); and (4) to cumulate this ladder in the form of a schedule showing the amount of money that can be invested to equal or better each of a series of rates of return. Figure 2.8 diagrams the resulting capital-demand schedule. In drawing such a schedule, the time span must be specified; for simplicity we shall assume the common one-year capital-planning period.

2.5.3.2 Yardsticks of Investment Worth. Measurement of the economic worth of capital proposals has two dimensions: (1) yardsticks

TABLE 2.3. *DEMAND SCHEDULE FOR CAPITAL**

(a) Prospective Rate of Return	(b) Volume of Proposed Investments	(c) Cumulative Demand
Over 100 per cent	2	2
50–100 per cent	38	40
25–50 per cent	200	240
15–25 per cent	1200	1440
5–15 per cent	3400	4840

* From Dean, *Managerial Economics*, p. 560.

of investment worth, and (2) applications of yardsticks by project analysis. Candidate yardsticks are (1) payback period, (2) book rate of return, and (3) discounted-cash-flow analysis.

In choosing among alternative yardsticks of investment worth, it is helpful to think first of the requirements of a good yardstick, such as:

1. *Accuracy:* measures productivity of capital correctly.
2. *Inclusiveness:* summarizes project merits in a single figure.
3. *Realism:* looks only at what happens to what is important—cash.
4. *Versatility:* makes different types of projects comparable.
5. *Parimutuelity:* reflects correctly the betting odds on getting the estimated earnings.
6. *Simplicity:* makes calculation easy.
7. *Screenability:* lines up with objective screening standards.

Payback Period. The payback-period yardstick is the number of years required for the earnings on the project to pay back the original outlay with no allowance for capital wastage.

For most projects, payback is an inadequate measure of investment worth. It is a cash concept, designed to answer the single question of how soon the cash outlay can be returned to the company's treasury. As such it fails in two important respects to provide a satisfactory yardstick for measuring the profit-producing power of candidate investments:

1. It ignores capital wastage. By confining analysis to the project's gross earnings (before depreciation), it takes no cognizance of its probable economic life.

2. It fails to consider the earnings of a project after the initial outlay has been paid back. Up to the end of the payback period the company has just got its bait back. How much longer the earnings will last is what determines the profitability of the investment.

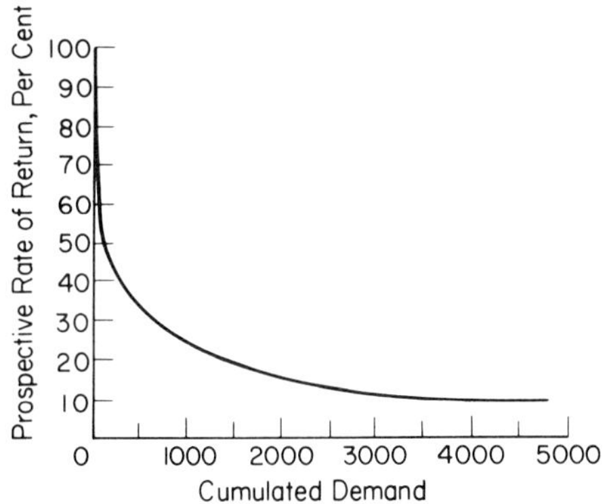

Fig. 2.8 *Demand for capital. From Dean,* Managerial Economics, *p. 561.*

Book Rate of Return. Any kind of rate-of-return yardstick is superior to payback in that it takes explicit account of capital wastage. It thus measures capital productivity by earnings over the whole life of the investment and permits rate-of-return rationing.

The book rate of return, though widely used, has the disturbing characteristic of having a host of variants. These depend upon how earnings are defined and how investment is measured, i.e., how questions like the following are answered:

1. Should the investment include expensed outlay?

2. Should the investment be deflated for prospective income taxes?

3. Is the investment the beginning amount or the average amount of capital tied up over the lifetime of the project?

4. Should earnings be gross or net of depreciation?

5. Should earnings be before or after income taxes?

Unless these questions are answered in a rigorously uniform manner throughout the corporation for all projects, the book method does not produce rates of return that are comparable among projects. If the answers to the questions are not uniform, a wide scatter of rates of return for the same project is obtained by the book method.

Although this book-return yardstick scores higher than does payback period, it is not brilliant in meeting the requirements of a good yardstick:

1. In *accuracy* it is only fair.
2. In *inclusiveness* it is deficient in not taking account of the time shape of either earnings or investments.
3. In *realism* it fails by not being confined to cash and not adequately reflecting taxes.
4. In *versatility* it comes off badly.
5. In *parimutuelity* it cannot reflect betting odds easily because it does not deal separately with each year.
6. In *simplicity* it is triumphant provided that ukases on exactly how to apply it are comprehensive and are obeyed.
7. In *screenability* it is acceptable though not quite perfect.

DISCOUNTED-CASH-FLOW (DCF) ANALYSIS. The discounted-cash-flow method is a relatively new approach to measuring the productivity and the cost of capital. In recent years the use of this method in industrial capital budgeting has grown rapidly. But it is this particular application that is new, not the principle. Time discounting has long been used in the insurance community for annuities, where accuracy and realism are indispensable.

Discounted-cash-flow analysis makes three contributions to top-management thinking: (1) An explicit recognition that *time* has economic value to the corporation; hence that near money is more valuable than distant money. (2) A recognition that *cash flows* are what matter; hence capitalization accounting and the resulting book depreciation are irrelevant for capital decisions except as they affect taxes. (3) A recognition that *income taxes* have such an important effect upon cash flow that their amount and timing must be explicitly figured into project worth.

Mechanics of method. Discounted-cash-flow analysis has two computational variants. The first is a rate-of-return computation that consists essentially of finding the interest rate that discounts gross future after-tax earnings of a project down to a present value equal to the project investment. This interest rate is the prospective rate of return on that particular investment. Examples of this variant are given in both Sec. 3, Engineering Economy, and Sec. 4, Capital Budgeting.

The second variant is a present-value computation that discounts gross future after-tax earnings of all projects at the same rate of interest. The rate of interest selected is, in effect, the company's minimum acceptable rate of return and should depend upon the company's cost of capital. Risk should be reflected by either deflating project earnings or adjusting the cutoff rate for projects of different categories of risk. The resulting present value is then compared with the project cost. If the present value exceeds it, the project is acceptable. If it falls below, it is rejected. In addition, projects can, by this variant, be ranked by various kinds of profitability indexes which reflect the amount or ratio of excess of present value over project cost. Section 3, Engineering Economy, contains a number of examples of calculations of present worth using an interest rate equal to a stipulated minimum attractive rate of return.

Both variants of the discounted-cash-flow approach require a timetable of after-tax cash flows of investment and of gross earnings that cover the entire economic life of the project. In practice the timetable can be simplified by grouping years in blocks. For projects for which investment is substantially instantaneous and gross earnings are level, simple computational charts and tables can be used to estimate the discounted-cash-flow rate of return directly from estimated economic life and after-tax payback. For projects with rising- or declining-earnings streams, this conversion is more complex.

Limitations of discounted cash flow. Both variants of the discounted-cash-flow approach have certain limitations. Because the approach is often unfamiliar in this application, its use requires persuasion and education. Moreover, it initially appears to be complex. But this appearance is deceptive; once the basic method is understood, it is actually simpler and quicker to use than the book method. Another deterrent is that it does not correspond precisely to accounting concepts about recording of costs and revenues. Consequently, special analysis is needed for "postmortem" audits of earnings on past investments.

Superiorities of discounted cash flow. The superiorities of the discounted-cash-flow method are impressive. It is immune to the distortions of capitalization policy. How much of the investment outlay is capitalized and how much is expensed differs from company to company and from project to project within a company. The true economic merit of a project stays the same. A discounted-cash-flow method reveals this; the book-return method hides it.

Divergent time patterns of earnings are correctly assessed by discounted cash flow and not by book return. The time shape of earnings differs greatly among projects. The economic significance of this disparity is underscored by the debates on the time pattern of capital wastage that have surrounded tax-law depreciation. The financial merit of a project depends on the time shape of earnings. The discounted-cash-flow method recognizes this dependence; the book method disregards it. The same principle applies to measuring the effect of end-of-life salvage value, which is so distant that at the high-earnings rates of many industrial projects it does not mean much. The accounts treat a distant-salvage dollar as having the same economic value as a near dollar, whereas the discounted-cash-flow method correctly says the values of the two dollars are quite different because of the economic value of time.

In terms of the requirements of a good yardstick, the discounted-cash-flow method makes an almost perfect score:

1. *Accuracy*. It measures the relevant concept of capital productivity with precision. It is not distorted by capitalization policy. It reflects only the effects of amortization policy upon income taxes and hence upon after-tax cash flows. It is correctly sensitive to the trend of earnings and to the timing of investment outlays, and it gives economically appropriate weight to ultimate salvage value.

2. *Inclusiveness*. It summarizes the merits of a project comprehensively in a single figure since it encompasses the project's whole lifetime of expected earnings and investment outlays measured relevantly and properly weighted for their timing.

3. *Realism*. It realistically confines the analysis to after-tax cash flows undistorted by inapplicable conventions of accountancy.

4. *Versatility*. It produces a measure of investment return which is precisely comparable among projects, regardless of the character and time shape of their receipts and of their investment outlays.

5. *Parimutuelity*. It is capable of reflecting the betting odds of actually getting project receipts of different types and in different years.

6. *Simplicity*. It is amazingly simple in spite of its apparent complexity. Once grasped, the method requires fewer rules to be mastered and uniformly interpreted. Moreover, the computation can be delegated to cheaper help because it does not require the use of trained accountants. Finally, computational graphs and tables can produce speed and economy in the calculations.

7. *Screenability*. It is conceptually compatible and therefore precisely comparable with a correctly determined company cost of capital. Hence it lines up with screening standards that can be logically and objectively justified.

2.5.3.3 Applying Yardstick to Capital Proposals. Two aspects of application of the yardstick to capital proposals will be discussed briefly: the economic dimensions of the project that need to be measured and the concepts of measurement to be applied to these dimensions.

Economic Dimensions of Capital Project. The investment worth of a capital project requires measurement and appraisal of at least four economic dimensions:

1. The amount and timing of added investment
2. The amount and timing of the added stream of earnings (net cash receipts)
3. The economic life, i.e., the duration of the earnings stream
4. The risks and uncertainties and the imponderable benefits associated with the project

The first three can generally be quantified with error margins that are tolerable for decision purposes. The fourth requires a high order of judgment.

Added investment. The appropriate investment base for calculating rate of return is the added outlay that will be occasioned by the adoption of a project as opposed to rejecting it, or adopting an alternative that requires less investment. For example, repairs that would be made whether or not the proposal is adopted should be excluded from the investment amount because they are not caused by it. The investment should include the entire amount of the original added outlay regardless of how it is treated in the books. Expensing certain items rather than capitalizing them may produce tax savings which should be reflected in estimating the investment. Any additional investment in working capital or other auxiliary facilities should be included in the investment amount, as should future research and promotional expenses caused by the proposal. Facilities transferred from other parts of the company should be included in the investment amount.

For the purpose of calculating prospective return, the items included in the investment amount should be valued at their economic rather than their accounting values. For cash outlays at the time of the investment decision, these are identical. For existing facilities, however, there can be a pronounced disparity. It is the present value of the earnings opportunities of such transferred facilities that is pertinent, and this is likely to differ from their book value. If the value of the foregone opportunity of continuing to use the facilities in the next best alternative way is lower than their disposal value, then it is their disposal value that should be used.

The timing of these added investments has an important effect upon the rate of return and should therefore be reflected in the rate-of-return computation.

Added earnings. The productivity of the capital tied up is determined by the increase in earnings or savings, i.e., net cash receipts, caused by making the investment as opposed to not making it. These earnings should be measured in terms of their after-tax cash or cash equivalent. Only costs and revenues that will be different as a result of the adoption of the proposal should be included. The concept of earnings should be broad enough to encompass intangible and often unquantifiable benefits. When these have to be omitted from the formal earnings estimates, they should be noted for subsequent appraisal of the project.

Economic life. Economic life of a project refers to the duration of the stream of benefits. Its length may be determined by physical deterioration, by obsolescence, or by the drying up of the source of earnings. Estimation of economic life is often the most difficult dimension of project value to quantify. But the problem cannot be ducked. Some estimate is better than none, and the depreciable life forecasted for bookkeeping purposes ought to be the best available forecast of economic life. For tax purposes this is not likely because estimates are made for a purpose different from making a capital-expenditure decision, namely, to minimize taxes.

Risks and imponderable benefits. Appraising the risks and uncertainties associated with a project requires such a high order of judgment that the problem should be explicitly faced and appraised by the collective wisdom of those best qualified to make the appraisal. Only disparities in risk among projects need to be allowed for, since the company's cost of capital reflects the overall risks. Only when the general character of the company's operations will be significantly altered by the investment will the risk that is reflected in the company's cost of capital be revalued in the market. In the process of measuring the probable rate of return on each project, the company may be successful in adjusting for the probability of getting the earnings of different types and of different years. If so, it is only the dispersion of possible outcomes that constitutes differential risk. For example, a labor-saving device might have a lower dispersion of outcomes than a new product. The chances of big, improbable gains or losses are smaller than for a new product. Though measurement of this sort of dispersion is difficult, some headway can sometimes be made by a necessarily arbitrary risk-ranking of candidate projects or categories of projects.

Most projects have some added benefits over and above the measurable ones. If excessive weight is given to these imponderables, there is danger that rate-of-return rationing will be displaced instead of being used to improve executive judgment in choosing among capital expenditures. When a low-rate-of-return project is preferred to a high one on the grounds of imponderable benefits, the burden of proof clearly rests on the imponderables.

CONCEPTS OF MEASUREMENT. For measuring these four dimensions of a project's return on investment, there are four key concepts:

1. *Futurity*. Future earnings and future outlays of the project are all that matter.
2. *Increments*. Added earnings and added investment of the project alone are material.
3. *Alternatives*. The proper benchmark for measuring added investment and the corresponding added earnings is the best alternative way to do it.
4. *Cash flows*. After-tax cash flows or their equivalents alone are significant for measuring capital productivity.

Futurity. The value of a proposed capital project depends on its future earnings. The past is irrelevant except as a benchmark for forecasting the future. Consequently, earnings estimates need to be based upon the best available projection of future volume, wage rates, price levels, etc. The earnings and ultimate salvage value corresponding to the proposed outlays need to be estimated year by year over the economic life of the proposed facilities and their time shape needs to be taken into account explicitly.

Increments. A correct estimate of earnings and investment must be based upon the simple principle that the earnings from the proposal are measured by the total added earnings or savings by making the investment as opposed to not making it, and that the same is true for the investment amount. Project costs should be unaffected by allocation of existing overheads, but they should reflect the changes in total overhead and other costs that are forecasted to result from the project. No costs or revenues that will be the same regardless of whether the proposal is accepted or rejected should be included, and the same goes for investment.

Alternatives. There is always an alternative to the proposed capital expenditure. The alternative may be so catastrophic that refined measurement is unnecessary to reject it. Often there are close alternatives so that an important problem of measuring capital productivity is to make a ladder of depth of alternative investments and measure the added earnings and added investment in a rate-of-return calculation for the ladder of alternatives to see how far it pays to go. The proper benchmark for the proposal is the next most profitable alternative way of doing it.

Cash flows. For economic realism, attention should be directed exclusively at the after-tax flows of cash or cash equivalents that will result from the adoption of the project. Book costs are confusing and immaterial.

2.5.4 Sources of Capital

The supply of capital for corporate investment comes ultimately from savings: of individuals, of corporations, and of governments. As to immediate sources, the corporation has two choices: its own savings (internal), or the savings of others (external).

2.5.4.1 Internal Sources. The main internal source of corporate savings is the generation of cash from operations. Corporate saving is the act of not paying it all out in dividends. This act has an accounting warrant and an economic warrant. The support from accountancy takes the form of depreciation and depletion charges, estimates of the wastage (consumption) of capital accompanying the generation of cash. These charges erect a cultural barrier to payment of dividends that accountants say are not "earned."

The economic warrant for corporate saving is quite different. Stockholders benefit from saving and internal investment of the corporation's after-tax cash generation when the rate of return on the investment is higher than the corporation's cost of capital. This is approximately the opportunity cost of capital of its stockholders, because the corporation usually has the option of buying its own stock. This economic sanction for corporate savings specifies the condition and the amounts that are economically warranted on the basis of the richness of the prospective return of internal investments. In so doing, it makes no distinction (as does the accounting sanction) between plowback of net income and reinvestment of depreciation, viewing gross after-tax cash generation as a single pool of capital, not separate puddles.

Gross cash earnings generated by operations constitute a common pool of future funds from which dividends will be paid and new investments financed. This pool should have corporate-wide availability for capital expenditures. To allow an operating division to reinvest its own depreciation charges regardless of the productivity of its investment proposals abrogates the principle of separability of sourcing and rationing of capital and destroys one of the major advantages of the multiproduct firm.

Another internal source of capital is disposal of assets. Each trade-in of a truck taps this source. Good capital sourcing should vigilantly ferret out opportunities to dispose of assets whose foregone earnings would in the future produce a rate of return on their disposal value lower than the corporation's combined cost of capital. Such disposals are a neglected source of capital in many corporations.

2.5.4.2 External Sources: Debt Capital. As external sources, there are two ways for a business to get savings from outsiders: borrow (debt capital) or partner (equity capital). Debt capital can be arbitrarily classified as either short-term or long-term.

SHORT-TERM DEBT. Short-term debt of the business enterprise can be of two sorts. The first is inadvertent borrowing which is a cultural by-product of normal operations. Mostly it is caused by lags between the time a service is performed and the time it is customary to pay for it. The corporation borrows a month's salary from its president. Although inadvertent, this borrowing is seldom avoidable.

The second sort of short-term borrowing is quite deliberate. The main source for most companies is the commercial bank. A *line of credit* is usually established, which sets a maximum amount of borrowing that the company may draw down to meet its needs.

Other explicit sources of short-term borrowing are: (1) loans from specialized *finance companies*, typically secured by accounts receivable or inventory; (2) *factoring* of accounts receivable, i.e., outright sale to a specialist who collects and absorbs bad-debt losses; and (3) sale of *commercial paper*, unsecured promissory notes generally of large corporations, sold publicly through the security markets.

LONG-TERM DEBT. Drawing the dividing line arbitrarily at one year, we classify intermediate-term debt capital as long-term. The *term loan* is a typical instrument. Its source may be a commercial bank, an insurance company, or a trust fund. Term loans result from direct (and usually unpublicized) negotiations between borrower and lender. They are privately placed as contrasted with public sale through the money market.

For longest-term borrowing, the usual loan contract is a *bond*. It may be placed privately by direct sale to a large financial institution or sold to the public, usually through an investment banker.

Long-term debt can be instructively subclassified in several ways: (1) nature of security (mortgage vs. debenture bonds); (2) directness of the obligation (direct debt vs. off-balance-sheet financing); and (3) degree of participation (pure debt vs. contingency debt). These three bases of classification overlap.

Nature of security. Borrowings may be secured by pledging specific assets (*mortgage bonds*) or secured only by the corporation's general credit (*debenture bonds*). The basic security for the debt of any corporation, however, is its uncommitted cash-generating ability. A bank does not want to run a blast furnace. Sale of a pledged asset is only a resort of desperation. The relevant economic measure of debt-carrying ability of a corporation is not the balance sheet but future generation of cash from operations, which can be derived from forecasted income statements.

Indirect debt. In addition to direct debt, a corporation can borrow indirectly by a variety of devices such as long-term leases, sale and lease backs, oil payments, royalty and throughput agreements. Regardless of legal status or accounting treatment, these forms of indirect obligation are the economic equivalent of direct debt. They are contractual obligations to make periodic payments for the use of capital and to repay the principal under specified conditions. These forms of off-balance-sheet borrowings have attractions. They can be tailored precisely to the borrower's needs as to amount and timing; they can be negotiated privately; and they can be left off the balance sheet. The price of these conveniences is usually a slightly higher interest cost.

Participation. Some debt has partnership features. For direct debt they are common and clear (e.g., for convertibles and for income bonds).

Convertible bonds are a cross between debt and equity financing. The bond holder has an option to convert his debt claim into shares of common stock at a predetermined price, which is equivalent to a long-term call on equity.

Income bonds are another hybrid, but a more disconsolate one. They participate only in the downside risk of equity. Interest payments of income bonds are obligatory only if covered by the pretax earnings during that current period.

Quasi-equity features are less clear but nevertheless present in much indirect debt (e.g., leasebacks and oil royalties). Most leasebacks and some oil-payment and oil-royalty arrangements give the lender options and residual values that have some attributes of equity.

2.5.4.3 External Sources: Equity Capital. Equity capital raised externally is of two main sorts: preferred stock, which is a kind of preferential but limited partnership; and common stock, which is full partnership.

PREFERRED STOCK. Dividends of preferred stock come ahead of common but are limited in amount. They are, however, not a contractual obligation and hence are not deductible as an expense for corporate income tax. In liquidation, the par value of the preferred comes ahead of common, though of course behind all debt obligations.

The limitations on the partnership of the preferred are sometimes relaxed. *Participating preferreds* share with the common stock some prearranged portions of earnings in excess of the minimum dividend. *Cumulative preferreds* get all previously omitted back dividends before any dividends can be paid on the common. *Convertible preferreds* have the option of swapping their limited partnership for the full partnership of common stock at a predetermined price.

Common Stock. Common stock is the economic equivalent of full partnership in the earnings and assets of the corporation. Dividends are discretionary and common stockholders get only the residual left after other payments. In liquidation also, only the amounts that remain after the company meets its obligations to lenders, suppliers, and preferred stockholders are distributed to common stockholders.

2.5.5 Cost of Capital

Saving, which is the ultimate source of capital funds for the corporation, is downright unpleasant. Therefore, we have to be paid to do it. That is the basic reason why it costs the corporation something to get capital. All resources command a price for their use. Capital is no exception. The financial manager's task is to blend the various capital sources available to him to achieve the lowest long-run cost for his company's total capital requirements.

This overall combined cost is the weighted average of the market cost of three main kinds of corporate capital: debt, preferred stock, and common equity. Capital from internal and indirect, hard-to-measure sources should be assigned the cost of its alternative direct source. The alternative for lease debt is direct borrowing for an equivalent term. The alternative for internal cash generation, whether labeled *earnings plowback* or *depreciation*, is flotation of common stock.

2.5.5.1 Cost of Debt Capital. That debt capital has a cost is undebatable. The use of money borrowed from outsiders has a price. This price is established by market forces and is knowable with considerable precision. This price the company must pay: the rate of interest.

The price of debt capital fluctuates with changes in supply and demand. At any time the price differs widely, depending on (1) the credit-worthiness of the borrower, (2) the duration and other terms of the loan, (3) the type of lending institution, and (4) the section of the country. The debt-cost range is wide, from 36 per cent for trade credit, if cash discounts are passed, down to the prime rate. The underlying causes of this wide disparity in the price of debt money are differences in risks and in costs of administration and collection. These two forces are sometimes opposed. For a short loan, the risk is less but the costs of launching and administration are proportionately higher. Other features affect the cost structure of debt capital; e.g., privacy has a price. Off-balance-sheet borrowing commands higher prices in the marketplace and all kinds of corporate private placements cost somewhat more than equivalent public debt.

Market imperfections distort the structure. Ignorance has a price: the astute borrower will try to borrow at the lowest rung of the debt-cost ladder that his credit-worthiness will permit. The competent lender will seek as comfortable a cushion of compensation for his risk differential as knowledge will warrant and competition permit. Thus, the twin forces of knowledge and competition continuously eat away at aberrations in the debt-cost structure caused by ignorance, timidity, sloth, and greed, making it come closer to the structure compatible with disparities in risk and in cost of administration and collection.

The true cost of debt capital is often higher than its nominal cost. The disparity is produced by many devices. Discounting (deducting interest in advance) increases true interest rate as does the requirement of minimum balances, since they pare down the amount really available to the lender. Ancillary charges for investigation, servicing, and insurance can also raise the effective cost of debt.

The concept of debt cost that is strictly pertinent is a forecast. What matters is the long-run average future cost of debt, not its past cost. But the past is, if averaged over a long period of time, a practical basis for estimating the future.

2.5.5.2 Cost of Preferred-Stock Capital. Preferred stock differs from debt in that there is no legal compulsion to pay. Dividends can be omitted without subjecting the company to bankruptcy proceedings. The principal, in most cases, never comes due, although the company may have the option to call in the issue. On the other hand, unlike common equity, the preferred stockholder has no claim to the residual profits of the company.

This limitation of the return to dividend payments is the crucial fact in computing the cost to the company of its preferred stock capital. The cost should be estimated in a manner similar to cost of debt, with the exception that there is no income tax adjustment (preferred-stock dividends are not allowed as a deductible expense). The company's maximum obligation is to pay dividends in the amounts negotiated at the time of issue.

2.5.5.3 Cost of Common-Stock Capital. The broad significance of the cost of equity capital needs to be understood if it is to be used intelligently. In a market economy any resource that is scarce, relative to the demand for it, commands a price, however difficult it may be to measure that price. This is true even though reimbursement for the suppliers of equity capital is a residual, that is, what is left after more tangible costs of operations, including interest on debt, are satisfied. In the long run, equity capital is not a discretionary cost. The fact that the rewards are residual and dividends are discretionary does not, as some think, mean that common-stock capital is costless. In the long run equity capital has a market price that is determined by investors' alternatives. Although owners of equity capital assume the ultimate risk by accepting as their compensation what is left, in most companies they have the option of selling at some price. They will not reinvest nor long leave their capital in any corporation that does not offer reasonable prospects of returns as high as those promised by alternative investments.

This residual is called *profit* by accounting convention. Profit is, however, a misnomer, disguising the fact that capital is a resource that commands a price for its use. Conventional accounting views equity capital as costless, that is, a free good. In an economic sense, however, there is no profit to the equity investor unless accounting profits exceed the cost of equity capital. Thus, despite its residual character, the cost of equity capital is a real cost to a business firm. It is what the free enterprise system has dictated and what a market economy has measured as the norm for the price of the use of equity capital in enterprises of this degree of risk.

Over any extended period, meeting the cost of equity capital at the corporate level is a necessity for maintaining the company's financial integrity, fulfilling its obligations to the financial community, and providing a continuously favorable climate for the capital formation that is needed to grow and prosper. To be sure, the earning power of the company at any given point in time is determined by the results of prior, essentially nonreversible business decisions and by the then-current business environment. Hence, current earnings may fall below cost of equity capital. But if they do so for a protracted period, they bring about an offsetting downward pricing of the stock designed to reestablish the balance with the then-anticipated future earnings and dividends.

Looking into the future, the investments made today must be capable of earning enough in the future to cover all costs, including the cost of equity capital, as well as providing for recovery of the capital invested during the economic life of the project. Only thus can the company maintain its financial integrity over the long run. For this reason, the policy of investment selection should establish investment standards that are based on the price it must pay for capital—the combined cost of equity capital and of debt capital—to assure that no investment will be taken on today that will be knowingly incapable of earning the cost of capital in the future.

Measurement. An obvious measurement of what common stockholders expect to receive is their share of the corporation's current earnings. A crude measure of the cost of equity capital, therefore, is the ratio of per-share accounting earnings to the then-current price of the stock.

A little reflection indicates that earnings-price ratios are not an adequate measurement of the cost of equity capital. There are four reasons for this: (1) the stockholder buys future earnings, not past earnings; (2) book earnings do not necessarily represent the true economic earnings of the corporation because of inadequacies and variations in accounting practice, especially in regard to allowances for inflation, for technological change, and for the recognition of changes in the value of mineral reserves; (3) current earnings have no simple relation to earnings growth; growth in earnings is largely the result of retention of earnings and the profitability of reinvestment; and (4) the stockholder cannot get his hands on his share of the earnings.

The problems associated with measuring the price of equity capital by the ratio of current accounting earnings per share to current price of stock can be avoided by casting the analysis in terms of what the investor can get his hands on. The price at which a share of stock is traded represents what the buyer is willing to pay now for a stream of future benefits. Basically, the stream of future benefits derives from future earnings per share, but the investor has no direct way of getting possession of them. Instead, he gets some combination of future dividends per share and capital appreciation.

If we knew the average expectation of future earnings, we could determine the rate at which they are being implicitly discounted to arrive at today's market price. This rate would be the cost of equity capital. But we cannot measure the investor's expectations of earnings performance. We can only look at past performance. If the time period is sufficiently long and a sufficiently large sample of companies is studied, it is probably a reasonable assumption that expectations, on the average, have been realized, and hence that actual returns can be used to represent expected returns. The argument is that there is no reason to suppose that expectations are subject to a consistent bias relative to outcomes. For any particular company or for any specific period of time, it is doubtful that outcomes would reflect expectations. But for a very long period of time or for a large sample of representative companies, errors of expectation ought roughly to cancel out.

Measuring cost of capital as the discounted-cash-flow rate of return on an investment in common equity is simple, straightforward, and easily understood as an accurate measure of returns actually received by the investor. At the outset of the analysis time period, our investor pays out a cost of x dollars for the stock. He receives a cash flow of dividends. He liquidates his investment by selling the stock at a price of y dollars at the end of the period. The rate of return that equates the present value of the investor's stream of returns (dividends plus liquidation price) with the purchase price of the stock at the start of the period is an estimate of the cost of equity capital.

The way to calculate the rate of discount that relates the price of the stock to the stream of dividends plus the appreciation of market price is illustrated in Table 2.4. The 1949 market price of *XYZ* stock was \$7.71 (adjusted for stock dividends and splits). It reflected the investors' anticipation of increases in earnings and dividends. When we discount the flow of dividends and the market-price appreciation to a present value of \$7.71, we find that the actual market yield was 20.2 per cent.

A second way to measure the cost of equity capital assumes that net cash flow from a corporation to the financial community as a whole consists only of a stream of dividends (since buying and selling transactions merely redistribute preexisting equity interest among individual investors). The cost of equity capital, then, is equal to the rate of discount of the future stream of dividends. This is estimated by taking an average dividend yield plus the

TABLE 2.4. *MEASURING COST OF EQUITY CAPITAL BY DISCOUNTED CASH FLOW (DCF), 1949–1960*

	Year	Cash Flow	Present Value at 25 per cent	Present Value at 20 per cent
Purchase price	1949	−\$7.71	−\$7.71	−\$7.71
Dividends	1950	nil	nil	nil
Dividends	1951	nil	nil	nil
Dividends	1952	nil	nil	nil
Dividends	1953	nil	nil	nil
Dividends	1954	+0.79	+0.26	+0.32
Dividends	1955	+1.30	+0.34	+0.44
Dividends	1956	+1.45	+0.30	+0.40
Dividends	1957	+1.52	+0.26	+0.35
Dividends	1958	+1.52	+0.20	+0.29
Dividends	1959	+1.52	+0.16	+0.25
Sale price	1960	+43.00	+3.70	+5.79
Net present value			−\$2.49	+\$0.13

DCF return = 20.2 per cent which is the cost of equity capital (not adjusted for flotation costs)

TABLE 2.5. *MEASURING COST OF EQUITY CAPITAL BY DIVIDEND-GROWTH GRADIENT METHOD, 1958–1965*

Year	Dividend per Share	Average Price of Stock	Dividend Yield
1958	$0.84	$21.38	4.0 per cent
1959	0.97	22.94	4.2
1960	1.00	20.63	4.9
1961	1.12	25.13	4.4
1962	1.15	24.56	4.7
1963	1.27	28.19	4.5
1964	1.33	37.94	3.5
1965	1.55	45.63	3.4

Dividend growth rate + average yield = cost of equity capital (not adjusted for flotation costs)
7.9 per cent + 4.2 per cent = 12.1 per cent

rate at which the dividend is expected to grow; this may be approximated by the past growth rate over a protracted period. The method is illustrated for a major oil company in Table 2.5. The current yield, 4.2 per cent, is added to the annual growth rate in dividends, 7.9 per cent, to produce an estimated cost of equity capital of 12.1 per cent.

Return on equity capital, then, is what investors received, measured in terms of either dividends plus growth in market value or dividends adjusted for growth. The return on the stock of a company during a specific time period is not necessarily the same thing as the cost of capital. Cost of capital is, by definition, what investors *require* as revealed by the relationship of the price they are willing to pay for the stock to the *future* economic benefits they anticipate (however hazily) through ownership. However, the rates of return experienced by investors in a group of economically similar common stocks should form a frequency distribution with a concentration around an average value that approximates the investors' required return for any given level of risk. The return actually provided to investors will vary among companies and over time, but should tend to concentrate around what is the investors' requirement.

The cost of capital obtained from internal cash generation is the same as the cost of equity capital, except no adjustment for flotation cost is needed. The cost is the same whether the saving is achieved by capital consumption allowances (depreciation and depletion), which may be viewed as recycling rather than capital formation, or from the reinvestment of net income (earnings plowback).

2.5.5.4 Combined Cost of Capital. The company's combined cost of capital will depend on how much of each kind of capital it obtains. The financial manager's task is to blend the numerous sources of capital available to him to obtain the lowest long-run cost of capital to his corporation.

Because debt capital is much cheaper after tax than common equity, one would expect a very high proportion of debt in the capital structure of well-managed companies. This, however, is not the case, primarily because as a company's depth of borrowing increases, approaching the theoretical limit of 100 per cent debt, the apparent (though not necessarily real) risk of bankruptcy rises. Presumably, lenders will exact higher and higher interest rates for this increasingly risky debt. Similarly, the hazards of common stockholders increase as debt goes deeper into this danger zone, and they, too, will be expected to require higher rates of return for equity capital (for example, the price of the common stock will fall). Hence, both the cost of debt and the cost of equity can be expected to rise as depth of debt increases. If they rise sharply enough, these higher costs will offset the decline in combined cost of capital that would otherwise result from a greater proportion of debt in the capital structure.

Table 2.6 illustrates part of the effect of capital structure on combined cost of capital. It is limited to the 0 to 50 per cent debt range observable in the petroleum industry. No increase in cost of equity capital is measurable over this range, whereas some increase in cost of debt is. Combined cost declines over this range, but should rise at some point beyond.

This minimum-cost point, where the increasing proportion of debt and the rising price of both debt and equity produce the lowest combined cost, is hard to locate empirically. It is hard because neither borrower nor lender will knowingly risk ruin in order to supply the needed data. Locating it is hard also because aversion for risk expresses itself in lending limits and in restrictive covenants rather than only in higher interest rates. The decision as to capital mix, therefore, requires a high order of mature judgment.

Despite some doctrine to the contrary, empirical evidence indicates that the combined cost of capital for a company with an actively

TABLE 2.6. *COMBINED COST OF CAPITAL (AFTER TAX) RELATED TO CAPITAL STRUCTURE*

Kind of Capital	10/90	20/80	30/70	40/60	50/50
Proportion of Debt	10 per cent	20 per cent	30 per cent	40 per cent	50 per cent
Proportion of Equity	90 per cent	80 per cent	70 per cent	60 per cent	50 per cent
Cost of debt	2.0 per cent	2.2 per cent	2.4 per cent	2.7 per cent	3.0 per cent
Cost of equity	11.0 per cent	11.0 per cent	11.0 per cent	11.0 per cent	11.0 per cent
Combined cost of capital	10.1 per cent	9.2 per cent	8.4 per cent	7.7 per cent	7.0 per cent

traded stock will decline if it gets more capital by borrowing, up to a limit of depth of debt which is beyond the depth that most managements would be willing to risk. Thus aversion for risk limits borrowing in most U.S. industrial companies far short of the point of lowest combined cost of capital. If instead of borrowing deeper the company gets growth capital by a mix of equity and debt such that its capital structure stays the same, then its cost of capital will be substantially the same over a wide supply range (capital price fluctuations aside).

2.5.6 Rationing of Capital

Two situations must be distinguished: (1) the autonomous financing firm that is determined to limit itself to internally generated funds only; and (2) the company that is willing to go outside for additional capital funds, either occasionally or regularly. The distinction in the budgeting problem in these two cases stems from the shape and behavior of the supply curves.

2.5.6.1 Autonomous Financing. The top panel of Fig. 2.9 diagrams the situation for the company that is limited to plowback earnings for its supply of capital. The demand schedule D_1 portrays for each prospective rate of return the amount of money that the firm can invest internally for earnings of at least that rate. At the point where the curve meets the supply, S_A^1, this firm can invest \$10 million at a rate of return of 20 per cent or better during the planning period.

The curve D_2 is the demand function in conditions when "needs" for capital expenditures are extremely pressing and profitable. If internal supply (mainly net cash generation from operations) remains at \$10 million (S_A^1), then an opportunity-cost of capital of 30 per cent is the indicated cutoff rate. Only if internal supply increases to S_A^2 will the cutoff stay at 20 per cent.

The internal supply curve is a vertical line because corporate saving is unaffected by the price of capital. Furthermore, it typically does not shift with shifts in prospective profitability of investment opportunities (e.g., from S_A^1 to S_A^2 when D_1 shifts to D_2). This unresponsiveness is because dividends are not (although they logically should be) inversely correlated with investment opportunity.

This model portrays a company that under no circumstances is willing to use outside money. In practice rich investment opportuni-

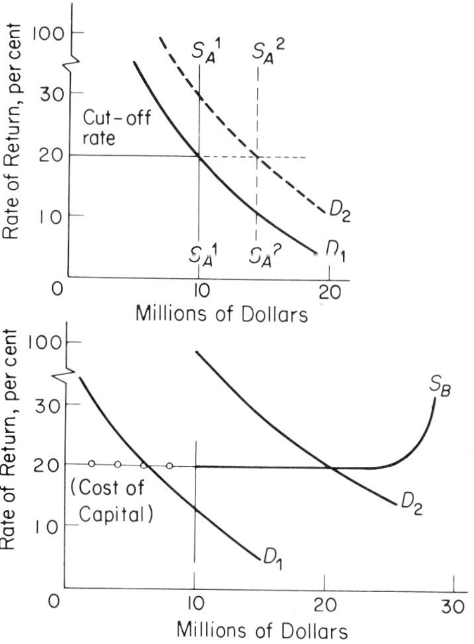

Fig. 2.9 *Theory of capital rationing: fluctuating effective rate. From Dean,* Managerial Economics, *p. 590.*

ties that drive the internal cutoff rate far above the costs of capital will act powerfully to undercut this policy and induce the company to borrow at least temporarily.

2.5.6.2 External Financing. The lower panel of Fig. 2.9 is a model for determination of the cutoff rate for a company willing to use outside money. In contrast to the autonomous supply case, the supply function extends to the right from the limit of retainable earnings to the point where the market balks at further financing. Combined cost of capital is constant up to this point because we assume the mix of debt and equity is unchanged during growth.

In the lower panel of Fig. 2.9, if the firm's investment-opportunity demand schedule shifts backward from D_2 to D_1 and its internal, net cash generation supply remains at $10 million, then the capital rationing cutoff should be established by the market cost of capital, 20 per cent. The decision rule is: Cut off at market cost of capital or at opportunity cost, whichever is higher.

2.5.6.3 Mechanics of Capital Rationing. Whether the cutoff criterion for rationing is determined by opportunity cost of capital (as in autonomous financing) or by market cost of external financing, the economic mechanics of capital rationing are about the same. The allocation of capital funds among rival investment projects should be based on the prospective profitability and risks of alternative capital proposals. Putting this principle to work requires three processes: foresight, screening, and follow-through.

FORESIGHT. Foresight is both a hallmark and an essential ingredient in good capital management. Foresight has two main dimensions: long-range projections of the firm's supply of and demand for capital, and short-range (usually one-year) capital budgets and cash budgets.

Five-year forecasts of a company's capital needs and capital sourcing are now common. Based on projections of future economic growth and sales of the industry and the company, some firms have prepared detailed plant-and-equipment targets toward which their entire capital expenditure program is oriented. Airlines, for example, were in 1969 planning the aircraft fleet for 1975.

In planning the sources of future capital funds, the long-term projections of probable capital outlays are stacked against predicted internal cash generation (cash receipts minus cash outlays) to forecast how much capital will have to be raised outside, e.g., by borrowing, by leasing, or by equity flotations.

Short-term budgets of capital outlays and working-capital sources are essential even though plans have wide margins of uncertainty. Budgets force advance study and early submission of capital proposals, which permit top management to screen them provisionally and weigh the need for cutbacks or the desirability of additional financing.

SCREENING. Screening of capital proposals is the cutting edge of capital rationing. With appropriate allowance for special risks and for imponderables, the criterion of minimum acceptable profitability should be the company's market cost of capital or its opportunity cost of capital, whichever is higher.

Making correct allowance for risks and for invisible benefits is where top management's judgment plays an important role in screening projects. Sometimes a big investment, e.g., an acquisition, changes the nature of the company's risk exposure enough to alter its ability and hence its costs of attracting capital. Thus an acquisition that is big enough and different enough in risk can modify the corporation's cost of capital. The general principle of rate-of-return rationing, nevertheless, still applies. Even when measurement of the project's profitability and riskiness is incomplete and subject to error, it provides a basis for improving executive judgment.

FOLLOW-THROUGH. Capital controls to assure that the program of capital rationing is actually carried out are of three main kinds: (1) outlay controls, i.e., making sure that the expenditures are in the amount and form authorized; (2) spending logs, i.e., comparing actual outlays with budgeted forecast outlays; and (3) postmortems, i.e., auditing the project earnings at early stages to keep the estimates honest and to learn from mistakes.

2.5.6.4 Economic Niche of Rate-of-Return Rationing. The intellectual validity of the discounted-cash-flow (DCF) approach to rate-of-return rationing that is sketched in this section is not seriously questioned. But it costs money and requires an enlightened management and a compatible overriding corporate goal. Consequently it is not appropriate for all capital-investing organizations and for all sorts of investment projects. What is its economic niche?

DCF capital rationing works best in enterprises whose overriding objective is maximizing profits in the long run. Governmental units, despite the current "cost-benefit analysis" fad, cannot really have economic capital rationing, because benefits cannot be calibrated and economic efficiency is not the overriding goal. Rate-of-return rationing contributes most to companies that are big, competitive, sophisticated, and primarily dedicated to the financial welfare of their owners. Projects should ideally compete for corporate capital on the basis of rate-of-return rivalry. DCF calibration is the best way to measure this capital productivity, providing that cost of superior measurement does not eat up benefits. It does, for some projects, including those (1) too profitable to require precise measurement (e.g., when the alternative is catastrophic), (2) too small to be worth the cost of calibration (simplistic shortcuts like MAPI formulas cut costs for homogeneous projects), (3) too vague and diffused in benefits to permit precise enough measurement (e.g., a new departure lounge for an airline), (4) too inextricably entwined in the total enterprise to measure separately, and (5) too risky to bother calibrating, i.e., risk so great and so unknowable it would swamp DCF measurements (e.g., the SST transport in 1969).

2.5.6.5 Alternatives to Rate-of-Return Rationing. Rate-of-return rationing by DCF analysis can serve as a standard of reference for appraising other methods of capital rationing. A common method, which is the antithesis of a rate-of-return system, is to let the determination of the total amount of capital expenditure and its allocation among projects be governed solely by the judgment of top executives who "consider each project on its merits" and tailor the total as best they can to the company's purse. This intuitive approach, when applied in large companies, burdens top management with a multitude of decisions that must be made without objective criteria. Hence the appraisal of an investment proposal is influenced by top management's appraisal of the executive who proposes it, and by his persuasiveness and persistence in presenting it.

Another method is to appraise and select individual investment proposals against an ideal of company balance and growth goals. For example, a petroleum company might seek, as a long-run objective, a 50 per cent growth in a decade and the attainment at the end of ten years of crude-oil production and refining facilities that equal its marketing demand. Such a goal may provide a criterion for approving and rejecting investment proposals, but the effects of this kind of plan upon the company's long-run rate of return are difficult to determine, because this approach views the company as a monolithic strategic investment. "Balance" usually means some form of vertical integration, and vertical integration has not proved universally profitable in all industries, nor is it sure to reduce the hazards of the enterprise.

PAUL T. NORTON, JR. *is a native of New Jersey, but moved to Philadelphia when he was twelve years old. After graduating from high school, he worked in various engineering jobs connected with the coal mining industry before deciding to seek an engineering degree. He received his engineering education at the University of Wisconsin, where he earned the Bachelor of Science degree in Electrical Engineering and the degree of Civil Engineer. Coincidentally, he and Eugene L. Grant, co-editor of this* Handbook, *were roommates at Wisconsin.*

Upon graduation from the University of Wisconsin, Mr. Norton enlisted in the Army Air Corps and served as an officer in Europe during the latter part of World War I. He returned to civilian life as a sales engineer with the Case Crane and Kilbourne Jacobs Company, Columbus, Ohio, and became vice president of the company in 1923. He remained in that position until 1926, at which time he joined the department of mechanics at the University of Wisconsin.

In 1929, Mr. Norton became the first head of the newly formed department of industrial engineering at the Virginia Polytechnic Institute. He developed that department into one of the leading ones in the country and remained there until 1948. In 1943, he was granted a leave of absence to become chief of the industrial processes branch of the Office of Production Research and Development, War Production Board. He also served as Treasurer of the Case Crane and Kilbourne Jacobs Company from 1943 until he joined the company on a full-time basis as vice president from 1948 to 1951. Later he was visiting professor of industrial engineering at Stanford University and lecturer in industrial engineering at the University of Florida.

An internationally recognized authority on depreciation and tax problems associated with depreciation matters, Mr. Norton has had published more than 60 papers on various phases of this and related subjects. He is co-author with Eugene L. Grant of Depreciation, *a leading text and reference book on the subject. He is a member of the American Society of Mechanical Engineers (Chairman, Publications Committee, 1953–4), American Society for Engineering Education (Council 1934–7), and the American Management Association.*

SECTION 3

ENGINEERING ECONOMY

PAUL T. NORTON, JR.

3.1 INTRODUCTION

It would be difficult to exaggerate the economic and social importance of engineering economy. The innumerable decisions that are made each day in this field in private industry determine whether proposals for investment in new plant and equipment are accepted or rejected. These decisions have far-reaching effects on our national standard of living.

It is a truism that a high standard of living depends upon the availability to the average person of an abundance of goods and services, plus the leisure necessary for the enjoyment of these goods and services. It is no accident that those countries possessing the highest degree of mechanization also possess the highest standard of living. Mechanization and the scientific management of facilities have caused the enormous increase in man-hour productivity in the industrialized areas of the world.

However, the same mechanization that has made possible a high standard of living has also been the cause of the greatly feared calamity that we call technological unemployment. It is small comfort to the individual worker who has lost his job, or to the community that has lost many jobs through the closing of a factory, to be told that in the long run mechanization has always meant greater total employment. For example, employment in the automobile industry is much greater than it was formerly in spite of the great decrease in the man-hours required to produce an automobile. Moreover, many other jobs in industries would never have developed at all if it had not been for the automobile and the motor truck.

Mechanization that is economically sound will widen markets by reducing the cost of the product, but reduction in total cost can occur only when the saving in operating cost (largely in labor cost) is greater than the increase in fixed charges that almost always follows an increase in mechanization. Since increased mechanization almost always reduces employment per unit of product, it can be justified economically and socially only if total cost, including fixed charges, is really reduced.

A thorough examination of the methods in general use for determining whether it will pay to make investments in productive equipment shows that many of them contain serious errors in principle. Some of these errors tend to encourage investments that should not be made because they reduce employment without reducing total cost. Other errors tend to prevent investments that really are desirable.

The principal purpose of this section is to develop methods for use in solving problems in engineering economy. Particular attention is devoted to problems that involve alternatives with differences in such factors as original investments, annual operating expenses, and lives. An attempt is made to introduce briefly the more important problems in this general field, but space limitations prevent the discussion of many problems of considerable importance to certain industries. Readers desiring a fuller discussion of the subject are referred to the bibliography at the end of the section.

3.1.1 Experience and Judgment Alone Are Inadequate Bases for Management Decisions

The great increase in the mechanization of our factories has created many new and serious problems for top management. Insufficient mechanization often reduces or eliminates profits by making competitive costs impossible to attain, but unwise mechanization sometimes leads to fixed charges that are excessive and that threaten the financial soundness of the company. Mechanization generally leads to investments that at best cannot be recovered for many years and that at worst may result in the loss of much of the investment. In addition, a correct decision among alternative ways for providing productive capacity often requires comparisons involving such factors as obsolescence. Where prospective obsolescence seems to be an important danger, it is necessary to weigh the increased chances for profit with a certain investment against the possible greater danger of loss with that particular investment because of rapid obsolescence.

When investments in fixed assets were a small part of the total investment in a business, an experienced manager with good business judgment could safely make decisions without formally considering what we are here calling engineering economy. Today the only safe way to handle alternatives involving investments in plant and equipment is to start with engineering economy studies. Good business judgment and experience are just as valuable as ever, but decisions involving investments in fixed assets should be based as far as possible on the tangible results of engineering economy studies.

3.1.2 Engineering Economy Formulas Are Not Always Safe

Formulas are useful in the solution of recurrent routine problems because a clerk who does not understand the derivation of a formula can still solve many problems by merely substituting in the formula. However, it should be clearly understood that no formula can be safely used unless someone in the organization understands its derivation and its limitations.

In many phases of engineering economy it is almost impossible to devise a formula that is both sound in principle and also simple in use. Moreover, a common objection to the use of even a sound and simple formula is the fact that it ordinarily gives merely a single value, with little or no information about the effect of even a slight deviation from that value. For example, minimum-cost formulas can often be devised that are both simple and economically sound, but their usefulness is limited by the fact that it is often the minimum-cost range that is really desired rather than the minimum-cost point, which is the only thing the formula can possibly give. What is even more significant, tabular methods that give all the desired information are often as simple and easy to use as the best possible formulas.

3.2 EQUIVALENCE

The typical engineering economy problem involves alternatives having differences in such factors as original investment, annual operating expenses, and lives. For example, the alternatives may be Machine *A*, with a first cost of $10,000, annual cash expenditures of $5,000, and an estimated life of 5 years, and Machine *B*, with a first cost of $15,000, annual cash expenditures of $4,000, and an estimated life of 6 years. It is obvious that the data as stated do not furnish sufficient information for comparing the two alternatives. It is necessary to make an economy study in which the first step is to convert the original data to other figures that are *equivalent* to the original data and that are in such shape as to be comparable.

In general, when alternatives such as Machines *A* and *B* have both initial investments and also cash expenditures spread over the period of the study, all payments must be converted either to an equivalent annual series or to an equivalent single amount at some specific date. In other words, it is necessary either to spread the investment uniformly over the study period and to add this amount to the estimated annual cash disbursements or else to convert the annual disbursements to a single payment at the time of the investment and to add it to the investment. The first method is commonly called the *annual-cost method* and the second the *present-worth method*. Both methods are explained later in this section.

Consider the $10,000 first cost of Machine *A* and the problem of converting this sum to a uniform annual figure for 5 years. It would be misleading merely to divide $10,000 by five and to assume that a payment of $2,000 a year for 5 years has the same effect as a payment of $10,000 at the start of the 5-year period. Because money has a time value, it is decidedly not a matter of indifference whether $10,000 is spent now or $2,000 is spent each year for 5 years. For equivalence calculations such as the one required to compare Machines *A* and *B*, it is necessary to select some particular interest rate that seems appropriate for the comparison to be made, all things considered.

Grant and Ireson describe the concept of equivalence as follows:*

> Given an interest rate, we may say that any payment or series of payments that will repay a present sum of money with interest at that rate

* E. L. Grant and W. G. Ireson, *Principles of Engineering Economy* (5th ed.) (New York: The Ronald Press Co., 1970), p. 30.

is equivalent to that present sum. Therefore, all future payments or series of payments that would repay the same present sum with interest at the stated rate are equivalent to each other.

3.2.1 Function of Interest in Equivalence

By means of interest calculations, we can determine an amount at any given time that is equivalent to a stated amount at some other time. For example, if an interest rate of 5 per cent per annum is used, \$1 now and \$1.05 a year from now may be said to be equivalent to each other.

Before attempting to determine an annual amount that is equivalent at 5 per cent per year over a 5-year period to the \$10,000 investment in Machine *A*, it is helpful to discuss three plans for the repayment of a loan of \$10,000 in 5 years with 5 per cent interest. These plans are shown in Table 3.1. All three are equivalent in the sense that in each plan the borrower pays and the lender receives 5 per cent per year on that part of the loan that has not already been repaid.

Table 3.1 shows three popular ways of repaying a \$10,000 loan in 5 years, with interest at 5 per cent. In Plan 1, the interest is paid at the end of each year, but no part of the principal is repaid until the end of the 5-year period, at which time the entire \$10,000 is repaid. In Plan 2, one-fifth of the principal is repaid at the end of each year, together with interest on the amount of the unrepaid principal during the year. In Plan 3, a uniform amount (partly interest and partly principal) is paid each year, this amount being just large enough so that all the principal, plus interest each year on the unrepaid principal during that year, will be repaid by means of these uniform annual payments of \$2,309.75 each. (The determination of the amount of the uniform annual payment will be explained in Art. 3.3.)

It will be noted that all three of these methods are equivalent to each other at 5 per cent interest because in each case the borrower pays back and the lender gets back the principal of the \$10,000 loan plus 5 per cent interest each year on the amount of the principal unrepaid during that particular year. These three series of year-end payments are equivalent to each other at 5 per cent and are also equivalent to \$10,000 at the beginning of the 5-year period.

The method of Plan 3 is the most useful method in most engineering economy problems since it makes possible the conversion of a present investment into an equivalent uniform annual cost of capital recovery, plus a return on the investment at the stated interest rate. In other words, if we assume that the \$10,000 investment in Machine *A*, described earlier, will

TABLE 3.1. *REPAYMENT OF \$10,000 IN 5 YEARS WITH INTEREST AT 5 PER CENT*

Plan	End of Year	Interest Due (5% of Money Owed at Start of Year)	Total Money Owed Before Year-End Payment	Year-End Payment	Money Owed After Year-End Payment
	0				\$10,000.00
	1	\$500.00	\$10,500.00	\$500.00	10,000.00
	2	500.00	10,500.00	500.00	10,000.00
1	3	500.00	10,500.00	500.00	10,000.00
	4	500.00	10,500.00	500.00	10,000.00
	5	500.00	10,500.00	10,500.00	0.00
	0				\$10,000.00
	1	\$500.00	\$10,500.00	\$2,500.00	8,000.00
	2	400.00	8,400.00	2,400.00	6,000.00
2	3	300.00	6,300.00	2,300.00	4,000.00
	4	200.00	4,200.00	2,200.00	2,000.00
	5	100.00	2,100.00	2,100.00	0.00
	0				\$10,000.00
	1	\$500.00	\$10,500.00	\$2,309.75	8,190.25
	2	409.51	8,599.76	2,309.75	6,290.01
3	3	314.50	6,604.51	2,309.75	4,294.76
	4	214.74	4,509.50	2,309.75	2,199.75
	5	109.99	2,309.74	2,309.74	0.00

be completely dissipated through use in the 5-year period (zero salvage value), the equivalent uniform annual cost of recovering this capital in 5 years, with 5 per cent return on the unrecovered balance during each year, will be $2,309.75 per year. Also the equivalent uniform annual cost of owning and operating this machine will be $7,309.75 per year, the figure that is obtained by adding the annual cash expenditures of $5,000 to the capital recovery cost of $2,309.75 per year. It is this equivalent uniform annual cost of $7,309.75 for Machine *A* that we can compare with a similarly obtained cost for Machine *B*, or any other available alternative.

3.2.2 Compound Interest Concept

All methods of repaying the $10,000 in 5 years with 5 per cent interest are equivalent, providing all payments of principal and interest are made at the end of some year and interest is charged each year at 5 per cent on the amount of principal remaining unrepaid during the year. This is true even if no payments of interest or principal are made until the end of the 5-year period. In that case, the interest will be added at the end of each year and in effect will become a part of a new principal. Thus, at the end of the first year, the interest will be $500, and when this is added to the original $10,000 principal, the principal during the second year becomes $10,500, and the interest at the end of the second year will be $525. Interest will be $551.25, $578.81, and $607.75 respectively in the third, fourth, and fifth years, and the amount to be repaid at the end of the fifth year will be $12,762.81. It will be recognized that this particular method employs the compounding of interest rather than the actual payment each year of the interest due.

It can easily be shown that the compound interest concept is the proper one for business loans and investments no matter how loans are repaid or investments recovered. This is true in actual loans from the viewpoint of both the borrower and the lender. Where interest is actually paid at the end of each interest period, it may be assumed that the lender could invest this amount and receive interest on it and also that the borrower could have invested the interest he actually paid if he had not paid it when he did.

It has been simpler to introduce the concept of equivalence by the use of a $10,000 loan for 5 years with 5 per cent interest, but in engineering economy most of the problems involve investments which it is hoped will be productive enough so that the amount of the investment will be recovered plus a satisfactory return.

Although the basic concept involved in the recovery of an investment (capital recovery with a return) is the same as that developed previously in this section for the repayment of a loan, the recovery of an investment in such things as plant and equipment is quite independent of the source of the funds that were used. The source and availability of investment funds may properly be considered when deciding whether certain types of investment should be made and in such matters as the setting of minimum attractive rates of return (interest rates). But care must be taken not to confuse the recovery of an investment with the repayment of a loan.

3.3 COMPOUND INTEREST FACTORS

The compound interest factors discussed in this article provide a convenient means for converting a single payment at some particular time into an equivalent single payment at some other time or of converting a single payment at some particular time into an equivalent uniform series of payments or of converting a uniform series of payments into an equivalent single payment at some particular time. By the use of two or more of these factors, various combinations of the three basic transformations just mentioned can be achieved, including such transformations as the conversion of an irregular series of payments into a uniform series of payments or a single payment at some particular time.

Compound interest tables giving in each case all six compound interest factors for a single interest rate are given in Tables 3.2 to 3.9, inclusive, for interest rates of 2, 3, 4, 5, 6, 7, 8, and 10 per cent, respectively. In addition, Table 3.10 gives capital-recovery factors for various interest rates from 6 to 50 per cent and Table 3.11 gives single-payment present-worth factors for the same interest rates.

The mathematical derivation of the six compound interest factors may be found in books on engineering economy or on the mathematics of finance. The mathematical expression for each factor is given at the head of each column

TABLE 3.2. *TWO PER CENT COMPOUND INTEREST FACTORS**

n	Single Payment		Uniform Payment Series				n
	Compound-Amount Factor	Present-Worth Factor	Compound-Amount Factor	Sinking-Fund Factor	Present-Worth Factor	Capital-Recovery Factor	
	Given P to Find S $(1+i)^n$	Given S to Find P $\frac{1}{(1+i)^n}$	Given R to Find S $\frac{(1+i)^n-1}{i}$	Given S to Find R $\frac{i}{(1+i)^n-1}$	Given R to Find P $\frac{(1+i)^n-1}{i(1+i)^n}$	Given P to Find R $\frac{i(1+i)^n}{(1+i)^n-1}$	
1	1.020	0.9804	1.000	1.00000	0.980	1.02000	**1**
2	1.040	0.9612	2.020	0.49505	1.942	0.51505	**2**
3	1.061	0.9423	3.060	0.32675	2.884	0.34675	**3**
4	1.082	0.9238	4.122	0.24262	3.808	0.26262	**4**
5	1.104	0.9057	5.204	0.19216	4.713	0.21216	**5**
6	1.126	0.8880	6.308	0.15853	5.601	0.17853	**6**
7	1.149	0.8706	7.434	0.13451	6.472	0.15451	**7**
8	1.172	0.8535	8.583	0.11651	7.325	0.13651	**8**
9	1.195	0.8368	9.755	0.10252	8.162	0.12252	**9**
10	1.219	0.8203	10.950	0.09133	8.983	0.11133	**10**
11	1.243	0.8043	12.169	0.08218	9.787	0.10218	**11**
12	1.268	0.7885	13.412	0.07456	10.575	0.09456	**12**
13	1.294	0.7730	14.680	0.06812	11.348	0.08812	**13**
14	1.319	0.7579	15.974	0.06260	12.106	0.08260	**14**
15	1.346	0.7430	17.293	0.05783	12.849	0.07783	**15**
16	1.373	0.7284	18.639	0.05365	13.578	0.07365	**16**
17	1.400	0.7142	20.012	0.04997	14.292	0.06997	**17**
18	1.428	0.7002	21.412	0.04670	14.992	0.06670	**18**
19	1.457	0.6864	22.841	0.04378	15.678	0.06378	**19**
20	1.486	0.6730	24.297	0.04116	16.351	0.06116	**20**
21	1.516	0.6598	25.783	0.03878	17.011	0.05878	**21**
22	1.546	0.6468	27.299	0.03663	17.658	0.05663	**22**
23	1.577	0.6342	28.845	0.03467	18.292	0.05467	**23**
24	1.608	0.6217	30.422	0.03287	18.914	0.05287	**24**
25	1.641	0.6095	32.030	0.03122	19.523	0.05122	**25**
26	1.673	0.5976	33.671	0.02970	20.121	0.04970	**26**
27	1.707	0.5859	35.344	0.02829	20.707	0.04829	**27**
28	1.741	0.5744	37.051	0.02699	21.281	0.04699	**28**
29	1.776	0.5631	38.792	0.02578	21.844	0.04578	**29**
30	1.811	0.5521	40.568	0.02465	22.396	0.04465	**30**
31	1.848	0.5412	42.379	0.02360	22.938	0.04360	**31**
32	1.885	0.5306	44.227	0.02261	23.468	0.04261	**32**
33	1.922	0.5202	46.112	0.02169	23.989	0.04169	**33**
34	1.961	0.5100	48.034	0.02082	24.499	0.04082	**34**
35	2.000	0.5000	49.994	0.02000	24.999	0.04000	**35**
40	2.208	0.4529	60.402	0.01656	27.355	0.03656	**40**
45	2.438	0.4102	71.893	0.01391	29.490	0.03391	**45**
50	2.692	0.3715	84.579	0.01182	31.424	0.03182	**50**
55	2.972	0.3365	98.587	0.01014	33.175	0.03014	**55**
60	3.281	0.3048	114.052	0.00877	34.761	0.02877	**60**
65	3.623	0.2761	131.126	0.00763	36.197	0.02763	**65**
70	4.000	0.2500	149.978	0.00667	37.499	0.02667	**70**
75	4.416	0.2265	170.792	0.00586	38.677	0.02586	**75**
80	4.875	0.2051	193.772	0.00516	39.745	0.02516	**80**
85	5.383	0.1858	219.144	0.00456	40.711	0.02456	**85**
90	5.943	0.1683	247.157	0.00405	41.587	0.02405	**90**
95	6.562	0.1524	278.085	0.00360	42.380	0.02360	**95**
100	7.245	0.1380	312.232	0.00320	43.098	0.02320	**100**

* Tables 3.2 to 3.10 from H. G. Thuesen, *Engineering Economy*. Copyright, 1950, by Prentice-Hall, Inc., Englewood Cliffs, N. J. pp. 483–490, 491.

TABLE 3.3. *THREE PER CENT COMPOUND INTEREST FACTORS*

n	Single Payment		Uniform Payment Series				n
	Compound-Amount Factor	Present-Worth Factor	Compound-Amount Factor	Sinking-Fund Factor	Present-Worth Factor	Capital-Recovery Factor	
	Given P to Find S $(1+i)^n$	Given S to Find P $\frac{1}{(1+i)^n}$	Given R to Find S $\frac{(1+i)^n-1}{i}$	Given S to Find R $\frac{i}{(1+i)^n-1}$	Given R to Find P $\frac{(1+i)^n-1}{i(1+i)^n}$	Given P to Find R $\frac{i(1+i)^n}{(1+i)^n-1}$	
1	1.030	0.9709	1.000	1.00000	0.971	1.03000	**1**
2	1.061	0.9426	2.030	0.49261	1.913	0.52261	**2**
3	1.093	0.9151	3.091	0.32353	2.829	0.35353	**3**
4	1.126	0.8885	4.184	0.23903	3.717	0.26903	**4**
5	1.159	0.8626	5.309	0.18835	4.580	0.21835	**5**
6	1.194	0.8375	6.468	0.15460	5.417	0.18460	**6**
7	1.230	0.8131	7.662	0.13051	6.230	0.16051	**7**
8	1.267	0.7894	8.892	0.11246	7.020	0.14246	**8**
9	1.305	0.7664	10.159	0.09843	7.786	0.12843	**9**
10	1.344	0.7441	11.464	0.08723	8.530	0.11723	**10**
11	1.384	0.7224	12.808	0.07808	9.253	0.10808	**11**
12	1.426	0.7014	14.192	0.07046	9.954	0.10046	**12**
13	1.469	0.6810	15.618	0.06403	10.635	0.09403	**13**
14	1.513	0.6611	17.086	0.05853	11.296	0.08853	**14**
15	1.558	0.6419	18.599	0.05377	11.938	0.08377	**15**
16	1.605	0.6232	20.157	0.04961	12.561	0.07961	**16**
17	1.653	0.6050	21.762	0.04595	13.166	0.07595	**17**
18	1.702	0.5874	23.414	0.04271	13.754	0.07271	**18**
19	1.754	0.5703	25.117	0.03981	14.324	0.06981	**19**
20	1.806	0.5537	26.870	0.03722	14.877	0.06722	**20**
21	1.860	0.5375	28.676	0.03487	15.415	0.06487	**21**
22	1.916	0.5219	30.537	0.03275	15.937	0.06275	**22**
23	1.974	0.5067	32.453	0.03081	16.444	0.06081	**23**
24	2.033	0.4919	34.426	0.02905	16.936	0.05905	**24**
25	2.094	0.4776	36.459	0.02743	17.413	0.05743	**25**
26	2.157	0.4637	38.553	0.02594	17.877	0.05594	**26**
27	2.221	0.4502	40.710	0.02456	18.327	0.05456	**27**
28	2.288	0.4371	42.931	0.02329	18.764	0.05329	**28**
29	2.357	0.4243	45.219	0.02211	19.188	0.05211	**29**
30	2.427	0.4120	47.575	0.02102	19.600	0.05102	**30**
31	2.500	0.4000	50.003	0.02000	20.000	0.05000	**31**
32	2.575	0.3883	52.503	0.01905	20.389	0.04905	**32**
33	2.652	0.3770	55.078	0.01816	20.766	0.04816	**33**
34	2.732	0.3660	57.730	0.01732	21.132	0.04732	**34**
35	2.814	0.3554	60.462	0.01654	21.487	0.04654	**35**
40	3.262	0.3066	75.401	0.01326	23.115	0.04326	**40**
45	3.782	0.2644	92.720	0.01079	24.519	0.04079	**45**
50	4.384	0.2281	112.797	0.00887	25.730	0.03887	**50**
55	5.082	0.1968	136.072	0.00735	26.774	0.03735	**55**
60	5.892	0.1697	163.053	0.00613	27.676	0.03613	**60**
65	6.830	0.1464	194.333	0.00515	28.453	0.03515	**65**
70	7.918	0.1263	230.594	0.00434	29.123	0.03434	**70**
75	9.179	0.1089	272.631	0.00367	29.702	0.03367	**75**
80	10.641	0.0940	321.363	0.00311	30.201	0.03311	**80**
85	12.336	0.0811	377.857	0.00265	30.631	0.03265	**85**
90	14.300	0.0699	443.349	0.00226	31.002	0.03226	**90**
95	16.578	0.0603	519.272	0.00193	31.323	0.03193	**95**
100	19.219	0.0520	607.288	0.00165	31.599	0.03165	**100**

TABLE 3.4. *FOUR PER CENT COMPOUND INTEREST FACTORS*

n	Single Payment		Uniform Payment Series				n
	Compound-Amount Factor	Present-Worth Factor	Compound-Amount Factor	Sinking-Fund Factor	Present-Worth Factor	Capital-Recovery Factor	
	Given P to Find S $(1+i)^n$	Given S to Find P $\frac{1}{(1+i)^n}$	Given R to Find S $\frac{(1+i)^n-1}{i}$	Given S to Find R $\frac{i}{(1+i)^n-1}$	Given R to Find P $\frac{(1+i)^n-1}{i(1+i)^n}$	Given P to Find R $\frac{i(1+i)^n}{(1+i)^n-1}$	
1	1.040	0.9615	1.000	1.00000	0.962	1.04000	**1**
2	1.082	0.9246	2.040	0.49020	1.886	0.53020	**2**
3	1.125	0.8890	3.122	0.32035	2.775	0.36035	**3**
4	1.170	0.8548	4.246	0.23549	3.630	0.27549	**4**
5	1.217	0.8219	5.416	0.18463	4.452	0.22463	**5**
6	1.265	0.7903	6.633	0.15076	5.242	0.19076	**6**
7	1.316	0.7599	7.898	0.12661	6.002	0.16661	**7**
8	1.369	0.7307	9.214	0.10853	6.733	0.14853	**8**
9	1.423	0.7026	10.583	0.09449	7.435	0.13449	**9**
10	1.480	0.6756	12.006	0.08329	8.111	0.12329	**10**
11	1.539	0.6496	13.486	0.07415	8.760	0.11415	**11**
12	1.601	0.6246	15.026	0.06655	9.385	0.10655	**12**
13	1.665	0.6006	16.627	0.06014	9.986	0.10014	**13**
14	1.732	0.5775	18.292	0.05467	10.563	0.09467	**14**
15	1.801	0.5553	20.024	0.04994	11.118	0.08994	**15**
16	1.873	0.5339	21.825	0.04582	11.652	0.08582	**16**
17	1.948	0.5134	23.698	0.04220	12.166	0.08220	**17**
18	2.026	0.4936	25.645	0.03899	12.659	0.07899	**18**
19	2.107	0.4746	27.671	0.03614	13.134	0.07614	**19**
20	2.191	0.4564	29.778	0.03358	13.590	0.07358	**20**
21	2.279	0.4388	31.969	0.03128	14.029	0.07128	**21**
22	2.370	0.4220	34.248	0.02920	14.451	0.06920	**22**
23	2.465	0.4057	36.618	0.02731	14.857	0.06731	**23**
24	2.563	0.3901	39.083	0.02559	15.247	0.06559	**24**
25	2.666	0.3751	41.646	0.02401	15.622	0.06401	**25**
26	2.772	0.3607	44.312	0.02257	15.983	0.06257	**26**
27	2.883	0.3468	47.084	0.02124	16.330	0.06124	**27**
28	2.999	0.3335	49.968	0.02001	16.663	0.06001	**28**
29	3.119	0.3207	52.966	0.01888	16.984	0.05888	**29**
30	3.243	0.3083	56.085	0.01783	17.292	0.05783	**30**
31	3.373	0.2965	59.328	0.01686	17.588	0.05686	**31**
32	3.508	0.2851	62.701	0.01595	17.874	0.05595	**32**
33	3.648	0.2741	66.210	0.01510	18.148	0.05510	**33**
34	3.794	0.2636	69.858	0.01431	18.411	0.05431	**34**
35	3.946	0.2534	73.652	0.01358	18.665	0.05358	**35**
40	4.801	0.2083	95.026	0.01052	19.793	0.05052	**40**
45	5.841	0.1712	121.029	0.00826	20.720	0.04826	**45**
50	7.107	0.1407	152.667	0.00655	21.482	0.04655	**50**
55	8.646	0.1157	191.159	0.00523	22.109	0.04523	**55**
60	10.520	0.0951	237.991	0.00420	22.623	0.04420	**60**
65	12.799	0.0781	294.968	0.00339	23.047	0.04339	**65**
70	15.572	0.0642	364.290	0.00275	23.395	0.04275	**70**
75	18.945	0.0528	448.631	0.00223	23.680	0.04223	**75**
80	23.050	0.0434	551.245	0.00181	23.915	0.04181	**80**
85	28.044	0.0357	676.090	0.00148	24.109	0.04148	**85**
90	34.119	0.0293	827.983	0.00121	24.267	0.04121	**90**
95	41.511	0.0241	1012.785	0.00099	24.398	0.04099	**95**
100	50.505	0.0198	1237.624	0.00081	24.505	0.04081	**100**

TABLE 3.5. *FIVE PER CENT COMPOUND INTEREST FACTORS*

n	Single payment		Uniform Payment Series				n
	Compound-Amount Factor	Present-Worth Factor	Compound-Amount Factor	Sinking-Fund Factor	Present-Worth Factor	Capital-Recovery Factor	
	Given P to Find S $(1+i)^n$	Given S to Find P $\frac{1}{(1+i)^n}$	Given R to Find S $\frac{(1+i)^n-1}{i}$	Given S to Find R $\frac{i}{(1+i)^n-1}$	Given R to Find P $\frac{(1+i)^n-1}{i(1+i)^n}$	Given P to Find R $\frac{i(1+i)^n}{(1+i)^n-1}$	
1	1.050	0.9524	1.000	1.00000	0.952	1.05000	**1**
2	1.103	0.9070	2.050	0.48780	1.859	0.53780	**2**
3	1.158	0.8638	3.153	0.31721	2.723	0.36721	**3**
4	1.216	0.8227	4.310	0.23201	3.546	0.28201	**4**
5	1.276	0.7835	5.526	0.18097	4.329	0.23097	**5**
6	1.340	0.7462	6.802	0.14702	5.076	0.19702	**6**
7	1.407	0.7107	8.142	0.12282	5.786	0.17282	**7**
8	1.477	0.6768	9.549	0.10472	6.463	0.15472	**8**
9	1.551	0.6446	11.027	0.09069	7.108	0.14069	**9**
10	1.629	0.6139	12.578	0.07950	7.722	0.12950	**10**
11	1.710	0.5847	14.207	0.07039	8.306	0.12039	**11**
12	1.796	0.5568	15.917	0.06283	8.863	0.11283	**12**
13	1.886	0.5303	17.713	0.05646	9.394	0.10646	**13**
14	1.980	0.5051	19.599	0.05102	9.899	0.10102	**14**
15	2.079	0.4810	21.579	0.04634	10.380	0.09634	**15**
16	2.183	0.4581	23.657	0.04227	10.838	0.09227	**16**
17	2.292	0.4363	25.840	0.03870	11.274	0.08870	**17**
18	2.407	0.4155	28.132	0.03555	11.690	0.08555	**18**
19	2.527	0.3957	30.539	0.03275	12.085	0.08275	**19**
20	2.653	0.3769	33.066	0.03024	12.462	0.08024	**20**
21	2.786	0.3589	35.719	0.02800	12.821	0.07800	**21**
22	2.925	0.3418	38.505	0.02597	13.163	0.07597	**22**
23	3.072	0.3256	41.430	0.02414	13.489	0.07414	**23**
24	3.225	0.3101	44.502	0.02247	13.799	0.07247	**24**
25	3.386	0.2953	47.727	0.02095	14.094	0.07095	**25**
26	3.556	0.2812	51.113	0.01956	14.375	0.06956	**26**
27	3.733	0.2678	54.669	0.01829	14.643	0.06829	**27**
28	3.920	0.2551	58.403	0.01712	14.898	0.06712	**28**
29	4.116	0.2429	62.323	0.01605	15.141	0.06605	**29**
30	4.322	0.2314	66.439	0.01505	15.372	0.06505	**30**
31	4.538	0.2204	70.761	0.01413	15.593	0.06413	**31**
32	4.765	0.2099	75.299	0.01328	15.803	0.06328	**32**
33	5.003	0.1999	80.064	0.01249	16.003	0.06249	**33**
34	5.253	0.1904	85.067	0.01176	16.193	0.06176	**34**
35	5.516	0.1813	90.320	0.01107	16.374	0.06107	**35**
40	7.040	0.1420	120.800	0.00828	17.159	0.05828	**40**
45	8.985	0.1113	159.700	0.00626	17.774	0.05626	**45**
50	11.467	0.0872	209.348	0.00478	18.256	0.05478	**50**
55	14.636	0.0683	272.713	0.00367	18.633	0.05367	**55**
60	18.679	0.0535	353.584	0.00283	18.929	0.05283	**60**
65	23.840	0.0419	456.798	0.00219	19.161	0.05219	**65**
70	30.426	0.0329	588.529	0.00170	19.343	0.05170	**70**
75	38.833	0.0258	756.654	0.00132	19.485	0.05132	**75**
80	49.561	0.0202	971.229	0.00103	19.596	0.05103	**80**
85	63.254	0.0158	1245.087	0.00080	19.684	0.05080	**85**
90	80.730	0.0124	1594.607	0.00063	19.752	0.05063	**90**
95	103.035	0.0097	2040.694	0.00049	19.806	0.05049	**95**
100	131.501	0.0076	2610.025	0.00038	19.848	0.05038	**100**

TABLE 3.6. *SIX PER CENT COMPOUND INTEREST FACTORS*

n	Single Payment		Uniform Payment Series				n
	Compound-Amount Factor	Present-Worth Factor	Compound-Amount Factor	Sinking-Fund Factor	Present-Worth Factor	Capital-Recovery Factor	
	Given P to Find S $(1+i)^n$	Given S to Find P $\frac{1}{(1+i)^n}$	Given R to Find S $\frac{(1+i)^n-1}{i}$	Given S to Find R $\frac{i}{(1+i)^n-1}$	Given R to Find P $\frac{(1+i)^n-1}{i(1+i)^n}$	Given P to Find R $\frac{i(1+i)^n}{(1+i)^n-1}$	
1	1.060	0.9434	1.000	1.00000	0.943	1.06000	**1**
2	1.124	0.8900	2.060	0.48544	1.833	0.54544	**2**
3	1.191	0.8396	3.184	0.31411	2.673	0.37411	**3**
4	1.262	0.7921	4.375	0.22859	3.465	0.28859	**4**
5	1.338	0.7473	5.637	0.17740	4.212	0.23740	**5**
6	1.419	0.7050	6.975	0.14336	4.917	0.20336	**6**
7	1.504	0.6651	8.394	0.11914	5.582	0.17914	**7**
8	1.594	0.6274	9.897	0.10104	6.210	0.16104	**8**
9	1.689	0.5919	11.491	0.08702	6.802	0.14702	**9**
10	1.791	0.5584	13.181	0.07587	7.360	0.13587	**10**
11	1.898	0.5268	14.972	0.06679	7.887	0.12679	**11**
12	2.012	0.4970	16.870	0.05928	8.384	0.11928	**12**
13	2.133	0.4688	18.882	0.05296	8.853	0.11296	**13**
14	2.261	0.4423	21.015	0.04758	9.295	0.10758	**14**
15	2.397	0.4173	23.276	0.04296	9.712	0.10296	**15**
16	2.540	0.3936	25.673	0.03895	10.106	0.09895	**16**
17	2.693	0.3714	28.213	0.03544	10.477	0.09544	**17**
18	2.854	0.3503	30.906	0.03236	10.828	0.09236	**18**
19	3.026	0.3305	33.760	0.02962	11.158	0.08962	**19**
20	3.207	0.3118	36.786	0.02718	11.470	0.08718	**20**
21	3.400	0.2942	39.993	0.02500	11.764	0.08500	**21**
22	3.604	0.2775	43.392	0.02305	12.042	0.08305	**22**
23	3.820	0.2618	46.996	0.02128	12.303	0.08128	**23**
24	4.049	0.2470	50.816	0.01968	12.550	0.07968	**24**
25	4.292	0.2330	54.865	0.01823	12.783	0.07823	**25**
26	4.549	0.2198	59.156	0.01690	13.003	0.07690	**26**
27	4.822	0.2074	63.706	0.01570	13.211	0.07570	**27**
28	5.112	0.1956	68.528	0.01459	13.406	0.07459	**28**
29	5.418	0.1846	73.640	0.01358	13.591	0.07358	**29**
30	5.743	0.1741	79.058	0.01265	13.765	0.07265	**30**
31	6.088	0.1643	84.802	0.01179	13.929	0.07179	**31**
32	6.453	0.1550	90.890	0.01100	14.084	0.07100	**32**
33	6.841	0.1462	97.343	0.01027	14.230	0.07027	**33**
34	7.251	0.1379	104.184	0.00960	14.368	0.06960	**34**
35	7.686	0.1301	111.435	0.00897	14.498	0.06897	**35**
40	10.286	0.0972	154.762	0.00646	15.046	0.06646	**40**
45	13.765	0.0727	212.744	0.00470	15.456	0.06470	**45**
50	18.420	0.0543	290.336	0.00344	15.762	0.06344	**50**
55	24.650	0.0406	394.172	0.00254	15.991	0.06254	**55**
60	32.988	0.0303	533.128	0.00188	16.161	0.06188	**60**
65	44.145	0.0227	719.083	0.00139	16.289	0.06139	**65**
70	59.076	0.0169	967.932	0.00103	16.385	0.06103	**70**
75	79.057	0.0126	1300.949	0.00077	16.456	0.06077	**75**
80	105.796	0.0095	1746.600	0.00057	16.509	0.06057	**80**
85	141.579	0.0071	2342.982	0.00043	16.549	0.06043	**85**
90	189.465	0.0053	3141.075	0.00032	16.579	0.06032	**90**
95	253.546	0.0039	4209.104	0.00024	16.601	0.06024	**95**
100	339.302	0.0029	5638.368	0.00018	16.618	0.06018	**100**

TABLE 3.7. *SEVEN PER CENT COMPOUND INTEREST FACTORS*

n	Single Payment		Uniform Payment Series				n
	Compound-Amount Factor	Present-Worth Factor	Compound-Amount Factor	Sinking-Fund Factor	Present-Worth Factor	Capital-Recovery Factor	
	Given P to Find S $(1+i)^n$	Given S to Find P $\frac{1}{(1+i)^n}$	Given R to Find S $\frac{(1+i)^n-1}{i}$	Given S to Find R $\frac{i}{(1+i)^n-1}$	Given R to Find P $\frac{(1+i)^n-1}{i(1+i)^n}$	Given P to Find R $\frac{i(1+i)^n}{(1+i)^n-1}$	
1	1.070	0.9346	1.000	1.00000	0.935	1.07000	**1**
2	1.145	0.8734	2.070	0.48309	1.808	0.55309	**2**
3	1.225	0.8163	3.215	0.31105	2.624	0.38105	**3**
4	1.311	0.7629	4.440	0.22523	3.387	0.29523	**4**
5	1.403	0.7130	5.751	0.17389	4.100	0.24389	**5**
6	1.501	0.6663	7.153	0.13980	4.767	0.20980	**6**
7	1.606	0.6227	8.654	0.11555	5.389	0.18555	**7**
8	1.718	0.5820	10.260	0.09747	5.971	0.16747	**8**
9	1.838	0.5439	11.978	0.08349	6.515	0.15349	**9**
10	1.967	0.5083	13.816	0.07238	7.024	0.14238	**10**
11	2.105	0.4751	15.784	0.06336	7.499	0.13336	**11**
12	2.252	0.4440	17.888	0.05590	7.943	0.12590	**12**
13	2.410	0.4150	20.141	0.04965	8.358	0.11965	**13**
14	2.579	0.3878	22.550	0.04434	8.745	0.11434	**14**
15	2.759	0.3624	25.129	0.03979	9.108	0.10979	**15**
16	2.952	0.3387	27.888	0.03586	9.447	0.10586	**16**
17	3.159	0.3166	30.840	0.03243	9.763	0.10243	**17**
18	3.380	0.2959	33.999	0.02941	10.059	0.09941	**18**
19	3.617	0.2765	37.379	0.02675	10.336	0.09675	**19**
20	3.870	0.2584	40.995	0.02439	10.594	0.09439	**20**
21	4.141	0.2415	44.865	0.02229	10.836	0.09229	**21**
22	4.430	0.2257	49.006	0.02041	11.061	0.09041	**22**
23	4.741	0.2109	53.436	0.01871	11.272	0.08871	**23**
24	5.072	0.1971	58.177	0.01719	11.469	0.08719	**24**
25	5.427	0.1842	63.249	0.01581	11.654	0.08581	**25**
26	5.807	0.1722	68.676	0.01456	11.826	0.08456	**26**
27	6.214	0.1609	74.484	0.01343	11.987	0.08343	**27**
28	6.649	0.1504	80.698	0.01239	12.137	0.08239	**28**
29	7.114	0.1406	87.347	0.01145	12.278	0.08145	**29**
30	7.612	0.1314	94.461	0.01059	12.409	0.08059	**30**
31	8.145	0.1228	102.073	0.00980	12.532	0.07980	**31**
32	8.715	0.1147	110.218	0.00907	12.647	0.07907	**32**
33	9.325	0.1072	118.933	0.00841	12.754	0.07841	**33**
34	9.978	0.1002	128.259	0.00780	12.854	0.07780	**34**
35	10.677	0.0937	138.237	0.00723	12.948	0.07723	**35**
40	14.974	0.0668	199.635	0.00501	13.332	0.07501	**40**
45	21.002	0.0476	285.749	0.00350	13.606	0.07350	**45**
50	29.457	0.0339	406.529	0.00246	13.801	0.07246	**50**
55	41.315	0.0242	575.929	0.00174	13.940	0.07174	**55**
60	57.946	0.0173	813.520	0.00123	14.039	0.07123	**60**
65	81.273	0.0123	1146.755	0.00087	14.110	0.07087	**65**
70	113.989	0.0088	1614.134	0.00062	14.160	0.07062	**70**
75	159.876	0.0063	2269.657	0.00044	14.196	0.07044	**75**
80	224.234	0.0045	3189.063	0.00031	14.222	0.07031	**80**
85	314.500	0.0032	4478.576	0.00022	14.240	0.07022	**85**
90	441.103	0.0023	6287.185	0.00016	14.253	0.07016	**90**
95	618.670	0.0016	8823.854	0.00011	14.263	0.07011	**95**
100	867.716	0.0012	12381.662	0.00008	14.269	0.07008	**100**

TABLE 3.8. *EIGHT PER CENT COMPOUND INTEREST FACTORS*

	Single Payment		Uniform Payment Series				
n	Compound-Amount Factor	Present-Worth Factor	Compound-Amount Factor	Sinking-Fund Factor	Present-Worth Factor	Capital-Recovery Factor	n
	Given P to Find S $(1+i)^n$	Given S to Find P $\frac{1}{(1+i)^n}$	Given R to Find S $\frac{(1+i)^n-1}{i}$	Given S to Find R $\frac{i}{(1+i)^n-1}$	Given R to Find P $\frac{(1+i)^n-1}{i(1+i)^n}$	Given P to Find R $\frac{i(1+i)^n}{(1+i)^n-1}$	
1	1.080	0.9259	1.000	1.00000	0.926	1.08000	**1**
2	1.166	0.8573	2.080	0.48077	1.783	0.56077	**2**
3	1.260	0.7938	3.246	0.30803	2.577	0.38803	**3**
4	1.360	0.7350	4.506	0.22192	3.312	0.30192	**4**
5	1.469	0.6806	5.867	0.17046	3.993	0.25046	**5**
6	1.587	0.6302	7.336	0.13632	4.623	0.21632	**6**
7	1.714	0.5835	8.923	0.11207	5.206	0.19207	**7**
8	1.851	0.5403	10.637	0.09401	5.747	0.17401	**8**
9	1.999	0.5002	12.488	0.08008	6.247	0.16008	**9**
10	2.159	0.4632	14.487	0.06903	6.710	0.14903	**10**
11	2.332	0.4289	16.645	0.06008	7.139	0.14008	**11**
12	2.518	0.3971	18.977	0.05270	7.536	0.13270	**12**
13	2.720	0.3677	21.495	0.04652	7.904	0.12652	**13**
14	2.937	0.3405	24.215	0.04130	8.244	0.12130	**14**
15	3.172	0.3152	27.152	0.03683	8.559	0.11683	**15**
16	3.426	0.2919	30.324	0.03298	8.851	0.11298	**16**
17	3.700	0.2703	33.750	0.02963	9.122	0.10963	**17**
18	3.996	0.2502	37.450	0.02670	9.372	0.10670	**18**
19	4.316	0.2317	41.446	0.02413	9.604	0.10413	**19**
20	4.661	0.2145	45.762	0.02185	9.818	0.10185	**20**
21	5.034	0.1987	50.423	0.01983	10.017	0.09983	**21**
22	5.437	0.1839	55.457	0.01803	10.201	0.09803	**22**
23	5.871	0.1703	60.893	0.01642	10.371	0.09642	**23**
24	6.341	0.1577	66.765	0.01498	10.529	0.09498	**24**
25	6.848	0.1460	73.106	0.01368	10.675	0.09368	**25**
26	7.396	0.1352	79.954	0.01251	10.810	0.09251	**26**
27	7.988	0.1252	87.351	0.01145	10.935	0.09145	**27**
28	8.627	0.1159	95.339	0.01049	11.051	0.09049	**28**
29	9.317	0.1073	103.966	0.00962	11.158	0.08962	**29**
30	10.063	0.0994	113.283	0.00883	11.258	0.08883	**30**
31	10.868	0.0920	123.346	0.00811	11.350	0.08811	**31**
32	11.737	0.0852	134.214	0.00745	11.435	0.08745	**32**
33	12.676	0.0789	145.951	0.00685	11.514	0.08685	**33**
34	13.690	0.0730	158.627	0.00630	11.587	0.08630	**34**
35	14.785	0.0676	172.317	0.00580	11.655	0.08580	**35**
40	21.725	0.0460	259.057	0.00386	11.925	0.08386	**40**
45	31.920	0.0313	386.506	0.00259	12.108	0.08259	**45**
50	46.902	0.0213	573.770	0.00174	12.233	0.08174	**50**
55	68.914	0.0145	848.923	0.00118	12.319	0.08118	**55**
60	101.257	0.0099	1253.213	0.00080	12.377	0.08080	**60**
65	148.780	0.0067	1847.248	0.00054	12.416	0.08054	**65**
70	218.606	0.0046	2720.080	0.00037	12.443	0.08037	**70**
75	321.205	0.0031	4002.557	0.00025	12.461	0.08025	**75**
80	471.955	0.0021	5886.935	0.00017	12.474	0.08017	**80**
85	693.456	0.0014	8655.706	0.00012	12.482	0.08012	**85**
90	1018.915	0.0010	12723.939	0.00008	12.488	0.08008	**90**
95	1497.121	0.0007	18701.507	0.00005	12.492	0.08005	**95**
100	2199.761	0.0005	27484.516	0.00004	12.494	0.08004	**100**

TABLE 3.9. *TEN PER CENT COMPOUND INTEREST FACTORS*

	Single Payment		Uniform Payment Series				
n	Compound-Amount Factor	Present-Worth Factor	Compound-Amount Factor	Sinking-Fund Factor	Present-Worth Factor	Capital-Recovery Factor	n
	Given P to Find S $(1+i)^n$	Given S to Find P $\frac{1}{(1+i)^n}$	Given R to Find S $\frac{(1+i)^n-1}{i}$	Given S to Find R $\frac{i}{(1+i)^n-1}$	Given R to Find P $\frac{(1+i)^n-1}{i(1+i)^n}$	Given P to Find R $\frac{i(1+i)^n}{(1+i)^n-1}$	
1	1.100	0.9091	1.000	1.00000	0.909	1.10000	**1**
2	1.210	0.8264	2.100	0.47619	1.736	0.57619	**2**
3	1.331	0.7513	3.310	0.30211	2.487	0.40211	**3**
4	1.464	0.6830	4.641	0.21547	3.170	0.31547	**4**
5	1.611	0.6209	6.105	0.16380	3.791	0.26380	**5**
6	1.772	0.5645	7.716	0.12961	4.355	0.22961	**6**
7	1.949	0.5132	9.487	0.10541	4.868	0.20541	**7**
8	2.144	0.4665	11.436	0.08744	5.335	0.18744	**8**
9	2.358	0.4241	13.579	0.07364	5.759	0.17364	**9**
10	2.594	0.3855	15.937	0.06275	6.144	0.16275	**10**
11	2.853	0.3505	18.531	0.05396	6.495	0.15396	**11**
12	3.138	0.3186	21.384	0.04676	6.814	0.14676	**12**
13	3.452	0.2897	24.523	0.04078	7.103	0.14078	**13**
14	3.797	0.2633	27.975	0.03575	7.367	0.13575	**14**
15	4.177	0.2394	31.772	0.03147	7.606	0.13147	**15**
16	4.595	0.2176	35.950	0.02782	7.824	0.12782	**16**
17	5.054	0.1978	40.545	0.02466	8.022	0.12466	**17**
18	5.560	0.1799	45.599	0.02193	8.201	0.12193	**18**
19	6.116	0.1635	51.159	0.01955	8.365	0.11955	**19**
20	6.727	0.1486	57.275	0.01746	8.514	0.11746	**20**
21	7.400	0.1351	64.002	0.01562	8.649	0.11562	**21**
22	8.140	0.1228	71.403	0.01401	8.772	0.11401	**22**
23	8.954	0.1117	79.543	0.01257	8.883	0.11257	**23**
24	9.850	0.1015	88.497	0.01130	8.985	0.11130	**24**
25	10.835	0.0923	98.347	0.01017	9.077	0.11017	**25**
26	11.918	0.0839	109.182	0.00916	9.161	0.10916	**26**
27	13.110	0.0763	121.100	0.00826	9.237	0.10826	**27**
28	14.421	0.0693	134.210	0.00745	9.307	0.10745	**28**
29	15.863	0.0630	148.631	0.00673	9.370	0.10673	**29**
30	17.449	0.0573	164.494	0.00608	9.427	0.10608	**30**
31	19.194	0.0521	181.943	0.00550	9.479	0.10550	**31**
32	21.114	0.0474	201.138	0.00497	9.526	0.10497	**32**
33	23.225	0.0431	222.252	0.00450	9.569	0.10450	**33**
34	25.548	0.0391	245.477	0.00407	9.609	0.10407	**34**
35	28.102	0.0356	271.024	0.00369	9.644	0.10369	**35**
40	45.259	0.0221	442.593	0.00226	9.779	0.10226	**40**
45	72.890	0.0137	718.905	0.00139	9.863	0.10139	**45**
50	117.391	0.0085	1163.909	0.00086	9.915	0.10086	**50**
55	189.059	0.0053	1880.591	0.00053	9.947	0.10053	**55**
60	304.482	0.0033	3034.816	0.00033	9.967	0.10033	**60**
65	490.371	0.0020	4893.707	0.00020	9.980	0.10020	**65**
70	789.747	0.0013	7887.470	0.00013	9.987	0.10013	**70**
75	1271.895	0.0008	12708.954	0.00008	9.992	0.10008	**75**
80	2048.400	0.0005	20474.002	0.00005	9.995	0.10005	**80**
85	3298.969	0.0003	32979.690	0.00003	9.997	0.10003	**85**
90	5313.023	0.0002	53120.226	0.00002	9.998	0.10002	**90**
95	8556.676	0.0001	85556.760	0.00001	9.999	0.10001	**95**
100	13780.612	0.0001	137796.123	0.00001	9.999	0.10001	**100**

in Tables 3.2 to 3.9. The symbols used in these mathematical expressions have the following meanings:

i = the interest rate per interest period

n = the number of interest periods

P = a present sum of money

R = a single end-of-period payment in a series of n equal payments made at uniform intervals, the entire series being equivalent to P

S = a sum of money n periods hence, which is equivalent to either P or R at interest rate i

In most problems in engineering economy, it is satisfactory to consider that interest is compounded annually, in which case i is the interest rate per year and n is the number of years. However, the tables may be used for any sort of compounding, such as semiannually, quarterly, or monthly. The rule is that i represents the interest rate per period and n represents the number of periods. Thus with quarterly compounding and a nominal interest rate of 8 per cent over a 12-year period, i would be 2 per cent and n would be 48.

It is advantageous to use mnemonic symbols to represent the six compound interest factors in problems and examples. The following symbols are used throughout this section:

Compound-amount factor (single payment)	(CA–i%–n)
Present-worth factor (single payment)	(PW–i%–n)
Sinking-fund factor	(SF–i%–n)
Capital-recovery factor	(CR–i%–n)

TABLE 3.10. *CAPITAL-RECOVERY FACTORS FOR INTEREST RATES FROM 6 PER CENT TO 50 PER CENT*

Given P, to Find R

$$\frac{i(1+i)^n}{(1+i)^n - 1}$$

n	i = 6%	8%	10%	12%	15%	20%	25%	30%	40%	50%	n
1	1.06000	1.08000	1.10000	1.12000	1.15000	1.20000	1.25000	1.30000	1.40000	1.50000	**1**
2	0.54544	0.56077	0.57619	0.59170	0.61512	0.65455	0.69444	0.73478	0.81667	0.90000	**2**
3	0.37411	0.38803	0.40211	0.41635	0.43798	0.47473	0.51230	0.55063	0.62936	0.71053	**3**
4	0.28859	0.30192	0.31547	0.32923	0.35027	0.38629	0.42344	0.46163	0.54077	0.62308	**4**
5	0.23740	0.25046	0.26380	0.27741	0.29832	0.33438	0.37184	0.41058	0.49136	0.57582	**5**
6	0.20336	0.21632	0.22961	0.24323	0.26424	0.30071	0.33882	0.37840	0.46126	0.54812	**6**
7	0.17914	0.19207	0.20541	0.21912	0.24036	0.27742	0.31634	0.35687	9.44192	0.53108	**7**
8	0.16104	0.17401	0.18744	0.20130	0.22285	0.26061	0.30040	0.34191	0.42804	0.52030	**8**
9	0.14702	0.16008	0.17364	0.18768	0.20957	0.24808	0.28876	0.33123	0.42034	0.51335	**9**
10	0.13587	0.14903	0.16275	0.17698	0.19925	0.23852	0.28007	0.32346	0.41432	0.50823	**10**
11	0.12679	0.14008	0.15396	0.16842	0.19107	0.23110	0.27349	0.31773	0.41013	0.50585	**11**
12	0.11928	0.13270	0.14676	0.16144	0.18448	0.22526	0.26845	0.31345	0.40718	0.50388	**12**
13	0.11296	0.12652	0.14078	0.15568	0.17911	0.22062	0.26454	0.31024	0.40510	0.50258	**13**
14	0.10758	0.12130	0.13575	0.15087	0.17469	0.21689	0.26150	0.30782	0.40363	0.50172	**14**
15	0.10296	0.11683	0.13147	0.14682	0.17102	0.21388	0.25912	0.30598	0.40259	0.50114	**15**
16	0.09895	0.11298	0.12782	0.14339	0.16795	0.21144	0.25724	0.30458	0.40185	0.50076	**16**
17	0.09544	0.10963	0.12466	0.14046	0.16537	0.20944	0.25576	0.30351	0.40132	0.50051	**17**
18	0.09236	0.10670	0.12193	0.13794	0.16319	0.20781	0.25459	0.30269	0.40094	0.50034	**18**
19	0.08962	0.10413	0.11955	0.13576	0.16134	0.20646	0.25366	0.30206	0.40067	0.50023	**19**
20	0.08718	0.10185	0.11746	0.13388	0.15976	0.20536	0.25292	0.30159	0.40048	0.50016	**20**
25	0.07823	0.09368	0.11017	0.12750	0.15470	0.20212	0.25095	0.30043	0.40009	0.50002	**25**
30	0.07265	0.08883	0.10608	0.12414	0.15230	0.20085	0.25031	0.30011	0.40002	0.50000	**30**
40	0.06646	0.08386	0.10226	0.12130	0.15056	0.20014	0.25003	0.30008	0.40001	0.50000	**40**
50	0.06344	0.08174	0.10086	0.12042	0.15014	0.20002	0.25000	0.30001	0.40000	0.50000	**50**
100	0.06018	0.08004	0.10001	0.12000	0.15000	0.20000	0.25000	0.30000	0.40000	0.50000	**100**
∞	0.06000	0.08000	0.10000	0.12000	0.15000	0.20000	0.25000	0.30000	0.40000	0.50000	∞

TABLE 3.11. *PRESENT-WORTH FACTORS FOR INTEREST RATES FROM 6 TO 50 PER CENT**

$\frac{1}{(1+i)^n}$ Given S, to Find P

n	i										n
	6%	8%	10%	12%	15%	20%	25%	30%	40%	50%	
1	0.9434	0.9259	0.9091	0.8929	0.8696	0.8333	0.8000	0.7692	0.7143	0.6667	**1**
2	0.8900	0.8573	0.8264	0.7972	0.7561	0.6944	0.6400	0.5917	0.5102	0.4444	**2**
3	0.8396	0.7938	0.7513	0.7118	0.6575	0.5787	0.5120	0.4552	0.3644	0.2963	**3**
4	0.7921	0.7350	0.6830	0.6355	0.5718	0.4823	0.4096	0.3501	0.2603	0.1975	**4**
5	0.7473	0.6806	0.6209	0.5674	0.4972	0.4019	0.3277	0.2693	0.1859	0.1317	**5**
6	0.7050	0.6302	0.5645	0.5066	0.4323	0.3349	0.2621	0.2072	0.1328	0.0878	**6**
7	0.6651	0.5835	0.5132	0.4523	0.3759	0.2791	0.2097	0.1594	0.0949	0.0585	**7**
8	0.6274	0.5403	0.4665	0.4039	0.3269	0.2326	0.1678	0.1226	0.0678	0.0390	**8**
9	0.5919	0.5002	0.4241	0.3606	0.2843	0.1938	0.1342	0.0943	0.0484	0.0260	**9**
10	0.5584	0.4632	0.3855	0.3220	0.2472	0.1615	0.1074	0.0725	0.0346	0.0173	**10**
11	0.5268	0.4289	0.3505	0.2875	0.2149	0.1346	0.0859	0.0558	0.0247	0.0116	**11**
12	0.4970	0.3971	0.3186	0.2567	0.1869	0.1122	0.0687	0.0429	0.0176	0.0077	**12**
13	0.4688	0.3677	0.2897	0.2292	0.1625	0.0935	0.0550	0.0330	0.0126	0.0051	**13**
14	0.4423	0.3405	0.2633	0.2046	0.1413	0.0779	0.0440	0.0254	0.0090	0.0034	**14**
15	0.4173	0.3152	0.2394	0.1827	0.1229	0.0649	0.0352	0.0195	0.0064	0.0023	**15**
16	0.3936	0.2919	0.2176	0.1631	0.1069	0.0541	0.0281	0.0150	0.0046	0.0015	**16**
17	0.3714	0.2703	0.1978	0.1456	0.0929	0.0451	0.0225	0.0116	0.0033	0.0010	**17**
18	0.3503	0.2502	0.1799	0.1301	0.0808	0.0376	0.0180	0.0089	0.0023	0.0007	**18**
19	0.3305	0.2317	0.1635	0.1161	0.0703	0.0313	0.0144	0.0068	0.0017	0.0005	**19**
20	0.3118	0.2145	0.1486	0.1037	0.0611	0.0261	0.0115	0.0053	0.0012	0.0003	**20**
25	0.2330	0.1460	0.0923	0.0588	0.0304	0.0105	0.0038	0.0014	0.0002	—	**25**
30	0.1741	0.0994	0.0573	0.0334	0.0151	0.0042	0.0012	0.0004	—	—	**30**
40	0.0972	0.0460	0.0221	0.0107	0.0037	0.0007	0.0001	—	—	—	**40**
50	0.0543	0.0213	0.0085	0.0035	0.0009	0.0001	—	—	—	—	**50**
100	0.0029	0.0005	0.0001	—	—	—	—	—	—	—	**100**

* Used by permission of H. G. Thuesen.

Compound-amount factor (uniform series) (SCA–i%–n)
Present-worth factor (uniform series) (SPW–i%–n)

The foregoing set of mnemonic symbols is one of two alternate sets that were suggested by the Committee on Standardization of Notation of the Engineering Economy Division of the American Society for Engineering Education.†

3.3.1 Compound-Amount Factor (Single-Payment)

It will be noted from the mathematical expression for each factor given at the head of each column of Tables 3.2 to 3.9 that $(1 + i)^n$ is the expression for the single-payment compound-amount factor, and also that the mathematical expression for each of the other factors contains this particular expression. Engineering handbooks and other nonfinancial handbooks often contain tables giving single-payment compound-amount factors but not the other five factors. In such cases, it is a simple matter to calculate any of the other five factors from the single-payment compound-amount factor for any desired value of i and n.

If tables are not available with the stipulated values of i or n or both, the expression $(1 + i)^n$ may be used to compute the desired factor. However, in most problems in engineering economy, it is good enough for practical purposes to interpolate between factors in available tables; there is no particular point in seeking accuracy in these factors that is greater than the

† In a report published in *The Engineering Economist*, XII, No. 1 (Oct.–Nov. 1966), 36–46.

accuracy of the data that are being used in the particular problem.

Problems that may be solved using this factor may be expressed in various ways, such as: If $5,000 is invested now at 7 per cent per annum, how much will it accumulate to in 8 years?

or

What is the compound amount of $5,000 for 8 years with interest at 7 per cent?

or

What must be the prospective cash receipt 8 years hence in order to justify a present cash disbursement of $5,000 if money is worth 7 per cent?

In all the foregoing questions, the cash flow by the prospective investor may be tabulated as follows:

Year	Cash Flow
0	–$5,000
8	$+S$

S = $5,000(CA–7%–8) = $5,000(1.718)

= $8,590

The required compound-amount factor, 1.718, is found in Table 3.7.

3.3.2 Present-Worth Factor (Single-Payment)

This factor, when multiplied by a future amount, will give the present worth of that future amount. It is obvious that the single-payment present-worth factor is the reciprocal of the single-payment compound-amount factor. Consider the questions: To get $2,000 in 14 years, how much must be invested now if interest is at 4 per cent?

or

What cash receipt now is acceptable in place of a prospective cash receipt of $2,000 in 14 years if interest is at 4 per cent?

or

What is the present worth of $2,000 in 14 years with interest at 4 per cent?

In all the foregoing questions, the cash flow by the prospective investor may be tabulated as follows:

Year	Cash Flow
0	P
14	$2,000

P = $2,000(PW – 4% – 14) = $2,000(0.5775)

= $1,155

The required present-worth factor, 0.5775, is found in Table 3.4.

3.3.3 Compound-Amount Factor (Uniform Series)

The compound-amount factor when multiplied by the amount of any payment of a uniform end-of-period series gives the equivalent total amount at the end of the series. The factor assumes that in the uniform series of payments, each payment is made at the end of its period. Consider the question: If $1,000 is invested at the end of each year for 4 years with interest at 5 per cent, what will be the amount of the accumulated fund at the end of the 4 years?

This is the cash flow by the prospective investor:

Year	Cash Flow
0	0
1	–$1,000
2	–$1,000
3	–$1,000
4	–$1,000 + S

S = $1,000(SCA – 5% – 4) = $1,000(4.310)

= $4,310

The required series compound-amount factor, 4.310, is found in Table 3.5.

3.3.4 Sinking-Fund Factor

This factor is used when it is desired to determine the magnitude of each of a series of equal end-of-period payments that will increase to some stipulated amount at the date of the final payment. Obviously, it is the reciprocal of the uniform-series compound-amount factor.

Consider the question: What annual end-of-year deposit must be made into a sinking fund to amount to $100,000 at the end of 6 years if interest is at 5 per cent?

This is the cash flow by the prospective investor:

Year	Cash Flow
0	0
1	$-R$
2	$-R$
3	$-R$
4	$-R$
5	$-R$
6	$-R$ + $100,000

$R = \$100{,}000(\text{SF} - 5\% - 6)$

$= \$100{,}000(0.14702) = \$14{,}702$

The required sinking-fund factor, 0.14702, is found in Table 3.5.

3.3.5 Present-Worth Factor (Uniform Series)

This factor when multiplied by one of a series of equal end-of-period payments gives the present worth at the start of the first period. Problems that may be solved using this factor may be expressed in various ways, such as: What investment is necessary to provide $200 at the end of each of the next 6 years, if interest is at 10 per cent?

or

What is the present worth of an end-of-year series of $200 a year for 6 years with interest at 10 per cent?

or

What immediate cash disbursement is justified in order to avoid cash disbursements of $200 at the end of each of the next 6 years if money is worth 10 per cent?

In all the foregoing questions this is the cash flow by the prospective investor:

YEAR	CASH FLOW
0	$-P$
1	+$200
2	+$200
3	+$200
4	+$200
5	+$200
6	+$200

(In the third statement of the question, the avoidance of an annual $200 cash disbursement has the same effect as an annual $200 cash receipt.)

$P = \$200(\text{SPW} - 10\% - 6) = \$200(4.355)$

$= \$871$

The required series present-worth factor, 4.355, is found in Table 3.9.

3.3.6 Capital-Recovery Factor

In engineering economy, this is often the most useful of the six compound-interest factors. By means of this factor one can distribute a present amount uniformly over a future interval of time by means of uniform end-of-period amounts. This concept may be expressed in many ways, one of them being the repayment of a loan through uniform annual end-of-year payments, such as the loan repayment in Plan 3 of Table 3.1. (CR–5%–5) is 0.230975. (Table 3.5, which gives only five significant figures, shows 0.23097.) The calculations for Plan 3 of Table 3.1 showed that $2309.75 exactly repays $10,000 in 5 years with interest at 5 per cent.

It is obvious that the capital-recovery factor is the reciprocal of the uniform series present-worth factor. Thus the six compound interest factors form three pairs in which the factors in each pair are reciprocals.

There is an extremely important relationship between the capital-recovery factor and the sinking-fund factor. It is not obvious from the mathematical expressions at the top of the columns for the capital-recovery factor and the sinking-fund factor, but these expressions may readily be rearranged so as to prove that the capital-recovery factor is always equal to the sinking-fund factor plus the interest rate. An examination of the values of these two factors in Tables 3.2 to 3.9 will show that this is true for every value of n in all these tables.

The reason the capital-recovery factor is equal to the sinking-fund factor plus the interest rate may easily be explained. It will be noted from Plans 1 and 3 of Table 3.1 that a loan of $10,000 may be repaid in 5 years with interest at 5 per cent by paying $500 interest at the end of each year plus $10,000 at the end of the 5-year period or by paying $2,309.75 (partly interest and partly principal) at the end of each year of the 5-year period. If the contract calls for payments according to Plan 1 but the borrower desires to pay off the principal of the loan in equal annual installments, he may make separate payments of $1,809.75 at the end of each year for 5 years and thus provide a sinking fund of $10,000 to pay off the principal of the loan when it becomes due. Thus the interest payments of $500 per year plus the sinking-fund payments of $1,809.75 per year are equal to the capital-recovery payments of $2,309.75 per year, and the capital-recovery factor is therefore equal to the sinking-fund factor plus the interest rate.

Problems requiring the use of the capital-recovery factor may be stated in various ways, such as: What 5-year annuity can be purchased for $10,000 with interest at 5 per cent?

or

If $10,000 is loaned for 5 years at 5 per cent

interest, what equal annual end-of-year payment would just repay the loan by the end of the 5 years? (This, of course, is the formulation of the problem illustrated in Plan 3, Table 3.1.)

or

What annual saving for 5 years must be anticipated in order to justify a present expenditure of $10,000, if money is worth 5 per cent?

or

What is the equivalent annual capital-recovery cost at 5 per cent interest of a machine having a first cost of $10,000, zero terminal salvage value, and a life of 5 years?

All the foregoing questions may be identified with cash flow by a prospective investor in this way:

YEAR	CASH FLOW
0	−$10,000
1	$+R$
2	$+R$
3	$+R$
4	$+R$
5	$+R$

$$R = \$10{,}000(\text{CR} - 5\% - 5)$$
$$= \$10{,}000(0.23097) = \$2{,}309.70$$

3.3.7 Using Several Factors in a Single Problem

Naturally, many problems can be solved to advantage with the aid of two or more of the foregoing six factors. For example, consider a proposal that involves an initial cash disbursement of $20,000, end-of-year cash disbursements of $1,000 for the first 6 years, a single cash disbursement of $10,000 at the end of the sixth year, end-of-year cash disbursements of $2,000 from the seventh to the fifteenth year, both inclusive, and a cash receipt of $5,000 at the end of the fifteenth year. It is desired to convert this irregular series of cash flows to an equivalent series of uniform disbursements for 15 years using an interest rate of 8 per cent. The purpose of this conversion is to make this proposal comparable with several alternative proposals that involve prospective cash flows that differ in their amount and timing.

The first step in converting such an irregular series to an equivalent uniform series should be to convert all cash flow either to its present worth at zero date or to its compound amount at the terminal date. The second step is to convert the algebraic sum of the present worths or compound amounts to an equivalent uniform series using the appropriate capital-recovery factor or sinking-fund factor. Conversion of this series to present worth can be done as follows:

$$\begin{aligned}\Sigma P = &- \$20{,}000 - \$1000(\text{SPW–8\%–6})\\ &- \$10{,}000(\text{PW–8\%–6})\\ &- \$2{,}000(\text{SPW–8\%–9})(\text{PW–8\%–6})\\ &+ \$5{,}000(\text{PW–8\%–15})\\ = &- \$20{,}000 - \$1{,}000(4.623)\\ &- \$10{,}000(0.6302)\\ &- \$2{,}000(6.247)(0.6302)\\ &+ \$5{,}000(0.3152)\\ = &- \$37{,}223\end{aligned}$$

Conversion to a uniform series of disbursements for 15 years is

$$\begin{aligned}R &= -\$37{,}223(\text{CR–8\%–15})\\ &= -\$37{,}223(0.11683) = -\$4349\end{aligned}$$

An alternate solution using compound amounts at the end of 15 years is

$$\begin{aligned}\Sigma S = &- \$20{,}000(\text{CA–8\%–15})\\ &- \$10{,}00(\text{SCA–8\%–6})(\text{CA–8\%–9})\\ &- \$10{,}000(\text{CA–8\%–9})\\ &- \$20{,}00(\text{SCA–8\%–9}) + \$5000\\ = &- \$20{,}000(3.172)\\ &- \$10{,}00(7.336)(1.999)\\ &- \$10{,}000(1.999) - \$2000(12.488)\\ &+ \$5{,}000\\ = &- \$118{,}071\end{aligned}$$

Conversion to a uniform series of disbursements for 15 years is

$$\begin{aligned}R &= -\$118{,}071(\text{SF–8\%–15})\\ &= -\$118{,}071(0.03683)\\ &= -\$4{,}349\end{aligned}$$

In most problems of the foregoing type, it is somewhat more convenient to use present worths rather than compound amounts. Moreover, in many circumstances all that is needed is the net present worth of a prospective cash flow, and there is no need to convert to an equivalent uniform annual series.

3.3.8 Gradient Factors

Consider a series of cash flows that occur in the form of an arithmetic progression, as follows:

End of Year	Cash Flow
0	0
1	0
2	G
3	$2G$
4	$3G$
—	—
$(n-1)$	$(n-2)G$
n	$(n-1)G$

In such a series, the annual change in cash flow G is called a gradient. The formulas for conversion of such a gradient series to an equivalent uniform annual end-of-year series R for the n years and to an equivalent present worth P are as follows:

$$R = G\left[\frac{1}{i} - \frac{n}{(1+i)^n - 1}\right]$$

$$= G\frac{n}{i}\left[\frac{1}{n} - \frac{i}{(1+i)^n - 1}\right]$$

$$P = G\left[\frac{1}{i} - \frac{n}{(1+i)^n - 1}\right]\left[\frac{(1+i)^n - 1}{i(1+i)^n}\right]$$

Most books on engineering economy published since about 1960 give the derivations of the foregoing formulas and give tables for the multipliers of G to convert a gradient series to an equivalent uniform series and to present worth. In the absence of such gradient tables, it is a fairly simple matter to compute any required uniform-series gradient factor using the sinking-fund factor $\frac{i}{(1+i)^n-1}$.

To compute a present-worth factor for a gradient series, it is also necessary to use the uniform-series present-worth factor $\frac{(1+i)^n-1}{i(1+i)^n}$.

Many economy studies involve estimated series of cash flows that increase or decrease by uniform amounts each year. Even though it is anticipated that actual receipts and disbursements will behave somewhat irregularly, the exact dates of the irregular variations may be unpredictable. Therefore, the best that can be done in making estimates may be to assume a uniform gradient wherever a fairly steady upward or downward trend is forecast. It follows that gradient factors may be helpful in many engineering economy studies.

3.3.9 Nominal and Effective Interest Rates

Consider a loan transaction in which interest is charged at 2 per cent per month. Such a transaction might conceivably be described as having an interest rate of 24 per cent per annum. However, in the technical language of compound interest, this should be referred to as a *nominal* 24 per cent per annum compounded monthly.

In such a loan transaction, the borrower pays more and the lender receives more than if interest of 24 per cent were paid at the end of each year. Consider that \$100 is borrowed with interest compounded at 2 per cent a month. The amount owed at the end of 12 months may be calculated using Table 3.2:

$$S = \$100(\text{CA–}2\%\text{–}12) = \$100(1.268)$$
$$= \$126.80$$

It is evident that borrowing (or investing) at 2 per cent a month has the same effect as borrowing (or investing) at 26.8 per cent per annum. The annual rate on the 2 per cent per month borrowing is described as an *effective* interest rate of 26.8 per cent per annum. The phrases *nominal interest rate* and *effective interest rate* may be defined in mathematical symbols as follows:

Assume that interest is compounded m times a year with an interest rate of r/m per compounding period.

The nominal interest rate per annum $= m(r/m) = r$

The effective interest rate per annum $= (1 + r/m)^m - 1$

3.4 COMPARING THE RELATIVE ECONOMY OF DESIGN ALTERNATIVES

In many economy studies relative to alternatives that need to be compared in the formulation of engineering projects, the prospective receipts from the sale of the product or service are the same regardless of which design is selected. The different methods of comparison of design alternatives can be illustrated to advantage by considering alternatives in pairs.

A typical situation is for one alternative—for example, Alternative A—to have a higher first cost than another alternative, Alternative B, but to have the advantage of lower annual disbursements and possibly also longer life or higher terminal salvage value or both. In judging whether such advantages are sufficient to justify the higher first cost, it is necessary to make calculations that will produce comparable figures for the two alternatives. Three common methods of securing comparable figures are:

1. Equivalent uniform annual cost. This is usually abbreviated to *annual cost*.
2. Present worth.
3. Rate of return on additional investment.

Before applying the annual-cost or present-worth method, it is necessary to make a decision on the minimum prospective rate of return deemed sufficient to justify an extra investment. All the examples in this article are based on the assumption that the minimum attractive rate of return before income taxes is 10 per cent. The examples assume that the design alternatives are compared before income taxes. The issues involved in deciding whether to make comparisons before or after prospective income taxes are discussed later in this section.

3.4.1 Annual Cost

Plan 3 of Table 3.1 may be used to illustrate the conversion of a first cost of a machine or structure into an equivalent uniform annual cost over the estimated life of an asset. In this plan, it was demonstrated that at the stated interest rate of 5 per cent, $10,000 at zero date was equivalent to $2309.75 a year for 5 years. Our discussion of compound interest tables and factors showed that the $2309.75 equivalent uniform annual figure was computed by multiplying the $10,000 by the capital-recovery factor for 5 per cent and 5 years (CR–5%–5).

In the following three examples, it is estimated that disbursements (for such things as operation, maintenance, property taxes, insurance) are uniform throughout the estimated lives of the machines and structures. (Actual future annual disbursements will no doubt be irregular rather than uniform; the assumption of uniformity for purposes of an economy study merely implies that the estimator has no basis for estimating any specific irregularities from year to year and that he therefore has no better estimate to make than one of uniformity.) Therefore, the only compound interest conversions required in these examples apply to the first costs of the assets and to their terminal salvage values, if any.

3.4.1.1 Example 1: Alternatives with Equal Lives and Zero Salvage Values. Machine *A* has a first cost of $15,000 and estimated annual cash disbursements during life of $6500. Machine *B* has a first cost of $25,000 and estimated annual cash disbursements during life of $4300. Each machine is estimated to have a life of 12 years, with zero salvage value at the end of life. Minimum attractive rate of return before income taxes is stipulated to be 10 per cent.

These are the equivalent uniform annual costs for comparative purposes:

Machine *A*

Cap. rec. cost	
= $15,000(CR–10%–12)	
= $15,000(0.14676)	= $2,201
Annual disbursements	= 6,500
Total annual cost	= $8,701

Machine *B*

Cap. rec. cost	
= $25,000(CR–10%–12)	
= $25,000(0.14676)	= $3,669
Annual disbursements	= 4,300
Total annual cost	= $7,969

The lower annual cost of Machine *B* indicates that its extra investment of $10,000 as compared to Machine *A* is sufficiently productive that it will earn more than the 10 per cent stipulated as the minimum attractive rate of return.

3.4.1.2 Example 2: Alternatives with Positive Salvage Values. When it is anticipated that there will be a positive salvage value at the end of the life of an asset, a satisfactory way to determine the annual cost of capital recovery is to separate the investment into two parts, one being the part that depreciates to zero and the other being the salvage value itself. It is obvious that the only capital cost for the salvage-value portion of the first cost is the amount of the salvage value multiplied by the interest rate.

Machine *C* has a first cost of $11,000, with anticipated salvage value of $2000 at the end of the estimated life of 8 years and prospective cash disbursements during life of $5420 per year. Machine *D* has a first cost of $19,000, with anticipated salvage value of $6000 at the end of the estimated life of 8 years and prospective cash disbursements during life of $4430 per year. Minimum attractive rate of return before income taxes is 10 per cent.

These are the equivalent uniform annual costs for comparative purposes:

Machine *C*

Cap. rec. cost	
= ($11,000 − $2000) (CR–10%–8)	
= ($9000) (0.18744)	= $1,687
Int. on salv. value	
= ($2000) (0.10)	= 200
Annual disbursements	= 5,420
Total annual cost	= $7,307

Machine *D*

Cap. rec. cost	
= ($19,000 − $6000) (CR–10%–8)	
= ($13,000) (0.18744)	= $2,437
Int. on salv. value	
= ($6000) (0.10)	= 600
Annual disbursements	= 4,430
Total annual cost	= $7,467

The lower annual cost of Machine *C* indicates that the $8000 of extra investment required for Machine *D* is not productive enough to earn the 10 per cent stipulated as the minimum attractive rate of return.

3.4.1.3 Example 3: Alternatives with Different Lives. Structure *E* has a first cost of $6,000, with anticipated zero salvage value at the end of the estimated life of 10 years and prospective cash disbursements during life of $1,600 per year. Structure *F* has a first cost of $20,000, with anticipated salvage value of $5,000 at the end of the estimated life of 20 years and prospective cash disbursements during life of $900 per year. Minimum attractive rate of return before income taxes is 10 per cent.

These are the equivalent uniform annual costs over the respective lives:

Structure *E*

Cap. rec. cost	
= $6,000(CR–10%–10)	
= $6,000(0.16275)	= $ 976
Annual disbursements	= 1,600
Total annual cost	= $2,576

Structure *F*

Cap. rec. cost	
= ($20,000 − $5,000) (CR–10%–20)	
= ($15,000) (0.11746)	= $1,762
Int. on salv. value	
= $50,00(0.10)	= 500
Annual disbursements	= 900
Total annual cost	= $3,162

Clearly, Structure *E* has the lower annual cost. However, an extra complication arises because of the difference in lives. The matter of the interpretation of these two annual costs as a basis for the decision between the two structures is discussed later in this article, after the same example has been analyzed by present-worth and rate-of-return methods.

3.4.2 Present Worth

The annual-cost method made alternatives comparable by converting estimated cash flows to equivalent uniform annual series throughout the estimated lives of the assets in question. The present-worth method makes alternatives comparable by converting each cash-flow series to an equivalent single figure at zero date. In both methods, the interest rate used in the equivalence conversion should be the stipulated minimum attractive rate of return. In both methods, the analysis provides a way to judge whether a proposed extra investment meets a stated standard of attractiveness, namely, the minimum attractive rate of return. The two methods naturally lead to the same conclusion about the relative merits of the alternatives being compared.

Whereas the problem in the annual-cost method was to convert first cost (and salvage value, if any) to an equivalent uniform annual series, the problem in the present-worth method is to convert annual disbursements (and salvage value, if any) to an equivalent figure at zero date. First cost is already at zero date and requires no conversion.

3.4.2.1 Example 1: Present-Worth Solution. A present-worth comparison of the costs of 12 years of service of Machines *A* and *B*, using 10 per cent interest, is as follows:

Machine *A*

First cost	= $15,000
PW of annual disb.	
= $6,500(SPW–10%–12)	
= $6,500(6.814)	= 44,290
Total PW of all disb. for 12 years	= $59,290

Machine *B*

First cost	= $25,000
PW of annual disb.	
= $4,300(SPW–10%–12)	
= $4,300(6.814)	= 29,300
Total PW of all disb. for 12 years	= $54,300

The conclusion is the same as in the annual cost comparison, namely, that the extra $10,000 investment in Machine *B* is economically justified under the given criterion of attractiveness. Moreover, the present worths are in the same proportion as the annual costs. That is, $59,290/$54,300 = $8,701/$7,969 = 1.092. Each total present worth multiplied by the appropriate capital-recovery factor gives the annual cost. Each annual cost multiplied by the appropriate series present-worth factor gives the total present worth.

3.4.2.2 Example 2: Present-Worth Solution. Machines *C* and *D* in this example had positive terminal salvage values at the end of their 8-year lives. Because a positive salvage value is a receipt, its present worth must be subtracted to find the total present worth of the net disbursements.

Machine *C*

First cost	=	$11,000
PW of annual disb.		
= $5420(SPW–10%–8)		
= $5420(5.335)	=	28,920
Total PW of disb. for 8 years	=	$39,920
Less: PW of receipt from salvage value		
= $2000(PW–10%–8)		
= $2000(0.4665)	=	930
PW of net disb. for 8 years	=	$38,990

Machine *D*

First cost	=	$19,000
PW of annual disb.		
= $4430(SPW–10%–8)		
= $4430(5.335)	=	23,630
Total PW of disb. for 8 years	=	$42,630
Less: PW of receipt from salvage value		
= $6000(PW–10%–8)		
= $6000(0.4665)	=	2,800
PW of net disb. for 8 years	=	$39,830

The comparison favors Machine *C* by a small margin, just as it did when annual costs were computed.

3.4.2.3 Example 3: Present-Worth Solution. When comparing alternatives by means of the present-worth method, it is necessary that the study period be the same for each alternative. For lives of 10 and 20 years, as in Example 3, it is necessary that the study period be the least common multiple of these lives (20 years) and also that definite assumptions be made as to the first cost and salvage value of each renewal of an asset and as to the annual disbursements for the entire study period. A common assumption is that the first costs and salvage values and the annual disbursements for renewals will be the same as for the asset being renewed. The following calculations for Example 3 make this assumption for the renewal of Structure *E*.

Structure *E*

First cost	=	$6,000
PW of renewal cost after 10 years		
= $6000(PW–10%–10)		
= $6000(0.3855)	=	2,310
PW of annual disb. for 20 years		
= $1600(SPW–10%–20)		
= $1600(8.514)	=	13,620
PW of net disb. for 20 years	=	$21,930

Structure *F*

First cost	=	$20,000
PW of annual disb. for 20 years		
= $900(SPW–10%–20)		
= $900(8.514)	=	7,660
PW of all disb. for 20 years	=	$27,660
Less: PW of salvage value		
= $5,000(PW–10%–20)		
= $5,000(0.1486)	=	740
PW of net disb. for 20 years	=	$26,920

3.4.3 Rate of Return on Additional Investment

Our comparisons of the alternatives in Examples 1, 2, and 3 by the annual-cost method and by the present-worth method have, in effect, answered the question "Will the extra investment in the alternative that has the higher first cost be recovered with a return (before income taxes) of at least our stipulated minimum attractive rate of 10 per cent?" In Example 1, the answer was yes. In Examples 2 and 3, the answer was no.

Another possible method of analysis is to find the expected rate of return on the additional investment and to compare this rate with the stated standard of attractiveness (10 per cent in the case of the three examples). Because a trial-and-error solution is necessary in many circumstances, this type of solution may be somewhat more time-consuming than one using annual costs or present worth. Nevertheless, for reasons discussed later in this article, there often may be definite advantages in using the rate-of-return method.

3.4.3.1 Example 1: Rate-of-Return Solution. Consider the respective cash flows for Machines *A* and *B* and the differences in year-by-year cash flows, as follows:

YEAR	MACHINE *A*	MACHINE *B*	*B–A*
0	–$15,000	–$25,000	–$10,000
1	–6,500	–4,300	+2,200
2	–6,500	–4,300	+2,200
3	–6,500	–4,300	+2,200
4	–6,500	–4,300	+2,200
5	–6,500	–4,300	+2,200
6	–6,500	–4,300	+2,200
7	–6,500	–4,300	+2,200
8	–6,500	–4,300	+2,200
9	–6,500	–4,300	+2,200
10	–6,500	–4,300	+2,200
11	–6,500	–4,300	+2,200
12	–6,500	–4,300	+2,200

The tabulation of differences in cash flow in the right-hand column clarifies the issue in the choice between these two alternatives. The selection of Machine *B* rather than Machine *A* causes an avoidable negative cash flow of $10,000 at zero date. However, this extra outlay of $10,000 will be responsible for positive cash flow of $2,200 a year for the next 12 years. (A reduction in annual disbursements from $6,500 to $4,300 has the same effect on cash as an annual receipt of $2,200.) Therefore, the question at issue is whether $2,200 a year for 12 years constitutes a high enough return on a $10,000 investment to be attractive, all things considered.

This simple case can be analyzed using the symbols employed in explaining compound interest formulas and tables. P is $10,000; R is $2,200; n is 12. The capital-recovery factor R/P = $2,200/$10,000 = 0.22. For an n of 12, Table 3.10 gives capital-recovery factors as follows:

i	CR
15 per cent	0.18448
20 per cent	0.22526

It is evident from inspection that a CR of 0.22 corresponds to an interest rate of a little more than 19 per cent. Interpolation between the two CR values gives approximately 19.4 per cent as the prospective rate of return on the extra investment in Machine *B*. Because this is in excess of the 10 per cent stipulated as the minimum attractive rate of return, Machine *B* appears to be the superior of the two stated alternatives.

3.4.3.2 Example 2: Rate-of-Return Solution. The difference in year-by-year cash flow between Machines *C* and *D* is slightly more complicated than in Example 1 because of the existence of terminal salvage values:

YEAR	MACHINE *C*	MACHINE *D*	*D* – *C*
0	–$11,000	–$19,000	–$8,000
1 to 8, inclusive	–5,420	–4,430	+990 per year
8	+2,000	+6,000	+4,000

Because of the terminal salvage values in this example, the unknown rate of return cannot be found by interpolation between tabulated capital-recovery factors as was done in the solution of Example 1. It is necessary to find the interest rate that makes the sum of the present worths of the differences in cash flow equal to zero. This requires a trial-and-error solution. In this instance, 7 per cent and 8 per cent are the tabulated interest rates for which the sum of the present worths should be computed.

At 7 per cent, the sum of the present worths is

$$\begin{aligned} &-\$8{,}000 + \$990(\text{SPW–}7\%\text{–}8) \\ &+ \$4{,}000(\text{PW–}7\%\text{–}8) \\ &= -\$8{,}000 + \$990(5.971) \\ &+ \$4{,}000(0.5820) \\ &= +\ \$239 \end{aligned}$$

A similar calculation at 8 per cent gives the sum of the present worths as –$94. Interpolation indicates that a decision to make the $8,000 extra investment in Machine *D* over Machine *C* is, in effect, a decision to make an avoidable investment that will yield about 7.7 per cent before income taxes. Because 7.7 per cent is less than the stated minimum attractive rate of 10 per cent, the extra investment in Machine *D* does not meet the given standard of attractiveness, and Machine *C* is therefore the economic choice.

3.4.3.3 Example 3: Rate-of-Return Solution. Like the present-worth solution for this example, a rate-of-return solution requires assumptions about the first cost, salvage value, and annual disbursements of the initial Structure *E*. If the assumption is the same one made in the present-worth solution, these are the prospective differences in cash flows between the two structures:

YEAR	STRUCTURE *E*	STRUCTURE *F*	*F*—*E*
0	−$6,000	−$20,000	−$14,000
1 to 20, incl.	−1,600	−900	+700 per year
10	−6,000	0	+6,000
20	0	+5,000	+5,000

The trial-and-error solution in this example requires calculations of the present worth of the differences in cash flows using the tabulated interest rates of 5 per cent and 6 per cent.

At 5 per cent the sum of the present worths is

$$- \$14,000 + \$700(\text{SPW–}5\%\text{–}20) + \$6,000(\text{PW–}5\%\text{–}10) + \$5,000(\text{PW–}5\%\text{–}20)$$

$$= - \$14,000 + \$700(12.462) + \$6,000(0.6139) + \$5,000(0.3769)$$

$$= + \$291$$

A similar calculation at 6 per cent gives the sum of the present worths as −$1,062. Interpolation between these two figures gives approximately 5.2 per cent as the prospective rate of return before income taxes on the extra investment in Structure *F*. Because this is less than the stipulated objective of at least a 10 per cent return, Structure F is not an economically attractive alternative to Structure *E*.

3.4.4 Some Aspects of Differences in Estimated Lives

Examples 1 and 2 dealt with alternatives that had equal lives. Certain relationships among the annual-cost method, present-worth method, and rate-of-return method could be brought out to good advantage by such examples. Nevertheless, a common condition is for design alternatives to have different lives.

In Example 3, Structure *E* has an estimated life of 10 years, whereas Structure *F* has an estimated life of 20 years. Consider the results of the annual-cost comparison stated as follows: The equivalent uniform-cost figure for Structure *E* is $2576 a year for 10 years. The equivalent uniform-cost figure for Structure *F* is $3,162 a year for 20 years.

Given the estimates of first cost, life, salvage value, and annual disbursements, and the stated interest rate of 10 per cent, the foregoing figures are unquestionably valid as equivalent uniform annual costs for the stated number of years. Clearly, Structure *E* has the lower annual cost. However, it is conceivable that even if all estimates should turn out to be correct, Structure *F* may be more economical in the long run. The question of long-run economy depends in part on what happens after the end of the life of the shorter-lived alternative.

In comparing a shorter-lived with a longer-lived alternative, an analyst may well anticipate certain changes that would favor the shorter-lived alternative and other changes that would favor the longer-lived one. For instance, the expectation that inflation will make the first cost of a replacement asset higher than the original cost of the asset to be replaced tends to favor longer-lived alternatives. On the other hand, the expectation that there will be technological progress tends to favor shorter-lived alternatives; presumably a replacement machine or structure will be superior to the original one and therefore will cause lower annual disbursements, higher annual receipts, or both.

The expectations about replacement assets do not necessarily have to be expressed in terms of cash flow by an analyst who uses the annual-cost method of comparison. He may merely view these expectations as matters to be given some weight in the final choice among the proposed alternatives, particularly in cases where the annual costs are fairly close together.

In contrast, as brought out in the solutions in Example 3, the present-worth method and the rate-of-return method require assumptions about cash flows for the same numbers of years for all alternatives. In that example, cash-flow estimates were made for 20 years, the least common multiple of the respective 10-year and 20-year lives of the two structures. It was assumed that the replacement for Structure *E* would have the same first cost, life, salvage value, and annual disbursements as the original structure that it would replace.

Although this simple assumption is made in many economy studies, an analyst is free to make some other assumption whenever it seems to him that a better one can be made. Often it is helpful to test the sensitivity of the conclusions of an economy study to specific changes in estimates of future cash flows. For instance, in Example 3, the question might be asked: "How would matters be changed if the replacement for Structure *E* should have a first cost of $15,000 rather than $6,000?" If no change in life, salvage value, or annual disbursements is assumed for the replacement structure, the present worth of 20 years service for Structure *E* is increased by ($15,000 − $6,000)(PW–10% –10) = ($9,000)(0.3855) = $3,470. The present worth of the net disbursements for Structure E for 20 years now becomes \$25,400, whereas the comparable present worth for Structure F remains at \$26,920. In this case, the comparison continues to favor Structure E in spite of the large increase in the estimated first cost of the renewal structure.

In this example, which uses an interest rate of 10 per cent, \$1 in 10 years is equivalent to less than 39¢ on the date of the economy study. In general, the farther into the future the date is made for an estimate of cash flow, the less influence the estimate ought to have on a present decision. Also, of course, the higher the minimum attractive rate of return (interest rate) used in an economy study, the smaller the present-worth factor applicable to any given future date and the less any changes in estimates of distant future cash flows will influence the conclusions of the economy study.

3.5 CALCULATING THE COST OF BORROWED MONEY

The same trial-and-error methods for finding an unknown interest rate that were illustrated in connection with rate of return are applicable to calculating the cost of long-term borrowing. In an investment in physical assets, there is usually an initial negative cash flow followed by a series of positive cash flows. In contrast, borrowing involves an initial positive cash flow followed by a series of negative cash flows.

Consider a long-term borrowing involving a \$10 million, 20-year bond issue with a stated interest rate of 5½ per cent per annum. Because interest is payable semiannually, the borrowing corporation pays interest of \$275,000 every 6 months; when the bonds mature at the end of 20 years, it pays the principal of \$10 million. The bonds are sold to an investment banking syndicate for \$9.7 million. This positive cash flow to the corporation on the date of issuance of the bonds is reduced by various outlays incident to the borrowing (for necessary accounting, engineering, financial reports, securities registration statement, printing of bonds, etc.) which total \$125,000. The semiannual disbursements throughout the life of the bond issue are increased above the actual interest payments by the need for various outlays associated with the bond issue (such as fees to the registrar and the trustee of the bonds and clerical costs of interest payments); these are estimated as totaling \$10,000. The resulting cash-flow series for time intervals of 6 months over the life of the bonds is as follows:

DATE	CASH FLOW
0	+\$9,700,000 − \$125,000 = +\$9,575,000
1 to 40 incl.	− \$275,000 − \$10,000 = − \$285,000 every 6 months
40	= −\$10,000,000

To find the unknown interest rate that makes the present worth of this cash-flow series equal to zero, the rate must be per stated time interval, namely, 6 months. A trial calculation with interest at 3 per cent is

$$\begin{aligned} PW &= +\$9{,}575{,}000 - \$285{,}000(\text{SPW}\text{–}3\%\text{–}40) - \$10{,}000{,}000(\text{PW}\text{–}3\%\text{–}40) \\ &= +\$9{,}575{,}000 - \$285{,}000(23.115) - \$10{,}000{,}000(0.3066) \\ &= -\$79{,}000 \end{aligned}$$

A similar calculation at 4 per cent gives the net present worth as +\$1,851,000. Interpolation gives 3.04 per cent as the interest rate per 6-month period. This corresponds to a nominal interest rate per annum of 2(3.04%) = 6.08%. The effective interest rate paid per annum is $(1.0304)^2 - 1 = 0.0617$ or 6.17 per cent.

The foregoing is a calculation of the cost of certain debt capital before income taxes. In the usual case where interest, debt discount, and the various costs associated with borrowing can be deducted from taxable income, either immediately or ultimately, the after-tax cost of debt capital can be considerably less than the before-tax cost, particularly for taxpayers who are taxed at relatively high rates.

If it is assumed that a taxpayer will be taxed throughout the entire period of a borrowing at an incremental rate t (expressed as a decimal), an approximate figure for the after-tax cost of borrowed money can be obtained by multiplying the before-tax cost by $(1 - t)$. Thus if the corporation that borrows at an effective before-tax rate of 6.17 per cent expects to be taxed at a 48 per cent rate throughout the life of the bond issue, its after-tax cost of borrowing will be approximately (6.17%)(1 − 0.48) = 3.21%.

3.6 CONTINUOUS COMPOUNDING AND THE UNIFORM-FLOW CONVENTION

The compound amount S of a principal sum P invested for n years with a nominal interest rate r and with m compounding periods a year is given by

$$S = P\left(1 + \frac{r}{m}\right)^{mn}$$

It may be shown that when the number of compounding periods per year m increases without limit and the nominal interest rate r remains constant, the foregoing expression becomes

$$S = Pe^{rn}$$

In this equation the symbol e refers to the base of natural, or Napierian, logarithms, approximately 2.7182818.

In continuous compounding, the single-payment compound-amount factor (CA–r%–n) is e^{rn} and the single-payment present-worth factor (PW–r%–n) is e^{rn}, where r is the nominal interest rate per annum.

The interest formulas and tables for uniform series presented earlier in this section all assume periodic cash flows; payments all occur at the ends of stated periods. Continuous compounding makes it possible to apply compound interest mathematics to the assumption of cash flow taking place uniformly throughout a period. With a nominal interest rate r per annum and continuous compounding, consider that a total of $1 flows continuously throughout a year. The formula for the present worth of this dollar at the start of the year is

$$P = \$1\,\frac{e^r - 1}{re^r}$$

3.6.1 Nominal and Effective Rates in Continuous Compounding

Writers on the mathematics of finance refer to r, the nominal interest rate used in continuous compounding, as the *force of interest*. In periodic compounding, the more often interest is compounded in a year, the greater the difference between the nominal and effective interest rates per annum. Moreover, the disparity between nominal and effective interest rates increases with the rate. Of course, this disparity is at its maximum value for any nominal rate when compounding is continuous.

The effective interest rate in continuous compounding is e^r. Grant and Ireson give Table 3.12, showing the relationship between effective and nominal rates in continuous compounding.

Tables of the various compound-interest factors similar to Tables 3.2 to 3.11 may be prepared assuming continuous compounding rather than periodic compounding. Such tables may apply to uniform cash flows as well as to periodic cash flows. Some of the published tables, such as those given by Grant and Ire-

TABLE 3.12. *FORCE OF INTEREST r TO BE USED IN INTEREST FORMULAS INVOLVING CONTINUOUS COMPOUNDING IN ORDER TO YIELD VARIOUS EFFECTIVE INTEREST RATES PER ANNUM**

Effective Rate per Annum	Force of Interest (i.e., Nominal Rate Compounded Continuously to Yield the Stated Effective Rate)	Effective Rate per Annum	Force of Interest (i.e., Nominal Rate Compounded Continuously to Yield the Stated Effective Rate)
1%	0.995033%	15%	13.976194%
2%	1.980263%	20%	18.232156%
3%	2.955880%	25%	22.314355%
4%	3.922071%	30%	26.236426%
5%	4.879016%	35%	30.010459%
6%	5.826891%	40%	33.647224%
7%	6.765865%	45%	37.156356%
8%	7.696104%	50%	40.546511%
10%	9.531018%		
12%	11.332869%		

* Grant and Ireson, *Principles of Engineering Economy*, p. 496.

son,* are based on effective interest rates. Other continuous compounding tables, such as the ones used in Sec. 4, 5 assume nominal interest rates. It is desirable for persons who make economy studies using either type of continuous compounding tables to be aware of the differences between nominal and effective rates shown here in Table 3.12.

3.6.2 The Uniform-Flow Convention

Some prospective cash flows are expected at specific predictable points in time. (An example was the payment of bond interest and principal illustrated in Art. 3.5.) However, many estimated cash flows are expected to be spread over stated periods of time, possibly in an irregular manner or possibly at a fairly uniform rate.

Usually the year is the most convenient unit of time to adopt in economy studies. With a full year as the minimum time unit, it is necessary to adopt some convention regarding cash flows expected to occur during each year. One common convention is to make an analysis as if such cash flows took place at the year's end. The various analyses of Examples 3.1 to 3.3 illustrated this end-of-year convention. Ordinary interest tables such as the ones given here as Tables 3.2 to 3.11 can be used with the end-of-year convention.

Another possible convention is to make an analysis as if cash flow took place uniformly throughout each year. A uniform-flow convention requires the use of interest tables that assume continuous compounding. The analysis form designated as Fig. 4.3 in Sec. 4 illustrates the use of continuous compounding factors applicable to uniform cash flows throughout each year. In this form, the analysis period extends from the third year before the zero point to the thirtieth year after that point. The form also shows continuous compounding factors for certain instantaneous cash flows associated with a proposed investment. In using this form, an analyst needs to recognize that certain types of investment may be assigned to a particular point in time whereas other investment outlays may be viewed as approximating uniform cash flows throughout a given year. All the continuous compounding factors in Fig. 4.3 are based on nominal interest rates.

Grant and Ireson state that it is their impression that the end-of-year convention is much more widely used than the uniform-flow convention and comment on the choice between these two conventions as follows:†

> The greater use of the end-of-year convention doubtless is based largely on grounds of its greater convenience. Standard interest tables and formulas may be used with this convention. In the numerous cases where annual cost comparisons are made, this convention has the advantage of being better adapted to such comparisons. Where compound interest methods require explanation in presenting the results of economy studies, it is easier to explain periodic compounding of interest than to explain continuous compounding.
>
> Compound interest conversions in economy studies constitute one of the steps in providing a rational basis for decisions among alternatives. In most cases the *decisions* will be the same regardless of the convention used. That is, both conventions will normally array a series of investment proposals in the same order.
>
> However, there are certain cases where the two conventions might lead to different recommendations for a decision between alternative investments and where the recommendation based on the uniform-flow convention will be sounder than the one based on the end-of-year convention. These cases occur particularly where an investment leading to positive cash flow concentrated in the near future is being compared with one leading to positive cash flow spread over a considerably longer period.

3.7 EVALUATION OF MAJOR INVESTMENT PROPOSALS

In this handbook the problems of evaluation of major investment proposals are discussed from slightly different viewpoints in Art. 5 of Sec. 2 and in Sec. 4. The writers of those sections also stress the usefulness of compound interest mathematics in project evaluation.

Examples 1, 2, and 3 in this section illustrate engineering economy techniques as applied to choices among design alternatives in the formulation of investment projects. It should be clear to the reader that the general principles and

* Grant and Ireson, *Principles of Engineering Economy* (5th ed.), pp. 622–27.

† Grant and Ireson, *Principles of Engineering Economy*, p. 542.

techniques needed in project formulation are generally applicable to project evaluation.

However, as explained in Sec. 2, decisions regarding competing proposals for major investments in a business enterprise are related to a variety of financial planning decisions, including such matters as dividend policy, new equity financing, and new debt financing. Also, as explained in Secs. 2 and 4, it is desirable to have systematic procedures that make possible sound project evaluation and comparison at top management levels. Because these matters are covered so well in Secs. 2 and 4, they are not discussed in this section on engineering economy.

3.8 ECONOMIC LOT SIZES IN MANUFACTURING

A correct understanding of the economic lot-size problem is desirable both because of the importance of the problem itself and because there are many other minimum-cost problems whose solutions are similar. The following discussion is adapted from that in a bulletin published in 1934.*

3.8.1 Importance of Problem

Many mass-production industries manufacture in advance of sale and ship articles as desired from finished stocks stored in warehouses. Modern manufacturing equipment ordinarily produces at a very rapid rate while operating but also has high preparation costs. Manufacturers know very well that under such circumstances the best way to reduce unit preparation costs is to produce in large lots. But manufacturers do not always understand to what an extent large lots increase unit charges incident to storage, such as rental of storage space, insurance and taxes on finished inventory, and increased risk of loss due to deterioration or obsolescence of inventory. The last-named factor may result in excessive amounts having to be sold at heavy discounts to get rid of styles that have not proved popular.

This rather general failure to pay sufficient attention to the unit charges incident to storage is undoubtedly due to the fact that cost-accounting methods may be perfectly satisfactory for the purposes for which they are designed but incapable of giving all the information required for establishing economic lot sizes. Unit preparation costs show up clearly on the cost-accounting records, whereas most of the costs incident to storage either do not appear at all in the accounting records or are not identified there as costs that vary with lot size. The unit charges incident to storage are nonetheless real and should be given full consideration. It will be shown presently that for the most economical size of lot the unit preparation costs are equal to the unit charges incident to storage. What should be aimed at is a size of lot which, by rapid turnover of capital invested in finished inventories, reduces the unit charges incident to storage to the point where they do not exceed the unit preparation costs.

3.8.2 Derivation of the Formula

Economic lot-size problems may be solved in several different ways. Many formulas have been developed for the purpose. Some are too complicated for practical use because their authors have tried to include every variable that has any effect. Others have limited usefulness because they do not include some variables of real importance. Many give lot sizes entirely too large because they use the interest rate on borrowed money instead of the minimum attractive rate of return when determining carrying charges on the investment in finished inventory. It is believed that the formula to be developed presently is simple enough to be understood and used by anyone and also that it includes all the factors necessary for a correct solution of the problem; a satisfactory turnover of capital invested in finished inventory is obtained by charging investment in finished inventory with the minimum attractive rate of return, based on the risks involved in the storage of this particular inventory.

Although formulas are useful in solving this problem, it should be understood that the economic lot size may be found just as well by tabular or graphical methods, which have the advantage that they give not only the economic lot size but also the economic range, something that no formula can possibly give. In addition, tabular and graphical methods show the effect on total unit charges of any deviation from the

* P. T. Norton, Jr., *Economic Lot Sizes in Manufacturing* (Blacksburg, Va.: Virginia Polytechnic Institute, 1934).

economic lot size—information that is often of great value.

In the formula to be developed, let

Q = the lot size (pieces per lot)

Q_e = the economic lot size (pieces per lot)

S = total preparation cost per lot (dollars), including cost of preparing manufacturing orders, cost of setting up machines, and any other similar costs that are independent of the number of pieces in the lot

P = pieces made per day

U = pieces used per day

N = days worked per year

C = material, direct labor, and factory overhead per piece (dollars)

A = cost of storing one piece for one year (dollars)

B = taxes, insurance, etc. (percentage per year on inventory)

I = minimum attractive rate of return on capital invested in finished inventory, risks considered (percentage per year)

V = total amount charged against each piece (dollars) for
- (a) preparation costs
- (b) material, direct labor, and factory overhead
- (c) storage charges, including return on invested capital

It can easily be proved that in the following derivation the preparation cost per piece may without error be neglected when figuring the capital tied up in finished inventory. The following derivation assumes that the total amount C will be charged from the moment production begins; other assumptions will be discussed later. As is customary, it will be assumed that the consumption rate is uniform; if the demand varies greatly between seasons, it may be desirable to compute the economic lot size for each season.

Under these assumptions, the maximum amount of capital tied up in finished inventory will be CQ dollars, and the average amount of capital thus tied up will be $CQ/2$ dollars. As there are NU pieces per year, the average amount of capital tied up *per piece* will be $CQ/2NU$ dollars.

Insurance and taxes must be paid on finished inventory, and the charge *per piece* for these items will be $BCQ/2NU$ dollars.

Cost accountants do not ordinarily include a charge for interest on capital tied up in inventory as part of the cost of a product. However, in studies of this sort, where different lot sizes result in different amounts of invested capital, it is necessary to consider this factor. It is only by making this charge at the minimum attractive rate of return that full consideration can be given to the important factor of capital turnover. On this basis, the charge *per piece* for return on invested capital will be $ICQ/2NU$ dollars.

It will be assumed that storage is in bins and that space must be reserved permanently for the largest number of pieces of each article that will ever be in storage. $(P - U)$ pieces will go into storage each day during production, which will continue for Q/P days. Hence the maximum number of pieces for which storage must be provided is

$$\frac{Q}{P}(P - U) \quad \text{or} \quad Q\left(1 - \frac{U}{P}\right)\text{pieces}$$

As it costs A dollars to store one piece for 1 year and there are NU pieces per year, the cost *per piece* for providing storage is

$$\frac{A\left(1 - \frac{U}{P}\right)Q}{NU} \text{ dollars.}$$

The total charges *per piece*, incident to storage, are the sum of the three expressions just derived, and may be written

$$\left[\frac{(B + I)C + 2A\left(1 - \frac{U}{P}\right)}{2NU}\right]Q \quad \text{or} \quad KQ$$

where K is a constant representing the terms within the brackets. The total charges *per piece* are

$$V = \frac{S}{Q} + C + KQ$$

This equation shows that the total charges against each piece are the sum of three parts: the preparation cost per piece S/Q, which varies inversely with size of lot; the approximately constant cost per piece C for material, direct labor, and factory overhead; and the charges per piece incident to storage KQ, which vary directly with size of lot.

The economic lot size is obviously the lot size for which V is a minimum. This size may easily be found by tabular or graphical methods, but the simplest way is to find the first derivative and to set it equal to zero.

$$\frac{dV}{dQ} = \frac{S}{Q^2} + 0 + K$$

and setting dV/dQ equal to zero,

$$Q_e = \sqrt{\frac{S}{K}}$$

The approximately constant unit cost of material, direct labor, and factory overhead C does not appear in the final formula, but it does affect the economic lot size because of its effect on the value of K.

The economic lot-size formula just derived may be written

$$\frac{S}{Q_e} = KQ_e$$

which proves that for the economic lot size, the unit preparation charges S/Q_e are equal to the unit charges incident to storage KQ_e. This relationship is useful when it is desired to solve a problem of this sort by tabular or graphical methods.

TABLE 3.13. *UNIT CHARGES FOR VARIOUS LOT SIZES OF EXAMPLE 4**

Q, Pieces	$\frac{S}{Q}$, Cents	C, Cents	KQ, Cents	V, Cents
1,000	1.000	10	0.041	11.041
2,000	0.500	10	0.082	10.582
3,000	0.333	10	0.123	10.456
4,000	0.250	10	0.164	10.414
4,500	0.222	10	0.184	10.406
4,940	0.202	10	0.202	10.404
5,500	0.182	10	0.225	10.407
6,000	0.167	10	0.246	10.413
7,000	0.143	10	0.287	10.430
8,000	0.125	10	0.328	10.453

* From Norton, *Economic Lot Sizes in Manufacturing*, Table 1, p. 15.

3.8.3 Example 4: Use of the Formula

In order to illustrate the use of the formula, the economic lot size will be determined under the following conditions:

$S = \$10$

$P = 1000$ pieces per day

$U = 100$ pieces per day

$N = 300$ days worked per year

$C = \$0.10$ per piece

$A = \$0.001$ per piece

$B = 3$ per cent per year

$I = 20$ per cent per year

Under these conditions, the value of K is

$$\frac{(0.03 + 0.20)0.10 + 2(0.001)\left(1 - \frac{100}{1000}\right)}{2(300)(100)} = \$0.00000041$$

and $$Q_e = \sqrt{\frac{10.00}{0.00000041}} = 4940 \text{ pieces per lot}$$

3.8.4 Tabular Methods May Be Used in Place of the Formula

By means of the formula, it has been a simple matter to determine that the economic lot size is 4940 pieces per lot. The formula cannot, however, give either the total unit charges or the difference in total unit charges with some other size of lot. The simplest way to obtain this valuable information is to tabulate the values as is done in Table 3.13.

Table 3.13 shows that the minimum total unit charges amount to \$0.10404 at the economic lot size of 4940 pieces per lot; also that the total unit charges are practically the same for lot sizes between 4500 pieces and 5500 pieces; and that for any lot size between 3000 pieces and 7000 pieces the total unit charges would not be more than one-half of 1 per cent more than at the economic lot size. Thus the lot size may ordinarily be varied to a considerable extent from the economic lot size without greatly increasing total unit charges. The fact that the lot size may ordinarily be reduced below the economic lot size without materially increasing the total unit charges is of great significance, as it enables a manufacturer to reduce his working capital requirements by reducing the amount of money tied up in finished inventory.

3.8.5 Importance of Using Minimum Attractive Rate of Return Instead of Interest Rate on Borrowed Money

Many economic lot-size formulas use the interest rate on borrowed money (cost of money) instead of the minimum attractive rate of return. If as high an interest rate as 6 per cent is used in this example, the formula will give a value of 7450 pieces per lot for what may well be called the "minimum cost lot size." However, this would not be a desirable lot size, as the turnover in capital invested in finished inventory would not be great enough to take care of the risks involved.

Figure 3.1 shows graphically the relationship between unit preparation charges and unit charges incident to storage for this example. The full lines are for the economic lot-size determination, using 20 per cent per year for *I*. The broken lines show the incorrect result that would be obtained if an interest rate of 6 per cent per year were used. (The horizontal line that would give the constant value of $0.10 for *C* in Fig. 3.1 is omitted because it would not change the shape of any of the curves of Fig. 3.1 or affect any of the information obtainable from Fig. 3.1.)

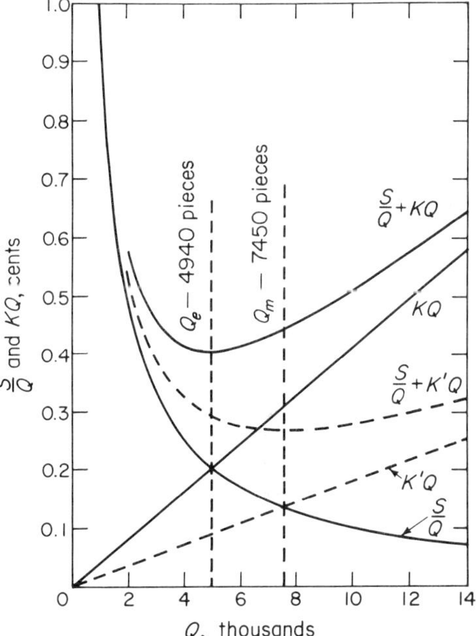

Fig. 3.1 Determination of economic lot size. From Norton, *Economic Lot Sizes in Manufacturing*, Figure 1, p. 18.

3.8.6 Modifications of Formula Factors

Factor *K* is a constant for any given set of conditions but may be varied to take care of several different situations.

In the derivation of *K* for use in Example 4, it was assumed that the total amount for material, direct labor, and factory overhead was to be charged at the moment production began. This is probably the situation in the majority of cases, but because certain manufacturing costs are not actually paid until some time after work is performed, it is often reasonable to assume that the investment in finished inventory is merely for that portion of a lot that actually goes into storage, in other words

$$Q\left(1 - \frac{U}{P}\right)$$

Under these circumstances, the *investment part* of *K* is

$$\frac{(B + I)\left(1 - \frac{U}{P}\right)C}{2NU}$$

In the derivation of *K* for use in Example 4, it was also assumed that storage space must be reserved permanently for the largest number of pieces of each article that ever was in storage. In many cases, storage space may be released for storing other articles whenever a shipment is made. Under these conditions, the total storage space requirements will be only half as much as in Example 4, and the storage cost part of *K* will be

$$\frac{A\left(1 - \frac{U}{P}\right)}{2NU}$$

It is now possible to state the four expressions for *K* that will permit one to use the formula for any of these assumptions for investment charges and storage costs.

If it is assumed that the entire cost for material, direct labor, and factory overhead must be charged at the moment production begins and that storage space must be reserved for the maximum number of articles ever in storage, then

$$K = \frac{(B + I)C + 2A\left(1 - \frac{U}{P}\right)}{2NU}$$

If it is assumed that the entire cost for material, direct labor, and factory overhead must be charged at the moment production begins but that any vacant storage space may be used for any article, then

$$K = \frac{(B + I)C + A\left(1 - \frac{U}{P}\right)}{2NU}$$

If it is assumed that the only investment that need be considered is that represented by articles actually in storage and that storage space must be reserved for the maximum number of articles ever in storage, then

$$K = \frac{[(B + I)C + 2A]\left(1 - \frac{U}{P}\right)}{2NU}$$

If it is assumed that the only investment that need be considered is that represented by articles actually in storage but that any vacant storage space may be used for any article, then:

$$K = \frac{[(B + I)C + A]\left(1 - \frac{U}{P}\right)}{2NU}$$

Several other relationships may be present in certain situations, but they are rarely given consideration in economic lot-size discussions. Some of these are important because when they are present they may reduce the time and effort required to make the studies.

The consumption rate U is one of the most important factors in the formula. It is standard practice to assume that this factor is a constant during the period under consideration. This is a sufficiently accurate procedure in most cases, but where the demand is quite seasonal, it may be desirable to determine a different lot size for each season, using the appropriate value of U in each case.

It will be noted that U appears in K in two places—in the denominator and also in the numerator as part of the expression $(1 - U/P)$. In many cases, U is so small when compared with P that $(1 - U/P)$ may be considered to be unity. Under these circumstances, it is evident that Q_e varies directly with $\sqrt{U}$. Where the assumption that $(1 - U/P)$ is unity is a reasonable assumption, it is a simple matter to modify the general formula to the form $Q_e = D\sqrt{U}$, where D is the value of Q_e found from the general formula $Q_e = \sqrt{S/K}$, when $U = 1$. The lot size for any value of U may then be easily determined by multiplying D by $\sqrt{U}$.

3.8.7 Reserve or Emergency Stock

Many manufacturers plan their production on the assumption that there will always be a certain number of pieces of an article in stock when the following order begins to enter storage. Many economic lot-size formulas attempt to include this reserve or emergency stock as a factor, but this is clearly not a sound procedure. If sales and production programs are carried out in accordance with the estimates on which the economic lot-size study was based, the use of a reserve stock will merely increase the number of pieces in stock at all times by the amount of the reserve stock. The added charges incident to the storage of the reserve stock, though real, do not in any way affect the economic lot size.

3.8.8 Summary of Economic Lot-Size Discussion

Lot sizes should not be made as large as possible in order to reduce unit preparation charges but should be made as small as possible in order to obtain a high rate of capital turnover and to reduce the risks incident to storage. However, if lot sizes are too small, the increase in the unit preparation charges will more than offset the advantages gained from lower charges incident to storage. All manufacturers are keenly aware of the reduction in unit preparation charges that is caused by large lots, but few manufacturers seem to understand how much large lots really cost them in charges incident to storage, including all the risks involved.

Total unit charges increase very little for considerable reductions in lot size below the economic lot size. Since the amount of capital tied up in finished inventory varies directly with the lot size, a manufacturer with limited working capital resources may often use to advantage lot sizes somewhat smaller than the economic lot size. By so doing, he will reduce the amount of capital tied up in inventory without materially increasing total unit charges.

Capital tied up in finished inventory should be charged with the minimum attractive rate of return on such capital and not merely with the simple interest rate paid on borrowed money. This procedure automatically takes care of the problems of capital turnover and such risks incident to storage as obsolescence and deterioration.

Formulas are useful in the routine determination of the economic lot size but do not show the difference in total unit charges that would result from using some other lot size. This information is important and may easily be obtained through tabular or graphical methods of solving the problem.

For the economic lot size, the unit preparation charges are equal to the unit charges incident to storage.

The economic lot size varies approximately as the square root of the consumption rate. This relationship makes it possible to redetermine quickly lot sizes for various consumption rates after the economic lot size has been determined for the normal consumption rate by means of the complete formula.

3.8.9 Other Minimum-Cost Problems

Most of the ideas that have been expressed in the foregoing discussion of the economic lot-size problem apply also to the many other minimum cost problems appearing in discussions of engineering economy. For example, in the derivation of the well-known Kelvin's law for the economical size of an electrical conductor, the factors are the same as in the derivation of the economic lot-size formula, with the exception that factor C does not appear directly in the Kelvin's law derivation. As this factor becomes zero in the first derivative, the Kelvin's law and economic lot-size formulas have the same final form.

In companies where the economic lot-size problem is important at all it is likely to be very important. This importance is usually due more to the fact that there are so many cases than to the great importance of any single case. Where these conditions apply, the formula approach may be better than the tabular or graphical approach in spite of the limitations in the kind of information that can be obtained from the formula solution. There are, however, many other minimum-cost problems that occur infrequently but that are very important when they do occur. In cases of this sort, the tabular or graphical method is certainly the better method because of the greater amount of useful information that may be secured.

3.9 THE BREAK-EVEN POINT CONCEPT

There are innumerable cases in engineering economy where one alternative is more economical under one set of circumstances and another alternative is more economical under other circumstances. For example, it was shown in Art. 3.4 that the rate of return on an additional investment may be ascertained by finding the interest rate that makes the annual costs or the present worths of the two alternatives equal; that interest rate is the break-even point.

Break-even charts have often been used in highly mechanized industries to determine the rate of production below which losses may be expected and above which profits may be expected. In its broadest sense, this problem is always a problem in engineering economy, but there is sometimes a lack of understanding of the great differences between the solutions of this problem for an existing plant and for a proposed plant. The fundamental basis for these differences is explained in Sec. 13, under the discussion of sunk costs.

An idea of the great variety of break-even point problems may be obtained from the following list of such problems discussed by Grant:*

The break-even point as a dimension.
The break-even point as an expected life.
The break-even point as a justifiable investment.
The break-even point as a capacity factor.

Many formulas have been proposed for determining break-even points, but in most cases it will be found that tabular or graphical

* E. L. Grant, *Principles of Engineering Economy* (3rd ed.) (New York: The Ronald Press Co., 1950), pp. 236–39.

methods give more and better information, without requiring much more time or effort.

3.10 IMMEDIATE VS. DEFERRED INVESTMENTS

When it seems reasonable to anticipate that growth of some sort will require additional capacity in the future, there is often the problem of deciding how much additional capacity it is economical to provide immediately. It is obvious that piecemeal construction, where additions are provided only when it is necessary to increase capacity, is not ordinarily so economical as a planned construction program. This observation is due to many factors, including the fact that larger units are likely to be more satisfactory than smaller units, from the viewpoint of both first cost and operating costs.

When making an engineering economy study to determine how great a present investment is economical, in comparison with the alternative of deferring at least part of the investment, present-worth calculations seem to have several advantages over annual-cost calculations. The selection of the interest rate is one of the most important decisions in studies of this sort.

Grant and Ireson make the following observations with respect to this problem:*

> Generally speaking, in any borderline case the irreducible factors are favorable to the alternative that involves a deferred investment rather than to an immediate investment with considerable excess capacity. This is particularly true if the interest rate used in the study has been the average cost of capital to an enterprise, without any increase to allow for a margin of safety. Unless the prospective rate of return is sufficiently higher than the cost of capital to justify a risk, that risk should not be undertaken.
>
> One irreducible factor favorable to the deferred investment is the possibility that the forecast growth may never materialize, and thus the excess capacity may never be needed. Another such factor in many situations may be the difficulty of securing investment funds with the resulting pressure to keep all investments to a minimum. Still another factor in some cases may be the possibility that anything installed in the future may be superior or somehow better adapted to the needs of the service than excess capacity installed at present in advance of such need.
>
> On the other hand there is a nuisance aspect to many deferred-investment plans that is favorable to immediate construction with considerable excess capacity. This is particularly true with regard to public utility services which must be placed underground in paved streets so that any addition to them requires cutting through the pavement and repaving.

* Grant and Ireson, *Principles of Engineering Economy* (4th ed.) (New York: The Ronald Press Co., 1960), pp. 255–56.

3.11 AMOUNT OF UTILIZATION OF A FIXED ASSET

Because the investment costs of such things as manufacturing equipment depend on time rather than on the degree of utilization, the greater the utilization the less the unit costs of product or services. This factor may have a significant effect on a comparison of alternatives that differ considerably in the relative importance of investment costs and annual disbursements.

For example, it may be economical in the long run to pay much more money for a high-efficiency motor, if the motor is to operate most of the time, but better to buy the cheapest available motor, with a very low efficiency, if the motor is to operate only occasionally. There are a very large number of seemingly different problems in engineering economy where this principle should be recognized when making a study.

An error frequently made when comparing several machines is to compare them on the basis of their unit costs for producing an article, with each machine operating at its capacity. What obviously should be done is to determine first what amount of product will be needed and next what it will cost to produce that amount with each machine. Any available capacity beyond the required amount is an irreducible in favor of the machine with the extra capacity but should not enter into the actual cost calculations of the study itself.

3.12 INCREMENT COSTS

In engineering economy studies, the prospective differences among alternatives are significant when making a decision. No one would be likely to dispute the statement made in the preceding sentence, but unfortunately many errors are made in practice because of a failure to understand just what are the prospective differences among alternatives in any given situation. A case in point has to do with what are usually called *increment costs*.

In engineering economy, the increment-cost concept often goes somewhat beyond a strict definition of the term, and in this discussion the use of the term will be explained rather than merely defined. Suppose, for example, that a certain domestic electric rate is as follows:

Service charge of $1 per month regardless of use.

4 ¢ per kilowatt-hour for first 100 kw-hrs per month.

2.5 ¢ per kilowatt-hour for the next 100 kw-hrs per month.

1.5 ¢ per kilowatt-hour for all over 200 kw-hrs per month.

A certain family uses an electric range for cooking, and the monthly electric bills have been varying from about $7 in July for 180 kw-hrs to about $9.30 in January for 320 kw-hrs, with an annual cost for electricity of about $100. It is estimated that during an average month the electric range consumes 120 kw-hrs. The monthly bills for electricity are considerably higher than for neighboring families who cook with gas. The present range must be replaced, and the householder is attempting to determine the operating cost of the electric range, which he suspects to be more than that of a gas range. It is obvious that what is desired is the increment cost of operating the electric range or, in other words, the cost in any month when using the electric range less the cost for that month if the electric range were not used.

Because of the structure of the rate schedule, the increment cost of using the electric range is quite different in different months, being $3.60 for July and only $1.80 for January, under the stated conditions. The only really safe way to determine what the cost of electricity is for operating this range is to calculate separately the monthly electric bill with and without the range. Calculations of this sort often give some surprising results for certain types of increment costs.

Much could be written about the many different kinds of increment costs, but that does not seem necessary here because most of these different situations will be discussed under some other characteristic. What is very important is the rule that it is the prospective differences between alternatives that are significant when a choice is to be made among the alternatives. Formulas and other similar methods are sometimes used to obtain directly the differences between the alternatives, but (as in the electric bill example previously stated in this article) the safest way to determine the differences among alternatives is to determine first the magnitudes of the various alternatives and then to get the differences by simple subtraction.

3.13 SUNK COSTS

Grant and Ireson explain the sunk-cost concept as follows:*

> Once the principle is recognized that it is the *difference* between alternatives that is relevant in their comparison, it follows that the only possible differences between alternatives for the future are differences in the future. The consequences of any decision regarding a course of action for the future cannot start before the moment of decision. Whatever has happened up to date has already happened and cannot be changed by any choice among alternatives for the future. This applies to past receipts and disbursements as well as to other matters in the past.
>
> From the viewpoint of an economy study, a past cost should be thought of as a *sunk cost*, irrelevant in the study except as its magnitude may somehow influence future receipts or disbursements or other future matters. Although this principle that a decision made now necessarily deals with the future seems simple enough, many people have difficulty in accepting the logical implications of the principle when they make decisions between alternatives. This seems particularly true when sunk costs are involved. Although some of the failures to recognize the irrelevance of sunk costs involve a misuse of accounting figures, these mental obstacles to clear reasoning are by no means restricted to people who have had contact with the principles and methods of accounting.

3.13.1 Example 5: Irrelevance of a Past Disbursement

Jones has decided to invest $1500 in a business venture and is faced with the problem of

---

* Grant and Ireson, *Principles of Engineering Economy* (5th ed.), p. 315.

raising the necessary cash. He has two alternatives: he owns twenty shares of common stock that he can sell at the present market value of $75 per share, or he can use a life insurance policy as the security for a 6 per cent loan. Jones estimates that the dividends on the stock will continue for an indefinite period at the present rate of $4.50 per share per year, so there will apparently be no difference between the interest cost of $90 per year on the loan and the $90 per year dividends on the stock that would be given up if the stock were sold.

Not being able to decide between the two alternatives, Jones seeks advice from a friend who tells him that his decision should be based on whether the price he paid for the stock was more or less than the present market value of $75 per share; that the stock should be sold if the sale would show a profit, but not if the sale would show a loss.

At first glance, this argument seems plausible enough, and there can be no doubt that this unwillingness to accept a loss is responsible for many of the misconceptions with respect to sunk costs and the irrelevance of past disbursements. Nevertheless, it is obvious that the cost of the stock should not in any way influence the decision on whether the stock should be sold or retained. The decision should be based solely on Jones' opinion as to future values for the stock. A book profit or loss should never influence a decision with regard to a sale. (Because profits are usually taxable and losses may within certain limitations reduce taxes, a decision concerning which assets to sell or whether to sell may depend on the tax consequences of the various alternatives, but this fact does not in any way nullify the rule that past disbursements are irrelevant.)

3.13.2 Example 6: Irrelevance of Book Value

An automatic lathe with a first cost of $4500 installed is being considered as a replacement for a 10-year-old turret lathe which originally cost $3000 and which has been depreciated at the rate of $120 per year, so that its present book value is $1800. The company would have no other use for the turret lathe if it were replaced in its present work, but it can be sold for $1500.

Many writers on the subject insist that where the book value of a present asset is greater than its realizable value, the difference between the two values should be added to the investment in the proposed asset, thus forcing the proposed asset to show savings that will recover both its own investment and also this so-called loss due to replacement. These persons claim that the proposed asset should be made to bear this added burden because the "loss" is due to the replacement and takes place at the time of the replacement.

It requires only a brief consideration of the subject to prove that the book value of the present asset should receive no consideration whatever in a replacement study. The first cost of $3000 for this present asset was spent 10 years ago, and no decision to retire or retain the asset can change either the amount or the time of this disbursement. The depreciation charge of $120 per year was merely a time allotment of the $3000 investment, which in depreciation accounting is considered to be a prepaid expense of service during the life of the asset. No attempt is made in depreciation accounting to have book values *during life* agree with any sort of values, realizable or otherwise. Further discussion of the differences in the aims of accounting allocations and engineering economy studies will be found in Art 3.14.

Further evidence that this $300 difference between book value and realizable value should not be considered in any way in the replacement study may be found in the fact that depreciation accounting does attempt to charge off during life the entire difference between first cost and salvage value, the latter being the realizable value of $1500 in this example. It is obvious that book value would always be realizable value at the time of a replacement if lives and salvage values could be estimated accurately at the beginning of life, which it is not possible to do.

As explained in the preceding paragraphs, there are several conclusive reasons why both the book value and any difference between book value and realizable value should be disregarded in making economy studies. But the really important reason is that in such studies only future receipts and disbursements are relevant. It is the realizable value of the present asset (the highest of such things as scrap value, second-hand value, or value to the owner for some other purpose) that should be used as the investment in the present asset when making a replacement study.

Some of the errors in theory with regard to

these matters are probably caused by an unfortunate use of words in defining both depreciation and book value. Grant and Norton discuss this as follows:*

> Modern writers on semantics have emphasized how readily one may be misled by words or phrases that have unfavorable associations. Two such phrases may be partly responsible for some of the confusion that exists in the interpretation of the depreciation accounts. One is the common definition of accounting depreciation as "loss in value." The other is the common description of unamortized cost as "unrecovered investment." "Loss" has an unpleasant sound that somehow makes depreciation seem worse than other operating expenses. . . . Depreciation accounting does not relate to value at all except as value may be defined in the neutral sense as any money amount that may be associated with property. "Unrecovered investment," despite its use by the Bureau of Internal Revenue, is an inaccurate description of unamortized cost. In some cases the investment in an asset actually may be recovered in its first few months of service, and in other cases it may never be recovered; neither fact has any relation to the depreciation that has been recorded on the books. (Even if unamortized cost really were unrecovered investment, it would not be relevant in a replacement economy study; the only differences between alternatives that properly enter into an economy study are future differences.)

3.14 INCORRECT INFERENCES FROM ACCOUNTING APPORTIONMENTS

Grant and Ireson explain the limitations of accounting as a basis for estimates in economy studies as follows:†

> Generally speaking, the accounts of an enterprise constitute the source of information that has the greatest potential value in making estimates for economy studies. Nevertheless, the uncritical use of accounting figures is responsible for many errors in such estimates. There are a number of important differences between the point of view of accounting and that which should be taken in an economy study.
>
> Accounting involves a recording of past receipts and expenditures. It deals only with what happened regarding policies actually followed and is not concerned with alternatives that might have been followed; it is concerned more with average costs than with differences in cost. It involves apportionment of past costs against future periods of time and apportionment of joint costs between various services or products. It does not involve consideration of the time value of money.
>
> Engineering economy, on the other hand, always involves alternatives; it deals with prospective differences between future alternatives. It is concerned with differences between costs rather than apportionments of costs. It does involve consideration of the time value of money.
>
> The statement is sometimes made that economy studies are merely a matter of accurate cost-accounting. Such a statement involves a failure to comprehend these fundamental differences in point of view. . . .
>
> The principle emphasized [here] is that it is always *differences* that are significant in economy studies and that the concept of cost, in order to be a useful guide to business decisions, must be related to specific alternatives to be compared. The diversity of alternatives that must be compared in business situations is such that no routine systematic procedure can be expected to give directly the "cost" figures that are needed for all comparisons.

3.14.1 Example 7: To Make or to Purchase

This example records an actual experience of the writer some years ago when he was being shown by the plant manager through a large mass-production furniture factory. This was one of several such factories operated by a very successful company with very progressive top management. There was a well-planned cost-accounting system employing the most modern techniques, which probably gave very good information concerning the cost of the various articles but which certainly was not adequate for the purpose for which its information was used in the decision recorded in this example.

In a lean-to off the machine room were four carving spindles which had evidently not been used for some time. In answer to a question, the plant manager stated that the same four men who had formerly operated these spindles on a time-wage basis were still producing the ornaments used by this company, utilizing second-hand rented spindles in a rented shed

* E. L. Grant and P. T. Norton, Jr., *Depreciation* (New York: The Ronald Press Co., 1955), p. 311.

† Grant and Ireson, *Principles of Engineering Economy* (4th ed.), pp. 22–23.

located in the same city. Further questioning developed the information that without working any harder than previously these four men were making more money, while the company was buying the ornaments cheaper than they had manufactured them. When asked how this could be, the plant manager stated that it was because they had a high overhead rate in this highly mechanized factory, whereas the four men had very little overhead in their operations with rented spindles in a rented shed.

Naturally, the writer then asked how the decision to discontinue manufacturing the ornaments had reduced to the slightest extent the *total overhead expense* of the factory. The factory manager replied that this had been bothering him, but that the head of the cost-accounting department had assured him that their method of figuring costs was correct and that they really were saving money by buying the ornaments. From further information furnished the writer by the plant manager, it was apparent that there had been no real decrease in total overhead expense as a result of this decision to buy the ornaments. Even the supervision had not been reduced to any great extent, because one of the four men had previously acted as a subforeman over the group. It seemed that the expense in the office incident to buying the ornaments was at least as great as the former expense in the production control department.

There was one feature of this cost-accounting system which, while satisfactory for the main purpose of the system, made it almost certain that any decision of this sort would be an incorrect decision. Because no repair parts had to be costed separately and all complete articles utilized the facilities of each department approximately to the same degree as their direct labor cost in each department, this company followed a rule quite common in the furniture industry of basing the charge for factory overhead on direct labor cost, using a single plant-wide overhead rate. This practice made the overhead cost of manufacturing the ornaments seem much greater than it actually was because the wage rates of the highly skilled men who operated these carving spindles were much higher than the average for the plant, while the overhead expense per man-hour for these inexpensive spindles was very much less than the overhead rate per man-hour for the many very expensive automatic machines that were used in this highly mechanized factory.

There could hardly be errors in the use of cost-accounting information more horrible than the one illustrated in this example, but the writer has found many cases that are nearly as bad in theory. This is the sort of error that could not possibly have been made by the men who founded these modern manufacturing companies. Most of them knew little or nothing about the principles that should govern either accounting procedures or engineering economy studies, but they did realize that the best and safest way to handle decisions of this sort was to determine as completely as possible the differences between the alternatives being considered. Those men were closer to the operations being performed than is the average theorist of today; it will be noted that it was the plant manager, who did not claim to know anything about cost-accounting, who was troubled about a decision made by a man who was an expert in cost-accounting but who probably did not know the difference between a carving spindle and a double-end tenoner. Modern industry has become so complicated that we simply must use these highly developed management procedures, but there is no safe way to apply their principles automatically or by formula.

3.15 APPRAISALS: MEASURING THE DISADVANTAGES OF OLD ASSETS AS COMPARED WITH NEW ONES

Appraisals are made for many different purposes and in many different ways. As is true of so many other problems in the general field of engineering economy, the purpose for which the appraisal is made governs to a large degree the procedure that should be used. For example, the conventional reproduction cost new of an identical property, less straight-line depreciation, seems to work quite well for fire insurance purposes but not at all well for most other purposes for which commercial appraisals are made.

In general, an industrial property does not have a value inherent in itself and separate from its use. If, therefore, it is desired to appraise an existing property, the first step should be to determine the most economical substitute property that could produce the desired service and then to find the value for the existing property that would make its annual costs equal to those of the most economical substitute.

Methods for making this sort of appraisal are explained by Grant and Norton, who also make the following statements with respect to some paradoxes in this field:*

Many paradoxes arise in the application of this equal annual cost viewpoint to replacement cost appraisals:

1. Often the appraised value of a plant as a whole should be less than the sum of the appraised values of its parts. The substitution of an entire new plant of radically different design may indicate many economies not possible in contemplating the most economical replacement asset for each part of the plant without any general change in the plant design or arrangement.

2. The more rapid the prospective future obsolescence, the more valuable may be the present obsolescent asset. Prospective improvements in design of new assets or prospective changes in service requirements may reduce the appraisal depreciation of existing old assets. For instance . . . the prospect of further design improvements would reduce the expected life of the new asset. . . . This in turn would increase the equivalent uniform annual costs during the life of the new asset and would therefore increase the appraised value of the old asset.

3. In some cases, as time goes on, the appraisal depreciation of an old asset may become less as compared to a specific new one. If engineering ingenuity permits its better adaptation to the present service, the value of an old asset may increase with time even without price-level changes. For instance, developments in materials handling by lift trucks . . . have served to decrease the value inferiority of . . . certain brick buildings to . . . mill-type steel-frame buildings. . . .

4. Appraisal depreciation may be negative; in other words, the appraised value of an old asset may be greater than the cost of the most economical substitute asset. For instance, if the most economical substitute is a new 18-in. pipe line, the lower pumping costs of an old 24-in. pipe line might give it a value superiority to the new pipe line.

5. Appraised values may be negative. The value inferiority of the old asset may be greater than the cost of the most economical substitute. . . . This simply means that it would pay to make an immediate replacement.

6. The less the contemplated service, the greater may be the appraised value. Although the most extreme instance of this seems to arise when appraising a plant used merely for stand-by purposes, it should be emphasized that the most economical plant for stand-by service may be quite different from that for regular operation.

* Grant and Norton, *Depreciation*, pp. 270–76. For other information on appraisals and the general concept of value, see J. C. Bonbright, *Valuation of Property* (New York: McGraw-Hill Book Company, Inc., 1937).

3.16 CONSIDERATION OF PROSPECTIVE INCOME TAXES IN ECONOMY STUDIES

Grant and Ireson comment on income taxes as follows:†

There is no room for debate on the point that income taxes need to be considered in choosing between alternatives in competitive industry; the only issue is *how* tax considerations can best be introduced into decision-making procedures. Two methods of considering income taxes in economy studies, both frequently used in industry [are] . . .

1. The first method . . . is to compare alternatives *before* income taxes. In this method, comparisons must be made using a minimum attractive rate of return before taxes that is high enough to yield a desired after-tax rate. . . .

2. The second method . . . is to compare alternatives *after* income taxes. The year-by-year differences in disbursements for taxes are estimated, and the analysis is made using a minimum attractive rate of return after taxes.

The first method has two major advantages. It is simpler in the sense of involving less calculation, and it can be applied by persons who are not familiar with the technicalities of income taxation. Nevertheless, the second method clearly is the one that is correct in principle. The first method is appropriate only where it is reasonable to expect that it will lead to the same *decisions* among alternatives that would be reached by applying the second method.

A rough generalization, subject to exceptions, is that the first method often is good enough for practical purposes at the level of decisions on design alternatives but that it will rarely be satisfactory in comparing alternative projects on the level of capital budgeting.

In this handbook, the consideration of income taxes at the capital-budgeting level is discussed and illustrated in Sec. 4.

† Grant and Ireson, *Principles of Engineering Economy* (5th ed.), pp. 374–75.

3.17 SUMMARY OF CONCEPTS IN ENGINEERING ECONOMY

Although the following summary of concepts was presented at a conference dealing with economy studies for certain public works, the concepts are general in their nature and apply equally well to economy studies in private industry:*

1. All decisions are among alternatives; it is desirable that alternatives be clearly defined and that all reasonable alternatives be considered.

2. Decision-making should be based on the expected consequences of the various alternatives. In comparing investment alternatives, it is desirable to make the consequences commensurable with the investments insofar as practicable. Money units are the only units that make consequences commensurable with investments.

3. Only the differences between alternatives are relevant in their comparison.

4. It is necessary to have a criterion for decision-making (or possibly several criteria). The criterion for investment decisions should recognize the time value of money and related problems of capital rationing.

5. In looking at the predicted consequences of various alternatives and in establishing criteria for decision-making, it is essential to decide whose viewpoint is to be adopted.

6. Insofar as possible, separable decisions should be made separately.

7. In organizing a plan of analysis to guide decisions, it is desirable to give weight to the relative degrees of uncertainty associated with various forecasts about consequences. In this connection, it is helpful to judge the sensitivity of the decision to changes in the different forecasts.

8. Decisions among investment alternatives should give weight to any expected differences in consequences that have not been reduced to money terms as well as to the consequences that have been expressed in terms of money.

9. Decisions among investment alternatives must be made at many different levels in an organization. The implementation of rules aimed at rational decision-making may appropriately be different at different levels.

* E. L. Grant, "Concepts and Applications of Engineering Economy," *HRB Special Report 56, Workshop Conference on Economic Analysis in Highway Programming, Location and Design* (Washington, D.C.: Highway Research Board, 1959).

BIBLIOGRAPHY

American Telephone and Telegraph Company, Engineering Department, *Engineering Economy* (2nd ed.) (New York, 1963).

Barish, N. N., *Economic Analysis for Engineering and Managerial Decision Making* (New York: McGraw-Hill Book Company, 1962).

Bonbright, J. C., *Valuation of Property* (New York: McGraw-Hill Book Company, 1937).

Bullinger, C. E., *Engineering Economy* (New York: McGraw-Hill Book Company, 1958).

Clark, J. M., *Studies in the Economics of Overhead Costs* (Chicago: University of Chicago Press, 1923).

Dean, Joel, *Capital Budgeting* (New York: Columbia University Press, 1951).

DeGarmo, E. P., *Engineering Economy* (4th ed.) (New York: The Macmillan Company, 1967).

Engineering Economist, The, a quarterly journal published at Stevens Institute of Technology, Hoboken, N. J., by the Engineering Economy Division of the American Society for Engineering Education; first issue was in 1955.

Fish, J. C. L., *Engineering Economics* (2nd ed.) (New York: McGraw-Hill Book Company, 1923).

Fleischer, G. A., *Capital Allocation Theory: The Study of Investment Decisions* (New York: Appleton-Century-Crofts, 1969).

Goetz, B. E., *Management Planning and Control* (New York: McGraw-Hill Book Company, 1949).

Grant, E. L., and L. F. Bell, *Basic Accounting and Cost Accounting* (2nd. ed.) (New York: McGraw-Hill Book Company, 1964).

Grant, E. L., and W. G. Ireson, *Principles of Engineering Economy* (5th ed.) (New York: The Ronald Press Co., 1970).

Grant, E. L., and P. T. Norton, Jr., *Depreciation* (New York: The Ronald Press Co., 1955).

Morris, W. T., *The Analysis of Management Decisions* (Homewood, Ill.: Richard D. Irwin, Inc., 1964).

Norton, P. T., Jr., *Economic Lot Sizes in Manufacturing*, Bulletin No. 31 (Blacksburg, Va.: Virginia Polytechnic Institute, 1934).

———, *The Selection and Replacement of Manufacturing Equipment*, Bulletin No. 32 (Blacksburg, Va.: Virginia Polytechnic Institute, 1934).

Schneider, Erich, *Wirtshaftlichkeits-rechnung* (Tübingen, Germany: J. C. B. Mohr, 1957).

Smith, G. W. *Engineering Economy: Analysis of Capital Expenditures* (Ames, Iowa: The Iowa State University Press, 1968).

Taylor, G. A., *Managerial and Engineering Economy* (Princeton, N. J.: D. Van Nostrand Company, Inc., 1964).

Terborgh, George, *Business Investment Management* (Washington, D.C.: Machinery and Allied Products Institute, 1967).

———, *Dynamic Equipment Policy* (New York: McGraw-Hill Book Company, 1949).

Thuesen, H. G., and W. J. Fabrycky, *Engineering Economy* (3rd ed.) (Englewood Cliffs, N. J.: Prentice-Hall, Inc., 1964).

Wellington, A. M., *The Economic Theory of Railway Location* (2nd ed.) (New York: John Wiley & Sons, Inc., 1887).

Zanobetti, Dino, *Economia Dell'Ingegneria* (Bologna, Italy: Casa Editrice—Professore Riccardo, Patron, 1966).

RAYMOND I. REUL *is coordinator of industrial engineering and systems engineering programs for FMC Corporation. He has been with FMC Corporation (or the former Westvaco Chlorine Products Corporation) since 1942 as assistant plant manager, staff industrial engineer, or in his present position. Prior to 1942 he was with Bethlehem Steel Company, the Permutit Company, and A. Hollander and Son.*

Mr. Reul has taught in many places as a lecturer and seminar leader for the Industrial Education Institute, Industrial & Commercial Techniques, Ltd., Irish Management Institute, Associazione Lombarda Dirigenti Aziende Industriali, Institutet For Lederskob Og Lonsombed, State Department Aid Programs, Rutgers University, University of Wisconsin, Fairleigh Dickinson University, University of North Carolina, Stanford University, and Bradford University in England. His seminars have been held in Canada, England, Ireland, Sweden, Denmark, Hawaii, Puerto Rico, Italy, Norway, Holland, Germany, France, Switzerland, and South Africa. He has served as a consultant and with in-plant training in over 200 major industrial firms in the United States and Europe.

Mr. Reul earned his B.S. degree in chemical engineering at Lehigh University and graduated from the A. H. Mogensen Lake Placid Work Simplification Course. He has published many articles, manuals, and sections for management manuals covering such subjects as measures of effectiveness for maintenance, profitability index, work simplification, materials handling, cost-reduction programs, equipment-replacement studies, rate-of-return calculations, and capital-investment analysis. He is a member of the American Chemical Society, American Institute of Industrial Engineers, American Association of Cost Engineers, American Society for Engineering Education, Manufacturing Chemists Association, and National Transportation Research Forum.

SECTION 4

CAPITAL BUDGETING

RAYMOND I. REUL

4.1 THE NATURE AND SCOPE OF CAPITAL BUDGETING

4.1.1 What It Is

The capital-budgeting process is an extremely complex and difficult assignment. It can probably be best described as the organized matching of prospective investment opportunities with anticipated financial resources and management capabilities.

Many and varied types of prospective investments must be explored, evaluated, and compared. At the same time, sources for funds must be established and the relative availability and costs of obtaining them determined. Then, taking into consideration existing or obtainable management capabilities, satisfactory candidates must be screened to select the most desirable projects to add to the firm's portfolio of operations.

4.1.2 Importance of the Function

Capital budgeting is one of the most important areas of management decision-making. This is because the conception, continuing existence, and growth of a successful business enterprise are entirely dependent upon the selection and implementation of sound, productive investments. The almost universal reluctance and often complete refusal of top-echelon management to delegate the authority for making investment decisions amply demonstrate the widespread recognition of this importance.

4.1.3 Historical Development

In the past, the majority of industrial enterprises tended to be small and relatively simple. They required unspecialized facilities easily convertible to other than the original purpose. Competition was limited. Compared with current expectations, most profit margins were large. Common sense, aided by a few simple rules of thumb, was usually adequate to assure the selection of productive opportunities for investment. Many enterprises were both owned and managed by a single proprietor. He provided the funds, made all the decisions, and reaped the rewards or suffered the consequences of his own decisions.

4.1.4 Evolution to Meet Changing Conditions

Within comparatively recent years, however, there has been a major change in the size, nature, and characteristics of most industrial enterprises. At the same time, the makeup and viewpoint of management have undergone a significant evolution. Both these factors have contributed to a rapidly expanding interest in capital budgeting and have also led to the development of many new techniques and procedures to expedite the performance of this function.

The change has been dramatic. Today's industrial enterprises are usually quite complex and require highly specialized facilities which can seldom be converted to other purposes. Financial commitments tend to be larger and more irrevocable. A greater number of factors and possible interactions, intensified competition with smaller margins, plus larger and more complicated income tax impacts are causing modern investment decision-making to become a far more critical process. The simple appraisal yardsticks of the past are seldom of sufficient accuracy or dependability to provide adequate bases for sound decisions. New, more universally applicable and more reliable approaches are becoming essential.

In addition, the owner-manager-proprietor has almost disappeared from the scene. Ownership is usually in the hands of a large group of stockholders, and the actual management is vested in a hired management team of professional decision-makers. In the majority of today's industrial complexes, this team has become so large that a common language for accurate appraisal, comparison, and reporting of prospective investment performance has a high priority.

4.1.5 Meeting Today's Needs

Because of these changed conditions, the intuitive "educated hunch" method of making decisions, based more on emotion than facts, has fallen into relative disrepute. The simple rules of thumb such as "years to pay back," "return on original investment," and the arbitrary indexes such as "average return on average book value" are being less relied upon to appraise the complex investment opportunities of today's industry. The more sophisticated net present worth, equivalent annual cost, and especially the true rate-of-return methods are now most frequently recommended and used by modern managements.

The growing acceptance of these more dependable appraisal methods should not be interpreted as the substitution of the scientific method for common sense or judgment. It must be recognized that the true objective of these new techniques is to provide more data of greater accuracy and validity in order to facilitate sounder and more dependable decisions.

4.2 THE CLASSIFICATION OF BUSINESS DISBURSEMENTS

4.2.1 Reasons for Segregation by Category

Business disbursements are made for a wide variety of purposes and exhibit many different patterns of behavior. The techniques used for their measurement and control must be matched to their behavior. Effective application of these techniques requires segregation of all disbursements into carefully designated categories. From the point of view of performance measurement and control, practically all business disbursements can be classified into four basic categories:

1. Operating costs
2. Maintenance expenses
3. Investment commitments
4. Research expenditures

4.2.2 Operating Costs

These are expenditures that can be directly related to results produced. Specific payments are exchanged for clearly definable products or services. The delay between the timing of expenditures and receipts tends to be very small and the risk of nonachievement minimal. Measurement and evaluation can be related to individual transactions with immediate control and corrective action performed on a continuing basis. Examples are: operating labor, raw materials, components, utilities, etc.

4.2.3 Maintenance Expenses

These are expenditures made to retain a status quo. Their objective may be the continuation of, or the return to, a previously existing economic status. Expenditures that produce new improvements in economic performance do not belong in this category. A direct relationship between individual expenditures and specific achievements can seldom be demonstrated. Performance can only be measured in terms of the amount of "continuation" produced. For this reason measurement, control, and corrective actions must be based upon trends in disbursements and results rather than individual transactions. Examples are: repairs, product or process improvements made to retain a competitive position, new facilities to maintain employee morale, etc.

4.2.4 Investment Commitments

These are expenditures that represent the deliberate risking of assets in anticipation of future receipt of a reward of greater value than that of commitment. A significant delay between the commitment of assets and the achievement of results, and a risk that the actual performance may not be as anticipated, are inherent in this type of transaction. In most cases, commitments are largely irrevocable, so to be effective, control and corrective action must take place before investment commitments are made—after is usually too late! Evaluations must be performed and decisions reached before commitments are made and must be based upon predictions. Examples are: new plant for new product, new machine for cost reduction, improved replacement for cost reduction, etc.

4.2.5 Research Expenditures

Expenditures in this category are not in themselves designed to produce a salable product or service. They are not expected to yield a profit. Their objective is to seek knowledge which will improve investment performance. Research expenditures are characteristically made over a considerable period of time and are monitored as spent. The risk of failure on individual projects is great. Failures usually outnumber successes. For this reason, yields on investments made possible by successful research projects must be sufficient to cover the losses of unsuccessful research efforts as well as the costs entailed by the successful efforts. Economic evaluation must be in terms of overall portfolio performance; physical success must be shown in relation to individual programs; and decisions must be made stepwise as the program proceeds. Control requires limitation of risk to assure reasonable probability of overall economic performance.

4.3 INVESTMENT APPRAISAL AND SELECTION

4.3.1 Defining a Valid Approach

Investment desirability can be demonstrated to be both a relative and a many-faceted function. A specific investment opportunity may be quite attractive to one prospective investor, yet totally unacceptable to another. This difference in desirability can result from each of the prospective investors having different commercial capabilities, capital availability or cost, alternative opportunities or a combination of them. Consequently, investment desirability does not seem a likely candidate for a positive selection technique or amenable to measurement by a single numerical criterion. Although the positive approach of rating comparative desirability does not appear practical, the reverse approach of screening out and rejecting undesirable proposals does appear workable. However, even rejection can be a relative matter, or it can involve relationships between two or more criteria. For this reason, evaluation yardsticks rather than decision techniques must be employed.

The concentration of effort on the screening out of undesirable opportunities has an additional advantage. Failure to recognize and implement a desirable proposal will seldom be serious. Other desirable opportunities will usually be discovered. But failure to recognize and reject an undesirable commitment can be serious, even fatal.

4.3.2 The Selection of Specific Criteria for Investment Desirability

The multitude of aspects that must be considered in rating the desirability of a prospective investment can be classified into three basic categories: functional feasibility, financial acceptability, and economic productivity.

4.3.3 Functional Feasibility

Functional feasibility is a measure of a proposed project's technical and commercial soundness. It includes consideration of the following three aspects.

4.3.3.1 Market Availability. Can the proposed product or service be sold at the estimated price and in anticipated volume? How big is the total market? What percentage of this market can be captured? Will this percentage change with time? Note: Neither improved quality nor better value can, by themselves, guarantee consumer acceptance. For instance:

1. Television sets providing much better pictures have long been technically possible but only at a price that few would be willing to pay. The result is negligible market availability.
2. Better automobiles could be produced and sold at a lower price, if styles were standardized and changed less often. But it is doubtful that many would buy such a product.

4.3.3.2 Facility Suitability. Is the proposed process or operation sufficiently efficient and dependable? For instance: Production of electricity from atomic energy has long been technically feasible. It has only been recently, however, that a facility of adequate safety producing a product of competitive cost has been developed. Will the selected location permit a satisfactory product or service to be delivered to the user at a competitive price? For instance, huge quantities of excellent coking coal have been discovered in Alaska. Unfortunately, due to a combination of distance from markets and very bad local weather conditions, no way has been found to make even the most modern efficient coking plant located in Alaska produce a product that could compete with a less efficiently produced product made elsewhere.

4.3.3.3 Availability of Management Staffing and Know-How. Has the proposed project been reviewed by experienced, technically sound, commercially competent people? Are qualified personnel available or obtainable to staff the proposed operation?

Poor investment decision-making and inefficient operation because of lack of proper staff and know-how have been one of the largest causes of business failures. Diversification is desirable, but it can be very hazardous if not preceded by acquisition of essential know-how and a properly experienced staff. For instance, a retail sales operation can be very profitable. To enter this field, however, without an experienced staff, thoroughly conversant with the special marketing problems of this kind of operation, can be fatal.

4.3.4 Financial Acceptability

Before any investment can be implemented, the funds required must be acquired. In any business enterprise, the amount of funds available or obtainable and the cost of acquiring them will vary with many different factors. The price that must be paid is called the financing cost.

The extent of past commitments, performance on past and current activities, the comparative magnitude of the prospective project, and the risks associated with its implementation are some of the more significant factors affecting this financing cost. Even if the required funds are already "in the hands" of the enterprise, there is the "lost opportunity" cost resulting from the decision not to accept alternative investment opportunities.

Financial acceptability is dependent upon a favorable comparison between the anticipated financing cost, the estimated earning rate (rate of return) of the prospective project, and the probability and consequences of project failure. To date, efforts to quantify these relationships have not met with signal success. But this is no excuse for not attempting to quantify the factors themselves to provide a means for improving the sharpness and reliability of management judgment.

4.3.4.1 Availability and Cost of Funds. Before one can assess the cost of obtaining funds, it must be decided from whose viewpoint this cost is to be estimated. On this question as applied to business enterprise, there are two widely divergent opinions:

1. One is that because the present stockholders or proprietors are the legal owners of the enterprise, the costs of new capital should be evaluated solely from their viewpoint. Under this assumption, funds obtained by the enterprise from other than the owner, at a cost different from the earning rate of the enterprise, exert leverage on the earning rate of the owner. In most cases, this cost of "borrowing" additional funds is less than the enterprise's earning rate and causes an inflation in the "equity earning rate" of the owner. Supporters of this point of view argue that each individual project ought to be evaluated with reference to an assumed scheme of financing of the project itself. They contend that where borrowing is assumed for a particular project, the individual project evaluation should recognize the enhancement of equity earnings because of the inflationary leverage of the assumed borrowing.

2. The other opinion is that the enterprise itself is an entity against which the weighted average cost of funds from all sources (including stockholders or proprietor) should be assessed. The acquisition of funds is viewed as a separate function. Supporters of this point of view argue that the management of a business enterprise have two, largely independent goals: the optimization of earning rate, which is project oriented; and the minimization of the cost of capital, which is enterprise oriented.

The latter of these approaches has two distinct advantages:

1. Permits direct comparison of the financial productivity of all different kinds of investment opportunities on a common basis.

2. Eliminates any possibility of confusion resulting from distortions of evaluations which can result from the building in of varying amounts of financing leverage.

These advantages make this approach both more reliable and easier to use. Its use is already widespread, with additional acceptance growing rapidly.

The funds required to finance an enterprise may be obtained in a wide variety of ways. The financing function of providing these funds is far too complicated to be detailed in a presentation of this scope, but the basic underlying principles are relatively simple, and an understanding of them is a prerequisite to effective capital budgeting.

While funds to finance industrial enterprises may be obtained in many different ways with all sorts of contingent obligations, practically all sources can be classified into two basic categories:

1. *Equity capital* from proprietors or stockholders who provide funds in exchange for ownership of the enterprise. The relationships between equity capital suppliers and the enterprise vary and may be quite complicated, but the eventual result of such ownership is a transfer of profits or losses of the enterprise to the owners. Although the potential for

profit to the owner may be great, there is no guarantee against losses or failure to recoup the original commitment. Consequently the risk may be great and the cost of obtaining equity capital can be high.

2. *Debt capital* loaned for a limited period of time at stipulated cost. The terms of such borrowing usually include a specific plan for return of the principal plus a fixed interest rate on the unamortized debt, plus a claim on all assets of enterprise to the extent of the indebtedness that takes precedence to the claims of the owners. However, except in the event of a default in agreed-upon payments, the suppliers of debt capital in no way share in the profits or losses of the enterprise.

The risk borne by the suppliers of debt capital is less than that borne by the suppliers of equity capital. Consequently the cost is lower. Conversely, while the cost of debt capital to the enterprise is lower, the risk to the enterprise is greater because any failure to meet continuing obligations can result in a takeover of the enterprise by the supplier of debt capital.

The cost of debt capital to the enterprise is also lower for another reason. Interest payments are considered business expenses, hence are deductible for income tax purposes, whereas profits, whether or not they are distributed to stockholders, are fully liable to income taxes. This means that as long as the enterprise is profitable and can utilize the deductions generated, the true cost of debt capital, as measured by the after-tax profit reduction, is equal to the interest rate times one minus the tax rate expressed as a decimal.

The objective of the financing function is to obtain the required funds at the optimum risk-cost relationship. Deciding what this optimum should be, how to obtain it, or how closely the true optimum is achieved are difficult and controversial subjects. But ascertaining a rough estimate of the actual cost of capital achieved is relatively simple. A calculation of the weighted-average (after-tax) cost of capital is illustrated by the following:

DATA

Income tax rate	40%
70% Equity capital @ cost of	12%
30% Debt capital @ cost of	6%

CALCULATION

$$(0.70 \times 0.12) = 0.0840$$
$$(0.30 \times 0.06) \times (1 - 0.4) = 0.0108$$
$$0.0948 = 9.48\%$$

The foregoing calculation assumes the existence of an acceptable measure of the cost of equity capital. In fact, the choice of such a measure is a controversial topic on which many different views have been expressed by writers on economics and finance.

To determine what the weighted average cost would be if the proportion of debt to equity capital were changed is not a matter of simple arithmetic. For instance, if the proportion of the cheaper debt capital is increased, the risk to the suppliers of both equity and debt capital will tend to increase with the result that both may seek to charge more, and the weighted-average cost of capital may actually increase rather than be reduced. The determination of and maintenance of the optimum debt vs. equity ratio is one of the prime functions of the financier.

4.3.4.2 Utility Functions. The acceptability of an investment opportunity increases with:

1. Increasing spread between anticipated profitability, as measured by rate of return, and cost of obtaining required funds.

2. Increasing probability of achieving anticipated performance. This factor is often expressed as a subjective estimate, but is probably better described by a composite distribution curve showing all possible results and their respective probabilities.

The acceptability of an investment opportunity decreases with:

1. Increasing exposure to risk of failure as measured by the time to "pay out" or time required to recover original investment if the project proceeds as anticipated.

2. Increasing extent of worst possible consequences as measured by determining the proportion of total resources placed in jeopardy.

3. Probability of failure to "break even." This can be a subjective estimate or be based upon a composite distribution curve of possible results.

The combination of the effect of all of these factors measures the utility of a proposed project to a particular prospective investor. The utility of the same project to another prospective investor may be quite different. Essentially, the utility function can be described as the attitude of an investor toward risk. Considerable empirical evidence has been developed that an individual's utility function represents a reproducible behavior pattern.

4.3.5 Economic Productivity

Economic productivity requires consideration of the following factors:

1. Most probable rate of return expressed as the equivalent average annual earning rate.
2. Sensitivity expressed either as the range of possible rates of return under foreseeable variations of input data or as a composite distribution curve based upon distribution curves of input data.
3. Comparison of most probable rate of return with comparable figures for mutually exclusive alternatives and with cost of capital.

All data must, of course, be expressed on a consistent basis.

4.4 MEASUREMENT OF ECONOMIC PERFORMANCE

4.4.1 Assignments of Management

A business enterprise may be viewed as a self-perpetuating economic entity designed to compete for its own continuing existence. It obtains the funds it requires by offering rewards for their use in the form of interest payments and dividends. These funds must be kept invested in profitable projects so that the promised rewards can be maintained. Thus, the management of a business enterprise actually has two major assignments: (a) caretaker or guardian of the performance of existing projects to which investments have already been committed; (b) originator and implementor of new investment commitments.

4.4.2 Difference Between Accounting and Engineering Economy Objectives

Both accounting and engineering economy are concerned with investment evaluation. But their assignments, points of view, and techniques are quite different.

4.4.2.1 The Caretaker Assignment of Accounting. The performance of projects to which assets have already been committed must be continuously monitored. The evaluations that do this job must be provided at frequent intervals so that appropriate actions can be taken promptly to ensure the continuing integrity and optimum profitability of the committed assets. The data provided by these evaluations are historical and require a variety of assumptions that need to be consistent but sometimes must be completely arbitrary. It is to this assignment that accounting techniques and procedures are addressed.

4.4.2.2 The Originator Assignment of Engineering Economy. The perpetuation of the business enterprise requires a continuing search for, selection of, and implementation of commitments to new investment opportunities with acceptable profitability and reasonable risk. These decisions must be based upon evaluations that are prepared in advance of any commitments and that predict performance over the entire life of the prospective project. Engineering economy techniques are designed to complete this assignment.

4.4.3 Accounting Techniques

Most business enterprises can be quite accurately described as a series of concurrent and sequential individual investments sponsored and controlled by a single management. Each of the individual investments usually has different patterns of disbursements and receipts which extend over a considerable period of time varying from a few to many years.

The accounting objective of measuring the total performance of a complete enterprise during a relatively short period of time must thus record a composite of the achievements of many different individual investments during an arbitrarily selected "accounting period." To accomplish this, a whole series of quite rational but completely arbitrary terms, definitions, and techniques has been developed.

4.4.3.1 The Balance Sheet. The purpose of the balance sheet is to record the financial condition of the business enterprise at a specific point in time—the end of each accounting period. It is based upon the following equation:

$$\text{owner's equity} = \text{enterprise's assets} - \text{liability of enterprise to others}$$

It achieves its objective by equating the enterprise's total assets against its total liabilities in accordance with the following rearrangement of the above equation.

$$\text{enterprise's assets} = \text{owner's equity} + \text{liability of enterprise to others}$$

In other words, the "owner's equity" is treated as part of the enterprise's total liabilities.

4.4.3.2 The Profit-and-Loss Statement. This analysis is also prepared at the end of each "accounting period." Its purpose is to measure the economic productivity of the enterprise during the preceding accounting period. It reports a gain in owner's equity as profit and decrease as a loss.

4.4.3.3 The Return on Investment. This is the ratio of one year's profit to some arbitrarily selected valuation of assets employed. The most commonly used denominator is the book value plus net working capital. But original cost and current replacement cost are frequently used in place of book value.

4.4.4 Accounting Terminology

The balance sheet, the profit-and-loss statement, and the return on investment are interim samples of a continuing operation. In order to make these analyses as representative as possible, accounting procedures employ a variety of standard assumptions in regard to the application of disbursements and receipts to individual accounting periods. The terminology of accounting is tied to these assumptions.

4.4.4.1 Capital. Disbursements made to acquire assets that are expected to be utilized over a period of longer than one year are "capitalized." That is, instead of charging the entire expenditure against the enterprise in the year it is spent, it is suspended in a special account called the "capital account." This amount is then prorated, at so much per accounting period, over the life of the asset as a measure of the consumption of the life of that asset during these time periods. This prorate is called "depreciation."

4.4.4.2 Depreciation. This may be defined as the arbitrary charging off of a prepaid expense. The portion of the capital account that has not been charged off is called the "book value." The depreciation charged is classed as an "expense" and treated as a period cost. There are many arbitrary patterns used to determine the amount to be charged off. The three most common patterns in the United States are called the straight-line method, the sum-of-the-year's-digits method, and the double-declining-balance method.

4.4.4.3 Book Value. This is the original cost of capitalized items less that portion which has been depreciated.

4.4.4.4 Expense. Although most expenses shown in the accounts for a fiscal year are equal to cash disbursements made during the same year, some expenses arise from prorating past disbursements (prepaid expenses, including depreciation) and others arise from prorating expected future disbursements (recognized in the accounts as accrued liabilities).

4.4.4.5 Inventory. Disbursements for items acquired in advance of need and held in storage are usually not charged as an expense to the enterprise until the accounting period when they are actually used. While in storage the amount expended is suspended in an account labeled "inventory."

4.4.4.6 Accounts Receivable or Payable. When a product or service is sold, accounting procedures treat the amount billed as an immediate receipt of the enterprise. But since no cash has actually been paid to the enterprise, a balancing amount is suspended in an "account receivable." In other words, the debt of the customer receives immediate recognition as an asset of the enterprise; hence, a debt of the customer is financed by the enterprise.

On the other side, the cost of materials received and billed to the enterprise but not yet paid for is considered a liability of the enterprise financed by the vendor. Between the time the amount is billed and the time payment is made, the amount due the vendor is suspended in the enterprise's account labeled "account payable."

4.4.5 Engineering Economy Approach

Each prospective project that is being considered for addition to the enterprise must be individually evaluated prior to its implementation.

The objective of the engineering economy approach is to determine the magnitude of the reward that can be expected to result from the implementation of specific prospective investments. This is measured by ascertaining the amount by which the anticipated receipts can be expected to exceed the prospective disbursements. The excess of cash receipts over cash disbursements can be expressed as an "earning rate," an "equivalent worth" at a specified time, or as an "annual equivalent" over the life of the project.

4.4.6 Problems in Establishing Inputs for Engineering Economy Calculations

There are two distinct areas where there are problems in establishing input data:

1. A great deal of source data must be obtained from accounting presentations that involve the use of arbitrary assumptions not appropriate for economy calculations.
2. All source data must be based upon predictions of future events, behavior, and results.

4.4.6.1 Conversion of Accounting Data to Economy Inputs. Accounting procedures record both actual cash flows and arbitrary prorated expenses, such as depreciation, accrued taxes, etc., as costs. When establishing cash flows for economy calculations, care must be taken to segregate and exclude all such prorated items.

In economy calculations, the amounts shown by accounting data to be suspended in "inventory" and "receivables" accounts must be handled in a special way as working funds. They must be treated as a disbursement at the start of the project and a receipt at the end of the project. Accounts payable represent negative working funds and must be treated as a receipt at the start of the project and as a disbursement at the end. Accounts payable and receivable must be valued at selling price; inventories, at value used in accounting procedure from which data are taken.

In economy calculations, the only difference between capitalized disbursements, expensed disbursements, and losses is the difference in the income tax implications.

4.4.6.2 The Reluctance to Predict the Future. A final judgment on the profitability of an investment is not possible until the consequences of the investment have terminated. Therefore, when one evaluates the profitability of a proposed investment, he must predict the complete future chain of cash flows—both receipts and disbursements—that he estimates will occur because of the commitment.

Because forecasts of future sales, costs, and technological developments can never be completely accurate, it has been argued that simple, approximate methods of making economy calculations should suffice. Although there is some justification for this attitude, an informed and unprejudiced assessment of the facts indicates that such an attitude is more likely to represent an alibi for continuation of the pleasure of making major decisions in a completely irresponsible manner.

4.5 INTEREST CONCEPTS AND COMPUTATIONS

4.5.1 Why Objections to the Use of Interest-based Methods Are Not Valid

It has been argued that interest calculations are complicated and difficult to understand. It is also claimed that the computations required are far too time-consuming to be employed in the making of management decisions.

Actually, this is not so. Properly explained, the principles involved in interest computations are quite simple and relatively easy to understand. And with the use of pretabulated calculation sheets and graphical interpolation, even rate-of-return calculations can be reduced to simple clerical procedures.

4.5.2 Basic Interest Principles

The following terms are used in connection with interest computations:

1. Interest rate
2. Earning rate
3. Rate of return
4. Equivalent worths
5. Principle of equivalence
6. Time value of money
7. Present worth
8. Future worth
9. Single payment factors
10. Discounting
11. Compounding
12. Trial-and-error solution

Perhaps the best way to become acquainted with the principles involved in interest calculations is to learn the meanings of these terms by observing how they apply to simple familiar transactions. For instance: If \$100 were deposited today in a bank paying an interest rate of 6 percent, compounded annually, one year from today the balance in this account (principal plus interest) would have grown to \$106. If this balance were then left in the bank another year at the same annually compounded

6 percent interest rate, the balance would grow to \$112.36. It is also possible to calculate that under these same conditions, \$94.35 would have had to be deposited in the bank one year ago in order for today's balance to be \$100. The difference between these balances can be tabulated as follows:

Differences	Balances
\$112.36 – \$106.00 =	\$6.36
\$106.00 – \$100.00 =	\$6.00
\$100.00 – \$ 94.35 =	\$5.65

These "differences" represent the earnings of the funds on deposit. Note that while the amounts differ, they are generated by a single interest rate. This interest rate is the rate at which the funds on deposit earn. From the point of view of the investor, this is the rate of return. Thus, the three terms—interest rate, earning rate, and rate of return—are identical in meaning.

Note that if this stipulated bank interest rate can be earned by funds on deposit, the \$94.35 a year ago can be stated to be equivalent to \$100 today which in turn is equivalent to \$106 one year later and this to \$112.36 the next year. Since things equivalent to the same thing must be equivalent to each other, it also can be stated that at this interest rate \$94.35 on any stipulated date must be equivalent to \$112.36 three years later. This is known as the principle of equivalence.

A word of caution: The differences tabulated above are also often referred to as the time value of money. This time value is not an inherent characteristic of money. The statement that a given sum of money received today is worth more than the same sum received at some future date is simply not valid. Money has time value only when it can be deposited or borrowed at a stipulated rate. This means that the principle of equivalence is applicable and can be used only when the stipulated interest rate has a real significance.

It is also possible to develop a common denominator by expressing all amounts in terms of their equivalent value now. Thus, the \$100 now is the "now" or present worth now of all the other amounts. Using this principle, past, present, and future cash flows can be expressed in terms of their present worth at the stipulated earning rate. The \$106 can be labeled the future worth at the end of one year of \$100 now at 6 per cent interest compounded annually.

To convert a given amount today to its future worth one year later using annual compounding, the amount is multiplied by $(1+i)$ where i is the interest rate. This calculation is called compounding forward. To compound further into the future, the multiplier is $(1+i)^n$ where n is the number of years. To find the present worth at a stipulated earlier year, we could merely divide by $(1+i)^n$. However, it is usually more convenient to multiply by the reciprocal $1/(1+i)^n$. This is called discounting back. The $(1+i)^n$ and $1/(1+i)^n$ are called single-payment compounding and discounting factors. They are called this because they permit the conversion of a specific amount of money at one point in time to a single equivalent amount at another point in time at a stipulated interest rate.

4.5.3 Selecting the Most Appropriate Type of Interest-based Evaluation

Fairly widespread agreement now exists that interest-based calculations are essential if accurate, dependable evaluations of economic productivity are to be obtained. There still is, however, considerable disagreement as to which of the many different promulgated methods is the most suitable.

Most of this disagreement appears to stem from either failure to realize that each of these methods measures a different aspect of economic productivity, or confusion as to just what aspect should be used in making decisions.

All valid interest-based methods utilize the principle of equivalence. In each of these methods, individual disbursements and receipts are converted to equivalents and compared. The equivalent to which the conversion is made determines what is measured by the answer. There are many of these interest-based methods, but they all fall into three basic categories, each measuring a different aspect of investment profitability. The three categories are:

1. Net equivalent worth at a specific time
2. Net level annual equivalent
3. Rate of return

4.5.3.1 Net Equivalent Worth at a Specific Time. In this method, a stipulated interest rate is used to convert and net out all cash flows to a stated point in time. When the stated point in time is the start of the project, the answer is called the "net present worth." When the stated point is the end of the project, the answer is called the "net terminal value."

These answers measure the results achieved by a given sum of money during an agreed-upon study period. They report the weighted average performance of the project and all other uses to which the funds are applied during the agreed-upon study period.

These answers are expressed as a single number which is in no way comparable to other answers when investment size or project life varies.

4.5.3.2 Net Level Annual Equivalent. In this method, all disbursements and receipts are converted to level annual equivalents at a stipulated interest rate and netted out. The answer measures the annual equivalent of the results of perpetual repetition of the prospective project.

This method also yields a single-number answer, but it is adjusted for investment size and project life so that direct comparisons between alternatives with such variations are feasible.

4.5.3.3 Rate of Return. This method requires no stipulations as to reference point in time, applicable interest rate, or study period. The answer obtained by this method is "the interest rate at which the proposed disbursements would have to be invested in an annuity fund, in order for that fund to be able to make payments equal to and at the same time as the receipts anticipated from the project." It measures the equivalent average earning rate of the project as an annual interest rate and permits direct comparisons between projects varying both in size and length of life.

4.5.3.4 Summary and Recommendations. The rate of return is the only one of these evaluation methods that gives an unequivocal answer to the question, "If the enterprise commits the required investment, what will the project it implements earn for the enterprise?" It is the only method that yields an answer which is directly comparable with the cost of borrowing the required funds. It is an evaluation method rather than a decision technique.

4.5.4 The Rate-of-Return Concept

To determine the interest rate (rate of return) of a single disbursement which results in a single receipt at a later date is a simple matter. To illustrate: Suppose it is anticipated that \$408 can be invested all at one time to yield an instantaneous receipt of \$800 exactly two years later. The relationship between these two figures and the applicable interest rate i is stated by the following equation:

$$408 \times (1+i)^2 = 800$$

Solving, we get:

$$(1 + i) = \sqrt{\frac{800}{408}} = \sqrt{1.96} = 1.40$$

$$i = 0.40 \Leftrightarrow 40 \text{ per cent}$$

The following calculations demonstrate that the answer of 40 per cent is the interest rate at which the \$408 would have to be deposited in order to achieve a balance sufficient to pay out the \$800 two years later as required by the rate-of-return concept.

$$\$408 \times 1.4 = \$571$$

$$\$571 \times 1.4 = \$800$$

Now suppose a project involving four separate pairs of anticipated disbursements and receipts, as in Table 4.1, is to be evaluated. (Disbursements are shown as negative cash flows, receipts as positive flows.)

Calculation of the rate of return exhibited by the additional pairs of disbursements and receipts yields the answers tabulated.

But suppose all of these disbursements and receipts are anticipated to occur in connection with a single project and the only data available are the overall cash flows shown in the last column. Could we not state the rate of return of the project to be the interest rate at which the two overall net disbursements would have to be invested in an annuity fund in order for the fund to be able to make payments equal to and at the same time as the three net overall receipts of the project?

The interest rate that satisfies conditions for the overall net cash flows tabulated in Table 4.1 is 20 per cent. The validity of this answer is demonstrated by Table 4.2. This table also makes two other facts clear:

1. If this project performs as anticipated, it will pay back the total of \$3000 invested plus 20 per cent interest on the funds in use. This interest rate is suitable for direct comparison with the cost of capital.

TABLE 4.1. *INDIVIDUAL CASH FLOWS*

TIMING	#1	#2	#3	#4	OVERALL NET FLOWS
Now	−$408	−$592			−$1000
1 year later			−$844	−$1156	− 2000
2 years later	+ 800				+ 800
3 years later					—
4 years later					—
5 years later		+ 954	+1046		+ 2000
6 years later				+3904	+ 3904
Individual interest rates	40%	10%	5%	28.5%	

TABLE 4.2. *TIMING*

Now	Deposit		$1000
		Balance	1000
End 1st Yr.	Plus interest @20%		+200
	Deposit		+2000
		Balance	$3200
End 2nd Yr.	Plus interest @20%		+640
		Less payment	−800
		Balance	$3040
End 3rd Yr.	Plus interest @20%		+608
		Balance	$3648
End 4th Yr.	Plus interest @20%		+730
		Balance	$4378
End 5th Yr.	Plus interest @20%		+876
		Less payment	−2000
		Balance	$3254
End 6th Yr.	Plus interest @20%		+650
		Less payment	−3904
		Balance	0

2. The rate-of-return concept as demonstrated involves neither explicit nor implicit reinvestment of proceeds.

4.5.5 The Calculation Algorithm

The foregoing answer of 20 per cent was demonstrated to be correct, but how can it be calculated? The overall net cash flows as previously tabulated provide an excellent demonstration problem illustrating many of the computational difficulties encountered in rate-of-return calculations. Note that despite the fact that there are only five figures, the following four difficulties are introduced:

1. Disbursements are made at more than one time.
2. Disbursements vary in amount.
3. Receipt timing is irregular.
4. Receipts vary in amount.

The first step in the solution is the selection of a reference or zero point. Any zero point can be used. The choice made will in no way affect the answer obtained, but it can greatly affect the ease of computation. In most cases, the selection of the beginning of the first year of receipts as the zero point will result in calculations of minimum complexity.

Now suppose the present worths of all net cash flows are calculated at a stipulated interest rate. The results of such a calculation using an annually compounded interest rate of 10 per cent are tabulated below:

			Present worths at 10 per cent		
Timing	Disbursements	Receipts	Factors	Disbursements	Receipts
−1	$1000		1.10	$1100	
0	2000		1.00	2000	
+1		$800	0.91		$728
+4		2000	0.68		1360
+5		3904	0.62		2420
				$3100	$4508

Note that the $800 receipt discounted back to zero time has a present worth of $728. This also means that if $728 were deposited in a bank at 10 per cent interest compounded annually, one year later the amount on deposit would be \$800 or just enough to make a payment equal to that anticipated from the project. In the same way, the \$1360 deposited at the same time would grow to \$2000 in four years and the \$2420 to \$3904 in five years. Thus, if the sum of \$728+\$1360+\$2420 or \$4508 were deposited at time zero, at the stipulated 10 per cent interest compounded annually, these deposits, plus the interest earned, would be just sufficient to make payments equal to and at the same time as the project. Note that this amount is larger than the present worth of the prospective investment (\$3100). Since the project's prospective investment is much smaller than the amount that would have to be invested in the bank at 10 per cent in order to achieve the same results, the prospective rate of return is demonstrated to be greater than 10 per cent.

Now note: If it is assumed that the bank will pay an interest rate higher than 10 per cent, the sum of the present worths of disbursements will increase and the sum of the present worths of receipts will decrease. Obviously, there must be some interest rate at which these two figures will be equal. This is the answer for which we must solve, for it is also by definition the interest rate at which deposits of identical amounts and timing would have to be made in order to provide exactly enough funds to make payments equal to and at the same time as receipts from the prospective project. In other words, we would have determined a bank interest rate exactly equivalent to the earning rate of the prospective project.

The only way this interest rate can be computed is by trial and error. As ordinarily performed, this requires repeated trials at different interest rates until the correct answer is determined. This can be a tedious and time-consuming operation. It has long tended to be a serious roadblock to the widespread acceptance and use of this method.

4.5.6 The Use of Pretabulated Worksheets and Graphical Interpolation

The "profitability index" approach utilizes pretabulated worksheets and graphical interpolation to reduce the necessary trial-and-error computations to routine clerical procedures. The use of these techniques to solve the demonstration problem is illustrated in Fig. 4.1.

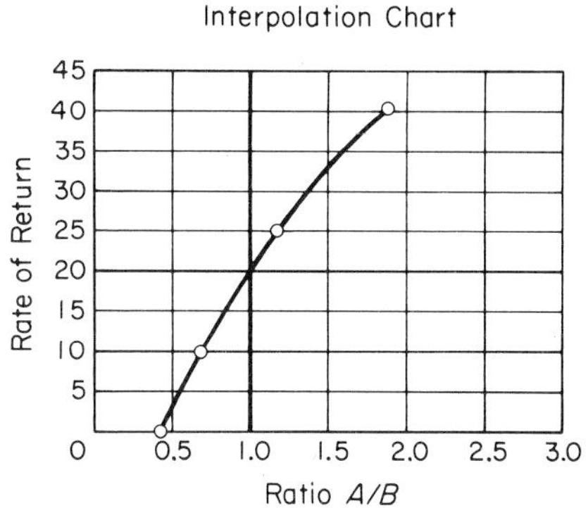

Fig. 4.1 *Use of pretabulated work sheet. Rate of return is 20 per cent.*

4.5.6.1 The Worksheet and Graphical Interpolation Chart (Fig. 4.1). Separate schedules are provided on the worksheet for disbursements and receipts. The zero points of both, however, are identical and are identified by a heavy line with a diamond at the left. The design of the form facilitates the assumption of the beginning of the first year of receipts as the zero point. Single-payment compounding and discounting factors are provided for three different interest rates together with adjacent blank columns for entering appropriate present worths. (On this demonstration form, the pretabulated present worth factors are based upon annual compounding and instantaneous cash flows. Conversion to the more frequently used convention of continuous compounding and during the year receipt is merely a matter of substituting different present worth factors.) The chart at the bottom of the page is for plotting the ratio (horizontal scale) of the sum of the present worth disbursements to the sum of the present worth of receipts at each of the stipulated interest rates (vertical scale).

4.5.6.2 Using the Worksheet and Chart. The use of this worksheet to find the rate of return of the demonstration problem is illustrated in Fig. 4.1.

The predicted disbursements and receipts are entered in the blank column headed "trial #1 at 0 per cent interest" and on the lines corresponding to the time the cash flows are anticipated to occur. The comparison of the totals of actual disbursements and receipts is equivalent to making a trial at zero per cent interest and provides one point on our interpolation chart.

"Trial #2 at 10 per cent" is made by multiplying each of the actual amounts by the adjacent present worth factor in the next column and entering the present worths in the next blank column. Similar trials are made at 25 and 40 per cent.

Then at each trial rate, including that at 0 per cent, the sum of the present worths of the disbursements is divided by the sum of the present worths of the receipts and the ratios entered on the line labeled ratios A/B.

The interest rate at which the sum of the present worths of receipts is exactly equal to the sum of the present worths of disbursements is determined by plotting on the chart in Fig. 4.1 each of the ratios obtained against the interest rate at which they were calculated, drawing a curve through these points and reading on the vertical scale the point where this curve intersects the unity ratio line. This shows, as it should, that the rate of return on this demonstration problem is 20 per cent.

4.5.6.3 Comments on Accuracy and Validity of This Algorithm. As long as this algorithm is applied to the type of problem for which it is designed (see Par. 4.5.7 for discussion of when algorithm cannot be used), the accuracy of the answers obtained by this graphical interpolation method will be as exact as the number of decimal points in the present worth factors and the legibility of the chart permit. In fact, because of the use of the most sensitive ratio, which is that at zero interest rate, as one point on the curve, this method offers the maximum accuracy of results obtainable with a given number of decimal places in the present worth factors.

In addition, this method is to a certain extent self-checking. Only three points are required to establish the curve. The fourth serves as a check. If a smooth curve cannot be drawn through all four points, an error in computation or plotting has been made.

4.5.7 Limitations of the Algorithm and How to Handle Them

The method of determining the earning rate or rate of return of a prospective investment by solving for the interest rate at which the sum of the present worths of receipts is equal to the sum of the present worths of disbursements has been labeled an "algorithm." This term has been applied because it is a trick method of limited application. It can be employed only when the problem to be solved is in the form for which the method was designed. Recognition of this limitation is extremely important. Improper use of this method can result in misconceptions and serious errors in evaluations.

The algorithm has been designed to compute the interest rate that relates one or more receipts to one or more *prior* disbursements. Valid answers will be obtained only if the problem can be reduced to this configuration. Specifically, if this method is misused to evaluate proposals in which all net disbursements do not precede all net receipts, overstated, or in some cases multiple overstated, answers may be obtained.

Fortunately, however, many of the problems to which this algorithm cannot be directly applied can be rearranged to a solvable form. A suggested method of handling two types of such cases is illustrated as follows:

CASE I: CONDITIONAL SALE (LEASE WITH OPTION TO BUY)

Year	Proposed payments	Anticipated receipts
1	$500	$300
2	300	300
3	300	300
4		300
5		300
Total	$1100	$1500

Direct application of the algorithm to the above data yields an answer of 38 per cent. This is not the earning rate on the $1100 commitment but actually the earning rate on the implied "equity" in accordance with the following net annual disbursements and receipt schedule.

Year	Proposed payments	Anticipated receipts
1	$200	—
2		—
3		—
4		$300
5		300
Total	$200	$600

It is suggested that the outright acquisition cost be obtained and two separate calculations be made. With the outright acquisition cost estimated to be $1000, the following calculations can be made:

THE TRUE EARNING RATE OF THE PROJECT

Year	Disbursements	Receipts
0	$1000	
1		$300
2		300
3		300
4		300
5		300

Answer given by algorithm: 15.2 per cent.

COST OF CAPITAL

(CHARGE FOR FINANCING PAYMENTS)

Year	Outright purchase	Conditional payments avoided
0	$1000	
1		$500
2		300
3		300

Answer given by algorithm: 6 per cent.

The true earning rate of the project, 15.2 per cent, provides a basis for determining whether the project is sufficiently profitable. The cost of capital, 6 per cent, provides a basis for determining whether the cost of financing via a conditional sale is acceptable.

CASE II: PRIOR RECEIPTS

Year	Disbursements	Receipts
−1		$50
0	$100	
+1		
+2		25
+3		25
+4		25

Direct application of the algorithm to this problem results in two answers: 35 per cent and 63 per cent. The successful trials at these interest rates are tabulated here:

Year	Actual cash flows	Present worths at 35 per cent: Fac-tors	Present worths at 35 per cent: Amounts	Present worths at 63 per cent: Fac-tors	Present worths at 63 per cent: Amounts
−1	+ 50	1.35	+ 68	1.63	+ 81
0	−100	1.00	−100	1.00	−100
+2	+ 25	0.55	+ 14	0.38	+ 5
+3	+ 25	0.41	+ 10	0.23	+ 6
+4	+ 25	0.30	+ 8	0.14	+ 4
Total net present worths			0		0

Neither of these answers represents a valid evaluation. Both overstate the earning rate by building in reinvestment of funds at the solution interest rate, which is higher than any reasonable expectation.

The best solution offered to date for this type of problem appears to be a compromise in which a stipulated, highly believable reinvestment rate is used to convert the unsolvable problem into a form to which the algorithm can be validly applied. The application of this "crutch" is illustrated in the two tables shown below. If a 10 per cent stipulated interest rate is used as a "crutch" the successful trial rate is 19 per cent. If a 6 per cent stipulated interest rate is used as a "crutch" the successful trial rate is 17 per cent.

On the basis of the foregoing calculations, it seems reasonable to consider this project

Year	Actual cash flow	Present worth: Factor at 10 per cent	Present worth: Factor at 19 per cent	Present worth: Amounts
−1	+ 50	1.10		+ 55
0	−100		1.00	−100
+2	+ 25		0.70	+ 18
+3	+ 25		0.59	+ 15
+4	+ 25		0.48	+ 12
				0

Year	Actual cash flow	Present worth: Factor at 6 per cent	Present worth: Factor at 17 per cent	Present worth: Amounts
−1	+ 50	1.06		+ 53
0	−100		1.00	−100
+2	+ 25		0.73	+ 18
+3	+ 25		0.62	+ 16
+4	+ 25		0.53	+ 13
				0

equivalent in profitability to other projects with a rate of return somewhere between 17 and 19 per cent.

4.6 APPRAISAL STRATEGIES WHEN ALTERNATIVES ARE MUTUALLY EXCLUSIVE

4.6.1 Decisions Are Between Alternatives

Every investment decision requires an evaluation of the difference between alternatives. Even when only one specific action is being proposed, there is always the alternative of "doing nothing." Therefore, for any specific proposal to "do something," the basic basis for appraisal must be a comparison between doing that "something" and doing "nothing."

When there are more than these two alternatives, valid appraisal requires comparison between pairs of alternatives. The selection of the alternatives to be compared and the factors considered can affect the validity of the appraisal.

4.6.2 "Do Nothing" Alternative with Implied Loss

The continuing profitability of an existing operation can be threatened in many different ways. Frequently, such threats can be eliminated or minimized by the expenditure of additional funds. Opportunities for such defensive moves generally fall into one of the following three categories:

1. Replacement of a portion of facilities which will no longer function satisfactorily. (Not complete failure which implies termination of the project and consideration of the origin of an entire new project.)
2. Addition of new facilities to rectify a competitive position that has been lost.
3. Addition of new facilities that were accidentally or intentionally omitted from the original installation.

Expenditures of these types neither add new profit nor extend the life of the original enterprise. They merely reestablish the previously enjoyed status quo. Frequently they must be repeated again and again just to hold the line.

In almost every case, both accounting practice and federal regulations require that such expenditures be capitalized. But actually, they are maintenance costs, not investments, and can only be validly evaluated and controlled by consideration of long-term trends. Hence, application of investment evaluation techniques cannot be expected to yield meaningful or useful results. For this reason, although accounting practice requires inclusion of such expenditures in the "capital budget," it is recommended that they be segregated and evaluated as continuing "profit-maintaining expenditures" rather than individual investment opportunities (Par. 4.9.1).

4.6.3 "Do Nothing" Alternative with Implied Investment

This type of alternative occurs in many forms. Usually there is an existing asset that can be retained as a source of continuing receipts or traded for one immediate receipt. A decision to retain the asset implies the investment of the immediate cash value of the asset to achieve the continuing receipts. For example: An owner of an enterprise, evaluating an offer to acquire his business, must consider the offered purchase price to be the implied investment required to continue as owner and realize the continuing receipts. Both the original purchase price and the present "book value" are sunk costs and irrelevant to the evaluation. The interest rate that relates the proposed purchase price to the continuing receipts represents the rate of return on the investment implied by not disposing of the asset.

4.6.4 Multiple Alternatives with Differing Investments

When there are more than two alternatives, they must be compared in pairs and in the proper sequence. For instance, suppose consideration is being given to three possible projects:

1. A small factory
2. A somewhat larger factory
3. Two separate factories at different locations

The probable cash flows are anticipated to be:

Year	Small factory	Larger factory	Two factories
1	−$1000	−$1000	−$2000
2	+ 475	+ 583	+ 913
3	+ 475	+ 583	+ 913
4	+ 475	+ 583	+ 913

(Note: Short lives, small amount of money, and uniform patterns have been and will continue to be used as a means of simplifying the illustration of basic principles. Negative cash flows represent "disbursements" and plus cash flows "receipts.")

The correct procedure is first to compare the alternative with the smallest investment with the alternative of "do nothing."

Year	Do nothing	Small factory	Difference
1	0	−$1000	−$1000
2	0	+ 475	+ 475
3	0	+ 475	+ 475
4	0	+ 475	+ 475

A true rate of return (based upon annual compounding and end-of-year convention) of 20 per cent on the $1000 investment will be found by application of the algorithm. This is probably an acceptable rate. If so, it means the small-factory alternative is acceptable.

Then, the larger-factory alternative with the next larger investment must be compared with the small-factory alternative.

Year	Small factory	Larger factory	Difference
1	−$1000	−$1300	−$300
2	+ 475	+ 583	+ 108
3	+ 475	+ 583	+ 108
4	+ 475	+ 583	+ 108

The true rate of return on the $300 additional investment resulting in the incremental receipts of $108 per year is 4 per cent. This is a very low earning rate and would, in most cases, probably be considered unacceptable. If this is so, it means that the larger-plant alternative itself is unacceptable and consideration of this alternative is eliminated.

The alternative with the next larger investment must now be compared with the last acceptable one, the small factory.

Year	Small factory	Two factories	Difference
1	−$1000	−$2000	−$1000
2	+ 475	+ 913	+ 438
3	+ 475	+ 913	+ 438
4	+ 475	+ 913	+ 438

In this case, application of the algorithm to the difference demonstrates that the true rate of return on the incremental investment of $1000 is 15 per cent. This is not so high as the 20 per cent obtainable on the small-factory alternative but is probably acceptable. If it is, it means that the two-factory alternative is acceptable.

4.6.5 Alternatives with Different Lives

The same approach must be applied. The answer is found by evaluating the difference between the alternatives. For instance, a decision between the following two alternatives can only be made by evaluating alternative *B* minus alternative *A*.

Year	Alternative *A*	Alternative *B*	Difference (*B* − *A*)
0	−1000	−1000	0
1	+ 475	+ 162.75	−312.25
2	+ 475	+ 162.75	−312.25
3	+ 475	+ 162.75	−312.25
4		+ 162.75	+162.75
5		+ 162.75	+162.75
6		+ 162.75	+162.75
7		+ 162.75	+162.75
8		+ 162.75	+162.75
9		+ 162.75	+162.75
10		+ 162.75	+162.75
Rate of return	20 per cent	10 per cent	4 per cent

The evaluation of the difference implies that if alternative *A* is available, acceptance of alternative *B* means the "giving up" or implied investment of $312.25 per year for three years in return for receipts of $162.75 per year for seven years. Application of the algorithm to these cash flows indicates a rate of return of only 4 per cent. This is probably too low to be acceptable. If so, alternative *B* is unacceptable. The rate of return of 10 per cent on *B* is the weighted average of 20 per cent on *A* and 4 per cent on *B* − *A*.

Summary: The principles detailed in this section are basic, universally valid, and relatively simple. However, it should be recognized that practical application of these approaches to the complexities of actual business situations can pose many difficulties. The analyst will frequently be forced to employ considerable ingenuity in arranging data so that they will yield valid results, yet be in a form that can be handled by the calculation algorithm.

4.7 INCOME TAX IMPACTS

Income taxes are a real cash cost. They are just as much a cost of doing business as wages paid, purchases, and property taxes. They are different, however, from all other costs in one respect—the amount of the tax to be paid depends on the amount of profit as determined by accounting procedures.

But taxable income is not necessarily the same as accounting income. Tax laws and regulations change from time to time and differ from one governmental unit to another. Income tax laws often contain provisions that constitute incentives for certain types of investment and deterrents for other types. Two examples of such incentives in the tax laws of the United States have been the percentage depletion allowance, a long-time feature of the tax laws, and the investment tax credit, introduced in 1962 and continued into 1970.

Unless a project is being evaluated for a government agency or nonprofit enterprise that pays no income tax, every evaluation in a capital budget should be based on estimated cash flows after consideration of income taxes. With high tax rates, complex rules, and laws providing tax incentives and tax deterrents, the impact of income taxes on projects competing for limited funds can be so large and so varied that evaluation without consideration of income taxes can no longer be considered a meaningful measure of performance.

4.7.1 Depreciation

When funds are deposited in a savings account, the interest paid is income and is taxable. If the deposited funds are gradually withdrawn over a period of years, the withdrawals clearly are a return of capital and are not income in the accounting sense. Neither are they subject to income taxation.

If the same funds had been used to purchase a piece of equipment intended to produce income, a portion of the cash receipts generated by the equipment should also be recognized as a return of capital and excluded from the income shown by the books of account. The depreciation charge in the accounts makes it possible for accounting statements to reflect this aspect of the return of capital. Nevertheless, the matter is not so simple as if funds had been withdrawn from a savings account. Orthodox depreciation accounting provides a number of ways in which the difference between the first cost of an asset and its estimated salvage value may be apportioned among its years of estimated life. Moreover, there frequently is a question about the appropriate estimate of the life of an asset and its terminal salvage value.

In 1954 and thereafter in the United States, the income tax laws permitted liberalized methods for computing depreciation charges for new depreciable assets. In general, these methods allowed a more rapid write-off in the early years of the assets' estimated lives than had previously been permissible. In 1962 and thereafter in the United States, revenue procedure 62–21 and its various revisions allowed the use for tax purposes of "guideline lives" for certain classes of depreciable assets; generally speaking, these lives were shorter than the ones that had commonly been used for tax purposes. Although the 1954 liberalized methods and the 1962 guideline lives apply only to the United States, many similar types of liberalization of depreciation allowances for tax purposes have been permitted in other countries that levy income taxes.

In the United States and elsewhere, many corporate taxpayers have taken advantage of the tax laws and regulations to use higher depreciation charges for tax purposes than they have used in their corporate accounts. As far as engineering economy evaluations are concerned, the depreciation methods adopted are relevant only in regard to their effect on the amount and timing of tax payments. Therefore, only the depreciation methods adopted for tax purposes require consideration in engineering economy studies.

Except for items placed on inventory, every asset acquired by an enterprise may be depreciable! That is, the cost of these assets may eventually be charged off against the profits of the enterprise, but sometimes these charges are not called or labeled "depreciation."

For instance, when an item is expensed, it can be thought of as a capitalized item depreciated 100 per cent at the time of purchase. Buildings and equipment are charged off over various periods of time by depreciation. Land and certain other items that are not usually considered to be depreciable can be thought of as being depreciated at the time of disposal to the extent that the net proceeds from the disposal are less than the original cost. What this means is that no economy evaluation can be considered valid unless it takes into account both aspects of an acquisition, namely, the cost of acquisition and the eventual tax deductions to be generated by that expenditure.

4.7.2 Depletion

Percentage depletion in the United States is a tax incentive to certain types of investment that applies only to the extractive industries. Taxpayers have been permitted to make deductions from taxable income for certain exhaustible resources equal to a specified percentage of the gross income from the property during the year. The highest percentage, 22 per cent, has applied to oil and gas wells; some minerals have received depletion allowances as low as 5 per cent. In all cases, the depletion allowance under the percentage method in any year could not exceed 50 per cent of the net income of the taxpayer from the property in that year as computed without the depletion allowance. Insofar as engineering economy evaluations are concerned, estimated percentage depletion must be recognized as a deduction that reduces the tax payable.

4.7.3 Investment Credits

The investment tax credit in the United States has been another type of tax incentive designed to stimulate certain kinds of investment. It has applied to tangible depreciable property, although there have been a number of types of exclusion. An important exclusion has been buildings and most of their components. Moreover, the credit was not available for the types of outlays that may be expensed for accounting and tax purposes.

The investment credit has taken the form of a reduction in income taxes ordinarily taken in the year of acquisition of the eligible assets. The tax reduction has been a stipulated percentage of the investment in the assets. The percentage has depended on the estimated useful life of each qualifying asset, with the maximum credit applicable to lives of eight years or more and no credit permitted where the estimated life was less than four years.

4.7.4 After-Tax Cash-Flow Calculations

The computation of each year's estimated cash flow after income taxes may be made as illustrated in the following tabulation for an assumed tax rate of 40 per cent:

Total receipts	\$10,000
Less: cash costs	4,000
Net cash receipts	\$ 6,000
Less: deductions*	2,000
Taxable profit	\$ 4,000
Less: tax @ 40 per cent	1,600
After-tax profit	\$ 2,400
Plus: deductions	2,000
Net after-tax cash flowback	\$ 4,400

$$\text{*deductions} = \text{depreciation} + \text{depletion} + \frac{\text{investment credit}}{\text{tax rate}}$$

The above after-tax cash flowback is the figure entered as the annual receipt in rate-of-return or profitability index calculations. The validity of this computation is demonstrated by the following calculation:

Net cash receipts (as above)	= \$6000
Taxes paid (as above)	= 1600
Net cash after-tax flowback	\$4400

While the above calculation is extremely simple, it cannot be used until a supplementary calculation is made to determine the amount of the tax. This returns the original complexity.

Another approach is to use the following computation:

Net cash receipts × (1 − tax rate)	= 6000 × (1 − 0.4)	= \$3600
Plus deduction × (tax rate)	= 2000 × (0.4)	= 800
Net after-tax cash flowback		\$4400

This calculation is based on an algebraic transformation of the usual calculations. It is recommended because it is much simpler and because it permits separate, independent valuation of after-tax cash flows from income receipts and tax deductions.

4.8 THE PROFITABILITY INDEX

The profitability index is a specific variation of the rate-of-return yardstick. It was first introduced in the FMC Corporation in 1954. Since that time, it has been incorporated as part of the FMC Corporation's standard procedures where it has proven a valuable tool for the surveillance of prospective investments. A great many other companies have adopted the same, or a very similar, approach.

4.8.1 Nature of the Yardstick

The "discounted cash flow,"* the "interest rate of return,† and the "internal rate of return" ‡ are names frequently used for the same or very similar methods. The profitability index, however, as used at FMC Corporation, involves some very specific assumptions and procedures. These are:

1. Continuous compounding
2. During-the-year receipt convention
3. Use of pretabulated worksheets
4. A standard zero point
5. Graphical interpolation

4.8.2 Continuous Compounding

Figure 4.2 illustrates the changes in effective interest rate with change in compounding period. Note that the more frequently compounding takes place, the greater the earnings, but with ever-diminishing gain. Yield can be said to approach a limiting value which will not be exceeded even if compounding takes place an infinite number of times per year (continuous compounding).

The assumption of continuous compounding introduces no mathematical complications. All that is necessary is to use present worth compounding or discounting factors as illustrated in Fig. 4.3. (Factors as shown also include adjustment to reflect during-the-year receipts.)

Compared with annual compounding, the use of continuous compounding tends to yield conservative evaluations which increase in conservatism as the reported yield increases. During-the-year receipts affect the answer in the opposite way and are appropriate when they are a more realistic simulation of real life.

* Joel Dean, "Measuring Productivity of Capital," *Harvard Business Review*, 1954.

† J. B. Weaver and R. J. Reilly, "Interest Rate of Return," *Chemical Engineering Progress*, 1956.

‡ Usually associated with University of Chicago School of Business.

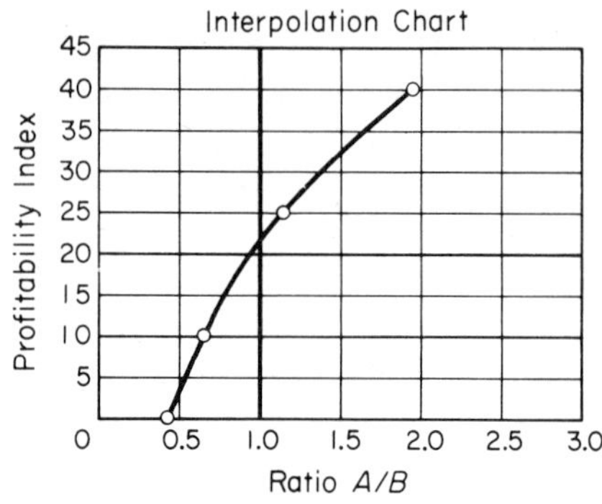

Fig. 4.3 *Pretabulated work sheet with continuous compounding. The P.I. is 21 per cent.*

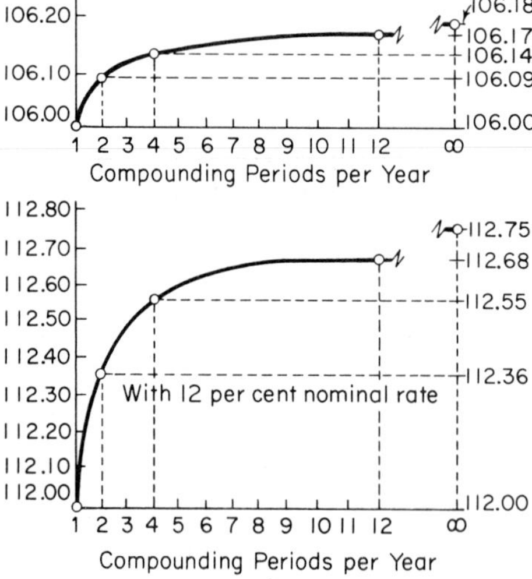

Fig. 4.2 *Change in effective interest rate with change in compounding period.*

4.8.3 During-the-Year Receipts, Cash Flow Convention

The profitability index calculation is based on a series of cash flows timed to better simulate the manner in which cash flows would actually be received from a prospective investment. The assumption of "during-the-year receipt" is certainly more representative of real life than the oversimplification of assuming instantaneous, end-of-year receipts usually made in textbook presentations.

The assumption of during-the-year receipts introduces no mathematical complications. All that is necessary is to revise the present worth compounding and discounting factors as shown in Fig. 4.3.

4.8.4 Importance of a Consistent Index

Consider the following prospective investment opportunity:

Timing	Cash flows
0	$ −1000
+1	+ 300
+2	+ 300
+3	+ 300
+4	+ 300
+5	+ 300

If we assume annual compounding, instantaneous end-of-year receipts as we did in our demonstration evaluation in Fig. 4.1, we would get a rate of return of 15.2 per cent. With the assumption of annual compounding and middle-of-the-year receipts (approximation of during-the-year receipts) the answer we would get would be 19.1 per cent. The profitability index method yields an answer of 17.5 per cent. The true earning rate is an absolute attribute, but like temperature, which is also an absolute attribute, it can be measured on different scales. Temperature can be measured in degrees fahrenheit, centigrade, kelvin, etc. The rate-of-return scales vary with the built-in assumptions. It is for this reason that the word "index" was included in the name profitability index as an indication that it is not an absolute value.

This explanation should also make it clear why individual names are applied to evaluation methods with specific sets of assumptions. It should also make it clear why it is imperative to use one standard method (scale) for comparing opportunities competing for an individual corporation's funds. Failure to do so is like comparing apples and oranges.

4.9 A COMPLETE SYSTEM FOR INVESTMENT APPRAISAL

The "authorization for expenditure" procedures and forms adopted by the Chemical Divisions of FMC Corporation sixteen years ago represent an organized program designed to:

1. Provide a uniform standard format for presenting requests for major expenditures.
2. Build into the format a requirement for sufficiently detailed disclosure of assumptions to allow adequate review of their reasonableness and reliability.
3. Specify and detail the computation of standard yardsticks of established accuracy, validity, and dependability.
4. Simplify the preparation of this material by providing standard forms for accumulation of data, performance of calculations, and summarizing of results.

These procedures are still in use in FMC, essentially unchanged, and they have been widely copied by many other companies. They are presented here as an example of a thoroughly tested and proven program.

A total of eight standard forms plus a letter of standard format are specified:

1. Cover sheet (summary of information)
2. Estimate of investment requirements
3. Depreciation and depletion worksheet
4. Time schedule of expenditures and receipts
5. Calculation sheet
6. Graphical interpolation chart
7. Calculation of payout and before-tax profit
8. Calculation of margin, breakeven, and turnover
9. The memo of justification

The use of the somewhat elaborate worksheets and detailed methods described here may seem unnecessarily tedious. This is because the proposed procedure has been designed for universal use and had to be made applicable to almost any type of investment proposal. The use of shortcuts, practical for many specific applications, has been avoided. This was done to obtain the much greater reliability that can be expected to result from the use of one standard, universally applicable method. Shortcut methods, even when used by qualified engineers or accountants, can lead to serious errors in evaluations.

4.9.1 The Summary Voucher (Form 3016 Fig. 4.4)

Purpose of form:

1. To provide a brief but official summary of each proposed project.
2. To serve as identifying voucher to receive and record authorizing signatures.
3. To be official means of transmitting notice of authorization.

Classification of justification:

DISPOSITION OF COPIES

Form 3016

A.F.E.
CHEMICAL
DIVISIONS
F.M.C.

NUMBER ______

LOCATION ______

DIVISION ______

DRAWN BY ______ DATE ______

REQUEST

AUTHORIZATION IS REQUESTED TO EXPEND $ ______ FOR:

EST. START ______ EST. COMPLETION ______

INVESTMENT DATA

ITEM	BUDGETED 19__ authorize	spend	EXPENDITURES total	19__	19__	19__
CAPITAL						
EXPENSE						
(other)						
TOTAL TO BE AUTHORIZED						
WORKING FUNDS REQUIRED						
FUTURE OBLIGATIONS						
TOTAL INVESTMENT TO BE JUSTIFIED						

ASSETS RETIREMENTS $______ ABANDONMENTS $______

JUSTIFICATION

☐ PROFIT ADDING P.I. ______ % PAY OUT ______ YRS.

☐ COST REDUCTION EXPECTED PROCESS LIFE ______ YRS.

☐ EXPANSION OF EXISTING OPERATION OR PRODUCT
INCREMENTAL MARGIN ______ % ADDED SALES $ TO BREAK EVEN ______

☐ EXPANSION WITH NEW OPERATION OR PRODUCT
TURNOVER ______ BREAKEVEN % CAPACITY
AVG. MARGIN ______ % ______ %

PREDICTED PERFORMANCE SCHEDULE

year	net sales return	profit at division level

☐ PROFIT MAINTAINING ☐ REPLACEMENT ☐ OTHER
AMT. PROFIT AT STAKE $______ ☐ NEW FACILITY ______

REVIEWS AND APPROVALS

STAFF REVIEW

DEPT.	SIGNATURE	DATE

EXECUTIVE APPROVALS

	SIGNATURE	DATE
LOCATION MGR.		
SALES MGR.		
DIV. PRES.		
CHEM. DIVS.		
CORPORATE		

Fig. 4.4 *Summary voucher form.*

Profit adding: Acquisition of assets to increase profit by:

a. Making possible new or expanded operations.
b. Improving output quality to raise prices or volume.
c. Reducing costs to improve profit margin or volume.

Profit maintaining: Acquisition of assets to hold on to existing profits by:

a. Replacing part of existing asset which has failed.
b. Improving existing facilities to circumvent competition.
c. Providing services accidentally or intentionally omitted in original installation.

Detailed instructions:

1. *Number:* The number entered here is a combination code and identification device. It is also to be used as the accounting charge for accumulating and reporting actual job costs.
   a. Code letter (C) indicates division of origin.
   b. Code number (1) designates location of origin.
   c. The second group of numbers (57) indicates year submitted.
   d. The third group of numbers (168) is the engineering estimate serial number.

2. *Location:* The plant, office, or laboratory at which request originates.

3. *Division:* Division authorizing submission of request.

4. *Drawn by:* Initials of individual preparing or delegating preparation of request. *Date:* Release date.

5. *Authorization requested:* Enter here the total expenditure substantiated by engineering estimate and for which authorization is requested. This figure also appears in "investment data." It is repeated here so that no possible confusion can exist as to the exact total of funds to be authorized. *For:* Give here a brief description of facilities proposed and the purpose for which they are requested.

6. *Estimated start:* Express as months or days after approval date.

7. *Estimated completion:* Express as months or days after approval date.

8. *Budgeted:* Every year a capital and major expense budget is submitted to corporate management by each division. These budgets contain a job-by-job prediction of expenditures (in each of these categories) to be *authorized* and to be *spent* during the ensuing year. Enter here the amounts budgeted, irrespective of amounts herein *requested,* for the specific facilities proposed. Working funds and future obligations are not budgeted, hence will not be shown here.

9. *Expenditures:* Enter here *detail of total investment* required in connection with this project. Show year-by-year schedule of expenditures predicted. These figures should include working funds and future obligations and agree with investment figures on FMC Form #3009 used in P.I. calculation (Fig. 4.5).

10. *Retirement:* Enter here the original cost of facilities which will be sold, given away, or scrapped as a result of this project and upon which no "*abandonment*" can be claimed.

11. *Abandonments:* Enter here the depreciated value of facilities upon which "abandonment" can be claimed because the retirement of these assets was due to a cause not contemplated in setting applicable depreciation rates. (See U.S. Treasury Dept. Cumulative Bulletin 1956–1; Section 1.167.)

Special Note: Under "justification" all investments are to be classified into *profit-adding* or *profit-maintaining*. Provision has also been made to break down profit-adding-type investments into three categories determined by the nature of the means for adding profit: (1) cost reduction, (2) expansion of existing operations, or (3) expansion via new operations or products. If the investment is to be justified on more than one of the foregoing bases, it should, if possible, be split up in a logically defensible manner and each part separately evaluated.

12. *Profit-adding:* Check here if the proposed project will result in *new, additional profit.* Check also the way or *ways of achieving* this added profit and complete the additional information requested in connection with each.

13. *P.I.:* The profitability index may be defined as the average (after federal income tax) rate of return (*or equivalent compound interest rate*) the proposed investment will earn. For a complete description of the method and forms to be used in the calculation of this figure, see Pars. 4.9.2 through 4.9.5.

14. *Payout:* The figure to be presented here may be defined as the *elapsed years and fraction thereof* from the date of the project's first operating income until the date when the project's cash flowbacks just equal the investment risked. The "investment risked" is defined as the "amount authorized." For method of computation, see Par. 4.9.6.1.

Predicted performance schedule:

15. *Year:* Enter in this column, in chronological order, the last two digits of the first ten consecutive years during which the project is predicted to provide a cash flowback.

16. *Net sales return:* Enter in this column year-by-year increased annual net sales income

as predicted in calculation of P.I. There will be no entry here, of course, in cost-reduction proposals.

17. *Profits at division level:* These are to be expressed before deduction of federal income taxes but after deduction of all current charges including: expensed investments, depreciation, abandonments, and inventory write-offs in

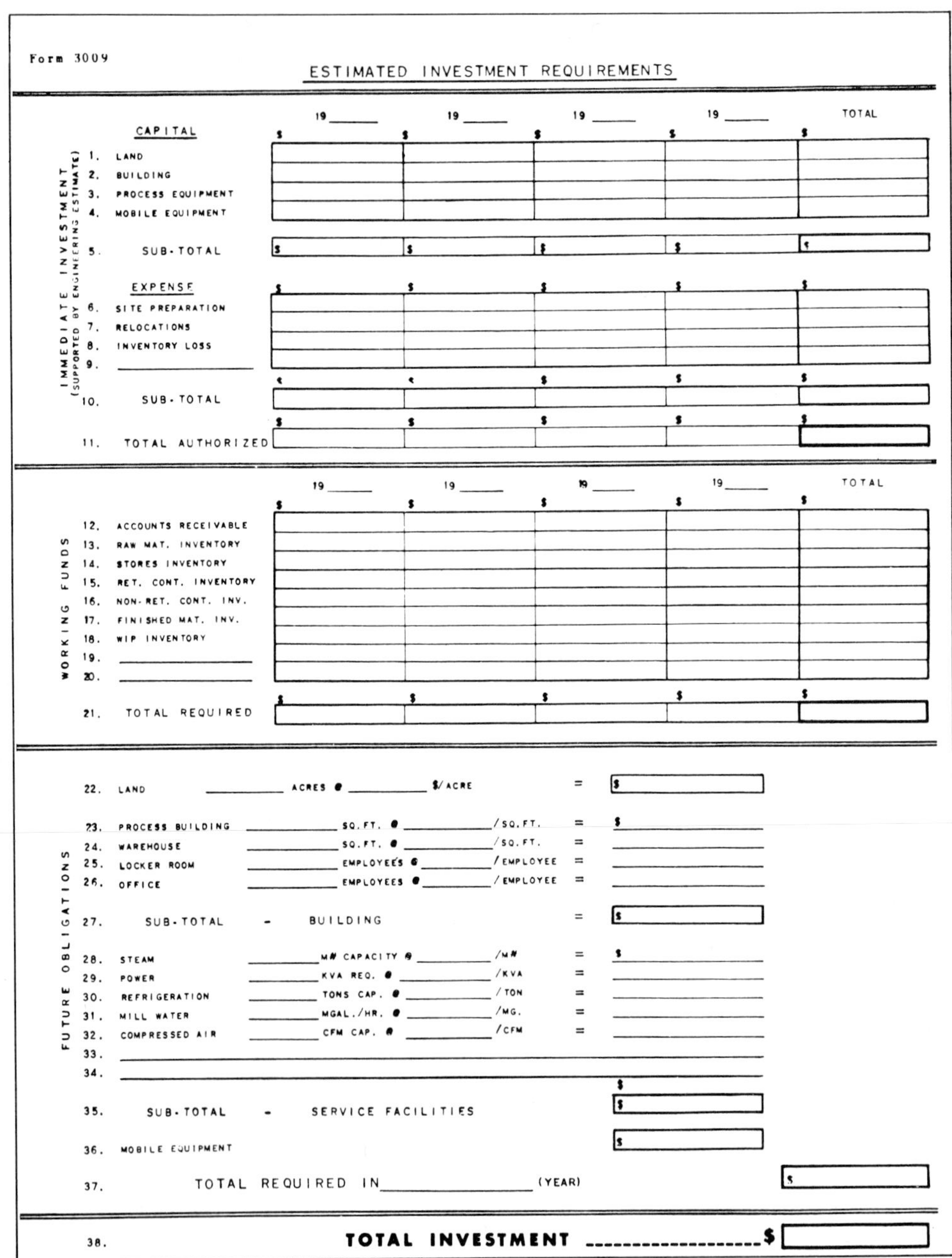

Form 3009

ESTIMATED INVESTMENT REQUIREMENTS

IMMEDIATE INVESTMENT (SUPPORTED BY ENGINEERING ESTIMATE)

	19 ___	19 ___	19 ___	19 ___	TOTAL
CAPITAL	$	$	$	$	$
1. LAND					
2. BUILDING					
3. PROCESS EQUIPMENT					
4. MOBILE EQUIPMENT					
5. SUB-TOTAL	$	$	$	$	$
EXPENSE	$	$	$	$	$
6. SITE PREPARATION					
7. RELOCATIONS					
8. INVENTORY LOSS					
9. ___					
10. SUB-TOTAL	$	$	$	$	$
11. TOTAL AUTHORIZED	$	$	$	$	$

WORKING FUNDS

	19 ___	19 ___	19 ___	19 ___	TOTAL
	$	$	$	$	$
12. ACCOUNTS RECEIVABLE					
13. RAW MAT. INVENTORY					
14. STORES INVENTORY					
15. RET. CONT. INVENTORY					
16. NON-RET. CONT. INV.					
17. FINISHED MAT. INV.					
18. WIP INVENTORY					
19. ___					
20. ___					
21. TOTAL REQUIRED	$	$	$	$	$

FUTURE OBLIGATIONS

22. LAND ___ ACRES @ ___ $/ACRE = $ ___

23. PROCESS BUILDING ___ SQ.FT. @ ___ /SQ.FT. = $ ___

24. WAREHOUSE ___ SQ.FT. @ ___ /SQ.FT. = ___

25. LOCKER ROOM ___ EMPLOYEES @ ___ /EMPLOYEE = ___

26. OFFICE ___ EMPLOYEES @ ___ /EMPLOYEE = ___

27. SUB-TOTAL - BUILDING = $ ___

28. STEAM ___ M# CAPACITY @ ___ /M# = $ ___

29. POWER ___ KVA REQ. @ ___ /KVA = ___

30. REFRIGERATION ___ TONS CAP. @ ___ /TON = ___

31. MILL WATER ___ MGAL./HR. @ ___ /MG. = ___

32. COMPRESSED AIR ___ CFM CAP. @ ___ /CFM = ___

33. ___

34. ___

35. SUB-TOTAL - SERVICE FACILITIES $ ___

36. MOBILE EQUIPMENT $ ___

37. TOTAL REQUIRED IN ___ (YEAR) $ ___

38. **TOTAL INVESTMENT** ------------------$ ___

Fig. 4.5 *Estimated investment requirements.*

accordance with usual accounting practice. For method and worksheet for computing these figures, see Par. 4.9.6.2.

18. *Cost reduction:* Check here if proposed project is predicted to result in reduced: operating, maintenance, administrative, or indirect costs.

19. *Expected process life:* Enter here the number of years the process (to which the facility is to be added) can be expected to continue. If this is not equal to or greater than the productive life claimed for the new *facility*, an explanation is due.

20. *Expansion of existing operation or process:* Check here if the proposed project is predicted to yield increased profit through expansion of existing facilities.

21. *Incremental margin percent:* (Net per cent margin on expanded production.) This is the difference between the project's average incremental income and incremental cost expressed as a percentage of the project's average incremental income. In making this calculation, the incremental costs and income used should be *based on normal performance* after start-up years. Annual depreciation should be charged on a *uniform basis* equal to the total amount (expense and capital) divided by the *productive life predicted* for these facilities. "Productive life" means actual project life, irrespective of depreciation rates used for accounting or tax purposes. See Par. 4.9.7 for computation.

22. *Added sales to breakeven:* This should be the *total annual sales*, based on the unit incremental income used above, to *just equal* the total annual *incremental costs* (fixed and variable) to produce this volume. For computation, see Par. 4.9.7.

23. *Expansion with new operation or product:* Check here if the proposed project envisions a new product, a new location, or a new facility.

24. *Turnover* (*sales/investment ratio*): Average annual *net sales* income (normal performance after start-up years) divided by the authorized *investment* (expressed as a decimal ratio). See Par. 4.9.7.

25. *Average percent margin:* This is the difference between *net income* and average costs charged (including average depreciation) expressed as a percentage of average net income. Figures used should be *normal performance* after start-up years. For computation, see Par. 4.9.7.

26. *Breakeven percent capacity:* The *volume of production*, expressed as a percentage of design capacity, at which the total *annual income* will just equal total *annual costs.* For computation, see Par. 4.9.7.

27. *Profit-maintaining:* Check here if justification of the proposed investment is based on *retention of existing profit* rather than *addition of new profit.*

28. *Amount of profit at stake:* Enter here the amount of existing annual profit (before tax) that will be jeopardized if the proposed investment is not implemented. If the entire plant operation is at stake, merely enter the word "plant." If a whole department, enter name of department, etc. It is recognized that in many cases a firm figure may be difficult to give and require some speculation.

29. *Nature:* Check replacement, new facility, or other. If "other," indicate nature on line below.

4.9.2 Estimation of Investment Requirements (Form 3009 Fig. 4.5)

There are three basically different categories into which investment expenditures can be classified. Since each of these is quite different in nature and risk, it is desirable to segregate them for estimation and presentation. Form 3009 is used for this purpose. You will note that provision has been made in all cases for timing of proposed expenditures.

1. *Immediate investment.*

Land: Funds expended for this purpose must always be capitalized. No depreciation is permitted during the life of the project, but a tax deduction equal to the original purchase cost is credited at the end of the project. Credit should also be given for resale price (salvage) at end of project.

Buildings and equipment: Funds expended for these facilities are usually capitalized. Federal income tax laws provide schedules for calculating depreciation allowances with schedules that vary with the type asset involved.

Expenses: In the case of certain facilities, site preparation, relocations, etc., the federal income tax law permits these expenditures to be charged off as current expenses. Since these are nonrepetitive, they are considered part of the investment. Calculation-wise, they are to be treated as a capital expenditure with 100 per cent depreciation the year of expenditure.

2. *Working funds.* A sizable sum of money is usually tied up for the duration of the project in carrying accounts receivable, inventories of

raw and finished material, containers, etc. These expenditures are of the nature of a revolving fund and involve a minimum of risk because they can usually be recovered upon fairly short notice. Since these expenditures are neither capital nor manufacturing expense, neither their initial expenditure nor eventual recovery have any income tax impact. Thus, in these evaluations, no depreciation is permitted on working funds, nor is their return (at project end) subject to income tax.

3. *Future obligations.* Many times a project will receive an unfair advantage, or "free ride" in its evaluation, by the use of existing land or surplus facilities. In other cases, we may recognize the need for certain facilities but choose to get along by utilizing existing equipment at a cost penalty until the need for more capacity allows us to install a unit of economic size. What we are actually doing is postponing an investment which we recognize as eventually necessary. Our evaluations would certainly not be correct or truly objective if we ignored these future obligations, yet including them as part of the current investment before they are actually going to be made does not seem realistic either.

Therefore, where surplus facilities are used and must eventually be replaced, or where acquisition of facilities is postponed, the full cost of these facilities is to be estimated and *included in the evaluation as a "future obligation"* and charged at the time the actual expenditures are anticipated to be made. To guard against accidental or deliberate over-optimism, it is further specified that assumption of deferment be limited to five years or one-third of the life of the proposed project, whichever is the lesser. In this way, these expenditures will be included in the total investment specified, but their impact on profitability is lessened, as it should be. The foregoing is not a perfect solution, but it does allow a "weighted" recognition of this type of obligation and should, on an average, tend to minimize unfair evaluations. Since the impact of these future obliga-

Form 3013

DEPRECIATION & DEPLETION WORK SHEET

INVESTMENT

CALENDAR YEAR	ANNUAL PERIOD	DEPRECIATION OF INDIVIDUAL ASSET GROUPS BY ______ METHOD							DEPLETION	TOTAL INCOME TAX DEDUCTION
		EXPENSE ● 100% AMT.____	BUILDINGS ● ____ AMT.____	PROCESS EQUIP. ● ____ AMT.____	MOBILE EQUIP. ● ____ AMT.____	AMT.____	AMT.____	AMT.____		
	1ST									
	2ND									
	3RD									
	4TH									
	5TH									
	6TH									
	7TH									
	8TH									
	9TH									
	10TH									
	11TH									
	12TH									
	13TH									
	14TH									
	15TH									
	16TH									
	17TH									
	18TH									
	19TH									
	20TH									
	21ST									
	22ND									
	23RD									
	24TH									
	25TH									
	26TH									
	27TH									
	28TH									
	29TH									
	30TH									

AFTER ZERO POINT

Fig. 4.6 *Depreciation and depletion work sheet.*

tions on the profitability will usually be comparatively small, estimation of these investments need not be very accurate. A set of factors which should be sufficiently accurate for estimating these items is easy to develop and simple to use.

4.9.3 Determination of Income Tax Deductions (Form 3013 Fig. 4.6)

Depreciation. Depreciation is not a current disbursement. It is merely an agreed upon method of charging off a prepaid (capitalized) expense. It enters P.I. and payout calculations only as an allowance which permits *income-tax-free* retention of *an equivalent amount of profit.*

Since 1954, the income tax laws of the United States have permitted taxpayers to choose *one* of three ways for calculating depreciation deductions:

1. Straight-line
2. Declining balance
3. Sum-of-the-year's-digits

Form 3013 shown in Fig. 4.6 provides for year-by-year statement of depreciation for each category of facilities involved. The separation into categories should be based upon the different legally accepted lives of the various types of facilities.

Depletion. Depletion is an allowance which permits retention of part of profits, tax free, to compensate for consumption of natural resources. The basis for, and calculation of, depletion allowances is a very involved subject. It is suggested that when proposals involve inclusion of depletion allowances, the help of the comptroller's department be sought to establish the exact amounts to be credited year by year.

Total deduction. Totaling the foregoing figures horizontally will give you the total deduction applicable each year.

4.9.4 Time Schedule of Expenditures and Receipts

Form 3010, Fig. 4.7, provides for the complete tabulation of all anticipated cash flows (both flowout and flowback). It also provides space for stating most of the sales volume, sales price, and cost assumptions made.

The first step in using this worksheet is to select a reference time or zero point in the time scale of the proposed project. Theoretically, any point may be chosen as the zero time, but the design of this and succeeding worksheets makes it decidedly advantageous to select as zero point *the beginning of* the first calendar year in which income is received. Once this point is chosen, it must be carefully adhered to throughout all following forms.

You will note that each increment of expenditure can be timed as occurring:

a. All at once at the beginning of the year
b. Continuously during the year

If we propose the purchase of a piece of equipment such as a truck, we can reasonably assume that we will pay for it in a single payment. This type of expenditure would fall under (a) above, all at once at the beginning of the year. A new building or process equipment, which might take months to build, during which time we can expect to expend our fund on a more or less continuous basis, would fall under (b) above, or continuously during the year. Cash flowback increments are all assumed to be received continuously during the year.

Definitions

Annual volume: express in lbs., tons, etc.

Plant net income: gross sales return less: freight absorption, commissions, divisional sales, administration and research expense.

Plant cost before depreciation: use total cost including all plant overheads and indirects. Exclude depreciation and federal income tax.

Profit before depreciation: plant net income less plant cost before depreciation.

Depreciation and depletion deduction: use annual totals from form 3013.

Cash flowbacks—income: obtained by multiplying profit before depreciation by (1 − tax rate) with the federal income tax rate expressed as a decimal.

Cash flowbacks—on deductions: obtained by multiplying depreciation and depletion deductions by the federal income tax rate expressed as a decimal.

4.9.5 Calculation of the Profitability Index

Form 3011, Fig. 4.8, is used to make the calculations required to determine the *profitability index*. Cash flows for each year, as shown in preceding form 3010, are entered in the column headed "Trial #1—0 per cent interest rate." These are the present worths at 0 per cent interest.

Each of these numbers is then multiplied by the factor in the adjacent subcolumn labeled

Form 3010

TIME SCHEDULE OF EXPENDITURES & RECEIPTS

INVESTMENT

CAL. YEAR	PERIOD	CAPITAL LAND	CAPITAL FACILITIES	EXPENSED	WORKING FUND	TOTAL
BEFORE ZERO POINT						
	3RD YR. AT START					
	3RD YR. DURING					
	2ND YR. AT START					
	2ND YR. DURING					
	1ST YR. AT START					
	1ST YR. DURING					
AFTER ZERO POINT	1ST YR. AT START					
	1ST YR. DURING					
	2ND YEAR DURING					
	3RD YEAR DURING					
	4TH YEAR DURING					
	5TH YEAR DURING					
	6TH YEAR DURING					
	7TH YEAR DURING					
	8TH YEAR DURING					
	9TH YEAR DURING					
	10TH YEAR DURING					
TOTALS						

RECEIPTS

CAL. YEAR	ANNUAL PERIOD	ANNUAL VOLUME IN ____	PLANT NET INCOME	PLANT COST BEFORE DEPRECIATION	PROFIT BEFORE DEPRECIATION	DEPRECIATION & DEPLETION DEDUCTION	CASH FLOW BACK AFTER ____% INCOME TAX ON INCOME	ON DEDUCTION
AFTER ZERO POINT	1ST							
	2ND							
	3RD							
	4TH							
	5TH							
	6TH							
	7TH							
	8TH							
	9TH							
	10TH							
	11TH							
	12TH							
	13TH							
	14TH							
	15TH							
	16TH							
	17TH							
	18TH							
	19TH							
	20TH							
	21ST							
	22ND							
	23RD							
	24TH							
	25TH							
	26TH							
	27TH							
	28TH							
	29TH							
	30TH							

Fig. 4.7 *Time schedule of expenditures and receipts.*

"factor" and the result entered in the sub-column headed "present worth." These are the present worths at 10 per cent interest rate. These calculations are repeated at 15, 25, and 40 per cent interest rates; in each case the present worth at each interest rate is computed by multiplying actual amounts by the present worth factors for the interest rate indicated.

After we have calculated the present worth of each cash flow at each of the trial interest rates, we add the columns to obtain the total present worth of expenditures and the total

Form 3011

CALCULATION OF PROFITABILITY INDEX

TIMING			TRIAL #1 0% INTEREST RATE	TRIAL #2 10% INTEREST RATE		TRIAL #3 15% INTEREST RATE		TRIAL #4 25% INTEREST RATE		TRIAL #5 40% INTEREST RATE	
	CAL. YEAR	PERIOD	ACTUAL AMOUNT OF DISBURSEMENTS	FACTOR	PRESENT WORTH	FACTOR	PRESENT WORTH	FACTOR	PRESENT WORTH	FACTOR	PRESENT WORTH
BEFORE ZERO POINT											
		3RD YR. AT ST.		1.350		1.568		2.117		3.320	
		3RD YR. DURING		1.285		1.456		1.873		2.736	
		2ND YR. AT ST.		1.221		1.350		1.649		2.225	
		2ND YR. DURING		1.162		1.253		1.459		1.834	
		1ST YR. AT ST.		1.105		1.162		1.284		1.492	
		1ST YR. DURING		1.052		1.079		1.136		1.230	
AFTER ZERO POINT		1ST YR. AT ST.		1.000		1.000		1.000		1.000	
		1ST YR. DURING		.952		.929		.885		.824	
		2ND YEAR DURING		.861		.799		.689		.553	
		3RD " "		.779		.688		.537		.370	
		4TH " "		.705		.592		.418		.248	
		5TH " "		.638		.510		.326		.166	
		6TH " "		.577		.439		.254		.112	
		7TH " "		.522		.378		.197		.075	
		8TH " "		.473		.325		.154		.050	
		9TH " "		.428		.280		.120		.034	
		10TH " "		.387		.241		.093		.023	
TOTALS (A)											

	CAL. YEAR	PERIOD	ACTUAL AMOUNT OF RECEIPTS	FACTOR	PRESENT WORTH	FACTOR	PRESENT WORTH	FACTOR	PRESENT WORTH	FACTOR	PRESENT WORTH
AFTER ZERO POINT											
		1ST YEAR DURING		.952		.929		.885		.824	
		2ND " "		.861		.799		.689		.553	
		3RD " "		.779		.688		.537		.370	
		4TH " "		.705		.592		.418		.248	
		5TH " "		.638		.510		.326		.166	
		6TH " "		.577		.439		.254		.112	
		7TH " "		.522		.378		.197		.075	
		8TH " "		.473		.325		.154		.050	
		9TH " "		.428		.280		.119		.034	
		10TH " "		.387		.241		.093		.023	
		11TH " "		.350		.207		.073		.015	
		12TH " "		.317		.178		.057		.010	
		13TH " "		.287		.154		.044		.007	
		14TH " "		.260		.133		.034		.005	
		15TH " "		.235		.114		.027		.003	
		16TH " "		.212		.098		.021		.002	
		17TH " "		.192		.084		.016		.001	
		18TH " "		.174		.073		.013		.001	
		19TH " "		.157		.062		.010		.001	
		20TH " "		.142		.054		.008			
		21ST " "		.129		.046		.006			
		22ND " "		.117		.040		.005			
		23RD " "		.105		.034		.004			
		24TH " "		.095		.029		.003			
		25TH " "		.086		.025		.002			
		26TH " "		.078		.022		.002			
		27TH " "		.071		.019		.001			
		28TH " "		.064		.016		.001			
		29TH " "		.058		.014		.001			
		30TH " "		.052		.012		.001			
TOTALS (B)											
RATIO A/B											

Fig. 4.8 *Calculation of profitability index.*

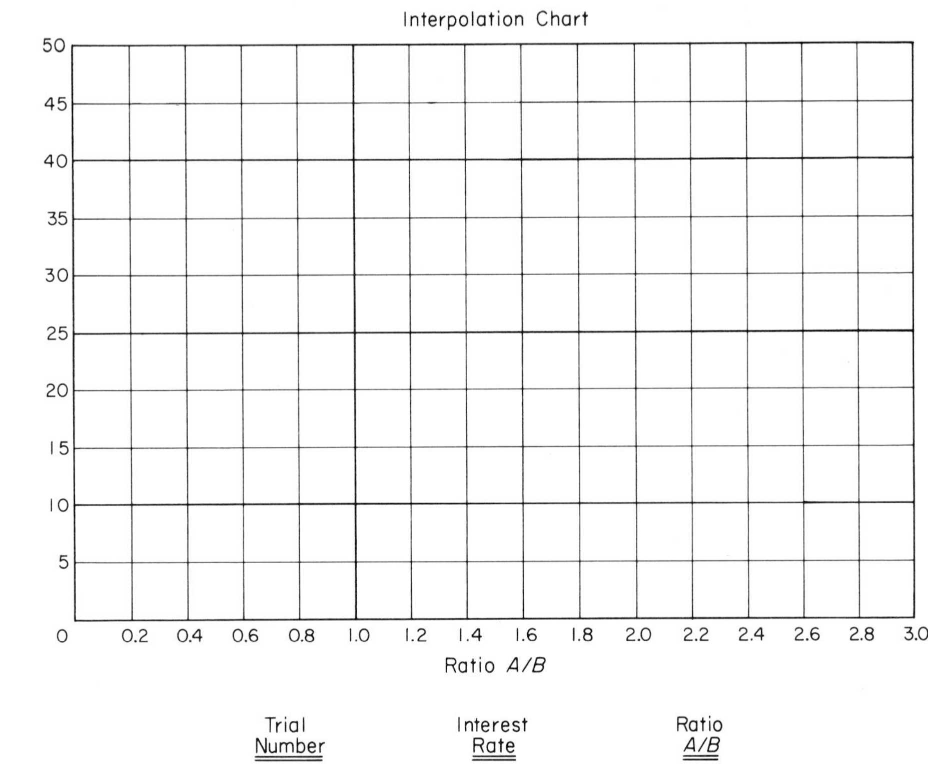

Trial Number	Interest Rate	Ratio A/B
1	0 per cent	____
2	10 per cent	____
3	15 per cent	____
4	25 per cent	____
5	40 per cent	____

Fig. 4.9 *Profitability index.*

present worth of receipts at each trial interest rate. At each trial interest rate, we divide the total present worth of expenditures by the total present worth of receipts, and enter in the space labeled "ratio A/B."

The P.I. is the interest rate at which the sum of the present worths of all receipts is equal to the sum of the present worths of all expenditures. This is the interest rate at which the "ratio A/B" is equal to unity.* While it is quite likely that the interest rates we have tried will bracket the answer we seek, we can scarcely hope that any one of them will be the exact answer. To find this exact answer, we must interpolate. Form 3012, Fig. 4.9, is provided for this purpose.

Graphical interpolation chart.

1. Enter the A/B ratios calculated on the preceding form 3011 in the spaces provided.

2. Plot these ratios against the interest rate at which they were calculated.

3. Draw a smooth curve through these points, including that corresponding to the ratio at zero interest rate. (Failure to obtain a smooth curve indicates that there is an error in calculation or plotting.)

4. Where the curve intersects the heavy 1.0 ratio line, read across horizontally to the P.I.

* Provided all net expenditures occur prior to all net receipts. Otherwise, "crutch" procedure must be used. See Par. 4.5.7.

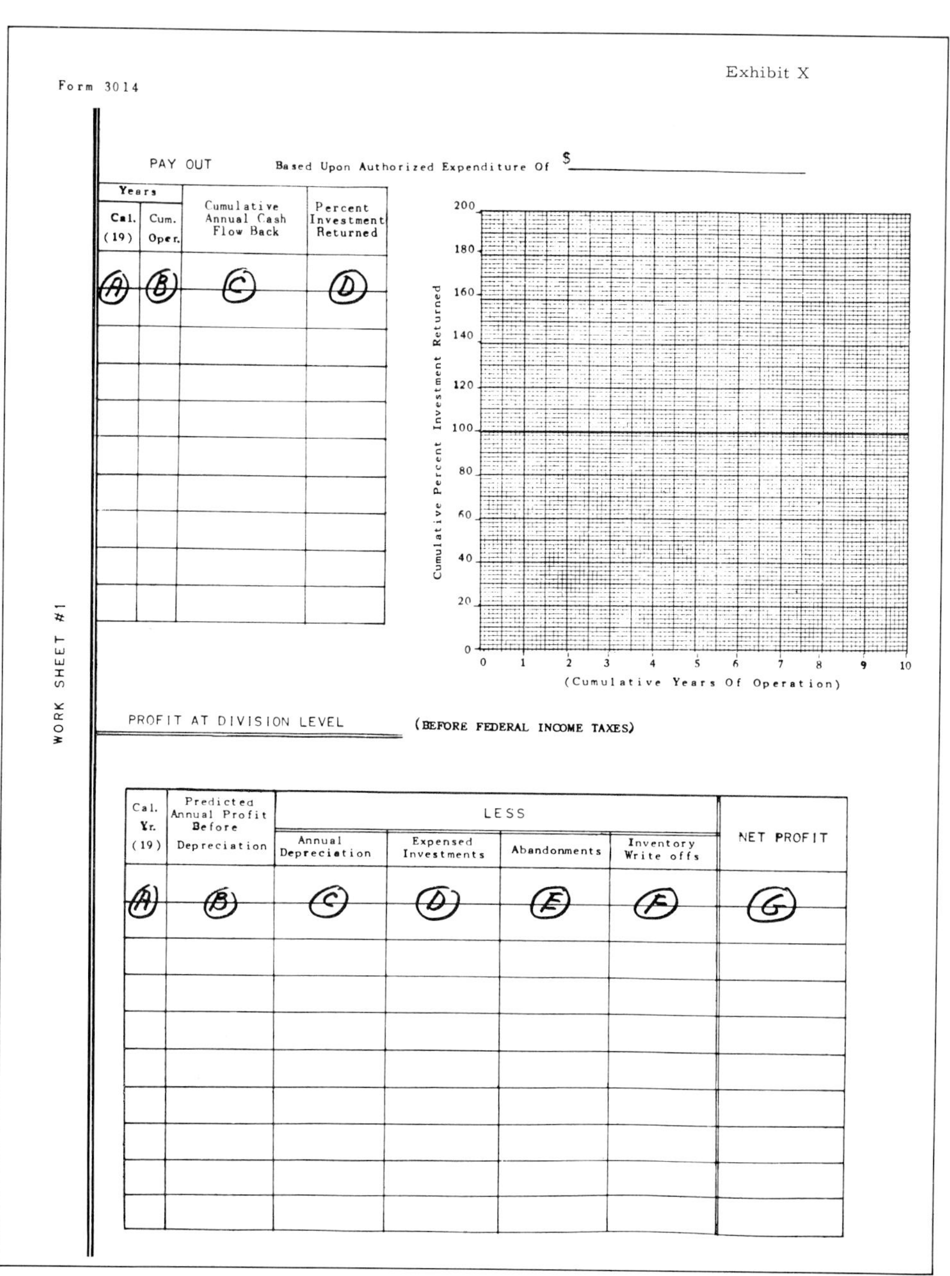

Form 3014

Exhibit X

WORK SHEET #1

PAY OUT Based Upon Authorized Expenditure Of $________

Years: Cal. (19)	Years: Cum. Oper.	Cumulative Annual Cash Flow Back	Percent Investment Returned
(A)	(B)	(C)	(D)

PROFIT AT DIVISION LEVEL (BEFORE FEDERAL INCOME TAXES)

Cal. Yr. (19)	Predicted Annual Profit Before Depreciation	LESS: Annual Depreciation	LESS: Expensed Investments	LESS: Abandonments	LESS: Inventory Write offs	NET PROFIT
(A)	(B)	(C)	(D)	(E)	(F)	(G)

Fig. 4.10 *Investment payout and proforma profit schedule.*

4.9.6 Calculation of Investment Payout and Proforma of a Profit Schedule (Form 3014 Fig. 4.10)

4.9.6.1 Payout

1. Enter authorized (immediate) investment.
2. Enter in column (A) the last two numerals of the first ten calendar years during which the project is predicted to provide a cash flowback.
3. Enter in column (B) (for each calendar year) the cumulative operating years and fraction thereof, beginning with the first month of operation.
4. Enter in column (C) the cumulative cash flowback (after deduction of federal income tax) for each calendar year. Calculate from annual flowbacks shown in P.I. computation on form 3011.
5. Divide each year's cumulative cash flowback by the authorized investment specified above to obtain for each year the percentage of investment returned by the end of that year and enter in column (D).
6. Plot cumulative per cent against cumulative operating years on chart. (Use straight lines between points).

4.9.6.2 Proforma of a Before-Tax Profit Schedule

1. Enter in column (A) the last two numerals of the first ten calendar years during which the project is predicted to provide a cash flowback.
2. Enter in column (B) the predicted annual profit, before depreciation and federal income taxes, for the above ten years (from P.I. calculation—FMC Form 3010).
3. Enter in column (C) total annual depreciation predicted to be booked for each year in accordance with accounting practice. (Omit depletion and expensed items).
4. Enter in column (D) all "expensed" investments (from P.I. calculation form 3010).
5. Enter in column (E) all abandonments (from P.I. calculation form 3010).
6. Enter in column (F) all inventory write-offs predicted as a result of this project and show in year scheduled.
7. Complete column (G) for each year by subtracting depreciation, expensed investments, abandonments and write-offs scheduled, from the profit before depreciation and taxes predicted for that year to obtain the "net profit" as reported by the usual accounting procedures.

4.9.7 Presentation of Return on Sales, Breakeven Analysis and Turnover

General Instructions. FMC Form 3015, Fig. 4.11, has been designed so that it can be used for evaluating predicted performance of investments involving either *completely new products and facilities* or *expansion of existing facilities*. The use to which it is put should be designated by *crossing out the nonapplicable part of the headings* (in parentheses).

1. *Expansion:* Cross out either (with new) or (of existing), leaving the term which correctly describes project being evaluated.
2. *Base year:* Select normal average year after start-up period.
3. *Predicted volume:* Volume of product estimated to be sold in base year.
4. *Predicted life:* Total productive life anticipated for the project (should be the same as used in P.I. calculations).
5. *Authorized investment:* The total capital and expense expenditure authorized. Exclude working funds and future obligations.
6. *Income and costs:* (average or incremental): Cross out the term which does not apply. (Average for new products or facilities, incremental for expansion of existing).
7. *Net income:* Enter all new income estimated to be received during base year. (Should be the same as the amount used in P.I. calculations for the selected base year). Net income is defined as total income less divisional: sales, administrative and research expenses, as well as freight absorptions and commissions to be paid. Enter total annual and unit amounts.
8. *Costs chargeable:* Enter all costs chargeable, including raw materials, direct manufacturing costs, plant overhead, taxes, insurance, etc. Enter average annual depreciation (total authorized investment divided by predicted life of project).
    a. Show basis for estimation of individual items.
    b. Show separately in appropriate column:
        (1) *Fixed annual cost:* Total annual cost of each item which *varies with time* rather than production. Estimate the amount on the assumption that the plant is set up and staffed to operate as predicted.
        (2) *Annual variable cost:* Total annual cost of each item which varies with *amount*

Form 3015

WORK SHEET #2

EXPANSION (With New or Of Existing) FACILITIES

BASE YEAR - 19 ________ PREDICTED ANNUAL VOL. ________

PREDICTED LIFE ________ YRS. AUTHORIZED INVESTMENT $ ________

(Average or Incremental) INCOME AND COSTS

	ITEM	BASIS	FIXED ANNUAL	ANNUAL VARIABLE	TOT./UNIT ——/.——
	PLANT NET INCOME				
LESS COSTS CHARGEABLE	DEPRECIATION	Author. Inv. ÷ Years Life			
	TOTALS				
				(Average or Incremental) UNIT MARGIN	

(Average or Incremental) Margin % = $\frac{\text{Unit Margin}}{\text{Unit Plant Net Income}}$ X 100 = ——— X 100 = ____ %

TOTAL COST

Fixed $ ________

Variable

Total $ ________

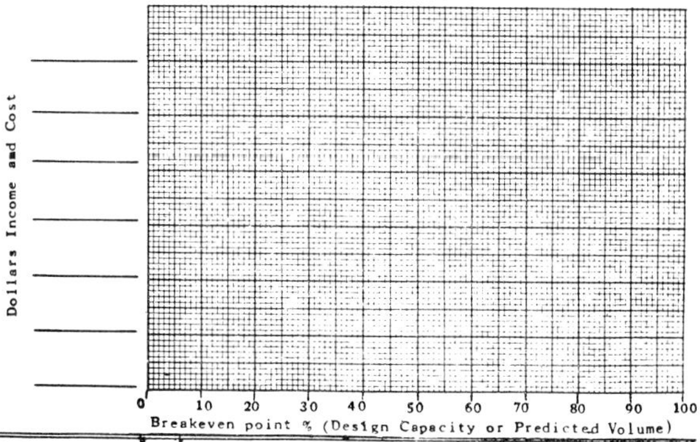

Dollars Income and Cost

0 10 20 30 40 50 60 70 80 90 100

Breakeven point % (Design Capacity or Predicted Volume)

EXISTING

ADDED SALES TO BREAKEVEN

$\frac{}{(\text{Predicted Income})}$ X $\frac{}{(\text{\% Predicted Volume})}$ = $ ____

NEW

TURNOVER

$\frac{(\text{Annual Plant Net Income})}{(\text{Author. Inv.})}$ = $\frac{\$}{\$}$ = ____

Fig. 4.11 *Presentation of return on sales, breakdown analysis, and turnover.*

produced. Estimate the annual amount on the assumption of the plant producing as predicted.

c. *Total unit cost per item (fill in unit basis—cents/lb. or dollars/ton)*: The sum of *fixed annual* costs and *annual variable* costs for each item, divided by the total production predicted for the base year.

4.9.7.1 Calculation of Per cent Margin. Margin (average or incremental): Cross out the term which does not apply. Add the individual unit costs to obtain *total unit cost* and subtract same from the *plant unit income* to obtain *unit margin.* Divide *unit margin* by *plant unit income* to obtain *per cent margin.*

4.9.7.2 Calculation of Turnover. Calculate only on expansion with new products or facilities. Divide total net income by total immediate (authorized) investment to obtain turnover, expressed as a ratio.

4.9.7.3 Use of Chart "F" to Calculate Breakeven Points. On expansion with new product or facilities:

a. Cross out "predicted volume" under chart. Enter total fixed and total variable costs from above and add to obtain *total cost.*

b. Enter suitable scale of dollars on ordinate (adequate to permit plotting of total income and total costs at predicted volume).

c. At zero per cent of "design capacity," place points to show:

1. Zero income

2. Total fixed costs as tabulated under total cost to left of chart

d. At the per cent of design capacity at which the new facilities are predicted to operate during average normal year, place points to show:

1. Total income at predicted level of operations

2. Total costs as shown in tabulation at left for predicted level of operations

e. Connect income points at zero per cent and at predicted per cent of design capacity with a straight line.

f. Connect cost points at zero per cent and at predicted per cent of design capacity with a straight line.

g. Drop a vertical line from where these two lines cross to the horizontal axis and read off breakeven volume as per cent of design capacity.

On expansion of existing facilities:

a. Cross out "design capacity" under chart. (Enter fixed and variable costs from above and add to obtain *total cost.*)

b. Insert suitable scale of dollars on ordinate (adequate to permit plotting of total income and total costs at predicted volume).

c. At zero per cent of "predicted volume," place points to show:

1. Zero income

2. Incremental fixed costs as tabulated under total cost to left of chart

d. At 100 per cent of "predicted incremental volume," place points to show:

1. Total incremental income at predicted volume

2. Total incremental cost as shown in tabulation at left of chart

e. Connect income points at zero per cent and 100 per cent predicted volume with straight line.

f. Connect cost points at zero and 100 per cent predicted incremental volume with straight line.

g. Drop a vertical line from where these two lines cross to the horizontal axis and read off *breakeven volume* as the per cent predicted incremental volume.

h. Obtain *added sales to breakeven* by multiplying *predicted gross incremental income* by *per cent predicted incremental volume.*

4.9.8 Memo of Justification

This document should contain the basic information required in a standardized sequence to facilitate review. The contents should enable any engineer or executive, even though not familiar with the plant or location, to understand fully the background and reasoning that lie behind the recommended investment. It may vary in length from one-half to several pages, depending upon the size and complexity of the project.

Most important, it should set forth the key assumptions, both commercial and technical, upon which the claimed profitability rests. Other essential ingredients are:

1. A description of the present situation

2. An outline of the proposed change or addition

3. The alternates considered

4. An explanation of the economic justification
5. The planned method of execution of the project

As attachments, the memo should include sufficient sketches, plot plans, flow sheets, photographs or vendor's literature to give the reader a clear understanding of what is proposed.

MARVIN E. MUNDEL *is now the principal in his own firm, M. E. Mundel and Associates, which is engaged in industrial engineering consulting work on an international scale. He established this firm in 1965 after almost two years as principal staff officer for industrial engineering and work measurement in the Executive Office of the President, Bureau of the Budget. Previously, Dr. Mundel served as industrial engineer for Tung-Sol Lamp Company, director of the Army Management Engineering Training Agency, Rock Island, Illinois; and as a private consultant.*

He has taught at Bradley University, Marquette University, University of Birmingham (England), Keio University (Japan) and Purdue University, and has been president and regional vice-president for Region VI of the American Institute of Industrial Engineering.

He is author of numerous books and articles, which have been published in six languages. In 1953, Dr. Mundel received the Gilbreth Medal from the Society for the Advancement of Management for his contributions to industrial engineering. During the centennial celebration of New York University in 1953, he was awarded a commendation as one of that university's distinguished engineering graduates. He is a member of Tau Beta Pi, Iota Alpha, and Sigma Xi, and a Fellow of the Society for the Advancement of Management and of the American Institute of Industrial Engineers.

He received the Bachelor of Science degree from New York University and the M. S. and Ph. D. degrees from The State University of Iowa. He is probably best known for his book, Motion and Time Study, Principles and Practices, *the fourth edition of which is now in process.*

SECTION 5

MOTION AND TIME STUDY*

MARVIN E. MUNDEL

5.1 DEFINITION OF MOTION AND TIME STUDY

5.1.1 General Definition

Motion and time study deals with (1) the scientific determination of preferable work methods, (2) the appraisal, in terms of time, of the value of work involving human activity, and (3) the development of material required to make practical use of these data.

5.1.2 Interrelationship of Motion and Time Study

It is difficult to separate these three phases, since a specified method is one of the conditions of time measurement, and time measurements often provide a basis for comparing alternative methods. In addition, method determination and time appraisal complement

*The material in this section is, in great part, adapted from M. E. Mundel, *Motion and Time Study: Principles and Practice*, 2nd ed. (Englewood Cliffs, N. J.: Prentice-Hall, Inc., 1955).

the utility of each other in application. The combined term—motion and time study—is used to denote all three phases of activity: method determination, time appraisal, and the development of material for the application of these data.

5.1.3 Universal Applicability

In any activity or occupation, motion and time study can help find a preferred way of doing the work and can provide measurements for controlling the activity. The motion and time study approach fits all human activities, including heavy or light factory work, storage or warehouse operation, farm work, housework, surgery, cafeteria work, department store or hotel work, and battle activities. What is accomplished by the work may vary from job to job, and the field of knowledge covering the raw material, product design, process, tools, equipment, and workplace may shift, but the human effort is always composed of the same basic acts. Consequently, the procedures for selecting a preferable method are essentially the same, and information relating to the economical use of human effort is universally applicable. Further, if managerial controls are to be effective, time study (or work measurement, as it is sometimes called) is always a basic requirement.

Motion and time study enables the various divisions of an organization to cooperate in selecting and planning the proper integration of materials, design of product or work achieved, process, tools, workplaces and equipment, and hand and body motions. Motion and time study techniques are aids for systematically performing many industrial engineering activities.

5.2 SHORT HISTORY

5.2.1 Early Beginnings

The prototype of modern time study appears to be the work begun in 1881 by Frederick W. Taylor, Chief Engineer of the Midvale Steel Co. The first widespread paper relating to this activity was entitled "Shop Management," Paper No. 1003, *American Society of Mechanical Engineers*, 1903, authored by Taylor. The prototype of modern motion study appears to be the work done during the late 1800s and early 1900s by Frank B. Gilbreth resulting in a chapter on "Motion Study" in *Bricklaying System*, F. B. Gilbreth, Myron C. Clark Publishing Company, Chicago, 1909. This chapter formed the basis of *Motion Study*, F. B. Gilbreth, D. Van Nostrand Co., New York, 1911. These early works were closely associated with an integrated approach to what was called "scientific management." Other than a few later highlights, space does not permit detailing the history beyond this point. Yost* has pointed out that the periodicals listed in the *Readers Guide* included 150 articles on the subject between 1911 and 1914, together with an even greater number in the technical and trade journals and newspapers for those years. During this later period, the full compatibility of the two approaches began to emerge.

5.2.2 Formalization of the Field

Gilbreth followed his early work with *Applied Motion Study*, Sturgis and Walton Co., New York, 1917; *Fatigue Study*, The Macmillan Company, New York, 1919; and *Motion Study for the Handicapped*, George Routledge, London, 1920. These later works were done in collaboration with his wife and coworker, Lillian M. Gilbreth. Also, to the main stream of published material was added the work of Jules Amar in France (available as *The Human Motor*, George Routledge, London, 1920), the work of Atzler in Germany, of Cathcart and Benedict published by the *Carnegie Institution of Washington, D. C.*, and the work of the *Industrial Fatigue Research Board* and the *Industrial Health Research Board* of Great Britain, beginning about the time of World War I. There were also numerous other contributors in many countries in the fields of industrial engineering, industrial psychology, and physiology. It is almost impossible to summarize this period without slighting (unintentionally) many who contributed greatly to this growing field. Considerable assistance was given to its growth by the number of articles published in the McGraw-Hill magazine *Factory*, under the editorship of L. C. Morrow.

---

*Edna Yost, *Frank and Lillian Gilbreth* (New Brunswick, N. J.: Rutgers University Press, 1949), p. 188.

5.2.3 Early Textbooks

Academic work in motion and time study was somewhat limited by the unavailability of suitable texts until the publication of *Time and Motion Study*, by S. M. Lowry, H. B. Maynard, and G. J. Stegemerten, McGraw-Hill, New York, 1927; *Common Sense Applied to Motion and Time Study*, A. H. Mogensen, McGraw-Hill, New York, 1932; and *Motion and Time Study*, R. M. Barnes, Wiley, New York, 1937, this last volume being an expansion of an earlier lithoprinted book by the same author. Although these early texts have largely been displaced by later editions or later works by other authors, they stand as historic landmarks in the subject.

5.2.4 Later Developments

By the 1940s the number of texts and related books began to grow to an imposing list and work in the area was firmly established on a scientific basis, on an international scale, and as a regular part of the normal industrial routine of the civilized world.

5.3 MOTION STUDY

5.3.1 Detailed Definition

Motion study is a scientific, analytical procedure for determining a preferable work method, considering:

a. The raw materials

b. The design of the product

c. The process or order of work

d. The tools, workplace, and equipment for each individual step in the process

e. The hand and body motions used in each step.

The criterion of preference is usually economy of money, but other criteria such as ease or economy of human effort, economy of work time, in-process time, material, skill, space, machine time, tools, etc., frequently may take precedence.

5.3.2 Uses

The basic use of motion study is implicit in the definition given in Par. 5.3.1. However, motion study is also applicable to:

a. Improving a work method by locating deficiencies in any of the five individual factors involved or in their interrelationships

b. Improving a work method by assisting in adjusting the pattern of interrelationships or any of the individual characteristics of the five factors involved to meet: (1) new economic conditions, (2) changes in criterion of preference, (3) new materials or equipment

c. Designing a work method through logical analysis and synthesis

d. Providing a schematic framework for organizing knowledge about the interrelationships and individual characteristics of the five factors involved.

5.3.3 Steps in the Application

The scientific method of solving problems involving the determination of a preferred way of doing a job, a preferred method of production, or a preferred method of doing new work requires the application of a logical procedure consisting of the following seven steps:

A. *Aim.* Determination of objective in terms of area of job to be changed, and establishment of criteria for evaluating the preferability or success of solutions. Innovations may be introduced in any one of the five areas that affect the performance of the job. These areas are:

1. *Hand and body motions.* The particular motions, their sequence, and their nature may be changed to ease or improve the task.

2. *Work station (tools, workplace layout, or equipment).* The design of any single work station or the equipment used for any part of the task may be modified.

3. *Process or work sequence.* The order or condition in which the various work stations receive the product may require change or the number of work stations may be modified.

4. *Product design, form of goods sold, or material or service produced.* The final form of the product, as it leaves the organization, may require slight modification to facilitate the attainment of the objectives of improvement.

5. *Incoming supplies or raw material.* The materials brought into the organization may require change in form, condition, or specification in order to allow the desired improvements to be made.

A change in any of these factors with numbers above 1 usually must be accompanied by changes in the areas with lower numbers in order to accommodate the change. Also, a change beginning in Area 3, 4, or 5 is usually evaluated with the assistance of an analysis of what happens to the product; a change beginning in Area 1 or 2 with an analysis of work performed by a man. Reference to these areas (usually referred to as "classes") may help clarify instructions given to the analysts.

B. *Analysis.* Analysis of the work method into subdivisions or steps possessing known characteristics, or concerning whose performance information is already available which is pertinent to the problem and is relatively unique or singular in nature, and graphic or tabular presentation of the actual or contemplated sequence of these steps. These analyses may be divided into two main groups: those for analyzing activity of the product and those for analyzing activity of people.

C. *Criticism.* Checking the analysis against basic data, check lists of desirable arrangements of the steps into a preferable work pattern, and information concerning desirable ways of performing each of the steps.

D. *Innovation* (*synthesis*). Formulating a new procedure for performing the work.

E. *Test.* Testing, by means of the data used in Step C, the desirability of the method formulated in Step D, in respect to the objectives set up in Step A.

F. *Trial.* Making a sample application of the method tested in Step E to ascertain whether all variables have been taken into account.

G. *Application.* Standardizing, installing, evaluating, and maintaining the improved work method.

5.3.4 Lists of Techniques

Motion-study techniques may be listed under the steps of the logical procedure which they implement. Techniques for the study of current activity are usually made by means of direct observation or by means of micromotion or memomotion study. (See Par. 5.3.5 for definitions.) In studying a contemplated activity, they may be used to detail the concept. In the following lists those marked by an * indicate analyses which may be made by micromotion or memomotion study. In some of these cases, analysis by direct observation is difficult, if not impossible.

5.3.4.1 For Determining Aim.

a. Possibility guide or list
b. Activity chart
c. Machine load chart
d. Functional form analysis chart

5.3.4.2 For Performing Analysis. These techniques are divided into two major groups, product and man, as explained in Par. 5.3.3B.

For Product Analysis

a. Process chart—product analysis
b. Flow diagram
c. Procedure analysis chart (in some cases this also shows personnel actions)

For Man Analysis

a.* Process chart—man analysis
b.* Man flow diagram
c. Workplace layout
d.* Operation chart
e.* Multiple-activity analysis
1. Man and machine (or multimachine) operation chart
2. Man and machine (or multimachine) operation time chart
3. Man and machine (or multimachine) process chart
4. Man and machine (or multimachine) process time chart
5. Multiman operation or process chart
6. Multiman operation time or process time chart
7. Multiman charts similar to 1, 2, 3, or 4 above
f.* Micromotion study
1. Simo (simultaneous motion cycle) chart
2. Memomotion charts (time sequence, flow or tabular)
g. Cyclegraphic or chronocyclegraphic records

5.3.4.3 For Criticism. Available check lists and basic data are arranged in suitable form for use with each of the techniques given under Pars. 5.3.4.1 and 5.3.4.2.

5.3.4.4 For Innovation. The techniques used to present the innovation are usually similar to those used to make the *Analysis* (see Par. 5.3.4.2).

5.3.4.5 For Test. A proposed solution may be tested against the basic data referred to in Par. 5.3.4.3 and against the criteria set up in determining the *Aim* (Par. 5.3.3). Summaries of what will be achieved are often needed in evaluating the utility of a proposed solution. The following forms are useful:

a. Methods proposal summary
b. Procedure proposal summary.

5.3.4.6 For Trial. Formal permission to conduct a trial should be obtained through routine channels and procedures.

5.3.4.7 For Application. This step usually requires the preparation of what is variously called a WSP (Written Standard Practice), SOP (Standard Operating Procedure), a layout (see Sec. 8 on Plant Layout), a tool, the design of a machine design or workplace (see any standard text on drafting practice), or an operator instruction sheet or procedure instruction (Sec. 1).

5.3.5 Definitions of Techniques

The techniques listed under Par. 5.3.4 are defined below. Again, most of them are applicable either to an existing job or process or to a projected job or process.

5.3.5.1 For Aim. These techniques are designed to help the analyst select the most feasible type of change. They help him select the appropriate analysis procedure for performing the next step of the motion study. This relationship is given in Table 5.1.

Possibility Guide. This is a systematic list of all possible changes suggested by the person familiar with the activity or product under

TABLE 5.1. *SELECTING THE ANALYSIS TECHNIQUE**

Class of Change Sought	Physical Characteristics of Job	Technique Recommended	Equipment Required	Notes
5, 4, 3	Several work stations in process.	Process chart-product analysis.	Paper and pencil.	Rapid analysis.
5, 4, 3	Flow of paper work or control form.	Procedure analysis chart.	Paper and pencil.	Rapid analysis of complex situations.
2 or 1	Person moves from place to place while doing work.	Process chart-man analysis.	Paper and pencil. Watch is sometimes used but is not necessary.	Rapid analysis.
2 or 1	Person does job at one place.	Operation chart.	Paper and pencil. Watch is sometimes used but is not necessary	Rapid analysis and widely applicable.
2 or 1	Man operates one or more machines. Machine time is an important factor.	Man and machine charts.	Paper and pencil. Watch is desirable. Film may be used.	Rapid analysis.
2 or 1	Several men work coordinately, as a crew with or without machine or machines.			
	A—Simple task.	A—Multiman or multiman and machine chart.	A—Paper and pencil. Watch is desirable.	A—Difficult analysis at times.
	B—Complex task.	B—Memomotion chart.	B — Micromotion equipment.	B—Very easy to analyze.
2, 1	Long-run short cycle or detailed training material is required.	Simo chart.	Micromotion equipment.	Analysis takes some time but is less costly for detailed analysis and where much methods work is done.

* Adapted from Mundel, *Motion and Time Study: Principles and Practice.*

scrutiny. It also shows the consequences of each suggestion.

Activity Chart. This is a chronological record of the various activities of an individual performing a variety of tasks. It includes a summary of the nature of each activity, the work-units produced, and the time spent at each activity. Several activity charts may be combined into a work distribution chart, described in Par. 5.3.5.1.

Work Distribution Chart. "A tabulation of the various tasks performed by the individuals in an organization classified in accordance with the major activities of that organization. Time spent by each individual on each task is indicated, so that a completed chart will show the total man-hours spent on each activity."*

Machine Load Chart. This is similar to an activity chart but records the activities of a machine rather than of a person.

Functional Form Analysis Chart. This is a tabular presentation of the activities of an organization and the forms used at each step in performing each activity. It indicates areas in which forms overlap, are inadequate, or are too numerous.

5.3.5.2 For Analysis. These techniques are described categorically in Par. 5.3.3B. Essentially, they provide a formal, graphic means of breaking a large problem down into smaller problems. They enable the analyst to develop or apply basic information and to synthesize satisfactory solutions.

For Product Analysis. These techniques are, for the most part, graphic.

Process chart—product analysis. This is a graphic presentation of the separable steps involved in performing the work required to modify a product from one stage of completion to another.

Flow diagram. This consists of a sketch of the part of the plant concerned in the study of a product with a line tracing the path followed by the product. In some cases, the symbols from the process chart are shown at the places they occur.

Procedure analysis chart. This is a graphic presentation of the separable steps of activities with forms or other paper work, showing the interrelationships of the various papers, forms, or copies involved and can be used also to indicate the work of related individuals and/or offices and/or products. Its primary use is to analyze paper-work routines.

For Man Analysis. These techniques also are primarily graphic.

Process chart—man analysis. This is a graphic presentation of the separable steps a person performs when doing a task that requires him to move from place to place in the course of his work, treating the person as a single unit.

Man-flow diagram. This is a sketch of the part of a plant connected with a man's work on which a line has been drawn indicating the path the operator travels in performing this work.

Workplace layout. This is a dimensioned sketch (usually showing both plan and profile views) of the place at which a man works. It helps the analyst to understand the body and eye motions required to perform the task. The path of motions may be added, as in a man-flow diagram.

Operation chart. This is a graphic presentation of the separable steps of a person's body members when he is performing a job that takes place essentially at one location and indicates the relative simultaneity of the movements of the various body members being considered. (An operation chart is an analysis of the work performed by a worker on any one operation on a process chart, either man, product, or combined analysis. It is a description of what the operator does.)

Multiple activity. These techniques are used to analyze situations in which a man works with one or more machines and in which the work of the machine is a controlling factor, and situations in which a group of men work coordinately with or without machines. To be useful, the analysis must show not only the sequence of steps, of the two or more items charted, but also the relative simultaneity of steps.

The man and machine operation chart is used to analyze the work of one operator and one machine in one location. It is similar to an operation chart, except that a third column is added for the machine. In this column, the activity of the machine is indicated by one of

*"Techniques of Work Simplification," Department of Army, *Pamphlet No. 20–300*, June 1951, 3.

two symbols: the symbol for suboperation when the machine is working, and the symbol for delay when it is idle. The machine is thus charted as if it were an additional body member with limited functions. If more than one machine is involved, an additional column is added for each machine.

The man and machine operation time chart is similar to a man and machine operation chart except that in place of a column of symbols, a plain vertical column and shadings of various types are used to indicate the nature of the activity, the length of each shading indicating the amount of time spent at that activity. The activities of the machine or machines are charted in a similar manner. The relative simultaneity of activity is thus indicated much more accurately.

The man and machine process chart is similar to a man and machine operation chart except that the person's body members are not charted separately. The person is treated as a single unit, as in a process chart—man analysis.

The man and machine process time chart is like a man and machine operation time chart except that the man's activities are charted as a single unit.

Multiman operation or process chart is a graphic presentation of the separable steps of the work of a crew and indicates the relationship between the work of each member when the work involves coordination of the crew. In the operation chart version, each pertinent body member of each man is charted separately. In the process chart version, each man is treated as a single unit.

Multiman operation or process time chart differs from a multiman operation or process chart only in that a shaded column instead of a column of symbols is used to represent each item being charted. Since the length of each shading is proportional to the amount of time spent on each step, the simultaneity of action is shown with high accuracy.

A multiman and multimachine chart may be made by adding a column for each machine to the two preceding types of charts.

*Micromotion study** refers to the analysis of a man-work method by using a motion-picture camera with a timing device in the field of view. A constant-speed drive incorporated in the camera may serve as the timing device. A video-tape recorder may be used in lieu of a motion-picture camera. Micromotion study involves three steps: filming, analysis of the film, and graphic presentation. Although the usual speed is 16 frames per second, or 1000 frames per minute, other speeds may be more convenient. For example, speeds of 50, 60, or 100 frames per minute are exceedingly useful in studying complex and varied tasks, long-cycle or crew activities, and long-period activities. Such speeds reduce the amount of film used, and consequently the amount of film that has to be analyzed, by as much as 95 per cent. Study conducted at these slower film speeds is frequently referred to as *memomotion study*.

A simo (*simultaneous motion cycle*) *chart* is a graphic presentation of the separable steps of each pertinent body member of the individual being charted and is normally accompanied by the *use* of a series of 17 analysis categories called therbligs†(see Par. 5.3.3B) together with a vertical bar for each body member charted. Since the number of separately identified categories (17) is too large for convenient use of shadings, colors are used in the vertical bars on the chart; one bar for each member is charted and the length of each colored section is made proportional to the amount of time involved. Normally, one bar is provided for each hand, but additional bars may be used for fingers, head, eyes, feet, knees, and so on. Several people, working as a crew, may be charted together.

A memomotion chart is a graphic presentation of the separable steps of a person's activity as taken from a memomotion film. Many of the previously described types of graphic presentations may be employed or in unusual cases a special set of categories may be developed to accommodate the requirements of the problem as suggested by the basic statement of Step B of Par. 5.3.3. Flow diagrams may also be employed or special tabular presentations resembling those described in Par. 5.3.5.1 may be used.

Cyclegraphic or chronocyclegraphic records‡ are photographic records of a body member's path of movement. A light is attached to the

* Originated by Frank and Lillian Gilbreth.

† After Gilbreth (spelled backwards), who developed these classifications.

‡ Developed by Frank and Lillian Gilbreth.

body member and the movement is photographed on still film. A long shutter opening equal to the time for a movement cycle is usually used, although the opening may be shorter or longer. In cyclegraphic records, a continuous light source is used. In chronocyclegraphic records, the light is caused to flash with a nonsymmetrical peak of brilliance* so that speed and direction may be measured later. The same result may be obtained by rotating a photographic wedge† in front of the lens in lieu of a flashing light. Other arrangements employ stereoptical photography or two cameras at right angles to each other. A similar record may be obtained by plotting the path of a definite point on a person's hand or body from successive frames of motion pictures. All these photographic techniques are exceedingly useful in studying skills.

---

* Developed by the Gilbreths.

† Developed by Professors Connolly and Wilshire, College of Aeronautics, Cranfield, England.

5.3.5.3 For Criticism. Check lists present desirable and undesirable features and facts common or pertinent to work of a given scope or nature. The lists are arranged in a fashion that facilitates comparison with an actual or contemplated job or process. Such a comparison helps the analyst to locate undesirable features and to insure conformance with desirable features. Basic data concerning human dimensions, capacities, and limitations should also be referred to.

5.3.5.4 For Test. Actual forms commonly used for summarizing are given in Art. 5.6. They are usually tabular presentations which include an organizational routing of the information and a listing of the criteria of success. They also contain pertinent facts relating to the important aspects of the proposal under consideration as well as an identification of the source of estimated data. They usually also summarize the proposal (see Par. 5.3.4.5).

TABLE 5.2. *SYMBOLS FOR PROCESS CHART—PRODUCT ANALYSIS**

Symbol	Name	Used to Represent
◯	Operation	Something done to the product at essentially one location.
□	Quantity inspection	A special form of operation involving the verification of the quantity of a product present against some record of the quantity that is supposed to be there.
◇	Quality inspection	A special form of operation involving the verification of some attribute or quality of a product against a standard.
○	Movement	A change in the location of a product which does not change it in any other way.
▽ (nested)	Temporary storage	The storage of a product under conditions such that it may be moved or withdrawn from storage without a requisition.
▽	Controlled storage	Storage of a product under controls such that a requisition or receipting is needed to withdraw it.

* From Mundel, *Motion and Time Study: Principles and Practice.*

TABLE 5.3. *SYMBOLS FOR PROCEDURE ANALYSIS CHART**

Symbol	Name	Used to Represent
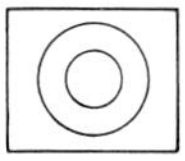	Origin of form	Form first being made out.
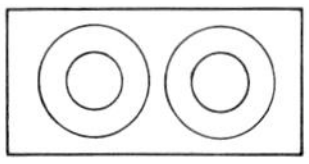	Origin of form	Form first being made out in duplicate.
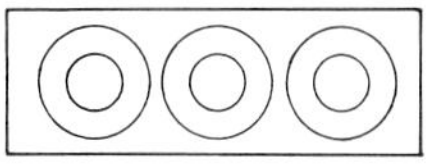	Origin of form	Form first being made out in triplicate, etc.
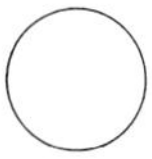	Operation	Work being done on form; computations or additional information added, etc.
	Movement	A change in location of form, not changing it.
	Delay	Forms waiting to be worked on, such as in desk basket.
	File	Forms in a file.
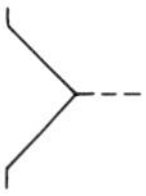	Information take-off	Information being taken off form for entry onto another or for use by someone. Point of line indicates symbol on other parallel chart where information is going. (Use – – – – broken line to indicate destination if destination appears on chart and line is aid to clarity.)
	Disposal	Form or copy destroyed.
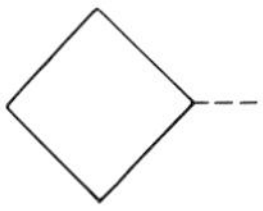	Inspection	Correctness of information on form checked by comparison with other source of information. (Use – – – – broken line drawn to other source if other source appears on chart and line is aid to clarity.)
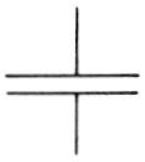	Item change	Change in item charted.
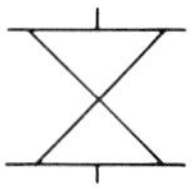	Gap	Activities not pertinent to study and hence not charted in detail.

* From Mundel, *Motion and Time Study: Principles and Practice.*

5.3.6 Symbols

To make it easier to prepare and interpret data, almost all the graphic techniques employed in motion study use symbols to represent the various separable steps identified. These symbols, and the categories they represent, have evolved over a considerable period of time. They are designed to divide the activity under consideration into a minimum number of groups while still obtaining the benefit of division into unique categories for which basic facts may be developed. Symbols, together with the scientific method, form the basis of motion study. The same symbols are used in systems analysis.

5.3.6.1 For Product Analysis Techniques. These symbols relate to identifiable steps of work done on a product or item.

Process chart—product analysis: Table 5.2.

Procedure analysis: Table 5.3.

5.3.6.2 For Man Analysis Techniques. These symbols relate to identifiable steps of work performed by a man and also to identifiable steps in the work of any machines or equipment used.

Process chart—man analysis: Table 5.4.

Operation chart: Table 5.5.

For multiple-activity analysis: These symbols are given separately for each type of chart.

Man and machine operation charts: Par. 5.3.5.2 explains the method of adapting the symbols shown in Table 5.5.

Man and machine operation time charts: Table 5.6.

Man and machine process charts: Par. 5.3.5.2 indicates the method of using the symbols shown in Tables 5.4 and 5.5.

Man and machine process time charts: Table 5.7.

Multiman operation or process charts: Tables 5.4 and 5.5.

Multiman time charts: Tables 5.6 and 5.7.

Multiman and machine charts: Tables 5.4, 5.5, 5.6, and 5.7.

5.3.6.3 For Micromotion Study. These symbols for identifiable steps are usually associated with film analysis. But, because of detail, they are useful for detailing the concept of a contemplated job.

Simo charts. These charts usually employ therbligs as their fundamental units. Defini-

TABLE 5.4. *SYMBOLS FOR PROCESS CHART—MAN ANALYSIS**

Symbol	Name	Used to Represent
○	Operation	The doing of something at one place.
□	Quantity determination	A special form of operation involving the person determining the quantity of an item present.
◇	Inspection	A special form of operation involving the person comparing an attribute of a product with a standard, or verifying the quantity present.
∘	Movement	A change in location; moving from one place to another.
▽	Delay	Idleness. Waiting or moving, provided the movement was not part of the job and the time could have been spent waiting.

* From Mundel, *Motion and Time Study: Principles and Practice.*

TABLE 5.5. *SYMBOLS FOR OPERATION CHART**

Symbol	Name	Used to Represent
	Suboperation	Body member doing something at one place, such as taking hold, lining up, assembling, etc.
	Movement	A movement of a body member toward an object or changing the location of an object.†
	Hold	Body member maintains an object in a fixed position so that work may be done with or on it at that location.
	Delay	Body member is idle or delaying for other body member.

* From Mundel, *Motion and Time Study: Principles and Practice.*

† On very long operations the analyst may combine some of these steps into larger steps using "get" in place of *reach for*, *take hold* of, and *bring* object to work area; "aside," meaning *move object* from work area, *let go* of object, and *return.* In such a case the chart is described as being made with "a gross breakdown" and is considerably shorter than when made with the usual steps.

TABLE 5.6. *SYMBOLS USED WITH MAN AND MACHINE OPERATION TIME CHART**

Symbol†	Name	With Man Activities Is Used to Represent	With Machine Activities Is Used to Represent
	Suboperation	Body member or operator doing something at one place.	Machine working ("on" time), machine paced.
	Suboperation	Not used.	Machine working ("on" time), operator paced.
	Movement	Body member or operator moving toward or with an object.	Not used.
	Hold	Body member maintaining an object in a fixed position.	Not used.
	Delay	Body member or operator is idle.	Machine is idle ("down" time).

* From Mundel, *Motion and Time Study: Principles and Practice.*

† The amount of shading is chosen to suggest automatically the general usefulness of the step. The less shading, the probable greater undesirability of the step.

TABLE 5.7. *SYMBOLS FOR MAN AND MACHINE PROCESS TIME CHART**

Symbol	Name	With Man Activities Is Used to Represent	With Machine Activities Is Used to Represent
	Operation	The doing of something at one place.	Machine working ("on" time), machine paced.
	Operation	Not used.	Machine working ("on" time), operator paced.
	Quantity determination	A special form of operation involving the person determining the quantity of an item present.	Not used.
	Inspection	A special form of operation involving the person comparing an attribute of a product with a standard, or verifying the quantity present.	Not used.
	Movement	A change in location; moving from one place to another.	Not used.
	Delay	Idleness. Waiting or moving, provided the movement was not part of the job and the time could have been spent waiting.	Machine is idle ("down" time).

* From Mundel, *Motion and Time Study: Principles and Practice.*

tions of these units and the colors used to represent them on charts are given in Table 5.8.

Memomotion charts. The symbols shown in Tables 5.4, 5.5, 5.6, 5.7, and 5.8 may be used in memomotion charts. However, special sets of steps and symbols designed for special problems are often very valuable. The series given in Table 5.9 was created for an analysis of air-crew activities* presented by means of a multiman and multimachine operation time chart. Special care should be exercised to interpret Par. 5.3.3B correctly so as to avoid the use of irrelevant subdivisions. Special categories may also make tabular presentations more effective. Table 5.10 gives the results of a memomotion analysis of a pharmacist's activity, together with the symbols used during the analysis. Note that Tables 5.9 and 5.10 merely suggest approaches to special problems rather than commonly used categories and symbols.

*M E. Mundel, "Motion Study Techniques Which Could Be Brought To Bear on Desirable Size of Aircraft Crews," *Scientific Methods for Use in the Investigation of Flight Crew Requirements* (Cambridge, Mass.: The Flight Safety Foundation, 1948).

5.3.7 Check Lists

An analyst will facilitate his activity if he will accumulate information on ways, peculiar to his industry or type of work, of improving the various type of steps concerned in each type

TABLE 5.8. *THERBLIG DEFINITIONS AND SYMBOLS**

Color Group and General Characteristics	Therblig	Symbol	Color	Eagle Pencil†	Dixon Thinex Pencil	Definition
Red–blue—terminal therbligs	Grasp	G	Lake red	744	369	Begins when hand or body member touches an object. Consists of gaining control of an object. Ends when control is gained.
	Position	P	Blue	741	376	Begins when hand or body member causes part to begin to line up or locate. Consists of hand or body member causing part to line up, orient, or change position. Ends when body member has part lined up.
	Pre-position	PP	Sky blue	740½	418	Same as position except used when line up is previous to use of part or tool in another place.
	Use	U	Purple	742½	396	Begins when hand or body member actually begins to manipulate tool or control. Consists of applying tool or manipulating control. Ends when hand or body member ceases manipulating tool or control.
	Assemble	A	Heavy violet	742	377	Begins when hand or body member causes parts to begin to go together. Consists of actual assembly of parts. Ends when hand or body member has caused parts to go together.
	Disassemble	DA	Light violet	742	422	Begins when hand or body member causes parts that were integral to begin to separate. Consists of taking objects apart. Ends when hand or body member has caused complete separation.
	Release load	RL	Carmine red	745	383	Begins when hand or body member begins to relax control of object. Consists of letting go of an object. Ends when hand or body member has lost contact with object.
Green—gross movement therbligs	Transport empty	TE	Olive green	739½	391	Begins when hand or body member begins to move without load. Consists of reaching for something.

TABLE 5.8 (*Continued*)

Color Group and General Characteristics	Therblig	Symbol	Color	Eagle Pencil†	Dixon Thinex Pencil	Definition
						Ends when hand or body member touches part or stops moving.
Green—gross movement therbligs	Transport loaded	TL	Grass green	738	416	Begins when hand or body member begins to move with an object. Consists of hand or body member changing location of an object. Ends when hand or body member carrying object arrives at general destination or movement ceases.
Gray–black—hesitant movement therbligs	Search	SH	Black	747	379	Begins when hand or body member gropes or hunts for part. Consists of attempting to find an object. Ends when hand or body member has found location of object.
	Select	ST	Light gray	734½	399	Begins when hand or body member touches several objects. Consists of locating an individual object from a group. Ends when the hand or body member has located individual object.
Yellow–orange—delay therbligs	Hold	H	Gold ochre	735	388	Begins when movement of part or object, which hand or body member has under control, ceases. Consists of holding an object in a fixed position and location. Ends with any movement.
	Unavoidable delay	UD	Yellow ochre	736	412	Begins when hand or body member is idle. Consists of a delay for other body member or machine when delay is part of method. Ends when the hand or body member begins any work.
	Avoidable delay	AD	Lemon yellow	735½	374	Begins when hand or body member deviates from standard method. Consists of some movement or idleness not part of method. Ends when hand or body member returns to standard routine.

TABLE 5.8 (*Continued*)

Color Group and General Characteristics	Therblig	Symbol	Color	Eagle Pencil†	Dixon Thinex Pencil	Definition
	Rest for over-coming fatigue	R	Orange	737	372	Begins when hand or body member is idle. Consists of idleness which is part of cycle and necessary to overcome fatigue from previous work. Ends when hand or body member is able to work again.
Brown—accompanied by thinking	Plan	PN	Brown	746	378	Begins when hand or body members are idle or making random movements while worker decides on course of action. Consists of determining a course of action. Ends when course of action is determined.
	Inspect	I	Burnt ochre	745½	398	Begins when hand or body member begins to feel or view an object. Consists of determining a quality of an object. Ends when hand or body member has felt or seen an object.

* From Mundel, *Motion and Time Study: Principles and Practice.*

† The colors of some of these pencils vary somewhat from the standard colors. They have been selected to match the standard as closely as commercial pencil colors allow.

of analysis—that is, specific ways of achieving the general suggestions implied by the check list questions contained in basic check lists. The following check lists are arranged in the same general order as the definitions of techniques. They are by no means exhaustive but should be used as guides to indicate the type of questioning the analyst should engage in.

5.3.7.1 For Aim. Check lists for determining Aim are given in Tables 5.11 and 5.12. In Table 5.11, which is designed to yield general ideas, the numders of the groups of questions refer to the five classes of change described in Par. 5.3.3A. Table 5.12 gives possible criteria of success. The check list questions as well as the following general statements and questions should be considered:

The particular class of change chosen as an objective is a function of many factors.

The higher classes of change often take longer to install, affect more people, and require higher authority. Hence, in addition to the desirability or apparent feasibility of the individual suggestions, the following economic and psychological factors affecting change must be carefully considered before a final decision is made:

1. How great is the actual or expected volume, and how often does the job occur?
2. How long will the job exist?
3. How much time, per unit, is spent on the job?
4. How much time is available to work up the change?
5. How much equipment is already invested in the job?
6. How much analysis time will be required?

TABLE 5.9. *SYMBOLS FOR USE WITH MAN AND MACHINE ANALYSIS OF AIRCRAFT FLIGHT CREW*

Category	Code	Color	Dixon Pencil Number	Remarks
Communications				
Listen	VL	Pink	381	Striped
Voice	VV	Vermillion	371	Solid
S/P phones	VP	Red	370	Solid
Intercom	VI	Lavender	424	Solid
Radio	VR	Purple	396	Solid
Mechanical (Servo)	VS	Violet	377	Solid
Gross movements				
Walk	WL	Olive green	391	Solid
Walk and supervise	WS	May green	416	Solid
Manipulations				
Adjust or operate in a special manner	MS	Sky blue	418	Striped
Operate in a normal manner	MN	Sky blue	418	Solid
Record or write	MR	Blue	376	Solid
(If desired, this group may be expanded into the normal 17 therbligs.)				
Considerations				
Form a decision	CD	Brown	378	Solid
Compute	CC	Burnt sienna	398	Solid
Inspect	CI	Gold ochre	388	Solid
Delays				
Unavoidable idleness	DU	Yellow	374	Solid
Avoidable idleness	DA	Yellow and black	374 379	Alternate stripes
Machines or instruments				
Normally operating	ON	None		Blank
Being manipulated	OM	Black	379	Solid
(Reading on instrument. Show actual reading in bar, at appropriate place.)				

7. How much loss of production or sales will be caused by the change?
8. How much retraining will be required?
9. What saving will the change effect?
10. What is the position of the analyst in the plant organization?
11. What personalities are involved?
12. What plant policies affect the problem?
13. Is the product for plant use or for customers?
14. What do other interested groups in the organization think of the feasibility of the suggestion?

The importance of each factor will vary from case to case. The order in which they are given here does not necessarily indicate their relative importance. The analyst must weigh and evaluate them for each situation.

5.3.7.2 For product analysis techniques. Check lists follow for the major techniques defined in Par. 5.3.5.2.

Process chart—product analysis: Table 5.13.

Procedure analysis: Table 5.14.

5.3.7.3. For man analysis techniques. The following check lists relate to individual techniques and to groups of techniques.

Process chart—man analysis: Table 5.15.

Operation chart: A check list for use with operation charts is given in Table 5.16. Related basic data are given in Fig. 5.1.

Multiple activity: Check lists for use with multiple-activity charts should be used in conjunction with the check lists relating to the basic charts of which they are an elaboration.

Man and machine charts: Table 5.17.

TABLE 5.10. *SYMBOLS AND RESULTS OF ANALYSIS OF PHARMACIST'S PRESCRIPTION FILLING TIME**

Description of Activity	Symbol	Rank Importance	Per Cent of Time Used
Work on labels or prescription blanks	L	1	23.3
Work wrapping	AWR	2	10.5
Work putting material into prescription containers or with containers	AM	3	10.1
Inspection of prescription blanks	IP	4	7.4
Work applying labels	AL	5	7.2
Travel to and from register	TR	6	5.4
Work counting items	AC	7	5.2
Travel to shelves or cupboards for material	TEM	8	5.0
Work with balance and accessories	AB	9	4.3
Work getting down items	AD	10	4.1
Work compounding	ACO	11	3.9
Work on drugs	ADR	12	3.3
Talking to customers	V	13	2.8
Work at cash register	CA	14	2.3
Inspection of shelves	IS	15	2.0
Inspection of drug containers or contents	IC	16	1.3
Travel to shelves or cupboards to put away	TLM	17	1.1
Work putting up items	U	18	0.6
Work with liquid measures	ALI	19	0.2
Total			100.0

* From Mundel, *Motion and Time Study: Principles and Practice.*

Multiman charts: Table 5.18.

Micromotion study. Check lists are available only for commonly used breakdowns. The analyst working with unusual breakdowns will have to organize his own data.

Therbligs and simo charts: Table 5.19.

Memomotion study. If appropriate check lists are available, they should be used. In most cases, however, new check lists, related to the criteria of success being used, must be created when new breakdowns are used. See Table 5.12.

5.3.8 Forms and Illustrations

Because of space requirements, a simple illustration has been given in each case. As the complexity of the problem increases, the advantage of the graphic technique as an aid to understanding also increases.

5.3.8.1 For Establishing Aim. The forms given are only samples. In practice, many versions exist.

Possibility list and guide. A possibility list for the manufacture of an armature arm (see Fig. 5.2) is given in Fig. 5.3. Here the possibilities and the classes of change affected are listed, but not the consequences. A form for detailing the consequences is shown in Fig. 5.4.

Activity chart: Fig. 5.5. Work distribution chart: Fig. 5.6.

Machine load chart: This type of chart is similar to that shown in Fig. 5.5, but the machine rather than the person is the subject.

Functional form analysis chart: Fig. 5.7.

5.3.8.2 For Analysis. Sample forms for both product and man analysis techniques are given. In practice, many variations exist.

For Product Analysis. Only simple cases are shown, in the form in which they would appear in practice.

Process chart—product analysis: A simple process chart—product analysis of a materials handling problem is shown in Fig. 5.8. An improved method is shown in Fig. 5.9. A more complex type of chart is shown in Fig. 5.10. This type is sometimes useful for the entire manufacture of a complex product. In

TABLE 5.11. *CHECK LIST FOR POSSIBILITY GUIDE**

5. Can a slightly different raw material be ordered or can the same material be ordered in a form that would be more advantageous? Can we change:
   a. Shape
   b. Size
   c. Packaging
   d. Quantity packaged together
   e. Material
   f. Amount of processing done by supplier
   g. Color
   h. Finish
   i. Any other specification
   j. The product so as to make any material or auxiliary material unnecessary
4. Can the product be made, sold, or sent out in a more advantageous form? Can we:
   a. Modify design
   b. Pack differently
   c. Change finish
   d. Change weight
   e. Change tolerances
3. a. Can we do the different jobs along the route between receiving and shipping in a different order?
   b. Is any step unnecessary?
      1. What does it accomplish?
      2. Why is it done?
      3. What would happen if it were not done?
   c. Can we combine any steps?
   d. Can we advantageously break any job into two or more separate operations?
2. a. Can any new tools or equipment or a change in the workplace make any job in the sequence easier? (This is almost always possible.)
   b. Can any tool or equipment be eliminated advantageously?
   c. Can any two tools be combined?
1. Can a new motion pattern make any job in the sequence easier? (This is almost invariably true.) Specific suggestions are usually more easily made after the method analyst is more familiar with the man-analysis techniques; hence, at this point, a mere list of the possible jobs that may be looked into or some rough suggestions will probably be all the student is capable of. (The analyst also tries to eliminate motions. In actual practice, as will be evident later, this part of the analysis may be done in more detailed fashion.)

* From Mundel, *Motion and Time Study: Principles and Practice.*

preparing such a chart, a large ruled sheet or even a blank sheet of paper would be used in preference to a form. In other cases, a preprinted form such as that shown in Fig. 5.11 may be used. In the illustration given, a paper-work problem is being followed. Although paper work is commonly analyzed in this manner, it is suggested that the method of Fig. 5.14 is preferable for paper work.

Flow diagram: The original flow diagram of the process charted in Fig. 5.8 is shown in Fig. 5.12, and the improved method is shown in Fig. 5.13.

Procedure analysis chart: A simple procedure analysis chart showing the activity with forms and of the personnel concerned with the forms is shown in Fig. 5.14. Although the charts for complex procedures may become rather large, they greatly facilitate understanding, study, discussion, and improvement. Indeed, without charting, some procedures defy understanding.

FOR MAN ANALYSIS. Samples of the various charts are given as they would appear in practice. In actual use, of course, many varieties appear.

Process chart—man analysis: An original process chart—man analysis is given in Fig. 5.15. The improved version shown in Fig.

TABLE 5.12. *LIST OF CRITERIA OF PREFERENCE**

1. Greater economy of operation:
   a. Through less time for the job.
   b. Through less effort required by the job. (These first two are often synonymous if only manual methods are altered.)
   c. Through less scrap.
   d. Through less material in the product.
   e. Through a change in amount of indirect labor.
   f. Through the use of less expensive equipment.
   g. Through the use of fewer people.
   h. Through the use of less critical skills.
   i. Through the use of less critical machines.
   j. Through the use of less space.
   k. Through the use of less in-production time.
2. Better product in respect to function or salability:
   This is a long-range aspect of number 1, although it may involve greater cost of operation on a particular job. It may, however, in many situations be accomplished with less cost and is then doubly desirable.
3. Better material control:
   This is also an economic objective as it relates to inventory cost, scheduling and control functions, and customer service.

* For industries or activities other than manufacturing these criteria may need to be restated; i.e., in department store work, as with other service occupations, customer service is of extreme importance; in medical work, patient safety assumes increased importance, etc. Adapted from Mundel, *Motion and Time Study: Principles and Practice.*

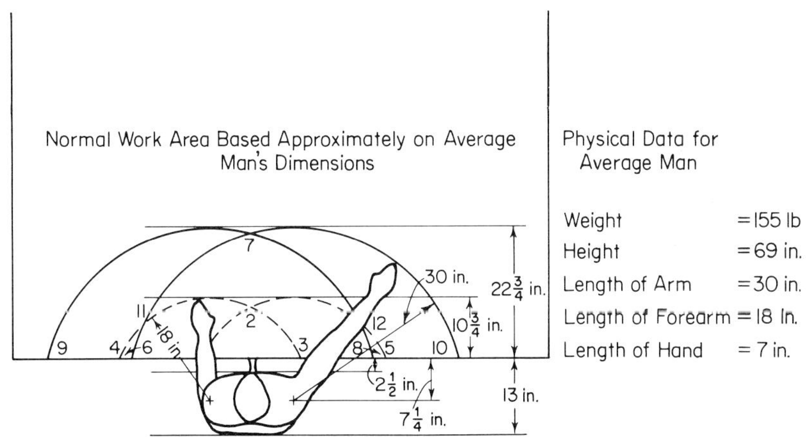

Fig. 5.1 *Human work areas. Preferred areas:*

1. *For work requiring visual direction:*
   Best: 1–2–3
   Next: 2–11–7–12
   Next: 6–7–8
   Next: 4–11–7–12–5
2. *For work with low visual requirements:*
   Best: 4–11–2–12–5
   Next: 9–7–10

In all cases, attempt to keep within 9–7–10 to avoid trunk movements.
From Mundel, Motion and Time Study: Principles and Practice, *p. 155.*

TABLE 5.13. *CHECK LIST FOR PROCESS CHART—PRODUCT ANALYSIS**

Basic Principles

A. Reduce number of steps
B. Arrange steps in best order
C. Make steps as economical as possible
D. Reduce handlings
E. Combine steps if economical
F. Shorten moves
G. Provide most economical means for moving
H. Cut in-process inventory to workable minimum
I. Use minimum number of control points at most advantageous places

1. Can any step be eliminated?
   a. as unnecessary (Ask: Why is it done?)
   b. by new equipment (Ask: Why is present equipment used?)
   c. by changing the place where it is done or kept (Ask: Why is it done there?)
   d. by changing the order of work (Ask: Why is it done in its present order?)
   e. by changing the product design (Ask: Why is it done as it is?)
   f. by changing the specifications of the incoming supply (Ask: Why is it ordered in its present form or used at all?)
2. Can any step be combined with another?
   Are there any possible changes that would make this feasible in
   a. workplace
   b. equipment
   c. order of steps
   d. product design
   e. specification of supply or any raw material
3. Can the steps be rearranged so as to make any shorter or easier?
4. Can any step be made easier?
   (If this looks like a possibility, make further detailed analysis of this step.)

* From Mundel, *Motion and Time Study: Principles and Practice.*

TABLE 5.14. *CHECK LIST FOR PROCEDURE ANALYSIS**

1. Each step should be necessary. If not, eliminate it.
2. Each step should have a reason for being by itself. Can it be combined?
3. Each step should have an ideal place in sequence. Where should it be?
4. Each step should be as easy as possible.
5. Each form should have a real purpose. Verify it. Is the form necessary? Can it be eliminated, combined with another form, or replaced by a copy of another form?
6. Files should have a purpose. Do they? Avoid duplication. Avoid excess files. File by subject used to enter files. Check on manner of use.
7. If form is finally destroyed, it should never, in some cases, have been originated.
8. Information going from one form to another suggests more copies in the first place. Are all information take-offs and readings necessary? If so, which are going to be given priority for design?
9. Are all copies getting equal use? Sharing the load may speed up the procedures.
10. Does someone sign all copies? How can this be avoided? Signers are busy.
11. Is there excess checking?
12. Where is the best place to check? Calculate the risk.
13. What would happen if a form was lost?
14. What equipment might help the job? (See commercial catalogs.)
15. Does one person have too much of the procedure?
16. Are as many steps as possible given to the lowest classification personnel applicable?
17. Can travel of forms be advantageously reduced?
18. Can the form be kept in action, out of file baskets?

* From Mundel, *Motion and Time Study: Principles and Practice.*

TABLE 5.15. *CHECK LIST FOR PROCESS CHART—MAN ANALYSIS**

Basic Principles

A. Eliminate all possible steps
B. Combine steps
C. Shorten steps
D. Place in best sequence
E. Make each step as economical as possible

1. Can any operation be eliminated, combined, shortened, or made easier?
   a. as unnecessary
   b. by changing the order of work
   c. by new or different equipment
   d. by changes in the layout; by grouping equipment better
   e. by changing the form of the product sent out
   f. by more knowledge on part of the worker
2. Can any movement be eliminated, combined, shortened, or made easier?
   a. by leaving out operations
   b. by changing the places where things are kept
   c. by shifting some operations to another job into which they fit more conveniently
   d. by changing the layout
   e. by changing equipment
   f. by changing the order of work
   g. by conveyors (Make sure they are economical.)
3. Can delays be eliminated, combined, or shortened?
   a. by changing the order of work
   b. by changing the layout
   c. by new or different equipment
4. Can countings or inspections be eliminated, combined, shortened, or made easier?
   a. Are they really necessary; what happens after they are done and the information obtained?
   b. Do they provide unnecessary duplication?
   c. Can they be performed more conveniently by another person?
   d. Are they done at the best point in the sequence?
   e. Can sample inspection or statistical control be used?
5. Can any step be made safer?
   a. by changing the order of work
   b. by new or different equipment
   c. by changing the layout

* From Mundel, *Motion and Time Study: Principles and Practice*.

5.16 resulted from applying the check list in Table 5.15.

Four general situations are encountered in applying process charts—man analysis.

1. *The work has a single repeated cycle.* Here a cycle is defined as all the steps necessary to bring a unit of product to the state of completion typical of the operation, or as all the steps typical of a single performance of the task. A single cycle is charted.

2. *The work is cyclic, but includes several subcycles performed with different frequencies.* For instance, the worker may perform subcycle A on each part and then subcycle B for ten parts together. The chart will show one performance of each subcycle and will also indicate its frequency.

3. *The work varies from cycle to cycle.*

A. Some variations result from operator habit rather than from the inherent nature of

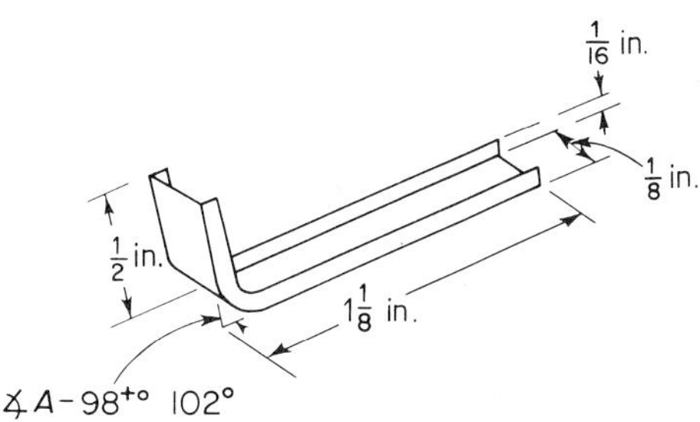

Fig. 5.2 *Armature arm. From Mundel,* Motion and Time Study: Principles and Practice, *p. 38.*

Type of Chart Preliminary Possibility Guide

Method	Present process	Machine No.	——
Operation	Mfg.	Operation No.	——
Criterion	Lower cost	Part No.	124 R
Part name	Armature Arm	Chart by	L. Edmond
Operator	——	Date charted	2/16/46

Suggestion Number	Class of Change	Description	Other classes affected
1	5	Use sheet stock	4, 3, 2, 1
2	5	Purchase formed to size and shape	3, 2, 1
3	4	Redesign relay to allow more tolerance in arm eliminating inspection	3, 2, 1
4	3	Inspect at punch press as part of operation	2, 1
5	2	Dual inspection fixture, electric response	1
6	1	Tip fixture to make easier, weigh to count at inspection	
7	1	Distribute work to both hands at inspection	

If more space is needed paste additional sheet on here.

Fig. 5.3 *Possibility list. From Mundel,* Motion and Time Study: Principles and Practice, *p. 39.*

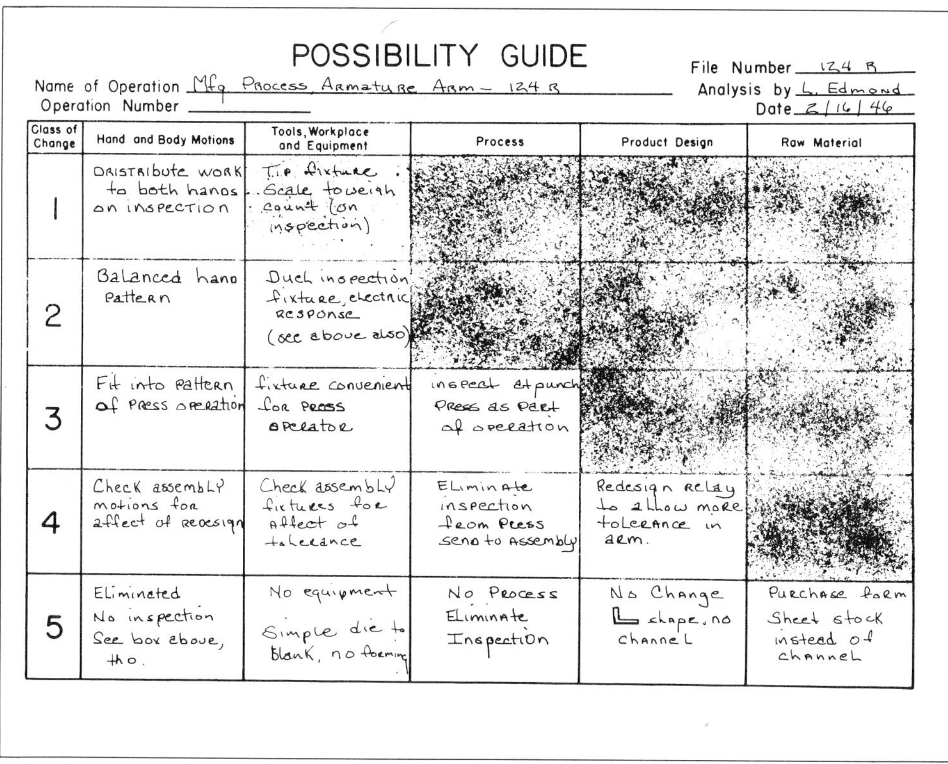

POSSIBILITY GUIDE

Name of Operation Mfg. Process, Armature Arm – 124 B
Operation Number
File Number 124 B
Analysis by L. Edmond
Date 2/16/46

Class of Change	Hand and Body Motions	Tools, Workplace and Equipment	Process	Product Design	Raw Material
1	Distribute work to both hands on inspection	Tip fixture : Scale to weigh count (on inspection)			
2	Balanced hand pattern	Dual inspection fixture, electric response (see above also)			
3	Fit into pattern of press operation	fixture convenient for press operator	inspect at punch press as part of operation		
4	Check assembly motions for affect of redesign	Check assembly fixtures for affect of tolerance	Eliminate inspection from press send to assembly	Redesign relay to allow more tolerance in arm.	
5	Eliminated No inspection See box above, tho.	No equipment Simple die to blank, no forming	No Process Eliminate Inspection	No Change L shape, no channel	Purchase form Sheet stock instead of channel

Fig. 5.4 *Possibility guide. From Mundel,* Motion and Time Study: Principles and Practice, *p. 40.*

the job (as in some maintenance work). Consequently, the analyst may have to plot several cycles in order to get enough material from which to develop a preferable work pattern.

B. Other variations are inherent in the job. Each performance may differ in detail but not in general pattern. In such a case, sample cycles are drawn to indicate the general pattern and the varying details. The factors controlling the variation are weighted approximately.

In either A or B, the application of the more complex memomotion study may be useful.

4. *The work is non-cyclic.* This is usually true of supervisory and similar activities. Here the chart may lead only to general suggestions for improved planning.

Man-flow diagram. A man-flow diagram corresponding to the job shown in Fig. 5.15 is given in Fig. 5.17.

Operation chart. A simple operation chart for the original method of drilling hinge channels is shown in Fig. 5.18. The workplace is shown in Fig. 5.19 and the hinge channel itself in Fig. 5.20.

As with process charts—man analysis, four types of jobs, with the following characteristics, may be encountered:

1. The work has a single repeated cycle.
2. The work is cyclic, but there are several subcycles performed with different frequency.
3. The work varies from cycle to cycle.
4. The task is such that there is no regular cycle.

The method of handling these different types of jobs is essentially similar to the procedure used with process charts—man analysis, except that the work of the individual body members is studied rather than the worker as a whole.

Multiple activity. Sample charts are shown for the major types. Other forms may appear

G. Shay — NAME
24 March 1953 — DATE
Mngt. Office — UNIT
GS-2 — GRADE

ACTIVITY CHART

Nature of activity	Remarks	Time began	Work units produced	Elap. time (min.)
Soundscriber transcription	Short letters	8:00	2 letters	30
" "	Type	8:30	1 report	120
Break		10:30		10
Run ditto	new stencils	10:40	2 stencils 50 each	10
Proof from Soundscriber	Report from 10:30	10:50	1 report	30
Type from handwrit. copy	Letters	11:20	2 letters	40
Lunch		12:00		60
Type from hand. copy	Letters	1:00	*	120
Break		3:00		10
Type from hand copy	Letters * (con't)	3:10	15 letters	110
Finish		5:00		
Summary				
1. Type letters from soundscriber			2	30
2. Type and proof report from soundscriber			1	150
3. Type letters from handwritten copy			17	270
4. Run ditto			2 stencils 50 each	10
5. Breaks			2	20
	TOTAL			480

Fig. 5.5 *Activity chart.*
From "Techniques of Work Simplification," Department of Army, Pamphlet 20–300, June 1951.

ORGANIZATIONAL UNIT CHARTED: Assignment Section

WORK DISTRIBUTION CHART

(DA Pamphlet 20-300)

APPROVED BY: | DATE:

[X] EXISTING ORGANIZATION [ ] RECOMMENDED ORGANIZATION

CHARTED BY: Major Wilson

				NAME: J. K. Wilson			NAME: Paul Smith			NAME: Wm. Chase			NAME: Miss Hall			NAME: Miss Brown			NAME: Miss Jones		
				POSITION: Section Chief	GRADE: Major		POSITION: Chief Clerk	GRADE: Sgt		POSITION: Strength Analyst	GRADE: GS 7		POSITION: Order Clerk	GRADE: GS 5		POSITION: Steno	GRADE: GS 4		POSITION: Clerk-typist	GRADE: GS 3	
ACTIVITY NUMBER	ACTIVITY	WORK[1] COUNT	HOURS PER WEEK	TASKS	WORK[1] COUNT	HOURS PER WEEK	TASKS	WORK[1] COUNT	HOURS PER WEEK	TASKS	WORK[1] COUNT	HOURS PER WEEK	TASKS	WORK[1] COUNT	HOURS PER WEEK	TASKS	WORK[1] COUNT	HOURS PER WEEK	TASKS	WORK[1] COUNT	HOURS PER WEEK
1.	Assignment of Enlisted Personnel	100	53	Allocates replacements to units Allocates levies from higher authority	100	7 5	Assembles strength and requirements data		14	Recommends allocations		7	Prepares & proofreads transfer orders	100	13	Proofreads orders	100	7			
2.	Assignment of Officer Personnel	50	49	Studies requirements against authorizations Review assignments allocations Approves orders	 50	4 5 3	Checks rosters and records for completeness		8	Recommends allocations		4	Prepares & proofreads transfer orders	50	9	Proofreads orders	50	5	Drafts transfer orders	50	11
3.	General Inquiry and Information Service		34	Conference with commanders and staff officers Review and sign letters of inquiry		3 1	Answers routine inquiries by phone	25	6	Presents studies to other branch officials		5	Routine inquiries on status of orders	10	5	Prepares records of branch conferences Dictation & transcription of inquiry letters		6 5	Answers routine inquiries	7	3
4.	Specific Case Studies & Requirements Analysis		48	Continuing survey of long-range requirements Review case studies		3 2	Reviews Case Studies		3	Statistical Analysis of Unit Strength Studies scarce MOS categories		12 6	Furnish data on personnel or orders		6	Dictation and transcription of case studies		6	Types summaries of requirements		8
5.	Administration and Supervision		39	Conference with branch personnel Prepare budget Staff Conference		1 4 2	Time and Attendance records Prepare Administrative reports	6 3	4 2	Staff conference		2	Staff conference Prepare Weekly Summaries of transfers		2 3	Records of Staff Conferences		4	Types Weekly transfer summaries Types Administrative reports		11 4
6.	Miscellaneous Activities		17				Analyzes reports from other sections		3	Security Inspections		4	Special Messenger Service		2	Phone Directory Office SOP, etc.		5	Community Chest, Blue Cross, etc.		3
	TOTALS (Manhours)		240			40			40			40			40			40			40

[1] Work Count data is optional.

ANALYSIS: WHAT TAKES THE MOST TIME?...IS THERE MISDIRECTED EFFORT?...ARE SKILLS USED PROPERLY?...ARE THERE TOO MANY UNRELATED TASKS?...ARE TASKS SPREAD TOO THINLY?...IS WORK DISTRIBUTED EVENLY?

NOTE: If additional columns are required, cut off required number from another blank form and attach by pasting to this edge.

DA AGO FORM 673, 1 MAY 51. REPLACES OCS FORM 511, WHICH MAY BE USED.

① Initial preparation

Figure 3. Work Distribution Chart.

949408 O - 51 (Face p. 6) No. 1

Fig. 5.6 *Work distribution chart.*
From "Techniques of Work Simplification," Department of Army, Pamphlet 20–300, June 1951.

TABLE 5.16. *CHECK LIST FOR OPERATION CHART**

Basic Principles

A. Reduce total steps to minimum
B. Arrange in best order
C. Combine steps where feasible
D. Make each step as easy as possible
E. Balance the work of the hands
F. Avoid the use of the hands for holding

1. Can a suboperation be eliminated?
   a. as unnecessary
   b. by a change in the order of work
   c. by a change of tools or equipment
   d. by a change of layout of the workplace
   e. by combining tools
   f. by a slight change of material
   g. by a slight change in product
   h. by a quick-acting clamp on jig, if jigs are used
2. Can a movement be eliminated?
   a. as unnecessary
   b. by a change in the order of work
   c. by combining tools
   d. by a change of tools or equipment
   e. by a drop disposal of finished material
      (The less exact the release requirements, the faster the release.)
3. Can a hold be eliminated? (Holding is extremely fatiguing.)
   a. as unnecessary
   b. by a simple holding device or fixture
4. Can a delay be eliminated or shortened?
   a. as unnecessary
   b. by a change in the work that each body member does
   c. by balancing the work between the body members
   d. by working simultaneously on two items
      (Slightly less than double production is possible with the typical person.)
   e. by alternating the work, each hand doing the same job, but out of phase
5. Can a suboperation be made easier?
   a. by better tools
      (Handles should allow maximum flesh contact without sharp corners for power; easy spin, small diameter for speed on light work.)
   b. by changing leverages
   c. by changing positions of controls or tools
      (Put into normal work area—Fig. 5.1)
   d. by better material containers
      (Bins that permit slide grasp of small parts are preferable to bins that must be dipped into.)
   e. by using inertia where possible
   f. by lessening visual requirements
   g. by better workplace heights
      (Keep workplace height below elbow.)
6. Can a movement be made easier?
   a. by a change of layout, shortening distances (see Fig. 5.1)
      (Place tools and equipment as near place of use and as nearly in position of use as possible.)
   b. by changing direction of movements
      (Optimum angle of workplace for light knobs, key switches, and hand-wheels is probably 30° and certainly between 0° to 45° to plane perpendicular to plane of front of operator's body.—Unpublished Purdue Research)
   c. by using different muscles
      Use the first muscle group in this list that is strong enough for the task:
      (See Fig. 5.1 for visual items that may affect this order.)

* From Mundel, *Motion and Time Study: Principles and Practice.*

TABLE 5.16. *(Continued)*

1. finger (not desirable for steady load or highly repetitive motions)
2. wrist
3. forearm
4. upper arm
5. trunk (for heavy loads shift to large leg muscles)

d. by making movements continuous rather than jerky (see Fig. 5.1)

7. Can a hold be made easier?
   a. by shortening its duration
   b. by using stronger muscle groups, such as the legs, with foot-operated vises

in practice. In extremely complex cases, large sheets of graph or tracing paper may be preferable to a ruled form.

Man and machine operation charts: A chart showing the original method of performing an operation on a centerless grinder is given in Fig. 5.21.

Man and machine operation time chart:

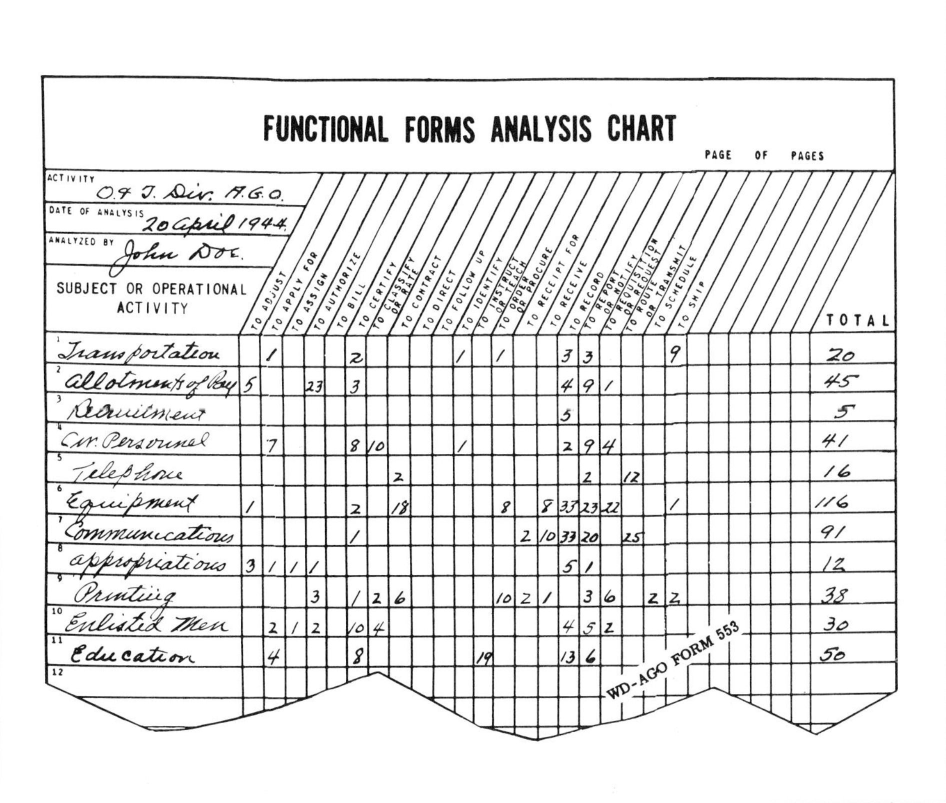

FUNCTIONAL FORMS ANALYSIS CHART

PAGE OF PAGES

ACTIVITY O. & T. Div. A.G.O.

DATE OF ANALYSIS 20 April 1944.

ANALYZED BY John Doe.

	SUBJECT OR OPERATIONAL ACTIVITY	TO ADJUST	TO APPLY FOR	TO ASSIGN	TO AUTHORIZE	TO BILL	TO CERTIFY	TO CLASSIFY OR RATE	TO CONTRACT	TO DIRECT	TO FOLLOW UP	TO IDENTIFY	TO INSTRUCT OR TEACH	TO ORDER OR PROCURE	TO RECEIPT FOR	TO RECEIVE	TO RECORD	TO REPORT OR NOTIFY	TO REQUISITION OR REQUEST	TO ROUTE OR TRANSMIT	TO SCHEDULE	TO SHIP							TOTAL
1	Transportation		1				2					1		1			3	3				9							20
2	Allotments of Pay	5			23		3										4	9	1										45
3	Recruitment																5												5
4	Civ. Personnel		7				8	10				1					2	9	4										41
5	Telephone								2									2		12									16
6	Equipment	1					2		18					8		8	33	23	22			1							116
7	Communications						1								2	10	33	20		25									91
8	appropriations	3	1	1	1												5	1											12
9	Printing				3		1	2	6					10	2	1		3	6		2	2							38
10	Enlisted Men		2	1	2		10	4									4	5	2										30
11	Education		4				8						19				13	6											50
12																													

WD-AGO FORM 553

Fig. 5.7 *Functional forms analysis chart.*
From "Techniques of Work Simplification," Department of Army, Pamphlet 20–300, June 1951.

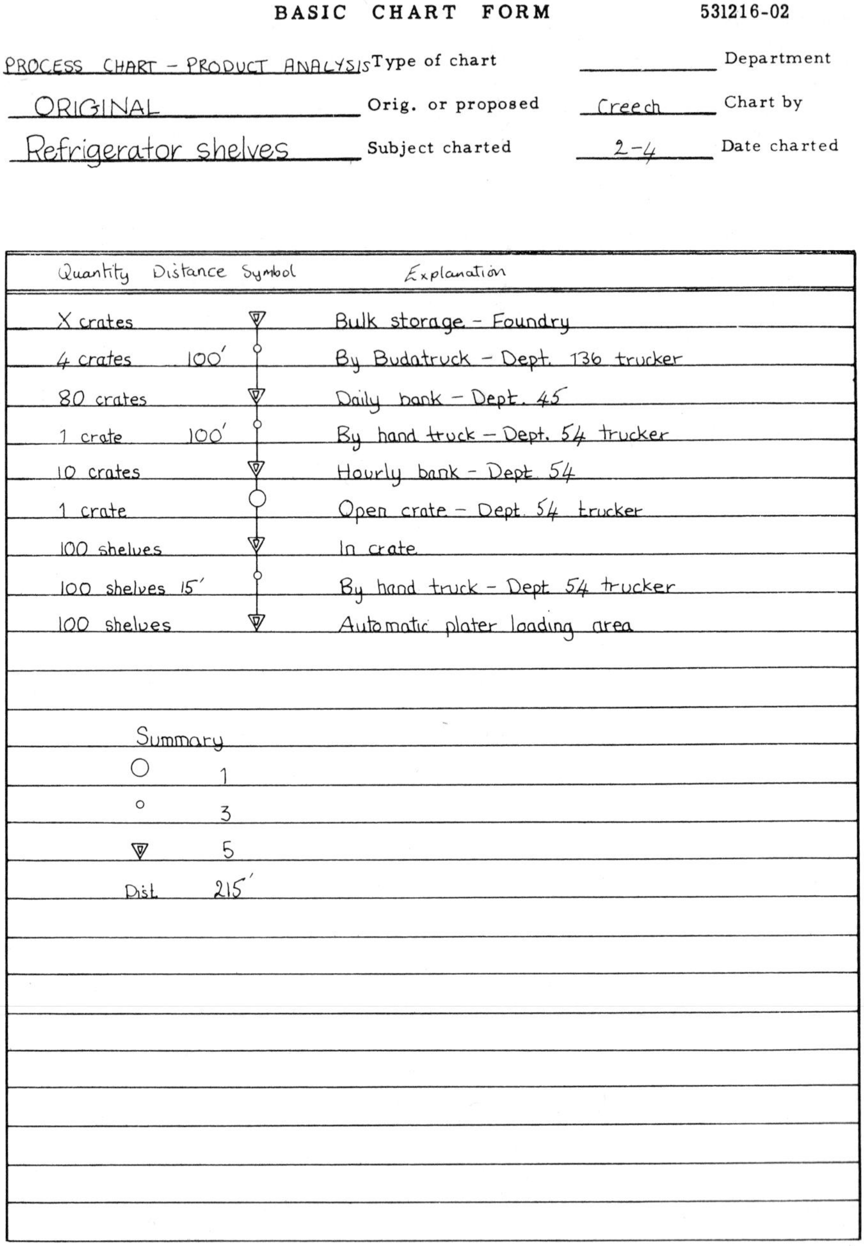

BASIC CHART FORM 531216-02

PROCESS CHART – PRODUCT ANALYSIS	Type of chart		Department
ORIGINAL	Orig. or proposed	Creech	Chart by
Refrigerator shelves	Subject charted	2-4	Date charted

Quantity	Distance	Symbol	Explanation
X crates		▽	Bulk storage – Foundry
4 crates	100′	○	By Budatruck – Dept. 136 trucker
80 crates		▽	Daily bank – Dept. 45
1 crate	100′	○	By hand truck – Dept. 54 trucker
10 crates		▽	Hourly bank – Dept. 54
1 crate		◯	Open crate – Dept. 54 trucker
100 shelves		▽	In crate
100 shelves	15′	○	By hand truck – Dept. 54 trucker
100 shelves		▽	Automatic plater loading area

Summary	
◯	1
○	3
▽	5
Dist.	215′

Fig. 5.8 *Process chart—product analysis for original method of handling refrigerator food shelves from bulk storage to plating department.*
From Mundel, Motion and Time Study: Principles and Practice, *p. 65.*

The chart in Fig. 5.22 shows an improved method of performing the same operation shown in Fig. 5.21.

Man and machine process time chart: The chart in Fig. 5.23 shows the operation of wind, stretch, chop, and peel radio-tube grids with the aid of a grid lathe.

Multiman and machine process chart: The

chart in Fig. 5.24 shows the original method for a three-man crew cutting studs on a cutoff saw.

Multiman and machine time charts: The chart in Fig. 5.25 shows an improved method for performing the same operation shown in Fig. 5.24.

For micromotion study. Many of the charts

BASIC CHART FORM 531216-02

PROCESS CHART-PRODUCT ANALYSIS	Type of chart		Department
Proposed	Orig. or proposed	Creech	Chart by
Refrigerator shelves	Subject charted	2-4	Date charted

Quantity	Distance	Symbol	Explanation
X crates		▽	Bulk storage – Foundry
1 crate	15′	○	By hand truck – Dept. 54 trucker
1 crate		◯	Open crate – Dept. 54 trucker
100 shelves		▽	In crate
100 shelves	60′	○	By hand truck – Dept 54 trucker
100 shelves		▽	Automatic plater loading area

Summary and Recap

	Original	Proposed	Saved
◯	1	1	0
○	3	2	1
▽	5	3	2
Dist.	215′	75	140

Fig. 5.9 *Process chart; product analysis for improved method of process of Fig. 5.8. From Mundel,* Motion and Time Study: Principles and Practice, *p. 67.*

Type of Chart PROCESS CHART – PRODUCT ANALYSIS

Method ORIGINAL Machine No. ~

Operation MANUFACTURE Operation No. ~

Part No. J.U. 5873

Part name UNIVERSAL JOINT Chart by J. Reich

Operator ~ Date charted Nov 19

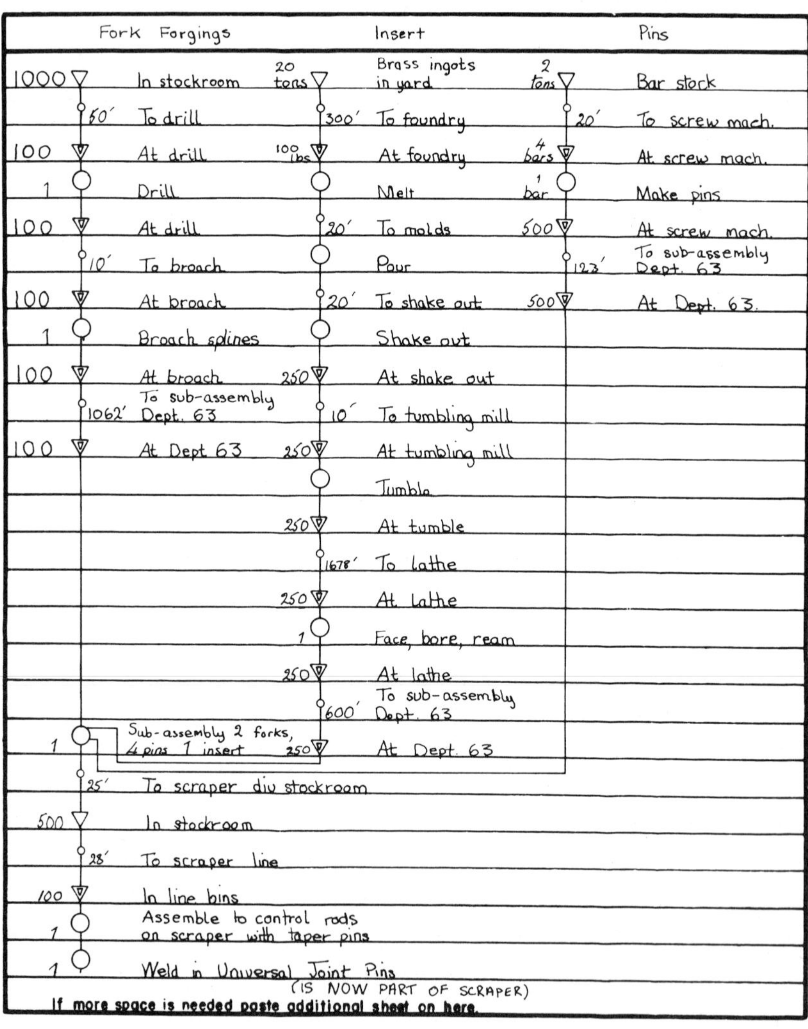

Fork Forgings		Insert		Pins	
1000	In stockroom	20 tons	Brass ingots in yard	2 tons	Bar stock
60′	To drill	300′	To foundry	20′	To screw mach.
100	At drill	100 lbs	At foundry	4 bars	At screw mach.
1	Drill		Melt	1 bar	Make pins
100	At drill	20′	To molds	500	At screw mach.
10′	To broach		Pour	123′	To sub-assembly Dept. 63
100	At broach	20′	To shake out	500	At Dept. 63.
1	Broach splines		Shake out		
100	At broach	250	At shake out		
1062′	To sub-assembly Dept. 63	10′	To tumbling mill		
100	At Dept 63	250	At tumbling mill		
			Tumble		
		250	At tumble		
		1678′	To lathe		
		250	At Lathe		
		1	Face, bore, ream		
		250	At lathe		
		600′	To sub-assembly Dept. 63		
1	Sub-assembly 2 forks, 4 pins 1 insert	250	At Dept. 63		
25′	To scraper div stockroom				
500	In stockroom				
28′	To scraper line				
100	In line bins				
1	Assemble to control rods on scraper with taper pins				
1	Weld in Universal Joint Pins (IS NOW PART OF SCRAPER)				

If more space is needed paste additional sheet on here.

Fig. 5.10 *Process chart—product analysis for proposed method of manufacture of universal joint. From Mundel,* Motion and Time Study: Principles and Practice, *p. 90.*

previously described may be made with the aid of micromotion study. Three steps are involved: filming, film analysis, and charting.

Simo charts: A simo chart, with therbligs, for a simple assembly operation, is shown in Fig. 5.26. The parts are shown in Fig. 5.27. The form used to record the film data, including the workplace layout, is shown in Fig.

FLOW PROCESS CHART — NUMBER 1 — PAGE NO. 1 — NO. OF PAGES 2

PROCESS: Requisition for OVM Kits

☐ MAN OR ☒ MATERIAL: Requisition form

CHART BEGINS: Receipt of requisition in Sup. off.

CHART ENDS: Dispatch of procurement action

CHARTED BY: Lt. John Jones

DATE: 25 Oct 1950

ORGANIZATION: Supply Office

SUMMARY						
ACTIONS	PRESENT		PROPOSED		DIFFERENCE	
	NO.	TIME	NO.	TIME	NO.	TIME
○ OPERATIONS	11	120				
⇨ TRANSPORTATIONS	9					
☐ INSPECTIONS	3	40				
D DELAYS	7	250				
▽ STORAGES	1					
DISTANCE TRAVELLED (Feet)	1015					

DETAILS OF ☒ PRESENT ☐ PROPOSED METHOD

	DETAILS	OPERATION / TRANSPORTATION / INSPECTION / DELAY / STORAGE	DISTANCE IN FEET	QUANTITY	TIME	WHAT?	WHERE?	WHEN?	WHO?	HOW?	NOTES	ELIMINATE	COMBINE	SEQUENCE	PLACE	PERSON	IMPROVE
1	Requisition for OVM Kits arrives at supply clerk's desk	○⇨☐D▽				x				x	Came from maintenance foreman						
2	Requisition entered in master register	○⇨☐D▽			10		x	x			Why?		x			x	
3	In outgoing basket of supply clerk	○⇨☐D▽			10												
4	Requisition sent to file clerk	○⇨☐D▽	75														
5	Entered in file register	○⇨☐D▽			5						Why?	x	x			x	
6	File searched for outstanding requisitions	○⇨☐D▽							x		This action was taken to make certain that the kits are not already on order						
7	Outstanding order removed from file	○⇨☐D▽			15												
8	Charge-out slip prepared for order file	○⇨☐D▽															
9	Requisition and previous order in outgoing basket	○⇨☐D▽			25												
10	To requisition clerk	○⇨☐D▽	120					x									
11	In incoming basket of requisition clerk	○⇨☐D▽			40												
12	Req. checked against authorized and on hand equipment	○⇨☐D▽			15			x		x	Could it be checked earlier?						
13	Requisition to correspondence clerk	○⇨☐D▽	90					x								x	
14	Acknowledgement prepared	○⇨☐D▽			20		x	x	x					x	x	x	
15	Acknowledgement and requisition file to supply officer	○⇨☐D▽	150								By messenger. May send work count here		x				
16	In incoming basket	○⇨☐D▽			60												
17	Acknowledgement signed	○⇨☐D▽			5						Why does he sign back?					x	x
18	Requisition to procurement clerk	○⇨☐D▽	160														
19	In incoming basket	○⇨☐D▽			15												
20	Procurement clerk prepares procurement order	○⇨☐D▽			45												
21	Procurement clerk checks source of supply	○⇨☐D▽			10			x	x								

DA AGO FORM 1 MAY 51 684 — REPLACES OCS FORM 391, 1 FEB 1951, WHICH MAY BE USED.

Fig. 5.11 *Process chart—product analysis using preprinted form.*
From "Techniques of Work Simplification," Department of Army, Pamphlet 20–300, June 1951.

TABLE 5.17. *CHECK LIST FOR MAN AND MACHINE CHARTS**

Basic Principles

A. Eliminate steps
B. Combine steps
C. Rearrange in best fashion
D. Make each step as easy as possible
E. Raise percentage of cycle of machine running time to maximum
F. Reduce machine loading and unloading to minimum
G. Raise machine speed to economic limit

(The first seven questions that follow are similar to those used with operation charts, where more detail was given; hence, the reader is also referred back to them. The bare questions are given here so as to provide, at one place, all of the check-list items to be used.)

1. Can a suboperation be eliminated?
   a. as unnecessary
   b. by a change in the order of work
   c. by a change of tools or equipment
   d. by a change in layout of the workplace
   e. by combining tools
   f. by a slight change of material
   g. by a slight change in product
   h. by a quick-acting clamp on the jigs or fixtures
2. Can a movement be eliminated?
   a. as unnecessary
   b. by a change in the order of work
   c. by combining tools
   d. by a change of tools or equipment
   e. by a drop disposal of finished material
3. Can a hold be eliminated? (Holding is extremely fatiguing.)
   a. as unnecessary
   b. by a simple holding device or fixture
4. Can a delay be eliminated or shortened?
   a. as unnecessary
   b. by a change in the work each body member does
   c. by balancing the work between the body members
   d. by working simultaneously on two items
   e. by alternating the work, each hand doing the same job, but out of phase
5. Can a suboperation be made easier?
   a. by better tools
   b. by changing leverages
   c. by changing positions of controls or tools
   d. by better material containers
   e. by using inertia where possible
   f. by lessening visual requirements
   g. by better workplace heights
6. Can a movement be made easier?
   a. by a change of layout, shortening distances
   b. by changing the direction of movements
   c. by using different muscles (see Fig. 5.1)
      Use the first muscle group in this list that is strong enough for the task:
      1. finger
      2. wrist
      3. forearm
      4. upper arm
      5. trunk
   d. making movements continuous rather than jerky
7. Can a hold be made easier?
   a. by shortening its duration
   b. by using stronger muscle groups, such as the legs, with foot-operated vises

* From Mundel, *Motion and Time Study: Principles and Practice.*

TABLE 5.17. (*Continued*)

8. Can the cycle be rearranged so that more of the handwork can be done during running time?
   a. by automatic feed
   b. by automatic supply of material
   c. by change of man and machine phase relationship
   d. by automatic power cut-off at completion of cut or in case of tool or material failure
9. Can the machine time be shortened?
   a. by better tools
   b. by combined tools
   c. by higher feeds or speeds

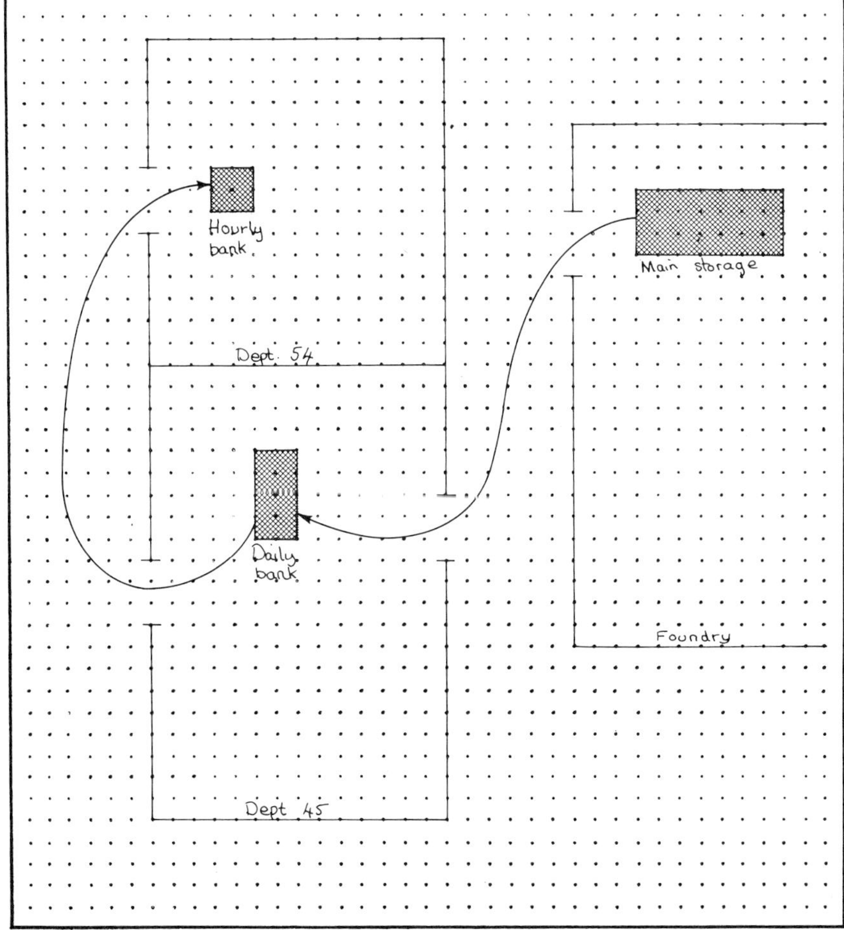

Fig. 5.12 *Flow diagram for process charted in Fig. 5.8.*
From Mundel, Motion and Time Study: Principles and Practice, *p. 66.*

Sketch of PROPOSED ROUTE USED FOR HANDLING REFRIGERATOR FOOD SHELVES

Scale – Each square = NO SCALE

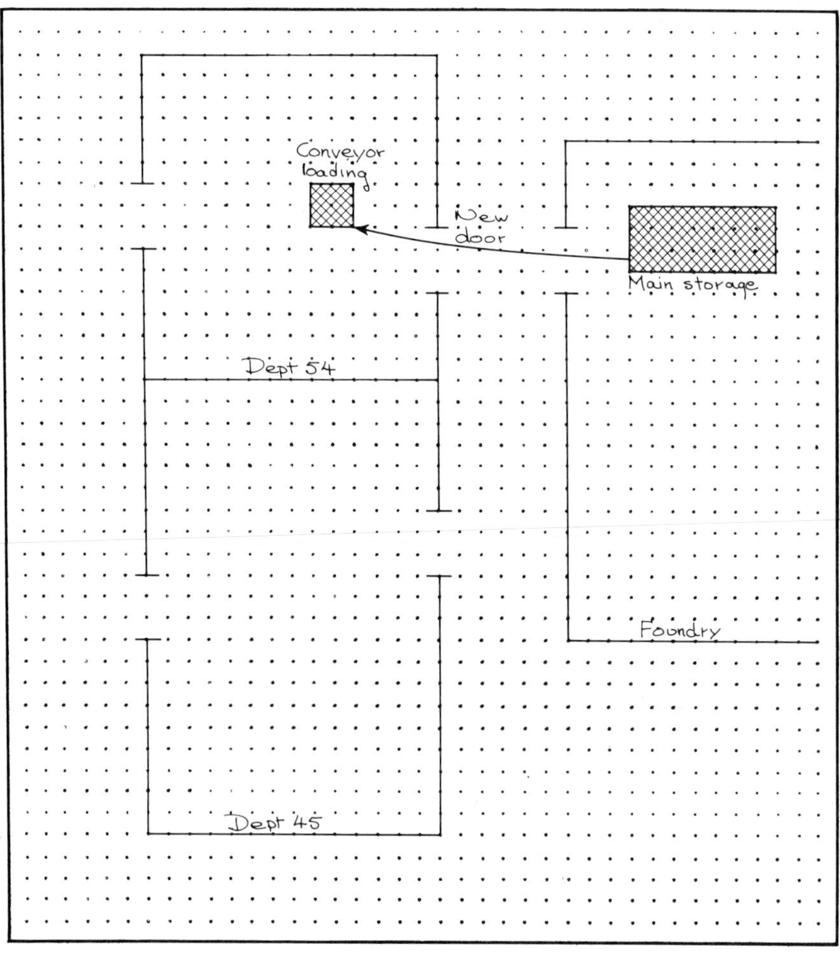

Fig. 5.13 *Flow diagram for process charted in Fig. 5.9.*
From Mundel, Motion and Time Study: Principles and Practice, *p. 68.*

5.28. An analysis of the film of one cycle of the operation from which the chart was made is shown in Fig. 5.29.

Memomotion chart: Sections of a memomotion chart (a process time chart—man analysis) of a hotel maid cleaning a hotel room are shown in Fig. 5.30.

5.3.9 Outlines of Procedures

Procedures for the major techniques are outlined below to provide a general guide to use. Since the same graphic techniques are used for analysis, innovation (synthesis), and test, the outline for each technique indicates how it is used at each step.

5.3.9.1 For Establishing Aim. The following outline indicates how a possibility guide is used. A similar procedure is followed with other related techniques.

1. Uses:

a. Aids in systematically listing possible changes and in collecting material from which to determine an objective.

b. Aids in determining a suitable analysis technique.

c. Indicates which divisions in the organization will be affected.

2. How made:

a. Either a form as shown in Fig. 5.3 or a blank sheet of paper may be used.

b. The analyst states his criterion of success.

c. The analyst determines roughly the degree of change that is warranted.

d. With the aid of a check list, the analyst lists the possibilities that occur to him, tempering his imagination with the decision reached in the previous step, identifies each possibility as to class of change and other areas affected, and gives each possibility an identifying number.

e. The analyst expands each possibility

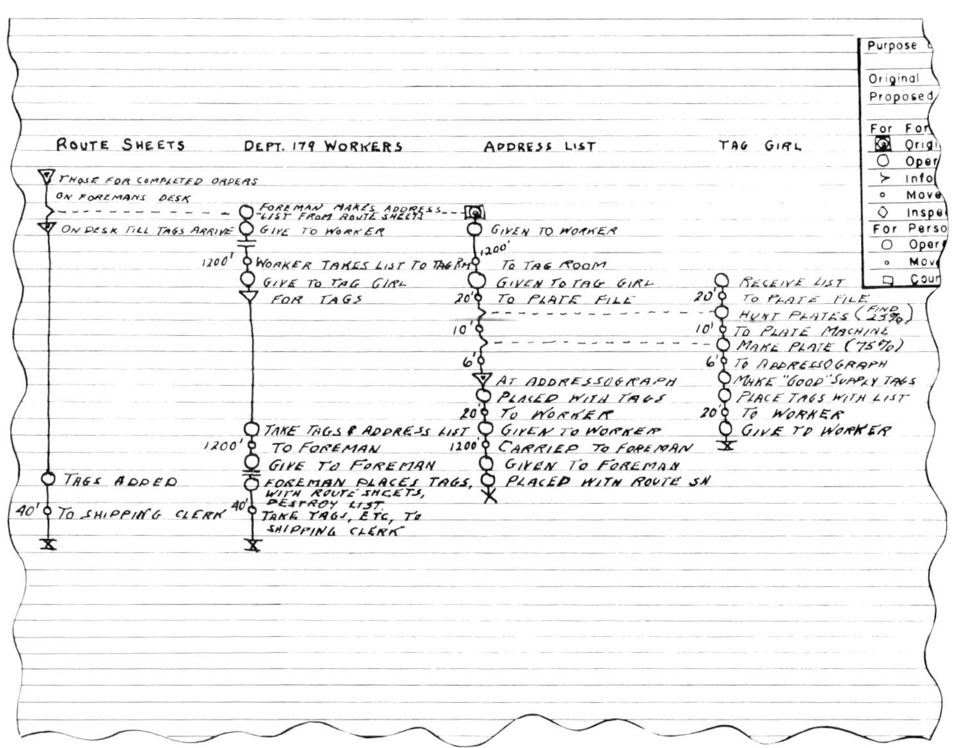

Fig. 5.14 *Procedure analysis chart (section).*

TABLE 5.18. *CHECK LIST FOR MULTIMAN AND MULTIMAN AND MACHINE PROCESS CHARTS**

Basic Principles

A. Balance the work of the crew
B. If a machine is involved, consider increasing percentage of use
C. Ease the job of the most-loaded man
D. Eliminate steps
E. Combine steps
F. Make steps as easy as possible

1. Can any operation be eliminated?
   a. as unnecessary
   b. by changing the order of work
   c. by new or different equipment
   d. by changes in the layout
2. Can any movement be eliminated?
   a. by leaving out operations
   b. by shifting some operations to another job into which they fit more conveniently
   c. by changing equipment
   d. by changing the layout
   e. by changing the order of work
   f. by conveyors (Make sure they are economical.)
3. Can delays be eliminated?
   a. by changing the order of work
   b. by changing the layout
   c. by new or different equipment
4. Can countings or inspections be eliminated?
   a. Are they really necessary; what happens after they are done and the information obtained?
   b. Do they give unnecessary duplication?
   c. Can they be performed more conveniently by another person?
   d. Are they done at the best point in the sequence?
5. Can operations be combined?
   a. by changing the order of work
   b. with new or different equipment
   c. by changing the layout
6. Can movements be combined?
   a. by changing the order of work
   b. by changing the layout
   c. by changing the quantity handled at one time
7. Can delays be combined?
   a. by changing the order of work
   b. by changing the layout
   c. by being grouped better, if they provide rest
8. Can countings or inspections be combined?
   a. by changing the order of work
   b. by changing the layout
9. Can any step be made safer?
   a. by changing the order of work
   b. by new or different equipment
   c. by changing the layout
10. Can any operation be made easier?
    a. by a better tool
    b. by changing positions of controls or tools
    c. by using better material containers or racks, bins, or trucks
    d. by using inertia where possible and avoiding it where worker must overcome it
    e. by lessening visual requirements (see Fig. 5.1)
    f. by better workplace heights

---

* From Mundel, *Motion and Time Study: Principles and Practice.*

TABLE 5.18. *(continued)*

- g. by using different muscles (see Fig. 5.1)
  Use the first muscle group in this list that is strong enough for the task:
  1. finger
  2. wrist
  3. elbow
  4. shoulder
  5. trunk
- h. by jigs or fixtures

11. Can any movement be made easier?
    - a. by a change in layout, shortening distances
    - b. by a change in the direction of movements
    - c. by changing its place in the sequence to one where the distance that must be traveled is shorter
12. Can any delay of one crew member, caused by another crew member, be eliminated?
    - a. by changing the number on the crew
    - b. by changing the number of machines that the crew uses
      (One must again bear in mind the following four possibilities, which were listed previously in connection with man and machine charts.)
      1. Reduction of operator delays to the minimum required for rest and personal time. There may be considerable machine delay.
      2. Reduction of machine delays to the minimum required to provide the operator with rest and personal time, at which times the machine is unattended. There may be considerable other operator delay.
      3. Reduction of machine and operator delays such that they will provide the most economical balance.
      4. Reduction of both operator and machine delays to the minimum required to provide the operator with rest and personal time.
    - c. by a redistribution of the work among the crew
    - d. by changing the order of work of the crew

on a detailed possibility guide form, or on a blank sheet of paper, so that he may examine the consequences of each.

3. How used:

a. Aided by his knowledge of business and manufacturing, and of the economic and psychological factors involved, the analyst selects the class of change or the possibility that appears most feasible. Several persons may participate in this step.

b. On the basis of the above decision and the nature of the job, the analyst selects an appropriate analysis technique from those listed in Table 5.1.

4. What then: The analyst is now ready for Step B (Analysis) of the logical approach. Although the trained analyst may not draw up a formal possibility guide, his thinking should follow its pattern. Experience suggests, however, that the formal approach is usually much more productive than the informal approach.

5.3.9.2 For Product Analysis Techniques. The procedures for using two different techniques are given.

Process Chart—Product Analysis.

1. Uses:

a. Provides a last check prior to a Class 1 or 2 change. (See Par. 5.3.3A.)

b. Supplies information leading to a Class 3, 4, or 5 change.

c. Aids in planning the manufacture of a product or the making of a plant layout.

2. How made:

a. A form may be used, or a blank sheet of paper may be substituted.

b. If possible, actually observe the process. If the item is not in process, or if direct observation is inconvenient, use a scale floor plan. Actual observation is more desirable, since discrepancies often exist between the supposed process and the actual process.

c. Pick a convenient starting place for the analysis.

d. Classify the first step according to the categories listed in Table 5.2.

Type of Chart PROCESS CHART – MAN ANALYSIS

Method	ORIGINAL	Machine No.	D-GROUP
Operation	INSPECT DRAWN COPPER WIRE	Operation No.	I-3
		Part No.	~
Part name	ALL DRAWN COPPER	Chart by	C.W. Mc.
Operator	SIMON GREEN	Date charted	6/4

Dist.	Symbol	Description
24′	∘	To next finished spool
	○	Take spool
24′	∘	To inspection bench
	○	Place spool on bench
	○	Strip and cut outer layer which is always damaged in drawing
15′	∘	To scrap container
	○	Dispose of scrap
15′	∘	To inspection table
	○	Cut 2′ from end, then 2″ sample from 2′ piece
	◇	O.D. with box micrometer
	○	Pick up spool
24′	∘	To machines' "inspected" rack
	○	Place spool

SUMMARY

Symbol	Count
○	7
□	0
◇	1
∘	5
▽	0

Dist. 102′ AVG.

If more space is needed paste additional sheet on here.

Fig. 5.15 *Process chart—man analysis for original method of inspecting production of copper wire. From Mundel,* Motion and Time Study: Principles and Practice, *p. 103.*

e. On the first line of the chart, enter the proper symbol and description. If the step is a movement, pace off the distance or measure it by some other means. In the quantity column, always show the amount handled as a unit on that step; if the step is a storage,

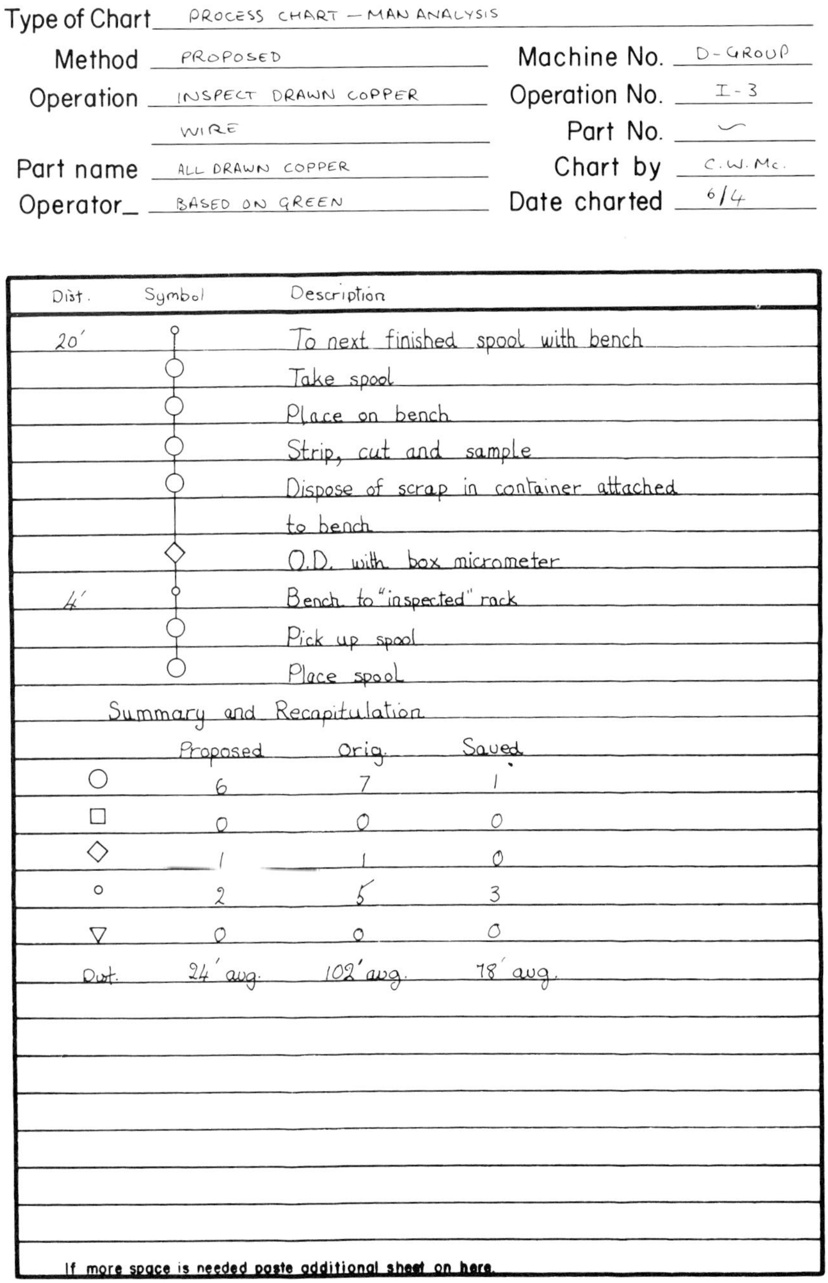

Type of Chart PROCESS CHART — MAN ANALYSIS

Method	PROPOSED	Machine No.	D-GROUP
Operation	INSPECT DRAWN COPPER WIRE	Operation No.	I-3
		Part No.	~
Part name	ALL DRAWN COPPER	Chart by	C.W.Mc.
Operator	BASED ON GREEN	Date charted	6/4

Dist.	Symbol	Description
20′	○ (small)	To next finished spool with bench
	○	Take spool
	○	Place on bench
	○	Strip, cut and sample
	○	Dispose of scrap in container attached to bench
	◇	O.D. with box micrometer
4′	○ (small)	Bench to "inspected" rack
	○	Pick up spool
	○	Place spool

Summary and Recapitulation

	Proposed	Orig.	Saved
○	6	7	1
□	0	0	0
◇	1	1	0
∘	2	5	3
▽	0	0	0
Dist.	24′ avg.	102′ avg.	78′ avg.

If more space is needed paste additional sheet on here.

Fig. 5.16 *Process chart—man analysis for improved method of job charted in Fig. 5.15. From Mundel,* Motion and Time Study: Principles and Practice, *p. 107.*

indicate the usual maximum quantity. If a time measurement is desired for the step, use an ordinary watch or a stop watch, or consult the file of standard times if it is available.

f. On the second line, enter the proper symbol for the second step, and so on.

g. Make a separate entry every time the item moves from one workplace to another, waits, is inspected, or is worked on. However, movements that occur at a given workplace are not usually noted separately.

3. How used: Each step of the process is

Sketch of ORIGINAL METHOD – MAN FLOW DIAGRAM
INSPECT DRAWN COPPER WIRE

Scale – Each square = 2 ft.

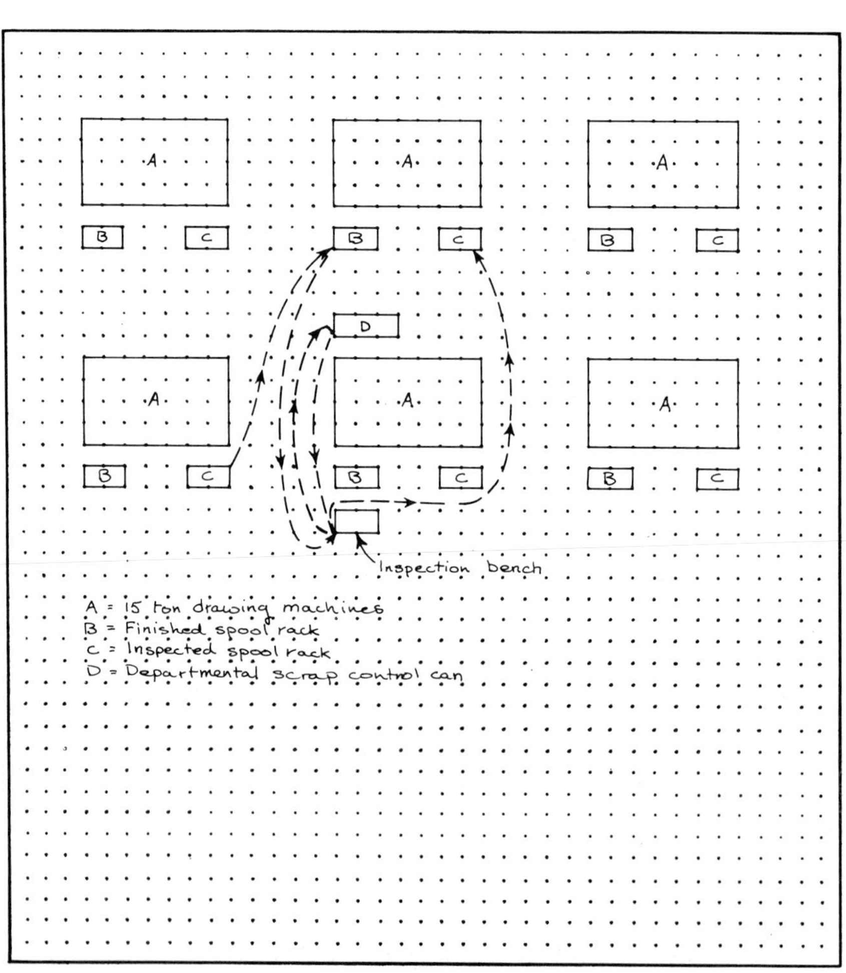

Fig. 5.17 *Man-flow diagram for original method of inspecting production of copper wire. From Mundel,* Motion and Time Study: Principles and Practice, *p. 104.*

checked against the questions in the check list being used.

4. What then: A process chart—product analysis is drawn for a resulting suggested im-

BASIC CHART FORM **531216-02**

OPERATION	Type of chart		Department
ORIGINAL	Orig. or proposed	H. Auxford	Chart by
Drill Door Hinge Channel	Subject charted	2-4	Date charted

Left-hand description	Symbols		Right-hand description
	▽	∘	To area A
		○	Pick up channel
		∘	To fixture
		○	Remove chips with channel
		○	Place in fixture
Lower spindle	○	▽·	In fixture
and pull		∘	To supply tub
		○	Pick up channel
		∘	To area A
		○	Preposition and place
		∘	To part in fixture
Raise spindle	○	▽·	In fixture
On spindle	▽	○	Remove channel
		○	Reverse ends
		○	Remove chips with channel
		○	Place in fixture
Lower spindle	○	▽·	In fixture
Raise spindle	○		
	▽	○	Remove from fixture
		∘	To finished tub
		○	Place in tub

Summary

	LH	RH	Both
○	9	11	20
∘	0	6	6
▽·	0	4	4
▽	12	0	12
Total	21	21	42

Fig. 5.18 *Original operation chart for drilling holes in hinge channel.*
From Mundel, Motion and Time Study: Principles and Practice, *p. 206.*

Fig. 5.19 *Workplace for drilling hinge channel.*
From Mundel, Motion and Time Study: Principles and Practice, *p. 207.*

proved method in order to permit a final check and to provide a means of describing the proposed new method.

Procedure Analysis Chart.

I. General goal (Follow A, B, or C as applicable.)

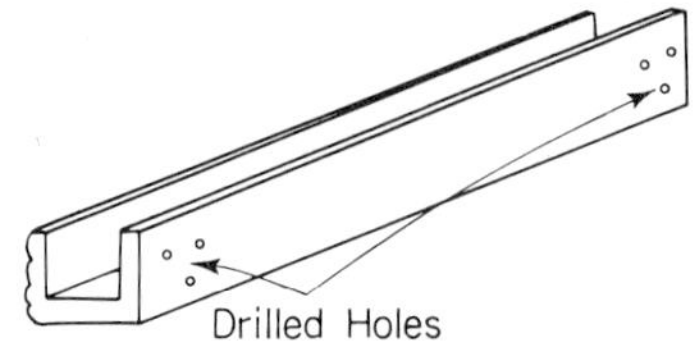

Fig. 5.20 *Hinge channel. From Mundel,* Motion Time Study: Principles and Practice, *p. 165.*

A. If problem is design of a new procedure and/or form(s):

1. Define purpose
2. Determine what organizational segments will be affected
3. Determine what personnel will be affected
4. Determine relationship to existing procedures and forms
5. Determine applicable regulations

B. If problem is unspecified, and if the problem area must be examined before the specific problem can be defined:

1. To analyze problem from viewpoint of forms and form load, use functional forms analysis chart. Designate all forms duplicating one function for detailed study.
2. To analyze problem from viewpoint

Type of Chart MAN AND MACHINE OPERATION CHART

Method	ORIGINAL	Machine No.	G-14
Operation	CENTERLESS GRIND	Operation No.	12
		Part No.	B-2
Part name	BEARING	Chart by	Wren
Operator	T. Silles.	Date charted	4/4

Left-hand description	Symbols		Right-hand description	Grinder
To bearing supply	∘	∘	To feed control	▽
Pick up bearing	○	○	Group control	
To grinder	∘	▽	On control	
Place in grinder	○			
	▽	○	Start hydraulic feed	
		▽	During most of grind	○
		○	Finish grind	
To finished bearing	∘	○	Back off	▽
Take bearing out	○	∘	To gage	
For gaging	▽ (hold)	○	Pick up gage	
		○	To bearing	
		○	Gage	
To finished bearing box	∘	∘	To bench with gage	
Place bearing in box	○	○	Lay on bench	

Summary

	LH	RH	BH	MACH
○	4	7	11	2
∘	4	4	8	—
▽ (hold)	3	0	3	—
▽	3	3	6	12
Total	14	14	28	14

If more space is needed paste additional sheet on here.

Fig. 5.21 *Man and machine operation chart for original method of centerless grind aircraft-engine bearing. From Mundel,* Motion and Time Study: Principles and Practice, *p. 218.*

TABLE 5.19. *CHECK LIST FOR THERBLIGS**

Basic Principles

1. Try to have both hands doing the same thing at the same time or balance the work of the two hands.
2. Try to avoid the use of the hands for holding.
3. Keep the work in the normal work area. (See Fig. 5.1.)
4. Relieve the hands of work whenever possible.
5. Eliminate as many therbligs or as much of a therblig as possible.
6. Arrange the therbligs in the most convenient order.
7. Combine therbligs when possible.
8. Standardize method and train worker.

Therblig	Examine				
	Design of Product	Tools	Jigs	Workplace Layout and Equipment	Motion Pattern
G	Easy to pick up No hazard	Combine Pre-position Assign place Design for grasp In holders	Easy to take parts from or self-ejecting If portable, design for grasp	Ejecting bins Lip bins Slide bins PP boxes No barriers to vision Tool holders Tweezers or tongs	Avoid hand to hand grasp PP parts Slide parts Use bins to advantage Use best type of grasp
P and PP	Less weight Maximum tolerances Bevel holes Round tops of pins Bevel screw ends Make parts for easy line-up Easy access Remove burrs	Self-guiding or locating Easy grip Good leverage Pre-position in holders	Hold parts at convenient angles Receive parts from convenient TL path Stops, guides, funnels Maximum tolerance in jig Large locking motion Self-locating for parts	Paint for seeing Maximum PP of tools and material Arrange for easy TL to place of P and PP	Natural, free motions with accuracy supplied by stops Combine P with TL Combine several P's into one

U, A, and DA	Minimum tool work Reduce screw lengths Easy to get at Combine parts Subassemble Remove burrs	Power Ratchets Combined tools PP tools Design for task Easy to use Best leverage Utilize momentum	Allow free action of tools Guide tools Bevel bushings Bullet top on locating pins Hold parts firmly Uniform type of fastening, preferably clamp levers At convenient height and angle Rotatable Few fastenings	Not in way of tools Tool holders Convenient height	Natural motions Lightest muscle group able to do job Proper leverage Proper posture Back brace on chair Combine U's and A's
RL	Droppable Easy to let go of	Suspended or in PP holder at all times	Easy to fit parts into Will automatically locate parts Kick, blow, drop, slide, or spring parts out	Chutes for RL near work area or in TL path Self-counting trays	As soon as possible Foot ejector As part of TL Without P
TE and TL	Fewer parts Less weight	Within easy reach Light Easy to hold Balanced Counterbalanced Self-returning Foot control	Near parts Chutes and drops Make following P less exacting Attach levers, wheels, and wrenches	Arrange parts for natural sequence Get parts and tools close to point of use	Use smooth continuous motions, circular paths, avoid backtracking Coordinate with use of eyes Use both hands systematically Use smallest amount of body required Two or more parts at once provided this does not interfere in subsequent P

TABLE 5.19. (*Continued*)

Therblig	Examine				
	Design of Product	Tools	Jigs	Workplace Layout and Equipment	Motion Pattern
SH and ST	Standardize parts Make nontangling Color code	Not tangle with other tools Minimum number Special eyeglasses Combine Paint in contrasting color Pre-position Definite location	Fixed in place Levers or wrenches attached Paint controls in contrasting color	Lip bins Definite places for tools and materials Label or color bins Bins contrast with parts Illuminate workplace Paint workplace for seeing	Use eyes to do work Use uniform motion pattern Use bins or trays of material systematically
H, UD, and AD	See basic rules. These therbligs are undesirable. Balance work with machine cycle if machine is used.				
R	If other therbligs are improved this will be reduced to a minimum. Rest is preferably provided by a rest pause rather than as a regular element in the cycle. If it occurs as part of a machine operation, it should take place during machine running time.				
PN	See basic rules. This therblig is undesirable. Balance work with machine cycle if machine is used.				
I	Easy reference points Minimum requirements	Easy reading Go-no-go Optical Rugged Combined gages	Minimum number of fastenings Uniform fastenings Light Built-in gages Easy reading	Good light, free of glare and flicker; of proper color, direction, and contrast	Fixed and definite pattern even for eyes Arrange so part is stationary when being viewed

* From Mundel, *Motion and Time Study: Principles and Practice*.

Min.	Left hand		Right hand		Grinder	
	To bearing	0.03	To supply	0.02		0.15
	Take out	0.02	Take one	0.03		
			To grinder	0.03		
0.1			Place in	0.03		
			To feed lever	0.02		
			Start feed	0.02		
			To gage	0.02		
			Pick up	0.01		
0.2	For gaging	0.25	To bearing	0.02		
			Gage	0.10		
0.3	To "finished" box	0.03	Gage to bench	0.03		
	Place in box	0.02	Lay on bench	0.02		
0.4						52
0.5				0.30		
		0.44				
0.6						
			To feed lever	0.02		
0.79			Finish grind	0.10		0.10
			Back off	0.02		0.02

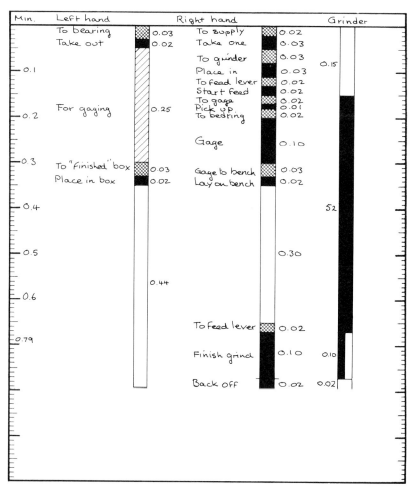

Fig. 5.22 *Man and machine operation time chart for improved method of centerless grind aircraft-engine bearing. From Mundel,* Motion and Time Study: Principles and Practice, *p. 228.*

of personnel and work load, use work activity analysis.

a. Supply personnel concerned with ruled sheets divided into three columns: Time, Activity, Remarks.

b. Request personnel to maintain log of their activity for adequate period.

c. Explain purpose in detail and assist in preparing standard terms for describing major activities to facilitate later summarizing.

d. Answer questions.

e. Assist and supervise collection of data.

f. Determine most pressing needs from above data and designate activities that need detailed study. (*Note:* A record may also be made of overloaded equipment.)

C. If problem is the improvement of a specific procedure and/or form(s):

________ MAN AND ________ MACHINE TIME CHART

Method ORIGINAL
Operation WIND, STRETCH, CHOP AND PEEL
Part name GRIND
Operator Kozoff

Machine No. G105
Operation No. 1
Part No. 2193
Chart by JK
Date charted 10/4

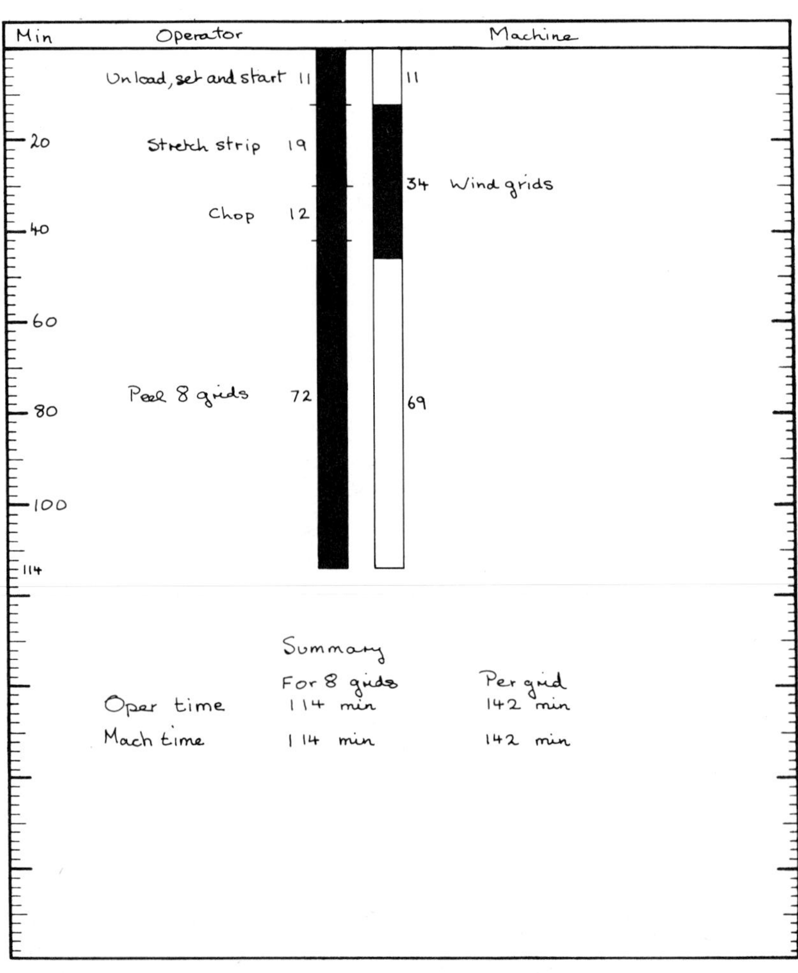

Fig. 5.23 *Man- and machine-process time chart for one-man crew on winding, stretching, chopping, and peeling radio-tube grids. From Mundel,* Motion and Time Study: Principles and Practice, *p. 234.*

1. Determine nature of specific problem and decide whether problem is:
   a. Redesigning a procedure
   b. Redesigning a form

II. Specific goal (Selecting an objective or guide. If problem concerns the initial design

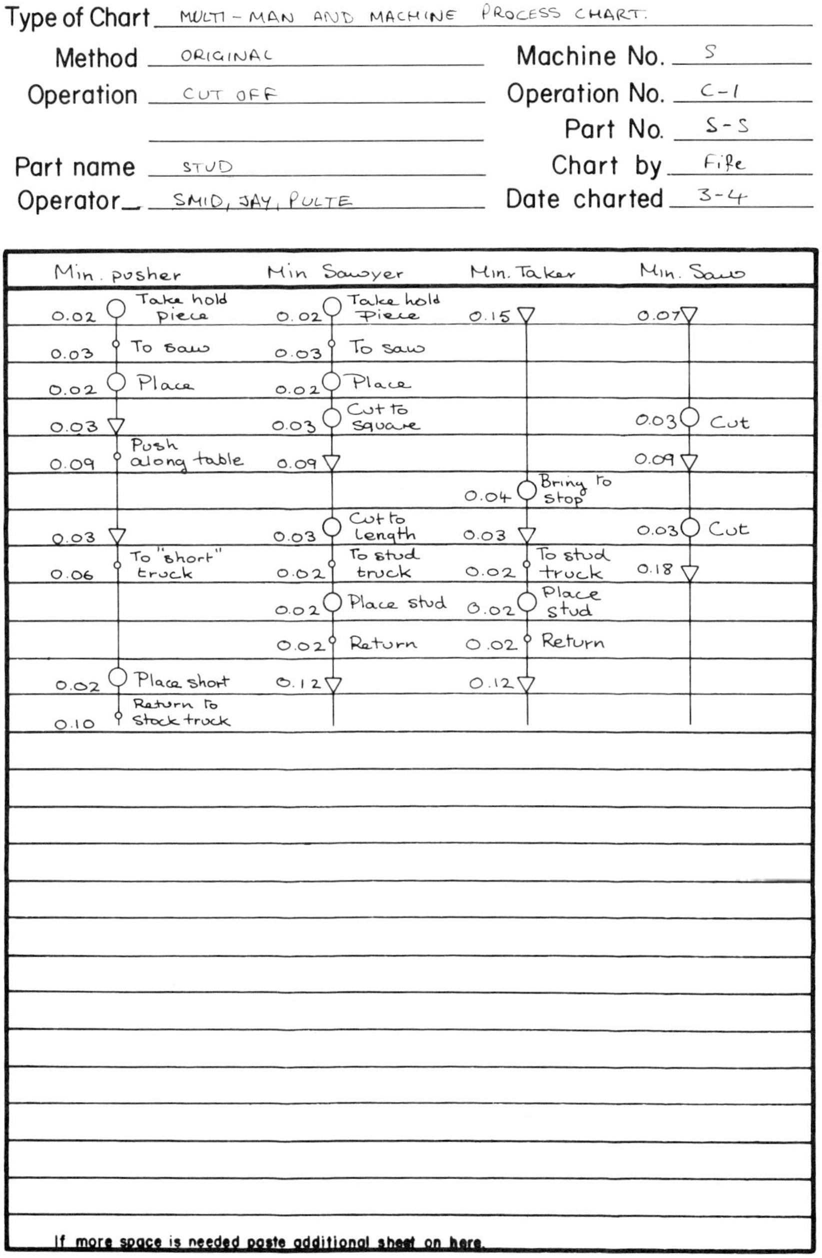

Type of Chart MULTI-MAN AND MACHINE PROCESS CHART.

Method	ORIGINAL	Machine No.	S
Operation	CUT OFF	Operation No.	C-1
		Part No.	S-S
Part name	STUD	Chart by	Fife
Operator	SMID, JAY, PULTE	Date charted	3-4

Min.	pusher	Min	Sawyer	Min.	Taker	Min.	Saw
0.02	Take hold piece	0.02	Take hold Piece	0.15		0.07	
0.03	To saw	0.03	To saw				
0.02	Place	0.02	Place				
0.03		0.03	Cut to square			0.03	Cut
0.09	Push along table	0.09				0.09	
				0.04	Bring to stop		
0.03		0.03	Cut to Length	0.03		0.03	Cut
0.06	To "short" truck	0.02	To stud truck	0.02	To stud truck	0.18	
		0.02	Place stud	0.02	Place stud		
		0.02	Return	0.02	Return		
0.02	Place short	0.12		0.12			
0.10	Return to Stock truck						

If more space is needed paste additional sheet on here.

Fig. 5.24 *Multiman and machine-process chart for original method of three-man crew cutting 8-ft length from 14′2″ × 4″ lumber. From Mundel,* Motion and Time Study: Principles and Practice, *p. 245.*

of a procedure or form, select most desirable objectives to optimize; note that it is difficult to optimize all factors at once.)

A. Consider the following list of possible objectives:

1. Fewer people
2. Fewer steps (in procedure or in using form)
3. Less time on a step or steps
4. Less time in production (for all-over procedure to be completed)
5. Less space

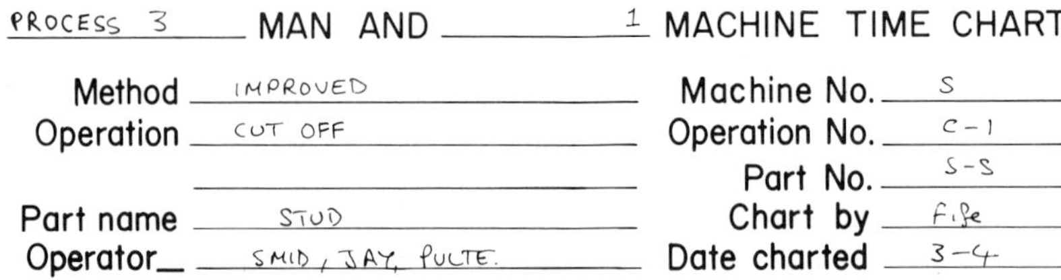

PROCESS 3 MAN AND 1 MACHINE TIME CHART

Method	IMPROVED	Machine No.	S
Operation	CUT OFF	Operation No.	C-1
		Part No.	S-S
Part name	STUD	Chart by	Fife
Operator	SMID, JAY, PULTE.	Date charted	3-4

Min.	Pusher		Sawyer		Taker		Saw
	Lift piece to saw	0.04	To "short" truck with "short"	0.02	Stud to stud truck	0.03	0.07
	Place	0.03	Place	0.03	Place	0.04	0.03
		0.03	To Saw	0.02	To saw table	0.03	0.07
0.1	Shove along table	0.03	Cut to square	0.03		0.03	0.03
	Return to "stock" truck	0.04		0.03	Place to stop	0.04	
0.2	Take hold piece	0.03	Place	0.04		0.03	
			Cut to length	0.03			

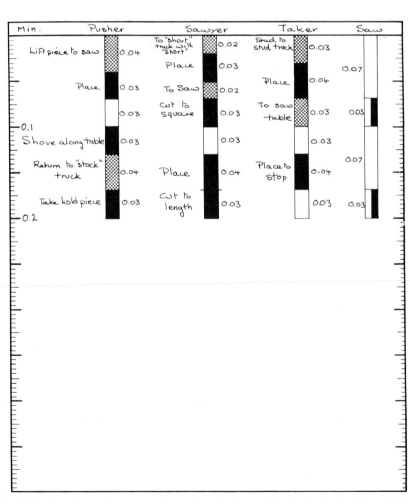

Fig. 5.25 *Multiman and machine-process time chart for improved method of job of Fig. 5.24. From Mundel,* Motion and Time Study: Principles and Practice, *p. 245.*

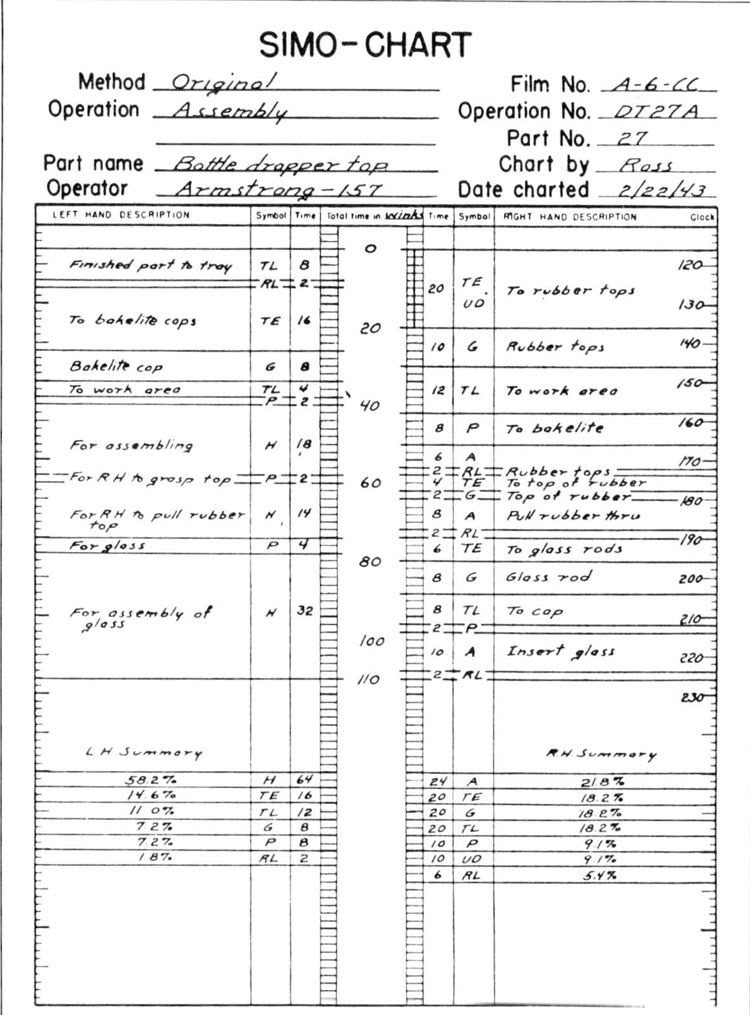

SIMO-CHART

Method: Original — Film No.: A-6-CC
Operation: Assembly — Operation No.: DT27A
Part No.: 27
Part name: Bottle dropper top — Chart by: Ross
Operator: Armstrong-157 — Date charted: 2/22/43

LEFT HAND DESCRIPTION	Symbol	Time
Finished part to tray	TL	8
	RL	2
To bakelite caps	TE	16
Bakelite cap	G	8
To work area	TL	4
	P	2
For assembling	H	18
For RH to grasp top	P	2
For RH to pull rubber top	H	14
For glass	P	4
For assembly of glass	H	32

Total time in winks: 0, 20, 40, 60, 80, 100, 110

Time	Symbol	RIGHT HAND DESCRIPTION
20	TE. UD	To rubber tops
10	G	Rubber tops
12	TL	To work area
8	P	To bakelite
6	A	Rubber tops
2	RL	
4	TE	To top of rubber
2	G	Top of rubber
8	A	Pull rubber thru
2	RL	
6	TE	To glass rods
8	G	Glass rod
8	TL	To cap
2	P	
10	A	Insert glass
2	RL	

Clock: 120, 130, 140, 150, 160, 170, 180, 190, 200, 210, 220, 230

LH Summary	Symbol	Time
58.2%	H	64
14.6%	TE	16
11.0%	TL	12
7.2%	G	8
7.2%	P	8
1.8%	RL	2

Time	Symbol	RH Summary
24	A	21.8%
20	TE	18.2%
20	G	18.2%
20	TL	18.2%
10	P	9.1%
10	UD	9.1%
6	RL	5.4%

Fig. 5.26 *Simochart for original method of assembling dropper bottle tops. From Mundel,* Motion and Time Study: Principles and Practice, *p. 293.*

6. Less time spent in using critical skills or critical personnel
7. Less time spent on critical equipment
8. Increased quality (accuracy)
9. Less cost
10. Less skill needed for various steps
11. Better control
12. Other (state)

B. Using experience, direction, or a balancing of needs to guide you, select one or more as an objective.

C. Check the objective selected with your superior or with affected personnel.

III. Preliminary analysis

A. Determine roughly the degree of change that is desirable or feasible

1. Consider the following factors:

a. How great is the actual or expected volume of the activity?

b. How long will it last?

c. How much time, per unit, is spent on the job?

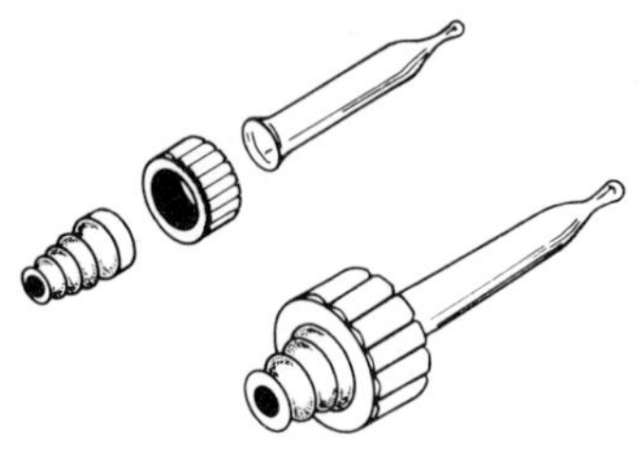

Fig. 5.27 *Parts of medicine-bottle dropper top and assembly. From Mundel,* Motion and Time Study: Principles and Practice, *p. 290.*

d. How much time is available to work up a change?

e. How much is already invested in equipment for the job?

f. How much analysis time will be required?

g. How much time would be lost in a change?

h. How pressing is the need?

i. How much retraining would be required?

j. What is the possible saving?

MICROMOTION STUDY DATA SHEET

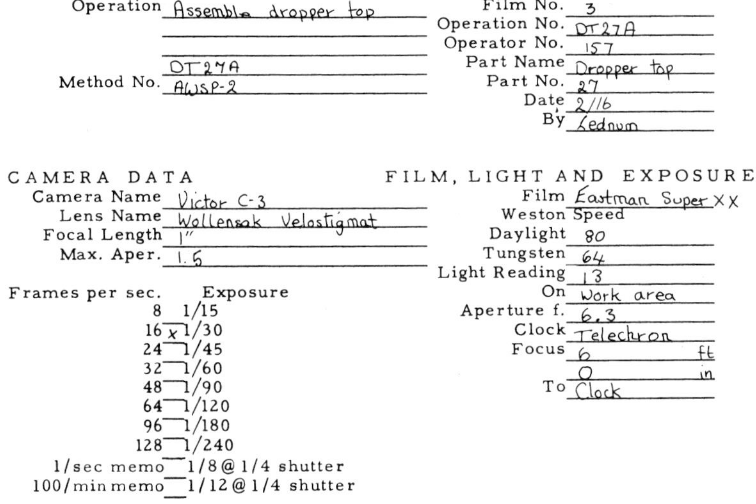

Operation Assemble dropper top

Method No. DT27A AWSP-2

Film No. 3
Operation No. DT27A
Operator No. 157
Part Name Dropper top
Part No. 27
Date 2/16
By Lednum

CAMERA DATA
Camera Name Victor C-3
Lens Name Wollensak Velostigmat
Focal Length 1"
Max. Aper. 1.5

Frames per sec. Exposure
8 1/15
16 x 1/30
24 1/45
32 1/60
48 1/90
64 1/120
96 1/180
128 1/240
1/sec memo 1/8 @ 1/4 shutter
100/min memo 1/12 @ 1/4 shutter

FILM, LIGHT AND EXPOSURE
Film Eastman Super XX
Weston Speed
Daylight 80
Tungsten 64
Light Reading 13
On Work area
Aperture f. 6.3
Clock Telechron
Focus 6 ft 0 in
To Clock

SKETCH OF CAMERA SET UP

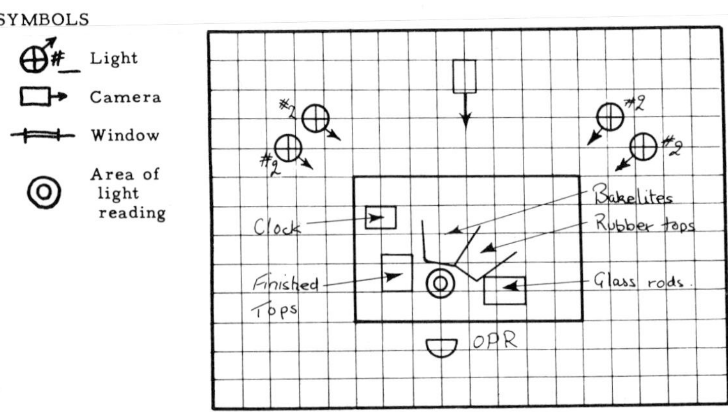

Fig. 5.28 *Film data sheet for film used for Fig. 5.26.*
From Mundel, Motion and Time Study: Principles and Practice, *p. 291.*

RECORD OF FILM ANALYSIS

Film No: 3 | Date filmed 2/16/43 | Analysis by J Ross | Date 2/22/43

Operation Assemble | Operator Armstrong-157 | Part name Bottle dropper top | Part No. 27

1 Sheet of 1 | Dept. Tops-92

Clock reading	Subtracted time	Therblig symbol	Left hand description	Clock reading	Subtracted time	Therblig symbol	Right hand description	Clock reading	Subtracted time	Notes
116	8	TL	Finished part to tray	116	20	TE, UD	To rubber tops			
124	2	RL	Dropper top into tray	136	10	G	Rubber top			
126	16	TE	To bakelite caps	146	12	TL	To work area			
142	8	G	Bakelite cap	158	8	P	To bakelite			
150	4	TL	To work area	166	6	A	" "			
154	2	P	For assembling	172	2	RL	Rubber top			
156	18	H	" "	174	4	TE	To top end of rubber			
174	2	P	For RH to grasp rubber	178	2	G	" " " "			
176	14	H	" " " pull "	180	8	A	Pull rubber thru			
190	4	P	To receive glass	188	2	RL	Rubber			
194	32	H	" " "	190	6	TE	To glass rods			
226	X	TL	Finished part to tray	196	8	G	Glass rod			
				204	8	TL	" " to cap			
116	110			212	2	P	" " " "			
110	/OK			214	10	A	" " " "			
				224	2	RL	" "			
				226	X	TE	To rubber tops			
				116	110					
				110	/OK					

Fig. 5.29 *Film analysis, with therbligs of film referred to in Fig. 5.28. From Mundel,* Motion and Time Study: Principles and Practice, *p. 292.*

k. What personalities are involved?

l. Does policy allow a change?

m. Is any aspect beyond control; if it is, who must be consulted?

B. Evaluate each of the above factors as well as possible at this initial stage.

C. On the basis of this evaluation, set rough limits for the analysis activity.

IV. Follow either A or B below, as applicable.

A. For the analysis of an existing procedure:

1. Using the symbols of Table 5.3, and a form such as Fig. 5.14, make charts of all the forms being used.

a. Discuss each form with individuals concerned. Follow a copy of form from inception of use to final disposal. Place remarks made by individuals on your notes if important.

b. Record separable activities of each form with symbols and explanations.

c. Check against standard instructions, if available.

2. Draw procedure chart, columnating by forms, persons, or offices as appears most desirable. Integrate material collected at Step 1 above so that a horizontal line drawn across chart at any point passes through items that are happening at the same time.

B. For the design of a procedure:

1. Using the symbols of Table 5.3 and a blank chart, lay out procedure. As you proceed, take into consideration the items of the check list and the applicable items listed under Step I (General goal).

2. Treat as you would an existing procedure and attempt to make improvements.

V. Check against the appropriate check list.

VI. Consult trade catalogs for alternatives and suggestions.

VII. Discuss suggestions from V and VI with affected personnel.

VIII. Draw up proposal, using procedure analysis chart.

IX. Consult with supervisor for interim advice and direction.

X. Treat your proposal with the items from Steps V, VI, and VII as if it were an original analysis.

XI. Discuss final result with all concerned as a "proposal." Work up through the affected segment, adjusting proposal if necessary. Con-

1 MAN AND 0 MACHINE TIME CHART

Method	ORIGINAL	Machine No.	~
Operation	CLEANING	Operation No.	~
		Part No.	1108
Part name	ROOM	Chart by	Osay
Operator	CALIDONIA	Date charted	MARCH 7

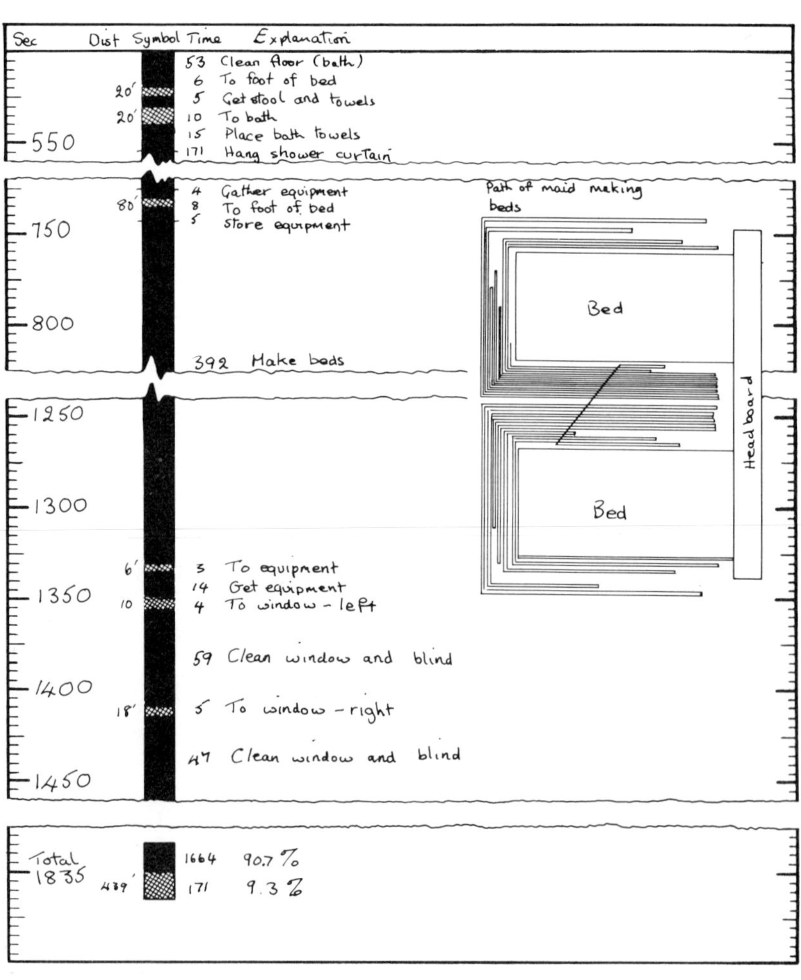

Fig. 5.30 *Sections of chart showing original method of maid cleaning hotel room. From Mundel,* Motion and Time Study: Principles and Practice, *p. 312.*

sult with supervisors and workers to obtain participation rather than just acquiescence.

XII. Submit proposal through channels.

XIII. Await authorization to proceed.

5.3.9.3 For Man Analysis Techniques. The uses of the major techniques are outlined below. A more detailed procedure such as that given in Par. 5.3.9.2 could be evolved for each technique.

PROCESS CHART—MAN ANALYSIS.

1. Use: Supplies information leading to a Class 1 or 2 change in a job that requires the worker to move from place to place. (*Note:* Higher classes of change are occasionally suggested by an analysis of this type and should not be overlooked).

2. How made:

a. Either a form such as appears in Fig. 5.11 or 5.15 or a blank sheet of paper may be used.

b. The chart may begin at any point in a cycle of work considering a cycle as the complete set of steps necessary to bring a unit of the goods to the degree of completion typical of the work. However, it is usually most convenient to begin with the first step connected with a particular unit and end on the last step before the next similar unit is worked on or a new routine is started.

c. The first step in the cycle should be carefully classified according to the categories in Table 5.4. The symbol, explanation, and, if the step is movement, the distance should be entered on the first line of the chart. The explanation should be as succinct as possible. The distance may be paced, estimated, measured on the actual floor plan, or scaled from a drawing.

d. If a time measurement for the step is desired, use an ordinary watch or stop watch.

e. Information for subsequent steps should be entered on subsequent lines, in such a way that the symbols, explanations, and so forth form separate columns for easy reading.

f. An entry should be made for every separable phase of the work, for every time the worker moves from one place to another, and for every time the worker works at a workplace. When two distinctly separate activities follow each other at a workplace without an intervening movement, two operation symbols, one after the other, may be entered.

g. The steps should represent the activities of the worker—i.e., what he does to the product, and where he goes—rather than what happens to the product.

h. If the job has several subcycles occurring at different frequencies, each subcycle should be charted separately, and its frequency of occurrence should be noted.

i. If the job has a variable cycle, classify and handle as indicated:

(1) Variations caused by operator habit. Chart several versions to insure adequate data.

(2) Variations inherent in job and each different. Plot general pattern, marking parts that are constant and parts that vary from cycle to cycle.

(3) No pattern or cycle. Plot a selected period of work as it occurs.

j. A man-flow diagram (see Par. 5.3.5.2) is often a useful adjunct to the process chart—man analysis.

3. How used: Each step of the process is checked against the questions in the check list being used.

4. What then: A process chart—man analysis is drawn for a resulting suggested improved method in order to permit a final check and to provide a means of describing the proposed new method.

OPERATION CHART.

1. Use: Supplies information leading to a Class 1 or 2 change in a job that takes place at one location and in which the operator rather than a machine controls the flow of work.

2. How made:

a. Operation charts are best prepared by actually observing the worker, although they may be prepared from a proposal for a method.

b. Either a form such as that in Fig. 5.18 or a blank sheet of paper may be used.

c. Although the chart may begin at any point in the work cycle, it is usually most convenient to begin with the hand that makes the first movement attributable to a unit of product.

d. The first and last cycle of a work period may be different from the other cycles. It is usually desirable to chart the most typical cycle.

e. The first step of the hand that begins the cycle should be classified according to the categories in Table 5.5. An entry should then be made on the first line, as in Fig. 5.18. The symbols are usually drawn freehand.

f. All the steps of this hand should then

be classified and plotted in order. Several cycles may have to be observed before this information can be recorded in complete form.

g. The steps of the other hand should then be plotted. Each step for this hand must be recorded opposite the step of the other hand during which it occurs.

h. Since the chart may become congested and confusing when all this information is coordinated, it is frequently necessary to redraw it in order to make it easily legible, as in Fig. 5.18. However, since the chart is mainly an aid to understanding, excess draftsmanship is a waste of time and should be avoided.

i. A summary should be placed at the bottom of the chart.

j. A sketch of the workplace is often a useful adjunct.

3. How used:

a. Each step of the operation is checked against the questions in the check list for operation charts (see Table 5.16).

b. The general principles listed at the top of the check list should serve as objectives.

4. What then:

a. An operation chart is drawn for the resulting suggested method in order to permit a final check and to provide a means of describing the proposed new method.

b. A revised sketch of the workplace is usually drawn, and drawings of the necessary tools, jigs, fixtures, bins, and so forth are prepared.

MULTIPLE-ACTIVITY CHARTS. For convenience, all procedures for using man and machine charts have been combined into one group and all procedures for multiman charts into another.

Man and machine charts

1. Uses:

a. Improve utilization of man or machine and improve integration of the two on jobs where the man works with one or more machines and where the work of the machine is a controlling factor.

b. Aid in selecting a man and machine work pattern appropriate to production requirements and cost conditions.

c. In the simpler forms, aid in determining where more detailed analysis techniques may be usefully applied.

2. How made:

a. Man and machine operation charts are constructed as follows:

(1) A right- and left-hand operation chart is constructed, as summarized in Par. 5.3.9.3.

(2) The machine is then charted by the use of a procedure similar to that used for placing the second hand on right- and left-hand operation charts, with the machine information being placed on the chart as shown in Fig. 5.21. Usually, only two symbols are used for the machine: the suboperation symbol when the machine is working and the delay symbol when it is idle.

(3) A summary is placed at the bottom of the chart.

b. Man and machine operation time charts are constructed as follows:

(1) A man and machine operation chart is drawn.

(2) Sample time values are obtained for each step, by means of a watch, film records, or standard time tables.

(3) A second chart is drawn either on the form shown in Fig. 5.22 or on a sheet of graph or cross-section paper, using the conventions established in Table 5.6. The length of the shading of the bar for each step on the chart is made proportional to the time involved. A time scale should be devised so that each line of the chart will represent the same time value and the chart will be of convenient size.

(4) A summary may be placed at the bottom of the chart, although the length of the chart is itself a rough summary, giving the total time for the operation.

c. Man and machine process time charts are constructed as follows:

(1) A process chart—man analysis is constructed, as summarized in Par. 5.3.9.3. The information is usually confined to the left half of the chart, and the right half is left blank.

(2) The machines used are charted on the same sheet, as with man and machine operation charts, and the steps of the machines are keyed into the process chart.

(3) By use of a watch, stop watch, camera, or table of time values, the time is determined for each item on the chart.

(4) By use of a form similar to that in Fig. 5.23 or a sheet of graph or cross-section paper, the chart is redrawn. The conventions of Table 5.7 are used, and the length of each item is made proportional to the time involved. A convenient scale should be chosen and a time value assigned to each line of the chart so

that the chart is of a convenient size. The scale should be the same throughout the chart.

(5) A summary may be placed at the bottom of the chart, although the length of the chart is itself a rough summary, giving the total operation time.

3. How used: Each step of the method analyzed should be checked against the questions included in suitable check lists.

With the more complex tasks, the desired type of solution should be stated as a criterion for selection.

The time chart may be used to locate sections of the task that need more intensive study.

4. What then: A chart of the proposed method is either constructed or selected from the available solutions for the subsequent steps; or more detailed analyses, for further study, may be made of sections of the task.

Multiman charts

1. Uses:

a. Arranging crew work for the best balance

b. Estimating the effect of varying the size of the crews

c. As a basis for instructing the crew

d. Locating sections of the task where a detailed right- and left-hand analysis is needed to develop an improvement

2. How made:

a. A process chart—man analysis is constructed for one of the crew members, with the information confined to one side of the chart. The classification of steps, symbols, and procedure is the same as with process charts—man analysis.

b. The other crew members are charted, one at a time, on the same sheet. Symbols that indicate simultaneous activities should be entered on the same line.

c. Next, machine or machines are analyzed in a similar manner on the same chart.

d. The time is obtained for each item on the chart. If watches are used, the times will be collected from several cycles; hence some adjustment may be necessary to obtain comparable values. If films are used, the same cycle of work can be used for all entries.

e. By the use of a form similar to Fig. 5.25 or a sheet of graph or cross-section paper, the information obtained in the previous steps is recharted, using the conventions of Table 5.7 and a convenient time scale.

f. A summary may be placed at the bottom of the chart, although the length of the chart is itself a rough summary, giving the total cycle time.

3. How used: Each step of the operation is checked against a suitable check list. The general principles listed at the top of the check list should serve as objectives.

4. What then: A new multiman chart is drawn for the resulting suggested method in order to permit a final check and to provide a means of describing the proposed new method. If effects of variations in crew size are being examined, several alternative charts may be drawn.

MICROMOTION STUDY. Summary procedures are given for fine and gross types of analysis.

Simo charts

1. Uses:

a. Primarily for short-cycle, repetitive jobs

b. For jobs involving high-order skills

c. For jobs representing a series where the changes involved may affect several jobs similarly. (*Note:* The general outline offered here can be used for any of the previously described graphic man-analysis techniques.)

2. How made:

a. A film is made of the operation, with a timing device in the field of view. As an alternative procedure a video-tape record may be made and used in lieu of a film.

b. If the film is analyzed with the therblig breakdown, it is desirable to use the following steps in connection with a form for film analysis such as that shown in Fig. 5.29 or a sheet of ruled columnar paper.

(1) To facilitate future reference, the data at the top of the sheet are obtained from the micromotion study film data sheet.

(2) The entire film is viewed, and a typical whole cycle is selected. A cycle is defined as the complete series of motions required to bring a unit of product to the degree of completion characteristic of the operation. As with the previous techniques, it is usually most convenient to select the cycle as starting with the first motion connected with the production of a unit and ending when the same motion is repeated with the next unit.

(3) The activities of one body member, usually the busiest, are recorded by noting, on the first line, the beginning time (the clock or counter reading for the first frame or picture in which the therblig appears), for the first

therblig of the cycle for that body member, the therblig letter symbol, and the explanation. Then the projector is turned till the next therblig appears; appropriate information is entered on the second line; and so on. The explanation should assume that the therblig is the verb and should tell what and where the action is. The actual therblig times may be obtained later by successive subtractions. It is customary practice to follow each body member through the entire cycle, and then to repeat the cycle for as many times as there are body members being studied. For therbligs such as RL, which may take place entirely between two successive pictures and not appear on the film at all, the time interval between the two pictures is usually arbitrarily assigned. At the normal speed of 16 frames, the interval will be approximately 0.001 minute. On the form given in Fig. 5.29 space is provided for the two hands; the column headed "Notes" is for the analysis of any other body member, such as feet, eyes, and so forth. This is the most common type of analysis. Special forms may be designed for more complex analyses.

c. A simo chart is drawn to assist in examining the interrelationships of the body members concerned.

3. How used: Each step of the operation is checked against a suitable check list. The general principles listed at the head of the check list should serve as objectives.

4. What then: A new simo chart is synthesized for the resulting suggested method in order to permit a final check and to provide a means of describing the proposed new method.

Memomotion study

Since the application of this technique varies greatly with the nature of the objective and the subject, a complete case is given as a guide for developing a procedure.

1. Uses:
   a. Long cycles
   b. Irregular cycles
   c. Crew activities
   d. Long-period studies

2. Sample case: *Three-man crew running heavy castings through a "Rotoblast."*

A possibility guide had indicated that a Class 1 or 2 change was desirable, and a study of the process chart—product analysis suggested that the operation, shot-blast casting, was necessary. This completed Step A of the logical approach, *Aim*. For convenience and accuracy, Step B, *Analysis*, was performed with a memomotion camera. Since sufficient light from the building monitor penetrated even the murky interior of the foundry, there was no need for supplementary lighting. The crew members, aware that they were being photographed, were filmed at work, and an analysis was made of the film in order to obtain data with which to construct a multiman and machine process time chart.

The analysis was made with a projector equipped with a frame-counter. A typical cycle was selected. Since the selected cycle was 439 seconds long (7.3 minutes), and there are 40 frames per foot on 16 mm. film, the selected section was only 11 feet long. At normal speeds, the cycle would have occupied almost 200 feet of film. (The analysis time saved by using the memomotion film should be obvious.) The selected cycle was identified for future examination by placing an ink spot on the beginning frame. The beginning counter reading, a process chart—man analysis symbol, and a description for one operator were entered on the first line of a ruled sheet. The projector was indexed until the first frame showing a new activity appeared, and the appropriate information was entered on the second line. This procedure continued until all the activities of the operator had been detailed for the whole cycle. The film was then returned to the starting point and the second operator's activities were analyzed.

In this case, the film was studied four times, once for each of three workers and once for the machine. Sections of the chart made for the original method are shown in Fig. 5.31. The sections on the chart marked "overblast" could not be determined from the film; they were worked out with the aid of experimental data on the minimum required shot-blasting time. In drawing the chart, these experimental data were used to denote the unnecessary blasting, which was labeled "overblast."

Although the chart could have been made without film, the use of film assured data on all men from the same cycle, was quicker, gave better accuracy, and provided a record to use in checking the analysis. Moreover, memomotion films possess a unique feature. When they are projected at normal speed, they exaggerate the movements made and call them to the analyst's attention for modification and improvement. They also permit a long period of activity to be reviewed in a short time. Since an hour's activity can be projected in

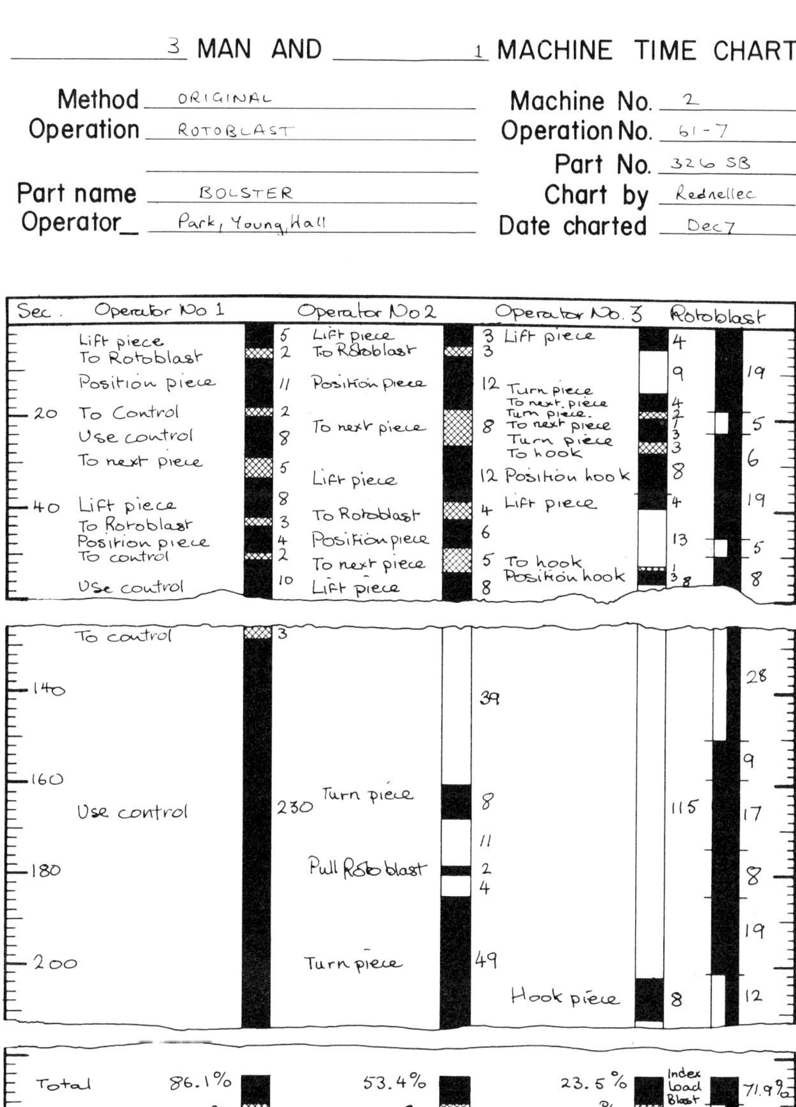

Fig. 5.31 *Sections of chart showing original method of rotoblast.*
From Mundel, Motion and Time Study: Principles and Practice, *p. 309.*

four minutes, a "bird's-eye" view of a task is presented that often aids in suggesting innovations. Memomotion films are also valuable in discussions with foremen and workers, since only a short time away from the job is needed to review a lengthy period of work.

The proper multiman check list was used on this foundry chart and the new method, parts of which are shown in Fig. 5.32, resulted. The improved method saved 35 per cent of the man-minutes of work per unit and increased hourly production by 2 per cent. It did not

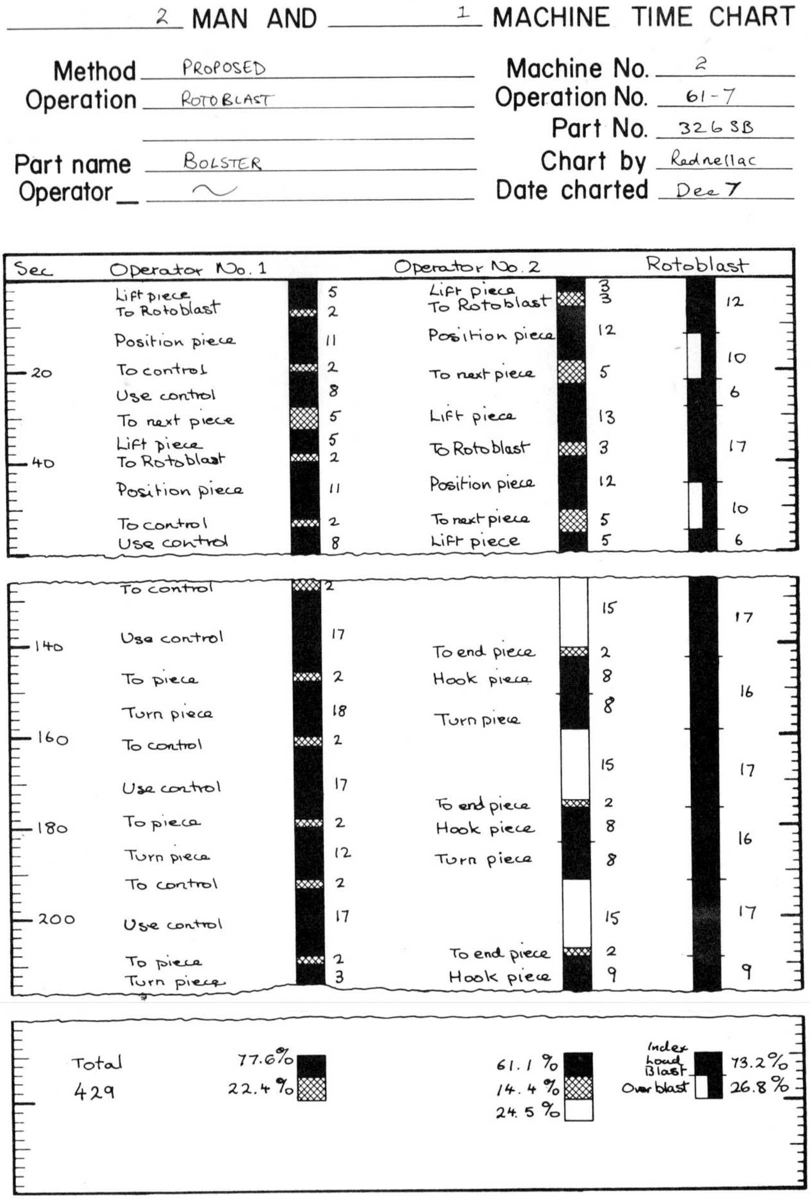

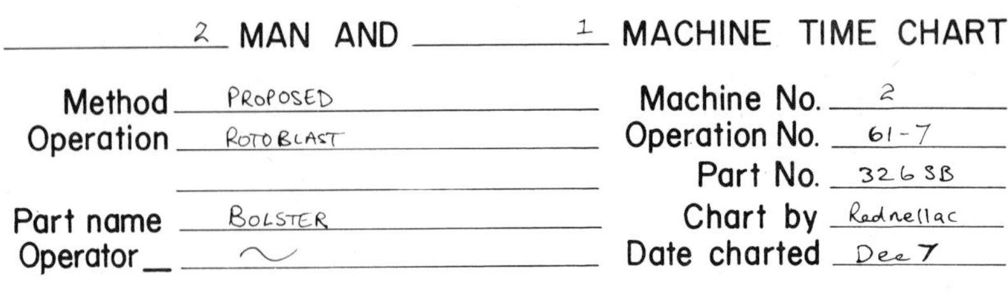

2 MAN AND 1 MACHINE TIME CHART

Method	PROPOSED	Machine No.	2
Operation	ROTOBLAST	Operation No.	61-7
		Part No.	326SB
Part name	BOLSTER	Chart by	Rednellac
Operator	~	Date charted	Dec 7

Sec	Operator No. 1		Operator No. 2		Rotoblast	
	Lift piece	5	Lift piece	3		12
	To Rotoblast	2	To Rotoblast	3		
	Position piece	11	Position piece	12		
20	To control	2	To next piece	5		10
	Use control	8				6
	To next piece	5	Lift piece	13		
	Lift piece	5	To Rotoblast	3		17
40	To Rotoblast	2				
	Position piece	11	Position piece	12		
	To control	2	To next piece	5		10
	Use control	8	Lift piece	5		6

Sec	Operator No. 1		Operator No. 2		Rotoblast	
	To control	2		15		17
140	Use control	17	To end piece	2		
	To piece	2	Hook piece	8		16
	Turn piece	18	Turn piece	8		
160	To control	2		15		17
	Use control	17	To end piece	2		
180	To piece	2	Hook piece	8		16
	Turn piece	12	Turn piece	8		
	To control	2				
200	Use control	17		15		17
	To piece	2	To end piece	2		
	Turn piece	3	Hook piece	9		9

Total	Operator No. 1	Operator No. 2	Rotoblast
429	77.6%	61.1%	Index Load Blast 73.2%
	22.4%	14.4%	Over blast 26.8%
		24.5%	

Fig. 5.32 *Sections of chart showing proposed method of rotoblast. From Mundel,* Motion and Time Study: Principles and Practice, *p. 310.*

require the crew to work faster, but it did make for a better distribution of the work and provided a more effective work pattern.

The memomotion film was also used in the discussions that preceded the installation of the new method.

5.4 WORK SIMPLIFICATION

To some, work simplification is synonymous with motion study. To others, it is a proprietary, simplified version of motion study that an individual can apply to his own work.

Perhaps it should be thought of more generally, as a participational motion-study approach suitable to the educational level of the foreman or worker and usually (but not necessarily) involving the use of only the simpler, pencil-and-paper techniques. Suitable parts of Art. 3 should be consulted.

5.5 TIME STUDY

5.5.1 Definition

Time study is the appraisal, in terms of time, of the value of work involving human effort. It produces a time standard for the performance of a series of acts by a man or group of men. To avoid chaos, a standard should be carefully defined so that consistent, reliable measurements may be made. A sample, specific definition of a standard is as follows: The standard time for a job will be 130/100 of the amount of time necessary to accomplish a unit of work, using a given method, under given conditions, by a worker who possesses sufficient skill to do the job properly, who is as physically fit for the job after adjustment to it as the average person who can be expected to be put on the job, and who works at the maximum pace that can be maintained on such a job, day after day, without harmful physical effects.* The term "time study" may also be used as synonymous with the term "work measurement."

5.5.2 Uses

Standard times may be used for the following purposes:

A. To set schedules

B. To determine supervisory objectives

C. To furnish a basis of comparison for determining operating effectiveness

D. To set labor standards for "satisfactory" performance

E. To determine the number of machines a person may run

F. To balance the work of crews or production lines

G. To compare alternative methods

H. To determine standard costs

I. To determine equipment and labor requirements

J. To determine basic times or standard data

K. To provide a basis for setting piece prices or incentive wages

*This limit is usually based on sociological acceptance rather than on physiological limits. The modern worker expects to leave work with sufficient energy to engage in leisure-time activities.

5.5.3 Techniques

In order to set up an adequate, reliable time standard, the following items must be specified: the internal content of the task (the contribution required of the individual), the unit of output involved, the type of individual for whom the standard is determined, and the degree of exertion required of the individual. Further, it should be obvious that for widespread use a set of standards should be consistent with each other in respect to difficulty of attainment.

5.5.3.1 List. Available techniques may be divided into three major groups, as follows:

A. Stop-watch (or camera) time study. First, an actual performance is studied and analyzed. Then the data obtained are synthesized into a standard. (The result is sometimes called an "engineered standard of performance.")

B. Synthesized standards. Data obtained from previous time studies, giving standards for parts of jobs, may be resynthesized into a standard for a job, in its totality, different from those previously studied. (Standards set in this manner are also referred to as "engineered.")

C. Statistical standards. Data obtained from a considerable number of people performing tasks over a considerable period of time are used on some arbitrary basis. Although such standards lack some of the features of engineered standards, they have many uses. Mathematical approaches such as linear programing and multiple regression may be used to set standards for work which may be unamenable to any direct analysis. As will be seen, such "standards," although useful, lack many of the requisites of true standards.

5.5.4 Steps in the Determination of a Standard

Each of the three basic groups contains numerous variants in method. However, the general steps remain the same; only the details vary.

5.5.4.1 Stop-watch (or camera) time study. The procedure usually involves the following five steps:

A. Defining the standard of measurement

B. Recording the standard method and identifying the unit of work

C. Observing the time taken by an actual worker

D. Rating or relating performance to standard

E. Applying allowances

DEFINING THE STANDARD OF MEASUREMENT. This need be done only once for all the standards in a plant. (See Par. 5.1.1 for an illustration.)

RECORDING THE STANDARD METHOD AND IDENTIFYING THE UNIT OF WORK. The essential criteria for the identification are:

A. Could the job be reproduced from it?

B. Does it contain everything the worker has to do?

C. Is the unit of work such as to allow convenient determination of production on the job?

The following items are usually included:

1. The department in which the job is performed

2. Job number

3. Product, material specifications, and identification as related to the operation

4. Workplace layout and dimensions

5. Description of equipment and its condition, if abnormal

6. Tool descriptions

7. Feeds and speeds of machines, welding currents used, and so forth

8. Surrounding environmental conditions

9. Services related to machine and tool maintenance, delivery, and material-handling rendered to, or required of, the worker

10. The unit of production to be counted

11. A description of the manual details of the job. These are usually broken down into so-called "elements," in accordance with the following criteria:

A. Easily detected and definite end points

B. As small as is convenient to time

C. As unified as possible

D. Hand time separated from machine time

E. Constant elements separated from variables (as related to several jobs)

F. Regular and irregular items separated

The terminology used in the description of the elements varies with the nature of the job. For heavy work involving movement from place to place, a description of the activities of the person as a whole is most suitable. For heavy work with crews, the activities of the individual crew members, and their coordination, must be indicated. For heavy and moderately heavy work done mainly at one place, the activity of the individual as a whole is sufficient, unless the coordination of the worker's body members is a critical factor, in which case the activities of each body member should be detailed and the coordination indicated. This latter procedure is also preferable for light work. Therblig terminology is also useful in descriptions, particularly for small handwork. Of course, if therblig time studies are shown to men who are unfamiliar with the terminology, they will be unintelligible and may arouse mistrust. However, since more and more plants are undertaking training programs in motion and time study for foremen and supervisors, therbligs are constantly coming into more common use. In any event, they furnish a shorthand for record-taking, and may be expanded for instructional use later on.

In some cases, a single time-study sheet is sufficient for recording the standard practice. In other cases, it is necessary to attach drawings of the tools and other equipment. In still others, the written standard practice may run to several pages. No one form has yet been devised that will handle the problems of all plants satisfactorily. A sample description for a bench job is shown in Fig. 5.33, and part of the description for a two-man job is shown in Fig. 5.34. Still another time-study form is shown in Fig. 5.35.

OBSERVING THE TIME TAKEN BY AN ACTUAL WORKER. Six different methods of recording observed times are in common use. Four of these involve the use of a stop watch; one, the use of a motion-picture camera; and one, the use of a special time-study machine. Under certain conditions, a sound recorder may be used advantageously.

Continuous timing. The watch runs continuously throughout the study. It is started at the beginning of the first element of the first cycle being timed, and is not stopped until the study has been completed. At the end of each element, the time is recorded. The individual element times are obtained by successive subtractions after the study has been completed.

Repetitive (or snap-back) timing. The watch

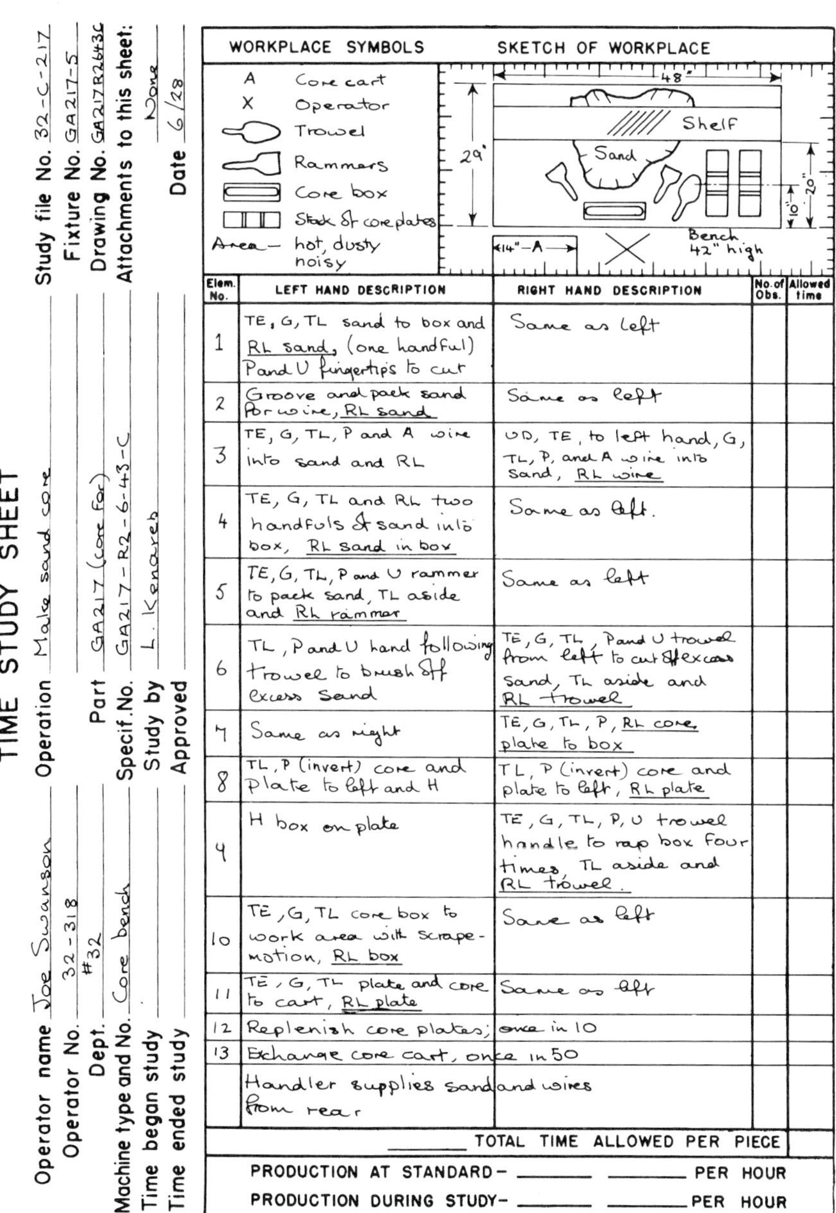

TIME STUDY SHEET

Operation Make sand core

Part GA217 (core for)

Specif. No. GA217-R2-6-43-C

Study by L. Kenareb

Approved

Operator name Joe Swanson

Operator No. 32-318

Dept. #32

Machine type and No. Core bench

Time began study

Time ended study

Study file No. 32-C-217

Fixture No. GA217-5

Drawing No. GA217R2643C

Attachments to this sheet: None

Date 6/28

WORKPLACE SYMBOLS

A Core cart
X Operator
Trowel
Rammers
Core box
Stack of core plates
Area — hot, dusty noisy

SKETCH OF WORKPLACE

Elem. No.	LEFT HAND DESCRIPTION	RIGHT HAND DESCRIPTION	No. of Obs.	Allowed time
1	TE, G, TL sand to box and RL sand, (one handful) P and U fingertips to cut	Same as left		
2	Groove and pack sand for wire, RL sand	Same as left		
3	TE, G, TL, P and A wire into sand and RL	UD, TE, to left hand, G, TL, P, and A wire into sand, RL wire		
4	TE, G, TL and RL two handfuls of sand into box, RL sand in box	Same as left.		
5	TE, G, TL, P and U rammer to pack sand, TL aside and RL rammer	Same as left		
6	TL, P and U hand following trowel to brush off excess sand	TE, G, TL, P and U trowel from left to cut off excess sand, TL aside and RL trowel		
7	Same as right	TE, G, TL, P, RL core plate to box		
8	TL, P (invert) core and plate to left and H	TL, P (invert) core and plate to left, RL plate		
9	H box on plate	TE, G, TL, P, U trowel handle to rap box four times, TL aside and RL trowel.		
10	TE, G, TL core box to work area with scrape-motion, RL box	Same as left		
11	TE, G, TL plate and core to cart, RL plate	Same as left		
12	Replenish core plates;	once in 10		
13	Exchange core cart, once in 50			
	Handler supplies sand and wires from rear			

______ TOTAL TIME ALLOWED PER PIECE

PRODUCTION AT STANDARD — ______ ______ PER HOUR

PRODUCTION DURING STUDY — ______ ______ PER HOUR

Fig. 5.33 *Front of time study sheet with method description in terms of therbligs for sand-core molding. From Mundel,* Motion and Time Study: Principles and Practice, *p. 351.*

STANDARD METHOD FOR BURR ROLL- "SUPER STRIP". 2 MAN CREW 1 sheet of 2
150 lb. COIL 3/4" WIDE 4 WIRE TIES

Element No.	Operator	Element No.	Helper
1	Stop machine and walk from control lever to coiler	1	Pick up 4 pieces of wire from holder, place 4 pieces of wire 90 degrees apart on top of finished coil on skid and step to coiler
2	Inspect last end of coil, pick up snips from machine framework, cut off last end of coil and lay end and snips on machine framework	2	Remove handwheel and front plate, pulls coils a little away from back plate and slip on coil clamp
3	Walk from coiler to spare reel		
4	Index swivel reel, pick up end of coil on spare reel and feed thru Burr Roller to feed rolls		
5	Walk from feed rolls to coiler		
5a	With helper lift coil off coiler and place on skid	3	With operator lift coil off coiler and place on skid
6	Walk from coil on skid to control lever	4	Remove core from coil on skid and place on coiler
7	Start machine and feed end of coil thru feed rolls and past coiler and stop machine	5	Delay
8	Walk to coiler		
9	Inspect first end of coil, pick up snips from machine framework, cut off first end of coil and lay end and snips on machine framework		
10	Walk from coiler to control lever	6	Rethread coiler, take up slack and replace front plate and handwheel
11	Delay		
12	Start machine and walk from control lever to spare reel	7	Step to coiler strip on skid
13	Unscrew reel handwheel from reel shaft and lay on floor out of way	8	Pick up one end of wire under finished coil on skid, make one wire tie around coil by hand, twists wire tight around coil with pliers, and bends ends flat on coil
14	Pull reel front plate from reel shaft, roll to support and lean front plate against support	9	Repeat No. 8
15	Pick up hook, pull coil about 8" off pile on skid and lay hook down	10	Repeat No. 8
		11	Repeat No. 8

Fig. 5.34 *Part of method description for two-man burr roll.*
From Mundel, Motion and Time Study: Principles and Practice, *p. 354.*

HISTORY DETAIL

Machine Data

	ELEMENT No. 1	2	3	4	5
Average Diameter of Part					
Spindle Speeds					
Cutting Speeds, Feet Per Min.					
Feed Per Minute					
Feed Per Revolution					

Type of Machine

Type of Chuck

Material Machined

Cutting Tools: Carbon ☐ High Speed ☐ T. C. C. ☐

Classification or Type of Tools

Moulding Data

Type of Machine

Metal Used

Type of Flask

Size of Flask: Length ... Width ... Height

Type of Pattern

Number of Pieces on Gate

Number of Cores

Dry Cores ... Green Cores

Number of Loose Inserts on Pattern

Number of Riser Blocks

Number of Men: Molders ... Helpers

Sketch:

Remarks:

Fig. 5.35 *A time study form adapted for machine and foundry operations.*

is started at the beginning of the first element of the first cycle being timed, and is simultaneously read and snapped back to zero at the completion of this, and each subsequent, element. This procedure allows the element times to be entered directly on the time-study sheet without the need for subtractions. Consistent over- and under-reading of the watch will cause errors that do not arise in the continuous method. Also, considerable manipulation of the watch is required. Many labor groups look upon the repetitive method as highly liable to error. With extremely short elements, any errors that occur may represent large percentages of the elements. Some time-study men use an additional watch to accumulate the total time as a check on these errors. However, in competent hands, the snap-back method can be used successfully.

Continuous accumulative timing. Two stop watches are mounted on a special holder with a mechanical linkage between the controls of the two watches. The push buttons on this holder are so arranged that when one watch is stopped the other is started, and vice versa. The watches are used alternately, each accumulating the time for half of the elements. This method allows each watch to be read while the hand is motionless. In obtaining the actual element times, alternate recordings are subtracted successively.

Repetitive accumulative timing. The same set-up is used as in continuous accumulative timing, except that another button is provided to return the stopped watch to zero before the next element ends. Hence, actual element times may be recorded directly. This is a reasonably accurate method. However, most time-study men regard the mechanism as troublesome, and accumulative methods are not widely used.

Motion-picture camera timing. Regular speeds may be used, although memomotion speeds are more useful. Regular film-analysis techniques are used to obtain the actual data.

Time-study machines and sound recorders. A

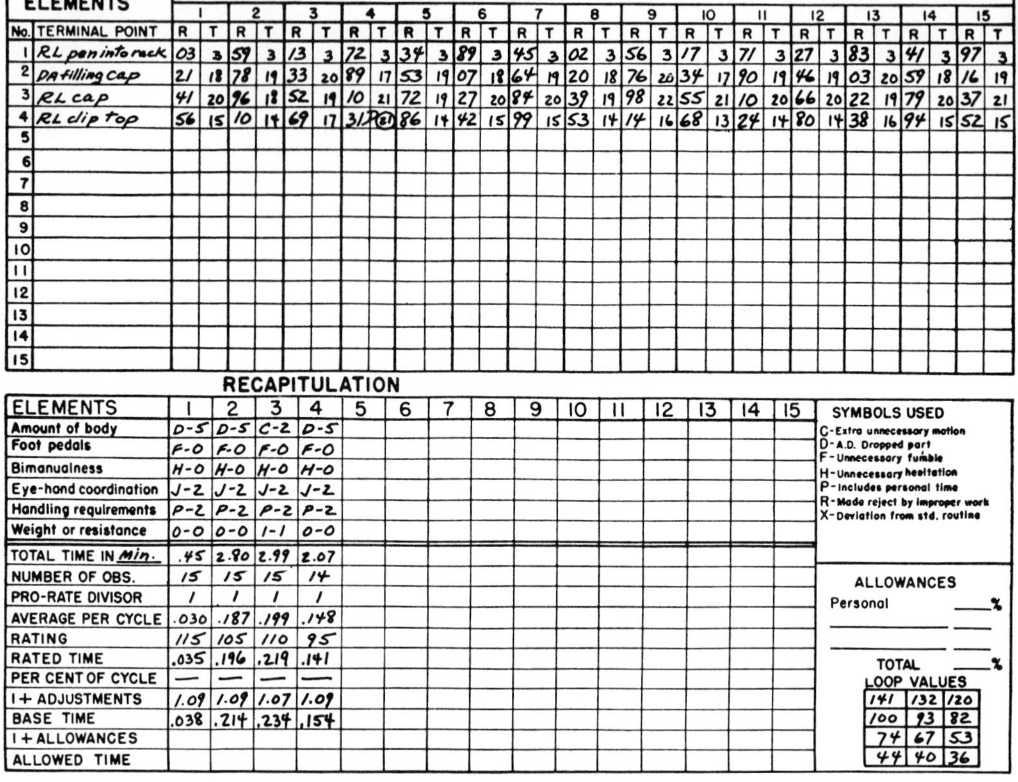

ELEMENTS		CYCLES 1		2		3		4		5		6		7		8	
No.	TERMINAL POINT	R	T	R	T	R	T	R	T	R	T	R	T	R	T	R	T
1	RL pen into rack	03	3	59	3	13	3	72	3	34	3	89	3	45	3	02	3
2	DA filling cap	21	18	78	19	33	20	89	17	53	19	07	18	64	19	20	18
3	RL cap	41	20	96	18	52	19	10	21	72	19	27	20	84	20	39	19
4	RL clip top	56	15	10	14	69	17	31 R	(21)	86	14	42	15	99	15	53	14
5																	
6																	
7																	
8																	
9																	
10																	
11																	
12																	
13																	
14																	
15																	

ELEMENTS		CYCLES 9		10		11		12		13		14		15	
No.	TERMINAL POINT	R	T	R	T	R	T	R	T	R	T	R	T	R	T
1	RL pen into rack	56	3	17	3	71	3	27	3	83	3	41	3	97	3
2	DA filling cap	76	20	34	17	90	19	46	19	03	20	59	18	16	19
3	RL cap	98	22	55	21	10	20	66	20	22	19	79	20	37	21
4	RL clip top	14	16	68	13	24	14	80	14	38	16	94	15	52	15
5															
6															
7															
8															
9															
10															
11															
12															
13															
14															
15															

RECAPITULATION

ELEMENTS	1	2	3	4	5	6	7	8	9	10	11	12	13	14	15
Amount of body	D-5	D-5	C-2	D-5											
Foot pedals	F-0	F-0	F-0	F-0											
Bimanualness	H-0	H-0	H-0	H-0											
Eye-hand coordination	J-2	J-2	J-2	J-2											
Handling requirements	P-2	P-2	P-2	P-2											
Weight or resistance	0-0	0-0	1-1	0-0											
TOTAL TIME IN Min.	.45	2.80	2.99	2.07											
NUMBER OF OBS.	15	15	15	14											
PRO-RATE DIVISOR	1	1	1	1											
AVERAGE PER CYCLE	.030	.187	.199	.148											
RATING	115	105	110	95											
RATED TIME	.035	.196	.219	.141											
PER CENT OF CYCLE	—	—	—	—											
1 + ADJUSTMENTS	1.09	1.09	1.07	1.09											
BASE TIME	.038	.214	.234	.154											
1 + ALLOWANCES															
ALLOWED TIME															

SYMBOLS USED
C-Extra unnecessary motion
D-A.D. Dropped part
F-Unnecessary fumble
H-Unnecessary hesitation
P-Includes personal time
R-Made reject by improper work
X-Deviation from std. routine

ALLOWANCES
Personal ____%
________ ____
________ ____
TOTAL ____%

LOOP VALUES

141	132	120
100	93	82
74	67	53
44	40	36

Fig. 5.36 *Continuous timing of an inspection of fountain pens.*
From Mundel, Motion and Time Study: Principles and Practice, *p. 373.*

common time-study machine employs a constant-speed paper tape with finger-actuated pencil markers. Accurate time measurements can be made by depressing the markers according to a predetermined scheme, and then measuring the distance between the marks with a calibrated ruler. If the task involves communications, a sound tape recording may be made and later transcribed and interpreted on a recording oscillograph. This record may be measured, as with the time-study machine. Also, special sound cues may be recorded on the tape while an operation is being observed, and these may be timed later either with a watch or a recording oscillograph. Under certain conditions, these procedures are ideal.

Time recording forms. For recording the time values measured with a stop watch, the form shown in Fig. 5.36 is convenient. Many variations of this form are used with equally good results. The form is often printed on the back of the form used to record standard practice. It should never be used without an adequate written standard practice. On the example shown, the end point of each element after which the time is to be recorded should be written in the small description column, as an aid in timing. Space is provided on this form for recording 15 cycles of 15 elements. If more elements or cycles are to be recorded, two or more sheets may be used. If the use of two sheets is a common occurrence, a larger form should be designed and printed. For each element of each cycle, two boxes are provided in columns labeled R and T. The R column is for "readings" when continuous timing is used. The subtracted times, which are the time values for each element, are placed in the T, or "time," column. Different-colored pencils keep the R and T values separate and facilitate correct calculations later. If repetitive timing is used, the values are entered directly in the T column, or else a form with T columns only may be drawn up. It is usual practice to time enough cycles to obtain a representative sample of performance.

Representative sample of times. Some variation almost always occurs from reading to reading for any element, even if the worker is not attempting to vary his pace. This variation is caused by the following factors, among others:

1. Random variations in operator movements and pace
2. Random variations in the positions of the parts worked with
3. Random variations in the position of the tools used
4. Random variations caused by slight errors in watch-reading

For any observed pace of performance, timings of 10 cycles will tend to produce a more stable average than timings of 5 cycles; the average of 15 timings will tend to be better than that of 10, and so forth. In practice, however, two sets of 15 readings on the same element will seldom, if ever, produce an identical average, even though the pace is the same in both cases.

A reasonable limit on the number of readings is to take enough to make the chances 95 out of 100 that the observed average will be within ± 5 per cent of the true average for the element for the pace at which it was performed. Some may prefer a looser criterion of 95 chances out of 100, ± 10 per cent. These odds may be restated as 68 out of 100, ± 5 per cent.

If the time studies are to be used for establishing incentive wages, either criterion is reliable enough, since ± 5 or 10 per cent usually approximates a bargainable increment in wages. Errors of more than this magnitude are to be avoided.*

Statisticians have developed methods for determining the probable accuracy of a sample. These methods may be adapted to this time-study problem.

The applicable sampling theories involve two simple formulas, Eqs. 5.1 and 5.2, which follow. Both are based on the assumption that chance or random causes are controlling the variation from reading to reading for a given element. In most cases, this is a tenable assumption.

Eq. 5.1 gives a measure of the variability of data about its average. The variability is represented by SD, the standard deviation, which is expressed as follows:†

$$\mathrm{SD} = \sqrt{\frac{\Sigma d^2}{N}} \tag{5.1}$$

d = $X - \bar{X}$ computed for each reading of the element separately before squaring and then summing
X = Individual readings of an element
$\bar{X}$ = Mean or average of all readings of an element
Σ = Sum of like items
N = Number of readings of an element

This equation may be expressed as follows for machine computation (Friden, Monroe, Marchant, etc.):

$$\mathrm{SD} = \sqrt{\frac{\Sigma X^2}{N} - \left(\frac{\Sigma X}{N}\right)^2} = \frac{1}{N}\sqrt{N\Sigma X^2 - (\Sigma X)^2}$$

Assuming that this represents the variability of a huge group of similar readings or the parent population (a commonly tenable assumption), another measure, $\mathrm{SD}_{\bar{x}}$, the standard error of the mean (or average), may be computed by Eq. 5.2, which indicates the probable variability of the averages of groups of N values of X about the obtained $\bar{X}$.

$$\mathrm{SD}_{\bar{x}} = \frac{\mathrm{SD}}{\sqrt{N}} \tag{5.2}$$

The property of this last measure is such that 95 per cent of the probable values of $\bar{X}$ (average for the element) will lie within $\pm 2\mathrm{SD}_{\bar{x}}$ of the true average.

Hence, if $2\mathrm{SD}_{\bar{x}}$ is equal to or less than 5 per cent of $\bar{X}$, we may say the chances are at least 95 out of 100 that our average for the element to which the rating will be applied is within ± 5 per cent of the true average representing the performance we observed. If the 10 per

*William Gomberg, *A Trade Union Analysis of Time Study* (Chicago: Science Research Associates, 1948), p. 14.

†SD is used here in place of the symbols σ or s, which are used in Sec. 13 and 14 for standard deviation. SD_{x} is used in place of σ_{x}. See Sec. 13 for a complete discussion of industrial statistical methods.

cent criterion is used, then the above may be restated with "10 per cent" in place of "5 per cent."

As was explained earlier, both criteria can reasonably be applied to time studies.

If the selected limiting condition is not met, we may work Eq. 5.2 backward, using the SD we first obtained, setting $2SD_{\bar{x}}$ equal to 5 per cent of $\bar{X}$, and solving for N', which will indicate the number of readings that will probably be needed.

Indeed, it is this last property that makes this test feasible, easy, convenient, and economical to use, after certain mathematical manipulations of the formulas have been made.

Combining Eqs. 5.1 and 5.2, we may state:

$$SD_{\bar{x}} = \frac{\frac{1}{N}\sqrt{N\Sigma X^2-(\Sigma X)^2}}{\sqrt{N'}}$$

Setting 5 per cent of $\bar{X}$ equal to $2SD_{\bar{x}}$, we get

$$0.05\bar{X} = \frac{\Sigma X}{20N} = 2\frac{\frac{1}{N}\sqrt{N\Sigma X^2-(\Sigma X)^2}}{\sqrt{N'}}$$

$$\frac{\Sigma X}{20} = \left(\frac{2\sqrt{N\Sigma X^2-(\Sigma X)^2}}{\sqrt{N'}}\right)$$

$$\text{and } N' = \left(\frac{40\sqrt{N\Sigma X^2-(\Sigma X)^2}}{\Sigma X}\right)^2 \quad (5.3)$$

where N' is the required number of readings.

This equation may be easily handled even by one who is not familiar with the mathematics of its derivation.

If the analyst prefers to set his limits as 95 chances out of 100 within ± 10 per cent, then:

$$N' = \left(\frac{20\sqrt{N\Sigma X^2-(\Sigma X)^2}}{\Sigma X}\right)^2 \quad (5.4)$$

Rating or Relating Performance to Standard. In the preceding section, techniques were described for timing a performance of an actual operator on a job. With luck, the operator studied might meet all the requirements of the standard type and might be working at the proper pace. Then the observed time values would require no further adjustment. In actual practice, however, the operator usually fails to perform ideally. Either he does not meet all the requirements of the standard operator, or fails to work at the pace required for standard performance, or both. Consequently, the analyst must ask himself (1) how to evaluate the performance observed as compared with the requirements given in the definition of standard used as the basis of the measurement, and (2) how to reduce this evaluation to a mathematical value that will allow the adjusting (if necessary) of the representative time values actually obtained, so as to determine a base for the standard time. These are the aims of rating.

Common rating procedures. Common time-study rating procedures can be divided into two main groups: mathematical procedures; procedures requiring judgment.

A. *Mathematical procedures.* The mathematical plans require a statistical sorting out, on the basis of the time recordings alone, of the effect of the operator's skill, aptitude, pace, relative rate of exertion, capriciousness, and so forth, from normal job variation in order to obtain a measure that would be relatively the same, regardless of whatever conditions of the above mentioned variables were in existence at the time the data were recorded. It is not surprising that this goal has never been realized. Any other mathematical method would require an outside reference point against which observed variations could be measured. The only reference point usable for such a purpose would be the standard time, but if this were known there would be no reason for the rating.

B. *Procedures requiring judgment.* Many rating systems involving judgment have been proposed and used. For the most part, they fit the following definition: "Rating is that process during which the time-study man compares the performance of the operator under observation with the observer's own concept of normal performance."*

Most of these procedures require the time-study man to perform the same two basic steps:

1. He must judge the difficulty of the job and form a mental concept of what the performance of the job under observation would look like if it met the requirements of standard performance as defined.

2. The observer must appraise the actual performance under observation as compared

*Society for the Advancement of Management, Committee on Rating of Time Studies, *Advanced Management*, Vol. 6, July-Sept. 1941, 110.

with the concept formed in Step 1 and place a numerical value on this appraisal.

The usual guide recommended for Step 1 is "experience."

Improved rating procedure. It may be seen that the evaluation or rating of performance (as is correctly done in many rating procedures) may be reduced to a judgment of not more than two items: (a) observed pace, and (b) job difficulty. In the typical time-study procedure, the time-study observer first judges (b) job difficulty and then judges (a) observed pace.

What makes a more reliable time-study rating procedure possible is, first of all, the realization that the difficulty of the job and its effect on maximum possible pace do not need to be judged but may be reduced to objective terms. These objective terms are based on observable phenomena, and may be reduced to tabular data as a function of strength required, amount of body used, degree of dexterity, and the like, and may be experimentally verified.

What is proposed is again a two-step rating procedure, but the steps are in the reverse of the conventional order. This is called *objective rating* and consists of the following steps:

1. Observed pace is rated against an objective pace standard, which is the same for all jobs. In this rating, no attention whatsoever is paid to job difficulty or to its effect on possible pace. Hence, a single pace standard may be used instead of a multiplicity of mental concepts.

2. A percentage increment is applied to adjust the resulting value subsequent to Step 1. This percentage increment is taken from experimentally determined tables of the effect of various observable factors on the exertion required at a given pace and is called "a secondary adjustment."*

It has been verified experimentally that this improved rating procedure gives a more consistent concept of standard time than do the conventional procedures. It is true that the procedure requires a certain amount of preliminary activity before any time studies can be made. However, this activity need be performed only once, and the subsequent details of making time studies are less complex than in conventional rating procedures. The first part of this preliminary activity may be done in any one of three convenient ways:

1. The time-study man, or department, may choose a simple job involving almost no skills or special aptitudes and then determine experimentally the pace on this job when performance meets the requirements of standard time. This plan may require some experimentation with several operators.

2. The time-study department may make a series of films of workers working at different paces on a simple job and then ask management to select the one that represents its concept of standard pace. True, judgment must be exercised here and some original error is possible, but neither factor is critical. At least a standard of unchanging pace is set up. Also, if management wants to assume the prerogative of defining acceptable pace, it should be assumed at a high enough organizational level. The pace selected may be jointly negotiated by labor and management, in which case the accuracy with which it represents a previously accepted definition of standard is of less consequence, although the discrepancy may be discovered later through experience with the use of the selected pace. At least the pace will be acceptable to both parties, and without such mutual acceptance joint agreements concerning money per hour appear inadequate.

3. A simple job may be shown to large groups of industrial engineers (or joint labor-management groups). Then values that have been corrected for concept of standard can be averaged and used as a basis for the standard pace.

The second step suggested in the preliminary activity being outlined here is the formal selection of a film showing the standard rate of activity for any one job. This film represents the unit of measurement, or the rate of activity, for 100 per cent standard pace.† Since the record is permanently preserved, the standard rate of activity may actually be included in the labor contract. At the very least, it is available for comparison and for use in discussing ratings on an objective basis. Such can never be the case when the standard represents merely the time-study man's unan-

*If a 60 base system is preferred, this can be identified as 60.

†Recently, "difficulty adjustments" has been used instead of "secondary adjustments," but since Dr. Mundel originally used "secondary adjustment," we have retained that term. Ed.

chored, mental concept of proper performance.

But a single film record is not enough for actual use. It is highly desirable to prepare *step films* that show step-by-step deviations from standard pace on the job. These records make it possible to establish markings on the scale of pace and facilitate the rating.* Once a standard pace has been selected by any one of these procedures, the step films or multi-image films can be easily prepared.† Experiments have shown‡ that about 6 per cent change in pace is the usual minimum detectable difference; hence, the steps on the film should approximate this magnitude. Although a considerable group of regular time-study engineers were used to obtain these data, it is possible that with further training the minimum detectable difference might be reduced. Therefore, this percentage should not be considered as an absolute value.

After the time-study man has completed this preliminary activity, he is ready to move on to the first step of the objective rating procedure. In practice, he may do one of the following:

A. Compare the observed job with the concept of the scale of standard pace that he has obtained through careful study of the step films or multi-image films.

B. Compare the observed job with the film by projecting the film into a shadow box near the job so that both may be viewed simultaneously.

C. Compare a film of the observed pace with the step films or multi-image films by projecting both films simultaneously.

Regardless of the method, the time-study man must only judge whether the job being studied (actual performance or film of performance) is being performed at a pace (rate of activity) equal to any one of the step films or steps on the multi-image film, or at a pace lying between any two of the steps, and then assign a rating as indicated by the predetermined values of the steps. He pays absolutely no attention to job difficulty and its effect on the possible pace for the task.

The next part of the preliminary activity prior to using the improved rating procedure is the determination of a table of secondary adjustments so that the time obtained by time study, after being adjusted to standard pace, may be further adjusted to represent the rate of exertion included in the definition of a standard for the actual job being studied. Such a table will enable the time-study observer to perform Step 2 of the two steps of the objective rating procedure.

Obviously, all jobs cannot be performed at the standard pace. Most jobs are more difficult than the job for which standard pace has been set, and some jobs are more difficult than others. For instance, some jobs involve heavier parts, closer visual work, and so forth. These variations set different limits on the pace possible for each job with a fixed rate of exertion. However, they may be evaluated objectively.

The method consists of determining the various factors that make for difficulty in the job, and evaluating their effect. This evaluation is applied as a *secondary adjustment* in computing the standard for the job, so that all the standards will be consistent in regard to attainability. These secondary adjustments may be set up as in Table 5.20.

The work of developing the values for these secondary adjustments is by no means complete, but even at the present stage of development it should be more satisfactory and reliable than leaving the adjustment to the mental evaluation of the time-study man, as is inherent in conventional rating procedures. Moreover, if inconsistencies appear when the secondary adjustments are applied, the source can be tracked down and lasting corrections can be made, a course of action that is not possible with the conventional approach. Reducing secondary adjustments to tabular form also eliminates other sources of subjectively caused variation.

Note that all the adjustments are indicated as positive increments of time above the time

*All the films may be in loop form—that is, the front end may be spliced to the back to permit continuous projection for any period of time. Also the frames may be divided into different areas, each area showing a different pace, so that a group of steps may be projected simultaneously. Such films are called *multi-image step films*.

†These films were first proposed in M. E. Mundel, *Systematic Motion and Time Study* (Englewood Cliffs, N. J.: Prentice-Hall, Inc., 1947).

‡M. E. Mundel, and R. N. Lehrer, "An Evaluation of Performance Rating," *Proceedings 12th Annual National Time and Motion Study Clinic*, Industrial Management Society, Chicago, 1948.

TABLE 5.20. *SECONDARY ADJUSTMENTS FOR TIME STUDIES**

Category Number	Description	Reference Letter	Condition	Per Cent Adjustment
1	Amount of body used	A	Fingers used loosely.	0
		B	Wrist and fingers.	1
		C	Elbow, wrist, and fingers.	2
		D	Arm, etc.	5
		E	Trunk, etc.	8
2	Foot pedals	F	No pedals or one pedal with fulcrum under foot.	0
		G	Pedal or pedals with fulcrum outside of foot.	5
3	Bimanualness	H	Hands help each other or alternate.	0
		H_2	Hands work simultaneously doing the same work on duplicate parts.	10
4	Eye-hand coordination†	I	Rough work, mainly feel.	0
		J	Moderate vision.	2
		K	Constant but not close.	4
		L	Watchful, fairly close.	7
		M	Within 1/64 inch.	10
5	Handling requirements†	N	Can be handled roughly.	0
		O	Only gross control.	1
		P	Must be controlled, but may be squeezed.	2
		Q	Handle carefully.	3
		R	Fragile.	5
6	Weight		Identify by the letter W followed by actual weight or resistance.	Use curve (Fig. 5.37)

* From Mundel, *Motion and Time Study: Principles and Practice.*
† Note: These scales could possibly go much higher in some cases.

required at the standard pace (film loop or concept of rate of activity). Hence, the film loop or concept should, at 100 per cent pace, represent the concept of standard time on an extremely simple operation. Also, these adjustments may be used only when all jobs have been rated against a single standard pace that does not take job difficulty into account. Essentially, the developed data of Table 5.20 are used as follows:

A. Separate adjustments are made for each element.

Assume that, on a given job, for 10 per cent of the time the operator pushes a lever with a 15-pound resistance. For the rest of the time, the operator loads very light parts into a fixture. Obviously, the operator's pace will vary from the first element to the second. Consequently, each element will be adjusted separately.

B. The total secondary adjustment for an element will be the simple sum of all the appropriate values from the scales for all the factors. As far as is known at present, these factors are additive. No complex interaction has yet been found at the element level.

C. The secondary adjustments are combined with the pace rating. If the rating is 90 per cent and the total secondary adjustment is 12 per cent, the observed time for that element will be multiplied by 0.90 × 1.12, or 1.01. This will give the same result that would have been obtained had the observed time first been multiplied by 0.90 to get the time required at the 100 per cent pace and then 12 per cent of the result added to the first product. The 0.12 and 0.90 cannot be added; otherwise the actual increment would vary with the pace observed, which would not be correct. (If a different numerical scale or scale base is used, the values used will change, but the essential procedure will not.)

D. The factors for which secondary adjustments should be added are:

1. Amount of body used
2. Foot pedals
3. Bimanualness
4. Eye-hand coordination
5. Handling or sensory requirements
6. Weight handled or resistance encountered

Allowances. The procedures presented thus far do not include provisions for three additional groups of adjustments that must commonly be made:

1. Allowance for personal time
2. Allowance for irregular occurrences that may not have been time studied or that cannot be prorated
3. Allowance for machine time

These allowances differ from the secondary adjustments in three respects: (1) usually they are applied similarly to every element in the task; (2) in some categories, they may actually represent a block of time that may be accumulated from a large number of cycles to eventually provide an interval in the work period during which the worker will not be working; and (3) they relate to factors external to the job.

1. *Allowance for personal time.* This category is not to be confused with the commonly used, catch-all term *fatigue allowance*, which, because of the many interpretations of the term *fatigue*, has led to a great deal of misunderstanding. After the time-study observations have been adjusted either by subjective ratings or objective ratings (including secondary adjustments), the result is supposedly a time value that will permit the operator to produce work in this time, or less, throughout the normal work period, so far as the internal work of the operation is concerned. However, note that no attention is paid to personal needs (among other items) or to the effect of external conditions upon personal needs.

A person usually cannot work through a normal industrial work period without attending to personal needs. The amount of time required will be affected by the conditions surrounding the work: less when the surroundings are comfortable and quiet, and more when they are hot, dusty, or noisy. Many plants have a set schedule to provide for personal time during the working day.

2. *Allowance for irregular occurrences.* It is desirable to study and rate all irregular occurrences and to prorate them so that they are properly apportioned to each cycle. However, it is extremely difficult to handle some occurrences in this manner. For example, let us assume a sewing-machine operator must clean his machine at the end of each day. Perhaps this takes five minutes. It is hardly proper to charge all this time to the last job of the day. Nor is it feasible to prorate it to a value that is added to each cycle, since the operator may work on several jobs each day and each job may have a different total cycle time. Yet the cost of these five minutes should somehow be distributed among the jobs. Consequently, some adjustment must be made that will permit the accumulation of five minutes during the day as part of the standards. Similar allowances may have to be made for tool and machine maintenance, and the like.

A unique method of determining and evaluating irregularly occurring elements is described as "ratio-delay" studies. These are made as follows:

The observer passes the work stations at random intervals. Each time he passes some predetermined spot, he notes on a check sheet what is occurring. The following precautions should be observed:

> 1. Only homogeneous groups should be combined in a single study; that is, similar observations on similar machines.
> 2. Fewer than 1000 observations are of little value. Best results are obtained in the range from 2500 to 3500 observations.
> 3. Observations must be at random intervals . . . and independent.
> 4. Observations should be made throughout all hours of a working day, and over a period of not less than two weeks.*

The percentage of occurrences of each type of activity in the sample of observations indicates the percentage of time spent at it. When these percentages are reduced to minutes per working day, they can be used as a basis for determining percentage allowances to be added to the standard time for all elements.

As an alternate procedure, a mechanical or electrical recorder is placed on the machine to provide an accurate time measurement of each delay. In an accompanying log, the foreman or worker notes the reason for each delay. These two records may later be analyzed to

---

*J. S. Petro, "Using Ratio-Delay Studies To Set Allowances," *Factory*, October 1948, 94.

provide a basis for allowances for irregular occurrences.

It is strongly suggested, however, that when the irregular work can be directly attributed to a specific job, and when a rate of occurrence during that job can be determined, the time value should be handled as an irregular element. Otherwise the allowances for irregular elements may get out of hand and may eventually lead to sizable inconsistencies in the comparative difficulty of standards for different jobs. Extended time studies can be made in which the work is divided only into productive work and delays. Either a stop watch or a memomotion camera can be used. With a camera, more than one machine can be observed at a time, with no loss of accuracy. Such studies provide a basis for determining accurate allowances.

3. *Allowance for machine time.* There are some who will feel that the adding of an additional allowance, so as to permit an equal production increment over standard on all jobs, to jobs where the machine controls in excess of 45 per cent of the cycle (the percentage at which the rest effect ceases to make up for the lack of production opportunity) is a "softening up" of the standard. This attitude is worth examining from two points of view: first, as if no incentives were to be applied; and second, with incentives applied.

In either case, production in excess of standard should be expected from the typical worker. The standard should be set below expected performance in order to provide a situation that leads to "attainment" rather than to "implied frustration."

When no incentives are used, an adjustment to make equal the excess attainment possible on all jobs will enhance the value of the standards for most uses, since they may be used similarly for scheduling and the men in the shop may inter-compare their performance with one another on a simple basis. This is their usual concept of equity and is hardly worth disturbing.

If incentives are in use, an incentive increment may be provided for machine jobs that is equal to the increment provided for jobs that involve only hand time. Or an even higher incentive increment may be established, since the actual burden rate is higher on the machine jobs. Certainly some adjustment must be made if the job hierarchy is not to be completely upset.

Table 5.21 gives a schedule of allowances, including a set of values for equalizing the over-standard production attainable on jobs that are partly machine-controlled. These values are based on a standard that makes 30 per cent over standard production typically expectable. The values of category three would have to be altered if any other percentage were introduced.

For tasks where the operator works during part of the machine cycle, further computations are necessary before the values from all three categories are added to obtain the allowance. Let us assume we have a task of which 60 per cent is controlled by the machine. However, since the operator prepares a load during part of this time, he actually works 70 per cent of the cycle. According to Table 5.21, an operator with a cycle of 70 per cent hand time would receive a 0 allowance; the implication is that he could work fast enough during the 70 per cent to exceed the standard by 30 per cent without any allowance. However, his excess productivity during 40 per cent of the cycle would show up. How fast he went during the remainder of the hand time would be of no value, since the machine controls the production after the 40 per cent for hand time is completed. Consequently, instead of gaining 30 per cent, he would gain only $30 \times 40/70 = 17$ per cent. An allowance of $30 - 17$, or 13 per cent, would have to be added to bring this standard to a level he could exceed by 30 per cent without greater exertion than a 100 per cent manually paced operator. Note that this allowance is larger, as it should be, than the 6 per cent allowance that would accompany a job involving 60 per cent machine time during all of which the operator rested. Other situations may be evaluated in a similar manner.

Table 5.21 is only suggestive. The nature of the allowance or the meagerness of data may require that some of the items be recomputed or negotiated before being used in an actual application.

SUMMARY. A time study complete with allowances, and with the final calculations made and transferred to the front, is shown in Fig. 5.38a and 5.38b. The two calculations at the bottom of the front of the sheet are designed to summarize the effect of the ratings and allowances for the time-study supervisor. The "production anticipated at standard" is computed on the basis of the allowed cycle time. The "production during study" is computed on

TABLE 5.21. *ALLOWANCE TABLE**

Category	Reference Letter	Condition	Time in Minutes in 8-Hour Work Spell†	Per Cent
1—Personal‡	S T U SP	Comfortable Warm or slightly disagreeable Hot, dusty, noisy, etc. Special or unusual	24 30 60 As required	
2—Irregular (cleanup, tool, etc.)	Use name	Evaluate	Suitable	

To reduce the total of time from categories 1 and 2 to a per cent value, use an equation in which: A = minutes per work day; B = minutes personal allowance per work day; C = minutes irregular allowance per work day.

$$\frac{B + C}{A - (B + C)} \times 100$$

Category	Reference Letter	Condition	Time in Minutes in 8-Hour Work Spell†	Per Cent
3—Per cent of base time of cycle controlled by the machine**	V followed by per cent controlled by machine	100 of cycle controlled by machine		30
		95		27
		90		24
		85		20
		80		17
		75		14
		70		11
		65		9
		60		6
		55		4
		50		3
		45 or less		0

* From Mundel, *Motion and Time Study: Principles and Practice.*

† Divided into two 4-hour work spells by a lunch period.

‡ If rest spells are used, as total time of rest pauses or, instead of this, if rest pauses are meant as substitutes for personal time taken at will.

** These per cent allowances are for standards with a 30 per cent anticipated typical production increment.

the basis of the total average time (with irregular elements prorated as necessary). These two values sum up the relationship between observed performance and anticipated performance; they are often of great aid in indicating when unusual or difficult situations may arise from the installation of the standard.

5.5.4.2 Synthesized Standards. Procedures for synthesizing standards may be divided into two main groups: (1) procedures that use standard elements and that are generally designed for use with a given type of work, and (2) procedures that are based on smaller action units to permit more general application. Each of these two types is discussed separately.

SYNTHESIZED STANDARDS FROM STANDARD ELEMENTS. To develop the basic element times from which standards may be synthesized, it is necessary to have a series of time studies of similar jobs. The time-study data must have the following characteristics:

A. Based on an adequate written standard practice with well-defined element end points

B. Broken down into similar elements

C. Similar methods used in the tasks observed

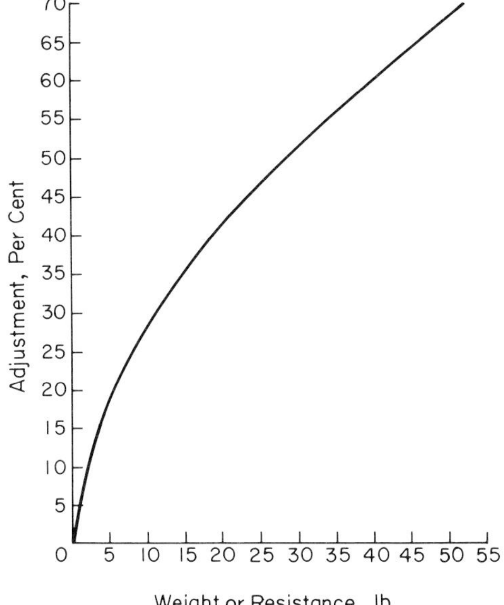

Fig. 5.37 *Adjustment for weight or resistance as a function of pounds (see Table 5.20).*

D. Similar equipment used in the tasks observed

E. Homogeneous elements

F. Rated to a uniform rate of activity

G. Comparable allowances used

In developing standard data for a group of jobs, the following types of elements may be encountered (and should be isolated):

A. Constant elements, identical from job to job

B. Variable elements, similar from job to job but varying in difficulty with the size, shape, and so forth, of the work. These may involve handwork or work with a machine, if the machine work is operator-paced.

C. Machine elements, mechanically controlled by the feed, speed, length of cut, welding current, thickness of material, and so forth

These three types of elements may exhibit any of the following three characteristics of occurrence:

1. Repeated the same number of times in each job

2. Repeated a different number of times in each job

3. Appear in some jobs and not in others

The data should be obtained and handled in the following manner:

1. Obtain time studies for as wide a range of jobs as possible, within a given group of similar jobs.

2. Be sure the time-study data meet the criteria previously stated.

3. Summarize the time studies on a summary from. (The columns usually are for given elements, and the lines for given jobs. It is customary to list also, on each line, all physical characteristics peculiar to the item handled on that job—i.e., weight, volume, etc.)

4. Ascertain which elements are constant and which are variable (a function of some characteristic of the product).

5. For the constant elements, determine the average standard time. Statistical techniques may be used to examine the significance of the variability of the data for such elements.

6. For the variable elements:

a. Determine, on a logical basis, what job characteristic or characteristics they are a function of.

b. Plot on a graph the time for each element against a variable or variables. (Note that sometimes the variables may interact or act in combination.)

c. Fit a smooth curve or curves to the points plotted.

This curve (or curves) may or may not pass close to the points. (If two variables control the values, a series of parallel curves may be necessary.) If the curve does not fit the points reasonably well, any of the following causes may be responsible:

1. The time studies were incorrectly or inconsistently rated.

2. Other variables as well as those plotted also affected the time.

3. The method varied.

4. An incorrect variable was used.

Each possibility should be investigated and evaluated. Note that even if the values fall along a smooth curve, it is still possible that two types of errors occurred that compensate for each other, although this is not likely. The variables used should be reasonable. When a reasonable fit with a smooth curve cannot be obtained, or when the reason for lack of fit cannot be explained with certainty, the investigations and study should continue until such can be done. Haphazard basic times are never justifiable. A good criterion is, "Could this be explained in a manner that would be acceptable to the typical worker?" This criterion rules out the use of complicated mathematics in the final statement, although the careful worker may choose to fit his curves mathematically. In many cases, only a straight line is justifiable, and it is often fitted by the method of least squares. This method, however, is designed to determine the straight line with the best fit and not to justify the line.

d. If the variable is discrete—that is, if it has only definite steps between two limits—prepare a table for the range of jobs covered. If the variable is continuous—that is, if it has an infinite number of possibilities between two limits—prepare a graph.

Some workers prepare a formula for the function. But if the formula is at all complex, it is often considered undesirable* because:

1. It may complicate the determination of standard. Of course, the analyst may prepare a very complex equation and use it himself either to plot the curve or to find the basic time for step-by-step values of the variable, and then prepare a detailed table for actual use. However, the questions may well be asked, "Are the data really as accurate as all this refinement would suggest? Could not acceptable values be read off the curve sketched for the data?" Also, since a table has a finite number of steps, the values nearest the actual condition of the variable in a specific application must be used as a reasonably good approximation; hence, extreme accuracy cannot be obtained.

2. It tends to befuddle most of the working groups to whom it is "explained."

3. It requires additional work that does not usually add utility to the basic times.

4. It may lead to extrapolation beyond the range covered by the data, and thus cause serious errors.

*This is not true of the common practice of using a simple mathematical formula to show how the elements are combined into a standard. Since such a formula usually involves no higher mathematics, it is often a helpful device.

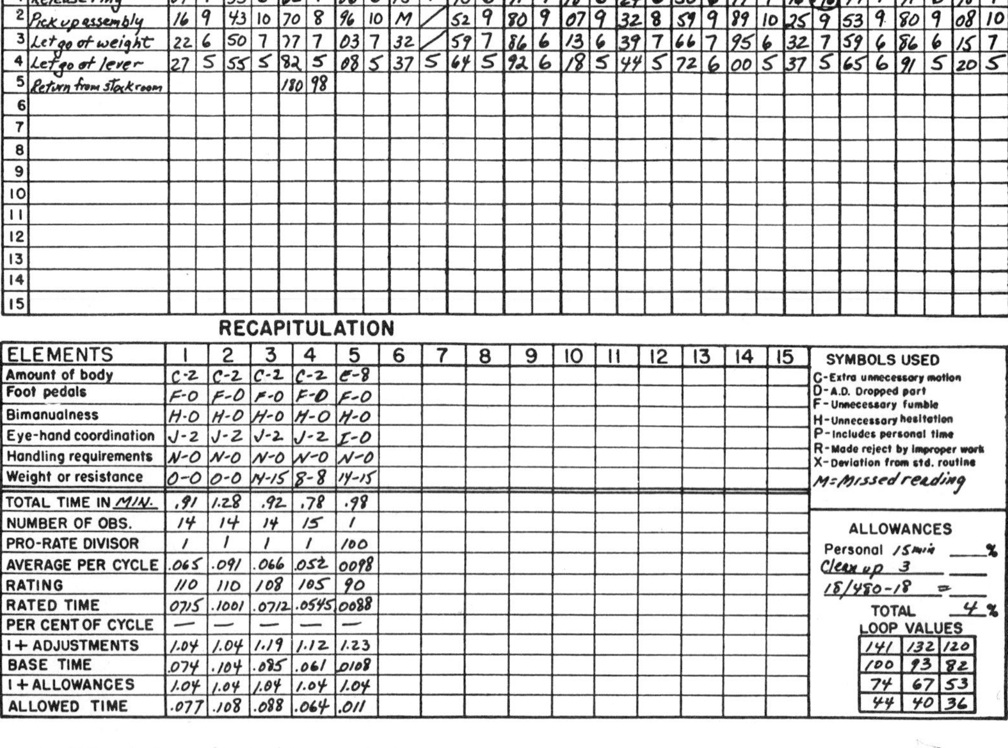

ELEMENTS		CYCLES																													
		1		2		3		4		5		6		7		8		9		10		11		12		13		14		15	
No.	TERMINAL POINT	R	T	R	T	R	T	R	T	R	T	R	T	R	T	R	T	R	T	R	T	R	T	R	T	R	T	R	T	R	T
1	Release ring	07	7	33	6	62	7	86	6	15	7	43	6	71	7	98	6	24	6	50	6	79	7	16	(F16)	44	7	71	6	98	7
2	Pick up assembly	16	9	43	10	70	8	96	10	M	/	52	9	80	9	07	9	32	8	59	9	89	10	25	9	53	9	80	9	08	10
3	Let go of weight	22	6	50	7	77	7	03	7	32	/	59	7	86	6	13	6	39	7	66	7	95	6	32	7	59	6	86	6	15	7
4	Let go of lever	27	5	55	5	82	5	08	5	37	5	64	5	92	6	18	5	44	5	72	6	00	5	37	5	65	6	91	5	20	5
5	Return from stockroom					180	98																								
6																															
7																															
8																															
9																															
10																															
11																															
12																															
13																															
14																															
15																															

RECAPITULATION

ELEMENTS	1	2	3	4	5	6	7	8	9	10	11	12	13	14	15
Amount of body	C-2	C-2	C-2	C-2	E-8										
Foot pedals	F-0	F-0	F-0	F-0	F-0										
Bimanualness	H-0	H-0	H-0	H-0	H-0										
Eye-hand coordination	J-2	J-2	J-2	J-2	I-0										
Handling requirements	N-0	N-0	N-0	N-0	N-0										
Weight or resistance	0-0	0-0	N-15	8-8	14-15										
TOTAL TIME IN MIN.	.91	1.28	.92	.78	.98										
NUMBER OF OBS.	14	14	14	15	1										
PRO-RATE DIVISOR	1	1	1	1	100										
AVERAGE PER CYCLE	.065	.091	.066	.052	0098										
RATING	110	110	108	105	90										
RATED TIME	0715	.1001	.0712	.0545	0088										
PER CENT OF CYCLE	—	—	—	—	—										
1 + ADJUSTMENTS	1.04	1.04	1.19	1.12	1.23										
BASE TIME	.074	.104	.085	.061	0108										
1 + ALLOWANCES	1.04	1.04	1.04	1.04	1.04										
ALLOWED TIME	.077	.108	.088	.064	.011										

SYMBOLS USED
C-Extra unnecessary motion
D-A.D. Dropped part
F-Unnecessary fumble
H-Unnecessary hesitation
P-Includes personal time
R-Made reject by improper work
X-Deviation from std. routine
M=Missed reading

ALLOWANCES
Personal 15 min ____ %
Clean up 3 ____
18/480-18 = ____
TOTAL 4 %

LOOP VALUES

141	132	120
100	93	82
74	67	53
44	40	36

Fig. 5.38 (*a*) *Rear of typical time study form.*

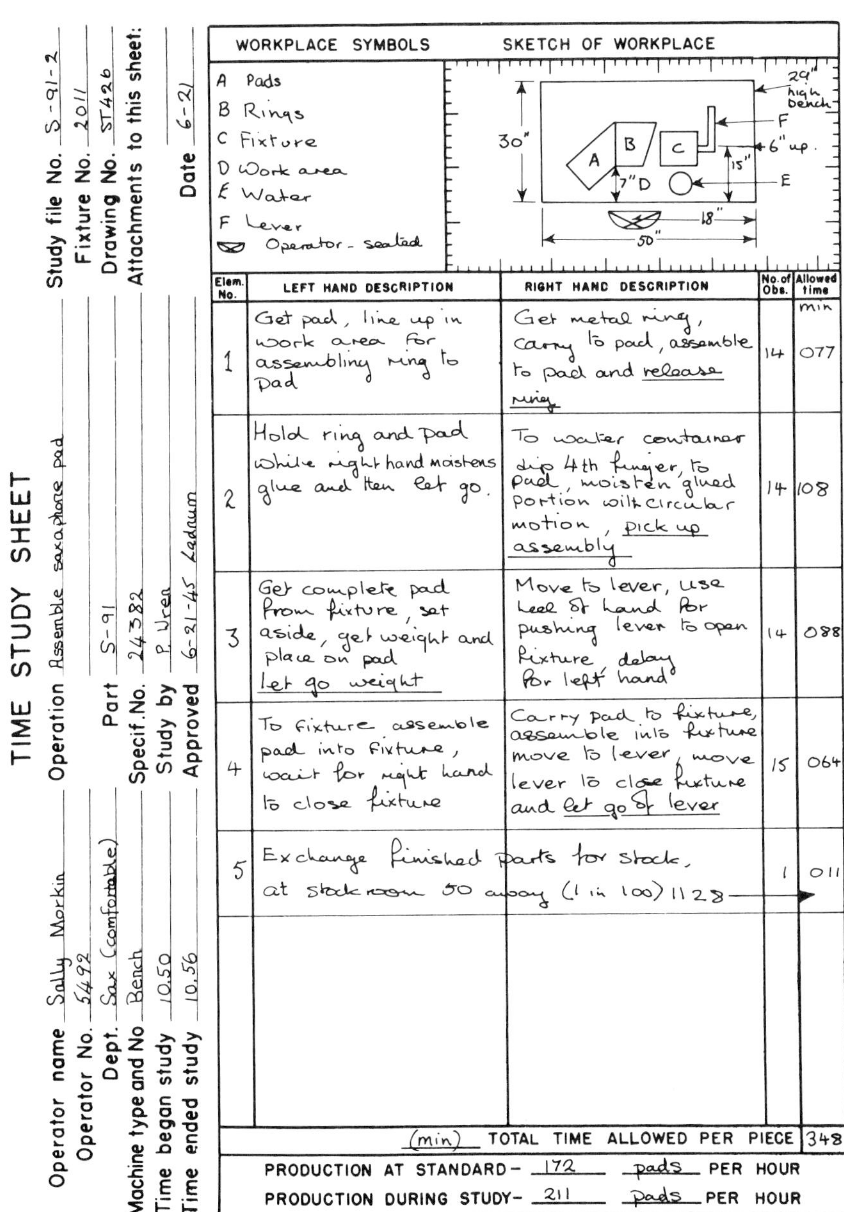

TIME STUDY SHEET

Operator name Sally Morkin — Operation Assemble saxophone pad — Study file No. S-91-2

Operator No. 5492 — Fixture No. 2011

Dept. Sax (comfortable) — Part S-91 — Drawing No. ST426

Machine type and No Bench — Specif. No. 24382 — Attachments to this sheet:

Time began study 10.50 — Study by P. Wren

Time ended study 10.56 — Approved 6-21-45 Ladnum — Date 6-21

WORKPLACE SYMBOLS

A Pads
B Rings
C Fixture
D Work area
E Water
F Lever
Operator - seated

SKETCH OF WORKPLACE

Elem. No.	LEFT HAND DESCRIPTION	RIGHT HAND DESCRIPTION	No. of Obs.	Allowed time min
1	Get pad, line up in work area for assembling ring to pad	Get metal ring, carry to pad, assemble to pad and release ring	14	077
2	Hold ring and pad while right hand moistens glue and then let go.	To water container dip 4th finger, to pad, moisten glued portion with circular motion, pick up assembly	14	108
3	Get complete pad from fixture, set aside, get weight and place on pad let go weight	Move to lever, use heel of hand for pushing lever to open fixture delay for left hand	14	088
4	To fixture assemble pad into fixture, wait for right hand to close fixture	Carry pad to fixture, assemble into fixture, move to lever, move lever to close fixture and let go of lever	15	064
5	Exchange finished parts for stock, at stockroom 50 away (1 in 100) 1128		1	011
	(min) TOTAL TIME ALLOWED PER PIECE			348

PRODUCTION AT STANDARD – 172 pads PER HOUR

PRODUCTION DURING STUDY – 211 pads PER HOUR

Fig. 5.38 (*b*) *Completed front of time study form for task of assembling saxophone pad. From Mundel,* Motion and Time Study: Principles and Practice, *pp. 483–84.*

SYNTHESIZED STANDARDS FROM BASIC MOTION TIMES. In some cases, it is either inconvenient or impossible to engineer a standard directly (the job may not be in operation), and studies of similar work from which applicable element times could be extracted or

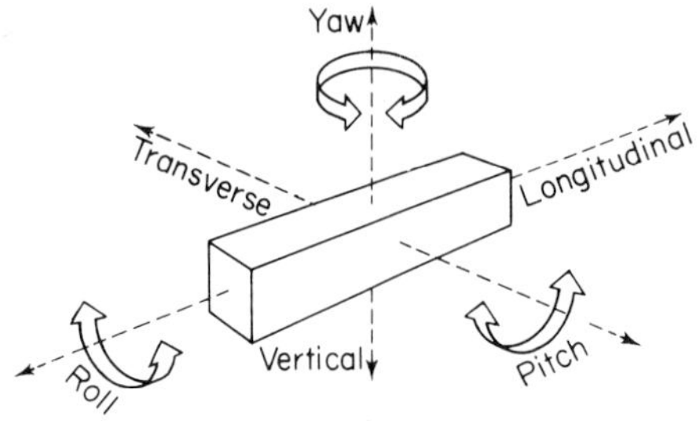

Fig. 5.39 *The six sources of restrictions possible with the therblig position.*

developed may not be available. Further, it may be desirable to develop a standard by another means as an independent basis of comparison for stop-watch standards. However, individual motion times for units as small as therbligs are not independent values; rather, they are affected by the pattern of motions in which they occur. Although in the method given here, corrections for this effect are made, the data given are not to be considered absolute values nor is the correction to be considered complete. Reference should be made to Par. 5.5.5 (Sources of error), which offers suggestions for the intelligent use of these data.

Source of the data. The data given in this section were developed by Dr. I. P. Lazarus and are reproduced with his permission. Using the camera timing technique described in Par. 5.5.4.1, he collected numerous time studies of industrial operations. These pictures were taken at 1000 frames per minute to permit later rating against the multi-image loops described in Par. 5.5.4.1. The reliability of the samples was checked using statistical methods. The rated therblig times were treated in the manner described in Par. 5.4.1, 6.

The data developed were tested by (1) reconstituting the original jobs, and (2) synthesizing a second group of new tasks and comparing the standards thus developed against standards set independently by stop-watch time-study and objective rating procedures. The differences between these latter pairs of values were in the order of the usual discrepancy between pairs of independently set time studies.

Variables evaluated. When all data were reduced to a common pace (see Par. 5.5.4.1), the therbligs *TE* and *TL* (see Table 5.8) were found to be an exponential continuous function of distance.

Grasp was a discrete function of condition of grasp. Four types were isolated:

1. *Contact:* control gained by mere contact
2. *Contact pinch:* control gained by contact with one or more fingers sliding the object into a position so that the thumb could oppose them
3. *Pinch:* control gained by maneuvering the hand so that the thumb and one or more fingers came into opposition on the object
4. *Wrap:* control gained by fingers and palm coming into opposition on the object

Position was a function of the number of degrees of positioning required. A degree was defined as a restriction in a dimension or orientation that had to be provided by the operator; there were six possible sources of restriction, but no more than five were found to occur at once. The six sources are shown in Fig. 5.39.

Assemble was found to be a function of the distance of restricted movement.

Disassemble and *release load* appeared to be constants. The interactions and interdependence of the therbligs were adjusted for by grouping them into elements and by applying secondary adjustments (see Table 5.20 and Fig. 5.37) to the elements as a whole. Consequently, it appears that the therblig times are a function of (at least) the variables given under each therblig, the variables covered by the secondary adjustments of Table 5.20, and the context of the element in which they occur.

Basic motion time values. The data determined by the procedure described above are given, in a form suitable for use in synthesizing a task, in Table 5.22.

Method of use. To develop a standard from the data given in Table 5.22 the following steps should be followed:

A. The actual or contemplated workplace or work area should be carefully dimensioned, as shown in Fig. 5.40, and the nature of what is to be accomplished should be summarized.

B. Using a form such as that shown in Fig. 5.41, each element should be detailed in terms of constituent therbligs. Elements of the magnitude of stop-watch time-study elements should be used to allow for therblig interactions. The conditions of each therblig should be carefully evaluated and noted in the proper column in terms of the variables used in Table 5.22.

C. Using Table 5.20, the secondary adjustment for each element should be computed using the proper columns of Fig. 5.41.

D. Using Table 5.22, the time for each

therblig, for each hand, should be entered on the computation form, as in Fig. 5.41. (The job shown in Fig. 5.40 was used.) Note that Table 5.22 gives basic times, the basic times plus 1 per cent increments to 1.10 of the basic time, and then successive 10 per cent steps from 0.10 to 0.90 of the basic time. If the secondary adjustment is 10 per cent or less, the therblig time may be entered on a form such as that in Fig. 5.41, with only one entry. If the secondary adjustment exceeds 10 per cent, two entries should be made; for a secondary adjustment of 23 per cent, for example, both the 1.03 multiple (the value from column E) and the multiple 0.20 (from column N) should be entered in the two time entry spaces for each therblig provided on the form in Fig. 5.41. Unavoidable delays, holds, and so forth take their value from the controlling hand. For *select*, which is basically a therblig with no controls, twice the *grasp* time should be used as an approximation.

E. The total element time is taken as that

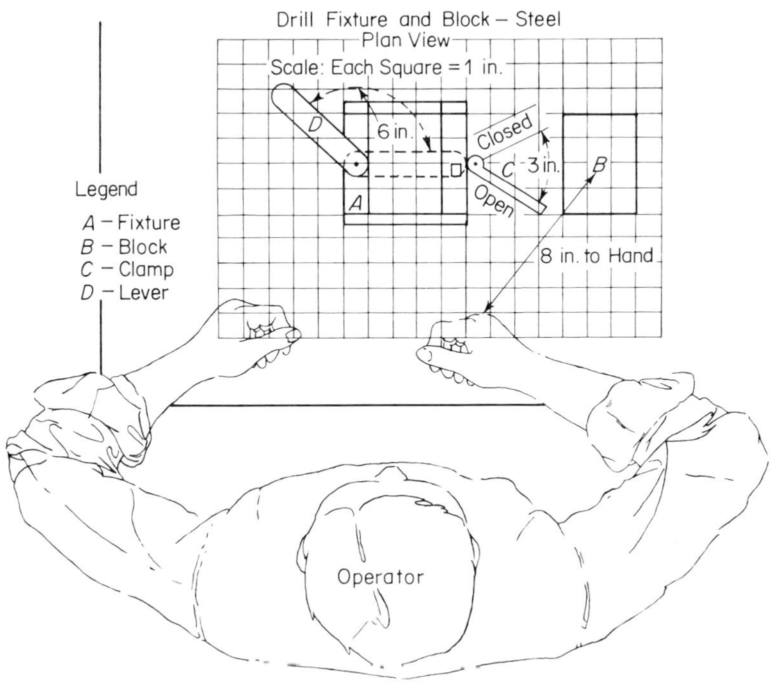

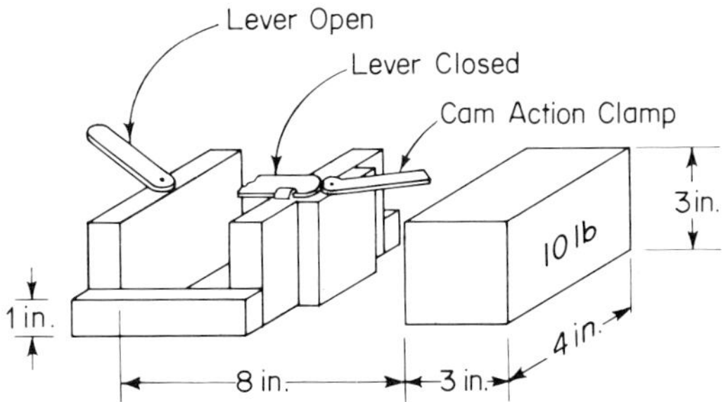

Fig. 5.40 *Sketch of workplace.*

TABLE 5.22. *BASIC TIME TABLES IN 1/100,000 MIN; (0.00001 MIN)**
Time for Transport Loaded, Transport Empty, and Walking

A	B	C	D	E	F	G	H	I	J	K	L	M	N	O	P	Q	R	S	T	U
Distance (inches)	Basic Time	Bx1.01	Bx1.02	Bx1.03	Bx1.04	Bx1.05	Bx1.06	Bx1.07	Bx1.08	Bx1.09	Bx1.10	Bx.10	Bx.20	Bx.30	Bx.40	Bx.50	Bx.60	Bx.70	Bx.80	Bx.90
1	172	174	175	177	179	181	182	184	186	187	189	17	34	52	69	86	103	120	138	155
2	243	245	248	250	253	255	258	260	262	265	267	24	49	73	97	122	146	170	194	219
3	297	300	303	306	309	312	315	318	321	324	327	30	59	89	119	149	178	208	238	267
4	343	346	350	353	357	360	364	367	370	374	377	34	69	103	137	172	206	240	274	309
5	384	388	392	396	399	403	407	411	415	419	422	38	77	115	154	192	230	269	307	346
6	420	424	428	433	437	441	445	449	454	458	462	42	84	126	168	210	252	294	336	378
7	454	459	463	468	472	477	481	486	490	495	499	45	91	136	182	227	272	318	363	409
8	485	490	495	500	504	509	514	519	524	529	534	49	97	146	194	243	291	340	388	437
9	515	520	525	530	536	541	546	551	556	561	567	52	103	155	206	258	309	361	412	464
10	543	548	554	559	565	570	576	581	586	592	597	54	109	163	217	272	326	380	434	489
11	569	575	580	586	592	597	603	609	615	620	626	57	114	171	228	285	341	398	455	512
12	594	600	606	612	618	624	630	636	642	647	653	59	119	178	238	297	356	416	475	535
13	619	625	631	638	644	650	656	662	669	675	681	62	124	186	248	310	371	433	495	557
14	642	648	655	661	668	674	681	687	693	700	706	64	128	193	257	321	385	449	514	578
15	665	672	678	685	692	698	705	712	718	725	732	67	133	200	266	333	399	466	532	599
16	686	693	700	707	713	720	727	734	741	748	755	69	137	206	274	343	412	480	549	617
17	707	714	721	728	735	742	749	756	764	771	778	71	141	212	283	354	424	495	566	636
18	728	735	743	750	757	764	772	779	786	794	801	73	146	218	291	364	437	510	582	655
19	748	755	763	770	778	785	793	800	808	815	823	75	150	224	299	374	449	524	598	673
20	767	775	782	790	798	805	813	821	828	836	844	77	153	230	307	384	460	537	614	690
21	786	794	802	810	817	825	833	841	849	857	865	79	157	236	314	393	472	550	629	707
22	805	813	821	829	837	845	853	861	869	877	886	81	161	242	322	403	483	564	644	725
23	823	831	839	848	856	864	872	881	889	897	905	82	165	247	329	412	494	576	658	741
24	841	849	858	866	875	883	891	900	908	917	925	84	168	252	336	421	505	589	673	757
25	858	867	875	884	892	901	909	918	927	935	944	86	172	257	343	429	515	601	686	772
26	875	884	893	901	910	919	928	936	945	954	963	88	175	263	350	438	525	613	700	788
27**	892	901	910	919	928	937	946	954	963	972	981	89	178	268	357	446	535	624	714	803
28	908	917	926	935	944	953	962	972	981	990	999	91	182	272	363	454	545	636	726	817
29	924	933	942	952	961	970	979	989	998	1007	1016	92	185	277	370	462	554	647	739	832
30	940	949	959	968	978	987	996	1006	1015	1025	1034	94	188	282	376	470	564	658	752	846
31	955	965	974	984	993	1003	1012	1022	1031	1041	1051	96	191	287	382	478	573	669	764	860

* Adapted from Irwin P. Lazarus, "A System of Predetermined Human Work Times," Ph.D. Thesis, Purdue Univ., 1952.
** Use 27″ for WALKING, per standard pace. For WALKING by distance, per foot, use line below item 48.

32	971	981	990	1000	1010	1020	1029	1039	1049	1058	1068	97	194	291	388	486	583	680	777	874
33	986	996	1006	1016	1025	1035	1045	1055	1065	1075	1085	99	197	296	394	493	592	690	789	887
34	1001	1011	1021	1031	1041	1051	1061	1071	1081	1091	1101	100	200	300	400	501	601	701	801	901
35	1015	1025	1035	1045	1056	1066	1076	1086	1092	1106	1117	102	203	305	406	508	609	711	812	914
36	1029	1039	1050	1060	1070	1080	1091	1101	1111	1122	1132	103	206	309	412	515	617	720	823	926
37	1044	1054	1065	1075	1086	1096	1107	1117	1128	1138	1148	104	209	313	418	522	626	731	835	940
38	1058	1069	1079	1090	1100	1111	1121	1132	1143	1153	1164	106	212	317	423	529	635	741	846	952
39	1072	1083	1093	1104	1115	1126	1136	1147	1158	1168	1179	107	214	322	429	536	643	750	858	965
40	1085	1096	1107	1118	1128	1139	1150	1161	1172	1183	1194	109	217	326	434	543	651	760	868	977
41	1099	1110	1121	1132	1143	1154	1165	1176	1187	1198	1209	110	220	330	440	550	659	769	879	989
42	1112	1123	1134	1145	1156	1168	1179	1190	1201	1212	1223	111	222	334	445	556	667	778	890	1001
43	1125	1136	1148	1159	1170	1181	1193	1204	1215	1226	1238	113	225	338	450	563	675	788	900	1013
44	1138	1149	1161	1172	1184	1195	1206	1218	1229	1240	1252	114	228	341	455	569	683	797	910	1024
45	1151	1163	1174	1186	1197	1209	1220	1232	1243	1255	1266	115	230	345	460	576	691	806	921	1036
46	1164	1176	1187	1199	1211	1222	1234	1245	1257	1269	1280	116	233	349	466	582	698	815	931	1048
47	1176	1188	1200	1211	1223	1235	1247	1258	1270	1282	1294	118	235	353	470	588	706	823	941	1058
48	1189	1201	1213	1225	1237	1248	1260	1272	1284	1296	1308	119	238	357	476	595	713	832	951	1070
per foot	396	400	404	408	412	416	420	424	428	432	436	40	79	119	158	198	238	277	317	356
Time for Grasp																				
Contact	174	176	177	179	181	183	184	186	188	190	191	17	35	52	70	87	104	122	139	157
Contact-pinch	278	281	284	286	289	292	295	297	300	303	306	28	56	83	111	139	167	195	222	250
Pinch	357	361	364	368	371	375	378	382	386	389	393	36	71	107	143	179	214	250	286	321
Wrap	566	572	577	583	589	594	600	606	611	617	623	57	113	170	226	283	340	396	453	509
Time for Position																				
First degree	200	202	204	206	208	210	212	214	216	218	220	20	40	60	80	100	120	140	160	180
Second degree	223	225	227	230	232	234	236	239	241	243	245	22	45	67	89	112	134	156	178	201
Third degree	311	314	317	320	323	327	330	333	336	339	342	31	62	93	124	156	187	218	249	280
Fourth degree	412	416	420	424	428	433	437	441	445	449	453	41	82	124	165	206	247	288	330	371
Fifth degree	556	562	567	573	578	584	589	595	600	606	612	56	111	167	222	278	334	389	445	500
Time for Assemble																				
1/4 inch†	00	00	00	00	00	00	00	00	00	00	00	00	00	00	00	00	00	00	00	00
3/8	24	24	24	25	25	25	25	26	26	26	26	2	5	7	10	12	14	17	19	22
1/2	89	90	91	92	93	93	94	95	96	97	98	9	18	27	36	45	53	62	71	80
5/8	154	156	157	159	160	162	163	165	166	168	169	15	31	46	62	77	92	108	123	139
3/4	219	221	223	226	228	230	232	234	237	239	241	22	44	66	88	110	131	153	175	197
7/8	284	287	290	293	295	298	301	304	307	310	312	28	57	85	114	142	170	199	227	256

† 1/4 inch and less of ASSEMBLE is containec in POSITION; at least it could not be evaluated as a separate value.

TABLE 5.22. *(Continued)*

A	B	C	D	E	F	G	H	I	J	K	L	M	N	O	P	Q	R	S	T	U
Distance (inches)	Basic Time	Bx1.01	Bx1.02	Bx1.03	Bx1.04	Bx1.05	Bx1.06	Bx1.07	Bx1.08	Bx1.09	Bx1.10	Bx.10	Bx.20	Bx.30	Bx.40	Bx.50	Bx.60	Bx.70	Bx.80	Bx.90
1	349	352	356	359	363	366	370	373	377	380	384	35	70	105	140	175	209	244	279	314
1 1/8	414	418	422	426	431	435	439	443	447	451	455	41	83	124	166	207	248	290	331	373
1 1/4	479	484	489	493	498	503	508	513	517	522	527	48	96	144	192	240	287	335	383	431
1 3/8	544	549	555	560	566	571	577	582	588	593	598	54	109	163	218	272	326	381	435	490
1 1/2	609	615	621	627	633	639	646	652	658	664	670	61	122	183	244	305	365	426	487	548
1 5/8	674	681	687	694	701	708	714	721	728	735	741	67	135	202	270	337	404	472	539	607
1 3/4	740	747	755	762	770	777	784	792	799	807	814	74	148	222	296	370	444	518	592	666
1 7/8	805	813	821	829	837	845	853	861	869	877	886	81	161	242	322	403	483	564	644	725
2	870	879	887	896	905	914	922	931	940	948	957	87	174	261	348	435	522	609	696	783
2 1/8	935	944	954	963	972	982	991	1000	1010	1019	1029	94	187	281	374	468	561	655	748	842
2 1/4	1000	1010	1020	1030	1040	1050	1060	1070	1080	1090	1100	100	200	300	400	500	600	700	800	900
2 3/8	1065	1076	1086	1097	1108	1118	1129	1140	1150	1161	1172	107	213	320	426	533	639	746	852	959
2 1/2	1130	1141	1153	1164	1175	1187	1198	1209	1220	1232	1243	113	226	339	452	565	678	791	904	1017
2 5/8	1195	1207	1219	1231	1243	1255	1267	1279	1291	1303	1315	120	239	359	478	598	717	837	956	1076
2 3/4	1260	1273	1285	1298	1310	1323	1336	1348	1361	1373	1386	126	252	378	504	630	756	882	1008	1134
2 7/8	1325	1338	1352	1365	1378	1391	1405	1418	1431	1444	1458	133	265	398	530	663	795	928	1060	1193
3	1391	1405	1419	1433	1447	1461	1474	1488	1502	1516	1530	139	278	417	556	696	835	974	1113	1252
Time for Disassemble All	476	481	486	490	495	500	505	509	514	519	524	48	95	143	190	238	286	333	381	428
Time for Release Load All	189	191	193	195	197	198	200	202	204	206	208	19	38	57	76	95	113	132	151	170

1 Sheet of ORDNANCE MANAGEMENT ENGINEERING TRAINING PROGRAM

3 Sheets WORK SHEET FOR PREDETERMINED APPROXIMATE PERFORMANCE TIME

2319 6 March 1953 Job number and date

Element number	LEFT HAND DETAILS What, from, to, etc.	Therblig	Condition of the therblig	Time value	RIGHT HAND DETAILS Time value	Condition of the therblig	Therblig	What, from, to, etc.	FOR ELEM. Adj. cat.	Adj. % and total
	From H to fixture	TE	8"	504 146	504 146	8"	TE	From H to block	C	2
	Fixture	G	P	371 107	371 107	P	G	Block	F H	0 0
	Fixture	H			504 146	8"	TL	To fixture with block	J P	2 2
1					578 167	5th	P	Block to fixture	W_{10}	28
					363 105	1"	A	Block into bottom of fixture		34
					197 57		RL	Block in fixture		
					3245					
	From fixture to lever	TE	4"	374 34	374 34	4"	TE	From block to clamp	C	2
	Lever D	G	C	190 17	617 57	W	G	Clamp C	F H	0 0
2	Lever to closed location	TL	6"	458 42			H		I O	0 1
	Lever	RL		206 19	258	*		* To balance	W_4	16
	While RH locks clamp	UD			324 30	3"	TL	Cam clamp to lock		19
					206 19		RL	Cam clamp		
					1919					

ORDBC 85AR ARMY-RIA

Fig. 5.41 *Element work sheet for use with predetermined approximate work times.* (*Filled in for job of Fig. 5.40.*)

required by the hand with the longest or controlling time.

F. On a form such as that shown in Fig. 5.42, the elements and job information are summarized, the allowances computed, and then the standard is computed. (The job of Fig. 5.40 was again used in this illustration.)

Since the original data were developed by the procedure described under Par. 5.5.4.1, the standard that results from the use of these data should be, within limits, comparable to the standard described in that article. This synthesizing procedure is particularly useful for planning new work or for comparing contemplated, alternative methods.

5.5.4.3 Statistical Standards. These are usually determined by the following procedure:

A. Defining the standard of measurement in terms of some time statistic

B. Selecting an apparently related work unit

C. Determining the relationship between A and B

Statistical standards are usually used for management control purposes rather than for detailed or incentive use.

Defining the Standard of Performance. The time statistic selected may be the mean or arithmetic average, one of the quartile boundaries, or some function of the mean and the standard deviation. Since all these measures are calculated from data on past performance (usually a considerable period and number of people are involved), the statistical standard may be said to be a function of past performance. For example, standard time for an activity may be defined as the average time previously taken, per work unit, over the past two months or over the best two months in the past year, and so forth.

Concerning statistical standards, W. R Vogel, Ordnance Manpower Control Specialist, has written:

ORDNANCE MANAGEMENT ENGINEERING TRAINING PROGRAM
SUMMARY SHEET - PREDETERMINED APPROXIMATE PERFORMANCE TIMES

Drill guide block 2319 — Job name and number

Guide block 31 — Part name and number

7 Number of elements
✓ Workplace sketch attached
✓ Machine data sheet attach.
C F Schneider — Analysis by
6 Mar 1953 — Date
1 Page of
6 Pages

ALLOWANCES
A - Personal 30 Min/day
B - Irregular — Min/day

ALLOWANCE CALCULATIONS
1 - Reduction of A and B to percent

$\frac{30}{480-30} = 6.7$

C- Machine or process details

2 - Total allow.(A,B,C) 7 percent

Elem no.	Element title	Adj.time Min.	Allowed time Min.	Occurs per cycle	Prorated time Min.
1	Place block in fixture	.03245	.03472	1	.03472
2	Close lever and lock clamp	.01919	.02053	1	.02053
3		.05164	.05164		.0553
4					
5					
6					
7					
8					
9					
10					
11					
12					
13					
14					
15					

Note: For illustrative purposes this sheet has been completed as if the two elements in figure 5.39 completed the cycle. As reference to the data at the top will show, there are actually 7 elements.

WORK UNIT Piece — TOTAL TIME PER CYCLE .0553 Min.

TOTAL TIME .922 hrs per 1000

ANTICIPATED PRODUCTION 1090 per Hr.

Fig. 5.42 *Summary form for developing standard time from data of Fig. 5.41.*

If we set a performance standard statistically using the arithmetic mean, we are defining the performance standard as the average amount of time it has taken to produce a work unit in the past. This definition makes no assumption of improvement.

If we set a performance standard using an upper quartile, we are setting a performance standard that is something higher than average past performance. On this, I have two comments.

First, in the light of the purposes for which this standard is to be used, I feel that by using the upper quartile method we are automatically introducing errors in scheduling work, making budget estimates, and arriving at manpower allowances. My reasoning on this is that the mere fact that we are setting a high performance standard is not at all indicative that we will meet this standard in other than the long run sense. If we do not meet this standard, obviously we are going to require more manpower, money, and equipment than the standard indicates, which will only complicate the budgeting, scheduling, and manpower activities.

Second, the upper quartile performance standard does not provide a uniform approach to the analysis in different work areas. In a work area where, due to the nature of the account and the work unit, the monthly fluctuations are very slight, the upper quartile will be very close to the average. In an account or work area where there is a great deal of dispersion in the data due to type of work being measured and the unit of measure, the upper quartile performance standard will be considerably higher than the average. Inasmuch as we have had a great deal of dispersion in our performance data in the past (which was only in part caused by fluctuations in productivity) we have no reason not to expect dispersion in the future simply because we have set a performance standard. The net result of using the upper quartile performance standard is that we are automatically setting the following criterion for how stiff the standard will be: The accounts or work areas with the greatest dispersion will have the stiffest standard. This appears to me to be a particularly poor criterion for adjusting manpower.

The ultimate effect of the application of the upper quartile would be to eliminate personnel in a very unrealistic and undesirable manner based on a defenseless criterion.

No matter what statistical method is used to set a performance standard, it should be clear that no basis is afforded by which we can say an operation is good or bad—highly effective or not. Statistically, we can validly make statements as to the changes in the effectiveness; not as to the degree of effectiveness attained.

Recording the Standard Method and Identifying the Unit of Work. The procedure indicated in Par. 5.5.4.1 can be followed; it needs no further detailing.

Great care must be exercised in selecting the work units for statistical standards, particularly with service activities. The work units should summarize all outputs without any double counting, be related to time required, and be such that a forecast of future work may be made. However, since statistical standards are usually applied as a gross control to departments, operating units, or even whole establishments, numerous methods on different tasks, in varying numbers of repetitions, enter as time-consumers. The statistical standard is usually expressed as total time for end-product units of work. In many cases, these work units are not easy to identify. Take, for example, a shipping department for a job shop plant. The work required is a function of the number of items packed, the packing used, the weight and size of the items, and so on. Keeping records of what is actually done, in terms of these numerous variables, might require extensive study; in many cases, such data are not available from past records. A convenient set of work units is usually selected and their suitability is determined by comparing time (man-hours or man-days) consumed in the past and work units produced. Statistical techniques such as curve-fitting, linear programing, and multiple regression are used to make this comparison.

Determining the Relationship Between Time and Work Units. Once a convenient and significant set of work units has been determined, time (man-hours or man-days) is stated as a function of the work units on the basis of the time statistic selected. Reference to Par. 5.5.5 (sources of error) will assist in the intelligent use of such standards.

5.5.5 Sources of Error

Considerable difficulty in working with time standards can be avoided if it is realized that they involve an extremely sensitive area of human relations, and that they are far from absolute measures.

5.5.5.1 Stop-watch or Camera Time Study. Even before the application of rating, several possible sources of error exist. The method observed may include deviations from that commonly used. Erroneous decisions may be made as to which observations are foreign elements, unallowable fumbles, or irregular elements. The sample of values observed may not be a random sample of the work required (by the usual product), or it may not be a truly random sample of performance. The application of statistical controls only increases the *probability* that the sample will be representative. Further, errors may be introduced by faulty reading of the stop watch, although these may be minimized by practice.

WITH COMMON RATING PROCEDURES. There is a considerable possibility that the time-study man will introduce error in establishing the mental concept with which the observed performance is to be compared. Contributing factors may be the relations among the time-study department, operators, and foremen, or past experience, or simply the time-study man's mood. Also, there is no way of preventing these mental concepts from drifting slowly over a period of time. This drift may lead to a biased, rather than to a random, error. Further, error may arise in the comparison of concept and performance. Also, since the application of allowances usually requires judgment as to which category is applicable, further inconsistencies may be introduced in the standards.

IMPROVED RATING PROCEDURE. Although the improved procedure removes some of the sources of error common to conventional rating procedures, error may still be introduced when performance is compared with standard pace. Further error may be introduced by a poor element selection, even if the secondary adjustments are correctly applied. Also, the judgment of selecting a pace, to which to compare an observed performance, is replaced by the judgment of selecting the proper secondary adjustments to apply. However, this is a simpler type of judgment, and, although it is still a possible source of error, it represents a source of lower probability and magnitude; further, the resulting error will be random rather than biased. (If biased error appears, it may be removed by altering the table of secondary adjustments.) In addition, with the improved rating procedure, the basis of judgment—both on pace comparison and secondary adjustments—remains constant over a period of time. And here again, the errors will be random rather than biased. The remarks made in respect to allowances apply equally to this method.

5.5.5.2 Synthesized Standards. Methods of synthesizing standards tend to hide but not eliminate some of the sources of error.

SYNTHESIZED STANDARDS FROM STANDARD ELEMENTS. Since these standards are usually based on data from numerous time studies, random errors included in the original studies are usually reduced in magnitude, whereas biasing errors are carried on through. Additional errors may be introduced in new jobs by applying elemental standards to elements that are not identical, or by reassembling element times into a new pattern in which the elements affect each other even though no allowance is made for such an effect. Further, if variable elements have been plotted against a fortuitous variable, any new value extracted from such a curve may be in error. If the curve is extrapolated, the implied relationship may no longer hold and a spurious value will result. In addition, the so-called constant elements may not be constant; here is another possible source of error in application. The remarks made previously concerning allowances are also applicable.

SYNTHESIZED STANDARDS FROM BASIC MOTION TIMES. In the first place, the original data may be of doubtful accuracy. Since the tables are based on average values, the data should be examined to determine possible inherent error. Secondly, the time for a therblig has been shown to be a function of:

a. Distance

b. Complexity of action

c. Amount of body involved

d. Bimanualness involved

e. Whether the use of the feet accompanies the action

f. The eye-hand coordination required

g. The sensory requirements

h. The weight or resistance involved

i. The preceding and following therbligs as well as the context of the whole pattern of the task

j. The direction of the movement

k. The place of the therblig in a motion pattern

l. Possible interactions of two variables

m. Several other variables as yet unidenti-

METHOD PROPOSAL SUMMARY

1. Date: ______________________ 1. Fewer people
2. Fewer steps
2. To: ______________________ 3. Less time on a step or steps
4. Less time in production
3. From: ______________________ 5. Less space
6. Less time for critical skills
4. Subject: ______________________ 7. Less time on critical equipment
8. Increased quality
______________________ 9. Less cost
10. Less skill on step or steps
______________________ 11. Better control

Improvement will (insert proper numbers or describe if not classifiable.)

5. ______________________

6. If this proposal is approved it will be necessary to (summarize):

____

____

Attachments: (Insert number of sheets in proper boxes;follow with page nos. on lines.

7. [ ] ________ Cost of change estimate detail sheet

8. [ ] ________ Original charts

9. [ ] ________ Proposed charts, including summary and comparison

10. [ ] ________ Proposed equipment list and details of placement

11. [ ] ________ Jig,fixture, workplace or layout sketches or drawings

12. [ ] ________ Job instruction sheets as required by proposal

13. [ ] ________ Additional attachments list.

14. Financial aspects of change and evaluation of change (summary)

a.______________________estimated the annual volume.

b.______________________estimated the fixed cost of tools etc.

c.______________________Time new method would take to pay for itself.

d.______________________Original labor cost and hours, annual.

e.______________________Proposed labor cost and hours, estimated annual.

f.______________________Cost of change

g.______________________Net savings, annual(estimate)

Fig. 5.43 *Method proposal summary form.*

fied. (Even when the effects of variables *a* through *k* are extracted, considerable variation in observed data still exists.)

In some available systems of predetermining standards, the existence of many of these variables is disregarded, although this is hardly a realistic procedure for eliminating them as sources of error. The system of data given in Par. 5.5.4.2 takes into account, to a degree, variables *a* through *i*. Error may still be introduced by an improper description of the task, a wrong classification of a therblig, a poor element grouping, and an incorrect selection of secondary adjustments. The remarks made in Par. 5.5.5.1 concerning allowances are also applicable.

5.5.5.3 Statistical Standards. Since statistical standards usually do not include a

PROCEDURE PROPOSAL SUMMARY

SPAS NO. I

1 DATE: ______

2 TO: ______

3 FROM: ______

4 SUBJECT: ______

5 ☐ NEW PROCEDURE

6 ☐ IMPROVED PROCEDURE

7 ☐ IMPROVED FORMS

1. FEWER PEOPLE
2. FEWER STEPS
3. LESS TIME ON A STEP OR STEPS
4. LESS TIME TO COMPLETE PROCEDURE
5. LESS SPACE
6. LESS TIME FOR CRITICAL PERSONNEL
7. LESS TIME ON CRITICAL EQUIPMENT
8. INCREASED ACCURACY
9. CHEAPER PROCEDURE
10. LESS SKILL ON STEP OR STEPS
11. BETTER CONTROL

8 Improvement will (insert proper numbers or describe if not classifiable): ______

9 If this proposal is approved, it will be necessary to (summarize) ______

10 Financial aspects of change and evaluation of change (summarize): ______

ATTACHMENTS (insert number of sheets in proper boxes; follow with page numbers on line)

11 ☐ ______ Original procedure chart

12 ☐ ______ Proposed procedure chart

13 ☐ ______ Proposed equipment list and details of placement

14 ☐ ______ Form designs and instructions for use, by forms

15 ☐ ______ Job instructions, by person

16 ☐ ______ Additional attachments list

ORDBC 43.1AR

Fig. 5.44 *Procedure analysis proposal summary form.*

description of the work content (in terms of actions called for) of the work unit, they lack this essential control and will be in error when the work content changes. Further, since they are based on past performances and provide no method of really appraising or evaluating these performances, they may be extremely inconsistent. Standards based on data from poor performances (poor because of equipment, work effort, worker ability, supervision, or intermittent activity) cannot be differentiated from standards based on good performances, nor can the reason for the differences be easily detected. Further, the larger the group, the more distantly the work unit used may actually be related to work or effort input. Indeed, there is some question whether the word "standard" has the same significance as in the techniques of Pars. 5.5.4.1 and 5.5.4.2. However, if one recognizes these possible deficiencies and uses the statistical values as guides, they may be of considerable value.

WRITTEN STANDARD PRACTICE

PROCESS EARTHWORM TRACTOR ·PISTON RINGS

STEP NO.	NO. DEPT.	MACHINE	OPERATION	METHOD STD.	STD. TIME
1	9 Foundry	Class L	Mold	FM-7	.28 minute per unit of 4
2	9 Foundry	On Conveyor C-1	Pour Mold	FM-8	.10 minute per tree of 40
3	9 Foundery	Bench and Hoist BH-3	Shake out	FM-9	.51 per tree of 40

Fig. 5.45 *Written standard for process.*
From Mundel, Motion and Time Study: Principles and Practice, *p. 573.*

ACE VALVE Co.-WORK STANDARD

DEPT:
OPERATION:
PART:

RATE

SET BY	APPROVED	STANDARD TIME (hours/pc)	PIECES PER HOUR	METHOD NO.	DATE

Fig. 5.46 *Sample time standard form for rate book.*
From Mundel, Motion and Time Study: Principles and Practice, *p. 575.*

WRITTEN STANDARD PRACTICE
The Perfect Circle Co.

SECOND INSPECTION

No. SI 14
Issue 2
Sheet 1 of 3
3-6-44

OPERATION: Gap Gage Earthworm Tractor 5 3/4" Rings

EQUIPMENT

1 Sizer Ring Gage
1 Set Feeler Gages

1 V Trough
1 V Trough (non-sectioned)

REQUIREMENTS

Operator will wear gloves when handling rings.

All pans of rings will be lifted by service men.

Pan ticket will remain in pocket on side of pan, except when being read by operator.

Operation will be performed by checker.

Ring gages will be checked at the beginning of each shift by a checker.

A ticket showing ring gage size, variation number, feeler gage number, operator's clock number and date will be filled out when ring gage is checked and will be kept with gage while it is in use.

Feeler gages will be kept at the Foreman's desk and will be returned to desk at end of working period.

SPECIFICATIONS

Gage Size: 5.747 Gap Clearance: .011 - .021

Rings will be gaged on a percentage basis. Operator will gage every 8th ring, beginning with bottom ring and ending with top.

If any ring is found out of limits, operator will check 8 rings above and 8 rings below place from which reject was removed until all rings not within limits have been rejected.

SECURE RINGS

1. HAVE SERVICE MAN PLACE ENTIRE TEST of rings at back of bench in a horizontal position.

CHECK SET-UP

2. CHECK PAN TICKET to see that previous operation has been completed. Check number of rings in pan against count shown on ticket. (Count remnant row and add to number in full rows). If counts do not agree, have floor clerk make necessary corrections.

Fig. 5.47 (*a*) *Written standard practice for operation.*

5.5.5.4 Summary. All methods of determining standards have at least several possible or probable sources of error. In using these methods, one should attempt, as far as possible, to control and reduce all controllable sources, while still recognizing their nature and the existence of other sources. Also, it should be recognized that the values produced are only approximations of the concept embodied in the statement of standard.

5.6 MOTION AND TIME-STUDY REPORTS

5.6.1 Types and Uses

If adequate use is to be made of the data developed through motion and time study, adequate reports should be submitted in connection with: (a) each project undertaken, and (b) total activity in these functional areas.

The individual project reports normally form a basis for acceptance or rejection. Adequate reports of total activity serve as: (a) a continual reminder of what can be done, (b) a basis for programing future activity, and (c) a basis for evaluating the worth of the activity.

5.6.2 Project Reports

Project reports may be submitted in the form of a method proposal, a procedure proposal, or a work standard.

5.6.2.1 Method Proposal. A form suitable

SECOND INSPECTION — WRITTEN STANDARD PRACTICE, The Perfect Circle Co. — No. SI 14, Issue 2, Sheet 2 of 3, 3-6-44

OPERATION: Gap Gage Earthworm Tractor 5 3/4" Rings

GAGE RINGS

3. REMOVE EVERY 8TH RING from first row of pan, beginning with bottom ring and ending with top.

4. PLACE RINGS ON BENCH in front of first row with gaps toward operator in the order removed from pan.

5. REMOVE SPECIFIED NUMBER OF RINGS from remaining two rows in the manner described in Steps #3 and #4.

6. MOVE GAGE to a position at left end of first pan.

7. PICK UP FEELERS AND HOLD them between thumb and index finger while gaging rings. (When placing ring in gage or removing ring from gage, hold feelers in palm of hand.)

8 PICK UP ONE RING from first stack on right approximately 1" on right of gap with thumb and index finger of right hand.

9. BRING RING TO A POSITION OVER GAGE.

10. GRASP RING on left of gap with left index finger and thumb, placing middle fingers on top of ring and thumbs on face and corners about ½" on either side of gap.

11. TILT RING SLIGHTLY AND SLIP BACK OF RING INTO GAGE.

12. INSERT REST OF RING INTO GAGE by squeezing thumbs together toward gap.

13. ROLL THUMBS OFF RING to release points slowly. Do not flip points into gage. Be sure back of rings is pressed firmly into gage.

14. TEST GAP by placing Go feeler between points, starting at inside diameter and bringing feeler toward operator through gap. Go feeler should pass between points freely. If feeler passes through gap with a slight drag, ring should be rejected.

15. PLACE NO GO FEELER BETWEEN POINTS, starting at inside diameter and bringing feeler toward operator through gap. The feeler should either refuse to pass or should fit very tightly. Do not force feeler between the points.

16. REMOVE RING FROM GAGE by closing gap with thumb and index fingers of each hand and lifting ring. Do not flip points out of gage.

17. PLACE IN NON-SECTIONED TROUGH ON LEFT OF GAGE AND AT SAME TIME REPEAT STEP #8, if ring is within limits.

 a. If ring is not within limits, place in sectioned trough on left of gage in section marked for that type reject.

Fig. 5.47 (*b*) *Continuation of form.* (*Space does not permit reproducing the rest of this*). *From Mundel,* Motion and Time Study: Principles and Practice, *pp. 395–96. Reproduced by courtesy of D. C. Parsons, manufacturing manager, Perfect Circle Co.*

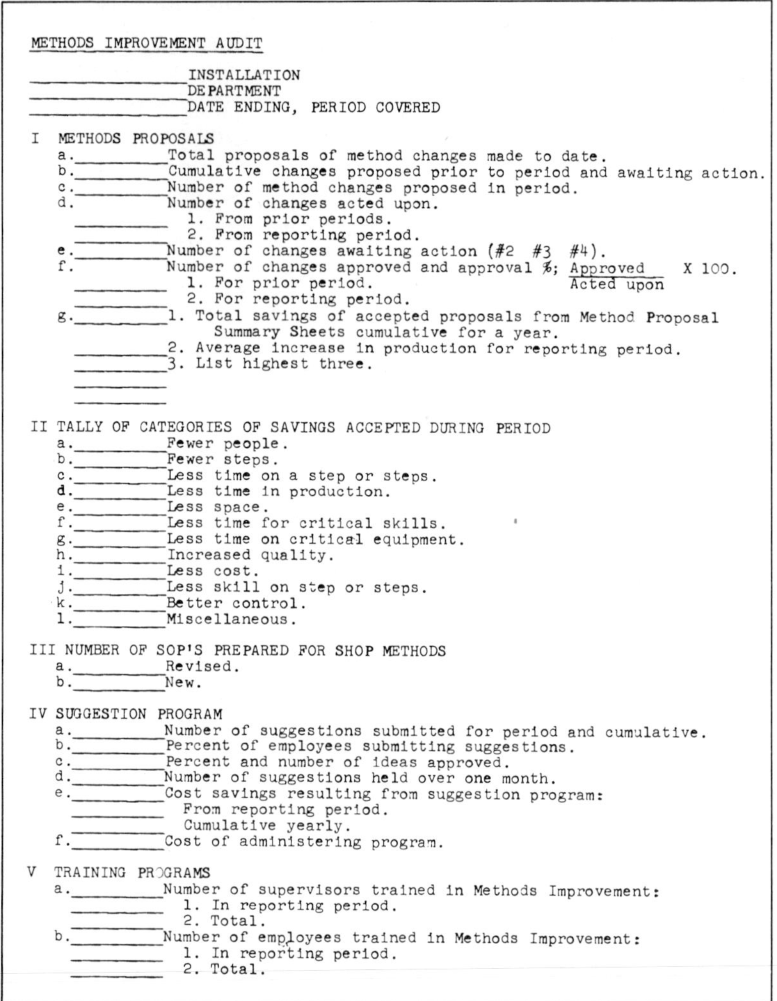

METHODS IMPROVEMENT AUDIT

_______ INSTALLATION
_______ DEPARTMENT
_______ DATE ENDING, PERIOD COVERED

I METHODS PROPOSALS
a.______ Total proposals of method changes made to date.
b.______ Cumulative changes proposed prior to period and awaiting action.
c.______ Number of method changes proposed in period.
d.______ Number of changes acted upon.
______ 1. From prior periods.
______ 2. From reporting period.
e.______ Number of changes awaiting action (#2 #3 #4).
f.______ Number of changes approved and approval %; Approved / Acted upon X 100.
______ 1. For prior period.
______ 2. For reporting period.
g.______ 1. Total savings of accepted proposals from Method Proposal Summary Sheets cumulative for a year.
______ 2. Average increase in production for reporting period.
______ 3. List highest three.
______
______

II TALLY OF CATEGORIES OF SAVINGS ACCEPTED DURING PERIOD
a.______ Fewer people.
b.______ Fewer steps.
c.______ Less time on a step or steps.
d.______ Less time in production.
e.______ Less space.
f.______ Less time for critical skills.
g.______ Less time on critical equipment.
h.______ Increased quality.
i.______ Less cost.
j.______ Less skill on step or steps.
k.______ Better control.
l.______ Miscellaneous.

III NUMBER OF SOP'S PREPARED FOR SHOP METHODS
a.______ Revised.
b.______ New.

IV SUGGESTION PROGRAM
a.______ Number of suggestions submitted for period and cumulative.
b.______ Percent of employees submitting suggestions.
c.______ Percent and number of ideas approved.
d.______ Number of suggestions held over one month.
e.______ Cost savings resulting from suggestion program:
______ From reporting period.
______ Cumulative yearly.
f.______ Cost of administering program.

V TRAINING PROGRAMS
a.______ Number of supervisors trained in Methods Improvement:
______ 1. In reporting period.
______ 2. Total.
b.______ Number of employees trained in Methods Improvement:
______ 1. In reporting period.
______ 2. Total.

Fig. 5.48 *Method improvement audit form.* (*Developed by Dr. I. Lazarus, Army Ordnance Management Engineering Training Program.*)

for summarizing the content of a method proposal is shown in Fig. 5.43. This would satisfactorily serve as: (a) a letter of transmittal, (b) a summary sheet, or (c) a source for summarizing the information into a periodic report.

5.6.2.2 Procedure Proposal. A form suitable for summarizing the results of a procedure proposal is shown in Fig. 5.44. This form serves the same three purposes served by Fig. 5.43.

5.6.2.3 Work Standard. The standards for a process may be reported in the form shown in Fig. 5.45. The standards for an operation may be reported on the form shown in Fig. 5.46. Either form would be backed up by time studies and operator instruction sheets. Part of an operator instruction sheet is shown in Fig. 5.47.

5.6.3 Activity Reports

Separate activity reports may be submitted for each of five functional areas:

a. Methods improvement
b. Procedure analysis
c. Labor standards development
d. Training (work simplification, etc.)
e. Suggestion plan operation

Only the first three of these are described in this section. Such reports are often referred to as "periodic audits of the performance of the function."

5.6.3.1 Methods Improvement. A form suitable for periodic reports on the performance of this function, as developed by Dr. I. Lazarus of the Army Ordnance Management Engineering Training Program, is shown in Fig. 5.48.

5.6.3.2 Procedure Analysis. A form suitable for periodic reports on the performance of this function, as developed by J. Moquin of the Army Ordnance Management Engineering Training Program, is shown in Fig. 5.49.

PROCEDURE ANALYSIS ACTIVITY AUDIT SHEET

________________ INSTALLATION

________________ DEPARTMENT

________________ DATE ENDING, PERIOD COVERED

I ACTIVITY

a. ________ Total number of man hours spent in procedure and forms design studies during report period.
b. ________ Reduction in man hours of work load per month as result of procedure and forms design changes during report period.
c. ________ Cumulative savings in man hours.

II PROCEDURE STUDIES

a. ________ Approximate number of procedures in installation.
b. ________ Total number of procedures recorded graphically at end of last report period.
c. ________ Number of procedures recorded graphically during this report period.
d. ________ Total number of procedures recorded graphically at end of this report period.
e. ________ Number of procedures proposals submitted during report period.
f. ________ Number of procedure proposals accepted during report period.

III FORMS DESIGN

a. ________ Total number of forms in use at end of last report period.
b. ________ New forms created not replacing old forms.
c. ________ Number of forms combined.
d. ________ Number of forms eliminated.
e. ________ Net change in number of forms in use.
f. ________ Total number of forms in use at end of present report period.
g. ________ Number of forms simplified.

Fig. 5.49 *Procedure analysis audit form. (Developed by Mr. J. Moquin, Army Ordnance Management Engineering Training Program.)*

LABOR STANDARD DEVELOPMENT AND APPLICATION AUDIT

__________ INSTALLATION

__________ DEPARTMENT

__________ DATE ENDING, PERIOD COVERED

I LABOR STANDARD DEVELOPMENT
- a. __________ Total number labor standards, present.
- b. __________ Total number labor standards, last report.
- c. __________ Number increase over last report.
- d. __________ Total number labor standards revised; method changes.
- e. __________ Total number labor standards revised; other reasons.

II STANDARD HOURS COVERAGE
- a. __________ Total elapsed man hours during report period covered by labor standards.
- b. __________ Total daywork man hours during report period.
- c. __________ Total elapsed man hours during report period.
- d. __________ Percentage of hours covered by labor standards.
- e. __________ Total number of operations performed during report period covered by labor standards.
- f. __________ Total number of operations performed during report period not covered by labor standards.
- g. __________ Percentage of operations performed covered by labor standards.

III PRODUCTIVITY ANALYSIS
- a. __________ Total standard man hours produced on work covered by standards.
- b. __________ Total elapsed man hours on work covered by standards.
- c. __________ Percentage productivity.

IV DELAY HOURS ANALYSIS
- a. __________ Total machine downtime delay man hours.
- b. __________ Percentage of machine downtime to total elapsed man hours.
- c. __________ Total set-up delay man hours.
- d. __________ Percentage of set-up time to total elapsed man hours.
- e. __________ Total miscellaneous delay man hours.
- f. __________ Percentage of miscellaneous delay to total elapsed man hours.
- g. __________ Total delay man hours.
- h. __________ Percentage total delay to total elapsed man hours.

Fig. 5.50 *Work measurement audit form.* (*Developed by Mr. C. Schneider, Army Ordnance Management Engineering Training Program.*)

5.6.3.3 Labor Standards Development. A form suitable for periodic reports on the performance of this function, as developed by C. Schneider of the Army Ordnance Management Engineering Training Program, is shown in Fig. 5.50.

5.7 MOTION AND TIME-STUDY POLICIES

5.7.1 Definition

Policies are statements of the procedures and aims to be followed by an organization in meeting recurring situations.

5.7.2 Areas to be Covered

If routine motion and time-study work is to be effective, the controlling policies should state the aims and procedures for at least the following seven items.

a. Who will determine the standard method?

b. How will the standard method be put into practice?

c. What does standard time represent?

d. Who will determine the standard time and how will it be determined?

e. Under what conditions may standard time be changed?

f. How will production and hours expended be reported?

g. How often and through what channels will reports on the performance of the motion and time-study functions be reported?

BIBLIOGRAPHY

Abruzzi, Adam, *Work, Workers and Work Measurement*. New York: Columbia University Press, 1956.

Aitken, H. G. J., *Taylorism at Watertown Arsenal.* Cambridge, Mass.: Harvard University Press, 1960.

American Institute of Industrial Engineers, *Proceedings of the Annual Conference*, annually since 1955. New York: American Institute of Industrial Engineers, 347 E. 47th Street.

American Society of Mechanical Engineers, *Fifty Years of Progress in Management*. New York: American Society of Mechanical Engineers, 347 E. 47th Street.

Bailey, G. B. and Ralph Presgrave, *Basic Motion Timestudy*. New York: McGraw-Hill Book Company, Inc., 1958.

Barnes, R. M., *Motion and Time Study: Design and Measurement of Work*, 5th ed. New York: John Wiley and Sons, Inc., 1963.

Barnes, R. M., *Work Sampling*, 2nd ed. New York: John Wiley and Sons, Inc., 1957.

Barnes, R. M. and Robert B. Andrews, *Performance Sampling*. Los Angeles: University of California, 1955.

Brouha, Lucien, *Physiology in Industry*. New York: Pergamon Press, 1960.

Buffa, E. S., *Models for Production and Operations Management*. New York: John Wiley and Sons, Inc., 1963.

Carson, Gordon B. (editor), *Production Handbook*, 2nd ed. New York: The Ronald Press Co., 1958.

Dreyfuss, Henry, *The Measurement of Man*. New York: Whitney Library of Design, 1960.

Friedman, Georges, *The Anatomy of Work*. New York: The Free Press of Glencoe, Inc., 1961.

Gilbreth, F. B., *Applied Motion Study*. New York: Sturgis and Walton Company, Inc., 1917.

Gilbreth, F. B., *Motion Study*. New York: D. Van Nostrand Company, Inc., 1911.

Gilbreth, F. B. and L. M. Gilbreth, *Motion Study for the Handicapped*. London: George Routledge and Sons, Ltd., 1920.

Gomberg, William, *A Trade Union Analysis of Time Study*. Chicago: Science Research Associates, 1948.

Gomberg, William, "Trade Unions and Industrial Engineering," Section 17 in *Handbook of Industrial Engineering and Management*. Englewood Cliffs, N. J.: Prentice-Hall, Inc., 1955.

Hadden, A. A. and V. K. Genger, *Handbook of Standard Time Data—For Machine Shops*. New York: The Ronald Press Co., 1954.

Hansen, B. L., *Work Sampling in Modern Management*. Englewood Cliffs, N. J.: Prentice-Hall, Inc,. 1960.

Karger, D. W. and F. H. Bayna, *Engineered Work Measurement*. New York: The Industrial Press, 1957.

Krick, E. V., *Methods Engineering*. New York: John Wiley and Sons, Inc., 1962.

Lehrer, R. N., *Work Simplification*. Englewood Cliffs, N. J.: Prentice-Hall, Inc., 1957.

Lowry, S. M., H. B. Maynard, and G. J. Stegemerten, *Time and Motion Study*, 3rd ed. New York: McGraw-Hill Book Company, Inc., 1940.

Maynard, H. B. (editor), *Top Management Handbook*. New York: McGraw-Hill Book Company, Inc., 1960.

Maynard, H. B., G. J. Stegemerten, and J. L. Schwab, *Methods-Time Measurement*. New York: McGraw-Hill Book Company, Inc., 1948.

Miles, L. D., *Techniques of Value Analysis and Engineering*. New York: McGraw-Hill Book Company, Inc., 1961.

Mundel, M. E., *Motion and Time Study*, 3rd ed. Englewood Cliffs, N. J.: Prentice-Hall, Inc., 1960.

Mundel, M. E. *Systematic Motion and Time Study*. Englewood Cliffs, N. J.: Prentice-Hall, Inc., 1947.

Nadler, Gerald, *Motion and Time Study*. New York: McGraw-Hill Book Company, Inc., 1955.

Nadler, Gerald, *Work Design*. Homewood, Ill.: Richard D. Irwin, 1963.

Niebel, B. W., *Motion and Time Study*, 3rd ed. Homewood, Ill.: Richard D. Irwin, 1962.

Presgrave, Ralph, *Dynamics of Time Study*. New York: McGraw-Hill Book Company, Inc., 1945.

Quick, J. H., J. H. Duncan, and J. A. Malcolm, Jr., *Work-Factor Time Standards*. New York: McGraw-Hill Book Company, Inc., 1962.

Shaw, A. G., *The Purpose and Practice of Motion Study*. London: Harlequin Press Company, Ltd., 1952.

Taylor, F. W., *The Principles of Scientific Management*. New York: Harper and Brothers, 1911.

UAW-CIO, *The UAW-CIO Looks at Time Study*. Detroit: International Union of Automobile, Aircraft and Agricultural Implements Workers of America, 1947.

WILLIAM GRANT IRESON *is currently professor and chairman of the industrial engineering department at Stanford University. From 1948 to 1951, before coming to Stanford, he was professor of industrial engineering at the Illinois Institute of Technology.*

Mr. Ireson is a native of Virginia and holds B.S. and M.S. degrees in industrial engineering from the Virginia Polytechnic Institute. Upon graduation from this institution, he joined the Wayne Manufacturing Corporation as an engineering trainee in production. Although serving the company in a number of capacities, he worked primarily in production control, where he was in charge of production schedules for the entire plant.

In 1941, he returned to V. P. I. as an instructor in industrial engineering. He advanced steadily through the ranks of assistant and associate professorships, becoming an acting professor and the acting head of the department in 1947.

Mr. Ireson has specialized in production, quality control, reliability, engineering economy, and industrial development. He is author of Factory Planning and Plant Layout *and coauthor, with Eugene L. Grant, of the fifth edition of* Principles of Engineering Economy. *He is editor of both the* Reliability Handbook *and the Prentice-Hall series of monographs in industrial engineering and management science. He has written many articles and technical reports.*

Mr. Ireson is a registered professional industrial engineer in California and a registered professional engineer in Illinois, and has served as a consultant to many government agencies, international organizations, and industrial firms. He is a member of the American Society of Mechanical Engineers, the Society for International Development, the American Society for Engineering Education, Sigma Xi, and the Institute of Management Science. He is also a Fellow of the American Institute of Industrial Engineers and the American Society for Quality Contol.

SECTION 6

FACTORY PLANNING AND MATERIALS HANDLING

W. GRANT IRESON

6.1 FACTORY PLANNING IS SYSTEM DESING

As the industrial world becomes more and more "automated," the "factory" becomes more and more like a single, complex machine made up of a large number of complex mechanical devices, computers, buildings, and persons, each having a specific set of functions to perform in a massive, integrated program. It is completely appropriate to describe a factory as a *productive system*; yet the word *system* requires explanation, and the system concept must be understood in order to develop an orderly, logical procedure in the design of an efficient factory.

In addition to describing a factory as a productive system, it is necessary to recognize that a productive system can produce things other than physical products. The product could be a service. The principles to be developed in this section will apply equally to hospitals, department stores, telephone systems, power generation, banks, and many other service institutions. Only the details of the applications differ among these service- and product-producing organizations. Even the measures of the effectiveness of the productive system design are similar and are based on the same assumptions and criteria. These ideas will be developed in later sections.

6.1.1 The System Concept

The word *system* now has such common usage that its meaning has become all-inclusive. Each writer seems to impart some special meaning to the word. In this section, the meaning will be the one commonly implied in the engineering profession. Essentially, a system consists of a processing unit that operates on the inputs in some describable fashion in order to produce an output. The inputs to the processing unit are usually matter, energy, and information. The processing unit is composed of buildings, materials, machines, and information-processing devices (which can be humans) that operate on the inputs in prescribed ways to obtain a result that contributes to the output. The output may be a product and/or service and/or information and (usually) waste. Obviously, persons must determine the desired outputs, the necessary inputs, and the nature and operation of the processing unit. Humans provide the control aspects of the processing system, determining the kind and rate of input and the processing functions, then evaluating the output and the corrective actions for the inputs and the processes.

Achieving the desired output in the process of system design requires specification of the internal components in the processing unit and their relationships. In some instances the relationships are so well known and understood that they can be expressed explicitly in mathematical formulations, and the design of the system is a matter of solving the appropriate set of equations. Most productive systems are much too complex to be designed in such a manner, but some process plants have been "designed by a computer." In other words, all possible combinations that might be used have been stored in a computer's memory; then the criteria of a successful or acceptable design have been written into a computer program, so the computer can effectively calculate every possible combination and compare its "cost" with the various criteria. The combination that meets all the criteria for output at the lowest "cost" is then printed out by the computer. This same fundamental process can be performed manually, but it is, of course, so time consuming that all possible alternatives cannot be examined, and a variety of methods must be employed to obtain close approximations to the optimum design specification. Much of this section will be devoted to presentations of methods by which the optimum can be approached, using manual methods that are within reasonable constraints of time and money.

The word *cost* was placed in quotation marks in the preceding paragraph because cost can be construed to mean other than monetary costs. Frequently the other kinds of costs can be translated into money units, as when space in buildings can be converted to an equivalent monetary cost per year. In some cases the cost is something that must be considered as a penalty in its own terms, as when increased weight in a motor vehicle reduces the performance of the vehicle regardless of the extra monetary cost of operating it. In productive systems design, the costs may be space requirements, floor loads in the buildings, numbers of persons for which personnel facilities must be provided, or even safety of personnel. Many such factors may be analyzed in the same way that monetary costs will be analyzed. One of the advantages of design by human beings over design by computers is that value judgments can be made in trading-off one such "cost" against another.

One additional concept of systems must be recognized: the relation of the system to the criteria that will be used in appraising system design proposals. A large system, such as an industrial plant for the production of lawn mowers or home laundry machines, is actually composed of a number of separate and identifiable subsystems. Choosing the most economical subsystem in a sequential design

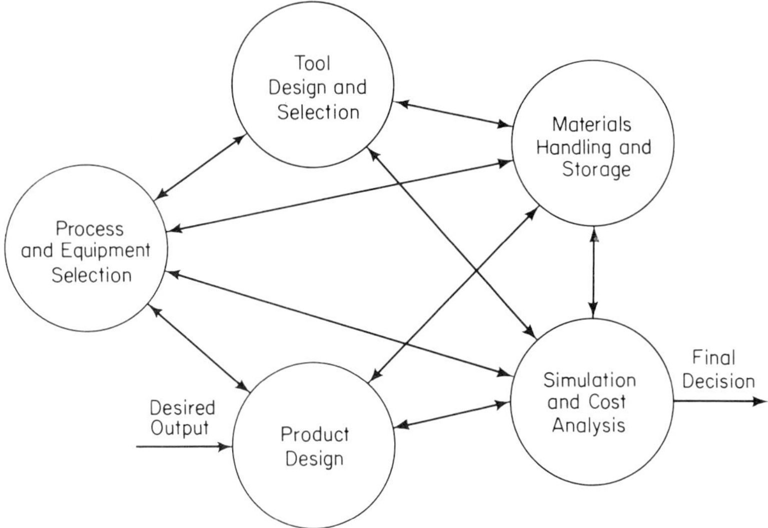

Iterative or Circular Design Process

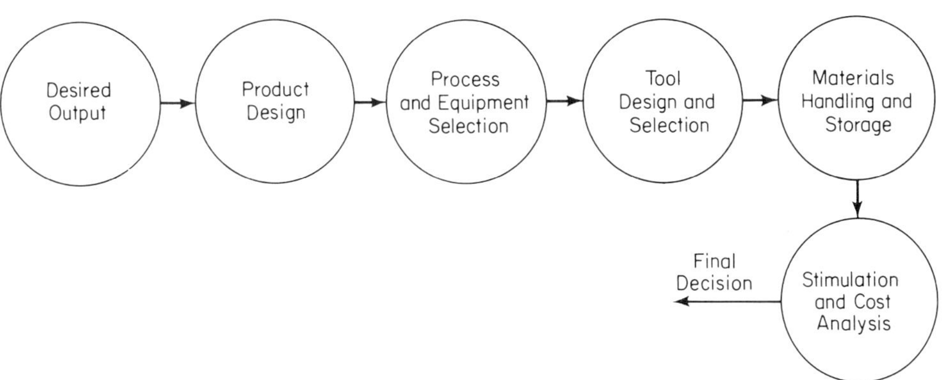

Sequential Design Process

Fig. 6.1 *Analogy of design process.*

process will not assure the optimal overall system any more than suboptimization will assure an optimal solution in typical problems analyzed by operations research techniques. And, just as we find it impossible in many operations research problems to devise an analytical technique that will assure an overall optimal solution, we will find that in many system design problems we are unable to proceed directly from output requirements and available inputs to an optimal design for the productive system. We are forced to use a series of trial and error approximations, repeating previous analyses in light of new findings, thus progressing in a sort of circular decision process to reach a close approximation to an optimal design for the proposed productive system* (Fig. 6.1).

*The reader is urged to refer to *System Engineering Handbook*, edited by Robert E. Machol (New York: McGraw-Hill Book Company, 1965) for very valuable information on systems concepts, and specific systems design problems and methods. Part V, System Techniques, is particularly valuable.

6.1.2 Factory Subsystems

There are several different ways by which a whole productive system (factory) can be broken up into subsystems for detailed analysis. Each of these subsystems can be identified and analyzed individually, but each interacts with other subsystems, and the interactions must be considered in the overall design if something approaching an optimal design is to be obtained. The usual subsystems are as follows:

6.1.2.1 Production Processes. The desired output or product and the selection of the input materials are the two primary factors that determine the alternative production methods that may be used. Many different kinds of machines, processes, and tools can be used to obtain the same output product, and the selection of the combination that will give all the desired characteristics at the lowest total cost in the long run is generally thought to be the best, or optimal, set. However, the choice of equipment may be influenced by other subsystems, such as materials handling, storage, inventory policies, etc.

6.1.2.2 Materials Handling System. All factories require means by which materials, parts, tools, and even machines can be moved into and out of storage, from process to process, and into and out of the processing equipment. Materials handling systems can be entirely manual or any combination of manual-mechanical, up to very sophisticated, automated systems requiring almost no human attention. The selection of the production processes and equipment will provide one set of bounds on the feasible handling system. The selection of a handling system or a production system without regard to interactions with the other system can lead to very uneconomical designs.

6.1.2.3 Support Systems. A number of support systems are necessary in every factory. These systems do not necessarily directly touch the product, but they are necessary in order to permit the production processes and the materials handling systems to convert the materials to finished goods or services. The support systems are usually man-machine type systems and most of them are primarily concerned with the collection, storage, analysis, and output of information for the purpose of directing or controlling other systems, including other support systems. Paper-work systems and procedures normally play a major role in the effectiveness of support systems. Many support systems require similar input information, and outputs from different systems frequently need to be integrated to supply the basis for decisions or actions. The support systems should be designed at the same time so that duplication of effort can be avoided and so that functions and activities can be assigned to the subsystem best equipped to handle them.

Computers and their collateral equipment usually play a central role in support systems. However, other machines, from office machines to repair and maintenance shop machines, are of concern to support systems. Selection of the appropriate mechanical equipment is just as important to the effectiveness of the support systems as to the production system.

The common support systems are the following, or some combinations of the following:

A. Production planning and control
B. Materials and inventory control
C. Quality assurance/reliability assurance
D. Purchasing
E. Cost accounting and cost control
F. Maintenance
G. Personnel (manpower planning)
H. Sales demand forecasting
I. Plant engineering
J. Communications

Each of these support systems can normally be broken down into several subsystems which can be treated as an entity but which must be designed in consonance with the other systems and subsystems. For example, the quality assurance/reliability assurance system can be broken down into such subsystems as vendor relations, incoming inspection, in-process inspection, functional test, life test, design review, data analysis, customer relations, or others. Each of these subsystems will have specifically designated duties and will need floor space, equipment, information-handling facilities, etc. The design of the man-machine system cannot be made without consideration of the relations with other subsystems, nature of the relations, flow of physical materials such as products, and the nature of the process to be performed. It is advisable to draw a chart that shows the gross relationships of all subsystems and then draw a series of charts to show the detailed relationships of each subsystem to all others. As will be pointed out later, the construction of such relationship charts will usually automatically force the determination of the information necessary to develop a good design for the subsystem. These relationship charts can take the form of block diagrams showing the flow of information, materials, parts, reports, drawings, etc., between the subsystem under study and all others. The functions of the subsystem are itemized as the subsystem determines what it does with each item in the flow diagram. Thus the work load and the requirements for the system can be developed with ease.

6.1.3 Definition of Terms

A few special terms need to be defined to establish the purpose and scope of this section. Other terms will be defined as they occur throughout the text.

Factory. Any place in which the factors of production, land, labor, capital, and enterprise are brought together for the creation of goods or services. The term *plant* is used synonymously with factory throughout this section.

Factory planning. The formulation of a complete plan for the creation of goods or services. The term embraces the determination of the location, production processes, equipment, physical arrangement, provisions for personnel, offices, and all functions that are necessary to the completion of the goods. It implies that a careful study has been made of the several alternatives at each phase of the process and that the course has been adopted that has the greatest likelihood of providing the required service most economically, considering the long run.

Plant layout. The analysis and proposal for the physical arrangement of the physical facilities after the decision on the site, production processes, and equipment has been made. This is a more restricted term than factory planning. It is not limited to the process of determining the arrangement of a given number of machines within a department by means of templates or models, as is often erroneously believed.

Continuous industry. An industry in which it is impossible to stop the production process on short notice without suffering considerable loss in partially processed materials, in damage to equipment, or in labor and materials required to clean out and recondition production equipment. This classification is useful in factory planning because the continuous industry's problems of factory planning are very different from those of the interruptible industry.

Interruptible industry. An industry in which the production process may be stopped on short notice without suffering any losses except those caused by idleness on the part of the workers and the equipment. Many factories of the interruptible type do operate on a twenty-four hour basis, seven days a week. Continuous industries have to work on this basis and take great precautions against breakdowns that might cause a rapid shutdown of the processes. A great many plants have both continuous and interruptible processes or divisions.

6.2 FACTORY PLANNING: MANAGEMENT CONSIDERATIONS

6.2.1 Management's Interest in Factory Planning

In most businesses, competition for the available market constantly forces the management of each concern to seek competitive advantages through such methods as: product improvement or new products, lower costs and lower selling prices for the same or better quality, better service to customers.

The selection and arrangement of physical facilities (the factory plan) can assist in achieving these advantages and the re-layout of facilities can frequently lead to relatively large annual savings. Some of the ways in which the factory plan can effect economies in operation are:

1. Lower labor costs, both direct and indirect
2. Reduction in work stoppages and down times; fewer interruptions of production
3. Reduction of the through-put time (the elapsed time between issue of raw materials to the movement of finished product to stores)
4. Reductions in the inventories of work-in-process, raw materials, and supplies
5. Greater production capacity without additional equipment, space, or employees

6. Simplification of production and material controls

7. Increased flexibility in the output, in terms of both the volume and the variety of products or services

8. Lower materials handling costs

9. Lower plant-maintenance and engineering costs

10. Lower storage costs

11. Provision for later expansion or contraction at a minimum cost and minimum interruption of production

12. Greater ease in effecting changes in layout

13. Better morale and lower turnover among employees

14. Reductions in employment and training costs

15. Faster service to customers

16. Fewer problems that require managerial attention as a result of better production facilities and better employee relations

The opportunities for these economies are not always evident to top management. It becomes the duty of the lower levels of management, which are closer to the problems, to be alert to the possibilities.

6.2.2 The Organization for Factory Planning

The factory-planning function is essentially a staff function, but current practice does not provide any real precedent for its inclusion in the organization chart as a separate functional group. Large and small industrial concerns vary considerably in their treatment of the planning function. Some assign it to the plant engineering group, some to the methods department, others to the production planning department, and still others to a separate department that may be in charge of factory planning, plant layout, materials handling, and other functions. In some concerns, the factory-planning group is a planning group only, leaving the execution of the plan to other divisions, such as plant engineering; in other concerns, the planning group is responsible for both the planning and the execution of the plan, for the installation of equipment, and for the establishment of production procedures and standards. Each organization must determine which plan of organization is best suited to the personalities involved, the abilities of the persons available, the availability of information to certain groups and not to others, and the existing division of functions.

Regardless of the place of the factory-planning group in the organizational chart, the group must have access to information on the demand for the several products; proposed changes in methods of production; product designs; sales promotion programs; availability of labor; labor rates; cost data on the past performance of materials handling, labor, and machines; and availability of funds. Since this information must come from a large number of different sources, it seems logical that a separate department, armed with authority to request the information, should be established to carry out the planning function. The very fact that such diverse kinds of information must be used in arriving at an economical plan for the plant indicates that a separate department should be responsible for handling it. The planning group should be free from the influence that might be exerted by its parent division.

The planning function should possess the authority to obtain necessary information from all departments. Such an organization does not relieve the planning department of the necessity for maintaining a cooperative and congenial working relationship with the other departments. It should consult with all interested groups and departments while it prepares plans and schedules of execution, so that the best interests of the company will be recognized and maintained.

In the smaller companies, where valid reasons exist for combining functions, certain plant activities might be combined into one department called the "Production Engineering Department." Such a department might be charged with the following functions: plant engineering, production planning and control, methods engineering, production and wage standards.

In continuous industries, the plan for a new plant is usually completed in great detail before construction begins. Once in operation, the continuous plant is usually not subject to further planning for several years. Large industrial concerns, which more or less continuously expand by building new plants or additions, usually maintain a staff of specialists in that particular type of industry. These specialists constantly search for new and better ways of producing the product and are responsible for planning the new plants, additions, and revisions. A smaller concern usually relies upon consultants and architects to design and plan its facilities, since it does not have a uniform and continuous need for the service (see Par. 6.2.5).

6.2.3 General Planning Procedure

The usual factory-planning project is divided up into a number of separate problems to be solved. It would be ideal to be able to design the entire system in one process, but limitations on interrelationships and the difficulty in obtaining necessary information more or less force a suboptimization approach. The procedure described in the following paragraphs makes no pretense at arriving at a perfect or optimal solution. How good a procedure will be depends upon the availability of information and the persistence of the design engineer in his attempts to relate all the significant factors to the overall goal.

By making use of all possible methods of analysis and synthesis, the one best solution for the given conditions can be determined with reasonable facility. Interdependence among the separate problems can be considered, and the solutions of the separate problems can be tempered to obtain the overall optimum solution for the entire factory.

Items that are variable in one problem may be fixed in another. It is important, then, to recognize and separate the variable factors from the fixed factors in each instance, so that the degrees of variation of variable factors will be considered only within the limits permitted by the fixed factors.

The development of a systematic procedure for the collection and analysis of data and for the formulation of the best possible plan is the objective of this section. Briefly, the procedure is:

1. Collect all factual data regarding the products, volumes, markets, facilities, financial resources, forecasts of future developments, labor supply, management policies, competition, and so on.
2. Separate the fixed factors from the variable factors and establish quantitative values, wherever possible, for each factor.
3. Working from the fixed factors, establish the practicable limits of variation for each variable factor and determine the results of allowing the variable factors to assume the different possible values.
4. Analyze the combined effects of selecting the value for each factor which will maximize the productivity and/or minimize costs. Later in the section simulation of the results of the plan as a means of evaluating the combined effects will be discussed.
5. Choose the combination that provides the optimum solution to the problem.

6.2.4 Cost Analysis

In any factory-planning problem it is assumed that certain facts will be known and that the plan will not be accepted unless certain criteria are met. For instance, it is assumed that the desired or necessary quality level is known and that no plan will be acceptable unless it will produce the quality required. Similarly, the expected demand establishes the criteria regarding volume capability for the product mix and the timing of the production. Flexibility to adjust to changing demands or changing mixes of product may be a primary requirement of the plan. Restrictions on the escape of objectionable fumes, dust, waste, etc., are known and either are or are not met by the plan. When the easily established requirements are known, one other criterion remains: cost. The plan that will meet all the minimum requirements at the lowest overall cost (monetary) will be the one that should be selected.

It will never be possible to predict exactly what demand, labor and material costs, and other relevant factors will be over a period of years. Therefore, the data for planning should not be treated as unchanging facts. On the other hand, the data should be examined to determine what reasonable variation might be expected to take place, and then to evaluate the proposed plans relative to the probable extreme values. The next subsection discusses this variables approach to estimation of future events.

6.2.4.1 Variables Approach. All factory planning is done for the *future*. The basis for most factory planning is the estimates of future production demands for the products, and the estimates of future costs of labor, materials, machines, tools, and so on. To consider probable variation from the estimates and to compute the effects of this probable variation on the final costs, or on the costs of each element, one should estimate probable maximum and minimum values for each factor. Then each cost item can be computed on the basis of both the maximum and minimum values as well as on the expected value. Analysis of the factor relative to changes in economic level, wage rates, public taxation, competition, military situations, employment, and population growth will indicate how the factor will react with each change. Then, assuming that certain changes may occur, either upward or downward, the effects on each cost factor can be estimated and the appropriate maximum or minimum value can be used in calculating the final total cost.

It is important to note that a given change from the expected in the future will cause some variables to be maximized and some to be minimized. The estimate of the total final costs, then, will not be made up of either the total of all the maximum cost estimates or the total of all the minimum cost estimates. Instead, it will be a mixture. The difference between the probable maximum cost and the probable minimum cost is not likely to be as great as the variation of individual factors indicates, for there are many compensating factors. Figure 6.2 is a simple illustration of the use of this concept.

In Fig. 6.2, probabilities of 0.90, 0.50, and 0.10 have been used for the probabilities that the various estimates will be exceeded. Obviously these are intuitive probabilities and the estimates are based almost entirely on the opinion of the persons making them. The author has frequently argued against the use of intuitive probabilities in decision processes,

ANALYSIS SHEET

Product: Thermostat bracket Part No.: B-475

Department: Press Analyst: K C. B.

Present Method: 3 Separate compound dies on 3 10-ton presses. Strip stock.

Proposed Method: Multistage progressive die on one 30-ton press. Automatic feed using coiled strip stock.

Purpose of Proposal: Reduce labor costs (No change in overhead costs)

Cost Factors	Present	Proposed Min.	Proposed Expected	Proposed Max. *
Annual Production (Units)		30,000	50,000	75,000
Die Cost ($)		6,000	8,000	10,000
Die Life (Units)		100,000	200,000	300,000
Labor Cost/unit (Includes setups)	$0.120	01	.03	.05
Material Cost/units ($)	.005	006	008	009

Cost Analysis		Best Possible	Expected	Worst Possible
Annual Production (units) Dies Cost ($)		75,000 6,000	50,000 8,000	30,000 10,000
Die Life (units) Die Life (Years)		300,000 4	200,000 4	100,000 3
Equivalent Annual Die Cost (including 10% return)		$1893	$2524	$4021
Unit Die Cost ($)	0.0120	0.0252	0.0505	0.1340
Unit Labor Cost ($)	0.1200	0.0100	0.0300	0.0500
Unit Material Cost ($)	0.0050	0.0060	0.0080	0.0090
Total Unit Cost ($)	$0.1370	$0.0412	$0.0885	$0.1930

*Probability of Estimate being exceeded: Min: 90; Expected: 50; Max: .10.

Fig. 6.2 *Illustration of concept of planning by minimum, expected, and maximum values.*

but in this case the accuracy of the probabilities is relatively unimportant. What is important is that the extreme values be made high enough or low enough that the probability of their actual occurrence is very small. Then, by the time several probabilities of extreme values are multiplied together to obtain the probability of a combination event, the probability is so small as to be almost impossible. For example, in Fig. 6–2 the unit die cost is the result of combining four factors, and the probability of either extreme value is 0.10^4, or 0.0001, an extremely small probability that the unit die cost will be as low as the low estimate or as high as the high estimate. Suppose the probabilities of 0.2 had been assigned to each of the four factors for the extreme estimates. The probability of the worst or best happening would be 0.2^4, or 0.0016, which is much larger than the former probability of the extremes of unit die cost, but still so small as to be very unlikely to occur. An engineer should still be very confident that his calculations of the worst possible and the best possible unit die costs are safe for planning purposes.

All estimates of future expenses and incomes will be in error to some extent, and it is impossible to predict the amount or the direction of the errors that will occur. However, future changes in such basic items as material prices, labor rates, and taxes are likely to be very similar regardless of the alternative chosen. As long as the proportions of labor, materials, and so on do not differ greatly between alternatives, the effects of future changes will be similar for all the alternatives. The better plan today is likely to be the better plan in the future, unless there are major differences in the amount and grade of labor or materials used. If such is the case, the analyst should carry out a sensitivity analysis to see how much change would need to take place to reverse his present decision. (See Sec. 3, Engineering Economy.)

6.2.4.2 Engineering Economy Approach. Many of the problems of factory planning are direct applications of the principles of engineering economy. Since an overwhelming portion of factory planning is the replanning of existing facilities and involves only one segment of a larger plant, the whole plan may be developed as a succession of small and relatively independent problems. In such cases, it is usual to find that a fairly limited number of practicable alternative solutions exists. The selection of one of the alternative solutions can be made by making a direct comparison of the alternatives by one of the engineering economy methods: rate of return on investment, equivalent annual cost, present worth, breakeven point, capitalized cost, or time required for the alternative to pay for itself. Articles 3.2 to 3.7 describe these methods completely.

The engineering economy methods are directly applicable to the following types of problems:

1. The selection of plant sites
2. The determination of the types of buildings and the number of floors in each
3. The selection of machinery and production equipment
4. The selection of materials handling systems and equipment
5. The arrangement of the producing departments within the buildings
6. The provision of factory services (immediate capacity and distribution of steam, gas, air, power, etc.)
7. The determination of the economic lot sizes to be produced

6.2.4.3 Planning for the Future. Each factory-planning problem must be considered from two viewpoints: immediate needs and future needs. The less stabilized the industry is, the more important it becomes to plan for future eventualities. Such factors as public acceptance of the product, rate of development of competing lines, financial conditions, amount of current research and development on the product, age of the industry, and research and development of production methods largely determine the length of time over which the present arrangement can be expected to remain unchanged and economical. If this expected time is short, there is a great incentive to plan the factory so that future changes can be made quickly and economically to accommodate changing conditions. This condition is described as the flexibility of the plant. Most of the noncontinuous industries must anticipate constantly changing conditions and must build this flexibility into the factory plan.

Flexibility is accomplished by providing in the original plans the following conditions:

1. Wide column spacing to permit freer arrangement of equipment

2. High floor-load capacities and structural strength

3. Adequately high ceilings or clearances for handling equipment, dust-collection equipment, air ducts, and the like

4. A network of exposed or conveniently enclosed service lines with frequent tapoffs: steam, gas, compressed air, electricity, water, etc.

5. Substantially larger capacity in service supply lines than immediately needed

6. Standardized materials handling systems that can be adapted readily to changes in size, weight, and nature of products

7. Provisions for expansion of production facilities and/or buildings without interrupting existing layouts. Location of trunk lines so that additions will not require their relocation

8. A long-range plan for growth by addition of buildings in the original plot plan

9. Movable walls or partitions

10. Adequate personnel facilities, strategically located

11. Extra land for expansion

6.2.5 Indicators of Need for Replanning

All management personnel, including foremen and supervisors, should constantly be sensitive to indications that existing layouts are adversely affecting the operations in a plant. Workmen frequently become aware of needed changes before the management personnel recognize the indicators. It is advisable to make frequent surveys, both by consciously looking for indications and by asking workmen for suggestions, in order to prevent a bad situation continuing indefinitely. The following items are common indicators that the layout needs to be restudied:

1. Changes in the product: new styling, new models

2. Addition of new products and deletion of old ones

3. Changes in the location or concentration of markets

4. Changes in the volume of demand

5. Worker complaints regarding working conditions (noise, light, temperature, etc.)

6. Worker complaints on availability of materials, tools, instructions, etc.

7. Sudden increase in absenteeism

8. Frequent accidents

9. High labor turnover

10. Large amount of, or increase in, idle machine time

11. Frequent failure to meet promised delivery dates or production schedules

12. Foremen's requests for additional equipment or workers

13. Foremen's complaints about delays in supplies from other departments

14. High or rapidly increasing maintenance on equipment and buildings (obsolescent facilities)

15. High percentage of rejected product by inspectors or customers

16. Increasing difficulty in finding suitable employees

17. Congestion in plant, lack of storage space, shortage of trucks, skids, etc.

18. Higher in-process inventory than usual.

The existence of any one of these conditions means that some opportunity for improvement exists, and there is a reasonable likelihood that factory planning is involved. An investigation should be initiated and the problem should then be turned over to the group or groups involved for a complete investigation and recommendations. Every group that is directly affected or that directly contributes to the condition should be consulted and invited to participate in the investigation.

A factory-planning study should be initiated whenever a condition exists that offers opportunities for improvement of productivity, improved morale, or lower production costs. However, the effectiveness of such a plan, if used exclusively, would be largely accidental. The factory-planning program should be more systematically organized in order to be most effective. The functions of factory planning should be definitely provided for in the organization chart, even if they constitute only part of the duties and responsibilities of one person.

Replanning programs that result only because some of the foregoing conditions exist are not as effective in maintaining the efficiency of the plant as are formal on-going programs. One type of program requires a review of every department or section on a periodic basis. A schedule is planned whereby each section will be reviewed and evaluated for possible improvement on a regular time scale. The size of the plant and the frequency of changes in product designs, volumes, etc., will determine the size of the factory-planning team required to carry out this periodic review. In small plants, the team may have other responsibilities that occupy most of the time, with the factory-planning reviews taking only a few weeks per year.

In large plants, where the factory-planning function can keep a sizable team busy all the time, the review and revaluation may be going on all the time. Specialization of labor may be practiced with considerable success, in that individuals can specialize in the planning of different types of facilities.

6.3 PLANT LOCATION

The problem of plant location exists whenever a plant expansion is contemplated. Presumably, there are always two or more alternatives open to the industrial concern. The first is not to expand, the second is to expand at the present location, and the successive alternatives are to expand at different locations. The nature of the alternative locations and the attractions and disadvantages of each are the determining factors in the final decision. The best location is one that will enable the company to produce and distribute its product with the greatest profit. It is commonly said that the best location is the one that permits the concern to produce and distribute its product (some assumed and fixed quantity) at the lowest overall cost. This is not necessarily true, because the location of the plant relative to markets will, in many instances, directly affect the volume of the product that can be sold. The greater volume at one location, even though it may result in a slightly higher unit cost, may still render a greater profit.

The three elements of cost for location analysis are cost of obtaining suitable raw materials and supplies, cost of conversion or production of the product ready for sale, and cost of distributing the product to the dealer or consumer. Each of these cost elements will be affected by the location of the plant relative to the sources of raw materials and supplies, labor and equipment markets, and markets for the finished product. In the following discussion of determining factors, each element will be present but will not be discussed at length. A little thought on the matter will reveal how each factor affects these cost elements.

The problem of plant location consists of two major divisions: the selection of the geographic region, and the specific site selection within the region.

6.3.1 Regional Site Selection

The factors and their effects on the total costs vary with the regions of the nation. Table 6.1 provides a check list of factors and examples of cost items that may be adversely affected by the region.

The information necessary to make a reliable estimate of the general costs for a given region is readily available. Comparative labor costs by cities and regions are available from the U.S. Department of Labor and other agencies.

Several journals frequently publish surveys of wage rates. Freight rates by classes of commodities can be obtained by regions from almost any railroad, and most railroads maintain special departments for the purpose of supplying information on many different cost items to prospective customers. Local building contractors and contractor associations can provide information about construction costs, and published information is available from many trade journals. State and local chambers of commerce will provide data on state and local laws, ordinances, codes, and regulations. Most states and communities realize the advantages of broadening the tax base by inducing industrial expansion. City councils, chambers of commerce, railroads, and merchants are eager to supply information, make contributions to desirable industries, and make tax concessions in order to attract companies. But herein lies a danger. Sometimes these immediate advantages appear to be more important than they really are in the long-run economy of the company, and the chosen site may ultimately prove to be a poor choice.

The industrial plant that is seeking a site today has much greater freedom than it has had in the past. Some of this freedom has come about as the result of technological advances, changes in worker attitudes, and labor legislation. Specifically, some of the developments that have increased the freedom of location are:

1. Improvements in transportation facilities and speed of service
2. Reduction in wage differentials between regions

TABLE 6.1. *FACTORS AFFECTING THE SELECTION OF A REGIONAL PLANT SITE*

Factor	For Each Alternative Region, Check the Effects of:
Raw material supply	Length of haul from source to region Freight rates on the commodity, both raw and finished Ratios of weights and volumes of finished products to weights and volumes of raw materials Labor supply and attractions or accommodations to induce migration of labor to source Availability of water, fuel, power, etc., at source Different sources of raw materials (They may be widely separated.) Availability of adequate transportation facilities
Labor supply in region	Adequacy of supply of desired type Competition for the existing supply Suitability of existing supply, by former work or training, for the intended type of work Union organization and strength Race relations Dependability of the type of labor available, and aptitude for factory work, training, and upgrading
Marketing	Market area to be served by the plant Concentration of market and stability of demand Effects of location on number of warehouses and inventories Freight rates to principal market areas Transportation facilities available to market areas Travel expense for salesmen and service personnel Risks of delays and damage to goods in shipment (customer relations) Competition for the market and relative location of competitors' plants
Factory services	Adequacy of supply of power, water, fuel, etc., for present and prospective plant size Availability of external plant services: sewage disposal system, repair and replacement parts, fire protection, public transportation for personnel, etc. Available supply of trained management personnel Attractions of region (cultural, climatic, etc.) for professional management personnel
Climatic conditions	Cost of construction to withstand forces of nature: earthquakes, winds, snow, etc. Heating or air-conditioning costs for either personnel comfort or process control Probability of absenteeism caused by weather Probability of work stoppages or interruption of supply of raw materials by weather Necessity for premium wages because of weather conditions Cost of maintenance, deterioration of products or raw materials, and rapid depreciation resulting from climatic conditions
Laws and codes	Laws limiting scope of work permitted "Fair Employment Practices Acts" Extra costs for unemployment insurance, workman's compensation, retirement insurance, and similar benefits Waste disposal, smoke abatement, and nuisance regulations Local tax laws on real property, corporate income, money on deposit, etc. State, county, or city building codes, and safety and health regulations Rulings of such bodies as the Interstate Commerce Commission, which may penalize certain types of industries in certain regions

3. Mobility of workers and management
4. Increased leisure time for workers, which places more importance on recreational facilities, cultural advantages, etc.
5. Improvements in construction methods and designs for plant buildings, which make them less expensive to build
6. Trend toward one-floor plans that require large land areas for expansion
7. Availability of automobiles for almost all workers, making large parking facilities a necessity for most plants and making more plant sites convenient to labor supply
8. Improvements in processing and machine designs that reduce the relative number of employees required for a given output
9. Equalization of freight rates between regions
10. Availability of economical air-conditioning equipment to counteract adverse climatic conditions for employees and processes
11. Expansion of markets for almost all goods and services so that more plants are needed to meet the demands

No plant site should be chosen without a careful investigation and economy study of all the factors involved. It is almost impossible for any plant management to select the most economical plant site by guess. What appear to be relatively small differences in costs between regions frequently turn out to be highly significant in the total cost.

6.3.2 Site Selection Within the Region

After the region for the industrial plant has been chosen, there remains the matter of selecting the specific site for the plant. There usually are as many alternative sites within the region as there are alternative regions, and the problem is frequently just as complicated and difficult to solve. Many of the same factors that are considered in the choice of a region are again considered relative to specific sites, but there are also other factors.

6.3.2.1 City, Suburban, or Country. A city site is one within the corporate limits of a city, and associated with the site are such conditions as heavily concentrated population, high land values, complete regulation of activities, and limited space. The suburban site is one outside the corporate limits of a city (usually a reasonably large city), and may be within the corporate limits of a smaller city or town. The country site is usually some distance from an established center of population. The choice among these is usually based upon an economy study of the effects of:

1. Land cost
2. Availability and cost of labor
3. Availability and cost of suitable transportation facilities to the site for raw materials, finished product, and personnel
4. Local ordinances, building codes, and restrictions
5. Available power, fuel, water, and sewage disposal
6. Union activities and strength in the area
7. Climatic conditions
8. Size of plant, immediately and in the future
9. Local property and other taxes
10. Nature of processes and hazards or nuisances produced

Some of the conditions that tend to dictate the choice of a country location are:

1. A need for a large amount of land for eventual development
2. A need to use processes that are normally considered objectionable within populated areas
3. A need for large volumes of relatively pure water, which can be obtained from streams, wells, or springs
4. A need for a favorable property tax
5. A need for protection against sabotage or the observation of processes or output

Suburban sites are chosen for such reasons as:

1. A need for reasonably large land areas not obtainable in the city, yet close to transportation and a large population center
2. A desire to be free from the more strict building codes and restrictions normally found in cities
3. A need for a large number of female employees, generally available in suburban areas because of a lack of other employment opportunities
4. A desire to escape high taxes
5. A desire to escape from highly unionized areas
6. A desire to locate nearer markets, transportation facilities, or employees' homes

The city location is frequently chosen for plants that require:

1. A large portion of employees to be highly skilled

2. Rapid transportation or quick contact with customers or suppliers

3. Relatively small total space, which can be contained in multistory buildings

4. A large variety of materials and supplies, but each in relatively small quantities

5. Public utilities, city water, electricity, gas, sewage disposal, police and fire protection, etc., at reasonable rates (The company may not be in a financial position to provide these facilities for its plant.)

6. The means of getting into production with the least possible investment in land, buildings, etc. (These facilities can more frequently be rented in a city.)

A recent development has radically changed the general problem of plant site selection for many young, small companies: the introduction of industrial parks or industrial estates. In the United States, as well as in the developing nations, land frequently is expensive, and the owners prefer to obtain an annual income from the land instead of selling it. Capital-gains taxes on highly appreciated land make selling an unattractive alternative for many owners. The expectation that land values will continue to climb rapidly suggests that many owners can profitably develop the land for industrial uses and lease it to plant operators under more favorable terms than they can sell it. The result is that many small- and medium-size firms (and in some circumstances the very large firms) find it advantageous to conserve limited capital by leasing as opposed to buying factory sites.

In areas of extreme land scarcity, such as in Hong Kong, industrial estates have been built into multistoried buildings, and considerable success has been realized in such ventures. In Kowloon, some of the industrial estates in such high-rise buildings not only house factories, which occupy only a portion of the total space available, but provide apartments, markets, and schools for the employees of the factories.

Few industrial concerns are fortunate enough to find a naturally ideal plant site. The objective of site selection studies is to find a site that can be developed and from which the operations can be conducted at the lowest overall costs in the long run. The investment in the development of a site can be amortized over a period of years along with the plant buildings. A site that requires a greater initial expenditure for development may prove to be more economical in the long run, because of consistent annual savings in other cost items, than what at first appears to be a more suitable site. An engineering economy study is a "must" for site selection. (See Sec. 3 for a complete discussion of these methods.)

6.3.3 Site Planning

The utilization of the site and the plan by which the ultimate factory will be arranged on the available land will have an important effect on the long-run economy of the plant. For a new plant, the tentative plans for the physical facilities, including immediate and prospective floor area, number and types of buildings, eventual employment, flow of materials, and auxiliary facilities, will have been prepared for use in the site selection. (See Art. 6.5.1 on the sequence of events in planning a factory.) The selected site conditions will in turn usually necessitate certain adjustments in the general plans. The site will first be described by a map showing the location of the site relative to highways; railway lines; rail sidings; power, gas, water, and sewer lines; towns; public transportation facilities; and any existing structures. A topographical map will be prepared to show the contours and elevations of all segments of the land area relative to the same items. Drawings will be made of all existing structures, showing building details, floors, floor loads, columns, piping, power circuits, walls, doors, windows, and so forth. Climatological data–wind velocities and directions, temperatures, rain and snowfall, humidity, sunshine and cloudiness, and probability of extremes of each factor by months or seasons—will be obtained.

With this accumulation of information, the planning group is ready to study the site with particular reference to the general plan for the factory. The group will attempt to place the several buildings on the site in such a way that full advantage can be taken of the site characteristics and existing conditions. Some of the things that the planners will attempt to do are:

1. Locate buildings that are to be served by rail lines near existing or extensions of existing rail sidings.

2. Locate areas to be served by highway trucks near existing highways or streets.

3. Locate employee parking facilities near

existing streets and convenient to work areas. Several parking areas may be required.

4. Arrange buildings, relative to each other, so that full advantage can be taken of the prevailing winds for ventilation or removal of unpleasant smoke or fumes, of sunlight for lighting and heating buildings (in winter), and of buildings to protect much-used outside traffic lanes from wind and rain.

5. Arrange building so that additions for expansion can be made later without disrupting production and so that roads, rail, power, water, sewer, and steam lines will not have to be relocated.

6. Avoid unnecessary scattering of buildings that will increase materials handling and other service costs.

7. Take advantage of any natural slopes of the terrain to accomplish drainage, waste and sewage disposal, and materials handling.

8. Locate sources of dirt, fumes, and obnoxious products to the leeward side of the site. (Site should have been chosen so that such products will not be blown over populated areas. Not to do so invites regulation and corrective action.) Provide for acceptable methods of waste disposal.

9. Locate office buildings near main street to facilitate visitors' entrance and exit, and plant protection.

10. Isolate especially hazardous processes and reduce the danger that fire will spread from building to building.

11. Make use of sound, existing structures, if their use will not increase other operating costs by an amount that will equal the amortization of the most economical substitute structure.

12. Plan for attractive appearance of plant, landscaping, lighting, etc., so that public and employee relations will be good.

It can be seen that the actual site serves to limit, and at the same time may assist, the planning group in arriving at the optimum factory plan. Advantages that were not expected when the general plan was drawn up may appear, and, also, adverse conditions may be present that were not expected. In any event, once the site has been selected, the planning engineers have a much greater amount of data that can be used directly in making the numerous decisions required. These data permit a more accurate evaluation of the effects or costs of each of the different alternatives. Decisions can be made with a greater confidence that they represent the "one best way."

6.3.4 One Plant or Several?

In the discussion of the selection of both the region and specific plant site, many factors were enumerated that must be considered in the selection. It was assumed that only one plant was to be located. Many growing industrial concerns are faced with the problem of deciding between a number of small plants and one large one. Where raw materials are more or less universally available, where no special skills are required, and where the market is represented by the general population, several plants, instead of one, usually represent the most economical solution. Each plant, however, must be designed for and operated at or near the optimum output. The cost of duplicating facilities which may not be used to capacity is one of the limitations that must be considered in the solution.

Soap, paper, gray cast iron, automobiles (assembly), synthetic fibers, millwork, furniture, plastics, fertilizer, and clothing are some products that have been decentralized with some success by individual companies.

Staley and Morse* have made a significant contribution to the literature of factory planning, especially in the analysis of plant location problems. They have classified the factors affecting factory planning for small industry into:

I. Locational influences
   - Factories that process a dispersed raw material
   - Products with local markets and relatively high transfer costs
   - Service industries

II. Process influences
   - Separable manufacturing operations
   - Craft or precision handwork
   - Simple assembly, mixing, or finishing operations

III. Market influences
   - Differentiated products having low-scale economies
   - Industries serving small total markets

A complete analysis of the influences of loca-

---

*Staley, Eugene and Richard Morse, *Modern Small Industry for Developing Countries* (New York: McGraw-Hill Book Company, 1965).

tion, process, and market, using their methods of classification, will lead to the establishment of the feasible alternative solutions to any problem concerning location, size, and number of plants. After the identification of the feasible alternatives, the planner can simulate the operation of the plant in each of the alternative solutions in order to arrive at reasonably complete knowledge of the prospective profitability of each. His simulation tools can be anything from simple pencil and paper to elaborate computers.

6.4 FACTORY BUILDINGS

The buildings housing the equipment and activities of any organization constitute one of the subsystems that must be considered in designing a factory system. The buildings, differing in numbers of floors, shapes, power and utility distribution systems, relative locations, ceiling heights, column spacings, floor load capacities, and many other specific features, provide the primary "environment" for the factory. As will be pointed out in the following articles, the ways by which these features are analyzed in a factory-planning program and the resulting plans depend very heavily upon the restrictions imposed by existing buildings or the freedom provided by the opportunity to design entirely new buildings.

By far the greater amount of factory planning is in connection with the re-layout of existing buildings, but each year a large number of new buildings is added to the total industrial structure of the country. Each re-layout problem presents a challenge to the planning engineers to devise an efficient production arrangement within certain limitations, but the design of new buildings is an equally important challenge. Each new building is designed to house a particular production process, but it in turn becomes an "existing" building henceforth, and the care and thought that go into its design determine the relative ease of its re-layout when that time comes.

6.4.1 Modern Concepts of Building Design

The tremendous construction programs for both war production and civilian products have provided many opportunities for architects and engineers to try new ideas and to accumulate results. A great amount of knowledge was gained about how to design factory buildings. It is impossible to review even a small portion of these trials and results in this volume, but a few generalities can be drawn from the experience to assist factory planners. A knowledge of these general principles will enable the factory-planning group to be of more specific assistance to the architects and structural engineers when faced with the problem of designing a new building. The following paragraphs briefly report these principles.

One of the most significant developments came through the realization that the building's principal function is to provide protection for machines, equipment, personnel, materials, products, or company secrets. Thus, if the equipment need not be protected, it need not be housed. Refinery and chemical plant equipment has long been erected without housing; a building is provided only for the protection of operating personnel, records, instruments, and the like. If the climate is mild the year round, and if the equipment or products cannot be affected by rain or sun, then proper fences and the usual plant-protection police force may provide all the protection that is needed for the production process. The building may be needed only for offices, packaging, inspection, and similar functions. One plant in Texas has been constructed without exterior walls. Not only did the plan reduce construction costs (and capital tied up in fixed assets), but the relatively steady winds help to prevent the accumulation of explosive dust in the building (see Art. 6.3.3).

It has been shown that in most regions, where sufficient land is available at reasonable prices, a one-floor plant is more economical than a multistory plant. Naturally, certain types of processes require a multistory arrangement, but, where there is a choice, the old idea that gravity materials handling would justify additional floors has been largely disproved. It is true that a sphere incloses the maximum volume per square foot of exterior surface, but it is not true that the cost of a square foot of wall, floor, or roof is the same in a cubical building as in a one-story building providing the same floor space. The relative ease of constructing single-story plants, the rapidity of construction, the unlimited floor load capacities, the elimination of stairways and elevators, the greater flexibility of layout (vertically or horizontally), and the ability to expand or contract operations are some of the advantages of the single-floor plant that tend to make it more economical.

Construction of plants for dynamic, rapidly changing industries has taught that industrial plants designed for one product can be economically converted to the production of very dissimilar products provided the original plant was planned for flexibility. Conversions were usually made through the complete removal of existing machines and equipment and the addition of new equipment, and the re-layout has not always been ideal because of limitations imposed by the original construction. On the other hand, it has been learned that no plant (except some continuous-type factories) can be expected to remain as planned and originally arranged for long periods of time. Changes in products, design, demand, and the like are inevitable, and efficient production will demand that rearrangements be made. Thus, a basic concept has been developed that plants for interruptible types of industry should be planned for great flexibility and ease of conversion or re-layout. (See Art. 6.2.5 for means of increasing flexibility.)

Techniques and construction practices are constantly changing and new materials of construction are constantly being employed. Speed of construction is important for several reasons (money tied up during construction period, immediate need for floor space, necessity to beat or meet competition, and so on), and where certain materials increase the speed of construction without increasing costs, there is a big incentive to use these materials. When costs of labor, freight, and lost time are considered, many of these materials are actually less expensive than conventional methods. Unfortunately, many of these newer materials and construction methods are not permitted under the building codes in some cities.

Some of these methods and materials are:

1. Use of precast walls, floors, and roofs (cast horizontally at one point and erected or "tilted up" at another)
2. Use of pre-stressed concrete
3. Use of light-weight aggregate
4. Prefabricated buildings or sections of buildings
5. Aluminum for structural members, wall covering, prefabricated panels, etc.
6. Insulated wall panels, with insulating core covered by metal or concrete
7. Nonmetallic insulating boards or tile for built-up roofs
8. Special treatments for concrete or wood floors to preserve and reduce wear.

Names and addresses of companies supplying such materials can be found in various trade journals.

Building plans once corresponded to the shapes of such letters as I, L, H, E, and C, but most of the recent, one-floor plants have been simple rectangles or squares. Use of natural light and natural ventilation was the principal reason for the odd shapes, but the advent of low-cost power, efficient light sources and fixtures, and forced ventilation has largely overcome any advantages the older plans may have had. Greater floor space inclosed per foot of wall, greater flexibility in layout, and greater ease of materials handling are some of the other advantages of the large, simple plans.

Figure 6.3 shows the standard types of factory buildings in common use. Although certain of these types have been popular for different industries, it is common to find two or more of these sections used in a single large

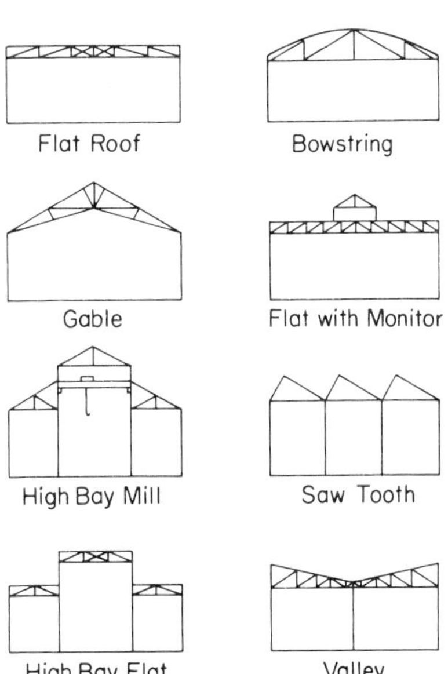

Fig. 6.3 *Most industrial plant buildings are made up of combinations of these standard types. From Ireson,* Factory Planning and Plant Layout, *p. 247.*

building, depending upon the processes to be housed in the different divisions. It is necessary to consider the process, equipment, materials handling systems, product, climate, and construction and maintenance costs in order to arrive at the best type for a given application.

6.4.2 Buildings for Continuous Industries

Another major concept of building design is that the production facilities should be completely planned, the ideal layout arranged, auxiliary facilities arranged relative to production facilities, and the building then designed around the complete plan. This procedure results in a very specialized building that usually will not be inherently flexible. This type of building design is usually employed only where there is an expectation that the original plan will remain unchanged for a long time. This concept is most often used by the continuous industries or to house certain highly specialized processes in what would otherwise be an interruptible industry.

Continuous industries plan for almost perfectly balanced production facilities, so that every division or department of the plant will have about the same capacity. The nature of the continuous industries is such that it is normally difficult to make changes after the plant is put into operation. The management goes to great lengths to assure that the equipment chosen, the layout, and the design of production facilities are the best currently available, so that the plant can enjoy a long period of operation without the danger that obsolescence will reduce the profit potential in a few years. The associated expenses of making alterations, adding new equipment, and changing processes make it difficult for a new piece of equipment to justify its installation on economic grounds. Thus a continuous industry expects its business to go on for many years, which really means that its building is unlikely ever to be used for any other purpose.

A long-established and financially sound company whose product is highly stabilized from the design viewpoint may choose this concept in order to obtain maximum efficiency and minimum costs. Thus, a large automobile manufacturer may build a highly specialized engine plant because there is little likelihood of radical changes in the design of auto engines for a long time. Minor changes can be easily accommodated in the specialized plant, and annual savings in operating expenses resulting therefrom may be sufficient to pay for the plant and earn a good return in a relatively short time.

6.4.3 Effects of Existing Buildings on Layout

Whenever the production facilities are to be arranged within an existing building, some of the freedom of arrangement is lost. Each building feature becomes a fixed factor, or else the feature must be changed. Some of the variables in the general factory-planning problem become constants and the systematic approach to the problem is altered accordingly.

The building design should be the last step in the procedure of factory planning when a new building is contemplated. This is true because it is only after other factors have been analyzed that the total floor space needed can be estimated and the conditions determined that will establish the features to be incorporated in the building. When an existing building must be used for the production facilities, the approach is altered so that the effects of the building on each different alternative can be considered as each separate decision is made. The effect is usually to eliminate certain alternatives and to impose limitations on others, so that the decision may be greatly simplified. However, it may necessarily be limited to the selection of the best of several undesirable alternative arrangements.

When an existing building is to be used, the planning group should obtain as soon as possible complete information on the building and building conditions. This information should include:

1. Up-to-date blueprints of each floor, showing columns, walls (fire, load-bearing, temporary), windows, doors, stairs, elevators, and other physical features
2. Accurate layout drawings of factory service piping and wiring, including sizes, capacity, and condition
3. Layout drawings, showing the location of existing equipment and machines, whether this equipment will be used in new layout or not (Note size, vertical height, and subfloor depth of each machine.)
4. Plot plans of surrounding area, specifically showing any other buildings that may

be used in connection with the one in question, driveways, rail sidings, contours of land, and elevation of this building relative to each of these items

5. Blueprints showing the elevation views of each different section of the building, ceiling or truss clearances, floor levels, ducts, piping, wiring

6. An engineering inspection report, giving:
   a. Estimated safe floor loads
   b. Estimated costs of rehabilitation needed
   c. Estimated costs of proposed or possible changes
   d. Estimated costs of strengthening or reinforcing to give some minimum safe floor load

Each of the preceding items limits the re-layout of the facilities and the alterations that might make the building more usable for the intended purposes. The planning engineers study these limitations and then establish the possible alternative arrangements, considering such factors as the space required for each department compared to that available, weights of machines and materials compared to safe floor loads, sequence of operations, materials handling facilities and costs, and special conditions or hazards associated with the several departments. Each of the possible alternative arrangements will be analyzed and compared by engineering economy methods.

6.5 THE LAYOUT OF PRODUCTION EQUIPMENT

6.5.1 Procedure for Plant Layout

Plant layout is just one limited phase of factory planning (see Art. 6.1). The layout of a factory generally consists of a number of separate problems of arrangement of facilities, and each problem is small enough to be manageable. This statement is not obviously true, especially if the plant involved happens to be a large and complex one. However, the procedures described in the following articles are designed to assist in the breakdown of a complex problem into a number of smaller problems that can be handled with relative ease. The more complex the product becomes, the more important it is to follow a systematic method to simplify the problem.

The procedure should first guide the planner's actions so that the probability of his arriving at an optimum arrangement is maximized, and, second, it should provide him with a basis for judging the relative merits of any one proposed layout versus others. In other words, the planner needs a road map to guide him to the best solution and a sign to tell him when he has arrived at the proper arrangement. At present, there are a number of general principles that help the planner find the best way, but little has been accomplished yet in developing a usable, foolproof mathematical model. The matter of criteria for judging the relative value of different proposals is being investigated in many places, but at this time the best measure is the prospective unit cost resulting from each proposed layout.

In spite of many attempts to formulate mathematical models that will direct the planner immediately to the best possible layout, there are so many factors to be considered that at best these attempts have only given us insights into the nature of the problems. The most effective method so far discovered is simulation. Massive, high-speed computers have made it possible to "test" proposed plans by simulation at relatively low cost. The simulation results not only tell fairly well which plans are best, but also point out the kinds of problems that can be expected due to random variance of the factors involved.

The following articles describe the procedure employed in the re-layout of existing facilities for the manufacture of a given product. It is assumed that the product, or a similar product, has been manufactured previously, and that the facilities must be arranged in an existing building. This is the most common layout problem, but special procedures to use when the product is entirely new, when new machines or equipment is to be used, and/or when a new building is to be used are given later. These procedures follow, in general, the pattern established in Art. 6.2.1.

6.5.1.1. Collect All Factual Data

1. Obtain or prepare an up-to-date print or drawing of the existing layout of the facilities to be rearranged.

2. Secure an accurate description (print and engineers' report) of the building and building features (see Art. 6.4.3).

3. Collect complete information on the product design and present manufacturing methods. This should include:

PARTS LIST

PRODUCT Brass Valve CATALOG NO 157-B

SIZES ½", ¾", 1" & 1¼" DRAWING NO 237-10

MANUFACTURE TO: STOCK ☑ SPECIFICATIONS All Brass

ORDER ☐ Std Pressure, see Drawings

PACKED 1 per Carton

12 Cartons per case

PART NO	PART NAME	DRAW NO	MATERIAL	QUAN PER UNIT	SOURCE
157-B1	Valve Body	237-11	Brass Casting	1	Make
" -B2	Gland Body	"	" "	1	"
" -B3	Stem	237-12	Brass Rod, Rd	1	"
" -B4	Gland Cap	237-11	Brass Hex Rod	1	"
" -B5	Handwheel	237-12	" "	1	Purchase
" -B6	Nut	" "	Brass	1	"
" -B7	Disc Ring	237-13	Brass Tubing	1	Make
" -B8	Disc	" "	Brass Casting	1	"
" -B9	Packing	237-11	Asbestos-Graphite	1	Purchase
" -B10	Tissue Paper	—	18" × 24"	1	"
" -B11	Carton	237-14	Cardboard	1	"
" B12	Case	237-14	Wood	1/12	"

Fig. 6.4 *Parts list for a brass valve. From Ireson,* Factory Planning and Plant Layout, *p. 18.*

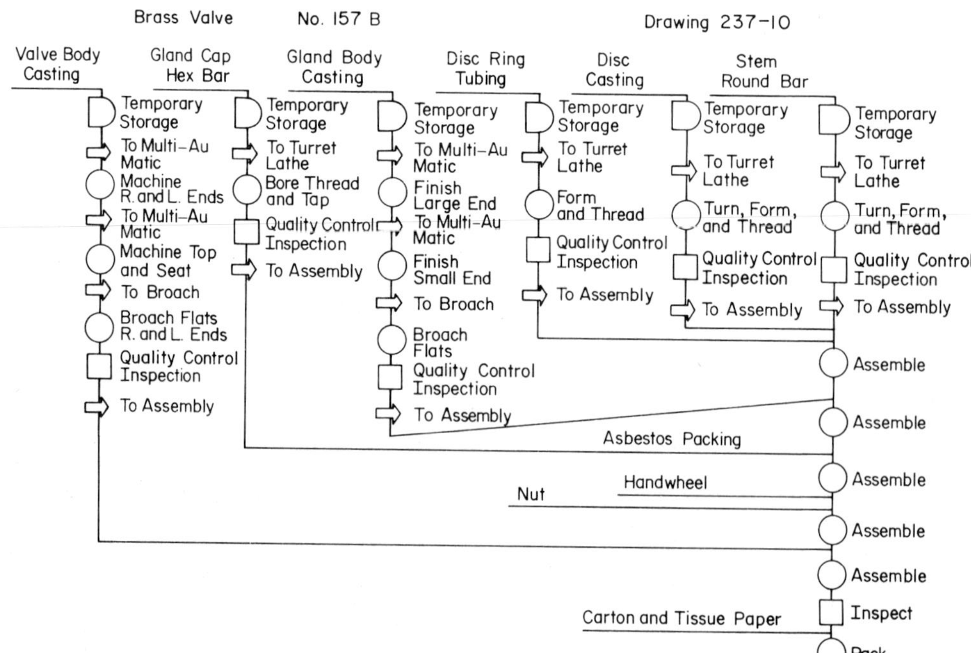

Fig. 6.5 *Master flow process chart for a brass valve. From Ireson,* Factory Planning and Plant Layout, *p. 19.*

a. List of parts and drawings for each part (see Fig. 6.4)

b. A flow process chart for each part by present method of manufacture

c. A master flow process chart for each part and the assembly of the final product (see Fig. 6.5)

d. Standard instruction sheet for each operation, showing tooling, machine, jigs, fixtures, etc., and the standard production time for the operation

4. From the sales department, obtain the results of the market survey and sales forecast for the product. Tabulate the prospective quantities by periods or years for the prospective life of the new layout. If the product is subject to seasonal fluctuation, prepare a chart showing the probable fluctuations by months.

5. Determine the budgetary limitations within which the new layout must be made, such as:

a. Amount of money that is available for the project

b. Criteria by which investment of funds in re-layout, new equipment, tools, or materials handling facilities must be justified (see Section 3.)

c. Managerial policies that will affect the freedom of choice in making the re-layout: May production be stopped or must it be maintained during the change? May re-layout extend beyond immediate problem if that will simplify the re-layout problem? Is management wholeheartedly behind the factory-planning group, or will each proposal have to be "sold" against active resistance?

6. Make a survey of existing materials handling facilities that affect this problem. This survey should be designed to determine the nature and effectiveness of the present facilities (see Art. 6.7.4). It should include:

a. Inventory of equipment, with an evaluation of its current condition and prospective life

b. Current materials handling costs (not only the actual operating costs, but also those resulting from interruptions in production caused by the system, damage to materials and parts, and cost of supervision and coordination)

c. Adaptability of existing facilities to work with other handling equipment and layouts

7. Secure a report on the labor supply available for present and future operations of this division. The report should provide information on the scarcity of unskilled, semi-skilled, and skilled labor, male or female, and going rates.

These factual data are generally available from the methods and standards, production planning and control, personnel, and sales or marketing departments and the factory management. Every effort should be made to obtain as complete and accurate information as possible, because lack of information or errors in data can cause the planning group to spend a great amount of time and effort on proposals or solutions that are impracticable or completely impossible.

6.5.1.2 Separate fixed and variable factors and attach quantities or numerical values wherever possible.

1. Analyze each item of information collected in the first stage (Art. 6.5.1.1) and determine whether or not it can or will be changed, or whether it is a factor that will vary with business activity, economic conditions, promotional programs, and the like. For example, if present machines must be used (possibly because no capital is available even if new machines might be more economical), the freedom to devise new and better operations and operation sequences is limited by the capabilities of the existing machines. On the other hand, the costs of different materials or different grades of labor may be variables that depend upon a number of factors, and it is necessary for the planning group to consider the probable variations in each in order to arrive at the combination of materials, operations, and kind of labor that will give the optimum plan.

2. Prepare a list of fixed factors. These fixed factors are conditions that must be satisfied if they are not to impose certain physical limitations on the proposed plans.

3. Prepare a list of variable factors and establish probable limits of variation for each. This step emphasizes the fact that many of the data are estimates of future activity and should not be treated as absolute quantities. The maximum capacity of output may be the basis for computing the number of machines and for balancing producing units, but the effects of operating at an output below this maximum must be considered in making the final layout and production plan.

6.5.1.3 Investigate all possible improvements that can be incorporated in the new plan. This is a study of manufacturing methods made in cooperation with other departments. It should cover such items as:

1. Machines and equipment: Are there other types of machines that can be used more economically for the operations?

2. Tools, jigs, fixtures: Will better accessories enable the operations to be performed more rapidly, more accurately, with less skilled labor, or to be combined with others? (See Section 9.)

3. Standard practices: Can the work place be improved, better procedures be devised, operations or movements be eliminated, simplified, or combined, and operation sequences be changed?

4. Materials handling: Can better use be made of existing materials handling equipment, better equipment be employed, movements be eliminated or shortened, transport and transfer be combined, or semi-automatic or automatic equipment be substituted? (See Art. 6.7.)

5. Materials: Can other materials be substituted for present materials in order to reduce materials, labor, or processing costs while maintaining or improving the product quality?

6. Personnel: How can personnel facilities be improved to reduce fatigue, improve morale, or improve efficiency?

This phase of the investigation is most important, for it provides the planning group with definite ideas for consideration in arranging the new layout. The planning engineer should discuss each phase of the manufacture with the individual workmen, foremen, methods engineers, designers, inspectors, and purchasers or buyers in order to uncover every possible difficulty that has been encountered in the past. At the same time, the engineer should seek suggestions from these persons and make careful notes of the suggestions and sources. He should also be so familiar with the several parts that he can anticipate difficulties and use direct questions to ascertain the facts regarding these ideas. It is a cardinal principle that the optimum alternative cannot be selected unless every possible alternative is recognized and investigated.

The planning engineer must also recognize that he is dependent upon the other departments in the plant for information and suggestions. Thus it pays him to cooperate fully with other departments in order to induce these departments to do as much of this investigation as possible. It is, however, his responsibility to talk with and stimulate the thinking of the members of the cooperating departments.

6.5.1.4 Determine the auxiliary departments or services that must be provided for this manufacturing setup. Certain auxiliary departments and services may be needed in the immediate vicinity of the manufacturing area, and provisions should be made for them in the initial planning of floor space utilization. Services that do not require floor space or special treatment, or departments that do not require physical proximity to the manufacturing facilities, need not be considered at this stage. Some of the items that may be necessary are:

1. Tool room, tool maintenance or die shop
2. Foremen's offices, clerical offices, etc. (see Art. 6.6)
3. Inspection rooms, gage storage, and maintenance
4. Materials handling facilities, such as conveyors, aisle widths to accommodate trucks or tractors, crane ways, etc.
5. Personnel facilities
6. Storage facilities for raw materials, supplies, work-in-process, and finished parts

6.5.1.5 Make basic decisions and establish feasible alternative methods of arranging the factory. The total number of different ways of arranging the production and service facilities is very large, approaching the permutation of the number of factors, machines, parts, and other contributing conditions. At this time it is not practicable to attempt to solve such a large set of conditions for the optimum by either intuition or mathematical models. In order to reduce the problem to a size that can be handled by the human mind, *it is necessary to eliminate whole sets of alternatives by analysis of past experience, records, and so forth in relation to the present problem.* Both inductive and deductive reasoning are employed to arrive at the conclusion that certain combinations are not worthy of future consideration, that certain factors or conditions naturally complement each other, and that the best possible solution must lie among a very limited number of combinations. This is a matter of analysis of facts and estimates, rather than the practice of an art. The reduction of the number of combinations makes it possible for the human mind to encompass the essentials and arrive at a reasonable solution. The following basic decisions help to accomplish this reduction and enable the planner to progress step by step toward the optimum solution:

1. What method of departmentalization will be used—product, process, or a combination for the several parts? (See Art. 6.5.3.)
2. How should the several departments be arranged within the building space, considering building characteristics, nature of processes, space requirements, etc.?
3. What materials handling facilities are most suitable and economical considering the nature of the parts, materials and products, type of movement required, processing methods, distances involved, etc.?
4. What method of determining the arrangement of machines and equipment within the departments will be most satisfactory from the viewpoint of time required to reach a solution, cost of making the layout, cost of the tools employed, and value of the tools in the future? (Sketching or drafting, two-dimensional templates, or three-dimensional models)

These basic decisions must be treated as tentative, for as the details are worked out it may become necessary to change the decisions partially or totally. Such a development is inconvenient, but should not be looked upon as a failure of the procedure. The decisions were made on the basis of their probability of being best, but they can be proved only by actual computation of their results.

6.5.1.6 Prepare and test the several alternative arrangements, choose the best, and prepare the report to management.

1. Block out the estimated department boundaries on the building floor plan.
2. Using one of the layout methods, arrange the machines within the department boundaries, making adjustments as necessary.
3. Analyze each arrangement to determine comparative costs of production, materials handling, supervision.
4. Have best plan reviewed by interested persons (workmen, foremen, and representatives of methods, production-control, personnel, materials handling, and design departments). Obtain criticisms and revise plan if necessary.
5. Prepare schedule for the installation.
6. Prepare final plan for management's approval.

6.5.1.7 Prepare complete instructions for the re-layout and supervise the installation.

6.5.1.8 Collect data on the operation, and locate and correct any faults that may have been incorporated. It is usual, especially in highly mechanized production, to find a number of "bugs" in a new layout, and the planning group is not relieved of responsibility until the plan is functioning satisfactorily.

6.5.2 Methods of Departmentalization

There are three recognized methods of organizing the production facilities of a factory by departments:

Product or line
Process or functional
Combination of the two.

Each of these methods of departmentalization has certain advantages and disadvantages. The decision on which method to use in a given situation depends upon the conditions that must be met.

No one method should be adopted without examining the possibility of using the other methods for some departments or products. Two or three methods are commonly used within a single building of a multibuilding plant. Many companies employ the combination method for the layout of fabrication and processing, line method for assembly and for some particular parts, and process layout for painting or finishing and packaging for shipment. The object is, of course, to minimize the cost of production in each phase of production by adopting the optimum method.

6.5.2.1 Line or Product Layout. In the line layout, all the equipment required for one part or product is grouped together in one department in the sequence of the operations performed, so that the part is completed there and does not have to be moved from department to department for processing. The line is "balanced," in that the output capacity of each different type of equipment (or process) is the same or as nearly so as practicable.

Ordinarily, only one product is processed by one line department. However, it is frequently possible to change the tooling on the machines and thereby use the same sequence of machines for the production of some similar part. Although it is quite common to use the same production line for manufacturing the same part in a series of sizes, entirely different parts, which happen to require the same or almost the same sequence of machine tools, may be produced on the line. This situation may make it possible to devise an efficient line layout for two or more products neither of which individually would justify the line.

A line layout is not usually a *straight* line. In fact, a straight line usually indicates inefficient use of floor space and greater intradepartmental handling costs. The ideal line is one in which the product of one machine automatically feeds into the next, and so on. Transfer machines, carrousel arrangements, automatic cleaning, dipping, spraying, and drying of paint by conveyorized movement of parts, and many other automatized production lines have been devised. If each machine must be attended by workmen, it is desirable to arrange the machines in such a way that each operator can conveniently pick up the part from the place where it was deposited by the previous operator. Conveyor belts, chutes, slides, and trays on short lengths of gravity roller conveyor are used when the size and shape of the machines make the direct arrangement inconvenient. The actual path of the part through the line department may be a straight line (in order to employ a flat belt conveyor), but it is more likely to be similar to the shapes of such letters as S, U, L, M, N, O, Y, or C, depending upon the number of machines involved, the shape and size of the floor space available, the size and weight of the product, and the number and duration of the operations performed on the part.

The following conditions tend to dictate the use of line-type layout:

1. The discovery that a slight reduction in the selling price of a product will produce a large increase in the demand
2. The discovery that two or more products have operations that are performed on the same types of machine and in the same sequence
3. Difficulty of maintaining production control if produced by a process layout
4. Difficulty in finding highly skilled employees for the existing functional layout
5. Complaints from foremen over crowded conditions, lack of storage, materials handling delays, lack of capacity, or overtime work

The line layout usually has certain unique characteristics:

1. Conveyorized, automatic, or semiautomatic materials handling, with most of the in-process inventory in temporary storage on the materials handling system
2. Small in-process inventory and short production cycle
3. Mechanical pacing, partially or completely
4. A labor force that is mostly unskilled or semiskilled
5. The operation of two or more machines by one operator
6. The utilization of more specialized tools, jigs, fixtures, and machines
7. Little or no flexibility in the volume of output (in a given time period) or in product design; little or no opportunity to use the machines for other products during slack period
8. The possibility that one breakdown will interrupt a whole line
9. A high investment in specialized equipment that may have little or no resale value

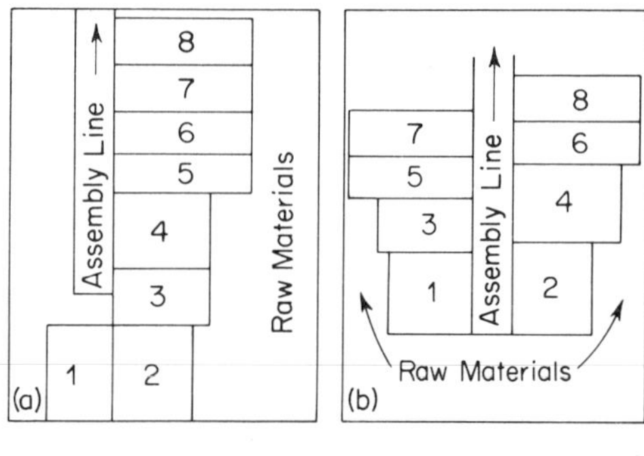

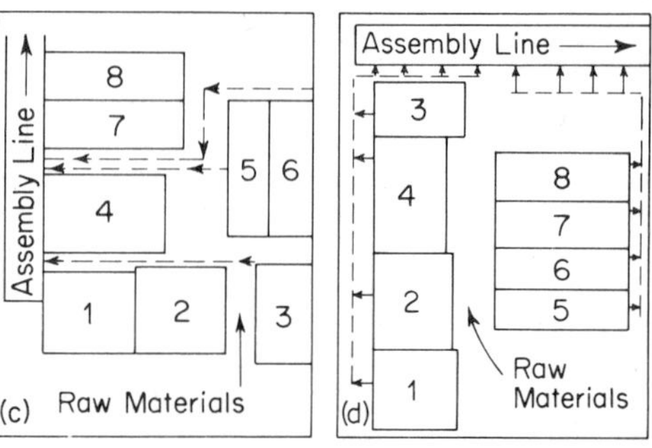

Fig. 6.6 *Four possible ways of arranging eight product lines feeding an assembly line. From Ireson,* Factory Planning and Plant Layout, *p. 58.*

10. Less necessity for written orders and records to obtain and maintain the desired production control; no detail scheduling for individual machines in the line

11. A need for supervisors to be familiar with widely diversified operations.

Assembly lines are customarily fed by a number of product departments that fabricate the individual parts. An attempt is usually made to arrange the product departments along the assembly line and in the sequence of assembly so that the parts are fed directly to it. For many reasons, this is not always feasible, and parts may be fabricated at a considerable distance from the point of assembly. The importance of having the feeder line near the assembly line decreases as the size and weight of the part decrease, as the ease of handling increases, or as the cost of making special arrangements near the assembly point goes up. Figure 6.6 shows a number of ways of arranging product-line departments in conjunction with an assembly line.

6.5.2.2. Process or Functional Layout. In the process or functional method of layout, the department is made up of machines, equipment, or processes that fall into one category, according to the functions performed. The product is fabricated by moving it from department to department according to the sequence of operations to be performed on it. The operations performed in each department are assigned to particular machines within the department according to the capacity required, availability of machines, precision required, and so forth. Functional layout is general-purpose layout, and provides for great flexibility in output, design of products, and methods of production. It minimizes the seriousness of a breakdown.

The functional layout is characterized by such conditions as:

1. The presence of a skilled labor force, capable of setting up the machine, reading blueprints, and determining the proper sequence, feeds, and speeds for efficient operation

2. Highly specialized supervision in each department

3. Many different production orders in process at the same time, resulting in a need for careful control and direction

4. Large storage space in each department for work-in-process and large in-process inventories

5. Extensive materials handling operations; frequent movement of small quantities over medium and long distances

6. Generalized materials handling equipment, requiring a large amount of labor and supervision

7. The need for large volumes of instructions, written and oral, to effect desired movements and operations at desired times

8. Frequent changes in workers' jobs and frequent instruction

9. No mechanical pacing of the work

10. A possibility of making greater use of machines and requiring less capital investment

6.5.2.3 The Combination Layout. The combination method of layout is feasible when a number of products require about the same sequence of functional operations but none enjoys sufficient volume to justify individual production lines. The principle of this method

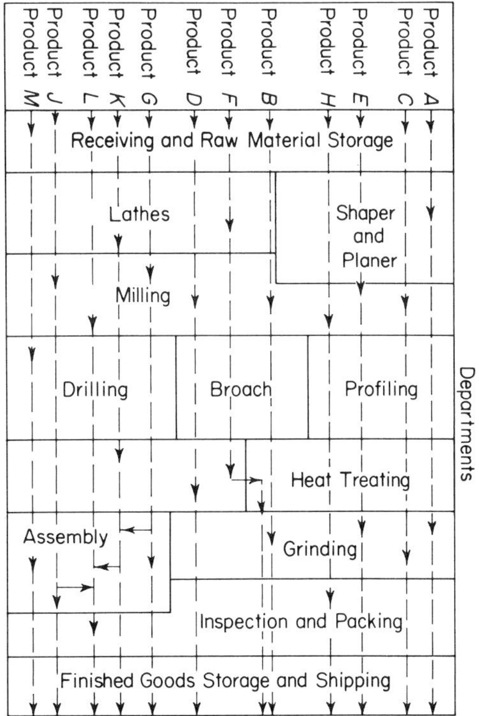

Fig. 6.7 *An illustration of the combination method of departmentalization. From Ireson,* Factory Planning and Plant Layout, *p. 45.*

lies in the arrangement of functional departments across the building at right angles to the flow of product and in the required sequence of operations. Particular sections of each department are assigned the different lines of products, but the sections can be adjusted as volumes change to accommodate larger or smaller orders. Figure 6.7 shows a typical combination layout.

Another example can be taken from the manufacture of steel office furniture or other hollow steel products. The sequence of operations for each component part is generally the same:

1. Shear or blank stock for part
2. Punch and notch blanks
3. Bend or form blanks
4. Deburr as required
5. Spot-weld or gas-weld component parts into subassemblies (drawers, legs, pedestals, tops)
6. Finish (clean, dip, sand, spray, bake, etc.)
7. Assemble final product
8. Install hardware (locks, drawer stops, hinges, etc.)
9. Pack and ship.

A company producing several different styles of desks, filing cabinets, chairs, cabinets, tables, and so forth can make profitable use of such a layout.

6.5.2.4 Analysis of Product and Process Layout. An example will serve to show the effects of layout method on the production cycle time and the in-process inventory. It also shows how the line layout is a natural development resulting from refinements of the process layout.

Assume that a certain part, weighing 75 pounds, has an annual volume of 400 pieces, produced in four lots of 100 pieces each. Three operations (others omitted for simplicity) are necessary, requiring 30 minutes, 20 minutes, and 45 minutes, respectively, on machines no. 1, 2, and 3. Two hours are allowed for the movement of a load of parts from one opera-

Production Schedule

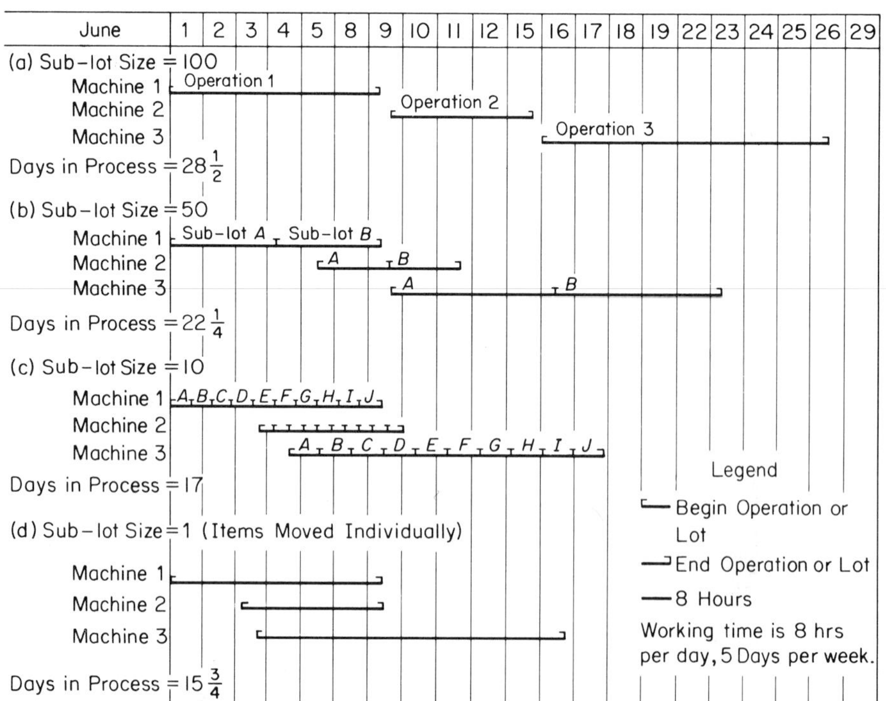

Fig. 6.8 *Effects of sub-lot size on over-all production times and schedules.*

tion to the next. Figure 6.8 shows several ways of scheduling this job, using sublots of different sizes as follows:

(a) 100
(b) 50
(c) 10
(d) 1

In each case, the production has been started on June 1, and 100 units have been completed as quickly as possible. Each sublot is scheduled to the next full hour except for (c) and (d). The schedule is also made in such a way that after a machine has been set up there will not be any idle time on it. Setup times have not been included.

It is obvious that it is not practical to try to handle 100 units in one "unit load" (see Art. 6.7), since that means that schedule (a) would have to be used, and the entire 100 units are tied up in-process for 28½ days. None can be packaged or shipped prior to June 29. If the lot is broken into two sublots (two unit loads), the first sublot is ready for packaging and shipment on June 16. Furthermore, the longest cycle time, sublot B, is just 20 days. If 10 sublots of 10 units each are used, the first 10 can be packaged on June 5, a total cycle time of 5 days, and the longest cycle time will be for sublot J, about 8 days.

Little is gained by reducing the sublot to a single unit (the assumption that each piece will be moved independently), but (d) shows what would happen if the three machines were placed along a roller conveyor. Machine no. 1 would build up a bank ahead of machine no. 2 before machine no. 2 starts working. As soon as no. 2 starts it reduces the bank or float more rapidly than machine no. 1 replenishes it. The two-hour move time here becomes a safety factor, so that when machine no. 1 finishes its last piece, machine no. 2 has two hours of work, or 6 pieces, ahead of it. When machine no. 2 started, it built up a bank of two hours' production (6 pieces) before machine no. 3 started, and continued to build up the bank faster than machine no. 3 used it at the rate of 45/20. The largest bank occurred when machine no. 2 finished its last piece. At that time, there were 62 pieces in the bank ahead of machine no. 3.

As the sublot size is increased—(c), (b), and (a)—the maximum bank or float increases, necessitating greater and greater storage space between each pair of operations. Such calculations as these provide a basis for deciding whether to employ a line or process layout for a particular product. They establish the data from which storage capacity, conveyor spacing, conveyor speeds, and production schedules can be computed. The ideal line is developed when the number of machines and operations is balanced perfectly, so that each one is taking the same amount of time, and so that the bank is the minimum size for safety from interruption.

6.5.2.5 Temporary Production Lines. Many companies have developed techniques for setting up product line layouts for short-run production. The basic principle behind this development is that it may be less costly to rearrange machines into a production line than to transport the materials from department to department of a functional layout. The necessary flexibility in layout is accomplished by:

1. Mounting light and medium-weight machines on platforms, skids, or casters so that they can be moved easily and quickly by a fork lift truck, an overhead crane, or a tractor. Machines must be stable without being fastened to the floor.
2. Locating heavy machines at such intervals that the necessary light machine can be arranged around them to form a production line under widely varying conditions.
3. Providing readily accessible services. The installation of a grid system of bus duct for power distribution and piping for compressed air, steam, water, and so forth, facilitates the rapid change-over between production runs.

Such a plan has disadvantages that should be carefully evaluated before embarking on the plan. It is highly probable that the company will have to have a larger number of machines than it can keep operating, but this may also be true for the company if it does custom work and employs the functional layout. This possibility requires that an extensive study be made of materials handling costs, and that standard data be developed that will enable the production-planning department to decide whether or not to set up a special line each time an order is received. If orders of any size are to be accepted, it will be necessary to maintain a functional layout to handle the small orders in addition to providing space and machines for the temporary line layouts. Unless all machines can be interchanged between the line and functional departments, a less efficient machine may have to be used on certain jobs because the better machine may be assigned to the other departments. This plan places great emphasis on scheduling, and poor control can render the plan almost useless.

6.5.2.6 The Equipment List. To arrive at the correct decision regarding the departmentalization method, it is necessary to tabulate the kinds of machines and equipment to be used and the amount of time each product will occupy each type. The flow process sheet is the logical place to accumulate the information on machine requirements for individual parts. An extra column can be added to show the product of the standard time per piece and the number of pieces to be run per week, month, or year. This is the machine-hour capacity required for that part. A second column can be added for the equivalent number of machines, which is the machine-hour capacity required divided by the standard number of working hours for the base period (week, month, or year). If flow process charts do not exist, and are not needed for other purposes, a form such as that shown in Fig. 6.9 can be used to show both the sequence of operations, the machines, and the machine capacity required. Such a form immediately shows the number of machines that would be required for a product line (rounding the values to the next larger whole integer) and points out the difficulties to be encountered in balancing the line. If the product is to be produced in a range of sizes, it may be possible to develop an efficient line by combining the times for the different sizes and allowing for the extra setups.

Figure 6.10 shows a summary sheet for a certain type of machine. On this sheet are summarized all the operations in the plant or division that require this kind of machine. A similar summary should be prepared for each type of machine.

MACHINE CAPACITY WORK SHEET

PART NAME Valve Body SIZE 1/2"

PART NUMBER 157-B1 QUANTITY PER Mo. 3000

SIZE AND WEIGHT OF PART 1" x 1" x 2 3/4", 6 oz., approx.

REMARKS: Operation Sequence No. 3

168 hrs. per month

OPER. NO.	OPERATION	MACHINE	OPER. TIME (MIN.)	OCCUR. PER PC	TOTAL MACH HRS. PER Mo.	NO. MACH. REQ.
1	Face, Bore and Tap end	Turret Lathe	1.30	2	130	.78
2	Face, Bore and Tap top	Turret Lathe	1.40	1	70	.42
3	Rough bore & Ream seat	2 sp Gang Drill	1.10	1	55	.33
4	Broach Hex Flats on end	Vertical Broach	1.00	2	100	.60

Fig. 6.9 *Machine capacity work sheet used in computing machine time required for each part. From Ireson,* Factory Planning and Layout, *p. 22.*

MACHINE CAPACITY REQUIREMENTS SUMMARY

MACHINE: Turret Lathe SIZE OR CAPACITY: 8' swing

NO. OWNED: 5 MINIMUM NO. REQUIRED: 8

SPECIAL FEATURES OR EQUIPMENT: Chuck for Castings, Air Operated, 3-Jaw

REMARKS: 168 hours per month

PART NO.	OPER. NO.	REQUIRED CAPACITY BY SIZES HOURS PER Mo.								TOTAL HRS. PER Mo.	ESTIM. SET-UP TIME
		1/2"	3/4"	1"	1 1/4"						
157-B1	1	130	145	85	80					440	12
157-B1	2	70	80	52	48					250	12
157-B2	1	65	73	43	40					221	12
157-B2	2	70	80	52	48					250	12
	$\frac{1209}{168} = 7.2$	TOTAL REQUIRED PER MONTH								1161	48

Fig. 6.10 *Summary sheet used in computing the minimum number of machines required. From Ireson,* Factory Planning and Plant Layout, *p. 23.*

The collection of summary sheets gives the *minimum* number of machines for the planned production by the *functional* layout method. It assumes continuous operation during the prescribed working hours and makes no allowance for scheduling difficulties or breakdowns. If any line-type layouts are to be used, and if they are not utilized for an equal percentage of the working time, additional machines will be needed. The consequences in terms of fixed costs can be computed for these extra machines and the decision on method of layout can be made on the basis of facts.

As the decisions regarding methods of departmentalization are made, an equipment list by departments should be compiled. Figure 6.11 shows a typical equipment list for a department. This example refers to an existing plant where all the equipment for the department is already owned. If new equipment is required, it would be added to the list and assigned a machine number. This form serves a number of purposes. It is very useful at a number of stages in the development and installation of the new plan. It provides the specifications for the department and will help to assure that the facts are considered each time a decision is to be made. The form is completed in the following order:

1. Insert department name and number.
2. List machines, machine numbers, present location, services required, size, and weight.
3. Tabulate the services required in the appropriate spaces in heading.
4. Compute power required.
5. Enter product description.
6. Indicate materials handling system, if known.

The space requirements are estimated (see Art. 6.5.4.3) and the location of the department is determined. (This may have been designated previously and the space may be a

EQUIPMENT LIST

DEPT. NO. P

DEPARTMENT Aluminum Specialties LOCATION: BLDG. #2

SPACE AVAILABLE Feb. 12,-19 PROD. TO START Feb. 22 FLOOR 1st

SERVICE REQ. STEAM No POWER: AC - 3 Ph. BAYS D-12,13,14

COMP AIR 100 p.s.i. CONNECTED LD. 25 KVA

GASES Natural gas VOLTAGE 110-220 DRAW. NO. L-145

WATER cold LIGHTING 110V - 6KW OTHER

PRODUCTS Misc. Alum. Sheet Parts VOL. PER HR. 150 CU. FT, WT. PER HR 300 LB.

MATERIALS HANDLING SYSTEM 4-wheel Hand truck and tote pans; Sheets by O-H Crane

EQUIPMENT	MACH NO.	PRES. LOC.			AVAILABLE			DEST. BAY	SERVICES REQ.	SIZE	WT.	MOVE SCHED.	MOVE COMP.
		BLDG	FL.	BAY	MO.	DAY	HR.						
Squaring Shear-8'	S-10	#1	1st	B-2	2	12	8A	D-12	220V-Com Air	5'x12'x6'	5000	2-12 A	✓
Punch Press-10ton	P-6	#1	1st	B-3	2	13	8A	D-12	220V- " "	3'x3'x7'	1200	2-13 A	✓
Punch Press-10ton	P-8	#1	1st	B-3	2	13	8A	D-12	220V- " "	3'x3'x7'	1200	2-13 A	✓
Press Brake-5'	P-2	#2	2nd	D-12	2	15	12N	D-13	220V-	7'x4'x8	3,000	2-15 P	2-16 P
Hyd. Press-200 ton	HP-4	#2	1st	D-13	2	12	8A	D-13	220V-N.Gas	—	—	No Move	✓
Press Brake-8'	P-10	#1	2nd	B-3	2	14	12N	D-13	220V-	11'x5'-9	5000	2-14 P	✓
Punch Press-Ind-25ton	P-20	#1	1st	B-4	2	13	8A	D-12	220V-Com Air	4'x4'x8	2500	2-13 P	2-14 A
Riveter - Auto	R-5	#2	2nd	C-5	2	12	8A	D-14	110V- " "	2'x3'x5	500	2-12 P	✓
Riveter - Auto	R-6	#2	2nd	C-5	2	12	8A	D-14	110V- " "	2'x3'x5	500	2-12 P	✓
Spot Welder-25KVA	W-10	#1	2nd	C-3	2	15	12N	D-14	220V- " "	3'x6'x5	600	2-16 A	2-15 P
Arc Welder-200amp	W-25	#1	2nd	C-3	2	15	12N	D-14	220V " "	4'x2'x3'	300	2-16 A	2-15 P

Fig. 6.11 *Typical departmental equipment list. From Ireson,* Factory Planning and Plant Layout, *p. 65.*

fixed amount.) After all tentative department locations have been decided:

7. Specify location by building number, floor, and bays in the heading.

After the machine arrangement has been determined within the department boundaries (see Art. 6.5.5):

8. Insert the drawing number and the destination by bay for each machine.

9. Determine when machine will be available for move, and insert available date.

10. Prepare move and installation schedule, and insert date of scheduled move.

11. Check off the completion of the move, or insert the actual move date if it differs from the schedule.

6.5.3 Department Location

6.5.3.1 Process Departments. The relative position of departments within the building(s) is very important to the efficient operation of the plant. Several factors enter into this decision regardless of the method of departmentalization. These factors are:

1. Sequence of departments entered by products and the amount of space needed for each department

2. Nature of the processes involved: Are they hazardous, offensive, dirty, noisy, etc.?

3. Weight, size, and characteristics of machines; floor load capacity, ceiling heights, space beneath floor

4. Weight, size, and nature of products: Are they large and bulky, fragile, flammable, explosive, etc.?

5. Materials handling costs: Does the arrangement minimize the movement distances, back-tracking, and crisscrossing?

6. Is this a service department that must serve a large area? Engineering, maintenance, production planning and control, inspection and quality control, raw stores, shipping, receiving, tool cribs, and maintenance

7. Quality characteristics of the building space: Is this space noisy, dirty, poorly lighted, or poorly ventilated? Does it provide necessary work conditions such as security, crane ways, freedom from vibrations, docks, or wide bays?

8. Contacts with the public, and internal communication: Is it necessary for members of certain departments to have frequent visitors (the personnel department, for example), or frequent person-to-person conferences with members of other departments?

Each of these factors affects the efficiency of operations within the plant and directly or indirectly affects the overall cost of producing a given volume of work. It is possible to reflect the effects of many of these factors on costs in mathematical equations of varying complexity. If the effects of all factors are shown in one equation, the equation usually becomes very complicated and is difficult to solve for the minimum value. Many attempts have been made to establish mathematical models for this and similar problems, but none can be classified as completely satisfactory at this time. Smaller problems have been attacked by industrial operations research teams with encouraging results. By limiting the departmental arrangement problem to include only a few factors, it is possible to obtain very good results.

For the present, the problems must be solved by a combination of cost data, logic, and good judgment. Analysis of fixed factors will eliminate a great many possible arrangements and point out those that remain. (Refer to Art. 6.5.1.2.) For instance, the fact that the second floor of a building has a limited floor load capacity will eliminate the possibility of placing certain departments there. The location of rail sidings and truck docks fairly well establishes the location of receiving, raw stores, finished goods storage, and shipping departments. Examination of a collection of process charts may reveal that parts always start in one or two departments, and that certain other departments always follow each other in a fixed order. For example, the finish grind of precision parts may always (or usually) follow heat treatment. The fact that most of the tools issued to individual workmen will be used in a small number of fabricating departments will help locate the tool room for economy of workmen's time.

By establishing such limiting factors, the number of satisfactory arrangements can be reduced to a number small enough to be handled and tested for the optimum solution. If necessary, several different arrangements can be drawn to a small scale and compared. Criticisms and suggestions can be obtained from all levels of employees to help assure that all the alternatives have been recognized.

6.5.3.2 Line Departments. The arrangement of line or product departments is commonly based upon the sequence of assembly of the parts, but this is not always practicable. Limitations, such as those enumerated in the previous paragraphs, may prevent such an arrangement. If the assembly line is to be fed by conveyors carrying the parts from the line departments, extra distance between the assembly and fabricating departments may not increase materials handling costs substantially (see Art. 6.7). Thus, the use of conveyors by companies that enjoy volumes of output that justify them increases the freedom of locating the line departments; line departments can be located where they make the best use of the space available.

6.5.3.3 Estimating Departmental Space Requirements. The space required for a department can be estimated by two different methods. The first method employs a ratio between the total space for the department and the space actually occupied by the machines and equipment. This ratio will vary with the type of layout, the type of industry, and the kind of machines involved. It commonly varies from 3 : 1 to 6 : 1. Each plant must conduct a study of its machines and operating conditions in order to arrive at a satisfactory ratio. This study can be made by surveying several existing departments that are considered to be typical of the plant. If the problem concerns a new plant, the planning engineers can prepare several layouts, using templates, and can compute the ratio.

The ratio method is most useful for estimating the space needed for line departments, since there will be several different kinds of machines in each department. The department estimate is the product of the actual space occupied by the machines, and the ratio.

A second method employs the principle of the production center. A production center is composed of a single machine plus all the equipment and space required for its proper functioning. Ordinarily, this means that the center includes space for the machine, the tool cabinet, worked and unworked parts, access to the aisle, operator space, and maintenance space around the machine. Figure 6.12 shows a typical production center. After the production center is designed, its space requirement is multiplied by the number of machines in order to obtain the departmental space estimate. It is obvious that this method is most useful in estimating the space for process departments. Furthermore, the dimensions of the production center provide a basis for establishing the most efficient dimensions for the department. Access aisles are usually included in the production-center estimate, but main traffic aisles are not, and must be provided when locating the department boundaries on the building floor plan.

There is more likelihood of errors in the space estimates from the ratio method than from the production-center method. Therefore, it may be necessary to make slight adjustments in the departmental boundaries

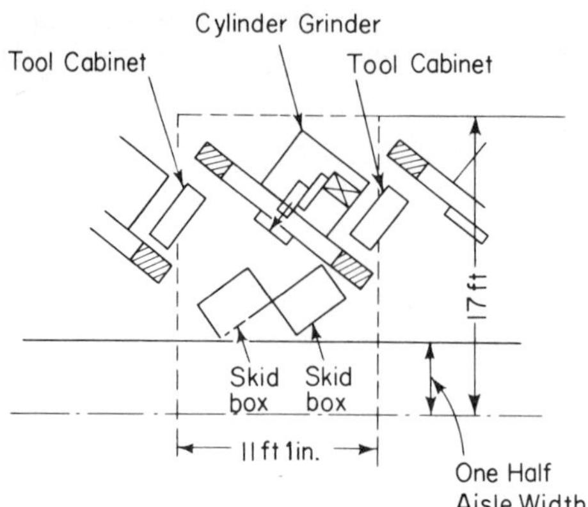

Fig. 6.12 *A typical production center, including space allowance for machine, tool storage, materials, maintenance and aisle. From Ireson,* Factory Planning and Plant Layout, *p. 67.*

when the machine layout is planned (see Art. 6.5.5). Also, if growth is expected in the immediate future, space for additional machines may be provided in the original plan, and the present machines may be arranged so that the new machines can be installed without disturbing the other machines. This space must be included in the original estimate.

6.5.4 Machine Layout

The arrangement of the machines within the departmental boundaries is made after the location of the department has been established. There are three common methods of planning this arrangement:

1. Drafting or sketching
2. Two-dimensional templates
3. Three-dimensional models

6.5.4.1 Drafting or Sketching Machine Arrangements. The preparation of a machine layout by this method usually requires a certain amount of preliminary work. The draftsman or planning engineer, working from the equipment list and on a scale drawing of the department space, establishes in his mind the basic pattern of the layout that he thinks will be satisfactory. He probably will make a few free-hand sketches, approximately to scale, showing the general position of the several machines, and obtain comments from his associates on the plan. After examining it from all viewpoints, he may prepare another free-hand sketch, or he may start drafting the layout on tracing paper or cloth. He actually makes a top view of each machine in its proper position relative to other machines, conveyors, columns, windows, and so on. He will usually work in pencil so that he can make changes with ease. The drawing is presented to management and other interested persons, and comments or approval is obtained. It is common for such a plan to be revised two or more times before it receives final approval.

It is obvious that such a method of determining machine layout can become very expensive if more than one or two revisions are required. There are instances, however, where this method is probably the most economical and perfectly satisfactory. The conditions that are favorable to this method are:

1. Process departmentalization
2. A large number of similar machines or pieces of equipment in each department
3. Ample floor space, with no major incentive to arrange the equipment in the smallest possible space

Since process departments are arranged to accommodate a variety of products, more space is usually allowed around the machines than in a line department. If a satisfactory production center has been designed for the type of machine to be used, the layout consists primarily of repeating this production-center layout as many times as is necessary. Thus, the real work is done in designing the production center. Once that has been approved, there is little likelihood that the departmental arrangement will not be approved. This means that the layout should be accomplished with few revisions.

The draftsman can simplify his work by preparing a cardboard or plastic template of the production center and using it as a guide for drawing the outlines of the machines, tables, conveyors, boxes, and so on.

The drafting method can usually be used quite satisfactorily for the layout of drafting rooms, clerical or typographical offices, insurance companies, banks, functional departments, packaging, inspection, and many others.

6.5.4.2 Two-Dimensional Templates. The two-dimensional template is simply a piece of some flat, stiff material on which is shown a projection of the machine and its overtravel and/or stock overhang to scale. It is then cut out of the material along the extreme outlines. The two-dimensional template is the most popular of the methods used and is the one that is most likely to give the best machine arrangement with the smallest overall cost of planning. The template method of layout has become so popular that the American Society of Mechanical Engineers has published a standard for templates and models.* This standard presents the most acceptable practice on material for templates, scale, representation of machine features, overhang and overtravel, and identification. A summary of the most important characteristics follows:

1. The template should be made of heavy cardboard, plastic, or fiber that can be used repeatedly without serious wear or soiling and that will lie flat.
2. The material of the template should take ink or some other convenient marking material.
3. The template should be a projection of the machine and *all* overhang and overtravel of machine parts, work, carriages, and so on; the machine base should be outlined on templates.
4. An agreed-upon set of symbols should show the location of motor, control levers, handwheels, and work.
5. The template should be capable of being photographed in black and white without losing any essential details.
6. The template must contain some form of identification.
7. The scale should be ¼″ = 1′0″.

A template of a milling machine is illustrated in Fig. 6.13.

Since the preparation of suitable templates by ordinary drafting methods is rather inefficient and costly, a number of companies have developed other methods of producing templates commercially and offer them for sale to using companies. Most of the commercial templates use clear plastic of various thicknesses as the basic material and provide means for attaching the templates to the tracing of the plant area. Some use pressure-sensitive cement and others use alnico magnets for this purpose. In some cases, the template is flexible and can be run through blueprint or Ozalid machines; others are stiff and the print paper must be exposed by placing strong lights above the layout on a flat surface.

The great advantage of clear plastic templates of all types over cardboard or fiber is that the internal lines, symbols, and identi-

**Plant Layout Templates and Models—1949* (New York: The American Society of Mechanical Engineers, 1949).

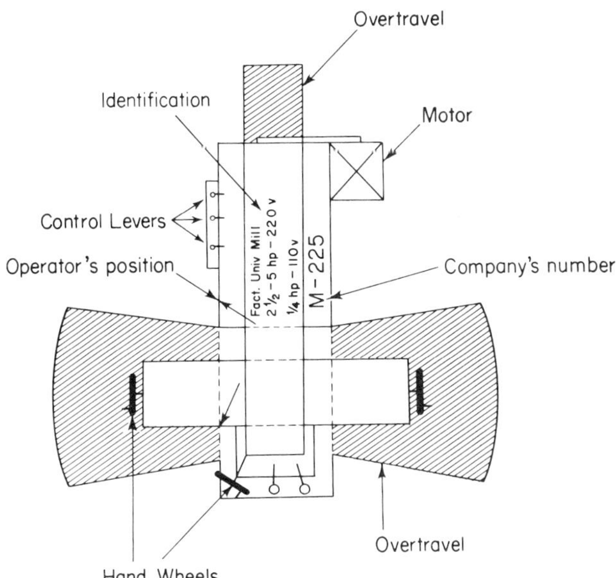

Fig. 6.13 *A typical template for plant layout purposes. Notes indicate the information desired and the symbols used. From Ireson,* Factory Planning and Plant Layout, *p. 81.*

fication information are recorded on the blueprint at the same time as the outlines are recorded. If opaque templates are used, the information must be lettered or typed within the machine outline after the print has been made. The ease with which the clear plastic templates can be removed and rearranged, and the ease with which a record of each trial can be made, encourage planning engineers to try several different layouts before reaching a decision. The results of the different layouts can be compared critically only if a record is made of each one.

Most of these templates will function properly on regular tracing paper, but a number of suppliers also provide thin plastic sheets to replace the tracing paper or cloth. These plastic sheets are printed or scored in quarter-inch squares so that the templates can be oriented with respect to columns, walls, and other machines simply by counting squares. The building outlines, columns, windows, conveyors, aisle boundaries, overhead monorails, stairs, flow lines, and so on are printed on translucent, pressure-sensitive tape and are applied directly to the plastic sheet to provide a complete drawing of the building area. Thus the complete layout and as many prints as desired can be made without drawing anything.*

6.5.4.3 Three-Dimensional Models. Three-dimensional models provide the most realistic means of representing the equipment of a plant, and in many ways make the machine arrangement easier to accomplish. These are scale models of the machines and equipment, with as much detail incorporated into them as their size will permit. The scale is customarily ¼″ = 1′0″, which corresponds to the most frequently used scale for two-dimensional templates, and permits the use of the same blueprints or tracings of the factory building. Models do not ordinarily show overtravel or overhang of work, but do show tables, carriages, and controls in their normal position.

Models should be made of a material that will withstand frequent handling and even dropping. They should be heavy enough to stand upright even in a light breeze. They should be painted according to a color code, such as that used by the Air Force or recommended by the Du Pont Company or the Pittsburgh Plate Glass Company. In addition to the usual colors for the machine base, points of danger, moving parts, and so forth, the machined surfaces are usually painted with an aluminum or white paint. The color code adds to the realism of the model, but, what is more important, the planning engineer can tell more about how to arrange the machines in order to provide for safe working conditions, ease of handling work, position relative to light source, and so on.

Although models are relatively expensive in first cost, they frequently save their cost in the first re-layout problem by saving time for everyone concerned, from the workmen through top management. The use of models makes it unnecessary to be able to read blueprints in order to understand a proposed layout. Anyone can grasp the essential characteristics of the proposed layout more quickly from seeing the three-dimensional model than from seeing a two-dimensional blueprint of the same layout. Thus the ideas of many persons can be brought to bear on the problem with a certain visual appeal, for most persons in a plant will stop and shift models around on the plan when they would not give a blueprint a second look. Even experienced layout men usually find it easier to visualize all the difficulties of a layout problem by using a model.

Just as with two-dimensional templates, it is important that a record be made of each different proposal so that all proposals can be compared objectively. Since the models do not lend themselves to blueprinting, the only practical method lies in photography. Figure 6.14 shows a photograph of a small section of a plant model. Obviously, it would be difficult for the millwrights to work from such a photograph in installing the machines. A grid on the floor of the plant model would help considerably. One solution of this problem lies in using both two-dimensional templates and three-dimensional models. In this method, the arrangement is made by using the models on the tracing or plastic sheet. When the arrangement has been perfected, each machine is replaced by a plastic template secured to the tracing in exactly the same position. One or more blueprints are then made, and another

---

*These products are available from such companies as: F. Ward Harman Associates, Halesite, L. I., New York; Repro-Templets, Inc., Oakmont (Allegheny Co.), Pennsylvania.

Fig. 6.14 *Plant layout using models of both equipment and buildings. From Ireson,* Factory Planning and Plant Layout, *p. 101.*

proposed arrangement can be set up. In this way, the advantages of both methods are obtained, but the costs of both methods are also involved.

Storage of three-dimensional models of plant buildings and machines presents a problem of major proportions for a large concern. The model of the building is the most difficult, for it may be so large that rearranging the machines in certain areas requires the layout men to work from a temporary bridge constructed over the model. This naturally increases the difficulty of keeping the model up-to-date. One solution dictates that sectional planning boards of some convenient size, such as 24 by 36 inches, be used in place of a plant model. The blueprint or tracing of the plant area is divided into pieces of the same size which are mounted on the boards. These are usually covered with a plastic sheet to provide a good working surface. These boards are identified along the edge and are stored horizontally in a closed cabinet. Two or more sections can be removed for study or re-layout when needed. If desirable, all boards can be placed in their proper position to show the complete plant plan, floor by floor. The models of the machines and equipment can be attached to the boards and stored as a permanent record of the plan and as a constantly available tool for use in replanning the area.

6.5.4.4 Aisles. Aisles are highways over which both materials and personnel must travel; the width and arrangement of aisles can be set only in relation to the kinds and volume of traffic that must be handled. The layout engineers can base the aisle width and location on the following variables:

1. Size and kinds of trucks and turning radii (power or manual)
2. Size of load to be carried by each type
3. Frequency of trips
4. Frequency and volume of personnel traffic
5. Boundary conditions (location of machines and other items along aisles)
6. Method of picking up and depositing load and position of load relative to aisle

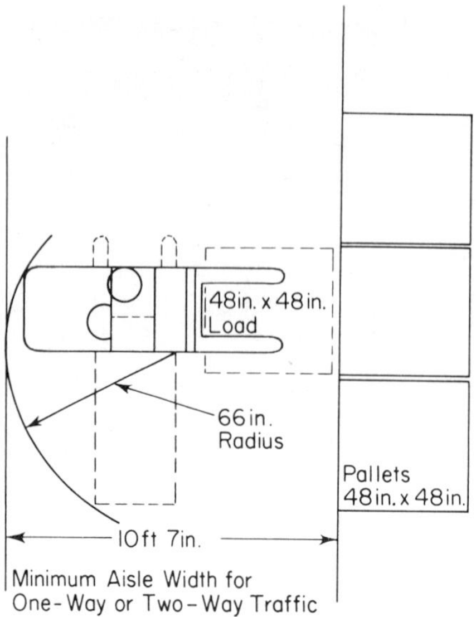

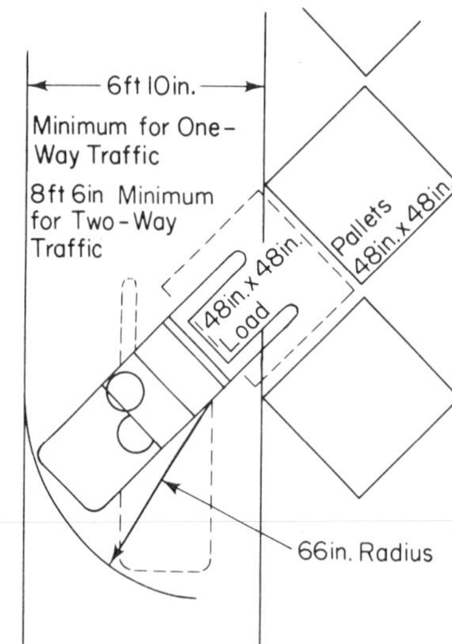

Fig. 6.15 *A study of the effects of pallet position on the width of aisles. From Ireson,* Factory Planning and Plant Layout, *p. 110.*

7. One-way or two-way traffic
8. Whether individual machines are to be served by aisles, or are only designated areas adjacent to a group of machines to be served.

Aisles should not be excessively wide, and should not have too many turns or obstructions. Turns, obstructions, and blind corners invite accidents, and excessively wide aisles invite use as temporary storage areas. Careful planning of aisles will reduce handling time, improve service and utilization of materials-handling equipment, and help prevent personnel injuries and damaged materials.

6.5.4.5 Storage Areas. A substantial portion of the floor space of every industrial plant is devoted to the storage of raw materials, tools, supplies, work-in-process, subassemblies, and finished goods. Careful planning of both the location and layout of storage facilities can reduce the time for controlling and issuing materials, costs of materials handling, loss or damage to materials, and direct labor time. Since each company's storage problems differ to some extent, no rules can be given that will apply in every case. However, the following principles can be used as guides in arriving at an economical arrangement:

1. Decentralize storage areas to conserve workmen's time spent in securing supplies, but keep each storeroom large enough to keep one stores clerk reasonably occupied.
2. Do not put bulky items (those that are not easily pilfered) in inclosed storerooms unless special storage conditions are required for safety or to prevent deterioration.
3. Try to store bulky items at the point of use to prevent rehandling.
4. Consider the cost of exerting complete control over stores against the probable cost of stolen or misplaced items and the inaccuracies that will occur in cost accounting.
5. Attempt to arrange storage areas according to classes of items in order to prevent unnecessarily large inventories through duplication.
6. If materials are stored at point of use, make the workmen responsible for notifying purchasing when stocks are reduced to some specified quantity.
7. Plan storage areas so that mechanical handling equipment can be used to the maximum advantage. Handle as little as possible by hand.
8. Classify all materials, tools, subassemblies, and so on according to some standard classification, and arrange storage facilities by classifications so that any item can be located quickly and easily.
9. Permanently allocate space, bins, racks, or shelves to standard and regularly stocked items, based upon probable maximum quantities of each item.

10. Select storage equipment for the specific storage problem.
11. Employ vertical space to conserve floor space by using stacking pallets, bins, skid boxes, and mezzanine floors.
12. Observe safety codes when planning storage facilities for hazardous materials.

6.5.4.6 Planning Personnel Facilities. It has been recognized in recent years that the attitudes of factory personnel are extremely important to successful operation. The physical facilities to care for personnel needs directly affect the workers' attitudes and should be planned as carefully as the production facilities. The worker's physical comfort, morale, and loyalty to the company are reflected in his productivity and affect the direct labor, materials, and burden costs of production. The planning engineers should make specific provisions for employee facilities at the same time as they are planning the production facilities. The most frequently provided personnel facilities include adequate provisions for the following:

1. Employment interviewing, testing, physical examinations, assignment to job, training, and upgrading
2. In-plant medical service, dispensaries, and first aid
3. Comfortable working conditions relative to heat, light, noise, humidity, dirt, dust, and fumes
4. Lockers, washrooms, and toilets
5. Food service
6. Recreation facilities and programs
7. Special services:
   a. Loans and banking
   b. Union activities
   c. Insurance programs
   d. Legal, income tax, and similar services

Almost all plants make some provisions for the first four items above. The community facilities, union demands, type of employees, number of employees, availability of suitable labor force, management attitudes, and employees' desires will largely determine the extent to which personnel facilities will be provided beyond the bare minimum. Modern management is in general far exceeding the bare minimum because it has recognized the advantages of a friendly, loyal, and cooperative work force. Companies that have employed more progressive programs, and that have thereby convinced the employees that management is genuinely concerned over their welfare, have enjoyed peaceful employee relations and long periods of uninterrupted production. Eastman Kodak Company and the Lincoln Electric Company are notable examples.

The space for personnel facilities should be determined according to the number of employees in each division or section of the plant. The principal services requiring space within the factory areas are those that are used frequently each day by the employees: washrooms, toilets, lockers, water fountains, and food services. First-aid and medical service, though not used daily by large numbers of employees, must be conveniently located within the factory in order properly to serve its function.

First-aid and medical service space depends upon the type of service to be rendered. Company policy regarding medical service beyond the minimum first aid required will dictate the number of offices for nurses, doctors, examinations, and treatments. As the medical service is extended to the families of employees, the facilities must be expanded even more. Some companies maintain complete hospitals for use by the employees and their families. Very few standards exist by which the space and facilities for a given service can be estimated.

The medical office should be readily accessible to the employees and also to the employment office.

Company-sponsored food service has become a standard practice in most industries. Even the very small plants usually provide some simple food service, such as hot coffee and cookies or doughnuts during the working hours and space for a good vendor to sell sandwiches and beverages during the lunch period. A sufficient quantity of nourishing food is essential to the well-being of the employees, and many companies subsidize the food service substantially in order to encourage the employees to eat wholesome meals. It is an accepted principle that company food service should provide comparable meals at lower cost to the employee than commercial establishments can provide. This means that the company must not try to recover from the sale of the food such items as rent on the space, power, heat and light, building maintenance, and, in some cases, even the cost of preparation.

The types of food service commonly

provided in plants of different sizes are:

1. A coffee bar
2. A rolling cafeteria
3. A small diner
4. A lunch counter
5. A cafeteria
6. A restaurant with table service.

The most important considerations in planning food service are:

1. Service must be fast so that employees will not have to wait and then eat rapidly.

2. The food service area should be located conveniently to the work areas of the majority of the employees.

3. The lunch period should be long enough to enable the employees to reach the service area, eat slowly, and return to work stations without rushing.

4. The eating habits of the employees should be considered in selecting kinds of food to serve. (Do not try to force employees to eat strange food.)

5. Variety of food should not be so great as to increase preparation expense unnecessarily or to increase the time required by the employee to make his choice.

6. The service should be planned to provide for continuous serving over a period of one and a half to two hours by staggering the times of the lunch periods.

7. An ample quantity of good quality is better than smaller portions of the best-quality foods. Preparation should be thorough.

8. Plan food service areas so that they can be used for other purposes when food is not being served (conference, recreation, or training space).

9. The food service area should be exceptionally clean and pleasant. (It should be light, colorful, relatively quiet, and lend an atmosphere of relaxation to the lunch period.)

Figure 6.16 shows a cafeteria planned to provide maximum seating capacity and service capacity within a limited space. It is arranged to keep the traffic flowing in two large circles with minimum interference.

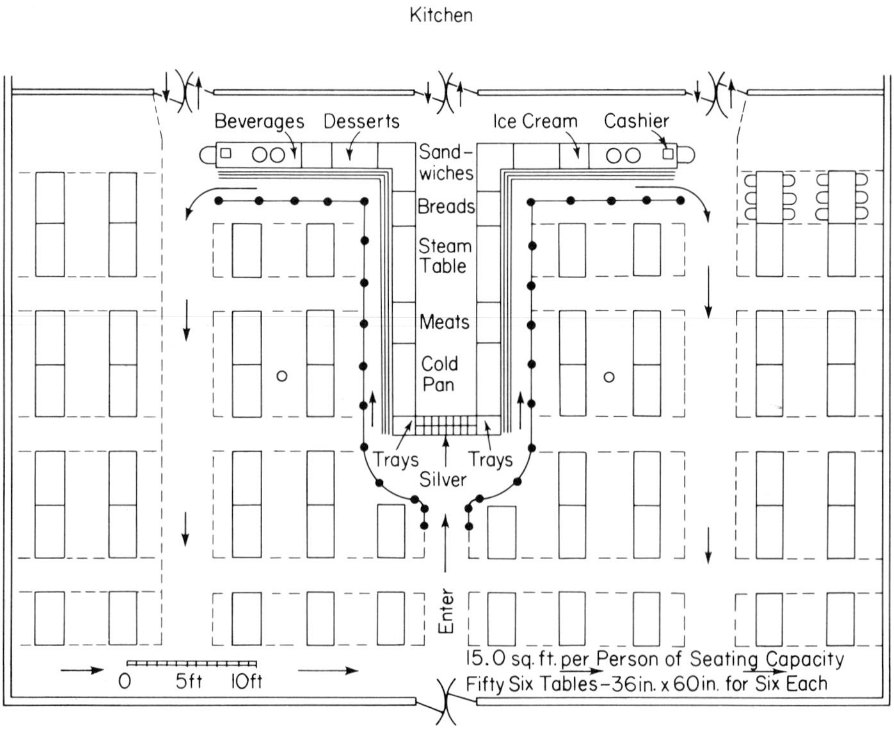

Fig. 6.16 *Cafeteria plan for feeding 200 to 300 employees simultaneously. From Ireson,* Factory Planning and Plant Layout, *p. 282.*

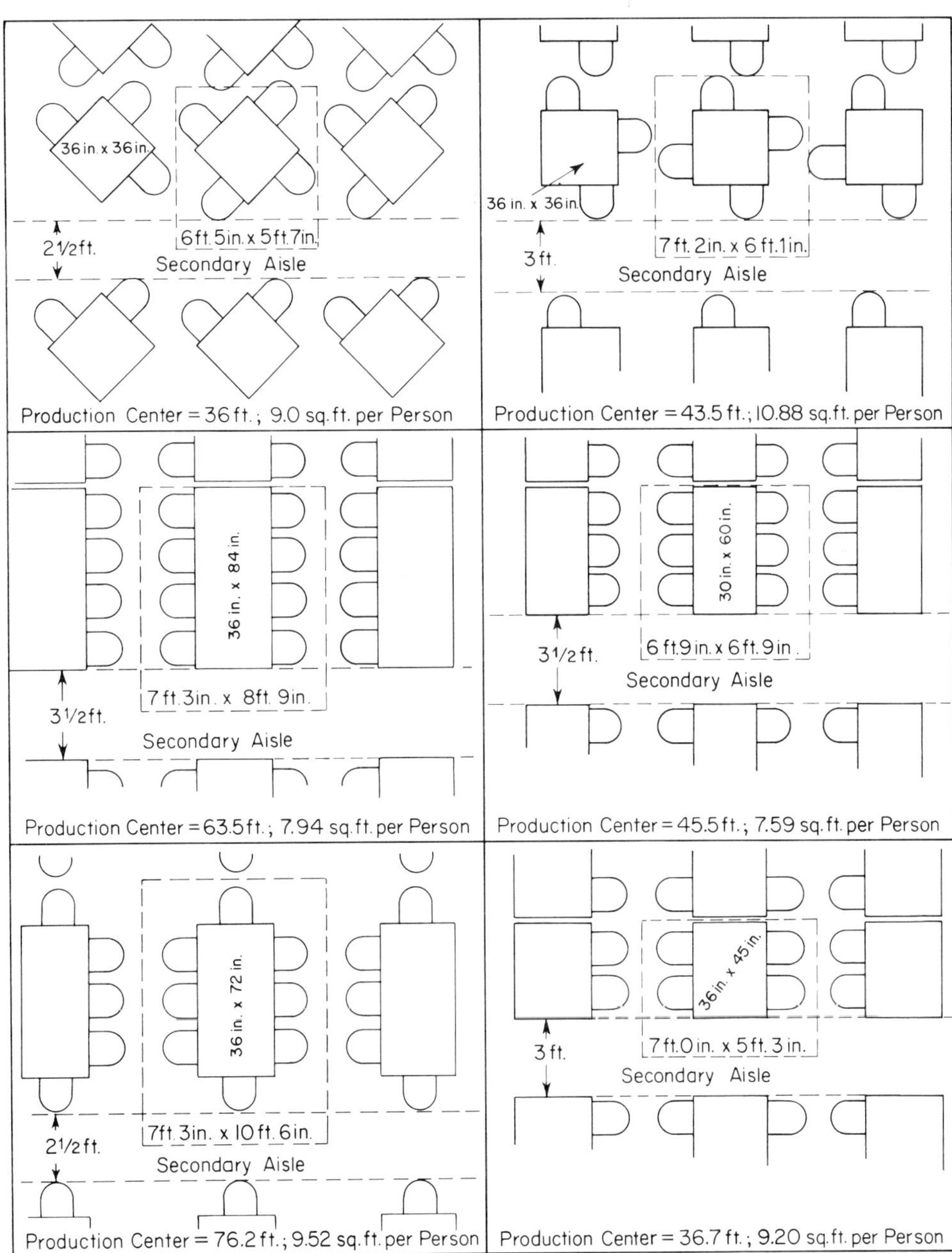

Fig. 6.17 *A study of table spacing for industrial restaurants by production center method. From Ireson,* Factory Planning and Plant Layout, *p. 283.*

Seating space is the principal space problem in food service. Figure 6.17 provides a graphical comparison of the space required to seat a person, using several sizes of tables and different arrangements. It should be noted that the values in square feet per person are only for the actual seating space; they do not include main aisles, serving facilities, or kitchen areas. Hotel and restaurant supply houses are always glad to provide professional assistance in planning the food service for a plant. They will provide complete listings of the kitchen, dish-washing, and serving equipment, and in some cases will actually prepare layouts to fit the specific space available.

Offices in which other personnel services are conducted should be located according to the amount of contact with the individual employees. Any services that require person-to-person communication with the employees should be conveniently located to the plant entrances or work areas. Such an arrangement will encourage the employees to use the service and will help to reduce the traffic through other office areas. Offices for services that are conducted by mail or intra-plant memo can be located without reference to the employee entrances. For example, payroll reductions for U.S. Bond purchases, life insurance, repayment of loans, or union dues can be made in the regular payroll offices.

State and city codes must be observed in determining what personnel facilities and services will be provided. Most states have labor codes covering the minimum number of washroom and toilet facilities, ventilation and cleanliness of facilities, inspection of food service facilities, first-aid and medical services, workmen's compensation, employer liability, and hours and conditions of employment for women and children. The planning engineers should become thoroughly familiar with the appropriate code before planning any personnel facilities and should be sure that all requirements are satisfied by the plans. In most cases, the codes should be interpreted as minimum requirements, and further investigation should be made to determine suitable facilities for both immediate and future needs. Provisions should be made for expansion of the facilities as growth occurs, even though the extra facilities may be just roughed-in now and completed later as needed.

6.5.4.7 Criteria of Effective Layout. There are no real *standards*, such as thermal efficiency, length, weight, or speed, by which the effectiveness of a plant layout can be measured. The criteria by necessity must be by comparison of costs of performing certain functions by different plans. Overall costs of producing a given quantity of goods of a certain quality and within a certain time period then become the means for determining the best plan of those proposed. How nearly the optimum plan is obtained depends upon the ingenuity of the planner and the thoroughness with which he performs the analyses of the data available. The objective of the previous articles of this section has been to point out ways by which factors that contribute to costs can be considered in arriving at a proposed plan.

To be of any real value, criteria must be capable of being applied to the proposed plans while still on paper and before an installation is made. The criteria must be reasonably accurate and easily applied, and must show some close relationship to the various costs of operation. Thus, the criteria will deal with three types of costs:

1. Direct labor
2. Direct material
3. Burden or overhead expenses

The effects of various proposed plans can be judged by:

1. Direct labor
   a. Reduction in task time
   b. Elimination or reduction in manual effort
   c. Reduction in distances to obtain tools, materials, and instruction
   d. Reduction in distances and waiting time to satisfy personal needs
   e. Improvement in working conditions: light, heat, noise, dirt, interruptions
   f. Improvement in employee morale or "will to work"
2. Direct material
   a. Elimination of damage to materials or parts in transit
   b. Means of controlling scrap and waste
   c. Use of less expensive materials (resulting from improved processing or better product design)
3. Burden expenses
   a. Better use of vertical space
   b. Better use of floor space
   c. Reduction of idle or down time on equipment
   d. Elimination of necessity for close supervision, written orders, move orders, etc.
   e. Substitution of mechanical for manual handling
   f. Elimination of unnecessary movements of materials
   g. Use of gravity for moving materials

h. Combining movement with processing (as drying on conveyors)

i. Elimination of temporary storage between operations

j. Elimination of need for more space and equipment by improvement in productivity and utilization

k. Improvement in flexibility of output (volume, design, variety of product, etc.)

Direct measures of the effectiveness of a plan have been proposed. Each of these measures is useful, but if used alone may be very misleading. It is advisable to use several of these measures to test each proposed plan and to select that plan which minimizes the largest number of the most significant of these criteria. Some of these direct measures are:

1. The expected direct labor cost per piece or unit
2. The indirect labor costs per piece or unit (should be based on actual cost items that vary among the alternatives rather than on some overhead allocation plan)
3. The materials handling cost per ton of shipped product
4. The ratio of direct labor to indirect labor per piece or unit
5. The distance traveled by each part, in feet
6. The number of handlings per piece
7.

$$\frac{\text{number of handlings} \times \text{weight at each stage}}{\text{shipping weight}}$$

8. The total direct labor hours per piece or unit

These measures become most valuable when they can be compared with a suitable standard or optimum value. A multiplant company will compare these measures for a proposed plant with the corresponding values from efficiently operated existing plants. If trade association journals publish such data from individual companies or from the industry as a whole, the individual concern can compare its values with the published data to find out if they are better or worse than its competitors'. Under such conditions, the direct measures become goals that can be reached or bettered. If they must be used only as comparative values among several alternative plans, they are useful in selecting the best of the alternatives but do not indicate whether or not the optimum plan has been devised.

6.5.4.8 Layout Check List. The following questions will provide a check list by which each layout can be checked in order to avoid the most common errors.

A. Space utilization and disposition

1. Is there sufficient space for the operator to perform all his tasks at the machine?
2. Is there sufficient space around the machine for easy maintenance?
3. Is the machine locked in by other machines, so that it cannot be moved without first moving other machines?
4. Is there space for the tools, auxiliary equipment, jigs, fixtures, tables, and tool cabinets for the proper operation of the machine?
5. Is there sufficient space for the worked and unworked parts or the "float"?
6. Is the machine accessible, so that the worker can get to and from his work station without danger of injury?
7. Is the machine located too close to the aisle or conveyor for the safety of the operator or others?
8. Is too much space allowed, so that the operator becomes inefficient?
9. Is vertical space used for storage and materials handling?
10. Are storage areas (tools, raw materials, in-process or finished goods) adequate for expected volumes?
11. Are personnel service areas adequate for the number of employees expected?

B. Machine location factors

1. Is the machine located at the best position or angle for the effective supply and removal of materials, and for effective use of floor space?
2. Is the machine in the best location for both natural and artificial lighting?
3. Does the location of the machine subject the operator to excessive heat, noise, drafts, or fumes from other machines or processes?
4. Is the location safe from flying particles, explosions, fire, moving trucks and cranes, and other hazards?
5. Is the machine located properly in relation to the sequence of operations?

C. Services and distribution of services

1. Is the distribution system for steam, gas, air, and power designed for ease of relayout and installation of equipment?

2. Has excess capacity been provided in the distribution system so that additional equipment can be installed?

3. Will re-layout or expansion necessitate relocation of main or trunk lines?

4. Are all pipe lines and conduits clearly labeled?

5. Are tap-offs, control panels, and valves provided at convenient intervals so that interruptions will be minimized?

6. Are service lines exposed for ease of maintenance?

7. Are service lines protected from freezing, damage by materials handling equipment, and machine operation?

8. Are protective devices, hoods, baffles, insulators, and the like provided to protect personnel and equipment?

9. Has too much space been allocated to aisles?

10. Are there too many aisles or are aisles too wide?

11. Are aisles clearly marked and do they have too many turns or obstructions?

12. Are aisles wide enough to provide for the volume of traffic expected and the manipulation of the expected loads?

D. Storage areas

1. Are tool cribs and storage areas at convenient locations?

2. Are storage areas frequented by the employees at excessive distances from their work areas?

3. Do the storage areas provide protection from theft or loss of materials of high value?

4. Are special storage facilities provided for paints, oils, acids, gas cylinders, chemicals, flammable and explosive substances?

5. Does the location of storage areas complicate the receiving and checking of incoming materials?

6. Does the location of storage areas increase the length of hauls of large volumes of materials?

7. Does the arrangement of storage areas permit the use of mechanical handling facilities?

8. Are there provisions for the inspection of incoming materials?

9. Are storage facilities, shelves, racks, bins, trays, etc., selected for ease of filling and issuing?

10. Are storage facilities centralized or decentralized according to the particular needs of the area served?

11. Are storage areas planned and equipped for systematic identification and location of items in the area?

12. Is proper use made of vertical space in storage areas by means of mezzanine floors, palletized loads, etc.?

13. Is enough space allocated to storage to care for peak loads?

14. Are in-process storage areas located for maximum efficiency of the materials handling system?

E. Personnel facilities

1. Are employee entrances at too great distances from work stations?

2. Are washrooms, lockers, and toilets located at frequent intervals (preferably within 200 feet of work stations)?

3. Has a sufficient number of toilets, lockers, and wash basins been provided on each floor and near each department?

4. Have water fountains been provided at frequent intervals (100 to 150 feet apart)?

5. Are first-aid rooms or dispensaries located conveniently to the work areas, and readily accessible?

6. Are employee services (counseling, savings and loan, legal, income tax, commissary, insurance programs) located conveniently to the employees' entrance?

7. Are working conditions (temperature, humidity, noise, light, cleanliness, and working position) planned to be as pleasant as the tasks will permit?

8. Is food service clean, economical, of good quality, adequate, and available conveniently to all employees?

9. Are all personnel facilities properly ventilated, maintained in extremely clean condition, and planned for ease of maintenance?

10. Are medical service, first-aid stations, and dispensaries adequate for the number of employees and the hazards present in the plant?

6.6 OFFICE LAYOUT

Office layout involves two problems: the location of the offices and the arrangement of the furniture and equipment within the offices.

6.6.1 Office Location

The object of office planning is to provide for the smoothest and most direct flow of communications, paper work, and information from department to department in the necessary sequence so that each department can perform its function as efficiently as possible. The functions performed by the several departments can be compared to the operations performed in the manufacture of a product, and the location of offices can be thought of as a special case of department location. Some offices will embody the principles of process or functional layout, in that a number of persons in each office or department perform similar work—i.e , accounts-receivable clerks in an accounting office. Other offices will embody the principles of line layout, in that a paper form will pass through the hands of a number of persons, each of whom performs some specialized function on it.

The location of the offices can be determined by analyzing the functions relative to the organization chart, communications, paperwork flow.

6.6.1.1 The Organization Chart. The organization chart of a company usually shows the lines of authority and the lines of response between and among the auxiliary and producing departments. The chart should show very clearly how the information flows from one division to another, the level of responsibility, where the information originates, and its ultimate destination. The organization chart will usually provide a guide for groupings of offices and will indicate what other offices or departments should be located nearby. The use of an organization chart, however, is seldom a practicable solution in that the building is not constructed like an organization chart and the chart does not necessarily indicate the amount of space that will be required by the different departments. The nature of the activity may be such that a location determined by this method would have serious disadvantages. For instance, the drafting department is tied closely to the factory, but it should be located in a clean, quiet area and out of the principal traffic patterns to reduce interruptions and disturbances.

6.6.1.2 Communications. The most important aspect of office location is that of communication with other departments, customers, vendors, and carriers. Since written and oral (telephone) communication is not seriously limited by distance, those offices whose principal flow of work is conducted by these means need not be located adjacent to or even near closely related departments. It is only when the communication must be conducted person-to-person that the location of offices becomes important. Thus, the plant manager, production planning and control, methods engineering, motion and time study, plant maintenance, and other activities that must perform part of their work in the factory area should be located conveniently to the factory. Offices that have to receive persons from outside the plant (salesmen, customers, advertising agencies, common carriers) should be located conveniently to the street or plant entrance so that visitors will not have to go into the factory area. These offices may be located in the business district of the city, a number of miles from the factory site. Offices that deal primarily with management functions should be close to those persons served.

Consideration of the communication problems helps to establish the general location of offices and auxiliary departments. Specific conditions, such as size and nature of space available at the different locations, orientation relative to light and prevailing winds, and the character of the work performed, will help to determine the specific location of the several departments.

6.6.1.3 Flow of Paper Work. Just as the process flow chart is the basic instrument in determining the location of machines, the process flow chart of paper work can be used as a basis for arranging the office functions. The sequence of operations on forms for any purpose can be shown by the chart, and the same sequence can be used for the location of offices or desks (for the particular functions) within an office. The process flow chart is particularly useful in arranging offices for banks, mail-order houses, insurance companies, and central offices for manufacturing concerns. The procedures and systems commonly used to accomplish the clerical work in sales, purchasing, production control, inspection, and accounting usually employ the principles of division of labor and specialization of labor. Thus, one form usually passes through a number of hands before all the information is recorded or noted. These operations can be lined up in sequence within the office in many cases. (See Sec. 1 for examples of such charts.)

6.6.1.4 Personal Considerations. It is not uncommon for the location of offices to be determined almost completely by personal preferences. A certain amount of prestige is involved in the location and assignment of space, and company policy may dictate that key individuals be given their choice of offices before general assignments are made. The wise planning engineer will make a thorough study of office needs and the most desirable locations from the standpoint of functions performed and then proceed diplomatically to solicit opinions and preferences. He should be prepared to point out difficulties that will arise if the personal preferences will result in serious errors.

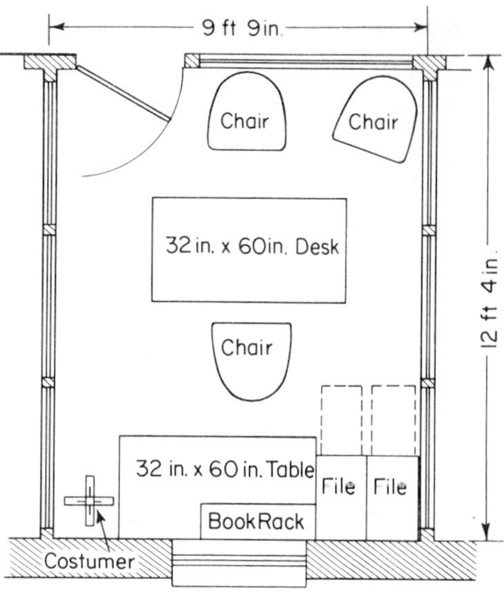

Fig. 6.19 *Typical private office designed for space economy with effective arrangement. From Ireson,* Factory Planning and Plant Layout, *p. 140.*

6.6.2 Space Estimation

The space requirements for offices usually bear some direct relationship to the number of employees to be accommodated. Some general rules have been devised for estimating space needs purely on the basis of the number of employees, but a more accurate estimate can be reached by making separate estimates of space for desks, filing, aisles, and private offices. Different functions within the same company require radically different ratios of desks, chairs, file cabinets, and other office equipment. It is advisable to prepare a typical production-center layout for each different function to be performed and to multiply the space for each production center by the number of persons to be employed in that function (see Art. 6.5.4.3). Figure 6.18 shows a production center for a function that requires an extra chair and a file cabinet, and with 40-inch-wide aisles between all desks. This function requires about 65 square feet.

Figure 6.19 shows a typical private office for a junior executive. Its space, including half of a 6-foot-wide aisle, is approximately 150 square feet.

The production-center area may be as low as 42 square feet, for a single desk and chair and half of a 4-foot-wide aisle, to around 100 square feet if wider aisles are used and if extra file cabinets are necessary.

Space for private offices may run from less than 100 square feet for functional offices to over 500 square feet for major executives. The trend is away from exceptionally large private offices to more modest offices and the

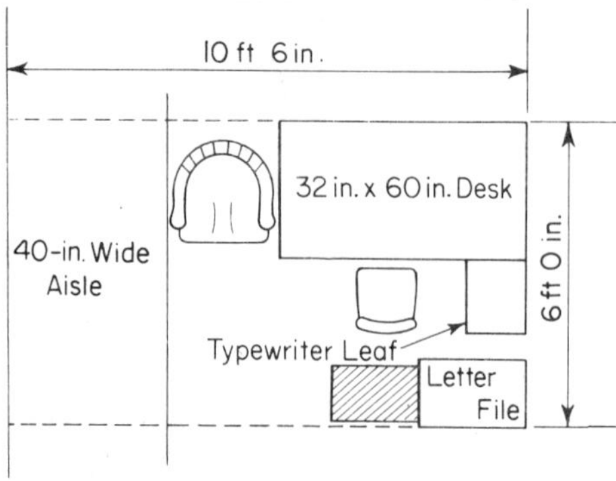

Fig. 6.18 *A typical production center layout for stenographic pool. From Ireson,* Factory Planning and Plant Layout, *p. 145.*

provision of one or more conference rooms to be used for meetings involving five or more persons.

An example of the space estimate for a certain department follows:

Private offices 2 10 ft × 12 ft	240 sq ft
General office	
10 desks (desk, 2 chairs) 6 ft × 8 ft	480 " "
Filing section	
1 vault 10 ft × 16 ft	160 " "
25 file cabinets, 6 sq ft each	150 " "
Lockers 16 4 sq ft each	64 " "
Aisles (not accounted for in previous estimates)	
Main 5 ft × 30 ft	150 " "
Cross 3 ft × 40 ft	120 " "
Total	1364 sq ft
Use area approximately 35 ft × 40 ft or 30 ft × 50 ft	

6.6.3 Office Layout

There are a few principles to guide the arrangement of the furniture and equipment in offices. Consideration of these principles will help to assure that offices will be efficient and comfortable. These principles are:

1. Provide sufficient aisles to prevent work interruption by persons entering and leaving the area. Provide aisle access for each desk.
2. Divide offices and functions so that distracting traffic will be minimized in each office.
3. Arrange desks and work tables so that both natural and artificial light fall on the desk from the same general angle.
4. Have the worker's back or side to the windows. Never have worker facing windows.
5. Arrange desks back to front (all in the same direction) unless the functions require constant collaboration of more than two persons.
6. Locate functions that require frequent visitors nearest the entrance of each office. (In most cases, this will be the chief person in each office group.)
7. Locate filing areas in the less desirable space, but convenient to those using them most frequently.
8. Do not locate file cabinets along heavily traveled aisles.
9. Locate functions requiring greatest concentration away from main aisles, entrances, conference areas, and other distracting influences.
10. Try to avoid placing the front of a desk against a wall.

6.7 MATERIALS HANDLING

6.7.1 Importance of Materials Handling

Materials handling is an inevitable part of every manufacturing plant. The magnitude of the problem is indicated by the fact that the cost of materials handling in all forms accounts for from 20 to 50 per cent of the total cost of converting the raw materials into the finished product. It is not uncommon for each part to be handled as many as 50 to 60 times before it is in final form and shipped. These facts alone are sufficient to prove that the materials handling system should be given exhaustive study whenever a new factory is being planned or an existing one is being remodeled.

The cost of materials handling arises from two sources: the cost of owning and maintaining equipment and the cost of operating the system. In general, higher investments in mechanical, semiautomatic, or automatic handling equipment are justified by the reduction of operating costs resulting from the better equipment. If the reduction in operating costs is greater than the increase in other costs, a net saving has been obtained. This is the usual attack on the problems of high handling costs in existing plants, but another attack, often overlooked, is available. That attack analyzes the selection and arrangement of the production facilities and attempts to eliminate the need for separate handling facilities.

The first objective, then, should be to select the most suitable production equipment and so arrange it within the factory buildings that the need for materials handling will be eliminated as far as practicable. The second objective should be to select the materials handling system that will accomplish the required handling at the lowest overall cost. It is obvious that successful factory planning requires the simultaneous consideration of material shandling problems, the selection of production equipment, and the arrangement of physical facilities. These factors are inseparable.

6.7.2 Terminology of Materials Handling

Certain terms will be used in the following sections that have special meanings with regard to materials handling. Some of these terms are commonly used in the industrial vocabulary, whereas others are subject to controversy.

Materials handling. Any movement of materials, vertically, horizontally, or both, manually or mechanically, in batches or one piece at a time.

*Transport.** The movement of a lot or batch of product from one production center or storage area to another production center or storage area. The purpose of transport is to locate the product for additional operations or storage. Transport is usually accomplished by "move men" and/or materials-handling equipment.

Transfer. The movement of pieces, singly or in small quantities, from a container to a machine where the operation is performed and back into another container. This handling is usually performed by the machine operator, and is not formally charged against the cost of materials handling.

Almost all materials are handled in either bulk or packaged form. This classification is useful in selecting a suitable handling system for the particular material.

Bulk materials. Any materials that are loose and handled in quantities without being contained in bags, boxes, barrels, and the like. Materials handling equipment for bulk materials supplies the temporary container, and the movement takes place by means of features such as pipes, buckets, belts, chutes, and tubes. The material is usually deposited in a bin, tank, elevator, or other container, from which it is fed into the processing equipment.

Packaged materials. Materials contained in convenient-sized packages, such as bags, boxes, crates, cartons, cans, barrels, or other types of containers that can be handled as individual pieces by the material shandling system.

Unit load. A certain number of packaged units mounted together on a skid, pallet, platform, or skid-box for movement as a single unit. The unit load may be made up of a certain number of individual pieces secured to a pallet or deposited in a box. It is most frequently used in connection with fork lift trucks and for purposes of warehousing, rail, and ship movement, but may be used with bulky items in manufacturing operations.

Rehandle. A term applied to the movement of a piece, package, or unit load. Rehandle consists of the pick-up, move, and set-down. The term is most frequently used in analyzing handling operations to discover unnecessary or excessive handlings that can be eliminated.

6.7.3 Principles of Materials Handling

Principles of materials handling—qualitative statements that are generally true—can be used to guide the analysis of and the decision regarding a handling problem. The purpose of the principles is to prevent serious mistakes resulting from oversights or incomplete examination of a problem. The principles do not give the solution; rather, they point out important considerations in the solution. In fact, some of the principles tend to be contradictory, but even these perform the useful function of emphasizing that the conditions unique to one problem must be analyzed from two or more entirely different viewpoints.

The following principles have been adapted from those contained in Stocker's *Materials Handling*† and from the author's own experience.

1. The objective of materials handling analysis should be the elimination of the necessity for handling as a separate function.

2. Methods studies should be used to determine the best equipment and handling methods for the particular conditions encountered.

3. Engineering economy principles should be employed to select the most economical alternative method or equipment when methods studies are not conclusive (see Sec. 3).

4. The prospective economy in materials handling will not be obtained unless the handling functions are coordinated with the other activities of the plant by efficient control systems (see Sec. 6).

5. Operating economy is obtained by:

a. Reducing idle or terminal time of handling equipment to the minimum

*It is believed that Preben Jessen, a consulting engineer in materials handling, originated the use of the terms *transport* and *transfer*, with the limitation of a movement of five feet on transfer.

†Harry E. Stocker, *Materials Handling*, 2nd ed. (Englewood Cliffs, N. J.: Prentice-Hall, Inc., 1951).

b. Increasing the size of the unit handled or the unit load

c. Employing powered, mechanical equipment to perform the moving task

d. Replacing obsolete equipment and methods with the most economical substitutes

e. Employing versatile equipment that can be used for several handling jobs

f. Selecting equipment of standard design instead of specially designed and built equipment

g. Reducing the ratio of the dead weight of the equipment to the live weight of the load to a minimum

h. Increasing the speed of the handling function without corresponding increases in other operating costs

i. Providing inspection and preventive maintenance to reduce failures and emergency repairs

j. Maintaining safe working conditions and training materials handlers in safe practices

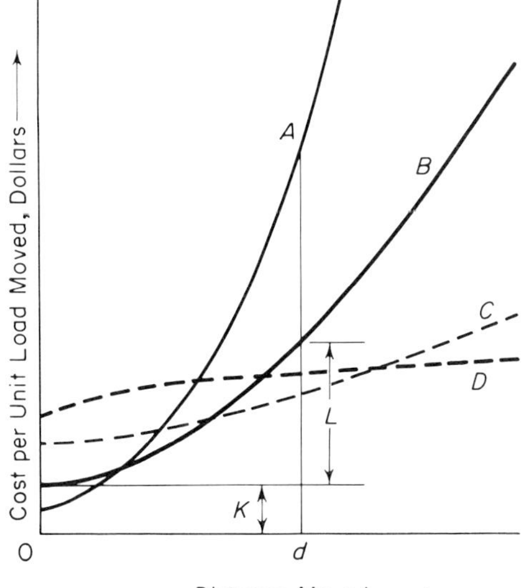

Fig. 6.20 *Types of costs curves for various materials-handling devices versus the distance for a predetermined unit load.*

k. Substituting mechanical controls, timers, and actuators for human operators

6. The economy of materials handling systems is measured by the cost of handling per ton or unit for a specific movement.

7. Materials handling costs increase with an increase in distance handled, but the rate of increase differs with the systems and is not necessarily uniform relative to distance (see Fig. 6.20).

8. Materials should be moved by gravity whenever the economies thus obtained are not counteracted by extra costs of factory facilities or operating labor.

9. The system selected should provide for flexibility to accommodate changes in output, products, or layout.

10. Whenever materials-handling methods affect production or factory costs, the combination that minimizes the total of such costs should be selected.

11. Whenever possible, the handling system should be combined with the production equipment so that transport and transfer are accomplished without human assistance and without process interruption.

6.7.4 Analysis of Materials Handling Problem

The solution of any materials handling problem requires the accumulation of a large amount of factual data and their analysis relative to the particular plant and production conditions. This analysis can best be performed when the collection and analysis of the data follow a logical pattern. Such a pattern will include the following items:

1. Plant factors
2. Methods factors
3. Products and materials
4. Present handling methods and equipment
5. Proposed handling methods and equipment
6. Cost data and economic analysis

6.7.4.1 Plant Factors. Plant factors refer to all the conditions of the building and the layout as they now exist or as they are proposed to be. The analyst needs to have data on the size and relative location of buildings and interior features such as column spacing; location of doors, elevators, stairs, and columns; floor load capacity; ceiling heights; aisles; piping and power circuits. These conditions establish the plant restrictions within which the materials handling system must function. They in turn become restrictions upon the types of handling systems and equipment that can be operated within the plant. It is important for the analyst to know which conditions must remain as they are and which can be altered to accommodate a more economical system.

6.7.4.2 Methods Factors. Methods factors encompass all the details of production methods, equipment, processes, sequence of operations, production plan (whether interruptible or continuous), temporary storages, volumes to be handled, and so forth. These methods factors describe the materials handling problems against a background of space, machines, weights and volumes of materials, and operators, so that the engineer can establish the specifications for a satisfactory materials handling system. (Par. 6.7.5 deals with the specifications for a system in terms of the features required.) The methods employed in production impose special restrictions on the handling system. These restrictions can be used to eliminate whole groups of handling systems, thereby simplifying the selection or indicating the methods that offer an opportunity for improvement or economy.

6.7.4.3 Products and Materials. The specific kinds of products and materials to be moved, and the volumes and distances of each move, provide the engineer with additional specifications for the system. The nature of the products, whether packaged or in bulk, fragile or rugged, dense and heavy or light and bulky, establishes the operating conditions to be met, and the quantities per period of time establish the load capacity and speed required of the system. Hazards or hazardous conditions are recognized, and protective equipment or measures are incorporated in the specifications.

Not only the existing products and materials, but also the possible changes in them in the future, must be considered if the materials handling system is to have the prospect of being economical over a reasonable period of time. Flexibility of the system is a guard against early obsolescence.

6.7.4.4 Present Handling Methods and Equipment. If the present materials handling problem is in an existing plant, it can be assumed that there is existing handling equipment and that it is either inadequate, expensive to operate, or unsuited to the handling job. Before existing equipment is scrapped and new equipment purchased, it is essential to realize that the problem may have come about as a result of poor utilization of the present equipment, or a plant layout that prevents its proper utilization, or both. Thus, the investigation of a materials handling problem should include a careful analysis of the effectiveness of the existing system. Points of failure, effectiveness of the coordination of handling with production, effects of layout on the utilization of the equipment and the coordination, and the results (in terms of cost) should be determined. With this information, the engineer may be able to point out ways by which the existing system can be used effectively through such actions as relocation or re-layout of certain departments or machines, the purchase of auxiliary equipment, the purchase of a few additional pieces of the same type, or the concentration of the existing equipment on jobs for which it is suited and the purchase of other types of equipment for the other handling tasks.

6.7.4.5 Proposed Handling Methods and Equipment. The fact that existing equipment can be used, apparently with satisfactory results, does not mean that this plan is necessarily the most economical. Only when the engineer has obtained all the data listed in the four preceding articles is he in a position to investigate the available materials handling systems to find those that will satisfy the needs of the plant. It is axiomatic that the most economical solution cannot be selected unless *all* systems that will satisfy the requirements have been identified as alternatives. Thus, having formulated the problem and established the specifications for the handling system, the engineer will proceed to use every possible source of information in order to obtain a complete catalog of functionally satisfactory systems, their first costs, operating characteristics, operating costs, prospective lives, and so on. He will then analyze the several systems and by preliminary economic analysis will discard all but the two or three most economical systems.

6.7.4.6 Cost Data and Economic Analysis. With the number of alternatives reduced to three or four systems, including the use of existing equipment as formulated in Art. 6.7.4.4, the analyst is ready to make exhaustive studies of the several systems. Complete cost information will be obtained from plants using similar equipment. Repair and maintenance histories will be accumulated. Firm bids on first cost and installation will be obtained. Irreducible advantages and disadvantages of each alternative will be enumerated. Finally, a complete engineering economy study will be made that will include all the pertinent information regarding the systems, and also the effects of taxes, insurance, and depreciation or amortization. The resulting amounts are comparable and, the irreducibles being equal, the system that gives the lowest prospective cost will be selected. Since this final selection is usually not made by the engineer but by the management, it is very important that the study be complete in every respect and present the proposals in clear, understandable language.

6.7.5 Features of Materials Handling Systems

The selection of a materials handling system involves two types of problems: technical problems and economic problems. Article 6.7.4 treated these two problems in general terms. The technical problems are involved in determining which handling systems will satisfactorily perform the required handling function. The solution of the technical problems provides the several alternatives that are then analyzed as economic problems. This article deals with the features of the handling system that are considered within the technical problems. The specifications for a technically satisfactory system can be almost completely prescribed in terms of the following features.

6.7.5.1 Flexibility. Flexibility of a materials handling system refers to its adaptability to changes in operating conditions. These changing conditions involve different products of varying size, nature, shape, and weight; changes in volumes to be handled; new layouts, machinery, and production processes. Seasonal and cyclical changes in business, development of new or complementary products, obsolescence of products, shifts in markets, and obsolescence of production equipment are some of the common causes of changes in the handling problem. The greater the probability that these changes will occur, the greater is the need for a *flexible* materials handling system.

Flexibility of handling systems may be illustrated by such characteristics as:

1. Ability to handle efficiently different sizes of packages simultaneously and at irregular intervals (rubber belt conveyors, roller conveyors, fork lift trucks, tractor-trailer trains, spiral chutes)
2. Ability to be rearranged with ease to accommodate a different path of movement (fork lift trucks, roller conveyors on portable stands, and sections of powered conveyors, as opposed to overhead monorails, under-floor drag lines, or elevators)
3. Ability of standard equipment to be fitted with special jigs or fixtures to do a specialized job that would otherwise require a specially designed system (monorail systems for finishing rooms with different types of hangers to accommodate widely varying sizes and shapes or products through a sequence of operations)

6.7.5.2 Space Requirements. All materials handling systems require space. Some require floor space constantly, whereas others use it intermittently. Some occupy vertical space that is otherwise unused. Trucks, trailers, and mobile equipment use aisle space that is also used for personnel traffic, but may require that the aisles be larger than would otherwise be necessary. The selection of a system that makes the most economical use of the available space, both floor and vertical, may eliminate the need for building expansion. The storage facilities are usually integrated with the handling system, and the use of pallets, stacking bins, or mezzanine floors may double or triple the storage capacity, provided the handling system is properly selected. Through such related economies in the utilization of space, the materials handling system can effectively expand the plant capacity without long-term commitment of capital funds in plant buildings.

6.7.5.3 Supervision. All materials handling systems require some direction, coordination, and supervision to accomplish effectively the assigned task. However, as the systems progress downward from the completely automatic, continuous ones to the system of individually dispatched mobile units, the amount of human effort required for supervision increases rapidly. Since the cost of human effort for supervision, coordination, and operation is a continuing annual expense, it is profitable to use the most automatic system practicable for the existing conditions. Conditions that tend to indicate automatic or semiautomatic systems are:

1. A stabilized line of products
2. Fairly uniform volumes or loads
3. Standardized tasks or fixed patterns of movement
4. A sufficient volume to justify the investment even if the several parts of the system are used only intermittently

In any semiautomatic or automatic systems, the probability of breakdowns and the consequences in lost production, cost of maintenance, and idle labor should be considered in making the final decision.

6.7.5.4 Speed. The rate of movement of the conveying equipment plays a substantial part in the determination of the most economical system. There are two points to be considered: Should speed be fixed or variable? Will the speed, in connection with the load capacity, provide the required volume of movement? Whenever the system is to be an integral part of the production system, as in mechanically paced work, it is almost essential that the speed be variable. Poor quality of materials, "green" workers, absenteeism, and variations in the production process are other reasons for selecting a variable-speed mechanism.

An increase in the speed of operation may permit the selection of a lower-capacity unit, with lower investment. However, the consequences of high speed, such as greater damage to materials, generation of heat, higher maintenance, and the hazards of a sudden breakdown, should be carefully weighed against the advantages.

6.7.5.5 Power. The nature and location of the movement may limit the handling equipment to the use of certain kinds of power. Mobile units with unlimited range must have self-contained power units, internal combustion engines, or batteries. Fixed systems, with limited areas of service, may use gravity, electricity (line), hydraulic power, or internal combustion engines. Both the initial investment and the cost of the power or fuel vary with the type of power used. Restrictions on the use of certain power sources (*i.e.*, the use of gasoline engines in enclosed space or areas containing explosive materials) may eliminate the possible use of the most economical or most satisfactory system.

6.7.5.6 Path of Movement. The path of movement can be classed as either fixed or variable, and, if variable, as limited or unlimited. Certain materials or parts follow the exact path through a plant and go through the same sequence of operations. The path of movement is fixed, and handling systems that provide fixed paths of movement, such as the roller conveyor or overhead chain-driven monorail, can be used for continuous or intermittent movement of materials over this path.

If the volumes are limited, and if the same handling units must serve many different materials or parts, the path of movement must be variable. The path is limited variable if the movement is to be confined to some relatively small area. Examples of the limited-variable path include the use of drag lines to handle bulk materials (as from a coal storage bin into a power plant), a bridge crane that provides completely variable movement within the limits of its runway, sectional conveyors that can be readily shifted about within a warehouse, and conveyor systems that incorporate a good switching system to permit a large number of different movements. Unlimited-variable paths require the use of mobile units, such as fork lift trucks, tractor-trailer trains, hand trucks, and mobile cranes.

Variable-path equipment generally requires greater supervision and coordination, in order to effect the desired movements on schedule, than does the fixed-path or even the limited-variable-path equipment. On the other hand, it is almost axiomatic that fixed-path equipment can only be used in connection with line or product layout.

At the same time that the path of movement is considered, it is important to analyze the possibilities of performing both the transport and transfer functions with the handling equipment. It is impossible to enumerate all the possible methods of accomplishing this objective. Some are very simple and others are elaborate and very costly. The nature of the product, the nature of the operation or process, the volumes to be handled, and relative time standards for successive operations are factors that enter into the analysis and the design of such a system.

Temporary storage of materials on the handling system is one way of eliminating rehandlings. This possibility should be studied along with the determination of the path of movement.

6.7.5.7 Load Capacity. The term *load capacity* refers to the ability of the equipment to carry a certain load. Unfortunately, however, all equipment is not rated on the same basis. Most conveyors are rated upon the safe load in pounds per foot of length. Fork lift trucks, trailers, and shop trucks are rated by the weight that they will carry. In the case of the fork lift trucks, this weight is specified at a certain distance from the face of the forks, but actually is determined by the distance from the center of the front wheels to the center of gravity of the load. Different types of fork trucks with the same rated capacity may have very dissimilar actual load capacities.

In the case of troughed belt conveyors or screw conveyors, the nature of the bulk material, size of particles, density, and angle of repose determine how much of the material can be moved under various conditions and by different sizes of conveyors.

It is the combination of the load capacity and the speed of the equipment that really defines the output capacity of the handling system.

6.7.6 Selecting the Handling System

There are literally hundreds of available materials handling systems of standard design from which to choose two or more technically satisfactory systems for complete economic analysis. Since it would be economically impracticable to make a complete study of every possible system, it is essential to eliminate a large number of systems quickly and easily without running any serious risk of eliminating one of the more economical ones. It is for this reason that Arts. 6.7.4 and 6.7.5 have emphasized, first, the collection of the necessary information to enable the engineer to completely understand the handling problems, and, second, the specification of the features that will make up a technically satisfactory handling system. By following these ideas and by using a check list such as that shown in Fig. 6.21, the engineer is able to locate quickly the types of handling equipment that satisfy the requirements. It will be noted that the features described in Art. 6.7.5 are repeated in the headings of Fig. 6.21 and that check marks identify the properties of each of the more common handling devices.

Since each type of equipment is manufactured by a number of different companies, the engineer is faced with the problem of selecting the equipment of a specific manufacturer after he has selected the basic types. Slight differences in design details, sizes, construction, materials, weight, available services, and so on account for the differences in initial cost of the same basic type from different manufacturers. As far as possible, the engineer should evaluate these differences in terms of initial cost, maintenance and repair, and operating costs, and should base his decision on the engineering economy study. However, many of these differences will not be reducible to monetary terms, and must be handled as irreducibles in the final analysis.

6.7.7 Materials Handling Equipment

It is impracticable to attempt to discuss in this volume the many types of handling equipment. Every different kind of equipment has certain merits relative to certain handling problems, and manufacturers are daily adding new designs, new features, and improvements in utility, and are deleting obsolete systems from their lines. The reader is urged to refer to recent publications for up-to-date information on specific handling systems. The periodicals devoted to materials handling problems,* the trade associations,† and periodicals devoted to factory management‡ are ready

**Mechanical Handling*, The Louis Cassier Co. Ltd., Dorset House, Stamford St., London, England; *Modern Materials Handling*, 795 Boylston St., Boston, Mass. 02116.; *Flow Magazine*, 1240 Ontario St., Cleveland, Ohio. 44113.

†American Materials Handling Society, Inc., 638 Phillips Ave., Toledo, Ohio 43612.; Conveyor Equipment Manufacturers Assoc., 1129 Vermont Ave. N.W., Washington, D. C. 20015.; Electric Industrial Truck Assoc., 3701 N. Broad St., Philadelphia, Pa. 19140.; Caster and Floor Truck Manufacturers Assoc., 7 W. Madison St., Chicago, Ill. 60602.; The Materials Handling Institute, Inc., 1108 Clark Building, Pittsburgh, Pa. 15222.

‡*Factory Management and Maintenance*, 330 W. 42nd St., New York, N. Y. 10018.; *Modern Industry*, 400 Madison Ave., New York, N. Y. 10017.; *Mill and Factory*, 205 E. 42nd St., New York, N. Y. 10017.

Materials–Handling Equipment Selection Chart

	Materials		Movement			Supervision Required			Path			Speed			Power				Load			Space		
	Bulk	Packaged	Vertical	Horizontal	Combination of Vertical and Horizontal	Close Supervision and Detailed Dispatch	Little Supervision and Detailed Dispatch	Automatic or Semiautomatic	Completely Variable	Fixed Path	Fixed Area	Variable	Fixed	Either Fixed or Variable	Electricity–Line	Electricity – Bat	Internal Combustion	Gravity	Unit Load with Limit	Continuous Loading, Maximum Load per Unit	Fixed Units by Spacing of Carriers	Minimum Width Aisles	Constant Utilization of Fixed Area	No Floor Space
Trucks Industrial																								
Manual																								
Four–Wheeled Platform	X	X		X		X			X			X			Manual Power				X			X		
Two–Wheeled Platform	X	X		X		X			X			X			Manual Power				X			X		
Two–Wheeled Platform (Barrel, etc.)	X	X		X		X			X			X			Manual Power				X			X		
Dollies		X		X		X			X			X			Manual Power				X			X		
Pallet Lift		X		X		X			X			X			Manual Power				X			X		
Powered																								
Driver–Walk																								
Pallet Lift		X		X		X			X			X				X	X		X			X		
Platform		X		X		X			X			X				X	X		X			X		
High–Lift Fork	X[1]	X		X	X	X			X			X				X	X		X			X		
Driver–Ride																								
Pallet Lift		X		X		X			X			X				X	X		X			X		
Platform		X		X		X			X			X				X	X		X			X		
Low–Lift Platform	X[1]	X		X		X			X			X				X	X		X			X		
High–Lift Platform	X[1]	X		X	X	X			X			X				X	X		X			X		
Telescoping Fork Lift	X[1]	X		X	X	X			X			X				X	X		X			X		
Tractors and Trailers																								
Industrial Tractors																								
Three–Wheeled																								
Four–Wheeled																								
Industrial Trailer																								
According to Wheel Arrangement	X[1]	X		X		X			X			X				X	X				X	X		
Cranes, Hoists, and Monorails																								
Crane																								
Overhead Bridge, Traveling	X[1]	X			X	X					X			X	X				X					X
Gantry	X[1]	X			X	X					X			X	X				X					X
Jib		X			X	X					X			X	X				X					X
Hoist																								
Chain Manual	X[1]	X	X			X					X	X			Manual Power				X					X
Electric Motor Drive	X	X	X		X	X					X	X			X				X					X
Pneumatic	X	X	X		X	X					X	X			X				X					X
Monorail																								
Carrier	X[1]	X		X	X		X			X					X				X				X	
Trolley		X		X	X		X			X					X								X	
Chain Trolley		X		X	X		X			X					X								X	
Conveyors																								
Roller, Gravity																								
Spiral		X	X		X			X		X		X						X		X			X	
Portable		X		X	X			X		X		X						X		X			X	
Fixed		X		X	X			X		X		X						X		X			X	
Roller, Live																								
Chain Drive		X		X	X			X		X				X	X					X			X	
Belt Drive		X		X	X			X		X				X	X					X			X	
Wheel, Gravity																								
Portable		X		X	X			X		X		X						X		X			X	
Fixed		X		X	X			X		X		X						X		X			X	
Belt																								
Flat	X	X	X	X	X			X		X				X	X		X			X			X	
Troughed	X		X	X	X			X		X				X	X		X			X			X	
Portable	X	X	X	X	X			X		X				X	X		X			X			X	
Fixed	X	X	X	X	X			X		X				X	X		X			X			X	
Drag																								
Pusher Bar		X	X	X	X			X		X			X		X					X			X	
Screw	X[1]		X	X	X			X		X			X		X					X			X	X
Floor Chain	X[1]	X		X	X		X			X			X		X					X	X	X		
Overhead Chain	X[1]	X		X	X		X			X			X		X					X	X	X		
Apron																								
Wood Slat	X	X	X	X	X			X		X			X	X	X		X			X			X	
Steel Slat	X	X	X	X	X			X		X			X	X	X		X			X			X	
Ball Transfer		X		X		X					X	X			Manual Power				X				X	
Bucket Conveyor	X[1]		X		X			X		X			X		X		X			X			X	
Continuous Chain Trolley	X[1]	X		X	X			X		X			X	X	X						X		X	
Slides and Chutes																								
Spiral or Straight																								
Wood	X	X	X		X		X			X			X					X		X			X	
Steel	X	X	X		X		X			X			X					X		X			X	
Vibrating	X		X		X		X	X		X			X		X			X		X			X	
Pneumatic Systems																								
Rigid Tube and Nozzle																								
Bulk Loaders	X				X		X			X			X		X					X				X
Bulk Unloaders	X				X		X			X			X		X					X				X
Rigid Tube																								
Cylindrical Carrier		X			X		X			X			X		X				X					X
Oval Carrier		X			X		X			X			X		X				X					X
Elevators																								
Freight																								
Electric	X[1]	X	X			X				X			X		X				X				X	
Hand	X[1]	X	X			X				X		X			Manual Power				X				X	
Hydraulic	X[1]	X	X			X				X			X		X				X				X	
Dumb Waiter		X																						
Continuous Lifts																								
Arm		X	X				X	X		X			X		X						X		X	
Tray		X	X				X	X		X			X		X						X		X	

[1]With scoop, bucket, etc.

Fig. 6.21 *Tabulation of principal types of materials-handling equipment, with characteristics and features. From Ireson,* Factory Planning and Plant Layout, *p. 171.*

sources of current information. Manufacturers' representatives are more than willing to give engineering assistance, cost data, and reference lists of equipment users.

BIBLIOGRAPHY

Apple, James M., *Plant Layout and Materials Handling*, 2nd ed. New York: The Ronald Press Company, 1963.

Coleman, J. R., Jr., S. Smidt, and R. York, "Optimum Plant Design for Seasonal Production," *Management Science*, *10*, No. 4 (July 1964), p. 778.

Conway, Richard W., "Some Tactical Problems in Digital Simulation," *Management Science*, *10*, No. 1 (Oct. 1963), p. 49.

Goode, H. P. and Sidney Saltzman, "Computing Optimum Shrinkage Allowances for Small Order Sizes," *Journal of Industrial Engineering*, *12*, No. 1 (Jan.–Feb. 1961).

Grant, Eugene L. and W. Grant Ireson, *Principles of Engineering Economy*, 4th ed. New York: The Ronald Press Company, 1960.

Ireson, W. Grant, *Factory Planning and Plant Layout*. Englewood Cliffs, N. J.: Prentice-Hall, Inc., 1952.

Lawless, Robert M. and Paul R. Haas, Jr., "How to Determine the Right Sized Plant," *Harvard Business Review*, *40*, No. 3 (May–June 1962).

Machol, Robert E., Editor, *System Engineering Handbook*. New York: The McGraw-Hill Book Company, 1965.

Mathews, Harold H., "Simulation Puts Conveyor System Through Mathematical Pilot Plant," *Modern Materials Handling*, *17*, No. 4 (April 1962), p. 72.

Maynard, H. B. Editor, *Industrial Engineering Handbook*, 2nd ed. New York: The McGraw-Hill Book Company, 1963.

Moore, James M., *Plant Layout and Design*. New York: The Macmillan Company, 1962.

Morris, William T., *Analysis for Materials Handling Management*. Homewood, Ill.: Richard D. Irwin, Inc., 1962.

Muther, Richard, *Systematic Layout Planning*. Boston: Industrial Education Institute, 1961.

Parson, James A., "You Can Simulate Systems with Paper and Pencil," *Modern Materials Handling*, *21*, No. 3 (March 1966), p. 50.

Starr, Martin K., *Production Management*. Englewood Cliffs, N. J.: Prentice-Hall, Inc., 1964.

Reed, Ruddell, Jr., *Plant Layout*. Homewood, Ill.: Richard D. Irwin, Inc., 1961.

Wilson, Richard C., "Evaluation of Special Relations and Empirical Plant Layout Criteria by Digital Computer," Ph.D. Dissertation, University of Michigan, 1961.

RAY K. LINSLEY, *joined the Stanford University faculty in 1950 as associate professor of hydraulic engineering. In 1956 he became professor and head of the department of civil engineering. Prior to his appointment at Stanford he was chief of hydrologic services for the United States Weather Bureau.*

Mr. Linsley, a native of Connecticut, received his B.S. degree in civil engineering from Worcester Polytechnic Institute. He is a Fellow of the American Society of Civil Engineers, a Fellow of the American Geophysical Union and past president of its section of hydrology, a professional member of the American Meteorological Society, and a member of the American Society for Engineering Education.

He is co-author, with Kohler and Paulhus, of Applied Hydrology *and* Hydrology for Engineers, *and with Franzini of* Water-Resources Engineering. *He has authored numerous technical papers and has served as consultant to UNESCO, the World Meteorological Organization, the U.S. Weather Bureau, and the governments of Israel and Venezuela. During the academic year 1957–58 he was a Fulbright lecturer at The Imperial College of Science and Technology in London, and during 1964–65 was water resources research coordinator in the Office of Science and Technology.*

Mr. Linsley has received the Meritorious Service Award of the U.S. Department of Commerce and the Collingwood Prize of the American Society of Civil Engineers.

SECTION 7

INDUSTRIAL CLIMATOLOGY

RAY K. LINSLEY

DEFINITIONS

Average. The arithmetic mean of a group of data. Commonly used in climatology to refer to the mean value of a weather element over a large area for a specified period.

Climate. The summation of weather at a given place over a long period of time.

Macro. A prefix combined with the words climate and variation to refer to the broad features of climate over an extensive area, such as a continent.

Mean. An arithmetic average of a group of data. Commonly used in climatology to refer to the average of a time series, such as the mean temperature for a day, month, or year.

Micro. A prefix combined with the words climate and variation to refer to the small-scale features of climate within a limited area, such as the microvariations of temperature within a city.

Normal. The arithmetic average of the values of a weather element for a specific day, month, season, or other portion of the year over a period of 30 years. In the United States the normal period is 1941 to 1970. This period will be advanced 10 years each decade. The World Meteorological Organization will use the period 1931 to 1960 until 1991, when a full, new, 30-year period will be available.

7.1 WEATHER AND MAN

Contrary to an oft-repeated saying, man has done a great deal about the weather. He has probably devoted more effort to combating weather than he has to any other purpose with the possible exception of waging war. Perhaps man will eventually learn how to actually control the weather. He has already exerted some measure of weather control over small areas by fog dispersal over airports (FIDO), orchard protection by smoke and heaters, and small-scale modification of clouds by cloud seeding. For the present, however, most of man's efforts against the weather can be summed up in two words—protection and avoidance.

7.2 WEATHER PROBLEMS IN INDUSTRY

7.2.1 Protection from the Weather

Buildings are intended almost exclusively to afford protection from weather. Starting with a natural cave, man has developed the art of shelter to the point where he can create enormous skyscrapers with artificial weather provided by heating and air conditioning. Space does not permit a detailed survey of efforts to provide protection against the weather, but irrigation farming, clothing, instrument landing aids for aircraft, lighthouses, ultra-high-frequency radio, the modern automobile, flood-control dams, and many other items may be cited as examples. Thus the concept of protection goes beyond mere physical shelter. It includes all devices designed to ward off the harmful effects of weather. The design of protective works must include consideration of physical resistance to water and wind, chemical resistance to radiation and airborne compounds, and numerous other problems associated with modern materials engineering.

Man is not inclined to seek protection from the weather until the need is great. Seasonal products such as snow tires, antifreeze, and air-conditioners find little demand during the off season. Thus it is necessary to schedule production, distribution, and sales of protective items to meet seasonal peaks of demand.

7.2.2 Avoidance of the Weather

If the weather of the world were invariant—if every location had the same weather at all times—the planning of protection against the weather would be simple. Unfortunately, weather is highly variable. In the interests of economy, man tries to plan his activities so that minimum protection is required. One way to do this is to locate an activity at a site where the climate is most favorable. Another is to schedule an operation outdoors during favorable weather.

7.3 WEATHER FORECASTS

If plans are to be made to avoid inclement weather for a specific operation, it is necessary to have some information concerning the weather to be expected. The ideal solution would be found in a detailed forecast of the weather for the times and places of interest. Highly successful utilization of special weather forecasts has been made in connection with frost warnings for citrus and cranberry growers, rain warnings for fruit drying, light-intensity forecasts for scheduling electric-power production, snow forecasts for highway and railroad snow-removal operations, and flood forecasts for emergency protection and evacuation operations of all types. For the most part, such forecasts are limited to periods not exceeding about 48 hours in advance. Forecasts for longer periods become, of necessity, less specific as to the time of occurrence of weather phenomena and less accurate in the estimates of the magnitude of the weather event. The limit of practical forecasts at the present time is about 30 days. Forecasts of this length indicate no more than that the average temperature and precipitation during the period will be near, above, or below normal.

Despite the limitations on range and accuracy of forecasts, an intelligent use of forecasts can provide many benefits to industry. Experience has shown that a trained meteorologist who devotes himself to forecasts for a specific purpose and who has a thorough understanding of the uses to which the forecast will be put can do a better job than one whose assignment is to prepare a general forecast for a large area. Consequently, the industrial meteorologist has taken his place on many industrial staffs to aid in advance planning against the weather over short periods.

7.4 CLIMATOLOGY

The lasting qualities of paint, the design of an industrial plant, the advance manufacture of seasonal merchandise, and solar heating are examples of operational problems requiring information about weather to be expected over a period extending beyond the range of successful weather forecasts. What is needed is information on *climate*, the statistical collective of weather elements over a long period of time. The climatological forecast is quite different from the usual weather forecast. The climatologist assumes that the weather of the future will be like the weather of the past and that statistical analysis of past weather records can provide a summary of the weather to be expected in the future. The climatological forecast may include a description of average weather; the possible range of values of a weather element; the probable number of times that a certain weather condition is likely to occur in a given period of time; or the probability that the weather of a given day, week, or month will be within a specified range of conditions. Climatology cannot predict the specific weather to be expected at a given place and time (unless its probability is 100 per cent) but only the probability or odds that certain weather conditions will occur. The climatological forecast provides the planner with the probabilities upon which he can base a calculated risk.

7.5 TYPICAL PROBLEMS IN CLIMATOLOGY

7.5.1 General

In the sections that follow, some examples of the application of climatology to industrial problems are presented. Most climatological problems involve numerous complications, and their solutions require extensive analysis of data. Although the list of examples is far from comprehensive, it is hoped that the reader will be encouraged to develop solutions to the specific problems that he encounters. A discussion of the sources of climatological data, how to interpret them, and how to treat them statistically is presented in Arts. 7.6 through 7.15.

7.5.2 General Information

Certain weather problems are best solved by compiling general information in a form that can be easily used. This type of problem is encountered where an operating official must make decisions when the controlling criteria depend upon the specific circumstances of each case. The pertinent data are summarized in tabular or graphic form so that the desired information can be readily obtained. The best presentation for a particular case depends on the nature of the problem and on the personal preferences of the user. The *Climatic Handbook for Washington, D. C.** and *Climatic Atlas of the United States* illustrate general summaries of data intended for a variety of users. The *Airway Meteorological Atlas for the U.S.*, *Atlas of Climatic Charts for the Oceans*, and *Summer Weather Data* represent compilations of data covering a large area but intended primarily for a specific class of problems.

Illustrative charts showing probabilities for wind speed, temperature, and cloudiness for a single station are shown in Fig. 7.1. Charts of this type are desirable when the variations in probability throughout the year are important. Isopleth charts (maps showing lines of equal values of a variable) are shown in Figs. 7.2, 7.12, 7.15, 7.17, 7.18, 7.19, and 7.20. Most such maps are suitable only for depicting the broad features of the climatic pattern. In the mountain regions, the small scale does not permit accurate representation of the variations in the

* See bibliography at the end of this section.

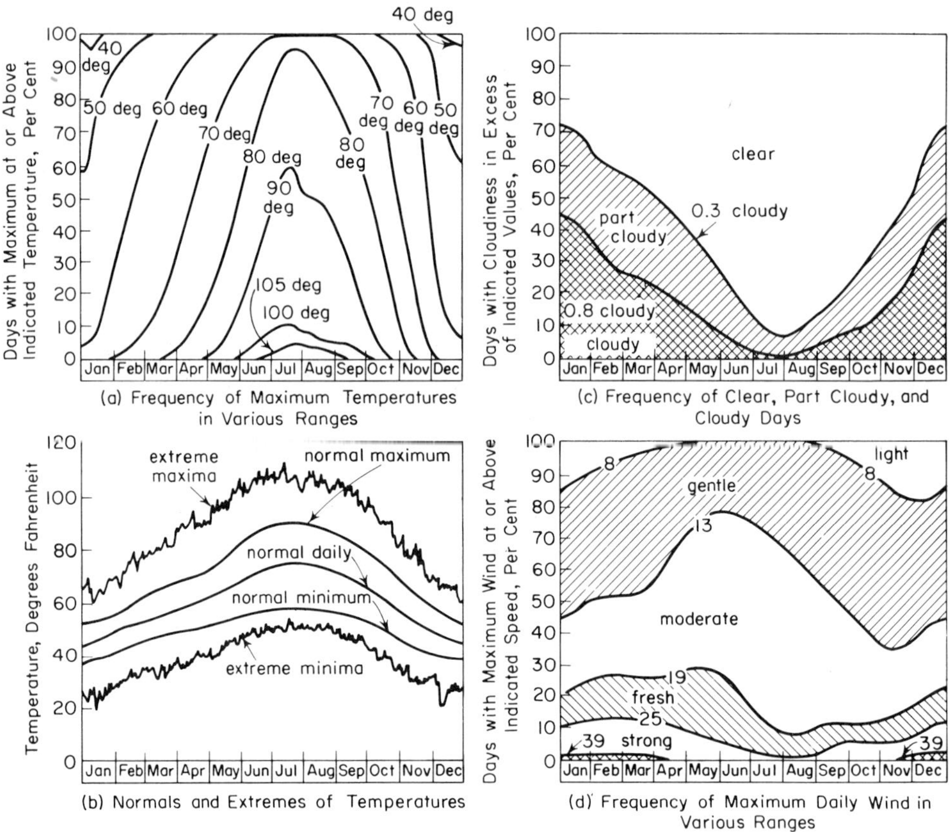

Fig. 7.1 *Examples of graphical presentation of climatic data for Sacramento, California (period 1877–1942). From Climate of Sacramento (Sacramento, Calif.) U. S. Weather Bureau Office 1944.*

element being pictured. A table of the data for the key stations used in plotting the map would be a more realistic presentation. Several other types of graphical presentation appear in Figs. 7.4, 7.5, 7.7, 7.8, 7.14, and 7.21. In each case, the data could be presented in several other pictorial or tabular forms.

7.5.3 Site Selection

Climate is rarely the only factor that governs the selection of a site for an industrial plant, but it is often quite important. Assume that an industry desires a site for a plant requiring constant temperature at 60°F. No location will afford such an ideal situation naturally, but a site that has a mean temperature of 60° and a minimum standard deviation of temperature will require the least amount of artificial heating and cooling. The climatic factor must be weighed with other economic aspects in the selection of a suitable site. Figure 7.3 shows the frequency distributions of mean monthly temperature for two cities having mean annual temperatures very close to 60°F as selected from Fig. 7.12. San Diego is clearly the preferable site. The example demonstrates how important it is to consider the frequency distribution of the weather element in question as well as its normal or average value. Note that if the distributions of Fig. 7.3 had been computed from daily temperature values, the differences between the two stations might have been more marked. Monthly or annual averages tend to suppress the extreme variations.

7.5.4 Site Layout

After an industrial site has been selected, weather factors may continue to be important

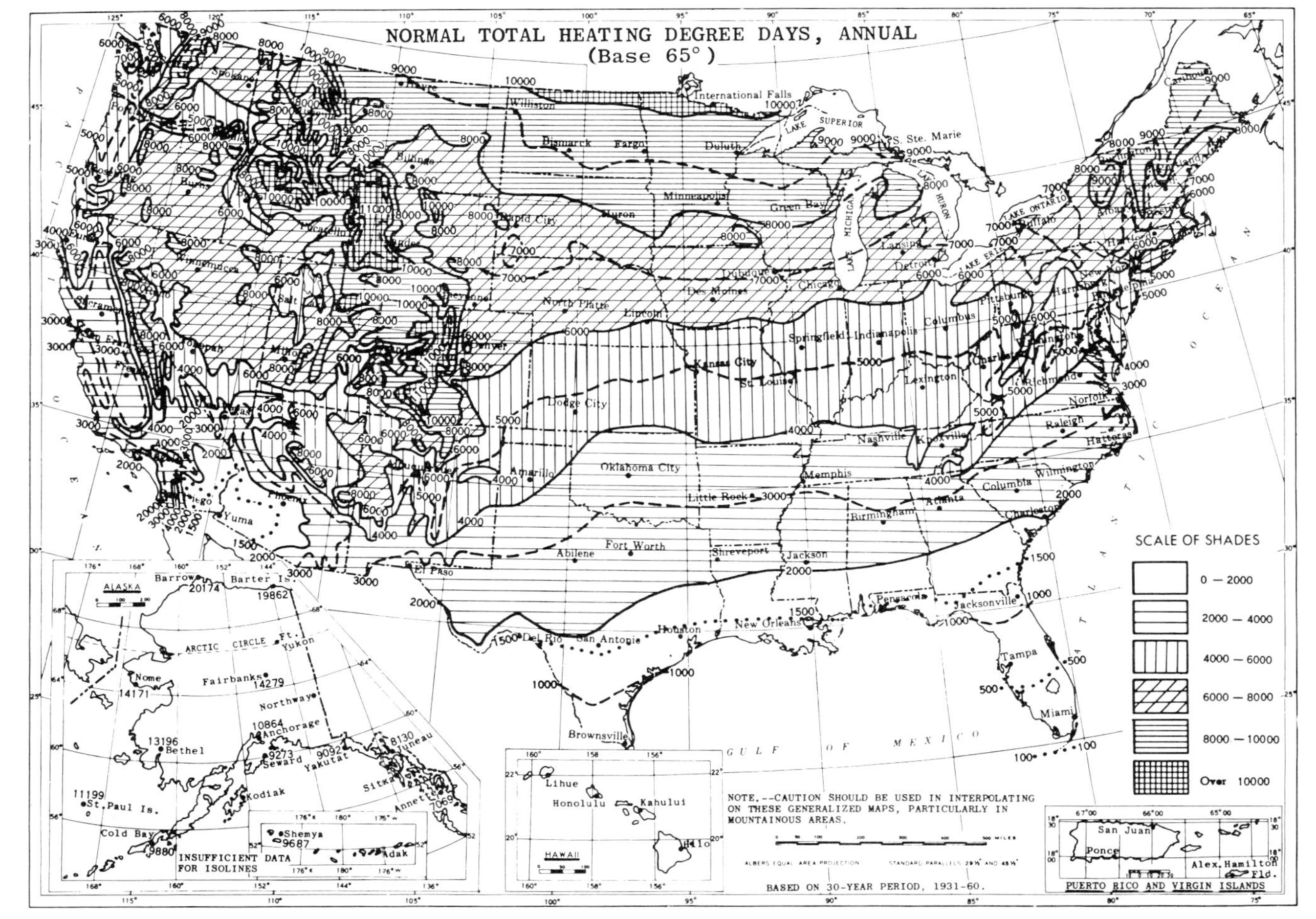

Fig. 7.2 *Normal degree days in the United States and their standard deviations. Courtesy U. S. Weather Bureau.*

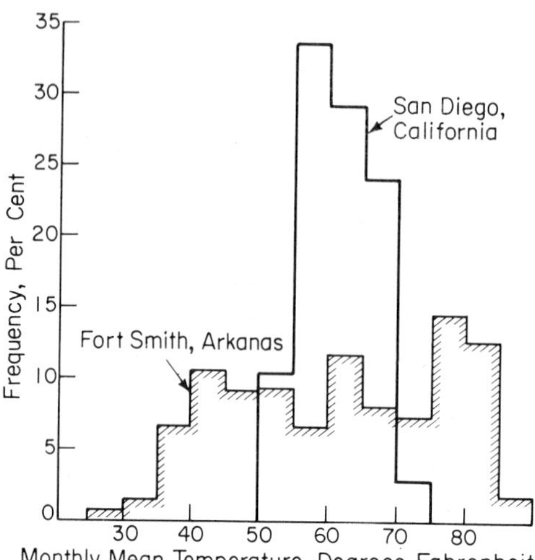

Fig. 7.3 *Comparison of temperature frequency distributions for two cities with the same normal annual temperature.*

in determining the plant layout. Assume a plant including an outdoor operation that should be located upwind of a plant unit that will produce considerable dust. In addition, the outdoor work area should be protected from wind for the comfort of the workers. Frequency distribution of wind direction is commonly represented by a *wind rose*, such as that shown in Fig. 7.4.

If this wind rose were applicable to the site under study, it would seem desirable to place the outdoor operation to the north of the other plant buildings for protection and southwest of the dustproducing unit to avoid the dust. The relatively light and infrequent east and northeast winds would be least harmful in transporting dust. Moreover, the wind roses for January and July suggest that the east and northeast winds are most common during the winter season when rain and snow might interfere with outdoor work. Careful consideration should be given to the possibility that construction of plant buildings might alter the local wind pattern and create undesirable eddies that would require a different solution.

7.5.5 Scheduling

Numerous operations must be scheduled to meet climatic deadlines. Sales are materially reduced unless shipments of antifreeze for cars are in dealers' hands before the first cold spell of the year. Chocolate shipped at temperatures over 87°F may be damaged by heat. Sales of soft drinks, citrus fruits, and ice cream are boosted by a heat wave. Figure 7.1(a) contains the information necessary to estimate the probability of high temperatures at any time during the year. Figures 7.5 and 7.21(i) summarize the

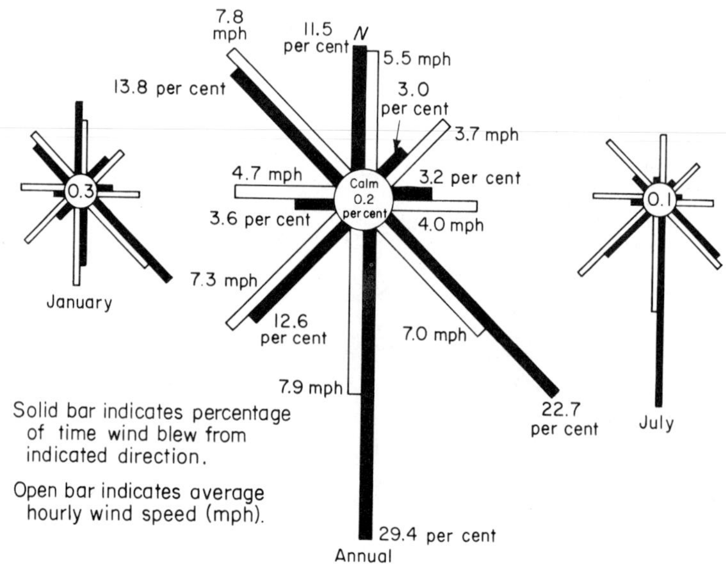

Fig. 7.4 *Wind roses for Sacramento, California.*

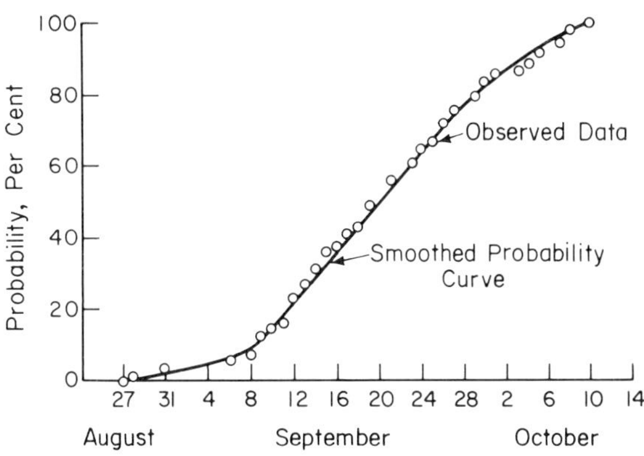

Fig. 7.5 *Cumulative probability of the occurrence of 32°F or less prior to any date in the fall for Bismarck, North Dakota.*

probabilities that temperatures at or below 32°F will occur on or before any given date in the fall. From charts such as these, the industrial planner can estimate the risk he is taking by scheduling delivery of products for a specific date. He can balance the risk of loss through late delivery against the cost of storage if delivery is too early.

Equally important in the distribution of seasonal products is the possible areal extent of a given weather condition. Figure 7.6 shows the area covered by temperatures over 90°F on August 10, 1944. New York City endured temperatures in excess of 90°F for 7 days during this heat wave. Products that would be in demand under such weather conditions must be available in sufficient quantity to serve the area and should be located at distribution points from which they can be delivered promptly to all points in the area.

7.5.6 Equipment Operation

Equipment must be selected to meet the conditions in the area where it is to be used. The wet-bulb depression (Par. 7.12.1) is a measure of the degree of cooling possible with evaporative-type air-conditioning equipment. Figure 7.7 shows the cumulative probability of various wet-bulb depressions at Shreveport, Louisiana. From a chart of this kind, the proper type of equipment could be determined for an installation in the area. Even more useful, however, would be a summary showing the wet-bulb depressions to be expected for various temperatures. Such a joint frequency is shown in Table 7.1, which lists the number of periods of consecutive cloudy days at San Jose, California, and the number of degree days below 65°F occurring during these periods. A table of this type would be helpful in designing solar-heating equipment since heat storage facilities or auxiliary heating must be supplied to assure adequate heat during cloudy days when solar heating is not satisfactory. Although single-city studies, such as those presented in Table 7.1 or Fig. 7.7, provide a valuable guide in selecting equipment for a given installation, studies for many points are more valuable for determining the best standard sizes for units to meet the diversified conditions that can be expected.

7.5.7 Precipitation

Precipitation (rain and snow) is an important factor in many industrial problems. Drainage ways for storm water are designed on the basis of precipitation intensity-frequency data.* Snowfall is a factor in roof design in many portions of the country.† Snowfall, temperature, humidity, and wind enter the design of road and walkway snow-melting systems.‡ Consider an industry involving certain operations that are interrupted by rain. A reserve of materials is necessary to permit the rest of the

---

* *Rainfall Frequency Atlas for the United States*, U. S. Weather Bureau Technical Paper 40 (Washington, D. C., May 1961).

† Housing and Home Finance Agency, *Snow Load Studies*, Housing Research Paper 19 (Washington, D. C., 1952).

‡ W. P. Chapman, "Design of Snow Melting Systems," *Heating and Ventilating*, April 1952, Reference Section, pp. 96–102.

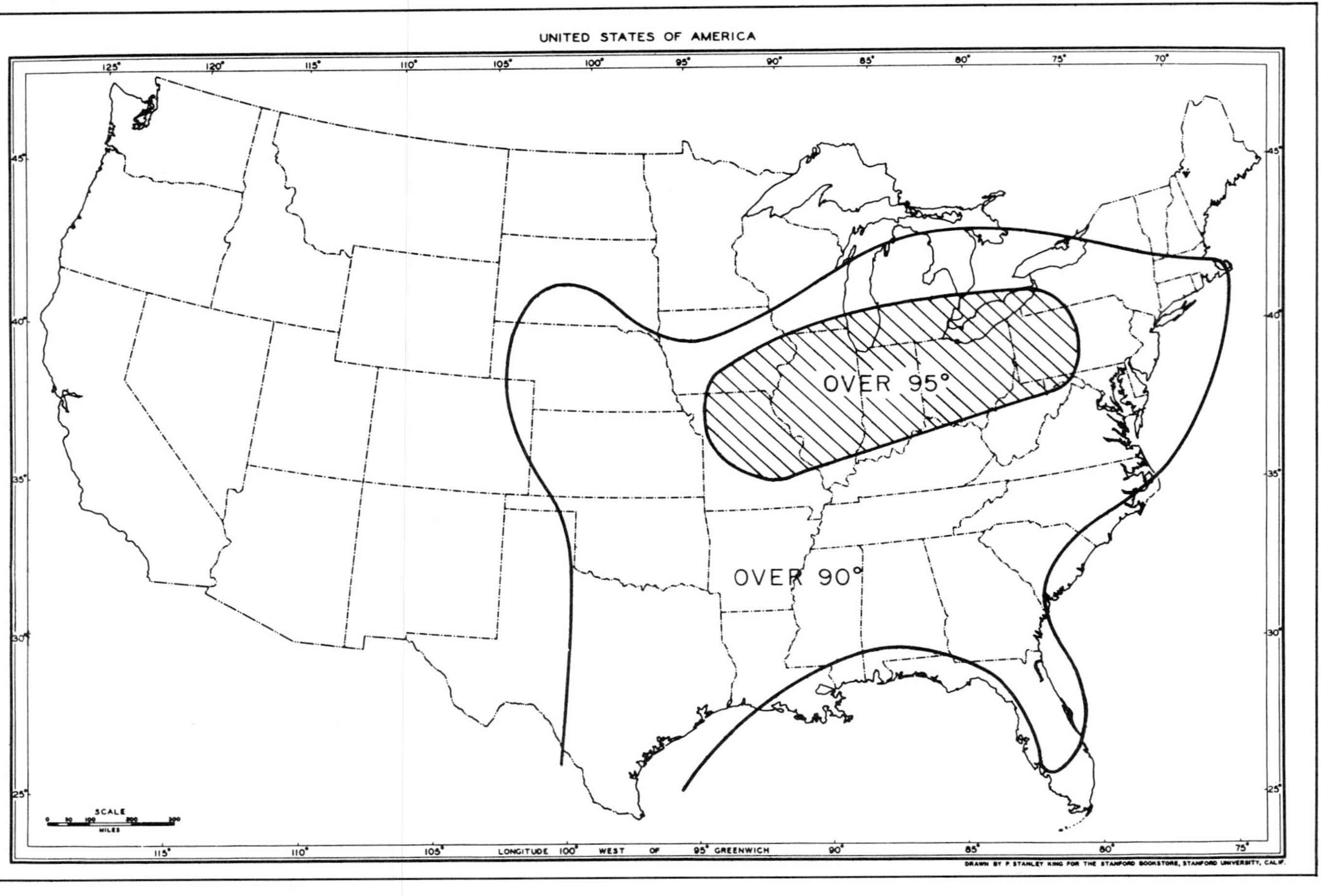

Fig. 7.6 *Area covered by the heat wave of August 10, 1944.*

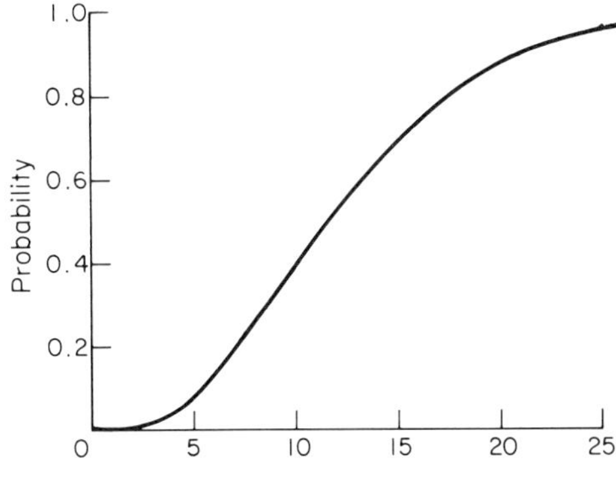

Fig. 7.7 *Cumulative probability of wet-bulb depression, Shreveport, Louisiana. Courtesy U. S. Weather Bureau.*

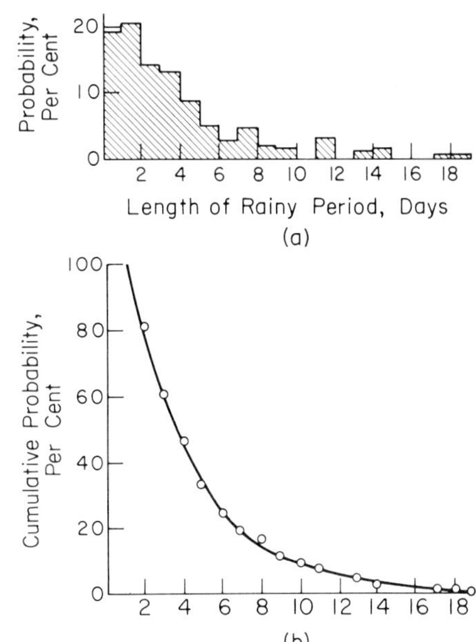

Fig. 7.8 *Probability of consecutive rainy days at San Francisco, California. (a) Probability that a rain day will be one of a series of various lengths. (b) Cumulative probability that a rain day will be one of a series equal to or greater than the indicated length.*

process to continue during the interruptions. A climatological analysis of the frequency of various numbers of consecutive days of rainfall indicates how big this material reserve should be. The same data, used in an economic analysis, might suggest whether protection should be provided for the interruptible operation. Such an analysis of rainfall frequency is shown in Fig. 7.8.

TABLE 7.1. *CONSECUTIVE CLOUDY DAYS AND CORRESPONDING HEATING DEGREE DAYS AT SAN JOSE, CALIF. (Based on 8 years of record.)*

Degree Days	Number of Days										Number of Cases	Per Cent Above Class Limit
	1	2	3	4	5	6	7	8	9	10		
1 to 9	85	15	4			1					105	100
10 to 19	46	18	5	3	1	1					74	67
20 to 29	3	25	4	2							34	43
30 to 39		11	11	12		1					35	32
40 to 49		6	7	6	4		1				24	21
50 to 59			2	4	2	2		1			11	13
60 to 69				4		2	1				7	10
70 to 79					2		1	6	1		10	8
80 to 89				2	2		2				6	4
90 to 99						3		1	1		5	3
100 to 109							1			1	2	1
110 to 119												0.3
120 to 130								1			1	0.3
Number of cases	134	75	33	33	11	10	6	9	2	1	314	
Number of days	134	150	99	132	55	60	42	72	18	10		
Per cent of heating days	6	7	5	6	3	3	2	4	1	1		
Cumulative per cent	38	32	25	20	14	11	8	6	2	1		

7.5.8 Air Pollution

Since meteorological factors play an important role in the movement of airborne pollutants, a study of the climate of an area will often suggest the problems that may result from a given industrial installation. Strong winds are normally indicative of satisfactory dispersion of contaminants, but if the winds are predominantly from one direction, a narrow strip of land downwind from the source may receive an excessive concentration. Variability in wind direction can be advantageous in spreading pollutant material over a wide area. However, mean monthly or annual wind data, such as were used to prepare the wind roses of Fig. 7.4, may conceal critical diurnal wind variations. Figure 7.9 shows the average hourly wind speed and direction for Santa Maria, California. As at many coastal stations, a land- and sea-breeze regime predominates, with light variable winds at night. Critical concentrations of pollutants could be expected at night, although the average July wind speed of 7 mph might at first inspection seem adequate for reasonable dispersion.

Under many conditions, dispersion of smoke and gases can be materially improved by increasing effective stack height or exit temperature. If an *inversion* is common in the area, however, the only solution may lie in the use of collection equipment at the stack or in redesign of the process to eliminate the source of nuisance. An inversion (Par. 7.8.3) is a layer of the atmosphere in which temperature increases with height. Vertical movement of air through such a layer is suppressed, and airborne pollutants collect beneath the layer in excessive concentrations. Meteorological analysis can often anticipate the existence of this problem in a new development or assess the magnitude of the problem in areas already faced with air-pollution problems.

7.5.9 Human Comfort

Human comfort and the effects of clothing are complex problems involving several weather factors. Much experimental work has been expended on, and numerous formulas and charts have been developed to try to solve, the problem of classifying climate with respect to comfort. One such chart is shown in Fig. 7.10, which utilizes temperature and humidity as parameters. An average wind velocity of 17 ft/sec or 11.6 mph is assumed. A joint frequency analysis of humidity and temperature would permit the climate of a given place to be classified with respect to comfort, as indicated on this chart. It is interesting to note that the range *AA* is considered by Huntington* to be the optimum range for mental work. Range *BB* is considered optimum by British standards, and *CC* is considered optimum in the United States. Climatologists generally agree that a considerable variation in weather is an important feature of a good climate. From the viewpoint of employee morale, comfort outside the plant is almost as important as working conditions in

* Ellsworth Huntington, *Civilization and Climate* (New Haven: Yale University Press, 1915), pp. 89–128.

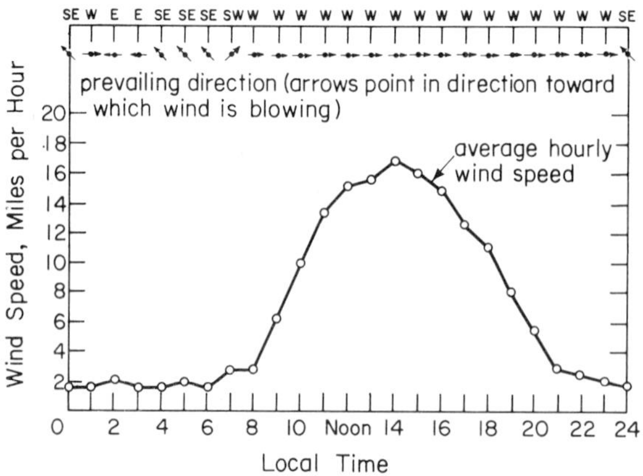

Fig. 7.9 *Average diurnal wind variation for July, Santa Maria, California.*

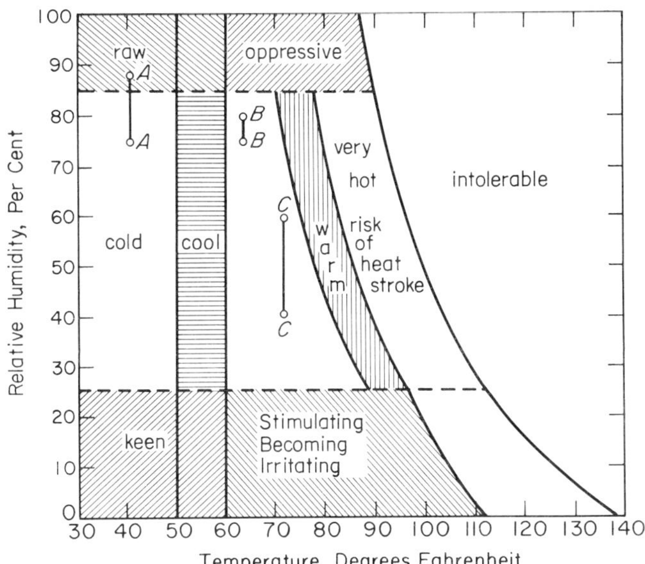

Fig. 7.10 *A tentative classification of climate as it affects human comfort. Assumed wind speed of 17 ft/sec. From D. Brunt, "The Relation of the Human Body to Its Physical Environment,"* Quarterly Journal of the Royal Meteorological Society, **69** April, 1943.

the plant. The difference between *BB* and *CC* (Fig. 7.10) suggests the differing opinions regarding comfort that result from a natural acclimatization of people exposed over a period of time to a given climate.

The *temperature-humidity index* (THI)* used by the U.S. Weather Bureau is given by the equation

$$\text{THI} = 0.4(T_d + T_w) + 15 \qquad (7.1)$$

where T_d is the dry-bulb temperature and T_w is the wet-bulb temperature. Some people begin to feel discomfort at an index of 70, about half the population is uncomfortable at an index of 75, and nearly everyone is uncomfortable at an index of 79. The constants in the equation are selected so that the THI will approximately equal the effective temperature of the American Society of Heating, Refrigerating and Air-Conditioning Engineers.†

* E. C. Thom, "The Discomfort Index," *Weatherwise*, 12 (Apr. 1959), 57–60.

† *American Society of Heating, Refrigerating and Air-Conditioning Engineers Guide and Data Book, 1961* (New York: American Society of Heating, Refrigerating and Air-Conditioning Engineers, 1961), p. 109.

7.6 SOURCES OF CLIMATOLOGICAL DATA

The main source of climatological data in the United States is the U.S. Weather Bureau, whose archives contain billions of items gathered, for the most part, since the last decade of the nineteenth century. Data are published monthly in *Climatological Data Bulletins* issued for each state. Files of these bulletins are to be found in many libraries and in Weather Bureau Offices in the principal cities (Fig. 7.11). Unpublished data may be obtained in the form of photostats, microfilm, or punched-card tabulations. Data for the years 1934–41 and 1948 to date are on punched cards. The Bureau is prepared to provide special services in the tabulation, processing, or reproduction of data at cost.

The most detailed weather records are obtained at approximately 300 Weather Bureau Offices, where hourly observations and automatic instruments provide comprehensive coverage. Volunteer observers at about 5000 additional stations make daily observations of precipitation and maximum and minimum temperatures. Other special stations bring the total to about 12,000 stations. These surface observations are supplemented by upper-air

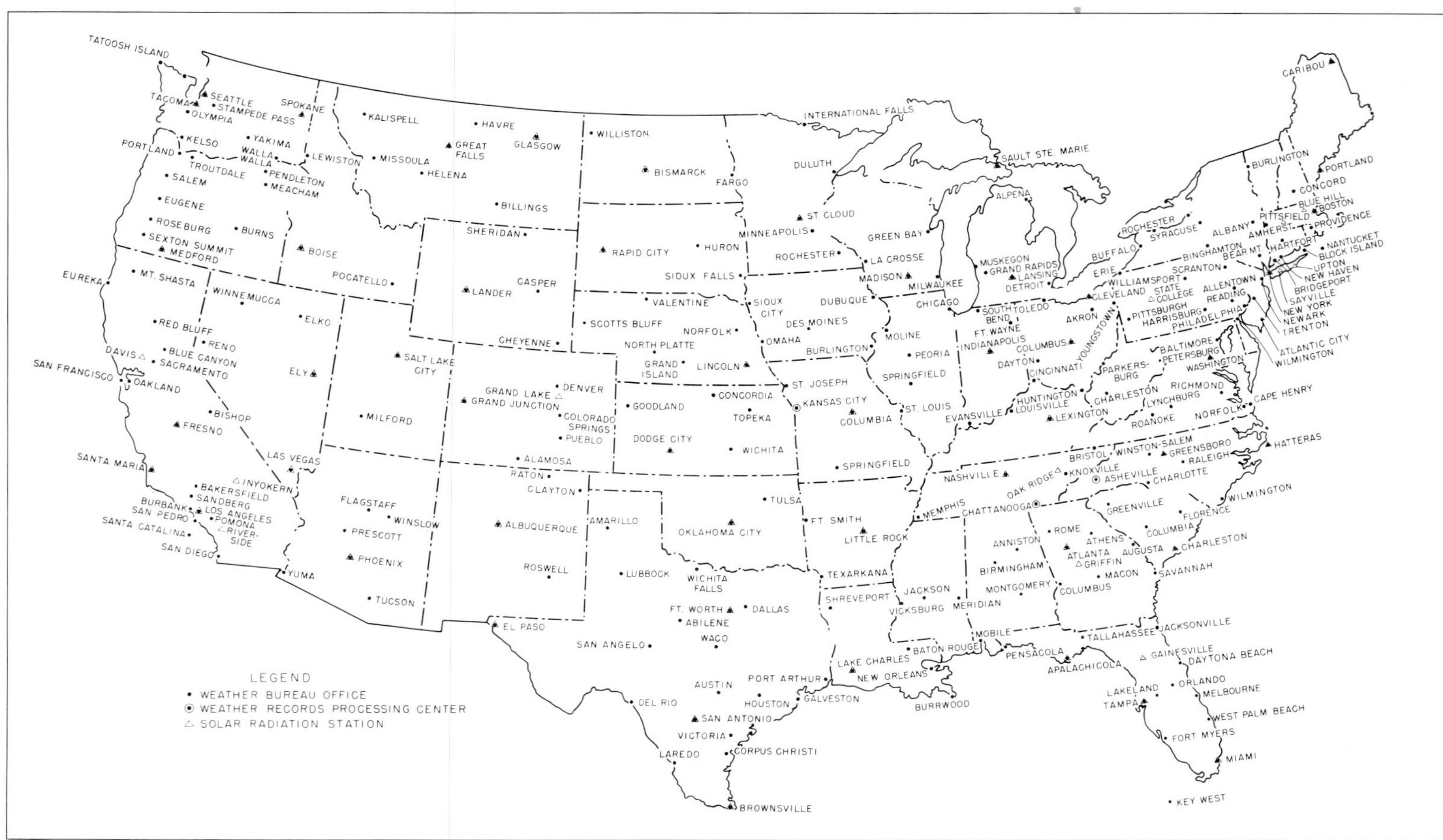

Fig. 7.11 *Weather Bureau Offices and Solar radiations stations in the United States.*

data from about 200 stations in the United States and Alaska.

An extensive summary of climatological data for the United States is given in *Climate and Man, Yearbook of Agriculture*.* A summary of monthly data for all available stations from the beginning of record to 1960 appears in the *Climatic Summary of the United States, Weather Bureau Bulletin W* and *Supplements*. Worldwide weather records are published annually in the *Réseau Mondial*. Clayton has summarized records at selected worldwide stations for the periods of record in *World Weather Records*.

7.7 CLIMATIC CHANGES

7.7.1 Fluctuations

Perhaps the most evident feature of weather is its variability. The various elements of weather at any point on the earth's surface are continuously fluctuating about a value called the normal. Since the basis of most climatological studies is the assumption that the weather of the past is a guide to the weather of the future, it is important to know whether this normal is fixed or variable.

Ahlmann† distinguishes between *climatic variations*, or trends measured in terms of geologic time, and *climatic fluctuations* observable in a period as short as the human life-span. Geological evidence clearly demonstrates the existence of climatic variations. Some parts of the world have experienced a change from near-polar climate to tropic climate over a period of some 50,000 years. This is a warming of about 0.001°F per year, an insignificant quantity in most climatological studies. Superimposed on these variations are fluctuations of shorter period and greater magnitude. Kincer‡ concludes that a definite upward trend in temperature persisted from about 1860 to 1940. At Washington, D.C., this change seems to be about 2°F, or about 0.02°F per year. Ahlmann finds somewhat larger fluctuations in Sweden. Precipitation records provide less conclusive evidence of fluctuation, and the lack of long records for other elements makes it difficult to study trends in humidity, wind, cloudiness, and so forth.

Changes in observation techniques may be responsible for at least part of the apparent trends that have been noted. It has been suggested that artificial heating in cities is responsible for apparent upward trends in temperature, although this seems to be discounted by the existence of similar trends at rural exposures. Increased height of buildings and expansion of cities increase frictional resistance to wind and lead to an apparent downward shift in wind speed in some cities. Average annual wind speed at Detroit, Michigan, decreased from 15 to 9 mph between 1909 and 1939.

It may be concluded that natural trends are too small to be of importance in industrial climatology but that changes caused by man's activities may be large enough to require special treatment. As a precaution, the data used in climatological studies should be the most recent available, and the record period used should be as short as possible consistent with requirements of an adequate statistical analysis. These limitations will minimize the effects of trends in the data, whatever their cause may be.

* See bibliography at the end of this section.

† H. W. Ahlmann, "The Present Climatic Fluctuation," *Geographical Journal*, 112 (June 1947), 165–95.

‡ J. B. Kincer, "Our Changing Climate," *American Geophysical Union Transactions*, 27 (1947), 342–47.

7.7.2 Cycles

A casual inspection of any series of meteorological data reveals fluctuations that seem to be roughly cyclical. A great deal of study has been given to determining the period and amplitude of such cycles and to explaining them through correlation with other cyclical phenomena such as sunspots and planetary movements. The results of these studies are not particularly conclusive. Fourier analysis carried to a sufficient number of terms will produce a reasonable fit for almost any series of climatic data. Serial correlation analysis suggests, however, that most climatic series are entirely random. Various investigators have identified nearly 100 "cycles" in climatic data, ranging from 1 year to 744 years in length.§ None of these cycles or combinations of cycles have proved useful in predicting future weather. In the light of present knowledge, it seems best to consider climatic data as random series.

§ Sir Napier Shaw, *Manual of Meteorology* (2nd ed.), Vol. 2 (London: Cambridge University Press, 1942), pp. 320–25.

The only clearly established cycles are the annual and diurnal ones, and even these are rarely identical. The earth-sun relationships responsible for daily fluctuations in temperature, wind, and humidity, and the annual march of almost all meteorological elements assure variations that are roughly similar from day to day and year to year.

7.7.3 Persistence

Persistence is a term applied to the tendency of a weather condition, once established, to continue for a period of time. Storms that bring rain are large air masses requiring time to move past a given point. Hence rain will persist until the system has moved from the area. Long-period persistence develops when a series of similar systems moves across an area; a given type of weather continues with minor interruptions for the time necessary for the series to pass. The circulation patterns that bring hot, dry weather to an area may be more common in some years than in others. The result is a dry year or drought. Since the reasons for the development of conditions of this type are only imperfectly understood, from the climatological viewpoint persistence is treated as a random occurrence—i.e., two rainy days in sequence will be common, three less frequent, four still less frequent, and so on (Fig. 7.8), much as would be expected for the occurrence of successive heads on the toss of a coin.

7.8 TEMPERATURE

7.8.1 Measurement

The most commonly observed temperatures are the daily maxima and minima measured about 5 ft above the ground in a white-painted, wooden shelter with louvered walls and double roof. Maximum temperature is observed by means of a mercury thermometer that functions like a clinical thermometer—i.e., the mercury rises freely but is restrained from falling by a constriction in the bore. Minimum temperature readings are obtained from an alcohol thermometer that has a small glass index in the bore. Surface tension of the alcohol draws the index down as temperature falls, but as temperature rises the alcohol flows past the index. The position of the upper end of the index indicates the minimum temperature since the last setting. Both thermometers are exposed in nearly horizontal positions. The shelter protects the thermometers from direct solar radiation so that they will record free-air temperature at a level approximately that of a person's head. Continuous records of temperature are obtained at Weather Bureau Offices from *thermographs*, in which a bimetallic or liquid-filled element actuates a pen on a clock-driven chart. Since these instruments are less accurate than the thermometers, their records are ordinarily adjusted to conform to the observed maximum and minimum temperatures.

7.8.2 Terminology

The following definitions cover the most commonly used temperature expressions and are in accord with U.S. Weather Bureau usage.*

Mean daily temperature is the average of the maximum and minimum temperatures for the day. This value is usually slightly high, but within a degree of the true average temperature for the day. The magnitude of the bias varies somewhat with the time of observation.†

Daily range is the difference between the maximum and minimum temperatures for a given day.

Mean monthly temperature is the average of the mean monthly maximum and minimum temperatures.

Absolute monthly range is the difference between the highest and lowest temperatures during the month.

Mean annual temperature is the average of the twelve monthly mean temperatures for the year.

Annual range is the difference between the mean temperatures of the warmest and coldest months of the year.

A *degree day* is a departure of one degree per day from a selected reference temperature. Most published degree-day data are based on departures below 65°F, a measure of heating required in buildings. A day with a mean daily temperature of 60°F represents five degree days below 65°F. If the daily temperature fluctuates above and below the reference value,

---

* A. H. Thiessen, *Weather Glossary* (Washington, D. C.: U. S. Weather Bureau, 1946).

† W. F. Rumbaugh, "The Effect of Time of Observation on the Mean Temperature," *Monthly Weather Review*, 62 (1934), 375–76.

degree days computed from the mean temperature will be less than those determined by summing the degree hours for the day and dividing by twenty-four.

7.8.3 Areal Variations in Temperature

Two types of areal variations in climatic elements may be noted: (1) those resulting from the conditions of exposure of instruments, and (2) point-to-point differences with uniform exposure. The standard thermometer shelter is intended to create a uniform exposure condition for the instruments. Different types of shelters (or no shelters at all) may result in differing temperatures on thermometers only a few feet apart. The temperature within a shelter is an *index* to temperatures near the shelter, but it is not intended to be a quantitative *measure* of these conditions. The temperature of a surface exposed to the direct rays of the sun will be substantially higher than the "official temperature." Temperatures in or near a building will differ from shelter temperatures, depending on the type of construction and the extent of artificial heating or cooling. Any climatological analysis of temperature must view the shelter temperature in the light of the actual temperature conditions at the point of interest. Published temperature data must either be treated as a statistical index of the temperature influence being studied or be adjusted for the differences in conditions of exposure.

Large variations in temperature may be observed when thermometers are exposed in standard shelters in different locations. Some of the macrovariations in normal annual temperature are depicted in Fig. 7.12. The most obvious variation is with latitude, as indicated by the general east-west alignment of isotherms (lines of equal temperature) in the eastern United States. Since large bodies of water exert a stabilizing influence on temperature, winter isotherms bend southward along the coasts and summer isotherms are diverted northward. The annual range of temperature is therefore lower along the coasts than in the interior. The effect of topography is evident in the greatly distorted pattern of the western states, with generally lower temperatures in regions of high elevation.

Microvariations in temperature are controlled by much the same factors as are the large scale variations. Large lakes modify the temperature regime of the immediate area.* Vegetation has a stabilizing effect, and maximum temperature, daily range, and annual range are usually lower at stations in a forest than at stations in open country. The artificial heat of cities and the shielding effect of haze and smoke result in higher temperatures than those in the immediately surrounding country. The difference is often as much as 2°F, resulting mainly from higher minimum temperatures in the cities.

Probably the most important factor controlling temperature is topography. If dry air is forced to rise, it expands because of lower pressure and therefore cools. Assuming an adiabatic process (no heat exchange between a rising air particle and its environment), free-air temperature decreases about 0.5°F per 100 ft of elevation. Air lifted above its condensation level becomes saturated and cools at about 0.3°F per 100 ft because of the latent heat of condensation released by the moisture. If the cooling and condensation of moisture in the air result in precipitation, the thermal process is not reversible, and when the air descends on the leeward slope of the mountain it is warmed at the dry adiabatic rate. Level for level, warmer temperatures are experienced on the leeward slopes of mountains than on the windward slopes. The *chinook*, or *foehn*, wind—a warm, dry, downslope wind—is the result of this process.

The average rate of temperature decrease with height assumed in the U.S. Standard Atmosphere (Table 7.2) is about 0.36°F per 100 ft near the ground. In general, we may expect a temperature decrease between 0.3° and 0.4°F per 100-ft increase in altitude. This condition will exist only when the wind is strong enough to keep the air well mixed. Under still-air conditions, radiative cooling of the ground results in a rapid lowering of the ground surface temperature at night, and an *inversion*, or increase of temperature with height, may develop. Under these conditions, frost may occur at the ground surface with above-freezing temperatures in the instrument shelter. An inversion may also form aloft,

---

* J. Leighly, "Effects of the Great Lakes on the Annual March of Temperature in Their Vicinity," *Papers of the Michigan Academy of Arts, Science, and Letters*, 27 (1941), 337–414.

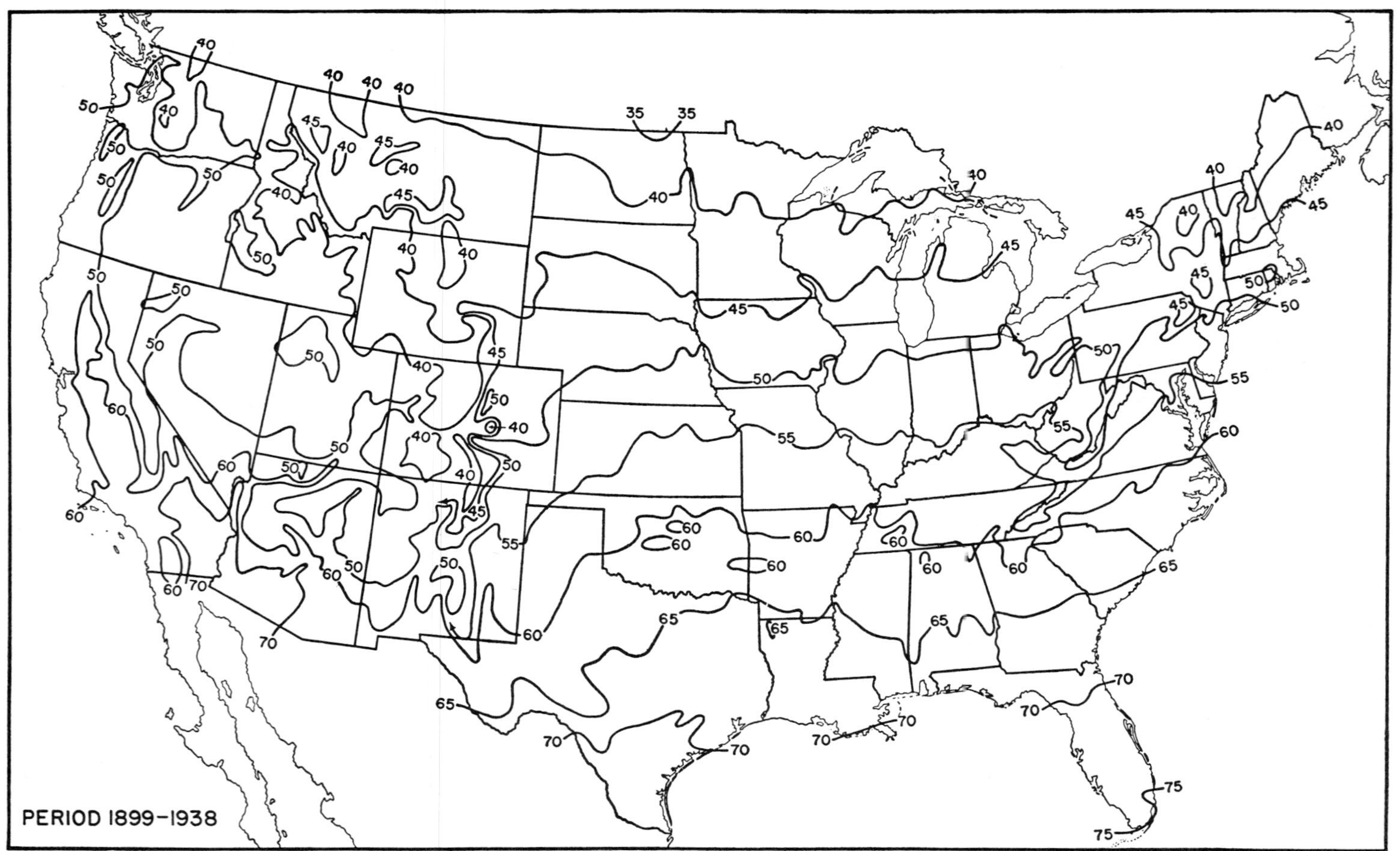

Fig. 7.12 *Normal annual temperature (degrees Fahrenheit) in the United States. Courtesy U. S. Weather Bureau.*

TABLE 7.2. *VARIATION OF PRESSURE, TEMPERATURE, AND BOILING POINT WITH ELEVATION**

Elevation (Feet Above Mean Sea Level)	Pressure		Air Temperature, Degrees Fahrenheit	Boiling Point, Degrees Fahrenheit
	Inches of Mercury	Millibars†		
−1000	31.02	1050.5	62.6	213.8
0	29.92	1013.2	59.0	212.0
1000	28.86	977.3	55.4	210.2
2000	27.82	942.1	51.8	208.4
3000	26.81	907.9	48.4	206.5
4000	25.84	875.0	44.8	204.7
5000	24.89	842.9	41.2	202.9
6000	23.98	812.1	37.6	201.1
7000	23.09	781.9	34.0	199.2
8000	22.22	752.5	30.6	197.4
9000	21.38	724.0	27.0	195.6
10,000	20.58	696.9	23.4	193.7
11,000	19.79	670.2	19.8	191.9
12,000	19.03	644.4	16.2	190.1
13,000	18.29	619.4	12.6	188.2
14,000	17.57	595.0	9.1	186.4

* Data represent average free-air conditions.
† The meteorological unit of pressure equivalent to 1000 dynes/cm^2

creating a thermal stratification beneath which smoke and dust tend to collect. Without strong winds, the cooling of the surface air may lead to *air drainage*, the flow of the cold air downslope. As a result, valley temperatures may be lower than those on the adjacent hills.

7.8.4 Adjustment of Temperature Records

The factors described in the preceding paragraphs mean that moving a temperature station may easily result in changes in observed temperatures. One method for detecting and correcting such changes is to plot the temperature data from the station in question against the average temperature at several nearby stations which are known to be unchanged (Fig. 7.13). A change in the conditions of observation will ordinarily be indicated by a shift in the relation. The adjustment may be different for maximum and minimum temperatures and may also vary during the year. A relation of this type may be used to estimate missing records or to extend the station record to the years prior to the actual beginning of observations. The importance of such an adjustment depends on the problem under study, but its magnitude should certainly be investigated for all stations known to have been moved. Since records of station moves during the early years of the Weather Bureau are rather inadequate, a test for all stations is preferable. Figure 7.14 indicates the magnitude and frequency of temperature differences that may be expected between two stations a short distance apart. The Sleepy Hollow station is about 6 miles SSW and about 350 ft higher than the Washington, D.C., city office. Note the difference between maximum and minimum temperatures and the seasonal variations.

7.9 PRECIPITATION

7.9.1 Measurement

Precipitation amounts are expressed in terms of the depth of liquid water that would accumulate on a horizontal surface. The official gage of the U.S. Weather Bureau consists of a funnel 8 in. in diameter which discharges into a measuring tube 2.53 in. in diameter. An outer container 8 in. in diameter serves as an overflow can. The ratio of the areas of the funnel and of the measuring tube is 10:1 so that the depth in the tube can be measured to the nearest 0.01 in. of precipitation on the 8-in. funnel with a scale graduated to 0.1 in. Snow is measured by removing the funnel and catching the precipitation in the overflow can. The snow is melted and poured into the measuring tube so that its water equivalent can be measured. Numerous gages of smaller diameter are in use. They are usually as accurate as the larger gage for meas-

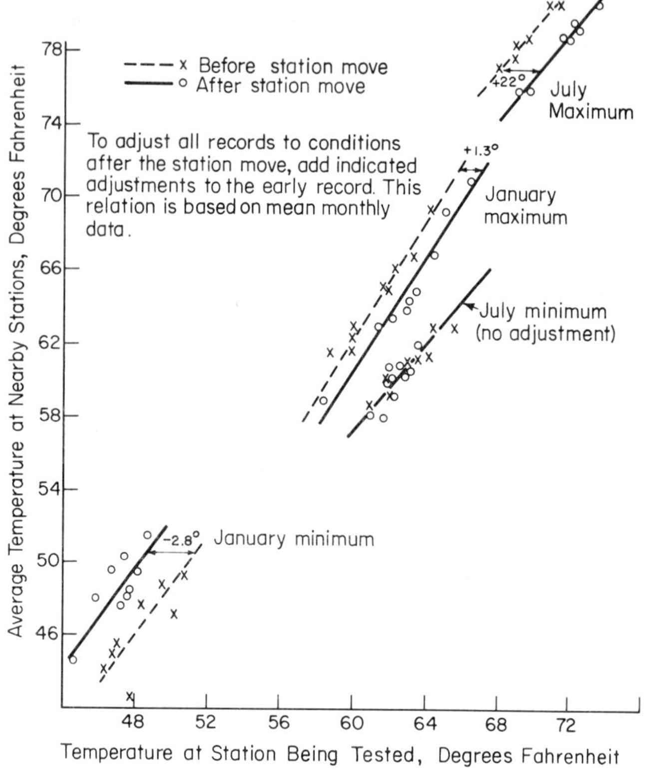

Fig. 7.13 *Relation for adjusting temperature records for effect of a station move.*

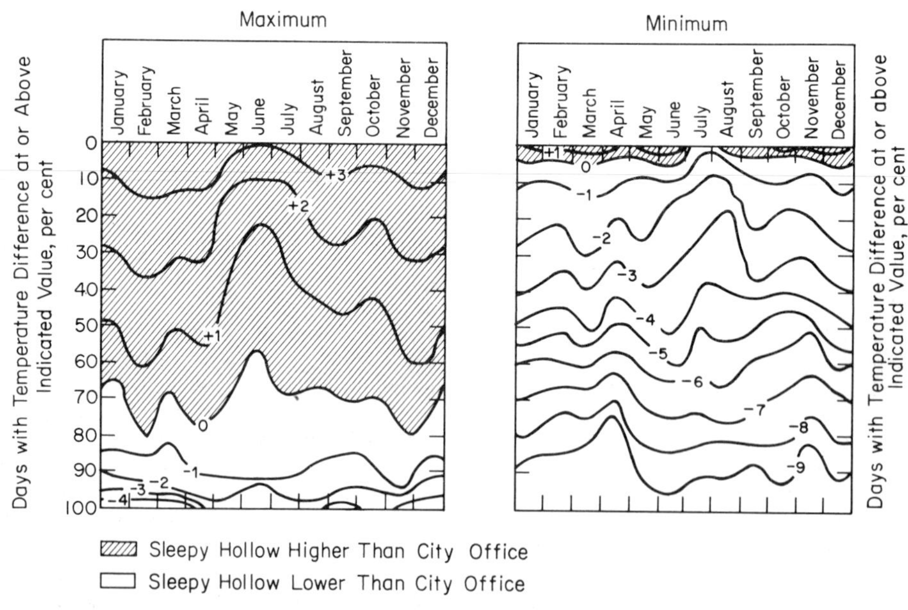

Fig. 7.14 *Frequency of daily maximum and minimum temperature differences between Washington, D.C. city office and Sleepy Hollow, Virginia, 1942–44. Courtesy U. S. Weather Bureau.*

uring rainfall but are unsatisfactory for measuring snowfall.

The most common recording gage used to obtain data on short-period rainfall intensity is the weighing gage. A collecting bucket is mounted on a platform supported by a spring or lever scale, and the increase in weight is recorded on a chart. The tipping-bucket gage consists of a pair of small buckets mounted under a funnel in such a way that, when one receives 0.01 in. of precipitation, it tips and discharges its contents and at the same time brings the other bucket under the funnel. Each tip of the buckets is recorded on a chart. The tipping-bucket records are usually more satisfactory for determining rainfall rates for periods of less than an hour, but the gage is not very suitable for measuring snowfall.

The exposure of the gage has more effect on the accuracy of rainfall records than does the type of gage used. The gage should be far enough from trees or buildings so that it is not sheltered from rainfall. On the other hand, strong winds tend to deflect a portion of the precipitation from the gage and result in catches that are too low. Some protection by low bushes, trees, or buildings about as far from the gage as their own height is desirable. No quantitative basis is available for suggesting the errors resulting from differing exposures, but in studies in which quantity of rain is important, records from severely windswept sites should be viewed with some suspicion unless a special windshield has been used on the gage.

7.9.2 Snow

Snow on the ground is measured in inches of depth with any convenient scale. To allow for variations resulting from local melting or drifting, the reported depth should be an average of several readings in the vicinity of the station. All cooperative observers of the Weather Bureau are asked to record snow depth, but these data are not published and must be obtained from the Bureau files.

7.9.3 Cause of Precipitation

Macrovariations of precipitation (Fig. 7.15) depend on nearness to a moisture source, topography, and latitude. Since the moisture content of the atmosphere is a function of temperature, there is a greater amount of water available for precipitation in the lower latitudes. With the oceans as the source of moisture for precipitation, it is also natural to find greater amounts of precipitable water along the coastlines than in the interior of continents. Evaporation from lakes and land surfaces and transpiration of moisture by plants are not major sources of atmospheric moisture.* And man-made lakes also do not materially affect the precipitation regime of an area. Moisture alone is not sufficient to bring heavy precipitation to an area. A mechanism capable of converting atmospheric water vapor to liquid water is essential. This is accomplished by cooling the air. The only means of cooling large air masses sufficiently to produce heavy precipitation is by lifting, which causes cooling by expansion (Par. 7.8.3). The three basic lifting processes are frontal, convective, and orographic.

A *front* is the boundary between two air masses of differing characteristics. A warm front is one in which the air behind the front is warmer than that preceding the front. A cold front is the reverse—i.e., the colder air follows the front. The warm, moist air following a warm front rises over the colder air in advance of the front. This lifting results in condensation and perhaps precipitation. Warm-front precipitation is usually gentle and continuous for a fairly long period. A cold front pushes under the warmer air in advance, lifting is more rapid, and precipitation is more intense but of shorter duration. Many cold fronts are accompanied by thunderstorms.

Convectional lifting takes place when an air mass is warmed from below and tends to rise through cooler air surrounding it. The thunderstorms of the eastern United States are convectional storms and result in very intense but localized precipitation. Convectional thunderstorms may bring rain to areas as small as 4 or 5 square miles while the surrounding country receives little or no rainfall.

Mountains form *orographic barriers* over which air masses must rise. If the air contains sufficient moisture, the mountains become regions of high precipitation. Because mountains are fixed barriers causing the rain to occur at the same place each time, rainfall is usually higher than in regions of frontal activity, where the moving front distributes the precipitation over large areas. The more rapidly air is lifted, the greater the amount of moisture condensed.

---

* G. S. Benton, R. T. Blackburn, and V. O. Snead, "The Role of the Atmosphere in the Hydrologic Cycle," *American Geophysical Union Transactions*, 31 (Feb. 1950), 61–73.

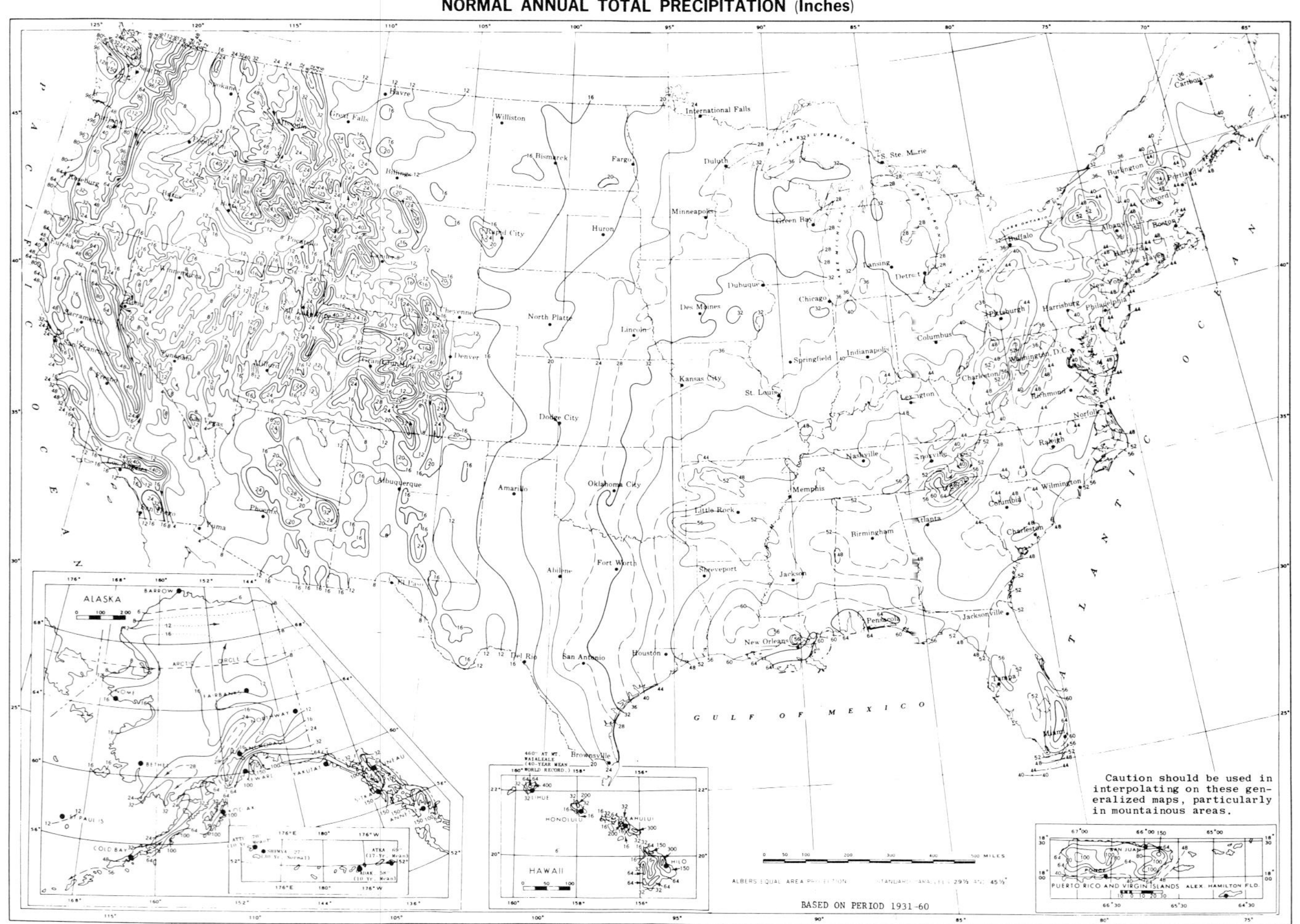

Fig. 7.15 *Normal annual precipitation (inches) in the United States. Courtesy U. S. Environmental Data Service.*

Therefore, the steeper the mountain slopes, the greater the precipitation. Orographic precipitation is distinguished by the regularity of the precipitation pattern. Areas of high and low precipitation remain reasonably stable from storm to storm. Because air is descending on the lee slopes of a mountain, a *rain shadow* or region of low precipitation is usually present on the slopes away from the prevailing wind. Rain drops do not fall vertically from their point of formation; rather, they describe a trajectory to leeward that is determined by wind speed and drop size. Hence leeward slopes immediately beyond a mountain crest may receive fairly high precipitation as a result of carry-over of rain formed during the ascent on the windward slope.

7.9.4 Areal Variations in Precipitation

Areal variations in precipitation are somewhat more difficult to rationalize than those in temperature. In regions of little relief, areal variations from storm to storm may be quite large, but, in the average over long periods of time, the differences from point to point are small because of the random location of the storm centers. In such regions one can interpolate linearly between rainfall stations to estimate normal monthly or annual rainfall at ungaged points. If there is substantial topographic relief, the situation is quite different. Large variations in precipitation may be observed over very short distances and linear interpolation may be quite inaccurate. Some success has been reported* in use of multiple correlations between precipitation and topographic factors as a basis for estimating normal annual rainfall in mountain regions. Vegetal types, erosion conditions, and other factors affected by rainfall offer further clues to the rainfall pattern in mountains.

7.9.5 Adjustment of Precipitation Records

Few precipitation stations have remained in exactly the same location throughout their period of record. In some instances, a station move has involved a large change in exposure conditions without a corresponding change in

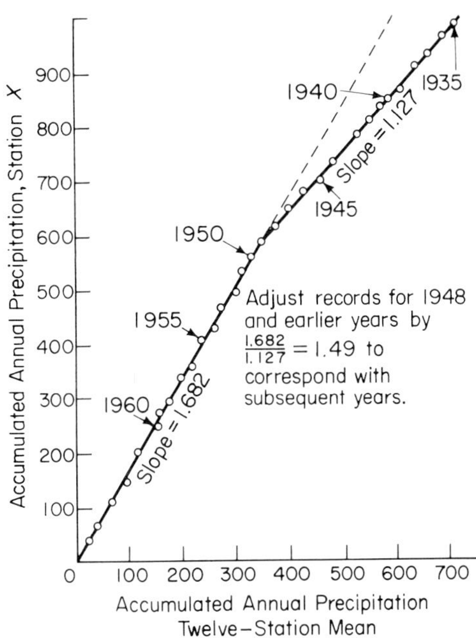

Fig. 7.16 *Double-mass curves for adjustment of precipitation record.*

name. Such records may be adjusted by use of the *double-mass curve.*† A double-mass curve is constructed by plotting the mass accumulation of precipitation at the station to be tested against a comparable mass accumulation for the same years at a group of control stations in the immediate area (Fig. 7.16). A change in the rainfall regime of the station is indicated by a change in slope of the double-mass curve. The station records may be adjusted by multiplying the older records by the ratio of the slopes of the double-mass curve before and after the change in slope. The method is somewhat subjective, and caution is necessary to avoid interpreting random variations about the double-mass curve line as critical changes in slope. Analysis of variance may be used to indicate the probable significance of an apparent change in slope if no record of an actual station move exists. Missing records at a station may be estimated by multiplying records from adjacent stations by the ratio of slopes of the respective mass

* W. C. Spreen, "A Determination of the Effect of Topography on Precipitation," *American Geophysical Union Transactions*, 28 (Apr. 1947), 285–90.

† M. A. Kohler, "Double-Mass Analysis for Testing the Consistency of Records and for Making Required Adjustments," *Bulletin of the American Meteorological Society*, 30 (1949), 188–89.

curves. The double-mass curve assumes the existence of a relation between precipitation amounts at nearby stations. Although this assumption is valid for annual and monthly precipitation, it may not be for daily records.

7.10 SUNSHINE AND CLOUDINESS

7.10.1 Observations

Three types of observations provide data on sunshine, solar radiation, and sky cover. The most common observations are those of state of the sky. All cooperative observers are asked to record the condition of the sky at observation time as clear, partly cloudy, or cloudy. These are subjective observations and are of limited value because they represent conditions only at one instant during the day. Regular offices of the Weather Bureau observe cloud cover hourly and express the observation in tenths of the sky covered by clouds. The observations are made by trained observers and are more reliable than those made by cooperative observers. The average of the hourly observations is the daily cloudiness. The rules governing these observations are such that a complete overcast of high cirrus clouds is reported as 10/10 cloudy, although much radiation reaches the ground. Quantitative estimates of radiation reaching the ground must be based on cloud type as well as cloudiness. Haurwitz* has developed an equation expressing insolation as a function of cloud type and density.

Sunshine-duration indicators in use at many stations consist of a mercury switch so arranged that incident radiation causes the mercury to expand and close a circuit; the circuit is also closed at 1-min intervals by another switch controlled by a time clock. If both switches are closed simultaneously, an indication is made on a chart. The time of sunshine and its total duration to the nearest minute are obtainable from the chart. The threshold radiation that will activate the indicator depends upon its adjustment. Records from two offices may not be exactly comparable because of differing threshold settings.

Solar radiation is measured with *pyrheliometers* at about fifty stations in the United States (Fig. 7.11). The Epply pyrheliometer is most common. This device consists of a series of concentric rings painted alternately black and white and exposed in a glass globe. The current developed by the temperature difference between the black and white rings is a measure of the incident radiation. The usual chart record shows instantaneous values of incident radiation and must be integrated to determine daily totals. The customary unit is the *langley* (*ly*), which is 1 g-cal/cm^2. One langley is equivalent to 3.69 Btu/ft^2.

7.10.2 Variations

Generally speaking, geographical variations of sunshine (or cloudiness) are not large. Except under special conditions, one can usually transpose records of sunshine or cloudiness over fairly large areas. Along coastlines, where the persistence of fog varies widely, the only safe basis for estimates of this condition is local observations. Similarly, local smoke and dust (smog) may reduce the sunshine received in cities. Figure 7.17 shows maps of average daily hours of sunshine in the United States during the winter and summer. Marked seasonal differences are evident. The amount of radiation reaching the ground depends both on the duration of sunshine and on the extent of interception of radiation by the atmosphere. Even 5000 ft of clear air will reduce the radiation received at the ground by as much as 10 per cent.† Because of the very few solar-radiation stations in the United States, the data for the map of Fig. 7.18 were obtained for the most part by computing radiation Q from

$$Q = Q_0(0.35 + 0.61S) \qquad (7.2)$$

where Q_0 is the cloudless-day radiation and S is the percentage of possible hours of sunshine.‡ The values on the map indicate radiation received on a horizontal plane at the earth's surface.

* B. Haurwitz, "Insolation in Relation to Cloudiness and Cloud Density," *Journal of Meteorology*, 2 (Sept. 1945), 154–66; 3 (Dec. 1946), 123–24; 5 (June 1948), 110–13.

† I. F. Hand, J. H. Conover, and W. A. Boland, "Simultaneous Pyrheliometric Measurements at Different Heights on Mt. Washington, N. H.," *Monthly Weather Review*, 71 (May 1943), 65–69.

‡ S. Fritz and T. H. MacDonald, "Average Solar Radiation in the United States," *Heating and Ventilating*, 46 (July 1949), 61–64.

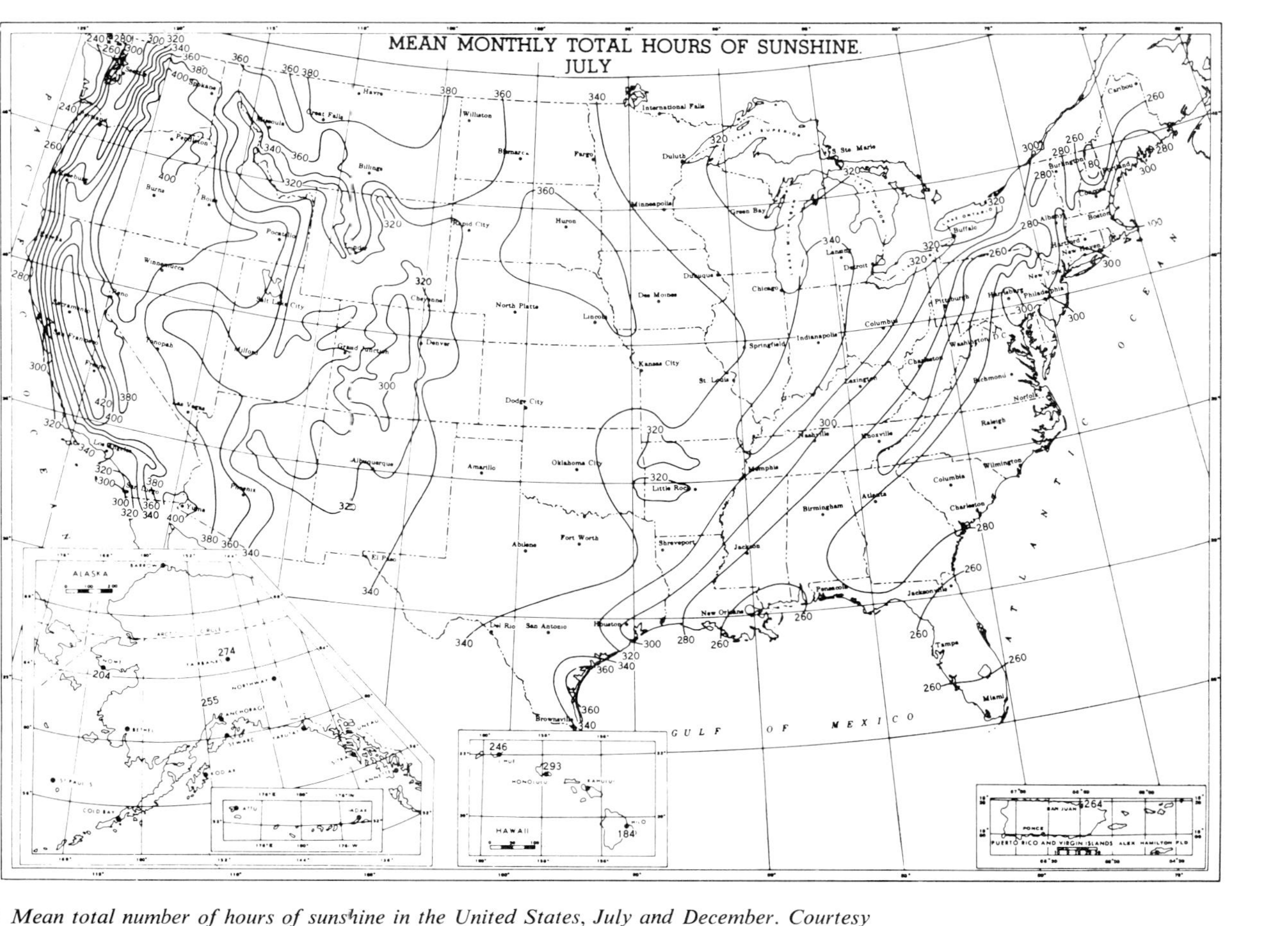

Fig. 7.17 (a) *Mean total number of hours of sunshine in the United States, July and December. Courtesy U. S. Environmental Data Service.*

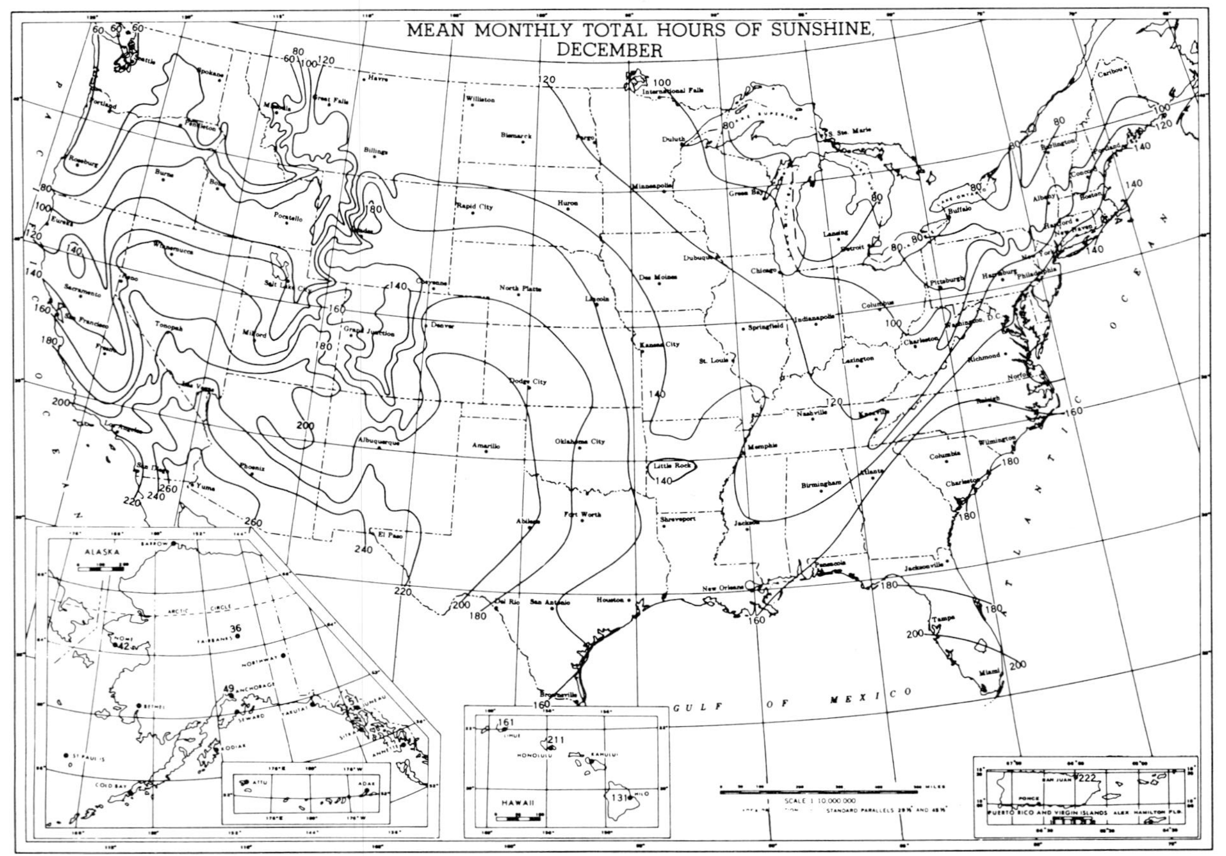

Fig. 7.17 (b) *(Continued)*

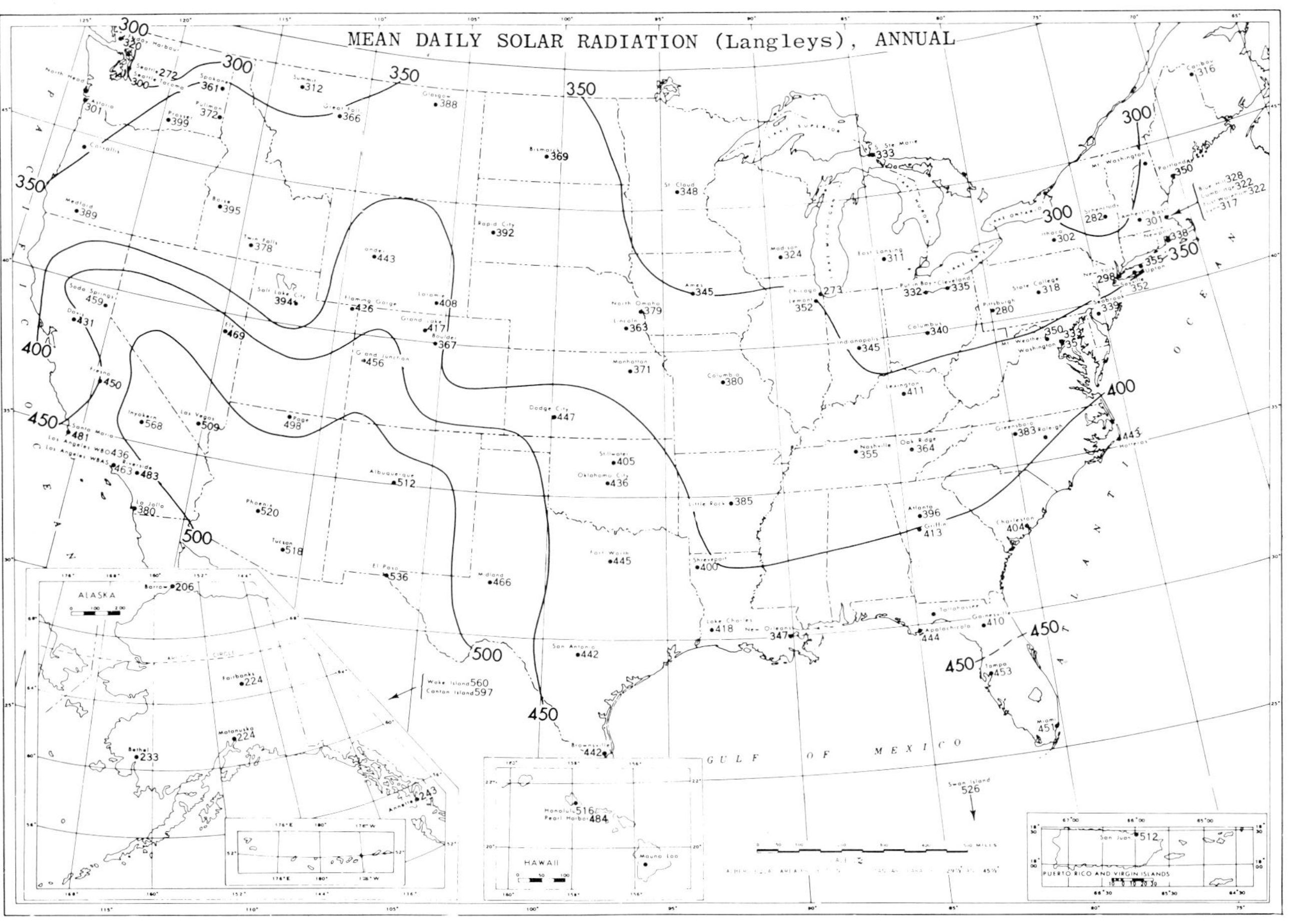

Fig. 7.18 *Average daily solar radiation (Langleys) in the United States. Courtesy U. S. Environmental Data Service.*

7.11 WIND SPEED AND DIRECTION

7.11.1 Observations of Wind Speed

Wind-speed records are obtained at almost all Weather Bureau Offices from instruments known as *anemometers*. The most common type of anemometer consists of three cups rotating about a vertical axis. Contacts are provided for each sixtieth mile of wind. By counting the number of contacts per minute, the average velocity during the minute (in miles per hour) is obtained. At many airports the only record is that of the hourly 1-min average velocity. The same type of anemometer can be geared to contact after each mile and can be connected to a recorder that indicates the time of each contact. The speed between contacts is determined by measuring the time interval.

The propeller anemometer consists of a two- or three-bladed propeller which rotates about a horizontal axis. A tail vane keeps the rotor pointing into the wind. An armature is rotated inside a coil by the propeller, and the current generated varies with the speed. A dial microammeter shows the instantaneous wind speed, or a recording ammeter can provide a continuous trace. This type of anemometer makes possible a better estimate of gust velocities, but the inertia of the assembly considerably dampens wind-speed fluctuations of very short periods.

A pressure-tube anemometer consists of a Pitot tube mounted with a tail vane to keep it pointed into the wind. Pressure variations are transmitted to a recording device, often a pen floating in a mercury manometer. The pressure-tube anemometer gives the most detailed record of wind-speed fluctuations if the manometer is sufficiently sensitive.

7.11.2 Observations of Wind Direction

Wind direction is observed with a wind vane. Most airport stations read the direction at hourly intervals from a direct-reading dial positioned by a selsyn transmitter at the vane. Some offices record direction each minute to eight compass points by means of an arm attached to the vane which contacts one of four directional segments. A clock-driven switch is closed at 1-min intervals, causing one of four pens to mark on a sheet depending on the momentary wind direction. Intermediate directions are recorded when the contact arm bridges between two segments and causes two pens to mark simultaneously. Hourly *prevailing-wind direction* is determined from the chart as the direction with the largest number of contacts during the hour. Daily prevailing direction is that direction with the largest number of hours prevailing. Note that prevailing direction is only the direction that occurred most frequently and may actually have occurred as little as 20 per cent of the time.

7.11.3 Variations in Wind

Wind is probably the most variable element of climate in terms of point-to-point differences. Local surface winds are greatly influenced by topography, vegetation, and structures. Obstruction by buildings usually results in lower wind velocities in cities than in the surrounding country (Par. 7.7.1). Valleys receive the same sort of shelter from surrounding hills. In open terrain with little relief, wind data are usually transposable from the anemometer location to a wide surrounding area. In mountains, wind velocity and direction may vary widely over very short distances. Local observations are the only sure check on such variations.

Since wind is fluid (air) in motion, its speed is greatly affected by friction with the earth's surface. The effect of friction is evident up to altitudes of about 2000 ft above the ground but is most marked immediately above the surface. Height of anemometer is therefore an important factor in interpreting wind data. The relation between wind speeds at different levels in the friction layer is given approximately by the equation

$$\frac{v_w}{v_{wo}} = \left(\frac{Z}{Z_0}\right)^k \tag{7.3}$$

where v_w is the wind speed and Z is height. The subscript 0 indicates the measured wind speed and anemometer elevation. The exponent k varies from $\frac{1}{2}$ for winds under 8 mph to $^1/_7$ for winds over 35 mph, with an average of about $^1/_5$.

7.11.4 Hurricanes and Tornadoes

Hurricanes reach maximum intensity over the oceans and dissipate rapidly as they move inland. Coastal wind data in hurricane areas should not be transposed more than a few miles inland. Tornadoes encompass such a small

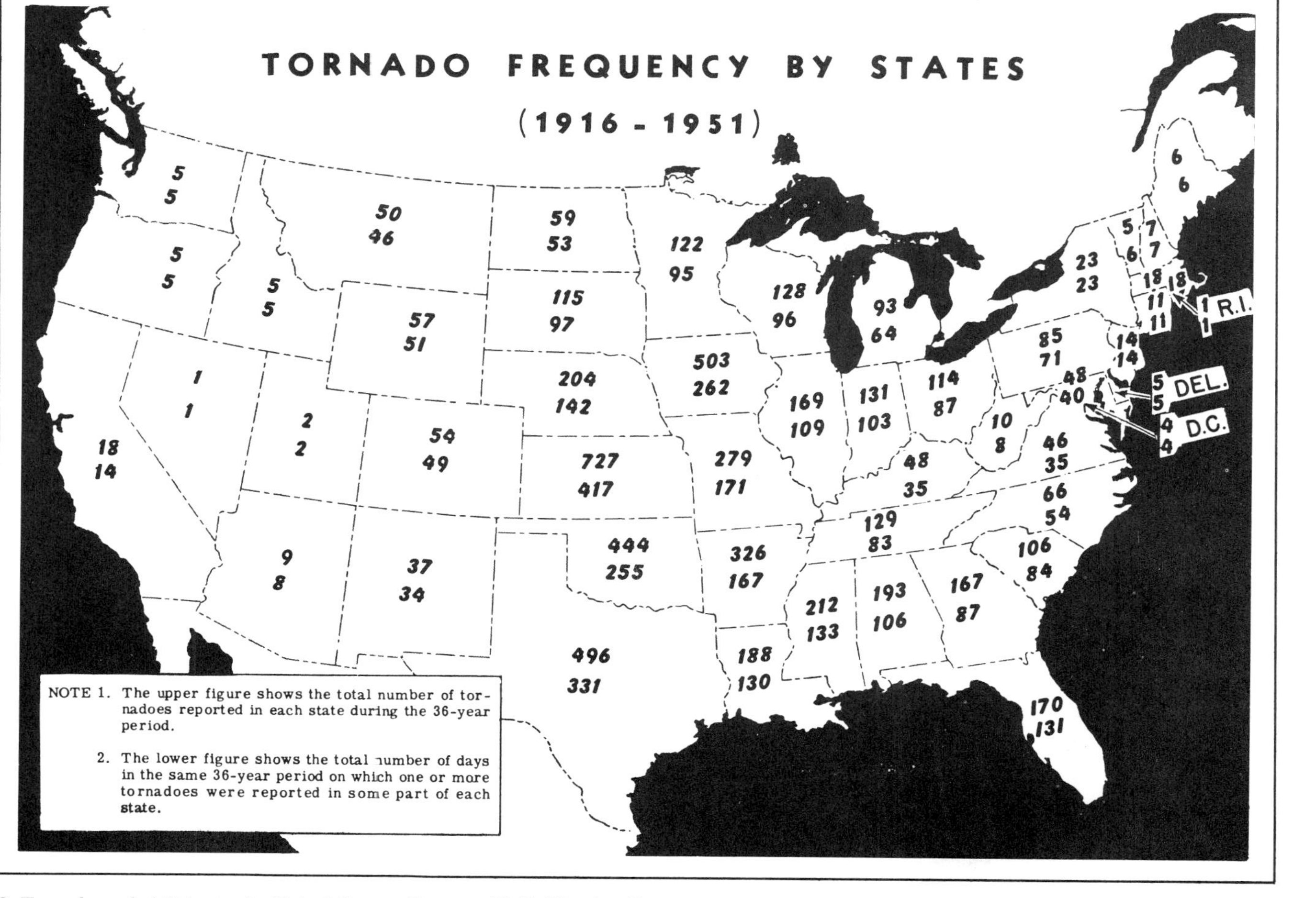

Fig. 7.19 *Tornado probabilities in the United States. Courtesy U. S. Weather Bureau.*

area and are so violent that no measurements of velocity are available. From the climatological viewpoint, tornadoes are treated as separate from the regular wind regime. Figure 7.19 shows tornado probabilities by states for the United States.

7.12 HUMIDITY

7.12.1 Observations

Since the moisture content of the air must be measured indirectly, it is one of the most difficult elements to measure. Most observations are made with *psychrometers*, consisting of a pair of thermometers, one of which has its bulb covered with a muslin wick moistened with water. When these thermometers are ventilated by whirling or by use of a fan, evaporation from the moistened wick depresses the wet-bulb temperature. An empirical correlation between the dry-bulb temperature and wet-bulb depression permits calculation of dew point or relative humidity.* Thermistors or thermocouples are sometimes used as elements in continuously recording psychrometers.

A more common humidity-recording device is the *hygrograph*, in which the variations in length of a strand of human hair actuate a pen which records relative humidity. Records from hygrographs may often be substantially in error.† The hairs are affected by temperature and show a lag that increases with decreasing temperature, becoming almost infinite at –40° F. Frequent exposure of the hair element to a wide range in humidity is necessary for good results. Continued exposure in a very dry or wet atmosphere may cause a drift in the instrument calibration. Records from hair hygrographs should be adjusted to conform to psychrometer observations near the daily maximum and minimum humidities.

7.12.2 Units of Humidity

Vapor pressure is the partial pressure exerted by water vapor in the air. It is most commonly expressed in millibars (1000 dynes/cm^2) and is computed from an empirical equation using wet- and dry-bulb temperatures and air pressure. The *dew-point temperature* is the temperature at which a given parcel of air would become saturated with water when cooled at constant pressure without change in moisture content. *Absolute humidity* is an expression of the mass of water vapor in a given space. It may be computed from the equation

$$a_h = 217(e/T_a) \tag{7.4}$$

where a_h is absolute humidity in grams per cubic meter, e is the vapor pressure, and T_a is the absolute temperature in degrees Kelvin ($0 = -273°$C).

Relative humidity is the percentage ratio of the moisture in a given space to that which the space could hold if saturated. It is given by the equation

$$f = 100(e/e_s) \tag{7.5}$$

where f is relative humidity in per cent, e is vapor pressure, and e_s is saturation vapor pressure. Relative humidity is widely used because it reflects the effect of humidity on human comfort, but it is an unsatisfactory measure for many purposes because it is a function of temperature. With a constant moisture content, relative humidity rises as air temperature falls, and vice versa. The other expressions for humidity remain constant unless there is an actual change in the moisture content of the air. Relative humidity data show a marked diurnal variation, but the other measures of humidity vary only slightly throughout the day.

7.12.3 Variations in Humidity

In contrast to the elements discussed in the previous sections, humidity is a relatively stable element. Dew-point temperature, vapor pressure, and absolute humidity are usually about the same over a fairly large area and their normal values are similar for large areas. Since relative humidity is a function of temperature, it is more variable because of the temperature differences imposed on the humidity differences.

In general, humidity is a function of distance from a source of moisture; it is maximum along coastlines and minimum in the interior of continents (Fig. 7.20). Since the maximum quantity of vapor the air can hold is a function of temperature, absolute humidity is greatest in the

---

* U. S. Weather Bureau, *Psychrometric Tables* (Washington, D. C., 1941).

† M. F. Mueller, "Characteristics of Hair-Element Humidity Instruments," *Instrumentation*, 22 (Sept. 1949), 798–99; and N. Sissenwine, "On Inaccuracies of the Hair Hygrograph Operated in Closed Tents," *Bulletin of the American Meteorological Society*, 28 (Apr. 1947), 192–96.

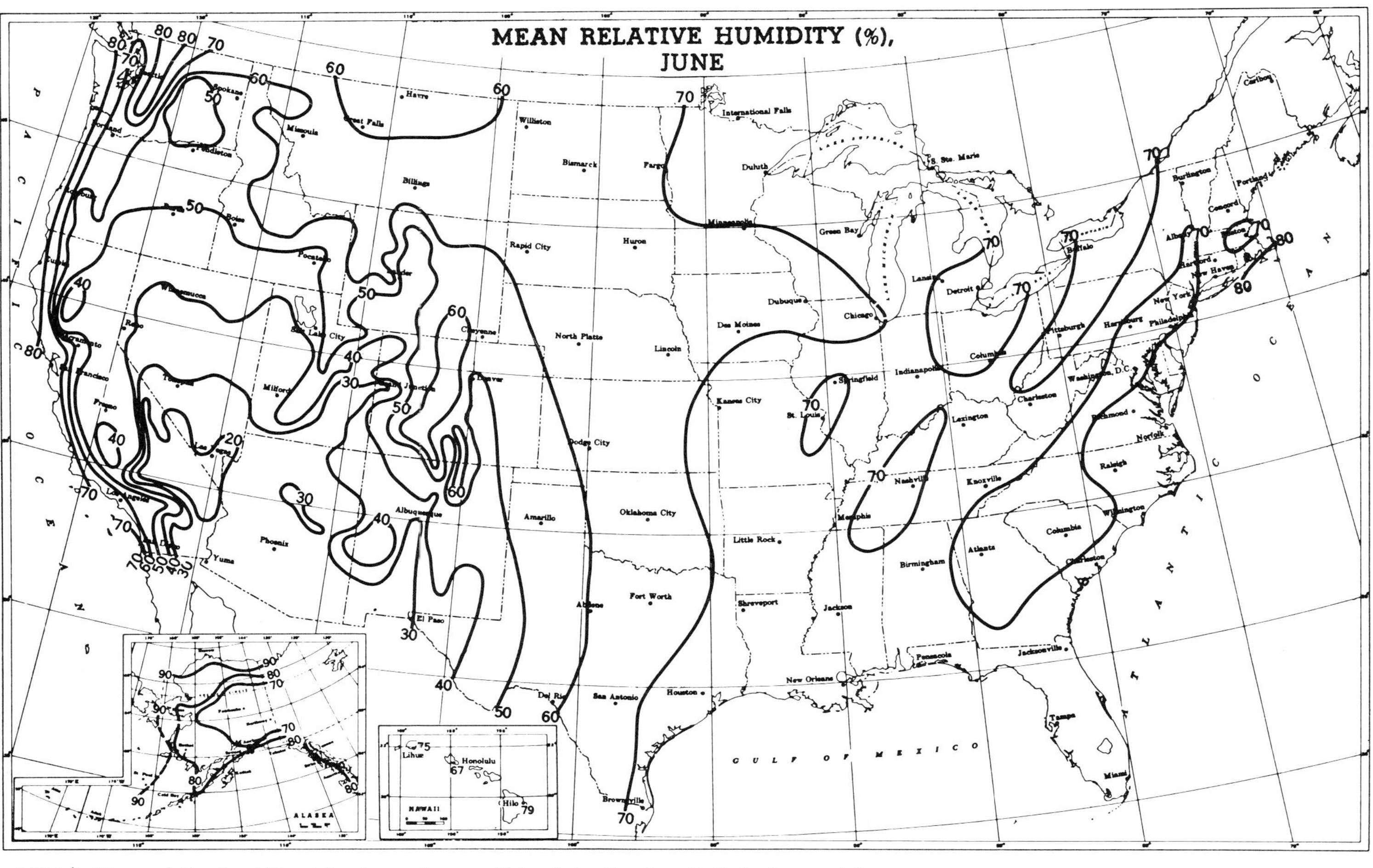

Fig. 7.20 (a) *Mean relative humidity at local noon (per cent) for June. Courtesy U. S. Environmental Data Service.*

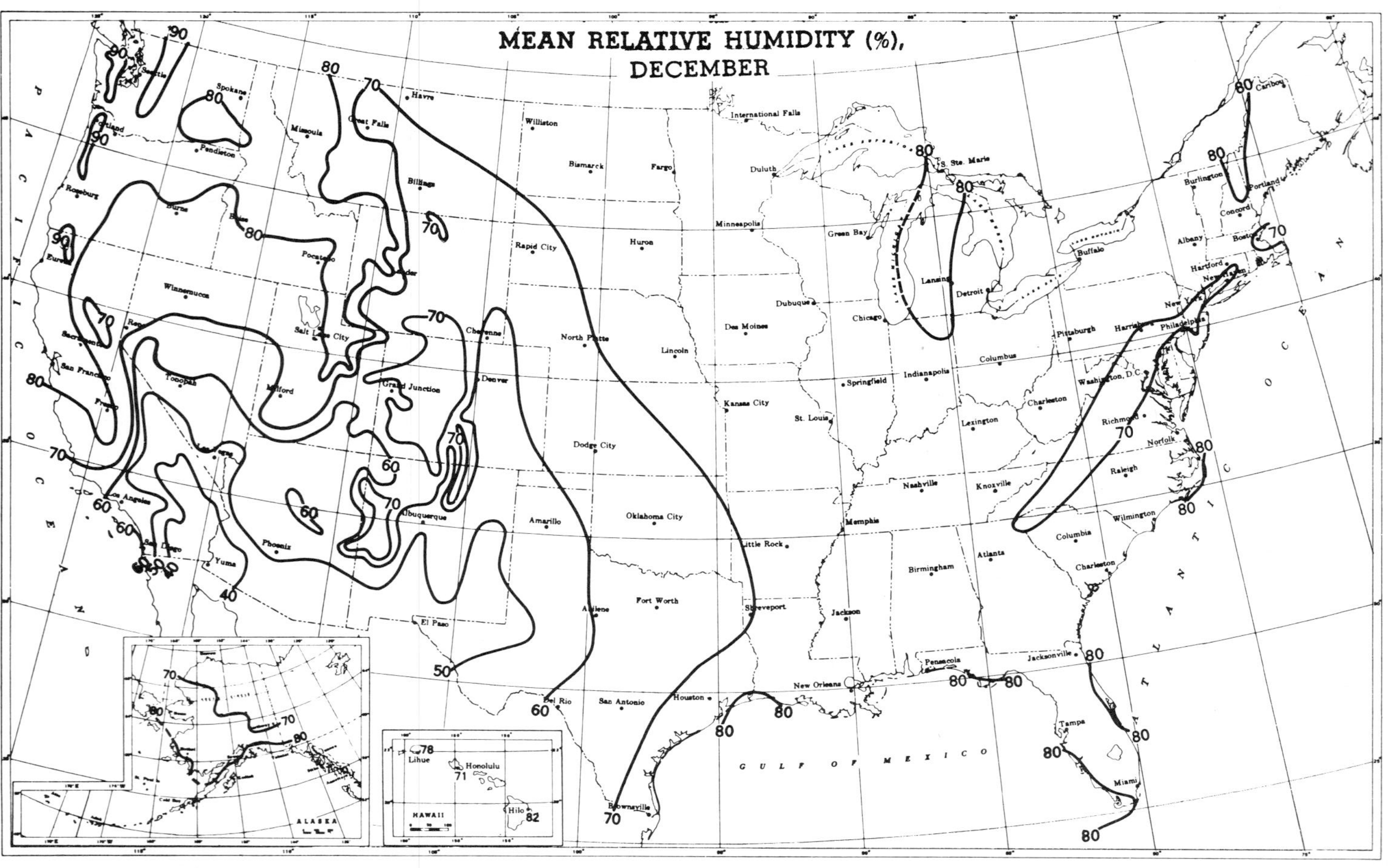

Fig. 7.20. (b) *Mean relative humidity at local noon (per cent) for December. Courtesy U. S. Environmental Data Service.*

low latitudes. Dew-point temperature cannot exceed the temperature of the source of atmospheric moisture. Hence coastal dew points approximate the temperature of the adjacent ocean waters. The maps of Fig. 7.20 show relative humidity at noon (local time). Since this is near the daily temperature maximum, relative humidity is near the diurnal minimum. The two maps also show a marked seasonal variation in humidity in the interior of the country but relatively little seasonal difference along the coast.

7.13 STATISTICAL METHODS IN CLIMATOLOGY

7.13.1 Frequency Distributions

Climatological analysis depends heavily on basic statistical methods. The climatologist's special role in an analysis lies in his understanding of the limitations of the data and the interrelations that may be expected. Methods of statistical analysis in climatology are discussed at some length by Conrad and Pollak and by Brooks and Carruthers.* Since climatology is concerned mainly with probability, the character of the frequency distributions encountered becomes important. With a large number of data, the events with moderately high probability can be predicted quite accurately from the raw data. If the data are limited in number or if interest centers on events of low probability, the results may be considerably enhanced by using a suitable theoretical distribution. A tentative estimate of the length of record required to obtain a stable frequency distribution for various elements and types of climate is shown in Table 7.3.

Climatologists have commonly assumed that the normal distribution is applicable to most climatological data. The adequacy of this assumption depends on the problem at hand. The distributions of several elements for Washington, D.C., are shown in Fig. 7.21. The smoothed curves are the normal distribution as fitted to the data. Monthly and annual means of temperature, wind speed, sunshine, and humidity conform to the assumption of normality. Precipitation for months or years is usually slightly skewed, but often it can be satisfactorily treated with the normal distribution in problems not concerned with extreme values. The

* See bibliography at the end of this section.

TABLE 7.3. *APPROXIMATE LENGTH OF RECORD IN YEARS REQUIRED FOR A STABLE FREQUENCY DISTRIBUTION**

Climatic Element	Type of Area			
	Islands	Coasts	Plains	Mountains
Temperature	10	15	15	25
Humidity	3	6	5	10
Cloudiness	4	4	8	12
Visibility	5	5	5	8
Precipitation	25	30	40	50

* From *Study of Length of Record Needed to Obtain Satisfactory Climatic Summaries for Various Meteorological Elements*, Army Air Forces Weather Information Branch Report 588 (Washington, D.C., Nov. 1943).

distribution of short-period values of almost all elements departs from the normal more than does that for long-period measurements. Hourly wind speed in New England appears to conform to the Pearson Type III distribution.† Distributions of daily and hourly precipitation are highly skewed. Cloudiness and wind direction often show bimodal distributions.

7.13.2 Extremes

Many climatic elements are restricted within very definite limits. Percentage of sunshine, cloudiness, and humidity can vary only between 0 and 100 per cent. Wind speed and precipitation are limited by zero at the lower end of their range, but for practical purposes can be assumed to have no physical upper limit. Temperature can vary between absolute zero and the boiling point, but actual temperatures are within such a limited portion of this range that temperature may be considered an unlimited variate.

For factors that can be considered as unlimited, the most satisfactory treatment of extreme values seems to be in the theory of extreme values.‡ The theory applies to a series made up from the extreme values of a group of

† P. C. Putnam, *Power from the Wind* (Princeton, N. J.: D. Van Nostrand Company, Inc., 1948).

‡ R. A. Fisher and L. H. C. Tippett, "Limiting Forms of the Frequency Distribution of the Largest or Smallest Member of a Sample," *Proceedings of the Cambridge Philosophical Society*, 24 (1928), 180–90; and E. J. Gumbel, "On the Frequency Distribution of Extreme Values in Meteorological Data," *Bulletin of the American Meteorological Society*, 23 (May 1942), 95–105.

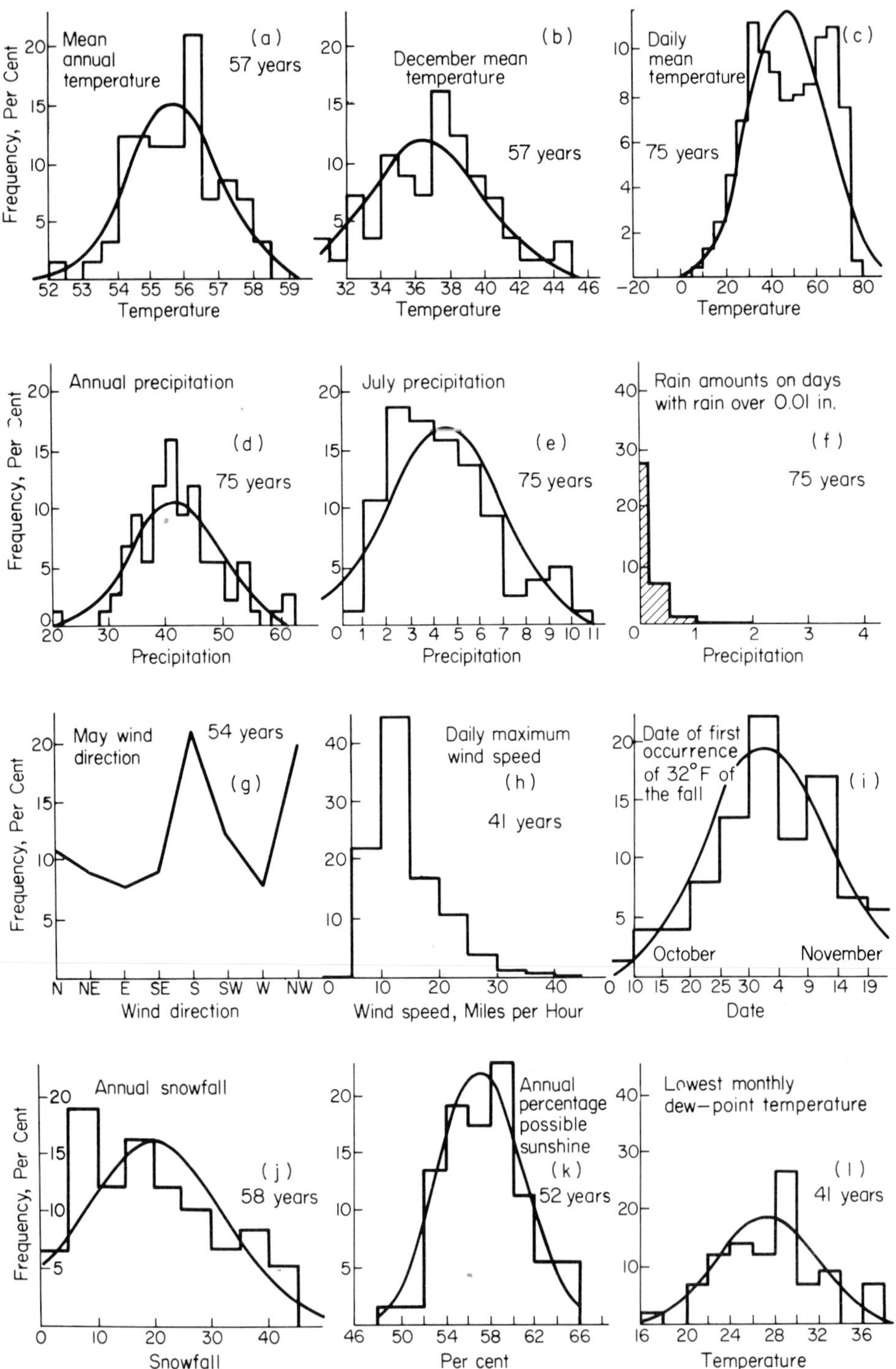

Fig. 7.21 *Some frequency distributions of climatic elements for Washington, D. C.*

TABLE 7.4. *PROBABILITY THAT AN EVENT OF GIVEN RETURN PERIOD WILL OCCUR WITHIN A SPECIFIED NUMBER OF YEARS*

Return Period, Years	Period, Years				
	1	10	50	100	500
1	1.0	1.0	1.0	1.0	1.0
5	0.2	0.89	0.99999	—	—
50	0.02	0.18	0.64	0.87	0.99996
100	0.01	0.10	0.40	0.74	0.993

series such as annual temperature maxima, annual precipitation maxima, and so forth. Analysis of extreme values of the annual total or average of a weather element is probably accomplished as well by the use of the normal frequency curve as by any other method. The annual averages of the limited factors such as percentage of sunshine vary so little that they can be considered unlimited variables in most cases. See Fig. 7.21(k).

Actually, there is no theory that can provide a reliable estimate of the return period for the maximum value of a series. If the true return period of an event is T years, the probability that it will occur in any year is $1/T$. From the principles of probability, the probability that an event that equals or exceeds the T-year event will occur in any series of n years is $1 - (1 - P)^n$ where $P = 1/T$. Table 7.4, which is computed from this expression, shows the probability that events of various return periods will be observed in a given interval of years. The table indicates, for example, that there is one chance in ten that the highest event in a 10-year period actually has a true return period of 100 years. However, there remains a slight probability (seven chances in 1000) that the largest event in 500 years is only the 100-year event. The use of a theoretical distribution as an aid in extrapolating frequency curves permits the inclusion of the lesser events whose return periods are better established and reduces the likelihood that the extremes will be given excessive weight in the extrapolation.

7.14 STATEMENT OF THE CLIMATOLOGICAL PROBLEM

The first step in the solution of a climatological problem is the rigorous statement of the conditions of the problem. Such a statement must come from a person who is intimately familiar with the application. Otherwise, an erroneous solution may develop. Aircraft spraying of insecticides is best accomplished under conditions of minimum air turbulence so that the dust will settle rapidly in the desired area. Although spraying of fungicides might seem to offer a similar problem, it is actually necessary

TABLE 7.5. *QUESTIONNAIRE FOR PLANNING CLIMATOLOGICAL ANALYSIS**

1. Where Will the Operation Take Place?
   a. at specified point or points?
   b. anywhere within an area limited by geographical, topographic, or economic factors?
   c. can a favorable location be selected?
2. When Will the Operation Occur?
   a. is the operation to be continuous?
   b. will operations be limited to a specific clock hour, season, month, etc.?
   c. can a favorable operation time be selected?
3. What Efficiency Is Required?
   a. is operation required under all possible conditions?
   b. is a calculated risk of inoperativeness acceptable?
      (1) Will the limitations on operation be on an areal, time, or economic basis?
4. What Weather Factors Are Involved?
   a. specifically how does each factor or combination of factors affect the proposed operation?

* Adapted from W. C. Jacobs and W. C. Spreen, *Some Climatological Problems Associated with the Assignment of Engineering Design Criteria* (Chicago: American Meteorological Society, Sept. 11, 1952).

to have a slight amount of turbulence so that the fungicide will reach the bottom as well as the top of the plants.

In stating his problem, the industrial user of weather data must be prepared to answer a questionnaire essentially like that shown in Table 7.5. The answers to this questionnaire define the area of interest, the controlling weather parameters and their tolerable limits, the operational efficiency required, and the time of the operations. Although some of this information is unimportant in specific cases, in general each item of information plays an important role in securing the correct solution. The design of searchlight cooling equipment would ordinarily be based on nighttime temperatures only, unless it is specified that the light may be used during the day. If an operational efficiency of 100 per cent is required, the climatologist searches for the probable upper limit of the controlling parameters. If an alternate location is permissible, a high efficiency may be obtained by planning operations at one site when the conditions are unfavorable in another location.

The exact answer to Item 4, Table 7.5, may be difficult to state because the effect of the weather factors is imperfectly understood. Deterioration of materials is a function of several weather factors, but until the nature of this function is determined experimentally, it is difficult for the climatologist to be helpful. An empirical equation for deterioration has been suggested* as follows:

$$A = ax^t + b\frac{(f-K)}{100}(1.054)^t(1 + cI)(1 + 0.067w) \quad (7.6)$$

where A is the deterioration rate, t is the temperature of the deteriorating surface, f is relative humidity, I is the concentration of effective impurities in the air, and a, b, c, x, and K are experimental constants. Such an equation defines precisely the weather factors of the problem. Although an equation of this type is not essential, simple least-squares correlation or analysis of variance may be helpful in selecting the important weather parameters.

Determination of the relation between climatological parameters and the operational function may become part of the climatological study. It may be necessary to collect data and develop a relation. In this process important nonclimatological parameters cannot be ignored. For example, in studying the relationship between temperature and attendance at a swimming pool, the differences between weekdays and weekends must be considered.

If positive values can be assigned to various operational results because of various weather conditions, the possibility of utilizing a formal decision process such as the maximin approach should be considered.

7.15 SITE TESTING

It is fortunate when a climatological problem develops for a point where a climatological station exists. But in most cases the problem involves a point at which no data have been collected. Transposition of record from a nearby station or interpolation between existing stations may prove satisfactory, but in many cases the accuracy of the analysis is materially improved by the collection of data at the actual site. If 3 or more years of record are collected, the techniques described in Pars. 7.8.4 and 7.9.5 will be helpful, but even a short period of record interpreted by a meteorologist in the light of the physical setting can be very useful. Very little research has been conducted to determine the length of record required to make the reduction of a short record to a long record possible. Putnam† found that for analysis of hourly wind data in connection with windpower studies a record of 10 days was adequate to estimate probable wind frequency at a site 10 miles from an established station, and 90 days at a site 38 miles from an existing station.

Obviously, the most desirable situation is that in which the necessary supplemental station is established as soon as the existence of a weather problem is recognized. The cost of instruments and operation may easily be less than $1000 for the first year, a cost almost certain to be negligible in comparison with project costs. In many cases, the instruments can be reused at other sites as subsequent weather problems develop.

* C. E. P. Brooks, "Climate and the Deterioration of Material," *Quarterly Journal of the Royal Meteorological Society*, 72 (1946), 87–92.

† *Power from the Wind.*

BIBLIOGRAPHY

Berry, F. A., E. Bollay, and N. R. Beers, *Handbook of Meteorology*. New York: McGraw-Hill Book Company, 1945.

Blair, T. A., *Climatology, General and Regional*. Englewood Cliffs, N. J.: Prentice-Hall, Inc., 1942.

———, *Weather Elements*, 5th. Edition. Englewood Cliffs, N. J.: Prentice-Hall, Inc., 1965.

Brooks, C. E. P. and N. Carruthers, *Handbook of Statistical Methods in Meteorology*. London: Her Majesty's Stationery Office, 1953.

Clayton, H. H., *World Weather Records*, **105**. Washington, D. C.: Smithsonian Miscellaneous Collection, 1947.

Conrad, V. and L. Pollak, *Methods in Climatology*, 2nd Edition. Cambridge, Mass.: Harvard University Press, 1950.

Haynes, B. C., *Techniques of Observing the Weather*. New York: John Wiley & Sons, Inc., 1947.

Jacobs, W. C., "Wartime Developments in Applied Climatology," *Meteorological Monographs*, **1** (1947).

Kincer, J. B., *Climate and Weather Data for the United States*, Yearbook of Agriculture, pp. 685–1228. Washington, D. C.: U. S. Department of Agriculture, 1941.

Linsley, R. K., M.A. Kohler, and J. L. H. Paulhus, *Applied Hydrology*. New York: McGraw-Hill Book Company, 1949.

———, *Hydrology for Engineers*. New York: McGraw-Hill Book Company, 1958.

Magill, Paul L., Francis R. Holden, and Charles Ackley, *Air Pollution Handbook*. New York: McGraw-Hill Book Company, 1956.

Malone, T. F. (editor), *Compendium of Meteorology*. Boston: American Meteorological Society, 1951.

Middleton, W. E. K. and A. Spilhaus, *Meteorological Instruments*, 3rd Edition. Toronto: University of Toronto Press, 1953.

Réseau Mondial. Geneva, Switzerland: World Meteorological Organization. Published annually.

Stern, Arthur C., *Air Pollution*. New York: Academic Press, Inc., 1962.

U.S Department of Agriculture, *Climate and Man, Yearbook of Agriculture*. Washington, D.C., 1941, 1248 pp.

U.S. Environmental Data Service, *Climatic Atlas of the United States*. Washington, D.C., 1968, 80 pp.

U.S. Weather Bureau, *Airway Meteorological Atlas for the United States*. Washington, D. C., 1944.

———, *The Climatic Handbook for Washington, D. C.*, Technical Paper 8. Washington, D. C., 1949.

———, *Atlas of Climatic Charts of the Oceans*. Washington, D. C., 1938.

———, *Climatic Summary of the United States, Bulletin* W. and *Supplements*. Washington, D. C., 1935.

LEO B. MOORE, *professor of management at the Sloan School of Management, Massachusetts Institute of Technology, is well known for his work in standardization, work simplification, and industrial engineering, as well as in management development, administration, and organization. He is the author of many articles in the field of standardization and had, for many years, a by-line column entitled "Standards Outlook" in* The Magazine of Standards, *published by the American Standards Association.*

Professor Moore has long been concerned with the contribution of standardization to industrial management, particularly through its impact as a tool of the manager. He initiated and directed a company standards program for several years and has been associated as a consultant with many leading company programs. As early as 1950 he introduced and developed an elective course in industrial standardization at M.I.T. He has been honored by the Standards Engineers Society both by the presentation of the combined ASTM-SES Award for his contributions to the field and by its designating its most distinguished award as the Leo B. Moore Medal.

In addition to his affiliations with several associations in management and engineering, Professor Moore is a fellow of the Standards Engineers Society and a member of the American Standards Association. He is a registered professional engineer in Massachusetts and was the recipient of the Gilbreth Medal in 1960 for his distinguished contributions to industrial engineering and scientific management.

SECTION 8

INDUSTRIAL STANDARDIZATION

LEO B. MOORE

8.1 WHAT IS STANDARDIZATION?

Standardization is a managerial concept which seeks effectiveness in operations through the acceptance and systematic employment of established and agreed upon procedures and practices. Management in its direction and control of enterprise uses the standardization process to draw upon past experience and proven engineering, economic, and administrative thought to organize, evaluate, and formulate the current practice that will most contribute to the success of its activities. The central theme of this process is consultation and consensus of all those with substantial interest in the resultant standard.

To this end, standardization is pursued not only in engineering and manufacturing but also in other functions of a business. The accepted standards in each of these areas enhance communication and improve understanding, particularly in interface problems. Management supports the use of such standards to reduce the variety and range of answers to recurrent problems, to conserve the enterprise resources, and to foster efficient and profitable operations.*

8.1.1 Definition of Standard

A standard is the decision that is approved and accepted as the best current solution to a

*For related view see R. D. Teece, "Standardization in Modern Industrial Management," *Standards Engineering*, **XVI**, No. 8 (Sept. 1964), 1–9.

problem situation; it remains the standard until a better solution is developed or created. For example, distance is a human concept of space that may be described in many ways. Down through human history the description of this concept has varied from the width of the thumb to the width of a ray of light, in accordance with the demands of man to measure distance. With each new demand has come a new decision (a new standard).

In its popular use, the word *standard* is most often associated with measurement, even in such expressions as decent standard of living, high standard of conduct, and fair standard of value, probably as a result of its initial development in the area of weights and measures, physical constants, and similar basic data. Such inadvertent confusion of the decision itself with one of its uses is not unusual since the formulation of the decision must always be succeeded by a representation of this decision and this representation often takes on more meaning than the decision itself. Thus, a yardstick means more to the user than does the scientific basis for its existence.*

8.1.2 Depictions of Standards

Both the kind of decision being reached and the way it will be used influence its form of representation, and this form typically becomes known as the standard. Standards of measurement are, for example, invariably converted into appropriate measuring devices and instruments. The expected method of application of the standard determines the exact final form and its variations since a decision is accepted only by its recurrent use. Thus, there are rulers and tape measures, micrometers and calipers, altimeters and depth gauges.

Other standards concerned with communication of information are depicted in an organized, symbolic way for easy use. Charts, graphs, calibrations, scales, tables, curves, and similar pictorial representations of standards information make their communication and understanding more effective. Standards that establish specifications, techniques, and practices may best be presented in an orderly format of sequential instructions using printed sheets, pictures, or detailed drawings. The effectiveness of the standard depends on the clarity of its formulation and of its depiction since these are essential to its intended use for communication, measurement, and judgment during its life.†

8.1.3 Types of Standards

Because there are so many different standards in industry, they are generally divided into two groups:

1. Technical standards—those standards that apply to the productive phases of the business, such as materials, parts, products, supplies, manufacturing practices, procedures and methods, test methods, drafting practices, nomenclature, abbreviations. These specify the what and the how of the business.

2. Managerial standards—those standards that apply to the administrative phases of the business, such as company policy, personnel procedure, accounting systems, expenditure controls, performance evaluation, safe practices, security regulations. These specify the who, when, and why of the business.

8.1.4 Descriptive Adjectives

Since a clear-cut division does not always exist between technical and managerial standards, some companies make no attempt to separate them. They merely label their standards with appropriate adjectives, such as material standards, safety standards, and drafting standards.

In the same vein, it is often the practice of an organization to group standards according to major functional areas, such as engineering standards, manufacturing standards, and administrative standards, and to subdivide these groupings in keeping with organizational elements and responsibilities. Thus, engineering standards would include those standards within the engineering area—for example, design standards, drafting standards, material standards. Manufacturing might include purchasing standards, safety standards, process standards, etc. Administrative standards would include those in personnel, office, finance, and similar areas. The extent to which this format

*"ISO Council Acts on Projects—Defines Standard," *The Magazine of Standards*, **33**, No. 10 (Oct. 1962), 301–2.

†Walter S. Brown, "The Birth and Maturing of a Standard," *Standards Engineering*, **XVI**, No. 11 (Dec. 1964), 9–10.

is followed varies according to managerial design and support of standards activities.

8.1.5 Definition of Standardization

Standardization is the process of establishing standards. It is the organized procedure we go through to decide what the standard will be and to have it accepted and used. Sometimes it involves merely writing down the rules that apply to a situation as established by authority; other times it may require discussion and ultimate agreement as established by group consent. The end result is the establishment of a standard. The fire department states that the fire doors on the spray room will be kept closed, and that is used as a mandatory managerial standard. The engineers agree after some discussion that CRS is cold-rolled, not corrosive-resistant, steel, and that is accepted as a voluntary, technical standard.*

8.1.6 Similar Words

In particular industries or activities, certain other words have come to be accepted as meaning generally the same thing as standard and standardization. Most of these terms, such as policy, practice, procedure, and regulation, are easy to identify if the above definitions and ideas are kept in mind. There are two, however, that warrant definition here: specification and simplification.

8.1.7 Specification

In many company activities—for example, purchasing, manufacturing, and marketing—the term specification has become acceptable. In all cases, a specification is a statement or description as complete and definite as possible of particular requirements, characteristics, or qualities that establish with certainty the terms of a contract, the details of construction, the composition of materials, the control of processes, the sequence in procedures, and the like. A specification often employs, through references, basic existing standards rather than spelling out this information specifically.†

In every sense, then, a specification used repeatedly to meet a recurrent problem is a standard. In industry, the term is applied to materials, jobs, products, and so forth, as a matter of custom and should in these instances be considered synonymous with standard. A specification for a house, ship, or other product that is being produced only once is, however, not a standard specification.

8.1.8 Simplification

Simplification is the reduction of variety by eliminating the unneeded. It is one of the procedures used in standardization and consists of removing from a list of items those that are deemed excessive, unnecessary, or redundant.

The technique of simplification is used to review materials, parts, components, products, forms, tools, equipment, or anything that is produced, purchased, and employed in a range of sizes, colors, designs, or other distinguishing characteristics. The items that survive the elimination process are referred to as standard items and the entire technique is properly referred to as standardization through simplification.‡

8.2 SOURCES OF STANDARDS

Industrial standards have so many varied origins that for purposes of general consideration they are classified according to their source and scope. Thus, we have four types or kinds of standards:

1. Company
2. Association and society
3. National
4. International

As we go down the list, it is apparent that the standard is effective in a larger and larger area. Consequently, its preparation and approval become bigger and bigger problems. For that reason, these groupings are frequently referred to as the *levels of standardization* in considering the field as a whole.

8.2.1 Company Activity

8.2.1.1 Informal Programs. Every company uses standards of some sort, but not every com-

*Andrew S. Schultz, Jr., "The Impact of Standardization," *The Magazine of Standards*, **34,** No. 11 (Nov. 1963), 340–43.

†For the relationship of specification to engineering and management see E. H. MacNeice, *Industrial Specifications* (New York: John Wiley & Sons, Inc., 1953).

‡S. David Hoffman, "Important Standards Ideas Defined," *The Magazine of Standards*, **34,** No. 8 (Aug. 1963), 246.

pany has a formal standards program. Those that do not have an organized program tend to inherit or absorb standards from around them and to adapt and employ those that fit their operations or are required by them. Most often this activity is carried on by individuals in the company as a simple matter of common sense, without fanfare or publicity. The companies in this group by far outnumber those that employ standardization on an organized basis. Some of these companies feel that they do not have the time, money, or personnel to spare from more pressing tasks. Many others have never given any consideration to establishing a formal standards program.

8.2.1.2 Formal Programs. Standardization is not for the exclusive benefit of the large company that can indulge it and indulge in it. A list of the companies active in standardization is not in any way a "Who's Who" of American industry. Whether the company is large or small, the executives have the same decision to make: "Do we organize our present use of standards and receive the additional benefits of such activity, or do we leave well enough alone?" The extent of the activity at the beginning and its subsequent growth are matters of policy for each company. To help answer these questions is the purpose of this section.

8.2.2 Association and Society

Much of the work of the company standards program extends beyond the organizational and geographical boundaries of the company itself. Many decisions require cooperation and coordination with other companies. The natural vehicles for such participation are the trade associations and the professional and technical societies, with their industrywide interest and membership. For that reason, many of these organizations maintain an active interest in standardization.

The usual approach to standardization used by associations and societies is to establish a standards committee that (1) reviews the opportunity for standardization, (2) appoints investigating subcommittees as needed, (3) presents the recommended standard to the membership for approval, and (4) publishes it for general use.

At least 450 of the 3000 associations and technical societies in the United States are actively concerned with matters of standardization. A description of these activities, as well as other pertinent information, has been prepared for many of these organizations and is available from the Superintendent of Documents, Washington, D. C.*

8.2.2.1 National Electrical Manufacturers Association. A brief account of the work of NEMA should be of interest since electrical products are so much a part of our everyday life. NEMA is a trade association of about 520 manufacturers of equipment and apparatus used for the generation, transmission, distribution, and utilization of electric power. NEMA may be considered an aggregation of Product Sections, each representing manufacturers of certain classes of products, such as motors and generators, steam and hydraulic turbines, transformers, wire and cable, switchgear, industrial controls, ranges, water heaters, and domestic appliances.

The standardization activities of NEMA are guided by policies established by the Board of Governors. The following general statement of NEMA policy appears as a foreword to every NEMA standard:

> National Electrical Manufacturers Association Standards are adopted in the public interest and are designed to eliminate misunderstandings between the manufacturer and the purchaser and to assist the purchaser in selecting and obtaining the proper product for its particular need. Existence of a National Electrical Manufacturers Association Standard does not in any respect preclude any member or nonmember from manufacturing or selling products not conforming to the standards.

The NEMA bylaws define a NEMA standard as follows:

> A Standard of the National Electrical Manufacturers Association defines a product, process, or procedure with reference to one or more of the following: nomenclature, composition, construction, dimensions, tolerances, safety, operating characteristics, performance, quality, rating, testing, and the service for which designed.

Correlating the standardization activities of NEMA is the Codes and Standards Committee, which supervises these activities by:

1. Reviewing for approval, or for other ap-

---

*Sherman F. Booth, *Standardization Activities in the United States*, Miscellaneous Publication 230 (Washington, D. C.: National Bureau of Standards, 1960).

propriate action, all proposed standards, reports, or technical documents which are to be issued as having official NEMA approval.

2. Designating the personnel and supervising the activities of NEMA representation on all technical committees working with other standardizing organizations, thereby assuring adequate representation and proper consideration of the diversified interests of the electrical manufacturer.*

In general, a NEMA standard originates in its Product Sections, where a committee develops it, and, after approval of the Section as a whole, it is referred to the Codes and Standards Committee for complete review and final approval.

NEMA has approximately 225 current publications covering standards on such varied items as brushes, terminals, sockets, lamp bases, wires, cables, insulators, transformers, circuit breakers, measuring instruments, electrical equipment, machinery and controls, and safety codes.

NEMA participates actively with seventy associations and organizations in national and international standardization projects.

8.2.2.2 American Society for Testing and Materials. A brief description of the work of ASTM should also be of interest since the standards of this organization have found their way into every company in the country. ASTM was organized in 1898 and formally incorporated in 1902 as a national technical society. Its specific purpose is "the promotion of knowledge of the materials of engineering and the standardization of specifications and the methods of testing." Over 3000 technical committees are responsible for the development of standard specifications and tests and for keeping them up to date. The Society's work is advanced by the presentation and subsequent publication of technical data in the form of papers, reports, and discussions. Its membership of over 12,000 includes (1) producers of raw materials and semifinished and finished products, (2) consumers of materials, and (3) a general-interest group comprised of engineers, scientists, educators, testing experts, research workers, and so forth. Membership is held by individuals, companies, associations and technical societies, governmental departments, schools, and libraries.

ASTM standardization applied to methods of testing and analysis recognizes that the properties shown by a material are dependent to a large extent upon the method of determination and, in order that comparative results may be secured in widely scattered laboratories, standardized procedures are essential. ASTM is splendidly equipped to develop standard test methods because on its 100 main technical and administrative committees and their hundreds of sections are outstanding experts in technical fields.

To develop specifications as well, the ASTM committee organization is very well adapted, for each committee is made up of producers who are familiar with the limitations of the manufacturing processes and of consumers who are fully acquainted with the requirements of the various materials. Before a standard is finally adopted by ASTM it has been given a rigorous examination in the committees, by the Society as a whole, and through actual use during the period of publication as a tentative standard since tentatives are recognized as embodying the latest thoughts and practices and are widely used throughout industry.

A great amount of authoritative data concerning materials is brought out by the Society, either as an adjunct to its standardization work or as the result of independent research. In fact, before any standardization is possible, some research is necessary. It is only after data from reliable tests are available that intelligent recommendations can be made in regard to the material in question. ASTM has over 100 separate research projects supporting its standardization work. The fact that the 3700 ASTM standards are issued in a book of thirty-two parts provides some idea of the breadth of its work.†

8.2.3 National Standardization

A centralizing force is essential to coordinate and correlate the standards work of the hundreds of companies, associations, and groups

*For treatment of the international standards problems of electrical manufacturers see W. A. McAdams, "CEE—Challenge or Opportunity for American Industry," *The Magazine of Standards*, **34**, No. 1 (Jan. 1963), 3–6.

†For new material concepts in the space age see Thomas A. Marshall, "The Care and Feeding of Standards," *The Magazine of Standards*, **34**, No. 1 (Jan. 1963), 8–12.

in the country; to prevent duplication of effort; and to bring together all interested parties. For the same reasons and in the same ways that companies seek the assistance of the association and society, a vehicle is needed to solve standards problems on a nationwide basis.

8.2.3.1 United States of America Standards Institute (USASI). The USASI, formerly the American Standards Association (ASA), was set up in 1918 (then called the American Engineering Standards Committee) to serve as a central, national clearinghouse for standardization work in the United States (see Fig. 8.1). It is a privately financed federation of over 135 national trade associations, professional societies, government and consumer organizations, with a membership of more than 2000 companies representing a cross section of commerce and industry.

The USA Standards Institute is structured to act in a judicial capacity in the approval of

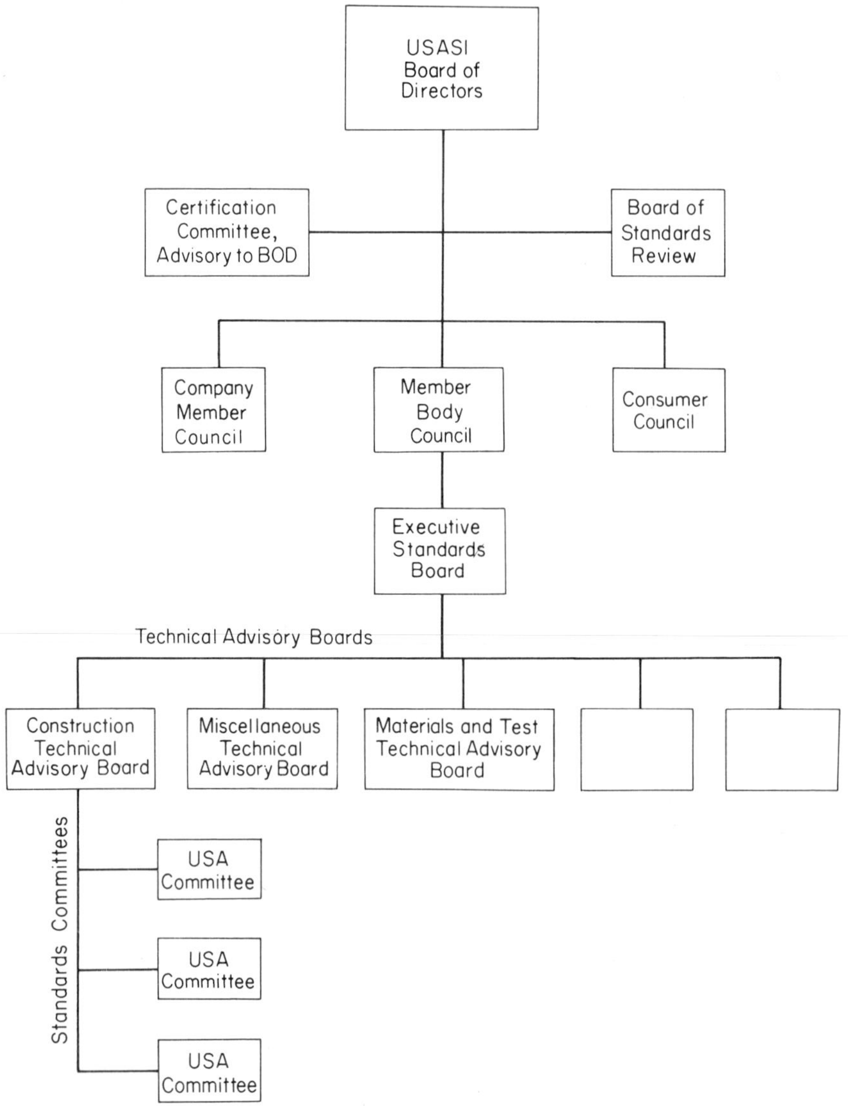

Fig. 8.1 *USA Standards Institute organization chart. Courtesy USASI.*

USA Standards through a Board of Directors and three operating councils—Member Body Council, Company Member Council, and Consumer Council. The Institute does not itself formulate standards. It depends upon the combined technical capability of its member bodies and other competent organizations to develop standards and to work through the procedures of the Institute for their approval as national or USA Standards. The Institute performs a coordinating function to ensure that the interests of all segments of the economy—consumers, labor, industry, government—are protected and represented in national voluntary standardization activities. It provides the means for determining the needs of these groups for new or revised standards and ensures that existing organizations competent to resolve these needs undertake the work. If an organization competent to develop standards and codes in a particular field does not exist, the Institute takes steps to establish a suitable group.

The Board of Directors is the governing body of the Institute, managing its affairs and determining its policies, as well as retaining final authority over all matters pertaining to procedures, including approval of USA Standards. The directors, besides the officers of the Institute, include significant representation from the three Councils, which have had certain responsibilities delegated to them. The Member Body Council is charged with the development and maintenance of all procedures relating to the preparation, approval, acceptance, and designation of standards, and the constitution of standards boards and committees. The Company Member Council is responsible for the development of programs to maintain liaison with, and represent the interests of, commerce and industry in the work of the Institute. The Consumer Council represents and protects the interests of the consuming public, serving as the Institute contact between the general public and industry in matters concerning standards affecting the consumer.*

The organization and procedures of the Institute permit alternative methods for the establishment of USA Standards. (See Fig. 8.2.) Essentially these choices build upon the standards work of competent organizations in the appropriate field by submitting their existing standards to the approval process or asking such organizations to prepare proposed standards for submittal. Where necessary, a USA Standards Committee is formed and performs the task of development under the administrative sponsorship of a suitable competent organization or one of the seventeen Standards Boards. Under these procedures more than 2500 USA Standards have been developed by representatives of industry, government, labor, and consumer organizations.

A critical part of the establishment process and fundamental to final acceptance and approval as a USA Standard is the spirit of voluntary standardization found in the consensus principle. The standards committee must agree on the scope and provisions of the project before them, accept the concept of voluntary standards, abide by the designated administrative sponsor of the project, and be truly representative of all interested and affected parties. Finally, standards produced by the committee are approved as USA Standards only when an impartial Standards Institute Board or other authority finds that they are backed by consensus—that those affected have in fact reached substantial agreement on their scope and provisions. Final adoption by the using organization is of course voluntary, with the trust in the process producing such unanimity as to encourage adoption. The ultimate aim is national acceptance of USA Standards.†

8.2.3.2 National Bureau of Standards. According to the provisions of the Constitution, Congress has authority over weights and measures. Responsibility for these standards basic to industry and commerce has been entrusted to the National Bureau of Standards, set up in 1901. Its functions outlined in the enabling act were restated more explicitly in Public Law 619, 81st Congress, as follows:

> Sec. 2. The Secretary of Commerce (hereinafter referred to as the "Secretary") is authorized to undertake the following functions:
>
> (a) The custody, maintenance, and development of the national standards of measurement, and the provision of means and methods for making measurements consistent with those standards, including the comparison of standards used in scientific investigations, engineering, manufacturing, commerce, and educational

*For greater detail see "How the USA Standards Institute Coordinates Standardization," *The Magazine of Standards*, **38**, No. 2 (Feb. 1967), 35–38.

†Francis K. McCune, "The Case for Voluntary Codes," *The Magazine of Standards*, **38**, No. 5 (May 1967), 131–32.

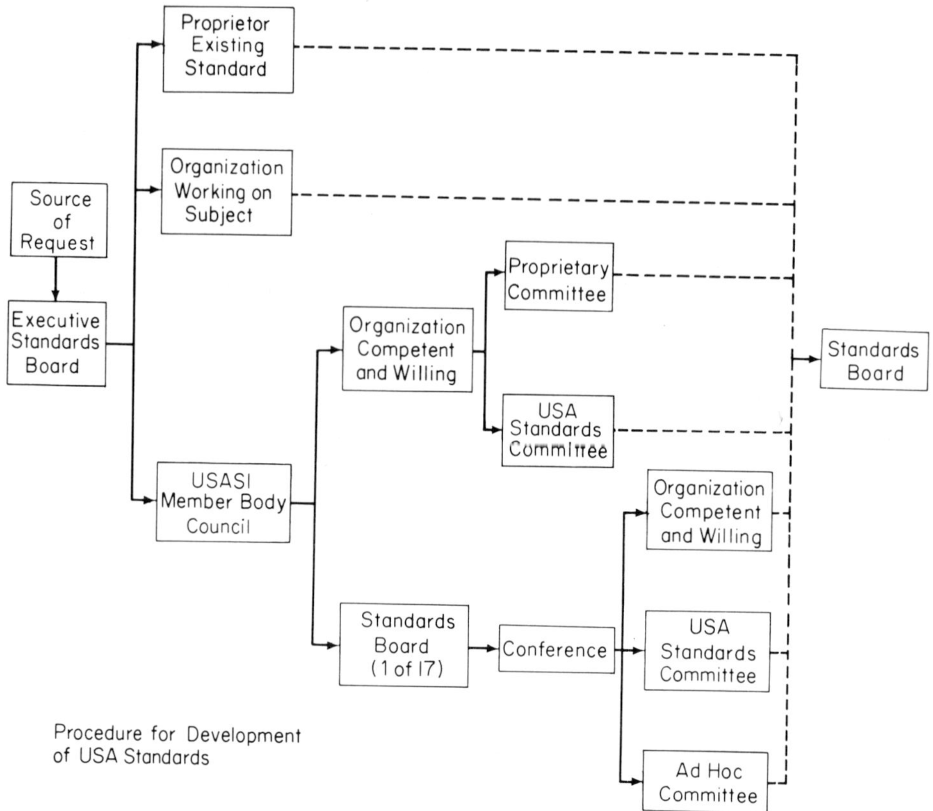

Fig. 8.2 *Flow chart, development of USA standards. Courtesy USASI.*

institutions with the standards adopted or recognized by the government.

(b) The determination of physical constants and properties of materials when such data are of great importance to scientific or manufacturing interests and are not to be obtained of sufficient accuracy elsewhere.

(c) The development of methods for testing materials, mechanisms, and structures, and the testing of materials, supplies, and equipment, including items purchased for use of government departments and independent establishments.

(d) Cooperation with other governmental agencies and with private organizations in the establishment of standard practices, incorporated in codes and specifications.

(e) Advisory service to government agencies on scientific and technical problems.

(f) Invention and development of devices to serve special needs of the government.

The Bureau is authorized to "exercise its functions for the government of the United States, for any state or municipal government within the United States, or for any scientific society, educational institution, firm, corporation, or individual within the United States engaged in manufacturing or other pursuits requiring the use of standards or standard measuring instruments." On this broad base of research and testing, the Bureau through the years has performed excellent service for the country. It is recognized as the principal agency of the federal government for research in physics, mathematics, chemistry, and engineering.

It is concerned not only with the fundamental standards of mass, length, and time but also with a wide variety of basic standards in electricity, optics, heat and power, atomic physics, chemistry, mechanics and sound, organic and fibrous materials, metallurgy, mineral products, building technology, applied mathematics, electronics, radio propagation, and data processing.

Although the Bureau has no regulatory power, it has gained a pre-eminent position in fundamental standards work through the high standard it has set for its work, its cooperation with individuals and groups throughout the country, and the publication of its findings.*

8.2.3.3 Department of Commerce. The Commodity Standards Division of the Office of Technical Services issues simplified practice recommendations and commercial standards as a service to industry and business and for general use by the public. As part of Herbert Hoover's program of waste elimination in industry,† the Division of Simplified Practice was established in 1921. The purpose of the Division was to establish procedures and facilities that would bring together all interested parties in a nationwide program to simplify lines of manufactured products, as, for example, an 84 per cent reduction in sizes and varieties of cans for fruits and vegetables, from 200 to thirty-two. This was a voluntary action on the part of all groups concerned—manufacturers, distributors, and consumers—with the Division acting as a center of assistance. The Division serves as publisher of the promulgated simplified practice recommendations. More than 245 such recommendations have been issued, covering products in many fields.

In 1927,commercial standards were instituted to assist industry in the establishment of standards of quality of manufactured goods. Commercial standards cover terminology, types, classifications, grades, sizes, and use characteristics of manufactured products, and describe standard methods of test, rating, certification, and labeling. Their purpose is to provide uni form bases for fair competition and to assist industry in maintaining quality that will command the respect of purchasers.

The function of the Department of Commerce in the establishment of commercial standards is that of a coordinating and fact-finding agency to assist in ascertaining the desires of all concerned. Commercial standards are developed only upon written request of interested parties. The promulgation of a standard developed through the resulting cooperative action is dependent upon written acceptances from the trade representing a satisfactory volume of business in the commodity covered. Adherence to commercial standards by the industry is entirely voluntary, but when guarantees of conformance are made by the manufacturer or seller, that guarantee is a part of the sales contract. A list of commercial standards currently in effect may be obtained on request from the U.S. Department of Commerce, Washington, D. C. 20225.

8.2.3.4 Department of Defense. The Standardization Division of the Defense Supply Agency has the responsibility for the DOD standardization program and activities. Prior to centralization, many segments of the military establishment had developed and issued their own specifications for required materials and supplies. Within the dozen or more separate series of specifications duplications or annoying variations created confusion and difficulty for contractors and others seeking to furnish defense items. To bring uniformity, order, and harmony to the extensive and complex purchasing for defense, the Standardization Division has issued over 20,000 military (MIL) specifications and cancelled many of the earlier series. In support of these specifications, it has developed and published over 3000 military standards in a wide range of technical and scientific areas.

The Department of Defense developed these specifications and standards to unify and coordinate its own operations in accordance with its recognized responsibilities. They are mandatory upon no one but the Department, but they represent knowledge and information of inestimable value to many who are influenced by them.‡

8.2.3.5 General Services Administration. In 1949, for the first time, the federal government, long concerned over its mounting expenditures and the efficiency of its purchasing activities, established (Public Law 152, 81st Congress) uniform controls over its supply and property management activities. Congress in that year

*For a fascinating account see John Perry, *The Story of Standards* (New York: Funk & Wagnalls Co., 1955).

†Federated American Engineering Societies, *Waste in Industry* (New York: McGraw-Hill Book Company, 1921).

‡For current information see *Standardization Policies, Procedures and Instructions*, Defense Standardization Manual M200A (Washington, D. C.: Defense Supply Agency, April 2, 1962).

set up the General Services Administration as the agency responsible for providing the government with an economical and efficient system for procurement and supply, including such necessary related activities as specifications and standards. The Standardization Division of the Federal Supply Service of GSA carries out the standardization program for materials, supplies, and equipment in common use by the federal government. This program includes the standardization of commodities, the development of federal specifications and standards, and the establishment of the federal catalog system.*

The federal catalog system is the result of efforts to establish and maintain uniformity in the classification, numbering, and description of commodities used by federal agencies, including the military departments. This monumental work, the consequence of extensive simplification and standardization, will be of increasing significance to American industry and commerce as well as to federal supply administration. The underlying philosophy of this gigantic program is to employ the typical production of industry, to seek the best technical knowledge and abilities in government and industry, and to reflect technological advance insofar as is possible and consistent with the aims set forth by Congress.

8.2.4 International Standardization

The highest standardization work is that accomplished on an international level. The recognition of the interdependence of nations and of the importance of their interrelationships urges such work upon us, but difficulties of national language, customs, and pride make for almost insurmountable obstacles. The first advances in this direction were made in scientific and technical areas.† In 1875 the Metric Convention established ways and means for international agreement on scientific data such as nomenclature, symbols, physical constants, temperature scales, and definitions, in addition to weights and measures.‡ The International Electrotechnical Commission, organized in 1906, has established standards for international symbols, resistance measurements, rules for rating machinery, nomenclature, voltages, heavy current systems, and motor specifications.§ The International Commission on Illumination, started in 1913, has made attempts at standardization of lighting systems and associated technical problems, including definitions of luminous intensity, fundamental quantities used in illumination and photometry, and uniform practices in lighting.

In spite of the limited success of the technical bodies in the international standardization of scientific data and in spite of the diverse problems of the different industrial nations, the national bodies of many countries have continually pressed for international cooperation and standardization of industrial activities. For many years countries linked by common language, boundaries, and commerce have engaged in joint standardization. In Europe, prior to World War II, the Dutch, Swiss, Swedish, Czechoslovakian, Austrian, and German standards bodies cooperated very closely. The same has been true of Canada and the United States, the United States and the Latin American countries, and the United Kingdom, Canada, and the United States. In 1948 the latter group reached an agreement for standard screw threads to apply in each of the three countries. During World War II, these countries worked closely on standardization matters and formed what was known as the United Nations Standards Coordinating Committee as a temporary body for international cooperation.‖

8.2.4.1 International Organization for Standardization (ISO). In 1947 in Zurich, Switzerland, the first general assembly was held of the International Organization for Standardization (ISO), with a membership of the national bodies of twenty-six countries. The organization agreed to have as its official languages English, French, and Russian, and in many

*"GSA Changes Standards Procedures," *The Magazine of Standards*, **34**, No. 2 (Feb. 1963), 43–45.

†Russell F. Holmes, "International Standards—Development, Effect," *Standards Engineering*, **XV**, No. 2 (Feb. 1963), 5–7.

‡"International Committee Acts on Weights and Measures," *The Magazine of Standards*, **34**, No. 3 (Mar. 1963), 74–75.

§"Electrical Industry Benefits from IEC Participation," *The Magazine of Standards*, **35**, No. 9 (Sept. 1964), 265–68.

‖For progress from regional to international see Russell F. Holmes, "Development of International Screw Thread Standards," *Standards Engineering*, **XV**, No. 9 (Sept. 1963), 8–9.

TABLE 8.1. *ISO MEMBER BODIES*

Australia	Finland	Korea (Democratic People's Republic of)	Romania
Austria	France	Korea (Republic of)	Singapore
Belgium	Germany	Lebanon	South Africa (Republic of)
Brazil	Ghana	Mexico	Spain
Bulgaria	Greece	Morocco	Sweden
Canada	Hungary	Netherlands	Switzerland
Central America	India	New Zealand	Thailand
Ceylon	Indonesia	Norway	Turkey
Chile	Iran	Pakistan	United Arab Republic
Colombia	Iraq	Peru	United Kingdom
Cuba	Ireland	Philippines	U.S.A.
Czechoslovakia	Israel	Poland	U.S.S.R.
Denmark	Italy	Portugal	Venezuela
	Japan		Yugoslavia

other respects to pattern its operation after the United Nations. The governing bodies of the ISO are (1) the General Assembly—a meeting of the delegates nominated by the member bodies—and (2) the Council, consisting of the president of the ISO and a representative from each of ten member bodies. At least five member nations must indicate their interest in a project before it is initiated. Acceptance of the resulting ISO recommendations is voluntary. At its inception, the ISO agreed to assume and resume the work of the prewar International Federation of National Standardizing Bodies (ISA), which had made by 1939 substantial progress through forty-seven projects toward securing international uniformity in such diverse fields as photographic and acoustical standards, documentation standards for libraries (including microfilm techniques), limit systems for machine parts to provide for interchangeable fits, inch-millimeter conversion ratio, preferred numbers, paper sizes, and film standardization.

This work, coupled with the programs already offered by the member nations, means that there are 116 projects before the ISO, ranging from automobile parts to textiles. Table 8.1 is a list of ISO member bodies. It indicates the extent of and potential interest in this work so essential to international trade and goodwill.*

8.3 COMPANY PROGRAM

Standardization starts and ends in the individual company and in its everyday operations. The extent of its use is determined partly by the size of the company, but mostly by the attitude of its personnel. And of all its personnel the top executives of the company exert the main influence—influence on its emphasis, organization, activity, and support. Consequently, there are as many programs as there are companies, and it would be foolhardy here to state specifically that any particular company should do this or that. Rather, the possibilities for its use within a given company should be explored and pursued.†

8.3.1 Starting Point

Every company has a set of standards, even though that term may not be applied to them. The purchasing department has some purchase specifications; the production group, some process instructions; the inspection section, some testing procedures; the accounting office, some paper routines; the sales department, some product catalogs. Every one of these and hundreds more, in even the smallest companies, are standards in every sense of the word. The program begins with an analysis and organization of these existing standards.

*For an account of the growth of international standardization see Cyril Ainsworth, "Standardization Abroad," *The Magazine of Standards*, **35**, No. 12 (Dec. 1964), 364–67.

†In this vein see John T. Milek, "The Role of Management in Company Standardization," *The Magazine of Standards*, **33**, No. 4 (Apr. 1962), 107–14.

Of all these opportunities for the standardization effort, the best by far for a given company is the place where it is needed. A company seldom enters into standardization deliberately simply because it is a sound principle of management. Standards work can always be traced to a need for which standardization was the answer. The purchasing department may discover an increasing raw-material inventory of special items in small lots. The manufacturing division opens a new plant a thousand miles away from home and finds that process techniques are stored in the heads of the employees at home. The sales department is saddled with a number of slow movers. The vendors are changing specifications until the engineers are swallowed up in confusion. Every one of these is a problem to the departments concerned, but it also is an opportunity for standardization. Once the need has been identified, standardization can be of service.*

A great help in starting standardization will be found outside the company. Visits to other companies active in standards work, not necessarily within a given industry, will always inspire ideas and opportunities. The work already done within an industry by other organizations, such as trade associations, technical societies, government groups, and the USA Standards Institute should be compiled early in the program. The experience of others is always valuable.

8.3.2 The Standards Engineer

The qualities of an effective standards engineer, based on what he has to do, indicate that he must be: (1) particularly strong in the ability to handle human relations problems, (2) a good engineer, versed in the technical and productive problems of his company, and (3) an able administrator, capable of making competent business decisions.†

8.3.2.1 Qualifications of a Standards Engineer.

A. Human relations skills

1. An abundant feeling for and understanding of people as individuals and in groups

2. A keen sensitivity to and appreciation of the attitudes of people

3. An ability to cope patiently with the common human characteristics of resistance to change and resentment of criticism

4. A capacity to alleviate conflicting views and opinions in a forthright manner

5. Consummate skill in obtaining agreeable cooperation from divergent individuals and groups

6. A natural quality for doing and saying the right thing at the right time

B. Technical knowledge

1. A well-rounded understanding of the fields of engineering basic to the operations of the company

2. A working knowledge of the specific technical problems of the company

3. An extensive familiarity with the equipment and processes used in the company

4. A practical skill in basic engineering terms, techniques, and troubles common to the industry

C. Business experience

1. A general understanding of the relationships of all phases of the company

2. A particular interest in the practices and operations of other companies in the industry

3. An extensive knowledge of the operations of individual departments in the company—*viz.* purchasing, production engineering, manufacturing, and industrial engineering

4. A working knowledge of other departments in the company—*viz.* sales, accounting and finance, design, research and development

5. A developed ability based on sufficient service in the company to make sound business decisions

6. A well-rounded capacity to accept responsibility, to display initiative, and to temper authority

8.3.2.2 Duties and Responsibilities of the Standards Engineer.

1. The standards engineer is charged with the total responsibility for the standards program, subject only to:

a. Higher authority on matters of policy, expenditure, and scope

b. Company divisions on matters of content, use, and revision

2. The higher authority is a formal committee on standardization policy composed of top executives.

---

*For a similar treatment see Madhu S. Gokhale, "The Importance of Company Standards," *Standards Engineering*, **XVI**, No. 2 (Feb. 1964), 1–8.

†See John L. Rayfield, "Training a Standards Engineer," *Standards Engineering*, **XIV**, No. 11 (Nov. 1962), 1–8.

3. The company divisions may be one or more of the sections of the enterprise that are expected to comply with the standard. Their compliance is to be contingent upon their acceptance, and they may reject a standard for cause.

4. The standards engineer establishes a program of activity for the company for a given period of time and submits it to higher authority for review, change, and comment.

5. The standards engineer breaks down his program into projects by company interests and has freedom to add or delete projects within the scope of the program approved by higher authority.

6. The standards engineer does all preliminary work on each project:

a. Gathers facts pertinent to the project both inside and outside the company.

b. Makes a tentative decision on the content of the proposed standard.

c. Obtains opinions and data supporting and refuting his proposal both inside and outside the company.

d. Sponsors research, investigation, or tests to establish the validity of his proposal.

e. Makes such changes or corrections as may be deemed necessary.

7. The standards engineer submits his proposed standard to all divisions of the company concerned for their approval and acceptance. Depending upon the scope of the standard, he may:

a. Submit it to each division head for his initials.

b. Submit it to a committee of representatives of these divisions for their approval.

8. The standards engineer revises any proposed standard not approved or any existing standard rejected by any division concerned with its provisions until it is approved.

9. The standards engineer keeps such records, minutes, or data in connection with each standard as are required to maintain a complete history of the standard.

10. The standards engineer is responsible for the publication, binding, and distribution of accepted standards.

11. The standards engineer is responsible for active participation in the standardization work of trade associations, professional groups, technical societies, and national associations.

These duties and responsibilities should not be accepted in detail by any company as here presented; they merely reflect good general practice. A specific company should establish its own method of operation.*

8.3.3 Method of Operation

The standards engineer works with problems that have a breadth of interest and a complication of detail that force him to use several means of accomplishment. There are, on the one hand, some purely routine activities vital to his program, but requiring decisions that rest exclusively with him. On the other hand, there will be decisions involving expenditures of money, obsolescence of inventory, scrapping of equipment, change of procedures, and sundry eddy effects throughout the company that by their nature no one man could engineer or install. These two opposite extremes of his activity naturally suggest two opposed methods of operating the standards program—the individual or expert approach versus the committee or participation approach. Making contacts with other companies and groups that will help him in his program and collecting standards peculiar to his industry, he can do himself. Converting his company to the new, standard screw threads or simplifying and standardizing the product line, however, require complete coordination and control, and full understanding and cooperation, which are best accomplished through group effort.†

8.3.3.1 Committee Approach. Some companies use the committee technique exclusively, with marked success. These committees range in responsibility all the way from policy-making down to working groups that prepare standards. There is much to be said for the committee since it serves several functions at once—it provides participation, promotes understanding, establishes agreement, and assures acceptance of the standards program. In a company that has the manpower and has highly developed and efficient committee-management relationships, it is only natural that standardi-

*See for contrast Donald E. Smith, "Streamlined Standardization," *Standards Engineering*, **XIII**, No. 1 (Feb.–Mar. 1961), 3–5.

†See Arnold M. Rosenwald, "Let's Talk Sense About Company Standards," *Standards Engineering*, **XIV**, No. 6 (June 1962), 3–6.

zation should receive the same treatment as other activities.*

Other companies which have tried to use the committee method have found it wanting. On the whole, committees are difficult to manage, particularly if they are expected to be productive. In those companies that have tried and failed, the problem always seems to be that of getting things done. The committee members are men already burdened with their own departmental duties and they react to standardization as another interference with their primary responsibilities. Many companies that have experienced this frustration are set against committees. Others have lost interest in standardization or have at least set up barriers to its extension. Sometimes the program is simply handed to the standards engineer, who is expected to go it alone.

8.3.3.2 Expert Approach. The standards engineer on his own is helpless. The very definition of standardization implies and demands that others be involved. Yet in too many companies the standards engineer is fighting time and tide, trying with all his might to put this standard in here and argue that one in there. He meets with limited success. He is the expert and is treated as such by everyone—the know-it-all a long way from home. Yet once the way has been cleared, he moves quickly to action and results.

8.3.3.3 Combined Approach. For the average company, the goal is to capture the swift movement of the lone man and the slow deliberation of the working committee. The decision of how much of each method the standards engineer should use depends primarily on what is to be accomplished and on the human relations problems involved.†

The standards engineer might in general be expected to gather the facts pertinent to a standards problem on his own and to prepare a recommendation that would be discussed by an appropriate group for its reaction and recommendation. Normally, he would be expected to make necessary changes and return to the group for approval. The final preparation, publication, and distribution would be within his domain.‡

8.3.4 Organizational Position

The decision to start a program of standardization should never be hampered by a fear of where to fit the standards engineer into the organization. On the basis of the specifications set down for this individual, he will probably be working in the company. The new standardization program should have a small beginning; the selected individual should probably be permitted to carry on a portion of his regular work at the same time as he works into standardization. He should have relative freedom in his program and should seek his own level in the organization. As the program grows and assumes increasing stature in the company, the problem of where to place the activity will disappear, for the answer will be self-evident.§

Obviously, the scope of the program is limited in time by its organizational assignment. The standards engineer who reports to the chief production engineer will never find opportunity to carry out standards work in the sales department. Consequently, the standards engineer, from the beginning, should be encouraged to cut across organizational lines somewhat to make his program companywide and not departmentalized. In a relatively short time, the universality of the standards effort should be apparent, and its worth to the company as a whole, completely clear. Thus, the standards engineer, if he is not in a high place in the organization, should in time certainly report to a high place.

8.3.5 Standards Department

When the amount of standards activity warrants, the formal establishment of a standards

---

*For this orientation see Roy P. Trowbridge, "The General Motors Standards Program," *The Magazine of Standards*, **34**, No. 3 (Mar. 1963), 67–72.

†See Leo B. Moore, "The Human Side of Standards," *Standards Engineering*, **X**, No. 4 (Aug.–Sept. 1958), 1–10.

‡For one approach see William D. Claus, "How Celanese Uses Standards," *The Magazine of Standards*, **35**, No. 12 (Dec. 1964), 355–59.

§To exemplify see "How Much Standardization Is Profitable?" 1. Forest H. Bump, The Small Standards Program, and 2. Arnold M. Rosenwald, The Medium Size Standards Program, *Standards Engineering*, **XV**, No. 1 (Jan. 1963), 7–10. Also see Henry A. Pohs, "Standards in a Multi-Plant Corporation," *The Magazine of Standards*, **33**, No. 12 (Dec. 1962), 361–66.

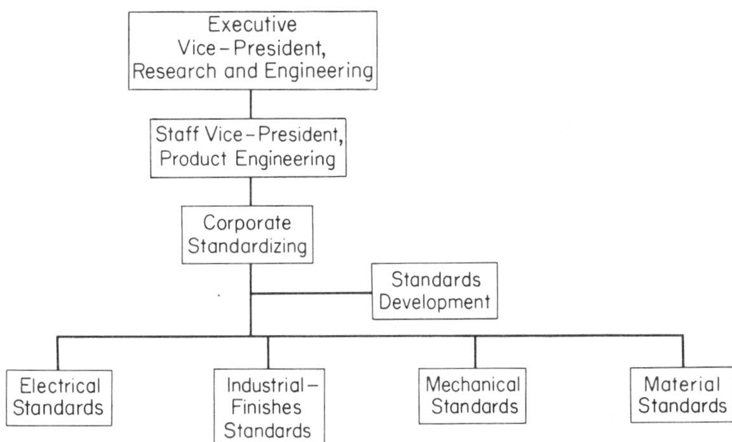

Fig. 8.3 *Standards organization—standards activity assigned to engineering. Courtesy RCA.*

department brings together the standards engineer(s) and appropriate support personnel and facilities as a responsible group. The organizational breakdown of this unit usually reflects the type of standards being pursued—for example, mechanical, electrical, and material—as well as such administrative activity as seems desirable. (See Fig. 8.3.)

Included in the realm of administration are the scheduling and control of standards projects, the setting up and management of standards committees, the publication and distribution of approved standards, and the updating and revision of existing standards—all of which are continuing special and unique responsibilities for the standards department. Frequently, the department must maintain, in addition to a complete file of its own company standards, a library of standards from other sources, a compilation of suppliers' catalogs, and general reference material. The standards department often becomes an information center of considerable value to the entire organization.*

8.3.6 Areas of Operation

The principles of standardization may be applied to every phase of a business. As a matter of fact, however, standards engineers concentrate on those areas that combine both need and return.

8.3.6.1 Materials. Almost every company has done standardization work in connection with materials. These material standards are incorporated in specifications, which are used not only by purchasing and by inventory control but also by manufacturing and by design engineering. The specification generally indicates the use of the material and covers a variety of characteristics, such as chemical, electrical, and physical properties, and usually test and rejection procedures. The basic principle is that at least all properties that are necessary will be specified. The effect of these standard specifications is to provide uniformity and agreement, as between vendor and purchaser. (See Fig. 8.4.) Within the company they provide, for receiving, an exact understanding of acceptable material, and, for purchasing, the possibility of alternate suppliers and quantity buying. The greatest advantage is the mutual understanding that results from a specification of the materials in all their particulars, with complete agreement on the meaning of terminology and on quantities and tolerances. Most companies have specifications for their raw materials, and many have extended this activity to their manufacturing, maintenance, and service supplies.

8.3.6.2 Parts. The next most extensive area in which standardization is practiced is in the

---

*See R. J. Abele, "A Standards Department—Organization and Operation," *Standards Engineering*, **XVI**, No. 3 (Mar. 1964), 1, 9–10. And in the same issue Sherwin Gardner, "A Standards Section—Function and Responsibility," 3–5.

PURCHASING SPECIFICATION

RADIO CORPORATION OF AMERICA
PRODUCT ENGINEERING CORPORATE STANDARDIZING CAMDEN, N.J.

2010024
PAGE - 1 OF 1
DATE - JUNE 1, 1963
(Formerly PS-24)

CODE IDENT NO. 49671
COMMODITY CODE 3211

SUBJECT SPRING STEEL STRIP, BLUED

1. Scope - This specification applies to hardened and blue tempered spring steel strip for use in the manufacture of high quality steel springs.

2. Composition - The material shall conform to the following requirements as to chemical composition:

Element	Percent
Carbon	0.89 to 1.04
Manganese	.30 to .50
Silicon	.10 to .20
Sulfur	.050 maximum
Phosphorus	.040 maximum
Iron	Remainder

3. Temper - The material shall conform to the following requirements as to temper.

3.1 Over .020 Inch Thick - Material over 0.020 inch in thickness shall be tested for hardness only. The hardness shall be 66 to 70, inclusive, Rockwell 30N Scale.

3.2 Up to .020 Inch Thick - Material 0.020 inch and under in thickness shall be tested for bending properties only. Strip of any width shall be bent 180 degrees between the jaws of a vise or testing machine, around pins of the diameters specified in Table 1, in accordance with Figure 1. The center line of the pin shall be approximately one inch from the outside of the bend.

The material shall either fracture transversely or remain intact in accordance with Table 1. When a break occurs, it shall approximate a straight line perpendicular to the longitudinal axis of the test strip.

TABLE 1 - Bend Test Requirements

Thickness, Inch	Diameter of Pin (D), Inch	
	Without Fracture	With Fracture
Up to .0015	.062	.010
Over .0015 to .003, incl.	.093	.035
Over .003 to .006, incl.	.156	.062
Over .006 to .009, incl.	.218	.093
Over .009 to .012, incl.	.312	.140
Over .012 to .015, incl.	.375	.187
Over .015 to .020, incl.	.438	.218

FIGURE 1 - Bend Test

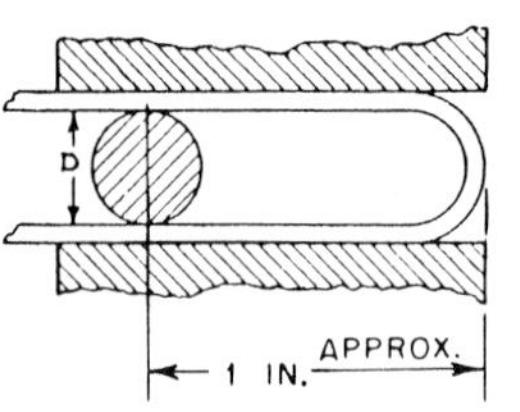

4. RCA Part Numbers - RCA part numbers for some thicknesses are given below. In earlier issues of this specification, material was identified as "PS-24" and the thickness was separately specified.

Nominal Thickness, Inch	RCA Part Number	Nominal Thickness, Inch	RCA Part Number
.001	2010024-1	.018	2010024-18
.002	2010024-2	.020	2010024-20
.003	2010024-3	.024	2010024-24
.004	2010024-4	.025	2010024-25
.005	2010024-5	.030	2010024-30
.006	2010024-6	.032	2010024-32
.007	2010024-7	.036	2010024-36
.008	2010024-8	.040	2010024-40
.009	2010024-9	.042	2010024-42
.010	2010024-10	.048	2010024-48
.012	2010024-12	.050	2010024-50
.014	2010024-14	.060	2010024-60
.016	2010024-16		

Assignment of an RCA part number does not necessarily indicate ready availability in any quantity. See RCA standards for guidance in selection of thickness.

5. Dimensions - Dimensions shall be as specified, in accordance with the tolerances listed below:

Nominal Thickness, Inch	Tolerance Plus or Minus
Up to .020, incl.	.0005
Over .020 to .040, incl.	.001
Over .040	.002

Width Plus or minus .005 inch

Strips shall be free from transverse bow, and the lateral bow or camber shall not exceed .38 inch in eight feet.

6. Form and Finish - Unless otherwise specified, material shall be supplied in coils of approximately equal length, in accordance with the usual commercial practice. The finish shall be the polished and blued grade.

The steel shall be uniform, clean, blue and smooth. It shall be free from rust, dirt, pits, seams, spots, cracks, slivers and other imperfections, in accordance with good commercial practice.

Edges shall be free from burrs, unless the order specifies that slit edges are acceptable.

7. Packing and Marking - Material shall be packed in such manner as to protect it from damage during transit or storage.

Each bundle or shipping unit shall be suitably marked or tagged with the RCA part number; the width, length and net weight of material; the Purchase Order number and the name of the supplier.

Similar Specifications -

AMS	5122B
American Iron and Steel Institute	C1095 (Chemical Composition)

PS-1
29/M

REAFFIRMATION OF ISSUE DATED DECEMBER 1, 1957; NO CHANGES

PRINTED IN U. S. A. ▌INDICATES CHANGE, ADDITION OR DELETION SUPERSEDES DECEMBER 1, 1957

Fig. 8.4 *Standard purchasing specification—an example of a standard that promotes communication and agreement in and outside the company. Courtesy RCA.*

MACHINE PARTS

SECTION C
Page 3

OUTLINE OF NUMBERING

February 6, 1962★

C1-C9 FASTENING DEVICES
- C1 Bolts and screws
  - C1.2♣ Screw and washer assembly
  - C1.3♣ Studs
  - C1.4♣ Weld screws
  - C1.5♣ Shoulder screws
  - C1.6♣ Wood screws
  - C1.7♣ Thumb screws
  - C1A Alloy steel
  - C1B Brass
  - C1C Medium carbon steel
  - C1D Chrome stainless steel
  - C1E Bronze
  - C1F Chrome-nickel stainless steel
  - C1G Galvanized
  - C1H Aluminum
  - C1L Low carbon steel
  - C1M Nickel-copper
- C2 Nuts
  - C2.1♣ Sheet metal
  - C2.2♣ Weld
  - C2.3♣ Wing
  - C2.4♣ Nut and washer assemblies
  - C2.5♣ Cap nuts
  - C2A Alloy steel
  - C2B Brass
  - C2D Chrome stainless steel
  - C2E Bronze
  - C2F Chrome-nickel stainless steel
  - C2G Galvanized
  - C2H Aluminum
  - C2L Low carbon steel
- C3 Rivets
  - C3.1 Eyelets
  - C3B Brass
  - C3F Chrome-nickel stainless steel
  - C3H Aluminum alloy
  - C3K Copper
  - C3L Low carbon steel
  - C3M Copper-nickel alloy
  - C3Y Miscellaneous
- C4 Washers
  - C4 1♣ Lockwashers
  - C4A Chrome-nickel stainless steel
  - C4B Copper alloy
  - C4G Galvanized
  - C4L Low carbon steel
  - C4N Mica
- C5 Pins
  - C5.1♣ Grooved straight
  - C5.2♣ Spring dowel
  - C5.3♣ Drive
  - C5.4♣ Plug button
  - C5.5♣ Threaded dowel
  - C5A Taper, dowel, cotter
- C9 Miscellaneous
  - C9.1♣ Spring retaining rings, precision
  - C9.2♣ Spring retaining rings, square or round

C1-C9 FASTENING DEVICES (Cont)
- C9 Miscellaneous (Cont)
  - C9.3♣ Spring retaining rings, spiral wound
  - C9.4♣ Wire inserts

C10-C19 CURRENT-CARRYING PARTS
- C10 Contacts
  - C10A Contact tips
- C11 Brushes
  - C11A Carbon brushes
- C12 Connectors
  - C12.0♣ General and index
  - C12.1♣ Brazed connectors
  - C12.10♣ Power connectors
  - C12.11♣ Power connectors
  - C12.12♣ Power connectors
  - C12.13♣ Power connectors
  - C12.14♣ Power connectors
  - C12.15♣ Power connectors
  - C12.16♣ Power connectors
  - C12.17♣ Power connectors
  - C12.20♣ Screw pressure - electrical
  - C12.21♣ Screw pressure - electrical
  - C12.22♣ Screw pressure - electrical
  - C12.23♣ Screw pressure - electrical
- C13.0 Terminal boards

C20-C29 BEARING DEVICES
- C20 Ball bearings
- C21 Roller bearings
- C22 Sleeve bearings and bushings
  - C22A Sintered
  - C22B Cast
  - C22C Wrought
  - ♣Page 1 oil rings
- C28 Lubricating fittings
  - C28.1♣ Lubricating fittings
- C29 Miscellaneous (meter jewels, etc)
  - C29A Balls

C30-C39 MOTION TRANSMITTING PARTS
- C30 Knobs, wheels, handles, etc
- C31 Pulleys, sprockets, etc
- C32 Gears
- C33 Springs
- C34 Keys
  - C34.1♣ Machine keys
  - C34A Woodrull keys
- C39 Miscellaneous

C52-C59 OTHER MECHANICAL PARTS
- C56 Seals and gaskets
- C57 Piping
  - C57.2♣ Screwed fittings
  - C57.3♣ Solder-joint fittings
  - C57A Pipe plugs
- C58 Shells, enclosures, etc
- C59 Nameplates
  - C59.1♣ Standard nameplates
  - C59B Sheet brass
  - C59E Cast bronze
  - C59F Stainless steel
  - C59H Aluminum

C60-C69 PRODUCT
- C60 Motor & generator parts
  - C60A Capacitors
  - C60B Vibration isolation mountings
  - C60C Commutator insulation rings
- C66 Electronic tube parts
  - C66A Lead wires
  - C66B Sleeves, cathodes, caps
  - C66C Magnet shapes
  - C66Y Material parts

C70-C79 FUNCTIONAL COMPONENTS
- C70 Protective and control components
  - C70A Resistors
  - C70B Capacitors
  - C70C Coils and inductors
  - C70D Switches
  - C70E Fuses
  - C70F Relays
- C71 Energy conversion components
- C72 Energy utilization components

C110♣ INSERTS FOR PLASTICS

♣ Parts identified solely by drawing numbers.

Courtesy General Electric.

SD46A GENERAL ELECTRIC ★Supersedes issue of May 9, 1957

Fig. 8.5 *Standard machine parts list. One page of the index of selected parts standardized for company use. Courtesy General Electric.*

field of parts, with particular emphasis on purchased parts. (See Fig. 8.5.) These specifications indicate the part by name and number and the various alternates of dimensions and shapes, including such other information as threads, finish, design data, and material. These standard-parts specifications are usually compiled in a standard-parts book and range from ball bearings to washers.

8.3.6.3 Engineering Procedures and Practices. The specification of materials and parts has naturally led to or has required standardization in many engineering procedures and practices. A standard numbering system might be developed for both parts and materials, and later extended to tools, jigs, fixtures, and equipment of all kinds—not only in the productive areas, but also in the service areas. A list of standard abbreviations, symbols, and terminology might be prepared, and from this will come the need for standard drafting procedures, including paper, lettering, and detailing. A further natural development would be the establishment of standard test methods. (See Table 8.2.) These test methods, including equipment and procedures, would be designed to assure the company that the supplier was complying with the standard material and part specifications. Many companies have gone a step further into the area of engineering design, including such items as limits, fits, tolerances, roughness, finishes, gaging, and similar design and mechanical data.

8.3.6.4 Manufacturing Processes. Many companies have done standardization work in what might be called manufacturing processes and methods. These standards are in the areas of tool and equipment standards, including

TABLE 8.2. *STANDARD TEST METHOD—AN EXAMPLE OF A TESTING PROCEDURE STANDARDIZED AND ACCEPTED FOR MEASURING COMPLIANCE WITH SPECIFICATION**

G–E Test Method E1B39 governs the storage stability test for pressure-sensitive tapes, as follows:

E1B39A—50 per cent relative humidity
E1B39B—90 per cent relative humidity

APPARATUS:
Constant-temperature oven
Desiccator

TEST SAMPLE:
Roll of tape 1 in. wide and at least 10 yd long.

E1B39A—50 Per Cent Relative Humidity

PROCEDURE:
Place the sample roll in a sealed desiccator over a solution which will give a relative humidity of 50 ± 5 per cent at a temperature of 150°F. (Suggested solutions are sulfuric acid with specific gravity of 1.339 or a glycerine solution having a refractive index between 1.4440 and 1.4486.)
Store the sealed desiccator in an oven maintained at 150° ± 5°F for a period of 30 or 72 hr as specified.
After aging, allow the tape to cool for 2 hr at room temperature.
Examine the tape to determine the condition of the adhesive and backing.

REPORT:
The report should include the purchase order number, the manufacturer's name, the G–E designation, and a statement of whether the tape meets the requirements of the specification.

E1B39A—90 Per Cent Relative Humidity

PROCEDURE:
Place the sample roll in a sealed desiccator over a solution which will give a relative humidity of 90 ± 1 per cent at a temperature of 150°F. (Suggested solutions are sulfuric acid with specific gravity of 1.1368 or a glycerine solution having a refractive index between 1.3773 and 1.3905.)
Store the sealed desiccator in an oven maintained at 150° ± 2°F for a period of 6 days.
After aging, allow the tape to cool for 2 hr at room temperature.
Examine the tape to determine the condition of the adhesive and backing.

REPORT:
The report should include the purchase order number, the manufacturer's name, the G–E designation and a statement of whether the tape meets the requirements of the specification.

* Courtesy General Electric.

TABLE 8.3. *STANDARD MANUFACTURING PROCESS—WRITTEN OPERATIONAL METHODS ASSURE UNDERSTANDING AND ENCOURAGE COMPLIANCE**

G-E Manufacturing Process P4C5 covers the procedure for electrolytic alkali cleaning of ferrous materials.

MATERIALS AND EQUIPMENT:

Electrolytic cleaning tank

Cleaning bath (1)

Alkaline cleaner .... 8–10 oz/gal

Water to make .... 1 gal

(1) Various proprietary cleaners such as Pennsalt or Clepo are available. Consult your laboratory for approved materials. Material substitutions with corresponding changes in concentrations should be made only with the approval of the laboratory.

SAFETY PRECAUTIONS:

Acid-resisting rubber gloves, rubber aprons, tight-fitting goggles or face shield must be worn.

If cleaning solution is spilled on the body, wash immediately with water.

If cleaning solution gets into eyes, a small stream of running water should be directed *gently under* the eyelids for several minutes before going to the dispensary.

An exhaust ventilation system with monel, acid-resisting coated wood, or galvanized iron ducts and scrubbed air is recommended.

PROCEDURE:

Surface preparation:

Preclean—Remove oil and dirt. Use G-E Manufacturing Process P4C2 (vapor degreasing) or/and G-E Process P4C4 (emulsion cleaning).

Electroclean—Immerse in electrolytic cleaning tank, making the parts cathodic for 90 sec and then anodic for 30 sec. The electrolytic cleaning bath operates at a temperature of 200°F with a current density of 50–60 amp/sq ft.

Rinse—Rinse thoroughly in running cold water.

Dry—If plating operations do not follow immediately, dip in hot water and dry by suitable means such as filtered air blast.

* Courtesy General Electric.

jigs and fixtures, and of operational standards, such as welding, heat treating, hardening, dehydrating, painting, cleaning, carbonizing, and other technical procedures. (See Table 8.3.) These manufacturing standards are particularly evident in the mechanical industry but are not restricted to it.

8.3.6.5 Additional Areas. Most companies experience a growth in standards activity that reflects increased need and value. The initial engineering standardization extends into the manufacturing area and then carries into essentially administrative functions. Safety standards, for example, are an important area of standards work. For some companies, personnel practices and procedures, security regulations, and even wage-incentive and bonus arrangements are standard methods of operation.

In the office area, the extension to written policy and procedure follows naturally in such fields as organization, accounting, branch plant and office operation, and other aspects of control of the enterprise. Within marketing, the treatment as standards problems of such matters as catalogs, installation and servicing instructions, replacement parts lists, and operating instructions establishes good practice and contributes to good customer relations.

The support that standards activity provides to ongoing programs in the company promotes increased opportunity for growing standards work. Quality control, reliability, value analysis, maintainability, configuration management, to mention but a few, all depend heavily upon a vital and expanding company standardization program.

8.3.7 Standardization Techniques

Even though standardization is generally considered an engineering activity, very few technical methods of attack on the problems of standardization have been developed.

8.3.7.1 Simplification. The easiest step to take in standardization is to use the technique known as simplification. This is the process of reducing variety by elimination. A company that makes 260 types of washers decides, by reviewing the cost and sales situation, not to make 150 of them. Another company, which uses twenty-six different types of lubricants, analyzes its needs and finds that ten will satisfy its requirements. This process is as simple as these examples suggest; the prime problem is to obtain some agreement on what will be eliminated. This work is the application on a company level of the principles and practices, described previously under the Department of Commerce, that resulted in simplified practice recommendations (Par. 8.2.3.3).

8.3.7.2 Deliberate Decision. Simplification consists of a review of current variety followed by a decision on what is acceptable and therefore standardized. This technique presupposes that the present variety has been the result of considered engineering thought and deliberate selection. Such is usually not the case. The result is, therefore, standardization by compromise. The more desirable engineering investigation to determine the needs of the situation has not been made. There is no technique that applies to this final accomplishment of standardization, referred to as *deliberate decision*, except the straightforward engineering approach coupled with patient participation.

8.3.7.3 Preferred Numbers. There is one tool of standardization—preferred numbers—that supports deliberate decision. These are numbers that have been selected and standardized in the form of a series of geometrical relationships, so that the relationship of the value of one number to the next is always the same. Thus, in the five series from 10 to 100 there are five numbers that are approximately 60 per cent apart from each other. These numbers (Table 8.4) have been adopted for use by designers in preference to arbitrary numbers in order to create uniformity and consequent interchangeability. They are applied to sizes, ratings, or characteristics, such as diameters, lengths, areas, volumes, weights, and capacities.*

*For an excellent treatment with examples see I. R. Weir, "A Standard Aid for Designers—Preferred Numbers," *The Magazine of Standards*, **33**, No. 3 (Mar. 1962), 72–77.

8.3.7.4 Numbering Systems. Standards engineers have developed administrative techniques that support and direct their programs. The most common of these is the standard numbering system. The type, form, and content of the system vary from company to company. The simplest system is to number from one on up; or the system may be more complex, employing combinations of letters and numbers. In any case, the purpose is to designate the material, part, process, procedure, or other standard to be specified. Through the numbering system, the standards group can extend its activities legitimately, for nonstandard items, being unnumbered, are plainly visible. In like manner, the group can control nonstandard practice in the company by requiring that a standard number be assigned to all materials, parts, and so forth.†

8.3.7.5 Standard Format. Unless detailed thought is given to the form and content of the prepared standard at the start, they present recurring problems. Most standards engineers follow the easiest course of being guided by examples from other companies or from other groups, like the USASI, and of accepting and incorporating the suggestions of their company executives. This means that experience gained in preparing standards tends to indicate necessary change in the form, content, wording, and presentation of the standard until one generally acceptable format, uniquely the company's, is developed.

Considerable help in developing a uniform format will be gained by consulting the USASI *Style Manual* for American Standards and by using its recommendations as a guide toward more readable and understandable standards.

8.3.7.6 Printed Standards. The next most consistent technique is the practice of putting standards in writing. The advantage of written standards in promoting understanding and compliance is evident. For the standards group, the decision on what the form and content of the standard should be requires a complete survey of the standard before its publication; in many cases, problems of agreement are

†For a thorough coverage see Slack Ulrich, "An Effective Engineering Drawing and Part Numbering System," *Standards Engineering*, **XV**, No. 7 (July 1963), 1–10.

TABLE 8.4. *BASIC PREFERRED NUMBERS—THE FIVE, TEN, TWENTY, AND FORTY DECIMAL SERIES (10 TO 100)**

Five Series (60 Per Cent Steps)	Ten Series (25 Per Cent Steps)	Twenty Series (12 Per Cent Steps)	Forty Series (6 Per Cent Steps)
10	10	10	10
			10.6
		11.2	11.2
			11.8
	12.5	12.5	12.5
			13.2
		14	14
			15
16	16	16	16
			17
		18	18
			19
	20	20	20
			21.2
		22.4	22.4
			23.6
25	25	25	25
			26.5
		28	28
			30
	31.5	31.5	31.5
			33.5
		35.5	35.5
			37.5
40	40	40	40
			42.5
		45	45
			47.5
	50	50	50
			53
		56	56
			60
63	63	63	63
			67
		71	71
			75
	80	80	80
			85
		90	90
			95

Preferred numbers below 10 are formed by dividing the numbers between 10 and 100 by 10, 100, etc.
Preferred numbers above 100 are formed by multiplying the numbers between 10 and 100 by 10, 100, etc.

Percentage steps in headings are approximate averages.

* Reproduced from *American Standard Preferred Numbers, Z17.1-1958* (New York: USA Standards Institute, 1958), p. 8, copies of which may be purchased from the USA Standards Institute, 1430 Broadway, New York, New York 10001.

introduced on control data that have never been considered by the company before. Thus, in the specification of a cleaning solution, prior to standardization, the brand name might have been the critical characteristic, whereas, after standardization, the quality of carbon tetrachloride by weight in the solution might be the critical characteristic. Obtaining agreement from all parties concerned in regard to really critical information provides the standards group with a broader base for its activities. The printed standards, collated and assembled, generally in loose-leaf binders, constitute for every company the heart of the standards program. These standards manuals, as they are called, are kept up to date by the standards department and are distributed to all interested sections of the company.*

8.3.7.7 Standards Control. Through standards manuals, the department gains standards control over company practices, such as in design, manufacturing, and purchasing. In turn, they furnish operating personnel with an opportunity to indicate additional areas for standardization work in the form of recommended standards projects. In exercising control, probably the best practice is to require that the standards group be informed whenever a nonstandard is employed. The purpose of this requirement is not to require compliance but to indicate the need for investigation of nonstandard practice. Thus, the purchasing department might be required to send a copy of a requisition for a nonstandard material to the standards department for the latter's information. Frequent use of nonstandard practice certainly would indicate the need for review of the situation and for action on the part of the standards engineer.†

8.3.7.8 Outside Contacts. The standards group should, and usually does, maintain an active liaison with other individuals, organizations, or groups that do standardization work. It is desirable that the company join the USA Standards Institute and that the standards engineer be a member of the Standards Engineers Society and be active in the standardization work of USASI, of trade associations, and of technical societies. In addition, the standards engineer should keep in close touch with the standards and specifications work of government agencies, particularly with respect to his own industry. It is good standardization practice to permit the standards promulgated by the USASI, trade associations, technical societies, and the government to form the basis for standards approved and accepted within the company.

8.4 BENEFITS FROM STANDARDIZATION

Around the world, rapid change in technological capability, greater employment of mass-production techniques, and increased complexity in communication and decision-making have provoked intensive activity at every level of standardization in order to gain its benefits and to tap its resources. At the international level, where the present problems of man and the future opportunities for his world are best explored, no force offers more potential for depth of understanding than the practical use of mutual agreement and consent, a basic principle of standardization. Consensus on matters technical and scientific, encourages fuller consideration of the economic and social, medical and ethical. The absolute dependence of effective world trade upon standards, for example, has long been the well-understood basis for continuing international cooperation in standardization, even under challenging and difficult conditions. The cohesive force that holds together various groups of nations, especially those in the common markets, is the growing strength of all types of standards that support mutual business and commerce.‡

On the national level, standardization continues to be the prime tool for the elimination of waste in industry. The standardization process, particularly critical thinking in the evaluation of alternatives, creates within every standard for materials, equipment, tools, products, and all things manufactured and produced the potential for increasingly efficient production. In like manner, mass production as an economic philosophy is supported by a solid foundation of standardization since production in large quantities is best accomplished under standardized conditions. The elimination of duplication and the reduction of variety

*J. R. Walgren, "Alcoa's Engineering Standards Manual," *The Magazine of Standards*, **33**, No. 8 (Aug. 1962), 241–43.

†For one aspect of control see R. C. Dahlin, "A Corporate Part Numbering System," *Standards Engineering*, **XII**, No. 6 (Dec. 1960–Jan. 1961), 3–7.

‡A. Y. Viatkine, "Standardization and the Development of Trade," *The Magazine of Standards*, **35**, No. 8 (Aug. 1964), 237.

wherever possible provide the same salutary effect through the conservation of the existing resources of the country and the more effective employment of those limited resources which are available.*

For the company that employs standardization even in a limited formal way the benefits that accrue from effective coordination are evident throughout the entire enterprise. Every standard represents a considered decision upon which dependable conduct can be based and the consequences predicted. On this basis alone, professional management can face its deeper problems with a sense of freedom from routine cares and concerns. In every company, standards implement the concept of the exception principle with full vigor and return. With the exception principle operating, the manager is assured that he will be made aware of any event that does not meet standard practice and of any circumstance that predicts difficulty in attaining standard practice in the future.

8.4.1 Basic Merit

The wide range of benefits that standardization provides can be explained only by the fact that standardization is a pervasive and underlying concept in the management of human affairs. For industrial management, standardization represents a mode of communication that provides knowledge and promotes commitment in support of its goals. The operating objectives of every manager are most effectively met by using standardization as a tool. When a manager rests his continuing technique upon the principles and practices of standardization, he will more consistently realize its basic merits.†

Each standard represents the codification of knowledge. The consensus method employed in the formulation of the standard provides a foundation for better understanding by bringing together all those with interest and experience in the project and encouraging intelligent discussion and deliberation. From this activity there emerges an expression of knowledge mutually agreed upon. Such knowledge provides the power not simply for reaching better decisions but also for making them consistently better because through standards we have a concrete way to talk about what we know and we can discuss more clearly what we do not know.

Commitment to a goal is best achieved through communication and understanding. The anticipation of the problems involved in the pursuit of any given goal invariably raises the question of interface situations. Nothing is more helpful in these cases than a set of standards that facilitate the required planning, direction, and control. With this advance knowledge, desired goals are reached with competent assurance in an atmosphere of conviction and determination.

The far-ranging merit of standardization results not only from the fact that a standard represents knowledge but also from the fact that it was agreed to and accepted by those concerned with its nature and intended employment. Thus it is not simple knowledge that a standard represents but useful knowledge to be employed in a worthy fashion. A standard, then, provides for management the kind of communication and commitment that serve the goals of a profitable, economic, and growing enterprise, and support a philosophy of abundant production for the good of more consumers around the world.‡

8.4.2 Functional Values

Although standards may be promulgated by the many organizations in the levels above the company, long-term values develop and grow within the ranks of the company. These values may specifically vary from company to company, but every company can expect to realize similar return on its investment in standards when it makes the philosophy of standardization an integral part of its operation.§

Purchasing depends upon standards and

*For the importance of standardization to national existence see Isidore Stern, "The Tempo of Standardization," *The Magazine of Standards*, **33**, No. 5 (May 1962), 139–41.

†For the service performed for management see W. E. Wall, "The Value of Standardization as a Tool of Management," *Standards Engineering*, **XV**, No. 2 (Feb. 1963), 1, 8, 9.

‡To understand this problem from another viewpoint, that of British industry, see "Standards for Productivity," *The Magazine of Standards*, **34**, No. 7 (July 1963), 216–17.

§Irving D. Miner, "Selling Standards to Management," *Standards Engineering*, **XV**, No. 8 (Aug. 1963), 3–6.

specifications to provide long-term effectiveness in its operations since it stands directly between the company and its suppliers. Standards provide the basis for buying only what is wanted on long-run contract terms in economical lot sizes and under truly competitive conditions. The reduction of special items and inventory levels, of consequent handling, storing, and record-keeping, and of the confusion of complicated communication results in the simplification and improvement of purchasing activities. Seeking more satisfactory sources of supply and opportunities for greater value through value analysis may then receive competent and careful attention.*

Engineering employs standardization to convey to all concerned its technical decisions regarding the various aspects of the product prepared for the marketplace. The vehicle for this communication is the drafting department as the typical source of the requisite engineering documentation. Design standards as well as those for performance, testing, and packaging, for example, represent the codification of engineering knowledge useful to other functions of the business. Through the processes of standardization, engineering accumulates specialized information in support of company operations while drawing upon current developments in knowledge and practice in its professional areas.†

Manufacturing finds its greatest resource for efficiency in standardization. Longer production runs of standard materials through uniform equipment with fewer changeovers and reduced in-process inventory and material handling contribute to low-cost operations. The establishment of standard methods, practices, and procedures simplifies control and schedule problems as well as reducing the impact of turnover and training.‡

Marketing expects of standardization an improvement in relations with the customer through more careful determination and more certain satisfaction of his needs and wants. The conversion of these needs and wants into standard products with a full line of engineered units composed of standard components gives evidence of real concern for consumer acceptance and approval through easier installation and service, replacement and repair. Such standards simplify sales-training programs, permit concentration on profitable items, and fortify continuing relations with the customer.§

Office management increasingly recognizes the need of applying standardization to its mounting problems of work loads and costs. Paper work itself wherever located in an enterprise represents a substantial and growing burden of business. The standardization of forms, equipment, and methods tends to eliminate unnecessary paper work, routinize office activities, and permit the measurement, control, and improvement of administrative work. The advent of the computer and automatic data-processing systems merely highlights the basic necessity for applying in the office the same techniques that have for so long contributed to the total effectiveness of other productive work.‖

8.4.3 Dollar Savings

The USASI undertook a survey representing 140 documented case studies, covering eighty-one industries and industrial products, for the purpose of compiling proof in terms of dollars or percentages of net savings attributable to standardization.#

8.4.3.1 Specific Cases. Some specific examples of savings given by these companies are listed below. In fairness, not all the best exam-

*For an example of a practical system in purchasing specifications and its significance see F. V. Kupchak, "Westinghouse Materials Standardization," *Standards Engineering*, **XII**, No. 2 (Apr.–May 1960), 15–18.

†N. M. Zizi, "The Standards Engineer's Role in Reliability, Value Engineering, Maintainability," *Standards Engineering*, **XVI**, No. 6 (June–July 1964), 1, 6.

‡For a concrete example of results of effective use of standardization in manufacturing see J. C. Hall, "Ford Manufacturing Standards," *Standards Engineering*, **XIV**, No. 4 (Apr. 1962), 6–9.

§For the role of standards in component product reliability see Paul S. Darnell, "Component and Product Reliability," *Standards Engineering*, **XV**, No. 5 (May 1963), 3–5.

‖Henry A. Pohs, "Part Numbers vs. Data Processing," *The Magazine of Standards*, **35**, No. 7 (July 1964), 202–6.

#For problems and approaches to evaluation see F. C. Ewert, "Savings from Standards," *The Magazine of Standards*, **33**, No. 9 (Sept. 1962), 273–75. Also, R. L. M. Rice, "Standards Evaluation for Profit," *Standards Engineering*, **XV**, No. 1 (Jan. 1963), 1, 11–13.

ples have been chosen. But the fact that savings exist and are known, no matter their extent, is helpful.

Figures show that in one year's output of American automobiles over $8 million was saved for the manufacturers, and of course in cost to the public, because of the large number of minor parts that have been standardized.

A large motor manufacturer reported that a record was kept of the number of hexagonal head screws used in the various models which they built. The reduction in cost after standardization was $52,758.96 in one year.

Ball bearing and electrical equipment manufacturers state that SAE standards have reduced production costs over 20 per cent.

A well-known case of a similar kind is that of an aircraft manufacturer who saved $268 per plane by substituting industry standards for bolts in the place of special company standards.

So many intangibles are involved that savings can only be "guesstimated." "Factory costs are reduced, probably, by 1 to 3 per cent" writes a large manufacturer of type-setting machinery.

Standards are really the lifeblood of motion pictures for without them there would be no motion-picture industry.

There is no doubt that standardization in the electric-utility business is responsible for many of the fine economies which we now enjoy. Also, it has made it possible for us to have more strict construction standards, which results in carrying a minimum of different kinds and types of materials in our storerooms.

We regret that we have no information which will allow us to make an appraisal of the benefits in terms of dollars, but we inherently know that they are substantial with us as with all companies employing the principles of standardization.

Standardization of stationery and printing by a small eastern railroad was reported, in 1944, to have reduced the cost from $110,000 to $65,000 a year, a saving of 41 per cent.

In a paper-manufacturing establishment an increase in "percentage of perfect" from 92 per cent to 96 per cent was achieved in 7 months as a result of the application of quality-control standards.

In a textile-machinery factory, stores inventories were reduced 9 per cent within a space of 4 months.

In a factory manufacturing cooking utensils, the reduction of storage and warehousing costs amounted to $184,716.*

*From "Dollar Savings Through Standards," *Standardization*, published by the American Standards Association (Oct. 1951), p. 306.

8.4.4 Popular Objections

It is not unusual for such a fundamentally useful managerial concept to cause concern among those who question its value. Most popular objections to standardization are based on the undue, unwise, or intensive pursuit of its principles and practices, and give little or no credit to counterforces. One objection is the prediction of drab and depressing sameness, with its regimentation and uniformity. This prediction is extended to include the suppression of creativity and innovation, with stultification of progress and growth.†

These and similar objections have no merit in a free economy characterized by market competition, enlightened management, and dynamic growth. Such forces as these lead to the regular review and revision of every standard to determine whether it continues to serve the overriding needs of the company and the economy. Thus, standardization provides the basis for quantity production and mass consumption for every nation on earth, while at the same time in no way interfering with the satisfaction of those who want something different, special, or new and are willing to pay for it. It should be noted that even special items are often constructed of standard parts and materials, utilizing standards knowledge and giving only the desired special aspect of individual treatment. The manager uses standardization as a tool wisely when he recognizes that the way the tool is employed is more important than the tool itself.‡

8.4.5 Standards Problems

Here are some of the perplexing situations that are likely to arise in any standardization program:

1. Gaining and maintaining top management's active, rather than passive, support
2. Deciding extent and limits of acceptance of standards throughout the company
3. Arbitrating requests for use of nonstandard conditions

†For the specific technical and human obstacles to standardization see I. R. Weir, "Obstacles in Standardization," *Standards Engineering*, **XII**, No. 2 (Apr.–May 1960), 7–10.

‡P. K. McElroy, "What Practical Standardization Is Not," *Standards Engineering*, **XIII**, No. 3 (June–July 1961), 14–17.

4. Timing the promulgation and adoption of standards

5. Exploiting the opportunity to extend the scope and activity of the standards group

6. Developing coordination in the total standards effort

7. Resisting the pressure of time and situation to formulate a standard before complete

8. Determining the purpose and coverage of a proposed standard

9. Deciding the method and extent of evaluation of the standards effort

10. Determining the nature and extent of participation in standards activities of outside organizations

These and many similar problems are the constant challenges of every company standards program, which must necessarily give continuing evidence of being well managed and productive. At the same time, they tend to reflect the value accorded standards activity by management and standards engineer alike. They symbolize the uncertainty, even indifference, on the part of both toward the full magnitude and true worth of a completely integrated standardization program. The continuing question that must be faced by both together is, "What are we losing by not standardizing?"

8.4.6 The Future of Standards

All signs point to increasing benefits from standardization in the future. Technological developments in their scope and complexity are expanding the horizons for the standards engineer both in the nature of the problems he must face and in the approach he must take to them.* His shift from the provincial view of the immediate demands of present standards efforts to the fuller perspective of the interrelationships of his objectives and those of the entire enterprise is already underway.† How his efforts integrate with the long-range requirements of the economy of his nation and of the world is the basic question he must continue to face for years ahead. Clearly, the reflection of his efforts in the broader mirror of national and international well-being must be one of creative good sense at work.

But creative breadth alone is not sufficient; an environment conducive to this viable potential is absolutely essential. Only when the management of an enterprise integrates the economic and social dimensions of standardization into the operating philosophy of the management group will there be continual conditions for standards contributions to that enterprise. In like manner, management must accept the future responsibility (1) for a broad educational program aimed at how the challenges of our times can be met through standards, (2) for the financial and leadership support of appropriate institutions in the standards field, and (3) for the encouragement of the use of standards principles and practices in the solution of managerial problems of every type. In the United States, there is an awakening realization of this need for breadth and strength in managerial concern, one evidence of which is the establishment of the USA Standards Institute for "USA Standards" to give status, organization, and coordination to the standards movement of the future.‡

In every respect current developments are expanding the future opportunities for the standards engineer and are challenging his ability to properly motivate his enterprise and to release his creativity for greater industrial ingenuity. The standards engineer who performs his function and the management that provides support for an integrated standards program are in the most strategic position to stimulate their industry to constant improvement as well as to greater productivity through standardization.

*Frank Philippbar, "Standards for the Space Age," *Standards Engineering*, **XV**, No. 4 (Apr. 1963), 1, 7.

†Capt. Jonathan A. Barker, USN, "Value, Quality-Reliability, Standards," *The Magazine of Standards*, **34**, No. 6 (June 1963), 175–76. Also, J. J. Conway, "Engineering Standards—A Systems Approach," *Standards Engineering*, **XII**, No. 3 (June–July 1960), 3–7.

‡"Report of Panel on Engineering and Commodity Standards," *The Magazine of Standards*, **36**, No. 4 (Apr. 1965), 115–26.

BIBLIOGRAPHY

Books

Brady, Robert A., *Industrial Standardization*. New York: National Industrial Conference Board, 1928.

Booth, Sherman F., *Standardization Activities in the United States*, Miscellaneous Publication

230. Washington, D.C.: National Bureau of Standards, 1960.

Chase, Stuart, *The Tragedy of Waste*. New York: The Macmillan Company, 1925.

Federated American Engineering Societies, *Waste in Industry*. New York: McGraw-Hill Book Company, 1921.

Gaillard, John, *Industrial Standardization: Its Principles and Application*. New York: H. W. Wilson Co., 1934.

Harriman, Norman F., *Standards and Standardization*. New York: McGraw-Hill Book Company, 1928.

Lansburgh, Richard H., *Standards in Industry*. Philadelphia: American Academy of Political and Social Science, 1928.

MacNeice, E. H., *Industrial Specifications*. New York: John Wiley & Sons, Inc., 1953.

Melnitsky, Benjamin, *Profiting from Industrial Standardization*. New York: Exposition Press, 1953.

Perry, John, *The Story of Standards*. New York: Funk & Wagnalls Co., 1955.

Reck, Dickson (Editor), *National Standards in a Modern Economy*. New York: Harper & Row, Publishers, 1956.

Pamphlets

Dickson, Paul W., *Industrial Standardization*. New York: National Industrial Conference Board, 1947.

Rogers, Jack, *Industrial Standardization*. New York: National Industrial Conference Board, 1957.

Magazines

The Magazine of Standards. New York: USA Standards Institute, published monthly.

Standards Engineering. Binghamton, New York: Standards Engineers Society, monthly.

Every practicing standards engineer should subscribe to the two magazines in the field and have for reference the two pamphlets prepared as Studies in Business Policy by the National Industrial Conference Board. Special publications by societies and associations on subjects pertinent to the standards problems of his company will supplement the general reference books. The standards field is dynamic and growing in importance; its activities are best reflected in current publications.

LAWRENCE E. DOYLE, *professor of mechanical engineering at the University of Illinois, teaches in both the mechanical engineering and industrial engineering areas. Much of his teaching and research deals with process and tool engineering, and he is recognized as one of the authorities in this field. He has written over fifteen papers on various aspects of this subject, and his book,* Tool Engineering: Analysis and Procedures, *is one of the classics. He has collaborated on the* Tool Engineers' Handbook *and* Manufacturing Processes and Materials for Engineers, *and is the author of* Metal Machining *and* Process Analysis and Control.

Professor Doyle earned his B.S. degree in mechanical engineering at Yale University and then worked for Dunn and Bradstreet and Lyon Furniture Merchandising Agency until he joined the Cincinnati Milling Machine Company. He then served as assistant supervisor of process and tool engineering for the Allison Division of General Motors Corporation until 1943, when he joined N. E. Miller and Associates in Detroit.

In 1946 he went to the University of Illinois and started teaching and working on his M.E. degree. After earning that degree he worked his way through the teaching ranks to his present position of full professor. He is a registered Professional Engineer.

Professor Doyle is a member of the American Society of Mechanical Engineers, the American Society for Metals, the Illinois Engineering Council (director 1954 to 1967), Pi Tau Sigma, and Sigma Xi. He has also served in many capacities in the American Society of Tool Engineers (now The Society of Manufacturing Engineers), having been chairman from 1948 to 1968 of the Chicago chapter Professional Engineering Committee and, from 1953 to 1955, chairman of the National Professional Engineering Committee. He received the Service Award of the Chicago Chapter in 1953 and the National Education Honor Award of the Society in 1961.

SECTION
9

TOOL AND MANUFACTURING ENGINEERING

LAWRENCE E. DOYLE

9.1 THE NATURE AND SCOPE OF TOOL AND MANUFACTURING ENGINEERING

9.1.1 Definitions

Tool and manufacturing engineering is concerned with planning the processes, supplying the tools and equipment, and coordinating the facilities for manufacturing a product in an optimum manner. Its main subdivisions are manufacturing analysis, process planning, tool designing, toolmaking, and tool controlling.

Manufacturing analysis consists essentially of long-range planning of manufacturing requirements and facilities; and feasibility studies and development of better processes, tools, machines, and other equipment and of new products and product improvements.

Process planning or *process engineering* is concerned with devising, selecting, and specifying processes, machines, tools, and other equipment to cut, form, and furbish materials, finish manufactured articles, and assemble products.

Tool designing or *tool design* involves the development and design of the special machinery, tools, and gages for production.

Toolmaking entails the fabrication and maintenance of special tools, dies, jigs, fixtures, and other accessories to production equipment.

Tool controlling or *tool control* covers the responsibilities of trying, correcting, and proving new tools, initial and periodic tool inspection, stocking and dispensing tools and gages, and studying tool use and performance.

9.1.2 The Place of Tool and Manufacturing Engineering in an Industrial Organization

Tool and manufacturing engineering is one of the staff functions that supply the necessary services and facilities to enable the productive core of an industrial organization to operate smoothly. In this respect, tool and manufacturing engineering is like purchasing, accounting, and plant maintenance. In a small plant the entire function may be performed by one person as part or all of his duties. In a large organization, engineering for manufacturing may be divided among several departments under the direction of a chief tool engineer, also variously known as the master mechanic, production engineer, or head manufacturing engineer, who reports to the factory manager. Typical department names are process engineering, tool design, gage design, tool room, tool inspection, tool trouble, and tool stores.

Typically, the product engineers send drawings, production-quantity releases, and models of a new or revised product to the tool and manufacturing engineers, who study the project, suggest changes to help production, and submit estimates of the cost of acquiring facilities and manufacturing the product to plant management, from which appropriations are obtained to prepare for production.

Process engineers write routings and issue

orders for machines, tools, and other equipment to establish the means to execute process plans. Toolmakers and the other specialists provide the special tools and machines. Orders for standard tools and machines go to the purchasing department for procurement of the physical facilities. The routings guide other staff functionaries and notify them about requirements: the purchasing department to obtain materials, the personnel department to hire and orientate workers, the production-control department to plan schedules, and factory management to direct production efforts.

9.2 PROCESS PLANNING: ROUTING AND TOOL ORDERS

A process plan is a complete concept of a process; it is expressed in various degrees of detail on several kinds of forms. One form almost always used is a *routing*, also known as a *route sheet*, *process sheet*, and *planning-operation sheet*. It identifies the process by giving the name and number of the article to be produced, the quantity and lot sizes required, etc. It lists the operations to be performed and the machines, tools, gages, and equipment to be used for each. It specifies performance expected in the form of estimated or standard cycle time per piece and setup time per lot. Figure 9.1 is one form of routing.

Routings are as brief as possible to save preparation and reading time; for example, departments and machines may be designated merely by numbers and symbols. Since a routing is used in many parts of a plant, a number of copies are generally issued. All copies contain the basic information, but some may have certain spaces reserved for and containing information of value only to their recipients; for instance, spaces for the accounting department to record certain standard costs.

A *tool order*, exemplified in Fig. 9.2, is a common supplement to the routing to convey specifications for designing or procuring a tool.

An *operator instruction sheet* or *operation sheet*, like the one in Fig. 9.3, describes an operation in more detail than is feasible on a route sheet and customarily includes a sketch of the work to be done in the operation. In some plants operation sheets are prepared for all operations to aid in operation setup, workplace layout, efficient supervision, and operator guidance.

9.3 PROCESS-PLANNING PROCEDURE

Process planning entails the following considerations.

9.3.1 The Requirements and Conditions of the Process

Before any problem can be solved, its requirements and conditions must be defined. The following must be determined first in planning a process:

1. The specifications of the finished product, which may be found for a mechanical device in part and assembly prints, an engineering release for production, a manufacturing order, and a list of parts and materials
2. The size, shape, and other properties of the raw material
3. The quantity or number of pieces of the product to be made and the date required for delivery

The process engineer should make a mental or written list of every item in the specifications to get a full grasp of the project. The items include the surfaces to be finished, quality of finished surfaces, individual dimensions and tolerances, geometric relationships among surfaces, physical properties of materials, surface painting or plating, assemblies, and other refinements. Figure 9.4 is one form for listing and analyzing the specifications for a workpiece.

9.3.2 Improving Specifications

A competent manufacturing engineer is always alert for ways to improve specifications to make production easier and cheaper. He is in close touch with the problems of production and thus in a position to point out changes, ranging from complete revisions to minor alterations in product design, to benefit production. The right to accept or reject changes belongs to the product designer responsible for the performance of the product. Figure 9.5 gives a few illustrations of typical changes.

The following is a suggestive outline of important types of changes to promote economy in processing:

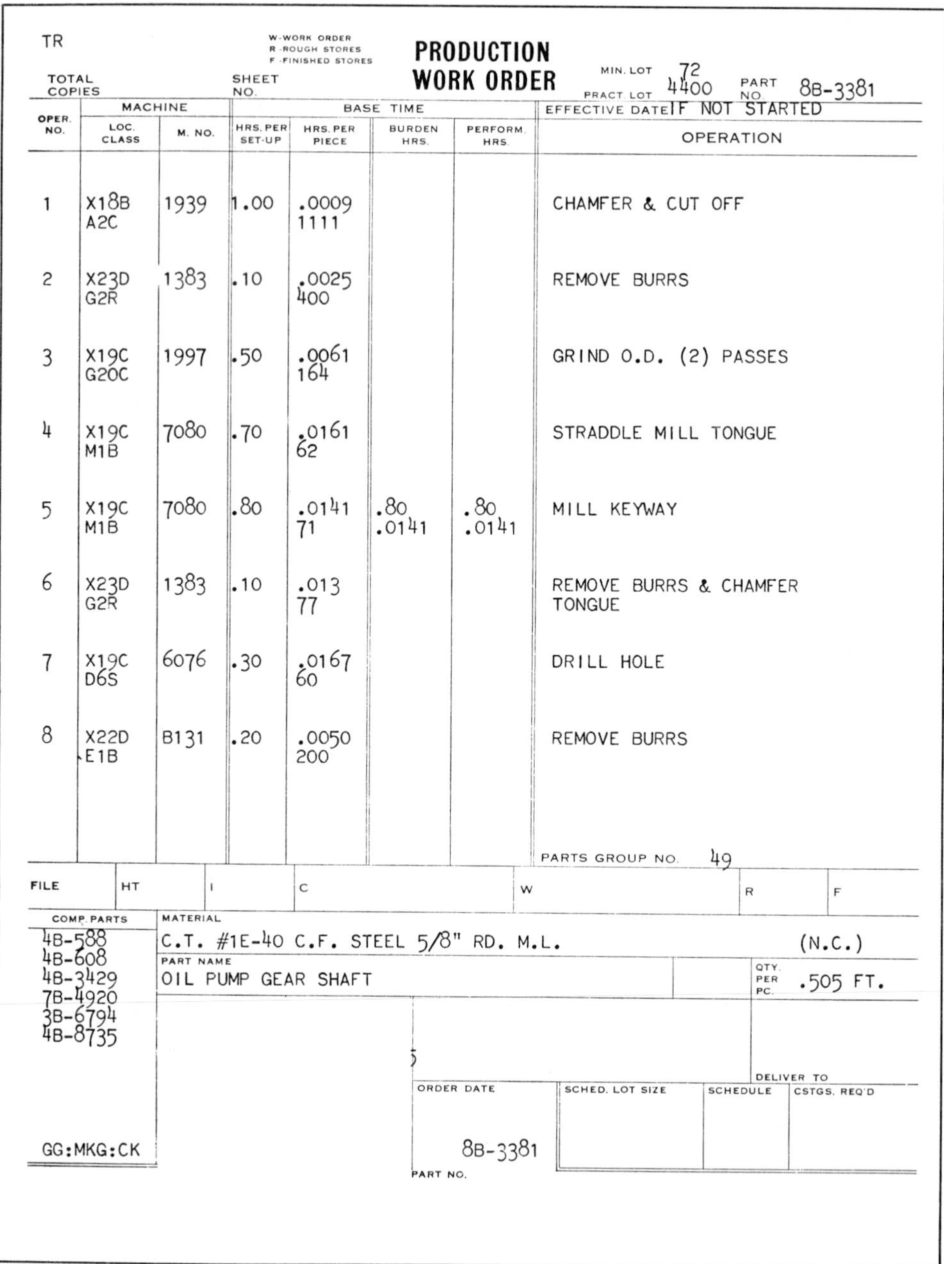

TR

W-WORK ORDER
R-ROUGH STORES
F-FINISHED STORES

PRODUCTION WORK ORDER

TOTAL COPIES — SHEET NO. — MIN. LOT 72 — PRACT. LOT 4400 — PART NO. 8B-3381

OPER. NO.	MACHINE LOC. CLASS	MACHINE M. NO.	BASE TIME HRS. PER SET-UP	BASE TIME HRS. PER PIECE	BURDEN HRS.	PERFORM. HRS.	EFFECTIVE DATE IF NOT STARTED — OPERATION
1	X18B A2C	1939	1.00	.0009 1111			CHAMFER & CUT OFF
2	X23D G2R	1383	.10	.0025 400			REMOVE BURRS
3	X19C G20C	1997	.50	.0061 164			GRIND O.D. (2) PASSES
4	X19C M1B	7080	.70	.0161 62			STRADDLE MILL TONGUE
5	X19C M1B	7080	.80	.0141 71	.80 .0141	.80 .0141	MILL KEYWAY
6	X23D G2R	1383	.10	.013 77			REMOVE BURRS & CHAMFER TONGUE
7	X19C D6S	6076	.30	.0167 60			DRILL HOLE
8	X22D E1B	B131	.20	.0050 200			REMOVE BURRS

PARTS GROUP NO. 49

FILE — HT — I — C — W — R — F

COMP. PARTS: 4B-588, 4B-608, 4B-3429, 7B-4920, 3B-6794, 4B-8735

MATERIAL: C.T. #1E-40 C.F. STEEL 5/8" RD. M.L. (N.C.)

PART NAME: OIL PUMP GEAR SHAFT — QTY. PER PC. .505 FT.

DELIVER TO

ORDER DATE — SCHED. LOT SIZE — SCHEDULE — CSTGS. REQ'D

GG:MKG:CK

PART NO. 8B-3381

Fig. 9.1 *One form of route sheet. Courtesy Caterpillar Tractor Co.*

1. Changes in assemblies to reduce the number of parts needed (as by combining parts) or to simplify parts (as by enlarging tolerances)
2. Additions to parts in the form of pads, ribs, fins, lugs, reliefs, etc., to control warpage, to reduce deflection, to add strength to resist processing forces, to aid in holding the workpiece (as by means of chucking rings), to aid in

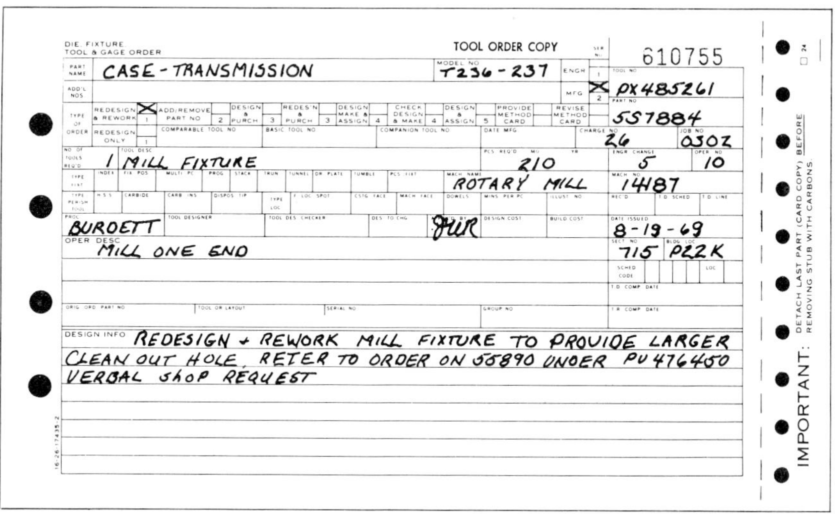

DIE, FIXTURE TOOL & GAGE ORDER — TOOL ORDER COPY — 610755

PART NAME	CASE-TRANSMISSION
MODEL NO.	T236-237
TOOL NO.	PX485261
PART NO.	5S7884
TYPE OF ORDER	REDESIGN & REWORK (X)
CHARGE NO.	26
JOB NO.	0302
NO. OF TOOLS REQ'D / TOOL DESC.	1 MILL FIXTURE
PCS. REQ'D	210
ENGR. CHANGE	5
OPER. NO.	10
MACH. NAME	ROTARY MILL
MACH. NO.	1487
PROC.	BURDETT
DATE ISSUED	8-19-69
OPER. DESC.	MILL ONE END
SECT. NO.	715
BLDG. LOC.	P22K

DESIGN INFO REDESIGN + REWORK MILL FIXTURE TO PROVIDE LARGER CLEAN OUT HOLE. REFER TO ORDER ON 5S890 UNDER PV476450 VERBAL SHOP REQUEST

IMPORTANT: DETACH LAST PART (CARD COPY) BEFORE REMOVING STUB WITH CARBONS.

Fig. 9.2 *A typical tool order. Courtesy Caterpillar Tractor Co.*

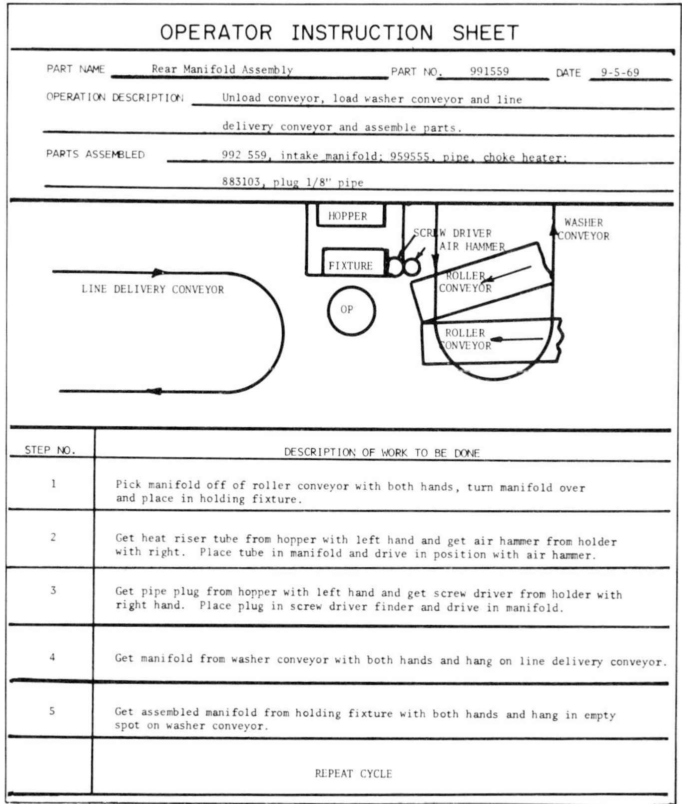

OPERATOR INSTRUCTION SHEET

PART NAME Rear Manifold Assembly PART NO. 991559 DATE 9-5-69

OPERATION DESCRIPTION Unload conveyor, load washer conveyor and line delivery conveyor and assemble parts.

PARTS ASSEMBLED 992 559, intake manifold; 959555, pipe, choke heater; 883103, plug 1/8" pipe

STEP NO.	DESCRIPTION OF WORK TO BE DONE
1	Pick manifold off of roller conveyor with both hands, turn manifold over and place in holding fixture.
2	Get heat riser tube from hopper with left hand and get air hammer from holder with right. Place tube in manifold and drive in position with air hammer.
3	Get pipe plug from hopper with left hand and get screw driver from holder with right hand. Place plug in screw driver finder and drive in manifold.
4	Get manifold from washer conveyor with both hands and hang on line delivery conveyor.
5	Get assembled manifold from holding fixture with both hands and hang in empty spot on washer conveyor.
	REPEAT CYCLE

Fig. 9.3 *An operator instruction sheet.*

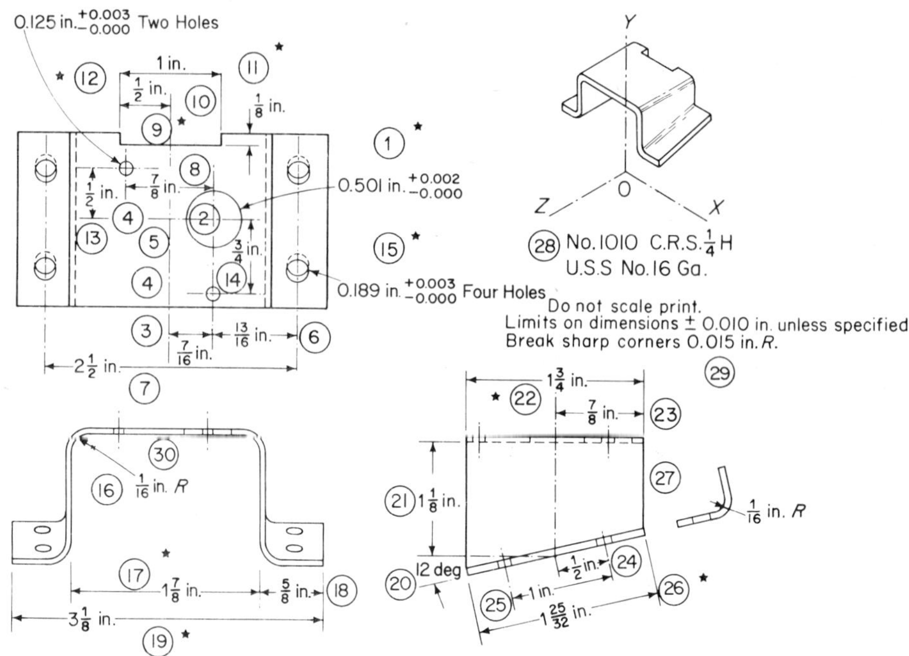

TABULAR ANALYSIS

PART NAME... *Cross shaft bracket* DATE... *Feb. 1, 1964*

PART NUMBER... *P-460 529* PLANNER... *E. A. Reed*

NO.	SPECIFICATIONS	BETWEEN SURFACES-DEPEND ON MATERIAL[1]	DIE[2]	PROCESSING[3]	REMARKS	OPERATIONAL REQUIREMENTS
1	0.501″ $^{+0.002}_{-0.000}$	✓	✓		Squareness implied 90°	★
2	℄ Part to ℄ 0.501″ hole			✓		
3	$\frac{7}{16}$″ to 0.501″ ℄ 0.125″ holes			✓	Possible limit stack with spec. No. 6	
4	℄ to ℄		✓	✓	Implied coincident	
5	℄ ⊥ ℄		✓	✓	Implied 90°	
6	$\frac{13}{16}$″	✓	✓	✓		
7	$2\frac{1}{2}$″	✓	✓	✓		
8	$\frac{7}{8}$″		✓	✓		
9	$\frac{1}{2}$″		✓	✓		
10	1″	✓	✓		Squareness implied 90°	★
11	$\frac{1}{8}$″	✓	✓	✓		
12	0.125″ $^{+0.003}_{0.000}$ (2) holes	✓	✓			★
13	$\frac{1}{2}$″ to 0.125″ hole		✓	✓		
14	$\frac{3}{4}$″ to 0.125″ hole		✓	✓		
15	0.189″ $^{+0.003}_{-0.001}$ (4) holes	✓	✓		Squareness implied 90°	★
16	$\frac{1}{16}$″ Radius (2)	✓	✓		Implied 90° bend	
17	$1\frac{7}{8}$″	✓	✓			★
18	$\frac{5}{8}$″	✓	✓		Possible limit stack with spec. No. 17	
19	$3\frac{1}{8}$″	✓	✓		Implied symmetrical	
20	12°		✓	✓	No tolerance on ∠	
21	$1\frac{1}{8}$″		✓	✓	No tolerance on flatness or parallelism	
22	$1\frac{3}{4}$″	✓	✓	✓		★
23	℄ to $\frac{7}{8}$″			✓		
24	℄ to $\frac{1}{2}$″			✓		
25	1″		✓	✓		
26	$1\frac{25}{32}$″	✓	✓	✓		
27	$\frac{1}{16}$″ Radius (2)	✓	✓		Implied 90° bend	★
28	1010 C.R.S. $\frac{1}{4}$ H U.S.S. No. 16 Ga.	✓			Tolerance on thickness	
29	Break sharp corners 0.015″ R	✓		✓	Question need	Not needed

Fig. 9.4 *A drawing for a stamped sheet-metal bracket and a tabular analysis sheet listing specifications. From American Society of Tool and Manufacturing Engineers.* ASTME Die Design Handbook (2nd ed.) (*New York: McGraw-Hill Book Company, 1965*), *pp. 3–5.*

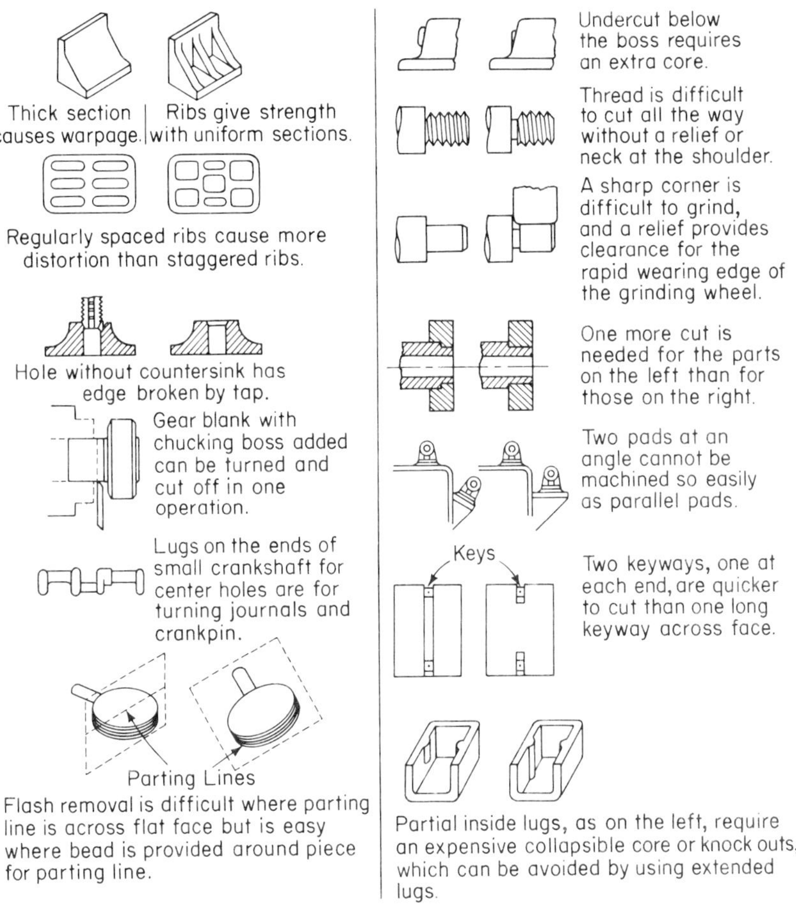

Fig. 9.5 *Typical product-design changes that have resulted in production economies.*

driving the workpiece, to aid in locating the workpiece, to prevent surface mutilation, to provide runout spaces for tools, and to provide foolproofing means

3. Changes in the shape, size, or material of a part to enable the part to be made by a more economical process, reduce tooling costs, decrease machining time, reduce the number of operations in a process, reduce material cost and waste, and aid in assembly

9.3.3 Principles of Dimensioning

The dimensions of a part specify the positions and conditions of its surfaces. In general, surfaces may be classified as *functional surfaces*, which enter into the operation or location of the part in a mechanism; *clearance surfaces*, which provide continuity in the part but have no functional role; and *atmospheric surfaces*, which are not near other surfaces in the mechanism. Dimensions between functional surfaces have relatively small tolerances, dimensions to clearance surfaces have larger tolerances, and dimensions to atmospheric surfaces have the largest. Base lines for dimensions frequently coincide with functional surfaces. The tolerances on the dimensions of a part drawing are indicative of the relative importance of the surfaces to which the dimensions apply.

The cost of holding a tolerance on a dimension increases as the tolerance decreases, in the manner indicated by the curve *cd* of Fig. 9.6. The best service is obtained from a particular part when a small tolerance is held; then the value of the product is high. With a larger tolerance, some pieces fail too soon and the

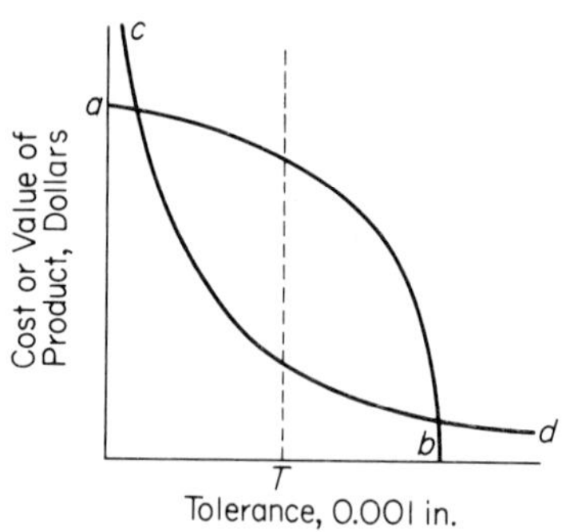

Fig. 9.6 *The factors that determine a desirable tolerance.*

average quality of the product is lowered, as indicated by the curve *ab* of Fig. 9.6. The most economical tolerance for the dimension is *T*, which gives the largest difference between product value and cost. Obviously, a larger tolerance may be as wasteful as a smaller tolerance.

The factors affecting every dimension are like those depicted by the curves of Fig. 9.6, but they do not have the same values for each dimension. A competent designer tries to assign a tolerance to each dimension corresponding to the tolerance *T*. In that way, the designer tells the tool engineer which surfaces are most important.

The clarity and exactness of the language of dimensioning can be upheld by observance of the following rules, which the tool engineer has a right to expect the product designer to follow:

1. Two points, lines, or surfaces on a drawing of a part should be connected only by one dimension or one set of dimensions.
2. Dimensions should be placed between the points, lines, or planes most closely related to each other.
3. Dimensions should be placed and tolerances assigned to reflect the functional requirements of the part.
4. A part must be dimensioned in three coordinate directions.

9.3.4 Critical Areas

The surfaces given prime attention and preferred for locating and gaging in manufacturing are called *critical areas.* Before a process can be planned, the process engineer must select the critical areas. To do this, he applies three *indicators* that point to the functional and clearance surfaces on the part because surfaces which serve the part functionally are likely to make the best critical areas for manufacturing. These are the indicators:

1. *The presence of a base line from which a number of dimensions are taken.* A base line must coincide with a real surface or an axis of a surface to be significant as a critical area.
2. *Close tolerances* on linear or geometric dimensions.
3. *Relatively fine surface finishes.* This may mean superfinished surfaces on a ground part or merely machined surfaces on an otherwise rough casting.

The indicators often point to several surfaces in each coordinate direction on a part as possible critical areas. For practical reasons, such as the size or position for location, a functional surface is not always a desirable critical area. Three *tests* are applied to determine which surfaces promise to serve best as critical areas from a practical point of view.

1. *A test for arithmetical superiority.* A surface that is the terminal point for the most dimensions is desirable as a critical area because it offers the best locating area to minimize tolerance accumulations.
2. *A test for geometrical superiority.* A surface least subject to runout or with points far apart is desirable as a critical area.
3. *A test for mechanical superiority.* A surface having a minimum susceptibility to deflection or mutilation is desirable as a critical area.

9.3.5 The Operations of a Process

A *major process* (machining, molding, forming, and assembling) is usually planned as an entity, but it is preceded by *basic processes*, such as forging or casting before machining, and may be interspersed with *auxiliary processes*, like heat treatment during a machining process. The operations that constitute a major process may be classified in the following way:

1. *Critical operations* primarily establish the critical areas but include all related areas treated economically at the same time.
2. *Qualifying operations* establish locating surfaces preliminary to the critical manufacturing operations or serve to reestablish critical areas after workpiece distortion such as may occur in heat treatment.
3. *Placement* or *auxiliary-process operations* complement the operations of the major process in order to meet specifications. Examples are heat treatment, straightening, or plating in a machining process; cleaning in a foundry process; or welding or staking in an assembly process.

4. *Tie-in* or *secondary operations* are those major process operations that occur after the critical operations and in relation to the placement operations in order to produce the dimensions, surface conditions, and geometrical relationships of the finished part. Examples for machining processes are gear cutting, finish grinding, and drilling operations.

5. *Supporting* or *enterprise-protection operations* control or expedite the manufacturing routine and protect the interests of the enterprise. Receiving, storage, and inspection are examples.

Each item in the list of requirements for a process can be assigned to one class of operations. The problem of forming the operations is simplified because the operations need to be formed only out of the items in one class at a time.

One or several operations may be formed to satisfy the items allotted to any one class. Some operations should obviously be done together. For instance, several adjacent concentric diameters would normally be turned in one operation. Other items require separate operations because of the techniques and equipment involved. For example, if hardening and plating a piece are two items in the placement-operation class, separate operations would be set up to take care of them.

The items in an operation class may be combined as groups into operations or treated separately in a number of ways; finding the best way is a matter of recognizing and comparing all the alternatives. Usually some alternatives are obviously undesirable; the rest must be analyzed and their costs compared to find the most economical. Process planning is at this stage an application of the techniques and principles of engineering economy discussed in Sec. 3.

The planning of an operation includes the comprehension of the elements or steps that must be followed to carry out the operation; selection of the machine tool, accessories, tools, and gages needed; calculation of the operating speeds, feeds, power requirements, and forces; and recognition of the tolerances that must be held and their feasibility.

9.3.6 The Sequence of Operations

The operations of a process can be relegated to a logical order in the manner depicted in Fig. 9.7. At the beginning and end the inspection operations pass the raw material from the

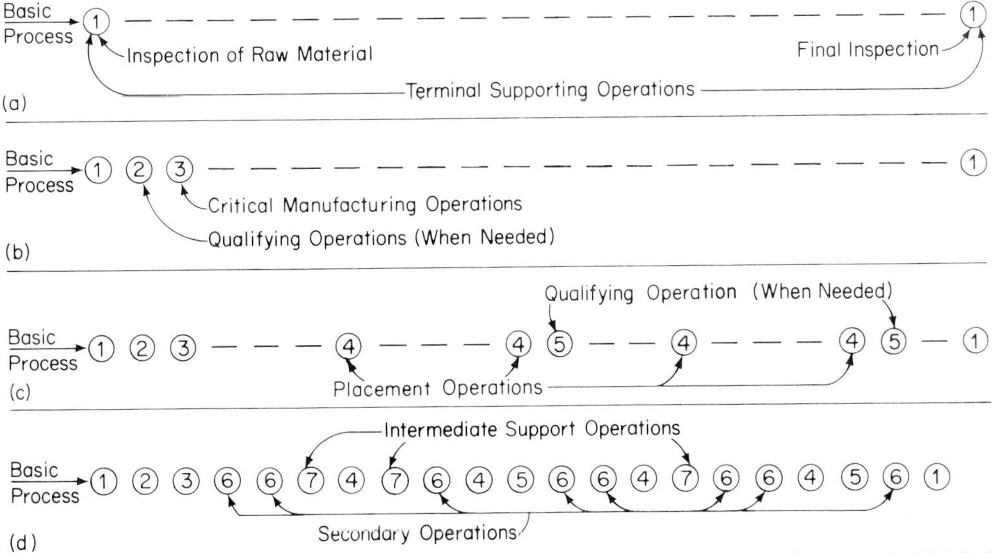

Fig. 9.7 *A diagram of the steps to establish the sequence of operations of a process.*

basic process and approve the finished product of the major process as in Fig. 9.7 (a).

A basic rule of processing is that the critical areas must be established as soon as possible to provide reliable surfaces for consistent location throughout the process; thus, the critical operations are put at the beginning as in Fig. 9.7(b). In some cases qualifying operations are needed to prepare rough locating spots such as on foundry pads. If placement operations are needed, they usually have an order dictated by their interactions. For example, heat treatment comes before final plating. They are inserted in the operation list with ample spaces around each as in Fig. 9.7(c). Necessary requalifying operations, like straightening after heat treatment, may be placed in line with the appropriate placement operations.

Each secondary or tie-in operation is usually related to placement operations, like drilling and gear cutting before hardening or finish grinding after heat treatment, or to other secondary operations. Thus they are interspersed in the manner depicted by Fig. 9.7(d). Where a number of secondary operations fall in a group, they are arranged with respect to each other in the most economical manner considering movement of materials, locations of machines, and the effect of each operation on the others.

Functional surfaces of a workpiece that are dimensioned with small tolerances and especially those that serve as base lines are commonly critical areas, but others are treated in secondary operations. Functional surfaces may impose exacting requirements, and their relationships with other operations are important in establishing operation sequence. If the requirements are for extra-fine surface finish, the necessary operations may come late in the process to avoid damaging the finish in other operations. Functional surfaces dimensioned with small tolerances may be machined as early as possible in a sequence for dimensional control and also to avoid wasting effort on pieces that may have to be scrapped in the difficult operations.

As a final step support operations go where they will best provide control, service, and protection throughout the process.

9.4 CUTTING SPEEDS AND FEEDS

Speed and feed recommendations for metal cutting are commonly given in tables, like Table 9.1, but in such form are really applicable to limited situations because many factors determine proper speed and feeds and their effects can be stipulated only in lengthy tables.

9.4.1 Cutting Speed

Cutting speed is the surface speed V in feet per minute at which either the work or cutter travels. Machine speed N is usually specified in revolutions per minute where a workpiece or cutter of diameter d in. is revolved. A practical relationship is $N = 4V/d$.

The following formula is like the one utilized in a popular analog computer for calculating cutting speed in feet per minute

$$V = ABCDEFGPQ^{0.2}/H^{1.72}R^{0.16}T^{n}f^{0.58}c^{0.2}$$

The factors are defined and evaluated in the following discussion. Figure 9.8 contains a simple Fortran program for solving this formula on digital computers.

A is an empirical constant depending on the tool material; Table 9.2 gives typical values.

B provides for the effect of a cutting fluid. Average values are 1.0 for dry cutting, 1.15 for cutting oils, and 1.25 for soluble oils and water emulsions. Cutting fluids are particularly beneficial with high-speed-steel tools in heavy cuts at low speeds, but they are not always of value at high speeds with cemented-carbide and sintered-oxide tools.

C represents the type of material; Table 9.1 displays feasible amounts for various materials.

D allows for different microstructures ranging from 0.7 for the austenitic structures of stainless steels; to 1.0 for commercial hot-rolled, cold-drawn, quenched and tempered, normalized, and annealed stock; to 1.4 for coarse, spheroidized structures.

E provides for the condition of the workpiece surface before cutting. A sand-cast surface is 0.7; cast and shot-blasted, 0.75; covered with heat-treatment scale, 0.80 to 0.95; clean, 1.0.

F accounts for the tool: 1.0 for single-point turning, facing, and boring tools, and most milling cutters; 0.7 for drills and weakened form cutters; 0.8 or less for reamers.

G is determined by several variables with the relationships shown in Table 9.3 for single-point tools. The factor W in the table is the depth of the cut divided by the nose radius of the tool, both in inches. The entering angle is that

TABLE 9.1. *METAL CUTTING SPEEDS AND FEEDS*

Reference Number	Work Material	Brinell Hardness Number	Cutting Speed, Feet per Minute[1] High-Speed Steel	Cemented Carbide	Feed,[2] Inches per Revolution	Factor C	Average[3] P_u HP/Cu. In. per Minute	Factor U	Factor z
1.	Carbon steel	170	160	660	0.010	0.8	0.75	6×10^{-2}	0.5
2.	Free-cutting steel	170	170	700	0.015	1.05	0.50	4×10^{-2}	0.5
3.	Alloy steel < 300 Bhn	200	160	660	0.010	1.10	0.70	5×10^{-2}	0.5
4.	Alloy steel > 300 Bhn	400	75	310	0.005	1.05	1.50	3×10^{-5}	1.8
5.	Cast steel	170	165	680	0.010	0.80	0.75	6×10^{-2}	0.5
6.	Cast iron	190	110	460	0.013	0.75	0.60	3.8×10^{-5}	1.85
7.	Ductile iron	190	130	540	0.013	0.90	0.60		
8.	Malleable iron	150	155	640	0.020	0.90	0.40		
9.	Stainless steel, Type 300	170	60	250	0.006	0.30			
10.	Stainless steel, free-cutting, Type 300	150	175	725	0.008	0.70			
11.	Stainless steel, Type 400	250	130	540	0.006	0.90	1.25	2.5×10^{-1}	0.3
12.	Stainless steel, free-cutting, Type 400	180	210	870	0.008	1.00			
13.	Monel metal, Type R	150	140	590	0.010	0.55	0.70		
14.	Monel metal, Type *K*	160	190	790	0.005	0.55	1.40		
15.	Titanium alloys	310	80	330	0.008	0.95	0.90	1.8×10^{-2}	0.7
16.	High-temperature alloys, nickel base	230	55	210	0.007	0.40	1.10		
17.	High-temperature alloys, cobalt base	310	25	105	0.005	0.25	1.50		
18.	Copper	80	150	620	0.008	1.25	0.90		
19.	Free-cutting brass	110	500	2100	0.025	2.00	0.30		
20.	Yellow brass	150	300	1120	0.020	2.00	0.40		
21.	Manganese bronze	185	170	700	0.022	1.25	0.35		
22.	Phosphor bronze	135	250	1050	0.020	1.25	0.30		
23.	Aluminum alloys	100[4]	170	700	0.050	0.85	0.15	3×10^{-2}	0.4
24.	Magnesium alloys	60[4]	340	1400	0.070	0.90	0.10	4×10^{-2}	0.3

[1] Cutting speed for 0.100-in. depth of cut at feed specified; 60-min tool life and 0.060-in. wear land for high-speed steel; 30-min tool life and 0.030-in. wear land for cemented carbide
[2] Feed for 5/8-in. sq. tool at 0.100-in. depth of cut and 1.5-in. overhang
[3] P_u for cut of 0.100-in. depth and 0.0125-in./tooth feed
[4] 500-kg load

between the normal to the workpiece surface and the side cutting edge of the tool; it is the same as the side-cutting-edge angle of a tool positioned with a 90 deg setting angle. Values of G for other tools: drills, 1.14; face mill (no chamfer) 0.8, (45 deg chamfer less than one-half depth of cut) 1.0, (45 deg chamfer more than one-half depth of cut) 1.2, (60 deg chamfer more than one-half depth of cut) 1.28; slab mill, 1.0; slot mill with sharp corner, 0.8.

H is the Brinell hardness number of the workpiece material.

P provides for differences in tool materials; Table 9.2 gives values.

Q is the actual amount of flank wear allowed on the tool. For small tolerances and good surface finishes, flank wear may be limited to 0.005 to 0.010 in.; with hard materials and thin workpiece sections, 0.015 in.; average practice with carbide tools allows up to 0.030 in.; and with high-speed-steel tools as much as 0.060 in.

R, the number of teeth in the cutter, is 1 for a single-point tool, the same as the number of flutes in a drill or reamer and the actual number of teeth in a milling or boring cutter. The more teeth in a cutter the greater the effects of chip crowding, heating, and vibration.

T is tool life, which should be between the economical tool life (the one that gives the lowest cost of metal removal) and the tool life

Cutting Speed

```
$          FORTRAN
         DIMENSION W(10)
1        RIT 7,2,(W(I),I=1,10)
2        FORMAT (10A6)
         RIT 7,3,A,B,C,D,E,F,G,H,P,Q,R,T,XN,SF,SC
3        FORMAT (F8.0,F6.3,F5.3,F4.2,F5.2,2F4.2,F6.1,F5.2,F6.4,F4.2,F6.1,F5
        1.4,F5.3,F6.3)
         V=(A*B*C*D*E*F*G*P*Q**.2)/(H**1.72*R**.165*T**XN*SF**.58*SC**.2)
         WOT 6,4,(W(I),I=1,10),V
4        FORMAT (1X,10A6,5X,F7.0)
         GO TO 1
         END
```

Horsepower

```
$          FORTRAN
         DIMENSION A(5)
1        RIT 7,2,(A(I),I=1,5),W,PU,U,H,Z,SF,AN,RN
2        FORMAT (5A6,6F7.3,F2.0,F4.3)
         IF (W) 3,15,3
3        IF (U) 4,11,4
4        IF (H) 5,11,5
5        IF (Z) 6,11,6
6        IF (SF) 7,11,7
7        IF (AN) 8,10,8
8        IF (RN) 9,10,9
9        P=(.456*U*H**Z*W*(1.25-(.015*AN))*(1.+RN))/SF**.18
         GO TO 13
10       P=(.456*U*H**Z*W)/SF**.18
         GO TO 13
11       IF (PU) 12,15,12
12       P=PU*W
13       WOT 6,14,(A(I),I=1,5),P
14       FORMAT (1X,5A6,F3.0)
         GO TO 1
15       WOT 6,16,(A(I),I=1,5)
16       FORMAT (5X,5A6,14HINCORRECT DATA)
         GO TO 1
         END
```

Fig. 9.8 *A FORTRAN program for calculating cutting speed and horsepower.*

for maximum production. The economical tool life for an operation is

$$T_e = [(1 - n)/n]\ [C_T/R_1].$$

The tool life for maximum production is

$$T_m = [(1 - n)/n]\ [C_c/R_1].$$

The factor C_c is the cost of changing and adjusting the tool on the machine; the factor C_T includes C_c and also the cost of sharpening the tool and the proportion of the initial cost of the tool consumed by each sharpening. The factor R_1 is the operation rate for direct and overhead costs in dollars per minute. Where r tools operate simultaneously, the economical or maximum production-tool lives are r times those for a single tool. The factor n is the same in all equations; Table 9.2 gives values for tools of various materials. Table 9.4 gives economical tool lives reflecting typical production-shop practices. (Values for *TM* are given in Fig. 9.9.) The economical life for a tool may differ from factory to factory because it depends on costs which vary. Specific tools may have economical tool lives quite different from those quoted; for instance, expensive broaches may

TABLE 9.2 *FACTORS THAT DEPEND ON TOOL MATERIAL*

Tool Material	Factor P	A	n
High-speed steel	1.0	144,000	0.125
Super-high-speed steel	1.2		
Cast nonferrous alloys	1.8	173,000	0.17
Cemented carbides			
(average)	5.0	268,000	0.250
Grade C–1	4.5		
C–2	4.8		
C–3	5.2		
C–4	5.4		
C–5	5.0		
C–50	4.0		
C–6	5.3		
C–7	5.5		
C–70	6.5		
C–8	8.0		
Sintered oxides	8.0	1,500,000	0.700

warrant tool lives of hundreds of hours. An estimate of the economical tool life for any tool can be made from the formulas once tool costs are ascertained.

The factors f for feed in inches per tooth and c for depth of cut in inches are discussed in detail in the next subsection. The exponents in the cutting-speed formula for these factors are average values for cemented carbides; averages for high-speed-steel are 0.50 for feed and 0.29 for depth of cut. Max Kronenberg has reported an exponent of 0 for depth of cut c, and exponents for feed f of 0.19 for cutting nonferrous metals with cemented carbides, and 0.45 for bronze and copper, 0.61 for brass, and 0.58 for aluminum cut with high-speed-steel.

The formula presented here for cutting speed takes into account the main factors that determine speed in most cases and assigns average values for them, but many other variables, like the shape and size of the workpiece, the condition and characteristics of the machine, and other tooling, influence performance to some extent in various cases. Actually no one formula can account for all the possibilities with the present state of knowledge of metal cutting. The same operation may show as much as 40 per cent difference in tool life from machine to machine; different lots of the same kind of steel may require 40 per cent or more difference in cutting speed for the same tool life. Niedzwiedzki* reported that ordinary differences in

* Antoni Niedzwiedzki, *Manual of Machinability and Tool Evaluation* (Cleveland, Ohio: Huebner Publications, Inc., 1960).

TABLE 9.3. *FACTOR FOR TOOL PROFILE*†

Entering Angle, Degrees	W: Sintered Oxide, Cemented Carbide 1000	10	2	1	0.5	0.2	Nonferrous Alloy High-Speed Steel 1000	10	2	1	0.5	0.2
0	0.80	0.83	1.00	1.15	1.35	1.41	0.71	0.78	1.00	1.22	1.45	1.57
15	1.02	1.07	1.15	1.23	1.35	1.41	0.90	1.02	1.15	1.29	1.45	1.57
30	1.13	1.20	1.27	1.32	1.36	1.41	1.08	1.18	1.29	1.36	1.45	1.57
45	1.21	1.27	1.33	1.37	1.39	1.43	1.26	1.34	1.41	1.46	1.50	1.57
60	1.28	1.33	1.39	1.42	1.44	1.47	1.40	1.45	1.49	1.52	1.56	1.60
75	1.32	1.36	1.42	1.45	1.47	1.50	1.50	1.54	1.57	1.59	1.61	1.64

† Adapted from recommendations for Carboloy machinability computer.

TABLE 9.4. *TYPICAL ECONOMIC TOOL LIVES, MINUTES*

Type of Tool	Tool Material: High-Speed Steel	Nonferrous Alloy	Cemented Carbide: Brazed	Cemented Carbide: Mechanical	Sintered Oxide
Single-point tools	60	30	30	15	10
Drills, cobores, and reamers	30		20		
Taps and dies	90				
Form tools	120		120		
Milling cutters	60 × *TM*‡	30 × *TM*‡	30 × *TM*‡	15 × *TM*‡	

‡ Values of *TM* are given in Fig. 9.9.

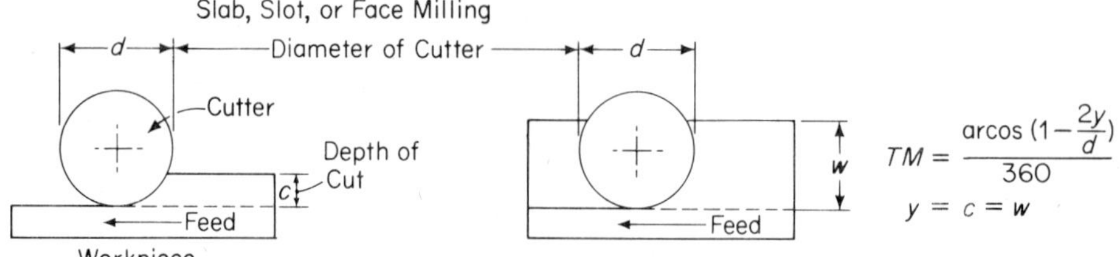

Face Milling with Centers of Cutter and Workpiece Coincident

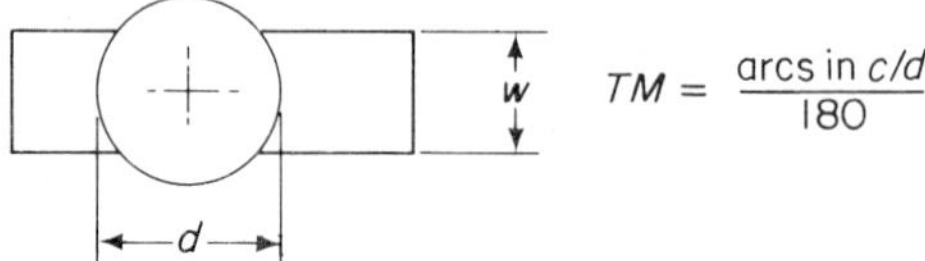

Fig. 9.9 *An explanation of the factor TM in Table 9.4.*

grinding tools in the shop resulted in variations of about 15 per cent in cutting speed for a 60-min tool life, and variations in the microstructure in the same bar caused the cutting speed for a 60-min tool life to vary in some cases more than 30 per cent.

When the cutting speed computed from the formula for a given tool life is not realized in the actual operation, the factors that appear to cause the discrepancy may be changed and the speed recalculated until the results conform with the practice. For instance, changing the value of factor *D* compensates for an unusual microstructure. Other factors may be inserted as they appear to be relevant. As more information is obtained from actual practice and more is learned about the correct interpretations of the factors, the accuracy of the results from the formula may be improved in line with the practices in any one plant.

9.4.2 Depth of Cut and Feed

Light cuts produce the best surface finishes. Depth of cut is usually less than $\frac{1}{16}$ in. for finish cutting and less than 0.010 in. for finish grind-

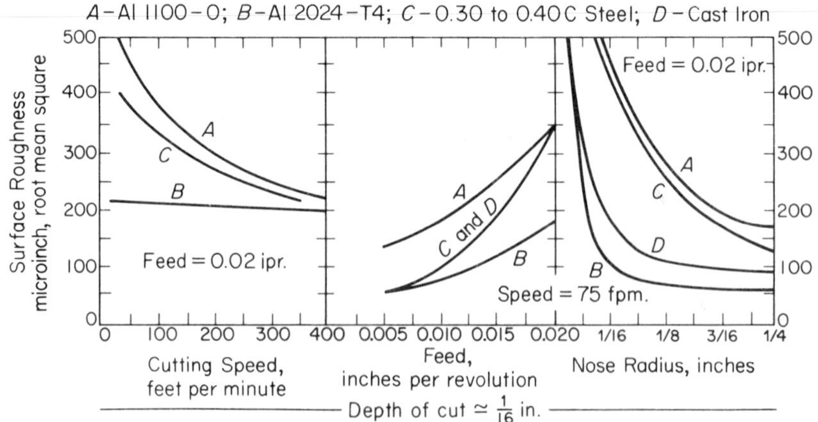

Fig. 9.10 *Average effects of cutting speed, feed, and nose radius on surface finish. From data in Manual on Cutting of Metals, (New York: The American Society of Mechanical Engineers, 1953).*

ing. Finish turning and shaping feeds range from 0.005 to 0.025 ipr or in./stroke. Figure 9.10 gives an indication of the finish to be expected from various feeds depending on the nose radius of the tool.

Rough cutting is done most efficiently with cuts as large as the workpiece, tool, and machine will stand. As deep a cut as the stock or tool allows with a moderate feed distributes the load over more cutting edge and is preferable to a shallow cut and heavy feed. Rough turning and shaping feeds range from 0.015 to 0.075 ipr or in./stroke. The horsepower formula of Par. 9.6.2 ascertains how much feed is possible with the power available in an operation.

A feed that induces an extreme fiber stress of 10,000 psi in a single-point tool is $f = bh^2/(210cLP_u)$, where b is the width and h the height of the tool shank, c the depth of cut, and L the tool overhang—all in inches. Table 9.1 gives values for the unit horsepower P_u for various materials. The feed may be increased proportionately for any other permissible extreme fiber stress.

For drills, the expression for feed in inches per revolution, $f_d = 0.010\sqrt{d}/P_u$, is based on drilling practice recommended by Orlan W. Boston; d is the diameter of the drill in inches. For a two-flute drill, f_d is divided by two to obtain the feed in inches per tooth for the velocity formula. The depth of cut c for a drill is one-half the diameter d.

The feed and speed should be reduced for deep-hole drilling without oil-hole or gun drills or frequent retraction of the drill from the hole. The feed in inches per revolution should be reduced to $f_r = f_d(1 - 0.02q)$ and the speed to $V_r = V(1 - 0.025q)$, where q is the number of diameters the hole is deep.

The numerical values of feeds in inches per revolution for various materials in Table 9.1 are also the feeds in inches per tooth for high-speed-steel face mills. The feed for a cemented-carbide or nonferrous-alloy face mill is approximately the feed for a high-speed-steel face mill times $0.1\sqrt{H}$, where H is the Brinell hardness number of the work material. According to the recommendations of the Cincinnati Milling Machine Co., the proper feed rates for helical mills are about 80 per cent of the feeds that face mills will stand, for slotting and side mills 60 per cent, for end mills 50 per cent, for form relieved cutters 30 per cent, and for circular saws 25 per cent. The proper dimension for feed for milling, designated f in the cutting-speed formula, is inches of radial chip thickness; $f = 2t\sqrt{cd - c^2}/d$, where t is the feed in inches per tooth in the direction the work is fed, c is the width or depth of cut perpendicular to the direction of feed and the cutter axis, and d is the diameter of the cutter. When $c \geq d/2$, $f = t$.

For rough broaching, the feed per tooth may be less than 0.003 in. for a weak broach to as much as 0.010 in. for a strong one. The cut per tooth is normally less than 0.001 in. in the burnishing or sizing section at the end of a broach and usually less than 0.0001 in. for the last few finishing teeth.

In traverse grinding, the wheel is fed from one-half to three-quarters of its width for each revolution of the workpiece.

9.5 CUTTING AND FORMING FORCES

The force and energy required are important considerations in the selection of a press for a cutting or forming operation. In metal cutting, the forces are sometimes necessary considerations in the design of tools and machines.

9.5.1 Blanking and Piercing Forces

The maximum force in pounds required to cut a piece with a perimeter L in. from material t in. thick with an ultimate shear strength of S_s lb/in.^{2} is $F_S = LtS_s$. Approximate strengths of common materials are given in Table 9.5.

The maximum force may be reduced by staggering several punches or by inclining the cutting edges of punch or die to provide shear. A punch needs to penetrate only a part of the thickness of a sheet equal to $p \times t$ to cut it. The per cent penetration p required for common materials is given in Table 9.5. If several punches are arranged in steps so that each acts only when the preceding one has penetrated a distance at least $p \times t$, the punches are said to be staggered, and the maximum force is only that imposed by the largest punch.

Shear on a punch distorts the blank or slug and on the die distorts the surrounding material. If the shear or height of inclination of v in. over the cutting edge is larger than $p \times t$, the force in pounds required to cut a piece is $F'_S = F_S tp/v$.

TABLE 9.5. *PROPERTIES OF SHEET METAL**

Material	Soft				Partly Cold-Worked or Heat-Treated			
	Ultimate Tensile Strength, Pounds per Inch Squared	Elongation, Per Cent	Shear Strength, Pounds per Inch Squared	Penetration, Per Cent	Ultimate Tensile Strength, Pounds per Inch Squared	Elongation, Per Cent	Shear Strength, Pounds per Inch Squared	Penetration, Per Cent
Aluminum 1100	13,000	35	9,000	30	18,000	9	11,000	30
3003	16,000	30	11,000	30	22,000	8	14,000	25
2024	27,000	20	18,000	25	70,000	18	41,000	15
7075	33,000	17	22,000	25	83,000	11	48,000	15
Brass 60–75 per cent Cu (yellow and cartridge)	50,000	65	33,000	50	60,000	8	40,000	20
Bronze, 90 per cent Cu (commercial)	37,000	45	28,000	25	52,000	11	35,000	15
Copper	32,000	45	22,000	55	42,000	14	26,000	30
Steel 0.06C	43,000	30	34,000	60	48,000	20		
0.20C	55,000	25	44,000	50	65,000	15	60,000	38
0.30C	68,000	20	52,000	33	76,000	12	67,000	22
1.00C	120,000	5	115,000	10				
Silicon steel			65,000	30			150,000	2
Stainless steel	85,000	50	60,000	50				
Titanium Ti–55A	73,000	29						
Titanium RC–70 75°F	125,000	20	100,000					
Titanium RC–70 800°F	90,000	30						

* Averages are given in the table; actual values vary with material content, heat treatment, and other conditions.

9.5.2 Bending Forces

For material t in. thick and of S psi tensile strength bent along a straight line of length W in. and with the unsupported stock L in. wide, Brozek[†] has reported the force in pounds to be $F_B = Wt^2S/2L$. A bend in a vee die with opening width of Y in. requires a force $F_B = Wt^2S/Y$. Sometimes the force is multiplied by 1.33 for a larger margin of safety. These are empirical formulas. (Figs. 9.35 and 9.41)

9.5.3 Drawing Forces

The force in pounds to draw a sheet-metal cup of diameter d from material of thickness t and ultimate tensile strength S may be estimated from $F_D = \pi tS(d - t)N$. Figure 9.11 gives values for N. Other forms may be approximated by considering straight sections deformed in bending and curved sections as segments of circular cups.

† John S. Brozek, "A Notebook on Die Design," *The Tool Engineer*, **XXVI**, No. 6 (June 1951), 49.

9.5.4 Press Ratings

About 10 per cent is usually added to the calculated forces to allow for losses in the press. As much as 25 per cent more may be added for springs, cushions, or buffers, especially for drawing, depending upon their sizes. The total force in pounds is converted to tons because presses are rated in tons. The tonnage capacity of a press may be found from the manufacturer's catalog. If that information is not available, the capacity in tons T may be calculated for a mechanical press with crankshaft diameter up to 7 in. For an end-wheel press with an overhanging crankpin of diameter D in., $T = 2\frac{1}{2}D^2$. For a press with crankshaft of diameter D supported on both sides of the crank (or double crank), $T = 3\frac{1}{2}D^2$. A press reaches its maximum capacity only when the ram is near the bottom of its stroke.

9.5.5 Metal-Cutting Forces

The three conventional components of forces acting on a single-point tool are the cutting or

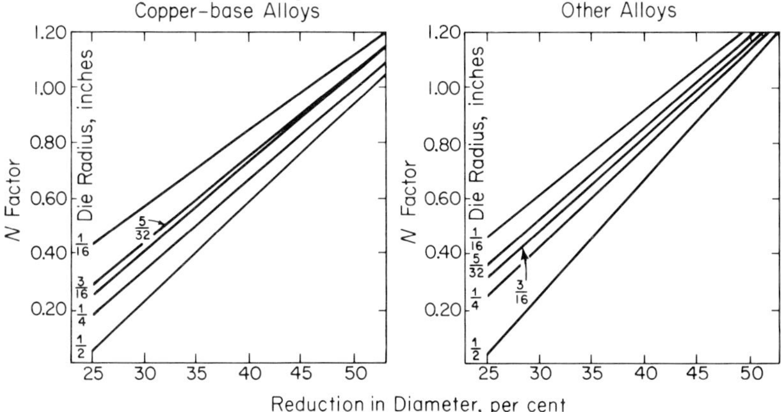

Fig. 9.11 *Values for factor* N *in cupping. From E. N. Ludington, "Prediction of Drawing Properties from Tensile Tests,"* Metal Progress *(Aug. 1956), p. 93.*

tangential force F_c in the direction of the cutting speed, a feed or longitudinal force F_L in the direction of feed and perpendicular to F_c, and a surfacing or radial force F_R at right angles to the other two.

Cutting force depends mostly upon the material cut and the cross-sectional area of cut, as designated in Table 9.6, and to a much less extent upon rake angles, cutting fluids, tool materials, and speeds. The force in an actual operation may be two to three times as high as that determined experimentally because dull tools may as much as double the force, material is not always uniform, and heavier cuts may be taken in the shop than those anticipated.

Longitudinal and radial forces are commonly less than the cutting forces in an operation but rise rapidly and approach the value of the cutting force when a tool becomes dull. Thus, the maximum values of F_L and F_R are likely to be close to the highest value of F_c.

Normally, over 95 per cent of the horsepower P_c in a metal-cutting operation is derived from the cutting force F_c lb and the cutting speed V ft/min, which act in the same direction. That is because the cutting speed is much larger than the feed rate. For many operations, the cutting force can be estimated from the relationship

$$F_c = 33{,}000 P_c / V$$

In a milling operation, F_c is the tangential force on the rotating cutter. For a drill, the torque T in.-lb is related to the speed N rpm by

$$T = 63{,}000 P_c / N$$

The thrust of a drill can be represented by

TABLE 9.6. *AVERAGE VALUES OF CUTTING FORCE F_c, POUNDS.**

Material	Cross-Sectional Area of Cut, Square Inches				
	0.001	0.010	0.020	0.030	0.040
Light alloys (50 Bhn)	35	305	580	860	1120
Brass	129	725	1215	1650	2020
Cast iron 100 Bhn	100	730	1330	1900	2440
150 Bhn	125	910	1660	2380	3050
200 Bhn	140	1020	1860	2660	3420
Cast steel 150 Bhn	268	1900	3400	4830	6170
Steel SAE 1015	300	1920	3360	4570	5800
SAE 1035	400	2560	4480	6070	7700
SAE 1060	520	3330	5820	7900	10000

* Based upon equation $F_c = C_p\ (1000\ A)^r$ and constants from Max Kronenberg, *Machining With Single Point Tools* (Cincinnati: The Cincinnati Milling Machine Co.).

$$B = Kf^x d^y$$

where f is the feed in inches per revolution and d is the diameter of the drill in inches. Experimentally determined values of K, x, and y are given in Table 9.7.

TABLE 9.7. *FACTORS FOR EQUATION* $B = Kf^x d^y$*

Material	K	x	y
Aluminum	50,000	1.1	1.2
Cast iron (163 Bhn)	14,720	0.6	1
Steel SAE 1020	40,000	0.78	1
SAE 1045	42,000	0.78	1
SAE 1095	69,000	0.78	1
SAE 6150 (187 Bhn)	53,400	0.78	1

* For values for other materials see O. W. Boston, *Metal Processing* (New York: John Wiley & Sons, Inc., 1951), p. 333.

9.6 ENERGY AND POWER

The energy and power that must be delivered by a machine determine the rate at which the machine can do a job or whether it can do the job at all.

9.6.1 Energy and Power in Press Work

The energy in foot-pounds per stroke to shear material of thickness t in. and percentage penetration p (see Table 9.5) under maximum force of F_s lb is $E = 1.16F_s pt/12$, which allows for normal losses from friction in machines and dies. E. V. Crane has reported the energy in a drawing operation to be $E = F_D hC/12$, where h is the length of draw in inches and C is a factor varying from 0.6 for light to 0.8 for heavy drawing.

Slowing down the flywheel of a mechanical press during a stroke supplies energy. Some makers prescribe energy ratings for their presses; one specifies that the permissible energy withdrawal during each stroke in inch-tons is equal to the press force rating in tons. A recognized practice allows a 10 per cent drop in flywheel speed during each stroke for continuous operation; for a flywheel weighing W lb and turning with a maximum speed of V ft/sec at its radius of gyration, the energy released is $E_F = 0.003WV^2$. A slowdown of 20 per cent is permitted for intermittent operation; the energy is $E_F = 0.0056WV^2$ in that case.

To restore the energy used during one stroke, the motor must be able to deliver, during the $1/R$ part of the stroke, horsepower $P = ENR/33{,}000$. N is the number of strokes of the press ram per minute, and R is usually between 1 and 2.

9.6.2 Power in Metal Cutting

The major factors that determine the power in a metal-cutting operation are included in the formula for horsepower $P_c = 0.465UH^zW[1.25 - 0.015(AN)]\,[1 + (RN)]/f^{0.18}$ The factors H, f, and c are the same as in the cutting-speed formula in Par. 9.4.1; U and z are specified for various materials in Table 9.1. W is defined in the next paragraph. The factor AN is the effective rake angle of the cutting tool in degrees, and RN is the nose radius in inches. If these are not known, an approximation within 10 per cent in most cases is obtainable from $P_c = 0.6\,UH^zW/f^{0.18}$. For materials for which U and z are not given, an estimate for horsepower is $P_c = P_u W$; P_u is given in Table 9.1.

For turning, the rate of metal removal in cubic inches per minute is $W = 12Vfc$; V is the cut speed in feet per minute. For drilling, $W = 0.7854D^2 f_a N$; D is the drill diameter in inches, f_a the feed in inches per revolution equal to $2f$, and N the speed in revolutions per minute. For milling, $W = 12cwfnN$; c and w are the width and depth of cut in inches, f the feed in inches per tooth, and n the number of teeth on the cutter.

The computer program of Fig. 9.8 gives a choice of horsepower formulas. The factors in these formulas are based on experiments with sharp tools. As tools normally become dull, unit power consumption increases up to 50 per cent; so do forces. Quite often a machine tool is loaded to capacity at the start of a run with sharp tools. As the tools wear, the machine becomes somewhat overloaded; for intermittent loading this can be done without harm up to 75 per cent over rated capacity when the time between cuts at least equals the cutting time. A machine tool should not be operated over its rated capacity for continuous service.

9.7 WORKPIECE LOCATION

The locating points and surfaces and clamping areas on a workpiece are selected when a process is planned. The actual locating and clamping devices are selected and proportioned when the tools are designed.

The conditions for full workpiece location are given by the 3-2-1 principle of location as follows: *To locate a piece fully, place and hold it against three points in a base plane, two points in a vertical plane, and one point in a plane square with the first two.* The three points for the base plane may be on parallel surfaces if occasion demands, as may be the two points for the second plane.

The three planes or three sets of planes for locations should be mutually perpendicular if possible. Such surfaces, or points on them, provide a boxlike arrangement in which a part can be held most securely. Also errors from foreign particles on the locators are minimized.

Points as far apart as possible on any surface should be selected to help minimize locating errors.

Fewer than the number of points specified by the 3-2-1 principle may be selected to locate a piece if partial location is sufficient, as it often is. For instance, a chuck normally locates a piece to be turned on the equivalent of four or five points. However, a free body has six degrees of freedom, and a locating point is necessary for each degree of freedom to be confined.

No more than the number of points prescribed by the 3-2-1 principle of location should be selected for locating a rough surface because more points provide unreliable location. However, more points than specified by the 3-2-1 principle, even whole surfaces, are often desirable for a finished surface. The extra points add nothing to location but do provide better support to the workpiece. On four buttons, a finished surface rocks if a chip is present on one of the buttons, but no evidence of the chip is given with three buttons.

A part should be located in an operation from surfaces most directly connected dimensionally to the surface or surfaces treated in the operation to minimize the accumulation of tolerances.

When the critical areas are machined on a rough workpiece, the locating surfaces should be selected to ensure (1) that a reasonable relationship exists between rough and finished surfaces on the finished part, (2) that thickness of sections is uniform, and (3) that adequate stock will be left on surfaces to be machined later.

In tie-in operations, a part should be located from critical areas. If these areas are distorted in placement operations, they should be reestablished by going back to the basic rough locating areas or to areas reliably related to the critical areas.

9.8 WORKPIECE CLAMPING

The purpose of clamping is to direct forces to seat a workpiece firmly against the locating points and surfaces and to hold it there securely against all disturbing forces.

Clamping forces must be applied to suit the locators. A clamping force must tend neither to upset the location of the workpiece nor to distort the workpiece. The best condition prevails when a force is directly opposed by a fixed locator through a heavy section of the workpiece. Otherwise, a clamping force should be counteracted directly by a jack support, especially if the workpiece is weak and likely to deflect. A clamp may dig into a rough surface but must not mar a finished surface to which it is applied.

Processing forces should be directed against fixed locators as much as possible, as indicated in Fig. 9.12. Clamping forces are applied to counteract processing forces not taken directly by fixed stops and to neutralize the moments set up by the processing forces. Initial clamping forces must be imposed to prevent even the slightest movement of a workpiece in an operation. To do that, the initial force must be larger than any reaction the clamp is likely to receive during the operation.

9.9 PRODUCTION TOLERANCES

9.9.1 Operation Tolerance

The natural tolerance or performance limits of an operation set up and conducted in a definite manner can be ascertained by the techniques of statistical quality control described in Secs. 13 and 14.

Definite tolerances must be held in each operation if a process is to produce finished parts within specified tolerances. As indicated in Fig. 9.6, a dimension can be held to any tolerance desired, but the cost increases as the tolerance is decreased. Usually, different kinds of operations are necessary for different tolerances. For instance, a diameter may be turned if its tolerance is more than 0.001 in. but must be ground for a smaller tolerance and must be lapped or superfinished for exceptional accuracy and finish.

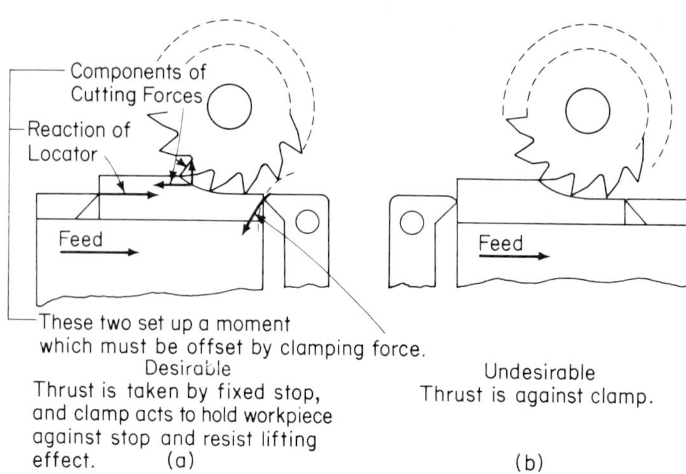

Fig. 9.12 *Forces opposed by fixed stops.*

The dimensional variations in an operation may result from variations in dimensions or material properties of rough workpieces; inaccuracies in tools and machines; wear, deflection, or thermal expansion of tool and machine details; dirt, chips, and burrs on slides and locating surfaces; and human errors. Common practice is to give an important tool dimension a tolerance of 5 to 20 per cent (most commonly 10 per cent) of the tolerance of the corresponding workpiece dimension. The aim of good process planning and tool designing is to control all the causes of errors, and principles to that end are cited throughout this chapter.

9.9.2 Tolerance Charts

A tolerance chart is a means of proving that the operations as planned will together produce the tolerance required for the product. It shows the dimensions, tolerances, and stock removal at all stages of manufacture in an easily understood form, saves time in making changes, and serves as a ready reference during discussions of a process.

The typical tolerance chart of Fig. 9.13 shows conventional designations, symbols, and notations. An X at one end of one dimension arrow in each operation shows the locating surface. The head of each arrow points to a line representing the surface cut. Lines without arrows represent resultant dimensions.

On the chart of Fig. 9.13, the tolerances on stock removal and resultant dimension are the sums of tolerances of working dimensions. For instance, the $\pm$ 0.002-in. tolerance of the 3.007-in. resultant dimension of Operation Number 20 is the sum of the $\pm$ 0.001-in. tolerances of the 0.118-in. and 3.125-in. working dimensions of Operation Numbers 5 and 20. If operations are subject to statistical quality control, with assurance that the deviations in the working dimensions are likely to occur at random within the specified working limits, the probable tolerances of the resultant dimensions and amounts of stock removal may be computed on the basis of $A = \sqrt{B^2 + C^2}$. The resultant tolerance is A, and the working tolerances are B and C. On this basis, the tolerance of the 3.007 resultant dimension of Operation Number 20 can be expected to be

$$\sqrt{0.002^2 + 0.002^2} = 0.0028 = \pm\ 0.0014 \text{ in.}$$

A tolerance chart must be realistic and take into consideration the following points:

1. Any condition contrary to good practice must not be allowed merely to satisfy the tolerance chart.

2. Tolerances must reflect economically attainable performance for each operation.

3. Tolerances specified from a surface which is neither a locator nor machined in the same operation must be large enough to absorb the accumulation of tolerances.

4. Dimensions must be chosen so they can be checked with a practicable gage, preferably both in the holding device and after release.

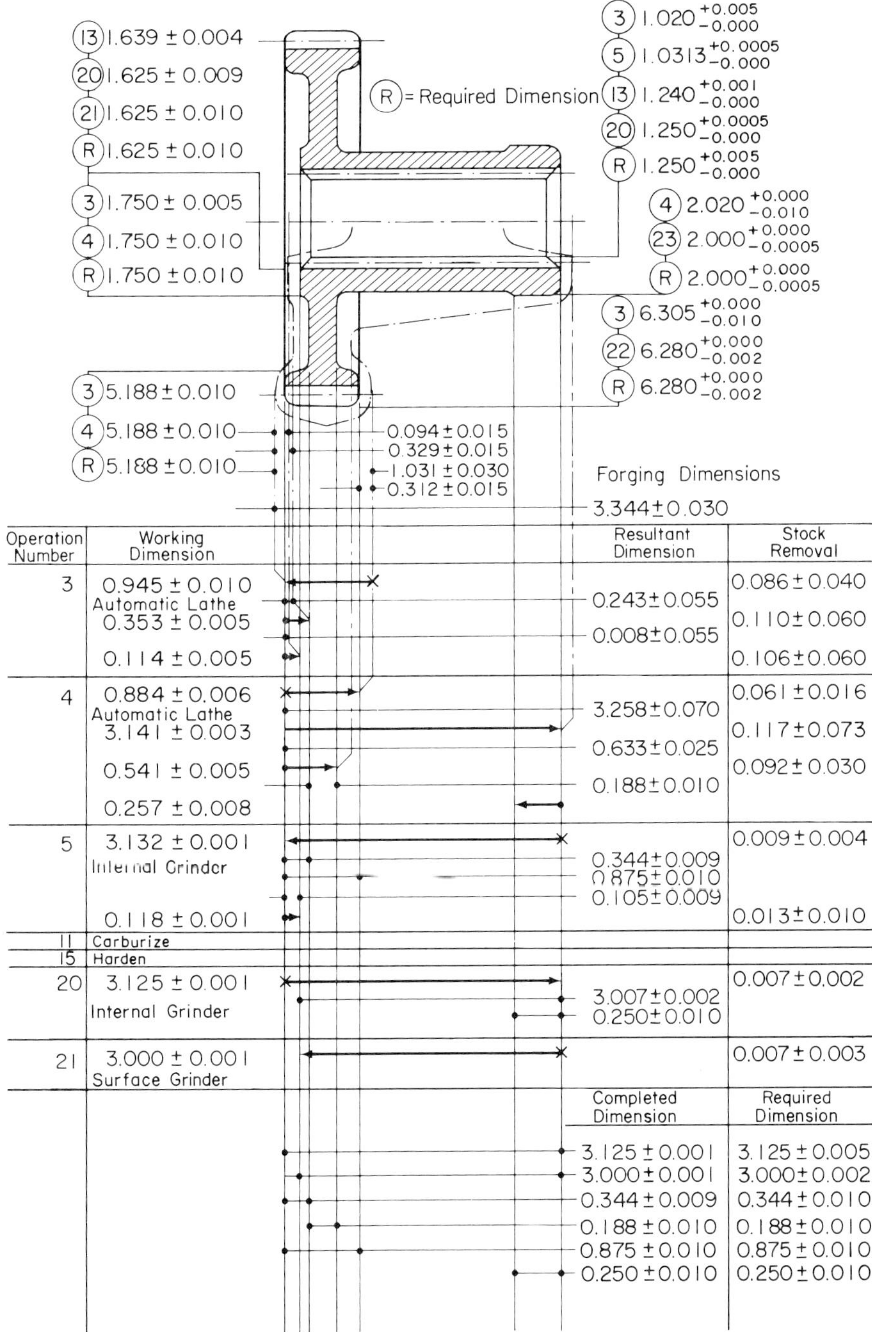

Fig. 9.13 *A typical tolerance chart.*

5. The least stock allowed must be enough to ensure cleanup of a surface, and the most stock must not be excessive.

If a tolerance chart shows that tolerance accumulations are likely to exceed specifications or that some operations must be held too closely, the following remedies may be presented:

1. Operations may be recast or rearranged.
2. Corrective operations may have to be added.
3. The sequence of operations may be changed.
4. The fact may be recognized that some part of the production may be outside of specifications.
5. Stock-allowance requirements may be waived to some extent.
6. An attempt may be made to have the tolerance of one or more dimensions increased on the part drawing. The tolerance chart may serve as proof of the impracticability of meeting specifications.

9.9.3 Stock Allowance

The material provided on a surface to be machined is called the stock allowance; it must be sufficient to allow for expected runout, for geometric and linear variations of the rough workpiece, and for complete removal of valleys, scratches, scale, inclusions, and burnt material, and to enable the tools to cut cleanly. On the other hand, definite upper limits for stock allowance must be observed. Good surface finishes and dimensional accuracy demand light cuts. Even in roughing, too much stock can impose excessive loads on tools or result in the removal of desirable surface material, as in the case of carburized steel parts.

Stock allowance often depends largely on the tolerances of basic processes. These may be ascertained from basic suppliers or, in some cases, from trade association standards.* In general, the larger a workpiece, the greater the stock allowance. The proper amount of stock allowance for any particular operation must be determined from the considerations applicable to the specific case. As a guide, Table 9.8 shows typical stock-allowance practices for common operations.

* For examples: *Steel Casting Design*, Section 3: "Tolerances" (Cleveland, Ohio: Steel Founders Society of America), and *Standard Practices and Tolerances for Impression Die Forgings* (Cleveland, Ohio: Drop Forging Association).

9.10 PLANNING AND TOOLING FOR LOW-COST PROCESSING

Low production costs are achieved by getting the best balance among direct, indirect, and fixed costs, which is a matter of engineering economy as explained in Sec. 3 and by making all costs as low as possible. The following subsections give principles that point the way to low costs.

9.10.1 Operation Analysis for Low Direct Costs

An operation is analyzed by being broken down into its elements. Each element is studied to find out how it can be done best. The tested elements are put together in the most efficient way, and facilities are provided to carry out the operation as planned.

9.10.2 Direct Handling Costs

The costs of handling elements can be minimized by motion economy; elimination of strain, fatigue, and heavy manual labor; conservation of skill; and combination of elements.

The principles of motion economy† of most importance to tool engineering are (1) eliminate all unnecessary motions, (2) shorten and simplify all necessary motions, (3) balance the work, (4) minimize use of the eyes, and (5) eliminate use of hands as holding devices. The subject of motion economy and techniques employed for operation analysis are discussed in Sec. 5.

Strain, fatigue, and heavy labor may be alleviated by providing (1) comfortable working conditions, (2) convenient and easy-to-operate controls and devices, (3) mechanical power to do work, and (4) means to reduce the weight or load an operator must bear.

Skill is conserved by transfer of skill and improvement of perception. Skill is transferred

† O. W. Habel and G. G. Kearful, "Machine Design and Motion Economy," *Mechanical Engineering*, **61**, No. 12 (Dec. 1939), 897. Also, *Motion Economy Rules to Guide Process and Tool Engineers* (Saginaw, Mich.: Saginaw Steering Gear Div., General Motors Corp., 1939).

TABLE 9.8. *TYPICAL STOCK ALLOWANCES FOR MACHINING OPERATIONS**

Condition of Rough Surface	Metal Removal Process	Dimension			
		Up to 2 in.	2 in. to 6 in.	6 in. to 12 in.	12 in. to 18 in.
Iron casting	Turning or milling	1/16	3/32	3/32	1/8
	Boring (diameter allowance)		1/8	1/8	3/16
Steel casting	Turning or milling	3/32	1/8	1/8	3/16
	Boring (diameter allowance)		3/16	3/16	1/4
Malleable casting	Turning or milling	1/32–1/16	1/16	3/32	5/32
	Boring (diameter allowance)		1/16	3/32	5/32
	Coining	0.15–0.030			
Drop forging	Turning or milling	1/64–3/32	3/32–1/8	1/16–1/8	
	Coining	0.015–0.030			
Die casting	Reaming (diameter allowance)	0.005–0.030			
	Diamond boring (diameter allowance)	0.008–0.025			
Rough-turned steel	Finish turning (diameter allowance)	1/64–1/32	1/32–3/64	3/64–1/16	1/16–3/32
Rough machined	Grinding (diameter allowance)	0.010–0.015	0.012–0.015	0.015–0.020	0.018–0.020
	Grinding (surface)	0.005–0.008	0.007–0.010	0.010–0.012	
Rough ground	Grinding (diameter allowance)	0.008	0.010	0.015	0.018
	Grinding (surface)	0.003–0.005	0.004–0.007	0.005–0.010	0.006–0.010
Finish ground	Microgrind (diameter allowance)	0.004	0.005	0.007	0.008
	Microgrind (surface)	0.002–0.003	0.002–0.003	0.003–0.004	0.004–0.005
Commercial grind	Superfinish	0.00025	0.0003		
Smooth grind	Superfinish	0.00015	0.0002		
Drilled hole	Boring (diameter allowance)	1/32–1/16			
	Reamed (diameter allowance)	0.010–0.035			
Bored hole	Precision boring or reaming (diameter allowance)	0.004–0.009			

* Specifications are for stock on surface or side unless otherwise stated.

when built into a machine or tool, such as a drill jig. Perception is improved by magnification of movements, as by indicators.

Elements are combined by simultation and integration. Simultation means the occurrence of two or more elements at the same time, such as when machine and operator are both working at once. Integration means the incorporation of two or more cuts or actions into one operation, even though they may not be done simultaneously. Machining as well as handling costs may be reduced by combining elements. Special machines are commonly built for high production to carry out these principles.

9.10.3 Direct Machining Costs

Machining time is influenced by the design of the workpiece, the machining method, the capacity of the machine, the design of the tools, the condition of the work material, and the conduct of the operation.

Design changes to reduce machining time are suggested in Par. 9.3.2 and Fig. 9.5.

Some machining processes are quicker than others under favorable circumstances. For instance, broaching is faster than milling but can be applied only to parts offering no obstruction to the broach and produced in sufficient quantities to justify the investment in broaching equipment.

The maximum attainable machining rate may depend upon the strength and power of the machine tool. Rugged and powerful machines are necessary to produce pieces in large quantities. With adequate machine capacity, carefully designed cutting tools are necessary to realize the most from an operation. As an example, a manufacturer found that the rate of a milling operation could be increased appreciably by increasing the number of teeth in the cutters over conventional designs in use for some time. Another requirement for fast machining rates is the use of heavy jigs and fixtures to give adequate support to the work.

A workpiece should usually be cut in the direction requiring the shortest travel or the fewest number of strokes. Tools should be moved to and from the work fast and over as short a distance as is safe. Most modern machine tools are provided with rapid traverse rates for this purpose.

9.10.4 Indirect and Fixed Costs

Ways of planning for low indirect costs include giving attention to the facilities needed for inspection, scrap disposal, setup, changeover, maintenance, and other off-line functions; thorough promotion of safety; coordination of operations with material-handling facilities between operations; and selection of equipment that is readily available.

The cost of equipment can be kept at a minimum by making it as simple as feasible to produce the required quantity and quality of product, by using toolmaking facilities knowingly and properly, by using versatile equipment, by making use of standard tool details, and by adapting standard machine units for construction of special machines.

9.11 PROGRAMING FOR NUMERICAL CONTROL

A program for a numerically controlled operation consists of a series of instruction sets, called blocks, each specifying a step in an operation. A block is made up of a number of words; each word contains a series of characters that designate the setting of a machine function. An example of a block of information is "003 TAB 01275 TAB 01562 TAB 2 TAB 6 TAB 3 EOB." The symbol EOB denotes the end of a block, and TAB the spaces between words. The first word is the block number, the second the x-coordinate setting for the table, next the y-coordinate setting, the feed rate, the speed, and the coolant-flow instruction. This is just one format; there are a number of different ones, and any one machine takes just one format. A series of blocks is punched in code on a paper tape to form the instructions for an operation.

Positioning or *point-to-point* N/C (numerically controlled) programs, exemplified by the drilling of several holes in a surface for which each instruction set specifies the worktable setting for a hole, are often simple and programed by hand, step by step. The more detailed programs, especially *contour* or *continuous-path* N/C programs, require many steps and often detailed calculations; they are economically set up only by digital computers.

An N/C source program for computer processing is generally written in a limited language made up of simple Englishlike words and numbers. An example of a couple of lines is:

```
GO PAST/LIN 3
TLRGT, GO FWD/CIRC 2
```

The rather obvious meaning is to move the tool past Line 3, and then, keeping the tool on the right of the workpiece, to go forward along Circle 2. Usually a line of such instructions culminates in a number of steps in the final program; e.g., a matrix of holes may be designated by a matrix symbol, the coordinates of a corner hole, and the specification of the number of holes and their coordinate spacings. The computer in that case calculates the coordinates of and sets up the instructions for each hole.

The computer program or compiler enables the computer to convert the source program into machine language, to make necessary calculations (such as finding the coordinates of a sufficient number of points along a specified curve), and to fill in the necessary instructions to constitute a complete program for an N/C machine. The output is called an *object program;* it may or may not be suitable for a particular N/C machine. A *post processor* is a supplementary computer program that modifies an object program to suit a specific N/C machine or control system. The following subsections briefly describe a number of leading computer programs for N/C programing.

9.11.1 Positioning Programs

AUTOPROPS (Automatic Program for Positioning System) produces two-axis, point-to-point programs from a relatively few, simple input statements including abbreviations of hole patterns, such as matrices and bolt circles. The outputs of several modifications are applicable to several widely used N/C drill and punch presses. It utilizes an IBM 1401 computer and provides a printed plot of the programed hole pattern for verification and cards or punched-tape (with attachment) output.

AUTOSPOT (Automatic System for Positioning Tools) produces three-axis, point-to-point programs with limited fourth-axis capability from a simple, English-like language. Three types of statements in the language are definitions of symbols, tool-data specifications, and machining statements that specify the unit operations. Features of the program are the manipulation of hole patterns and arrangements for tool offset and face and pocket milling, even along arcs and slopes. AUTOSPOT requires a post-processor program but has been adapted to many makes of N/C drill presses, jig boring machines, and Milwaukee-Matic machining centers. It utilizes an IBM 1620 computer and usually takes several passes through the machine.

CAMP I (Compiler for Automatic Machine Programing) is a three-axis, point-to-point program written by the Westinghouse Electric Corporation for the LGP 30 computer and for the Milwaukee-Matic machining center, for which it does not require a post processor. The program directs tool changes and arc and slope milling, drilling, boring, reaming, and tapping operations. It can transpose coordinate systems and select proper cutting speed, feed, and depth of cut to suit a tool and work-material combination. Limitations of machine tool and fixture (such as imposed by obstructions) are observed. A later version, CAMP II for large computers with a generalized post processor readily adaptable to almost any N/C system, has been called possibly the most powerful point-to-point system. It has all of the features of CAMP I plus pattern-expansion and subroutine-generation capability and complete symbolic labeling. It can handle all point-to-point machining operations and a certain amount of pocket- and connected-path milling.

PRONTO (Program for Numerical Tools) is a three-axis, point-to-point program written by the General Electric Company originally for large computers; it has been individual-machine oriented to several N/C control systems but is readily adaptable by means of a generalized post processor to any N/C machine. The PRONTO vocabulary consists of fifty words of three to six letters each. Subroutines supply the complete cycles for drilling shallow and deep holes, boring and counterboring, and tapping; hole pattern manipulation is feasible for matrices and bolt circle holes, and coordinate rotation and translation are available—all from minimal instructions.

SNAP (Simplified Numerical Automatic Programer) is a two-axis, point-to-point program specifically designed for the Brown and Sharpe Turr-E-Tape drill through the IBM 1401 system. Instructions are relatively few and simple but include means for defining linear, matrix, and bolt-circle-hole patterns. A plot of the output is given for verification.

9.11.2 Continuous-Path Programs

APT (Automatically Programed Tools), one of the most extensive programs, is the result of over 100 man-years of direct effort and the support of scores of cooperating user concerns. It is a general composite of computer programs that cover as many aspects of N/C programing as possible and even include auxiliary functions such as cost estimating and production scheduling in some applications. Data-processing for APT requires the largest computers. APT can be instructed through a large vocabulary of Englishlike words and symbols, which in many respects are like those of Fortran. It directs tool paths along straight lines, points, circles, ellipses, parabolas, and even curves faired through prescribed points, with necessary tool offset. The APT program translates the instructions it receives, performs all necessary calculations, and issues the numerically coded list of instructions for the machine tool. Postprocessors are available for a large number of N/C systems. APT has gone through several stages of development. Originally sponsored by the Air Force, it is managed as APT III and IV by the Illinois Institute of Technology and is available to subscribers.

ADAPT (Air Materiel Command Developed APT) is a three-axis, continuous-path program prepared by IBM for moderate-size computers. The third-axis capability is limited in some respects (for example, it is not able to describe solid geometric figures), but the program can render limited cylindrical and conical work, figures in canted or ramp planes, and simple copy logic and repeated patterns and instructions (macros). Postprocessors have been developed for a number of machines, and various computer manufacturers have been extending the capabilities of the program for their respective systems.

AUTOMAP (Automatic Machining Program) is a two-axis, continuous-path program for contours of lines, arcs, and slopes in two planes and third-axis nonsimultaneous positioning. Source programs are written in a language of about 40 APT words to describe the workpiece in geometric terms for the IBM 1620 computer, which makes necessary calculations, such as for arcs, and formulates generalized machining instructions. A second pass postprocesses for a specific N/C machine. Postprocessors have been reported written for an Airco flame cutter and a G and L boring mill.

AUTOPROMPT (Automatic Programed Tools) is a five-axis, three-dimensional contouring program written for large computers. Although the vocabulary is different, AUTOPROMPT does about what APT does but in a somewhat more sophisticated manner. The programer needs to specify only the surfaces and boundaries of the part and the regions where multiple passes must be made. The program then ascertains what paths the tool must follow; the programer does not have to define the cutter paths. APT postprocessors can translate AUTOPROMPT output for specific N/C machines. Steps have been taken to integrate AUTOPROMPT into the APT system.

SPLIT (Sundstrand Processing Language Internally Translated) is a three- to five-axis, continuous-path program adaptable to medium and large size computers, specifically for all Sundstrand machine tools with Bendix, GE, and TRW controls without postprocessors. The language is Englishlike and employs sixteen to eighteen major and about forty minor word statements. The program processes macroinstructions (preconstructed cycles) for ordinary and deep-hole drilling, boring, and tapping operations, locations of matrices, and rotation and translation of coordinate axes.

SYMPAC (Symbolic Program for Automatic Controls) is a three-axis, continuous-path program for the Sperry Rand Univac computers. Input is in shortened English statements in a fixed-field format. Each line of entry may contain sequencing, part-number coding, and programing notes. Geometric combinations of points, straight lines, circles, cubes, cones, and curves fitted to points in the x-y plane may be processed in horizontal as well as canted planes. The cutter path in the third direction may be added in the form of cutting instructions. SYMPAC consists of four parts. The preprocessor fits cubics to point arrays. A geometry compiler, called USCAN, calculates and compiles the parameters specified by the geometric input. The intermediate processor converts the results of the first two sections to output statements defining the tool path. The fourth part, the postprocessor, encodes the output for a specific N/C machine, including accelerations, decelerations, feed rates, and point coordinates.

9.12 COST ESTIMATING

A cost estimate is an attempt to ascertain the expenses that must be incurred to carry through a project, i.e., to manufacture a product in industry. The actual factors and costs differ from one case to another, but the principles and techniques of making estimates are the same for all. Essentially there are two ways to estimate the cost of an item as an entity, whether it be a complete manufacturing plant or an element of an operation. One way is to multiply some magnitude of the item by a factor. The cost of a chemical plant, for instance, may be estimated roughly by multiplying its expected output in tons by a factor representing the dollar investment per ton of output found applicable for similar plants. A second way is to adopt the value of a similar item of known cost. Thus, the cost of performing a certain operation may be estimated to be the same as or some proportion of the cost of performing a like operation in the past.

The cost of a complete project or item as an entity may be estimated by one of the two ways just explained. More accurate results for a large project often are obtainable by dividing it into parts, estimating the cost of each part, and adding the costs of the elements together to arrive at the total estimate. The smaller elements are simpler, more likely to have comparable counterparts, and easier to appraise, and errors in estimating them tend to compensate for each other when added together. The number of elements into which a project should be divided for the lowest total cost is

$$n = b(P_1C/c)^q.^*$$

Here b and q depend upon the estimating system; b is usually less than $\frac{1}{2}$, and the exponent q less than 1. In a given system P_1 represents the average percentage error in estimates with the information available; if information is sparse and P_1 is large, n should be large, and vice versa. The factor C is the total value of the project; with more at stake, a more detailed breakdown is warranted. The average cost of making an estimate of each element is c; the higher the cost, the fewer the number of divisions justified.

9.12.1 Estimating Practices

Method 1 in Fig. 9.14 involves a detailed breakdown with full information available. This is a final cost estimate; its preparation cost is high, but the error is likely to be minimal. At other stages shown, less information is available or sought, the methods are less thorough, preparation costs are less, and margins of error are likely to be proportionately higher.

Among major costs of a project are those for design and development, preparation of production facilities, testing and tryout, and fabrication—each estimated separately. Of the fabrication costs, direct material and labor are the most readily (and often the only) identifiable ones for the parts of a project. Other costs commonly are grouped into overhead or indirect cost factors.

9.12.2 Estimating Material Costs

The material chargeable to a piece is that in the rough state. The volume of a piece is estimated or calculated and multiplied by the density of the material (0.26 $lb/in.^3$ for cast iron, 0.28 for steel, 0.092 for aluminum, and 0.30 for brass) to find the weight. An irregular piece is divided into regular parts, and each part then estimated separately. Experienced estimators are able to judge the weight of intricate pieces rather closely by comparing them with pieces of known weights.

The length of bar stock required for a piece is the sum of the length of the finished piece plus 1/32 in. to 1/16 in. on each faced end plus the width of the cutoff tool. The last item may be fairly estimated as $w = 0.06 + 0.25P_ud$ for various materials in Table 9.1.

The dimensions of a blank developed for a formed piece, as explained in Par. 9.16.13, are sufficiently accurate for estimating purposes. The scrap allowance between successive blanks in a strip should be 1/32 in. for stock less than 1/32 in. thick, an amount equal to the stock thickness for stock between 1/32 in. and 3/16 in., and a maximum of 3/16 in. for greater thicknesses. The scrap allowance on each side of a blank at the edges of a strip is equal to the stock thickness plus the product of 0.015 times the diameter or width of blank.

From 1 to 12 per cent of material is normally lost in processing in scrapped pieces, butt ends, droppings, and so forth, depending upon conditions. An average allowance of 5 per cent is commonly made for such bulk losses.

The weight of a piece multiplied by the unit cost of material gives the material cost per piece. If the unit cost covers only the purchase price of the material, the material cost is multiplied by one or more additional factors to account for bulk losses, purchasing, and handling costs. The unit-cost figure frequently is larger than the purchase price and accounts for the other costs, in some cases for a charge for heat treatment. Material unit costs generally decrease for larger amounts of material. Scrap from a few materials, such as brass, has appreciable salvage value, and that may be deducted from the material cost.

Unit material costs need to be checked at intervals, on the average about every 6 months, and corrected.

9.12.3 Estimating Direct Labor Costs

The total direct labor required to make a part may be estimated by reference to that required on similar jobs. A more refined method is to divide the job into operations, which is the same as prescribing the process.

The basis for estimating operation time is previous performance on similar operations. Some estimators with backgrounds as toolmakers, mechanics, or foremen are able to

---

* L. E. Doyle, "An Analysis of Cost Estimating Principles and Practices," Paper T–9 (Detroit: American Society of Tool Engineers, 1952).

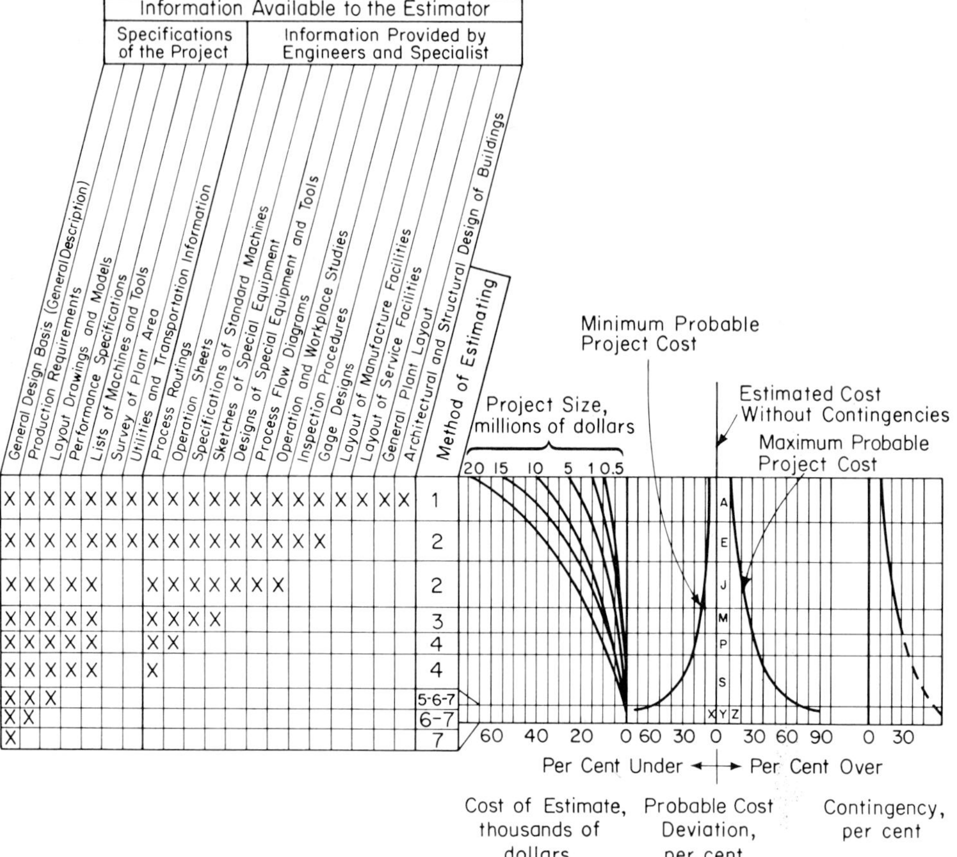

Fig. 9.14 *Cost and reliability of various estimating methods. From L. E. Doyle, "Cost Estimating, How to Minimize the Dangers of Chance,"* The Tool Engineer, *(June 15, 1949) p. 82.*

judge operation time values closely. Much tool estimating and some product estimating are done in this way.

Some estimators have found rules that give sufficiently accurate and quick estimates of operation times. The time for a weld may be estimated by multiplying the length by an ascertained average time to weld 1 in. The cutting time may be calculated for an operation and multiplied by a factor that studies in a particular plant have shown gives the total time for that particular type of operation. Such factors must be used with discretion.

Those without sufficient shop experience may have to refer to records for operation-time values. Cost-accounting records may be consulted for specific operations. The estimate sheets of Fig. 9.15 have a column to enter the actual time for each operation after the job has been completed. After a person has worked with such records for a while, he remembers the times of typical operations and need not refer to the files for every case.

Tables* or charts, prepared to show the time values for particular typical operations performed on various shapes and sizes of pieces of various materials, and with various amounts of stock removal, are helpful for quick and accurate estimating.

When cost-accounting or shop records are used, care must be taken that the information

* W. N. Nordquist, *Die Designing and Estimating* (Cleveland, Ohio: Huebner Publications, Inc., 1960). Also, Leonard Nelson, "How to Estimate Dies, Jigs, and Fixtures from a Part Print," *American Machinist*, Special Report No. 510 (Sept. 4, 1961).

is reliable. Because of equipment breakdowns, carelessness, or intention, time may not be recorded properly in the shop on some jobs. An operator may be inexperienced, or proper tools may not be available for the first run of an operation.

9.12.4 Breakdown of Operations

Where work methods are standardized, operations can be divided into standard elements. The most accurate estimates of operation time are obtained by evaluating operation elements and adding them together.

The time needed for almost any fabrication operation can be divided into (1) setup time, (2) man or handling time, (3) machine time, and (4) lost time or downtime.

Machine time differs from the others because it can be calculated. The machine time for an operation is $T = L/F$. The length of cut L in inches includes the length of the surface cut

TOOL GROUP CODE 804
WORK SHEET PREPARED ☒ YES ☐ NO

JOB ESTIMATE SUMMARY

EST. NO. B 03348
DATE 11/19/51

CUSTOMER Perfection Mfg. Co.
TOOL NO. 97625
DESCRIPTION: Boring bar with wear strips - 17½ × 2¼

INQ'Y NO.
QUOT. NO. 175432
ORD. NO.
JOB NO.
EST. DEL'Y 5/1/52

NO. PCS. 1
UNIT PRICE

MATERIAL REQUIREMENT

DESCRIPTION	AMOUNT PER UNIT	RESERVED FOR ACTUAL COST
Rough material	3 50	
Standard parts	58 50	
Make bar and assemble with standard # XR 765 wear strips, cutters, etc.		
TOTAL MATERIAL	62 00	

ENGINEERING TIME

HOURS	RATE	TOTAL	AMOUNT PER UNIT
3	4.50	13.50	13 50

PRODUCTION UNIT TIME

DESCRIPTION OF OPERATION OR DETAILS	MACH. HOURS PER UNIT	RATE	AMOUNT
Make bar	11.55	6.10	70 50
OUTSIDE LABOR			
HEAT TREATING			6 25
TOTAL ENGINEERING AND PRODUCTION UNIT COST			90 25
PLUS MARKETING AND PROFIT RATIO %			90 25

COST EST. BY J.B.S. DATE 11/19/51 UNIT SELLING PRICE 152 25

REMARKS:

DATE COMPLETED BY COST DEPT.

DESIGN TIME	90 DAYS	ENG. EST. BY C.B.L.	DATE 11/19/51
METHOD TIME	3 DAYS		
SHOP TIME	14 WEEKS	DEL'Y EST. BY A.H.D.	DATE 11/19/51

ENGINEERING DEPARTMENT

FORM BJ-603-8-48-R-7-51

UARCO INC. CHICAGO

Fig. 9.15(A) *A cost-estimate summary sheet.*

JOB ESTIMATE WORK SHEET

REF. TO EST. NO. B03348

CUSTOMER: Perfection Mfg. Co.

JOB NO.

DESCRIPTION OF MATERIAL AND OPERATIONS	OUTSIDE LABOR AND HEAT TREATING	MATERIAL COST	UNIT TIME 1st Est.	UNIT TIME 2nd Est.	ACTUAL UNIT TIME PER COST DEPT.
Body - 30# H.T. & Material x 1.15	6.25	3.50			
Operations					
Turning				2.20	
Mill slot				1.50	
Mill tang				0.75	
Bore				1.00	
Drill & tap holes				0.80	
Grind O.D.'s.				2.50	
Assemble				0.50	
Strips XR765 - 4 Req'd. x 1.15		45.00			
Cutters AY625 - 2 Req'd. x 1.15		9.00			
Nuts LN1630 - 2 Req'd. x 1.15		1.50			
Screws, et x 1.15		3.00			
SUB-TOTAL	XXXX	XXXX	XXXX	9.25	XXXX
ESTIMATING CONTINGENCY 25 %	XXXX	XXXX	XXXX	2.30	XXXX
TOTALS	6.25	62.00	XXXX	11.55	XXXX

Fig. 9.15(B) *A cost-estimate work sheet.*

plus the distance traveled by the cutter at feed rate in approaching the surface plus the overtravel of the cutter. The feed F is expressed in inches per minute. Charts like Table 9.9 and special types of slide rules are commonly utilized to ascertain machine time quickly. The unit time in minutes per inch obtained from the chart of Table 9.9 is multiplied by the length of cut to obtain the machine or cutting time.

Setup and man time elements are evaluated by time study. Average values are tabulated in convenient ways in many plants for estimating purposes. An example of standard elements for lathe operations is given in Fig. 9.16. If standards are not available in a particular factory, reference may be made to published tables* of standard elemental times for common operations.†

Setup time is normally applied once to each lot of pieces, but for estimating purposes may

* W. A. Nordhoff, *Machine Shop Estimating* (New York: McGraw-Hill Book Company, 1960). Also, C. W. S. Parsons, *Estimating Machining Costs* (New York: McGraw-Hill Book Company, 1957).

† See Sec. 5 for estimation of direct labor time by means of synthetic time standards.

be prorated among the pieces in a lot to give a unit setup time. A prorated unit setup cost applies to only one lot size and must be clearly specified as such so that the figures will not be used for other jobs with erroneous results.

Allowances for personal needs, fatigue, and other justifiable items are added to elements measured by time-study methods. Tables prepared for estimating purposes may show elemental times that include necessary allowances or may show the allowances separately.

Lost time or downtime resulting from equipment breakdowns, parts that do not fit, tools that do not work properly, defective material, and so forth is accounted for by a performance factor applied to estimated operation time. Just what the lost time may be for a specific operation cannot be predicted, and the performance factor reflects average losses. It may be derived by dividing the total actual time for all jobs done in a plant over a period of several months to a year by the sum of the times estimated for the same jobs. This factor must be checked from time to time. Studies in various plants have shown performance factors from 1.10 to 1.70.

A performance factor may include corrections for average errors in estimating, such as the items overlooked in making a rough estimate without the benefit of detailed part drawings or the tendency of a particular estimator to be low. The estimating contingency in Fig. 9.15b is a performance factor.

9.12.5 Estimating Indirect Costs

Indirect costs are commonly distributed for estimating purposes by rates determined by cost-accounting methods. The number of hours of direct labor estimated for a job is generally multiplied by a rate to account for factory overhead. Often a single factor is used, including both the direct labor and overhead rates. For most tool estimating and for some product estimating, one overhead rate is applied to all operations in an estimate. However, to reflect difference in methods and equipment, overhead should be broken down to an extent corresponding to the detail to which direct costs are estimated. Thus, different overhead rates may be used to estimate costs for different departments, work centers, or types of operations in a plant.

As a rule, general administrative and selling costs are charged in proportion to the number of dollars of total manufacturing cost. This may be done by a separate factor, or the charges may be included with direct labor and factory overhead in one rate, as in the example of Fig. 9.15. In that case, the $6.10/hr rate against production unit time covers direct labor, factory overhead, general overhead, and marketing costs, and expected profit. However, general overhead is not always related to factory overhead. The space designated "Plus Marketing and Profit Ratio" in Fig. 9.15 is used to enter general overhead, marketing costs, and profit against outside labor which is not charged with factory overhead and against which the $6.10 rate is not applicable.

Standard purchased parts and material are not generally required to carry factory and general overhead, although practice in this respect is not uniform. A large markup puts such merchandise in a poor competitive position, but charges for out-of-pocket expenses for shipping, purchasing, and handling are justifiable.

An overhead rate based upon the practical capacity of a plant may be desirable from the accountant's standpoint to establish standard costs. The estimator needs a more realistic rate based upon an actual current or anticipated level of activity to forecast costs as they are likely to be. The estimator must understand the principles of the cost-accounting system that furnishes him information to determine the relevancy and adequacy of the data. The estimator must be careful not to duplicate items by estimating them individually if they are already included in the overhead rates furnished by the cost accountant.

When costs are estimated to set a selling or contract price, two figures may be helpful, especially as terminal points for negotiating the price. The first is an estimate of costs directly attributable to the project, including material, labor, and other direct costs plus that part of overhead that varies with the activity and that will be incurred only if the job is undertaken. The second figure is the sum of the first plus the proportion of fixed costs that do not depend on the specific job but toward which the job should contribute as much as possible.

9.12.6 Estimating the Costs of Auxiliary Services

As a rule, most of the auxiliary services of production are included in overhead for estimating purposes. However, when such items are sizable and do not vary in proportion to

TABLE 9.9. *CHART FOR QUICKLY COMPUTING TIME IN TURNING*

Diameter, Inches

½	⅝	¾	⅞	1	1⅛	1¼	1⅜	1½	1⅝	1¾	1⅞	2	2⅛	2¼	2⅜	2½	2⅝	2¾	2⅞	3	rpm
0.13	0.16	0.20	0.23	0.26	0.29	0.33	0.36	0.39	0.43	0.46	0.49	0.52	0.56	0.59	0.62	0.65	0.69	0.72	0.75	0.79	1
																					6
																					9
																					12
																					15
																					19
																	17	18	19	20	25
												16	17	18	19	20	21	22	23	24	31
									17	19	20	21	23	24	25	27	28	30	31	32	41
						15	19	20	22	24	26	27	29	31	32	34	36	37	39	41	52
				18	20	22	24	27	29	31	33	35	38	40	42	45	47	49	51	53	68
		18	19	22	25	27	30	33	36	38	41	44	47	49	52	55	58	60	63	66	84
	17	22	25	29	32	36	40	43	47	50	54	58	61	65	68	72	76	79	83	86	110
	20	25	29	33	37	41	45	49	53	57	61	65	70	74	78	82	86	90	94	98	125
18	23	27	32	37	41	46	50	55	60	64	68	73	78	82	87	92	96	101	105	110	140
21	26	31	37	42	47	52	58	63	68	73	78	84	89	94	99	105	110	115	120	126	160
24	29	35	41	47	53	59	65	71	77	82	88	94	100	106	112	118	123	130	135	141	180
26	33	39	46	52	59	65	72	79	85	92	98	105	111	118	124	131	137	144	150	157	200
29	37	44	52	59	66	74	81	88	96	103	110	118	125	133	140	147	155	162	169	177	225
33	41	49	57	65	74	82	90	98	106	115	123	131	139	147	155	164	172	180	188	196	250
38	48	57	67	76	86	96	105	115	124	134	143	153	162	172	182	191	201	210	220	229	292
43	54	65	76	86	97	108	119	130	140	151	162	173	184	194	205	216	227	238	248	259	330
49	61	74	86	98	110	123	135	147	160	172	184	196	209	221	233	245	258	270	282	295	375
56	70	83	97	111	125	139	153	167	181	195	209	223	236	250	264	278	292	306	320	334	425
64	80	96	112	127	143	159	175	191	207	223	239	255	271	287	303	319	335	351	367	382	487
72	90	108	126	144	162	180	198	216	234	252	270	288	306	324	342	360	378	396	414	432	550
82	102	123	143	164	184	205	225	245	266	286	307	327	348	368	389	409	430	450	470	491	625
91	112	140	161	182	203	231	252	273	301	322	343	364	392	413	434	455	483	504	525	553	700
104	128	160	184	208	232	264	288	312	344	368	392	416	448	472	496	520	552				800
117	144	180	207	234	261	297	324	351	387	414	441	468	504	531	558						900
130	160	200	230	260	290	330	360	390	430	460	490	520									1000
143	176	220	253	286	319	363	396	429	473	506	539										1100
156	192	240	276	312	348	396	432	468	516												1200
169	208	260	299	338	377	429	468	507	559												1300
182	224	280	322	364	406	462	504	546													1400
195	240	300	345	390	435	495	540														1500

Peripheral Speed, Feet per Minute

TABLE 9.9. (Cont'd.)

rpm	Feed per Revolution, Thousandths										
	0.002	0.004	0.007	0.010	0.015	0.020	0.025	0.031	0.046	0.060	0.090
1	500.	250.	143.	100.	67.	50.	40.	32.	22.	17.	11.
6	83.30	41.65	23.82	16.66	11.16	8.33	6.66	5.23	3.66	2.83	1.83
9	55.55	27.78	15.89	11.11	7.44	5.56	4.44	3.56	2.44	1.89	1.22
12	41.65	20.83	11.91	8.33	5.58	4.17	3.33	2.67	1.83	1.42	0.91
15	33.30	16.65	9.52	6.66	4.46	3.33	2.66	2.13	1.47	1.13	0.73
19	26.30	13.15	7.52	5.26	3.52	2.63	2.10	1.68	1.16	0.89	0.58
25	20.00	10.00	5.72	4.00	2.68	2.00	1.60	1.28	0.88	0.68	0.44
31	16.10	8.05	4.60	3.22	2.16	1.61	1.29	1.03	0.71	0.55	0.35
41	12.15	6.08	3.47	2.43	1.63	1.22	0.97	0.78	0.53	0.41	0.27
52	9.60	4.80	2.75	1.92	1.29	0.96	0.71	0.61	0.42	0.32	0.21
68	7.35	3.68	2.10	1.47	0.98	0.74	0.59	0.47	0.32	0.25	0.16
84	5.95	2.98	1.70	1.19	0.80	0.60	0.48	0.38	0.26	0.20	0.13
110	4.55	2.28	1.30	0.91	0.61	0.46	0.36	0.29	0.20	0.15	0.10
125	4.00	2.00	1.14	0.80	0.54	0.40	0.32	0.26	0.18	0.14	0.09
140	3.55	1.78	1.02	0.71	0.48	0.36	0.28	0.23	0.16	0.12	0.08
160	3.15	1.58	0.90	0.63	0.42	0.32	0.25	0.20	0.14	0.11	0.07
180	2.75	1.38	0.79	0.55	0.37	0.28	0.22	0.18	0.12	0.09	0.06
200	2.50	1.25	0.72	0.50	0.34	0.25	0.20	0.16	0.11	0.08	0.05
225	2.20	1.10	0.63	0.44	0.29	0.22	0.18	0.14	0.10	0.07	0.05
250	2.00	1.00	0.57	0.40	0.27	0.20	0.16	0.13	0.09	0.07	0.04
292	1.70	0.85	0.49	0.34	0.23	0.17	0.14	0.11	0.07	0.06	0.04
330	1.50	0.75	0.43	0.30	0.20	0.15	0.12	0.10	0.07	0.05	0.03
375	1.30	0.65	0.37	0.26	0.17	0.13	0.10	0.08	0.06	0.04	0.03
425	1.15	0.58	0.33	0.23	0.15	0.12	0.09	0.07	0.05	0.04	0.03
487	1.00	0.50	0.29	0.20	0.13	0.10	0.08	0.06	0.04	0.03	0.02
550	0.90	0.45	0.26	0.18	0.12	0.09	0.07	0.06	0.04	0.03	0.02
625	0.80	0.40	0.23	0.16	0.10	0.08	0.06	0.05	0.04	0.03	0.02
700	0.71	0.36	0.20	0.14	0.09	0.07	0.06	0.05	0.03	0.02	0.01
800	0.62	0.31	0.18	0.12	0.08	0.06	0.05	0.04	0.03	0.02	0.01
900	0.55	0.28	0.16	0.11	0.07	0.05	0.04	0.03	0.02	0.02	0.01
1000	0.50	0.25	0.14	0.10	0.07	0.05	0.04	0.03	0.02	0.02	0.01
1100	0.45	0.23	0.13	0.09	0.06	0.04	0.04	0.03	0.02	0.01	0.01
1200	0.42	0.21	0.12	0.08	0.05	0.04	0.03	0.03	0.02	0.01	0.009
1300	0.38	0.19	0.11	0.08	0.05	0.04	0.03	0.02	0.02	0.01	0.008
1400	0.36	0.18	0.10	0.07	0.05	0.03	0.02	0.02	0.01	0.01	0.008
1500	0.33	0.16	0.09	0.07	0.04	0.03	0.02	0.02	0.01	0.01	0.007
	Time, Minutes per Inch										

STANDARD OPERATION SHEET DEPT. "L"

DATE — JOB NO.

WORK ELEMENTS

NO.	A. WORK ELEMENTS	STD.	OCC.	TOTAL
1	WORK UP #			
2	IN 4 JAW CHUCK			
A.	TRUE WORK RGH.			
B.	TRUE WORK ACCUR.			
3	IN 3 JAW CHUCK			
A.	TRUE WORK RGH.			
B.	TRUE WORK ACCUR.			
4	IN TAILSTOCK CTR.	.20		
5	DRIVE DOG	.26		
6	BETWEEN CTRS.	.38		
7	IN STEADY REST	.37		
8	MOVE STEADY REST	1.95		
9	ON TAPER ARBOR	.20		
10	PRESS ON ARBOR	.50		
11	ON MANDREL W/NUT			
12	IN SPINDLE SOCKET	.55		
13	IN DRAW COLLET	.27		
14	IN SLEEVE	.20		
15	SET SCREW (EACH)	.20		
16	PLUG STR. OR TAPER	.25		
17	TOOL IN TAILSTOCK	.40		
18	TOOL IN POST & FIND SIZE	2.50		
19	S. POINT THRD. TOOL & FIND SIZE	3.00		
20	ROLL LOCK CHUCK	.70		
21	FIXT. F/SQ. TOOL BIT			
22	SCR. WRK. SHORT STUD	.52		
23	INDEX SQUARE	.08		
24	ROLL LOCK ARBOR	.45		
25	ECCENTRIC HOLDER			
26	CONNECT & DISCON. TAPER ATT.			
	TOTAL WORK ELEMENTS			

Fig. 9.16 *Handling time elements for engine lathe operations. Courtesy Scully-Jones & Co.*

direct manufacturing costs, it is preferable to estimate them individually. For instance, the cost of entering and following up an order may be the same for a small as for a large order. Engineering development and design may involve only slight modifications for one job but may be quite extensive for another.

Services that involve creative work or the solution of unforeseeable problems are not amenable to standardization. The amount of time to design a mechanical device may be estimated on the basis of what has been found necessary for a similar job or on the basis of how much time the designer thinks he will need. Whenever a sizable amount is at stake, it is always well for an estimator to seek the opinions of those who will have to fulfill the estimate in performance.

The time estimated for a service such as engineering is multiplied by an hourly rate to obtain the cost of the service. If several grades of designers, draftsmen, or other workers are employed on a job, the rate may be different for the time expended by each group. The hourly rate covers direct labor and also charges for administration, selling, supervision, housing, light, heat, and so forth.

9.12.7 Manufacturing-Progress Curves

Manufacturing-progress curves, also called learning curves, manufacturing-time forecasting curves, improvement curves, etc., are empirical log-log lines which approximately reflect the improvement in performance that results from doing a job again and again. The basic mathematical model has the formula

$$Y = KX^{-n},$$

where Y is the cumulative average hours expended in producing X number of units, and K is the number of hours required to make the first unit. A well-known characteristic of this relationship is that for a given n, Y is reduced a certain constant amount whenever X is increased by a constant percentage. Most commonly X doubles and Y is reduced according to the ratio $Y_{2X}/Y_X = \frac{1}{2}^n$, called the percentage of the curve. Figure 9.17 is an 80 per cent manufacturing curve. When the amount of output doubles, the ratio of second to first cost is 80 per cent. The upper line represents the cumulative average time Y_c, and the lower line the unit time $Y_u \simeq (1 - n) Y_c$, which is an approximation and applies only for more than ten units.

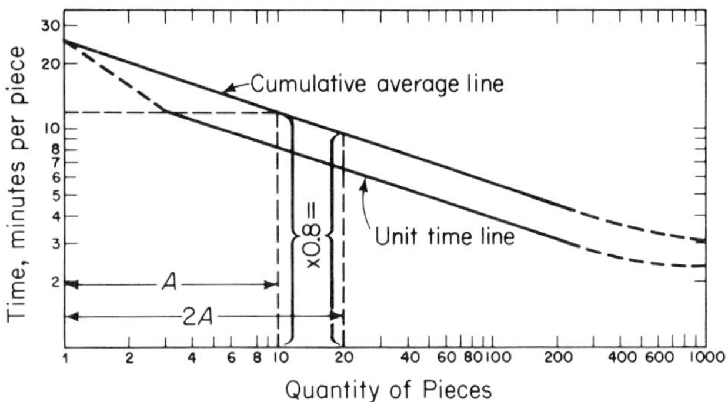

Fig. 9.17 *An 80% manufacturing progress or learning curve. From L. E. Doyle, "Cost Estimating, How to Minimize the Dangers of Chance,"* The Tool Engineer (*June 15, 1959.*)

Information may be obtained from a program curve by formulas. Besides those already given, the formula for the cumulative total time through unit X is $T_X = KX^{1-n}$. The individual unit time at unit X is

$$t_X = K\,[X^{1-n} - (X - 1)^{1-n}].$$

Graphical construction on log-log coordinates is fast and convenient, and commercial templates are available for those who work often with the curves.

Improvement curves are applicable in two contexts. One is that of the individual workman who learns to do a job more efficiently as he acquires experience while the method stays the same. Actually improvement is seldom limitless; this fact may be depicted by the curve deviating from a straight line and asymptotically approaching a minimum or standard time, as indicated by the lower dashed lines of Fig. 9.17. The practice of some is to cut off the line at the estimated standard time; others use a larger percentage (less slope) later in the curve than at the start.

The second context is that of the whole production system in a plant, where the manufacturing improvement results not only from the increased skill of the workmen but also from better methods, tools, designs, machines, etc. This is evidenced by the fact that the continuing output of a plant, a company, or a whole industry generally conforms to a manufacturing-progress curve. Although their usefulness has spread to many industries, manufacturing-improvement curves were first applied in the aircraft industry, for which studies show that an 80 per cent curve is average. Performance varies from plant to plant, however. Records of a number of aircraft companies producing various models of planes during World War II revealed plant curves of 65.3, 69.5, 70.8, 71.8, 75.0, 75.0, 76.4, 76.7, 77.4, 79.0, 80.5, 89.0, and 98.0.

The aspect of learning appears in the close relationship that exists between the improvement-curve ratio (or percentage) and manual ratio (the ratio of manual to machine work) of a task. Under certain conditions it has been possible to ascertain learning-curve percentages within 2½ per cent from measured manual ratios. Typical curve percentages are less than 80 per cent for all manual work, about 90 per cent for a 75 per cent manual-to-25 per cent machine-time ratio, 93 per cent for 50–50, 97 per cent for 25–75, and 100 per cent for all-machine time. Another observation that has been made is that the more mental effort required for a task, the greater the rate of learning.

The percentage of a curve for an aggregate of

tasks is $C_A = P_1C_1 + P_2C_2 \cdots + P_nC_n$ (not mathematically exact), where P_i represents the percentage of the individual task in the aggregate and C_i the percentage of the curve for the individual task.

Care must be exercised in making observations of manufacturing improvement. Some common errors are:

1. Giving credit for false labor savings that result when raw material is purchased in improved or finished form
2. Changing the usage of indirect labor to add or subtract from the efficiency of direct labor (assignment of functions from direct to indirect labor or vice versa without actually altering the work done)
3. Misinterpreting savings in labor input that actually result from increasing the volume of production or from reorganizing the work force in a more efficient manner
4. Overlooking the fact that a sufficiently long break between two periods of attention to a task may cause the learning curve to revert approximately to the starting point when the second period is begun

Sometimes manufacturing progress follows what is called an arched curve rather than a straight line on log-log coordinates. A straight line is approximated when a job is started from scratch; an arched curve occurs when those engaged in the task have the advantage of some previous related experience, which may have been acquired on similar jobs. Figure 9.18 gives an example. It is assumed that those starting the job have the proficiency acquired from producing five units and are really starting at the level of a sixth unit at Point A. This original advantage is the B factor. In this case they will complete their first unit in the time A'. Their second unit is actually equivalent to the seventh unit for a true beginner and is represented by B' corresponding to B in time. The third unit is C' corresponding to C, etc. Thus in the example, progress proceeds along the arched curve that becomes asymptotic to the original curve. Manufacturing progress on a redesigned part may follow an arched curve if the change in design does not materially change the process that has been used to make the parts of the old design. For a first-piece time Y'_A for point A' in Fig. 9.18, the cumulative average time on the arched curve for X pieces is

$$Y' = Y'_A(1 + B)^n/(X + B)^n,$$

where n is the slope of the original curve.

Some of the applications of progress curves are predicting direct labor costs and therefore accurately pricing proposed increases in contracts or additional contracts, setting a standard before an operator has fully mastered a job, projecting labor loads to plan schedules, determining efficiency in case of order cutbacks, and planning future requirements, production rates, and budgets.

9.12.8 Statistical Evaluation of Cost Estimates

Figure 9.14 shows that in almost all cases a discrepancy exists between a cost estimate and actual performance. The rougher the estimate, the larger the error may be. To find what the probable degree of error may be for an estimate, it is necessary to ascertain the standard deviations of the distributions of errors for the

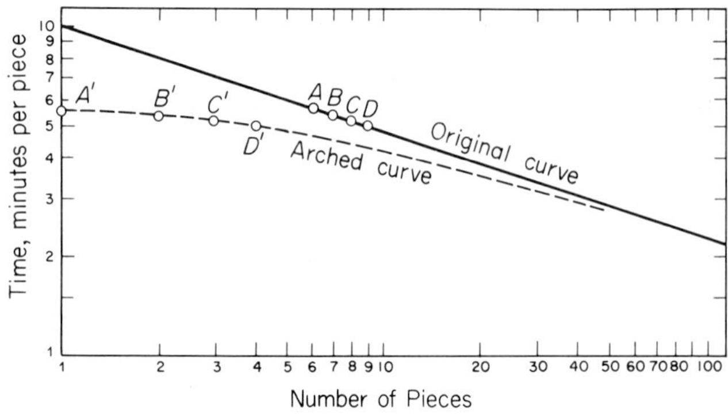

Fig. 9.18 *An illustration of the effect of the* B *factor on the learning curve.*

TABLE 9.10. *A SUMMARY OF COST ESTIMATES TO ILLUSTRATE HOW THE STANDARD DEVIATION FOR THE TOTAL COST MAY BE ASCERTAINED FROM THE STANDARD DEVIATIONS OF THE ELEMENTS**

Using Process *A*

Item	Mean Value	Standard Deviation
Annual production	35,000 units	$(40{,}000 - 30{,}000)/6 = 1667$
Selling price	\$75	$(78 - 72)/6 = 1$
Total annual income	\$2,625,000	$\sqrt{75^2(1667)^2 + (35{,}000)^2 1^2} = 1.295(10)^5 = 129{,}500$
Direct cost per unit	\$28	$(31 - 25)/6 = 1$
Annual direct costs	\$980,000	$\sqrt{(28)^2(1667)^2 + (35{,}000)^2(1)^2} = 58{,}400$
Annual overhead	\$880,000	$(900{,}000 - 860{,}000)/6 = 6667$
Selling cost per unit	\$4.60	$(5 - 4.20)/6 = 0.133$
Annual selling cost	\$161,000	$\sqrt{(4.6)^2(1667)^2 + (35{,}000)^2(0.133)^2} = 1470$
Total annual cost	\$2,021,000	$\sqrt{(5.84)^2(10)^8 + (6.667)^2 10^6 + (1.47)^2(10)^6} = 58{,}800$
Profit	\$604,000	$\sqrt{[(12.95)^2 + (5.88)^2](10)^8 - 2(75)(1667)^2[(28)^2 + (4.6)^2]^{1/2}}$ $= 91{,}700$†
Fixed investment	\$4,729,000	40,900 from estimates
Working capital	\$400,000	5000
Return on total investment	11.8%	$\sqrt{\dfrac{[(9.17)^2 + (4.09)^2 - (0.5)^2]10^6}{(5.129)^2(10)^{12}}} = 1.96$ per cent

† Adjustment for the correlation between annual sales and production. When sales are up, production must be high, and vice versa.

Using Process *B*

Item	Mean Value	Standard Deviation
Total annual cost, dollars	1,734,000	46,400
Profit, dollars	891,000	101,600
Total investment, dollars	7,200,000	48,400
Return on total investment, per cent	12.4	1.56

* From L.E. Doyle, "Cost Estimating, How to Minimize the Dangers of Chance," *The Tool Engineer* (June 15, 1959), p. 91.

elements that make up the estimate. Section 5 under Motion and Time Study explains how this is done for the elements of an operation. For complete operations, material prices, overhead, etc., statistical quality-control techniques described in Sec. 13 may be applied to variations in costs instead of dimensions.

Table 9.10 illustrates how the standard deviation of the error of a total estimate is derived from formulas of basic statistical theory (explained in Sec. 13) for adding, dividing, and multiplying the standard deviations of the elements together. The results are more meaningful than a single figure for each estimate. For instance, a profit of \$604,000 and return of 11.8 per cent are estimated for Process *A*, but only as a mean. Suppose a minimum return of 10 per cent is required. By normal probability, there is an 18 per cent chance that the return on the investment will be less than 10 per cent.

The probability is about 52 per cent that the profit for Process *A* will be \$600,000 or more per year and about 87 per cent that it will be over \$500,000 per year. There is one chance in ten that the profit will be lower than \$485,000 and the rate of return less than 9.2 per cent. If a more accurate estimate is desired, the estimator can readily see what items offer the most possibility for improvement.

Table 9.10 also gives an abbreviated summary for making the same product by Process *B* to show how statistical techniques are helpful in making comparisons. If most of the elements of both estimates vary in cost because of chance and if their distributions are near normal, Process *A* will give a larger return than *B* about 50 per cent of the time. The probability is about 20 per cent that the difference in return shown for the two processes could have come from chance alone, and thus there is no significant difference between the returns from the two different processes.

9.13 TOOL DESIGN

In the broadest sense all the machines and equipment of production are tools, but tool engineering in manufacturing is concerned primarily with the specialized adjuncts of machine tools—namely, cutting tools and their adapters, jigs, fixtures, dies, and gages. The principles for designing such tools are the subjects of this subsection.

9.13.1 Classes of Tools

Tools differ in quality and consequently in cost. A durable, accurate, and rapid-acting but expensive tool is economical for producing a large number of parts. A relatively cheap tool is adequate for few parts. In this respect, tools may be classified as follows:

Class I or *A* tools are of the highest quality to produce large quantities of precision products, generally in excess of 10,000 to 100,000 pieces, depending on the type of tool and plant practices. They must have long lives, be accurate throughout their lives, be easy to maintain, and be able to produce at a high rate. They are the most expensive to build. All details subject to wear are made of hard, wear-resistant materials and are readily replaceable, as are also any necessarily fragile details. The main members of such tools are heavy and rugged. Provision is made for easy chip removal and for application of adequate cooling and lubricating fluid. Mechanisms must be quick-acting, often automatic, and foolproof. Dies are designed in progressive, combination, or compound forms to minimize work handling. Molds are made with multiple cavities, knockouts, and chrome plating for long life and are adaptable to semiautomatic and transfer presses. Cutting tools are usually cemented carbide. Casting patterns with aluminum matchplates and multiple prints and metal core boxes and dryers are used.

Class II or *B* tools are designed to produce interchangeable products but in medium quantities, a few thousand or so. They are simple, made of cheaper materials, have lighter members, cost less than Class I tools, and made to last only long enough to produce definite quantities of parts. Mechanisms are simple, slow acting, and require operator attention. For this class, several simple dies are built instead of one complex die, but they are steel dies of conventional construction. Single- and double-cavity molds are used on hand presses. Many cutting tools in this class, especially form tools, are of high-speed steel.

Class III or *C* tools are the cheapest that can be used for low production, a few hundred to a few thousand pieces, without excessive operating cost. They are simple, rarely combine operations, and require skilled operators. Wear plates, blocks, and so forth may be provided but are seldom hardened. Standard measuring instruments are preferred to gages. Conventional machine tools with standard attachments and accessories are used. Continental dies are characteristic of this class. For castings, aluminum matchplates but wooden patterns and core boxes are common.

Class IV, *D*, or *temporary* tools are made to produce only a very few parts, often on an experimental basis. They are the lowest cost tools that will do a job at all. Standard holding devices, cutting tools, and basic machine tools are used almost entirely. Where a chuck or vise is not suitable, the work is clamped to the machine table. Dies frequently are not made for this class of work, but, where used, they are often made from soft metal or nonmetallic materials, and pieces may have to be finished by extra operations. Wood patterns are the rule.

If the cost of a Class I tool is represented by 4, then the approximate cost of a Class II tool is 3, a Class III tool is 2, and a Class IV tool is 1.

9.13.2 Tool-Design Procedures

A process plan informs the tool designer what a tool must do, under what conditions it must operate, and the type of tool required. The designer must select the necessary details and arrange and proportion the device to perform the required task efficiently.

By the *project method* of tool design, found in small plants, one designer is assigned to designing all the tools for one part. By the *group method*, each designer specializes in a particular type of tool, such as jigs, fixtures, or dies. In large plants, a method having the advantages of both these methods has each project directed by a senior designer or group leader with assistants who specialize.

Some tool designers first draw two or more views of the workpiece for fixtures, jigs, and dies, then suitable locating details at the prescribed locating points, clamping details to apply the necessary forces, and finally the body or frame to contain and tie together the locating and clamping details with the cutter setting or guiding details, and the adjuncts which position or fasten the tool in place for operation.

9.13.3 Tool Drawings

Practice in one plant, typical of others, recognizes three classes of tool-design drawings. *Class A* is the ultimate in drafting. The workpiece is drawn in phantom, commonly in red, in its proper position with respect to the tool during the operation, on the assembly or layout sheet. As many views are shown as are necessary for complete clarity. Small auxiliary sections are added where helpful. A stock list on the layout sheet specifies the detail number, quantity required, description or stock size, and material or specifications for each detail. The layout carries assembly dimensions and general notes for the guidance of the toolmaker. All parts with the exception of standard screws, dowels, etc., are completely detailed, usually on sheets other than the layout. Class *A* drawings are warranted for complex tools where a large amount of information must be presented.

Class B drawings permit detailing in assembly, where clarity is not impaired, and separate detailing on the assembly sheet. Small details, such as screws of one size, may only be shown once even though they occur in several places. Center lines indicate their positions. Class *B* drawings are adequate for most tools. They must not be confused, however, with the class of tool; a Class *B* drawing may well be suitable for a Class *A* tool.

Class C drawings may be partial or incomplete drawings or freehand sketches sufficient to convey a minimum of information. They are useful for alterations or repair work, where there is close contact between the designer and the toolmaker.

9.13.4 Tool-Drawing Dimensions

Tool dimensions that are related to the workpiece, machine, or other tools should have tolerances. For instance, the diameter of an arbor or locating plug on which a workpiece must fit must be given a definite tolerance. However, interchangeability of tool details is seldom advantageous, and the dimensions of mating details within a tool may not have tolerances specified, depending upon the system for making the tools. By one system, practically all the work on a single tool is done by a skilled toolmaker or small group, who can be depended upon to get the best results from the means available. For instance, a tolerance is not assigned to the diameter of a plug press fitted into a hole under such conditions, but the nominal diameter and a note calling for a press fit are specified. The toolmaker reams the hole and grinds the plug to fit. A second system of toolmaking employs specialists for the machining operations assigned to them by a toolmaker who assembles the details. The specialists are not aware of the specific purposes of the details they make, and detail drawings for their use should have tolerances specified on all dimensions.

9.14 CUTTING TOOLS

9.14.1 Cutting-Tool Materials

Carbon tool steel contains from 0.90 to 1.20 per cent carbon and nominally no alloying elements, but often moderate and sometimes large amounts of these elements are present. It is hard and wear resistant at low temperatures and takes a keen edge, but it softens at temperatures above 400°F and is limited to slow speeds. Carbon tool steel is not often used for production cutting, but it is cheap and easy to shape and harden. It is advantageous in operations that must be done slowly, for easy-cutting brass and other soft metals, and for seldom-used tools, like odd-size drills, that warrant only a minimum investment.

High-speed steel is highly alloyed and is made in many analyses and brands. Two main standard classes are tungsten and molybdenum high-speed steels. The commonest in the first class is designated T-1, is known as 18-4-1, and contains about 0.55 to 0.75 per cent carbon, 18 per cent tungsten, 4 per cent chromium, and 1 per cent vanadium. The most widely used "moly" high-speed steel is M-2, containing 0.85 per cent carbon, 5 per cent molybdenum, 4 per cent chromium, 2 per cent vanadium, and 6 per cent tungsten. These and types M-1 and M-10 account for over 90 per cent of the usage of high-speed steel. So-called "super high-speed steels" for severe operations may contain cobalt to improve red hardness or larger than usual amounts of vanadium for resistance to abrasion. High-speed steel loses its hardness at red heat—i.e., above 1100°F—and is capable of only moderate speeds. It is tough and can be fabricated and ground into many forms fairly easily but requires care in heat treatment. Its cost is moderate.

Cast, nonferrous alloys contain no iron, cannot be softened and cut or worked, and must be cast to shape and ground. Various alloys give a number of combinations of hardness and toughness for different purposes. A typical alloy with the trade name Stellite may contain from 43 to 48 per cent cobalt, 17 to 19 per cent tungsten, 30 to 35 per cent chromium, and about 2 per cent carbon. Such alloys retain hardness and become tougher at red heat, although they are much less shock resistant than high-speed steel. They have a hard skin that withstands the abrasive action of cast iron, malleable iron, and bronze.

Cemented carbides, sintered carbides, or plain *carbides* consist of very hard particles joined together in a metallic matrix. The primary ingredient in most commercial tools is tungsten carbide (particularly for cutting cast iron and nonferrous alloys), often mixed with various amounts of tantalum, titanium, and columbium carbides to lower the coefficient of friction and reduce cratering (of particular benefit in cutting steel). The binder for the particles is usually cobalt. Titanium carbide with a binder of nickel or molybdenum withstands higher temperatures than other carbides and therefore can be operated at high speeds, but it is quite brittle.

Carbide ingredients are mixed together in powdered form, compressed in a mold, presintered, and cut to desired shapes. The blanks are sintered at high temperatures and finished by grinding.

Cemented carbides stay hard up to relatively high temperatures and resist the large pressures of cutting because of a large modulus of elasticity (two to three times that of steel). They have almost no plastic flow at stresses up to 500,000 psi and have low thermal expansion and high thermal conductivity. Carbides are brittle and must be used carefully. They are more expensive than high-speed-steel tools but can be operated at higher speeds for good surface finishes; they remove metal faster and in most cases offer the most economical means of machining parts in quantities.

Different grades of cemented carbides are made by varying the kinds, proportions, and particle sizes of the ingredients to obtain properties which meet various requirements. At one extreme are the hardest but most brittle carbides with high resistance to wear and suitable for high-speed finishing cuts and for abrasive and hard materials. Other grades sacrifice hardness to various degrees for strength and shock resistance, for crater or edge-wear resistance, and for heavy cuts and tough materials. Each manufacturer identifies his grades by a code that usually corresponds to a general industry code. Carbides for cast iron, nonferrous, and nonmetallic materials are designated C-1 for roughing, C-2 for general-purpose work, C-3 for finishing, and C-4 for precision finishing; carbides for steel and its alloys are designated C-5 for roughing, C-50 for roughing alloy steel in particular, C-6 for general-purpose work, C-7 for finishing, C-70 for finishing alloy steel in particular, and C-8 for precision finishing.

Sintered oxides, ceramics, or *oxide tools* are basically compressed and sintered aluminum oxide powder, usually in the form of mechanically held inserts but sometimes of inserts fastened by adhesives to steel shanks. The material is quite hard, has a compressive strength of up to 400,000 psi, and has a transverse rupture strength of about 100,000 psi, enabling it to withstand normal chip loads with adequate support. Sintered oxides have poor thermal conductivity and do not absorb the heat of cutting readily, retain their hardness and strength up to 2000°F, resist abrasive wear, and perform well on most materials at speeds of 400 to 1400 fpm (in some cases up to 3000 fpm). However, tensile strength is relatively low, and the tools are brittle, easily chipped, and susceptible to breakage from shocks or loads varying in size or direction. As a result, use of oxide tools has been limited mostly to light finishing cuts, quite often on nonferrous and nonmetallic materials. They have not been accepted in many cases because of the extra care they require; the brittle tools must be mounted on rigid machine tools resistant to vibration and are not suitable for interrupted cuts. Some success has been reported in milling soft cast iron but not steel. Machine tools must be extra powerful to utilize sintered oxides at capacity with high speeds and heavy cuts.

Diamond is the hardest known material, can be run at cutting speeds up to 5000 fpm, cuts hard material, and produces fine finishes; but it is expensive, brittle, a poor heat conductor, and limited to light cuts. Typical applications are precision boring of holes and truing grinding wheels.

9.14.2 Elements and Angles of Cutting Tools

The elements, angles, and standard sizes of single-point tools are defined in the USA Standard B5.22–1950, *Single Point Tools and Tool Posts*. The chief elements and angles are depicted in Fig. 9.19, together with the conventional ways of specifying the size and angles of a tool.

Rake makes a tool cut easier but takes metal from behind and weakens the cutting edge. A practical rake angle, like those indicated in Table 9.12, is a compromise between easier cutting and longer tool life. In general, a small rake angle is desirable for cutting hard materials. If a tool is set other than "on center," its angles must be modified to give the proper effect in the particular position.

The *side rake angle* helps to direct the chip away from the tool holder, reduces side deflection of the tool, reduces feeding force, and weakens the tool less than the back rake angle.

Negative rake angles that slope upward from the cutting edge have been found to give good results with carbide tools under certain conditions, such as when the tools are subjected to shocks from interrupted cuts. A tool with a negative rake angle receives impact behind the cutting edge and has extra material backing up its edge. A negative rake angle increases cutting forces at low speeds, but forces drop off at the high speeds possible with carbides; negative rake angles are usually from 2 deg to 10 deg, with the other rake angle on the tool positive and 2 deg to 5 deg greater in magnitude.

A *relief angle* keeps the tool from rubbing on the work but takes some support from below the cutting edge. For turning, side relief must be greater than the helix angle of the cut and usually is larger than end relief. Relief angles be-

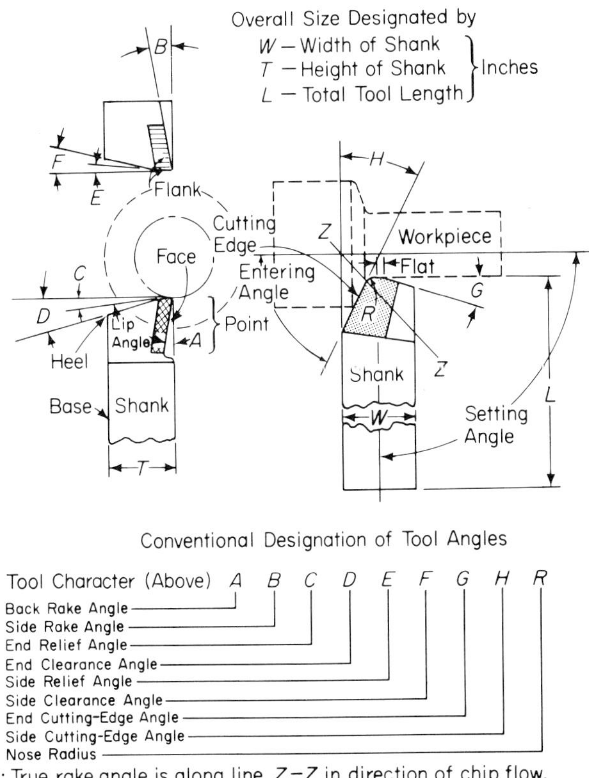

Fig. 9.19 *The elements and angles of a single-point tipped tool.*

TABLE 9.11. *A COMPARISON OF CUTTING-TOOL MATERIALS*

Material	Relative Tool Cost*	Typical Cut Speed	Cutting Cost, † Dollars per Cubic Inch of Metal Removed
Carbon tool steel	0.10	40	0.25
High-speed steel	0.50	90	0.13
Nonferrous alloy	2.00	150	0.06
Cemented carbide	5.25	500	0.04
Sintered oxide	12.00	800	0.02

* Prices are approximate for a $\frac{3}{8}$ in^2. tool bit or equivalent but vary with quantity, grade, and market conditions.

† Estimates based on performance under ideal conditions, cutting cold-drawn SAE 4140 steel, with labor and overhead rate of $6/hr.

TABLE 9.12. *RAKE ANGLES FOR SINGLE-POINT TOOLS, DEGREES*

Work material	Bhn	B–back Rake[4] A–Side	High-Speed Steel	Cast Non-ferrous Alloys	Cemented carbides Brazed[5]–Inserts–Mechanical		Sintered Oxides
Aluminum alloys	40–110[1]	B	15 to 35[3]	20	0 to 30[3]	0	
		A	15 to 35[3]	15	15 to 30[3]	5	
Copper alloys	120–185	B	0 to 15[3]		0	0	
		A	−5[2] to 10		0 to 15[3]	5	
High-temperature alloys	260–320	B	5		0	0	
		A	6		6	5	
Magnesium alloys	30–80[1]	B	20		0 to 30[3]	0	
		A	15		15 to 30[3]	5	
Titanium alloys	280–360	B	0		0	0	
		A	5		6	5	
Gray or flake-graphite		B	5	0	0	−5	
cast iron	140	A	10	5	5 to 15	−5	
Nodular or ductile		B	3	0	0	−5	
cast iron	180	A	8	5	5 to 15	−5	
Malleable cast iron	220	B	0	0	−5 to 0	−5	
		A	5	5	−5 to 5	−5	
Free-machining and		B	10		0	−5	−5 to −20[6]
plain-carbon steel	180	A	12		5 to 15	−5	−5 to −20
Free-machining alloy		B	8		0	−5	−10 to −30
steel	250	A	10		6	−5	−10 to −30
Alloy steel and cast		B	0	8	0	−5	−10 to −30
steel	350	A	8	8	6	−5	−10 to −30
Hot-work die steels		B	0		−5 to 0	−5	
and tool steels	500	A	5		−5 to 0	−5	
Stainless steel	180–220	B	5	10	0	0	
		A	8	15	6	5	
Precipitation-hardening		B	0	0	0	0	
stainless steel	220–320	A	5	5	6	5	

[1] 500-kg load

[2] Negative angles are sometimes necessary to prevent gouging in brass.

[3] Larger angles for softer materials

[4] Designations are for side-cutting tools; reverse B and A for end-cutting tools.

[5] Reduce rake angle 5 deg to 10 deg for interrupted or scaly cut.

[6] A maximum of −5 deg to −10 deg is obtainable with mechanical insert holders for reasonable relief angles.

tween 5 deg and 12 deg are usually ground on lathe tools. The larger angles are suitable for soft materials like aluminum. Brittle cutting edges of carbide and oxide tools need support and usually have small relief angles. Shaper or planer tools are subject to shock and may have relief angles as small as 4 deg. The tip on a tool usually overhangs the shank, with a secondary relief or clearance angle from 2 deg to 5 deg larger than the relief angle to make the tool easy to grind.

An *end cutting-edge angle* reduces drag on the front of the tool that tends to cause chatter. In most cases, 8 deg to 15 deg are sufficient behind a flat 1/16 to 5/16 in. long adjoining the nose radius for wiping action. Cutoff and other end cutting tools have no end cutting-edge angles.

A *side cutting-edge* or *lead angle* increases tool life because it permits initial contact behind the tip when a tool enters a cut, gradual emergence from a cut, and an increased length of cutting edge in action. This angle is usually 5 deg to 20 deg, but sometimes is as high as 30 deg to 40 deg. Too large an angle promotes chatter. No side cutting-edge angle is desirable for cutting a forging or casting with a hard skin, under which the tool should get to crumble the hard material, or for cutting up to a square shoulder. The actual effect of the side cutting-edge angle depends upon the setting angle.

A *nose radius* increases tool life, but too much radius promotes chatter. An empirical formula for nose radius is $R = 0.05c + 1.75f$, where c is the depth of cut in inches and f the feed in inches per revolution. Figure 9.10 indicates the nose radius for a required surface finish and feed.

One tool manufacturer calculates the thickness of an *inserted tip* for continuous cuts as $T = 2.4f^{0.5}c^{0.35}$, where f is the feed in inches per revolution and c the depth of cut in inches. A tip is one-third thicker for an interrupted cut.

The *shank* dimensions of a single-point tool in inches are width b and height h, proportioned in the relationship $bh^2 = 210fcLP_u$, where f is the feed in inches per revolution, c the depth of cut in inches, L the overhang of the tool in inches, and P_u the specific power to cut the work material in horsepower per cubic inch per minute from Table 9.1.

Chip breakers break up troublesome, long, and stringy chips, particularly with ductile materials like steel. Figure 9.20 shows examples of common types. The radius r should be about twice the amount of feed for feeds less than 0.015 ipr and from 1/32 to 1/16 in. for larger feeds. The step h should be about the same as the feed in inches per revolution at which the tool is to be operated. The width w may vary from 1/16 in. for feed of 0.010 ipr and depth of cut of 1/32 in., to 1/4 in. for feeds up to 1/16 ipr and depths of cut over 3/4 in. A groove chip breaker has a land between the groove and cutting edge equal to one to one and a half

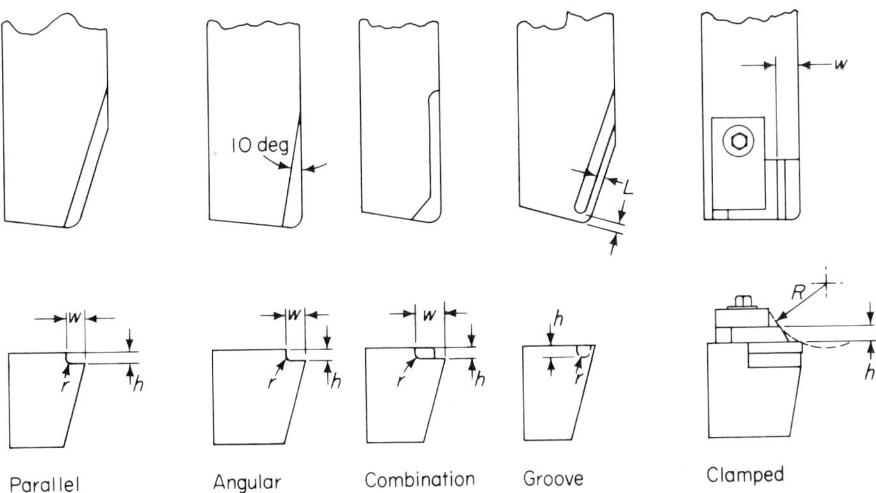

Fig. 9.20 *Types of chip breakers.*

times the feed. Commercial, mechanically held insert tools commonly have clamped or mechanical chip breakers.

Henriksen* showed that the radius to which a chip is bent is $R=(w^2/2h)+(h/2)$ as dimensioned in Fig. 9.20. The chips will be broken efficiently when R is within a certain range depending mainly on the work material and feed. Chip breakers should be designed for the following ranges. For free-cutting steel, R should have a maximum value for feeds up to 0.018 ipr of $2300f^{1.76}$, an average value for feeds from 0.010 to 0.030 ipr of $230f^{1.55}$, and a minimum for feeds of 0.010 to 0.040 ipr of $135f^{1.64}$. For plain carbon steel at feeds from 0.005 to 0.016 ipr, maximum $R = 490f^{1.54}$; at feeds from 0.010 to 0.022 ipr, average $R = 2224f^{2.03}$; and at feeds from 0.010 to 0.030 ipr, minimum $R = 3.625f^{2.34}$. For alloy steel (annealed 4140) at feeds from 0.005 to 0.020 ipr, maximum $R = 840f^{1.74}$; at feeds from 0.010 to 0.030 ipr, average $R = 1788f^{2.07}$; and at feeds from 0.010 to 0.035 ipr, minimum $R = 67f^{1.39}$. Cutting speed, depth of cut, and side cutting-edge angle have minor influences upon the bend-radius effect in breaking chips.

9.14.3 Types of Single-Point Cutting Tools

A *solid tool* has the full section of its cutting end made of the cutting material. The shank may be of the same material as the end or may be of soft steel welded to the end. A *tipped tool* has a small piece or tip of cutting material attached to the end on top of the soft shank by brazing, welding, or clamping. A *tool bit* is a relatively small cutting tool clamped in a *tool holder*, and the assembly is called a *bit tool*. A *radial tool* acts with the axis of its shank substantially in a radial position with respect to the work, as in Fig. 9.19. A *tangential tool* is held with its shank substantially tangent to the work surface. A *roughing tool* is designed to remove large amounts of stock with a maximum tool life and for that purpose may have negative rake, a large side cutting-edge angle, or a rounded nose, as required. A *finishing tool* has a keen edge, usually with more rake than a roughing tool. The keen edge and greater rake produce a good finish, and often the tool has a specific shape to suit a given job.

Common shapes of cutting tools are depicted in Fig. 9.21. A *recessing tool* is like a boring tool but is designed to be fed in radially, to undercut a form within a hole. A *knurling tool* has hardened steel serrated rollers on the end of a shank and is pressed against a revolving workpiece to raise a nonslip pattern on the surface. A *radius tool* has a cutting edge in the form of an arc of a circle, concave or convex, of uniform radius. A *form tool* has a cutting edge with a profile or contour to be reproduced on a workpiece. Some form tools are of the radial type, but most of those for screw machines are tangential. A *skiving tool* is fed tangentially over or under a revolving workpiece for finish forming.

9.14.4 Multiple-Point Tools

Most drills, countersinks, counterbores, spotfacers, milling cutters, reamers, and taps for manufacturing are standard commercial items. The tool engineer's main consideration is selecting them from manufacturers' catalogs in which they are described. Even where special tools of these kinds must be designed, the problem is usually one of modifying a standard tool to meet a specific situation. A multiple-point tool is basically an arrangement of single-point tools, each of which has angles and other features corresponding to those described for single-point tools.

Specifications for standard, multiple-point tools are given by the USA Standards Institute, Bulletins ASA B5.3–1960 *Milling Cutters;* USAS B94.9–1967 *Taps;* B94.11–1967 *Twist Drills;* B94.2–1962 *Reamers;* and B94.8–1967, *Inserted Blade Milling Cutter Bodies.*

9.15 FIXTURES AND JIGS

A *fixture* holds and locates a workpiece during an inspection or manufacturing operation. A *jig* also guides, controls, or limits one or more cutting tools.

9.15.1 Design of Fixtures and Jigs

An efficient tool design results from an analysis of the project. A primary investigation gives a complete understanding of the process

---

* Eric K. Henriksen, "Balanced Design Will Fit the Chip Breaker to the Job," Special Report No. 360, *American Machinist*, **98**, No. 9 (April 26, 1954), p. 117. Also, Eric K. Henriksen, "Clamped Chip Breakers, Their Advantages and Design," *Tool Engineer*, **XXXIV**, No. 3 (March 1955), p. 103.

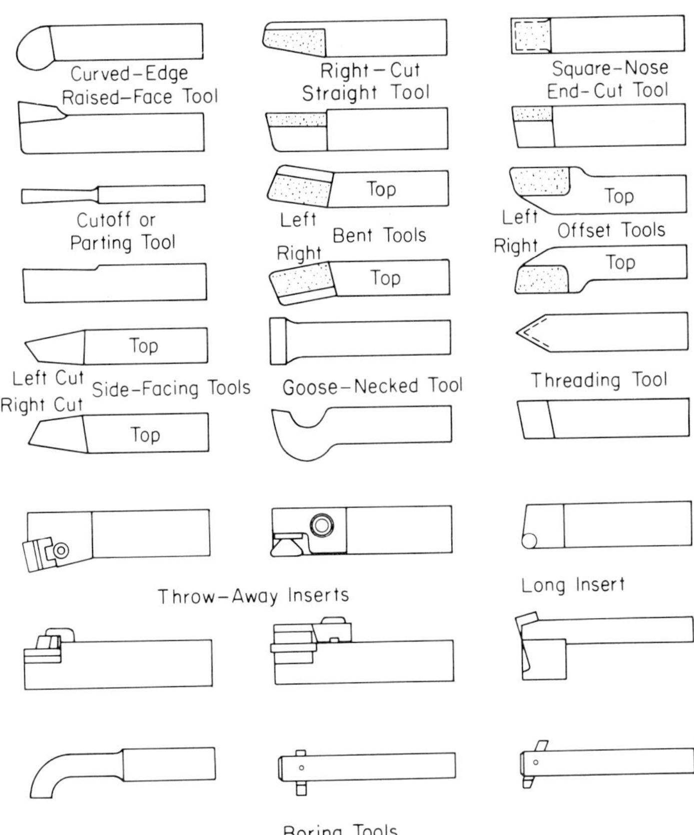

Fig. 9.21 *Single-point cutting tool shapes.*

and the interrelationships of operations. This primary investigation includes (1) making a survey of all operations performed on the part, (2) ascertaining the purpose of each operation, the type of material required, and its properties, (3) recognizing the tolerances and other factors of accuracy required, and (4) discerning the necessary machine tools, equipment, tools, workplace layout, and working conditions.

A detailed study is made of the operation for which each specific tool is required to determine the most economical disposition of the operation. Among the factors investigated are the total quantity of pieces to be produced, the rate of production and sizes of lots, the elements of the operation, the specific cuts to be made and their dimensions, the capabilities and dimensions of the machine tool, the desired work-cycle time, and a motion study of the operation routine. A major aim is to achieve the most economical arrangement for operation of a fixture or jig. The principles of efficient motion planning described in Sec. 5 and of low-cost processing discussed in Art. 9.10 should be understood and applied to the design of each tool.

A common low-cost fixture consists of a set of special jaws for a standard vise. Figure 9.22 indicates several examples. Manufacturers of tooling components offer a variety of commercial universal fixture and jig bodies. Special locating and supporting details may be added to one of these to make a low-cost fixture or jig for a particular job. Standard fixture and jig components are economical because they cost less than special details and are sturdy, accurate, easily converted from job to job, and well designed. Figure 9.23 gives examples.

To fulfill its purpose, a fixture or jig should have details (1) to locate, (2) to clamp and support the workpiece, (3) to tie the other details together into a workable whole with a

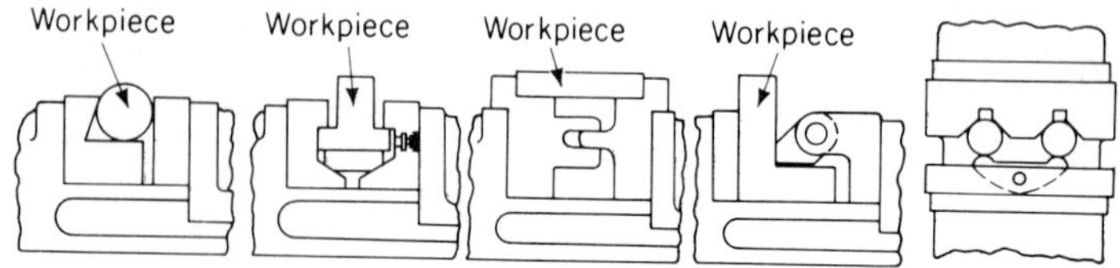

Fig. 9.22 *Special vise jaws.*

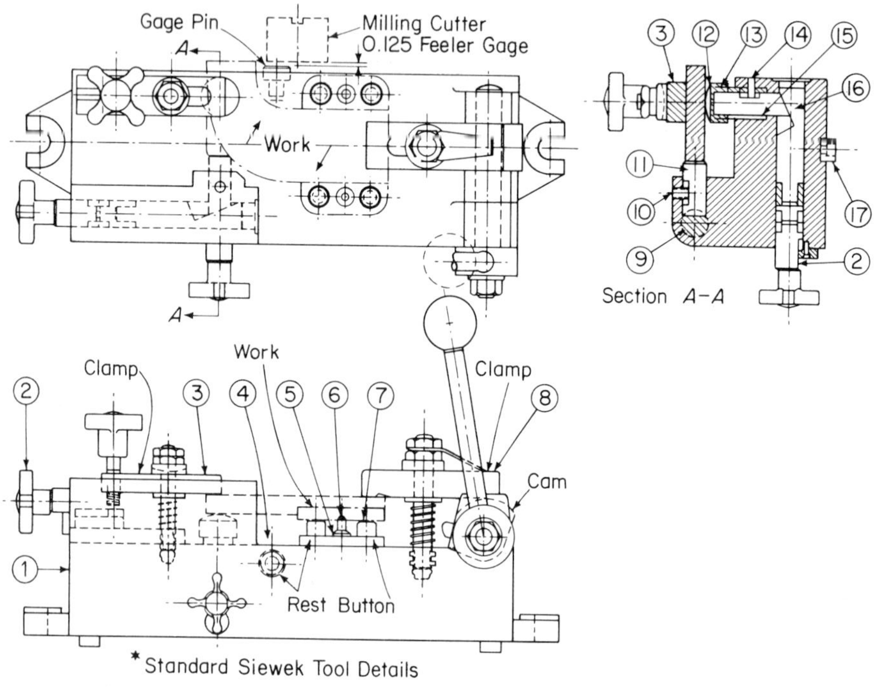

Det. No.	No. Req	Description	Stock.
1	1	Pattern K-216-1	C.I.
*2	2	Siewek Jack Lock	10,101
*3	1	Siewek Clamp	10,021
*4	1	Siewek Rest Button	10,226
5	2	Retainer	Standard
6	2	Locator	Standard
*7	4	Siewek Rest Button	10,225
*8	1	Siewek Cam Clamp	10,081
9	1	7/8" Dia. x 4"	M.S.
10	1	3/8" - 16 x 1/2"	Screw
11	1	3/4" Dia. x 2"	M.S.
12	1	1¼" Dia. x 3"	M.S.
13	1	1¼" Dia. x ½"	M.S.
14	1	3/8 - 16 x 3/8"	Screw
15	1	1" Dia. x 1 3/8"	M.S.
16	1	7/8" Dia. x 3 3/8"	M.S.
*17	2	Siewek Fixture Key	10,703
SIEWEK TOOL COMPANY			
		MILLING FIXTURE	T-1945

Fig. 9.23 *A layout drawing of a milling fixture with standard details. Courtesy Siewek Tool Company, Inc.*

body, base, or frame, (4) to guide cutting tools (jig) or to set cutters (fixture), and (5) to position or fasten the tool to the machine, bench, or equipment on which it is used. Some components or details may serve more than one purpose.

9.15.2 Locating Devices

Figure 9.24 illustrates common locators. *Sight location* by means of scribed lines, points, or sight holes is slow but is applicable for rough workpieces varying considerably in size and produced in small quantities.

Round pins or *plugs* for holes may have a bullet nose or large chamfer to facilitate insertion. If the fit is tight, initial jamming may occur but may be avoided by designs like those in Fig. 9.25*. A plug of diameter d extending from a shoulder, as in Fig. 9.25(a), will not stick in a hole of diameter D if the plug length

$$l = \sqrt{2WC},$$

where W is the minimum outside dimension of the workpiece and $C = (D - d)$. A projection of diameter d on a workpiece will not stick in

* L. E. Doyle, B. R. Better, and B. T. Chao, "A Design for Non-Sticking Plug and Ring Gages and Locators," *University of Illinois Engineering Experiment Station Bulletin* 433, **53**, No. 38 (Jan. 1956).

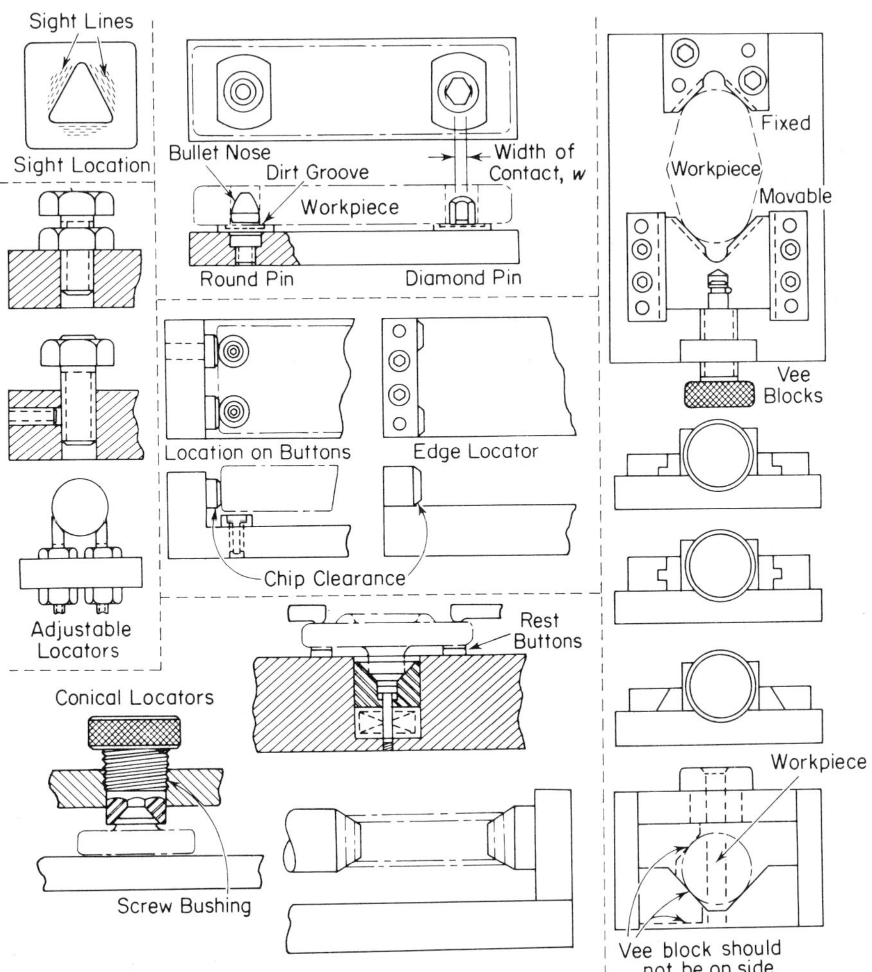

Fig. 9.24 *Locating devices.*

a locating hole of diameter D, as in Fig. 9.25(b), if the hole is countersunk or relieved to leave a length of engagement between hole and workpiece of only $E = l - m \leq \sqrt{WC}$. A conventional way of avoiding sticking is to relieve a plug by removing three equally spaced segments as in Fig. 9.25(c). A common value for angle a is 15 deg, which makes $m = 0.35d$. Such a plug will not stick if its length $l = [2.4(2a + 0.85d)(D - d)]^{1/2}$; but this method allows extra displacement of the workpiece in three directions in the amount of $0.207(D - d)$, which represents a loss of about 20 per cent in locating precision. An aligning groove on the end of a long plug of diameter d, as in Fig. 9.25(d), obviates sticking. Workable dimensions are $l_1 = (2AC_2)^{1/2}$, $l = \mu d$, and groove diameter $B = 0.95d$, where C_2 is the clearance between the hole and plug pilot, or $D - A$. The coefficient of friction μ usually is between 0.15 and 0.25 for steel. The pilot diameter A may be any convenient size between d and $[(2d^2/D) - d]$. A simple form of nonsticking hole has a pilot diameter $D_1 = d \sec T$ and groove as in Fig. 9.25(e), where T is the angle of friction. The distance

$$l = \mu d(D_1 - d)/2(D - d).$$

In many cases, since l must be excessively long, the design is not satisfactory.

A *diamond pin* used with a full round pin to locate in two holes, as in Fig. 9.24, compensates for the tolerances—S_1 on the dimension between hole centers and S_2 between pin centers, with $S = S_1 + S_2$. The width of contact W illustrated also in Fig. 9.25(f), must be as small as possible so that d may be as large as possible for a close fit in the hole D. However, W must not be so small that the pin will wear or fail rapidly; a lower limit for W is commonly $D/8$ but no smaller than $\frac{1}{64}$ to $\frac{1}{32}$ in. The pin diameter may then be

$$d = D - [S(W + S/2)/D]$$

and the pin thickness

$$T = 0.866w + (\sqrt{D^2 - W^2})/2.$$

After the diameter of a diamond pin has been determined, the maximum locating error that it provides is found. This may be expressed as a function of the maximum angle of displacement that may occur between the center line through both holes and the center line through both pins, in the form of

$$\tan a = [(D_a - d_a) + (D_b - d_b)]/X,$$

where D_a and D_b are the maximum sizes of the two holes, d_a and d_b the minimum sizes of round pin and diamond pin respectively, and X the distance between the pin axes. If the possible error in location is too great, the pin diameters may be increased as much as the minimum-hole diameters allow, but it must be recognized that any increase in diamond-pin diameter necessitates a decrease in contact width W and sacrifice of wearing surface.

Hardened *rest buttons* give good wear and have relatively small exposed surfaces. Figures 9.23 and 9.24 give examples. Thrust is better taken endwise than sideways by a button or pin. Large locating blocks should be relieved, and ample space provided for burr or chip clearance. Locators should be made easy to clean by being as small as is consistent with durability; they should be accessible; they should be higher than surrounding areas so that chips can fall off rather than onto them; and they should have sharp edges where suitable to scrape loose particles from the work.

Vee blocks are widely used for locating round pieces. A 90-deg vee angle is best for most applications, but care must be taken that the axis is placed so that variations in workpiece location cause the least errors. Opposed vee blocks in pairs, with one fixed and the other movable but guided, have the effect of locating three points on a workpiece.

Internal and *external conical locators* at the ends of diameters allow for wide variations in workpiece length or diameters. The centers on a lathe are examples.

Adjustable locators often are needed for castings, forgings, or welded assemblies that vary appreciably in size, especially from lot to lot. The head of a capscrew or bolt, hardened if subject to wear, is one form of adjustable locator; the shank is screwed into a threaded hole and tightened with a jamb nut.

A *centralizer* locates the center line of two or more surfaces or points as exemplified in Fig. 9.26. Conical locators and adjustable vee blocks act as centralizers. A centralizer may perform a clamping as well as a locating function.

9.15.3 Clamping Devices

The conditions that clamping devices must satisfy are described in Art. 9.8.

Two examples of *strap clamps* are seen on the

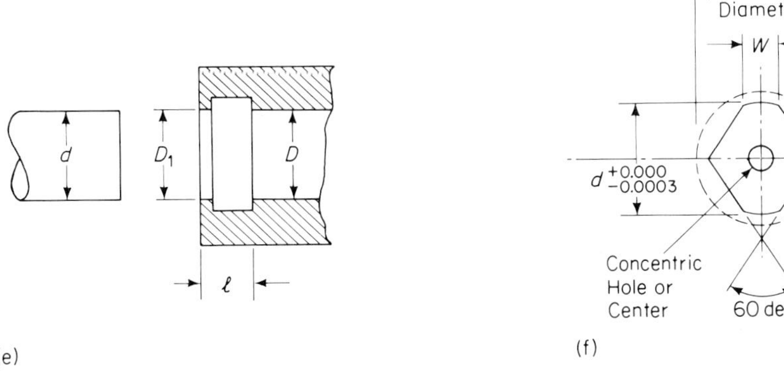

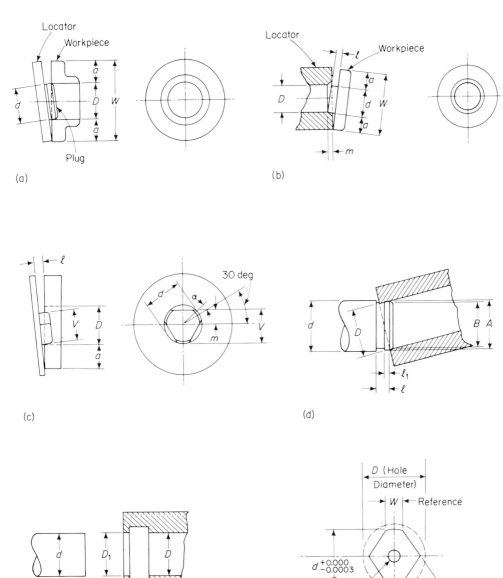

Fig. 9.25 *Dimensions of plug and pin locators.*

fixture of Fig. 9.23. One is operated by a hand knob and screw and the other by a cam and lever. Both are held down by a stud, nuts, and spherical washer between the ends. A strap clamp may be arranged with the force applied to it from above at the stud and may have its back end bearing on a block or on another workpiece. Other typical clamping devices are shown in Fig. 9.27.

Most clamps employ some variation of the

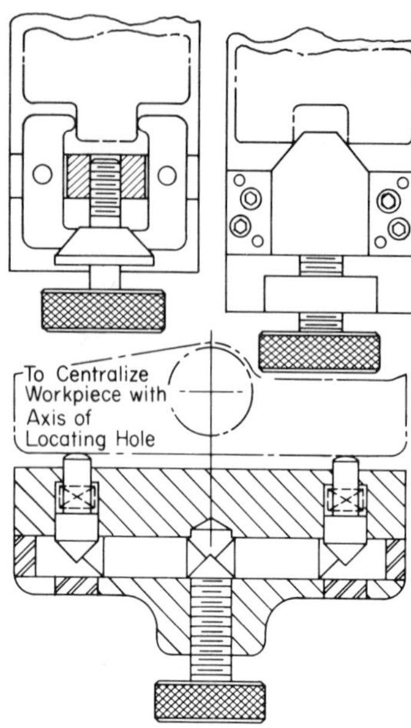

Fig. 9.26 *Centralizers.*

wedge, screw, cam, or toggle, with one or more levers, and are made irreversible. Screw clamps are simple, cheap, and versatile and can be made to act over a large range but are relatively slow. Cam and toggle clamps are faster but more expensive and have limited ranges. Thus, they may not be suitable for clamping rough workpieces that vary considerably in size.

In general, the higher the rate of production, the more worthwhile it is to make a clamping device quick and easy to operate. Air- or hydraulic-actuated clamping devices are much more expensive but take less time and effort for operation than manual ones.

Even with simple devices, much can be done to promote clamping efficiency. Each strap clamp of Fig. 9.23 is supported by a spring around the stud to keep the strap from falling down when withdrawn from the work. Such springs are partly enclosed on some clamps to keep chips from clogging them. The straps are slotted so they can be slid away from the work. A pin projecting from the cam withdraws and returns the right-hand strap when the cam is loosened and tightened. The tightening screw of the left-hand clamp is guided in a groove to control the strap when it is moved back. Stops are helpful to control swinging clamps.

9.15.4 Mechanics of Clamps

The force exerted in tightening a clamp manually depends upon the strength of the operator and the effort he is willing to make. From 50 to 100 in.-lb may be exerted on a hand knob. Small, knurled screw heads are not consistent because even a little wetness or oiliness makes them rather slippery. The forces in Table 9.13 are those that studies have indicated are exerted on various types of levers. A clamp is designed to deliver the required clamping force when the input is low but to be able to withstand the highest force likely to be imposed on it.

The force exerted by an air or hydraulic cylinder is $F_c = 0.7854(D^2 - d^2)P$. The diameter of the piston is D in., and of the rod d in. Most factories have air pressure P of 80 to 100 lb/in.^{2}, but it may drop to as low as 40 to 50 lb/in.^{2} in remote areas. Commercial hydraulic units are available to supply almost any pressure up to 1200 lb/in.^{2} A clamp is less likely to stick in closed position if closed by pressure applied to the rod side of the piston. Danger of a clamp opening as a result of pressure failure is eliminated by transmitting the clamping force through a locking mechanism, such as a cam.

Fatigue strength is the design criterion for clamp screws as well as for other tool details subjected to repeated stresses. The endurance limit in tension alone may be one and a half times the endurance limit for completely reversed stresses commonly given in tables of material properties. The endurance limit in compression may be 40 to 50 per cent higher than in tension, provided the yield strength is not exceeded. Stress-concentration factors of 4 for heat-treated alloy steels and 3 for mild steel have been reported for standard screw threads. Other factors need consideration, too; a practical working stress is one-sixth the endurance limit of a material. The nominal diameter of a clamping screw is

$$d = 1.35\sqrt{W/S},$$

where W is the axial load on the screw in pounds. The torque in inch-pounds that must be exerted on an N/C threaded screw or nut is $T = 0.2dW$, within 5 per cent.

The slot in a strap clamp should be $\frac{1}{16}$ in. wider than the bolt or stud passing through it. The width of the strap should be

$$W = 2.3d + \tfrac{1}{16},$$

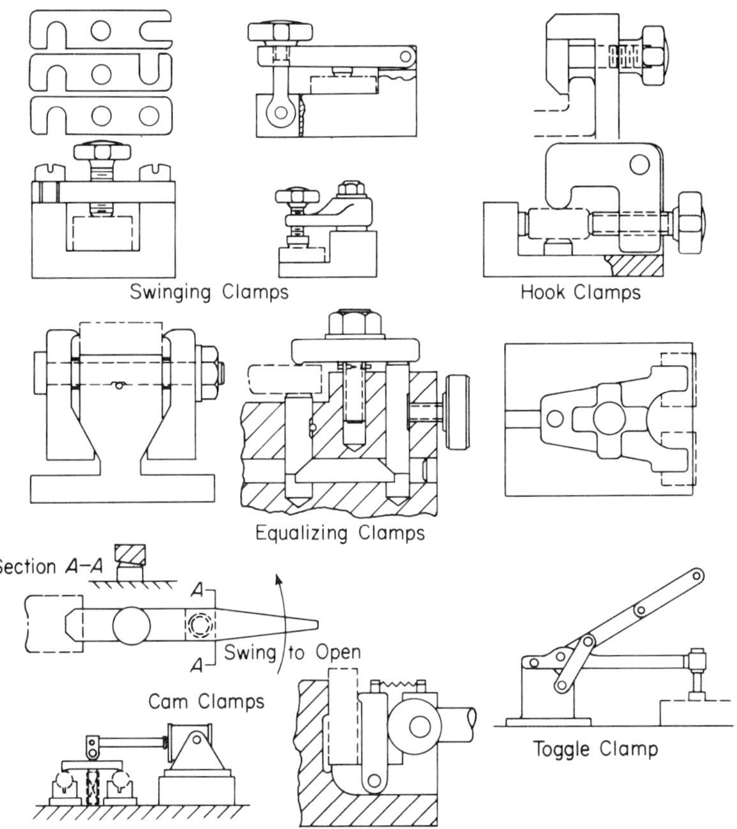

Fig. 9.27 *Common clamps.*

TABLE 9.13. *MANUAL FORCES EXERTED ON VARIOUS TYPES OF LEVERS*

Type of Lever	Force, Pounds Low	Average	High
Single lever			
Push or pull vertically from about 20 to 35 in. above floor level	50	95	140
From about 35 to 50 in. above floor level	30	65	100
Push or pull horizontally	30	65	100
Cross bars			
Push or pull vertically or horizontally from convenient position	80	160	240
Handwheel			
Vertical and parallel to body	60	125	190
Vertical and perpendicular to body	80	160	240
Horizontal	70	140	210

only slightly more than the diameter of the washer for the stud of diameter d in. With the nut and washer bearing halfway between the ends of the strap of length L in., a reasonable thickness for the strap is $h = 0.46\sqrt{dL}$.

An eccentric circular cam clamp is cheaper but must have 1.6 times as large a radius to stay locked and therefore has proportionately less mechanical advantage than a spiral cam clamp (one with a spiral curve generated around its axis). A cam gives a rise H in. of the cam surface when it is turned through an angle C deg, called its throw. The curve of a spiral cam must have a lead $l = 360H/\theta$. The minimum locking radius in inches of a spiral cam is $R = l/(2\pi f) = 57H/\theta f$, where f is the coefficient of friction on the cam face (considered 0.1 or less for security). The eccentricity of a circular cam must be $e = H/(1 - \cos\theta)$. It must have a minimum radius $R = H/f$ to stay locked in the most unfavorable position. The torque in inch-pounds needed for a spiral cam to exert a clamping force of F lb is approximately $T = 2rfF + r_1f_1F$, where r is the largest acting radius of the cam, f is 0.2 or more, r_1 is the radius and f_1 the coefficient of friction of the cam pivot, and F is the force rendered by the cam. The torque applied to a circular cam is

$$T = 2RfF + r_1f_1F.$$

A cam bearing on a plane surface should have a thickness in inches of $t = 132F/H^{2.19}r$, where F is the normal force on the cam surface, H the Brinell hardness number of the softer of the two surfaces, and r the maximum radius of the cam surface.

A simple *toggle clamp* like that in Fig. 9.28 has a pin radius $r = 0.47\sqrt{F/S_s}$, where S_s is the allowable shear stress. The force to actuate the toggle to deliver a force F is approximately $F_c = 2rfF(m + n)/ln$; the factor f is the coefficient of friction at the joints.

9.15.5 Selection of Clamping Device

Screw clamps are simple and cheap, especially when made of commercial screws, and have long ranges of action. For practical purposes, a cam is limited to a rise of about $\frac{1}{8}$ in. for its full force, and a quick-acting toggle clamp to $\frac{1}{16}$ in. or less.

Vibration tends to reduce the effectiveness of friction in keeping a clamp locked. Cam and wedge clamps are more susceptible to vibration than screw clamps and should be avoided where vibration is likely to be excessive.

Average operating times in minutes for typical devices are: socket-head screw—0.22; bar nut, screw jack, tail knob, and hexagonal nut—0.16; spoke nut—0.12; handwheel—0.11; star knob—0.09; hand cam—0.07; C-washer—0.06; quick-acting toggle—0.04; and sliding-strap clamp—0.01. A ratio of 4 to 1 has been reported between the times of manual- and air-operated chucks. Hydraulic is close to air operation in speed.

9.15.6 Other Fixture and Jig Details

Jack supports are necessary under rough workpieces for support other than that provided by the locators. (More than the number of locators prescribed by the 3–2–1 principle cannot be applied to rough surfaces.) A screw makes the simplest jack support, but with it the operator can distort the work if he is not careful. One commercial form of jack screw has a ratchet head that limits the force that can be applied. Figures 9.23 and 9.29 show other types of jack supports. Operating time is saved when several jack supports are locked or operated from one control.

Set blocks, *gage buttons*, or *gage pins* commonly are put on fixtures; each has one or more

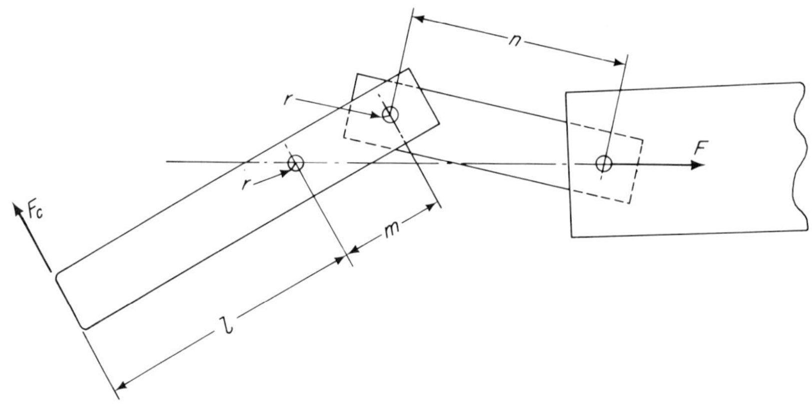

Fig. 9.28 *A simple toggle clamp.*

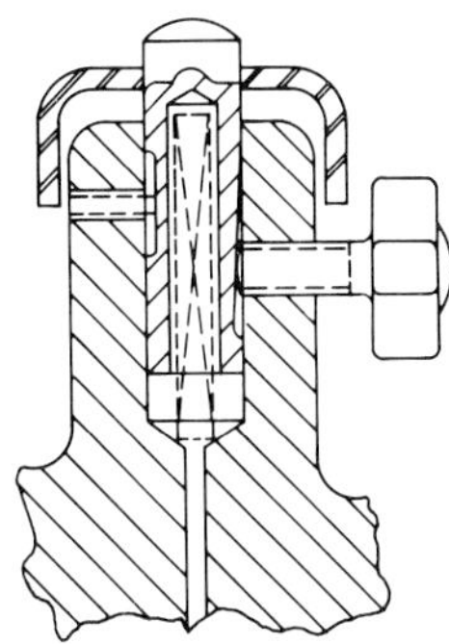

Fig. 9.29 *Adjustable jack support.*

reference surfaces to which cutters are set by means of machine adjustments, as in Fig. 9.23. A cutter thus positioned produces surfaces to the required dimensions on pieces located in the fixture. Sometimes the setting surfaces may be on details used for other purposes also, such as on locating blocks.

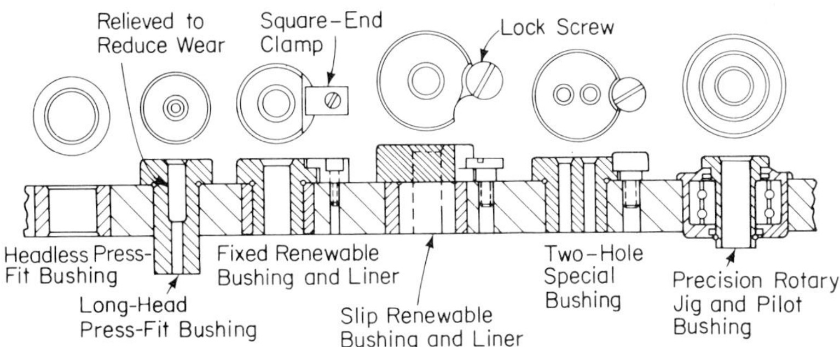

Fig. 9.30 *Jig bushings.*

Typical *jig bushings* are depicted in Fig. 9.30. *Headless bushings* are pressed into place, may be used alone for low production, and serve as liner bushings. A head keeps a bushing from being pushed through the plate, provides a place for clamping, helps protect the soft bushing plate from misdirected cutting tools, can serve as a bearing for a depth stop collar on a tool, and is the place where a slip bushing is grasped. *Fixed* or *plain-head bushings* are pressed in place to remain until worn out or no longer needed. If expected to be replaced when worn, a bushing is inserted in a liner with or without a head. A clamp or lock screw keeps a bushing from being pushed out by chips or from turning with the tool and wearing the hole in which it fits. A *slip renewable bushing* can be turned back slightly and withdrawn from the hole. Slip bushings can be used to guide tools of different sizes, such as a drill and reamer, in succession in one hole, or one bushing may be used for several holes in a workpiece where production is low. *Screw bushings*, like the one in Fig. 9.24, act as locators or clamps as well as tool guides.

Jig bushings may be purchased in standard sizes listed in manufacturers' catalogs and in USAS B5.6–1962, *Jig Bushings*, based upon jig-plate thicknesses of $\frac{5}{16}$, $\frac{1}{2}$, $\frac{3}{4}$, 1, $1\frac{3}{8}$, and $1\frac{3}{4}$ in. They are available ground to size or $\frac{1}{64}$ in. oversize for fitting. Special bushings are made to suit unusual conditions, such as when extra length is required or two holes are too close together for separate bushings.

A tool is guided best when a bushing is close to the work surface, but chips must pass out through the bushing and cause wear. That arrangement gets the chips out of the jig, which is particularly helpful where long and thin chips may become tangled in the jig. Wear is reduced by keeping the bushing away from the work surface a distance equal to about one and a half times the tool diameter. Thus, both positions have merits, and the one to adopt depends upon conditions in each case. However, compromise is not desirable because it has the disadvantages of both extremes.

Positioning and *fastening details* in a number of forms adapt fixtures and jigs to the machine tools or workplaces where they are used. Milling and similar fixtures may have bolt slots for fastening and keys for aligning, as in Fig. 9.23. Many jigs are moved about on machine tables and have four legs or pads for each position in which they are set. A jig may have legs that can be swung into different positions, or it may be placed on angle plates for drilling holes at angles.

In addition to the basic details, others may be added to fixtures and jigs to provide various refinements. In some cases, *ejection pins* with an operating mechanism may be supplied to make removal of work easy. *Foolproofing details* make it impossible to load a workpiece except in its correct position. That is commonly done by placing blocks or pins so that the workpiece will clear them only if it is positioned correctly.

9.16 PUNCHES AND DIES

9.16.1 Applications

Punches and dies are used in many industries for metal cutting, forming, stretching, and squeezing. Metal-cutting operations include blanking, piercing, perforating, shearing, parting, cutting off, trimming, slitting, notching, shaving, and broaching. In *blanking*, the piece wanted is punched from the stock. Holes are punched in a part by *piercing* or *punching;* small closely spaced holes are punched by *perforating*. Parts are separated from each other or from the parent stock and cut to length by *shearing*, *parting*, and *cutting off*. *Trimming* is the shearing of the flanges of drawn shells or forgings. *Shaving* removes 0.001 in. or so from the edge of a blank to make it accurate and square and to give it a good finish.

Metal forming beyond the elastic limit by bending or shaping, curling, wiring, stamping, embossing, and beading operations is not severe. *Bending* stresses a part in tension on one side and in compression on the other side of its neutral axis to form one or more angles. *Curling*, *wiring*, or *false wiring* consists of bending over the edge of a cup or bucket with a curl to form a smooth rim. The wire inside the curl is usually omitted in modern practice. The thickness of the material is not changed appreciably in *stamping*, *embossing*, or *beading* to raise or depress a pattern or rib on thin metal. Embossing is somewhat more severe and produces a sharper impression than stamping.

Drawing, redrawing, reducing, bulging, and ironing are stretching operations for producing relatively deep cavities by working or forming materials severely. In *drawing*, the metal is made to flow from a flat blank into a cup or shell with a continuous wall. Several draws may be necessary to produce a part, and operations after the first are called *redrawing* or *reducing*. *Bulging* is commonly performed to bulge out the contour of a cup between the rim and bottom. *Ironing* thins the walls by redrawing and at the same time squeezing a cup between the punch and die opening.

Stretch forming or *contour forming* is done by pulling and stretching sheets, roll or brake formed sections, or extrusions, while at the same time wrapping or drawing them around a die, punch, or form.

Squeezing operations are coining, sizing, swaging, forging, upsetting, and extruding. *Coining* is a severe high-pressure operation in a closed die causing the material to flow into sharp designs on the surfaces of a piece. In *sizing*, a part is squeezed to a definite thickness. Tolerances of 0.005 in. or less are commonly held in sizing parts in production. *Swaging*, *forging*, and *upsetting* are operations for squeezing slugs into definite shapes in production. Forging is done hot or cold and is called upsetting when done lengthwise on the end of a piece. *Extruding* is the method of compressing metal and causing it to flow rapidly through an orifice with a definite cross section that is imparted to the product.

9.16.2 Types of Dies

A *punch* is a detail of a press tool that enters the opening in another detail called the *die* or *die block*. The complete assembly or press tool is also called a *die*. A die that performs a single operation is named for that operation, such as a blanking die, bending die, or coining die.

A *progressive* or *follow die* has a series of stations. Each piece in a strip passes from station to station and through a series of operations, such as piercing, forming, and blanking, as in Fig. 9.31. At each stroke, a piece or pieces at the end of the strip are finished, and the strip is moved forward.

A *compound die* has a single station in which blanking and piercing or perforating are done on one piece at each stroke. The blanking punch contains the piercing die openings and is usually on the bottom, with the blanking die and piercing punches on top.

A *combination die* performs cutting and forming or drawing operations simultaneously in one station, as exemplified in Fig. 9.32. Because all operations are performed while a piece is in one station, combination and compound dies produce more accurate parts than

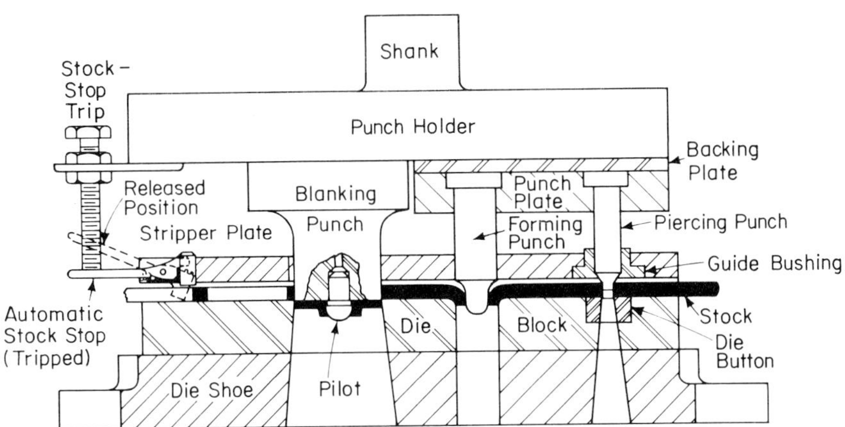

Fig. 9.31 *A piercing, forming, and blanking progressive die.*

do progressive dies but as a rule are more expensive to make and maintain.

A *sectional die* is composed of assembled sections instead of having a solid die block. *Cam die* is the name applied to the die often used for forming and side-piercing operations; it has horizontal or angular slides carrying punches actuated by cams on the descending punch holder.

Dies may be *push-through*—where the work-

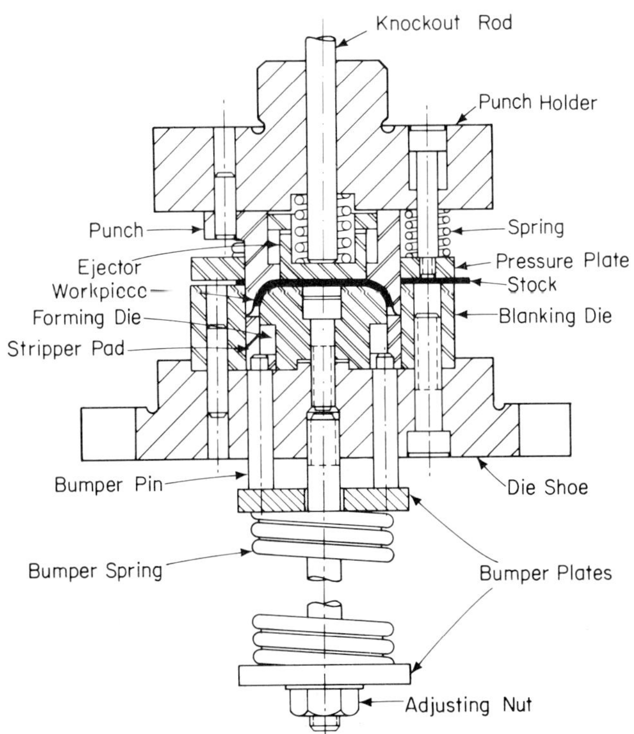

Fig. 9.32 *A combination blanking and drawing die.*

pieces or slugs pass through the die block, shoe, etc., as in Fig. 9.31—or *return-blank*, like that in Fig. 9.32.

The dies corresponding to the classes of tools described in Par. 9.13.1 are permanent or production dies for Classes *A* and *B*, short-run dies for Class *C*, and experimental, developmental, and temporary dies for Class *D*. *Experimental* or *developmental dies* are for experimentation with new model parts, materials, or shapes. *Temporary dies* may last for definitely limited short runs or until permanent tooling becomes available, but, in either event, they are discarded after serving the single purpose. *Short-run dies* are for repeated production of small lots, or they may have universal features to accommodate similar parts in small quantities. *Permanent* or *production dies* are for moderate or large quantities in long or continuous runs.

Several types of dies have cost-saving features for temporary or small-quantity production. A *hinge-type die* has its upper member or punch holder hinged to the die shoe for compactness and ease of setup. A *continental* or *push-through die* has a loose punch set on top of the work material and pushed through the die block for each piece. A *magnetic die* has its punch and die block, readily interchangeable with others, held by a magnetic die set for temporary operation. A *cutting-rule die*, constructed with sharp-edged steel rule material standing on one edge and bent to the shape to be blanked, is capable of short and moderate quantity runs on thin or soft metal but normally is used for blanking cork, paper, and fibrous materials. An *adjustable perforating die* with *universal tooling* consists of standard, commercial punching units bolted to the members of a die set. Assorted jobs can be tooled easily by changing the positions of the punching units. As a rule, temporary and short-run dies are likely to be single-operation dies, constructed as simply as possible without die sets, trip fingers, automatic stops, and the like. The more permanent dies tend to be more complex and to have multiple stations.

9.16.3 Die Details

The details of dies may be classified according to functions similar to those performed by jigs, fixtures, and other tools. The cutting and forming tools are the punch and die block. Stops, pilots, nests, and stripper plates locate the stock or workpieces; pressure plates, pads, and pins do the clamping. The die set serves as the body, base, or frame of the die and provides the elements for fastening and positioning the die on the press. Slender punches may be guided in stripper-plate bushings much like drill bushings, and blocks in some dies set the correct amount of punch travel.

9.16.4 Die Sets

A production die is conventionally constructed upon a *die set* that consists of a punch holder and die shoe kept in permanent alignment by two or more vertical leader or guide pins. The guide pins are fastened to the die shoe, and the punch holder, often bushed, slides over them. The die shoe has ledges or bolt slots for fastening it to the bolster plate of the press. The punch holder is bolted to or has its shank clamped in the press ram. The punch holder and die shoe are cast iron or semisteel for ordinary purposes and steel for highest quality. The punches and die block are positioned in a die set so that the resultant of all vertical forces acting during a stroke passes through the center of the ram. Die sets are available commercially in a large variety of styles and sizes.

9.16.5 Punches, Plates, and Pads

A large blanking or forming punch usually is made with a flange by which it is bolted and doweled to the punch holder after being aligned with the die block in assembly.

The length of the shank of a blanking punch for quantity production should include $\frac{1}{8}$ to $\frac{1}{4}$ in. for grinding to resharpen it during its life and enough more to reach the die block through the stripper plate and to shear the stock after the punch has been ground down. For stability, the width of the flange should not be less than the height of the punch. The nose of a drawing punch should have a smooth radius all around of four to six times the stock thickness.

Piercing punches are pressed in a punch plate, as shown in Fig. 9.31, and the plate is bolted and doweled to the punch holder. Head-type punches are widely used but require that the plate be removed to remove broken punches, which may have to be done frequently with small punches. Commercial punches are available with locking devices that make punch replacement quick and easy. A punch that is not round must be kept from turning by a pin through its flange and the punch

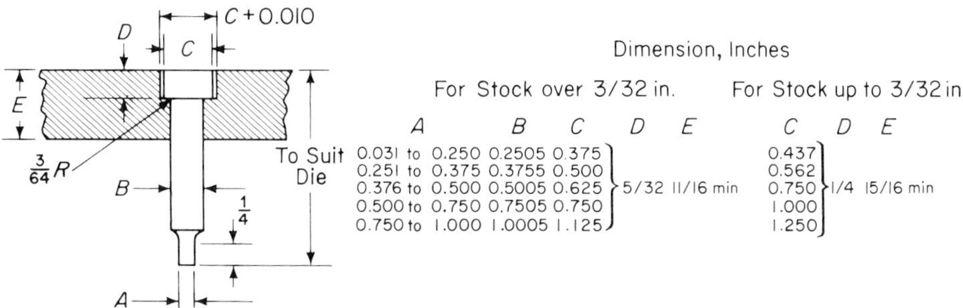

Fig. 9.33 *Piercing-punch and punch-plate dimensions.*

plate. Small punches are often made shorter than large punches by the depth of penetration to shear the stock to avoid crowding and breakage. Proportions of piercing punches and plates are suggested in Fig. 9.33. It is generally difficult to pierce a hole smaller in diameter than the thickness of the stock.

A hardened *backing plate* or *punch pad* may be inserted behind small punches, as shown in Fig. 9.31, to keep them from digging into the soft punch holder and working loose.

A popular technique, especially for temporary or short-run blanking dies, is to saw out the die-block opening at an angle, undersize on one side and oversize on the other. The opening on the small side is filed and stoned to proper size, leaving ample taper at the other end. The slug is filed on the large side to a proper clearance fit in the die opening and becomes the punch.

9.16.6 Die Blocks

Figure 9.34 indicates average thicknesses of production die blocks for blanking. The thickness should be increased $\frac{1}{8}$ in. for hard or tough materials like lamination stock and may be decreased $\frac{1}{8}$ in. for soft materials like brass or plastics. Temporary die blocks usually are thinner. A rule observed by some die designers is that the minimum distance from a cutting edge to the outside of the die block should not be less than the thickness of the die block to avoid cracking in heat treatment and service. A more conservative specification by one authority* is that the minimum distance from the cutting edge to the outside edge

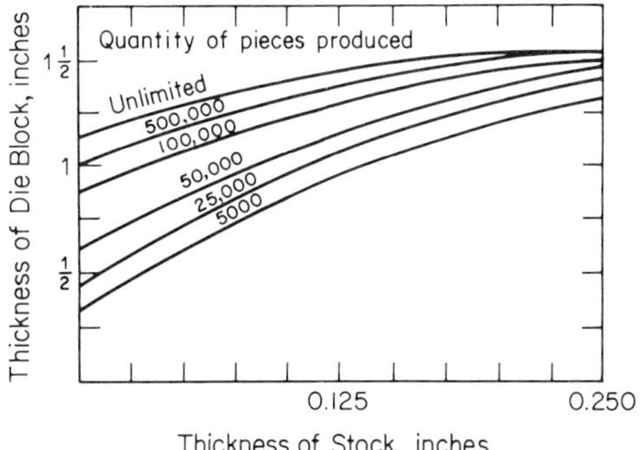

Fig. 9.34 *Average die-block thicknesses.*

should be $W = ft^{0.25}$, and the minimum cross-sectional area between the cutting opening and outside edge should be $A = 0.03T^{0.9}$. The factor f varies from 1.5 to 2 for small dies and from 2 to 3 for large dies; t is the die-block thickness in inches; and T is the punching force in tons.

A die opening for blanking or piercing is straight for a short distance below the top of the block and then is tapered $\frac{1}{4}$ deg to $\frac{3}{4}$ deg on a side on through the die block and shoe to prevent blanks and slugs from sticking. Satisfactory practice is to make the straight portion $\frac{1}{8}$ in. long for stock less than $\frac{1}{8}$ in. thick, and equal to the stock thickness when the stock is more than $\frac{1}{8}$ in. thick.

The width of the opening for a simple bending die as in Fig. 9.35 is $Y = AR + 2t$. Values of A from 2 to 3, with an average of 2.5, have been found acceptable.†

* F. Strasser, "Should Die Thickness Be Calculated?", *American Machinist*, **95**, No. 4 (Feb. 19, 1951), p. 146.

† W. W. Wood *et al.*, *Theoretical Formability* **II** (Washington, D. C.: Office of Technical Services, U.S. Department of Commerce, 1961), p. I-1.

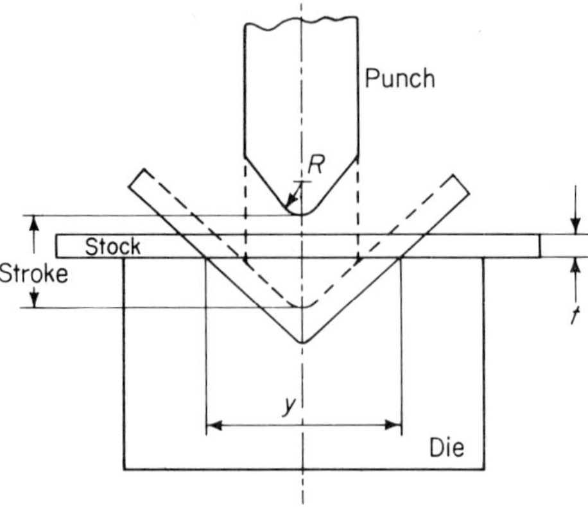

Fig. 9.35 *Dimensions of a simple bending die.*

The edge of a die opening over which metal is drawn has a radius four to six times the metal thickness and is highly polished. Characteristic of drawing dies, especially for irregular or unsymmetrical shapes, are drawing beads and grooves, as Fig. 9.36 illustrates. They may or may not be continuous around the die; they restrain and stretch the metal to control its movement into the die cavity as needed, at various places around the die. Experience and trial and error usually determine bead placement and size.

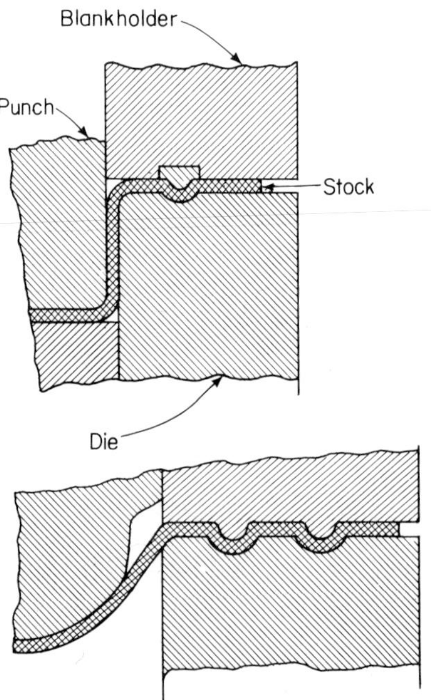

Fig. 9.36 *Drawing beads and grooves.*

Although most dies are made from solid blocks, *sectional dies* (depicted in Fig. 9.37) are often more economical to construct because all surfaces of the sections are external. Cutting or forming forces tend to separate the sections of the assembled die in use, but heavy dowel pins or keys or sinking the block into a slot in the die shoe can prevent this.

Die buttons, like the one in Fig. 9.31, are commonly inserted for piercing holes in long-run die blocks because they can be replaced if breakage or wear occurs. The bushing should extend about three-quarters of the way through the block to allow for grinding and also for ample bearing in the block.

Most punches and die blocks are of tool and die steel. The large variety of steel available suits many purposes. Large-production punches and dies, such as auto-body panel dies, are commonly cast iron with hard steel or other material inserts at critical places. Nonferrous-alloy and cemented-carbide punch and die-block inserts, tips for small punches, even solid punches, have proven successful for severe service and large-quantity production. Cemented-carbide dies are reported to last twenty to forty times longer than steel dies for cutting, and ten to thirty times longer for forming. For low production, especially but not entirely for soft work materials, dies are of rubber, plastic materials, Masonite, and low-melting-temperature alloys, which are quicker and easier to fabricate and are sometimes more available than steel.

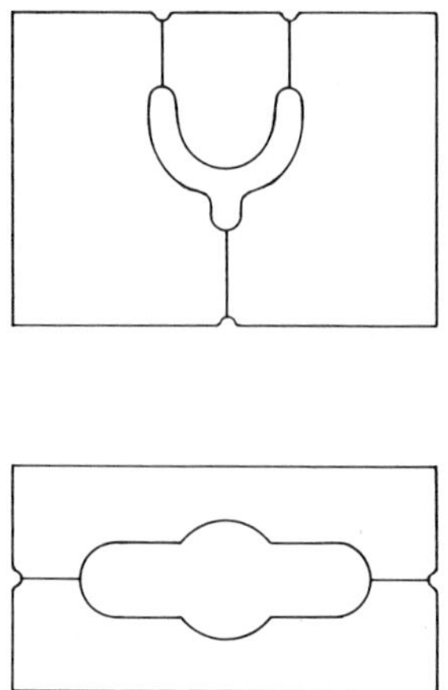

Fig. 9.37 *Top views of two sectional dies.*

9.16.7 Die Clearance

A uniform space or clearance between a punch and die produces a clean cut and a low cutting force. Figure 9.38 specifies the amounts of clearance for various materials for average work, but clearances may range from 50 per cent of those specified for small and close-tolerance work to 150 per cent for large, heavy-gage work.

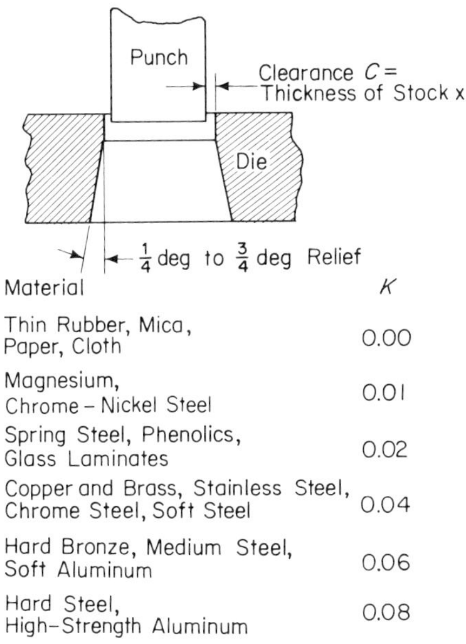

Punch
Clearance C =
Thickness of Stock x K
Die
$\frac{1}{4}$ deg to $\frac{3}{4}$ deg Relief
Material
K
Thin Rubber, Mica, Paper, Cloth 0.00
Magnesium, Chrome-Nickel Steel 0.01
Spring Steel, Phenolics, Glass Laminates 0.02
Copper and Brass, Stainless Steel, Chrome Steel, Soft Steel 0.04
Hard Bronze, Medium Steel, Soft Aluminum 0.06
Hard Steel, High-Strength Aluminum 0.08

Fig. 9.38 *Die clearance.*

A hole punched in a piece is the size of the punch; the slug or blank removed is the size of the die opening. Thus, for piercing the punch is the size of the hole desired, and the die opening is larger than the punch by the amount of the clearance on two sides. For blanking the die opening is the size of the blank required, and the punch is smaller.

The clearance between a drawing punch and die usually is more than the thickness of the stock, as much as 50 per cent more for the first draw and 25 per cent for subsequent draws when thickening of the material is not objectionable. Ironing clearance is less than the original thickness of the stock, with a theoretical maximum reduction of 50 per cent in wall thickness, although as much as 64 per cent has been reported.

9.16.8 Stripper Plates

The two kinds of *stripper plates*, *fixed* and *spring*, serve to strip material from punches, hold down material for forming or drawing, and guide punches. A spring stripper that acts as a pressure plate is shown in Fig. 9.32. A spring stripper that guides piercing punches must be accurately held in line by guide posts. A stripper that does not guide the punches may have clearance of 0.010 to 0.015 in. around small punches and 0.001 to 0.005 in. around sturdy punches.

A fixed stripper plate, as on a progressive die, has a lengthwise slot to guide the stocks. Proportions found adequate for stripper-plate slots by a large manufacturer are given in Table 9.14. A safe thickness in inches for a stripper plate above the slot is $T = W/30 + 2t$, where W is the stock width in inches and t is the stock thickness in inches. The screws that hold down the stripper plate and die block must have sufficient strength to resist the stripping force, which, with dull tools, may amount to 15 per cent of the cutting forces.

Stripper plates are made of machine steel for ordinary dies and of unhardened tool steel for good-quality dies and hardened for severe service. Guide bushings, like that in Fig. 9.31, similar to jig bushings, are inserted in soft stripper plates to inhibit wear in guiding punches.

A knockout supplementing the springs actuating the ejector in Fig. 9.32 is depressed by a positive stop on the press.

TABLE 9.14. *DIMENSIONS OF SLOTS IN FIXED STRIPPER PLATES*

Width of Stripper Slot		Depth of Stripper Slot	
Thickness of Stock, Inches	Slot Wider than Stock by —— Inches	Thickness of Stock, Inches (T_1)	Depth of Slot, Inches
0 to 0.040	5/64	0 to 0.0625	$0.025 + 2T$
0.041 to 0.080	3/32	Over 0.0625	$0.090 + T$
0.081 to 0.120	7/64	(for silicon steel the	
Over 0.120	$\frac{1}{8}$	minimum depth is $\frac{1}{8}$ in.)	

9.16.9 Stops, Pilots, and Stock Guides

Finger or *starting stops*, as in Fig. 9.39, position the stock in one station after another when a new strip is started through a die and retract when not needed. *Stock-aligning pins* are spring loaded to press the stock against one side of the stripper-plate slot. After a strip has been started through a die, it is located for one piece after another by a *stock stop* that usually registers against the scrap left between the blank openings. A simple stop is a fixed pin. A *swinging stop* is shown in Fig. 9.39. The strip is moved under the stop, which swings out of the way, and then is brought back to bear against the stop. This stop is slower than an *automatic stop*, one version of which is shown in Fig. 9.31. The pin that locates the stock is a loose fit in a hole in the stripper plate. A trip on the punch holder pushes the handle of the stop down and raises the stop pin, which is pressed by a spring against the right side of the hole and falls on top of the stock when the ram ascends. When the stock is moved forward, the pin falls into the next blank opening. The side of the opening that comes up next pushes the pin against the right side of the hole in the stripper plate, and the stock is stopped and positioned.

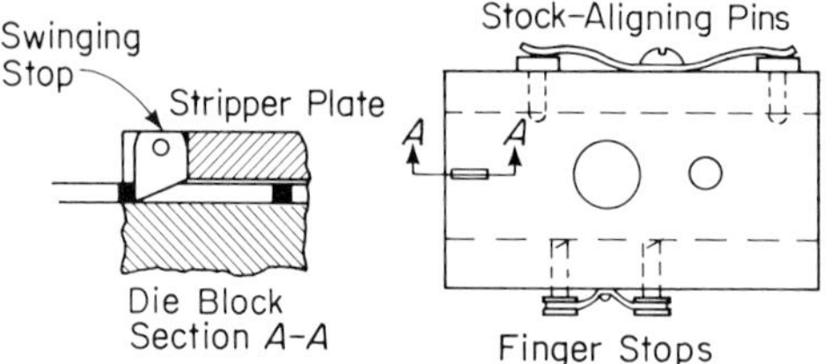

Fig. 9.39 *Stock stops.*

To produce accurate parts, a progressive die has pilots that register in holes pierced in the first stations to align them with the blank. If a part does not have holes in itself suitable for piloting, holes for that purpose may be pierced in the scrap. A simple form of pilot is shown in Fig. 9.31. Other types and proportions are given in Fig. 9.40. If misfeed occurs, a spring-loaded pilot retracts without causing damage. A die should be arranged so that when the pilots register they draw the stock a few thousandths of an inch away from the stops and side of the stripper plate.

Parts fed individually into a die are commonly located in a *nest* that should be three to four times as high as the blank thickness and should have well-rounded edges on top.

A *shedder*, *spring* or *kick pin type*, as shown in Fig. 9.40, dislodges blanks or slugs that tend to adhere to the face of a punch.

Stop blocks about $1\frac{1}{2}$ in. in diameter may be placed in a die set within the slide area of the press to help set up by indicating when the die is closed and to prevent the die from closing in storage.

A punch that cuts, shaves, or forms mostly or entirely on one side must be backed up to prevent excessive deflection. A backing-up surface may be provided on the die block in some cases. Otherwise, the punches are reinforced by bearing against backing blocks that often have wear plates of hardened and ground tool steel or bronze, with provision for lubrication and adjustments to compensate for wear. The heel of the punch must make contact with the backing-up surface before the punch acts.

9.16.10 Press Dimensions

An important dimension of a press that must be considered in the design of a die is the

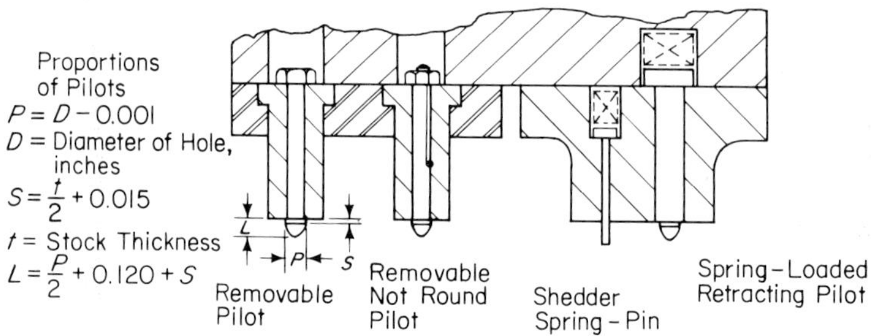

Fig. 9.40 *Pilots.*

shut height, which is the distance from the top of the bolster plate to the bottom of the ram when the ram is at the bottom of its stroke and its adjustment is all the way up, resulting in the maximum vertical space the die can occupy when closed. Other dimensions to be considered are the length of stroke, ram adjustment, diameter of hole in ram for punch-holder shank, bolt holes for clamping punch holder to ram, distance from center line of shank opening to press frame, relation of axis of shank opening to bolster plate, distance between ways if ram goes up between ways, sizes of openings in bolster plate and press bed, and sizes and positions of bolster-plate holes for bolts. If a roll-feed device is to be used on the press, the direction of feed and the distance from the top of the bolster plate to the center of the roll feed must be ascertained.

9.16.11 Stock Layout

Planning the size of stock and arrangement of the pieces for least scrap loss and most economy is one of the most important steps in the design of a die. This leads to the arrangement of the punches in the die. Templates of the blank may be cut out and tried in all possible arrangements to find the one requiring the smallest total amount of material per piece. The parts should be arranged to utilize standard commercial widths of stock if possible. A part to be bent later must be blanked in proper relation to the grain of the material. Bending is done best at right angles to the grain.

The side of a blank impinged upon by the punch has a square and sharp edge, often with a burr, and is called the flat side. The other side has a slightly rounded edge and is called the round side. The opposite is true of the stock from which a slug is punched. The placement of these sides is important for some parts.

9.16.12 Bending

The radius of bend below which the stock is split is called the *minimum bend radius*. Nearly all materials can be bent to a radius R equal to the stock thickness t and some to a radius equal to half the stock thickness. The limiting ratio R/t increases with length of bend (W in Fig. 9.41) up to a length $W = 8t$, after which the ratio remains almost constant. The minimum bend radius depends upon the material and its properties, the shape of the part, the

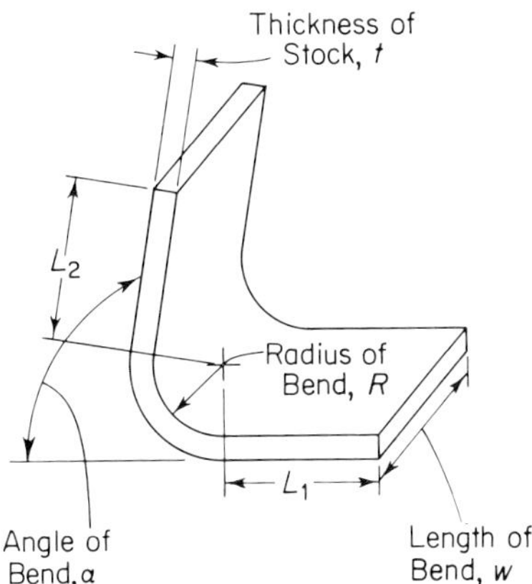

Fig. 9.41 *Elements of a bend.*

state of stress and strain gradients in the bend, the edge condition, and the angle of bend; it is given for any particular case by complex relationships and experimental factors.*

When a bend is made in a die to an angle α_1 and radius R_1 and the pressure released, the piece returns slightly to its prebent condition, to an angle α and radius R (Fig. 9.41). This is *springback*, which depends upon the material, its hardness, it stress-strain properties, the R/t ratio, and type of bend. Springback is expressed by a factor

$$K = \alpha/\alpha_1 = (R_1 + t/2)/(R + t/2)$$

This factor varies from 0.99 for soft steel or aluminum to 0.90 for harder grades when the R/t ratio is one. K generally decreases as R/t increases; K as small as 0.5 is reported for stainless steel for $R/t = 20$. Some of the springback effect may be eliminated by pinching the bend in a die, but is usually eliminated by overbending a part so that it springs back to the desired angle. The angle to which the die should be made may be predicted from experience, which may have been tabulated, but often is determined by trial and error during the construction of the die.

The length of stock for a piece bent as in Fig. 9.41 is

$$L = L_1 + L_2 + 0.017\alpha(R + kt)$$

The factor k varies from $\frac{1}{3}$ when $R < 2t$ to $\frac{1}{2}$ for $R > 2t$, as an approximation.

---

* Wood, *et al., Theoretical Formability.*

9.16.13 Drawing

Because drawing is a severe operation, it must often be done in stages and a series of tools must be provided accordingly. As a rule, a part requires more than one draw if the ratio of its height to diameter exceeds $\frac{1}{4}$ to $\frac{3}{4}$; more than two draws if the ratio exceeds $\frac{5}{8}$ to $1\frac{1}{8}$; more than three draws if the ratio exceeds $1\frac{1}{8}$ to 2; and so forth, depending upon the material ductility and condition and the nature of the operation. A corner radius in a shell should exceed four times the stock thickness if extra operations are to be avoided. Considerations are more complex for drawing rectangular and irregular parts, such as auto-body parts.*

Since metal is almost incompressible, the volume of the material in a drawn shell before being trimmed is the same as that of the original blank. The exact area of a blank may be ascertained by dividing the volume of material in the finished piece by the stock thickness. For thin stock it is sufficient to ascertain the area of the shell. One way is by using the theorem of Pappus and Guldinus that the area of a surface of revolution is equal to the length of its profile times the length of the path of its center of gravity. Often a profile may be divided into geometric components like the example of Fig. 9.42. For each part, L represents the length of line and R the radius to the center of gravity. The total area of the shell is $A = 2\pi(R_1L_1 + R_2L_2 + R_3L_3 + R_4L_4 + R_5L_5)$.

A round cup of outside diameter d, metal thickness t the same as the blank thickness, height h, and corner radius r between wall and bottom requires a blank of diameter

* E. V. Crane, *Plastic Working of Metals and Non-Metallic Materials in Presses (3rd ed.)* (New York: John Wiley & Sons, Inc., 1944), Chap. VIII.

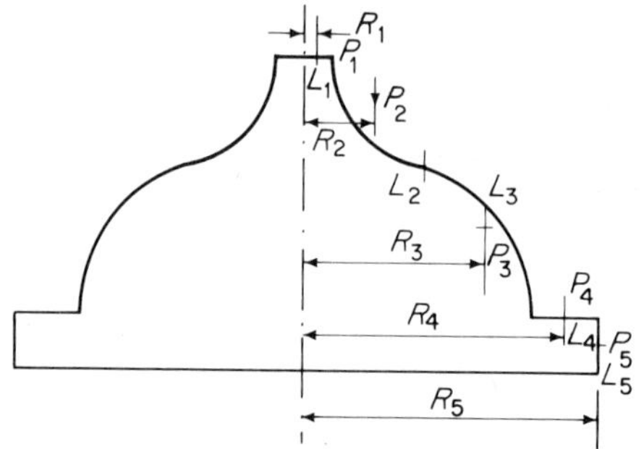

Fig. 9.42 *A cross section of a drawn shell.*

$$D = [(d - 2r - 2t)^2 + 4(d - t)(h - r) + 2\pi(r + 0.4t)(d - 0.7r - 0.3t)]^{1/2}$$

Stock of about 13 gage or heavier may be drawn in a simple shape without pressure applied to the blank. Under other conditions, the blank must be confined between pads under pressure while being drawn. With a single-action press, the pressure may be supplied by springs, as in Fig. 9.32, rubber pads, or hydraulic or air cylinders. With a double-action press, the blank is held mechanically by the first action of the press while being drawn by the second action. The actual pressure required is difficult to specify for any particular case and is ascertained experimentally.

9.17 GAGES†

9.17.1 Types of Gages

A *gage* is a device for determining whether one or more dimensions of a manufactured part are within specified limits. A *working gage* is used by an operator on a machine to check the work he is producing. *Inspection gages* are used to check finished pieces. *Reference* or *master gages* are reserved for checking other gages.

Figure 9.43 illustrates typical *fixed gages.* A gage that checks the high and low limits of a dimension is called a *limit gage* or a go and not-go gage. A *double-end gage* has a go member at one end and a not-go member at the other end and must be applied twice to a workpiece. A *progressive gage* is quicker, with its go and not-go members next to each other, but its action is sometimes limited. For instance, it is not suitable for probing the full depth of a blind hole. A *solid gage* is fixed for one set of limits. An *adjustable gage* can be set and locked to a predetermined size within its range.

An *indicating gage* reflects differences in dimensions by greatly magnifying them on a visual scale; for example, divisions 1/16 in. apart may represent 0.001 in. on one circular scale, 0.0001 in. on another, and 0.00001 in. on still another. Magnification is achieved by mechanical, pneumatic, or electrical means on various makes and models of indicating gages. A master customarily sets such a gage to a dimension, and the gage measures discrepancies of dimensions from that on the master for in-

† See Sec. 14 for gages in quality control and inspection.

dividual parts. The procedure is commonly followed to obtain data for statistical quality control of operations. A *combination gage* checks more than one dimension on a workpiece at one time.

9.17.2 Gage Standards

The dimensions of standard gages are specified in *Gage Blanks*, Commercial Standard CS8–51, U.S. Department of Commerce.

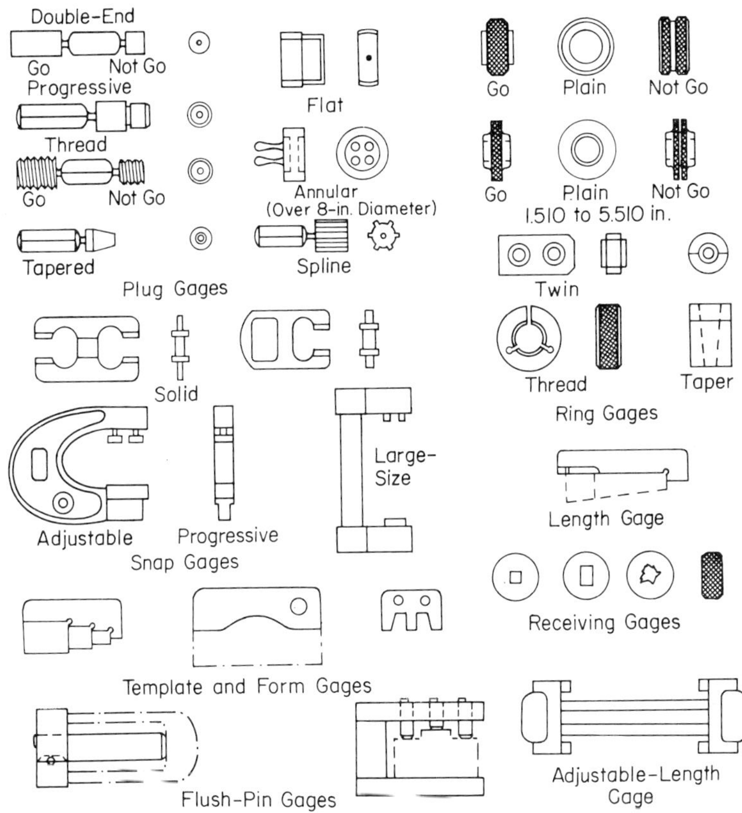

Fig. 9.43 *Typical gages.*

TABLE 9.15 *STANDARD GAGE TOLERANCES FOR PLAIN CYLINDRICAL PLUGS AND RINGS**

Nominal Size of Dimension, Inches		Gagemakers' Tolerance, Inches Class of Gage				
Above	To and Including	XX	X	Y	Z	ZZ
0.029	0.825	0.00002	0.00004	0.00007	0.00010	0.0020
0.0825	1.510	0.00003	0.00006	0.00009	0.00012	0.0024
1.510	2.510	0.00004	0.00008	0.00012	0.00016	0.0032
2.510	4.510	0.00005	0.00010	0.00015	0.00020	0.0040
4.510	6.510	0.000065	0.00013	0.00019	0.00025	0.0050
6.510	9.010	0.00008	0.00016	0.00024	0.00032	0.0064
9.010	12.010	0.00010	0.00020	0.00030	0.00040	0.0080

* From *Screw Thread Standards for Federal Services*, Handbook H–28 (Washington, D. C.: U. S. Department of Commerce, 1957), Part 1, p. 119.

Tolerances for plain cylindrical plug and ring gages have been standardized as shown in Table 9.15. Class XX gages are precision lapped for the highest degree of accuracy practicable and should be used only for the final inspection of close-tolerance dimensions and as reference gages. Class X gages are also precision lapped and are usually applied to close-tolerance inspection. Classes X to Z have progressively larger tolerances; they are used to inspect dimensions with correspondingly larger tolerances. Class Z is the regular commercial class of ring gage furnished when specifications do not call for exceptional accuracy. Class ZZ gages are ground but not lapped, are the most inexpensive of ring gages, and are suitable only for gaging small quantities of pieces with liberal tolerances.

A gage tolerance is commonly limited to 10 per cent of the tolerance on the part dimension and often to 5 per cent for inspection gages. A closer tolerance than necessary should not be specified for a gage because the gage cost rises rapidly as the tolerance is reduced. If the tolerance of a part dimension is large—for example 0.010 in.—the gagemakers' tolerance may be selected from the most liberal class and a substantial allowance may be made for gage wear.

9.17.3 Gage Policy

A number of systems and many variations of each are in use for assigning tolerances to gages. A policy developed by ordnance engineers and especially well suited for liberal part tolerances is depicted in Fig. 9.44. This system provides for wear of go gages but not of not-go gages, always allows the gagemaker a tolerance, prevents conflict between working and inspection gages, and always remains within the workpiece limits.

Three classes of thread-gage tolerances and their applications are specified in *Screw Thread Standards for Federal Services*, Handbook H–28.

BIBLIOGRAPHY

American Society of Tool and Manufacturing Engineers, *Fundamentals of Tool Design*. Englewood Cliffs, N.J.: Prentice-Hall, Inc., 1962.

———, *ASTME Die Design Handbook*, 2nd Edition. New York: McGraw-Hill Book Co., 1965.

Boston, Orlan W., *Metal Processing*. New York: John Wiley & Sons, Inc., 1951.

Brozek, John S., "A Notebook on Die Design," *The Tool Engineer*, **XXVI**, No. 6 (Feb.–June 1951), 49.

Carboloy Application Data. Detroit: Carboloy Department of General Electric Co., 1961.

Carlson, Richard F., *Metal Stamping Design*. Englewood Cliffs, N.J.: Prentice-Hall, Inc., 1961.

Chase, Herbert, *Handbook on Designing for Quantity Production*. New York: McGraw-Hill Book Co., 1950.

Crane, E. V., *Plastic Working of Metals and Non-Metallic Materials in Presses*, 3rd Edition. New York: John Wiley & Sons, Inc., 1944.

Die Engineering Manual. Detroit: Carboloy Department of General Electric Co.

Donaldson, C. and G. H. LeCain, *Tool Design*, 2nd Edition. New York: McGraw-Hill Book Co., 1957.

Doyle, L. E., "An Analysis of Cost Estimating Principles and Practices," *The Tool Engineer*, **XXVIII**, No. 5 (May–Aug. 1952), 37.

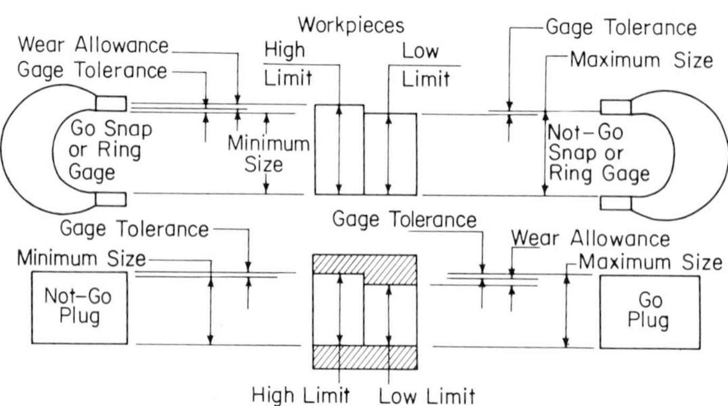

Fig. 9.44 *Gage-tolerance policy.*

———, "Cost Estimating—How to Minimize the Dangers of Chance," *The Tool Engineer*, **XXXXII**, No. 7 (June 15, 1959), 82.

———, "The Mechanics of Clamping Devices," *The Tool Engineer*, **XXV**, No. 1 (July–Oct. 1950), 17.

———, *Tool Engineering: Analysis and Procedure*. Englewood Cliffs, N.J.: Prentice-Hall, Inc., 1950.

Doyle, L. E., B. R. Better, and B. T. Chao, "A Design for Non-Sticking Plug and Ring Gages and Locators," *University of Illinois Engineering Experiment Station Bulletin 433*, **53**, No. 38 (Jan. 1956).

Doyle, L. E. *et al.*, *Manufacturing Processes and Materials for Engineers*. Englewood Cliffs, N.J.: Prentice-Hall, Inc., 1969.

Eary, Donald F. and G. E. Johnson, *Process Engineering for Manufacturing*. Englewood Cliffs, N.J.: Prentice-Hall, Inc., 1962.

Eary, Donald F. and Edward A. Reed, *Techniques of Pressworking Sheet Metal*. Englewood Cliffs, N. J.: Prentice-Hall, Inc., 1958.

Gadzala, John L., *Dimensional Control in Precision Manufacturing*. New York: McGraw-Hill Book Co., 1959.

Gage Blanks, Commercial Standard CS8. Washington, D.C.: U.S. Department of Commerce, 1951.

Hinman, C. W., *Die Engineering Layouts and Formulas*. New York: McGraw-Hill Book Co., 1943.

———, *Practical Designs for Drilling, Millings, and Tapping Tools*. New York: McGraw-Hill Book Co., 1946.

Jergens, John J., *Tool Design and Tool Engineering Handbook*. Cleveland, Ohio: 18107 Invermere Ave., 1959.

Kronenberg, Max, *Grundzüge der Zerspanungslehre*. Berlin: Springer-Verlag, 1954.

———, *Machining Science and Applications*. New York: Pergamon Press, 1966.

Machining Data Handbook. Cincinnati, Ohio: Metcut Research Associates, Inc., 1966.

Manual on Cutting of Metals. New York: The American Society of Mechanical Engineers, 1953.

Nelson, Leonard, "How to Estimate Dies, Jigs, and Fixtures from a Part Print," *American Machinist*, Special Report No. 510 (Sept. 4, 1961), p. 111.

Niedzwiedski, A., *Manual of Machinability and Tool Evaluation*. Cleveland, Ohio: Huebner Publications, Inc., 1960.

Nordhoff, W. A., *Machine Shop Estimating*. New York: McGraw-Hill Book Co., 1960.

Nordquist, W. N., *Die Designing and Estimating*. Cleveland, Ohio: Huebner Publications, Inc., 1960.

Parsons, C. W. S., *Estimating Machining Costs*. New York: McGraw-Hill Book Co., 1957.

Reichert, David I., "Estimating for Short Run Production of Electronic Systems," ASTME Paper MM66–701, Dearborn, Mich., March 7, 1966.

Screw Thread Standards for Federal Services, Handbook H–28. Washington, D.C.: U.S. Department of Commerce, 1963.

Thomas, Richard A., "The Language of Tape," *American Machinist*, **108**, No. 1 (Jan. 6, 1964), 73.

Treatise on Milling and Milling Machines. Cincinnati, Ohio: The Cincinnati Milling Machine Co., 1951.

Tucker, S. A., *Cost-Estimating and Pricing with Machine Hour Rates*. Englewood Cliffs, N.J.: Prentice-Hall, Inc., 1962.

U.S. Air Force Machinability Reports. Wood Ridge, N.J.: Curtiss-Wright Corp., 1950–60.

USA Standards Institute, New York City.

B4.1-1966 *Preferred Limits and Fits for Cylindrical Parts.*

B5.3-1960 *Milling Cutters.*

B9.49-1967 *Taps.*

B5.6-1962 *Jig Bushings.*

B5.8-1959 *Chucks and Chuck Jaws.*

B5.9-1960 *Spindle Noses.*

B5.10-1963 *Machine Tapers.*

B5.11-1954 *Spindle Noses and Adjustable Adapters for Multiple Spindle Drilling Heads.*

B94.11-1967 *Twist Drills.*

B94.2-1962 *Reamers.*

B5.18-1960 *Spindle Noses and Arbors for Milling Machine Noses.*

B5.21-1949 *Straight Cut-Off Blades for Lathes and Screw Machines.*

B5.22-1950 *Single Point Tools and Tool Posts.*

B94.8-1967 *Inserted Blade Milling Cutter Bodies.*

B94.10-1967 *High-Speed Steel and Cast Nonferrous Single-Point Tools and Toolholders.*

B5.27-1959 *Punch and Die Sets for Two-Post Punch Press Tools.*

Wage, Herbert W., *Manufacturing Engineering*. New York: McGraw-Hill Book Co., 1963.

Wilson, F. W. and P. D. Harvey, *Manufacturing Planning and Estimating Handbook*. New York: McGraw-Hill Book Co., 1963.

Wilson, F. W. and John M. Holt, Jr., *Handbook of Fixture Design*. New York: McGraw-Hill Book Co., 1962.

Wilson, F. W. *et al.*, *Machining with Carbides and Oxides*. New York: McGraw-Hill Book Co., 1962.

Wilson, F. W. *et al.*, *Numerical Control in Manufacturing*. New York: McGraw-Hill Book Co., 1963.

Wood, W. W. *et al.*, *Final Report on Sheet Metal Forming Technology*, **I** and **II**, Washington, D.C.: Office of Technical Services, U.S. Department of Commerce, 1963.

Wood, W. W. *et al.*, *Theoretical Formability*, **I** and **II**. Washington, D.C.: Office of Technical Services, U.S. Department of Commerce, 1961.

DAN H. BARBER *is manager of the Safety Division, Corporation Level, of the Standard Oil Company of California. He has spent the last twenty-two years in safety engineering work, the last seven of these in his present position.*

He earned his engineering degree from the University of Southern California and is a registered professional mechanical engineer in California. Most of Mr. Barber's professional career has been with the Standard Oil Company of California and he has made many contributions to the outstanding safety record compiled by that company.

Mr. Barber is a member of the American Petroleum Institute Committee on Safety and Fire Protection. He is a member of the Institute's reporting committees on manufacturing safety (chairman), production safety, safety training, and awards and statistics, and of the legislative committee. He is president of the San Francisco Chapter of the American Society of Safety Engineers. He is also a member of Veterans of Safety and Northern California Industrial Safety Society.

SECTION 10

INDUSTRIAL SAFETY

DAN H. BARBER
AND
ROBERT E. DONOVAN*

10.1 INDUSTRIAL-ACCIDENT EXPERIENCE AND COST

Industry pays a substantial bill each year for the treatment and cure of workmen disabled by on-the-job accidents. The National Safety Council reports that the direct out-of-pocket costs to American industry for these injuries amount to billions of dollars annually. Indirect costs, representing the money value of damaged equipment and materials, production delays, and time losses of other workers not

* Deceased.

involved in the accidents, involve more billions per year. The direct out-of-pocket costs include medical and hospital expenses, death benefits, and workmen's disability compensation as provided for by various state and federal laws. These costs detract from profits since the responsibility is placed by law on the employer. Whether the employer insures his risk with an insurance carrier or self-insures the industrial-injury liability, he pays for medical attention required by the injured employee and the compensation the employee receives during his disability.

In addition to industry, the injured employee suffers financially because injury-disability compensation never equals his earnings. Although some companies have full-pay benefit plans, there is a limit on the amount an injured worker can receive. Furthermore, many injuries leave the worker permanently disabled and may prevent him from earning at the same level he did prior to the injury. The National Safety Council estimates that wage losses because of temporary inability to work, lower wages after returning to work because of permanent impairment, and present value of future earnings lost by those totally incapacitated or killed exceed billions of dollars per year.

After workmen's compensation laws were enacted starting in 1910, informed and progressive industrial management realized that the payment of insurance premiums or costs under a self-insurer's permit represented a large drain on profits. Steel, mining, and oil companies were the first to make progress in the prevention of accidents and to profit by the lower injury costs secured. Through a simple program of preventing contact with moving machinery and installing guard rails at elevated workplaces to prevent falls, substantial reductions in the incidence of injury were made. However, when a level about 25 per cent below the old rate was reached, it was found impossible to reduce the rate further by means of physical safeguards. The reason was, and still is, that the human factors have the greatest influence on safety in industrial operations. These human factors include, in order of importance, the attitude of management regarding safety in company operations, the quality of supervision the worker receives, the skill or training that regulates the worker's action on the job, and the worker's attitude toward his employing company and its employee-relations policies and practices.

As time went on, the compensation benefits provided for the injured worker were increased by law, and medical and hospital expenses doubled and trebled. Progressive management, faced with continued increases in workmen injury costs in spite of all the safeguarding that was done, realized that more planning, better organization, and more intelligent research were needed if accidents were to be prevented. The larger companies have been most aware of this need. Consequently, as reported by the U.S. Department of Labor, the accident rates and injury costs in the larger companies are substantially lower than in the medium-sized and smaller companies.

10.2 NEED FOR ORGANIZING AND PLANNING THE PROGRAM

The need for planning and for intelligent administration of a safety program and the need for an adequate organization to carry on the many parts of that program are greater today than they were early in the history of the organized accident-prevention programs. There are many reasons for this. There has been a tremendous industrial expansion, and production processes are more complicated and diversified. Almost all states have adopted restrictive safety orders and laws regulating design, type, and operation of industrial equipment. Organized labor has entered the field of safety and has demanded better protection for its members while on the job.

In any industry, the program to prevent accidents or to reduce them to a reasonable minimum and the organization to carry on that program must be intelligently planned. Even in the smallest companies, where the responsibilities of developing production practices, maintaining equipment, purchasing, employee relations, and safety rest in the hands of a single man, the safety program must be planned. The pattern differs from that of a large company mainly in the number of people needed to carry it on. With the average cost of compensation insurance premiums for operating and maintenance personnel ranging from about 2 per cent of payroll in manufacturing industries to 10 per cent and higher in more hazardous industries, there should be no question of the value of a program that will reduce the accident experience and the cost of insurance premiums. Since the inception of organized safety in

American industry, no safety program intelligently planned and diligently executed has failed to produce a good accident experience. This result, however, has been secured only after consideration of the many factors and actions outlined below.

Safety is not an entity separate from production. It is not a field of industry set apart for the accident-prevention specialist. A safety program must be coordinated and integrated with the production program. The same executives who are responsible for low production costs and high production efficiency are responsible for carrying on the accident-prevention program. If the firm employs a safety engineer or a corps of safety engineers, they should be in a staff capacity, and their only authority should be to advise the executives and lower supervisors on problems encountered in providing safe working places and promoting safe practices. They should have no authority to issue orders to workmen or supervisors. A safety engineer is responsible only for the quality and the reliability of the service, advice, and assistance that he gives to line management. As in the case of other staff specialists, such as design engineers, doctors, and purchasing agents, a safety engineer must be sure that the advice he gives is sound and is based on the best known and most thoroughly explored theories and practices.

10.3 FUNDAMENTALS OF SUCCESSFUL ACCIDENT PREVENTION

Success in preventing needless accidents and their attendant personal injuries requires neither a complicated system of policing and controls nor involved scientific knowledge. There are, however, certain premises or fundamentals that must be accepted by industrial management and that, once accepted, must be continually emphasized and filtered down from the top of the organization to every level of management and supervision and to every job performed. The degree of success is dependent upon the degree to which these principles are sincerely accepted and actively promoted by management.

The following are the principles that must be accepted to keep accidents in industry at a reasonable level; they are accepted by people in charge of American industry who have been successful in accident-prevention work.

1. There must be a sincere desire by management to have an effective safety program. The attitude of management will be reflected down through supervision to every job.

2. Participation must be so evident as to leave no doubt in the minds of subordinates about the attitude of management. There is a vast difference in the accomplishments of an accident-control program in which management is visibly active and one in which there is passive acceptance of the idea but no active participation readily apparent to the working force.

3. The fact must be accepted that to prevent accidents money must be spent to provide safeguards, to plan and design safe operating processes and procedures, and to staff an adequate organization to assist line supervisors in carrying out the safety policies that management has established. Experience has proved that every dollar intelligently spent on safety has returned big dividends. This type of spending should be given the same serious consideration that is given requests for funds to replace inefficient or outmoded production equipment and processes. Haphazard procedures and obvious tolerance of unsafe practices are also outmoded and inefficient in modern industry. It should be realized that supplying money to provide mechanical safeguards, good lighting, good ventilation, good safety training, and instruction for workers and supervisors is not so much an indication of leniency by management as it is a necessity—a must in efficient industrial-production practices.

4. There must be an acceptance of the principle that the safety program is necessarily a part of the production-process program. Therefore, the same executives who are charged with the responsibilities of developing and directing efficient production methods must be held responsible for the administration of the safety program. The safety engineers are their staff advisers who assist them in meeting their accident-control responsibilities.

5. Top management should put the accident-control program and the safety-engineer staff on the same plane in the organization of the company as other staff functions, such as engineering, medical, industrial-relations, accounting, purchasing, and cost-control. As a matter of fact, with the exception of the accounting function, the accident-control program is an adjunct to the activities of all the other staff units mentioned and assists them in carrying out their functions.

10.4 ENGINEERING DESIGN OF EQUIPMENT AND PROCESSES

The basic goal of the design engineer during any project is to originate a structurally sound piece of equipment or process with operational perfection and economic justification. But there is also an ultimate goal. An engineering project must not be considered complete unless the following specifications are fulfilled in addition to the basic goal.

1. Provision of sufficient room and proper equipment for the safe storage and handling of materials
2. Location of machines and equipment not only from an operating standpoint but also to allow a maximum degree of safety to the installation and personnel during maintenance operations
3. Location and physical guarding of installations to reduce or avoid hazards
4. Installation of facilities, including walkways, stairways, platforms, and so forth, to provide safe access to otherwise inaccessible locations during operation, maintenance, and emergency conditions
5. Consideration of fire-prevention facilities
6. Consideration of employee facilities in the form of lunchrooms, washrooms, change rooms, toilet facilities, and offices, including the adequacy of lighting, heating, cooling, and ventilating. (See Sec. 6.)

It is evident then that a project design cannot be completed with only the basic goal in mind. The above specifications must be included during the design stage because they are economically justified and they fulfill a design engineer's professional obligation.

The safety engineer or representative can play an important role in liaison with the project engineer during the design stage of a project. In the performance of this work, the functions of the safety engineer may be summarized as follows.

1. Advise project engineers of personnel and equipment safety requirements as early as possible in the course of design work to avoid costly revisions later in the design program or during construction.
2. Assist project engineers in complying with applicable personnel safety codes and regulations pertinent to current project work.
3. Be thoroughly informed concerning current designs in progress in the engineering department so that all safety questions that arise may be comprehensively and quickly answered.
4. Assist project engineers in obtaining operating organizations' approval involving safety in designs as early as practicable in the course of a project to avoid costly revisions. This may be accomplished by assisting the engineer to interpret codes and regulations as they relate to specific designs and by discussing current designs with representatives of the operating organization to assure agreement of interpretations.
5. Assist project engineers by making analyses to determine economic justification of changes or inclusions in designs or construction recommended for safety reasons.

In carrying out these functions, the safety engineer should act only in an advisory-staff category and should not assume any of the basic responsibilities for design. On his own initiative, the safety engineer should spend sufficient time in the engineering organization to keep abreast of design progress on current projects. In effect, he serves as liaison between the engineering organization project engineers, representatives of the operating organizations affected, and his own organization. The safety engineer should therefore continually work with operating organizations to become thoroughly acquainted with operating techniques that may affect design.

The project engineer should then magnify his basic goal to include the responsibility for personnel safety requirements in design. He should seek the advice and assistance of the safety engineer or representative. The safety department may help him by preparing a manual covering safety in designs. The purpose of this manual should be to assemble in convenient form information and pertinent designs so that preferred practices can be incorporated in all installations.

10.5 INCLUSION OF SAFETY IN WRITTEN OPERATING STANDARDS AND INSTRUCTIONS

Although continued vigilance and educational programs, along with safety in designs, are necessary to ensure a successful safety program, we cannot be assured that human error

TABLE 10.1. *RECOMMENDED FORM FOR OPERATING STANDARDS*

	Operating standard no.———
	Operation of————Power Plant
Purpose	1. This standard prescribes the general procedure to be followed in the power generation at ———— Power Plant.
	Safety
General precautions	2. The detailed instructions contained in this standard are designed to promote safety of personnel and equipment and economy of operation in ———— Power Plant. It is important to prevent sudden changes in boiler operating or electric generating conditions which will seriously affect their capacity to carry loads and maintain a constant supply of steam or electricity to meet demands. To prevent such conditions, follow the instructions provided in this standard.
Avoid hazards	3. ————
Rescue work	4. ————
Fire-fighting equipment	5. ————
Respiratory protection	6. ————
Spills	7. ————
Eye protection	8. ————
Clothing	9. Care around rotating machinery, etc.
Hot lines	10. ————
Hot water	11. ————
Steam leaks	12. ————
Electrical	13. Avoid contact, etc.
	Steam Generation
Boiler water	14. Description of process, etc.
High pressure steam	15. ————
Exhaust steam	16. ————
	Starting cold boiler
Inspection	17. ————
Set lines and valves	18. ————
Filling with water	19. ————
Hydrostatic tests	20. ————
	Firing the boiler
Set lines	21. ————
Check gas supply	22. ————

will not interfere and cause serious consequence to both personnel and equipment. This is especially true of complex operations in plants or processes.

Airline and air-force pilots, though highly trained, use check lists and written operating procedures so that they will not have to rely on memory or instinct. Similarly, in the operation of complex plants or installations, where the slightest error in operating procedure may cause extensive damage to equipment and serious injury to personnel, operating standards or instructions should be a necessary part of routine operation. Operating personnel, maintenance personnel, and all others connected with the operation should be thoroughly familiar with these standards and instructions and should be required to use them.

Although the standard or instruction should explain the operation in its entirety, including charts, check lists, and so forth, specific safety items should be included to assure protection of personnel and equipment from designated hazards. These items may include hazardous chemicals and their use, electrical hazards, rotating equipment, rescue procedures, protective clothing and equipment, and fire-prevention and fire-fighting procedures.

Table 10.1 is a recommended form for operating standards. A power plant is used as the subject material since this is a necessary part of many plants or installations, large and small.

In addition to a complete explanation and running check list as outlined in the table, charts and graphs for quick reference should be included for operators' use. For a power plant, such a chart might list the safe limits of temperature and pressure for all equipment.

This summary is merely an example of the type of material that can be used as an aid to operating personnel and maintenance crews in the safe performance of their duties. There are many ways to accomplish this purpose, but it is essential to fortify other phases of the safety effort with a scheme of safety instructions and check charts to be included in operating standards that are guides to personnel in all plants, especially those with complex processes.

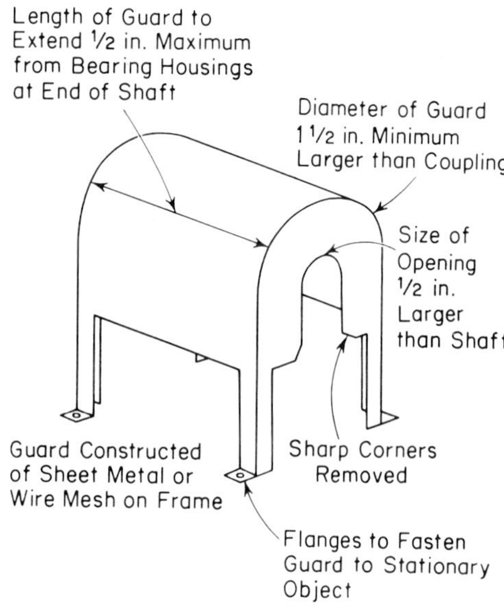

Fig. 10.1 *Typical coupling guard.*

10.6 SPECIFICATIONS FOR SAFEGUARDING

Safeguarding of machines and workplaces in general is an important phase of accident prevention. Manufacturers of machines are now more cognizant of the practice of guarding moving parts. On many machines, self-lubricating or automatic lubricating devices reduce the hazard of personnel contacting moving parts. However, it is essential that the user take steps to assure complete protection for his personnel. Working areas which are by location hazardous should be guarded to eliminate the dangers involved.

Specification for safeguarding all conditions is directly related to the design stage of a project and to the engineer responsible. (Refer to Specification 3, Art. 10.4.) In placing orders for machinery, equipment, and other material, the engineer can gain economical and operational, as well as mechanical, advantages by specifying that it be guarded in compliance with codes and standards.

10.6.1 Types of Guards

When dangerous moving parts of machinery are remote from working areas or are located with reference to a structure so that accidental contact is not possible, they can be considered locationally guarded.

Otherwise, shield guards can be used to eliminate the possibility of accidental contact with the moving parts. For example, Fig. 10.1 shows a typical guard to protect against rotating couplings and keyways on shafts. Figure 10.2 depicts a typical guard for a belt-and-pulley installation that allows for lubrication of parts without removal of guard. This is extremely important since many types of equipment must be lubricated in motion. Another typical belt-and-pulley guard is shown in Fig. 10.3. For larger installations, the guard must extend to at least 7 ft above the ground or platform level to give complete protection. Figure 10.4 shows one type of guard for gear mechanisms. The hinged installation provides easy access for lubrication and assures that the guard will always be in place.

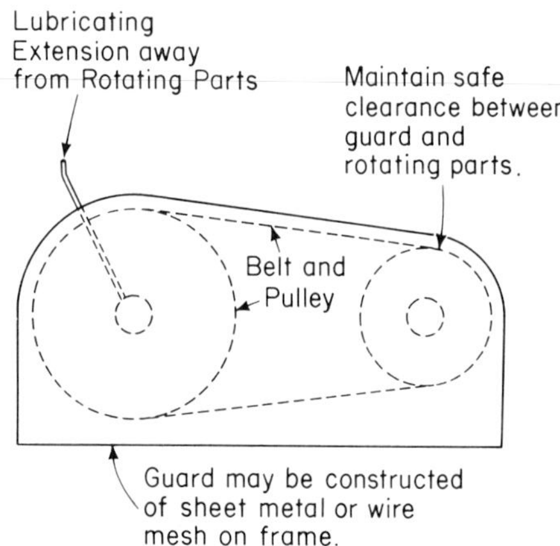

Fig. 10.2 *Typical belt-and-pulley guard.*

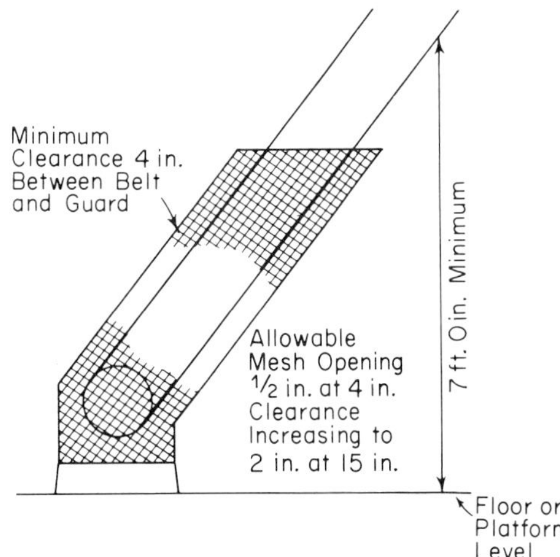

Fig. 10.3 *Typical inclined belt-and-pulley guard.*

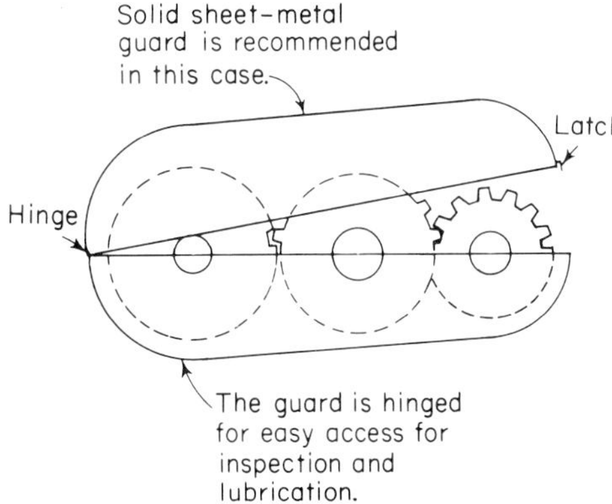

Fig. 10.4 *Typical guard for gear mechanisms.*

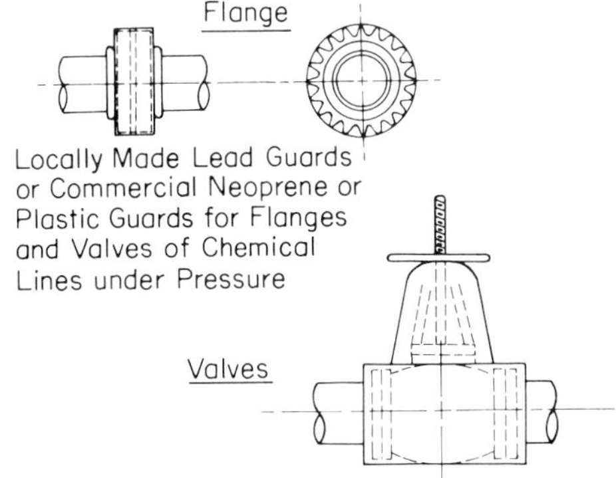

Fig. 10.5 *Typical chemical guard for flanges and valves.*

There are many other types of mechanisms involving hazardous moving parts that it is necessary to safeguard. Such installations as power conveyors, fans on internal-combustion engines, counterweights, and flywheels are a few hazards to consider. Surfaces hot enough to burn animal tissue on momentary contact should be guarded by insulation or made inaccessible by barricade. Adequate guards should be placed around fittings, such as flanges and valves, in which corrosive chemicals are handled under pressure. Figure 10.5 shows such a guard.

10.6.2 Means of Access

Operating and maintenance areas that are located in permanently inaccessible locations and that require ordinary operating attention and/or frequent repair, adjustment, or handling should be provided with platforms, maintenance runways, walkways, or ramps accessible by means of ladders or, if possible, by stairways. In choosing between a ladder and a stairway, the following points should be considered.

1. Number of times used per day
2. Necessity for carrying tools or equipment when ascending or descending
3. Age and physical condition of personnel who will use the installation

Every stairway with four or more treads should have a handrail. Figure 10.6 shows a typical stair railing with recommended dimensions. Note that the stair railing ties in smoothly with the platform or landing handrail. Figure 10.7 depicts the typical handrail for platforms, walkways, runways, or ramps which are 4 ft or more above the ground. The toeboard is not required for elevations under 6 ft.

The preferred slope of stairways is shown in Fig 10.8. All stairways on any one installation should have the same slope.

Ramps should be used as little as possible if stairways and a straight run of walkway can be used instead. Figure 10.8 indicates preferred slopes of ramps, and Fig. 10.9 shows a typical ramp with recommended dimensions.

Where stairways are not possible or practicable, fixed ladders should be used. Vertical ladders are recommended. Figure 10.10 shows a typical fixed ladder with opening to platform. This opening should be guarded with a drop bar or a chain guard attached to the handrail.

Vertical fixed ladders over 30 ft in length or

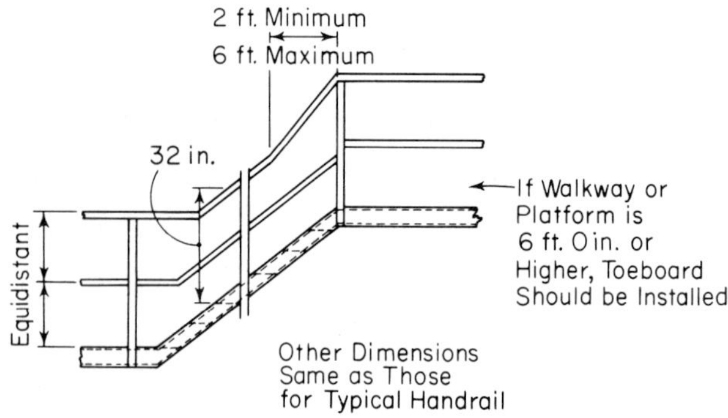

Fig. 10.6 *Typical stair railing.*

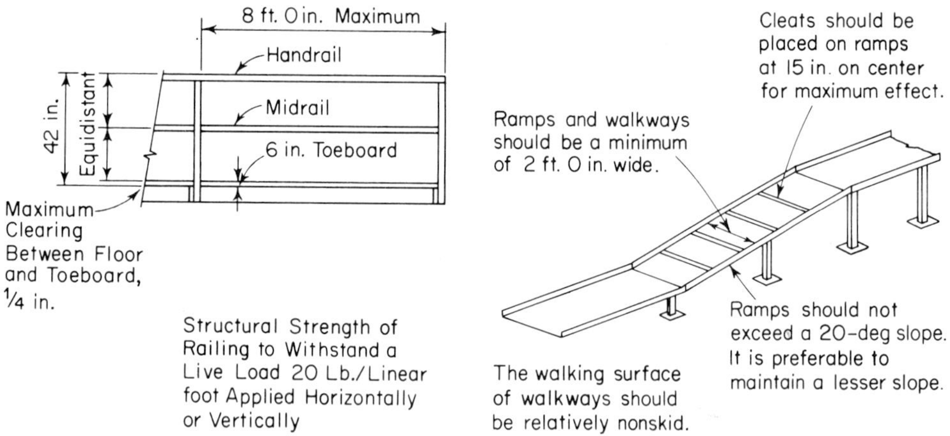

Fig. 10.7 *Typical handrail.*

Fig. 10.9 *Typical ramp.*

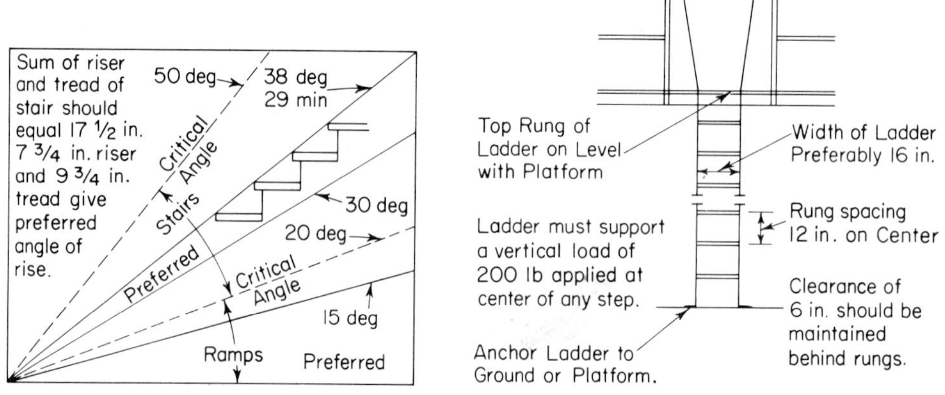

Fig. 10.8 *Preferred slopes for stairs and ramps.*

Fig. 10.10 *Typical fixed ladder.*

30 ft or more above the ground should be supplied with a backscreen or ladder cage. Figure 10.11 depicts a typical ladder cage.

In addition to the structural phase of designing walkways, stairways, and platforms, the nonskid qualities of walking surfaces must be considered. Nonskid surfaces are especially important around equipment or installations exposed to oil, water, ice, or other slippery conditions. Any material that may become a hazard through normal wear and tear should be avoided for use in walkways.

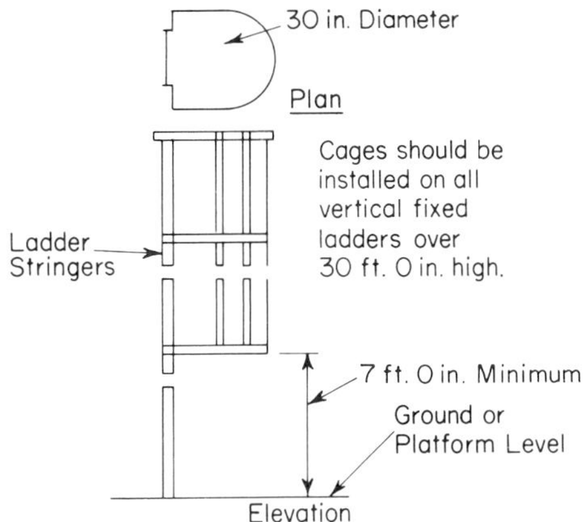

Fig. 10.11 *Typical ladder cage.*

10.7 ADDITIONAL ENGINEERING FACTORS IN SAFETY

Many safety-in-design features have been discussed that deal with the more common problems encountered in project design and operating and maintenance practices. The safety aspects of several other conditions require special handling. This section deals with these special conditions.

10.7.1 Electrical

Only the basic requirements for safety in electrical designs are presented here. They must be supplemented by city, county, state, and national codes in force where designs are originated. Where the codes are not stringent, these basic requirements should govern.

All electrical materials, devices, and appliances used in electrical installations should be of an approved type listed or labeled by the Underwriters' Laboratories, U.S. Bureau of Standards, U.S. Bureau of Mines, or other institutions of recognized standing. When questions arise concerning material or equipment not approved by a specific organization, the safety engineer should be contacted for advice.

Bonding and grounding techniques should follow good practice to prevent hazard from static electricity, short circuits, and stray currents. The following practices are especially important.

1. Frames and all exposed metal parts of portable electric hand tools which do not carry current should be grounded.

2. Exposed, non-current-carrying metal parts of all fixed electrical equipment, such as motors, generators, and control equipment, should be permanently grounded when in hazardous locations; when equipment is within reach of a person who can make contact with any grounded surface or object; when equipment is supplied by metal-clad wiring; and when equipment is operated at more than 150 volts to ground.

3. The metal guards of portable lamps must be grounded with a three-conductor cord and two-pole, three-wire receptacle unless the base and handle of the lamp are made of a nonconducting material.

There should be enough space around electrical equipment to allow ready and safe operation. This work space should not be less than 2½ ft. Around equipment such as switchboards and control panels, it should be increased as voltage ratings increase. Applicable codes should be closely followed in these instances.

Another important factor in the selection of equipment for electrical installations is the location involved. Locations can be divided into three classifications that should be included in the plant general-engineering specifications for electrical equipment.

Type 1: Locations where hazardous concentrations of flammable vapors or gases, volatile liquids, continuously suspended combustible dust, or easily ignitible fibers or materials are anticipated during normal operations

Equipment required: Use equipment with *all* electrical parts in an enclosure capable of withstanding an internal explosion without igniting flammable vapors outside.

Type 2: Locations where these conditions

TABLE 10.2. *CLEARANCES FOR SUPPLY CONDUCTORS*

Nature of Clearance	Wire or conductor concerned					
	A Span Wires (Other Than Trolley Span Wires), Overhead Guys, and Messengers	B Communication Conductors (Including Open Wire, Cables, and Service Drops), Supply Service Drops of 0–750 v	C Trolley Contact Feeder and Span Wires of 0–5000 v	D Supply Conductors of 0–750 v and Supply Cables	E Supply Conductors and Supply Cables of 750–20,000 v	F Supply Conductors and Supply Cables of More Than 20,000 v
1. Crossing above tracks of railroads which transport or propose to transport freight cars (maximum height 15 ft 1 in.) where not operated by overhead contact wires	25 ft	25 ft	22 ft	25 ft	28 ft	34 ft
2. Crossing or paralleling above tracks of railroads operated by overhead trolleys	26 ft	26 ft	19 ft	27 ft	30 ft	34 ft
3. Crossing or along thoroughfares in urban districts or crossing thoroughfares in rural districts	18 ft	18 ft	19 ft	20 ft	25 ft	30 ft
4. Above ground along thoroughfares in rural districts or across other areas capable of being traversed by vehicles or agricultural equipment	15 ft	15 ft	19 ft	16 ft	25 ft	30 ft
5. Above ground in areas accessible to pedestrians only	7 ft	10 ft	19 ft	12 ft	17 ft	25 ft
6. Vertical clearance above buildings and bridges (or other structures which do not ordinarily support conductors and on which men can walk) whether attached or unattached	8 ft	8 ft	8 ft	8 ft	12 ft	12 ft
7. Horizontal clearance of conductor from buildings (except generating and substations), bridges, or other structures (upon which men may work) where such conductor is not attached thereto	—	3 ft	3 ft	3 ft	6 ft	6 ft
8. Distance of conductor from center line of pole, whether attached or unattached	—	15 in.	15 in.	15 in.	15 or 18 in.	18 in.
9. Distance of conductor from surface of pole, crossarm, or other overhead line structure upon which it is supported, providing it complies with Item 8 above	—	3 in.	3 in.	3 in.	3 in.	—

may occur only as a result of abnormal conditions or equipment failure

Equipment required: Equipment with arcing or sparking contacts must be in an enclosure capable of withstanding an internal explosion without igniting flammable vapors outside. Equipment which produces no sparks that will ignite explosive vapors under normal operating conditions can have general-purpose enclosures.

Type 3: Locations where there is little or no hazard from flammable vapors

Equipment required: There are no restrictions regarding the use of sparking equipment in nonhazardous locations.

Clearances for supply conductors, guy wires, messenger wires, and trolley wires from the top surface of rails, buildings, thoroughfares, and other objects are essential for safe installations. It is recommended that Table 10.2 be followed in this case.

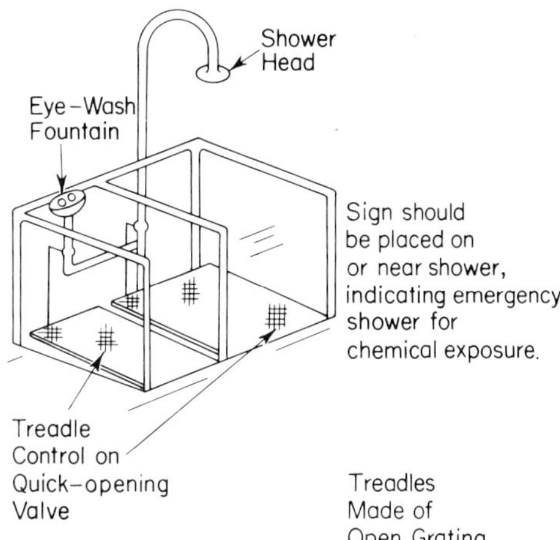

Fig. 10.12 *Typical emergency shower and eye-wash fountain.*

10.7.2 Chemical

Although some chemicals are harmless, most of them are hazards to personnel safety and must be closely guarded. An injury caused by chemical exposure can be extremely serious.

Protection against chemical exposure or contact by personnel can be accomplished to a great degree in the design stage of an installation. Further protection in the form of protective clothing, respiratory protective devices, or emergency devices is essential. (See Fig. 10.12.)

Proper guarding of equipment will eliminate most accidental contacts with hazardous chemicals. In preparing the specifications for equipment that will handle corrosive chemicals, the following factors should be thoroughly considered.

1. Equipment or pipelines that are to handle corrosive chemicals should be resistant to the corrosive action.

2. Where it is necessary to provide valves, flanges, or other fittings in lines handling corrosive chemicals, lead guards locally fabricated or commercial neoprene or plastic guards should be considered for installation on the fittings. Where possible, use tongue and groove fittings. See Fig. 10.5 for typical chemical guard installations.

3. Pumps, siphons, or other equipment should be guarded. The liquid end of such pumps should be entirely enclosed.

4. Special chemical-resistant packings must be used in valves, pump glands, and so forth, as added protection.

5. Sampling devices for removing samples of corrosive chemicals from closed systems should be of special design to prevent splashing and dripping.

Identification of chemicals in equipment is an essential part of providing safe operating facilities in plants or installations. All vessels or equipment in which corrosive chemicals are stored or used should be identified with durable signs that give the name of the chemical and a warning, such as "Caution." This not only aids operators in their performance but also provides additional warning to maintenance personnel working with this equipment. Pipelines that carry corrosive chemicals should be identified by stenciling the name of the commodity and the direction of flow on the line or by tagging the line. Since yellow is generally considered a caution color, a yellow background with black lettering is recommended for these identifying signs. Marking pipelines at the plot limits of plants or installations assures correct identification by operators and maintenance personnel of where the pipeline enters the operating area.

Storage of corrosive chemicals must be carried out with systematic care. All containers must be plainly labeled, and storage areas must be plainly designated by signs. Small containers of corrosive chemicals should not be stored on high storage shelves, and all storage areas should be cool, well ventilated, and remote from acute fire hazards. Incompatible chemicals must never be stored and handled so that contact is possible. For example:

1. Never store acetone next to concentrated nitric and sulfuric acid mixtures.

2. Never store copper near acetylene or hydrogen peroxide.

3. Isolate fluorine from everything.

4. Never store sulfuric acid near chlorates, perchlorates, or permanganates.

In designing storage facilities for corrosive chemicals, these problems should be thoroughly considered. In plants or installations where corrosive chemicals are used to a great degree, protective clothing is an essential part of the operating and maintenance personnel equipment. Such equipment must be maintained in perfect condition and should be cleaned and sterilized after each wearing.

10.7.3 Air Pollution

When dusts, mists, fumes, gases, or vapors are produced in quantities that may be harmful to personnel, control should be maintained by exhaust ventilation systems at the point of generation. These are essential when general ventilation proves ineffective or when elimination or prevention of the source of pollution is not possible or practicable.

The exhaust system must be designed and operated to maintain a volume and velocity of exhaust air to convey all impurities to points of safe disposal. The system must not draw the impurities through the breathing zone of workmen and must not exhaust them into working places where they can cause harmful exposure.

Stacks from equipment that produces quantities of smoke or harmful vapors should be located so that harmful quantities are not spread over working areas and public places. Where it is impossible to eliminate this condition by location, smoke eliminators should be designed into the equipment.

Where air-pollution conditions cannot be entirely eliminated from working areas, personnel should be required to wear respiratory protective equipment. Information on the maximum safe concentration of air contaminants can be obtained from any good publication on noxious gases and vapors.

10.7.4 Welding

The many types of welding operations create numerous hazards to personnel and equipment unless proper safeguards are taken. For instance, the harmful light rays produced by welding flames and arcs may seriously injure the eyes or burn the skin. Similarly, poisonous fumes and gases produced in welding operations may cause serious illness. Because of the many possibilities for injury to personnel and for damage to equipment in welding operations, it is essential that safe practices, regulations, and explicit operating standards be issued and be strictly observed.

Of primary importance is the strict adherence to a standard that permits only authorized personnel to perform welding operations and that requires all personnel involved to use suitable personal protective clothing, such as gloves and aprons or shields. Goggles must be carefully selected on the basis of lens shade to give proper protection for the welding operation involved.

In areas designated as welding-operating locations, ventilation must be adequate to carry away harmful concentrations of fumes that may be generated. In most cases, local exhaust ventilation at the point of welding is necessary, especially in confined spaces.

Welding-operating areas must not be located near flammable or explosive substances. All flammable material should be removed from the immediate vicinity and wooden floors or walls should be avoided or adequately protected by fire-resistant shields. Tanks, cylinders, or other containers must never be welded until it has been assured that they do not contain, or have not contained, flammable or explosive substances. Since the heat of welding may cause a sufficient increase in pressure to rupture or explode an unvented container, sealed containers must be vented.

In *oxyacetylene welding* the oxygen, the acetylene, and the flame require special handling to prevent injury to personnel or damage to equipment. The following precautions should be a required standard.

1. Grease or oil must not be allowed to come in contact with welding equipment.

2. Acetylene should not be used from cylinders at pressures in excess of 15 psi.

3. All cylinders should be provided with protective caps when not in use.

4. Cylinder valves must be immediately accessible when cylinders are in use to permit quick shut-off in emergency.

5. Hose must be protected from mechanical damage and must be periodically inspected for leaks by being submerged in water.

6. Torch must not be left burning when not in use, and its valve should be shut when not in use.

Electric arc welding presents the additional hazard of electrical contact and fire from electrical sources. The precautions necessary for electric welding are as follows:

1. Shields of curtains or screens must be provided around arc-welding locations to prevent injury to eyes of personnel in surrounding area.
2. Circuits must be checked only when system is dead.
3. The polarity and rotary switches must not be operated while equipment is under load.
4. Motor-generator and other electric-welding apparatus must be provided with adequate ground.
5. Cables must be protected from mechanical damage and any loose connections or defective electrode holders must be replaced.
6. Repairs must be made on this type of equipment by qualified electrical maintenance personnel.

10.7.5 General

There are many other engineering factors related to safety. All must be considered in an active accident-prevention program. In general, when all phases of a design or operation are considered (including safety), operating perfection will be assured and the plant will be safely operated and maintained.

10.8 SPECIFICATIONS FOR PERSONAL PROTECTIVE EQUIPMENT

Many hazards to workmen cannot be entirely eliminated. To avoid injury in such cases, the workmen must use personal protective equipment such as goggles, respirators, and safety hats depending on the hazard. Management must have this equipment available at all times, and provide strict supervision to assure use of the equipment by workmen.

Many brands of personal protective equipment are available, but not all of them are made in accordance with the standards set by such agencies as the Bureau of Standards, Bureau of Mines, and American Standards Association. These organizations determine the specifications to which protective equipment should be made and, in some instances, test the devices and issue specific approvals. Only personal protective equipment that meets the requirements of these standards should be used.

In order to determine which equipment is satisfactory, the purchasing agents, those who initiate requisitions, should be provided with a guide. This guide should be issued by the safety department and should be kept up to date in order to take advantage of new devices that are made available. The safety department should determine the proper standard and should list several brands or styles of equipment that meet the standard. The purchasing agent should buy only the approved types or brands.

10.8.1 Eye Protection

Protection for the eyes is one of the most important of the personal safety devices. There are many types of goggles available, and care should be taken to assure that the right type is used for a particular hazard. Hardened lenses are made in accordance with standards set up by the National Bureau of Standards.

Cup-type goggles protect the eyes from all angles and should be used where complete protection from heavy flying particles is needed—for example, in riveting, chipping, grinding, and heavy hammering. They are made in two general styles, one of which is for use as a cover goggle worn over corrective spectacles. The frames are shaped and vented to accommodate such hazards as dust, heavy particles, or chemicals.

Spectacle safety goggles can be used by shop men and other workers who require protection from light flying particles. These goggles provide protection from particles coming from the front and can be obtained with side shields to protect against particles coming from the side. They come in a full range of sizes and can be fitted in much the same manner as corrective spectacles. Proper fitting is extremely important in the use of this equipment.

For those who must wear corrective spectacles and need eye protection during their work, it is possible to have the individual's prescription ground in hardened safety lenses. These provide full-time protection and in many jobs eliminate the need for the heavier type of goggle normally used over prescription spectacles. Many employers pay the full cost of the spectacles, and others assist the employees by paying a portion of the cost, but in either case they seldom pay the cost of the refraction.

Both gas and electric welding require special eye protection to guard against the effects of the ultraviolet and infrared rays. The lenses of the goggles or welding helmets are made in various shades required for protection against rays. The filtering qualities for each shade are determined by the specifications set up by the National Bureau of Standards.

10.8.2 Respiratory Protection

Various types of protection against the breathing of harmful dust, fumes, or gases are available. The use of respiratory equipment should be regarded in most cases as emergency protection against casual or relatively brief exposure. Certain work, such as that done by sand blasters, may require continuous use of the equipment. However, every effort should be made to reduce the concentration of gases, vapors, or dust to such a level that the air is safe to breathe for 8-hr periods.

When the hazard cannot be completely eliminated, one of the following types of respiratory equipment should be used. The U.S. Bureau of Mines has set up rigid standards and tests for respiratory protective equipment and issues certificates of approval to manufacturers meeting these standards. Use of such approved equipment assures full protection when the equipment is properly used for the hazard involved.

Mechanical-filter respirators are designed to remove "particulate" matter such as dust, fumes, and mists from inhaled air by mechanical filtration. They will not provide protection against vapors or an atmosphere deficient in oxygen.

Chemical-cartridge respirators are designed for respiratory protection against organic vapors which are not immediately dangerous to life but which may produce discomfort, a chronic type of affliction or poisoning after repeated exposure, or mild acute symptoms after prolonged exposure. The Bureau of Mines approves this type of respirator for protection in atmospheres containing not more than $^1/_{10}$ of 1 per cent organic vapors by volume (1000 parts per million) which are not immediately dangerous to life. A partial list of such vapors includes acetone, alcohol, amyl acetate, benzine, benzol, carbon disulphide, carbon tetrachloride, chloroform, creosote, ether, hexane, lacquer, naphtha, pentane, phenol, toluene, trichlorethylene, and turpentine.

Canister gas masks are used for protection against harmful gases, vapors, fumes, mists, dusts, or smokes. All the air is drawn through a canister which filters, absorbs, neutralizes, or catalyzes limited amounts of harmful materials. Since the canister serves only to remove gases from the air and will not supply oxygen, this type of gas mask will not protect against suffocation if worn in atmospheres deficient in oxygen. Canisters are provided for use as protection against specific materials in concentrations of not more than 2 per cent (3 per cent in case of ammonia). Canister gas masks are intended for use in the open and should not be worn as protection inside tanks or other confined places where higher concentrations may be encountered.

Supplied-air respirators provide protection by conveying respirable air to the wearer through hoses, tubes, piping systems, air holders, or combinations of these. Various hose masks, airline respirators, compressed-air respirators, self-contained breathing apparatus, sand-blast helmets, and hoods are used for protection against specific hazards. When the air is supplied by mechanical compressors, purifiers should be provided in the line to remove carbon monoxide, which may be generated if the compressors are not in good mechanical condition or if they have their air intakes near engine exhausts. Only hose masks with positive pressure blowers or self-contained breathing apparatus should be used in immediately harmful atmospheres or in atmospheres from which the wearer could not escape without the aid of the equipment.

10.8.3 Head and Face Protection

Safety hats have prevented many injuries, particularly in the construction industry. They should be worn by all workers exposed to falling objects. The objects need not be large to cause serious injury. Small bolts and rivets have been known to fracture unprotected skulls.

Safety hats are made of metal and of glass, plastic, and other nonmetallic materials. So long as the hats comply with the test specifications of the National Bureau of Standards, the type to be used can be determined by factors of cost, weight, headband suspension, color, and so forth. In some work, such as electrical work, the hats should be of nonmetallic material. Safety hats equipped with plastic visors provide eye and face protection as well as head protection.

Face shields provide satisfactory protection

for certain hazards, such as small, flying particles or splashes of liquids; however, they are not acceptable for most uses where safety goggles are required for eye protection since the plastic shields will not withstand the impact of heavy, fast-flying objects. The shields are made of plastic of various thicknesses and degrees of scratch resistance. The shield should have a minimum thickness of 0.04 in. Face shields are also available in green and other shades for protection against excessive light. The plastic material must not be of the type that is easily ignited or that burns rapidly.

For full protection of the head from such hazards as extreme heat and spray painting, hoods that cover the head are best. Many types are available for various kinds and degrees of hazard. For use in abrasive blasting work, the hoods must provide respirable air to the wearer while protecting his head and shoulders against impact and abrasion by the rebounding sand or shot used in the process. An air-supplied hood, an airline respirator, or a mechanical-filter respirator may be used, depending on the degree of hazard. When sand-blasting operations are conducted in a room or enclosure, the operator and helpers should be protected with supplied-air equipment.

10.8.4 Protective Clothing

Workers who handle hot objects, rough materials, or chemicals should wear gloves made for the specific hazard involved. Usually, gloves are worn along with other body protection, such as aprons, coveralls, hoods, goggles, and shields, since the hazard often involves other parts of the body.

The American National Standards Institute publishes specifications for certain types of protective clothing, and these should be used whenever applicable. If no standard applies, or if the effect of the hazard on the protective material is not known, trials and tests should be made to ascertain that the material will actually provide satisfactory protection.

Safety shoes with protective toe caps provide excellent protection against toe and foot injuries. They should be worn in any occupation where there is danger of objects falling or being dropped on the workers' toes. In order to encourage their use, many employers make these shoes available—sometimes at prices below cost. Safety shoes are made in many styles, from dress shoes to heavy-duty boots.

10.9 SAFETY INSPECTIONS

Full success of a planned accident-prevention program cannot be realized unless sincere and determined efforts are made to discover hazardous situations through a positive safety-inspection system. Safety inspections are too often confined to detecting unsafe physical conditions, such as lack of safeguards, ungrounded electrical equipment, and the like. Inspectors should be trained to observe work practices as well as work conditions during an inspection tour because 90 per cent of all accidents involve an unsafe personal act—some act that should not have occurred, or some action that should have been taken but was not.

Safety inspections should not become a police action. Inspection plans that point a finger at someone are doomed to failure; once this type of program has been used, considerable time and effort are required to overcome the bad effects produced. A safety inspection should be a cooperative effort. It should use the various personal talents available and should be aimed at improving the accident situation by finding and correcting unsafe conditions and educating supervisors and workmen in safe work practices.

The line supervisor should make frequent inspections of his area and should welcome inspections from others since it is difficult for one person to see or notice everything about an area. He is responsible for the safe working conditions in his area or department. When safety inspections are made by someone other than the supervisor of the area, as they sometimes are, the line supervisor must not feel that the inspection party has assumed his responsibilities or that his prerogatives have in any way been challenged since inspections help the line supervisor carry out his responsibilities in this area.

Several different methods are used to set up procedures or guides for conducting safety inspections. One plan that has been successful is outlined in the following paragraphs. In considering safety inspections, basic safety can be reduced to eight simple general items: five concerning conditions to be checked and three concerning actions that can be taken to remedy the hazardous conditions found.

The five general categories to be inspected for unsafe conditions include the work area, the material handled, the hand tools used, the machines used, and the personal protective equipment required. The brief breakdown of the five categories that follows can be enlarged upon as supervisors gain experience in planned inspections. Printed on a check list, they serve as effective reminders for members of an inspection party. This breakdown lists the items to be observed for unsafe physical conditions and unsafe practices or personal acts.

Work area
1. Housekeeping
2. Job layout
3. Aisles and exits
4. Stairs, ramps, and ladders
5. Floors and other working levels

Materials handled
1. Piling and storage
2. Loading and unloading
3. Racks, platforms, and bins
4. Hand signals
5. Cranes, chains, hoists, slings, etc.

Hand tools
1. Condition of hand tools
2. Improper use of hand tools
3. Inspection and maintenance of hand tools
4. Difficulties, if any, in securing hand tools
5. Housekeeping as related to hand tools

Machines used
1. Lack of physical safeguards
2. Condition of physical safeguards
3. Actual protection afforded by safeguards
4. Location of controls
5. Mechanical condition of machinery and equipment as it affects safety

Personal protective equipment
1. Goggles and other eye protection
2. Respirators, fresh-air hoses, masks, etc.
3. Equipment provided but not used
4. Shoes, hats, gloves, special clothing
5. Difficulties, if any, in securing equipment

The three actions that can be taken when an unsafe personal act or unsafe condition is noted are:

1. Remove the hazard.
2. Protect the worker against the hazard.
3. Develop safe practices to avoid injury from the hazard.

Safety inspections do not accomplish their purpose unless immediate and positive action is taken to correct the unsafe conditions or practices revealed. If the individual making the inspection has the authority, he should take immediate action personally, or if the situation requires higher authority he should pass the information on for immediate disposition.

10.10 INVESTIGATION OF ACCIDENTS

10.10.1 Purpose and Organization

Successful accident-prevention programs include three fundamental activities: inspection of working areas, thorough study of all operating methods and practices, and education of employees to minimize human errors. Because of our human failings, accidents often occur even when these three functions have been stressed heavily. It is then that a fourth activity, although a secondary defense against accidents, becomes a vitally important function of the accident-prevention program. This activity is a thorough, impartial, and honest investigation of all the circumstances leading up to an accident.

The primary purpose of accident investigations is to determine the true basic cause of the accident for the express purpose of taking remedial action to prevent a recurrence and to remedy the weakness in one or more of the safety-program activities. Never should investigations be conducted for the purpose of placing blame. Investigation held for this purpose will do irreparable damage to the program. With the discovery of weak spots in the three previously mentioned fundamental activities, action can be taken to bolster this phase of the program.

In order to obtain an accurate picture of what occurred, investigations should be held without delay. Delay often results in failure to detect the true basic cause of the accident. Immediate investigations allow for first-hand observance of the condition of the work area and tools, and, most important, they help eliminate colored stories told by witnesses. A witness can see just so much as an accident occurs, and he is likely to become curious about what else happened. Being human, he has a natural desire to get the complete story —and he usually gets his information from others. Then he draws his own conclusions on the cause of the accident. In delayed investigations, the witness sometimes gives information based not wholly on what he saw but rather on the conclusion he has made on the cause. Such evidence could lead to the selection of an improper basic cause that would reduce the value of the investigation.

The line supervisor is best qualified to investigate accidents because he knows the situation that resulted in the accident. He is closest to all the job conditions, such as the purpose of

EMPLOYER'S REPORT OF INDUSTRIAL INJURY GO-42

DO NOT WRITE HERE

DO NOT WRITE HERE
No.

Every work injury to an employee which causes disability lasting longer than the day of the injury or which requires medical services other than first aid treatment must be reported within five days after the injury. If the injury results in death, a report must be made by telephone or telegraph to the Home Office not later than 24 hours after death.

I. EMPLOYER

1. Name (Give name under which concern does business)
2. Office address (No. and Street) (City or Town)
3. Nature of business

II. INJURED EMPLOYEE

4. **Name** Last First Middle Service Date **Yrs. Srvc.** Immediate Supervisor
5. Address (No. and Street) (City or Town)
6. **Age** 7. Sex: Check (✓) **Male** **Female** 8. Check (✓) **Married** **Single**
9. Number of hours worked per day; per week Number of days worked per week
10. Wages: $ per Average weekly wages: $
11. If board, lodging or commissions, furnished in addition to wages, give estimated value per month: $

III. ACCIDENT

12. (a) Place of accident (No and Street) (City or Town) (County)
    (b) If injured in connection with Joint Operations, give name of operation
13. On employer's premises (Yes or No) If answer is yes, give specific location
14. Department
15. **Date of accident** 16. (a) **Hour of day** **A.M./P.M.** 16. (b) **Day of Week**
17. Did injury result in disability beyond day of accident? (Yes or No) 18. If yes, give date last worked
19. Was injured paid in full for this **day**? (Yes or No) 20. Not in a mine

    Give names of witnesses

IV. CAUSE OF ACCIDENT

21. (a) **Occupation (job title)** 21. (b) **Hrs. on Duty**
22. **How long employed by you at this occupation? Check (✓) less than 3 mos.** **3 mos.-6 mos.** **6 mos.-1 yr.** **1 yr.-2 yrs.** **2 yrs.-5 yrs.** **5 yrs.-10 yrs.** **10 yrs.-15 yrs.** **15 yrs-20 yrs.** **20 yrs.-25 yrs.** **over 25 yrs.**
23. What was employee doing when accident occurred? (Describe briefly, such as loading truck, operating drill press, shoveling dirt, walking down stairs, etc.)
24. How did the accident happen? (Describe fully, stating whether the injured person fell, was struck, etc.; give all factors contributing to accident. Use other side of report for additional space.)
25. What machine, tool, substance, or object was most closely connected with the accident? (Name the specific machine, tool, appliance gas, liquid, etc., involved.)
26. If mechanical apparatus or vehicle, what part of it? (State if gears, pulley, motc., etc.)
27. Were mechanical guards, or other safeguards provided? (Yes or No) 28. Was injured using them? (Yes or No)
29. What do you recommend for preventing this type of accident? (State the specific preventive measures that can be taken by employer and workers. Do not say "By being more careful." Specify what should or should not be done.)

V. NATURE OF INJURY AND PART OF BODY AFFECTED

30. (Describe in detail the nature of the injury and the part of the body affected. For example: amputation of right index finger at second joint, fracture of ribs, lead poisoning, dermatitis of left hand, etc.)
31. Name and address of Physician
32. Name and address of hospital
33. Has employee returned to work? (Yes or No) 34. If yes, give date 35. At what wage? $ per
36. Did injury result in death? (Yes or No) 37. If yes, give date
38. In case of death, give name and address of nearest relative

DEPT'S. USE

Signed by
General Superintendent

Div. or Dist.

Date

SAFETY DIV. USE

Signed by

Official Position

Date

Filing of this report is not an admission of liability. ". . . No report of injury required to be filed by an employer or an insurer by this chapter shall be admissible as evidence in any adversary proceeding before the Industrial Accident Commission." Labor Code, Section 6413.

(SUPERVISOR MUST COMPLETE REVERSE SIDE OF FORM FOR ALL ACCIDENTS)

CODE	DO NOT WRITE IN THIS COLUMN
	Case No.
DEPT	Employer No.
UNIT	Industry
	Age
	Sex and Marital Status
	Weekly Wage
	County
	Accident Date
NATURE WORK	Occupation
TYPE ACCID.	Accident Type
EVENT	Agency
IMM CAUSE	Agency Part
UNDER REASON	Mech. Defect
	Unsafe Act
	Personal Defect
NATURE INJURY	Nature of Injury
BODY PART	Location
DEGREE DISAB	Extent of Injury
TAB	Insurance Carrier
	Report Lag
CODED BY	Coded By

Fig. 10.13 *Employer's report of industrial injury.*

the job, correct job procedures, equipment and materials used, and working conditions, and, of vital importance, he knows the employees. This knowledge fits him to arrive at the correct basic cause of the accident. Other people, such as a safety engineer, can help him conduct the investigation, but the line supervisor should not feel that his responsibility has been assumed by the other party. Committee investigations often prove to be of benefit. The line supervisor should, of course, be a member of the committee. This assignment proves to be a medium of education for the foreman, especially if his superior is also a committeeman. Having higher management represented on the investigating committee is beneficial in two ways: It allows management a chance to become better acquainted with conditions in the field, and it is visual proof to the employees that management has their interest at heart.

ACCIDENT ANALYSIS

The Supervisor for whom the injured worked shall investigate the accident at once and complete this side of the Form.

VI. Cause (Check Only One Immediate Cause)

A—Unsafe Condition

☐ 1. Equipment ineffectively guarded or unguarded.
☐ 2. Improper type or poor design of tools, equipment or materials.
☐ 3. Defective tools, equipment or materials.
☐ 4. Defective motor vehicle equipment.
☐ 5. Faulty plant facilities. (Improper or no emergency exits, poor light or ventilation, poor layout.)
☐ 6. Poor housekeeping. (Congestion, improper piling and storing, scattered tools, equipment and material.)
☐ 7. Slippery surfaces.
☐ 8. Improper clothing.
☐ 9. Physical disability.

B—Improper Action

☐ 1. Short-cut, improper route, path or walk.
☐ 2. Proper mechanical aids such as barrel trucks, skids, hoists, etc., not used.
☐ 3. Proper personal protective equipment such as goggles, respirators, masks, etc., not used.
☐ 4. Improper tools or equipment used.
☐ 5. Horseplay or fooling.
☐ 6. Improper posture for lifting or straining.
☐ 7. Improper position in relation to work (including improper placement of hands and feet).
☐ 8. Improper operation of motor vehicle, power driven equipment or bicycle.
☐ 9. Improper method or procedure.

VII. Proposed Correction (Check the One Correction that you feel will do most to remedy the Cause selected.)

C—By Direct Supervision

☐ 1. Give adequate or complete job instruction.
☐ 2. Enforce rules, standards or instructions closely.
☐ 3. Provide sufficient or better personal safety devices (goggles, safety belts, masks, respirators, etc.)
☐ 4. Provide correct or safe tools or equipment.
☐ 5. Provide safe plant facilities or equipment.
☐ 6. Review and correct inspection procedure.
☐ 7. Review and correct job planning.
☐ 8. Regulate pace of job.

D—By Counselling and/or Placement

☐ 1. Change employee's duties or assignment to be more compatible with his physical and/or mental abilities.
☐ 2. Adequately train employee.
☐ 3. Counsel employee on haste with which he works.
☐ 4. Reprimand or take other disciplinary action.
☐ 5. Counsel employee on attention to detail of job.
☐ IF CORRECTION APPLIES TO FELLOW EMPLOYEE, CHECK HERE ALSO.

E—☐ Other

☐ 1. Contact contractor, customer or other member of public to effect correction.

VIII. State Why the Above Cause was Selected.

IX. Elaborate on Corrective Action Taken. Point Out Any Contributing Causes and What Has Been Done Regarding Them.

X. .... SUPERINTENDENT **REPORT PREPARED BY** .... SUPERVISOR **DATE** .... 19......

GO-42 (50M-8-62)
PRINTED IN U.S.A.

Fig. 10.14 *Reverse side of employer's report of industrial injury.*

10.10.2 Investigation Procedures

Of the various procedures in use, one that has proved itself is briefly submitted in the following paragraphs for use as a guide.

1. *Care for the injured.* After getting safely to the accident scene, the investigator's first concern should be to see that the injured person is properly cared for. He should apply first aid if necessary and should arrange for immediate medical and ambulance service.

2. *Interview the injured.* If the injured person's physical and mental condition permits, he should be interviewed on the details of what he was doing and the sequence of events leading up to the accident. If he is not in condition

to be questioned, he should be contacted later as soon as his condition permits.

3. *Interview the witnesses.* Each witness should be interviewed in detail to obtain his version of all events leading up to the accident. Any discrepancies found in comparing witnesses' accounts of what happened should be immediately investigated to find the true facts.

4. *Notice the details of the work area.* All details of the work area, such as position of tools, lines, equipment, housekeeping, and unusual conditions, should be recorded.

5. *Make detailed sketches if necessary.* If details are complicated, sketches often make the picture of what happened stand out much better. For reports, sketches paint the picture more easily than words. Photographs often help in this respect.

6. *Draw conclusions.* From the detailed information produced by the investigation, conclusions can be reached. If they are derived in the proper manner, the basic cause of the accident can be selected without guesswork.

7. *Make recommendations.* From the basic cause and the facts obtained from the investigation, recommendations should be made on how to prevent a recurrence of a similar accident and on the action to be taken to strengthen the weak points in the three fundamental activities of the accident-prevention program. For educational purposes, a review of the accident, along with conclusions and recommendations, should be made in detail for all members of the crew involved in the accident and for any other crews doing similar work.

8. *Submit written report.* Reports will vary in makeup with the legal requirements and the desires of higher management. One type of report that has proved effective (see Figs. 10.13 and 10.14) gives a brief description of the accident, along with other information on the front side to meet legal requirements. The reverse side of the report provides space for the basic cause selected, reasons for its selection, and recommendations, which should be filled in by the line supervisor.

This report provides a compact statement of details for review by higher executives as it passes up through channels. Experience has proved that this type of form, especially the requirement that the supervisor select and explain a basic cause, invites the supervisor to give more thought to the real causes of accidents instead of passing them off as an act of carelessness on the part of the injured worker. Also, because a review of the list of supervisory responsibilities suggests proper actions to the supervisor, he is likely to give more thought to things he could have done before the accident occurred.

When an accident results in very serious injuries or death, an additional detailed report including sketches, witnesses' statements, and other detailed information proves valuable. This should be made out for use by the employer in furthering his accident-prevention program.

10.11 SURVEYS OF SPECIFIC CONDITIONS, PROCESSES, SPECIAL HAZARDS, AND HEALTH CONDITIONS

Industrial processes produce various dusts, gases, and vapors. If such products or by-products are not controlled in closed systems or by removal, they may create a health hazard to workmen. When it is known or suspected that the concentration of air contaminants around process equipment is sufficient to create a hazard, a thorough survey of the equipment and operating practices should be made. Surveys are instituted whenever any condition in a plant develops an unexpected health or accident hazard.

A successful survey requires considerable preliminary work in order that the individuals making it will be thoroughly familiar with the general problem and that the employees involved will know why the condition is being studied. This latter factor is extremely important since it eliminates speculation. When information is to be obtained from the workers, it is necessary to be completely above board with them to ensure that they will give factual information.

The safety engineer is well qualified to conduct surveys of specific hazards and conditions that may be detrimental to health or that may lead to accidents. Many companies employ industrial hygienists or toxicologists assigned to the safety staff to do this work. Since the workers and supervisors know that these specialists are interested in their safety and health, they usually cooperate during a survey, whereas unknown persons might meet opposition. Furthermore, these specialists are in a position to seek advice and assistance from the medical, chemical, operating, and engineering staffs.

When the industrial hygienist or toxicologist

is not available it may be desirable to call in an outside agency, such as the state Department of Health, U.S. Bureau of Mines, or an industrial agency that handles specific problems. The advantage of this method of surveying a problem is that such agencies are thoroughly familiar with specific hazards, have the necessary equipment for measurements and determinations, and have background and experience in handling similar problems. When an outside agency is used, it is important that someone, preferably the safety engineer, act as contact between the agency and the operating personnel.

Before any field work is done, the individual making a survey should become thoroughly familiar with the known information concerning the hazardous material, process, or equipment under study. This will keep him from doing unnecessary work that may already have been done, point the best way to investigate the problem, provide discussion information for talking to workers and supervisors, and instill confidence in the workers and supervisors that he will eventually reach the correct solution to the problem. The procedure to be followed in gathering information should be determined in advance and should be thoroughly planned in order that the necessary data can be obtained in the minimum time. Equipment for making measurements, analyses, or determinations should be fully understood and tested in advance, so that the operator is familiar with all the details of its use. All employees are strongly interested in any conditions that may be detrimental to their personal health, and they should be informed of the reason for any survey that is made of potentially hazardous conditions. This precaution will alleviate their fears concerning possible effects on their health, will reduce speculation on why someone is checking their working conditions, will encourage them to cooperate in providing information, and will enlist their support and use of corrective measures.

While collecting data, the person making the survey should ascertain that normal conditions exist. Sufficient time should be spent on the job so that all conditions of exposure are considered, and all concerned should be satisfied that the problem has been thoroughly investigated and that all necessary information has been obtained. Frequently, workers and supervisors have been familiar with the hazardous condition for a long period and may have excellent suggestions on ways of correcting or overcoming it. Their suggestions and advice should be sought and considered.

After all the data have been collected, the information should be assembled into a complete report. The report should include a statement of the problem, the procedure of investigating, the technique in the use of measuring devices, a compilation of the data, summary statement of the conditions found, and recommendations for correction. Before the final report is presented for consideration, it is advisable to review the results with the supervisors involved. This procedure will assure that the recommendations are practical and will enlist the supervisors' assistance and cooperation in correcting the hazards. Credit should be given for any specific suggestions used that were made by the workers or supervisors. Although reports of surveys are prepared in detail, when transmitted to management a brief summary should be included that covers the problem, conditions found, and recommendations made.

10.12 EDUCATIONAL ACTIVITIES FOR ACCIDENT CONTROL

10.12.1 Education Defined

Education for accident prevention is the systematic development and cultivation of natural powers by instruction, training, and example. The goal is accident-free production.

10.12.2 Importance of Psychological Factors

All accidents, whether directly caused by unsafe acts of workers or by the presence of hazardous conditions, may, if traced to their origin, be attributed to people either from the worker groups or management. Psychological influences must be considered in the selection, placement, induction, and training of workmen. Accident-prone workers are those whose characteristics and behavior are such as to make them considerably more likely to be injured than the average person. Recklessness, stubbornness, excitability, low intelligence, lack of interest in the job, or dislike of the foreman makes these workers difficult to train and leads them to unsafe acts, many times deliberately.

Psychology is important in arousing and maintaining enthusiasm for safety through the usual media of meetings, posters, first-aid training, and so forth. However, it is difficult to reach every worker through these means because messages must be addressed to whole groups of employees. Appeal cannot be made to each man's individual desires or traits but must be made to those most commonly felt by the largest number of employees. These will strike a responsive chord in some individuals but leave others untouched. Therefore, the supervisor is the key man in any induction and training program that considers psychological factors as they relate to safe performance. The foreman must apply practical psychology. He is in close contact with the employee and is in the best position to judge abilities and evaluate work performance, including safety factors. He must familiarize himself with the employee's likes and dislikes, his family situation, his financial troubles, and his background, which may influence his attitudes and on-the-job performance. If the foreman can adapt his training methods to appeal to the worker's individual desires and traits, there is a good chance that the worker will perform his job in a safe manner. It is important that consideration of the psychological factors be continued not only during the training period of an employee but also throughout his employment.

10.12.3 Instruction of Employees

Instruction should start when an employee reports for work. Most phases of his work should be outlined to him until he is fully apprised of every detail. Because habits (which are perfectly learned performances) are developed through repetition, work methods should be constantly observed and improper methods should be immediately corrected. Instruction implies a personal relationship between the supervisor and the employees as individuals. Since this relationship is established from the beginning of employment, the supervisor is the prime factor in its success or failure.

Whenever work methods are to be changed, employees should be prepared in advance, and adequate instructions should be given to assure a smooth transition from the old way to the new. Assuming that workers become self-proficient with the change invites serious problems.

10.12.4 Committees and Meetings

Management safety committees in many companies formulate policy regarding the accident-prevention program. These committees are effective in maintaining management responsibility for safety in their organizations. The committee structure often carries down to the various levels of supervision to facilitate dissemination of the safety program to the worker level. The safety engineer should be the adviser to the safety committees and can act as secretary in order to maintain a permanent record of the proceedings of committee meetings and activities. Employee safety committees are beneficial. Management should appoint members of these committees, and the members should work through their supervisors. Membership should be on a rotating basis, the period served being 6 months to 1 year. All committees should gear their work toward the satisfaction of objectives set at the beginning.

A plant always needs good safety meetings to bolster its overall program. To maintain a level of interest, they must appeal to the employees and should be well prepared. An agenda along the line of the following may be of assistance:

1. Start meeting on time.
2. Make opening remarks brief.
3. Review status of safety suggestions.
4. Collect new safety suggestions.
5. Have very brief discussion of accident statistics.
6. Introduce main feature of the meeting.
7. Request comments from visitors.
8. Express thanks to participants and those attending.
9. Close meeting on time.

Sources of material for program features are numerous. Whatever is selected should have a definite application to problems. Some of the following may stimulate other good ideas.

1. A safety subject dealing with some phase of accident experience that is giving trouble locally
2. A talk by some outside expert—e.g., a doctor or an equipment expert
3. A motion picture or slide film pertinent to the situation
4. Statistical reports
5. Articles or professional papers from trade journals, accident-prevention bulletins from such agencies as U.S. Bureau of Mines and National Fire Protection Association

10.12.5 Publications and Visual Aids

Periodicals such as safety magazines and house publications are good media for the dissemination of safety messages. Great care should be taken that the editorial content is readable. It should appeal to the intellectual level of those to whom it is directed.

Special bulletins, with adequate illustrations, may be put out to cover procedures in combating hazards. Where more detailed information is required, a manual may be issued —e.g., for investigating procedures in industrial injury cases, automotive accidents, designing for safety, and so forth.

Motion pictures and other visual aids are of great value in training programs. Care should be exercised to preview content. Visual aids that are poorly edited, inappropriate, lack continuity, or show improper work practices can do more harm than good. Indexes of safety motion-picture films and slide films may be procured from business and industrial photography magazines.

10.12.6 First-Aid Courses

Since first aid has an off-the-job as well as on-the-job application, it offers an opportunity to sell employees on management's interest in their welfare at all times. Industrial associations, U.S. Bureau of Mines, and American Red Cross offer good courses. Many successful safety programs have been launched through courses in first aid for all employees. Men are thus informed that the company has or is starting a formal program.

10.12.7 Conferences

Periodic seminars or directed conferences for supervisory personnel should be a cornerstone in the safety-training program. Since the supervisor is responsible for the efficiency of his segment of production, such training is essential to help him do his job. Production must be safe to be efficient. Such courses should embrace employee working relationships, industrial psychology, and allied information. Material for such meetings is available from many corporations and universities.

Since most states require a number of forms in the event of an industrial injury, forum meetings on administration of workmen's compensation are valuable. Information exchanged in these sessions enables supervisory personnel to understand the need for prompt, accurate reporting. Good reports facilitate administration and help in the preparation of valid statistical analyses.

10.12.8 Motor-Vehicle Driver Training

Where fleet operation is part of a business, motor-vehicle accident prevention is essential. Driver-training activity, coupled with eye examinations, can help management control this situation. A meeting or a series of meetings can be arranged to present the basic causes for accidents and an explanation of the physical, physiological, and psychological forces that culminate in accident-producing situations. Stress should be placed on the need for individual responsibility in adjusting to traffic situations and in avoiding accidents.

10.12.9 Management Attitude and Example

Instruction and training will be of little value without the third component—example. The employees will be influenced by the example set for them since the supervisor represents management and management's thinking. If he supports a safety-education program because he has to or if he gives it only lip service, he will get the same sort of response from his people. But enthusiasm for making safety a vital part of the everyday work activity will lead to a reduction of injuries in each phase of operation.

10.13 STATISTICAL REPORTS

Statistical reports are essential as a source of periodic information on progress or retrogression. They can quickly and effectively pinpoint weak areas where more effort needs to be expended on accident control. Such reports may be compiled for establishing:

1. Industrial injury frequency rates (disabling injuries per 1 million man-hours worked)
2. Industrial injury severity rates (days lost per 1 million man-hours worked)
3. Analyses of basic causes, nature of work, and types of accidents
4. Motor-vehicle accident frequency rates (recordable accidents per 1 million miles traveled)
5. Costs
6. Any other pertinent subject, such as vision-testing results and preemployment testing

Such reports may be compiled monthly, quarterly, semiannually, or annually. Once the report system is in operation, it provides a means of comparing any given period with another—e.g., last year vs. a base period, say the previous 5-year or 10-year averages.

TABLE 10.3. *TYPICAL LAYOUT FOR STATISTICAL REPORT*

ABC Manufacturing Company: Disabling Industrial Injuries									
Department	Number of Disabling Injuries			Frequency Rate (Injuries per Million Hours Worked)			Severity Rate (Days Lost per Million Hours Worked)		
	This Year	Last Year	Average, Previous 5 Years	This Year	Last Year	Average, Previous 5 Years	This Year	Last Year	Average, Previous 5 Years
Production	4*	3	4	4.1	2.9	11.2	6,250	300	3,220
Assembling	2	9	6	3.5	14.5	10.4	270	180	590
Warehousing	37	29	36	19.9	14.5	26.2	460	220	1,100
Maintenance	7	8	9	6.3	6.9	16.2	110	210	440
Total	50	39	55	8.5	9.7	16.0	1,770	230	1,340

* Each asterisk indicates a fatality.

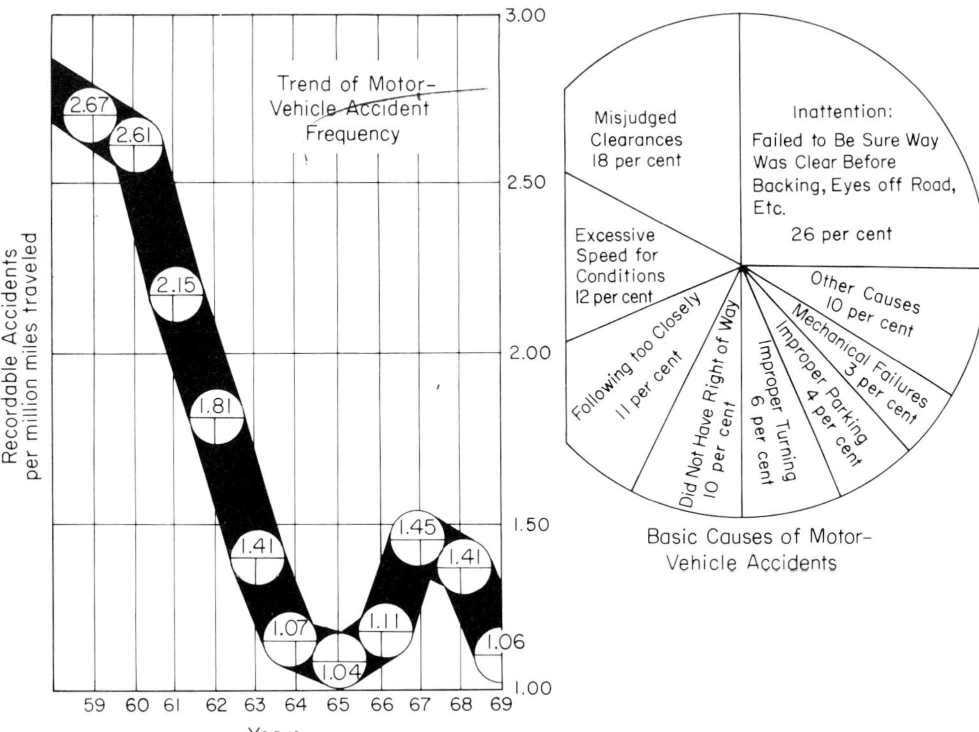

Fig. 10.15 *Frequency polygon and pie chart.*

The information should be concise and easily understood. Voluminous statistics are boring and often confusing. Consideration must be given to those who will read the reports.

A simple layout such as that in Table 10.3 may be used for the most frequently issued reports. Figure 10.15 shows an effective combination of a graph and a pie chart. The form of chart should be varied to keep them from becoming stereotyped.

Properly prepared, these reports may rapidly and graphically tell management in what direction the organization is headed as far as accidents are concerned. They provide an educational tool for use in the safety program. They should be easily interpreted and be as brief as possible.

DAVID A. HEMMES *has been teaching evening data-processing classes at Foothill College in Los Altos, California, for the past six years. A teacher at heart, he claims that to supplement his meager teaching salary he must "daylight" at Lockheed Missile and Space Company in Sunnyvale, California, where he is manager of the Data Processing Laboratory.*

His data-processing experience dates back to 1949 and includes work in programing research and in development of programing languages for IBM as well as various managerial positions in data processing during his twelve years with Lockheed.

SECTION
11

COMPUTERS AND DATA PROCESSING

DAVID A. HEMMES

11.1 ADVENT OF ELECTRONIC DATA-PROCESSING

The development of digital computers was an inevitable result of the need for speed and accuracy in calculation and data-processing. This section outlines some of the basic considerations for understanding the use of computers as electronic data-processing systems.

11.1.1 Historical Highlights

One of the earliest mathematical devices in common use was the abacus, which consists of movable beads on parallel wires. The abacus is a digital counter in the sense that the rightmost wire represents the units position of a number, the next wire to the left represents the tens position, and so forth. The user positions beads on each wire to represent the value of each digit. Although the Chinese were undoubtedly responsible for the abacus as we know it today, Gerbert, who learned of it from the Arabs in Spain, first introduced it in Europe. It is still the commonest calculating aid in Asia.

Credit for devising the first simple calculating machine goes to Blaise Pascal, who in 1642, at the age of 19, built a hand-operated machine which could add and subtract. Pascal's machine used figure wheels (each bearing the numbers 0 to 9) turned by hand with a small peg. This device, with a slight modification for the method of carrying, is still widely sold today as a pocket calculator.

Subsequent efforts to design and build mechanical aids for calculation produced a variety of unique and interesting devices too numerous to document here; however, no history of computing machinery would be complete without mentioning the efforts of the English mathematician Charles Babbage. Babbage's first attempt to design computing machinery resulted in his Difference Engine, which he built a small model of in 1822. He designed this machine to compute tabular functions, and a larger version of it was used in 1863 for calculating life tables for rating insurance. In 1833, while development of the Difference Engine was suspended because of misunderstandings and disputes, Babbage conceived the idea for his Analytical Engine, a machine which could be programed to perform a sequence of arithmetic operations while storing intermediate results in an internal memory. That Babbage could conceive of this Analytical Engine demonstrates that he clearly understood all the fundamental principles of modern digital computers. Although he devoted the rest of his life to its development, however, he never completed building it. His blueprints and assemblies now reside in the South Kensington Science Museum in London, and his ideas reside in thousands of modern computers.

The first modern digital computer was completed in 1944 by Howard Aiken, Harvard University, and IBM after 7 years of combined effort. This MARK I, officially known as the Automatic Sequence Controlled Calculator, was an electromechanical machine and was capable of operational speeds of $^3/_{10}$ sec for addition or subtraction.

The first all-electronic computer was completed in 1945; J. W. Mauchly and J. Presper Eckert, at the Moore School of Engineering of the University of Pennsylvania, developed it. Known as the ENIAC, Electronic Numerical Integrator and Calculator, this machine could execute 5000 additions or subtractions per second.

Development of the ENIAC led to the design of the EDVAC, Electronic Discrete Variable Computer. This machine, the forerunner of the commercial electronic data-processing machine, was the first to use the concept of the stored program; that is, it stored instructions as well as data in its memory. All modern large-scale computers now store programs.

11.1.2 Impact on Science and Management

11.1.2.1 Science. The speed and accuracy of computers have freed scientists from the time-consuming drudgery of computation. This freedom has enabled them not only to spend their time doing creative rather than routine work but also to undertake problems that would be impossible to solve without a computer as more than a lifetime of calculation would be involved.

11.1.2.2 Management. The use of a computer in a company has more implications for management than merely the increased speed and accuracy of processing data that benefit the scientist. Management, which is concerned with budgets, schedules, personnel, and making a profit, finds that the computer can be expensive, can turn out wrong answers at the same high rate of speed that it turns out right answers, and is often the center of politics and empire-building. As a result, management must learn how to ensure that the computer pays the proper return on the investment instead of becoming a costly monster.

In addition to cost considerations, the problem of fitting a computer or data-processing division organizationally into a company is one that has not been solved to popular satisfaction. In addressing this problem, management must keep in mind that a computer does not exist for its own sake but rather is a tool for functions which require data-processing to accomplish their objectives. The introduction of electronic computer processing has already created a whole new level of management in some companies and has eliminated a middle level of management in others. With increasing frequency, introduction of electronic data-processing systems is forcing a realignment of company organization.

11.2 A DIGITAL COMPUTER AS A DATA-PROCESSING SYSTEM

A system for processing data must have the ability to read, remember or recall, do arithmetic, move or rearrange data, and write. In addition, it must have the ability to be taught, or programed, to perform these basic functions as a series of operations following the set of rules, or algorithm, for processing the data. And it must have the ability to be programed to follow different logical paths through the algorithm based on tests performed on the data during processing. Although the word *compute* implies *do arithmetic*, a modern computer has the ability, although limited, to be a data-processing system. Figure 11.1 diagrams how a computer acts as such a system. The components in the figure are described below.

11.2.1 Functional Components for Data-Processing

11.2.1.1 Input Devices for Reading. A digital computer reads (or senses) data with input devices, of which the principal ones are *card readers*, which sense data recorded as holes punched in cards; *magnetic tape units*, which read data recorded on magnetic tape; and *terminals*, the most common of which reads data manually keyed in through a typewriter-like keyboard.

11.2.1.2 Storage Units for Remembering. A computer has, to use an unapproved analogy, a memory. Formally known as internal storage and informally as working memory, or core, this feature separates the computer from a calculator. Data or instructions are read by an input device into the memory, which is segmented into numbered cells or locations, each capable of holding either one character or a group of characters known as a *word*. The number of the location is its *address*. Data can be remembered or retrieved from the memory by calling for the contents of a specified address. In addition to the working memory a computer may have auxiliary storage units which can store millions of words. Physically these units are magnetic drums, magnetic discs, and specialized units. These extensions of the memory differ from the working memory in that they do not communicate directly with the input and output units and do not present instructions to the central processing unit for execution.

11.2.1.3 Central Processing Unit

For arithmetic. Performance of arithmetic is one of the three main functions of the central processing unit (CPU). The arithmetic (limited to addition, subtraction, multiplication, and division) is performed by transferring data from the working memory to the CPU for the operation and then storing the result in the memory. The CPU performs arithmetic through the use of its registers, which like memory locations are devices capable of receiving information, storing it, and transferring it as directed by control circuits. For an addition operation, the first number is transferred or copied from its memory location to the register, then adder circuitry adds the contents of the register to the contents of a second memory location, with the sum forming in the register. This sum is then stored in the memory, freeing the CPU for the next operation.

For logical decisions. The CPU contains the "logical ability" of a computer. It can compare the contents of one of its registers with the contents of a memory location or another register and set a switch indicating whether the contents of the register are equal to, greater than, or less than the contents of the memory location or register. In addition, it can set a switch indicating whether the contents of one of its registers are equal to, greater than, or less than zero. These switch settings are then used to control the proper sequence of subsequent operations.

For execution of the program. The CPU contains the circuitry for controlling both input-output units and the arithmetic and logical elements and for moving data between memory locations and registers in the CPU or between memory locations and input-output devices. Data-processing is accomplished by executing in the CPU the series of instructions known as the *program*. Each instruction in the program controls the circuitry for execution of its specified operation. Prior to processing, the program is stored in the memory. Processing occurs as the instructions are transmitted one by one to the CPU for execution.

11.2.1.4 Output Devices for Writing. A digital computer writes by transferring data from its memory to an output device for recording. Principal output devices are *magnetic-tape units*, which record data on magnetic tape; *line printers*, which print data a line at a time; and *terminals*, which can be line printers, type-

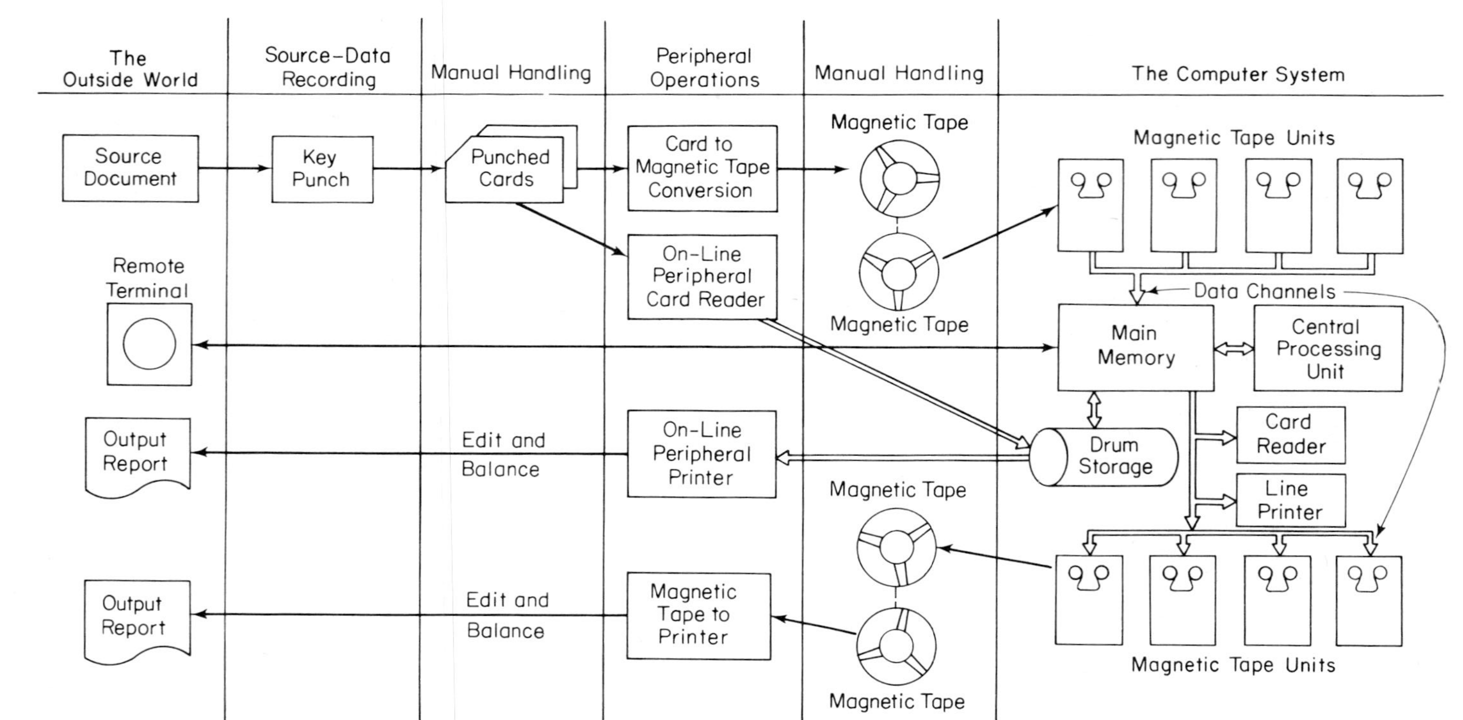

Fig. 11.1 *A digital computer as a data-processing system.*

writers, teletype, video (utilizing a television tube [CRT]), or specialized devices.

An interesting feature of input-output terminals is that they need not be physically located near the computer. Remote terminals are commonly used and are linked to the computer through the use of telephone lines, wide band lines, and microwave transmission.

11.2.1.5 Data Channels. Input-output units and the CPU are linked to the memory unit by data channels, which serve as paths over which data may be moved from one component to another.

11.2.1.6 System Configuration. The arrangement and number of components of a computer are known as the system configuration. A typical configuration could have eight magnetic-tape units on one channel, eight more tape units plus a card reader and a line printer on another channel, one main memory, a CPU, and 50 or 100 terminals. Since channels can operate independently of one another, such a configuration allows simultaneous use of input-output units without interrupting the data exchange between the memory and the CPU.

11.2.2 Peripheral Operations

Certain peripheral operations enable the computer to communicate with the outside world and still not to have its speed of processing curtailed by its limited ability to read and write. Peripheral operations allow the computer to read and write on that medium which provides the fastest input and output. In most cases, the computer reads and writes only magnetic tape. (This is not intended to minimize use of remote terminals where speed is not so important.)

11.2.2.1 Off-Line Peripheral Operations. The most common method of recording source data on a machine-sensible medium is punching the information into cards; and the most popular medium for presenting computer output data to people is a printed report. Therefore, the most common off-line peripheral operations are *card to tape*, which consists of reading the information from cards and writing it on magnetic tape, and *tape to print*, which consists of reading the information on magnetic tape and printing it as a report. When computer output needs to be retained in a machine-sensible medium but still needs some manual handling, a third peripheral operation is occasionally employed—*tape to card*, which consists of reading magnetic tape and punching the information into cards.

These peripheral operations are performed by a small computer which has the same general characteristics as the main computer, but which, because of its smaller size, is less expensive to operate and therefore can be used for such chores as reading cards and printing. In a data-processing center, one or more small peripheral computers, used for card-reading and for printing, make it entirely unnecessary for the main computer to use magnetic tape and remote terminals for input or output of data.

11.2.2.2 On-Line Peripheral Operations. The object of on-line peripheral operations is the same as that of off-line—to avoid tying up the main processor with the relatively slow cardreading and printing tasks. With on-line peripheral operations, a data channel links the peripheral computer to the main computer system. A large storage unit, similar in characteristics to the computer memory, serves as a data buffer and replaces the magnetic tape. A magnetic drum storage is commonly used for this purpose.

When cards are read by an on-line peripheral machine, the information is transmitted to the buffer storage with no interruption of the main processor. When all the information has been read into the buffer storage, it can then be transferred at a relatively high rate of speed into the main computer memory for processing. For printing, output data are transferred from the main memory to the buffer storage. The data can then be printed from the buffer storage without interrupting the central processor. Figure 11.1 shows how the computer uses peripheral operation for communication with the outside world.

11.3 DATA REPRESENTATION FOR ELECTRONIC PROCESSING

11.3.1 Internal Representation

Electronic elements with two states represent data in the computer memory and in the registers. For example, transistors are maintained either conducting or nonconducting;

TABLE 11.1. *BINARY REPRESENTATION OF DECIMAL VALUES 0 TO 17*

	Binary Numbers					Decimal Value
Power of base	2^4	2^3	2^2	2^1	2^0	
Place value	16	8	4	2	1	
	0	0	0	0	0	0
	0	0	0	0	1	1
	0	0	0	1	0	2
	0	0	0	1	1	3
	0	0	1	0	0	4
	0	0	1	0	1	5
	0	0	1	1	0	6
	0	0	1	1	1	7
	0	1	0	0	0	8
	0	1	0	0	1	9
	0	1	0	1	0	10
	0	1	0	1	1	11
	0	1	1	0	0	12
	0	1	1	0	1	13
	0	1	1	1	0	14
	0	1	1	1	1	15
	1	0	0	0	0	16
	1	0	0	0	1	17

ferromagnetic elements are magnetized with a plus polarity or a minus polarity; specific voltage potentials are present or absent. The values of these two-state elements are graphically represented by a 1 in one state and a 0 in the other; thus data can be considered to be all ones and zeros within a computer. This data representation implies a binary numbering system, and within this system each two-state device represents one bit.

11.3.1.1 Binary Mode. In some computers the internal data representation leads naturally to the use of the binary numbering system for numeric notation and arithmetic since this base 2 system contains only the symbols 0 and 1.

Definition of a binary number. Any number N to a base B is defined as

$$N_B = \cdots C_3 \times B^3 + C_2 \times B^2 + C_1 \times B^1 + C_0 \times B^0 + C_{-1} \times B^{-1} + C_{-2} \times B^{-2} \cdots\cdot \uparrow$$

radix point

Here is an example in decimals:

$$532.34_{10} = 5 \times 10^2 + 3 \times 10^1 + 2 \times 10^0 + 3 \times 10^{-1} + 4 \times 10^{-2}$$

$$\text{or } 532.34_{10} = 5 \times 100 + 3 \times 10 + 2 \times 1 + 3 \times {}^1/_{10} + 4 \times {}^1/_{100}$$

$$\text{or } 532.34_{10} = 500 + 30 + 2 + 0.3 + 0.04$$

Therefore a binary number is defined as

$$N_2 = \cdots C_3 \times 2^3 + C_2 \times 2^2 + C_1 \times 2^1 + C_0 \times 2^0 + C_{-1} \times 2^{-1} + C_{-2} \times 2^{-2} \cdots\cdot \uparrow$$

binary point

$$\text{or } N_2 = \cdots C_3 \times 8 + C_2 \times 4 + C_1 \times 2 + C_0 \times 1 + C_1 \times \tfrac{1}{2} + C_{-2} \times \tfrac{1}{4} \cdots\cdot$$

Binary notation. The above calculation leads to the valid conclusion that, when numbers are represented in binary notation, each bit has a place value which is a power of 2, and the decimal value of the number represented is the sum of these place values. Table 11.1 gives an example of this system.

Binary arithmetic. Binary arithmetic involves simple calculations which can be performed easily by computer circuitry. These are the rules.

Addition: $0 + 0 = 0$; $0 + 1 = 1$; $1 + 0 = 1$; $1 + 1 = 10$ (or 0 with a carry to the left).

Subtraction: $0 - 0 = 0$; $1 - 0 = 1$; $1 - 1 = 0$; $0 - 1 = -1$(or 1 with a borrow from the left).

Multiplication: $1 \times 1 = 1$; $1 \times 0 = 0$; $0 \times 1 = 0$; $0 \times 0 = 0$.

Division: $1 \div 1 = 1$; $0 \div 1 = 0$. Division by zero is never performed. Borrowing and carrying follow basically the same rules as those

TABLE 11.2. *BINARY ARITHMETIC OPERATIONS*

```
      1 1             0 1              1 0 1
      1 1 1         1 1 0 0          × 0 1 1
    +     1       −       1          -------
(a) -------   (b) ---------            1 0 1
    1 0 0 0         1 0 1 1    (c)   1 0 1
                                   ---------
                                     1 1 1 1

             1 1
     1 0 0 ) 1 1 0 0
             1 0 0
             -------
               1 0 0
               1 0 0
(d)            -----
               0 0 0
```

in decimal arithmetic. Table 11.2 illustrates binary representation of the following arithmetic operations: (a) $7 + 1 = 8$, (b) $12 - 1 = 11$, (c) $3 \times 5 = 15$, (d) $12 \div 4 = 3$.

Octal representation of binary numbers. Since conversion from binary to octal and octal to binary can be performed by inspection, the octal, or base 8, numbering system is commonly used to represent binary numbers. The octal system is not used by computers but by people who find it more convenient to say or write, for example, "octal 542" instead of "101100010."

Octal numbering system. The octal numbering system contains eight characters: 0, 1, 2, 3, 4, 5, 6, 7. An octal number is defined as

$$N_8 = \cdots C_3 \times 8^3 + C_2 \times 8^2 + C_1 \times 8^1 + C_0 \times 8^0 + C_{-1} \times 8^{-1} + C_{-2} \times 8^{-2} \cdots$$

↑ octal point

Conversion from binary to octal. Examination of Table 11.3 shows that three binary bits are represented by one octal character. Conversion is performed by grouping the binary bits in threes from right to left starting with the binary point and simply writing the octal equivalent for each group of three. For example, the binary number 10110111010101, when grouped in threes, becomes 010 110 111 010 101. Converting each group to octal gives 26725. (See Table 11.3.)

Conversion from octal to binary. Conversion from octal to binary requires only the writing of the three-bit binary equivalent for each octal character. For example, the octal number 72531 is converted in this way.

Octal	7	2	5	3	1
Binary	111	010	101	011	001

Hexadecimal representation of binary numbers. Many computers handle data internally in groups of four bits and eight bits. With these machines it is more convenient to use hexadecimal (base 16) notation to represent binary numbers.

Hexadecimal numbering system. The hexadecimal numbering system contains sixteen characters: 0, 1, 2, 3, 4, 5, 6, 7, 8, 9, A, B, C, D, E, F. (The graphics A, B, C, D, E, F are commonly used to represent those numbers between 9 and 10.) A hexadecimal number is defined as

$$N_{16} = \cdots C_3 \times 16^3 + C_2 \times 16^2 + C_1 \times 16^1 + C^0 \times 16^0 + C_0 \times 16^0 + C_{-1} \times 16_{-1} + C_{-2} \times 16^{-2} \cdots$$

↑ hexadecimal point

Conversion from binary to hexadecimal. Similar to conversion from octal, conversion from hexadecimal to binary can be performed by inspection. Conversion is performed by grouping the binary bits in fours from right to left, starting with the hexadecimal point and simply writing the hexadecimal equivalent for each group of four. For example, the binary number 101101111101, when grouped in fours, becomes 1011 0111 1101. Converting each group to hexadecimal gives B7D. (See Table 11.3.)

Conversion from hexadecimal to binary. Conversion from hexadecimal to binary requires the writing of only the four-bit binary equivalents for each hexadecimal character. For example the hexadecimal number 7F32C is converted in this way

hex	7	F	3	2	C
binary	0111	1111	0011	0010	1100

11.3.1.2 Character Mode. This mode, often called the *decimal mode* or *BCD* (binary coded decimal) *mode*, permits the handling of nonnumeric characters as well as numeric informa-

TABLE 11.3. *DECIMAL, BINARY, OCTAL, AND HEXADECIMAL EQUIVALENTS*

Decimal	Binary	Octal	Hexadecimal
0	000	0	0
1	001	1	1
2	010	2	2
3	011	3	3
4	100	4	4
5	101	5	5
6	110	6	6
7	111	7	7
8	001000	10	8
9	001001	11	9
10	001010	12	A
11	001011	13	B
12	001100	14	C
13	001101	15	D
14	001110	16	E
15	001111	17	F
16	010000	20	10
17	010001	21	11
Etc.			

tion in decimal notation. In character mode, a group of bits represents each character. These groups are fixed in length and encoded so that every character has a unique bit pattern. (Representing characters with coded groups is not limited to computers—Morse code is a well-known example of a coded character set using two states: a dot and a dash.) While some computer systems can operate only in character mode, others, under control of the program, can operate in either character or binary mode.

Character set. A character set is the total collection of characters recognizable by the computer system. The size of the character set depends upon the number of bits allotted to represent each character. A six-bit code has 64 (2^6) possible bit combinations and thus can represent 64 different characters. An eight-bit code can represent 256 characters. Character sets may vary from computer to computer. A typical set includes at least these characters.

Alphabetic characters	ABCDEFGHIJKLMNOP QRSTUVWXYZ
Numeric characters	0 1 2 3 4 5 6 7 8 9
Special characters	($*)−+/.,=

Character sets such as the above can be used to write a lot of language and represent a lot of data; however the lack of such characters as lower-case alphabetics, a full range of punctuation marks, and mathematical symbols such as subscripts, superscripts, $\neq$, $>$, Σ, etc., limits the amount of meaning the set can convey. Users have brought pressure to bear upon computer manufacturers for machines capable of handling a seven- or eight-bit code, thus allowing a set of 128 or 256 characters. One computer company has coined the word *byte* to mean a group of eight bits (plus a ninth bit for parity check). The eight data bits in a byte can represent one alphameric character, two packed (four-bit) decimal digits, or a portion of a binary word.

The byte must not be confused with the eight-bit representation used on some computer systems capable of handling alphameric characters only (no binary representation). On these computers six bits are data bits, one bit is a flag to mark either the first or last character of a word, and one bit is a parity bit (explained below). Some smaller systems use a six-bit code with four data bits, one flag, and one parity bit. This system, of course, limits the representation to numeric characters. Alphabetic characters must be represented by two numeric characters.

Collation sequence. All character sets must have a collation sequence, which is the order of all characters with respect to one another. It could be called alphabetical order, but it is extended to include all characters in the set. A collation sequence in a data-processing system enables the ordering and sorting of data.

11.3.2 External Representation

An output device or a manual operation from a source data document records externally stored data on machine-sensible media.

11.3.2.1 Punched Cards. The standard punched card is about 7⅜ in. long and 3¼ in. wide and contains a matrix of eighty columns and twelve rows. This matrix has 960 elements, or positions where a small rectangular hole can be punched. The presence or absence of a hole in each position provides 960 two-state devices for recording data. Input devices read cards by sensing the hole or no-hole condition in each position. Computer output devices (card punches) or manually operated card-punch machines punch data into cards.

11.3.2.2 Magnetic Tape

Physical characteristics. A reel of magnetic tape is normally ½ in. wide and about 2400 ft long. The tape is plastic and is coated on one side with a metallic oxide. Like a card it has rows and columns, the number of columns being limited only by the physical length of the tape. A row on tape is referred to as a *track*, and a column as a *frame*. Data are recorded as very small magnetized spots along each track. Each magnetized spot is a small magnet with a north and south pole. Since these poles are reversible, each magnet is in effect a two-state device for recording data. N→S can represent a 1, and S→N can represent a 0. A typical magnetic tape has seven or nine tracks, and thus each column, or frame, has seven or nine bits. Because the density of the columns is normally about 800 per inch, if one column is used to represent one character, data can be recorded on magnetic tape at a density of 800 characters per inch.

On-line-to-the-computer input-output devices known as *magnetic-tape units* read and write magnetic tape. These units move tape under the read-write heads at 100 to 150 in./sec. Each track has its own read-write heads. This high-speed tape movement and the high density of data on the tape make magnetic tape the most efficient medium for reading data into and recording data from a computer system. Some advanced tape units, reading tape with more than seven tracks, can read and write data at a rate of up to 340,000 numerical characters per second.

Magnetic-tape codes. Data can be represented on magnetic tape either in binary or character mode. In both modes, six of the seven bits or eight of the nine bits in each column record data. The remaining bit is a parity bit, which causes the number of bits in each column to be either even or odd. Using an internal parity check when reading or writing data eliminates transcription and reading errors. When data are recorded in character mode, a six-bit code or an eight-bit code is used. It is similar to the code used for internal representation.

11.3.2.3 Punched Paper Tape. Originally developed for transmitting data over telegraph lines, punched paper tape is one of the lesser used data-recording media. Its characteristics are the same as those of magnetic tape, except that a hole is actually punched into the tape to represent a bit. Paper tapes normally have either five or eight tracks, or channels, of information.

11.4 COMPONENT PARTS OF A DATA-PROCESSING SYSTEM

11.4.1 Internal Storage Devices

All internal storage is divided into locations or cells, each of which stores data in the form of bits. When data are sent to a location for storage, the data replace the previous contents of that location. When data are taken from a storage location, the contents of that location remain unchanged.

Storage devices used in a computer system are classified as main memory or auxiliary storage. Main memory is in line between input-output devices and the CPU. Its function is to store instructions as well as data. Auxiliary storage augments main memory to create additional internal storage. It differs from main memory in that it is not directly accessible to either the CPU or the input-output devices.

11.4.1.1 Magnetic-Core Storage. The most commonly used storage device for main memory is the magnetic-core storage. It is fast and provides random access, but it is expensive. The length of time it takes to transfer data from the memory to the CPU (known as access time) or to store data in the memory is very short. Access time in modern computers is 200 μs or less. With random access there is no significant or even measurable difference between the access times of different memory locations; thus there is no problem of choosing optimum locations to store data in order to reduce access time.

A magnetic core is a small ring or doughnut of ferromagnetic material which can be magnetized in a few billionths of a second. Core memories are made of core planes, which can be compared to a very small screen door with a magnetic core strung at every intersection of wire. Thus every core has two wires running at right angles through its center. If half the electrical current needed to magnetize a core is sent down through one wire and the other half of the current is sent across another wire, then all the current needed to magnetize a core coincides at the intersection of these wires, and magnetism is induced in one and only one core in the plane. When the same amount of current is sent through the same two wires but in the opposite direction, the magnetism, or the polarity, of the core is reversed. Thus a core has two states and can represent a bit with a 0 in one state and a 1 in the other.

Each intersection of wires, or each core, is one memory location. In order to have memory locations capable of holding more than one bit, core planes are placed one on top of the other to form a stack. Data, in the form of bits, are stored with one bit in each plane. The planes in the stack act in parallel, and the number of planes in the stack determines the size in bits of the memory location. Core stacks vary from six to seventy-two planes. In a typical core stack, each plane has 4096 cores, and a computer system could have enough stacks to provide memory sizes approaching 1 million words.

11.4.1.2 Magnetic Drums. Magnetic drums are used for both main memory and auxiliary storage, principally the latter. A magnetic drum is a rotating cylinder whose outside surface passes under read-write heads which record and read information in the form of bits. The surface of the drum is divided logically into bands. Each band forms a loop around the drum and has its own read-write heads. It has the same characteristics as an endless loop of magnetic tape.

11.4.1.3 Magnetic Disc. Not so fast as core or drum, the magnetic-disc storage is used when very large amounts of data must be stored internally as in the data-bank concept, where the user must maintain an up-to-date file and make frequent direct reference to the file.

A magnetic disc is a thin metal disc coated on both sides with a magnetic material. Data in the form of bits are recorded on the disc in concentric tracks. A disc memory consists of twenty or more discs mounted on a vertical or horizontal shaft. The shaft revolves, spinning the discs. Access arms, each with a read-write head, move in and out between the discs to read and write data on the proper track. The number of characters of data that can be stored in disc memories is in the tens of millions.

11.4.1.4 Other Bulk Storage. Since the ability of computers to process data is limited by the amount of data that can be stored internally, much research is being conducted toward developing large, fast, and inexpensive bulk memory devices. Devices such as magnetic cards and endless tape loops are already in common use for auxiliary storage, and other developments will follow. The magnetic core will likely remain the most popular main memory-storage device.

11.4.2 Central Processing Unit

11.4.2.1 Arithmetic and Logical Section. This section of the CPU performs all arithmetic and logical operations. Arithmetic is usually limited to addition, subtraction, multiplication, and division. (Strictly speaking, the CPU of a typical computer contains circuitry only for addition. Multiplication is performed by successive additions; subtraction, by addition of complements; division, by successive additions of complements.) All mathematical operations performed by a digital computer must therefore necessarily be performed by numerical methods.

The logical operations performed in the CPU consist of setting a switch to indicate whether the contents of a register are equal to, greater than, or less than the contents of a memory location or another register. Logical operations also consist of setting a switch to indicate whether the contents of a register are equal to, greater than, or less than zero. This switch setting is used by the control section of the CPU for determining the sequence of subsequent operations.

11.4.2.2 Control Section. The control section of the CPU controls the entire computer system. The circuitry for this control must be triggered by instructions which are transferred to the registers of the control section from the computer memory. A single instruction may do any of the following operations.

1. Cause data to be brought into storage from some external source, such as a card reader.
2. Cause a specified arithmetic operation to be performed on selected numbers.
3. Institute a logical test to determine what part of the program should be performed next.
4. Cause results to be sent from storage to a recording device, such as a magnetic-tape unit.

11.4.3 Input-Output Devices

11.4.3.1 Card Readers. Card readers sense the holes in punched cards as the cards are mechanically moved under reading brushes or photoelectric cells. Although card readers are normally too slow for reading data, they are used on line for reading occasional control cards.

11.4.3.2 Card Punches. A card punch moves blank cards one at a time under punch dies, which punch data received from the memory. Card punches are normally too slow to be used on line or directly connected to the computer; therefore most card punching is performed by an off-line device with the information punched being read from a magnetic tape.

11.4.3.3 Magnetic-Tape Units. Magnetic-tape units both read and write magnetic tape and serve as the principal input and output devices of a computer system. Sixteen, eighteen, and twenty tape units are common in a single computer system.

11.4.3.4 Paper-Tape Readers. These devices sense punched holes in paper tape and transmit the data to main memory. Too slow to be used as on-line input devices, paper-tape readers normally are used to copy the information from paper tape onto magnetic tape.

11.4.3.5 Paper-Tape Punches. As an output device, a paper-tape punch records data transmitted from main memory as punched holes in paper tape. As in the case of the paper-tape reader, a paper-tape punch is commonly used off line for copying information from magnetic tape onto paper tape.

11.4.3.6 Line Printers. Strictly an output device, a line printer prints one complete line of characters at a time. Line printers operate at speeds varying from 150 to over 1000 lines per minute and normally print either 120 or 132 characters in each line. When used on line, a line printer prints data transmitted from main memory. When used off line, it prints data from magnetic tape.

11.4.3.7 Typewriters. Used on line as both an input and output device, a typewriter is normally employed by an operator to control the computer by typing coded instructions which start and stop various programs. Diagnostic feedback is typed automatically and is used as information by the operator. Typewriters are also used as remote terminals, and in this role they are for data input and output.

11.4.3.8 Display Consoles. Display consoles normally are output devices and remote terminals. Some models allow limited manual input ability. The commonest example of a display console is the cathode-ray tube found in television sets. This output device, used on line, displays bits transmitted from main memory in the form of data. When properly positioned on the face of the tube these bits can represent a curve, a simple picture, or a graphic such as A, B, or C. More deluxe display consoles contain a plate etched with a character set. An electron passing through one of the characters etched on this plate is shaped and thus is displayed as a character, the plate working the same as a cookie cutter. Manual input to the CPU from a display console is through a key set on the console or through a light pencil (which is like a small flashlight) pointed into the face of the tube.

11.4.3.9 Control Consoles. For manual input all computer systems have a control console, which may or may not be augmented with a typewriter. It contains start, stop, and program-loading keys as well as lights for visual output of the contents of registers or of a manually selected memory location.

11.4.3.10 Others. A large variety of other devices, too numerous to document, may be used for special applications.

11.5 COMPUTER PROGRAMING CONCEPTS

A series of instructions which when executed direct the computer components to process data is known as a *program*. The best way to describe a program is by example.

In order to read two numbers, add them, and write their sum, an electronic computer executes the following instructions:

Instruction No. 1: Read a data word from magnetic tape into memory location 100.

Instruction No. 2: Read a data word from magnetic tape into memory location 101.

Instruction No. 3: Clear a register in the CPU to zero and then add the contents of memory location 100 to the contents of that register.

Instruction No. 4: Add the contents of memory location 101 to the contents of the register.

Instruction No. 5: Store the contents of the register in memory location 103.

Instruction No. 6: Write the contents of memory location 103 as a data word on magnetic tape.

The above six instructions constitute a computer program.

Since instructions reside in the computer memory prior to their execution, they must be represented in the memory as if they were data. Each instruction is cryptically encoded and occupies perhaps one location in computer memory. During execution of the program, each instruction is transferred or copied from its memory location into the instruction register of the CPU, which decodes the instruction and sets the proper switches in the circuitry for execution.

11.5.1 Execution of Instructions

11.5.1.1 Instructions. A machine-language instruction has at least two main parts: an *operation code*, which specifies the operation to be performed, and an *address* of the data or device needed for the operation specified by the code.

11.5.1.2 Machine Cycles. The CPU cycles between *instruction time* and *execution time*. Instructions are fetched from the memory during instruction time and executed during execution time. The CPU then returns to instruction time and fetches the next instruction to be executed.

11.5.1.3 Instruction Address Register. This register in the CPU holds in the memory the address of the next instruction to be executed. During instruction time the contents of the memory location specified by the instruction address register are transferred to the instruction register. The contents of the instruction address register are then incremented by one.

11.5.1.4 Instruction Register. This register in the CPU separates the instruction into the operation code and the address. The circuitry of the CPU, triggered by the bits in the instruction register, then fetches the data from the memory location specified by the address portion of the instruction and performs the operation specified by the code.

11.5.2 Stored Program

Since instructions are stored in the computer memory, nothing prevents an instruction from being brought into the CPU during execution time. When this happens, the instruction is treated as if it were data. Thus a stored-program machine has the ability to process its own instructions. This ability makes program modification during execution possible, a factor which adds greatly to the flexibility of the computer system and also contributes to its logical ability.

11.6 THE CONCEPT OF SOFTWARE

The word *software* was coined in order to identify a computer program as the nonhardware part of a data-processing system. It originally meant all computer programs; it now generally describes just those that complement the hardware to form a system. (Computer programs that find answers and process data are called *application programs.*)

Software falls into two main classes: *programing systems*, which consist of a language for writing programs and a translator, which encodes the programs to machine language; and *operating systems*, which take complete control of the entire computer system. They call application programs for execution, control input-output devices, call translator programs, provide program debugging aids, account for machine utilization, and perform other functions as required. In short, operating systems are programs that monitor and control other programs and take charge of computer operation. The human operator no longer operates the computer; he makes requests of the operating system.

11.6.1 Programing Systems

Programing systems allow a programer to write his source program in a language which, although rather stilted and limited to the character set of a given computer or class of computers, is designed for ease of expression.

11.6.1.1 Translators. A translator, or processor, encodes the source program to the machine-language program, known as the *object program.* The translator (a computer program itself) in addition to translating usually includes some programing aids such as automatic storage assignment and checks for grammatical errors. Diagnostic feedback generated during translation can be presented to the programer as an aid to his program debugging.

The output of the translator is the machine-language object program, which is always retained on some machine-sensible medium such as punched cards or magnetic tape. If the diagnostic feedback indicates no programing errors, the object program is then loaded into the computer memory for execution or further testing.

Translators are known as assemblers, compilers, generators, and interpretive systems, depending upon how far removed the source language is from the object language and on the method and amount of translation.

11.6.1.2 Languages

Machine-oriented languages. Known as an MOL, a machine-oriented language is one whose structure resembles the language of the computer. Just as in machine language, instructions have two parts—an operation code and an address; however an MOL allows a programer to write in symbolic mnemonic codes. For example, machine-language code for an addition might be 000101000001 (octal 0501), but in the MOL the code would be ADD. This and other typical operation codes such as READ, SUB, DIV lighten the programing chore, and their translation to machine language is a relatively simple process.

Another typical feature of an MOL is the use of symbolic names for storage assignments and variables. These can be as mnemonic as the programer makes them.

Although there is normally a one-to-one correspondence between the symbolic operation codes of a machine-oriented programing language and the operation codes of the machine language, most machine-oriented programing systems provide for additional operation codes known as *pseudooperations* and *macrooperations*. Pseudooperations translate to nothing at all. They are used to direct the translation process when necessary. Macrooperations translate to more than one machine-language operation.

Translators for machine-oriented programing languages are usually known as assemblers, assembly programs, or assembly systems.

Procedure-oriented languages. The need for powerful programing languages and the desire for easy exchange of source programs among the programing community have led to the extensive development of procedure-oriented programing languages (POLs). These languages differ from MOLs in that they describe a problem, an algebraic expression, or an algorithm rather than direct operation of computer components. The goal of developers of POLs is to create a system that is so machine independent that a programer does not need to know or even to care which computer will be used to execute his program. He does not need to learn how to program for a specific computer; he only needs to learn the POL, a process advertised as being simple. Another major benefit is that no reprograming is necessary if the computer is replaced by one with different characteristics. All that is necessary is a retranslation from the original source program to the machine language of the new computer. Furthermore, the use of a common procedure-oriented language implies standardization, with its many and well-known advantages. This goal has yet to be achieved in its entirety.

Some obstacles are that, while languages are fairly easy to specify, the writing of translators is slow and expensive and, to compound this problem, new versions (or dialects) of the languages are produced almost as rapidly as new models of automobiles. As a matter of fact, some languages are identified with their year of development, just as automobiles are.

In spite of a certain amount of resistance against them, POLs have gained widespread acceptance. Such languages as FORTRAN, COBOL, and PL1 (as well as numerous others) have become the programing standard in most data-processing organizations. It is common for a programing contract to specify that applications programs be written in such a language.

11.6.2 Operating Systems

Sometimes called *system software*, operating systems are an extension of the hardware; their purpose is to supervise and control the entire computer system. Operating systems are a complex hierarchy of computer programs which reside in whole or in part in main memory. (Operating systems are normally the only programs that are ever resident programs.) Designed to improve the overall effectiveness of the computer, the operating system allows the operator to stack jobs for continuous processing and has the ability to call in required programs or data, to schedule jobs by a priority algorithm, to allocate resources such as memory and input-output unit assignments, and to act as supervisor or monitor for the whole system. The operating system partitions, or divides, computer memory into separate logical areas, each with its own application and each unaware of the other, and then schedules the interface of each partition with the CPU, thus giving the user, effectively, as many computers as he has partitions. This feature takes advantage of the high speed of the CPU and never allows it to be idle while waiting for some slower component. Instead of having the CPU waiting, the operating system switches it to another partition. Regarded as bordering on the black arts by the uninitiated, this concept would allow, for example, one partition to be handling remote terminals, another to be performing an in-house application, and a third to be performing batch processing in what is called the *background* (partitions are not limited to three). Each of these would not process any faster (at least not significantly faster) if the others went away.

Theoretically, when a computer is installed, the operating system is generated and loaded into memory and it stays there forever. Actually, operating systems are changed, are modified, and are inadvertently erased. This is more of a nuisance than a problem as the task of reloading the operating system is straightforward but time consuming. Some of the programs found in an operating system are listed below.

11.6.2.1 Supervisor Program. Normally the only program that is always resident in memory, the supervisor is used to call applications programs for execution. Control is then transferred to the new program, which, after execution, returns control to the supervisor. The supervisor may also handle "interrupts" from devices or from the application program and may schedule which memory partition is active.

11.6.2.2 Input-Output Control. Input-output control is designed to relieve the programer of the often tedious task of writing detailed input-output instructions. He needs to give only a description of the files and of what data are to be read or written. The input-output control program will control physical input-output, block and unblock logical records, overlap input-output areas in memory for faster reading and writing, monitor end-of-file and end-of-reel conditions, and may even handle telecommunications activities.

11.6.2.3 Scheduling Program. Working with the supervisor, this program schedules the next job to be executed either on a first-in–first-out basis or according to some priority algorithm.

11.6.2.4 Language Translators. Sometimes called compilers, these translators convert source programs written in a programing language to the machine language.

11.6.2.5 Service and Utility Programs. As an aid to programers, an operation system provides a complete library of service and utility programs for such mundane and repetitive tasks as sorting, merging, loading, copying tape to disc, copying disc to tape, and formating for printer, as well as for most of the commonly required mathematical functions and peculiar user programs.

11.6.2.6 Accounting Programs. A necessary function of computer operations is accounting for the time used. An operating system has programs for keeping records of the jobs processed and of the lengths of their execution times.

11.7 DATA-PROCESSING CONCEPTS

11.7.1 Records and Files

A record with respect to electronic data-processing is a group of related facts or fields of information treated as a unit. A file is an organized set of records directed toward some purpose. The records in a file may or may not be sequenced according to a key contained in each record. Any collection of records is called a file; even a deck of punched cards may be correctly called a file of cards. Files being read are called input files and files being written are called output files.

11.7.1.1 Master Files. A file used as an information source for answering interrogations and generating reports is called a master file. An example is an employee-status file, which has a record for each employee. Each record contains such information as (Field 1) employee name, (Field 2) employee number, (Field 3) Social Security number, (Field 4) rate of pay, (Field 5) earnings in year to date, etc.

11.7.1.2 Transaction Files. The records in a transaction file contain information to change, update, or modify records in a master file. A transaction file is sometimes called a *change file*.

11.7.1.3 File Organization. In order to be used by a computer system, a file must be organized in such a way that the computer can identify it, recognize its beginning and end, and find records within it. If the file is on magnetic tape, this organization can be accomplished by dividing the file into three parts: the beginning of file label, the text, and the end of file label. In the case of master files, a directory or dictionary is needed to describe the format and contents of the records within the text. This directory should be machine sensible and is either physically a part of the master file or a part of any computer program for processing the file. The directory may also describe the format and content of the records in a transaction file used to update the master file.

In addition, the records within the text must be ordered in such a way that they are easily found without searching. The simplest way to do this is to order or sequence the records on some key field, such as an employee number. To find a record, the computer system looks at the key field and follows the same logic a human being follows when looking up a word in a dictionary.

11.7.2 File-Processing and Report Generation

Most data-processing consists of keeping a master file up to date so that it can be used as an information source for generating reports.

11.7.2.1 File-Processing. The term *file-processing* is used here to mean all the processes necessary to have an up-to-date, orderly file of information, and thus it includes updating and housekeeping.

Updating a master file. A transaction file, which contains information on all transactions or changes which have occurred since the last master-file updating, is readied on a computer input unit, and the master file to be updated is readied on another input unit if it is on an external storage medium. As the computer reads each transaction-file record, it finds the matching master-file record, updates it with the changed information, and writes a new master-file record into a newly created master file. The old master file remains unchanged.

Housekeeping. Sometimes called *file maintenance*, housekeeping of files is necessary to keep them up to date. Housekeeping includes such functions as removing no longer needed records and correcting information known to be filed incorrectly.

11.7.2.2 Report Generation. Report generation is normally performed independently of file maintenance. It must, of course, be performed with an up-to-date master file. Report generation involves reading the master file, selecting that information to be reported, arranging the information for output (including punctuation), and writing the arranged information on an output file. The output file can be printed as a peripheral operation.

11.7.3 Sequential Processing vs. Random Access

The use of sequential processing or random access is determined by whether the file is recorded on an external storage medium that must be read sequentially or on an on-line random-access file such as a magnetic drum or disc.

11.7.3.1 Sequential Processing. Magnetic tape is an example of a medium that must be read sequentially. The tape is wound on a reel and passed under a reading head of a tape unit. There is no way to go directly to a record in the middle of the tape, and once read, the tape has to be rewound before it can be read again. Records must be ordered in the text of the file on some key field to eliminate going back and forth on the tape. In order to update a master file, the transaction file must be sorted in the same order as the master file. Updating is performed by reading a record from each file and then looking at the key field on each to see whether they match; if so the master record is updated; if not another master record is read to see whether it matches and so on.

11.7.3.2 Random-Access Processing. When a master file is recorded on a medium such as a disc, a drum, or other bulk memory, the CPU can access any record in the file in the same length of time. Also the file never gets physically disturbed, thus it never needs any equivalent of rewinding a tape. Random-access processing is faster and more flexible than sequential processing, but it is also more expensive since bulk memory devices are more expensive than reels of tape. In random-access processing change records do not have to be ordered; thus random access lends itself well to real-time and on-line processing, where change information is sent directly to the processor as the change occurs and the file is immediately updated.

11.7.4 Batch Processing vs. On-Line Processing

11.7.4.1 Batch Processing. When change information is saved until there is a batch of it large enough to make a computer pass worthwhile, the procedure is called batch processing. Data processed sequentially usually have the change data batched. This batched changed data comprise the transaction file.

11.7.4.2 On-Line Processing. On-line file updating is accomplished by having data-gathering devices (key sets, token readers, terminals, etc.) on line, or connected to the central processor. As change data are recorded, rather than being batched in some machine-sensible medium, they are transmitted immediately and directly to the central processor.

Upon receipt of a transaction record the central processor updates the master file. This procedure is said to be a real-time one as the master file always contains up-to-date information with no lag for batching between data-gathering and updating. Real-time processing is too impractical if the master file is maintained on magnetic tape because of the sequential nature of the tape. Transaction records in an on-line system arrive at the central processor as the transaction occurs, which is in random order; thus the file must be stored in a random-access storage. Because the main memory is too small and too expensive, bulk memory or auxiliary memory is used.

11.7.5 Advantages and Limitations

11.7.5.1 Batch Processing: The Assembly Circle. Properly used, an electronic computer provides a means of putting the data-processing of a company on an automated, assembly-line basis; however this assembly-line concept differs from a manufacturing assembly line in that instead of raw data entering the line to be processed and assembled with final products coming out the end of the line, data-processing is more of a circle or loop, with finished products in the form of reports dropping off at various stages and a final product in the form of files being fed back into the book to be merged and processed with new raw data.

The economic advantages of mass producing on an assembly-line basis are obvious and well known, and so are the disadvantages. These disadvantages exist in the data-processing assembly circle as well as in the manufacturing assembly line. A long lead time is necessary for design of the circle. More time is necessary for the data-processing equivalent of tooling, which for the computer is programing. Still more time is necessary for training personnel who will operate or will be affected by the circle. And all of this time can be translated into money. Once a manufacturing assembly line or a data-processing assembly circle is operating, it becomes noticeably inflexible, and any effort to make changes or to use the line to produce a unique or a one-of-a-kind product can be prohibitively costly.

11.7.5.2 On-Line Processing: The Data Bank. The other concept of data-processing is more like a spoked wheel than a circle. At the hub of the wheel is the data bank—a file containing all the up-to-date information about a given operation. At the end of each spoke is a remote terminal for "deposits" and "withdrawals" of information. The remote terminals are located physically in the various operating units or departments of a company and are used to record data from transactions and to enter this data into the bank. Any operating department of the company with a remote terminal can also receive information from the bank by using its remote terminal as an inquiry station.

This data-bank concept has the advantage of a central file that is updated in real time and thus contains only current information. A remote terminal on line to the bank allows an operating department the flexibility of receiving only those data it wants, needs, and will use.

From an economic standpoint, a trade-off between the cost of the bank (which must be an electronic storage device) and the value of having current information readily accessible must be considered.

11.7.5.3 Time-Sharing. The use of random-access devices and concepts has led naturally to time-sharing of the CPU by many remote terminals. The advantage of time-sharing is that each terminal user is completely unaware and unhampered by the other users and has the realistic illusion of having the whole computer to himself. The concept of time-sharing is not new; for example, spark plugs for years have been time-sharing the use of the battery and the coil in an automobile ignition system. With computers, remote terminals are used, each not only working independently of the others but also doing a completely different and unrelated task. The vast difference between the internal speed of the computer and the speed of a manually operated remote terminal makes the handling of 50 or 100 terminals by one computer possible. Indeed, the main limitations on the number of terminals are the data channels and communication lines to link them up, not the ability or speed of the computer.

11.8 A MANAGEMENT VIEW OF THE COMPUTER

Management's concern with a computer should be primarily how to make the computer benefit the company and how to account for computer usage. Management need not concern itself with what bits go into what register in the CPU.

11.8.1 The Computer as a Management Tool

11.8.1.1 Management Decisions. The idea that a computer can make a management decision should be completely discarded. Management sets policy and makes decisions within the framework of that policy. A computer can be used effectively as a tool only if management sets up the criterion for the computer to follow.

11.8.1.2 Management-Information Control System. The main purpose of a management-information control system is to provide information to control an operation. However, a computer should not be used simply to prepare reports; this procedure leaves more of a burden on management than there need be, particularly in the area of selecting information for making a decision from the volumes of data a computer can output. A step toward a management-information control system is to program the computer to select only the information needed. This involves setting a target or expected result and then establishing an acceptable range above and below the target. Any item falling outside of this range is an exception and should be reported on in detail so that the manager responsible for its control will have enough information to take corrective action. This concept of *exception reporting* can be carried a step further by programing the computer to take appropriate action when an exception is encountered. This appropriate action may take the form of printed instructions or even mechanical preparation of such items as a purchase order. It must be remembered that this does not mean that the computer is making a decision. The computer only implements the decision. In this type of control system all decisions must be made beforehand and must be in the form of "This is what action should be taken under these circumstances and that is what action should be taken under those circumstances," etc. The management-information control system is completed when it produces just enough information or indicators so that management may monitor the system.

11.8.1.3 Review Techniques. Management review techniques, such as PERT (Programed Evaluation Review Technique) are examples of the use of a computer as a management tool. Everything in the PERT technique can be performed by manual methods; but to gather, process, and present all the necessary data by manual methods would take too long for the information to be of any value. The point is, similar to the scientist's using a computer to free himself from a lifetime of calculation, management uses the high speed of a computer to evaluate and review techniques that would be impossible or impractical to implement by hand.

11.8.2 Comments Concerning Costs

A company which uses larger-scale computers may have a data-processing budget higher than that of any other operating unit. A large-scale computer, if leased, can cost $100,000 per month for equipment rental alone, and if purchased the monthly depreciation approaches that figure. Operating costs for salaries for programers and operators, for overhead, and for replacement of perishable items often equal the equipment rental costs.

11.8.2.1 Lease vs. Purchase. The question of lease vs. purchase of equipment must always be under review; however, no general statement can be made as to whether leasing or purchasing is more economical or advantageous. Factors to be considered include availability of funds, cost of money, expected useful life of the equipment, stability of data-processing requirements, and ratio of rental price to sales price.

11.8.2.2 Third-Party Lease, or Lease-Back. Leasing computers from a third party which in turn has purchased the machine from the manufacturer, as opposed to purchase or lease directly from the manufacturer, has become increasingly popular as a means of reducing computer equipment costs. Such arrangements, which were in existence long before computer people began using them, are usually of two types: (1) the full payout, or financial lease, and (2) the nonpayout, or operating lease. The full payout, which is a method of financing a capital asset, places the risks and rights of a purchaser on the lessee; reasons for using this plan are entirely financial. The nonpayout is comparable to rental, the difference being the length of the commitment. The advantage to the lessee is that rentals are 10 to 20 per cent lower than the manufacturer's rentals. In this plan the lessor assumes the risks of obsolescence and must recover his costs on the residual value of the equipment and his ability to rent it again or must sell it at the end of the original lease term.

11.8.2.3 Cost of Equipment Usage. All plans for accounting for the cost of data-processing must include a technique for measuring the cost per hour of using the equipment and for allocating these costs to the using departments.

Clocks and meters. Clocks are used to measure the length of time a job, such as production control, payroll, or inventory control, runs on the computer. The job is essentially clocked on and clocked off. A common practice is to use a time clock and to have the computer operator punch a job card. Using a time clock becomes impractical with very high-speed computers because a stack of jobs can run faster than the operator can manually insert the job cards into the time clock. The time clock becomes impossible when, with multiprocessing, more than one job is being processed at the same time.

A solution is to use the internal clock of the machine. This clock is based on machine cycles and can be programed to measure time in hundredths of an hour. A good operating software system can cause the start time and finish time of a job to be printed or punched on line. Both an internal clock and an external clock must be wired to the program-control trigger of the computer so that the clock runs only when the computer runs.

When computing equipment is rented, meters are usually installed on all major components. These meters, like the clocks, run when the equipment runs, and rental billing is based on meter readings. Unlike the clocks, the meters cannot determine start and end times of individual jobs.

Rates. In order to properly allocate costs for machine usage, a computer rate in terms of dollars per hour of use must be established. When the computer is leased, one method for establishing the rate is as follows: A rental agreement usually has a flat price for the first 176 hr of use per month. This is known as the *base shift rental.* Every hour per month over the first 176 costs 10 or 20 per cent (depending on the rental agreement) of the base shift rental divided by 176. This is known as the *extra shift rental.* The total rental bill equals the base shift rental plus the extra shift rental. With third-party lease, there is often no charge for extra shift rental.

The total cost of computer usage equals the rental bill plus overhead and salaries of operators. Dividing the total cost by the number of hours of use gives a rate per hour, which can be used internally to allocate computer usage costs. The rate is inversely proportional to the usage (or utilization), and lower rates result from high computer utilization.

When the computer is purchased, the cost is the allowable depreciation plus maintenance costs plus overhead and operators' salaries. This total amount divided by hours of use is the rate for purchased equipment.

11.8.3 Basic Economic Considerations —Making the Computer Pay

A digital computer can process data faster, cheaper, and with more accuracy than human beings can. Using a digital computer to process data requires an initial investment in facilities and systems design and a continuing dollar outlay for equipment rental (or maintenance if the equipment is purchased) and for salaries of computer operators and programers.

11.8.3.1 False Economy

Rate of return on investments. The fact that computers work more cheaply than human beings do can mislead management into believing that, if they replace clerical personnel with computers, they will realize a cost saving. What they often fail to take into consideration is that proper money management requires that an investment in an electronic data-processing system have a return equal to the demand rate of return on any other good investment.

Lack of foresight. Another pitfall is that, if the rate of return is low enough that the business environment changes before the breakeven point on the initial investment in the data-processing system is reached, then more investment is needed in order to have the system reflect the changed conditions, and the breakeven point fades elusively into the future. If this process continues, the breakeven point will never be reached.

Nonmeasurable factors. Tunnel-vision systems design can cause any paper cost savings realized to be drained off by nonmeasurable, even nondetectable, factors. This drain occurs when, even though audits show an electronic data-processing system is costing less to operate than the old manual system, the nonclerical operating units of the company have inadequate operating information. This lack often leads heads of the operating units into keeping their own records (known as cheat sheets) by hand as a defense against the system. This manual record-keeping, surreptitiously bootlegged, is never detected in audits and can amount to a complete duplication of the electronic system.

11.8.3.2 Real Economy. Real economy can best be achieved by using digital computers to lower the operating costs of the entire company, rather than limiting them to the task of reducing the data-processing budget. Often if the data-processing budget is increased, other operating costs go down by a greater amount. In other words, the computer should be used for more than just replacing the existing manual data-processing. It should be used for a broad range of applications, such as production scheduling and inventory control. Successful economic use of a computer returns, typically, for every dollar invested less than a dollar in lower data-processing costs but returns an additional amount from gains in operating efficiency, such as improved production scheduling and lower inventories, so that the total return is greater than a dollar. The point is to use the computer as much as possible for as many things as possible.

11.8.3.3 Economy vs. Economy. Some applications, such as space-vehicle tracking and control, require a computer for the data-processing and computing, and there is no substitute. The only economic consideration in such cases is in determining which computer will be best for the job and still not cause a budget overrun.

11.8.4 Legal Considerations

The time is approaching when management must concern itself with the legal considerations of computer use.

11.8.4.1 Forced Use of Computers. A company could possibly be forced to use computers for data-processing. Stockholders could declare that management is not keeping pace with competitors who have installed electronic data-processing systems. Proper audits of accounts could demand machine-processing of data in order to ensure accuracy. Standards of government agencies could require that reports, statements, and tax information be submitted on a machine-sensible medium such as magnetic tape.

11.8.4.2 Liability. Members of the legal profession are already discussing hypothetical cases such as the following: A digital computer is used for real-time control of switching of railroad trains. A computer malfunction causes a wreck with resulting loss of life and property. The railroad management had taken all reasonable precautions for proper computer operation. In the subsequent lawsuit, who is liable and for what? The railroad management, the computer manufacturer, the programers? No one has yet set a precedent for suing a computer.

JAMES A. BAKER *is head of the Mathematics and Computing Group of the Lawrence Radiation Laboratory, University of California, Berkeley. Except for a period of about two years, Mr. Baker has been at the University of California since 1947. He first served as a teaching assistant in the Department of Mathematics while doing graduate studies in mathematics. Then he became a lecturer in the Departments of Industrial Engineering and Operations Research and of Electrical Engineering and Computer Science and in the Division of Transportation Engineering. He joined the Lawrence Radiation Laboratory in 1952 and continued his association with it until the present, except for two years during which he was associate director of the Data Systems Division of the Broadview Research Corporation.*

Mr. Baker earned his A.B. in mathematics at Pomona College. His primary interests are in the application of automatic digital computers to the solution of problems in engineering and science. He devised the first system for automatic reduction of data from high-energy physics experiments. He also directed the design and construction of an early program for the production of definitive orbits for near-earth satellites. Presently he directs the operation of a computer center employing sixty-seven professional and sixty-six subprofessional persons. The center provides operational and programing services for two CDC 6600 computers.

He is a consultant to the National Academy of Science Committee on the Uses of Computers. His teaching and research are directed at computer-assisted design, design automation, simulation, and network theory.

SECTION 12

CRITICAL PATH METHODS

JAMES A. BAKER

12.1 INTRODUCTION

Critical-path methods, CPM, PERT, PEP, and PERT-COST are names for a set of related activities which employ network techniques to study and solve certain problems in the management of large, complex projects.

12.1.1 Planning

One of the most important of these problems is that of project planning and scheduling. Critical-path methods have been used very successfully in estimating the cost and duration of large projects. Many government agencies (the Department of Defense in particular) now require critical-path analyses to accompany proposals for large projects.

Critical-path methods allow the planner of a large project to see in graphic form the relationships among various parts of a project; they also allow him to assess quickly and reasonably accurately the consequence of changing the scope of a project. These methods are also effective when used in estimating and scheduling the resources to complete a large project.

12.1.2 Control

Critical-path methods are also extensively used as a dynamic control tool in the management of large projects. These methods give the project manager a comprehensive picture of project status at any time. With the information from a critical-path analysis, he can pinpoint activities which are behind schedule. He can also use this information to allocate his resources most effectively in order to regain lost time.

12.2 HISTORICAL BACKGROUND

The Remington-Rand Division of the Sperry Rand Corporation devised the critical-path method in 1957 in work done for duPont. DuPont was concerned with scheduling and controlling maintenance shut-downs of large chemical process plants. Because such shutdowns are extremely expensive in terms of lost productivity, they hoped that this Remington-Rand contract would result in a method which would minimize shut-down length. James E. Kelley, then of Remington-Rand, and Morgan R. Walker of duPont devised the CPM* system under this contract.

In 1958, under the sponsorship of the U.S. Navy Special Projects Office, the PERT system was devised. PERT is an acronym for Program Evaluation and Review Technique. At that time, the navy was just starting development of the fleet-ballistic-missile, or Polaris, system. This project was of very large magnitude and complexity, and the navy wanted a management tool which would enable them to control the project effectively. It has been reported that the use of PERT in the Polaris project reduced the elapsed time by 18 months.

* M. R. Walker and J. S. Sayer, *Project Planning and Scheduling*, Report 6959 (Wilmington, Del.: E. I. duPont de Nemours and Co., Inc., March 1959). J. E. Kelley, Jr., and Morgan R. Walker, "Critical Path Planning and Scheduling," *Proceedings of Eastern Joint Computer Conference* (Dec. 1–3, 1959), pp. 160–73.

12.3 THE PROJECT

When developing an analytical tool, such as CPM, one must identify and isolate the features of the system that need to be modeled. The system that CPM is concerned with is the project. The tasks that are to be performed with CPM are planning, forecasting, and control. We describe below the features of a project involved in the performance of these tasks.

12.3.1 Activities

A project may be subdivided into activities—that is, time-consuming tasks or subprojects. Each activity in a project should be under the direction of a single individual. Other criteria often used in the identification and definition of activities are that they should be performed by a single craft or that they should be performed in a single shop.

If, for example, the project under consideration is the construction of a house, typical activities might be: excavating foundation, building forms, pouring concrete, framing, applying siding, wiring, rough plumbing, finish plumbing, etc.

When a project is modeled through the use of network techniques, the activity is the basic building block.

12.3.2 Events

Because time is an important parameter in the scheduling and control of projects, any model which is to assist in the performance of these tasks must take account of time. As a result, the project feature called an *event* is introduced into the network model.

An event is defined to be an instant in time. In a project, an event may mark the initiation of an activity, the completion of an activity, or a time after which an activity may be initiated.

An important function of the critical-path method is to assign times to the events in a project. Two distinguished events in any project are the beginning event and the ending event. The elapsed time between these two events is the total project time.

12.3.3 Precedence Relations

In most projects, the start of certain activities must await the completion of other activities. In order to construct a schedule for a project, one must have complete knowledge of these precedence relations among activities, and any useful project model must take these relations into account.

Examples of precedence relations in the construction of a house are that the foundation must be excavated before forms can be built, that forms must be built before concrete can be poured, etc.

12.4 THE ARROW DIAGRAM

In network modeling of projects, the arrow diagram is of primary importance. In this section, the arrow diagram will be defined, and the correspondence between its parts and the project will be explained.

12.4.1 Nodes

Nodes in an arrow diagram correspond to events in a project. Nodes are represented by circles, as in Fig. 12.1.

Fig. 12.1 *Representation of Node 17.*

Fig. 12.2 *An arrow representing an activity whose estimated duration is 16.2 weeks.*

12.4.2 Arrows

Arrows are used in the arrow diagram to represent activities. An arrow is drawn as a straight-line segment with a point at one end, as in Fig. 12.2. Quite often the estimated duration of an activity is printed along the top of the arrow for that activity. The pointed end of an arrow is referred to as its *head*. The nonpointed end is called the *tail*. The length of the arrow is arbitrary and is not associated with the duration of the activity it represents.

12.4.3 Connectivity

The connection of arrows and nodes represents the precedence relations among the

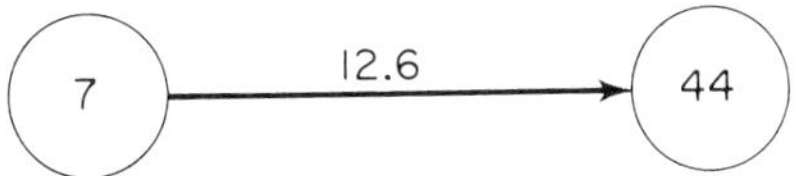

Fig. 12.3 *Two nodes connected by an arrow.*

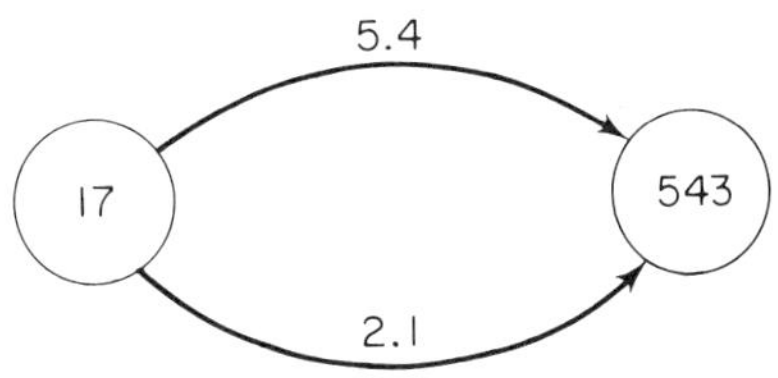

Fig. 12.4 *Two nodes connected by two arrows.*

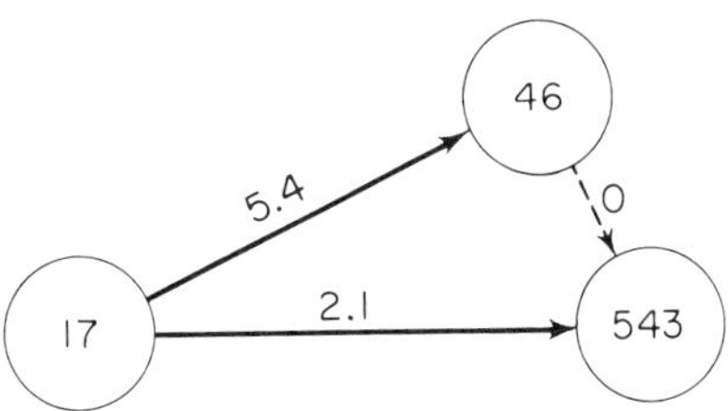

Fig. 12.5 *A dummy activity.*

activities in a project. Each arrow connects two nodes, as in Fig. 12.3. The node at the tail of the arrow is called the *preceding node* or event, and the node at the head of the arrow is called the *following node* or event. In Fig. 12.3, the arrow represents an activity whose estimated duration is 12.6 time units. The preceding event is Event 7. The following event is Event 44.

In most critical-path systems, one arrow at most connects each pair of nodes. We prohibit the situation illustrated in Fig. 12.4. In this figure, we have two arrows, representing activities of estimated durations 2.1 and 5.4, going from Node 17 to Node 543. Even though this practice corresponds to a real situation in the project, its prohibition makes the arrow diagram easier to talk about and to analyze than it would otherwise be.

In order to represent the situation pictured in Fig. 12.4, we introduce the concept of a *dummy* activity. Such activities have zero duration and are represented by dashed arrows, as in Fig. 12.5. In this figure, a new node, 46, has been introduced. The activity with duration 5.4 now has 17 as a preceding node and 46 as a following node. A dummy activity goes from Node 46 to Node 543.

With the convention that a pair of nodes may be connected by, at most, one arrow, we can identify each arrow uniquely by its preceding and following nodes. For example, in Fig. 12.5, the arrow with estimated duration 2.1 is represented by (17, 543) since 17 is its preceding node and 543 is its following node. Similarly, we refer to the arrow with estimated duration 5.4 as (17, 46), and to the dummy as (46, 543). In general, we refer to an arbitrary activity as (i, j), where i is the number of the preceding node and j is the number of the following node.

In order to make clear the relationship between the arrow diagram and the project which it represents, we employ the following rules:

Rule 1. No event may occur until all the activities which have it as a following event (node) have been completed.

Rule 2. No activity may be initiated until its preceding event has occurred.

These rules give us a method of representing in our arrow diagram the precedence relations among the activities in our project. For example, if Activity A must precede Activity B, we may make the following node of A the preceding node of B.

12.4.4 An Example of an Arrow Diagram

We shall consider here a simple project, changing a tire, and construct an arrow diagram for it. The first task is to break the project into activities and to list them. Table 12.1 is such a list. Each activity in it, except (1, 2) and (11, 12), is a task that can be performed continuously by one man.

After the activities in the project are listed, we must determine the precedence relations among them. A good first step is to search the list for initial activities—activities which may be performed first. The activity called "remove coats" is the only one in the list that has this property. We assign a preceding node number of 1 to it and place the number in the leftmost column of the table. The next step is to search the list for ending activities. In this case, the activity called "clean up and put on coats" is the only one. We assign to it Nodes 11 and 12. The choice of these event numbers is quite arbitrary.

Now we search backward for activities that may immediately precede an activity already listed. We find three of them: "put flat away," "put jack away," and "final tighten and replace hubcap." To each of these, we assign following node number 11 since each of them can immediately precede Activity (11, 12). We assign preceding node numbers 7, 9, and 10 to them. We now continue to work backward, always looking for activities which can immediately precede ones already assigned.

As we go through this process, we find it necessary to create two dummy activities: (4, 5) and (9, 10). We create (9, 10) because activities "lower car" and "tighten lug nuts" may be performed concurrently, but they both immediately follow "place spare" and both im-

TABLE 12.1. *ACTIVITY LIST FOR THE TIRE-CHANGING PROJECT*

Preceding Node	Following Node	Estimated Duration, Minutes (T_{ij})	Description of Activity
1	2	1.0	Remove coats
2	3	1.0	Procure jack and lug wrench
2	7	2.0	Procure spare
3	4	1.5	Remove hubcap and break lug nuts loose
3	5	0.5	Position jack
4	5	0	Dummy
4	6	1.0	Remove lug nuts
5	6	2.0	Jack up car
6	7	0.5	Remove flat
7	8	0.5	Place spare
7	11	2.0	Put flat away
8	9	1.0	Lower car
8	10	1.5	Tighten lug nuts
9	10	0	Dummy
9	11	1.0	Put jack away
10	11	2.0	Final tighten and replace hubcap
11	12	2.0	Clean up and put on coats

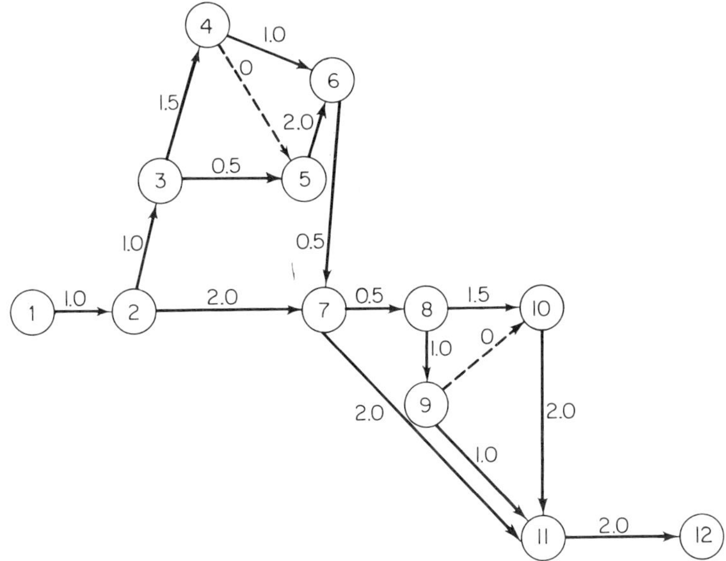

Fig. 12.6 *Arrow diagram for the tire-changing project.*

mediately precede "final tighten and replace hubcap." We must insert the dummy to avoid having "lower car" done concurrently with "final tighten and replace hubcap." Similarly, the dummy (4, 5) is inserted to ensure that (3, 4) precedes (5, 6).

When we have assigned node numbers to all events, we can draw the arrow diagram (Fig. 12.6) itself. Two or three attempts are generally necessary to obtain a diagram in which the arrows do not cross one another.

Notice that it was possible to express all the precedence relations in the project and to draw the arrow diagram without ever referring to the duration of an activity.

The next step then is to estimate the duration of each activity. One should make these estimates for each activity independently; in other words, one should assume that all resources required for an activity are available when it is started and will remain available until it is completed. The estimated duration of activity (i, j) is called T_{ij}. Estimates for this example are given in the column headed T_{ij} in Table 12.1.

12.5 THE CONSTRUCTION OF ARROW DIAGRAMS

In this section, we consider some practical problems associated with the use of network methods in the planning and management of large projects.

12.5.1 Who Should Construct Diagrams?

Although many firms now assign the construction of arrow diagrams to a separate operations-research or CPM group, this practice leads to ineffective use of this powerful tool. The construction of arrow diagrams is properly a line-management function. Only persons directly concerned with the management of a project have sufficient knowledge of the relationships within it to construct a proper model of it. Moreover, construction of a network model of a project often provides valuable insights into its structure.

A manager working with a network devised by others is more reluctant to modify it than he would be if he were working with his own network. Activities described as dependent in a diagram often turn out in practice to be independent and can, hence, be undertaken concurrently. The project manager is more likely to detect such situations than anyone else.

A special CPM group can, however, take the responsibility for routine maintenance and updating of arrow diagrams. Such a group can perform valuable services by relieving management of a variety of tedious bookkeeping tasks.

12.5.2 Who Should Make Time Estimates?

The proper person to estimate the duration of an activity is the person who will supervise it. Management should review his estimates, however, since there is a tendency to insert liberal fudge factors. An easy way for a supervisor to become a hero is to give a liberal time estimate for an activity and then to finish ahead of schedule. Such heroism is, of course, not good project planning or estimating.

After a project is under way, all time estimates for uncompleted activities should be reviewed periodically because unforeseen circumstances which affect activity durations almost invariably arise.

12.5.3 The Physical Construction of Diagrams

When constructing the diagram of a large project, it is often convenient to divide the project into a number of subprojects. This dividing should be done so that it minimizes the number of interconnections between pairs of subprojects. The diagram for each subproject may then be constructed on a separate sheet of paper. This procedure tends to minimize the effort of maintaining a diagram during the lifetime of a project.

The numbering of nodes in a diagram is often troublesome. In the past, some network analysis systems required that nodes be numbered so that the preceding node of an activity had a smaller number than the following node. In other words, if (i, j) were an activity, then i had to be less than j. Such restrictions are not now common.

A block of numbers should be assigned to each subproject, and nodes within the subproject should be assigned numbers from that block. A current ordered listing of all numbers used to date should be maintained, and any new node numbers assigned should be checked against the list. It is extremely important to avoid assigning the same number to two different nodes.

Another easy blunder to make is to introduce loops into a network. A loop is a sequence of activities of the form (i, j), (j, k), ... (l, m), (m, i). An example is (6, 17), (17, 24), (24, 6). Our rules say that Activity (6, 17) may not start until Event 6 has occurred; Activity (17, 24) cannot start until Activity (6, 17) has been completed; Activity (24, 6) cannot start until Activity (17, 24) has been completed; but Event 6 cannot occur until Activity (24, 6) has been completed. A pictorial presentation of this problem is given in Fig. 12.7.

If you use the convention mentioned above that (i, j) can be an activity only when $i < j$, then you avoid the problem of loops. This convention, however, is not recommended. When it is used, modifying, maintaining, and updating a network become extremely time consuming and difficult.

Most modern computer programs used in the analysis of networks contain provisions for the automatic detection and listing of loops. The responsibility for correcting these errors still lies with the network builder, however.

12.6 THE ANALYSIS OF ARROW DIAGRAMS

12.6.1 Notation

12.6.1.1 Earliest Start Time. The earliest time that an event can occur is designated by E_i, where i is the event number. The earliest time for a beginning event is taken, by convention, to be zero. Hence, if Event 6 is a beginning event, then $E_6 = 0$. If i is an event other than a beginning event, then E_i is the length of the longest path (or chain) of activities going from the beginning event to Event i. The length of a

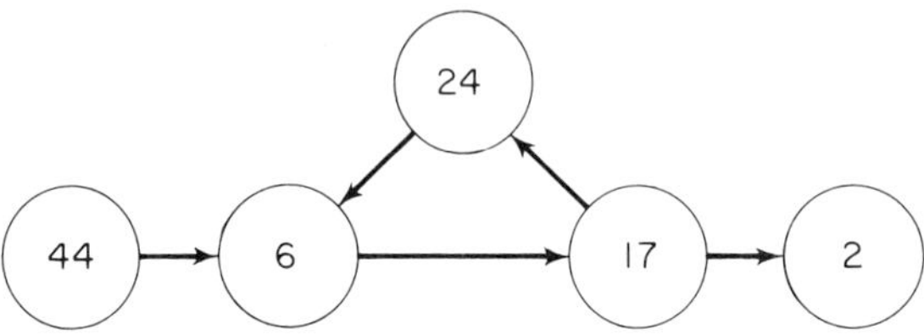

Fig. 12.7 *An arrow diagram containing a loop.*

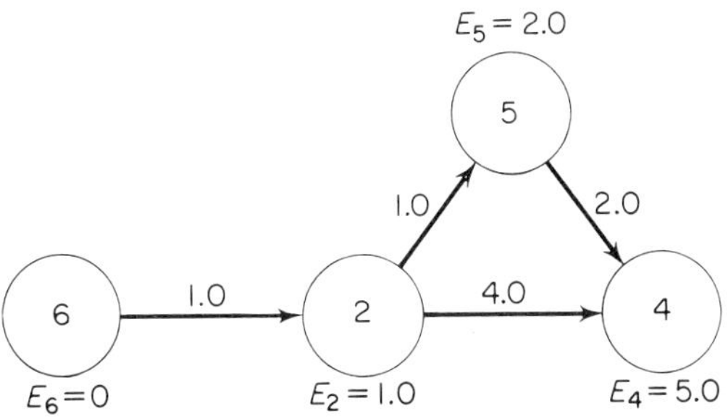

Fig. 12.8 *Example of the calculation of earliest event times.*

path of activities is the sum of the durations of all the activities on the path. The notation T_{ij} is used for the duration of activity (i, j).

An example of the computation of earliest event times is given in Fig. 12.8. In this example, Event 6 is the starting event; hence, by convention, $E_6 = 0$. There is only one path from the starting event (6) to Event 2, namely the path containing the single activity (6, 2). The length of this path is 1.0; hence $E_2 = 1.0$. Again, only one path goes from Event 6 to Event 5, namely the path containing the two activities (6, 2) and (2, 5). The length of this path is $T_{62} + T_{25} = 1.0 + 1.0 = 2.0$; hence, $E_5 = 2.0$.

Two paths go from Event 6 to Event 4, one containing Activities (6, 2), (2, 5), (5, 4), and the other containing two activities—(6, 2) and (2, 4). The length of the first of these paths is $T_{62} + T_{25} + T_{54} = 1.0 + 1.0 + 2.0 = 4.0$. The length of the second path is $T_{62} + T_{24} = 1.0 + 4.0 = 5.0$. Because the longest path is the second (containing two activities), $E_4 = 5.0$.

The earliest starting time for Activity (i, j) is designated by E_{ij} and is equal to the earliest occurrence time for its predecessor event—that is, $E_{ij} = E_i$.

12.6.1.2 Latest Start Time. If n is the number of the unique finishing event in a network, then E_n, the earliest event time for Event n, is the shortest elasped time in which the entire project can be completed. Planners generally assume that it would be desirable to complete the project in that time.

The latest time of occurrence for Event i is designated by L_i. L_n, if n is the finishing event, is defined to equal E_n. In other words, the latest time that the last event can occur is defined to be the earliest time at which it can occur!

If i is an event other than the finishing event, L_i is defined to be equal to L_n minus the length of the longest path from Event i to Event n. This definition implies that the latest time of occurrence of an event is the latest time that the event can occur and allow the project to be completed in elapsed time E_n.

It should be noted that if k is the number of the starting event, then $L_k = 0$. If L_k were different from zero, project elapsed time different from E_n would be implied.

An example of the computation of latest event time is given in Fig. 12.9. The diagram here is identical to the diagram of Fig. 12.8. We

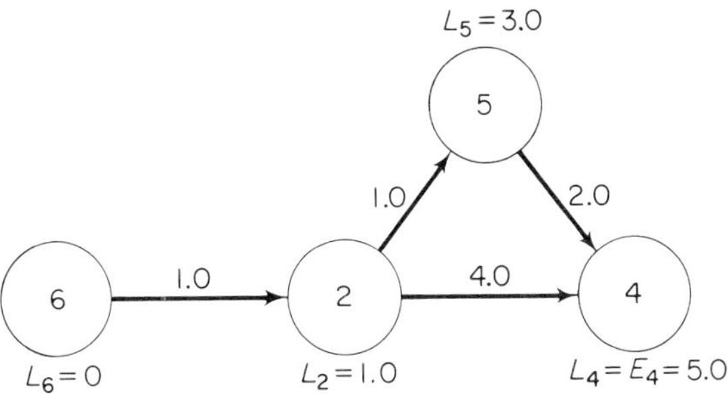

Fig. 12.9 *Example of the calculation of latest event times.*

observe that in this diagram $L_i = E_i$ except for $i = 5$. Since Event 4 is the finishing event, $L_4 = E_4 = 5.0$ by definition. The longest path from 5 to 4 is of length 2.0, and hence $L_5 = 5.0 - 2.0 = 3.0$. The longest path from 2 to 4 contains just one arrow and is of length 4.0. Hence, $L_2 = 5.0 - 4.0 = 1.0$.

The latest start time for an activity is designated by L_{ij}, and $L_{ij} = L_i$.

TABLE 12.2. *SLACK COMPUTATION FOR THE NETWORK OF FIG.* **12.10**

Event i	E_i	L_i	Slack $S_i = L_i - E_i$
1	0	0	0
2	1	2	1
3	3	3	0
4	4	4	0
5	5	5	0

12.6.1.3 Float. An important parameter in planning and scheduling a project is the amount of freedom one has in selecting a start time for each activity. Such freedom is, of course, available only within the constraint of finishing the entire project on schedule. The concepts of float and slack model this amount of freedom in critical-path systems. We talk about slack in events and float in activities.

The slack S_i in Event i is simply the latest occurrence time for that event minus its earliest occurrence time:

$$S_i = L_i - E_i$$

There are three kinds of float in activities. The first and most common is called *total float*. It is the measure of the time interval within which an activity may be started providing all preceding activities are completed as early as possible and all following activities are completed as late as possible. The total float in activity (i, j) is designated by TF_{ij}.

$$TF_{ij} = L_j - E_i - T_{ij}$$

The *free float* FF_{ij} in activity (i, j) is the free time in scheduling that activity providing that both its predecessor and its successor events take place as early as possible.

$$FF_{ij} = E_j - E_i - T_{ij}$$

The final and rarest float is called *independent float*. It is computed on the assumptions that the predecessor event takes place as late as possible and that the successor event takes place as early as possible.

$$IF_{ij} = E_j - L_i - T_{ij}$$

Consider the example network given in Fig. 12.10. Table 12.2 lists the earliest and latest event times and the slack for all events in this network. Notice that only Event 2 has nonzero slack. The floats for all the activities in Fig. 12.10 are computed in Table 12.3. Notice that three of the activities—(1, 2), (2, 3), and (3, 5)—have positive total float. Two of the activities—(2, 3) and (3, 5)—have free float; and only one—(3, 5)—has independent float. Thus we have the most freedom in scheduling

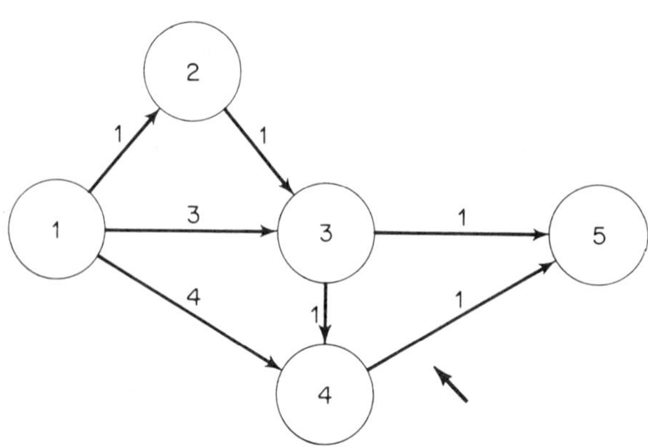

Fig. 12.10 *Example of float and slack.*

TABLE 12.3. *FLOAT COMPUTATION FOR THE NETWORK OF FIG.* **12.10**

Predecessor Event i	Successor Event j	E_i	L_i	E_j	L_j	Estimated Duration T_{ij}	Total Float $(L_j - E_i - T_{ij})$ TF_{ij}	Free Float $(E_j - E_i - T_{ij})$ FF_{ij}	Independent Float $(E_j - L_i - T_{ij})$ IF_{ij}
1	2	0	0	1	2	1	1	0	0
1	3	0	0	3	3	3	0	0	0
1	4	0	0	4	4	4	0	0	0
2	3	1	2	3	3	1	1	1	0
3	4	3	3	4	4	1	0	0	0
3	5	3	3	5	5	1	1	1	1
4	5	4	4	5	5	1	0	0	0

(3, 5), some freedom in (2, 3), a little in (1, 2), and none in all the rest.

An important output of any method for the analysis of arrow diagrams is information about slack and float.

12.6.1.4 The Critical Path. A critical path in a network is a sequence of activities, going from the starting event to the finishing event which has the property that increasing the length of any activity on the path delays the completion time of the project. Every network has at least one critical path.

A critical event is defined to be an event which has zero slack. Thus, Event i is critical if and only if $E_i = L_i$.

A critical activity is one which has zero total float. Thus, it also has zero free float and zero independent float. Moreover, its predecessor and successor events are critical.

A critical path is therefore a path of critical activities going from the beginning event to the ending event. There are two critical paths in Fig. 12.10: (1, 4), (4, 5) and (1, 3), (3, 4), (4, 5).

Increasing the duration of any critical activity increases (by the same amount) the duration of the project. However, shortening a critical activity does not necessarily shorten the project completion time. Observe, for example, the effect in the network of Fig. 12.10 of changing the duration of (critical) Activity (1, 3) from 3 to 2. The total project time is still 5, as it was before. Activity (1, 3) is no longer critical, and the sequence (1, 3), (3, 4), (4, 5) is no longer a critical path.

Shortening any noncritical activity will, of course, have no effect on total project time.

Identification of critical activities is an important part of any network-analysis system.

12.6.1.5 Loops. Another important output of network-analysis systems is the identification of loops. In Art. 12.5 we noted that loops in networks representing projects are mistakes. Uncovering loops by manual methods is often time consuming and difficult; hence, a system which performs this function automatically is desirable.

12.6.2 Analysis Methods Based on Sorting

In Art. 12.5 we noted that some early network-analysis systems required events to be numbered so that activities went from nodes with smaller numbers to nodes with larger numbers. In other words, if (i, j) is an event, then $i < j$.

There are two reasons for making this demand. One is that, if the user observes this rule rigorously, the possibility of introducing loops into the network is eliminated. The other and more important reason is that analyzing networks in which this rule is observed is very much simpler than analyzing those in which it is not.

Networks in which activities go only from smaller numbered nodes to larger numbered nodes are called *sorted networks.* Every network with no loops may be numbered so that it is a sorted network. Many network-analysis systems renumber the nodes to obtain a sorted network, perform the appropriate analysis, and then restore the original node numbers. The user is never aware that the renumbering has taken place. This node-renumbering pro-

TABLE 12.4. *AN ACTIVITY FORM SORTED FOR THE DETERMINATION OF EARLIEST EVENT TIMES*

i	j	T_{ij}	E_j
1	2	1	
1	3	3	
2	3	1	
1	4	4	
3	4	1	
3	5	1	
4	5	1	

cess to obtain a sorted network is called *topological sorting*.

In the remainder of this section, we assume that we are dealing with a sorted network. We assume, moreover, that the network contains exactly n nodes, that Node 1 is the starting node, and that Node n is the ending node.

The immediate goal of the analysis is to compute, for each event, its earliest and latest occurrence time. Two forms should be used in the calculation: an event form and an activity form. An event form is illustrated in Table 12.2, and an activity form is illustrated in Table 12.4. Initially the only information on the event form is a list of event numbers. The only information on the activity form is the predecessor and successor event numbers (Columns i and j) and the duration of each activity (Column T_{ij}). It is convenient to use separate activity forms for the calculation of earliest and latest event times.

Let us start with the determination of earliest event time. Sort the activities so that activity (k, l) follows activity (i, j) in case $l > j$. If $l = j$, then activity (k, l) follows activity (i, j) in case $k > i$. The activities in the network of Fig. 12.10 are sorted in this way in Table 12.4. In this table we make just one entry in the column headed E_j for each group of activities whose successor event is j. We use the following procedure for calculating E_j.

1. Set $E_1 = 0$.
2. Set $j = 2$.
3. Let $E_j = \max\ (E_i + T_{ij})$ for $i < j$ and (i, j) an event.
4. If $j = n$, stop; otherwise add 1 to j and go back to step 3.

Step 1 merely says that, by agreement $E_1 = 0$, that is the project starts at time zero. Step 3 is the crux of the procedure. Let us consider it in more detail. It says that we can calculate the earliest time at which Event j can occur by considering all the activities which have j as a successor event. The earliest time that event j can occur is the earliest time that the latest of these activities finishes. Consider one of these activities—(i, j) for example; it can finish, at the earliest, at time $E_i + T_{ij}$. Now, if we consider all activities whose successor event is j and take the maximum of their earliest finishing times, we shall have calculated E_j. That is exactly what Step 3 does.

Note that the E_i in $E_i + T_{ij}$ is always available when we are calculating E_j because $i < j$. Let us calculate the E_is for the network of Fig. 12.10, using Table 12.4 as a working space. We first set $E_1 = 0$. Then we set $j = 2$, and calculate E_2 according to the formula: $E_2 = \max\ (E_i + T_{i2})$. $i < 2$, and $(i, 2)$ is an event.

There is only one i, namely $i = 1$, such that $i < 2$ and $(i, 2)$ is an event. Hence, our calculation is reduced to

$$E_2 = \max\ (E_1 + T_{12}) = E_1 + T_{12}$$
$$= 0 + 1 = 1$$

So we set $E_2 = 1$ and enter this quantity in Column E_j in Table 12.4 opposite activity (1, 2).

Since j is not yet equal to n (which is 5 in this example), we add 1 to j giving $j = 3$ and proceed to calculate E_3 by the formula: $E_3 = \max\ (E_i + T_{i3})$. $i < 3$, and $(i, 3)$ is an activity. Only two values of i, namely $i = 1$ and $i = 2$, satisfy the conditions that $i < 3$ and that $(i, 3)$ is an activity, and our calculation is reduced to

$$E_3 = \max\ (E_1 + T_{13},\ E_2 + T_{23}) = \max$$
$$(0 + 3, 1 + 1) = \max\ (3, 2) = 3$$

So we enter $E_3 = 3$ opposite activity (1, 3) in Table 12.4. We calculate E_4 and E_5 similarly. We then determine the L_is in a manner analogous to the one in which the E_js are calculated. A suitable form for this calculation is in Table 12.5, which has been filled in with data from the network of Fig. 12.10. Activities are sorted so that activity (k, l) follows activity (i, j) in case $k > i$. If $k = i$ then (k, l) follows (i, j) in case $l > j$. The following procedure is used to calculate the L_is:

1. Set $L_n = E_n$.
2. Set $i = n - 1$.
3. Calculate L_i from the formula: $L_i = \min\ (L_j - T_{ij})$. $j > i$, and (i, j) is an activity.
4. If $i = 1$, stop; otherwise decrease i by 1 and go back to Step 3.

A good check on the accuracy of the com-

TABLE 12.5. *AN ACTIVITY FORM USED IN THE DETERMINATION OF LATEST EVENT TIMES*

i	j	T_{ij}	L_i
1	2	1	0
1	3	3	
1	4	4	
2	3	1	2
3	4	1	3
3	5	1	
4	5	1	4

putation of the L_is is that L_1 should be equal to zero.

The calculations described here for E_i and L_i may conveniently be performed manually for networks containing up to 200 activities.

When the earliest and the latest occurrence times have been calculated, floats may be calculated using the methods described in Par. 12.6.1.3.

12.6.3 Direct Analysis of Unsorted Networks

The analysis method described above in Par. 12.6.2 is applicable only to sorted networks. In this section, we describe a method which may be used directly on unsorted networks. Before that, however, it is necessary to develop some mathematical formalism.

We wish to define two new types of matrix operations called $+_1$ and $\times_1$. If A and B are $n \times n$ matrices of real numbers, then $C = A +_1 B$ and $D = A \times_1 B$ are defined to be $n \times n$ matrices whose elements are given by the formulas

$$C_{ij} = \max(A_{ij}, B_{ij})$$

$$D_{ij} = \max(A_{ik} + B_{kj})$$

$$k = 1, \ldots, n$$

In this section A^2 will mean $A \times_1 A$ and A^m will mean $A^{m-1} \times_1 A$. Similarly $\sum_{i=1}^{m} A^{(i)}$ means $A^{(1)} +_1 A^{(2)} +_1 \cdots +_1 A^{(m)}$.

Now, consider the matrix T whose element T_{ij} is the estimated duration of activity (i, j). If (i, j) is an activity, then T_{ij} may be zero if (i, j) is a dummy activity. If (i, j) is not an activity, we now define T_{ij} to be equal to minus infinity. (When this application is implemented on a computer, we simply set T_{ij} equal to some large, negative number.) We represent minus infinity by the sign $-\infty$.

The rules for calculating with minus infinity are given below, where a is understood to be any real number except plus infinity.

$$a + (-\infty) = -\infty + a = -\infty$$

The T matrix for the arrow diagram of Fig. 12.10 is in Table 12.6. The entry T_{ij} in this table is found in Row i and Column j. For example T_{35} (which is equal to 1) is in Row 3 and Column 5. By glancing at this matrix, we can quickly identify starting events and ending events. An event K is a starting event only if it is not the successor event for any activity. This means that, for each i, (i, K) is not an activity; i.e., that $T_{iK} = -\infty$. Hence, if we find a column, column K for example, that contains only minus infinities, then we know that event K is a starting event. Similarly, we observe that K is an ending event only if, for every j, $T_{Kj} = -\infty$. Hence, if we find a row, Row K for example, which contains nothing but minus infinities, then K must be an ending event. In Table 12.5, we observe that Event 1 is the only starting event and Event 5 is the only ending event.

We now interpret the meaning of the matrices T^2, T^3, etc. A path from Event i to Event j is a sequence of activities beginning at i and ending at j. Such a sequence might look like this:

$$(i, K_1), (K_1, K_2), (K_2, j)$$

This path contains three activities. The predecessor event of the first activity in the path must always be i and the successor event of the last activity on the path must always be j. The length of a path is defined to be the sum of the durations of all the activities in the path. The length of the path in the above example is

$$T_{iK_1} + T_{K_1K_2} + T_{K_2j}$$

E_i, the earliest time of occurrence of Event i, is simply the length of the longest path from the starting event to Event i. The minimum

TABLE 12.6. *T (DURATION) MATRIX FOR THE DIAGRAM OF FIG.* **12.10**

$$\begin{pmatrix} -\infty & 1 & 3 & 4 & -\infty \\ -\infty & -\infty & 1 & -\infty & -\infty \\ -\infty & -\infty & -\infty & 1 & 1 \\ -\infty & -\infty & -\infty & -\infty & 1 \\ -\infty & -\infty & -\infty & -\infty & -\infty \end{pmatrix}$$

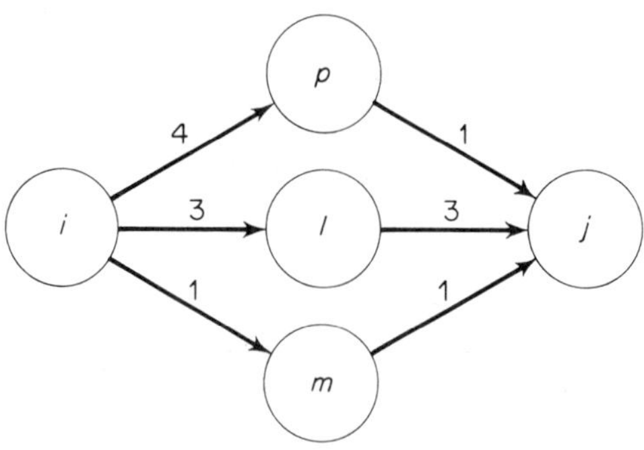

Fig. 12.11 *Calculation of the longest path from* i *to* j *containing exactly two activities.*

total project time is the length of the longest path from the starting event to the ending event.

The matrices T^2, T^3, etc., are simple tools for the calculation of the lengths of these longest paths. For example, T^2_{ij} is the length of the longest path from i to j containing exactly two arrows. Consider the situation illustrated in Fig. 12.11. Here we consider all the paths from i to j containing exactly two arrows, and we suppose that there are just three such paths: those whose central events are p, l, and m. For any other event, s for example, either (i, s) is not an activity or (s, j) is not an activity. This means that either $T_{is} = -\infty$ or $T_{sj} = -\infty$, and that $T_{is} + T_{sj} = -\infty$.

In order to compute T^2_{ij} we use the formula

$$T^2_{ij} = (T \times_1 T)_{ij} = \max_{k = 1, \ldots, n} (T_{ik} + T_{kj})$$

where n is the number of the highest numbered event in the network. There are n numbers, $T_{ik} + T_{kj}$ over which this maximum is taken. In our example (Fig. 12.11) only three of these numbers are different from minus infinity. These three numbers are

$$T_{ip} + T_{pj} = 4 + 1 = 5$$

$$T_{il} + T_{lj} = 3 + 3 = 6$$

$$T_{im} + T_{mj} = 1 + 1 = 2$$

Since any real number is always greater than $-\infty$, the maximum must be one of the above candidates. It is obviously 6. The longest path containing two activities going from i to j is $(i, l), (l, j)$; its length is 6. By similar arguments we can easily show that T^r_{ij} is the length of the longest path from i to j containing exactly r arrows.

We now define a new sequence of matrices $D^{(m)}$ as follows:

$$D^{(1)} = T$$

$$D^{(2)} = D^{(1)} +_1 T^2 = T +_1 T^2$$

$$D^{(3)} = D^{(2)} +_1 T^3 = T +_1 T^2 +_1 T^3$$

$$D^{(m)} = D^{(m-1)} + T^m = \sum_{j=1}^{m} T^j$$

$D^{(m)}_{ij}$ is the length of the longest path from i to j containing at most m arrows. This follows from the definition of $+_1$ which says that

$$(A +_1 B)_{ij} = \max (A_{ij}, B_{ij})$$

Hence

$$D^{(m)}_{ij} = \left(\sum_{k=1}^{m} T^k\right)_{ij} = \max (T_{ij}, T^2_{ij}, \ldots, T^m_{ij})$$

If our network contains n activities, then (since loops are forbidden) the maximum number of activities in any path through the network is $n - 1$. Each path contains one fewer activities than events. Thus, in constructing the $D^{(m)}$s we need to consider only ms as large as $n - 1$. We define

$$D = D^{(n-1)}$$

D_{ij} is the length of the longest path from i to j. Since E_i is the length of the longest path from the starting event to Event i, $E_i = D_{1i}$, for $i \neq 1$, assuming that Event 1 is the starting event.

If Event n is the ending event, then

$$L_n = E_n$$

$$L_i = E_n - D_{in} \quad \text{for} \quad i \neq n$$

That is, L_i, the latest occurrence of Event i, is the total project time E_n minus the length of the longest path from Event i to the ending event. When the E_is and L_is are known, the calculation of floats proceeds as described in Par. 12.6.1.3.

A sample calculation of the matrix D for the simple network illustrated in Fig. 12.12 is in Table 12.7. The starting event is 3 and the ending event is 2. Also, $D^{(2)} = D^{(3)}$. Only one path contains three activities in the example network, namely (3, 1), (1, 4), (4, 2), and it is

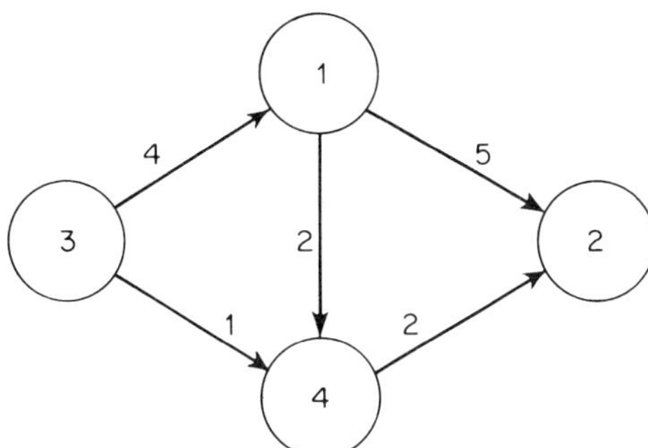

Fig. 12.12 *Network used to illustrate the nonsort analysis method.*

TABLE 12.7. *NONSORT ANALYSIS OF THE NETWORK IN FIG.* **12.12**

$$T=\begin{pmatrix} -\infty & 5 & -\infty & 2 \\ -\infty & -\infty & -\infty & -\infty \\ 4 & -\infty & -\infty & 1 \\ -\infty & 2 & -\infty & -\infty \end{pmatrix} \quad D^{(1)}=\begin{pmatrix} -\infty & 5 & -\infty & 2 \\ -\infty & -\infty & -\infty & -\infty \\ 4 & -\infty & -\infty & 1 \\ -\infty & 2 & -\infty & -\infty \end{pmatrix}$$

$$T^2=\begin{pmatrix} -\infty & 4 & -\infty & -\infty \\ -\infty & -\infty & -\infty & -\infty \\ -\infty & 9 & 6 & 6 \\ -\infty & -\infty & -\infty & -\infty \end{pmatrix} \quad D^{(2)}=\begin{pmatrix} -\infty & 5 & -\infty & 2 \\ -\infty & -\infty & -\infty & -\infty \\ 4 & 9 & -\infty & 6 \\ -\infty & 2 & -\infty & -\infty \end{pmatrix}$$

$$T^3=\begin{pmatrix} -\infty & -\infty & -\infty & -\infty \\ -\infty & -\infty & -\infty & -\infty \\ -\infty & 8 & -\infty & -\infty \\ -\infty & -\infty & -\infty & -\infty \end{pmatrix} \quad D^{(3)}=\begin{pmatrix} -\infty & 5 & -\infty & 2 \\ -\infty & -\infty & -\infty & -\infty \\ 4 & 9 & -\infty & 6 \\ -\infty & 2 & -\infty & -\infty \end{pmatrix}$$

shorter than the path (3, 1), (1, 2). The earliest start times in this network are given by $E_3 = 0$ and $E_i = D_{3i}$ for $i \neq 3$. Hence $E_2 = D_{32} = 9$, $E_1 = D_{31} = 4$, and $E_4 = D_{34} = 6$.

The latest start times are given by $L_2 = E_2 = 9$ and $L_i = E_2 - D_{i2}$ for $i \neq 2$. Hence $L_1 = E_2 - D_{12} = 9 - 5 = 4$, $L_3 = E_2 - D_{32} = 9 - 9 = 0$, $L_4 = E_2 - D_{42} = 9 - 2 = 7$.

12.6.4 The Detection of Loops

An important role for any network-analysis system is the detection of loops. The presence of loops indicates that errors were made in the preparation of the network.

Systems which work only on sorted networks need check only each input activity (i, j) to make sure that $i < j$. This type of error-checking is very simple to perform and produces results which enable the user to correct his mistakes quickly.

In systems which accept unsorted networks, perform a sorting step, and then analyze the resulting sorted network, loop detection must be accomplished during the sorting phase. If a network contains loops, it is not possible to rearrange it into a sorted network. Every sort routine must be written to cope with networks containing loops and to report their presence and their identity back to the user.

Loop detection is quite simple in the analysis method described in Par. 12.6.3. As the T^ks are calculated, their diagonal terms are examined. If, for example $T^k_{ii} \neq -\infty$, there is a loop containing k activities in which Event i is present. All other events involved in the loop also generate nonnegative entries along the diagonal of T^k. In adapting this analysis system to perform loop-detection functions, one must compute T^n when considering an n-node network because there may be a loop containing exactly n activities.

12.6.5 Networks with Multiple Starting or Ending Events

Many projects contain multiple starting events or multiple ending events or both. Often two or more activities may be started independently of one another at the beginning of a project. Two such activities in a construction project, for example, are the excavation of the foundation and the ordering of framing lumber. It is possible either to handle these networks directly or to modify them so as to eliminate the multiple beginning and ending events.

Consider the network in Fig. 12.13. It contains two beginning events and two ending events. We could eliminate this situation by inserting dummy activities (1, 2) and (4, 5). If we do this, Event 1 becomes the only starting

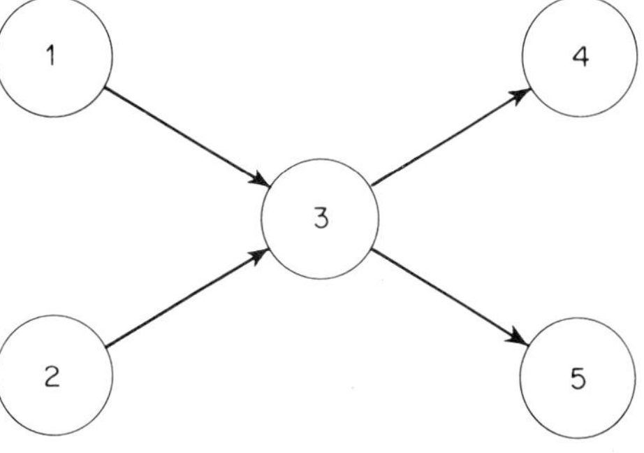

Fig. 12.13 *A network containing multiple starting and ending events.*

event, and Event 5 becomes the only ending event.

To analyze a network containing multiple starting or ending events or both directly, we may assign E_s arbitrarily to each starting event s. Often during the planning stage all E_ss are set equal to zero. If we use in-sort analysis, we then proceed exactly in the manner described in Par. 12.6.2 in the calculation of the E_is.

In calculating the L_i using in-sort analysis, for each ending event e we set

$$L_e = \max_{f \text{ an ending event}} (E_f)$$

In other words, the latest time of occurrence for any ending event is equal to the total project time. We then calculate the other L_is just as before.

If we are using the nonsort-analysis method of Par. 12.6.3, we calculate the Matrix D just as before. We then assign all E_ss for starting events s. Then, if i is not a starting event,

$$E_i = \max_{s \text{ a starting event}} (E_s + D_{si})$$

To calculate the L_is, we set, for each ending event e

$$L_e = \max_{f \text{ an ending event}} (E_f)$$

just as we did for the in-sort calculation. If i is not an ending event, we set

$$L_i = \min_{e \text{ an ending event}} (L_e - D_{ie})$$

12.6.6 Network Analysis During the Control Phase

The types of analysis discussed this far are directly applicable to the planning and scheduling phases of a project. When it is under way, one wishes to use network techniques for control.

To control a project effectively one must have information about its current status. From this information one can learn (via a network model) how well the project schedule is being adhered to and where effort should be expended to remedy any slippages.

Calculations performed during the control phase of the project are very similar to those performed during the planning phase. Calculation of the L_is, the latest occurrence times, is identical—and, in fact, the values of L_i previously calculated should be used in the control phase. However, new values should be calculated for the E_is.

This new calculation of the E_is is done by replacing the original planning network with a network which reflects the current status of the project. This current network is obtained by discarding all events which are predecessor events of only completed activities. The network which results from this process is, in general, one with multiple starting events. Each starting event s in this current network should be assigned its actual time of occurrence as E_s. Calculation of the remaining E_is then proceeds as described in Par. 12.6.5. Notice that E_n (assuming that n is the number of the ending event) may turn out to be different from L_n.

Floats are now calculated as before, using the new E_is and the old (planning) L_is. If the critical activities in the project have slipped, negative floats occur in some activities. If critical activities have been finished ahead of schedule, some activities which were formerly critical may no longer be so.

An example of a control-phase–network calculation is in Fig. 12.14 and Table 12.8. It is assumed that this calculation takes place 5 weeks after the project was started. In Fig. 12.14 the activities which have been completed are cross-hatched, and their actual durations are written in parentheses beside their estimated durations. Activities (1, 2) and (1, 3)

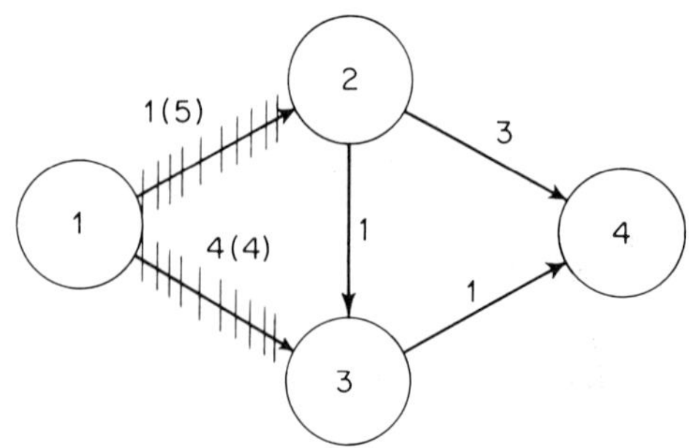

Fig. 12.14 *Example of a control-phase network.*

TABLE 12.8. *EXAMPLE OF A CONTROL PHASE NETWORK CALCULATION*

Event Table

Event Number i	Planned E_i	Planned L_i	Current E_i
1	0	0	0
2	1	3	5
3	4	4	6
4	5	5	8

Activity Table

Activity i	j	Estimated Duration T_{ij}	Actual Duration	Current Total Float TF_{ij}	Planned Total Float
1	2	1	5	—	—
1	3	4	4	—	—
2	3	1	—	−2	2
2	4	3	—	−3	1
3	4	1	—	−2	0

have been completed. Activity (1, 2) required 5 weeks to finish instead of the estimated 1 week. Activity (1, 3) was completed in 4 weeks, as estimated. Event 1 is eliminated from the network since it is the predecessor of only completed activities. The resulting network has only one starting event, namely Event 2. We assign $E_2 = 5$ since Event 2 actually occurred at Time 5. We then proceed to calculate $E_3 = 6$ and $E_4 = 8$. Noting that the entire project has slipped by 3 weeks, we now estimate that it will require a total of 8 weeks instead of the originally estimated 5.There is now only one critical activity, namely Activity (2, 4). Its total float is −3. If we were managing this project, we would direct our efforts toward shortening Activity (2, 4). If we could reduce its duration from 3 weeks to 2, we would shorten the total project time by 1 week.

12.7 VARIATIONS OF THE BASIC NETWORK MODEL

Since 1958, when network methods were first used in project planning and control, various modifications of the basic techniques described in the above sections have been introduced. Each of these modifications has been a response to a requirement generated by project management.

PERT (Program Evaluation and Review Techniques) was designed to enable management to obtain more realistic estimates of activity duration. In addition, it provides a model which reflects some of the statistical uncertainties involved in project scheduling and control.

PERT-COST is an extension of PERT which contains provisions for financial estimating and planning and provides links between traditional accounting structures and the network model.

CPM (the Critical Path Method) provides a model for studying the trade-offs between time and costs.

Resource-allocation models, such as RAMPS (Resource Allocation and Multiple Project Scheduling) and JAM (Joint Allocation Method) attack the problem of allocating finite resources to activities in one or more projects in a network frame.

12.7.1 PERT

PERT is the network scheduling and control model devised in 1958 by the U.S. Navy Special Projects Office. In PERT the user is asked to make three estimates for the duration of each activity: an optimistic estimate a, a pessimistic estimate b, and a most probable estimate m. These estimates are fitted to a beta distribution and the mean t_e of this distribution is calculated.

$$t_e = (a + 4m + b)/6$$

This mean or expected time takes the place of

the estimated duration T_{ij} in most subsequent calculations.

It was hoped that PERT systems could answer questions of the form: "With what probability will this project be completed in 18 weeks?" However, because of the (necessarily) qualitative nature of the estimating process, the answers to such questions have little validity.

The PERT multiple-estimate feature does have some advantages aside from its statistical one. In single-estimate systems potential activity managers tend to insert fat into their estimates. The result is that estimates obtained in such systems are often unrealistic. The PERT three-estimate system helps to discourage this practice. When a work leader is asked to make three estimates, he becomes aware that management is not asking him for an ironclad figure. The result is generally a fairly rational estimate.

The processing of three-estimate data is very little more expensive than standard single-estimate processing.

12.7.2 PERT-COST

In the PERT-COST system a sequence of accounting functions has been associated with the PERT system described in Par. 12.7.1. The network-analysis, time-oriented features of the system are unchanged.

The user of PERT-COST must supply certain inputs in addition to the usual arrow diagram. These include the assignment of each activity to an account number, estimates of the effort required in each activity, and estimates of current rates for each type of effort in the project.

Outputs include extensions of costs for each account, schedules of cost accumulation, and summaries of costs by effort type. These outputs are available during both the planning and the control phases of the project.

The use of the PERT-COST system is a requirement in proposals submitted in response to Defense Department requests for proposal for large-scale development contracts. Its use as a control tool is also mandatory in most Defense Department contracts.

12.7.3 The Critical-Path Method

The critical-path method (CPM) is an adaptation of network-scheduling methods which allows the user to study the problem of time-cost trade-offs. This system is designed to supply answers to questions of the following form: "What is the minimum cost increase involved in shortening the project by 2 days? To what activities should this extra cost be allocated?"

When using the CPM system, the user must supply two estimates for the duration of each activity: T_{ij}, the normal completion time; and C_{ij}, the crash completion time. The normal completion time is that in which the activity can be reasonably completed without the use of overtime and without the mobilization of extra resources. The crash time, as its name implies, is the minimum time required for completion of the activity through the use of overtime, extra shifts, added resources, etc.

Together with these two time estimates, the user must supply an estimate A_{ij} of the added cost involved in completing activity (i, j) in time C_{ij} instead of T_{ij}. This is an incremental cost, not a total cost. In the analysis it is assumed that this added cost goes linearly between the normal time and the crash time.

The cost slope S_{ij} is computed for each activity:

$$S_{ij} = A_{ij}/(T_{ij} - C_{ij})$$

The cost involved in shortening an activity by a time x ($x < T_{ij} - C_{ij}$) is given by xS_{ij}. The usual network analysis is made, employing the time estimates T_{ij}. Then a minimum time-cost curve is computed. For each time x, it gives the added cost $c(x)$ of completing the

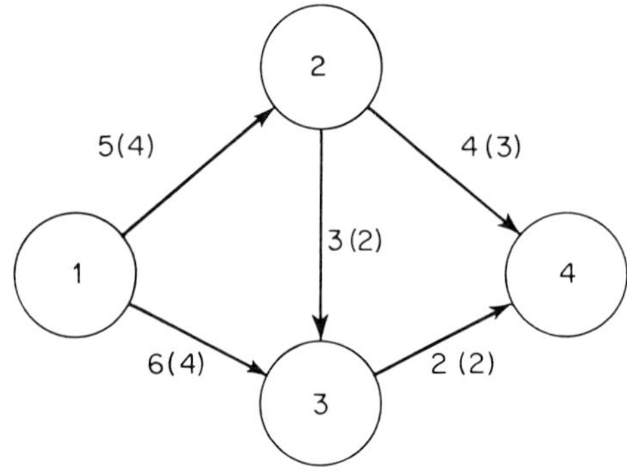

Fig. 12.15 *Example of a CPM network. Crash time estimates* (C_{ij}) *are in parentheses.*

TABLE 12.9. *ACTIVITY LIST FOR THE CPM NETWORK SHOWN IN FIG.* **12.15**

Predecessor Event i	Successor Event j	Normal Time T_{ij}	Crash Time C_{ij}	Added Cost A_{ij}	Cost Slope S_{ij}
1	2	5	4	\$ 6	\$ 6 per day
1	3	6	4	\$20	\$10 per day
2	3	3	2	\$ 3	\$ 3 per day
2	4	4	3	\$ 5	\$ 5 per day
3	4	2	2	\$ 0	—

TABLE 12.10. *ANALYSIS OF THE CPM NETWORK OF FIG.* **12.15** *USING NORMAL ACTIVITY TIMES*

Event Number i	Earliest Time E_i	Latest time L_i
1	0	0
2	5	5
3	8	8
4	10	10

project in time x. If n is the ending event in the project, $c(E_n) = 0$.

The computation of the function c involves looking at the original arrow diagram, selecting a combination of parallel critical activities (the shortening of which will shorten the entire project at minimum cost), shortening them until some additional activity becomes critical, and then repeating the whole process.

An illustration of CPM is in Fig. 12.15. Normal-time estimates T_{ij} are given as usual. Crash-time estimates C_{ij} are enclosed in parentheses. In Table 12.9 is a complete activity list for this network, including added cost A_{ij} and cost slope S_{ij}. Note that for Activity (3, 4) the normal time and the crash time are identical; each is 2 days.

In Table 12.10 the analysis of this network using normal activity times is displayed. Observe that the total project time E_4 is equal to 10 days.

There is only one critical path, and it contains activities (1, 2), (2, 3), and (3, 4). By shortening any of these activities, we can shorten the total project time. We wish, however, to shorten the critical-path activity which has the minimum cost slope. This is Activity (2, 3), and its cost slope is \$3 per day. (Notice that we cannot shorten Activity (3, 4).) We may shorten

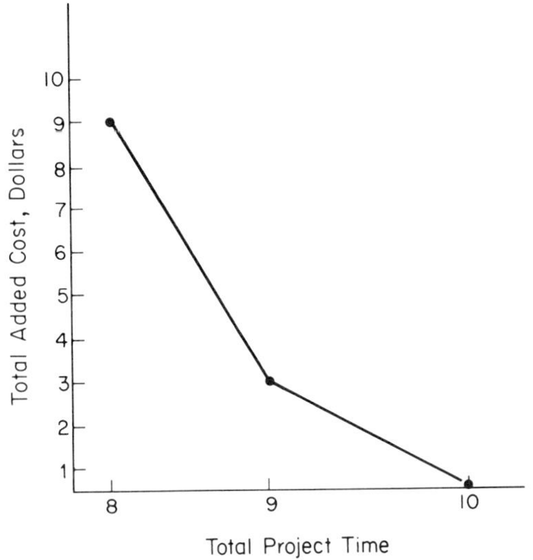

Fig. 12.16 *Minimum cost curve for the CPM network of Fig. 12.15.*

Activity (2, 3) to 2 days, its crash time, at a cost of $3. Thus we have shortened the total project time to 9 days, and Activity (2, 4) becomes critical. We may shorten Activity (1, 2) by 1 day at an added cost of $6. We have now shortened the total project time to 8 days at a total added cost of $9. Activity (1, 3) has now become critical. No further shortening of total project time is possible since, at their crash times, Activities (1, 2), (2, 3), and (3, 4) are all critical.

The minimum cost curve *c* for this project is illustrated in Fig. 12.16. The characteristic of the minimum cost curve is that it is flattest at the right-hand end and becomes increasingly steep toward the left. In other words, the shorter we make the project, the more it costs per day.

Analysis of CPM networks containing more than ten or fifteen events is best undertaken on a computer.

12.7.4 Resource-Allocation Methods

None of the project models discussed thus far attacks the problem of allocating finite resources. In computing start times for activities we have always assumed that the resources for their accomplishment would be available whenever we wanted them. This situation does not generally obtain.

When we construct a project schedule, we must be sure that the resources required by each activity are available at its scheduled start time. The achievement of this goal is the purpose of several resource-allocation models, most of which are based on network-analysis methods. They provide algorithms which answer the question: "When two activities are competing for a single resource, which one should use it first?"

Probably the most extensively used resource-allocation model is a proprietary system called RAMPS (Resource Allocation and Multiple Project Scheduling system) developed by CEIR. Another scheme, devised by Angus McDonald and Bruce Goeller, is called JAM (Joint Allocation Method).

The problem of finding a truly optimal schedule—one that minimizes total project time and uses only available resources—has not been completely solved. That is, no computational algorithm which invariably produces an optimum schedule in a reasonable time has yet been found. The systems cited above, however, do generally produce a near optimum schedule at a reasonable cost in computation.

12.8 COMPUTER PROGRAMS FOR THE ANALYSIS OF CRITICAL-PATH NETWORKS

During the past several years there has been a great proliferation of computer programs for use in network analysis. All major computer manufacturers provide systems for use on their machines. The most generally implemented systems are PERT-TIME and PERT-COST.

Most of these systems have features which make them convenient for controlling a project in progress. These features include the capability to store the basic network on tape and to update or modify it from cards.

Output from these systems is often flexible. It is generally possible to output activities sorted in order of increasing predecessor event number or in order of increasing float. Critical activities are generally represented distinctively by underlining, overprinting, or indentation. In some systems, a graphic output capability is provided.

Manuals for these systems may be obtained through the manufacturers' local sales offices.

An excellent bibliography of many of these systems is provided in the publication by Phillips and Beek (Operations Research, Inc.) listed in the bibliography below. This publication contains a summary of the characteristics of each system listed. The U.S. Air Force has produced a series of publications on PERT and PERT-COST (see reference in bibliography below). These publications describe systems implemented by the Air Force for use by Defense Department contractors.

BIBLIOGRAPHY

Kadet, Jordan, and Bruce H. Frank, "PERT for the Engineer," *IEEE Spectrum*, **1**, No. 11 (Nov. 1964), 131–37.

Lambourn, S., "Resource Allocation and Multi-Project Scheduling (RAMPS)—A New Tool in Planning and Control," *Computer Journal*, **5**, No. 4 (Jan. 1963), 300–304.

Phillips, C. R., and C. R. Beek, *Computer Programs for PERT and CPM*. Silver Springs, Md.: Operations Research, Incorporated, 1963.

United States Air Force, *ASAF PERT*. Washington, D. C., 1963.

DR. GERALD J. LIEBERMAN *is professor of Operations Research and Statistics and chairman of the Operations Research Department at Stanford University. He was graduated from Cooper Union with a bachelor's degree in mechanical engineering in 1948, then entered Columbia University and earned the master's degree in mathematical statistics. In 1953 he received the Ph.D. in statistics with a minor in industrial engineering.*

Before joining the Stanford faculty, Dr. Lieberman worked for the National Bureau of Standards.

He is a fellow in the American Society for Quality Control, the American Statistical Association, the Institute of Mathematical Statistics, and the American Association for the Advancement of Science. He is a member of the Institute of Management Science, Operations Research Society of America, and the American Society for Engineering Education. He served as national director and national executive director of ASQC, national treasurer of IMS and vice president and chairman of the Physical and Engineering Sciences section of ASA.

He has contributed many articles to Industrial Quality Control, Technometrics, Journal of the American Statistical Association, Biometrika *and other scientific journals. He is co-author of four books:* Engineering Statistics, *with A. H. Bowker;* Handbook of Industrial Statistics, *with A. H. Bowker*, Tables of the Hypergeometric Probability Distribution, *with D. B. Owen; and* Introduction to Operations Research, *with F. E. Hillier.*

DR. ALBERT BOWKER *chancellor of the City University of New York since 1963, received his Ph.D. degree from Columbia University after having earned a B.S. at Massachusetts Institute of Technology. He joined the Stanford University faculty as assistant professor of mathematical statistics in 1947, and successively became associate professor, director of the Applied Mathematics and Statistics Laboratories, professor of mathematics and statistics, and dean of the Graduate Division.*

Dr. Bowker is a fellow and past president of the Institute of Mathematical Statistics and of the American Statistical Association and a fellow of the American Association for the Advancement of Science and of the American Society of Quality Control. He is a member of the Biometric Society, the Operations Research Society of America, the Society for Industrial and Applied Mathematics, Sigma XI, and Stanford University advisory committees on computer science and the School of Education.

Dr. Bowker has written many articles for the scientific journals and has contributed chapters or sections in a number of books and handbooks. He is co-author with Henry P. Goode of Sampling Inspection by Variables, *with Gerald J. Lieberman of the* Handbook of Industrial Statistics and Engineering Statistics.

He is a trustee of the Dalton School, the Hall of Science of the City of New York, the Institute for Educational Development, the Institute of International Education, the Massachusetts Institute of Technology, and the Mount Sinai School of Medicine. For his leadership in expanding educational opportunities, the New York Urban League presented Dr. Bowker with the 1969 Frederick Douglass Award.

SECTION 13

INDUSTRIAL STATISTICS

ALBERT H. BOWKER
AND
GERALD J. LIEBERMAN

13.1 BASIC STATISTICAL CONCEPTS

13.1.1 Introduction

The science of statistics deals with drawing conclusions from observed data; the popular conception of statistics is that it involves large masses of data and concerns itself with percentages, averages, or presentation of data in tables or charts. This represents only a small part of the field today and is of less interest to industrial engineers than are other aspects of statistics—e.g., quality control, sampling inspection, and the design and analysis of experiments. These latter topics are the major ones presented in this section.

Most scientific investigations, whether concerned with the effect of a new drug on polio, methods of allaying traffic congestion in a big city, consumer reaction to a new product, or the quality of manufactured products, depend on observations, even if they are of a rudimentary sort. In scientific and industrial experimentation, these observations are taken to study the effect of variation of certain factors or the relation between certain factors. One may wish to study the quality of a raw material from a new supplier, the relation between tensile strength and hardness for a particular alloy, or the optimum combination of conditions in a manufacturing process. Ultimately, these observations are to be used for making decisions; the remainder of this section deals with providing procedures for making decisions with preassigned risks on the basis of the limited information in samples. These procedures are illustrated by examples. Many of these examples contain published data with references to the source.

13.1.2 Empirical Distributions and Histograms

A basic notion of statistics is the notion of variation. For example, there is no single figure

TABLE 13.1. *ITEM LIFETIMES FOR INCANDESCENT LAMPS*

Date	Item Lifetimes										Average of Sample
1–2–47	1067	919	1196	785	1126	936	918	1156	920	948	997
1–9–47	855	1092	1162	1170	929	950	905	972	1035	1045	1012
1–16–47	1157	1195	1195	1340	1122	938	970	1237	956	1102	1121
1–23–47	1022	978	832	1009	1157	1151	1009	765	958	902	978
1–30–47	923	1333	811	1217	1085	896	958	1311	1037	702	1027
2–6–47	521	933	928	1153	946	858	1071	1069	830	1063	937
2–13–47	930	807	954	1063	1002	909	1077	1021	1062	1157	998
2–20–47	999	932	1035	944	1049	940	1122	1115	833	1320	1029
2–27–47	901	1324	818	1250	1203	1078	890	1303	1011	1102	1088
3–6–47	996	780	900	1106	704	621	854	1178	1138	951	923
3–13–47	1187	1067	1118	1037	958	760	1101	949	992	966	1014
3–20–47	824	653	980	935	878	934	910	1058	730	980	888
3–27–47	844	814	1103	1000	788	1143	935	1069	1170	1067	993
4–3–47	1037	1151	863	990	1035	1112	931	970	932	904	993
4–10–47	1026	1147	883	867	990	1258	1192	922	1150	1091	1053
4–17–47	1039	1083	1040	1289	699	1083	880	1029	658	912	971
4–23–47	1023	984	856	924	801	1122	1292	1116	880	1173	1017
5–1–47	1134	932	938	1078	1180	1106	1184	954	824	529	986
5–8–47	998	996	1133	765	775	1105	1081	1171	705	1425	1015
5–15–47	610	916	1001	895	709	860	1110	1149	972	1002	922
5–22–47	990	1141	1127	1181	856	716	1308	943	1272	917	1045
5–29–47	1069	976	1187	1107	1230	836	1034	1248	1061	1550	1130
6–5–47	1240	932	1165	1303	1085	813	1340	1137	773	787	1058
6–12–47	1438	1009	1002	1061	1277	892	900	1384	1148	—	1123
6–19–47	1117	1225	1176	709	1485	1225	1011	1028	1227	1277	1148
6–26–47	1222	912	885	1562	1118	1197	976	1080	924	1233	1111
7–3–47	1135	623	983	883	1088	1029	1201	898	970	1058	987
7–10–47	1160	831	1023	1354	1218	1121	1172	1169	1113	1308	1147
7–17–47	1166	1470	1635	1141	1555	1054	1461	1057	1228	1187	1295
8–7–47	1016	744	1197	1122	666	1022	964	1085	612	1003	943
8–14–47	1235	942	1055	893	1235	1056	968	1056	1014	1096	1055
8–21–47	1013	889	1430	926	1297	1033	1024	1103	1385	—	1122
8–28–47	1077	813	1121	960	1156	1033	1255	225	525	675	884
9–4–47	1211	995	924	732	935	1173	1024	1254	1014	—	1029
9–11–47	798	1080	862	1220	1024	1170	1120	898	918	1086	1018
9–18–47	1028	1122	872	826	1337	965	1297	1096	1068	943	1055
9–25–47	1490	918	609	985	1233	985	985	1075	1240	985	1051
10–2–47	1105	1243	1204	1203	1310	1262	1234	1104	1303	1185	1215
10–9–47	759	1404	944	1343	932	1055	1381	816	1067	1252	1095
10–16–47	1248	1324	1000	984	1220	972	1022	956	1093	1358	1118
10–23–47	1024	1240	1157	1415	1385	824	1690	1302	1233	1331	1260
10–30–47	1109	827	1209	1202	1229	1079	1176	1173	769	905	1068

for the life of all incandescent lamps produced under certain conditions; some may last many times as long as others. Consider, for example, the data in Table 13.1 on the lifetimes in hours of 417 40–w 110–v internally frosted incandescent lamps taken from forced life tests.* The life varies from a low of 225 hr to a maximum of 1690. Many factors including variations in raw materials, workmanship, function of automatic machinery, and, in fact, test conditions may account for these differences; some of these factors may be controlled carefully but some pattern of variation is inherent in all observational data. Perhaps no one would expect all light bulbs to have exactly the same life, but even the results of the most carefully

* D. J. Davis, "An Analysis of Some Failure Data," *Journal of the American Statistical Association*, Vol. 47, 1953.

TABLE 13.2. *FREQUENCY TABLE FOR LENGTH OF LIFE OF INCANDESCENT LAMPS*

Class Interval, 100 Hr	Frequency, f	Mid-Point of Class Interval, m	Deviation from Arbitrary Origin (1050) in Class Intervals, d	fd	fd^2
200–299	1	250	−8	−8	64
300–399	—	350	−7	—	—
400–499	—	450	−6	—	—
500–599	3	550	−5	−15	75
600–699	10	650	−4	−40	160
700–799	21	750	−3	−63	189
800–899	43	850	−2	−86	172
900–999	91	950	−1	−91	91
1000–1099	87	1050	0	0	0
1100–1199	79	1150	1	79	79
1200–1299	44	1250	2	88	176
1300–1399	24	1350	3	72	216
1400–1499	9	1450	4	36	144
1500–1599	3	1550	5	15	75
1600–1699	2	1650	6	12	72
Totals	417			−1	1513

controlled experiments, such as those designed to measure the velocity of light, exhibit variation due to experimental error. Returning to the light bulbs, the data in Table 13.1 recorded serially do not present a clear picture of the nature of the variation, and it is more useful to present the data in a frequency table (Table 13.2), grouping adjacent observations into classes, which are usually called class intervals or cells.

Calculations

1. Raw data
   a. Mean:

$$\bar{x} = \frac{\sum_{i=1}^{N} x_i}{N} = \frac{435{,}921}{417} = 1{,}045.37$$

   b. Standard deviation:

$$s^2 = \frac{\sum(x_i - \bar{x})^2}{N-1} = \frac{\sum x_i^2 - (\sum x_i)^2/N}{N-1}$$
$$= \frac{470{,}808{,}333 - 190{,}027{,}118{,}241/417}{416}$$
$$= 36{,}316.85$$
$$s = 190.57$$

2. Grouped data
   a. Mean:

$\bar{x}$ in class intervals from arbitrary origin of 1050

$$= \frac{\sum fd}{N}$$
$$= -\,0.002$$

$\bar{x}$ in original units $=$ arbitrary origin

$$+ \frac{\sum fd}{N}\text{(class interval)}$$
$$= 1050 - 0.002(100) = 1050$$

   b. Standard deviation:

s in class interval units

$$= \sqrt{\frac{\sum fd^2 - (\sum fd)^2/N}{N-1}}$$
$$\sqrt{\frac{1513 - (1)^2/417}{416}} = 1.91$$

s in original units $=$ s in class interval units (class interval) $=$ 1.91

The most common method of presenting graphically such data as Table 13.1 is in terms of a histogram (Fig. 13.1), which consists of a number of columns with sides equal to the class interval boundaries and height proportional to the frequency.

For comparative or summary purposes, it is often useful to describe an observed frequency distribution by one or two numbers; the most common method is to use the arithmetic mean,

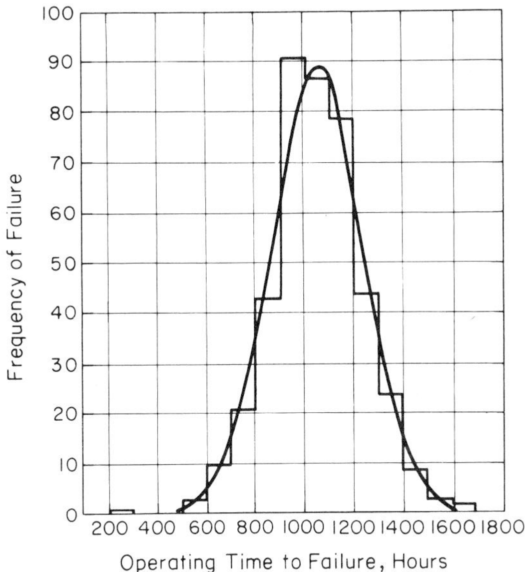

Fig. 13.1 *Life length histogram for incandescent lamps.*

which is a measure of central tendency, and the sample standard deviation, which is a measure of dispersion around the mean. The calculation of the mean and standard deviation from both the raw data and the grouped data were illustrated in Table 13.2.† For small numbers of observations, grouped data computations are not usually advantageous.

13.1.3 Theoretical Distributions

In the preceding article we described the concept of variables and the presentation of a large volume of data in the form of a histogram. Many decisions from observed data are based on small samples that are assumed to be drawn at random from a larger source, a so-called parent universe or population. To make valid inferences on the basis of small samples it is necessary to make some assumptions about the form of this population.

It is worthwhile to introduce here an important notion about these models; the constants that characterize these populations are called parameters and should always be clearly distinguished from the quantities we calculate from the observations, i.e., statistics. Thus the statistic, the arithmetic mean of a sample, may differ from the parameter, the true mean value* of the population.

13.1.3.1 Normal Distribution. The most important distribution in statistics is the normal distribution. This distribution has a symmetric bell-shaped form and tends to infinity in both directions (see Fig. 13.2). One of the classical theorems of probability states in essence that if observed quantities can be considered to be the result of large numbers of additive chance effects, the distribution of these quantities should be approximately normal. This theory, plus a mass of empirical evidence, indicates that the normal distribution may be assumed

† Throughout this section terms such as $x_1+x_2+x_3+x_4+x_5$ will be represented as $\sum_{i=1}^{5} x_i$. When no confusion exists, this may instead be written as $\sum x_i$. Terms such as $x_{11}+x_{12}+x_{13}+x_{14}+x_{21}+x_{22}+x_{23}+x_{24}+x_{31}+x_{32}+x_{33}+x_{34}$ will be represented as $\sum_{i=1}^{3}\sum_{j=1}^{4} x_{ij}$.

* If $f(x)$ represents the equation for the probability distribution, the true mean or population mean given by $\mu = \int_{-\infty}^{+\infty} xf(x)\,dx$ is a parameter.

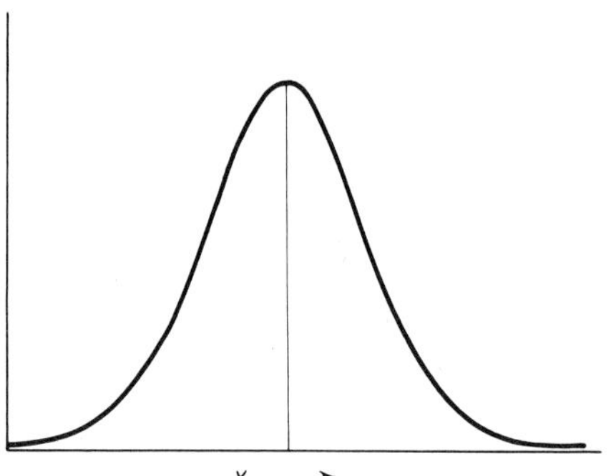

Fig. 13.2 *Normal distribution.*

as the underlying population for a large number of industrial problems. The equation for the normal curve is

$$\frac{1}{\sqrt{2\pi}\sigma} e^{-(x-\mu)^2/2\sigma^2}$$

The area under the curve is 1. The curve is determined by the two parameters μ, the population mean, and σ, the population standard deviation.*

The normal curve in standard form has area equal to 1, but it is sometimes desirable to construct a normal distribution which has the same area as a histogram. In this case, we write the equation

$$\frac{Nw}{\sqrt{2\pi}\sigma} e^{-(x-\mu)^2/2\sigma^2}$$

where N is the total number of observations and w is the width of the class interval. Areas of the normal distribution may be found in Table 13.3.

Most of the techniques of analysis presented in the remaining articles depend on the distribution of statistics based on random samples from a normal distribution. The sample mean has a normal distribution; in fact, for large samples the distribution of the sample mean from any distribution will be approximately normal. Other distributions, the chi-square, the t distribution, and the F distribution, are the distributions of various functions of means and variances of samples from a normal distribution. The explicit forms of these distributions need not concern us.

13.1.3.2 The Binomial Distribution. If we have a series of n independent trials and if at each trial p is the probability that the event will occur, the probability that r events occur in the n trials is $\binom{n}{r}p^r q^{n-r}$, where $q = 1 - p$, and $\binom{n}{r}$ is the number of combinations of n things taken r at a time.

$$\binom{n}{r} = \frac{n!}{r!(n-r)!} \quad \text{and}$$
$$n! = n(n-1)(n-2)\ldots(3)(2)(1)$$

The most common application in industrial work is lot-by-lot acceptance inspection, where the lot is large compared with the sample size; p is the fraction defective in the lot, n is the size of a sample drawn at random from the lot, and r is the observed number of defectives.

The observed number of defectives is some number from 0 to n, and hence

$$\sum_{r=0}^{n} \binom{n}{r} p^r q^{n-r} = 1$$

For example, suppose $n = 18, p = 0.10$.

r	Probability of r defectives
0	0.150
1	0.300
2	0.284
3	0.168
4	0.070
5	0.022
6	0.005
7	0.001
8	0.000
—	—
—	—
—	—
18	0.000

The distribution function of the above binomial distribution function is plotted in Fig. 13.3.

13.2 STATISTICAL QUALITY CONTROL: CONTROL CHARTS

13.2.1 Introduction

In his book *Statistical Quality Control*, Eugene L. Grant† states: "Measured quality of

---

$$* \quad \frac{1}{\sqrt{2\pi}\,\sigma}\int_{-\infty}^{+\infty} e^{-(x-\mu)^2/2\sigma^2}\,dx = 1$$
$$\frac{1}{\sqrt{2\pi}\,\sigma}\int_{-\infty}^{+\infty} xe^{-(x-\mu)^2/2\sigma^2}\,dx = \mu;$$
$$\frac{1}{\sqrt{2\pi}\,\sigma}\int_{-\infty}^{+\infty} (x-\mu)^2 e^{-(x-\mu)^2/2\sigma^2}\,dx = \sigma^2$$

† *Statistical Quality Control*, 3rd ed. (New York: McGraw-Hill Book Company Inc., 1964), p. 3.

TABLE 13.3. *AREAS UNDER THE NORMAL CURVE FROM K_α TO ∞*†

$$\int_{K_\alpha}^{\infty} \frac{1}{\sqrt{2\pi}} e^{-x^2/2}\, dx = \alpha$$

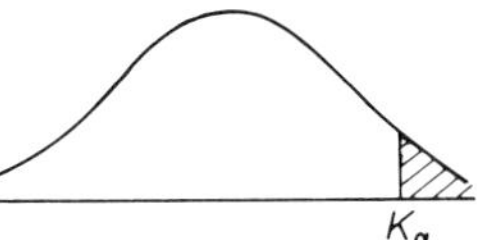

K_α	0.00	0.01	0.02	0.03	0.04	0.05	0.06	0.07	0.08	0.09
0.0	0.5000	0.4960	0.4920	0.4880	0.4840	0.4801	0.4761	0.4721	0.4681	0.4641
0.1	0.4602	0.4562	0.4522	0.4483	0.4443	0.4404	0.4364	0.4325	0.4286	0.4247
0.2	0.4207	0.4168	0.4129	0.4090	0.4052	0.4013	0.3974	0.3936	0.3897	0.3859
0.3	0.3821	0.3783	0.3745	0.3707	0.3669	0.3632	0.3594	0.3557	0.3520	0.3483
0.4	0.3446	0.3409	0.3372	0.3336	0.3300	0.3264	0.3228	0.3192	0.3156	0.3121
0.5	0.3085	0.3050	0.3015	0.2981	0.2946	0.2912	0.2877	0.2843	0.2810	0.2776
0.6	0.2743	0.2709	0.2676	0.2643	0.2611	0.2578	0.2546	0.2514	0.2483	0.2451
0.7	0.2420	0.2389	0.2358	0.2327	0.2296	0.2266	0.2236	0.2206	0.2177	0.2148
0.8	0.2119	0.2090	0.2061	0.2033	0.2005	0.1977	0.1949	0.1922	0.1894	0.1867
0.9	0.1841	0.1814	0.1788	0.1762	0.1736	0.1711	0.1685	0.1660	0.1635	0.1611
1.0	0.1587	0.1562	0.1539	0.1515	0.1492	0.1469	0.1446	0.1423	0.1401	0.1379
1.1	0.1357	0.1335	0.1314	0.1292	0.1271	0.1251	0.1230	0.1210	0.1190	0.1170
1.2	0.1151	0.1131	0.1112	0.1093	0.1075	0.1056	0.1038	0.1020	0.1003	0.0985
1.3	0.0968	0.0951	0.0934	0.0918	0.0901	0.0885	0.0869	0.0853	0.0838	0.0823
1.4	0.0808	0.0793	0.0778	0.0764	0.0749	0.0735	0.0721	0.0708	0.0694	0.0681
1.5	0.0668	0.0655	0.0643	0.0630	0.0618	0.0606	0.0594	0.0582	0.0571	0.0559
1.6	0.0548	0.0537	0.0526	0.0516	0.0505	0.0495	0.0485	0.0475	0.0465	0.0455
1.7	0.0446	0.0436	0.0427	0.0418	0.0409	0.0401	0.0392	0.0384	0.0375	0.0367
1.8	0.0359	0.0351	0.0344	0.0336	0.0329	0.0322	0.0314	0.0307	0.0301	0.0294
1.9	0.0287	0.0281	0.0274	0.0268	0.0262	0.0256	0.0250	0.0244	0.0239	0.0233
2.0	0.0228	0.0222	0.0217	0.0212	0.0207	0.0202	0.0197	0.0192	0.0188	0.0183
2.1	0.0179	0.0174	0.0170	0.0166	0.0162	0.0158	0.0154	0.0150	0.0146	0.0143
2.2	0.0139	0.0136	0.0132	0.0129	0.0125	0.0122	0.0119	0.0116	0.0113	0.0110
2.3	0.0107	0.0104	0.0102	0.00990	0.00964	0.00939	0.00914	0.00889	0.00866	0.00842
2.4	0.00820	0.00798	0.00776	0.00755	0.00734	0.00714	0.00695	0.00676	0.00657	0.00639
2.5	0.00621	0.00604	0.00587	0.00570	0.00554	0.00539	0.00523	0.00508	0.00494	0.00480
2.6	0.00466	0.00453	0.00440	0.00427	0.00415	0.00402	0.00391	0.00379	0.00368	0.00357
2.7	0.00347	0.00336	0.00326	0.00317	0.00307	0.00298	0.00289	0.00280	0.00272	0.00264
2.8	0.00256	0.00248	0.00240	0.00233	0.00226	0.00219	0.00212	0.00205	0.00199	0.00193
2.9	0.00187	0.00181	0.00175	0.00169	0.00164	0.00159	0.00154	0.00149	0.00144	0.00139

K_α	0.0	0.1	0.2	0.3	0.4	0.5	0.6	0.7	0.8	0.9
3	0.00135	$0.0^{3}968$	$0.0^{3}687$	$0.0^{3}483$	$0.0^{3}337$	$0.0^{3}233$	$0.0^{3}159$	$0.0^{3}108$	$0.0^{4}723$	$0.0^{4}481$
4	$0.0^{4}317$	$0.0^{4}207$	$0.0^{4}133$	$0.0^{5}854$	$0.0^{5}541$	$0.0^{5}340$	$0.0^{5}211$	$0.0^{5}130$	$0.0^{6}793$	$0.0^{6}479$
5	$0.0^{6}287$	$0.0^{6}170$	$0.0^{7}996$	$0.0^{7}579$	$0.0^{7}333$	$0.0^{7}190$	$0.0^{7}107$	$0.0^{8}599$	$0.0^{8}332$	$0.0^{8}182$
6	$0.0^{9}987$	$0.0^{9}530$	$0.0^{9}282$	$0.0^{9}149$	$0.0^{10}777$	$0.0^{10}402$	$0.0^{10}206$	$0.0^{10}104$	$0.0^{11}523$	$0.0^{11}260$

† From *Tables of Areas in Two Tails and in One Tail of the Normal Curve*, by Frederick E. Croxton. Copyright 1949 by Prentice-Hall, Inc. Permission is given to reproduce this table provided credit is given to the author and the Prentice-Hall copyright line is included.

manufactured product is always subject to a certain amount of variation as a result of chance. Some stable 'system of chance causes' is inherent in any particular scheme of production and inspection. Variation within this stable pattern is inevitable. The reasons for variation outside this stable pattern may be discovered and corrected. . . ." These variations outside the stable pattern are known as *assignable causes* of quality variation. A process

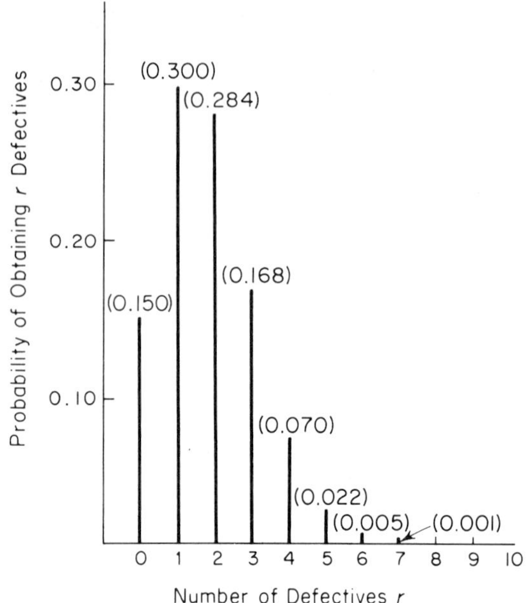

Fig. 13.3 *Distribution function for the binomial distribution with* n = *18* and p = *0.10.*

operating in the absence of any assignable causes of erratic fluctuations is said to be in *statistical control.*

To the manufacturer, the primary purpose of the control chart is to provide a basis for *action.* The introduction of a control chart aids in determining the capabilities of the production process. Action is taken when these estimated capabilities are unsatisfactory in relation to the design specifications. Furthermore, once the process capabilities have been determined, and are satisfactory, action is taken only when the control chart indicates that the process has fallen out of control; e.g., assignable causes of variation have entered.

13.2.2 Obtaining Data from Rational Subgroups

The essential feature of the control chart method is the drawing of inferences about the production process on the basis of samples drawn from the production line. The success of the technique depends upon grouping observations under consideration into subgroups or samples, within which a stable system of chance causes is operating, and between which the variations may be due to assignable causes whose presence is suspected or considered possible. Order of production is one of the more commonly used bases for obtaining rational subgroups. If items are coming from more than one source, the source may be a basis for rational subgrouping.

The size of the subgroup or sample usually is not less than 4. In industry 5 seems to be the most common.* It is preferable that all samples be of equal size.

13.2.3 Control Charts for Variables: $\bar{x}$ Charts

13.2.3.1 Statistical Concepts. If observations on an item are normally distributed, with mean $\bar{x}'$ and standard deviation σ', it is possible to find the probability that an observation will lie in an interval by referring to Table 13.3. Furthermore, if we denote the variable by x, the probability that x will be within the interval $\bar{x}' \pm 3\sigma'$ is 0.9973. In other words, if a plot such as that in Fig. 13.4 is made, on the average only 27 observations out of 10,000 will be outside the above interval, if the *population mean is* $\bar{x}'$

* When characteristics are classified into two groups, those containing defects and those not containing defects, and no other measurement is recorded, the sample size is usually much larger.

and the population standard deviation is σ'.* If the underlying population is normal with mean $\bar{x}'$ and standard deviation σ', the population of averages of subgroups of n drawn from the above population is also normal with mean $\bar{x}'$ and standard deviation $\sigma'_{\bar{x}} = \sigma'/\sqrt{n}$. Denoting the n observations by $x_1, x_2, \ldots, x_n$, the average of these n observations, $\bar{x}$, is defined as

$$\bar{x} = \frac{x_1 + x_2 + \cdots + x_n}{n}$$

If instead of plotting individual values as in Fig. 13.4, averages of samples of n are plotted, it is expected that on the average only 27 in 10,000 values of these averages will fall outside the interval

$$\bar{x}' \pm 3\sigma'_{\bar{x}} = \bar{x}' \pm \frac{3\sigma'}{\sqrt{n}}$$

* The notation in the articles on quality control is that recommended by the American Society for Quality Control. In the remaining articles, the usual notation found in texts on industrial statistics will be used—i.e., μ will represent the population mean, and σ will represent the population standard deviation.

The plot in Fig. 13.5 is referred to as a control chart for $\bar{x}$. Here $\bar{x}' + 3\sigma'/\sqrt{n}$ is referred to as the upper control limit (*UCL*) while $\bar{x}' - 3\sigma'/\sqrt{n}$ is referred to as the lower control limit (*LCL*).

In all the above discussion it was assumed that the underlying distribution is normal. Although in actual practice many distributions of observations are "nearly" normally distributed, many others do not resemble a normal distribution at all. However, it can be shown that under fairly weak restrictions, the average of samples of n tend toward normality as n gets large. This theorem is called the central limit theorem, and a proof can be found in most standard texts in mathematical statistics.

From a practical point of view n need not be too large before the results of the theorem begin to apply. In Figs. 13.6 and 13.7, it is demonstrated that even when the underlying distribution is from a rectangular or a triangular distribution, the distribution of $\bar{x}$ values from samples of 4 is approximately normal.

It is evident, then, that the central limit theorem is one of the keys to the success of the control chart for $\bar{x}$. No matter what the underlying distribution may be (there are certain weak conditions that must be satisfied), the above

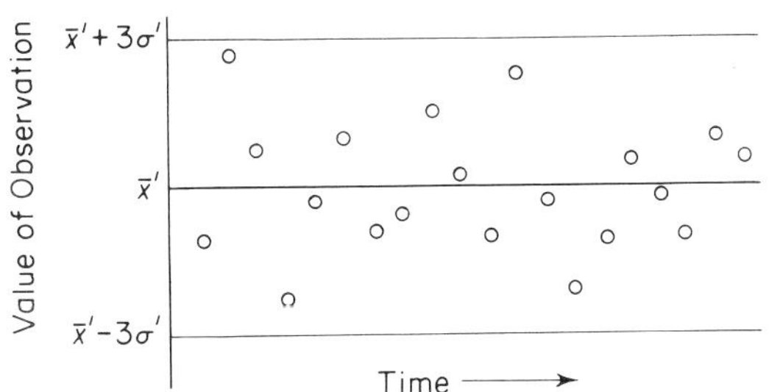

Fig. 13.4 *Plot of individual observations.*

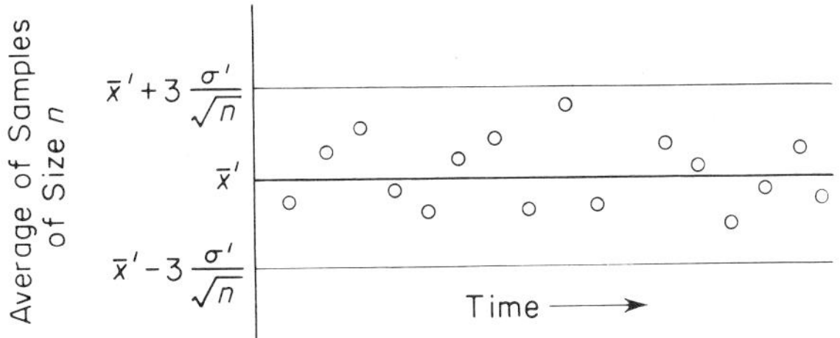

Fig. 13.5 *Control chart for* $\bar{x}$.

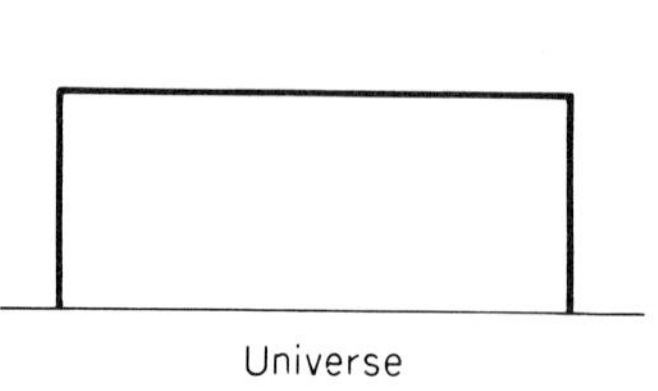

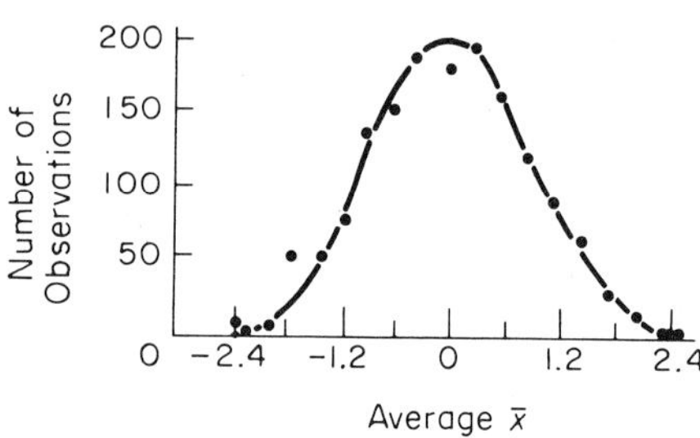

Fig. 13.6 *Distribution of the sample average from a rectangular universe.* Economic Control of Quality of Manufactured Product, *W. A. Shewhart, Bell Telephone Laboratories, Inc., Copyright 1931, D. Van Nostrand Co., Inc., Princeton, N.J.*

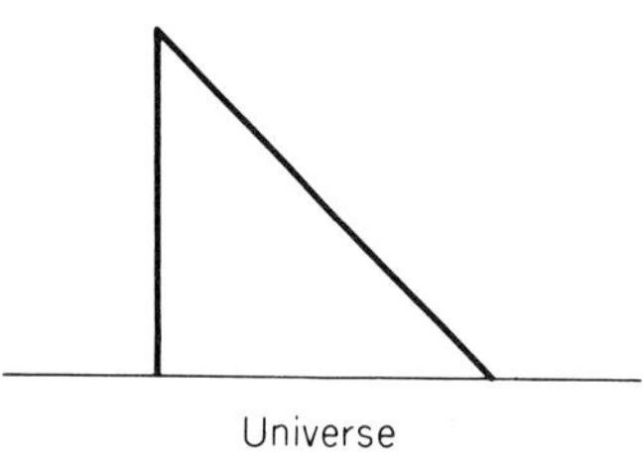

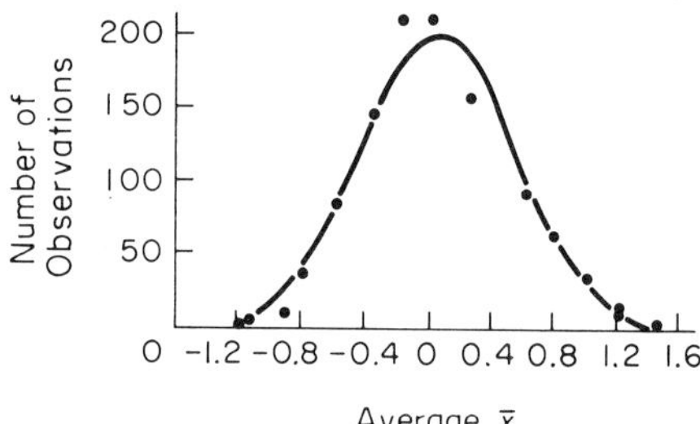

Fig. 13.7 *Distribution of the sample average from a right triangular universe.* Economic Control of Quality of Manufactured Product, *W. A. Shewart, Bell Telephone Laboratories, Inc., Copyright 1931, D. Van Nostrand Co., Inc.*

theorem states that the theory about the properties of the normal distribution is applicable, provided, of course, sample averages are considered as the variable plotted on the control chart.

13.2.3.2 Estimates of $\bar{x}'$. In Fig. 13.5, a control chart for $\bar{x}$ was shown, where the control limits are drawn in as functions of $\bar{x}'$ and σ'. In most practical applications $\bar{x}'$ and σ' are not known, and consequently estimates of these parameters must be obtained. It is desirable that these estimates be based on at least 25 subgroups of n observations. Naturally, the larger the number of subgroups, the better the estimates of the population parameters, provided the process is in control. Suppose the estimates are based upon k subgroups of size n. Denoting the averages of the subgroups by $\bar{x}_1, \bar{x}_2, \ldots, \bar{x}_k$, the best estimate of $\bar{x}'$, the population mean, is

$$\bar{\bar{x}} = \frac{\bar{x}_1 + \bar{x}_2 + \cdots + \bar{x}_k}{k}$$

where $\bar{\bar{x}}$ is also the average of all the nk observations.

13.2.3.3 Estimate of σ' by $\bar{\sigma}$. As was pointed out in the previous article, it is usually necessary to estimate σ', the population standard deviation. Define, for each subgroup:

$$\sigma = \sqrt{\frac{(x_1 - \bar{x})^2 + (x_2 - \bar{x})^2 + \cdots + (x_n - \bar{x})^2}{n}}$$
$$= \sqrt{\frac{x_1^2 + x_2^2 + \cdots + x_n^2 - n\bar{x}^2}{n}}$$

Let $\bar{\sigma}$ be the average of the standard deviations of the k subgroups, i.e.,

$$\bar{\sigma} = \frac{\sigma_1 + \sigma_2 + \cdots + \sigma_k}{k}$$

An estimate of σ' is then $\bar{\sigma}/c_2$, where values of c_2 for different n can be found in Table 13.4.

To summarize, the estimated central line for the control chart for $\bar{x}$ is $\bar{\bar{x}}$, and the estimated control limits are $\bar{\bar{x}} \pm 3\bar{\sigma}/\sqrt{n}c_2$. Values of 3/

$(\sqrt{n}c_2) = A_1$ can be found in Table 13.4 so that the estimated control limits can be written as $\bar{\bar{x}} \pm A_1\bar{\sigma}$.

13.2.3.4 Estimate of σ' by $\bar{R}$. Define, as the range (R) of a subgroup of n observations, the difference between the largest and smallest value. Let $\bar{R}$ be the average of the ranges of the k subgroups, i.e.,

$$\bar{R} = \frac{R_1 + R_2 + \cdots + R_k}{k}$$

Another estimate of σ' is $\bar{R}/d_2$. Values of d_2 for different values of n can be found in Table 13.4.

In estimating the control limits for $\bar{x}$ with $\bar{R}$ as an estimate of σ', the limits become $\bar{\bar{x}} \pm 3\bar{R}/\sqrt{n}d_2$. Values of $3/(\sqrt{n}d_2) = A_2$ can be found in Table 13.4, so that the estimated control limits can be written as $\bar{\bar{x}} \pm A_2\bar{R}$.

It must be pointed out that although $\bar{R}$ is simpler to calculate than $\bar{\sigma}$, the estimate of σ' based upon $\bar{\sigma}$ is a better estimate in the sense that on the average $\bar{\sigma}/c_2$ will lie nearer to σ' than $\bar{R}/d_2$ will.

13.2.3.5 Starting a Control Chart for $\bar{x}$. It has been pointed out that observations should be classified into rational subgroups, and each subgroup should contain at least 4 observations. The determination of the minimum number of subgroups is a compromise between obtaining the guidance of the control chart as quickly as possible, and the desire for the guidance to be reliable. Usually, at least 25 subgroups should be chosen.

With the data at hand, trial control limits can be calculated. Estimates of $\bar{x}'$ and σ' can be obtained in the manner described previously. The trial control limits are then $\bar{\bar{x}} \pm A_1\bar{\sigma}$ or $\bar{\bar{x}} \pm A_2\bar{R}$, depending upon which estimate of σ' is chosen. If σ' and/or $\bar{x}'$ are known, the control limits should be calculated, using the known values; e.g., if both are known, $\bar{x}' \pm 3\sigma'/\sqrt{n}$. Values of $3/\sqrt{n}$ can be found in Table 13.4, so that the control limits can be written $\bar{x}' \pm A\sigma'$.

Returning to the case where the parameters $\bar{x}'$ and σ' are unknown, and estimates have to be computed, some modifications may have to be made. Lack of control is usually indicated by points falling outside the limits. If all the points fall within the limits, the process is said to be in control. However, there is no assurance that assignable causes of variation are not present, and that this is a constant cause system. It merely means that for practical purposes it pays to act as if no assignable causes of variation are present, although realizing that an error of judgment is quite possible.

The trial control limits serve the purpose of determining whether past operations are in control. To continue using these limits as a basis for action on future production may require a revision of the trial limits, especially if a lack of control is exhibited by points falling outside the trial control limits. If the process is not in control, all the points do not come from a stable population. However, it is evident that future control limits should be based upon data from a controlled process. As a practical rule, then, those points falling outside the trial control limits are eliminated, and new trial control limits are computed using the remaining points. This procedure may be continued until all points fall within the control limits. There is no theoretical justification for this, other than the fact that those points falling outside the trial control limits are more likely to belong to another population, i.e., may be due to some assignable cause.

The control chart when used as a basis for action on future production may be set, using aimed-at values of $\bar{x}'$ and/or σ'. In this case, the control limits are modified by using the aimed-at values as if they were the known values in the computations. If this is done, and the control chart exhibits a lack of control in the future, the trouble may be that the process is not in control at the aimed-at values, but is in control at some other values. For example, suppose that an item is being produced at a level such that the mean $\bar{x}'$ is equal to 20. If the aimed-at value is $\bar{x}' = 25$, the control chart based on $\bar{x}' = 25$ will exhibit a lack of control. Suppose the mean can be shifted by a change in machine setting. The interpretation, then, is that there is an assignable cause present, namely, the machine setting, preventing the production process from operating at the aimed-at value. Yet, with the actual machine setting, the process is in control at $\bar{x}' = 20$. The term "state of control" may have to be interpreted in the above light. Aimed-at values of $\bar{x}'$ or σ' are often used when they can be achieved by a simple adjustment of the machine.

13.2.3.6 Relation Between Control Limits and Specification Limits. After ascertaining at what level the process is in control by means of

TABLE 13.4. *FACTORS FOR COMPUTING CONTROL CHART LINES*†

Number of Observations in Sample, n	Chart for Averages			Chart for Standard Deviations						Chart for Ranges						
	Factors for Control Limits			Factors for Central Line		Factors for Control Limits				Factors for Central Line		Factors for Control Limits				
	A	A_1	A_2	c_2	$1/c_2$	B_1	B_2	B_3	B_4	d_2	$1/d_2$	d_3	D_1	D_2	D_3	D_4
2	2.121	3.760	1.880	0.5642	1.7725	0	1.843	0	3.267	1.128	0.8865	0.853	0	3.686	0	3.276
3	1.732	2.394	1.023	0.7236	1.3820	0	1.858	0	2.568	1.693	0.5907	0.888	0	4.358	0	2.575
4	1.501	1.880	0.729	0.7979	1.2533	0	1.808	0	2.266	2.059	0.4857	0.880	0	4.698	0	2.282
5	1.342	1.596	0.577	0.8407	1.1894	0	1.756	0	2.089	2.326	0.4299	0.864	0	4.918	0	2.115
6	1.225	1.410	0.483	0.8686	1.1512	0.026	1.711	0.030	1.970	2.534	0.3946	0.848	0	5.078	0	2.004
7	1.134	1.277	0.419	0.8882	1.1259	0.105	1.672	0.118	1.882	2.704	0.3698	0.833	0.205	5.203	0.076	1.924
8	1.061	1.175	0.373	0.9027	1.1078	0.167	1.638	0.185	1.815	2.847	0.3512	0.820	0.387	5.307	0.136	1.864
9	1.000	1.094	0.337	0.9139	1.0942	0.219	1.609	0.239	1.761	2.970	0.3367	0.808	0.546	5.394	0.184	1.816
10	0.949	1.028	0.308	0.9227	1.0837	0.262	1.584	0.284	1.716	3.078	0.3249	0.797	0.687	5.469	0.223	1.777
11	0.905	0.973	0.285	0.9300	1.0753	0.299	1.561	0.321	1.679	3.173	0.3152	0.787	0.812	5.534	0.256	1.744
12	0.866	0.925	0.266	0.9359	1.0684	0.331	1.541	0.354	1.646	3.258	0.3069	0.778	0.924	5.592	0.284	1.719
13	0.832	0.884	0.249	0.9410	1.0627	0.359	1.523	0.382	1.618	3.336	0.2998	0.770	1.026	5.646	0.308	1.692
14	0.802	0.848	0.235	0.9453	1.0579	0.384	1.507	0.406	1.594	3.407	0.2935	0.762	1.121	5.693	0.329	1.671
15	0.775	0.816	0.223	0.9490	1.0537	0.406	1.492	0.428	1.572	3.472	0.2880	0.755	1.207	5.737	0.348	1.652
16	0.750	0.788	0.212	0.9523	1.0501	0.427	1.478	0.448	1.552	3.532	0.2831	0.749	1.285	5.779	0.364	1.636
17	0.728	0.762	0.203	0.9551	1.0470	0.445	1.465	0.466	1.534	3.588	0.2787	0.743	1.359	5.817	0.379	1.621
18	0.707	0.738	0.194	0.9576	1.0442	0.461	1.454	0.482	1.518	3.640	0.2747	0.738	1.426	5.854	0.392	1.608
19	0.688	0.717	0.187	0.9599	1.0418	0.477	1.443	0.497	1.503	3.689	0.2711	0.733	1.490	5.888	0.404	1.596
20	0.671	0.697	0.180	0.9619	1.0396	0.491	1.433	0.510	1.490	3.735	0.2677	0.729	1.548	5.922	0.414	1.586
21	0.655	0.679	0.173	0.9638	1.0376	0.504	1.424	0.523	1.477	3.778	0.2647	0.724	1.606	5.950	0.425	1.575
22	0.640	0.662	0.167	0.9655	1.0358	0.516	1.415	0.534	1.466	3.819	0.2618	0.720	1.659	5.979	0.434	1.566
23	0.626	0.647	0.162	0.9670	1.0342	0.527	1.407	0.545	1.455	3.858	0.2592	0.716	1.710	6.006	0.443	1.557
24	0.612	0.632	0.157	0.9684	1.0327	0.538	1.399	0.555	1.445	3.895	0.2567	0.712	1.759	6.031	0.452	1.548
25	0.600	0.619	0.153	0.9696	1.0313	0.548	1.392	0.565	1.435	3.931	0.2544	0.709	1.804	6.058	0.459	1.541
Over 25	$\frac{3}{\sqrt{n}}$	$\frac{3}{\sqrt{n}}$	—	—	—	‡	§	‡	§	—	—	—	—	—	—	—

† Reproduced by permission from *ASTM Manual on Quality Control of Materials,* American Society for Testing Materials, Philadelphia, Pa., 1951.

‡ $1 - \frac{3}{\sqrt{2n}}$ § $1 + \frac{3}{\sqrt{2n}}$

TABLE 13.5. *FORMULAS FOR CENTRAL LINES AND CONTROL LIMITS*

Statistic	Standards Given		Analysis of Past Data	
	Central Line	Limits	Central Line	Limits
Average, using σ'	$\bar{x}'$	$\bar{x}' \pm A\sigma'$	$\bar{\bar{x}}$	$\bar{\bar{x}} \pm A_1\bar{\sigma}$
Average, using R	—	—	$\bar{\bar{x}}$	$\bar{\bar{x}} \pm A_2\bar{R}$
Standard deviation	$c_2\sigma'$	$B_1\sigma'$, $B_2\sigma'$	$\bar{\sigma}$	$B_3\bar{\sigma}$, $B_4\bar{\sigma}$
Range	$d_2\sigma'$	$D_1\sigma'$, $D_2\sigma'$	$\bar{R}$	$D_3\bar{R}$, $D_4\bar{R}$

the control chart, it remains to determine whether or not the process can meet the specification limits set for the item. In Pars. 13.2.3.3 and 13.2.3.4, it was shown that σ' can be estimated from $\bar{\sigma}$ or $\bar{R}$, i.e., estimate of $\sigma' = \bar{\sigma}/c_2$ or $\bar{R}/d_2$. Furthermore, $\bar{x}'$ can be estimated from $\bar{\bar{x}}$. Consequently, it is known that if $\bar{x}'$ and σ' are really equal to the estimates, almost all the individual items will lie between $\bar{x}' \pm 3\sigma'$, i.e., on the average 9973 out of 10,000.

Specification limits are usually specified by giving an aimed-at value plus an interval surrounding this value. These specification limits *should not* be interpreted as meaning that all the items produced will fall within these limits, but rather *almost* all. If the designer is pressed, he will admit to allowing, say 1 in 1000, to fall outside. Once the number is ascertained, the corresponding probability limits can be obtained from the estimated $\bar{x}'$ and σ'.* For example, suppose the designer specifies specification limits of 20 ± 2, and he states that only 1 in 1000 should fall outside these limits. If the estimate of $\bar{x}'$ is 20.40 and $\sigma' = 1$, from the normal tables it follows that on that average, 999 out of 1000 items will fall between

$$[\bar{x}' - 3.29\,\sigma', \bar{x}' + 3.29\,\sigma'] = [17.11, 23.69]$$

The specification limits allow only [18, 22]. Consequently, the above process cannot meet the specification limits, and either the process must be changed or the specifications revised. To summarize, the specification limits must first be properly interpreted. As a practical rule, they are often interpreted to mean the allowance of 27 in 10,000 to fall outside. This corresponds to $3\sigma'$ limits. Then $\bar{x}' \pm 3\sigma'$* is calculated and compared with the specification limits. If the specifications fall outside the probability limits, either the process must be altered, or the specifications revised.

* Assuming the estimated values of $\bar{x}'$ and σ' are equal to $\bar{x}'$ and σ', respectively.

13.2.3.7 Interpretation of Control Charts for $\bar{x}$. As long as the process is in control, almost all the values of $\bar{x}$ will fall within the control limits. In this case, no action need be taken. A point falling outside these limits is a signal to hunt for trouble. In some instances, production may be stopped until a source of trouble has been discovered. Occasionally, approximately 27 in 10,000 times, an error will be made in that trouble will be sought even though nothing has gone wrong with the process. On the other hand, if something has gone wrong with the process, points will begin to fall outside the control limits. For example, a shift in the mean $\bar{x}'$ (σ' remaining constant) will result in points falling outside the control limits. This is usually indicated by points falling above *or* below the limits (but never both), depending on whether the shift is in the positive or negative direction. On the other hand, an increase in σ' ($\bar{x}'$ remaining constant) will result in points falling above *and* below the control limits. In addition to examining the process, when points fall outside the control limits the control limits themselves should be reexamined and perhaps brought up to date.

It has been pointed out that an error may be committed when action is taken after a point falls outside the control limits. There is a small probability that a point will fall outside the control limits even though the process is in control; i.e., if $3\sigma'_{\bar{x}}$ limits are used the probability is 0.0027. This probability of falling outside the control limits when the process is in control is known as the probability of committing an *error of the first type* or *type 1 error*. Similarly, if the process goes out of control, there is a probability greater than 0, that a point will fall within the control limits. This is known as the probability of an *error of the second type* or *type 2 error*. These two errors are related. A decrease in one results in an increase in the other. An increase in one results in a decrease in the other. The use of $3\sigma'_{\bar{x}}$ limits implies a probability of a type 1 error of

0.0027. If the process jumps out of control, and the new level is specified, the probability of a type 2 error can be computed. If $2\sigma'_{\bar{x}}$ limits were used instead of $3\sigma'_{\bar{x}}$, the probability of a type 1 error is increased to about 0.0455, but the probability of a type 2 error is decreased.

13.2.3.8 Example. Suppose $\bar{x}'$ is 25, and $\sigma' = 1$, and there are 4 observations in each subgroup. The control limits are then $25 \pm \frac{3}{2}$. The probability of a type 1 error is 0.0027. If the mean shifts to 27, the probability of a point falling inside the control limits of $25 \pm \frac{3}{2}$ is 0.1587. On the other hand, if $2\sigma'_{\bar{x}}$ is used as control limits (25 ± 1) the probability of a type 1 error is increased to 0.0455, whereas the probability of committing a type 2 error is only 0.0228.

It is evident that type 1 errors or type 2 errors (but not both) can be made as small as is desirable, at the expense of increasing the other type error. It is also evident that using $3\sigma'_{\bar{x}}$ limits is conservative from the point of view of considering the type 1 error. A point falling outside the control limits is almost sure evidence that the process is no longer in control. On the other hand, one cannot be assured that a small shift has not taken place even if the points fall within the control limits.

13.2.4 *R* Charts and σ Charts

13.2.4.1 Statistical Concepts. Although both R and σ *do not* have normal distributions, both of these functions are chance variables, each having a distribution. Furthermore, the population mean (of these chance variables) plus and minus three standard deviations (of these variables) contain almost all of the underlying population. Consequently, for control chart purposes, it is necessary to calculate both the population mean value and standard deviation of these chance variables.

The population average of σ is $c_2\sigma'$. The standard deviation of σ is

$$\sigma_\sigma = [2(n-1) - 2nc_2^2]^{1/2}\frac{\sigma'}{\sqrt{2n}}$$

The control limits are then $c_2\sigma' \pm 3\sigma_\sigma$ which can be written

$$\sigma'\left(c_2 \pm \frac{3}{\sqrt{2n}}[2(n-1) - 2nc_2^2]^{1/2}\right)$$

The factors

$$B_2 = c_2 + \frac{3}{\sqrt{2n}}[2(n-1) - 2nc_2^2]^{1/2}$$

$$B_1 = c_2 - \frac{3}{\sqrt{2n}}[2(n-1) - 2nc_2^2]^{1/2}$$

can be obtained from Table 13.4. Thus

$$UCL_\sigma = B_2\sigma' \qquad LCL_\sigma = B_1\sigma'$$

When σ' is unknown, it can be estimated from $\bar{\sigma}/C_2$, so that the estimated control limits are written

$$\bar{\sigma}\left\{1 \pm \frac{3}{\sqrt{2n}\,c_2}[2(n-1) - 2nc_2^2]^{1/2}\right\}$$

The factors

$$B_4 = 1 + \frac{3}{\sqrt{2n}\,c_2}[2(n-1) - 2nc_2^2]^{1/2}$$

$$B_3 = 1 - \frac{3}{\sqrt{2n}\,c_2}[2(n-1) - 2nc_2^2]^{1/2}$$

can be obtained from Table 13.4. The estimated control limits are then

$$UCL_\sigma = B_4\bar{\sigma} \qquad LCL_\sigma = B_3\bar{\sigma}$$

Note that the center line is $c_2\sigma'$ if σ' is known, and $\bar{\sigma}$ if σ' is unknown.

The control limits for R are obtained in a similar manner. The population average of R is $d_2\sigma'$. The standard deviation of R is $\sigma_R = d_3\sigma'$. The control limits are then $d_2\sigma' \pm 3\sigma_R$, which can be written $\sigma'(d_2 \pm 3d_3)$. The factors

$$D_2 = d_2 + 3d_3 \qquad \text{and} \qquad D_1 = d_2 - 3d_3$$

can be obtained from Table 13.4. Thus

$$UCL_R = D_2\sigma' \qquad LCL_R = D_1\sigma'$$

When σ' is unknown, it can be estimated from $\bar{R}/d_2$ so that the estimated control limits are written

$$\bar{R}\left[1 \pm 3\frac{d_3}{d_2}\right]$$

The factors

$$D_4 = 1 + 3\frac{d_3}{d_2} \qquad \text{and} \qquad D_3 = 1 - 3\frac{d_3}{d_2}$$

can be obtained from Table 13.4. The estimated control limits are then

$$UCL_R = D_4\bar{R} \qquad LCL_R = D_3\bar{R}$$

The center line is $d_2\sigma'$ if σ' is known, and $\bar{R}$ if σ' is unknown.

13.2.4.2 Setting Up a Control Chart for R or σ. The range of each subgroup should be obtained and $\bar{R}$ calculated from these values. The control limits are obtained from

$$UCL_R = D_4\bar{R} \qquad LCL_R = D_3\bar{R}$$

If a σ chart is desired, similar calculations are made in accordance with the rules described above. If all the points fall inside the control limits, no modification is made unless it is desired to reduce the process dispersion. In this case, an aimed-at value of σ' should be used in the calculations. When the $\bar{R}$ (or σ) chart indicates a possible lack of control by points falling outside the control limits, it is desirable to estimate the value of σ' that might be attained if the dispersion were brought into control. A method of estimation is to eliminate those points out of control (only those above the UCL_R if points fall both above and below) and recompute the values of the control limits based only on the remaining observations. If more points fall out of control, the procedure is carried out once again.

The final revised values of $\bar{R}, \bar{\sigma}$, or σ', whichever is used, may also be used to obtain new control limits for $\bar{x}$. This has the effect of tightening the limits on the $\bar{x}$ chart, making them consistent with a σ' that may be estimated from the revised $\bar{R}$ or $\bar{\sigma}$.

Control limits should be revised from time to time as additional data are accumulated.

13.2.5 Example of $\bar{x}$ and R Chart

The following are the $\bar{x}$ and R values for 20 subgroups of five readings.

The specifications for this product characteristic are 0.4037 ± 0.0010. The values given are the last two figures of the dimension reading, i.e., 31.6 should be 0.40316.

Subgroup	$\bar{x}$	R	Subgroup	$\bar{x}$	R
1	34.0	4	11	35.8	4
2	31.6	4	12	38.4	4
3	30.8	2	13	34.0	14
4	33.0	3	14	35.0	4
5	35.0	5	15	33.8	7
6	32.2	2	16	31.6	5
7	33.0	5	17	33.0	5
8	32.6	13	18	28.2	3
9	33.8	19	19	31.8	9
10	37.8	6	20	35.6	6

$\sum \bar{x} = 671.0,\ \bar{\bar{x}} = 33.6\ ;\ \sum R = 124,\ \bar{R} = 6.20$

The trial control limits are computed from

$$\begin{aligned} &\bar{\bar{x}} \pm A_2\bar{R} && UCL_{\bar{x}} = 37.2 \\ &\bar{\bar{x}} \pm (0.577)(6.20) && LCL_{\bar{x}} = 30.0 \\ &\bar{\bar{x}} \pm 3.58 && UCL_R = (2.115)(6.20) \\ &&& = 13.1 \\ &&& LCL_R = 0 \end{aligned}$$

The control charts for $\bar{x}$ and R with trial limits are shown in Fig. 13.8. Since some points fall outside the control limits, the process is assumed to be out of control. Eliminating these points, new control limits are computed.

$$\begin{aligned} \textstyle\sum \bar{x} &= 5660 \qquad \bar{\bar{x}} = 33.3 \\ \textstyle\sum R &= 91 \qquad \bar{R} = 5.06 \\ UCL_{\bar{x}} &= 33.3 + (0.577)(5.06) \\ &= 33.3 + 2.92 = 36.2 \\ LCL_{\bar{x}} &= 33.3 - (0.577)(5.06) \\ &= 33.3 - 2.92 = 30.4 \\ UCL_R &= (2.115)(5.06) = 10.7 \\ LCL_R &= (2.115)(0) = 0 \end{aligned}$$

None of the remaining points fall outside the limits. If it is now assumed that the process can be brought into control at this level we find

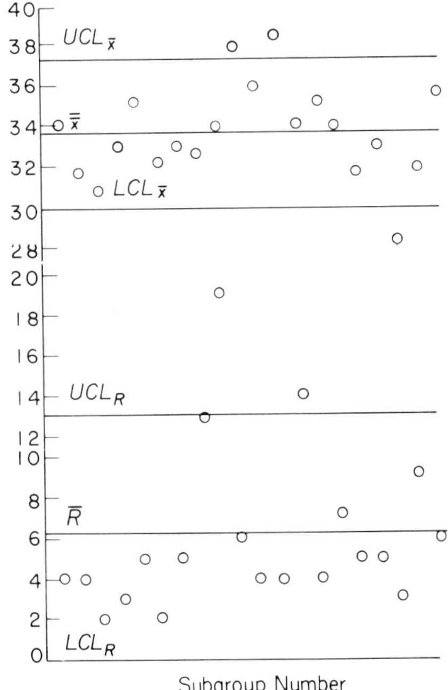

Fig. 13.8 *Control chart for $\bar{x}$ and* R.

$\bar{x}' = 33.3$ and $\sigma' = 2.175$. The value of σ' is obtained from

$$d_2 = \bar{R}/\sigma'$$
$$\sigma' = \bar{R}/d_2 = 5.06/2.326 = 2.175$$
$$[\bar{x}' - 3\sigma', \bar{x}' + 3\sigma'] = [\bar{x}' - 6.525, \bar{x}' + 6.525] = [26.8, 39.8]$$

In terms of the actual data, $[\bar{x}' - 3\sigma', \bar{x}' + 3\sigma'] = [0.4027, 0.4040]$. The specifications are

$$[0.4037 - 0.0010, 0.4037 + 0.0010] = [0.4027, 0.4047]$$

so that this production process is able to meet these specifications even though the process is not centered at the nominal value of 0.4037 provided the process remains in control at the above level.

13.2.6 Control Chart for Fraction Defective

13.2.6.1 Relation Between Control Charts Based on Variables Data and Charts Based on Attribute Data. The $\bar{x}$ and R control charts are charts for variables, i.e., for quality characteristics that can be measured and expressed in numbers. However, many quality characteristics can be observed only as attributes, i.e., by classifying the item into one of two classes, usually defective or nondefective. Furthermore, with the existing techniques, an $\bar{x}$ and R chart can be used for only one measurable characteristic at a time. For example, if an item consists of 10,000 measurable characteristics, each characteristic is a candidate for an $\bar{x}$ and R chart. However, it would be impossible to have 10,000 such charts, and only the most important and troublesome would be charted. As an alternative to $\bar{x}$ and R charts, and as a substitute when characteristics are measured only by attributes, a control chart based on the fraction defective can be used. This is known as a p chart. A p chart can be applied to quality characteristics that are actually observed as attributes even though they may have been measured as variables. The cost of obtaining attribute data is usually less than that for obtaining variables data. The cost of computing and charting may also be less, since one p chart can apply to any number of characteristics.

Basically, the p chart has the same objective as an $\bar{x}$ and R chart. It discloses the presence of assignable causes of variation, although it is not so sensitive as an $\bar{x}$ and R chart.

13.2.6.2 Statistical Theory. The statistical theory of the binomial distribution discussed in Par. 13.1.3 is applicable here. Let the probability of an item being defective be given by p', so that the probability of obtaining exactly d defectives in a sample of n is

$$\binom{n}{d} p'^{d}(1 - p')^{n-d}$$

The probability of obtaining a defectives or less in a sample of n is

$$\sum_{d=0}^{a} \binom{n}{d} p'^{d}(1 - p')^{n-d} = \binom{n}{0} p'^{0}(1 - p')^{n} + \binom{n}{1} p'^{1}(1 - p')^{n-1} + \cdots + \binom{n}{a} p'^{a}(1 - p')^{n-a}$$

Furthermore, the mean value of the total number of defectives in a sample of n is np', and the standard deviation is $\sqrt{np'(1 - p')}$.

If the fraction defective p is defined as the ratio of the number of defectives d to the total number of items in the sample n—i.e., d/n—the mean value of the fraction defective is p' and the standard deviation is $\sqrt{p'(1 - p')/n}$.

As a working rule, the control limits for the fraction defective are defined as

$$UCL_p = p' + 3\sqrt{\frac{p'(1 - p')}{n}}$$
$$LCL_p = p' - 3\sqrt{\frac{p'(1 - p')}{n}}$$

It is important to note that the probability of d/n falling within these limits depends on the value of p', even if the process is in control.* However, almost all the d/n will fall within the above limits provided the process is in control. Furthermore, the above limits result in a simple empirical rule and for large n, result in an accurate approximation.

If p' is not known, it is usually estimated from past data. The rules are similar to those used in estimating the parameters for the $\bar{x}$ charts. The estimate in this case is d_T/n_T where d_T is the total number of defectives found in the past data of size n_T.

It often happens in control charts for fraction

---

* This is quite different from the $\bar{x}$ chart, where the probability of $\bar{x}$ falling between $\bar{x}' \pm 3\sigma'/\sqrt{n}$ is 0.9973 for any values of $\bar{x}'$ and σ' provided the process is in control.

defective that the size of the subgroup varies. In this case, three possible solutions to the problem are as follows: (1) Compute control limits for every subgroup and show these fluctuating limits on the p chart. (2) Estimate the average subgroup size, and compute one set of limits for this average and draw them on the control chart. This method is approximate and is appropriate only when the subgroup sizes are not too variable. Points near the limits may have to be reexamined in accordance with (1). (3) Draw several sets of control limits on the chart corresponding to different subgroup sizes. This method is also approximate and is actually a cross between (1) and (2). Again, points falling near the limits should be reexamined in accordance with (1).

13.2.6.3 Starting the Control Chart. The subgroup size is usually large compared with that used for $\bar{x}$ and R charts. The main reason for this is that if p' is very small, and n is small, the expected number of defectives in a subgroup will be very close to 0.

For each subgroup compute p, where

$$p = \frac{\text{number of defectives in the subgroup}}{\text{number inspected in the subgroup}}$$

Whenever practicable, no fewer than 25 subgroups should be used to compute trial control limits. These limits are computed from the data by finding the average fraction defective $\bar{p}$, where

$$\bar{p} = \frac{\text{total number of defectives during period}}{\text{total number inspected during period}}$$

and substituting in the relations

$$UCL_p = \bar{p} + \frac{3\sqrt{\bar{p}(1-\bar{p})}}{\sqrt{n}}$$

$$LCL_p = \bar{p} - \frac{3\sqrt{\bar{p}(1-\bar{p})}}{\sqrt{n}}$$

Inferences about the existence of control or lack of control can be drawn in a manner similar to that described for the $\bar{x}$ and R chart.

13.2.6.4 Continuing the p Chart. The preliminary plot reveals two facts, namely, whether or not the process is in an apparent state of control, and the quality level. If the process appears to be in control, it may be in control at too high a level; i.e., the estimate of p' is higher than can be tolerated. In this case, the production process must be examined, and possible major changes made. On the other hand, the estimate of p' may indicate a good quality level even though the process may appear to be out of control. In this case assignable causes should be sought and eliminated. Of course, control limits can be determined on the basis of an aimed-at value of p', but an indicated lack of control must be interpreted with this in mind.

At first glance, falling below the lower control limit may appear to be desirable. However, this may be attributed to a poor estimate of p', although it may possibly indicate a change for the better in the quality level; and tracking down the assignable cause may enable an improvement to be made in the production process. In either case, points falling below the lower control limit call for a reexamination of the control chart or the production process or both.

13.2.6.5 Example. A sample of 50 pieces is drawn from the production of the last two hours from a single spindle automatic screw machine, and each item is checked by go and no-go gauges for several possible sources of defectives. The number of defective items found in 25 such successive samples was:

Subgroup	d	p	Subgroup	d	p
1	1	0.02	14	1	0.02
2	2	0.04	15	0	0.00
3	5	0.10	16	2	0.04
4	6	0.12	17	1	0.02
5	3	0.06	18	0	0.00
6	5	0.10	19	0	0.00
7	2	0.04	20	1	0.02
8	1	0.02	21	1	0.02
9	1	0.02	22	0	0.00
10	0	0.00	23	0	0.00
11	0	0.00	24	1	0.02
12	1	0.02	25	0	0.00
13	0	0.00			

$$\bar{p} = \frac{34}{1250} = 0.0272$$

$$UCL = 0.0272 + \frac{3\sqrt{(0.0272)(0.9728)}}{7.071}$$

$$= 0.0963$$

$$LCL = 0$$

The control chart for p with trial limits is shown in Fig. 13.9. Many points fall outside the control limits, and it appears that an assignable cause of variation was present when the first samples were taken that was not present when the later samples were taken. Recomputing the control limits starting with the seventh sample we find

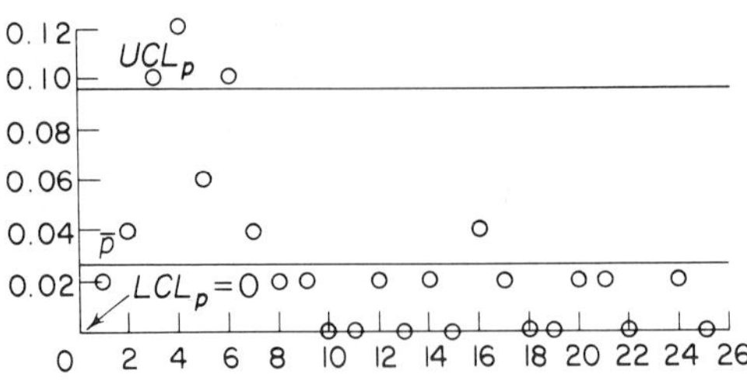

Fig. 13.9 *Control chart for* p.

$$\bar{p} = \frac{12}{950} = 0.0126$$

$$UCL = 0.0126 + \frac{3\sqrt{(0.0126)(0.9874)}}{7.071}$$

$$= 0.0599$$

$$LCL = 0$$

No points fall outside these limits and hence we use these as the control limits for future production.

13.2.7 Control Charts for Defects

13.2.7.1 Difference Between a Defect and a Defective. An item is considered to be defective if it fails to conform to the specifications in any of the characteristics. Each characteristic that does not meet the specifications is a defect. An item is defective if it contains at least one defect.

The c chart is a control chart for defects per unit. The unit considered may be a single item, a group of items, part of an item, etc. The unit is examined and the number of defects found is recorded on the c chart. If, for each unit, there are numerous opportunities for defects to occur, and if the probability of a defect occurring in a particular spot is small, the statistical theory for the c chart is based on the Poisson distribution. Some examples where c charts are applicable are in counting the number of defective rivets on an airplane wing, the number of imperfections in a piece of cloth, etc.

13.2.7.2 Statistical Theory. The probability of finding c defects in an item where the number of defects follows the Poisson law is

$$c'^{c}e^{-c'}/c!$$

where c' is the *average* number of defects per unit, and the standard deviation is $\sqrt{c'}$.

As a working rule, control limits for the c chart are defined as

$$UCL_c = c' + 3\sqrt{c'} \qquad LCL_c = c' - 3\sqrt{c'}$$

As in the case of the control chart for the fraction defective, these limits do not guarantee that a fixed percentage of the population will be between them for all values of c' even if the process is in control; i.e., the probability of c falling between these limits depends on the value of c'. However, for practical purposes, it can be assumed that "almost all the values" will fall between the control limits provided the process is in control.

If c' is not known, it is usually estimated from past data. The rules are similar to those used in estimating the parameters from the $\bar{x}$ charts. The estimate in this case is $\bar{c} = c_T/N_T$ where c_T is the total number of defects found in N_T units.

13.2.7.3 Starting and Continuing the Control. The discussion on starting and continuing the control chart for fraction defective given in Arts. 13.2.6.3 and 13.2.6.4 is pertinent to the c chart.

13.2.7.4 Example. The following table gives the number of missing rivets noted at final inspection on 25 airplanes. Read column downward and from left to right.

8	15	23	9	10
16	8	16	9	22
14	11	9	14	7
19	21	25	11	28
11	12	15	9	9

$$\bar{c} = \frac{351}{25} = 14.04$$

$$UCL_c = 14.04 + 3\sqrt{14.04} = 25.26$$

$$LCL_c = 14.04 - 3\sqrt{14.04} = 2.82$$

The control chart for c with trial limits is shown in Fig. 13.10. However, the next to the last plane falls outside the control limits, and it would be best to base future calculations on a $\bar{c}$ which did not involve this airplane.

$$\bar{c} = \frac{323}{24} = 13.46$$

$$UCL_c = 13.46 + 3\sqrt{13.46} = 24.47$$

$$LCL_c = 13.46 - 3\sqrt{13.46} = 2.45$$

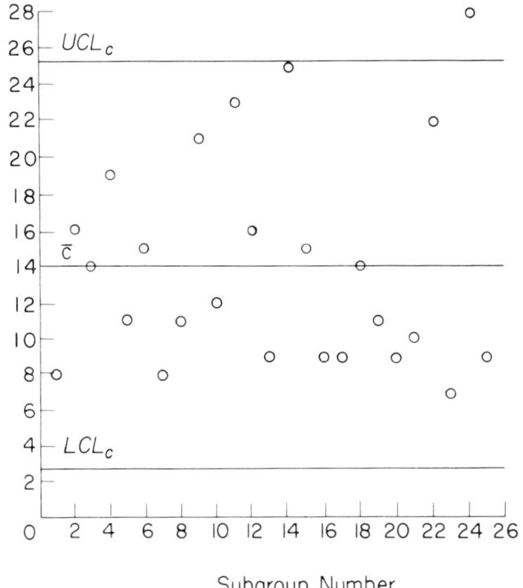

Fig. 13.10 *Control chart for* c.

The fourteenth airplane now falls outside the control limits, so that $\bar{c}$ is again recomputed without the results for this plane.

$$\bar{c} = \frac{298}{23} = 12.96$$

$$UCL_c = 12.96 + 3\sqrt{12.96} = 23.76$$

$$LCL_c = 12.96 - 3\sqrt{12.96} = 2.16$$

13.3 SAMPLING INSPECTION

13.3.1 Introduction

Sampling inspection is of two kinds, namely, lot-by-lot sampling inspection and continuous sampling inspection. In the former, items are formed into lots, a sample is drawn from the lot, and the lot is either accepted or rejected on the basis of the quality of the sample. This is most appropriate for acceptance inspection. In continuous sampling inspection, current inspection results are used to determine whether sampling inspection or screening inspection is to be used for the next articles to be inspected. Sampling plans are further classified depending on whether the quality characteristics are measured and expressed in numbers, i.e., variables inspection; or whether articles are classified only as defective or nondefective, i.e., attributes inspection.

An alternative to sampling inspection is to screen every item. The cost of such a scheme is prohibitive, with perfect quality rarely achieved. Still another advantage of sampling inspection schemes—if properly designed—is that they create more effective pressure for quality improvement, and therefore result in the submission of better quality product for inspection.

Although the above factors were recognized somewhat prior to World War II, and a few sampling inspection schemes were then in use, it was not until the outbreak of hostilities that modern sampling inspection received its impetus. Good sampling plans replaced bad ones, sometimes at the expense of an increase in the total amount of inspection but almost always resulting indirectly in better quality being produced.

Any lot-by-lot sampling plan has as its primary purpose the acceptance of good lots and the rejection of bad lots. It is important to define what is meant by a good lot. Naturally, the consumer would like all of his accepted lots to be free of defectives. On the other hand, the manufacturer will usually consider this to be an unreasonable request since some defectives are bound to appear in the manufacturing process. If the manufacturer screens the lot a few times he may get rid of all the defectives, but at the prohibitive cost of screening. This cost will naturally be reflected in his price to the consumer. Ordinarily, the consumer can really tolerate some defectives in his lot, provided the number is not too large. Consequently, the manufacturer and the consumer get together and agree on what constitutes good quality. If lots are submitted at this quality or better, the lot should be accepted; if otherwise, rejected. Again this is an imposing task and can be accomplished only at the expense of screening. It is at this point that sampling inspection, with its corresponding advantage of reduced inspection costs, can be instituted. This advantage should not be minimized. Few manufacturers or consumers, whichever has to bear the cost of inspection, can stay in business very long if all lots are screened.

13.3.2 Drawing the Sample

A decision must be made on what shall constitute a lot for acceptance purposes, for each lot must be identified. Each lot should represent, as nearly as possible, the output of one machine or process during one interval of time, so that all parts or products in the lot have been turned out under essentially the same conditions. Wherever practicable, parts from different sources or different conditions should not be mixed into one lot. The power of the sampling plans to distinguish between good and bad lots is dependent on the variation from lot to lot, and provision should be made to maintain the identity of, and prevent the mixing of, the different lots.

A sample from each lot supplies the information on which the decision to accept or reject the lot is based. Therefore it is essential that the sample be drawn from each lot in a random manner. A sample is random if every piece in the lot has an equal chance of being selected. This may be accomplished by using a table of random numbers, a deck of cards, or some such chance method, to ensure the equal chance for each piece. Human attempts to randomize a sample without such aids often result in biased samples.

13.3.3 Lot-by-Lot Sampling Inspection by Attributes: Single Sampling

A single sampling procedure can be characterized by the following: One sample of n items is drawn from a lot of N items; the lot is accepted if the number of defectives d in the sample does not exceed c. Here c is referred to as the acceptance number.

Certain risks must be taken if sampling inspection is to be used. A graph of these risks plotted as a function of the incoming lot quality (p') is known as an operating characteristic (OC) curve and is illustrated in Fig.13.11.

If quality is good, it is desirable to have the probability of acceptance high. On the other hand, if quality is bad, it is desirable to have the probability of acceptance small. Note that if p' is 0, the lot contains no defectives and hence the lot will always be accepted, i.e., $L(p') = 1$. If p' is 1 the entire lot is defective and hence the sample will always contain all defectives so that the lot will always be rejected, i.e., $L(p') = 0$.

Assume that a quality standard p_g is established and that all lots better than this standard are considered to be "good" and all lots worse are considered to be "bad." An "ideal" OC curve would then be of the form shown in Fig. 13.12, with all good lots accepted and all bad ones rejected. Obviously, no sampling plan can have such a curve. The degree of approximation to the ideal curve depends on n and c.

If c is held constant, and n is increased, the slope becomes steeper. On the other hand, holding n constant and changing the acceptance number has the effect of shifting the OC curve to the left or right. These concepts are illustrated in Figs. 13.13 and 13.14.

If the lot size N is large compared with the sample size n, the OC curve is essentially independent of the lot size. In other words, if the OC curve of a sampling plan defined by a sample of size n, a lot of size N, and an acceptance number of c is compared with the OC curve of a sampling plan defined by a sample of size n, a lot of size $N_1 (N_1 > N)$, and an acceptance number of c, the difference is negligible, if the sample size is small compared with the lot size. Since this is the situation in most industrial applications, we will make this assumption throughout the rest of this section. As a result, a single sampling plan will now be defined by

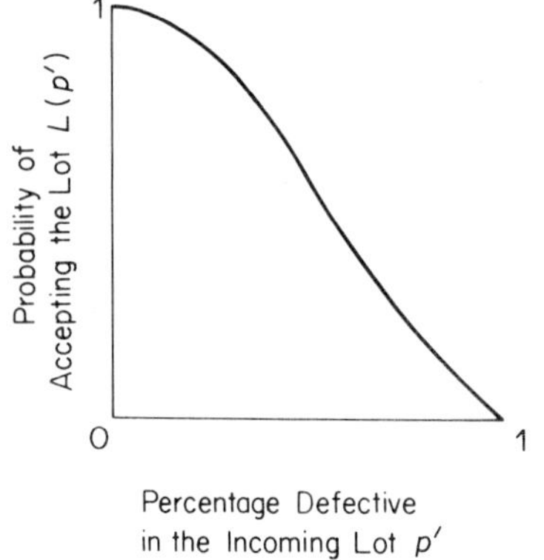

Fig. 13.11 *Operating characteristic curve.*

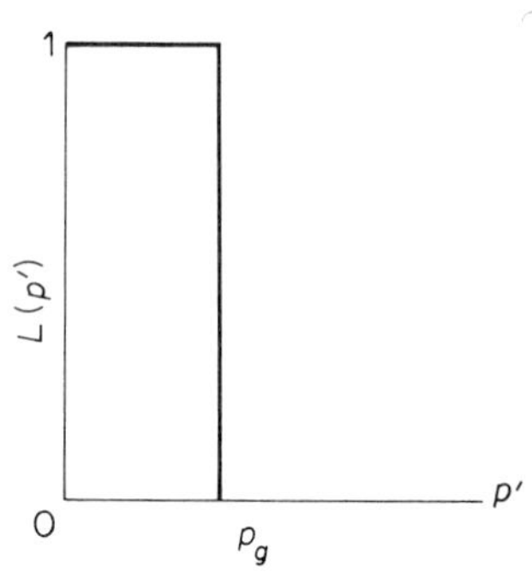

Fig. 13.12 *Ideal operating characteristic curve.*

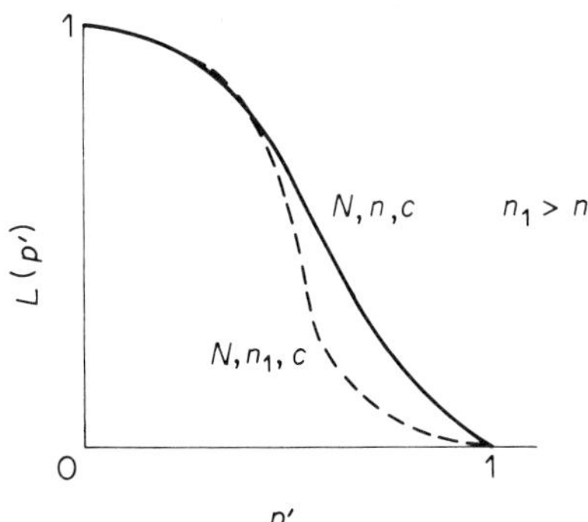

Fig. 13.13 *Effect on OC of varying* n.

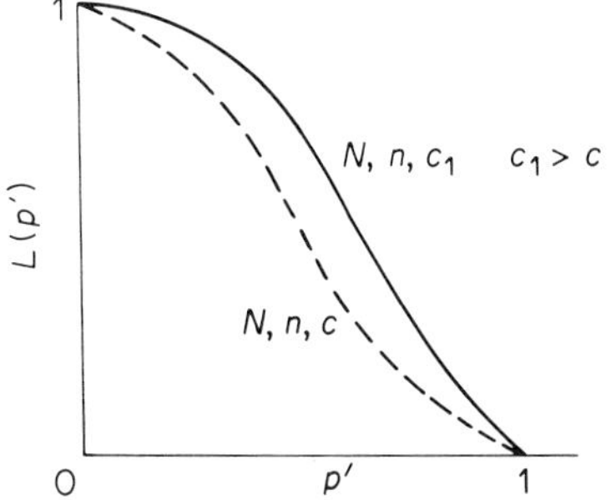

Fig. 13.14 *Effect on OC of varying* c.

only two numbers, the sample size n and acceptance number c.*

The single sampling procedure can be viewed from another side. By drawing a sample of n, and looking at the number of defectives d present, the consumer is essentially estimating the percentage defective, i.e., $\hat{p} = d/n$ is an estimate of the incoming quality p'. A lot is rejected if this estimate is too high, i.e., if $\hat{p} > c/n = p^*$. Therefore, a procedure which says to reject a lot if the number of defectives d is greater than c in a sample of size n is equivalent to rejecting a lot if the estimated percentage defective, $d/n = \hat{p}$, is greater than $c/n = p^*$. This viewpoint will be useful later in the discussion of sampling inspection by variables.

13.3.3.1 Choosing a Sampling Plan. The consumer has at his disposal the choice of one of many OC curves. The one chosen should reflect his views as to the cost of making wrong decisions. By varying n and c, he can always find an OC curve which will pass through two preassigned points. In other words, by specifying the points $[p'_1, L(p'_1)]$ and $[p'_2\ L(p'_2)]$, an n and a c can be found such that the OC curve passes through these two points. Before using sampling inspection, then, the consumer can locate two points p'_1 and p'_2 such that if quality is submitted better than p'_1, he will accept the lot with probability greater than $L(p'_1) = 1 - \alpha$; and if quality is submitted worse than p'_2 he will accept the lot with probability less than $L(p'_2) = \beta$. Here α is known as the producer's risk and β is known as the consumer's risk. The long-run average of submitted quality is known as the process average.

13.3.3.2 Calculation of OC Curves for Single Sampling Plans. It was pointed out previously that a sampling plan is defined by two numbers: n, the sample size, and c, the acceptance number. Thus the sampling procedure is to accept a lot if the number of defectives d in a sample of n does not exceed c. The OC curve is defined by this procedure. If lot quality is submitted at p' per cent defective, the probability of accepting the lot is $L(p')$.

$$L(p') = \sum_{d=0}^{c} \binom{n}{d}(p')^d(1 - p')^{n-d}$$

$$= (1 - p')^n + \binom{n}{1}(p')(1 - p')^{n-1} + \binom{n}{2}(p')^2(1 - p')^{n-2} + \cdots + \binom{n}{c}(p')^c$$

where $\binom{n}{d} = \dfrac{n!}{d!(n - d)!}$ and $0!$ is defined to be equal to 1.

Such calculations as the above are often cumbersome, and an approximation is desirable. Approximate answers may be obtained very rapidly by the use of the Poisson distribution, which is tabulated in Table 13.6.† The larger the n and smaller the p, the closer the approximate answer is to the true probability.

13.3.3.3 Example. What is the probability of accepting a lot whose incoming quality is 4.0 per cent defective, using a sample of size 30 and an acceptance number $c = 1$? (*page 569.*)

* From a mathematical viewpoint, this implies that the binomial distribution can be used as an approximation to the hypergeometric distribution.

† Here $L(p')$ is approximately the value read out of Table 13.6 with entries c and np'.

TABLE 13.6. *SUMMATION OF TERMS OF POISSON'S EXPONENTIAL BINOMIAL LIMIT*†‡

c' or np' \ c	0	1	2	3	4	5	6	7	8	9
0.02	980	1,000								
0.04	961	999	1,000							
0.06	942	998	1,000							
0.08	923	997	1,000							
0.10	905	995	1,000							
0.15	861	990	999	1,000						
0.20	819	982	999	1,000						
0.25	779	974	998	1,000						
0.30	741	963	996	1,000						
0.35	705	951	994	1,000						
0.40	670	938	992	999	1,000					
0.45	638	925	989	999	1,000					
0.50	607	910	986	998	1,000					
0.55	577	894	982	998	1,000					
0.60	549	878	977	997	1,000					
0.65	522	861	972	996	999	1,000				
0.70	497	844	966	994	999	1,000				
0.75	472	827	959	993	999	1,000				
0.80	449	809	953	991	999	1,000				
0.85	427	791	945	989	998	1,000				
0.90	407	772	937	987	998	1,000				
0.95	387	754	929	984	997	1,000				
1.00	368	736	920	981	996	999	1,000			
1.1	333	699	900	974	995	999	1,000			
1.2	301	663	879	966	992	998	1,000			
1.3	273	627	857	957	989	998	1,000			
1.4	247	592	833	946	986	997	999	1,000		
1.5	223	558	809	934	981	996	999	1,000		
1.6	202	525	783	921	976	994	999	1,000		
1.7	183	493	757	907	970	992	998	1,000		
1.8	165	463	731	891	964	990	997	999	1,000	
1.9	150	434	704	875	956	987	997	999	1,000	
2.0	135	406	677	857	947	983	995	999	1,000	
2.2	111	355	623	819	928	975	993	998	1,000	
2.4	091	308	570	779	904	964	988	997	999	1,000
2.6	074	267	518	736	877	951	983	995	999	1,000
2.8	061	231	469	692	848	935	976	992	998	999
3.0	050	199	423	647	815	916	966	988	996	999
3.2	041	171	380	603	781	895	955	983	994	998
3.4	033	147	340	558	744	871	942	977	992	997
3.6	027	126	303	515	706	844	927	969	988	996
3.8	022	107	269	473	668	816	909	960	984	994
4.0	018	092	238	433	629	785	889	949	979	992
4.2	015	078	210	395	590	753	867	936	972	989

† Reprinted by permission from *Statistical Quality Control*, 2nd ed., by E. L. Grant. Copyright 1952. McGraw-Hill Book Company, Inc.

‡ 1,000 × probability of c or less occurrences of event that has average number of occurrences equal to c' or np'.

TABLE 13.6. *Cont.*

c' or np' \ c	0	1	2	3	4	5	6	7	8	9
4.4	012	066	185	359	551	720	844	921	964	985
4.6	010	056	163	326	513	686	818	905	955	980
4.8	008	048	143	294	476	651	791	887	944	975
5.0	007	040	125	265	440	616	762	867	932	968
5.2	006	034	109	238	406	581	732	845	918	960
5.4	005	029	095	213	373	546	702	822	903	951
5.6	004	024	082	191	342	512	670	797	886	941
5.8	003	021	072	170	313	478	638	771	867	929
6.0	002	017	062	151	285	446	606	744	847	916

	10	11	12	13	14	15	16
2.8	1,000						
3.0	1,000						
3.2	1,000						
3.4	999	1,000					
3.6	999	1,000					
3.8	998	999	1,000				
4.0	997	999	1,000				
4.2	996	999	1,000				
4.4	994	998	999	1,000			
4.6	992	997	999	1,000			
4.8	990	996	999	1,000			
5.0	986	995	998	999	1,000		
5.2	982	993	997	999	1,000		
5.4	977	990	996	999	1,000		
5.6	972	988	995	998	999	1,000	
5.8	965	984	993	997	999	1,000	
6.0	957	980	991	996	999	999	1,000

c' or np' \ c	0	1	2	3	4	5	6	7	8	9
6.2	002	015	054	134	259	414	574	716	826	902
6.4	002	012	046	119	235	384	542	687	803	886
6.6	001	010	040	105	213	355	511	658	780	869
6.8	001	009	034	093	192	327	480	628	755	850
7.0	001	007	030	082	173	301	450	599	729	830
7.2	001	006	025	072	156	276	420	569	703	810
7.4	001	005	022	063	140	253	392	539	676	788
7.6	001	004	019	055	125	231	365	510	648	765
7.8	000	004	016	048	112	210	338	481	620	741
8.0	000	003	014	042	100	191	313	453	593	717
8.5	000	002	009	030	074	150	256	386	523	653
9.0	000	001	006	021	055	116	207	324	456	587
9.5	000	001	004	015	040	089	165	269	392	522
10.0	000	000	003	010	029	067	130	220	333	458

	10	11	12	13	14	15	16	17	18	19
6.2	949	975	989	995	998	999	1,000			
6.4	939	969	986	994	997	999	1,000			
6.6	927	963	982	992	997	999	999	1,000		
6.8	915	955	978	990	996	998	999	1,000		
7.0	901	947	973	987	994	998	999	1,000		

TABLE 13.6. *Cont.*

c' or np' \ c	10	11	12	13	14	15	16	17	18	19
7.2	887	937	967	984	993	997	999	999	1,000	
7.4	871	926	961	980	991	996	998	999	1,000	
7.6	854	915	954	976	989	995	998	999	1,000	
7.8	835	902	945	971	986	993	997	999	1,000	
8.0	816	888	936	966	983	992	996	998	999	1,000
8.5	763	849	909	949	973	986	993	997	999	999
9.0	706	803	876	926	959	978	989	995	998	999
9.5	645	752	836	898	940	967	982	991	996	998
10.0	583	697	792	864	917	951	973	986	993	997

	20	21	22
8.5	1,000		
9.0	1,000		
9.5	999	1,000	
10.0	998	999	1,000

c' or np' \ c	0	1	2	3	4	5	6	7	8	9
10.5	000	000	002	007	021	050	102	179	279	397
11.0	000	000	001	005	015	038	079	143	232	341
11.5	000	000	001	003	011	028	060	114	191	289
12.0	000	000	001	002	008	020	046	090	155	242
12.5	000	000	000	002	005	015	035	070	125	201
13.0	000	000	000	001	004	011	026	054	100	166
13.5	000	000	000	001	003	008	019	041	079	135
14.0	000	000	000	000	002	006	014	032	062	109
14.5	000	000	000	000	001	004	010	024	048	088
15.0	000	000	000	000	001	003	008	018	037	070

	10	11	12	13	14	15	16	17	18	19
10.5	521	639	742	825	888	932	960	978	988	994
11.0	460	579	689	781	854	907	944	968	982	991
11.5	402	520	633	733	815	878	924	954	974	986
12.0	347	462	576	682	772	844	899	937	963	979
12.5	297	406	519	628	725	806	869	916	948	969
13.0	252	353	463	573	675	764	835	890	930	957
13.5	211	304	409	518	623	718	798	861	908	942
14.0	176	260	358	464	570	669	756	827	883	923
14.5	145	220	311	413	518	619	711	790	853	901
15.0	118	185	268	363	466	568	664	749	819	875

	20	21	22	23	24	25	26	27	28	29
10.5	997	999	999	1,000						
11.0	995	998	999	1,000						
11.5	992	996	998	999	1,000					
12.0	988	994	997	999	999	1,000				
12.5	983	991	995	998	999	999	1,000			
13.0	975	986	992	996	998	999	1,000			
13.5	965	980	989	994	997	998	999	1,000		
14.0	952	971	983	991	995	997	999	999	1,000	
14.5	936	960	976	986	992	996	998	999	999	1,000
15.0	917	947	967	981	989	994	997	998	999	1,000

TABLE 13.6. *Cont.*

'or np' \ c	4	5	6	7	8	9	10	11	12	13
16	000	001	004	010	022	043	077	127	193	275
17	000	001	002	005	013	026	049	085	135	201
18	000	000	001	003	007	015	030	055	092	143
19	000	000	001	002	004	009	018	035	061	098
20	000	000	000	001	002	005	011	021	039	066
21	000	000	000	000	001	003	006	013	025	043
22	000	000	000	000	001	002	004	008	015	028
23	000	000	000	000	000	001	002	004	009	017
24	000	000	000	000	000	000	001	003	005	011
25	000	000	000	000	000	000	001	001	003	006

	14	15	16	17	18	19	20	21	22	23
16	368	467	566	659	742	812	868	911	942	963
17	281	371	468	564	655	736	805	861	905	937
18	208	287	375	469	562	651	731	799	855	899
19	150	215	292	378	469	561	647	725	793	849
20	105	157	221	297	381	470	559	644	721	787
21	072	111	163	227	302	384	471	558	640	716
22	048	077	117	169	232	306	387	472	556	637
23	031	052	082	123	175	238	310	389	472	555
24	020	034	056	087	128	180	243	314	392	473
25	012	022	038	060	092	134	185	247	318	394

	24	25	26	27	28	29	30	31	32	33
16	978	987	993	996	998	999	999	1,000		
17	959	975	985	991	995	997	999	999	1,000	
18	932	955	972	983	990	994	997	998	999	1,000
19	893	927	951	969	980	988	993	996	998	999
20	843	888	922	948	966	978	987	992	995	997
21	782	838	883	917	944	963	976	985	991	994
22	712	777	832	877	913	940	959	973	983	989
23	635	708	772	827	873	908	936	956	971	981
24	554	632	704	768	823	868	904	932	963	969
25	473	553	629	700	763	818	863	900	929	950

	34	35	36	37	38	39	40	41	42	43
19	999	1,000								
20	999	999	1,000							
21	997	998	999	999	1,000					
22	994	996	998	999	999	1,000				
23	988	993	996	997	999	999	1,000			
24	979	987	992	995	997	998	999	999	1,000	
25	966	978	985	991	994	997	998	999	999	1,000

$$L(0.04) = \sum_{d=0}^{1} \binom{30}{d}(0.04)^d(0.96)^{30-d}$$

$$= (0.96)^{30} + \frac{(30)!}{(29)!(1)!}0.04(0.96)^{29}$$

$$= 0.661$$

Using the Poisson approximation,

$$np' = 30 \times 0.04 = 1.20 \qquad c = 1$$
$$L(0.04) = 0.663$$

which, in this case, is in good agreement with the correct value.

13.3.4 Double Sampling Plans

A double sampling procedure can be characterized by the following: A sample of n_1 items are drawn from a lot; the lot is accepted if there are no more than c_1 defective items. The lot is rejected if the number of defectives is greater than or equal to r_1. If there are between $c_1 + 1$ and $r_1 - 1$ defective items, a second sample of size n_2 is drawn; the lot is accepted if there are no more than c_2 defectives in the combined sample of $n_1 + n_2$; the lot is rejected if there are $r_2 = c_2 + 1$ or more defective items in the combined sample of $n_1 + n_2$.

A lot will be accepted on the first sample if the incoming quality is very good. Similarly, a lot will be rejected on the first sample if the incoming quality is very bad. If the lot is of intermediate quality, a second sample may have to be taken. Double sampling plans have the psychological advantage of giving a second chance to doubtful lots. Furthermore, they can have the additional advantage of requiring fewer total inspections, on the average, than single sample plans for any given quality protection. On the other side of the ledger, double sampling plans can have the following disadvantages: (1) more complex administrative duties; (2) inspectors will have variable inspection loads depending upon whether one or two samples are taken; (3) the maximum amount of inspection exceeds that for single sampling plans (which is constant).

13.3.4.1 OC Curves for Double Sampling Plans. A lot will be accepted if, and only if, (1) the number of defectives d_1 in the first sample of n_1 does not exceed c_1, i.e., $d_1 \leq c_1$; (2) there are more than c_1 defectives in the first sample but less than r_1 defectives, and the total number of defectives $d_1 + d_2$ in the combined sample of $n_1 + n_2$ does not exceed c_2, $c_2 \geq r_1 - 1$, i.e., if $c_1 < d_1 \leq r_1 - 1$ and $d_1 + d_2 \leq c_2$.

Thus for any incoming quality p', the probability of accepting the lot is

$$L(p') = Pr(d_1 \leq c_1 | p') + Pr(c_1 < d_1 \leq r_1 - 1 \text{ and } d_1 + d_2 \leq c_2 | p')$$

where the symbol | reads "given" and $Pr(d_1 \leq c_1 | p')$ reads "the probability that d_1 is less than or equal to c_1, given p'."

$$\begin{aligned} L(p') &= \sum_{d_1=0}^{c_1} \binom{n_1}{d_1} (p')^{d_1}(1 - p')^{n_1-d_1} \\ &\quad + \sum_{d_1=c_1+1}^{r_1-1} \sum_{d_2=0}^{c_2-d_1} \binom{n_1}{d_1} (p')^{d_1} (1 - p')^{n_1-d_1} \binom{n_2}{d_2} (p')^{d_2} (1 - p')^{n_2-d_2} \\ &= \sum_{d_1=0}^{c_1} \binom{n_1}{d_1} (p')^{d_1} (1 - p')^{n_1-d_1} \\ &\quad + \sum_{d_1=c_1+1}^{r_1-1} \left[\binom{n_1}{d_1} (p')^{d_1} (1 - p')^{n_1-d_1} \sum_{d_2=0}^{c_2-d_1} \binom{n_2}{d_2} (p')^{d_2} (1 - p')^{n_2-d_2}\right] \end{aligned}$$

13.3.4.2 Example. A large lot of 3/4 inch screws is submitted for sampling inspection by means of double sampling. If the first sample of 20 contains no defectives, the lot is to be accepted. If it contains three or more defectives, the lot is to be rejected. Otherwise a second sample of 20 is to be drawn, and the lot is to be accepted if the total number of defectives in both samples does not exceed three. This plan can be characterized by the entries in Table 13.7. Find the probability of accepting the lot if $p' = 3.5$ per cent.

$$\sum_{d_1=0}^{0} \binom{n_1}{d_1} (p')^{d_1} (1 - p')^{n_1-d_1} = (1 - p')^{20}$$

$$= (0.965)^{20} = 0.490 \qquad (13.1)$$

$$\binom{n_1}{1} (p') (1 - p')^{n_1-1} \sum_{d_2=0}^{2} \binom{n_2}{d_2} (p')^{d_2}$$

TABLE 13.7. *A DOUBLE SAMPLING PLAN*

Sample	Sample Size	Combined Samples		
		Size	Acceptance Number	Rejection Number
First	20	20	0	3
Second	20	40	3	4

$$(1 - p')^{n_2 - d_2}$$
$$= \binom{20}{1}(0.035)(0.965)\sum_{d_2=0}^{2}\binom{20}{d_2}(0.035)^{d_2}$$
$$(0.965)^{20-d_2}$$
$$= 0.345 \tag{13.2}$$
$$\binom{n_1}{2}(p')^2(1-p')^{n_1-2}$$
$$\sum_{d_2=0}^{1}\binom{n_2}{d_2}(p')^{d_2}(1-p')^{n_2-d_2}$$
$$= \binom{20}{2}(0.035)^2(0.965)^{18}$$
$$\sum_{d_2=0}^{1}\binom{20}{d_2}(0.035)^{d_2}(0.965)^{20-d_2}$$
$$= 0.104 \tag{13.3}$$

Adding the three expressions results in $L(p') = 0.939$. Using the Poisson approximation we have

$$0.497 \tag{13.4}$$
$$(0.844 - 0.497)(0.966) = 0.335 \tag{13.5†}$$
$$(0.966 - 0.844)(0.844) = 0.103 \tag{13.6†}$$
$$L(p') = 0.935$$

13.3.5 Multiple Sampling Plans

A multiple sampling procedure can be represented in a table such as Table 13.8.

A first sample of n_1 is drawn, the lot is accepted if there are no more than c_1 defectives, the lot is rejected if there are more than r_1 defectives. Otherwise a second sample of n_2 is drawn. The lot is accepted if there are no more than c_2 defectives in the combined sample of $n_1 + n_2$. The lot is rejected if there are more than r_2 defectives in the combined sample of $n_1 + n_2$. The procedure is continued in accordance with Table 13.8. If, by the end of the sixth sample, the lot is neither accepted nor rejected, a sample of n_7 is drawn. The lot is accepted if the number of defectives in the combined sample of $n_1 + n_2 + n_3 + n_4 + n_5 + n_6 + n_7$ does not exceed c_7. Otherwise, the lot is rejected. Note that $c_1 < c_2 < \ldots < c_7$ and $c_i < r_i$ for all i. Of course, plans can be devised which permit any number of samples before a decision is reached. A multiple sampling plan will generally involve less inspection, on the average, than the corresponding single or double sampling plan guaranteeing the same protection. The advantage is quite important, since inspection costs are directly related to sample sizes. On the other hand, some of the disadvantages of multiple sampling plans are: (1) they usually require higher administrative costs than single or double sampling plans; (2) the variability of inspection load introduces difficulties in scheduling inspection time; (3) higher caliber inspection personnel may be necessary to guarantee proper use of the plans; (4) adequate storage facilities must be provided for the lot while multiple sampling is being carried on.

13.3.6 Classification of Sampling Plans

Published tables of sampling plans have been classified into four categories, namely, by acceptable quality level (AQL), by lot tolerance per cent defective (LTPD), by point of control, and by average outgoing quality limit (AOQL).

13.3.6.1 Classification by AQL. Acceptance procedures based on the acceptable quality level (AQL) generally make use of the process average to determine the sampling plan to be used. The AQL may be viewed as the highest per cent defective that is acceptable as a process average. In normal sampling, a lot at AQL quality will have a high probability of acceptance. The probability of acceptance, $1 - \alpha$, at the AQL is usually set near the 95 per cent point, and α is known as the producer's risk. Thus a producer has good protection against rejection of submitted lots from a process that is at the AQL or better. On the other hand, this type of classification does not specify anything about the protection the consumer has against the acceptance of a lot worse than the AQL.

---

†The Poisson table is set up to give $\sum_{d=0}^{c}\binom{n}{d}p'^d(1-p')^{n-d}$. To get an individual term, say $\binom{n}{c}p'^c(1-p')^{n-c}$, we take

$$\sum_{d=0}^{c}\binom{n}{d}p'^d(1-p')^{n-d}$$
$$-\sum_{d=0}^{c-1}\binom{n}{d}p'^d(1-p')^{n-d}$$
$$= \binom{n}{c}p'^c(1-p')^{n-c}$$

TABLE 13.8. *MULTIPLE SAMPLING PLAN*

Sample	Sample Size	Combined Samples		
		Size	Acceptance Number	Rejection Number
First	n_1	n_1	c_1	r_1
Second	n_2	$n_1 + n_2$	c_2	r_2
Third	n_3	$n_1 + n_2 + n_3$	c_3	r_3
Fourth	n_4	$n_1 + n_2 + n_3 + n_4$	c_4	r_4
Fifth	n_5	$n_1 + n_2 + n_3 + n_4 + n_5$	c_5	r_5
Sixth	n_6	$n_1 + n_2 + n_3 + n_4 + n_5 + n_6$	c_6	r_6
Seventh	n_7	$n_1 + n_2 + n_3 + n_4 + n_5 + n_6 + n_7$	c_7	$c_7 + 1$

13.3.6.2 Classification by LTPD. The lot tolerance per cent defective (LTPD) usually refers to that incoming quality above which there is a small chance that a lot will be accepted. This probability is usually taken to be near the 10 per cent point. Thus a consumer, inspecting lots submitted from a process that is at the LTPD or worse, has a small probability of accepting such lots. This probability is known as the consumer's risk. On the other hand, this type of classification does not specify anything about the protection the producer has against the rejection of lots better than the LTPD.

13.3.6.3 Classification by Point of Control. The point of control is the lot quality for which the probability of acceptance is 0.50. The concept here is one of splitting the risk between producer and consumer. Lots submitted from processes whose quality is better than the point of control have a probability of acceptance that is higher than 50 per cent. Lots submitted from processes whose quality is worse than the point of control have a probability of acceptance smaller than 50 per cent.

13.3.6.4 Classification by AOOL. The average outgoing quality limit (AOQL) does not refer to a point on the OC curve, but rather to the upper limit on outgoing quality that may be expected in the long run when all rejected lots are subjected to 100 per cent inspection, with all defective articles removed and replaced by good articles. The average outgoing quality (AOQ) can be computed from the formulas

$$\text{AOQ} = p'L(p') \qquad \text{AOQL} = \max_{p'} \text{AOQ}$$

where $\max_{p'}$ AOQ is the maximum AOQ over the range $p' = 0$ to $p' = 1$.

Figure 13.15 shows a graphic comparison of these four methods of indexing acceptance plans in relation to a stated quality standard. The four curves shown in this figure are OC curves for single sampling plans that are indexed in their respective tables by the 1.0 per cent defective figure. The most lenient plan (curve IV) is that classified by the AQL; in this plan with an AQL of 1.0 per cent, 4 defectives are permitted in a sample of 150. The strictest plan (curve I) is the LTPD plan permitting 0 defectives in a sample of 225. The point of control plan (curve II) permits 1 defective in a sample of 150; here the probability of acceptance of a 1.0 per cent defective lot is 0.50. The 1.0 per cent AOQL plan allows 1 defective in a

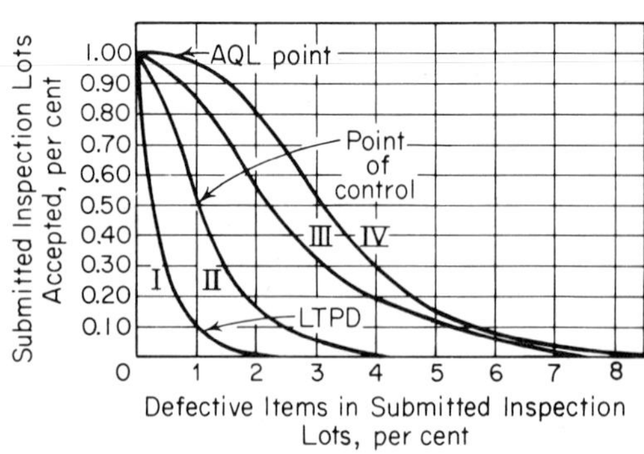

Fig. 13.15 *Single sampling plans classified on: I. 1 per cent LTPD:* n = *225;* c = *50. II. Point of control* = *1 per cent:* n = *50;* c = *1. III. 1 per cent AOQL:* n = *75;* c = *1. IV. AQL* = *1 per cent:* n = *150;* c = *4.*

sample of 75; the OC curve (curve III) is intermediate between the AQL 1.0 per cent and the point of control 1.0 per cent plans.

13.3.7 Dodge-Romig Tables

In 1944, H. F. Dodge and H. G. Romig published a volume of attribute sampling tables called *Sampling Inspection Tables.** These tables originated in the Bell Telephone Laboratories and reflect many years of experience with acceptance sampling in the Bell Telephone System. It should be pointed out that these tables were originally prepared for use within the Bell Telephone System. They were designed primarily to minimize the total amount of inspection, considering total sampling inspection and screening inspection of rejected lots. The Dodge-Romig volume contains four sets of tables, as follows:

(1) Single sampling lot tolerance tables
(2) Double sampling lot tolerance tables
(3) Single sampling AOQL tables
(4) Double sampling AOQL tables

13.3.7.1 Single Sampling Lot Tolerance Tables. This set indexes plans according to the following lot tolerance per cent defectives (LTPD) with the consumer's risk = 0.10.

0.5 %	2.0 %	4.0 %	7.0 %
1.0 %	3.0 %	5.0 %	10.0 %

Table 13.9 presents a typical table from Dodge-Romig single sampling LTPD. All the sampling plans in this table have the same LTPD. Furthermore, if rejected lots are screened, the table gives the AOQL values for each plan. For any given lot size there is a choice of six plans, each for a different value of the process average. If the plan chosen corresponds to the true process average of the production process, and rejected lots are screened, the plan will guarantee that the average amount of inspection, considering inspection of samples and screening of rejected lots, will be smaller than for any of the remaining five plans.† The process average is usually estimated from past data by taking the ratio of the total number of defectives to the total number inspected. If there are no past data to estimate the process average, the plan should be selected from the right-hand column of the table. This gives good lots a better chance of acceptance.

It must be emphasized that if screening of rejected lots does not take place, these plans still guarantee a stated LTPD.

13.3.7.2 Double Sampling Lot Tolerance Tables. This set indexes plans according to the same LTPD's as for the single sampling LTPD plans. A typical double sampling LTPD table is shown in Table 13.10.

This table has the same features as the single sampling LTPD tables with the exception of using double sampling instead of single sampling. The first sample is always smaller than the corresponding single sampling plan, but if a second sample is required, the combined sample size exceeds the sample size of the single sampling plan. However, the average amount of sampling inspection is generally smaller with the double sampling plan. Since the second sample is not always taken (depending on the submitted quality), the process average is usually estimated from the first sample only.

13.3.7.3 Single Sampling AOQL Tables. This set indexes plans according to the following average outgoing quality limits (AOQL).

1.0 %	1.5 %	4.0 %
0.25 %	2.0 %	5.0 %
0.5 %	2.5 %	7.0 %
0.75 %	3.0 %	10.0 %
1.0 %		

A typical single sampling AOQL table is shown in Table 13.11 A.

All the sampling plans in this table have the same AOQL, assuming all rejected lots are screened. Furthermore, the table gives the LTPD values for each plan. Like all the Dodge-Romig plans these plans have the property that the average amount of inspection is smallest for that plan which belongs to the class under whose heading the true process average falls.

13.3.7.4 Double Sampling AOQL Tables. This set indexes plans according to the same AOQL's as for the single sampling AOQL plans. A typical double sampling AOQL table is shown in Table 13.11 B. This table has the same features as the single sampling AOQL

---

* H. F. Dodge and H. G. Romig, *Sampling Inspection Tables* (New York: John Wiley and Sons, Inc., 1944).

† The average amount of inspection (AOI) is given by the formula

$$AOI = nL(p') + N[1 - L(p')].$$

TABLE 13.9. *EXAMPLE OF DODGE-ROMIG SINGLE SAMPLING LOT TOLERANCE TABLES*: Lot tolerance per cent defective = 5.0 per cent; consumer's risk = 0.10.†

Process Average, %	0–0.05			0.06–0.50			0.51–1.00			1.01–1.50			1.51–2.00			2.01–2.50		
Lot Size	n	c	AOQL, %	n	c	AOQL, %	n	c	AOQL, %	n	c	AOQL, %	n	c	AOQL, %	n	c	AOQL, %
1–30	All	0	0	All	0	0	All	0	0	All	0	0	All	0	0	All	0	0
31–50	30	0	0.49	30	0	0.49	30	0	0.49	30	0	0.49	30	0	0.49	30	0	0.49
51–100	37	0	0.63	37	0	0.63	37	0	0.63	37	0	0.63	37	0	0.63	37	0	0.63
101–200	40	0	0.74	40	0	0.74	40	0	0.74	40	0	0.74	40	0	0.74	40	0	0.74
201–300	43	0	0.74	43	0	0.74	70	1	0.92	70	1	0.92	95	2	0.99	95	2	0.99
301–400	44	0	0.74	44	0	0.74	70	1	0.99	100	2	1.0	120	3	1.1	145	4	1.1
401–500	45	0	0.75	75	1	0.95	100	2	1.1	100	2	1.1	125	3	1.2	150	4	1.2
501–600	45	0	0.76	75	1	0.98	100	2	1.1	125	3	1.2	150	4	1.3	175	5	1.3
601–800	45	0	0.77	75	1	1.0	100	2	1.2	130	3	1.2	175	5	1.4	200	6	1.4
801–1,000	45	0	0.78	75	1	1.0	105	2	1.2	155	4	1.4	180	5	1.4	225	7	1.5
1,001–2,000	45	0	0.80	75	1	1.0	130	3	1.4	180	5	1.6	230	7	1.7	280	9	1.8
2,001–3,000	75	1	1.1	105	2	1.3	135	3	1.4	210	6	1.7	280	9	1.9	370	13	2.1
3,001–4,000	75	1	1.1	105	2	1.3	160	4	1.5	210	6	1.7	305	10	2.0	420	15	2.2
4,001–5,000	75	1	1.1	105	2	1.3	160	4	1.5	235	7	1.8	330	11	2.0	440	16	2.2
5,001–7,000	75	1	1.1	105	2	1.3	185	5	1.7	260	8	1.9	350	12	2.2	490	18	2.4
7,001–10,000	75	1	1.1	105	2	1.3	185	5	1.7	260	8	1.9	380	13	2.2	535	20	2.5
10,001–20,000	75	1	1.1	135	3	1.4	210	6	1.8	285	9	2.0	425	15	2.3	610	23	2.6
20,001–50,000	75	1	1.1	135	3	1.4	235	7	1.9	305	10	2.1	470	17	2.4	700	27	2.7
50,001–100,000	75	1	1.1	160	4	1.6	235	7	1.9	355	12	2.2	515	19	2.5	770	30	2.8

† These tables reprinted with permission from H. F. Dodge and H. G. Romig, *Sampling Inspection Tables*. Copyright 1944, John Wiley & Sons, Inc.

tables with the exception of using double sampling instead of single sampling.

13.3.8 Military Standard 105D

Various sets of sampling tables using the AQL as an index have been published. The first set of tables was developed in 1943 for Army Ordnance, and with some changes and extensions became the Army Service Forces tables used during the final years of World War II.

An Administration Manual, *Standard Sampling Inspection Procedures*, was issued by the Navy in October 1945. This manual, including tables and OC (operating characteristic) curves for all the acceptance sampling plans given, had been prepared by the Statistical Research Group of Columbia University during World War II. Although the tables and procedures in this manual were similar in many respects to those of the Army Service Forces, sequential (multiple) sampling plans were included in addition to single and double sampling plans, and there were several other important points of difference. The actual tables from this manual were issued by the Navy Department in April 1946 as Appendix X to *General Specifications for Inspection of Material*. After the unification of the armed services, Appendix X was adopted by the Department of Defense in February 1949 as JAN–STD–105.

In 1949, a book entitled *Sampling Inspection*, edited by H. A. Freeman, M. Friedman, F. Mosteller, and W. A. Wallis‡ was published. This book was prepared by the Statistical

‡ H. A. Freeman, M. Friedman, F. Mosteller, and W. A. Wallis, *Sampling Inspection* (New York: McGraw-Hill Book Company, Inc., 1949).

TABLE 13.10. *EXAMPLE OF DODGE-ROMIG DOUBLE SAMPLING LOT TOLERANCE TABLES*

Lot tolerance per cent defective = 5.0 per cent; consumer's risk = 0.10.

Process Average, %	0–0.05						0.06–0.50						2.01–2.50					
	Trial 1		Trial 2			AOQL in %	Trial 1		Trial 2			AOQL in %	Trial 1		Trial 2			AOQL in %
Lot Size	n_1	c_1	n_2	n_1+n_2	c_2		n_1	c_1	n_2	n_1+n_2	c_2		n_1	c_1	n_2	n_1+n_2	c_2	
1–30	All	0	—	—	—	0	All	0	—	—	—	0	All	0	—	—	—	0
31–50	30	0	—	—	—	0.49	30	0	—	—	—	0.49	30	0	—	—	—	0.49
51–75	38	0	—	—	—	0.59	38	0	—	—	—	0.59	38	0	—	—	—	0.59
76–100	44	0	21	65	1	0.64	44	0	21	65	1	0.64	44	0	21	65	1	0.64
101–200	49	0	26	75	1	0.84	49	0	26	75	1	0.84	49	0	51	100	2	0.91
201–300	50	0	30	80	1	0.91	50	0	30	80	1	0.91	50	0	100	150	4	1.1
301–400	55	0	30	85	1	0.92	55	0	55	110	2	1.1	85	0	105	190	6	1.3
401–500	55	0	30	85	1	0.93	55	0	55	110	2	1.1	85	1	140	225	7	1.4
501–600	55	0	30	85	1	0.94	55	0	60	115	2	1.1	85	1	165	250	8	1.5
601–800	55	0	35	90	1	0.95	55	0	65	120	2	1.1	120	2	185	305	10	1.6
801–1,000	55	0	35	90	1	0.96	55	0	65	120	2	1.1	120	2	210	330	11	1.7
1,001–2,000	55	0	35	90	1	0.98	55	0	95	150	3	1.3	175	4	260	435	15	2.0
2,001–3,000	55	0	65	120	2	1.2	55	0	95	150	3	1.3	205	5	375	580	21	2.3
3,001–4,000	55	0	65	120	2	1.2	55	0	95	150	3	1.3	230	6	420	650	24	2.4
4,001–5,000	55	0	65	120	2	1.2	55	0	95	150	3	1.4	255	7	445	700	26	2.5
5,001–7,000	55	0	65	120	2	1.2	55	0	95	150	3	1.4	255	7	495	750	28	2.6
7,001–10,000	55	0	65	120	2	1.2	55	0	120	175	4	1.5	280	8	540	820	31	2.7
10,001–20,000	55	0	65	120	2	1.2	55	0	120	175	4	1.5	280	8	660	940	36	2.8
20,001–50,000	55	0	65	120	2	1.2	55	0	150	205	5	1.7	305	9	745	1050	41	2.9
50,001–100,000	55	0	65	120	2	1.2	55	0	150	205	5	1.7	330	10	810	1140	45	3.0

TABLE 13.11.A *EXAMPLE OF DODGE-ROMIG AOQL TABLES—SINGLE SAMPLING*
Average outgoing quality limit = 2.0 per cent.

Process, Average %	0–0.04			0.05–0.40			0.41–0.80			0.81–1.20			1.21–1.60			1.61–2.00		
Lot Size	n	c	p_t, %	n	c	p_t, %	n	c	p_t, %	n	c	p_t, %	n	c	p_t, %	n	c	p_t, %
1–15	All	0	—	All	0	—	All	0	—	All	0	—	All	0	—	All	0	—
16–50	14	0	13.6	14	0	13.6	14	0	13.6	14	0	13.6	14	0	13.6	14	0	13.6
51–100	16	0	12.4	16	0	12.4	16	0	12.4	16	0	12.4	16	0	12.4	16	0	12.4
101–200	17	0	12.2	17	0	12.2	17	0	12.2	17	0	12.2	35	1	10.5	35	1	10.5
201–300	17	0	12.3	17	0	12.3	17	0	12.3	37	1	10.2	37	1	10.2	37	1	10.2
301–400	18	0	11.8	18	0	11.8	38	1	10.0	38	1	10.0	38	1	10.0	60	2	8.5
401–500	18	0	11.9	18	0	11.9	39	1	9.8	39	1	9.8	60	2	8.6	60	2	8.6
501–600	18	0	11.9	18	0	11.9	39	1	9.8	39	1	9.8	60	2	8.6	60	2	8.6
601–800	18	0	11.9	40	1	9.6	40	1	9.6	65	2	8.0	65	2	8.0	85	3	7.5
801–1,000	18	0	12.0	40	1	9.6	40	1	9.6	65	2	8.1	65	2	8.1	90	3	7.4
1,001–2,000	18	0	12.0	41	1	9.4	65	2	8.2	65	2	8.2	95	3	7.0	120	4	6.5
2,001–3,000	18	0	12.0	41	1	9.4	65	2	8.2	95	3	7.0	120	4	6.5	180	6	5.8
3,001–4,000	18	0	12.0	41	1	9.3	65	2	8.2	95	3	7.0	155	5	6.0	210	7	5.5
4,001–5,000	18	0	12.0	42	1	9.3	70	2	7.5	125	4	6.4	155	5	6.0	245	8	5.3
5,001–7,000	18	0	12.0	42	1	9.3	95	3	7.0	125	4	6.4	185	6	5.6	280	9	5.1
7,001–10,000	42	1	9.3	70	2	7.5	95	3	7.0	155	5	6.0	220	7	5.4	350	11	4.8
10,001–20,000	42	1	9.3	70	2	7.6	95	3	7.0	190	6	5.6	290	9	4.9	460	14	4.4
20,001–50,000	42	1	9.3	70	2	7.6	125	4	6.4	220	7	5.4	395	12	4.5	720	21	3.9
50,001–100,000	42	1	9.3	95	3	7.6	160	5	5.9	290	9	4.9	505	15	4.2	955	27	3.7

TABLE 13.11.B *EXAMPLE OF DODGE-ROMIG AOQL TABLES—DOUBLE SAMPLING*

Average outgoing quality limit = 2.0 per cent.

Process Average, %	0–0.04						0.05–0.40						1.61–2.00					
	Trial 1		Trial 2			p_t, %	Trial 1		Trial 2			p_t, %	Trial 1		Trial 2			p_t, %
Lot Size	n_1	c_1	n_2	n_1+n_2	c_2		n_1	c_1	n_2	n_1+n_2	c_2		n_1	c_1	n_2	n_1+n_2	c_2	
1–15	All	0	—	—	—	—	All	0	—	—	—	—	All	0	—	—	—	—
16–50	14	0	—	—	—	13.6	14	0	—	—	—	13.6	14	0	—	—	—	13.6
51–100	21	0	12	33	1	11.7	21	0	12	33	1	11.7	23	0	23	46	2	10.9
101–200	24	0	13	37	1	11.0	24	0	13	37	1	11.0	27	0	28	55	2	9.6
201–300	26	0	15	41	1	10.4	26	0	15	41	1	10.4	32	0	48	80	3	8.4
301–400	26	0	16	42	1	10.3	26	0	16	42	1	10.3	36	0	69	105	4	7.6
401–500	27	0	16	43	1	10.3	30	0	35	65	2	9.0	60	1	90	150	6	7.0
501–600	27	0	16	43	1	10.3	31	0	34	65	2	8.9	65	1	95	160	6	6.8
601–800	27	0	17	44	1	10.2	31	0	39	70	2	8.8	70	1	120	190	7	6.4
801–1,000	27	0	17	44	1	10.2	32	0	38	70	2	8.7	70	1	145	215	8	6.2
1,001–2,000	33	0	37	70	2	8.5	33	0	37	70	2	8.5	110	2	205	315	11	5.5
2,001–3,000	34	0	41	75	2	8.2	34	0	41	75	2	8.2	160	3	310	470	15	4.7
3,001–4,000	34	0	41	75	2	8.2	38	0	62	100	3	7.3	235	5	415	650	20	4.3
4,001–5,000	34	0	41	75	2	8.2	38	0	62	100	3	7.3	275	6	475	750	23	4.2
5,001–7,000	35	0	40	75	2	8.1	38	0	62	100	3	7.3	280	6	575	855	26	4.1
7,001–10,000	35	0	40	75	2	8.1	38	0	62	100	3	7.3	320	7	645	965	29	4.0
10,001–20,000	35	0	40	75	2	8.1	39	0	66	105	3	7.2	395	9	835	1,230	37	3.9
20,001–50,000	35	0	40	75	2	8.1	43	0	92	135	4	6.6	480	11	1,090	1,570	46	3.7
50,001–100,000	35	0	45	80	2	8.0	43	0	92	135	4	6.6	580	13	1,460	2,040	58	3.5

Research Group of Columbia University, and with a few minor differences, the tables in *Sampling Inspection* are identical with JAN–STD–105.

JAN–STD–105 was superseded by MIL–STD–105A in 1950. Minor revisions resulted in MIL–STD–105B in 1958 and MIL–STD–105C in 1961. In 1963, MIL–STD–105D was adopted and is currently in use. It is evident that this new standard is likely to be used in the acceptance inspection of many billions of dollars worth of items purchased by the Army, Navy, and Air Force. In fact, this standard was developed by a joint working group from the military agencies of the United States, Great Britain, and Canada (known as the ABC Working Group) and hence will have impact throughout the world. Moreover, like its predecessors, MIL–STD–105D will doubtless have a great influence on acceptance procedures used in industry.

MIL–STD–105D is available for private use and is for sale by the Superintendent of Documents, U.S. Government Printing Office, Washington, D. C. 10025. Most of the remaining paragraphs in this section on sampling inspection by attributes will be concerned with a discussion of this document.

13.3.8.1 Classification of Defects. Defects are grouped into three classes: critical, major, and minor. A critical defect "is a defect that judgment and experience indicate is likely to result in hazardous or unsafe conditions for individuals using, maintaining, or depending upon the product; or a defect that judgment and experience indicate is likely to prevent performance of the tactical function of a major end item such as a ship, aircraft, tank, missile, or space vehicle." A major defect "is a defect, other than critical, that is likely to result in a failure, or to reduce materially the usability of the unit of product for its intended purpose." A minor defect "is a defect that is not likely to reduce materially the usability of the unit of product for its intended purpose, or is a departure from established standards having little bearing on the effective use or operation of the unit."

13.3.8.2 Acceptable Quality Levels. The acceptable quality level (AQL) "is the maximum per cent defective (or the maximum number of defects per hundred units) that, for purposes of sampling inspection, can be considered satisfactory as a process average." The distinction between a defect and a defective is readily seen if we define a defective item as one containing one or more defects. If the percentage defective is relatively small, more than one defect will occur on an item infrequently. Consequently, in this case, the mathematical theory using defects is essentially equivalent to using defectives. The values of the AQL's are

0.010	0.10	1.0	10	100
0.015	0.15	1.5	15	150
0.025	0.25	2.5	25	250
0.040	0.40	4.0	40	400
0.065	0.65	6.5	65	650
				1000

Here AQL values of 10.0 or less are expressed either in per cent defective or in defects per hundred units; those over 10.0 are expressed in defects per hundred units only. These points are approximately in geometric progression and multiples of 1, 1.5, 2.5, 4.0, and 6.5. The probabilities of acceptance of submitted lots having these AQL's range from about 0.88 for the smallest sample sizes to about 0.99 for the largest sample sizes. The particular AQL values to be used for a given product "will be designated in the contract or by the responsible authority. Different AQL's may be designated for groups of defects considered collectively, or for individual defects. An AQL for a group of defects may be designated in addition to AQL's for individual defects, or subgroups, within that group."

13.3.8.3 Normal, Tightened, and Reduced Inspection. Normal inspection is to be used at the start of inspection unless otherwise directed by the responsible authority. During the course of a contract, normal, tightened, or reduced inspection is continued unchanged for each class of defects or defectives on successive lots except where the switching procedures given as follows require change (a switch must then be made).

When normal inspection is in effect, tightened inspection is instituted when two out of five consecutive lots are rejected on original inspection. When tightened inspection is in effect, normal inspection is instituted where five consecutive lots have been considered acceptable on original inspection. Under tightened inspection, the producer's risk is increased, whereas the consumer's risk is decreased. In other words, the probability of accepting bad lots (as well as good lots) is decreased. This is accomplished by reducing the acceptance numbers while keeping the sample size fixed. In most cases, the acceptance numbers are identical with those for the next *lower* AQL class, although there are many exceptions to this rule.

When normal inspection is in effect, reduced inspection is instituted provided that all of the following conditions are satisfied:

a. The preceding ten lots (or more if noted) have been on normal inspection and none have been rejected on original inspection; and

b. The total number of defectives (or defects) in the samples from the preceding ten lots (or more if noted) is equal to or less than the applicable number given in Table 13.12. If double or multiple sampling is in use, all samples inspected should be used, not "first" samples only; and

c. Production is at a steady rate; and

d. Reduced inspection is considered desirable by the responsible authority.

Under reduced inspection, the sampling procedure may terminate without either acceptance or rejection criteria having been met. (The rejection number need not be one greater than the acceptance number.) In these circumstances, the lot will be considered acceptable.

When reduced inspection is in effect, normal inspection is instituted if any of the following occur on original inspection:

a. A lot is rejected; or

b. The sampling procedure for a lot terminates without either acceptance or rejection criteria having been met (although the particular lot is then accepted); or

c. Production becomes irregular or delayed; or

d. Other conditions warrant that normal inspection shall be instituted.

Under reduced inspection, the sample size is decreased and the rejection numbers modified so that the consumer's risk is increased and the producer's risk decreased. However, there is a very substantial saving in inspection costs for the consumer.

The use of tightened inspection is an important feature in the success of MIL–STD–105D. If quality is submitted close to but above the AQL, there is a relatively large probability that it will be accepted. Although this appears to be harmful on the surface, in reality it is not too damaging. In the first place, rejected lots cost the manufacturer a great deal of money. In fact, too many rejected lots can easily force a manufacturer out of business. For example, few manufacturers can tolerate even a 20 per cent rejection rate for their lots. Second, if lots are constantly being submitted above the AQL, this will eventually result in two of five successive lots being rejected so that tightened inspection will be instituted. This will cause even more of the manufacturer's lots to be rejected. Consequently, the manufacturer is actually forced to keep his submitted quality no worse than the agreed-upon level.

13.3.8.4 Sampling Plans. Sample sizes are designated by code letters from A to R (although code letter S appears for tightened inspection). The sample size code letter depends on the inspection level and the lot size. There are three inspection levels, I, II, and III, given for general use. Unless otherwise specified, inspection level II is used. However, inspection level I may be specified when less discrimination is needed, or level III may be specified for greater discrimination. Four additional special levels, S–1, S–2, S–3, and S–4, are also given for use where relatively small sample sizes are necessary and large sampling risks can or must be tolerated. The sample size code letter applicable to the specified inspection level and for lots of given size is obtained from Table 13.13.

Since it has been pointed out that the OC curves are essentially independent of lot size (for large lots), it may appear incongruous that one of the entries in the table is lot size. However, a moment's reflection will reveal that, to the inspection agency, the acceptance of a bad lot is much more serious when the lot size is large compared with when it is small. Consequently, better protection in the form of steeper OC curves is desired for large lots.

The appropriate master sampling table is selected as follows.

For Single Sampling:
- Normal inspection-Table 13.14
- Tightened inspection-Table 13.14
- Reduced inspection-Table 13.14

For Double Sampling:
- Normal inspection-Table III–A of MIL–STD–105D (which is not reproduced here)
- Tightened inspection-Table III–B of MIL–STD–105D (which is not reproduced here)
- Reduced inspection-Table III–C of MIL–STD–105D (which is not reproduced here)

For Multiple Sampling:
- Normal inspection-Table IV–A of MIL–

TABLE 13.12. *LIMIT NUMBERS FOR REDUCED INSPECTION* (Table VIII of MIL-STD 105D)

Number of Sample Units from Last Ten Lots or Batches	Acceptable Quality Level																									
	0.010	0.015	0.025	0.040	0.065	0.10	0.15	0.25	0.40	0.65	1.0	1.5	2.5	4.0	6.5	10	15	25	40	65	100	150	250	400	650	1000
20–29	†	†	†	†	†	†	†	†	†	†	†	†	†	†	†	0	0	2	4	8	14	22	40	68	115	181
30–49	†	†	†	†	†	†	†	†	†	†	†	†	†	†	0	0	1	3	7	13	22	36	63	105	178	277
50–79	†	†	†	†	†	†	†	†	†	†	†	†	†	0	0	2	3	7	14	25	40	63	110	181	301	
80–129	†	†	†	†	†	†	†	†	†	†	†	†	0	0	2	4	7	14	24	42	68	105	181	297		
130–199	†	†	†	†	†	†	†	†	†	†	†	0	0	2	4	7	13	25	42	72	115	177	301	490		
200–319	†	†	†	†	†	†	†	†	†	†	0	0	2	4	8	14	22	40	68	115	181	277	471			
320–499	†	†	†	†	†	†	†	†	†	0	0	1	4	8	14	24	39	68	113	189						
500–799	†	†	†	†	†	†	†	†	0	0	2	3	7	14	25	40	63	110	181							
800–1249	†	†	†	†	†	†	†	0	0	2	4	7	14	24	42	68	105	181								
1250–1999	†	†	†	†	†	†	0	0	2	4	7	13	24	40	69	110	169									
2000–3149	†	†	†	†	†	0	0	2	4	8	14	22	40	68	115	181										
3150–4999	†	†	†	†	0	0	1	4	8	14	24	38	67	111	186											
5000–7999	†	†	†	0	0	2	3	7	14	25	40	63	110	181												
8000–12499	†	†	0	0	2	4	7	14	24	42	68	105	181													
12500–19999	†	0	0	2	4	7	13	24	40	69	110	169														
20000–31499	0	0	2	4	8	14	22	40	68	115	181															
31500–49999	0	1	4	8	14	24	38	67	111	186																
50000 & Over	2	3	7	14	25	40	63	110	181	301																

† Denotes that the number of sample units from the last ten lots or batches is not sufficient for reduced inspection for this AQL. In this instance more than ten lots or batches may be used for the calculation, provided that the lots or batches used are the most recent ones in sequence, that they have all been on normal inspection, and that none has been rejected while on original inspection.

TABLE 13.13. *SAMPLE SIZE CODE LETTERS* (Table I of MIL-STD-105D)

Lot or Batch Size			Special Inspection Levels				General Inspection Levels		
			S–1	S–2	S–3	S–4	I	II	III
2	to	8	A	A	A	A	A	A	B
9	to	15	A	A	A	A	A	B	C
16	to	25	A	A	B	B	B	C	D
26	to	50	A	B	B	C	C	D	E
51	to	90	B	B	C	C	C	E	F
91	to	150	B	B	C	D	D	F	G
151	to	280	B	C	D	E	E	G	H
281	to	500	B	C	D	E	F	H	J
501	to	1200	C	C	E	F	G	J	K
1201	to	3200	C	D	E	G	H	K	L
3201	to	10000	C	D	F	G	J	L	M
10001	to	35000	C	D	F	H	K	M	N
35001	to	150000	D	E	G	J	L	N	P
150001	to	500000	D	E	G	J	M	P	Q
500001	and	over	D	E	H	K	N	Q	R

STD–105D (which is not reproduced here)
Tightened inspection-Table IV–B of MIL–STD–105D (which is not reproduced here)
Reduced inspection-Table IV–C of MIL–STD–105D (which is not reproduced here)

The double sampling plans are characterized by having the first and second samples equal and approximately equal to $0.631n$ where n is the equivalent single sample size.

Other than for the usual restrictions, the acceptance and rejection numbers are arbitrary, although they are chosen so that the OC curve for the double sampling plan matches that for the corresponding single sampling plan, and the average amount of inspection is always smaller than that for the single sampling plan. This can be seen from Fig. 13.16.

The multiple sampling plans have seven steps, all of which have the same sample size which is chosen equal to approximately $0.25n$, where n is the equivalent single sample size. The OC curve for these plans also matches that of the corresponding single sampling plan, and for almost all plans the average amount of inspection is smaller than that for the double sampling plan. This can also be seen from Fig. 13.16.

The Standard contains the OC curve for each of the plans given. Each OC curve corresponds to a single sampling plan, double sampling plan, and multiple sampling plan. The Standard also contains tables of the average outgoing quality limit (AOQL) for each of the single sampling plans (normal and tightened) based on the assumption that rejected lots are screened without error. Except for those plans where $c = 0$, the AOQL values in tightened inspection are close to the AQL values. Tables are given for the quality corresponding to probabilities of acceptance of 5 per cent and 10 per cent, called the "limiting quality," for each of the single sampling plans presented.

13.3.9 Designing Your Own Attribute Plan

In many cases, it will not be desirable to select a plan according to the standard procedure of MIL–STD–105D, implying, as it does, a rather arbitrary relation between lot size and sample size. If the characteristics of an OC curve are specified in advance, it is possible to determine the sample size and acceptance number accordingly. This article will deal with procedures for determining sampling plans when two points are specified, and with finding the OC curve of an arbitrary sampling plan.

13.3.9.1 Computing the OC Curve of a Single Sampling Plan. In Par. 13.3.3.2 a method for computing the OC curve of a single sampling plan with sample size n and acceptance number

TABLE 13.14. *SINGLE SAMPLING PLANS*

(Table II–A of MIL–STD–105B) For normal inspection (Master table)

Sample Size Code Letter	Sample Size	Acceptable Quality Levels (Normal Inspection)											
		0.010	0.015	0.025	0.040	0.065	0.10	0.15	0.25	0.40	0.65	1.0	1.5
		Ac Re	Ac Re	Ac Re	Ac Re	Ac Re	Ac Re	Ac Re	Ac Re	Ac Re	Ac Re	Ac Re	Ac Re
A	2	↓	↓	↓	↓	↓	↓	↓	↓	↓	↓	↓	↓
B	3	↓	↓	↓	↓	↓	↓	↓	↓	↓	↓	↓	↓
C	5	↓	↓	↓	↓	↓	↓	↓	↓	↓	↓	↓	↓
D	8	↓	↓	↓	↓	↓	↓	↓	↓	↓	↓	↓	0 1
E	13	↓	↓	↓	↓	↓	↓	↓	↓	↓	↓	0 1	↑
F	20	↓	↓	↓	↓	↓	↓	↓	↓	↓	0 1	↑	↓
G	32	↓	↓	↓	↓	↓	↓	↓	↓	0 1	↑	↓	1 2
H	50	↓	↓	↓	↓	↓	↓	↓	0 1	↑	↓	1 2	2 3
J	80	↓	↓	↓	↓	↓	↓	0 1	↑	↓	1 2	2 3	3 4
K	125	↓	↓	↓	↓	↓	0 1	↑	↓	1 2	2 3	3 4	5 6
L	200	↓	↓	↓	↓	0 1	↑	↓	1 2	2 3	3 4	5 6	7 8
M	315	↓	↓	↓	0 1	↑	↓	1 2	2 3	3 4	5 6	7 8	10 11
N	500	↓	↓	0 1	↑	↓	1 2	2 3	3 4	5 6	7 8	10 11	14 15
P	800	↓	0 1	↑	↓	1 2	2 3	3 4	5 6	7 8	10 11	14 15	21 22
Q	1250	0 1	↑	↓	1 2	2 3	3 4	5 6	7 8	10 11	14 15	21 22	↑
R	2000	↑	↑	1 2	2 3	3 4	5 6	7 8	10 11	14 15	21 22	↑	↑

↓ = Use first sampling plan below arrow. If sample size equals or exceeds lot or batch size, do 100 per cent inspection.
↑ = Use first sampling plan above arrow.
Ac = Acceptance number.
Re = Rejection number.

TABLE 13.14. *Cont.*

Sample Size Code Letter	Sample Size	Acceptable Quality Levels (Normal Inspection) 2.5 Ac Re	4.0 Ac Re	6.5 Ac Re	10 Ac Re	15 Ac Re	25 Ac Re	40 Ac Re	65 Ac Re	100 Ac Re	150 Ac Re	250 Ac Re	400 Ac Re	650 Ac Re	1000 Ac Re
A	2	↓	↓	0 1	↓	↓	1 2	2 3	3 4	5 6	7 8	10 11	14 15	21 22	30 31
B	3		0 1	↑		1 2	2 3	3 4	5 6	7 8	10 11	14 15	21 22	30 31	44 45
C	5	0 1	↑	↓	1 2	2 3	3 4	5 6	7 8	10 11	14 15	21 22	30 31	44 45	↑
D	8	↑	↓	1 2	2 3	3 4	5 6	7 8	10 11	14 15	21 22	30 31	44 45	↑	
E	13	↓	1 2	2 3	3 4	5 6	7 8	10 11	14 15	21 22	30 31	44 45	↑		
F	20	1 2	2 3	3 4	5 6	7 8	10 11	14 15	21 22	↑	↑	↑			
G	32	2 3	3 4	5 6	7 8	10 11	14 15	21 22	↑						
H	50	3 4	5 6	7 8	10 11	14 15	21 22	↑							
J	80	5 6	7 8	10 11	14 15	21 22	↑								
K	125	7 8	10 11	14 15	21 22	↑									
L	200	10 11	14 15	21 22	↑										
M	315	14 15	21 22	↑											
N	500	21 22	↑												
P	800	↑													
Q	1250														
R	2000														

TABLE 13.14. *Cont.*

(Table II–B of MIL–STD–105D) For tightened inspection (Master Table)

Sample Size Code Letter	Sample Size	Acceptable Quality Levels (Tightened Inspection)											
		0.010 Ac Re	0.015 Ac Re	0.025 Ac Re	0.040 Ac Re	0.065 Ac Re	0.10 Ac Re	0.15 Ac Re	0.25 Ac Re	0.40 Ac Re	0.65 Ac Re	1.0 Ac Re	1.5 Ac Re
A	2	↓	↓	↓	↓	↓	↓	↓	↓	↓	↓	↓	↓
B	3	↓	↓	↓	↓	↓	↓	↓	↓	↓	↓	↓	↓
C	5	↓	↓	↓	↓	↓	↓	↓	↓	↓	↓	↓	↓
D	8	↓	↓	↓	↓	↓	↓	↓	↓	↓	↓	↓	↓
E	13	↓	↓	↓	↓	↓	↓	↓	↓	↓	↓	↓	0 1
F	20	↓	↓	↓	↓	↓	↓	↓	↓	↓	↓	0 1	↓
G	32	↓	↓	↓	↓	↓	↓	↓	↓	↓	0 1	↓	↓
H	50	↓	↓	↓	↓	↓	↓	↓	↓	0 1	↓	↓	1 2
J	80	↓	↓	↓	↓	↓	↓	↓	0 1	↓	↓	1 2	2 3
K	125	↓	↓	↓	↓	↓	↓	0 1	↓	↓	1 2	2 3	3 4
L	200	↓	↓	↓	↓	↓	0 1	↓	↓	1 2	2 3	3 4	5 6
M	315	↓	↓	↓	↓	0 1	↓	↓	1 2	2 3	3 4	5 6	8 9
N	500	↓	↓	↓	0 1	↓	↓	1 2	2 3	3 4	5 6	8 9	12 13
P	800	↓	↓	0 1	↓	↓	1 2	2 3	3 4	5 6	8 9	12 13	18 19
Q	1250	↓	0 1	↓	↓	1 2	2 3	3 4	5 6	8 9	12 13	18 19	
R	2000	0 1	↑	↓	1 2	2 3	3 4	5 6	8 9	12 13	18 19		
S	3150		↑	1 2									

↓ = Use first sampling plan below arrow. If sample size equals or exceeds lot or batch size, do 100 per cent inspection.
↑ = Use first sampling plan above arrow.
Ac = Acceptance number.
Re = Rejection number.

TABLE 13.14. *Cont.*

Sample Size Code Letter	Sample Size	Acceptable Quality Levels (Tightened Inspection)													
		2.5 Ac Re	4.0 Ac Re	6.5 Ac Re	10 Ac Re	15 Ac Re	25 Ac Re	40 Ac Re	65 Ac Re	100 Ac Re	150 Ac Re	250 Ac Re	400 Ac Re	650 Ac Re	1000 Ac Re
A	2	↓	↓	↓	↓	↓	↓	1 2	2 3	3 4	5 6	8 9	12 13	18 19	27 28
B	3	↓	↓	0 1	↓	↓	1 2	2 3	3 4	5 6	8 9	12 13	18 19	27 28	41 42
C	5	↓	0 1	↓	↓	1 2	2 3	3 4	5 6	8 9	12 13	18 19	27 28	41 42	↑
D	8	0 1	↓	↓	1 2	2 3	3 4	5 6	8 9	12 13	18 19	27 28	41 42	↑	↑
E	13	↓	↓	1 2	2 3	3 4	5 6	8 9	12 13	18 19	27 28	41 42	↑	↑	↑
F	20	↓	1 2	2 3	3 4	5 6	8 9	12 13	18 19	↑	↑	↑	↑	↑	↑
G	32	1 2	2 3	3 4	5 6	8 9	12 13	18 19	↑	↑	↑	↑	↑	↑	↑
H	50	2 3	3 4	5 6	8 9	12 13	18 19	↑	↑	↑	↑	↑	↑	↑	↑
J	80	3 4	5 6	8 9	12 13	18 19	↑	↑	↑	↑	↑	↑	↑	↑	↑
K	125	5 6	8 9	12 13	18 19	↑	↑	↑	↑	↑	↑	↑	↑	↑	↑
L	200	8 9	12 13	18 19	↑	↑	↑	↑	↑	↑	↑	↑	↑	↑	↑
M	315	12 13	18 19	↑	↑	↑	↑	↑	↑	↑	↑	↑	↑	↑	↑
N	500	18 19	↑	↑	↑	↑	↑	↑	↑	↑	↑	↑	↑	↑	↑
P	800	↑	↑	↑	↑	↑	↑	↑	↑	↑	↑	↑	↑	↑	↑
Q	1250	↑	↑	↑	↑	↑	↑	↑	↑	↑	↑	↑	↑	↑	↑
R	2000	↑	↑	↑	↑	↑	↑	↑	↑	↑	↑	↑	↑	↑	↑
S	3150	↑	↑	↑	↑	↑	↑	↑	↑	↑	↑	↑	↑	↑	↑

TABLE 13.14. *Cont.*

(Table II–C of MIL–STD–105D) For reduced inspection (Master Table)

Sample Size Code Letter	Sample Size	Acceptable Quality Levels (Reduced Inspection)†											
		0.010 Ac Re	0.015 Ac Re	0.025 Ac Re	0.040 Ac Re	0.065 Ac Re	0.10 Ac Re	0.15 Ac Re	0.25 Ac Re	0.40 Ac Re	0.65 Ac Re	1.0 Ac Re	1.5 Ac Re
A	2	↓	↓	↓	↓	↓	↓	↓	↓	↓	↓	↓	↓
B	2	↓	↓	↓	↓	↓	↓	↓	↓	↓	↓	↓	↓
C	2	↓	↓	↓	↓	↓	↓	↓	↓	↓	↓	↓	↓
D	3	↓	↓	↓	↓	↓	↓	↓	↓	↓	↓	↓	0 1
E	5	↓	↓	↓	↓	↓	↓	↓	↓	↓	↓	0 1	↑
F	8	↓	↓	↓	↓	↓	↓	↓	↓	↓	0 1	↑	↓
G	13	↓	↓	↓	↓	↓	↓	↓	↓	0 1	↑	↓	0 2
H	20	↓	↓	↓	↓	↓	↓	↓	0 1	↑	↓	0 2	1 3
J	32	↓	↓	↓	↓	↓	↓	0 1	↑	↓	0 2	1 3	1 4
K	50	↓	↓	↓	↓	↓	0 1	↑	↓	0 2	1 3	1 4	2 5
L	80	↓	↓	↓	↓	0 1	↑	↓	0 2	1 3	1 4	2 5	3 6
M	125	↓	↓	↓	0 1	↑	↓	0 2	1 3	1 4	2 5	3 6	5 8
N	200	↓	↓	0 1	↑	↓	0 2	1 3	1 4	2 5	3 6	5 8	7 10
P	315	↓	0 1	↑	↓	0 2	1 3	1 4	2 5	3 6	5 8	7 10	10 13
Q	500	0 1	↑	↓	0 2	1 3	1 4	2 5	3 6	5 8	7 10	10 13	
R	800	↑	↑	0 2	1 3	1 4	2 5	3 6	5 8	7 10	10 13	↑	↑

↓ = Use first sampling plan below arrow. If sample size equals or exceeds lot or batch size, do 100 per cent inspection.
↑ = Use first sampling plan above arrow.
Ac = Acceptance number.
Re = Rejection number.
† = If the acceptance number has been exceeded but the rejection number has not been reached, accept the lot but reinstate normal inspection (see 10.1.4).

TABLE 13.14. *Cont.*

Sample Size Code Letter	Sample Size	Acceptable Quality Levels (Reduced Inspection) 2.5 Ac Re	4.0 Ac Re	6.5 Ac Re	10 Ac Re	15 Ac Re	25 Ac Re	40 Ac Re	65 Ac Re	100 Ac Re	150 Ac Re	250 Ac Re	400 Ac Re	650 Rc Re	1000 Ac Re
A	2	↓	↓	0 1	↓	↓	1 2	2 3	3 4	5 6	7 8	10 11	14 15	21 22	30 31
B	2	↓	0 1	↑	↓	0 2	1 3	2 4	3 5	5 6	7 8	10 11	14 15	21 22	30 31
C	2	0 1	↑	↓	0 2	1 3	1 4	2 5	3 6	5 8	7 10	10 13	14 17	21 24	↑
D	3	↑	↓	0 2	1 3	1 4	2 5	3 6	5 8	7 10	10 13	14 17	21 24	↑	↑
E	5	↓	0 2	1 3	1 4	2 5	3 6	5 8	7 10	10 13	14 17	21 24	↑	↑	↑
F	8	0 2	1 3	1 4	2 5	3 6	5 8	7 10	10 13	↑	↑	↑	↑	↑	↑
G	13	1 3	1 4	2 5	3 6	5 8	7 10	10 13	↑	↑	↑	↑	↑	↑	↑
H	20	1 4	2 5	3 6	5 8	7 10	10 13	↑	↑	↑	↑	↑	↑	↑	↑
J	32	2 5	3 6	5 8	7 10	10 13	↑	↑	↑	↑	↑	↑	↑	↑	↑
K	50	3 6	5 8	7 10	10 13	↑	↑	↑	↑	↑	↑	↑	↑	↑	↑
L	80	5 8	7 10	10 13	↑	↑	↑	↑	↑	↑	↑	↑	↑	↑	↑
M	125	7 10	10 13	↑	↑	↑	↑	↑	↑	↑	↑	↑	↑	↑	↑
N	200	10 13	↑	↑	↑	↑	↑	↑	↑	↑	↑	↑	↑	↑	↑
P	315	↑	↑	↑	↑	↑	↑	↑	↑	↑	↑	↑	↑	↑	↑
Q	500	↑	↑	↑	↑	↑	↑	↑	↑	↑	↑	↑	↑	↑	↑
R	800	↑	↑	↑	↑	↑	↑	↑	↑	↑	↑	↑	↑	↑	↑

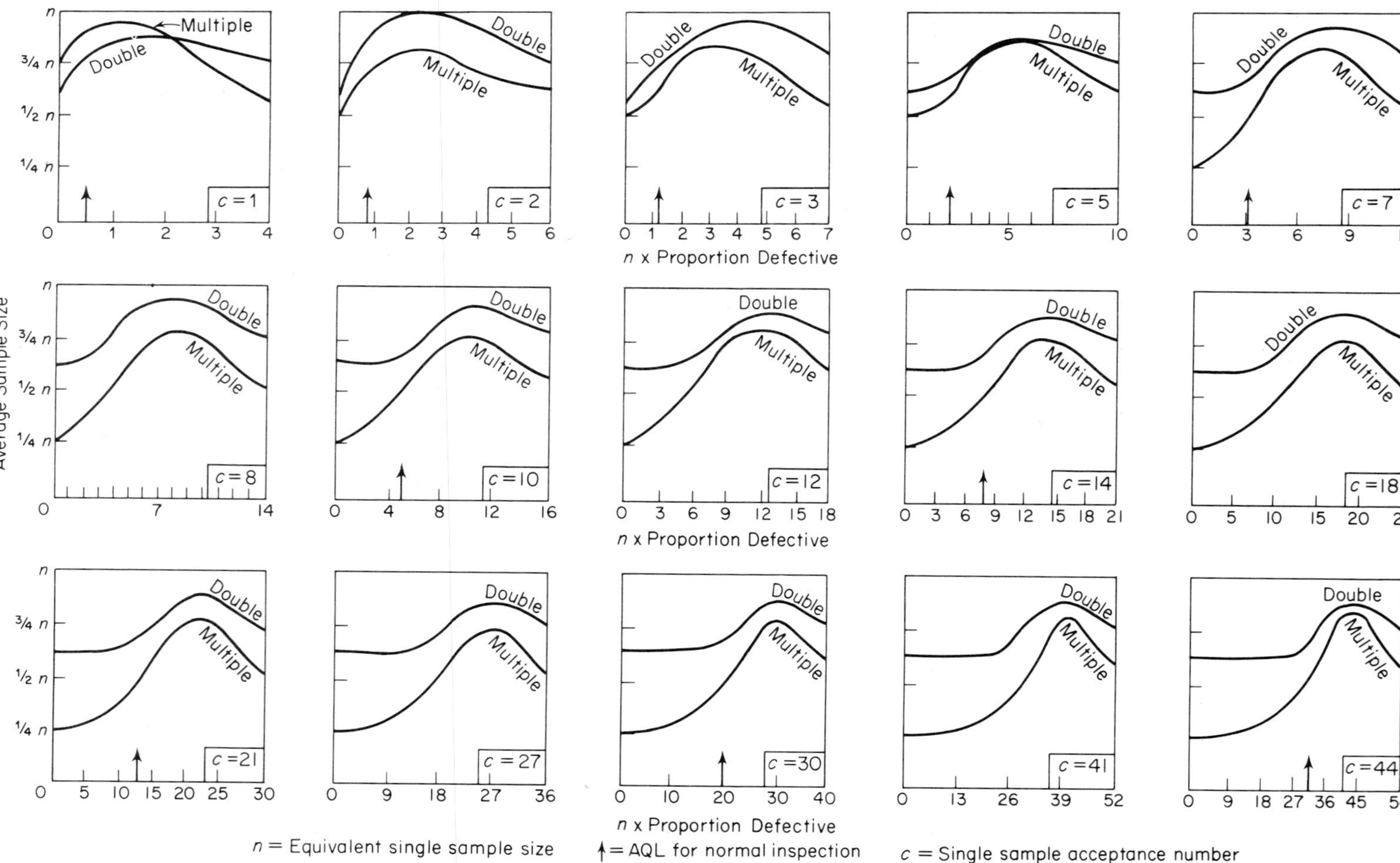

Fig. 13.16 *Average sample size curves for double and multiple sampling* (*normal and tightened inspection*) (*Table IX of MIL–STD–105D*).

c was presented. The OC curve may also be constructed from Table 13.15 by dividing each entry in the row for the given c by the sample size. The fraction defective p', for which the probability of acceptance is shown in the column heading, is obtained.

Example. For $n = 75$, $c = 4$; find the OC curve.

Dividing the numbers in the row $c = 4$ by 75 we find the following 13 points on the OC curve.

Probability of Acceptance	Fraction Defective	Probability of Acceptance	Fraction Defective
0.995	$\frac{1.078}{75} = 0.01437$	0.250	$\frac{6.274}{75} = 0.08365$
0.990	$\frac{1.279}{75} = 0.01705$	0.100	$\frac{7.994}{75} = 0.1066$
0.975	$\frac{1.623}{75} = 0.02164$	0.050	$\frac{9.154}{75} = 0.1221$
0.950	$\frac{1.970}{75} = 0.02627$	0.025	$\frac{10.242}{75} = 0.1372$
0.900	$\frac{2.433}{75} = 0.03244$	0.010	$\frac{11.605}{75} = 0.1547$
0.750	$\frac{3.369}{75} = 0.04492$	0.005	$\frac{12.594}{75} = 0.1679$
0.500	$\frac{4.671}{75} = 0.06228$		

13.3.9.2 Finding a Sampling Plan Whose OC Curve Passes Through Two Points. As pointed out in Par. 13.3.3, it is often useful to specify two points on an OC curve; p'_1 and p'_2 such that $L(p'_1) = 1 - \alpha$ and $L(p'_2) = \beta$. Here p'_1 is usually denoted as acceptable quality (quality we want to accept) and p'_2 usually represents unacceptable quality (quality we want to reject). Hence α and β are usually taken as small numbers, 0.01, 0.05, or 0.10. For these values, sampling plans may be found in Table 13.16. To construct a plan for a given p'_1, α and p'_2, β, calculate the ratio p'_2/p'_1 and find the entry in the appropriate α, β column which is equal to or just greater than the desired ratio. The acceptance number is read off directly and the sample size is determined by dividing np'_1 by p'_1.

Example. $p'_1 = 0.02$, $p'_2 = 0.04$, $\alpha = 0.05$, $\beta = 0.05$. Find n and c.

$$\frac{p'_2}{p'_1} = \frac{0.04}{0.02} = 2$$

The value in the table just greater than 2 is 2.030. Hence, $c = 21$, $n = 14.894/0.02 = 745$.

13.3.9.3 Design of Item-by-Item Sequential Plans. It is not possible to give simple formulas for the acceptance and rejection numbers in double or multiple sampling plans; however, simple formulas do exist for item-by-item sequential plans, i.e., plans in which the decision to accept, reject, or continue sampling is made after each observation. The plan can be represented graphically as a pair of parallel lines (Fig. 13.17). The plan is determined by the acceptance line $a_m = -h_1 + ms$ and the rejection line $r_m = h_2 + ms$. To calculate h_1, h_2, and s, we introduce the quantities

$$b = \ln \frac{1-\alpha}{\beta} \qquad a = \ln \frac{1-\beta}{\alpha}$$

$$g_1 = \ln \frac{p'_2}{p'_1} \qquad g_2 = \ln \frac{1-p'_1}{1-p'_2}$$

Then

$$h_1 = \frac{b}{g_1 + g_2} \qquad h_2 = \frac{a}{g_1 + g_2}$$

$$s = \frac{g_2}{g_1 + g_2}$$

Example. Suppose we want $p'_1 = 0.01$, $p'_2 = 0.10$, $\alpha = 0.05$, $\beta = 0.20$; we calculate

$$b = \ln \frac{0.95}{0.20} = 1.55815$$

$$a = \ln \frac{0.80}{0.05} = 2.77259$$

$$g_1 = \ln \frac{0.10}{0.01} = 2.30258$$

$$g_2 = \ln \frac{0.99}{0.90} = 0.09531$$

$$h_1 = 0.649796 \qquad h_2 = 1.156407$$

$$s = 0.039747$$

and get the equations

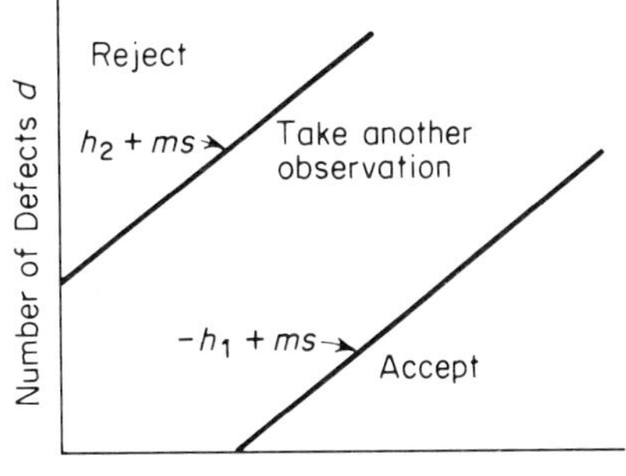

Fig. 13.17 *Graphic procedure for item by item sequential sampling plan.*

TABLE 13.15. *VALUES OF* np'_1 *FOR WHICH THE PROBABILITY OF ACCEPTANCE OF* C *OR FEWER DEFECTIVES IN A SAMPLE OF* n *IS* P(A)†‡

c	$P(A) =$ 0.995	$P(A) =$ 0.990	$P(A) =$ 0.975	$P(A) =$ 0.950	$P(A) =$ 0.900	$P(A) =$ 0.750	$P(A) =$ 0.500	$P(A) =$ 0.250	$P(A) =$ 0.100	$P(A) =$ 0.050	$P(A) =$ 0.025	$P(A) =$ 0.010	$P(A) =$ 0.005
0	0.00501	0.0101	0.0253	0.0513	0.105	0.288	0.693	1.386	2.303	2.996	3.689	4.605	5.298
1	0.103	0.149	0.242	0.355	0.532	0.961	1.678	2.693	3.890	4.744	5.572	6.638	7.430
2	0.338	0.436	0.619	0.818	1.102	1.727	2.674	3.920	5.322	6.296	7.224	8.406	9.274
3	0.672	0.823	1.090	1.366	1.745	2.535	3.672	5.109	6.681	7.754	8.768	10.045	10.978
4	1.078	1.279	1.623	1.970	2.433	3.369	4.671	6.274	7.994	9.154	10.242	11.605	12.594
5	1.537	1.785	2.202	2.613	3.152	4.219	5.670	7.423	9.275	10.513	11.668	13.108	14.150
6	2.037	2.330	2.814	3.286	3.895	5.083	6.670	8.558	10.532	11.842	13.060	14.571	15.660
7	2.571	2.906	3.454	3.981	4.656	5.956	7.669	9.684	11.771	13.148	14.422	16.000	17.134
8	3.132	3.507	4.115	4.695	5.432	6.838	8.669	10.802	12.995	14.434	15.763	17.403	18.578
9	3.717	4.130	4.795	5.426	6.221	7.726	9.669	11.914	14.206	15.705	17.085	18.783	19.998
10	4.321	4.771	5.491	6.169	7.021	8.620	10.668	13.020	15.407	16.962	18.390	20.145	21.398
11	4.943	5.428	6.201	6.924	7.829	9.519	11.668	14.121	16.598	18.208	19.682	21.490	22.779
12	5.580	6.099	6.922	7.690	8.646	10.422	12.668	15.217	17.782	19.442	20.962	22.821	24.145
13	6.231	6.782	7.654	8.464	9.470	11.329	13.668	16.310	18.958	20.668	22.230	24.139	25.496
14	6.893	7.477	8.396	9.246	10.300	12.239	14.668	17.400	20.128	21.886	23.490	25.446	26.836
15	7.566	8.181	9.144	10.035	11.135	13.152	15.668	18.486	21.292	23.098	24.741	26.743	28.166
16	8.249	8.895	9.902	10.831	11.976	14.068	16.668	19.570	22.452	24.302	25.984	28.031	29.484
17	8.942	9.616	10.666	11.633	12.822	14.986	17.668	20.652	23.606	25.500	27.220	29.310	30.792
18	9.644	10.346	11.438	12.442	13.672	15.907	18.668	21.731	24.756	26.692	28.448	30.581	32.092
19	10.353	11.082	12.216	13.254	14.525	16.830	19.668	22.808	25.902	27.879	29.671	31.845	33.383
20	11.069	11.825	12.999	14.072	15.383	17.755	20.668	23.883	27.045	29.062	30.888	33.103	34.668
21	11.791	12.574	13.787	14.894	16.244	18.682	21.668	24.956	28.184	30.241	32.102	34.355	35.947
22	12.520	13.329	14.580	15.719	17.108	19.610	22.668	26.028	29.320	31.416	33.309	35.601	37.219
23	13.255	14.088	15.377	16.548	17.975	20.540	23.668	27.098	30.453	32.586	34.512	36.841	38.485
24	13.995	14.853	16.178	17.382	18.844	21.471	24.668	28.167	31.584	33.752	35.710	38.077	39.745

25	14.740	15.623	16.984	18.218	19.717	22.404	25.667	29.234	32.711	34.916	36.905	39.308	41.000
26	15.490	16.397	17.793	19.058	20.592	23.338	26.667	30.300	33.836	36.077	38.096	40.535	42.252
27	16.245	17.175	18.606	19.900	21.469	24.273	27.667	31.365	34.959	37.234	39.284	41.757	43.497
28	17.004	17.957	19.422	20.746	22.348	25.209	28.667	32.428	36.080	38.389	40.468	42.975	44.738
29	17.767	18.742	20.241	21.594	23.229	26.147	29.667	33.491	37.198	39.541	41.649	44.190	45.976
30	18.534	19.532	21.063	22.444	24.113	27.086	30.667	34.552	38.315	40.690	42.827	45.401	47.210
31	19.305	20.324	21.888	23.298	24.998	28.025	31.667	35.613	39.430	41.838	44.002	46.609	48.440
32	20.079	21.120	22.716	24.152	25.885	28.966	32.667	36.672	40.543	42.982	45.174	47.813	49.666
33	20.856	21.919	23.546	25.010	26.774	29.907	33.667	37.731	41.654	44.125	46.344	49.015	50.888
34	21.638	22.721	24.379	25.870	27.664	30.849	34.667	38.788	42.764	45.266	47.512	50.213	52.108
35	22.422	23.525	25.214	26.731	28.556	31.792	35.667	39.845	43.872	46.404	48.676	51.409	53.324
36	23.208	24.333	26.052	27.594	29.450	32.736	36.667	40.901	44.978	47.540	49.840	52.601	54.538
37	23.998	25.143	26.891	28.460	30.345	33.681	37.667	41.957	46.083	48.676	51.000	53.791	55.748
38	24.791	25.955	27.733	29.327	31.241	34.626	38.667	43.011	47.187	49.808	52.158	54.979	56.956
39	25.586	26.770	28.576	30.196	32.139	35.572	39.667	44.065	48.289	50.940	53.314	56.164	58.160
40	26.384	27.587	29.422	31.066	33.038	36.519	40.667	45.118	49.390	52.069	54.469	57.347	59.363
41	27.184	28.406	30.270	31.938	33.938	37.466	41.667	46.171	50.490	53.197	55.622	58.528	60.563
42	27.986	29.228	31.120	32.812	34.839	38.414	42.667	47.223	51.589	54.324	56.772	59.717	61.761
43	28.791	30.051	31.970	33.686	35.742	39,363	43.667	48.274	42.686	55.449	57.921	60.884	62.956
44	29.598	30.877	32.824	34.563	36.646	40.312	44.667	49.325	53.782	56.572	59.068	62.059	64.150
45	30.408	31.704	33.678	35.441	37.550	41.262	45.667	50.375	54.878	57.695	60.214	63.231	65.340
46	31.219	32.534	34.534	36.320	38.456	42.212	46.667	51.425	55.972	58.816	61.358	64.402	66.529
47	32.032	33.365	35.392	37.200	39.363	43.163	47.667	52.474	57.065	59.936	62.500	65.571	67.716
48	32.848	34.198	36.250	38.082	40.270	44.115	48.667	53.522	58.158	61.054	63.641	66.738	68.901
49	33.664	35.032	37.111	38.965	41.179	45.067	49.667	54.571	59.249	62.171	64.780	67,903	70.084

† To find the fraction defective p', corresponding to a probability of acceptance $P(A)$ in a single sampling plan with sample size n and acceptance number c, divide by n the entry in the row for the given c and the column for the given $P(A)$.

‡ Reprinted by permission from J. M. Cameron, "Tables for Constructing and for Computing the Operating Characteristics of Single-Sampling Plans," *Industrial Quality Control*, July 1952, pp. 37–39.

TABLE 13.16. *VALUES OF* np'_1 *AND* C *FOR CONSTRUCTING SINGLE SAMPLING PLANS WHOSE OC CURVE IS REQUIRED TO PASS THROUGH THE TWO POINTS* $(p_1, 1 - \alpha)$ *AND* (p_2, β)†‡

	Values of p'_2/p'_1 for:					Values of p'_2/p'_1 for:			
c	$\alpha = 0.05$ $\beta = 0.10$	$\alpha = 0.05$ $\beta = 0.05$	$\alpha = 0.05$ $\beta = 0.01$	np_1	c	$\alpha = 0.01$ $\beta = 0.10$	$\alpha = 0.01$ $\beta = 0.05$	$\alpha = 0.01$ $\beta = 0.01$	np_1
0	44.890	58.404	89.781	0.052	0	229.105	298.073	458.210	0.010
1	10.946	13.349	18.681	0.355	1	26.184	31.933	44.686	0.149
2	6.509	7.699	10.280	0.818	2	12.206	14.439	19.278	0.436
3	4.890	5.675	7.352	1.366	3	8.115	9.418	12.202	0.823
4	4.057	4.646	5.890	1.970	4	6.249	7.156	9.072	1.279
5	3.549	4.023	5.017	2.613	5	5.195	5.889	7.343	1.785
6	3.206	3.604	4.435	3.286	6	4.520	5.082	6.253	2.330
7	2.957	3.303	4.019	3.981	7	4.050	4.524	5.506	2.906
8	2.768	3.074	3.707	4.695	8	3.705	4.115	4.962	3.507
9	2.618	2.895	3.462	5.426	9	3.440	3.803	4.548	4.130
10	2.497	2.750	3.265	6.169	10	3.229	3.555	4.222	4.771
11	2.397	2.630	3.104	6.924	11	3.058	3.354	3.959	5.428
12	2.312	2.528	2.968	7.690	12	2.915	3.188	3.742	6.099
13	2.240	2.442	2.852	8.464	13	2.795	3.047	3.559	6.782
14	2.177	2.367	2.752	9.246	14	2.692	2.927	3.403	7.477
15	2.122	2.302	2.665	10.035	15	2.603	2.823	3.269	8.181
16	2.073	2.244	2.588	10.831	16	2.524	2.732	3.151	8.895
17	2.029	2.192	2.520	11.633	17	2.455	2.652	3.048	9.616
18	1.990	2.145	2.458	12.442	18	2.393	2.580	2.956	10.346
19	1.954	2.103	2.403	13.254	19	2.337	2.516	2.874	11.082
20	1.922	2.065	2.352	14.072	20	2.287	2.458	2.799	11.825
21	1.892	2.030	2.307	14.894	21	2.241	2.405	2.733	12.574
22	1.865	1.999	2.265	15.719	22	2.200	2.357	2.671	13.329
23	1.840	1.969	2.226	16.548	23	2.162	2.313	2.615	14.088
24	1.817	1.942	2.191	17.382	24	2.126	2.272	2.564	14.853
25	1.795	1.917	2.158	18.218	25	2.094	2.235	2.516	15.623
26	1.775	1.893	2.127	19.058	26	2.064	2.200	2.472	16.397
27	1.757	1.871	2.098	19.900	27	2.035	2.168	2.431	17.175
28	1.739	1.850	2.071	20.746	28	2.009	2.138	2.393	17.957
29	1.723	1.831	2.046	21.594	29	1.985	2.110	2.358	18.742
30	1.707	1.813	2.023	22.444	30	1.962	2.083	2.324	19.532
31	1.692	1.796	2.001	23.298	31	1.940	2.059	2.293	20.324
32	1.679	1.780	1.980	24.152	32	1.920	2.035	2.264	21.120
33	1.665	1.764	1.960	25.010	33	1.900	2.013	2.236	21.919
34	1.653	1.750	1.941	25.870	34	1.882	1.992	2.210	22.721
35	1.641	1.736	1.923	26.731	35	1.865	1.973	2.185	23.525
36	1.630	1.723	1.906	27.594	36	1.848	1.954	2.162	24.333
37	1.619	1.710	1.890	28.460	37	1.833	1.936	2.139	25.143
38	1.609	1.698	1.875	29.327	38	1.818	1.920	2.118	25.955
39	1.599	1.687	1.860	30.196	39	1.804	1.903	2.098	26.770
40	1.590	1.676	1.846	31.066	40	1.790	1.887	2.079	27.587
41	1.581	1.666	1.833	31.938	41	1.777	1.873	2.060	28.406
42	1.572	1.656	1.820	32.812	42	1.765	1.859	2.043	29.228
43	1.564	1.646	1.807	33.686	43	1.753	1.845	2.026	30.051
44	1.556	1.637	1.796	34.563	44	1.742	1.832	2.010	30.877
45	1.548	1.628	1.784	35.441	45	1.731	1.820	1.994	31.704
46	1.541	1.619	1.773	36.320	46	1.720	1.808	1.980	32.534
47	1.534	1.611	1.763	37.200	47	1.710	1.796	1.965	33.365
48	1.527	1.603	1.752	38.082	48	1.701	1.785	1.952	34.198
49	1.521	1.596	1.743	38.965	49	1.691	1.775	1.938	35.032

† Here p'_1 is the fraction defective for which the risk of rejection is to be α and p'_2 is the fraction defective for which the risk of acceptance is to be β. To construct the plan, find the tabular value of p'_2/p'_1 in the column for the given α and β which is equal to or just less than the given value of the ratio. The sample size is found by dividing the np'_1 corresponding to the selected ratio by p'_1. The acceptance number is the value of c corresponding to the selected value of the ratio.

‡ Reprinted by permission from "Tables for Constructing and for Computing the Operating Characteristics of Single-Sampling Plans," *Industrial Quality Control*, by J. M. Cameron, July 1952.

$$a_m = -0.649796 + 0.039747m$$
$$r_m = 1.156407 + 0.039747m$$

Usually the sequential plan is applied in a tabular rather than graphical form; a_m and r_m are computed for every m and listed in Table 13.17. Note that a_m is rounded down to the nearest integer and r_m is rounded up to the nearest integer.

Simple formulas may be found for the probability of acceptance and the expected number of observations from five values of the fraction defective including the two points defining the plan.

$$p' = 0 \qquad L(p') = 1$$
$$p' = p_1' \qquad L(p') = 1 - \alpha$$
$$p' = s \qquad L(p') = \frac{h_2}{h_1 + h_2}$$
$$p' = p_2' \qquad L(p') = \beta$$
$$p' = 1 \qquad L(p') = 0$$
$$\bar{n}_0 = \frac{h_1}{s}$$
$$\bar{n}_{p'1} = \frac{(1-\alpha)h_1 - \alpha h_2}{s - p_1'}$$
$$\bar{n}_s = \frac{h_1 h_2}{s(1-s)}$$
$$\bar{n}_{p'2} = \frac{(1-\beta)h_2 - \beta h_1}{p_2' - s}$$
$$\bar{n}_1 = \frac{h_2}{1-s}$$

In the above example,

$$p' = 0 \qquad L(0) = 1$$
$$\bar{n}_0 = \frac{0.649796}{0.039747} = 16.348$$
$$p' = 0.01 \qquad L(0.01) = 0.95$$
$$\bar{n}_{.01} = \frac{(0.95)(0.649796) - (0.05)(1.156407)}{0.029747} = 18.808$$
$$p' = 0.039747$$
$$L(0.039747) = \frac{1.156407}{0.649796 + 1.156407} = 0.64024$$
$$\bar{n}_{(.039747)} = \frac{(0.649796)(1.156407)}{(0.039747)(0.960253)} = 19.688$$
$$p' = 0.10 \qquad L(0.10) = 0.20$$
$$\bar{n}_{.10} = \frac{0.80(1.156407) - 0.20(0.649796)}{0.10 - 0.039747} = 13.197$$
$$p' = 1.00 \qquad L(1.00) = 0$$
$$\bar{n}_{1.00} = \frac{1.156407}{0.960253} = 1.2043$$

Usually these five points may be used to make an adequate sketch of the OC and ASN curves such as the ones in Figs. 13.18 and 13.19.

If more points are needed, they may be found

TABLE 13.17. *TABULAR PROCEDURE FOR ITEM-BY-ITEM SEQUENTIAL SAMPLING PLAN*

Number of Observations m	Acceptance Number a_m	Rejection Number r_m	Observed Number of Defects d
1	—	—	
2	—	—	
3	—	2	
—	—	—	
—	—	—	
—	—	—	
17	0	2	
—	—	—	
—	—	—	
—	—	—	
22	0	3	
—	—	—	
—	—	—	
—	—	—	
42	0	3	
—	—	—	
—	—	—	
—	—	—	
47	1	4	
—	—	—	
—	—	—	
—	—	—	
67	2	4	
—	—	—	
—	—	—	
—	—	—	
72	2	5	
—	—	—	
—	—	—	
—	—	—	
97	3	6	
—	—	—	
—	—	—	
—	—	—	
117	4	6	
—	—	—	
—	—	—	
—	—	—	
122	4	7	
—	—	—	
—	—	—	
—	—	—	
143	5	7	
—	—	—	
—	—	—	
—	—	—	
148	5	8	

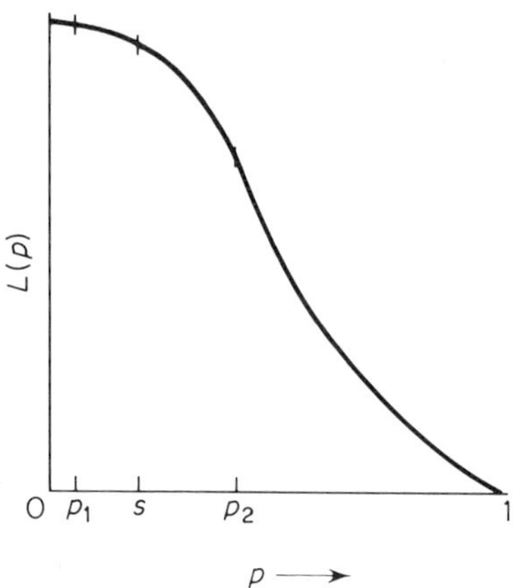

Fig. 13.18 *OC curve of sequential plan sketched from five points.*

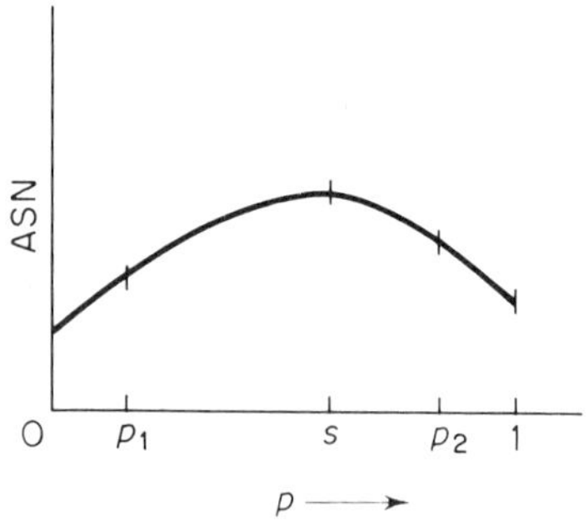

Fig. 13.19 *ASN curve of sequential plan sketched from five points.*

by substituting values between 1 and ∞ for x in the equations

$$L(p') = \frac{x^{h_1+h_2} - x^{h_1}}{x^{h_1+h_2} - 1} \qquad p' = \frac{x^s - 1}{x - 1}$$

The point $[p', L(p')]$ has the conjugate $[p'_c, L(p'_c)]$ given by

$$L(p'_c) = \frac{L(p')}{x^{h_1}} \qquad p'_c = p'(x^{1-s})$$

The expected number of observations is

$$\bar{n}'_p = \frac{L(p')\,(h_1 + h_2) - h_2}{s - p'}$$

13.3.10 Lot-by-Lot Sampling Inspection by Variables: Introduction*

Inspection procedures by variables are based upon the outcomes of a quality characteristic, measured on a continuous scale, and the decision to accept or to reject the lot is a function of these measurements. Variables inspection is applicable whenever the testing of individual items involves measurement on a continuous scale and the form of the distribution is known. Since inspection by variables makes greater use of the information concerning the lot than does inspection by attributes, variables plans require smaller sample sizes than attributes plans furnishing the same protection. Variables pertain to a single quality characteristic, and it is usually assumed that measurements of this quality characteristic are *independent, identically distributed normal random variables* having mean $\bar{x}'$ and standard deviation σ' either known or unknown.

Associated with each inspection characteristic are the design specifications. If only an upper specification limit U is given, the item is considered defective if its measurement exceeds U; the per cent defective in the entire lot p'_U is represented by the shaded area to the right of U in Fig. 13.20. If only a lower specification limit, L, is given, the item is considered defective if its measurement is smaller than L; the per cent defective in the entire lot, p'_L, is represented by the shaded area to the left of L in Fig. 13.20. Whenever both upper and lower limits are specified, the item is considered defective if its

* Much of the material on Sampling Inspection by Variables is taken from A. H. Bowker and G. J. Lieberman, *Engineering Statistics* (Englewood Cliffs, N. J.: Prentice-Hall, 1959).

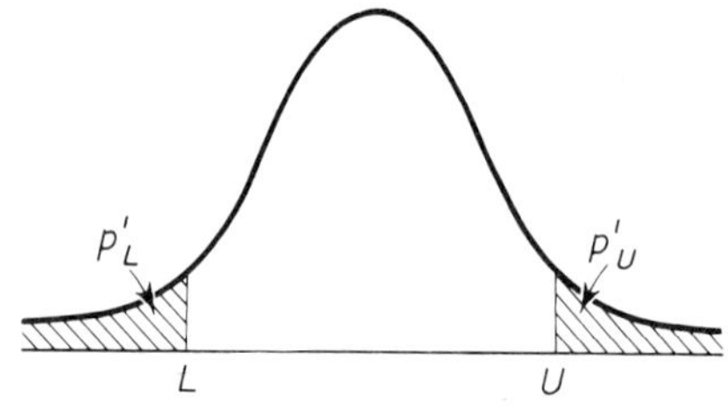

Fig. 13.20 *The shaded areas represent the fraction defective below the lower specification limit and above the upper specification limit.*

measurement either exceeds U or is smaller than L; the per cent defective in the entire lot, $p'_U + p'_L$, is represented by the entire shaded area in Fig. 13.20. If the mathematical model is realistic, i.e., the assumption of a normal distribution is valid, the ratio of the total number of defective units in the lot to the lot size should approximate the area outside of the specification limits.

If the per cent defective of a submitted lot is known, i.e., p'_L and/or p'_U are given (whichever is appropriate), sampling inspection is unnecessary to determine whether or not the lot is to be accepted. If the per cent defective is sufficiently small, the lot is accepted; otherwise, it is rejected. Since such knowledge about the incoming quality is rare, a logical procedure is to *estimate* the per cent defective from a sample, and to accept or reject the lot on the basis of this estimate. A sampling plan is then described as consisting of the sample size n; a method for estimating the per cent defective with the estimate being denoted by p_U, p_L, or $p_U + p_L$, whichever is appropriate; and a maximum allowable per cent defective M. If only any upper specification limit U is given, the estimate of the percentage above this limit p_U is obtained from the sample of size n. If $p_U \leq M$, the lot is accepted. If only a lower specification limit L is given, the estimate of the percentage below this limit p_L is obtained from the sample of size n. If $p_L \leq M$, the lot is accepted. If a double specification limit is given, both p_U and p_L are computed. If $p_U + p_L \leq M$, the lot is accepted.

13.3.11 Variables Acceptance Procedures: Standard Deviation Method

For an upper specification limit U, the "optimum"* estimate of p'_U is a function of $Q_U = [(U - \bar{x})/s]$ and is denoted by p_U, where $\bar{x}$ is the sample mean and s is the sample standard deviation, i.e.,

$$s = \sqrt{\sum(x_i - \bar{x})^2/(n-1)}$$

For a lower specification limit, L, the "optimum" estimate of p'_L is a function of $Q_L = [(\bar{x} - L)/s]$ and is denoted by p_L. The form of the functions p_U and p_L are rather complicated, and have been extensively tabulated.† A graph for these estimates is given in Fig. 13.21. Therefore, in order to obtain the quantity p_U or p_L, Fig. 13.21 is entered with the sample

* Optimality is defined as the minimum variance unbiased estimate.

† Such tables appear in MIL-STD-414, *Sampling Procedures and Tables for Inspection by Variables for Per Cent Defective.*

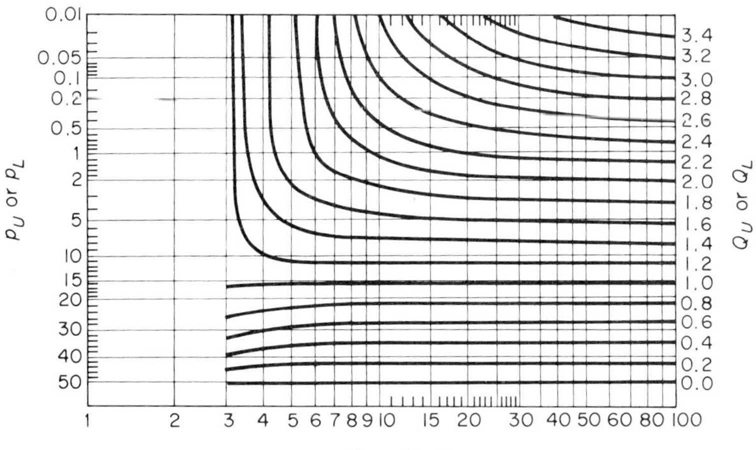

Fig. 13.21 *Chart for determining p_U or p_L from Q_U or Q_L by standard deviation method. Reproduced by permission from A. J. Duncan,* Quality Control and Industrial Statistics, *copyright 1959, Richard D. Irwin, Inc., Homewood, Illinois.*

size and the quality index Q_U or the quality index Q_L, whichever is appropriate; the required estimate, in per cent, is read from the graph. For two-sided specification limits, the "optimum" estimate of the lot per cent defective is given by $p_U + p_L$, where these quantities are also read from Fig. 13.21.

For an upper specification limit, the lot is accepted whenever $p_U \leq M$, where M is a preassigned constant. Alternatively, it can be shown that $p_U \leq M$ if, and only if, $[(U - \bar{x})/s] \geq k$, where k is a constant related to M. Hence, an equivalent procedure is to accept the lot if $[(U - \bar{x})/s] \geq k$.

For a lower specification limit, the lot is accepted whenever $p_L \leq M$. Alternatively, it can be shown that $p_L \leq M$ if, and only if, $[(\bar{x} - L)/s] \geq k$. Hence, an equivalent procedure is to accept the lot if $[(\bar{x} - L)/s] \geq k$.

For two-sided specification limits, the lot is accepted whenever $p_U + p_L \leq M$. There is no equivalent "k" procedure for this situation.

13.3.12 Variables Acceptance Procedures: Average Range Method

For an upper specification limit U, an "approximately optimum"† estimate of p'_U is a function of $Q_U = [(U - \bar{x})c/\bar{R}]$ and is denoted by p_U, where $\bar{x}$ is the sample mean, $\bar{R}$ is the average range of subgroup ranges (each subgroup usually consists of five measurements), and c is a constant depending upon the sample size. For a lower specification limit L, the estimate of p'_L is a function of $Q_L = [(\bar{x} - L)c/\bar{R}]$ and is denoted by p_L. The form of the functions p_U and p_L are rather complicated, and have been extensively tabulated.‡ A graph for these estimates is given in Fig. 13.22. Values of c for various sample sizes are given in Table 13.20 later in this text. Therefore, in order to obtain the quantity p_U or p_L, Fig. 13.22 is entered with the sample size and the quality index Q_U, or the quality index Q_L, whichever is appropriate, the required estimate, in per cent, is read from the graph. For two-sided specification limits, the estimate of the lot per cent defective is given by $p_U + p_L$, where these quantities are also read from Fig. 13.22.

For an upper specification limit, the lot is accepted whenever $p_U \leq M$, where M is a preassigned constant. Alternatively, it can be shown that $p_U \leq M$ if, and only if, $[(U - \bar{x})/\bar{R}] \geq k$, where k is a constant related to M. Hence, an equivalent procedure is to accept the lot if $[(U - \bar{x})/\bar{R}] \geq k$.

† No minimum variance unbiased estimate exists which is a function of $\bar{R}$.

‡ Such tables appear in MIL-STD-414, *Sampling Procedures and Tables for Inspection by Variables for Per Cent Defective.*

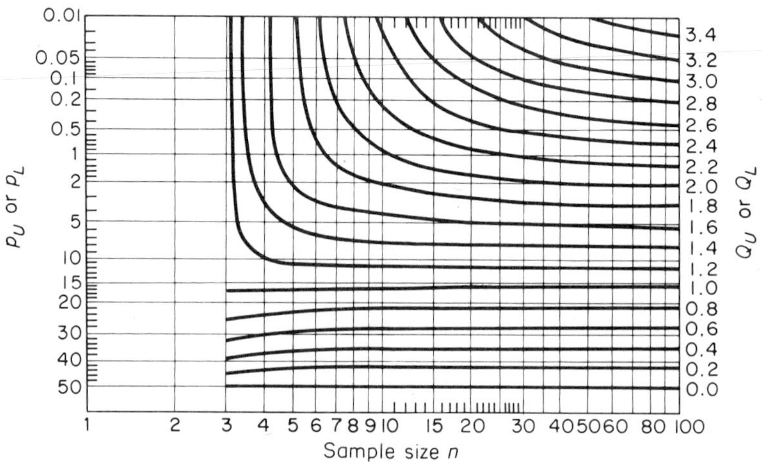

Fig. 13.22 *Chart for determining p_U or p_L from Q_U or Q_L by average range method. Reproduced by permission from A. J. Duncan,* Quality Control and Industrial Statistics, *copyright 1959, Richard D. Irwin, Inc., Homewood, Illinois.*

For a lower specification limit, the lot is accepted whenever $p_L \leq M$. Alternatively, it can be shown that $p_L \leq M$ if, and only if, $[(\bar{x} - L)/\bar{R}] \geq k$. Hence, an equivalent procedure is to accept the lot if $[(\bar{x} - L)/\bar{R}] \geq k$.

For two-sided specification limits, the lot is accepted whenever $p_U + p_L \leq M$. There is no equivalent "k" procedure for this situation.

13.3.13 Variables Acceptance Procedures: Known Standard Deviation Method

For an upper specification limit, U, the "optimum" estimate of p'_U is a function of

$$Q_U = \left(\frac{U - \bar{x}}{\sigma'}\right)\sqrt{\frac{n}{n-1}}$$

and is given by

$$P_U = \int_{Q_U}^{\infty} \frac{1}{\sqrt{2\pi}} e^{-z^2/2}\, dz$$

For a lower specification limit, L, the "optimum" estimate of p'_L is a function of

$$Q_L = \left(\frac{\bar{x} - L}{\sigma'}\right)\sqrt{\frac{n}{n-1}}$$

and is given by

$$P_L = \int_{\infty}^{-Q_L} \frac{1}{\sqrt{2\pi}} e^{-z^2/2}\, dz$$

It is evident that these estimates can be obtained directly from Table 13.3, "Areas Under the Normal Curve."

For two-sided specification limits, the "optimum" estimate of the lot per cent defective is given by $p_U + p_L$.

For an upper specification limit, the lot is accepted whenever $p_U \leq M$, where M is a preassigned constant. Alternatively, it is evident that $p_U \leq M$ if, and only if, $[(U - \bar{x})/\sigma'] \geq k$, where k is a constant given by

$$k = \sqrt{\frac{n-1}{n}}\, K_M$$

and K_M is the M percentage point of the normal distribution. Hence, an equivalent procedure is to accept the lot if $[(U - \bar{x})/\sigma'] \geq k$.

For a lower specification limit, the lot is accepted whenever $p_L \leq M$. Alternatively, it is evident that $p_L \leq M$ if, and only if, $[(\bar{x} - L)/\sigma'] \geq k$, where k is as defined above. Hence, an equivalent procedure is to accept the lot if $[(\bar{x} - L)/\sigma'] \geq k$.

For two-sided specification limits, the lot is accepted whenever $p_U + p_L \leq M$. There is no equivalent "k" procedure for this situation.

13.3.14 Comparison of Variables Procedures with M and k

It was indicated that single specification limit procedures based on estimates of the per cent defective are equivalent to single specification limit procedures involving k. Any advantages of one over the other are independent of OC curve arguments since they both result in the same OC curve, provided M and k are chosen properly. The estimation procedure has several advantages over the k procedure. Like attribute plans, variables plans based on the estimation procedure can involve measurement of lot quality by the per cent defective. The k procedure involves measurement of lot quality by the average and variability of the measurements. Thus, with the estimation procedure, it is unnecessary to shift to attributes sampling whenever per cent defective is the appropriate measure. Second, the estimation procedure has intuitive appeal because it is a logical one, whereas relating a measurement of lot quality based on the average and variability of the measurements to a practical criterion is difficult. In other words, rejecting a lot because the estimated per cent defective is too large is understandable whereas rejecting a lot because $(U - \bar{x})/s$ is too large is difficult to explain. Finally, the estimation procedure leaves the vendor and the consumer with an estimate of lot quality whether the lot is accepted or rejected. On the other hand the k procedure has the advantage of requiring one less step in carrying through the procedure. It does not require a table or graph look-up for the estimate of the lot quality before deciding on acceptance or rejection.

All variables sampling procedures for two-sided specification limits suffer from the fact that the probability of accepting a submitted lot with given per cent defective p' does not depend on p' alone but on the division of p' into two components, the per cent lying above the upper specification limit p'_U and the per cent lying below the lower specification limit p'_L. For this reason a two-sided procedure does not have a unique OC curve but rather a band of curves, each curve within the band representing a possible division of p'. Using the double specification procedure based on the estimate of the lot quality, this band is so narrow as to be, for all practical purposes, a single curve. Since the OC curve for the single specification limit is contained within this narrow band (corresponding to all the defectives outside one limit and none outside the other), it is used as the OC curve for the double specification procedure. No procedure based upon k has this desirable property.

13.3.15 The Military Standard for Inspection by Variables, MIL-STD-414

13.3.15.1 Introduction. In recent years there has been a recognition of the importance of inspection by variables for per cent defective as a test procedure to evaluate product quality. Both government and industry prepared sets of standard sampling plans designed for their own needs. The use of sampling inspection by variables increased to such an extent that the need to prepare a military standard that would serve as an alternative to MIL–STD–105B for inspection by attributes became apparent. In the latter part of 1957, the first military standard for inspection by variables, MIL–STD–414, was issued.

The variables standard is divided into four sections, namely:

Section A—General Description of Sampling Plans

Section B—Variability Unknown—Standard Deviation Method

Section C—Variability Unknown—Range Method

Section D—Variability Known

Section A will always be used in conjunction with the other sections since it provides general concepts and definitions needed for sampling inspection by variables. Sections B, C, and D consist of three parts each:

1. Sampling plans for the single specification limit case
2. Sampling plans for the double specification limit case
3. Procedures for estimation of process average and criteria for tightened and reduced inspection

For the single specification limit case, the acceptability criterion is given in two forms, namely, the k procedure (form 1), e.g., accept the lot if $[(U - \bar{x})/s] \geq k$; and the estimation procedure based upon M (form 2), e.g., accept the lot if $p_U \leq M$. Either of the forms may be used since they require the same sample size and have identical OC curves. Form 2 is required for the estimation of the process average, if such an estimate is desired. The estimation procedure (form 2) is required for double specification limits.

13.3.15.2 Section A: General Description of Sampling Plans. The variables standard has many features that are similar to the attributes standard. Defects are grouped into three classes, critical, major, and minor, with the definition of each similar to that in MIL–STD–105D. Like the attributes standard, the sampling plans are indexed by AQL, inspection level, and lot size. The AQL is defined as "a nominal value expressed in terms of per cent defective specified for a single quality characteristic." The AQL values given, in per cent, are

0.04	0.40	4.0
0.065	0.65	6.5
0.10	1.0	10.0
0.15	1.5	15.0
0.25	2.5	

The probabilities of acceptance of submitted lots having these AQL range from about 0.89 for the smallest sample sizes to about 0.99 for the largest sample sizes. The particular AQL value to be used for a single quality characteristic of a given product must be specified. In the case of a double specification limit, either an AQL value is specified for the total per cent defective outside both upper and lower specification limits or two AQL values are specified, one for the upper limit and another for the lower limit.

Sample sizes are designated by code letters B to Q. The sample size code letter depends on the inspection level and the lot size. There are five inspection levels, I, II, III, IV, V. Unless otherwise specified, inspection level IV is used. The sample size code letter applicable to the specified inspection level and for lots of given size is obtained from Table 13.18 (Table A–2 of MIL–STD–414).

After the AQL, inspection level, and sample code letter are chosen, the appropriate sampling plan is chosen from a master table in Section B, C, or D of the standard, whichever is appropriate. It is interesting to note that the variables standard states that "unless otherwise specified, unknown variability, standard deviation method sampling plans and the accept-

TABLE 13.18. *SAMPLE SIZE CODE LETTERS* (Table A–2 of MIL–STD–414)†

Lot Size	Inspection Levels I	II	III	IV	V
3 to 8	B	B	B	B	C
9 to 15	B	B	B	B	D
16 to 25	B	B	B	C	E
26 to 40	B	B	B	D	F
41 to 65	B	B	C	E	G
66 to 110	B	B	D	F	H
111 to 180	B	C	E	G	I
181 to 300	B	D	F	H	J
301 to 500	C	E	G	I	K
501 to 800	D	F	H	J	L
801 to 1,300	E	G	I	K	L
1,301 to 3,200	F	H	J	L	M
3,201 to 8,000	G	I	L	M	N
8,001 to 22,000	H	J	M	N	O
22,001 to 110,000	I	K	N	O	P
110,001 to 550,000	I	K	O	P	Q
550,001 and over	I	K	P	Q	Q

† Sample size code letters given in body of table are applicable when the indicated inspection levels are to be used.

ability criterion of form 2 (for the single specification limit case) shall be used." This refers to Section B and the estimation procedure.

Section A also contains all the OC curves. These OC curves represent the plans in Section B exactly. The OC curves of the corresponding plans in Sections C and D were matched as well as possible and are extremely close in most cases. Any discrepancies are due to the requirement of a constant sample size (multiple of five) for a code letter for the range plans and integer values of the sample size for the known standard deviation plans. Hence, for a given lot size, inspection level, and AQL, the user may select a plan from any of Sections B, C, or D and be assured of the same probability of accepting or rejecting material for any given quality. These OC curves are reproduced in Fig. 13.13.

13.3.15.3 Section B: Variability Unknown; Standard Deviation Method. When there is a single specification limit and form 1 is used (k procedure), the appropriate master sampling table is selected as follows:*

For normal or tightened inspection, Table 13.19 (Table B–1 of MIL–STD–414).

For reduced inspection, Table B–2 of MIL–STD–414 (which is not reproduced here).

The sample size, n, and acceptance criterion, k, are read from these tables. The lot is accepted if $[(U - \bar{x})/s] \geq k$ or $[(\bar{x} - L)/s] \geq k$, whichever is appropriate.

When there is a single specification limit or a double specification limit (with the AQL applying to both limits) and form 2 is used (estimation procedure), the appropriate master sampling table is selected as follows:

For normal or tightened inspection, Table 13.19 (Table B–3 of MIL–STD–414).

For reduced inspection, Table B–4 of MIL–STD–414 (which is not reproduced here).

The sample size n and the acceptance criterion M are read from these tables. The quality indices $Q_U = [(U - \bar{x})/s]$ and/or $Q_L = [(\bar{x} - L)/s]$ are computed. Figure 13.21 (Table B–5 of MIL–STD–414) is entered with n and Q_U and/or Q_L, whichever is appropriate; the estimates p_U and/or p_L, in per cent, are read from the graph. For an upper specification limit, the lot is accepted if $p_U \leq M$; for a lower specification limit, the lot is accepted if $p_L \leq M$; and for both upper and lower specification limits, the lot is accepted if $p_U + p_L \leq M$. Procedures are also given for two-sided specification limits when one AQL value is specified for the upper limit and another for the lower limit.

As in MIL–STD–105D, the variables standard provides for normal, tightened, and reduced inspection. Procedures for going on tightened and reduced inspection are described in detail in MIL–STD–414.

13.3.15.4 Section C: Variability Unknown; Range Method. When there is a single specification limit and form 1 is used (k procedure), the appropriate master sampling table is selected as follows:

For normal or tightened inspection, Table 13.20 (Table C–1 of MIL–STD–414).

For reduced inspection, Table C–2 of MIL–STD–414 (which is not reproduced here).

The sample size n and acceptance criterion k are read from these tables. The lot is accepted if $[(U - \bar{x})/\bar{R}] \geq k$ or $[(\bar{x} - L)/\bar{R}] \geq k$, whichever is appropriate.

When there is a single specification limit or a double specification limit (with the AQL applying to both limits) and form 2 is used (estimation procedure), the appropriate master sampling table is selected as follows:

For normal or tightened inspection, Table 13.20 (Table C–3 of MIL–STD–414).

* The k procedure cannot be used for double specification limits.

TABLE 13.19. *MASTER TABLE FOR NORMAL AND TIGHTENED INSPECTION FOR PLANS BASED ON VARIABILITY UNKNOWN, STANDARD DEVIATION METHOD*

(Single Specification Limit—Form 1; Table B–1 of MIL-STD-414)

Sample Size Code Letter	Sample Size	Acceptable Quality Levels (Normal Inspection)													
		0.04 *k*	0.065 *k*	0.10 *k*	0.15 *k*	0.25 *k*	0.40 *k*	0.65 *k*	1.00 *k*	1.50 *k*	2.50 *k*	4.00 *k*	6.50 *k*	10.00 *k*	15.00 *k*
B	3	↓	↓	↓	↓	↓	↓	↓	↓	↓	1.12	0.958	0.765	0.566	0.341
C	4	↓	↓	↓	↓	↓	↓	↓	1.45	1.34	1.17	1.01	0.814	0.617	0.393
D	5	↓	↓	↓	↓	↓	↓	1.65	1.53	1.40	1.24	1.07	0.874	0.675	0.455
E	7	↓	↓	↓	↓	2.00	1.88	1.75	1.62	1.50	1.33	1.15	0.955	0.755	0.536
F	10	↓	↓	↓	2.24	2.11	1.98	1.84	1.72	1.58	1.41	1.23	1.03	0.828	0.611
G	15	2.64	2.53	2.42	2.32	2.20	2.06	1.91	1.79	1.65	1.47	1.30	1.09	0.886	0.664
H	20	2.69	2.58	2.47	2.36	2.24	2.11	1.96	1.82	1.69	1.51	1.33	1.12	0.917	0.695
I	25	2.72	2.61	2.50	2.40	2.26	2.14	1.98	1.85	1.72	1.53	1.35	1.14	0.936	0.712
J	30	2.73	2.61	2.51	2.41	2.28	2.15	2.00	1.86	1.73	1.55	1.36	1.15	0.946	0.723
K	35	2.77	2.65	2.54	2.45	2.31	2.18	2.03	1.89	1.76	1.57	1.39	1.18	0.969	0.745
L	40	2.77	2.66	2.55	2.44	2.31	2.18	2.03	1.89	1.76	1.58	1.39	1.18	0.971	0.746
M	50	2.83	2.71	2.60	2.50	2.35	2.22	2.08	1.93	1.80	1.61	1.42	1.21	1.00	0.774
N	75	2.90	2.77	2.66	2.55	2.41	2.27	2.12	1.98	1.84	1.65	1.46	1.24	1.03	0.804
O	100	2.92	2.80	2.69	2.58	2.43	2.29	2.14	2.00	1.86	1.67	1.48	1.26	1.05	0.819
P	150	2.96	2.84	2.73	2.61	2.47	2.33	2.18	2.03	1.89	1.70	1.51	1.29	1.07	0.841
Q	200	2.97	2.85	2.73	2.62	2.47	2.33	2.18	2.04	1.89	1.70	1.51	1.29	1.07	0.845
		0.065	0.10	0.15	0.25	0.40	0.65	1.00	1.50	2.50	4.00	6.50	10.00	15.00	
		Acceptable Quality Levels (Tightened Inspection)													

All AQL values are in per cent defective.

Use first sampling plan below arrow, that is, both sample size as well as *k* value. When sample size equals or exceeds lot size, every item in the lot must be inspected.

(Double Specification Limit and Form 2—Single Specification Limit; Table B-3 of MIL-STD-414)

Sample Size Code Letter	Sample Size	Acceptable Quality Levels (Normal Inspection)													
		0.04 *M*	0.065 *M*	0.10 *M*	0.15 *M*	0.25 *M*	0.40 *M*	0.65 *M*	1.00 *M*	1.50 *M*	2.50 *M*	4.00 *M*	6.50 *M*	10.00 *M*	15.00 *M*
B	3	↓	↓	↓	↓	↓	↓	↓	↓	↓	7.59	18.86	26.94	33.69	40.47
C	4	↓	↓	↓	↓	↓	↓	↓	1.53	5.50	10.92	16.45	22.86	29.45	36.90
D	5	↓	↓	↓	↓	↓	↓	1.33	3.32	5.83	9.80	14.39	20.19	26.56	33.99
E	7	↓	↓	↓	↓	0.422	1.06	2.14	3.55	5.35	8.40	12.20	17.35	23.29	30.50
F	10	↓	↓	↓	0.349	0.716	1.30	2.17	3.26	4.77	7.29	10.54	15.17	20.74	27.57
G	15	0.099	0.186	0.312	0.503	0.818	1.31	2.11	3.05	4.31	6.56	9.46	13.71	18.94	25.61
H	20	0.135	0.228	0.365	0.544	0.846	1.29	2.05	2.95	4.09	6.17	8.92	12.99	18.03	24.53
I	25	0.155	0.250	0.380	0.551	0.877	1.29	2.00	2.86	3.97	5.97	8.63	12.57	17.51	23.79
J	30	0.179	0.280	0.413	0.581	0.879	1.29	1.98	2.83	3.91	5.86	8.47	12.36	17.24	23.58
K	35	0.170	0.264	0.388	0.535	0.847	1.23	1.87	2.68	3.70	5.57	8.10	11.87	16.65	22.91
L	40	0.179	0.275	0.401	0.566	0.873	1.26	1.88	2.71	3.72	5.58	8.09	11.85	16.61	22.86
M	50	0.163	0.250	0.363	0.503	0.789	1.17	1.71	2.49	3.45	5.20	7.61	11.23	15.87	22.00
N	75	0.147	0.228	0.330	0.467	0.720	1.07	1.60	2.29	3.20	4.87	7.15	10.63	15.13	21.11
O	100	0.145	0.220	0.317	0.447	0.689	1.02	1.53	2.20	3.07	4.69	6.91	10.32	14.75	20.66
P	150	0.134	0.203	0.293	0.413	0.638	0.949	1.43	2.05	2.89	4.43	6.57	9.88	14.20	20.02
Q	200	0.135	0.204	0.294	0.414	0.637	0.945	1.42	2.04	2.87	4.40	6.53	9.81	14.12	19.92
		0.065	0.10	0.15	0.25	0.40	0.65	1.00	1.50	2.50	4.00	6.50	10.00	15.00	
		Acceptable Quality Levels (Tightened Inspection)													

All AQL and table values are in per cent defective.

Use first sampling plan below arrow, that is, both sample size as well as *M* value. When sample size equals or exceeds lot size, every item in the lot must be inspected.

TABLE 13.20. *MASTER TABLE FOR NORMAL AND TIGHTENED INSPECTION FOR PLANS BASED ON VARIABILITY UNKNOWN, RANGE METHOD*

(Single Specification Limit—Form 1; Table C–1 of MIL-STD-414)

Sample Size Code Letter	Sample Size	Acceptable Quality Levels (Normal Inspection)													
		0.04 k	0.065 k	0.10 k	0.15 k	0.25 k	0.40 k	0.65 k	1.00 k	1.50 k	2.50 k	4.00 k	6.50 k	10.00 k	15.00 k
B	3	↓	↓	↓	↓	↓	↓	↓	↓	↓	0.587	0.502	0.401	0.296	0.178
C	4								0.651	0.598	0.525	0.450	0.364	0.276	0.176
D	5					↓	↓	0.663	0.614	0.565	0.498	0.431	0.352	0.272	0.184
E	7				↓	0.702	0.659	0.613	0.569	0.525	0.465	0.405	0.336	0.266	0.189
F	10	↓	↓	↓	0.916	0.863	0.811	0.755	0.703	0.650	0.579	0.507	0.424	0.341	0.252
G	15	1.09	1.04	0.999	0.958	0.903	0.850	0.792	0.738	0.684	0.610	0.536	0.452	0.368	0.276
H	25	1.14	1.10	1.05	1.01	0.951	0.896	0.835	0.779	0.723	0.647	0.571	0.484	0.398	0.305
I	30	1.15	1.10	1.06	1.02	0.959	0.904	0.843	0.787	0.730	0.654	0.577	0.490	0.403	0.310
J	35	1.16	1.11	1.07	1.02	0.964	0.908	0.848	0.791	0.734	0.658	0.581	0.494	0.406	0.313
K	40	1.18	1.13	1.08	1.04	0.978	0.921	0.860	0.803	0.746	0.668	0.591	0.503	0.415	0.321
L	50	1.19	1.14	1.09	1.05	0.988	0.931	0.893	0.812	0.754	0.676	0.598	0.510	0.421	0.327
M	60	1.21	1.16	1.11	1.06	1.00	0.948	0.885	0.826	0.768	0.689	0.610	0.521	0.432	0.336
N	85	1.23	1.17	1.13	1.08	1.02	0.962	0.899	0.839	0.780	0.701	0.621	0.530	0.441	0.345
O	115	1.24	1.19	1.14	1.09	1.03	0.975	0.911	0.851	0.791	0.711	0.631	0.539	0.449	0.353
P	175	1.26	1.21	1.16	1.11	1.05	0.994	0.929	0.868	0.807	0.726	0.644	0.552	0.460	0.363
Q	230	1.27	1.21	1.16	1.12	1.06	0.996	0.931	0.870	0.809	0.728	0.646	0.553	0.462	0.364
		0.065	0.10	0.15	0.25	0.40	0.65	1.00	1.50	2.50	4.00	6.50	10.00	15.00	
		Acceptable Quality Levels (Tightened Inspection)													

All AQL values are in percent defective.

Use first sampling plan below arrow, that is, both sample size as well as k value. When sample size equals or exceeds lot size, every item in the lot must be inspected.

(Double Specification Limit and Form 2—Single Specification Limit; Table C–3 of MIL-STD-414)

Sample Size Code Letter	Sample Size	c Factor	Acceptable Quality Levels (Normal Inspection)													
			0.04 M	0.065 M	0.10 M	0.15 M	0.25 M	0.40 M	0.65 M	1.00 M	1.50 M	2.50 M	4.00 M	6.50 M	10.00 M	15.00 M
B	3	1.910	↓	↓	↓	↓	↓	↓	↓	↓	↓	7.59	18.86	26.94	33.69	40.47
C	4	2.234	↓	↓	↓	↓	↓	↓	↓	1.53	5.50	10.92	16.45	22.86	29.45	36.90
D	5	2.474	↓	↓	↓	↓	↓	↓	1.42	3.44	5.93	9.90	14.47	20.27	26.59	33.95
E	7	2.830	↓	↓	↓	↓	0.28	0.89	1.99	3.46	5.32	8.47	12.35	17.54	23.50	30.66
F	10	2.405	↓	↓	↓	0.23	0.58	1.14	2.05	3.23	4.77	7.42	10.79	15.49	21.06	27.90
G	15	2.379	0.061	0.136	0.253	0.430	0.786	1.30	2.10	3.11	4.44	6.76	9.76	14.09	19.30	25.92
H	25	2.358	0.125	0.214	0.336	0.506	0.827	1.27	1.95	2.82	3.96	5.98	8.65	12.59	17.48	23.79
I	30	2.353	0.147	0.240	0.366	0.537	0.856	1.29	1.96	2.81	3.92	5.88	8.50	12.36	17.19	23.42
J	35	2.349	0.165	0.261	0.391	0.564	0.883	1.33	1.98	2.82	3.90	5.85	8.42	12.24	17.03	23.21
K	40	2.346	0.160	0.252	0.375	0.539	0.842	1.25	1.88	2.69	3.73	5.61	8.11	11.84	16.55	22.38
L	50	2.342	0.169	0.261	0.381	0.542	0.838	1.25	1.60	2.63	3.64	5.47	7.91	11.57	16.20	22.26
M	60	2.339	0.158	0.244	0.356	0.504	0.781	1.16	1.74	2.47	3.44	5.17	7.54	11.10	15.64	21.63
N	85	2.335	0.156	0.242	0.350	0.493	0.755	1.12	1.67	2.37	3.30	4.97	7.27	10.73	15.17	21.05
O	115	2.333	0.153	0.230	0.333	0.468	0.718	1.06	1.58	2.25	3.14	4.76	6.99	10.37	14.74	20.57
P	175	2.333	0.139	0.210	0.303	0.427	0.655	0.972	1.46	2.08	2.93	4.47	6.60	9.89	14.15	19.88
Q	230	2.333	0.142	0.215	0.308	0.432	0.661	0.976	1.47	2.08	2.92	4.46	6.57	9.84	14.10	19.82
			0.065	0.10	0.15	0.25	0.40	0.65	1.00	1.50	2.50	4.00	6.50	10.00	15.00	
			Acceptable Quality Level (Tightened Inspection)													

All AQL and table values are in per cent defective.

Use first sampling plan below arrow, that is, both sample size as well as M value. When sample size equals or exceeds lot size, every item in the lot must be inspected.

For reduced inspection, Table C–4 of MIL–STD–414 (which is not reproduced here). The sample size n and the acceptance criterion M are read from these tables. The quality indices $Q_U = (U - \bar{x})c/\bar{R}$ and/or $Q_L = (\bar{x} - L)c/\bar{R}$ are computed; c is obtained from Table 13.20. Figure 13.22 (Table C–5 of MIL–STD–414) is entered with n and Q_U and/or Q_L, whichever is appropriate; the estimates p_U and/or p_L in per cent are read from the graph. For an upper specification limit, the lot is accepted if $p_U \leq M$; for a lower specification limit, the lot is accepted if $p_L \leq M$; and for both upper and lower specification limits, the lot is accepted if $p_U + p_L \leq M$. Again, procedures for going on tightened and reduced inspection are described in detail in MIL–STD–414.

It is interesting to note that the sample size required using a range plan will always be greater than or equal to the sample size using the sample standard deviation plan. If a sample standard deviation plan requires a small n, the comparable range plan requires a not too different sample size; this difference increases as the sample size of the sample standard deviation plan increases.

13.3.15.5 Section D: Variability Known. This section assumes that the standard deviation σ' of the normal distribution is known.*

When there is a single specification limit and form 1 is used (k procedure), the appropriate master sampling table is selected as follows:

For normal or tightened inspection, Table D–1 of MIL–STD–414 (which is not reproduced here)

For reduced inspection, Table D–2 of MIL–STD–414 (which is not reproduced here) The sample size n and acceptance criterion k are read from these tables. The lot is accepted if $[(U - \bar{x})/\sigma'] \geq k$ or $[(\bar{x} - L)/\sigma'] \geq k$, whichever is appropriate.

When there is a single specification limit or a double specification limit (with the AQL applying to both limits) and form 2 is used (estimation procedure), the appropriate master sampling table is selected as follows:

For normal or tightened inspection, Table D–3 of MIL–STD–414 (which is not reproduced here)

For reduced inspection, Table D–4 of MIL–STD–414 (which is not reproduced here) The sample size n and the acceptance criterion M are read from these tables. The quality indices

$$Q_U = \frac{(U - \bar{x})}{\sigma'}\sqrt{\frac{n}{n-1}} \quad \text{and/or}$$

$$Q_L = \frac{(\bar{x} - L)}{\sigma'}\sqrt{\frac{n}{n-1}}$$

are computed. The quantity

$$\sqrt{\frac{n}{n-1}}$$

is denoted by v and is given in the master table. Table 13.3 is entered with Q_U and/or Q_L, whichever is appropriate; the estimates p_U and/or p_L are read from the table and converted to readings in per cent by multiplying by one hundred. For an upper specification limit, the lot is accepted if $p_U \leq M$; for a lower specification limit, the lot is accepted if $p_L \leq M$; and for both upper and lower specification limits, the lot is accepted if $p_U + p_L \leq M$. Again, procedures for going on tightened and reduced inspection are described in the Standard.

It is evident that the sample sizes required using the known standard deviation plan will always be smaller than the sample size using the sample standard deviation plan.

13.3.16 Continuous Sampling Inspection: Introduction

The preceding articles on sampling inspection have dealt with lot-by-lot inspection when items are classified according to attribute data or variables data. In the following articles, inspection procedures will be given for items that are not broken into lots and that are sampled by attributes. In these continuous sampling procedures, current inspection results are used to determine whether sampling inspection or screening inspection is to be used for the next items to be inspected.

13.3.17 Dodge Continuous Sampling Plan (CSP-1)

In 1943, Dodge published a continuous sampling plan in the *Annals of Mathematical*

* The known standard deviation is denoted by σ in MIL-STD-414; σ' will be used in this section so that the notation conforms to the notation of the American Society for Quality Control.

Statistics.* The procedure, as stated by Dodge, is as follows:

a. At the outset, inspect 100 per cent of the units consecutively as produced and continue such inspection until i units in succession are found clear of defects.

b. When i units in succession are found clear of defects, discontinue 100 per cent inspection, and inspect only a fraction f of the units, selecting individual sample units one at a time from the flow of product in such a manner as to assure an unbiased sample.

c. If a sample unit is found defective, revert immediately to a 100 per cent inspection of succeeding units and continue until again i units in succession are found clear of defects, as in paragraph (a).

d. Correct or replace with good units all defective units found.

In his paper, Dodge studied the properties of this plan, and presented equations and charts for determining the average outgoing quality limit (AOQL) as functions of the parameters f and i, under the assumption that the process is in a state of statistical control. A production process is said to be in statistical control if there is a positive constant $p \leq 1$ such that, for every item produced, the probability that it is defective is p, and is independent of the state (defective or nondefective) of all the other items produced. Figure 13.23 presents the necessary chart for the selection of the appropriate plan.

13.3.17.1 Example. Suppose the desired AOQL is 4 per cent and f is chosen to be $^1/_5$. From Fig. 13.23, i is found to be 17. The plan is as follows:

Inspect all the units as produced and continue such inspection until 17 units in succession are found clear of defects.

When 17 units in succession are found clear of defects, discontinue 100 per cent inspection, and inspect only one out of five units. As soon as a defective is found, revert to 100 per cent inspection of succeeding units and continue until again 17 units in succession are found clear of defects. Then resume sampling inspection.

13.3.17.2 Further Results with the Dodge Procedure. Although the Dodge procedure is always applicable, it only guarantees the specified AOQL provided the process is in a state of statistical control. Lieberman† has shown that the Dodge procedure guarantees an AOQL whether or not the process is in a state of statistical control. In fact, without the assumption of control, and for a given f and i,

* H. F. Dodge, "A Sampling Inspection Plan for Continuous Production," *Annals of Mathematical Statistics*, Vol. 14, September 1943.

† G. J. Lieberman, "A Note on Dodge's Continuous Inspection Plan," *Annals of Mathematical Statistics*, Vol. 24, September 1953.

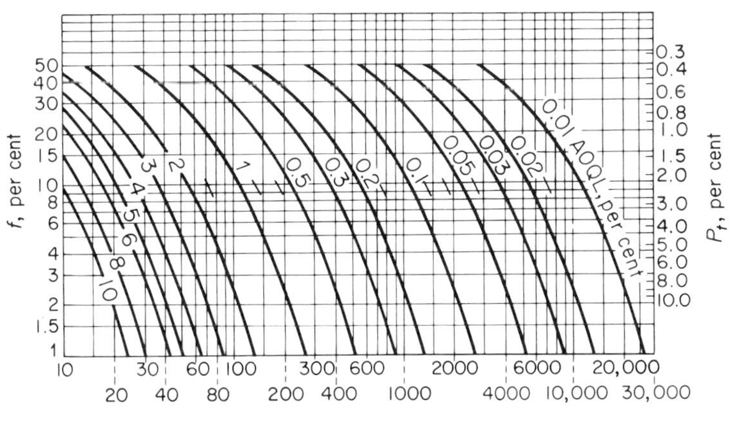

Fig. 13.23 *Curves for determining values of* f *and* i *for a given value of AOQL in Dodge's plan for continuous production. Reproduced by permission from "A Sampling Inspection Plan for Continuous Production," by H. F. Dodge.* Annals of Mathematical Statistics, *Vol. 14, September 1943.*

$$\text{AOQL} = \frac{1/f - 1}{1/f + i}$$

In the above example, if the assumption of control is *not* made and an AOQL of 4 per cent is desired with $f = 1/5$, it is necessary to take $i = 95$.

In 1951, H. F. Dodge and Miss M. N. Torrey* developed two modifications of CSP–1. They are referred to as CSP–2 and CSP–3. These plans remove the feature of reverting to 100 per cent inspection as soon as one defect is found; instead, 100 per cent inspection is reinstated only when one defect falls too closely on the heels of another during sampling inspection, that is, when their separation is smaller than a prescribed minimum spacing. Naturally, the probability of accepting a short run of product of poor quality is higher in these plans than in the original one. These last two plans differ in that one provides for the inspection of four additional units whenever a defect is found under conditions that do not require reinstating 100 per cent inspection, thereby providing protection against short runs of poor product.

Although no military standard for continuous sampling has evolved, there exists a U.S. Department of Defense document entitled "Inspection and Quality Control Handbook (Interim) H107, Single Level Continuous Sampling Procedures and Tables for Inspection by Attributes" which gives a set of continuous sampling plans based upon the Dodge procedures.

13.3.18 Multilevel Sampling Plans

A plan that allows for smooth transition between sampling and 100 per cent inspection, requiring 100 per cent inspection only when quality is quite inferior, and that allows for the inspection to continue to reduce when quality is definitely good, is a multilevel sampling plan. This plan allows for any number of sampling levels subject to the provisions that transitions can occur only between adjacent levels. A particular multilevel plan has been analyzed in considerable detail by Lieberman and Solomon†. Their plan is called MLP and is as follows:

As with the Dodge CSP–1 plan, inspect 100 per cent of the units consecutively as produced and continue until i units in succession are found clear of defects. When i units are found clear of defects, discontinue 100 per cent inspection and inspect only a fraction f. If the next i units inspected are nondefective, then proceed to sampling at rate f^2. If a defective is found, revert to inspection at the next lowest level, etc. This plan can be used with any number of levels, but from two to four levels seem to be the greatest practical interest. The first Dodge plan, CSP–1, is easily recognized as a special case containing only one sampling level. Lieberman and Solomon present curves of constant AOQL as a function of f and i for two, three, four, and an infinite number of levels. There have been numerous modifications of these plans. One of the most useful plans of this type was prepared for the U.S. Air Force by the Department of Industrial Engineering at Stanford University. With minor changes, this set of plans was later published by the U.S. Department of Defense as "Inspection and Quality Control Handbook (Interim) H106, Multi-level Continuous Sampling Procedures and Tables for Inspection by Attributes." As in the Lieberman-Solomon plans, at the outset 100 per cent of the product is inspected until a specified number of consecutive items i is found free of defects. At that time, the inspection is reduced to a fraction f of the items presented. If i successively inspected items are found free of defects, the inspection rate is reduced to a new fraction f^2 of the items presented. This procedure is continued through the several levels for the plan. However, if at any level a defective item is found before i successively inspected items are passed, the plan provides a special procedure to determine if sampling should continue at the present rate or revert back to the previous sampling rate.

The special procedure, after a defective item has been found, consists of inspecting the next four consecutive items. If no defectives are found in the four consecutive items, sampling is resumed at the same rate as when the defective was found. If no more defectives are found

* H. F. Dodge and M. N. Torrey, "Additional Continuous Sampling Inspection Plans," *Industrial Quality Control*, Vol. 7, March 1951.

† G. J. Lieberman and H. Solomon, "Multi-Level Continuous Sampling Plans," *Annals of Mathematical Statistics*, **26** (1955), 686–704.

before a total (including the four consecutive items) of i successively inspected items are determined to be acceptable, the sampling rate is reduced to the next smaller fraction (the next level). If a second defective is found, either in the four consecutive items or after resuming the sampling rate, revert to the former level (next larger sampling rate) and inspect four consecutive items, repeating the procedure just described. Thus, the sampling rate may be reduced or increased by specific steps as the quality of the presented items varies up and down. At the same time, the plans provide a check to differentiate between the random occurrence of a defective, which can occur even though the overall quality level is acceptable, and inferior or spotty quality.

13.3.19 Girshick Continuous Sampling Plan

In 1948, M. A. Girshick* presented a continuous sampling plan. It is defined by the three integers m, N, and K. The plan operates as follows:

The units of product in the production sequence are divided into segments of size K. Inspection begins by selecting at random one item from each consecutive segment of K items. The items are inspected in sequence and the number of defectives found, as well as the number of items examined, are cumulated. This procedure is continued until the cumulative number of defectives reaches m. At this point, the size of the sample n is compared with the integer N. If $n \geq N$, the product that has passed through inspection is considered acceptable and the inspection procedure is repeated on the new incoming product. If, on the other hand, $n < N$, the following actions are taken: (a) the next $N - n$ segments are inspected 100 per cent; and (b) after that, the inspection procedure is repeated. This procedure always guarantees that the AOQL cannot exceed

$$(K - 1)\, m/KN$$

13.3.19.1 Example. If an AOQL of 4 per cent is desired with $K = 5$ and $m = 3$, the value of N for the Girshick plan is found by solving the equation $0.04 = (4 \times 3)/(5N)$ for N, i.e., $N = 60$.

Inspection begins by selecting at random one item from each consecutive segment of five items. This procedure is continued until the cumulative number of defectives reaches three. At this point, the size of the sample n is compared with 60. If $n \geq 60$, the product that has passed through inspection is considered acceptable, and the inspection procedure is repeated on the new incoming product. If, on the other hand, $n < 60$, the next $60 - n$ segments are inspected 100 per cent. After that, the initial inspection procedure is repeated.

* M. A. Girshick, "A Sequential Inspection Plan for Quality Control," Technical Report No. 16, Applied Mathematics and Statistics Laboratory, Stanford University, 1954.

13.4 COMMON SIGNIFICANCE TESTS

13.4.1 Introduction

Significance tests are important tools in scientific and industrial experimentation; they are used for making decisions on the basis of the limited information available in samples and provide procedures for making these decisions with preassigned risks. For example, an experimenter may wish to determine whether a new method of sealing vacuum tubes will increase their life, whether a new alloy will have an increased breaking strength, whether a new source of raw material has resulted in a change in the quality of output, or whether a special treatment of concrete will have an effect on breaking strength.

Consider the breaking strength example: We are really concerned with two distributions, one the distribution of breaking strength of concrete made by the standard method, and the other the distribution of concrete receiving the special treatment. Our problem is to decide whether these two distributions are the same or different, and in particular to decide if their mean breaking strengths are the same. We make the hypothesis that there is no difference in mean breaking strength, and will accept or reject this hypothesis on the basis of experimental evidence. A number of test specimens of both the special and standard treatment will be made up and the compressive strength of each test specimen determined. A statistic will be computed from these measurements.

Ideally, before the data are at hand, the experimenter should decide on his rule of rejection and on the number of items to include in the sample. As explained in the detailed instructions for the various cases included in this article, the test for the equality of two population means is based on the difference of the sample means (say h = mean of special treatment minus mean of standard treatment) divided by some measure of their sampling variability. In this case, we will keep the standard method of treatment unless the special method actually increases the breaking strength, and we are interested in rejecting the hypothesis of equality only in this case. Hence, we will reject the hypothesis only if the sample mean of the new material is substantially greater than the sample mean of the old, i.e., h is positive. This kind of test that rejects when differences are in one direction only is called a one-tailed test, since the region of rejection consists of one tail of the distribution of the test statistic. Often we are interested in rejecting the hypothesis of equality of means when the true means differ in either direction, as when we want to see if a new source of raw materials has changed the quality of output. In this case, we reject when the difference of the sample means is either too large or too small; such tests are called two-tailed tests.

13.4.2 OC Curves of the Test

Suppose we let Δ equal the true difference in true means, i.e., Δ = mean of special treatment population minus mean of standard treatment population, and we are interested in the one-tailed test. Now we have indicated that we want a test that will have a high probability of rejecting the hypothesis of equality of means for Δ positive and a high probability of accepting the hypothesis (low probability of rejecting) if $\Delta \leq 0$. Of course, in repeated sampling from populations with a given Δ we will accept some of the time and reject the rest of the time; the percentage of times we accept the hypothesis is the probability of acceptance. A plot of the probability of acceptance as a function of Δ is called the operating characteristic curve of the test; just as the protection of an acceptance sampling plan is characterized by the OC curve (the probability of accepting the lot as a function of presented quality) so the risk of making the wrong decision in any significance test is characterized by the OC curve.

An operating characteristic curve for a one-sided test will look something like Fig. 13.24. An operating characteristic curve for a two-sided test will look something like Fig. 13.25. The probability of rejecting the hypothesis ($\Delta = 0$ in this case) when it is true (1 − probability of acceptance as given by the OC curve) is often called the level of significance or size of the test. The letter α will be used to denote the level of significance. Standard practice is to take α as 0.05 or 0.01 depending on how serious it is from a practical point of view to reject the hypothesis when it is true.

A family of OC curves of the one-tailed test for various values of the sample size is given in Fig. 13.26. As in the case of sampling inspection, increasing the number of observations increases the slope of the OC curve; the probability of accepting the hypothesis $\Delta = 0$ if Δ is really 0 remains the same; i.e., the significance level can be achieved for any sample size. What changes is the probability of deciding that Δ =

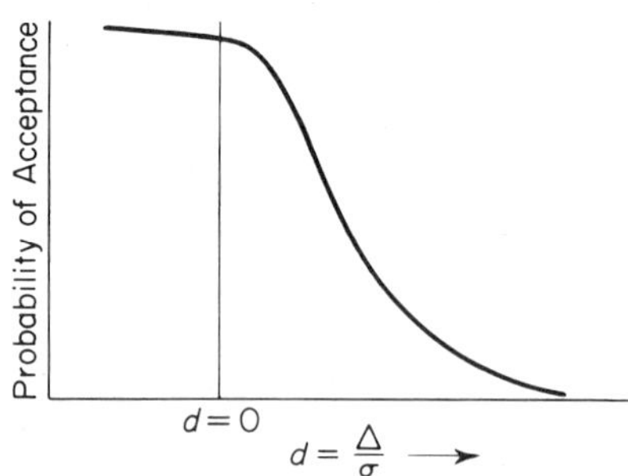

Fig. 13.24 *OC curve of the one-sided test for the equality of two means.*

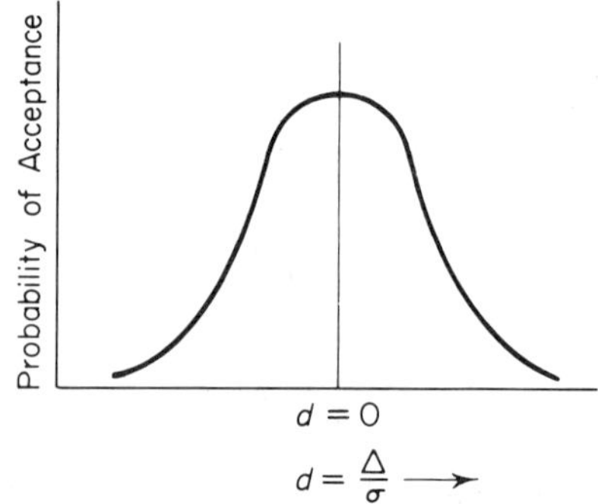

Fig. 13.25 *OC curve of the two-sided test for the equality of two means.*

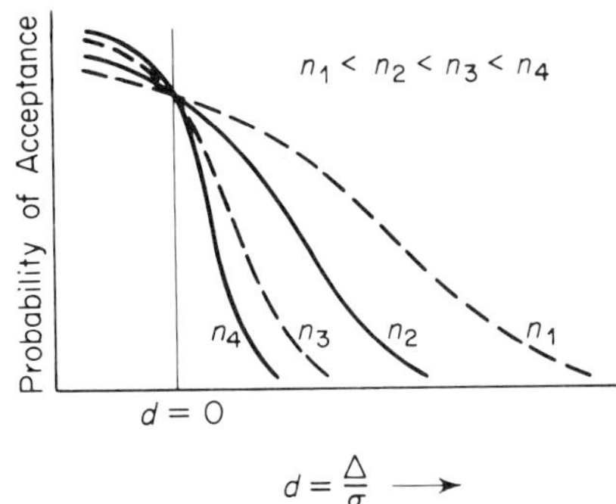

Fig. 13.26 *Family of OC curves of the one-tailed test for the equality of means.*

0 if the true situation is such that the means actually differ; for any given $\Delta > 0$ the probability of accepting will decrease as n increases. To pick a sample size, the experimentalist must pick a difference Δ that is important from the practical point of view and then choose the number of observations which gives him small chance of accepting the hypothesis if the true difference is as great as the practical difference. Very often he must weigh the cost of taking additional observations against the advantage in decreasing the probability of accepting the null hypothesis when it is false.

In practice, there are two major limitations to the use of OC curves. Very often, the number of observations is fixed in advance, by custom, limitation of testing equipment, or indeed the statistical analysis may be of secondary importance based on data taken for another purpose. Even in this case, a look at the OC curve is important, since it gives an idea of the type of differences you are likely to detect and hence an indication of the sensitivity of the analysis. In the second place, it often happens that the OC curve depends on parameters in which you are not interested, since the OC curve for the test of equality of two means depends on the ratio of the difference of the means to the standard deviation. However, even if only the general magnitude of the standard deviation is known, the OC curves are useful in designing experiments. The experimentalist must realize that whenever he picks a sample size he is implicitly picking an OC curve, and the more information he has available in making this decision the better. The OC curves for most of the standard significance tests are presented in this chapter.

All the tests described in this article are based on the assumption that the populations from which we draw samples are normal distributions. This assumption, which underlies a good deal of statistical technique today, is, of course, never exactly true, but has been shown to be approximately true by both empirical and theoretical studies, and the techniques derived on this assumption are adequate for most practical problems. Care should be taken, however, not to apply one-sided tests to very skew distributions.

A summary of the tests discussed in this section can be found in Tables 13.21 and 13.22.

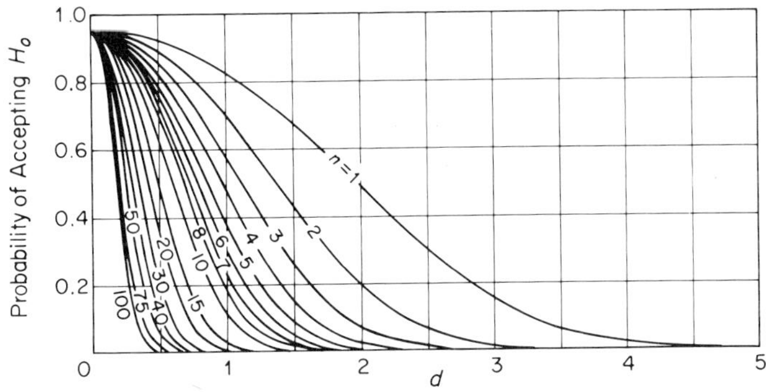

Fig. 13.27 *Operating characteristics of the two-sided normal test for a level of significance equal to 0.05. Reproduced by permission from "Operating Characteristics for the Common Statistical Tests of Significance" by Charles D. Ferris, Frank E. Grubbs, Chalmers L. Weaver,* Annals of Mathematical Statistics, *June, 1946.*

TABLE 13.21. *SUMMARY OF SIGNIFICANCE TESTS: TESTING FOR THE VALUE OF A SPECIFIED PARAMETER*

Hypothesis	Notation for Hypothesis	Qualifying Conditions	Statistic Used in Test	Reference to Detailed Explanation	Rule of Rejection	Tables	Choice of Sample Size
The mean of a normal distribution has a specified value	$\mu = \mu_0$	known σ	$U = \frac{\sqrt{n}(\bar{x} - \mu_0)}{\sigma}$	Normal dist. Art. 13.4.4	$U \geq K_\alpha$ if we wish to reject when the true mean exceeds μ_0.	13.3	Figs. 13.29 and 13.30
					$U \leq -K_\alpha$ if we wish to reject when the true mean is less than μ_0.	13.3	Figs. 13.29 and 13.30
					$\lvert U \rvert \geq K_{\alpha/2}$ if we wish to reject when the true mean departs in either direction from μ_0.	13.3	Figs. 13.27 and 13.28
	$\mu = \mu_0$	unknown σ	$t = \frac{\sqrt{n}(\bar{x} - \mu_0)}{s}$	t Art. 13.4.5	$t \geq t_{\alpha;n-1}$ if we wish to reject when the true mean exceeds μ_0.	13.25	Figs. 13.33 and 13.34
					$t \leq -t_{\alpha;n-1}$ if we wish to reject when the true mean is less than μ_0.	13.25	Figs. 13.33 and 13.34
					$\lvert t \rvert \geq t_{\alpha/2;n-1}$ if we wish to reject when the true mean departs in either direction from μ_0.	13.25	Figs. 13.31 and 13.32
The standard deviation (or variance) of a normal distribution has a specified value	$\sigma = \sigma_0$		$\chi^2 = \frac{\sum_{i=1}^{n} (x_i - \bar{x})^2}{\sigma_0^2}$	χ^2 Art. 13.4.6	$\chi^2 \geq \chi^2_{\alpha;n-1}$ if we wish to reject when the true standard deviation exceeds σ_0.	13.27	Figs. 13.37 and 13.38
					$\chi^2 \leq \chi^2_{1-\alpha;n-1}$ if we wish to reject when the true standard deviation is less than σ_0.	13.27	Figs. 13.39 and 13.40
					$\chi^2 \leq \chi^2_{1-\alpha/2;n-1}$ or $\chi^2 \geq \chi^2_{\alpha/2;n-1}$ if we wish to reject when the standard deviation departs in either direction from σ_0.	13.27	Figs. 13.35 and 13.36

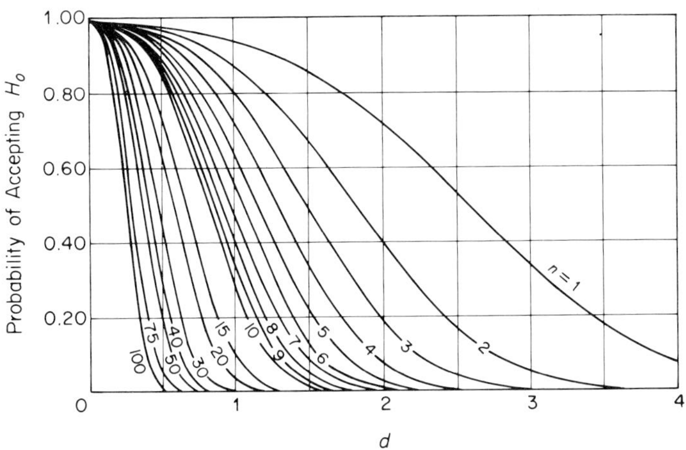

Fig. 13.28 *Operating characteristics of the two-sided normal test for a level of significance equal to 0.01.*

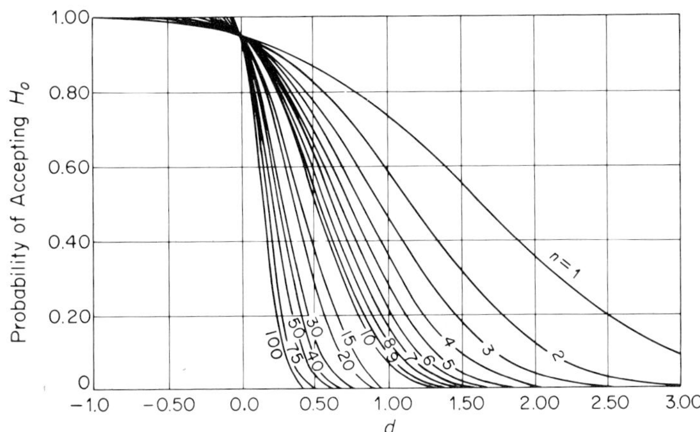

Fig. 13.29 *Operating characteristics of the one-sided normal test for a level of significance equal to 0.05.*

13.4.3 Notation

The notation used in the following discussion will be the standard statistical notation found in most statistics texts. The observations x_1, x_2, x_3, . . . , x_n will be assumed to be drawn from a normal distribution with mean μ and variance σ^2.

$$\bar{x} = \frac{x_1 + x_2 + \cdots + x_n}{n} = \frac{\sum_{i=1}^{n} x_i}{n}$$

is an estimate of μ, and

$$s^2 = \sum_{i=1}^{n} \frac{(x_i - \bar{x})^2}{n - 1}$$

$$= \frac{(x_1 - \bar{x})^2 + (x_2 - \bar{x})^2 + \cdots + (x_n - \bar{x})^2}{n - 1}$$

is an estimate of σ^2. The best computational form for s^2 is given by

$$s^2 = \frac{x_1^2 + x_2^2 + \cdots + x_n^2 - n\bar{x}^2}{n - 1}$$

Here α is the level of significance, and K_α, $t_{\alpha;\upsilon}$, $\chi^2{}_{\alpha;\upsilon}$, and $F_{\alpha;\upsilon 1,\upsilon 2}$ are respectively the α

TABLE 13.22. *SUMMARY OF SIGNIFICANCE TESTS: COMPARISON OF TWO POPULATIONS*

Hypothesis	Notation for Hypothesis	Qualifying Conditions	Statistic Used in Test
The means of two normal distributions are equal	$\mu_x = \mu_y$	Known standard deviations σ_x, σ_y	$U = \dfrac{\bar{x} - \bar{y}}{\sqrt{\sigma_x^2/n_x + \sigma_y^2/n_y}}$
The means of two normal distributions are equal	$\mu_x = \mu_y$	Unknown standard deviations with $\sigma_x = \sigma_y$	$t = \dfrac{\bar{x} - \bar{y}}{\sqrt{\dfrac{1}{n_x} + \dfrac{1}{n_y}}\sqrt{\dfrac{\sum_{i=1}^{n_x} (x_i - \bar{x})^2 + \sum_{i=1}^{n_y} (y_i - \bar{y})^2}{n_x + n_y - 2}}}$
The means of two normal distributions are equal	$\mu_x = \mu_y$	Unknown standard deviations with σ_x and σ_y not necessarily equal	$t' = \dfrac{\bar{x} - \bar{y}}{\sqrt{s_x^2/n_x + s_y^2/n_y}}$
The variance of two normal distributions are equal	$\sigma_x^2 = \sigma_y^2$		$F = \dfrac{s_x^2}{s_y^2} = \dfrac{\sum_{i=1}^{n_x} (x_i - \bar{x})^2/(n_x - 1)}{\sum_{i=1}^{n_y} (y_1 - \bar{y})^2/(n_y - 1)}$ In a two-sided test, put the larger mean square in the numerator

percentage points of the normal distribution, the α percentage point of student's t distribution with ν degrees of freedom, the α percentage point of the chi-square distribution with ν degrees of freedom, and the α percentage point of the F distribution with ν_1 and ν_2 degrees of freedom.

13.4.4. Test of the Hypothesis That the Mean of a Normal Distribution Has a Specified Value When the Standard Deviation Is Known (Table 13.23)

Percentage points of the normal distribution

Level of Significance	For One-Sided Tests	For Two-Sided Tests
$\alpha = 0.05$	$K_{0.05} = 1.645$	$K_{0.025} = 1.960$
$\alpha = 0.01$	$K_{0.01} = 2.326$	$K_{0.005} = 2.576$

13.4.4.1 Example. A manufacturer produces a special alloy steel with an average tensile strength of 125,800 psi. A change in the composition of the alloy is said to increase the breaking strength. A sample of 20 items of the new material is tested, and the average tensile strength is found to be 127,000. The standard deviation of the tensile strength is known to be 300 psi. To decide whether to adopt the new alloy, we test the hypothesis that the breaking strength is unchanged, i.e., that the new value could have arisen by chance from a population with true mean of 125,800. The statistic U is calculated:

$$U = \frac{\sqrt{n}\,(\bar{x} - \mu_0)}{\sigma}$$

$$= \frac{\sqrt{20}\,(127{,}000 - 125{,}800)}{300} = 17.888$$

Since we are interested in rejecting the hypothesis that the breaking strength is unchanged only if it is increased, we use a one-tailed test. As our observed value of U exceeds the critical value for the 1 per cent level $K_{.01} = 2.326$, we reject

TABLE 13.22. *Cont.*

Reference to Detailed Explanation	Rule of Rejection	Table	Choice of Sample Size
Art. 13.4.7	$U \geq K_\alpha$ if we wish to reject when $\mu_x > \mu_y$.	13.3	Figs. 13.29 and 13.30
	$U \leq -K_\alpha$ if we wish to reject when $\mu_x < \mu_y$.	13.3	Figs. 13.29 and 13.30
	$\lvert U\rvert \geq K_{\alpha/2}$ if we wish to reject whenever the means differ.	13.3	Figs. 13.27 and 13.28
Art.13.4.8	$t \geq t_{\alpha;n+n_y-2}$ if we wish to reject when $\mu_x > \mu_y$.	13.25	Figs. 13.33 and 13.34
	$t \leq -t_{\alpha;n_x+n_y-2}$ if we wish to reject when $\mu_x < \mu_y$.	13.25	Figs. 13.33 and 13.34
	$\lvert t\rvert \geq t_{\alpha/2;n_x+n_y-2}$ if we wish to reject whenever $\mu_x \neq \mu_y$.	13.25	Figs. 13.31 and 13.32
Art.13.4.9	$t' \geq t_{\alpha;v}$ if we wish to reject when $\mu_x > \mu_y$. $t' \leq -t_{\alpha;v}$ if we wish to reject when $\mu_x < \mu_y$. $\lvert t'\rvert \geq t_{\alpha/2;v}$ when we wish to reject when $\mu_x \neq \mu_y$. where the degrees of freedom v is given by the closest integer to $v = -2 + \left(\frac{s_x^2}{n_x} + \frac{s_y^2}{n_x}\right)^2 / [(s_x^2/n_x)^2/(n_x+1) + (s_y^2/n_y)^2/(n_y+1)]$.	13.25 13.25 13.25	There are no results available
Art. 13.4.10	$F \geq F_{\alpha;n_x-1,n_y-1}$ if we wish to reject when $\sigma_x > \sigma_y$.	13.32	Figs. 13.43 and 13.44
	$F \geq F_{\alpha/2;n_x-1,n_y-1}$ if $s_x^2 > s_y^2$ and we wish to reject whenever $\sigma_x \neq \sigma_y$.	13.32	Figs. 13.41 and 13.42
	$F \geq F_{\alpha/2;n_y-1,n_x-1}$ if $s_x^2 < s_y^2$ and we wish to reject whenever $\sigma_x \neq \sigma_y$.	13.32	Figs. 13.41 and 13.42

TABLE 13.23. *OUTLINE OF TEST*

Notation for Hypothesis	Test Statistic
$\mu = \mu_0$	$U = \sqrt{n}\,(\bar{x} - \mu_0)/\sigma$
Rule of Rejection	Rule for Choosing Sample Size
$U \geq K_\alpha$ if we wish to reject when the true mean exceeds μ_0.	Select a value $\mu_1 > \mu_0$ for which we want to reject the null hypothesis with high probability, and calculate $d = (\mu_1 - \mu_0)/\sigma$, and select n from the OC curve of Fig. 13.9 or 13.30.
$U \leq -K_\alpha$ if we wish to reject when the true mean is less than μ_0.	Choose a value $\mu_1 < \mu_0$ which we want to reject and enter Fig. 13.29 or 13.30 with $d = (\mu_0 - \mu_1)/\sigma$.
$\lvert U\rvert \geq K_{\alpha/2}$ if we wish to reject when the true mean departs from μ_0 in either direction.	Select a value Δ that is a departure from μ_0 we want to reject with high probability and enter Fig. 13.27 or 13.28 with $d = \lvert\Delta\rvert/\sigma$.

the hypothesis and conclude that the new alloy produces greater tensile strength than the old.

In this case, we based our decision on a sample of 20 observations; the sample size was picked arbitrarily. However, the operating characteristic curves of the U test in Fig. 13.30 (for 0.01 level of significance) provide an objective basis for the selection of a sample

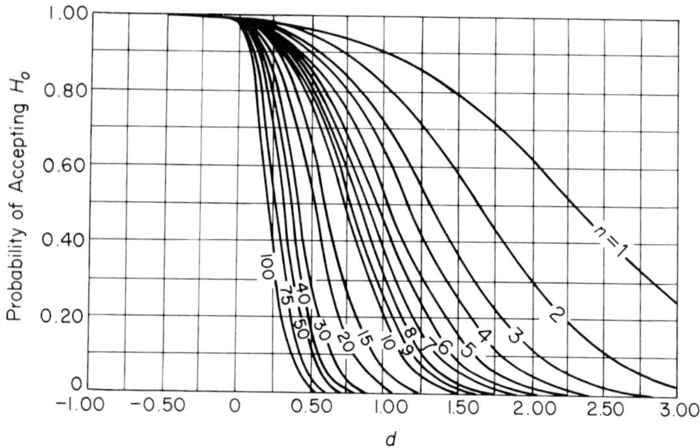

Fig. 13.30 *Operating characteristics of the one-sided normal test for a level of significance equal to 0.01.*

size. These curves give the probability of accepting the hypothesis as a function of the difference between the true mean and the hypothesized one, in σ units; e.g., for a sample size of 20 the probability of accepting is about 0.10 for a difference of $0.80\sigma = 240$. Ordinarily, we would use this graph before running the experiment and select the sample size which will have a high probability of detecting a difference that is important from a practical point of view.

13.4.5 Test of the Hypothesis That the Mean of a Normal Distribution Has a Specified Value When the Standard Deviation Is Unknown (Table 13.24)

13.4.5.1 Example. In the manufacture of a food product, the label states that the box contains 10 lb. The boxes are filled by machine, and it is of interest to determine whether or not the machine is set properly. Previous experience has indicated that the standard deviation is approximately 0.05 lb, although this is not known precisely. The company feels that the present setting is unsatisfactory if the machine fills the boxes so that the mean weight differs from 10 lb by more than 0.1 lb. From Fig. 13.31 and $d = 0.1/0.05 = 2$, a sample size of 6 is sufficient to ensure rejection with probability 0.95 if the mean weight differs from 10 lb by more than 0.1 lb. The data are: 9.99 lb, 9.99 lb, 10.00 lb, 10.11 lb, 10.09 lb, 9.95 lb.

$$\bar{x} = 10.021667^*$$

$$s^2 = \frac{\sum(x_i - \bar{x})^2}{n-1} = \frac{\sum x_i^2 - n\bar{x}^2}{n-1}$$

$$= \frac{602.6229 - 602.6029}{5} = 0.0040$$

$$s = 0.0632$$

$$\frac{\sqrt{n}(\bar{x} - 10)}{s} = \frac{\sqrt{6}(10.0217 - 10)}{0.0632} = 0.84$$

$t_{.025;5} = 2.571$ for 5 degrees of freedom (from Table 13.25)

Thus $|t| \leq t_{.025;5}$ so that it is unnecessary to change the machine setting.

13.4.6 Test of the Hypothesis That the Variance of a Normal Distribution Has a Specified Value (Table 13.26)

Percentage points of the chi-square distribution can be found in Table 13.27.

13.4.6.1 Example. The standard deviation of a dimension of standard product is $\sigma =$

---

* It is important to compute $\bar{x}$ to more places than appears necessary for the significance of the final result. The reason becomes clear upon looking at the calculation of s^2 where a subtraction operation is involved.

TABLE 13.24. *OUTLINE FOR TEST*

Notation for Hypothesis	Test Statistic
$\mu = \mu_0$	$t = \sqrt{n}\,(\bar{x} - \mu_0)/s$
Rule of Rejection	**Rule for Choosing Sample Sizes**
$t \geq t_{\alpha;n-1}$ if we wish to reject when the true mean exceeds μ_0.	Choose a value of $d = (\mu_1 - \mu_0)/\sigma$† for which we wish to reject the null hypothesis with high probability and enter Fig. 13.33 or 13.34 to find the necessary sample size.
$t \leq -t_{\alpha;n-1}$ if we wish to reject when the true mean is less than μ_0.	Choose a value of $d = (\mu_0 - \mu_1)/\sigma$ for which we wish to reject the null hypothesis with high probability and enter Fig. 13.33 or 13.34 to find the necessary sample size.
$\lvert t \rvert \geq t_{\alpha/2;n-1}$ if we wish to reject when the true mean differs from μ_0 in either direction.	Choose a value of $d = (\mu_1 - \mu_0)/\sigma$ for which we wish to reject the null hypothesis with high probability and enter Fig. 13.31 or 13.32 to find the necessary sample size.

† The operating characteristic curve of the t test, unlike the U test, depends on the ratio of $\mu_0 - \mu_1$ to σ, although the test procedure is independent of σ, as is the level of significance. Usually a rough idea of the magnitude of σ is sufficient for choosing the necessary sample size.

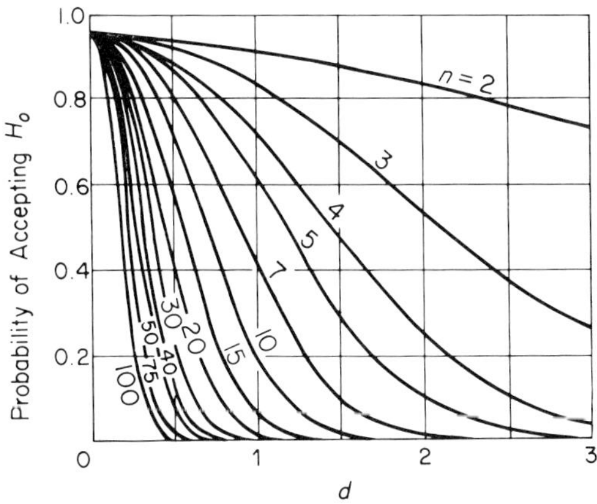

Fig. 13.31 *Operating characteristics of the two-sided* t *test for a level of significance equal to 0.05. Reproduced by permission from "Operating Characteristics for the Common Statistical Tests of Significance" by Charles D. Ferris, Frank E. Grubbs, Chalmers L. Weaver,* Annals of Mathematical Statistics, *June, 1946.*

0.1225 in. A new product is under consideration and will be adopted if the variability of this dimension is not larger. The decision about variability will be based on a sample of items of the new product. If σ is as large as 0.2450 in., we want to be quite sure to reject. From Fig. 13.38 it appears that a sample size of 25 will give a probability of acceptance of 0.01 for $\lambda = 0.2450/0.1225 = 2$ if we make the test at the 1 per cent level. A sample of twenty-five items is drawn and s^2 is found to be 0.0384. Hence $\chi^2 = (n-1)s^2/\sigma_0^2 = 61.4$, which exceeds $\chi^2_{.01;24} = 43.0$. The new product is not adopted.

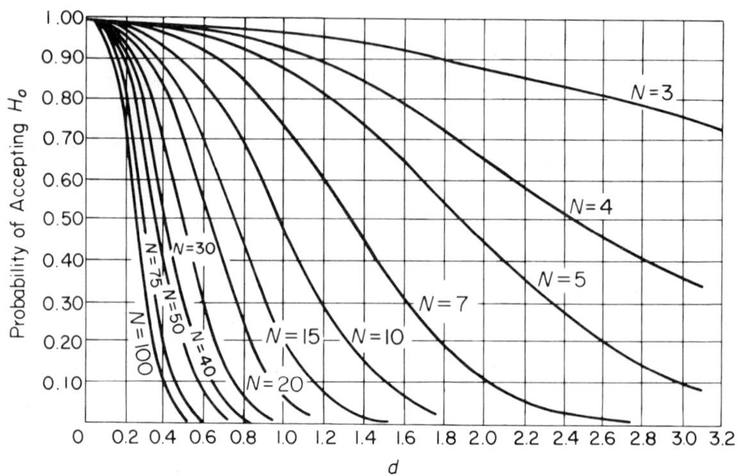

Fig. 13.32 *Operating characteristics of the two-sided* t *test for a level of significance equal to 0.01.*

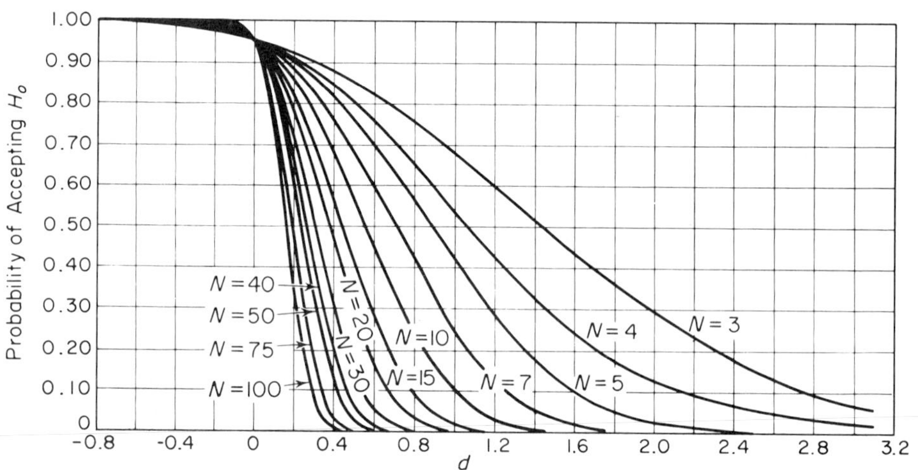

Fig. 13.33 *Operating characteristics of the one-sided* t *test for a level of significance equal to 0.05.*

13.4.7 Test of the Hypothesis That the Means of Two Normal Distributions Are Equal When Both Standard Deviations Are Known (Table 13.28)

Percentage points of the normal distribution

Level of Significance	For One-Sided Tests	For Two-Sided Tests
$\alpha = 0.05$	$K_{0.05} = 1.645$	$K_{0.025} = 1.960$
$\alpha = 0.01$	$K_{0.01} = 2.326$	$K_{0.005} = 2.576$

13.4.7.1 Example. There is evidence that surface finish has an effect on the endurance limit (reversed bending) of steel. An experiment is performed on 0.4 per cent carbon steel, using both unpolished and polished smooth-turned specimens. The finish on the smooth-turned polished specimens was obtained by polishing with No. 0 and 00 emery cloth.

For 0.4 per cent carbon steel, it has been estimated that an average increase in the endurance limit of approximately 7500 psi should be detected on the polished specimens. Furthermore, polishing should not have any effect on the standard deviation of the endurance limit, which is known from the performance of

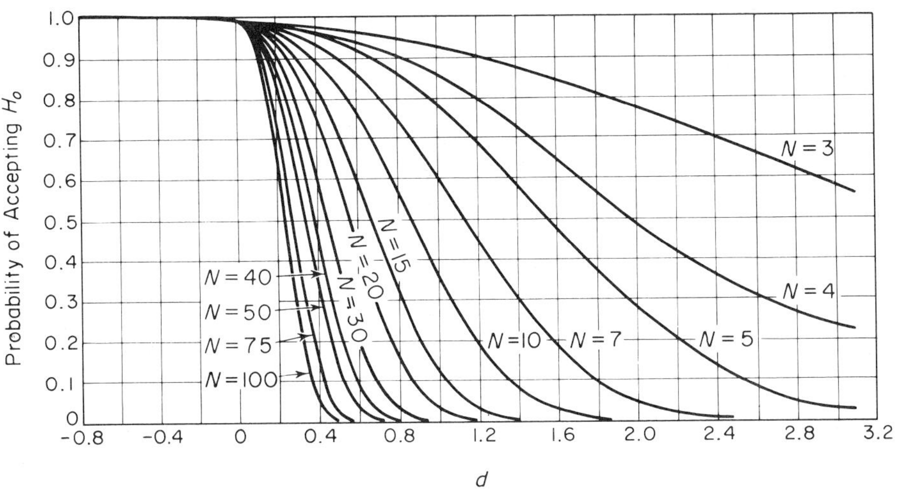

Fig. 13.34 *Operating characteristics of the one-sided* t *test for a level of significance equal to 0.01.*

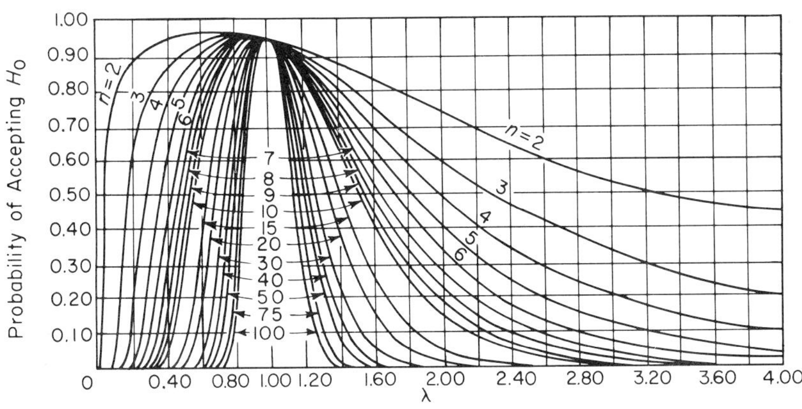

Fig. 13.35 *Operating characteristics of the two-sided chi-square test for a level of significance equal to 0.05.*

numerous endurance limit experiments to be 4000 psi. From Fig. 13.28 (1 per cent level of significance) with

$$d = 7500/\sqrt{(4000)^2 + (4000)^2} = 1.33$$

we find that about eight observations on each group of specimens are required to have a probability of 0.9 of detecting a difference as large as 7500 psi. The data are as follows:

Endurance Limit for Polished 0.4 per cent Carbon Steel x	Endurance Limit for Unpolished 0.4 per cent Carbon Steel y
86,500	82,600
91,900	82,400
89,400	81,700
84,000	79,500
89,900	79,400
78,700	69,800
87,500	79,900
83,100	83,400
$\bar{x}$=86,375	$\bar{y}$=79,838

$$U = \frac{6537}{\sqrt{(4000)^2/8 + (4000)^2/8}} = 3.27$$

$$K_{.01} = 2.326$$

Hence we reject the hypothesis at the 1 per cent significance level that surface polish has no effect on the endurance limit, and conclude that

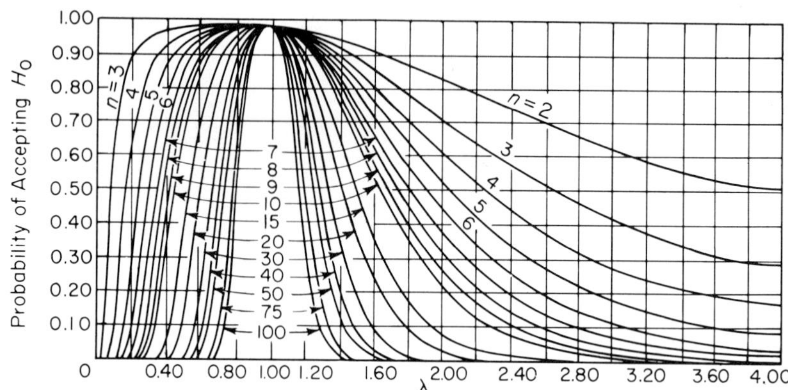

Fig. 13.36 *Operating characteristics of the two-sided chi-square test for a level of significance equal to 0.01.*

polished 0.4 per cent carbon steel specimens have a higher mean endurance limit than unpolished ones.

13.4.8 Test of the Hypothesis That the Means of Two Normal Distributions Are Equal Assuming That the Standard Deviations Are Unknown But Equal; Two-Sample t Test (Table 13.29)

Example. A manufacturer of electric irons produces these items in two plants. Both plants have the same suppliers of small parts. A saving can be made by purchasing thermostats for plant B from a local supplier. A single lot was purchased from the local supplier and it was desired to test whether or not these new thermostats were as accurate as the old. It was decided to test the irons on the 550°F setting, and the actual temperatures were to be read to the nearest 0.1 deg with a thermocouple. With the old supplier, very few complaints were received, and the manufacturer feels that the switch should not be made if the average temperature changes by more than 10.5°. The order of magnitude of the standard deviation is roughly 10° for the old supplier, and there is no reason to suspect that it will be different for the new supplier. For $d = 10.5/(2 \times 10) = 0.525$, and from Fig. 13.31, $n' = 45$, corresponding to a probability of 0.9 of detecting a change of 10.5 deg or more. Hence $n = 23$. The data are:

New Supplier x	(Degrees F)
530.3	559.1
559.3	555.0
549.4	538.6
544.0	551.1
551.7	565.4
566.3	554.9
549.9	550.0
556.9	554.9
536.7	544.7
558.8	536.1
538.8	569.1
	543.3

Old Supplier y	(Degrees F)
559.7	564.6
534.7	554.5
554.8	553.0
545.0	538.4
544.6	548.3
538.0	552.9
550.7	535.1
563.1	555.0
551.1	544.8
553.8	558.4
538.8	548.7
	560.3

$$\sum x = 12{,}684.3 \qquad \sum y = 12{,}648.3$$
$$\bar{x} = 551.491304 \qquad \bar{y} = 549.926086$$
$$\sum x^2 = 6{,}997{,}436.07 \qquad \sum y^2 = 6{,}957{,}333.23$$
$$\sum(x_i - \bar{x})^2 = \sum x_i^2 - n\bar{x}^2 = 2{,}154.930$$
$$\sum(y_i - \bar{y})^2 = \sum y_i^2 - n\bar{y}^2 = 1{,}703.130$$
$$\frac{\sum(x_i - \bar{x})^2 + \sum(y_i - \bar{y})^2}{n_1 + n_2 - 2} = \frac{3858.06}{44}$$
$$= 87.68318$$
$$|t| = \frac{551.491304 - 549.926086}{\sqrt{2/23}\sqrt{87.68318}} = 0.567$$
$$\leq t_{\alpha/2} = 2.02$$

Therefore we accept the hypothesis at the 5 per cent level of significance that there is no difference in the mean temperatures of the two suppliers.

13.4.8.1 Procedure Based on Correlated Samples. An important special case of the t test arises when the observations are "paired," each pair being taken under the same experimental conditions, with the conditions varying from pair to pair. In this case, the test is made on the differences of the observations, and is carried out as a one sample problem, testing the hypothesis that the mean of the difference is zero according to the procedure of Par. 13.4.5, where the observations now consist of the differences.

Example. A research laboratory is embarking on an experiment to determine whether or not outside temperature has any effect on the accuracy of a Bourdon gage. The experiment consists of placing the gage in a steam line which is completely insulated from room temperature. The line passes through two adjacent rooms. One room always remains at 32°F, whereas the temperature in the other can be varied. Since there is a time lag in changing the temperature of the controlled room, it is likely that the steam pressure will fluctuate during this time interval. Consequently, no effort is made to maintain the same pressure for a period of time. It can be assumed that the pressure is the same at the stations in each room and will be kept at approximately 60 psi. The experiment should be designed such that a difference as great as 0.75 psi should be picked up with high probability, say 0.9. The order of magnitude of the standard deviation is 0.4 psi. This is obtained as follows: The manufacturer usually specifies that these gages are good to ± 2 per cent. If we interpret this to mean that most of the observations will lie between ± 2 per cent of the true value, and in particular, if "most" means 0.9987, we have true value ± 2 per cent $=$ true value $\pm 3\sigma$. If the true value is near 60 psi, we have $3\sigma = 1.2$ or $\sigma = 0.4$. Hence the standard deviation of the differences is $(\sqrt{2})(0.4) = 0.57$. It must be emphasized that this is just an estimate of the order of magnitude of the standard deviation and should be used only for determining sample size. Referring to Fig. 13.31, $d = (0.75)/(\sqrt{2})(0.4) = 1.3$, so that eight observations are required. The data are:

Pressure Read at Room Temperature = 32 degrees x	Pressure Read at Other Room Temperature y	$x-y$	$(x-y)^2$
60.3	61.0	−0.7	0.49
59.9	60.0	−0.1	0.01
59.5	60.9	−1.4	1.96
58.5	58.2	0.3	0.09
59.6	59.9	−0.3	0.09
59.0	60.4	−1.4	1.96
60.1	59.6	0.5	0.25
59.4	59.9	−0.5	0.25

$$s^2 = \frac{5.10 - 8(0.450)^2}{7} = 0.497$$

$$\sum(x-y) = -3.6$$

$$\sum(x-y)^2 = 5.10$$

$$s = 0.705 \qquad |t| = \left|\frac{-0.45\sqrt{8}}{0.705}\right|$$

$$= \frac{1.406}{0.705} = 1.99 \le t_{\alpha/2} = 2.365$$

Therefore the data do not reveal any significant difference between gage readings when the temperature changes.

13.4.9 Test of the Hypothesis That the Means of Two Normal Distributions Are Equal When the Standard Deviations Are Unknown and Not Necessarily Equal (Table 13.30)

13.4.9.1 Example. A manufacturer of automobile crankshafts was troubled with the problem of bend in the final shaft. Bend may be caused by the length and weight of the shaft, by improper setup of machine tools, by the heat-treating process, or by some combination of these causes. It is suspected that nitriding the shaft, i.e., the process that hardens the surface of the shaft by a heat and nitrous oxide reaction with the steel, is the main cause for bend. Twenty-five shafts were measured before nitriding at the front main center journal by a dial indicator gage accurate to the 0.0001 in. Similarly, another 25 shafts were measured at the same spot after the nitriding operation. It cannot be assumed that the variability in the bend is the same before and after nitriding. We test the hypothesis that the mean value of the bend is the same before and after nitriding, wishing to reject if the average bend after nitriding is larger. Denoting by x the values before nitriding and by y the values after nitriding, the following results were obtained:

$$\sum_{i=1}^{25} x = 203 \times 10^{-4}$$

$$\bar{x} = 0.0008120$$

$$\sum_{i=1}^{25} x^2 = 2{,}933 \times 10^{-8}$$

$$\sum_{i=1}^{25} y = 453 \times 10^{-4}$$

$$\bar{y} = 0.001812$$

$$\sum_{i=1}^{25} y^2 = 14{,}362 \times 10^{-8}$$

$$s_x^2 = \frac{2{,}933 \times 10^{-8} - 25(0.0008120)^2}{24}$$

$$= 53.5 \times 10^{-8}$$

$$\frac{s_x^2}{n_x} = 2.14 \times 10^{-8}$$

$$s_y^2 = \frac{14{,}362 \times 10^{-8} - 25(0.001812)^2}{24}$$

$$= 256 \times 10^{-8}$$

$$\frac{s_y^2}{n_y} = 10.24 \times 10^{-8}$$

$$\nu = \frac{[(2.14)10^{-8} + (10.24)10^{-8}]^2}{(2.14)^2(10^{-16})/26 + (10.24)^2(10^{-16})/26} - 2 = 34$$

$$t' = \frac{0.0008120 - 0.001812}{\sqrt{(2.14)10^{-8} + (10.24)10^{-8}}}$$

$$= -\frac{0.001}{(3.52)10^{-4}} = -2.84$$

$$t_{.05;34} = 1.69$$

Since $t' \leq t_{\alpha;\upsilon}$ we reject the hypothesis that at the 5 per cent significance level the means are the same before and after nitriding, and we conclude that the average bend is larger after nitriding.

TABLE 13.25. *PERCENTAGE POINTS OF THE* t *DISTRIBUTION.*† *TABLE OF* $t_{\alpha;\nu}$*—THE 100* α *PERCENTAGE POINT OF THE* t *DISTRIBUTION FOR* ν *DEGREES OF FREEDOM.*

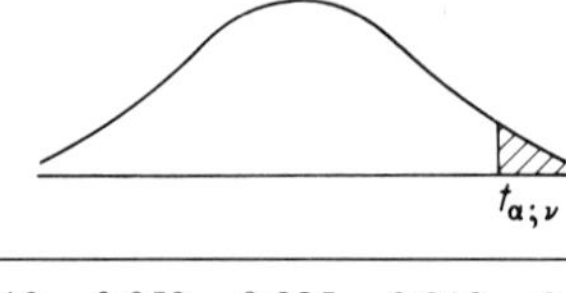

ν \ α	0.10	0.050	0.025	0.010	0.005	α / ν
1	3.078	6.314	12.706	31.821	63.657	1
2	1.886	2.920	4.303	6.965	9.925	2
3	1.638	2.353	3.182	4.541	5.841	3
4	1.533	2.132	2.776	3.747	4.604	4
5	1.476	2.015	2.571	3.365	4.032	5
6	1.440	1.943	2.447	3.143	3.707	6
7	1.415	1.895	2.365	2.998	3.499	7
8	1.397	1.860	2.306	2.896	3.355	8
9	1.383	1.833	2.262	2.821	3.250	9
10	1.372	1.812	2.228	2.764	3.169	10
11	1.363	1.796	2.201	2.718	3.106	11
12	1.356	1.782	2.179	2.681	3.055	12
13	1.350	1.771	2.160	2.650	3.012	13
14	1.345	1.761	2.145	2.624	2.977	14
15	1.341	1.753	2.131	2.602	2.947	15
16	1.337	1.746	2.120	2.583	2.921	16
17	1.333	1.740	2.110	2.567	2.898	17
18	1.330	1.734	2.101	2.552	2.878	18
19	1.328	1.729	2.093	2.539	2.861	19
20	1.325	1.725	2.086	2.528	2.845	20
21	1.323	1.721	2.080	2.518	2.831	21
22	1.321	1.717	2.074	2.508	2.819	22
23	1.319	1.714	2.069	2.500	2.807	23
24	1.318	1.711	2.064	2.492	2.797	24
25	1.316	1.708	2.060	2.485	2.787	25
26	1.315	1.706	2.056	2.479	2.779	26
27	1.314	1.703	2.052	2.473	2.771	27
28	1.313	1.701	2.048	2.467	2.763	28
29	1.311	1.699	2.045	2.462	2.756	29
Inf.	1.282	1.645	1.960	2.326	2.576	Inf.

† Reproduced from J. E. Freund, *Modern Elementary Statistics* (New York: Prentice-Hall, Inc., 1952), Table II, p. 390, Control Values of *t*. This table is abridged from Table IV of R. A. Fisher, *Statistical Methods for Research Workers*, published by Oliver & Boyd Ltd., Edinburgh, by permission of the author and publishers.

13.4.10 Test of the Hypothesis That the Variances of Two Normal Distributions Are Equal (Table 13.31)

Percentage points of the F distribution may be found in Tables 13.32–13.37.

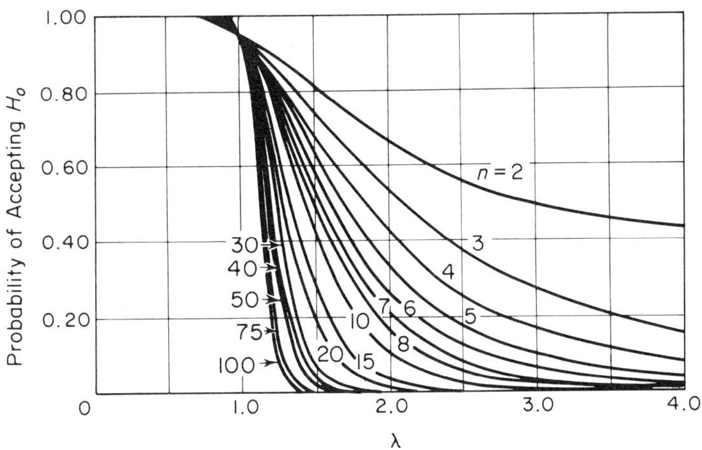

Fig. 13.37 *Operating characteristics of the one-sided (upper-tail) chi-square test for a level of significance equal to 0.05. Reproduced by permission from "Operating Characteristics for the Common Statistical Tests of Significance" by Charles D. Ferris, Frank E. Grubbs, Chalmers L. Weaver,* Annals of Mathematical Statistics, *June, 1946.*

13.4.10.1 Example. The standard deviation of a particular dimension of a metal component is small enough so that it is satisfactory in subsequent assembly; a new supplier of metal plate is under consideration and will be preferred if the standard deviation of his product is not larger, as the cost of his product is lower than that of the present supplier. From Fig. 13.43, it is decided to base this decision on a sample of 100 items from each supplier, since it is desired to ensure that the probability of switching to the new product is less than 0.02 if $\lambda \geq 1.5$; data on dimensions are relatively easy to obtain, and small numbers of observations give relatively little protection against erroneous decisions. The following data are obtained:

New supplier: $s^2 = 0.00041$
Current supplier: $s^2 = 0.00057$

$$F = \frac{(0.00041)}{(0.00057)} < F_{.05;99,99}$$

TABLE 13.26. *OUTLINE OF TEST*

Notation for Hypothesis	Test Statistic
$\sigma = \sigma_0$	$\chi^2 = (n-1)s^2/\sigma_0^2$
Rule of Rejection	Rule for Choosing Sample Sizes
$\chi^2 \geq \chi^2_{\alpha;n-1}$ if we wish to reject when the true standard deviation exceeds σ_0.	Select a value $\sigma_1 > \sigma_0$ for which we wish to reject the null hypothesis with high probability. Calculate $\lambda = \sigma_1/\sigma_0$ and enter Fig. 13.37 or 13.38 to find the required sample size.
$\chi^2 \leq \chi^2_{1-\alpha;n-1}$ if we wish to reject when the true standard deviation is less than σ_0.	Select a value $\sigma_1 < \sigma_0$ for which we wish to reject the null hypothesis with high probability. Calculate $\lambda = \sigma_1/\sigma_0$ and enter Fig. 13.39 or 13.40 to find the required sample size.
$\chi^2 \geq \chi^2_{\alpha/2;n-1}$ or $\chi^2 \leq \chi^2_{1-\alpha/2;n-1}$ if we wish to reject when the true standard deviation differs from σ_0 in either direction.	Select a value of the relative error σ_1/σ_0 for which we wish to reject the null hypothesis with high probability. Enter Fig. 13.35 or 13.36 to find the required sample size.

TABLE 13.27. *PERCENTAGE POINTS OF THE χ^2 DISTRIBUTION, TABLE OF $\chi^2_{\alpha;\nu}$—THE 100 α PERCENTAGE POINT OF THE χ^2 DISTRIBUTION FOR ν DEGREES OF FREEDOM.*

ν \ α	0.995	0.99	0.98	0.975	0.95	0.90	0.80	0.75	0.70	0.50
1	0.0^4393	0.0^3157	0.0^3628	0.0^3982	0.00393	0.0158	0.0642	0.102	0.148	0.455
2	0.0100	0.0201	0.0404	0.0506	0.103	0.211	0.446	0.575	0.713	1.386
3	0.0717	0.115	0.185	0.216	0.352	0.584	1.005	1.213	1.424	2.366
4	0.207	0.297	0.429	0.484	0.711	1.064	1.649	1.923	2.195	3.357
5	0.412	0.554	0.752	0.831	1.145	1.610	2.343	2.675	3.000	4.351
6	0.676	0.872	1.134	1.237	1.635	2.204	3.070	3.455	3.828	5.348
7	0.989	1.239	1.564	1.690	2.167	2.833	3.822	4.255	4.671	6.346
8	1.344	1.646	2.032	2.180	2.733	3.490	4.594	5.071	5.527	7.344
9	1.735	2.088	2.532	2.700	3.325	4.168	5.380	5.899	6.393	8.343
10	2.156	2.558	3.059	3.247	3.940	4.865	6.179	6.737	7.267	9.342
11	2.603	3.053	3.609	3.816	4.575	5.578	6.989	7.584	8.148	10.341
12	3.074	3.571	4.178	4.404	5.226	6.304	7.807	8.438	9.034	11.340
13	3.565	4.107	4.765	5.009	5.892	7.042	8.634	9.299	9.926	12.340
14	4.075	4.660	5.368	5.629	6.571	7.790	9.467	10.165	10.821	13.339
15	4.601	5.229	5.985	6.262	7.261	8.547	10.307	11.036	11.721	14.339
16	5.142	5.812	6.614	6.908	7.962	9.312	11.152	11.912	12.624	15.338
17	5.697	6.408	7.255	7.564	8.672	10.085	12.002	12.792	13.531	16.338
18	6.265	7.015	7.906	8.231	9.390	10.865	12.857	13.675	14.440	17.338
19	6.844	7.633	8.567	8.907	10.117	11.651	13.716	14.562	15.352	18.338
20	7.434	8.260	9.237	9.591	10.851	12.443	14.578	15.452	16.266	19.337
21	8.034	8.897	9.915	10.283	11.591	13.240	15.445	16.344	17.182	20.337
22	8.643	9.542	10.600	10.982	12.338	14.041	16.314	17.240	18.101	21.337
23	9.260	10.196	11.293	11.688	13.091	14.848	17.187	18.137	19.021	22.337
24	9.886	10.856	11.992	12.401	13.848	15.659	18.062	19.037	19.943	23.337
25	10.520	11.524	12.697	13.120	14.611	16.473	18.940	19.939	20.867	24.337
26	11.160	12.198	13.409	13.844	15.379	17.292	19.820	20.843	21.792	25.336
27	11.808	12.879	14.125	14.573	16.151	18.114	20.703	21.749	22.719	26.336
28	12.461	13.565	14.847	15.308	16.928	18.939	21.588	22.657	23.647	27.336
29	13.121	14.256	15.574	16.047	17.708	19.768	22.475	23.567	24.577	28.336
30	13.787	14.953	16.306	16.791	18.493	20.599	23.364	24.478	25.508	29.336

For values of $\nu > 30$, approximate values for χ^2 may be obtained from the expression $\nu\left[1 - \frac{2}{9\nu} \pm \frac{x}{\sigma}\sqrt{\frac{2}{9\nu}}\right]^3$ where $\frac{x}{\sigma}$ is the normal deviate cutting off the corresponding tails of a normal distribution. If $\frac{x}{\sigma}$ is taken at the 0.02 level, so that 0.01 of the normal distribution is in each tail, the expression yields χ^2 at the 0.99 and 0.01 points. For very large values of ν, it is sufficiently accurate to compute $\sqrt{2\chi^2}$, the distribution of which is approximately normal around a mean of $\sqrt{2\nu - 1}$ and with a standard deviation of 1.

This table is reprinted from *Statistical Tables for Biological, Agricultural, and Medical Research*, by R. A. Fisher and F. Gates, published by Oliver and Boyd, Edinburgh, and from *Biometrika*, Vol. **32**, Part II, October 1941, pp. 187–191, "Table of Percentage Points of the χ^2 Distribution," by Catherine M. Thompson. Reproduced in Croxton, *Elementary Statistics with Applications in Medicine*, Appendix VI, pp. 328–329, Values of χ^2.

TABLE 13.27. *Cont.*

0.30	0.25	0.20	0.10	0.05	0.025	0.02	0.01	0.005	0.001	α / ν
1.074	1.323	1.642	2.706	3.841	5.024	5.412	6.635	7.879	10.827	1
2.408	2.773	3.219	4.605	5.991	7.378	7.824	9.210	10.597	13.815	2
3.665	4.108	4.642	6.251	7.815	9.348	9.837	11.345	12.838	16.268	3
4.878	5.385	5.989	7.779	9.488	11.143	11.668	13.277	14.860	18.465	4
6.064	6.626	7.289	9.236	11.070	12.832	13.388	15.086	16.750	20.517	5
7.231	7.841	8.558	10.645	12.592	14.449	15.033	16.812	18.548	22.457	6
8.383	9.037	9.803	12.017	14.067	16.013	16.622	18.475	20.278	24.322	7
9.524	10.219	11.030	13.362	15.507	17.535	18.168	20.090	21.955	26.125	8
10.656	11.389	12.242	14.684	16.919	19.023	19.679	21.666	23.589	27.877	9
11.781	12.549	13.442	15.987	18.307	20.483	21.161	23.209	25.188	29.588	10
12.899	13.701	14.631	17.275	19.675	21.920	22.618	24.725	26.757	31.264	11
14.011	14.845	15.812	18.549	21.026	23.337	24.054	26.217	28.300	32.909	12
15.119	15.984	16.985	19.812	22.362	24.736	25.472	27.688	29.819	34.528	13
16.222	17.117	18.151	21.064	23.685	26.119	26.873	29.141	31.319	36.123	14
17.322	18.245	19.311	22.307	24.996	27.488	28.259	30.578	32.801	37.697	15
18.418	19.369	20.465	23.542	26.296	28.845	29.633	32.000	34.267	39.252	16
19.511	20.489	21.615	24.769	27.587	30.191	30.995	33.409	35.718	40.790	17
20.601	21.605	22.760	25.989	28.869	31.526	32.346	34.805	37.156	42.312	18
21.689	22.718	23.900	27.204	30.144	32.852	33.687	36.191	38.582	43.820	19
22.775	23.828	25.038	28.412	31.410	34.170	35.020	37.566	39.997	45.315	20
23.858	24.935	26.171	29.615	32.671	35.479	36.343	38.932	41.401	46.797	21
24.939	26.039	27.301	30.813	33.924	36.781	37.659	40.289	42.796	48.268	22
26.018	27.141	28.429	32.007	35.172	38.076	38.968	41.638	44.181	49.728	23
27.096	28.241	29.553	33.196	36.415	39.364	40.270	42.980	45.558	51.179	24
28.172	29.339	30.675	34.382	37.652	40.646	41.566	44.314	46.928	52.620	25
29.246	30.434	31.795	35.563	38.885	41.923	42.856	45.642	48.290	54.052	26
30.319	31.528	32.912	36.741	40.113	43.194	44.140	46.963	49.645	55.476	27
31.391	32.620	34.027	37.916	41.337	44.461	45.419	48.278	50.993	56.893	28
32.461	33.711	35.139	39.087	42.557	45.722	46.693	49.588	52.336	58.302	29
33.530	34.800	36.250	40.256	43.773	46.979	47.962	50.892	53.672	59.703	30

Since the value of F does not exceed $F_{.05;99,99}$ the hypothesis of equality of variances is adopted.

13.4.11 Problems of Estimation: Confidence Intervals

A common statistical problem is to estimate parameters on the basis of a sample. A single number (e.g., a sample mean) is occasionally adequate but usually it is desirable to give an indication of the reliability of these estimates and present an interval that may be said to be "likely" to contain the true value. More precisely, an interval is said to be a confidence interval of size $(1 - \alpha)$ if in the long run $100(1 - \alpha)$ per cent of intervals constructed in this way contain the true value.

For example, suppose we are interested in estimating the mean breaking strength of 1040

TABLE 13.28. *OUTLINE OF TEST*

Notation for Hypothesis	Test Statistic
$\mu_x = \mu_y$	$U = (\bar{x} - \bar{y})/\sqrt{\sigma_x^2/n_x + \sigma_y^2/n_y}$
Rule of Rejection	Rule for Choosing Sample Sizes, Assuming $n_x = n_y = n$
$U \geq K_\alpha$ if we wish to reject when $\mu_x > \mu_y$.	Select a value $\mu_x - \mu_y > 0$ for which we want to reject the null hypothesis with high probability. Calculate† $d = (\mu_x - \mu_y)/\sqrt{\sigma_x^2 + \sigma_y^2}$ and select n from the OC curve of Fig. 13.29 or 13.30.
$U \leq -K_\alpha$ if we wish to reject when $\mu_x < \mu_y$.	Select a value $\mu_x - \mu_y < 0$ for which we want to reject the null hypothesis with high probability. Calculate† $d = (\mu_y - \mu_x)/\sqrt{\sigma_x^2 + \sigma_y^2}$ and select n from the OC curve of Fig. 13.29 or 13.30.
$\lvert U \rvert \geq K_{\alpha/2}$ if we wish to reject when the true difference in means is either positive or negative.	Select a value $\lvert \mu_x - \mu_y \rvert > 0$ for which we want to reject the null hypothesis with high probability. Calculate† $d = \left\lvert \dfrac{\mu_x - \mu_y}{\sqrt{\sigma_x^2 + \sigma_y^2}} \right\rvert$ and select n from the OC curve of Fig. 13.27 or 13.28.

† In order to find the required sample size for the OC curves given, it is necessary to choose $n = n_x = n_y$. However, if n_x and n_y are fixed in advance, $n_x \neq n_y$, the resulting protection using the above rule can be obtained by entering the curves using d and $n = (\sigma_x^2 + \sigma_y^2)/[(\sigma_x^2/n_x) + (\sigma_y^2/n_y)]$.

TABLE 13.29. *OUTLINE OF TEST*

Notation for the Hypothesis	Test Statistic
$\mu_x = \mu_y$	$t = \dfrac{\bar{x} - \bar{y}}{\sqrt{\dfrac{1}{n_x} + \dfrac{1}{n_y}}\sqrt{\dfrac{\sum(x_i - \bar{x})^2 + \sum(y_i - \bar{y})^2}{(n_x + n_y - 2)}}}$
Rule of Rejection	Rule for Choosing Sample Size Assuming $n_x = n_y = n$
$t \geq t_{\alpha;n_x+n_y-2}$ if we wish to reject when $\mu_x > \mu_y$.	The OC curve depends on $d = (\mu_x - \mu_y)/2\sigma$. Estimate that value of d for which we want to be sure to reject the hypothesis and pick n' from Fig. 13.33 or 13.34. The required sample size is $(n' + 1)/2$.
$t \leq -t_{\alpha;n_x+n_y-2}$ if we wish to reject when $\mu_x < \mu_y$.	Estimate the $d = (\mu_y - \mu_x)/2\sigma$ for which we want to be sure to reject, and pick n' from Fig. 13.33 or 13.34. The required sample size is $(n' + 1)/2$.
$\lvert t \rvert \geq t_{\alpha/2;n_x+n_y-2}$ if we wish to reject whenever μ_x is not equal to μ_y.	Estimate the $d = \lvert(\mu_x - \mu_y)/2\sigma\rvert$ for which we want to be sure to reject, and pick n' from Fig. 13.30 or 13.32. The required sample size is $(n' + 1)/2$.

† A test for the equality of standard deviations is given in Par. 13.4.10.

carbon steel; ten specimens are tested. One possibility would be to present the average of these numbers as an estimate of the mean breaking strength, and this is the best estimate if a single number is desired. However, this single number may by chance depart substantially from the true mean value, and no indication of the magnitude of the departure is available. For many purposes, a more useful estimate is an interval which contains the true value with high probability.

Procedures for calculating confidence inter-

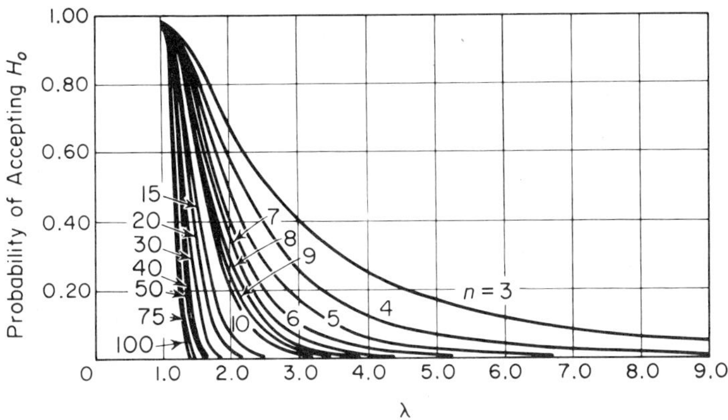

Fig. 13.38 *Operating characteristics of the one-sided (upper-tail) chi-square test for a level of significance equal to 0.01.*

TABLE 13.30. *OUTLINE OF TEST*

Notation for the Hypothesis	Test Statistic
$\mu_x = \mu_y$	$t' = \dfrac{\bar{x}-\bar{y}}{\sqrt{s_x^2/n_x + s_y^2/n_y}}$
Rule of Rejection	Formula for Obtaining the Degrees of Freedom υ
$t' \geq t_{\alpha;\nu}$ if we wish to reject when $\mu_x > \mu_y$.	$\upsilon = \dfrac{(s_x^2/n_x + s_y^2/n_y)^2}{(s_x^2/n_x)^2/(n_x+1) + (s_y^2/n_y)^2/(n_y+1)} - 2$
$t' \leq - t_{\alpha;\nu}$ if we wish to reject when $\mu_x < \mu_y$.	
$\|t'\| \geq t_{\alpha/2;\nu}$ if we wish to reject whenever μ_x is not equal to μ_y.	

vals most likely to arise in practical work are given in Table 13.38; these intervals usually involve percentage points of the standard distributions that are presented in Tables 13.3, 13.25, and 13.27 through 13.37. The notation for the percentage points is explained in Table 13.38.

13.5 CURVE FITTING

13.5.1 Introduction

It is common practice for engineers to represent relationships by means of graphs. Although these graphs are depicted as smooth curves, rarely do the experimental points lie on the fitted curves. This is usually explained by the fact that the experimental points are chance variables. The usual procedures are to plot the points and (1) connect all the points as in Fig. 13.45; (2) fair the best line in by eye as in Fig. 13.46; (3) fair the best curve in by eye as in Fig. 13.47. It is useful to discuss these three methods when there is a priori knowledge that the plotted relation is linear and when the plotted relation is curvilinear.

If there is an underlying linear relationship, procedure (1) has serious implications. In the first place, the problem is really to obtain the best estimate of the true underlying relationship (the word estimate is used because on the basis of a sample the true relationship is never obtained). The procedure pictured in Fig. 13.45 does not represent a linear relationship, and it is intuitively evident that the best estimate of a straight line should be a straight line. Furthermore, the performance of another experiment with the same variables would lead to a completely different graph. Very little confidence can be placed in curves that will differ radically with the performance of another experiment using the same variables. Frequently, graphs are used for prediction purposes, and no statements about the chances of being right or wrong can be made if procedure (1) is used.

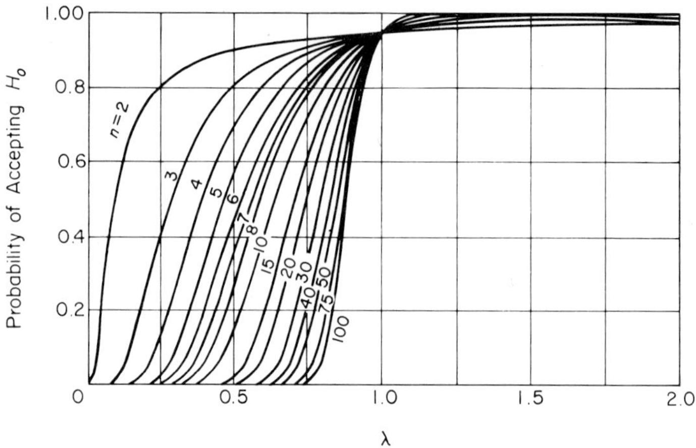

Fig. 13.39 *Operating characteristics of the one-sided (lower-tail) chi-square test for a level of significance equal to 0.05. Reproduced by permission from "Operating Characteristics for the Common Statistical Tests of Significance" by Charles D. Ferris, Frank E. Grubbs, Chalmers L. Weaver,* Annals of Mathematical Statistics, *June, 1946.*

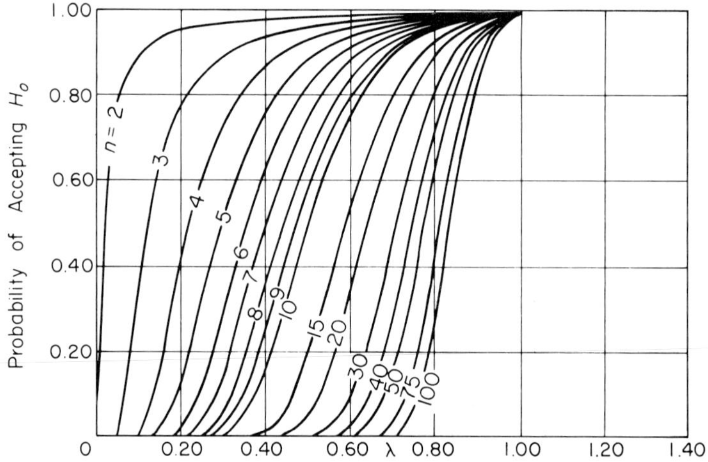

Fig. 13.40 *Operating characteristics of the one-sided (lower-tail) chi-square test for a level of significance equal to 0.01.*

The second procedure depicted in Fig. 13.46 is probably most commonly used. This has two serious disadvantages. Two people confronted with the same data will draw different lines through the points, so that there is not a single interpretation of the data. If these graphs are used for prediction purposes, the best point estimates based on the faired line will differ for different people. Furthermore, since fairing in a line is not an objective procedure, probability statements about the slope and intercept cannot be made.

The third procedure depicted in Fig. 13.47 can be dismissed for essentially the same reasons as those for procedures (1) and (2). In addition, it is again intuitively evident that the best estimate of a true linear relationship should be a straight line. If the underlying

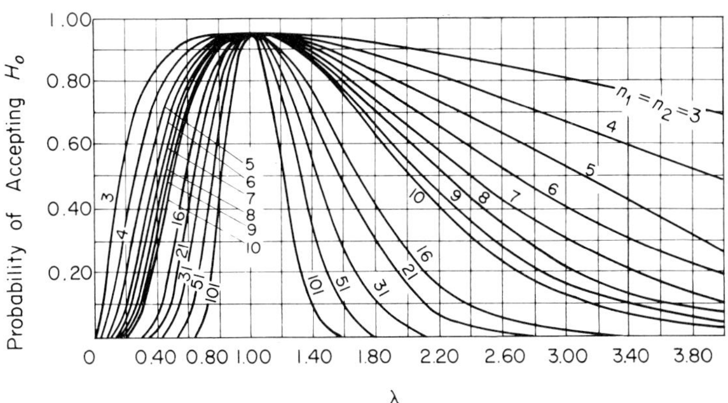

Fig. 13.41 *Operating characteristics of the two-sided F test for a level of significance equal to 0.05.*

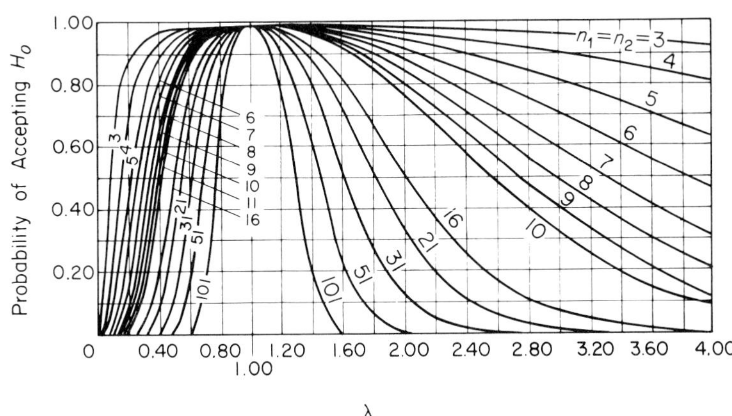

Fig. 13.42 *Operating characteristics of the two-sided F test for a level of significance equal to 0.01.*

TABLE 13.31. *OUTLINE FOR TEST*

Notation for the Hypothesis	Test Statistic
$\sigma_x^2 = \sigma_y^2$ or $\sigma_x = \sigma_y$	$F = \dfrac{[\sum (x_i - \bar{x})^2]/(n_x - 1)}{[\sum (y_i - \bar{y})^2]/(n_y - 1)} = \dfrac{s_x^2}{s_y^2}$
Rule of Rejection	Rule for Choosing Sample Sizes
$F \geq F_{\alpha; n_x-1, n_y-1}$ if we wish to reject when $\sigma_x > \sigma_y$. The notation x and y is arbitrary in a one-sided test; let x be the symbol for the variable with possible larger variance.	The OC curve depends on $\lambda = \sigma_x/\sigma_y$, and n_x and n_y. The OC curve for $n_x = n_y = n$ is given in Figs. 13.43 and 13.44. Pick a λ for which we want to reject and choose n such that the probability of accepting for that λ is sufficiently small.
In a two-sided test, let s_x^2 be the larger sample variance. Reject if $F \geq F_{\alpha/2; n_x-1, n_y-1}$.	The OC curve depends on $\lambda = \sigma_x/\sigma_y$ and is given in Figs. 13.41 and 13.42 for the case $n_x = n_y = n$. Pick the n for a specified λ such that the probability of accepting is sufficiently small.

TABLE 13.32. *PERCENTAGE POINT TEN OF THE F DISTRIBUTION,† TABLE OF* $F_{.10;\nu_1,\nu_2}$

Degrees of Freedom for the Denominator ν_2	Degrees of Freedom for the Numerator ν_1																	
	1	2	3	4	5	6	7	8	9	10	15	20	30	50	100	200	500	∞
1	39.9	49.5	53.6	55.8	57.2	58.2	58.9	59.4	59.9	60.2	61.2	61.7	62.3	62.7	63.0	63.2	63.3	63.3
2	8.53	9.00	9.16	9.24	9.29	9.33	9.35	9.37	9.38	9.39	9.42	9.44	9.46	9.47	9.48	9.49	9.49	9.49
3	5.54	5.46	5.39	5.34	5.31	5.28	5.27	5.25	5.24	5.23	5.20	5.18	5.17	5.15	5.14	5.14	5.14	5.13
4	4.54	4.32	4.19	4.11	4.05	4.01	3.98	3.95	3.94	3.92	3.87	3.84	3.82	3.80	3.78	3.77	3.76	3.76
5	4.06	3.78	3.62	3.52	3.45	3.40	3.37	3.34	3.32	3.30	3.24	3.21	3.17	3.15	3.13	3.12	3.11	3.10
6	3.78	3.46	3.29	3.18	3.11	3.05	3.01	2.98	2.96	2.94	2.87	2.84	2.80	2.77	2.75	2.73	2.73	2.72
7	3.59	3.26	3.07	2.96	2.88	2.83	2.78	2.75	2.72	2.70	2.63	2.59	2.56	2.52	2.50	2.48	2.48	2.47
8	3.46	3.11	2.92	2.81	2.73	2.67	2.62	2.59	2.56	2.54	2.46	2.42	2.38	2.35	2.32	2.31	2.30	2.29
9	3.36	3.01	2.81	2.69	2.61	2.55	2.51	2.47	2.44	2.42	2.34	2.30	2.25	2.22	2.19	2.17	2.17	2.16
10	3.28	2.92	2.73	2.61	2.52	2.46	2.41	2.38	2.35	2.32	2.24	2.20	2.16	2.12	2.09	2.07	2.06	2.06
11	3.23	2.86	2.66	2.54	2.45	2.39	2.34	2.30	2.27	2.25	2.17	2.12	2.08	2.04	2.00	1.99	1.98	1.97
12	3.18	2.81	2.61	2.48	2.39	2.33	2.28	2.24	2.21	2.19	2.10	2.06	2.01	1.97	1.94	1.92	1.91	1.90
13	3.14	2.76	2.56	2.43	2.35	2.28	2.23	2.20	2.16	2.14	2.05	2.01	1.96	1.92	1.88	1.86	1.85	1.85
14	3.10	2.73	2.52	2.39	2.31	2.24	2.19	2.15	2.12	2.10	2.01	1.96	1.91	1.87	1.83	1.82	1.80	1.80
15	3.07	2.70	2.49	2.36	2.27	2.21	2.16	2.12	2.09	2.06	1.97	1.92	1.87	1.83	1.79	1.77	1.76	1.76
16	3.05	2.67	2.46	2.33	2.24	2.18	2.13	2.09	2.06	2.03	1.94	1.89	1.84	1.79	1.76	1.74	1.73	1.72
17	3.03	2.64	2.44	2.31	2.22	2.15	2.10	2.06	2.03	2.00	1.91	1.86	1.81	1.76	1.73	1.71	1.69	1.69
18	3.01	2.62	2.42	2.29	2.20	2.13	2.08	2.04	2.00	1.98	1.89	1.84	1.78	1.74	1.70	1.68	1.67	1.66
19	2.99	2.61	2.40	2.27	2.18	2.11	2.06	2.02	1.98	1.96	1.86	1.81	1.76	1.71	1.67	1.65	1.64	1.63
20	2.97	2.59	2.38	2.25	2.16	2.09	2.04	2.00	1.96	1.94	1.84	1.79	1.74	1.69	1.65	1.63	1.62	1.61
22	2.95	2.56	2.35	2.22	2.13	2.06	2.01	1.97	1.93	1.90	1.81	1.76	1.70	1.65	1.61	1.59	1.58	1.57
24	2.93	2.54	2.33	2.19	2.10	2.04	1.98	1.94	1.91	1.88	1.78	1.73	1.67	1.62	1.58	1.56	1.54	1.53
26	2.91	2.52	2.31	2.17	2.08	2.01	1.96	1.92	1.88	1.86	1.76	1.71	1.65	1.59	1.55	1.53	1.51	1.50
28	2.89	2.50	2.29	2.16	2.06	2.00	1.94	1.90	1.87	1.84	1.74	1.69	1.63	1.57	1.53	1.50	1.49	1.48
30	2.88	2.49	2.28	2.14	2.05	1.98	1.93	1.88	1.85	1.82	1.72	1.67	1.61	1.55	1.51	1.48	1.47	1.46
40	2.84	2.44	2.23	2.09	2.00	1.93	1.87	1.83	1.79	1.76	1.66	1.61	1.54	1.48	1.43	1.41	1.39	1.38
50	2.81	2.41	2.20	2.06	1.97	1.90	1.84	1.80	1.76	1.73	1.63	1.57	1.50	1.44	1.39	1.36	1.34	1.33
60	2.79	2.39	2.18	2.04	1.95	1.87	1.82	1.77	1.74	1.71	1.60	1.54	1.48	1.41	1.36	1.33	1.31	1.29

80	2.77	2.37	2.15	2.02	1.92	1.85	1.79	1.75	1.71	1.68	1.57	1.51	1.44	1.38	1.32	1.28	1.26	1.24
100	2.76	2.36	2.14	2.00	1.91	1.83	1.78	1.73	1.70	1.66	1.56	1.49	1.42	1.35	1.29	1.26	1.23	1.21
200	2.73	2.33	2.11	1.97	1.88	1.80	1.75	1.70	1.66	1.63	1.52	1.46	1.38	1.31	1.24	1.20	1.17	1.14
500	2.72	2.31	2.10	1.96	1.86	1.79	1.73	1.68	1.64	1.61	1.50	1.44	1.36	1.28	1.21	1.16	1.12	1.09
∞	2.71	2.30	2.08	1.94	1.85	1.77	1.72	1.67	1.63	1.60	1.49	1.42	1.34	1.26	1.18	1.13	1.08	1.00

Example: $P\{F_{.10;8,20} < 2.00\} = 90$ per cent.

$F_{.90;\nu_1,\nu_2} = 1/F_{.10;\nu_2,\nu_1}$. Example: $F_{.90;8,20} = 1/2.42 = 0.413$.

Approximate formula for ν_1 and ν_2 larger than 30: $\log_{10} F_{.10;\nu_1,\nu_2} \simeq \dfrac{1.1131}{\sqrt{h} - 0.77} - 0.527\left(\dfrac{1}{\nu_1} - \dfrac{1}{\nu_2}\right)$,

where $\dfrac{1}{h} = \dfrac{1}{2}\left(\dfrac{1}{\nu_1} + \dfrac{1}{\nu_2}\right)$.

† The following tables of the F distribution are abridged from *Statistical Tables and Formulas*, by A. Hald, published by John Wiley & Sons, Inc., a major part of which has been abridged from "Tables of Percentage Points of the Inverted Beta (F) Distribution," ccmputed by M. Merrington and C. M. Thompson, *Biometrika* Vol. 33, 1943, 73–88, by permission of the proprietors, or reproduced from Table V of R. A. Fisher and F. Yates: *Statistical Methods for Research Workers*, Oliver and Boyd, Edinburgh, by permission of the authors and the publishers.

relationship is curvilinear, the three procedures can be dismissed for essentially the same reasons as discussed above.

The method commonly used for fitting linear and curvilinear relations is the method of least squares. Besides the advantage of giving a unique solution, i.e., a single line, this procedure has several optimum statistical properties. Furthermore, with some knowledge about the underlying distribution, confidence statements about the parameters can be made.

13.5.2 Simple Linear Curve Fitting

Two cases will be distinguished, namely, where there exists an underlying physical linear relationship, and where there exists a degree of association between the variables.

13.5.2.1 An Underlying Physical Linear Relationship. In this case, there is a mathematical relationship of the form $y = B_0 + B_1x$, which relates the variables. Examples of this case are the equations of physics:

$$\text{velocity} = \text{acceleration} \times \text{time} + \text{initial velocity}$$
$$\text{force} = \text{mass} \times \text{acceleration}$$

where the intercept, in the second example, is 0. Upon accumulating data for such relations, all the observed points $y_1, y_2, \ldots, y_n$ do not lie on a straight line because of experimental errors. However, it is reasonable to assume that the distribution of errors is such that for a fixed value of x, the population mean value of the y's is $B_0 + B_1x$, where B_0 and B_1 are the intercept and slope to be estimated from a sample $(x_1, y_1), (x_2, y_2), \ldots, (x_n, y_n)$. Here x is the independent variable and is *assumed* to be known without error. From a practical point of view, it is sufficient to have the error in x small compared with the variability in y. In this section it will be assumed that $y_1, y_2, \ldots, y_n$ are independent chance variables having a normal distribution* with mean $B_0 + B_1x$ and standard deviation σ (constant for all x_i).

13.5.2.2 A Degree of Association Between

* The equations for the least square estimates of B_0 and B_1 do not depend on the assumption of normality. However, tests of hypothesis and confidence statements presented later do depend on this assumption.

TABLE 13.33. *PERCENTAGE POINT FIVE OF THE F DISTRIBUTION, TABLE OF $F_{.05;\nu_1,\nu_2}$.*

Degrees of Freedom for the Denominator ν_2	Degrees of Freedom for the Numerator ν_1																	
	1	2	3	4	5	6	7	8	9	10	11	12	13	14	15	16	17	18
1	161	200	216	225	230	234	237	239	241	242	243	244	245	245	246	246	247	247
2	18.5	19.0	19.2	19.2	19.3	19.3	19.4	19.4	19.4	19.4	19.4	19.4	19.4	19.4	19.4	19.4	19.4	19.4
3	10.1	9.55	9.28	9.12	9.01	8.94	8.89	8.85	8.81	8.79	8.76	8.74	8.73	8.71	8.70	8.69	8.68	8.67
4	7.71	6.94	6.59	6.39	6.26	6.16	6.09	6.04	6.00	5.96	5.94	5.91	5.89	5.87	5.86	5.84	5.83	5.82
5	6.61	5.79	5.41	5.19	5.05	4.95	4.88	4.82	4.77	4.74	4.70	4.68	4.66	4.64	4.62	4.60	4.59	4.58
6	5.99	5.14	4.76	4.53	4.39	4.28	4.21	4.15	4.10	4.06	4.03	4.00	3.98	3.96	3.94	3.92	3.91	3.90
7	5.59	4.74	4.35	4.12	3.97	3.87	3.79	3.73	3.68	3.64	3.60	3.57	3.55	3.53	3.51	3.49	3.48	3.47
8	5.32	4.46	4.07	3.84	3.69	3.58	3.50	3.44	3.39	3.35	3.31	3.28	3.26	3.24	3.22	3.20	3.19	3.17
9	5.12	4.26	3.86	3.63	3.48	3.37	3.29	3.23	3.18	3.14	3.10	3.07	3.05	3.03	3.01	2.99	2.97	2.96
10	4.96	4.10	3.71	3.48	3.33	3.22	3.14	3.07	3.02	2.98	2.94	2.91	2.89	2.86	2.85	2.83	2.81	2.80
11	4.84	3.98	3.59	3.36	3.20	3.09	3.01	2.95	2.90	2.85	2.82	2.79	2.76	2.74	2.72	2.70	2.69	2.67
12	4.75	3.89	3.49	3.26	3.11	3.00	2.91	2.85	2.80	2.75	2.72	2.69	2.66	2.64	2.62	2.60	2.58	2.57
13	4.67	3.81	3.41	3.18	3.03	2.92	2.83	2.77	2.71	2.67	2.63	2.60	2.58	2.55	2.53	2.51	2.50	2.48
14	4.60	3.74	3.34	3.11	2.96	2.85	2.76	2.70	2.65	2.60	2.57	2.53	2.51	2.48	2.46	2.44	2.43	2.41
15	4.54	3.68	3.29	3.06	2.90	2.79	2.71	2.64	2.59	2.54	2.51	2.48	2.45	2.42	2.40	2.38	2.37	2.35
16	4.49	3.63	3.24	3.01	2.85	2.74	2.66	2.59	2.54	2.49	2.46	2.42	2.40	2.37	2.35	2.33	2.32	2.30
17	4.45	3.59	3.20	2.96	2.81	2.70	2.61	2.55	2.49	2.45	2.41	2.38	2.35	2.33	2.31	2.29	2.27	2.26
18	4.41	3.55	3.16	2.93	2.77	2.66	2.58	2.51	2.46	2.41	2.37	2.34	2.31	2.29	2.27	2.25	2.23	2.22
19	4.38	3.52	3.13	2.90	2.74	2.63	2.54	2.48	2.42	2.38	2.34	2.31	2.28	2.26	2.23	2.21	2.20	2.18
20	4.35	3.49	3.10	2.87	2.71	2.60	2.51	2.45	2.39	2.35	2.31	2.28	2.25	2.22	2.20	2.18	2.17	2.15
21	4.32	3.47	3.07	2.84	2.68	2.57	2.49	2.42	2.37	2.32	2.28	2.25	2.22	2.20	2.18	2.16	2.14	2.12
22	4.30	3.44	3.05	2.82	2.66	2.55	2.46	2.40	2.34	2.30	2.26	2.23	2.20	2.17	2.15	2.13	2.11	2.10
23	4.28	3.42	3.03	2.80	2.64	2.53	2.44	2.37	2.32	2.27	2.23	2.20	2.18	2.15	2.13	2.11	2.09	2.07
24	4.26	3.40	3.01	2.78	2.62	2.51	2.42	2.36	2.30	2.25	2.21	2.18	2.15	2.13	2.11	2.09	2.07	2.05
25	4.24	3.39	2.99	2.76	2.60	2.49	2.40	2.34	2.28	2.24	2.20	2.16	2.14	2.11	2.09	2.07	2.05	2.04
26	4.23	3.37	2.98	2.74	2.59	2.47	2.39	2.32	2.27	2.22	2.18	2.15	2.12	2.09	2.07	2.05	2.03	2.02
27	4.21	3.35	2.96	2.73	2.57	2.46	2.37	2.31	2.25	2.20	2.17	2.13	2.10	2.08	2.06	2.04	2.02	2.00
28	4.20	3.34	2.95	2.71	2.56	2.45	2.36	2.29	2.24	2.19	2.15	2.12	2.09	2.06	2.04	2.02	2.00	1.99

Degrees of Freedom for the Denominator ν_2	Degrees of Freedom for the Numerator ν_1																	
	1	2	3	4	5	6	7	8	9	10	11	12	13	14	15	16	17	18
29	4.18	3.33	2.93	2.70	2.55	2.43	2.35	2.28	2.22	2.18	2.14	2.10	2.08	2.05	2.03	2.01	1.99	1.97
30	4.17	3.32	2.92	2.69	2.53	2.42	2.33	2.27	2.21	2.16	2.13	2.09	2.06	2.04	2.01	1.99	1.98	1.96
32	4.15	3.29	2.90	2.67	2.51	2.40	2.31	2.24	2.19	2.14	2.10	2.07	2.04	2.01	1.99	1.97	1.95	1.94
34	4.13	3.28	2.88	2.65	2.49	2.38	2.29	2.23	2.17	2.12	2.08	2.05	2.02	1.99	1.97	1.95	1.93	1.92
36	4.11	3.26	2.87	2.63	2.48	2.36	2.28	2.21	2.15	2.11	2.07	2.03	2.00	1.98	1.95	1.93	1.92	1.90
38	4.10	3.24	2.85	2.62	2.46	2.35	2.26	2.19	2.14	2.09	2.05	2.02	1.99	1.96	1.94	1.92	1.90	1.88
40	4.08	3.23	2.84	2.61	2.45	2.34	2.25	2.18	2.12	2.08	2.04	2.00	1.97	1.95	1.92	1.90	1.89	1.87
42	4.07	3.22	2.83	2.59	2.44	2.32	2.24	2.17	2.11	2.06	2.03	1.99	1.96	1.93	1.91	1.89	1.87	1.86
44	4.06	3.21	2.82	2.58	2.43	2.31	2.23	2.16	2.10	2.05	2.01	1.98	1.95	1.92	1.90	1.88	1.86	1.84
46	4.05	3.20	2.81	2.57	2.42	2.30	2.22	2.15	2.09	2.04	2.00	1.97	1.94	1.91	1.89	1.87	1.85	1.83
48	4.04	3.19	2.80	2.57	2.41	2.29	2.21	2.14	2.08	2.03	1.99	1.96	1.93	1.90	1.88	1.86	1.84	1.82
50	4.03	3.18	2.79	2.56	2.40	2.29	2.20	2.13	2.07	2.03	1.99	1.95	1.92	1.89	1.87	1.85	1.83	1.81
55	4.02	3.16	2.77	2.54	2.38	2.27	2.18	2.11	2.06	2.01	1.97	1.93	1.90	1.88	1.85	1.83	1.81	1.79
60	4.00	3.15	2.76	2.53	2.37	2.25	2.17	2.10	2.04	1.99	1.95	1.92	1.89	1.86	1.84	1.82	1.80	1.78
65	3.99	3.14	2.75	2.51	2.36	2.24	2.15	2.08	2.03	1.98	1.94	1.90	1.87	1.85	1.82	1.80	1.78	1.76
70	3.98	3.13	2.74	2.50	2.35	2.23	2.14	2.07	2.02	1.97	1.93	1.89	1.86	1.84	1.81	1.79	1.77	1.75
80	3.96	3.11	2.72	2.49	2.33	2.21	2.13	2.06	2.00	1.95	1.91	1.88	1.84	1.82	1.79	1.77	1.75	1.73
90	3.95	3.10	2.71	2.47	2.32	2.20	2.11	2.04	1.99	1.94	1.90	1.86	1.83	1.80	1.78	1.76	1.74	1.72
100	3.94	3.09	2.70	2.46	2.31	2.19	2.10	2.03	1.97	1.93	1.89	1.85	1.82	1.79	1.77	1.75	1.73	1.71
125	3.92	3.07	2.68	2.44	2.29	2.17	2.08	2.01	1.96	1.91	1.87	1.83	1.80	1.77	1.75	1.72	1.70	1.69
150	3.90	3.06	2.66	2.43	2.27	2.16	2.07	2.00	1.74	1.89	1.85	1.82	1.79	1.76	1.73	1.71	1.69	1.67
200	3.89	3.04	2.65	2.42	2.26	2.14	2.06	1.98	1.93	1.88	1.84	1.80	1.77	1.74	1.72	1.69	1.67	1.66
300	3.87	3.03	2.63	2.40	2.24	2.13	2.04	1.97	1.91	1.86	1.82	1.78	1.75	1.72	1.70	1.68	1.66	1.64
500	3.86	3.01	2.62	2.39	2.23	2.12	2.03	1.96	1.90	1.85	1.81	1.77	1.74	1.71	1.69	1.66	1.64	1.62
1000	3.85	3.00	2.61	2.38	2.22	2.11	2.02	1.95	1.89	1.84	1.80	1.76	1.73	1.70	1.68	1.65	1.63	1.61
∞	3.84	3.00	2.60	2.37	2.21	2.10	2.01	1.94	1.88	1.83	1.79	1.75	1.72	1.69	1.67	1.64	1.62	1.60

Example: $P\{F_{05;8,20} < 2.45\} = 95$ per cent.

$F_{.95;\nu_1,\nu_2} = 1/F_{.05;\nu_2,\nu_1}$. Example: $F_{.95;8,20} = 1/F_{.05;20,8} = 1/3.15 = 0.317$.

TABLE 13.33. *Cont.*

Degrees of Freedom for the Numerator v_1																	Degrees of Freedom for the Denominator v_2
19	20	22	24	26	28	30	35	40	45	50	60	80	100	200	500	∞	
248	248	249	249	249	250	250	251	251	251	252	252	252	253	254	254	254	1
19.4	19.4	19.5	19.5	19.5	19.5	19.5	19.5	19.5	19.5	19.5	19.5	19.5	19.5	19.5	19.5	19.5	2
8.67	8.66	8.65	8.64	8.63	8.62	8.62	8.60	8.59	8.59	8.58	8.57	8.56	8.55	8.54	8.53	8.53	3
5.81	5.80	5.79	5.77	5.76	5.75	5.75	5.73	5.72	5.71	5.70	5.69	5.67	5.66	5.65	5.64	5.63	4
4.57	4.56	4.54	4.53	4.52	4.50	4.50	4.48	4.46	4.45	4.44	4.43	4.41	4.41	4.39	4.37	4.37	5
3.88	3.87	3.86	3.84	3.83	3.82	3.81	3.79	3.77	3.76	3.75	3.74	3.72	3.71	3.69	3.68	3.67	6
3.46	3.44	3.43	3.41	3.40	3.39	3.38	3.36	3.34	3.33	3.32	3.30	3.29	3.27	3.25	3.24	3.23	7
3.16	3.15	3.13	3.12	3.10	3.09	3.08	3.06	3.04	3.03	3.02	3.01	2.99	2.97	2.95	2.94	2.93	8
2.95	2.94	2.92	2.90	2.89	2.87	2.86	2.84	2.83	2.81	2.80	2.79	2.77	2.76	2.73	2.72	2.71	9
2.78	2.77	2.75	2.74	2.72	2.71	2.70	2.68	2.66	2.65	2.64	2.62	2.60	2.59	2.56	2.55	2.54	10
2.66	2.65	2.63	2.61	2.59	2.58	2.57	2.55	2.53	2.52	2.51	2.49	2.47	2.46	2.43	2.42	2.40	11
2.56	2.54	2.52	2.51	2.49	2.48	2.47	2.44	2.43	2.41	2.40	2.38	2.36	2.35	2.32	2.31	2.30	12
2.47	2.46	2.44	2.42	2.41	2.39	2.38	2.36	2.34	2.33	2.31	2.30	2.27	2.26	2.23	2.22	2.21	13
2.40	2.39	2.37	2.35	2.33	2.32	2.31	2.28	2.27	2.25	2.24	2.22	2.20	2.19	2.16	2.14	2.13	14
2.34	2.33	2.31	2.29	2.27	2.26	2.25	2.22	2.20	2.19	2.18	2.16	2.14	2.12	2.10	2.08	2.07	15
2.29	2.28	2.25	2.24	2.22	2.21	2.19	2.17	2.15	2.14	2.12	2.11	2.08	2.07	2.04	2.02	2.01	16
2.24	2.23	2.21	2.19	2.17	2.16	2.15	2.12	2.10	2.09	2.08	2.06	2.03	2.02	1.99	1.97	1.96	17
2.20	2.19	2.17	2.15	2.13	2.12	2.11	2.08	2.06	2.05	2.04	2.02	1.99	1.98	1.95	1.93	1.92	18
2.17	2.16	2.13	2.11	2.10	2.08	2.07	2.05	2.03	2.01	2.00	1.98	1.96	1.94	1.91	1.89	1.88	19
2.14	2.12	2.10	2.08	2.07	2.05	2.04	2.01	1.99	1.98	1.97	1.95	1.92	1.91	1.88	1.86	1.84	20
2.11	2.10	2.07	2.05	2.04	2.02	2.01	1.98	1.96	1.95	1.94	1.92	1.89	1.88	1.84	1.82	1.81	21
2.08	2.07	2.05	2.03	2.01	2.00	1.98	1.96	1.94	1.92	1.91	1.89	1.86	1.85	1.82	1.80	1.78	22
2.06	2.05	2.02	2.00	1.99	1.97	1.96	1.93	1.91	1.90	1.88	1.86	1.84	1.82	1.79	1.77	1.76	23
2.04	2.03	2.00	1.98	1.97	1.95	1.94	1.91	1.89	1.88	1.86	1.84	1.82	1.80	1.77	1.75	1.73	24
2.02	2.01	1.98	1.96	1.95	1.93	1.92	1.89	1.87	1.86	1.84	1.82	1.80	1.78	1.75	1.73	1.71	25
2.00	1.99	1.97	1.95	1.93	1.91	1.90	1.87	1.85	1.84	1.82	1.80	1.78	1.76	1.73	1.71	1.69	26
1.99	1.97	1.95	1.93	1.91	1.90	1.88	1.86	1.84	1.82	1.81	1.79	1.76	1.74	1.71	1.69	1.67	27
1.97	1.96	1.93	1.91	1.90	1.88	1.87	1.84	1.82	1.80	1.79	1.77	1.74	1.73	1.69	1.67	1.65	28

Degrees of Freedom for the Numerator ν_1																	Degrees of Freedom for the Denominator ν_2
19	20	22	24	26	28	30	35	40	45	50	60	80	100	200	500	∞	
1.96	1.94	1.92	1.90	1.88	1.87	1.85	1.83	1.81	1.79	1.77	1.75	1.73	1.71	1.67	1.65	1.64	29
1.95	1.93	1.91	1.89	1.87	1.85	1.84	1.81	1.79	1.77	1.76	1.74	1.71	1.70	1.66	1.64	1.62	30
1.92	1.91	1.88	1.86	1.85	1.83	1.82	1.79	1.77	1.75	1.74	1.71	1.69	1.67	1.63	1.61	1.59	32
1.90	1.89	1.86	1.84	1.82	1.80	1.80	1.77	1.75	1.73	1.71	1.69	1.66	1.65	1.61	1.59	1.57	34
1.88	1.87	1.85	1.82	1.81	1.79	1.78	1.75	1.73	1.71	1.69	1.67	1.64	1.62	1.59	1.56	1.55	36
1.87	1.85	1.83	1.81	1.79	1.77	1.76	1.73	1.71	1.69	1.68	1.65	1.62	1.61	1.57	1.54	1.53	38
1.85	1.84	1.81	1.79	1.77	1.76	1.74	1.72	1.69	1.67	1.66	1.64	1.61	1.59	1.55	1.53	1.51	40
1.84	1.83	1.80	1.78	1.76	1.74	1.73	1.70	1.68	1.66	1.65	1.62	1.59	1.57	1.53	1.51	1.49	42
1.83	1.81	1.79	1.77	1.75	1.73	1.72	1.69	1.67	1.65	1.63	1.61	1.58	1.56	1.52	1.49	1.48	44
1.82	1.80	1.78	1.76	1.74	1.72	1.71	1.68	1.65	1.64	1.62	1.60	1.57	1.55	1.51	1.48	1.46	46
1.81	1.79	1.77	1.75	1.73	1.71	1.70	1.67	1.64	1.62	1.61	1.59	1.56	1.54	1.49	1.47	1.45	48
1.80	1.78	1.76	1.74	1.72	1.70	1.69	1.66	1.63	1.61	1.60	1.58	1.54	1.52	1.48	1.46	1.44	50
1.78	1.76	1.74	1.72	1.70	1.68	1.67	1.64	1.61	1.59	1.58	1.55	1.52	1.50	1.46	1.43	1.41	55
1.76	1.75	1.72	1.70	1.68	1.66	1.65	1.62	1.59	1.57	1.56	1.53	1.50	1.48	1.44	1.41	1.39	60
1.75	1.73	1.71	1.69	1.67	1.65	1.63	1.60	1.58	1.56	1.54	1.52	1.49	1.46	1.42	1.39	1.37	65
1.74	1.72	1.70	1.67	1.65	1.64	1.62	1.59	1.57	1.55	1.53	1.50	1.47	1.45	1.40	1.37	1.35	70
1.72	1.70	1.68	1.65	1.63	1.62	1.60	1.57	1.54	1.52	1.51	1.48	1.45	1.43	1.38	1.35	1.32	80
1.70	1.69	1.66	1.64	1.62	1.60	1.59	1.55	1.53	1.51	1.49	1.46	1.43	1.41	1.36	1.32	1.30	90
1.69	1.68	1.65	1.63	1.61	1.59	1.57	1.54	1.52	1.49	1.48	1.45	1.41	1.39	1.34	1.31	1.28	100
1.67	1.65	1.63	1.60	1.58	1.57	1.55	1.52	1.49	1.47	1.45	1.42	1.39	1.36	1.31	1.27	1.25	125
1.66	1.64	1.61	1.59	1.57	1.55	1.53	1.50	1.48	1.45	1.44	1.41	1.37	1.34	1.29	1.25	1.22	150
1.64	1.62	1.60	1.57	1.55	1.53	1.52	1.48	1.46	1.43	1.41	1.39	1.35	1.32	1.26	1.22	1.19	200
1.62	1.61	1.58	1.55	1.53	1.51	1.50	1.46	1.43	1.41	1.39	1.36	1.32	1.30	1.23	1.19	1.15	300
1.61	1.59	1.56	1.54	1.52	1.50	1.48	1.45	1.42	1.40	1.38	1.34	1.30	1.28	1.21	1.16	1.11	500
1.60	1.58	1.55	1.53	1.51	1.49	1.47	1.44	1.41	1.38	1.36	1.33	1.29	1.26	1.19	1.13	1.08	1000
1.59	1.57	1.54	1.52	1.50	1.48	1.46	1.42	1.39	1.37	1.35	1.32	1.27	1.24	1.17	1.11	1.00	∞

Approximate formula for ν_1 and ν_2 larger than 30: $\log_{10} F_{.05;\nu_1,\nu_2} \simeq \dfrac{1.4287}{\sqrt{h - 0.95}} - 0.681\left(\dfrac{1}{\nu_1} - \dfrac{1}{\nu_2}\right)$, where $\dfrac{1}{h} = \dfrac{1}{2}\left(\dfrac{1}{\nu_1} + \dfrac{1}{\nu_2}\right)$.

TABLE 13.34. *PERCENTAGE POINT 2.5 OF THE F DISTRIBUTION. TABLE OF* $F_{.025;\nu_1\nu_2}$.

Degrees of Freedom for the Denominator ν_2	Degrees of Freedom for the Numerator ν_1																	
	1	2	3	4	5	6	7	8	9	10	11	12	13	14	15	16	17	18
1	648	800	864	900	922	937	948	957	963	969	973	977	980	983	985	987	989	990
2	38.5	39.0	39.2	39.2	39.3	39.3	39.4	39.4	39.4	39.4	39.4	39.4	39.4	39.4	39.4	39.4	39.4	39.4
3	17.4	16.0	15.4	15.1	14.9	1.47	14.6	14.5	14.5	14.4	14.4	14.3	14.3	14.3	14.3	14.2	14.2	14.2
4	12.2	10.6	9.98	9.60	9.36	9.20	9.07	8.98	8.90	8.84	8.79	8.75	8.72	8.69	8.66	8.64	8.62	8.60
5	10.0	8.43	7.76	7.39	7.15	6.98	6.85	6.76	6.68	6.62	6.57	6.52	6.49	6.46	6.43	6.41	6.39	6.37
6	8.81	7.26	6.60	6.23	5.99	5.82	5.70	5.60	5.52	5.46	5.41	5.37	5.33	5.30	5.27	5.25	5.23	5.21
7	8.07	6.54	5.89	5.52	5.29	5.12	4.99	4.90	4.82	4.76	4.71	4.67	4.63	4.60	4.57	4.54	4.52	4.50
8	7.57	6.06	5.42	5.05	4.82	4.65	4.53	4.43	4.36	4.30	4.24	4.20	4.16	4.13	4.10	4.08	4.05	4.03
9	7.21	5.71	5.08	4.72	4.48	4.32	4.20	4.10	4.03	3.96	3.91	3.87	3.83	3.80	3.77	3.74	3.72	3.70
10	6.94	5.46	4.83	4.47	4.24	4.07	3.95	3.85	3.78	3.72	3.66	3.62	3.58	3.55	3.52	3.50	3.47	3.45
11	6.72	5.26	4.63	4.28	4.04	3.88	3.76	3.66	3.59	3.53	3.47	3.43	3.39	3.36	3.33	3.30	3.28	3.26
12	6.55	5.10	4.47	4.12	3.89	3.73	3.61	3.51	3.44	3.37	3.32	3.28	3.24	3.21	3.18	3.15	3.13	3.11
13	6.41	4.97	4.35	4.00	3.77	3.60	3.48	3.39	3.31	3.25	3.20	3.15	3.12	3.08	3.05	3.03	3.00	2.98
14	6.30	4.86	4.24	3.89	3.66	3.50	3.38	3.29	3.21	3.15	3.09	3.05	3.01	2.98	2.95	2.92	2.90	2.88
15	6.20	4.76	4.15	3.80	3.58	3.41	3.29	3.20	3.12	3.06	3.01	2.96	2.92	2.89	2.86	2.84	2.81	2.79
16	6.12	4.69	4.08	3.73	3.50	3.34	3.22	3.12	3.05	2.99	2.93	2.89	2.85	2.82	2.79	2.76	2.74	2.72
17	6.04	4.62	4.01	3.66	3.44	3.28	3.16	3.06	2.98	2.92	2.87	2.82	2.79	2.75	2.72	2.70	2.67	2.65
18	5.98	4.56	3.95	3.61	3.38	3.22	3.10	3.01	2.93	2.87	2.81	2.77	2.73	2.70	2.67	2.64	2.62	2.60
19	5.92	4.51	3.90	3.56	3.33	3.17	3.05	2.96	2.88	2.82	2.76	2.72	2.68	2.65	2.62	2.59	2.57	2.55
20	5.87	4.46	3.86	3.51	3.29	3.13	3.01	2.91	2.84	2.77	2.72	2.68	2.64	2.60	2.57	2.55	2.52	2.50
21	5.83	4.42	3.82	3.48	3.25	3.09	2.97	2.87	2.80	2.73	2.68	2.64	2.60	2.56	2.53	2.51	2.48	2.46
22	5.79	4.38	3.78	3.44	3.22	3.05	2.93	2.84	2.76	2.70	2.65	2.60	2.56	2.53	2.50	2.47	2.45	2.43
23	5.75	4.35	3.75	3.41	3.18	3.02	2.90	2.81	2.73	2.67	2.62	2.57	2.53	2.50	2.47	2.44	2.42	2.39
24	5.72	4.32	3.72	3.38	3.15	2.99	2.87	2.78	2.70	2.64	2.59	2.54	2.50	2.47	2.44	2.41	2.39	2.36
25	5.69	4.29	3.69	3.35	3.13	2.97	2.85	2.75	2.68	2.61	2.56	2.51	2.48	2.44	2.41	2.38	2.36	2.34
26	5.66	4.27	3.67	3.33	3.10	2.94	2.82	2.73	2.65	2.59	2.54	2.49	2.45	2.42	2.39	2.36	2.34	2.31
27	5.63	4.24	3.65	3.31	3.08	2.92	2.80	2.71	2.63	2.57	2.51	2.47	2.43	2.39	2.36	2.34	2.31	2.29
28	5.61	4.22	3.63	3.29	3.06	2.90	2.78	2.69	2.61	2.55	2.49	2.45	2.41	2.37	2.34	2.32	2.29	2.27

Degrees of Freedom for the Denominator ν_2	Degrees of Freedom for the Numerator ν_1																	
	1	2	3	4	5	6	7	8	9	10	11	12	13	14	15	16	17	18
29	5.59	4.20	3.61	3.27	3.04	2.88	2.76	2.67	2.59	2.53	2.48	2.43	2.39	2.36	2.32	2.30	2.27	2.25
30	5.57	4.18	3.59	3.25	3.03	2.87	2.75	2.65	2.57	2.51	2.46	2.41	2.37	2.34	2.31	2.28	2.26	2.23
32	5.53	4.15	3.56	3.22	3.00	2.84	2.72	2.62	2.54	2.48	2.43	2.38	2.34	2.31	2.28	2.25	2.22	2.20
34	5.50	4.12	3.53	3.19	2.97	2.81	2.69	2.59	2.52	2.45	2.40	2.35	2.31	2.28	2.25	2.22	2.19	2.17
36	5.47	4.09	3.51	3.17	2.94	2.79	2.66	2.57	2.49	2.43	2.37	2.33	2.29	2.25	2.22	2.20	2.17	2.15
38	5.45	4.07	3.48	3.15	2.92	2.76	2.64	2.55	2.47	2.41	2.35	2.31	2.27	2.23	2.20	2.17	2.15	2.13
40	5.42	4.05	3.46	3.13	2.90	2.74	2.62	2.53	2.45	2.39	2.33	2.29	2.25	2.21	2.18	2.15	2.13	2.11
42	5.40	4.03	3.45	3.11	2.89	2.73	2.61	2.51	2.44	2.37	2.32	2.27	2.23	2.20	2.16	2.14	2.11	2.09
44	5.39	4.02	3.43	3.09	2.87	2.71	2.59	2.50	2.42	2.36	2.30	2.26	2.21	2.18	2.15	2.12	2.10	2.07
46	5.37	4.00	3.42	3.08	2.86	2.70	2.58	2.48	2.41	2.34	2.29	2.24	2.20	2.17	2.13	2.11	2.08	2.06
48	5.35	3.99	3.40	3.07	2.84	2.69	2.57	2.47	2.39	2.33	2.27	2.23	2.19	2.15	2.12	2.09	2.07	2.05
50	5.34	3.98	3.39	3.06	2.83	2.67	2.55	2.46	2.38	2.32	2.26	2.22	2.18	2.14	2.11	2.08	2.06	2.03
55	5.31	3.95	3.36	3.03	2.81	2.65	2.53	2.43	2.36	2.29	2.24	2.19	2.15	2.11	2.08	2.05	2.03	2.01
60	5.29	3.93	3.34	3.01	2.79	2.63	2.51	2.41	2.33	2.27	2.22	2.17	2.13	2.09	2.06	2.03	2.01	1.98
65	5.27	3.91	3.32	2.99	2.77	2.61	2.49	2.39	2.32	2.25	2.20	2.15	2.11	2.07	2.04	2.01	1.99	1.97
70	5.25	3.89	3.31	2.98	2.75	2.60	2.48	2.38	2.30	2.24	2.18	2.14	2.10	2.06	2.03	2.00	1.97	1.95
80	5.22	3.86	3.28	2.95	2.73	2.57	2.45	2.36	2.28	2.21	2.16	2.11	2.07	2.03	2.00	1.97	1.95	1.93
90	5.20	3.84	3.27	2.93	2.71	2.55	2.43	2.34	2.26	2.19	2.14	2.09	2.05	2.02	1.98	1.95	1.93	1.91
100	5.18	3.83	3.25	2.92	2.70	2.54	2.42	2.32	2.24	2.18	2.12	2.08	2.04	2.00	1.97	1.94	1.91	1.89
125	5.15	3.80	3.22	2.89	2.67	2.51	2.39	2.30	2.22	2.15	2.10	2.05	2.01	1.97	1.94	1.91	1.89	1.86
150	5.13	3.78	3.20	2.87	2.65	2.49	2.37	2.28	2.20	2.13	2.08	2.03	1.99	1.95	1.92	1.89	1.87	1.84
200	5.10	3.76	3.18	2.85	2.63	2.47	2.35	2.26	2.18	2.11	2.06	2.01	1.97	1.93	1.90	1.87	1.84	1.82
300	5.08	3.74	3.16	2.83	2.61	2.45	2.33	2.23	2.16	2.09	2.04	1.99	1.95	1.91	1.88	1.85	1.82	1.80
500	5.05	3.72	3.14	2.81	2.59	2.43	2.31	2.22	2.14	2.07	2.02	1.97	1.93	1.89	1.86	1.83	1.80	1.78
1000	5.04	3.70	3.13	2.80	2.58	2.42	2.30	2.20	2.13	2.06	2.01	1.96	1.92	1.88	1.85	1.82	1.79	1.77
∞	5.02	3.69	3.12	2.79	2.57	2.41	2.29	2.19	2.11	2.05	1.99	1.94	1.90	1.87	1.83	1.80	1.78	1.75

Example: $P\{F_{.025;8,20} < 2.91\} = 97.5$ per cent.

$F_{.975;\nu_1,\nu_2} = 1/F_{.025;\nu_2,\nu_1}$. Example: $F_{.975;8,20} = 1/F_{.025;20,8} = 1/4.00 = 0.250$.

TABLE 13.34. *Cont.*

Degrees of Freedom for the Numerator ν_1																	Degrees of Freedom for the Denominator ν_2
19	20	22	24	26	28	30	35	40	45	50	60	80	100	200	500	∞	
992	993	995	997	999	1000	1001	1004	1006	1007	1008	1010	1012	1013	1016	1017	1018	1
39.4	39.4	39.5	39.5	39.5	39.5	39.5	39.5	39.5	39.5	39.5	39.5	39.5	39.5	39.5	39.5	39.5	2
14.2	14.2	14.1	14.1	14.1	14.1	14.1	14.1	14.0	14.0	14.0	14.0	14.0	14.0	13.9	13.9	13.9	3
8.58	8.56	8.53	8.51	8.49	8.48	8.46	8.44	8.41	8.39	8.38	8.36	8.33	8.32	8.29	8.27	8.26	4
6.35	6.33	6.30	6.28	6.26	6.24	6.23	6.20	6.18	6.16	6.14	6.12	6.10	6.08	6.05	6.03	6.02	5
5.19	5.17	5.14	5.12	5.10	5.08	5.07	5.04	5.01	4.99	4.98	4.96	4.93	4.92	4.88	4.86	4.85	6
4.48	4.47	4.44	4.42	4.39	4.38	4.36	4.33	4.31	4.29	4.28	4.25	4.23	4.21	4.18	4.16	4.14	7
4.02	4.00	3.97	3.95	3.93	3.91	3.89	3.86	3.84	3.82	3.81	3.78	3.76	3.74	3.70	3.68	3.67	8
3.68	3.67	3.64	3.61	3.59	3.58	3.56	3.53	3.51	3.49	3.47	3.45	3.42	3.40	3.37	3.35	3.33	9
3.44	3.42	3.39	3.37	3.34	3.33	3.31	3.28	3.26	3.24	3.22	3.20	3.17	3.15	3.12	3.09	3.08	10
3.24	3.23	3.20	3.17	3.15	3.13	3.12	3.09	3.06	3.04	3.03	3.00	2.97	2.96	2.92	2.90	2.88	11
3.09	3.07	3.04	3.02	3.00	2.98	2.96	2.93	2.91	2.89	2.87	2.85	2.82	2.80	2.76	2.74	2.72	12
2.96	2.95	2.92	2.89	2.87	2.85	2.84	2.80	2.78	2.76	2.74	2.72	2.69	2.67	2.63	2.61	2.60	13
2.86	2.84	2.81	2.79	2.77	2.75	2.73	2.70	2.67	2.65	2.64	2.61	2.58	2.56	2.53	2.50	2.49	14
2.77	2.76	2.73	2.70	2.68	2.66	2.64	2.61	2.58	2.56	2.55	2.52	2.49	2.47	2.44	2.41	2.40	15
2.70	2.68	2.65	2.63	2.60	2.58	2.57	2.53	2.51	2.49	2.47	2.45	2.42	2.40	2.36	2.33	2.32	16
2.63	2.62	2.59	2.56	2.54	2.52	2.50	2.47	2.44	2.42	2.41	2.38	2.35	2.33	2.29	2.26	2.25	17
2.58	2.56	2.53	2.50	2.48	2.46	2.44	2.41	2.38	2.36	2.35	2.32	2.29	2.27	2.23	2.20	2.19	18
2.53	2.51	2.48	2.45	2.43	2.41	2.39	2.36	2.33	2.31	2.30	2.27	2.24	2.22	2.18	2.15	2.13	19
2.48	2.46	2.43	2.41	2.39	2.37	2.35	2.31	2.29	2.27	2.25	2.22	2.19	2.17	2.13	2.10	2.09	20
2.44	2.42	2.39	2.37	2.34	2.33	2.31	2.27	2.25	2.23	2.21	2.18	2.15	2.13	2.09	2.06	2.04	21
2.41	2.39	2.36	2.33	2.31	2.29	2.27	2.24	2.21	2.19	2.17	2.14	2.11	2.09	2.05	2.02	2.00	22
2.37	2.36	2.33	2.30	2.28	2.26	2.24	2.20	2.18	2.15	2.14	2.11	2.08	2.06	2.01	1.99	1.97	23
2.35	2.33	2.30	2.27	2.25	2.23	2.21	2.17	2.15	2.12	2.11	2.08	2.05	2.02	1.98	1.95	1.94	24
2.32	2.30	2.27	2.24	2.22	2.20	2.18	2.15	2.12	2.10	2.08	2.05	2.02	2.00	1.95	1.92	1.91	25
2.29	2.28	2.24	2.22	2.19	2.17	2.16	2.12	2.09	2.07	2.05	2.03	1.99	1.97	1.92	1.90	1.88	26
2.27	2.25	2.22	2.19	2.17	2.15	2.13	2.10	2.07	2.05	2.03	2.00	1.97	1.94	1.90	1.87	1.85	27
2.25	2.23	2.20	2.17	2.15	2.13	2.11	2.08	2.05	2.03	2.01	1.98	1.94	1.92	1.88	1.85	1.83	28

Degrees of Freedom for the Numerator ν_1																	Degrees of Freedom for the Denominator ν_2
19	20	22	24	26	28	30	35	30	45	50	60	80	100	200	500	∞	
2.23	2.21	2.18	2.15	2.13	2.11	2.09	2.06	2.03	2.01	1.99	1.96	1.92	1.90	1.86	1.83	1.81	29
2.21	2.20	2.16	2.14	2.11	2.09	2.07	2.04	2.01	1.99	1.97	1.94	1.90	1.88	1.84	1.81	1.79	30
2.18	2.16	2.13	2.10	2.08	2.06	2.04	2.00	1.98	1.95	1.93	1.91	1.87	1.85	1.80	1.77	1.75	32
2.15	2.13	2.10	2.07	2.05	2.03	2.01	1.97	1.95	1.92	1.90	1.88	1.84	1.82	1.77	1.74	1.72	34
2.13	2.11	2.08	2.05	2.03	2.00	1.99	1.95	1.92	1.90	1.88	1.85	1.81	1.79	1.74	1.71	1.69	36
2.11	2.09	2.05	2.03	2.00	1.98	1.96	1.93	1.90	1.87	1.85	1.82	1.79	1.76	1.71	1.68	1.66	38
2.09	2.07	2.03	2.01	1.98	1.96	1.94	1.90	1.88	1.85	1.83	1.80	1.76	1.74	1.69	1.66	1.64	40
2.07	2.05	2.02	1.99	1.96	1.94	1.92	1.89	1.86	1.83	1.81	1.78	1.74	1.72	1.67	1.64	1.62	42
2.05	2.03	2.00	1.97	1.95	1.93	1.91	1.87	1.84	1.82	1.80	1.77	1.73	1.70	1.65	1.62	1.60	44
2.04	2.02	1.99	1.96	1.93	1.91	1.89	1.85	1.82	1.80	1.78	1.75	1.71	1.69	1.63	1.60	1.58	46
2.02	2.01	1.97	1.94	1.92	1.90	1.88	1.84	1.81	1.79	1.77	1.73	1.69	1.67	1.62	1.58	1.56	48
2.01	1.99	1.96	1.93	1.91	1.88	1.87	1.83	1.80	1.77	1.75	1.72	1.68	1.66	1.60	1.57	1.55	50
1.99	1.97	1.93	1.90	1.88	1.86	1.84	1.80	1.77	1.74	1.72	1.69	1.65	1.62	1.57	1.54	1.51	55
1.96	1.94	1.91	1.88	1.86	1.83	1.82	1.78	1.74	1.72	1.70	1.67	1.62	1.60	1.54	1.51	1.48	60
1.95	1.93	1.89	1.86	1.84	1.82	1.80	1.76	1.72	1.70	1.68	1.65	1.60	1.58	1.52	1.48	1.46	65
1.93	1.91	1.88	1.85	1.82	1.80	1.78	1.74	1.71	1.68	1.66	1.63	1.58	1.56	1.50	1.46	1.44	70
1.90	1.88	1.85	1.82	1.79	1.77	1.75	1.71	1.68	1.65	1.63	1.60	1.55	1.53	1.47	1.43	1.40	80
1.88	1.86	1.83	1.80	1.77	1.75	1.73	1.69	1.66	1.63	1.61	1.58	1.53	1.50	1.44	1.40	1.37	90
1.87	1.85	1.81	1.78	1.76	1.74	1.71	1.67	1.64	1.61	1.59	1.56	1.51	1.48	1.42	1.38	1.35	100
1.84	1.82	1.79	1.75	1.73	1.71	1.68	1.64	1.61	1.58	1.56	1.52	1.48	1.45	1.38	1.34	1.30	125
1.82	1.80	1.77	1.74	1.71	1.69	1.67	1.62	1.59	1.56	1.54	1.50	1.45	1.42	1.35	1.31	1.27	150
1.80	1.78	1.74	1.71	1.68	1.66	1.64	1.60	1.56	1.53	1.51	1.47	1.42	1.39	1.32	1.27	1.23	200
1.77	1.75	1.72	1.69	1.66	1.64	1.62	1.57	1.54	1.51	1.48	1.45	1.39	1.36	1.28	1.23	1.18	300
1.76	1.75	1.70	1.67	1.64	1.62	1.60	1.55	1.51	1.49	1.46	1.42	1.37	1.34	1.25	1.19	1.14	500
1.74	1.72	1.69	1.65	1.63	1.60	1.58	1.54	1.50	1.47	1.44	1.41	1.35	1.32	1.23	1.16	1.09	1000
1.73	1.71	1.67	1.64	1.61	1.59	1.57	1.52	1.48	1.45	1.43	1.39	1.33	1.30	1.21	1.13	1.00	∞

Approximate formula for ν_1 and ν_2 larger than 30: $\log_{10} F_{.025;\nu_1,\nu_2} \simeq \dfrac{1.7023}{\sqrt{h - 1.14}} - 0.846\left(\dfrac{1}{\nu_1} - \dfrac{1}{\nu_2}\right)$, where $\dfrac{1}{h} = \dfrac{1}{2}\left(\dfrac{1}{\nu_1} + \dfrac{1}{\nu_2}\right)$.

TABLE 13.35. *PERCENTAGE POINT ONE OF THE F DISTRIBUTION. TABLE OF* $F_{.01;\nu_1,\nu_2}$.

Degrees of Freedom for the Denominator ν_2	Degrees of Freedom for the Numerator ν_1																	
	1	2	3	4	5	6	7	8	9	10	11	12	13	14	15	16	17	18
	Multiply the Numbers of the First Row ($\nu_2 = 1$) by 10																	
1	*405*	*500*	*540*	*563*	*576*	*586*	*593*	*598*	*602*	*606*	*608*	*611*	*613*	*614*	*616*	*617*	*618*	*619*
2	98.5	99.0	99.2	99.2	99.3	99.3	99.4	99.4	99.4	99.4	99.4	99.4	99.4	99.4	99.4	99.4	99.4	99.4
3	34.1	30.8	19.5	28.7	28.2	27.9	27.7	27.5	27.3	27.2	27.1	27.1	27.0	26.9	26.9	26.8	26.8	26.8
4	21.2	18.0	16.7	16.0	15.5	15.2	15.0	14.8	14.7	14.5	14.4	14.4	14.3	14.2	14.2	14.2	14.1	14.1
5	16.3	13.3	12.1	11.4	11.0	10.7	10.5	10.3	10.2	10.1	9.96	9.89	9.82	9.77	9.72	9.68	9.64	9.61
6	13.7	10.9	9.78	9.15	8.75	8.47	8.26	8.10	7.98	7.87	7.79	7.72	7.66	7.60	7.56	7.52	7.48	7.45
7	12.2	9.55	8.45	7.85	7.46	7.19	6.99	6.84	6.72	6.62	6.54	6.47	6.41	6.36	6.31	6.27	6.24	6.21
8	11.3	8.65	7.59	7.01	6.63	6.37	6.18	6.03	5.91	5.81	5.73	5.67	5.61	5.56	5.52	5.48	5.44	5.41
9	10.6	8.02	6.99	6.42	6.06	5.80	5.61	5.47	5.35	5.26	5.18	5.11	5.05	5.00	4.96	4.92	4.89	4.86
10	10.0	7.56	6.55	5.99	5.64	5.39	5.20	5.06	4.94	4.85	4.77	4.71	4.65	4.60	4.56	4.52	4.49	4.46
11	9.65	7.21	6.22	5.67	5.32	5.07	4.89	4.74	4.63	4.54	4.46	4.40	4.34	4.29	4.25	4.21	4.18	4.15
12	9.33	6.93	5.95	5.41	5.06	4.82	4.64	4.50	4.39	4.30	4.22	4.16	4.10	4.05	4.01	3.97	3.94	3.91
13	9.07	6.70	5.74	5.21	4.86	4.62	4.44	4.30	4.19	4.10	4.02	3.96	3.91	3.86	3.82	3.78	3.75	3.72
14	8.86	6.51	5.56	5.04	4.70	4.46	4.28	4.14	4.03	3.94	3.86	3.80	3.75	3.70	3.66	3.62	3.59	3.56
15	8.68	6.36	5.42	4.89	4.56	4.32	4.14	4.00	3.89	3.80	3.73	3.67	3.61	3.56	3.52	3.49	3.45	3.42
16	8.53	6.23	5.29	4.77	4.44	4.20	4.03	3.89	3.78	3.69	3.62	3.55	3.50	3.45	3.41	3.37	3.34	3.31
17	8.40	6.11	5.18	4.67	4.34	4.10	3.93	3.79	3.68	3.59	3.52	3.46	3.40	3.35	3.31	3.27	3.24	3.21
18	8.29	6.01	5.09	4.58	4.25	4.01	3.84	3.71	3.60	3.51	3.43	3.37	3.32	3.27	3.23	3.19	3.16	3.13
19	8.18	5.93	5.01	4.50	4.17	3.94	3.77	3.63	3.52	3.43	3.36	3.30	3.24	3.19	3.15	3.12	3.08	3.05
20	8.10	5.85	4.94	4.43	4.10	3.87	3.70	3.56	3.46	3.37	3.29	3.23	3.18	3.13	3.09	3.05	3.02	2.99
21	8.02	5.78	4.87	4.37	4.04	3.81	3.64	3.51	3.40	3.31	3.24	3.17	3.12	3.07	3.03	2.99	2.96	2.93
22	7.95	5.72	4.82	4.31	3.99	3.76	3.59	3.45	3.35	3.26	3.18	3.12	3.07	3.02	2.98	2.94	2.91	2.88
23	7.88	5.66	4.76	4.26	3.94	3.71	3.54	3.41	3.30	3.21	3.14	3.07	3.02	2.97	2.93	2.89	2.86	2.83
24	7.82	5.61	4.72	4.22	3.90	3.67	3.50	3.36	3.26	3.17	3.09	3.03	2.98	2.93	2.89	2.85	2.82	2.79
25	7.77	5.57	4.68	4.18	3.86	3.63	3.46	3.32	3.22	3.13	3.06	2.99	2.94	2.89	2.85	2.81	2.78	2.75
26	7.72	5.53	4.64	4.14	3.82	3.59	3.42	3.29	3.18	3.09	3.02	2.96	2.90	2.86	2.82	2.78	2.74	2.72
27	7.68	5.49	4.60	4.11	3.78	3.56	3.39	3.26	3.15	3.06	2.99	2.93	2.87	2.82	2.78	2.75	2.71	2.68

Degrees of Freedom for the Denominator ν_2	Degrees of Freedom for the Numerator ν_1																	
	1	2	3	4	5	6	7	8	9	10	11	12	13	14	15	16	17	18
28	7.64	5.45	4.57	4.07	3.75	3.53	3.36	3.23	3.12	3.03	2.96	2.90	2.84	2.79	2.75	2.72	2.68	2.65
29	7.60	5.42	4.54	4.04	3.73	3.50	3.33	3.20	3.09	3.00	2.93	2.87	2.81	2.77	2.73	2.69	2.66	2.63
30	7.56	5.39	4.51	4.02	3.70	3.47	3.30	3.17	3.07	2.98	2.91	2.84	2.79	2.74	2.70	2.66	2.63	2.60
32	7.50	5.34	4.46	3.97	3.65	3.43	3.26	3.13	3.02	2.93	2.86	2.80	2.74	2.70	2.66	2.62	2.58	2.55
34	7.44	5.29	4.42	3.93	3.61	3.39	3.22	3.09	2.98	2.89	2.82	2.76	2.70	2.66	2.62	2.58	2.55	2.51
36	7.40	5.25	4.38	3.89	3.57	3.35	3.18	3.05	2.95	2.86	2.79	2.72	2.67	2.62	2.58	2.54	2.51	2.48
38	7.35	5.21	4.34	3.86	3.54	3.32	3.15	3.02	2.92	2.83	2.75	2.69	2.64	2.59	2.55	2.51	2.48	2.45
40	7.31	5.18	4.31	3.83	3.51	3.29	3.12	2.99	2.89	2.80	2.73	2.66	2.61	2.56	2.52	2.48	2.45	2.42
42	7.28	5.15	4.29	3.80	3.49	3.27	3.10	2.97	2.86	2.78	2.70	2.64	2.59	2.54	2.50	2.46	2.43	2.40
44	7.25	5.12	4.26	3.78	3.47	3.24	3.08	2.95	2.84	2.75	2.68	2.62	2.56	2.52	2.47	2.44	2.40	2.37
46	7.22	5.10	4.24	3.76	3.44	3.22	3.06	2.93	2.82	2.73	2.66	2.60	2.54	2.50	2.45	2.42	2.38	2.35
48	7.19	5.08	4.22	3.74	3.43	3.20	3.04	2.91	2.80	2.72	2.64	2.58	2.53	2.48	2.44	2.40	2.37	2.33
50	7.17	5.06	4.20	3.72	3.41	3.19	3.02	2.89	2.79	2.70	2.63	2.56	2.51	2.46	2.42	2.38	2.35	2.32
55	7.12	5.01	4.16	3.68	3.37	3.15	2.98	2.85	2.75	2.66	2.59	2.53	2.47	2.42	2.38	2.34	2.31	2.28
60	7.08	4.98	4.13	3.65	3.34	3.12	2.95	2.82	2.72	2.63	2.56	2.50	2.44	2.39	2.35	2.31	2.28	2.25
65	7.04	4.95	4.10	3.62	3.31	3.09	2.93	2.80	2.69	2.61	2.53	2.47	2.42	2.37	2.33	2.29	2.26	2.23
70	7.01	4.92	4.08	3.60	3.29	3.07	2.91	2.78	2.67	2.59	2.51	2.45	2.40	2.35	2.31	2.27	2.23	2.20
80	6.96	4.88	4.04	3.56	3.26	3.04	2.87	2.74	2.64	2.55	2.48	2.42	2.36	2.31	2.27	2.23	2.20	2.17
90	6.93	4.85	4.01	3.54	3.23	3.01	2.84	2.72	2.61	2.52	2.45	2.39	2.33	2.29	2.24	2.21	2.17	2.14
100	6.90	4.82	3.98	3.51	3.21	2.99	2.82	2.69	2.59	2.50	2.43	2.37	2.31	2.26	2.22	2.19	2.15	2.12
125	6.84	4.78	3.94	3.47	3.17	2.95	2.79	2.66	2.55	2.47	2.39	2.33	2.28	2.23	2.19	2.15	2.11	2.08
150	6.81	4.75	3.92	3.45	3.14	2.92	2.76	2.63	2.53	2.44	2.37	2.31	2.25	2.20	2.16	2.12	2.09	2.06
200	6.76	4.71	3.88	3.41	3.11	2.89	2.73	2.60	2.50	2.41	2.34	2.27	2.22	2.17	2.13	2.09	2.06	2.02
300	6.72	4.68	3.85	3.38	3.08	2.86	2.70	2.57	2.47	2.38	2.31	2.24	2.19	2.14	2.10	2.06	2.03	1.99
500	6.69	4.65	3.82	3.36	3.05	2.84	2.68	2.55	2.44	2.36	2.28	2.22	2.17	2.12	2.07	2.04	2.00	1.97
1000	6.66	4.63	3.80	3.34	3.04	2.82	2.66	2.53	2.43	2.34	2.27	2.20	2.15	2.10	2.06	2.02	1.98	1.95
∞	6.63	4.61	3.78	3.32	3.02	2.80	2.64	2.51	2.41	2.32	2.25	2.18	2.13	2.08	2.04	2.00	1.97	1.93

Example: $P\{F_{.01;8,20} < 3.56\} = 99$ per cent.

$F_{.99;\nu_1,\nu_2} = 1/F_{.01;\nu_2,\nu_1}$. Example: $F_{.99;8,20} = 1/F_{.01;20,8} = 1/5.36 = 0.187$.

TABLE 13.35. *Cont.*

Degrees of Freedom for the Numerator ν_1																	Degrees of Freedom for the Denominator ν_2
19	20	22	24	26	28	30	35	40	45	50	60	80	100	200	500	∞	
Multiply the Numbers of the First Row ($\nu_2 = 1$) by 10																	
620	*621*	*622*	*623*	*624*	*625*	*626*	*628*	*629*	*630*	*630*	*631*	*633*	*633*	*635*	*636*	*637*	1
99.4	99.4	99.5	99.5	99.5	99.5	99.5	99.5	99.5	99.5	99.5	99.5	99.5	99.5	99.5	99.5	99.5	2
26.7	26.7	26.6	26.6	26.6	26.5	26.5	26.5	26.4	26.4	26.4	26.3	26.3	26.2	26.2	26.1	26.1	3
14.0	14.0	14.0	13.9	13.9	13.9	13.8	13.8	13.7	13.7	13.7	13.7	13.6	13.6	13.5	13.5	13.5	4
9.58	9.55	9.51	9.47	9.43	9.40	9.38	9.33	9.29	9.26	9.24	9.20	9.16	9.13	9.08	9.04	9.02	5
7.42	7.40	7.35	7.31	7.28	7.25	7.23	7.18	7.14	7.11	7.09	7.06	7.01	6.99	6.93	6.90	6.88	6
6.18	6.16	6.11	6.07	6.04	6.02	5.99	5.94	5.91	5.88	5.86	5.82	5.78	5.75	5.70	5.67	5.65	7
5.38	5.36	5.32	5.28	5.25	5.22	5.20	5.15	5.12	5.09	5.07	5.03	4.99	4.96	4.91	4.88	4.86	8
4.83	4.81	4.77	4.73	4.70	4.67	4.65	4.60	4.57	4.54	4.52	4.48	4.44	4.42	4.36	4.33	4.31	9
4.43	4.41	4.36	4.33	4.30	4.27	4.25	4.20	4.17	4.14	4.12	4.08	4.04	4.01	3.96	3.93	3.91	10
4.12	4.10	4.06	4.02	3.99	3.96	3.94	3.89	3.86	3.83	3.81	3.78	3.73	3.71	3.66	3.62	3.60	11
3.88	3.86	3.82	3.78	3.75	3.72	3.70	3.65	3.62	3.59	3.57	3.54	3.49	3.47	3.41	3.38	3.36	12
3.69	3.66	3.62	3.59	3.56	3.53	3.51	3.46	3.43	3.40	3.38	3.34	3.30	3.27	3.22	3.19	3.17	13
3.53	3.51	3.46	3.43	3.40	3.37	3.35	3.30	3.27	3.24	3.32	3.18	3.14	3.11	3.06	3.03	3.00	14
3.40	3.37	3.33	3.29	3.26	3.24	3.21	3.17	3.13	3.10	3.08	3.05	3.00	2.98	2.92	2.89	2.87	15
3.28	3.26	3.22	3.18	3.15	3.12	3.10	3.05	3.02	2.99	2.97	2.93	2.89	2.86	2.81	2.78	2.75	16
3.18	3.16	3.12	3.08	3.05	3.03	3.00	2.96	2.92	2.89	2.87	2.83	2.79	2.76	2.71	2.68	2.65	17
3.10	3.08	3.03	3.00	2.97	2.94	2.92	2.87	2.84	2.81	2.78	2.75	2.70	2.68	2.62	2.59	2.57	18
3.03	3.00	2.96	2.92	2.89	2.87	2.84	2.80	2.76	2.73	2.71	2.67	2.63	2.60	2.55	2.51	2.49	19
2.96	2.94	2.90	2.86	2.83	2.80	2.78	2.73	2.69	2.67	2.64	2.61	2.56	2.54	2.48	2.44	2.42	20
2.90	2.88	2.84	2.80	2.77	2.74	2.72	2.67	2.64	2.61	2.58	2.55	2.50	2.48	2.42	2.38	2.36	21
2.85	2.83	2.78	2.75	2.72	2.69	2.67	2.62	2.58	2.55	2.53	2.50	2.45	2.42	2.36	2.33	2.31	22
2.80	2.78	2.74	2.70	2.67	2.64	2.62	2.57	2.54	2.51	2.48	2.45	2.40	2.37	2.32	2.28	2.26	23
2.76	2.74	2.70	2.66	2.63	2.60	2.58	2.53	2.49	2.46	2.44	2.40	2.36	2.33	2.27	2.24	2.21	24
2.72	2.70	2.66	2.62	2.59	2.56	2.54	2.49	2.45	2.42	2.40	2.36	2.32	2.29	2.23	2.19	2.17	25
2.69	2.66	2.62	2.58	2.55	2.53	2.50	2.45	2.42	2.39	2.36	2.33	2.28	2.25	2.19	2.16	2.13	26
2.66	2.63	2.59	2.55	2.52	2.49	2.47	2.42	2.38	2.35	2.33	2.29	2.25	2.22	2.16	2.12	2.10	27
2.63	2.60	2.56	2.52	2.49	2.46	2.44	2.39	2.35	2.32	2.30	2.26	2.22	2.19	2.13	2.09	2.06	28

Degrees of Freedom for the Numerator ν_1

19	20	22	24	26	28	30	35	40	45	50	60	80	100	200	500	∞	ν_2
2.60	2.57	2.53	2.49	2.46	2.44	2.41	2.36	2.33	2.30	2.27	2.23	2.19	2.16	2.10	2.06	2.03	29
2.57	2.55	2.51	2.47	2.44	2.41	2.39	2.34	2.30	2.27	2.25	2.21	2.16	2.13	2.07	2.03	2.01	30
2.53	2.50	2.46	2.42	2.39	2.36	2.34	2.29	2.25	2.22	2.20	2.16	2.11	2.08	2.02	1.98	1.96	32
2.49	2.46	2.42	2.38	2.35	2.32	2.30	2.25	2.21	2.18	2.16	2.12	2.07	2.04	1.98	1.94	1.91	34
2.45	2.43	2.38	2.35	2.32	2.29	2.26	2.21	2.17	2.14	2.12	2.08	2.03	2.00	1.94	1.90	1.87	36
2.42	2.40	2.35	2.32	2.28	2.26	2.23	2.18	2.14	2.11	2.09	2.05	2.00	1.97	1.90	1.86	1.84	38
2.39	2.37	2.33	2.29	2.26	2.23	2.20	2.15	2.11	2.08	2.06	2.02	1.97	1.94	1.87	1.83	1.80	40
2.37	2.34	2.30	2.26	2.23	2.20	2.18	2.13	2.09	2.06	2.03	1.99	1.94	1.91	1.85	1.80	1.78	42
2.35	2.32	2.28	2.24	2.21	2.18	2.15	2.10	2.06	2.03	2.01	1.97	1.92	1.89	1.82	1.78	1.75	44
2.33	2.30	2.26	2.22	2.19	2.16	2.13	2.08	2.04	2.01	1.99	1.95	1.90	1.86	1.80	1.75	1.73	46
2.31	2.28	2.24	2.20	2.17	2.14	2.12	2.06	2.02	1.99	1.97	1.93	1.88	1.84	1.78	1.73	1.70	48
2.29	2.27	2.22	2.18	2.15	2.12	2.10	2.05	2.01	1.97	1.95	1.91	1.86	1.82	1.76	1.71	1.68	50
2.25	2.23	2.18	2.15	2.11	2.08	2.06	2.01	1.97	1.93	1.91	1.87	1.81	1.78	1.71	1.67	1.64	55
2.22	2.20	2.15	2.12	2.08	2.05	2.03	1.98	1.94	1.90	1.88	1.84	1.78	1.75	1.68	1.63	1.60	60
2.20	2.17	2.13	2.09	2.06	2.03	2.00	1.95	1.91	1.88	1.85	1.81	1.75	1.72	1.65	1.60	1.57	65
2.18	2.15	2.11	2.07	2.03	2.01	1.98	1.93	1.89	1.85	1.83	1.78	1.73	1.70	1.62	1.57	1.54	70
2.14	2.12	2.07	2.03	2.00	1.97	1.94	1.89	1.85	1.81	1.79	1.75	1.69	1.66	1.58	1.53	1.49	80
2.11	2.09	2.04	2.00	1.97	1.94	1.92	1.86	1.82	1.79	1.76	1.72	1.66	1.62	1.54	1.49	1.46	90
2.09	2.07	2.02	1.98	1.94	1.92	1.89	1.84	1.80	1.76	1.73	1.69	1.63	1.60	1.52	1.47	1.43	100
2.05	2.03	1.98	1.94	1.91	1.88	1.85	1.80	1.76	1.72	1.69	1.65	1.59	1.55	1.47	1.41	1.37	125
2.03	2.00	1.96	1.92	1.88	1.85	1.83	1.77	1.73	1.69	1.66	1.62	1.56	1.52	1.43	1.38	1.33	150
2.00	1.97	1.93	1.89	1.85	1.82	1.79	1.74	1.69	1.66	1.63	1.58	1.52	1.48	1.39	1.33	1.28	200
1.97	1.94	1.89	1.85	1.82	1.79	1.76	1.71	1.66	1.62	1.59	1.55	1.48	1.44	1.35	1.28	1.22	300
1.94	1.92	1.87	1.83	1.79	1.76	1.74	1.68	1.63	1.60	1.56	1.52	1.45	1.41	1.31	1.23	1.16	500
1.92	1.90	1.85	1.81	1.77	1.74	1.72	1.66	1.61	1.57	1.54	1.50	1.43	1.38	1.28	1.19	1.11	1000
1.90	1.88	1.83	1.79	1.76	1.72	1.70	1.64	1.59	1.55	1.52	1.47	1.40	1.36	1.25	1.15	1.00	∞

Degrees of Freedom for the Denominator ν_2

Approximate formula for ν_1 and ν_2 larger than 30: $\log_{10} F_{.01;\nu_1,\nu_2} \simeq \dfrac{2.0206}{\sqrt{h} - 1.40} - 1.073\left(\dfrac{1}{\nu_1} - \dfrac{1}{\nu_2}\right)$, where $\dfrac{1}{h} = \dfrac{1}{2}\left(\dfrac{1}{\nu_1} + \dfrac{1}{\nu_2}\right)$.

TABLE 13.36. *PERCENTAGE POINT 0.5 OF THE F DISTRIBUTION. TABLE OF* $F_{.005;\nu_1,\nu_2}$.

Degrees of Freedom for the Denominator ν_2	Degrees of Freedom for the Numerator ν_1: 1	2	3	4	5	6	7	8	9	10	11	12	13	14	15	16	17	18
	Multiply the Numbers of the First Row ($\nu_2 = 1$) by 100																	
1	*162*	*200*	*216*	*225*	*231*	*234*	*237*	*239*	*241*	*242*	*243*	*244*	*245*	*246*	*246*	*247*	*247*	*248*
2	198	199	199	199	199	199	199	199	199	199	199	199	199	199	199	199	199	199
3	55.6	49.8	47.5	46.2	45.4	44.8	44.4	44.1	43.9	43.7	43.5	43.4	43.3	43.2	43.1	43.0	42.9	42.9
4	31.3	26.3	24.3	23.2	22.5	22.0	21.6	21.4	21.1	21.0	20.8	20.7	20.6	20.5	20.4	20.4	20.3	20.3
5	22.8	18.3	16.5	15.6	14.9	14.5	14.2	14.0	13.8	13.6	13.5	13.4	13.3	13.2	13.1	13.1	13.0	13.0
6	18.6	14.5	12.9	12.0	11.5	11.1	10.8	10.6	10.4	10.2	10.1	10.0	9.95	9.88	9.81	9.76	9.71	9.66
7	16.2	12.4	10.9	10.0	9.52	9.16	8.89	8.68	8.51	8.38	8.27	8.18	8.10	8.03	7.97	7.93	7.87	7.83
8	14.7	11.0	9.60	8.81	8.30	7.95	7.69	7.50	7.34	7.21	7.10	7.01	6.94	6.87	6.81	6.76	6.72	6.68
9	13.6	10.1	8.72	7.96	7.47	7.13	6.88	6.69	6.54	6.42	6.31	6.23	6.15	6.09	6.03	5.98	5.94	5.90
10	12.8	9.43	8.08	7.34	6.87	6.54	6.30	6.12	5.97	5.85	5.75	5.66	5.59	5.53	5.47	5.42	5.38	5.34
11	12.2	8.91	7.60	6.88	6.42	6.10	5.86	5.68	5.54	5.42	5.32	5.24	5.16	5.10	5.05	5.00	4.96	4.92
12	11.8	8.51	7.23	6.52	6.07	5.76	5.52	5.35	5.20	5.09	4.99	4.91	4.84	4.77	4.72	4.67	4.63	4.59
13	11.4	8.19	6.93	6.23	5.79	5.48	5.25	5.08	4.94	4.82	4.72	4.64	4.57	4.51	4.46	4.41	4.37	4.33
14	11.1	7.92	6.68	6.00	5.56	5.26	5.03	4.86	4.72	4.60	4.51	4.43	4.36	4.30	4.25	4.20	4.16	4.12
15	10.8	7.70	6.48	5.80	5.37	5.07	4.85	4.67	4.54	4.42	4.33	4.25	4.18	4.12	4.07	4.02	3.98	3.95
16	10.6	7.51	6.30	5.64	5.21	4.91	4.69	4.52	4.38	4.27	4.18	4.10	4.03	3.97	3.92	3.87	3.83	3.80
17	10.4	7.35	6.16	5.50	5.07	4.78	4.56	4.39	4.25	4.14	4.05	3.97	3.90	3.84	3.79	3.75	3.71	3.67
18	10.2	7.21	6.03	5.37	4.96	4.66	4.44	4.28	4.14	4.03	3.94	3.86	3.79	3.73	3.68	3.64	3.60	3.56
19	10.1	7.09	5.92	5.27	4.85	4.56	4.34	4.18	4.04	3.93	3.84	3.76	3.70	3.64	3.59	3.54	3.50	3.46
20	9.94	6.99	5.82	5.17	4.76	4.47	4.26	4.09	3.96	3.85	3.76	3.68	3.61	3.55	3.50	3.46	3.42	3.38
21	9.83	6.89	5.73	5.09	4.68	4.39	4.18	4.01	3.88	3.77	3.68	3.60	3.54	3.48	3.43	3.38	3.34	3.31
22	9.73	6.81	5.65	5.02	4.61	4.32	4.11	3.94	3.81	3.70	3.61	3.54	3.47	3.41	3.36	3.31	3.27	3.24
23	9.63	6.73	5.58	4.95	4.54	4.26	4.05	3.88	3.75	3.64	3.55	3.47	3.41	3.35	3.30	3.25	3.21	3.18
24	9.55	6.66	5.52	4.89	4.49	4.20	3.99	3.83	3.69	3.59	3.50	3.42	3.35	3.30	3.25	3.20	3.16	3.12
25	9.48	6.60	5.46	4.84	4.43	4.15	3.94	3.78	3.64	3.54	3.45	3.37	3.30	3.25	3.20	3.15	3.11	3.08
26	9.41	6.54	5.41	4.79	4.38	4.10	3.89	3.73	3.60	3.49	3.40	3.33	3.26	3.20	3.15	3.11	3.07	3.03
27	9.34	6.49	5.36	4.74	4.34	4.06	3.85	3.69	3.56	3.45	3.36	3.28	3.22	3.16	3.11	3.07	3.03	2.99

Degrees of Freedom for the Denominator ν_2	Degrees of Freedom for the Numerator ν_1																	
	1	2	3	4	5	6	7	8	9	10	11	12	13	14	15	16	17	18
28	9.28	6.44	5.32	4.70	4.30	4.02	3.81	3.65	3.52	3.41	3.32	3.25	3.18	3.12	3.07	3.03	2.99	2.95
29	9.23	6.40	5.28	4.66	4.26	3.98	3.77	3.61	3.48	3.38	3.29	3.21	3.15	3.09	3.04	2.99	2.95	2.92
30	9.18	6.35	5.24	4.62	4.23	3.95	3.74	3.58	3.45	3.34	3.25	3.18	3.11	3.06	3.01	2.96	2.92	2.89
32	9.09	6.28	5.17	4.56	4.17	3.89	3.68	3.52	3.39	3.29	3.20	3.12	3.06	3.00	2.95	2.90	2.86	2.83
34	9.01	6.22	5.11	4.50	4.11	3.84	3.63	3.47	3.34	3.24	3.15	3.07	3.01	2.95	2.90	2.85	2.81	2.78
36	8.94	6.16	5.06	4.46	4.06	3.79	3.58	3.42	3.30	3.19	3.10	3.03	2.96	2.90	2.85	2.81	2.77	2.73
38	8.88	6.11	5.02	4.41	4.02	3.75	3.54	3.39	3.25	3.15	3.06	2.99	2.92	2.87	2.82	2.77	2.73	2.70
40	8.83	6.07	4.98	4.37	3.99	3.71	3.51	3.35	3.22	3.12	3.03	2.95	2.89	2.83	2.78	2.74	2.70	2.66
42	8.78	6.03	4.94	4.34	3.95	3.68	3.48	3.32	3.19	3.09	3.00	2.92	2.86	2.80	2.75	2.71	2.67	2.63
44	8.74	5.99	4.91	4.31	3.92	3.65	3.45	3.29	3.16	3.06	2.97	2.89	2.83	2.77	2.72	2.68	2.64	2.60
46	8.70	5.96	4.88	4.28	3.90	3.62	3.42	3.26	3.14	3.03	2.94	2.87	2.80	2.75	2.70	2.65	2.61	2.58
48	8.66	5.93	4.85	4.25	3.87	3.60	3.40	3.24	3.11	3.01	2.92	2.85	2.78	2.72	2.67	2.63	2.59	2.55
50	8.63	5.90	4.83	4.23	3.85	3.58	3.38	3.22	3.09	2.99	2.90	2.82	2.76	2.70	2.65	2.61	2.57	2.53
55	8.55	5.84	4.77	4.18	3.80	3.53	3.33	3.17	3.05	2.94	2.85	2.78	2.71	2.66	2.61	2.56	2.52	2.49
60	8.49	5.80	4.73	4.14	3.76	3.49	3.29	3.13	3.01	2.90	2.82	2.74	2.68	2.62	2.57	2.53	2.49	2.45
65	8.44	5.75	4.68	4.11	3.73	3.46	3.26	3.10	2.98	2.87	2.79	2.71	2.65	2.59	2.54	2.49	2.45	2.42
70	8.40	5.72	4.65	4.08	3.70	3.43	3.23	3.08	2.95	2.85	2.76	2.68	2.62	2.56	2.51	2.47	2.43	2.39
80	8.33	5.67	4.61	4.03	3.65	3.39	3.19	3.03	2.91	2.80	2.72	2.64	2.58	2.52	2.47	2.43	2.39	2.35
90	8.28	5.62	4.57	3.99	3.62	3.35	3.15	3.00	2.87	2.77	2.68	2.61	2.54	2.49	2.44	2.39	2.35	2.32
100	8.24	5.59	4.54	3.96	3.59	3.33	3.13	2.97	2.85	2.74	2.66	2.58	2.52	2.46	2.41	2.37	2.33	2.29
125	8.17	5.53	4.49	3.91	3.54	3.28	3.08	2.93	2.80	2.70	2.61	2.54	2.47	2.42	2.37	2.32	2.28	2.24
150	8.12	5.49	4.45	3.88	3.51	3.25	3.05	2.89	2.77	2.67	2.58	2.51	2.44	2.38	2.33	2.29	2.25	2.21
200	8.06	5.44	4.41	3.84	3.47	3.21	3.01	2.85	2.73	2.63	2.54	2.47	2.40	2.35	2.30	2.25	2.21	2.18
300	8.00	5.39	4.37	3.80	3.43	3.17	2.97	2.81	2.69	2.59	2.51	2.43	2.37	2.31	2.26	2.21	2.17	2.14
500	7.95	5.36	4.33	3.76	3.40	3.14	2.94	2.79	2.66	2.56	2.48	2.40	2.34	2.28	2.23	2.19	2.14	2.11
1000	7.92	5.33	4.31	3.74	3.37	3.11	2.92	2.77	2.64	2.54	2.45	2.38	2.32	2.26	2.21	2.16	2.12	2.09
∞	7.88	5.30	4.28	3.72	3.35	3.09	2.90	2.74	2.62	2.52	2.43	2.36	2.29	2.24	2.19	2.14	2.10	2.06

Example: $P\{F_{.005;8,20} < 4.09\} = 99.5$ per cent.

$F_{.995;\nu_1,\nu_2} = 1/F_{.005;\nu_2,\nu_1}$. Example: $F_{.995;8,20} = 1/F_{.005;20,8} = 1/6.61 = 0.151$.

TABLE 13.36. *Cont.*

Degrees of Freedom for the Numerator ν_1																	
19	20	22	24	26	28	30	35	40	45	50	60	80	100	200	500	∞	Degrees of Freedom for the Denominator ν_2
Multiply the Numbers of the First Row ($\nu = 1$) by 100																	
248	*248*	*249*	*249*	*250*	*250*	*250*	*251*	*251*	*252*	*252*	*253*	*253*	*253*	*254*	*254*	*255*	1
199	199	199	199	199	199	199	199	199	199	199	199	199	199	199	200	200	2
42.8	42.8	42.7	42.6	42.6	42.5	42.5	42.4	42.3	42.3	42.2	42.1	42.1	42.0	41.9	41.9	41.8	3
20.2	20.2	20.1	20.0	20.0	19.9	19.9	19.8	19.8	19.7	19.7	19.6	19.5	19.5	19.4	19.4	19.3	4
12.9	12.9	12.8	12.8	12.7	12.7	12.7	12.6	12.5	12.5	12.5	12.4	12.3	12.3	12.2	12.2	12.1	5
9.62	9.59	9.53	9.47	9.43	9.39	9.36	9.29	9.24	9.20	9.17	9.12	9.06	9.03	8.95	8.91	8.88	6
7.79	7.75	7.69	7.64	7.60	7.57	7.53	7.47	7.42	7.38	7.35	7.31	7.25	7.22	7.15	7.10	7.08	7
6.64	6.61	6.55	6.50	6.46	6.43	6.40	6.33	6.29	6.25	6.22	6.18	6.12	6.09	6.02	5.98	5.95	8
5.86	5.83	5.78	5.73	5.69	5.65	5.62	5.56	5.52	5.48	5.45	5.41	5.36	5.32	5.26	5.21	5.19	9
5.30	5.27	5.22	5.17	5.13	5.10	5.07	5.01	4.97	4.93	4.90	4.86	4.80	4.77	4.71	4.67	4.64	10
4.89	4.86	4.80	4.76	4.72	4.68	4.65	4.60	4.55	4.52	4.49	4.44	4.39	4.36	4.29	4.25	4.23	11
4.56	4.53	4.48	4.43	4.39	4.36	4.33	4.27	4.23	4.19	4.17	4.12	4.07	4.04	3.97	3.93	3.90	12
4.30	4.27	4.22	4.17	4.13	4.10	4.07	4.01	3.97	3.94	3.91	3.87	3.81	3.78	3.71	3.67	3.65	13
4.09	4.06	4.01	3.96	3.92	3.89	3.86	3.80	3.76	3.73	3.70	3.66	3.60	3.57	3.50	3.46	3.44	14
3.91	3.88	3.83	3.79	3.75	3.72	3.69	3.63	3.58	3.55	3.52	3.48	3.43	3.39	3.33	3.29	3.26	15
3.76	3.73	3.68	3.64	3.60	3.57	3.54	3.48	3.44	3.40	3.37	3.33	3.28	3.25	3.18	3.14	3.11	16
3.64	3.61	3.56	3.51	3.47	3.44	3.41	3.35	3.31	3.28	3.25	3.21	3.15	3.12	3.05	3.01	2.98	17
3.53	3.50	3.45	3.40	3.36	3.33	3.30	3.25	3.20	3.17	3.14	3.10	3.04	3.01	2.94	2.90	2.87	18
3.43	3.40	3.35	3.31	3.27	3.24	3.21	3.15	3.11	3.07	3.04	3.00	2.95	2.91	2.85	2.80	2.78	19
3.35	3.32	3.27	3.22	3.18	3.15	3.12	3.07	3.02	2.99	2.96	2.92	2.86	2.83	2.76	2.72	2.69	20
3.27	3.24	3.19	3.15	3.11	3.08	3.05	2.99	2.95	2.91	2.88	2.84	2.78	2.75	2.68	2.64	2.61	21
3.20	3.18	3.12	3.08	3.04	3.01	2.98	2.92	2.88	2.84	2.82	2.77	2.72	2.69	2.62	2.57	2.55	22
3.15	3.12	3.06	3.02	2.98	2.95	2.92	2.86	2.82	2.78	2.76	2.71	2.66	2.62	2.56	2.51	2.48	23
3.09	3.06	3.01	2.97	2.93	2.90	2.87	2.81	2.77	2.73	2.70	2.66	2.60	2.57	2.50	2.46	2.43	24
3.04	3.01	2.96	2.92	2.88	2.85	2.82	2.76	2.72	2.68	2.65	2.61	2.55	2.52	2.45	2.41	2.38	25
3.00	2.97	2.92	2.87	2.83	2.80	2.77	2.72	2.67	2.64	2.61	2.56	2.51	2.47	2.40	2.36	2.33	26
2.96	2.93	2.88	2.83	2.79	2.76	2.73	2.67	2.63	2.59	2.57	2.52	2.47	2.43	2.36	2.32	2.29	27
2.92	2.89	2.84	2.79	2.76	2.72	2.69	2.64	2.59	2.56	2.53	2.48	2.43	2.39	2.32	2.28	2.25	28

Degrees of Freedom for the Numerator ν_1																	Degrees of Freedom for the Denominator ν_2
19	20	22	24	26	28	30	35	40	45	50	60	80	100	200	500	∞	
2.88	2.86	2.80	2.76	2.72	2.59	2.66	2.60	2.56	2.52	2.49	2.45	2.39	2.36	2.28	2.24	2.21	29
2.85	2.82	2.77	2.73	2.69	2.66	2.63	2.57	2.52	2.49	2.46	2.42	2.36	2.32	2.25	2.21	2.18	30
2.80	2.77	2.71	2.67	2.63	2.60	2.57	2.51	2.47	2.43	2.40	2.36	2.30	2.26	2.19	2.15	2.11	32
2.75	2.72	2.66	2.62	2.58	2.55	2.52	2.46	2.42	2.38	2.35	2.30	2.25	2.21	2.14	2.09	2.06	34
2.70	2.67	2.62	2.58	2.54	2.50	2.48	2.42	2.37	2.33	2.30	2.26	2.20	2.17	2.09	2.04	2.01	36
2.66	2.63	2.58	2.54	2.50	2.47	2.44	2.38	2.33	2.29	2.27	2.22	2.16	2.12	2.05	2.00	1.97	38
2.63	2.60	2.55	2.50	2.46	2.43	2.40	2.34	2.30	2.26	2.23	2.18	2.12	2.09	2.01	1.96	1.93	40
2.60	2.57	2.52	2.47	2.43	2.40	2.37	2.31	2.26	2.23	2.20	2.15	2.09	2.06	1.98	1.93	1.90	42
2.57	2.54	2.49	2.44	2.40	2.37	2.34	2.28	2.24	2.20	2.17	2.12	2.06	2.03	1.95	1.90	1.87	44
2.54	2.51	2.46	2.42	2.38	2.34	2.32	2.26	2.21	2.17	2.14	2.10	2.04	2.00	1.92	1.87	1.84	46
2.52	2.49	2.44	2.39	2.36	2.32	2.29	2.23	2.19	2.15	2.12	2.07	2.01	1.97	1.89	1.84	1.81	48
2.50	2.47	2.42	2.37	2.33	2.30	2.27	2.21	2.16	2.13	2.10	2.05	1.99	1.95	1.87	1.82	1.79	50
2.45	2.42	2.37	2.33	2.29	2.26	2.23	2.16	2.12	2.08	2.05	2.00	1.94	1.90	1.82	1.77	1.73	55
2.42	2.39	2.33	2.29	2.25	2.22	2.19	2.13	2.08	2.04	2.01	1.96	1.90	1.86	1.78	1.73	1.69	60
2.39	2.36	2.30	2.26	2.22	2.19	2.16	2.09	2.05	2.01	1.98	1.93	1.87	1.83	1.74	1.69	1.65	65
2.36	2.33	2.28	2.23	2.19	2.16	2.13	2.07	2.02	1.98	1.95	1.90	1.84	1.80	1.71	1.66	1.62	70
2.32	2.29	2.23	2.19	2.15	2.11	2.08	2.02	1.97	1.93	1.90	1.85	1.79	1.75	1.66	1.60	1.56	80
2.28	2.25	2.20	2.15	2.12	2.08	2.05	1.99	1.94	1.90	1.87	1.82	1.75	1.71	1.62	1.56	1.52	90
2.26	2.23	2.17	2.13	2.09	2.05	2.02	1.96	1.91	1.87	1.84	1.79	1.72	1.68	1.59	1.53	1.49	100
2.21	2.18	2.13	2.08	2.04	2.01	1.98	1.91	1.86	1.82	1.79	1.74	1.67	1.63	1.53	1.47	1.42	125
2.18	2.15	2.10	2.05	2.01	1.98	1.94	1.88	1.83	1.79	1.76	1.70	1.63	1.59	1.49	1.42	1.37	150
2.14	2.11	2.06	2.01	1.97	1.94	1.91	1.84	1.79	1.75	1.71	1.66	1.59	1.54	1.44	1.37	1.31	200
2.10	2.07	2.02	1.97	1.93	1.90	1.87	1.80	1.75	1.71	1.67	1.61	1.54	1.50	1.39	1.31	1.25	300
2.07	2.04	1.99	1.94	1.90	1.87	1.84	1.77	1.72	1.67	1.64	1.58	1.51	1.46	1.35	1.26	1.18	500
2.05	2.02	1.97	1.92	1.88	1.84	1.81	1.75	1.69	1.65	1.61	1.56	1.48	1.43	1.31	1.22	1.13	1000
2.03	2.00	1.95	1.90	1.86	1.82	1.79	1.72	1.67	1.63	1.59	1.53	1.45	1.40	1.28	1.17	1.00	∞

Approximate formula for ν_1 and ν_2 larger than 30: $\log_{10} F_{.005;\nu_1,\nu_2} \simeq \dfrac{2.2373}{\sqrt{h} - 1.61} - 1.250\left(\dfrac{1}{\nu_1} - \dfrac{1}{\nu_2}\right)$, where $\dfrac{1}{h} = \dfrac{1}{2}\left(\dfrac{1}{\nu_1} + \dfrac{1}{\nu_2}\right)$.

TABLE 13.37. *PERCENTAGE POINT 0.1 OF THE F DISTRIBUTION. TABLE OF* $F_{.001;\nu_1,\nu_2}$.

Degrees of Freedom for the Denominator ν_2	Degrees of Freedom for the Numerator ν_1																	
	1	2	3	4	5	6	7	8	9	10	15	20	30	50	100	200	500	∞
	Multiply the Numbers of the First Row ($\nu_2 = 1$) by 1000																	
1	*405*	*500*	*540*	*562*	*576*	*586*	*593*	*598*	*602*	*606*	*616*	*621*	*626*	*630*	*633*	*635*	*636*	*637*
2	998	999	999	999	999	999	999	999	999	999	999	999	999	999	999	999	999	999
3	168	148	141	137	135	133	132	131	130	129	127	126	125	125	124	124	124	124
4	74.1	61.2	56.2	53.4	51.7	50.5	49.7	49.0	48.5	48.0	46.8	46.1	45.4	44.9	44.5	44.3	44.1	44.0
5	47.0	36.6	33.2	31.1	29.8	28.8	28.2	27.6	27.2	26.9	25.9	25.4	24.9	24.4	24.1	23.9	23.8	23.8
6	35.5	27.0	23.7	21.9	20.8	20.0	19.5	19.0	18.7	18.4	17.6	17.1	16.7	16.3	16.0	15.9	15.8	15.8
7	29.2	21.7	18.8	17.2	16.2	15.5	15.0	14.6	14.3	14.1	13.3	12.9	12.5	12.2	11.9	11.8	11.7	11.7
8	25.4	18.5	15.8	14.4	13.5	12.9	12.4	12.0	11.8	11.5	10.8	10.5	10.1	9.80	9.57	9.46	9.39	9.34
9	22.9	16.4	13.9	12.6	11.7	11.1	10.7	10.4	10.1	9.89	9.24	8.90	8.55	8.26	8.04	7.93	7.86	7.81
10	21.0	14.9	12.6	11.3	10.5	9.92	9.52	9.20	8.96	8.75	8.13	7.80	7.47	7.19	6.98	6.87	6.81	6.76
11	19.7	13.8	11.6	10.4	9.58	9.05	8.66	8.35	8.12	7.92	7.32	7.01	6.68	6.41	6.21	6.10	6.04	6.00
12	18.6	13.0	10.8	9.63	8.89	8.38	8.00	7.71	7.48	7.29	6.71	6.40	6.09	5.83	5.63	5.52	5.46	5.42
13	17.8	12.3	10.2	9.07	8.35	7.86	7.49	7.21	6.98	6.80	6.23	5.93	5.62	5.37	5.17	5.07	5.01	4.97
14	17.1	11.8	9.73	8.62	7.92	7.43	7.08	6.80	6.58	6.40	5.85	5.56	5.25	5.00	4.80	4.70	4.64	4.60
15	16.6	11.3	9.34	8.25	7.57	7.09	6.74	6.47	6.26	6.08	5.53	5.25	4.95	4.70	4.51	4.41	4.35	4.31
16	16.1	11.0	9.00	7.94	7.27	6.81	6.46	6.19	5.98	5.81	5.27	4.99	4.70	4.45	4.26	4.16	4.10	4.06
17	15.7	10.7	8.73	7.68	7.02	6.56	6.22	5.96	5.75	5.58	5.05	4.78	4.48	4.24	4.05	3.95	3.89	3.85
18	15.4	10.4	8.49	7.46	6.81	6.35	6.02	5.76	5.56	5.39	4.87	4.59	4.30	4.06	3.87	3.77	3.71	3.67
19	15.1	10.2	8.28	7.26	6.61	6.18	5.84	5.59	5.39	5.22	4.70	4.43	4.14	3.90	3.71	3.61	3.55	3.51
20	14.8	9.95	8.10	7.10	6.46	6.02	5.69	5.44	5.24	5.08	4.56	4.29	4.01	3.77	3.58	3.48	3.42	3.38
22	14.4	9.61	7.80	6.81	6.19	5.76	5.44	5.19	4.99	4.83	4.32	4.06	3.77	3.53	3.34	3.25	3.19	3.15
24	14.0	9.34	7.55	6.59	5.98	5.55	5.23	4.99	4.80	4.64	4.14	3.87	3.59	3.35	3.16	3.07	3.01	2.97
26	13.7	9.12	7.36	6.41	5.80	5.38	5.07	4.83	4.64	4.48	3.99	3.72	3.45	3.20	3.01	2.92	2.86	2.82
28	13.5	8.93	7.19	6.25	5.66	5.24	4.93	4.69	4.50	4.35	3.86	3.60	3.32	3.08	2.89	2.79	2.73	2.70
30	13.3	8.77	7.05	6.12	5.53	5.12	4.82	4.58	4.39	4.24	3.75	3.49	3.22	2.98	2.79	2.69	2.63	2.59
40	12.6	8.25	6.60	5.70	5.13	4.73	4.43	4.21	4.02	3.87	3.40	3.15	2.87	2.64	2.44	2.34	2.28	2.23
50	12.2	7.95	6.34	5.46	4.90	4.51	4.22	4.00	3.82	3.67	3.20	2.95	2.68	2.44	2.24	2.14	2.07	2.03

Degrees of Freedom for the Denominator ν_2	Degrees of Freedom for the Numerator ν_1																	
	1	2	3	4	5	6	7	8	9	10	15	20	30	50	100	200	500	∞
60	12.0	7.76	6.17	5.31	4.76	4.37	4.09	3.87	3.69	3.54	3.08	2.83	2.56	2.31	2.11	2.01	1.93	1.89
80	11.7	7.54	5.97	5.13	4.58	4.21	3.92	3.70	3.53	3.39	2.93	2.68	2.40	2.16	1.95	1.84	1.77	1.72
100	11.5	7.41	5.85	5.01	4.48	4.11	3.83	3.61	3.44	3.30	2.84	2.59	2.32	2.07	1.87	1.75	1.68	1.62
200	11.2	7.15	5.64	4.81	4.29	3.92	3.65	3.43	3.26	3.12	2.67	2.42	2.15	1.90	1.68	1.55	1.46	1.39
500	11.0	7.01	5.51	4.69	4.18	3.82	3.54	3.33	3.16	3.02	2.58	2.33	2.05	1.80	1.57	1.43	1.32	1.23
∞	10.8	6.91	5.42	4.62	4.10	3.74	3.47	3.27	3.10	2.96	2.51	2.27	1.99	1.73	1.49	1.34	1.21	1.00

Example: $P\{F_{.001;8,20} < 5.44\} = 99.9$ per cent.

$F_{.999;\nu_1,\nu_2} = 1/F_{.001;\nu_2,\nu_1}$ Example: $F_{.999;8,20} = 1/F_{.001;20,8} = 1/10.5 = 0.095$.

Approximate formula for ν_1 and ν_2 larger than 30: $\log_{10} F_{.001;\nu_1,\nu_2} \simeq \dfrac{2.6841}{\sqrt{h} - 2.09} - 1.672\left(\dfrac{1}{\nu_1} - \dfrac{1}{\nu_2}\right)$, where $\dfrac{1}{h} = \dfrac{1}{2}\left(\dfrac{1}{\nu_1} + \dfrac{1}{\nu_2}\right)$.

Chance Variables x and y. Measurements on one variable, say x, may be relatively inexpensive compared with measurements on y. Consequently, one would like to predict y from a knowledge of x. For example, determining abrasion loss is difficult, whereas measuring hardness by means of a Rockwell hardness machine is relatively simple. There exists a degree of association between abrasion loss and hardness. Similarly, rainfall in one area is associated with rainfall in a surrounding area. From measurements of the amount of rainfall in one area, one would like to predict rainfall in the other area.

A linear relationship exists between x and y if for a fixed x, the population mean value of y, given x, is of the form $B_0 + B_1x$. A sufficient condition for the existence of a linear relationship is that both x and y have a bivariate normal distribution.* The model implies that the actual observations $y_1, y_2, \ldots, y_n$ should not be on a straight line, but for fixed x, the population mean value of y should fall on the line $B_0 + B_1x$, where B_0 and B_1 are the intercept and slope, respectively, to be estimated from a sample $(x_1, y_1), (x_2, y_2), \ldots, (x_n, y_n)$.In this section it will be assumed that the y's are at least independently normally distributed† with mean $B_0 + B_1x$ and standard deviation σ (for all x_i).‡ Thus, returning to the abrasion loss-hardness example, if it is assumed that abrasion loss y and hardness x have a bivariate normal distribution, the average value of abrasion loss, given hardness, is $B_0 + B_1$ (hardness).

13.5.3 Least Square Estimates of B_0 and B_1

It should be noted that the cases presented in Pars. 13.5.2.1 and 13.5.2.2 fall into the general model of the following: For fixed x, the $y_1, y_2, \ldots, y_n$ are independently normally dis-

* If x and y are related such that x is normally distributed, the mean of y, given x, is a linear function of x, and the variance of y, given x, is constant, x and y are said to have a bivariate normal distribution.

† The equations for the least square estimates of B_0 and B_1 do not depend on the assumption of normality. However, tests of hypothesis and confidence statements presented later do depend on this assumption.

‡ That x and y have a bivariate normal distribution implies this condition.

tributed with mean value $B_0 + B_1x$ and standard deviation σ (for all x_i).

The problem is to estimate B_0 and B_1 by the method of least squares. Denote by b_0 and b_1 the least squares estimates of B_0 and B_1, respectively. The line is of the form $\tilde{y} = b_0 + b_1x$ and is usually referred to in statistics as a regression line. The least squares estimates of B_0 and B_1 are obtained by minimizing the sum of squares of the deviations about the estimated line with respect to b_0 and b_1, i.e., minimizing $\sum_{i=1}^{n} (y_i - \tilde{y}_i)^2$. The results are

$$b_1 = \frac{\sum_{i=1}^{n} (y_i - \bar{y})(x_i - \bar{x})}{\sum_{i=1}^{n} (x_i - \bar{x})^2} = \frac{\sum_{i=1}^{n} x_i y_i - n\overline{xy}}{\sum_{i=1}^{n} x_i^2 - n\bar{x}^2}$$

$$b_0 = \bar{y} - b_1\bar{x}$$

13.5.4 Significance Test for B_0 and B_1

In Art. 13.4 various significance tests for the mean and standard deviation of a normal distribution are given. In a similar manner, significance tests can be made for B_0 and B_1, and these are outlined in Table 13.39. These tests depend on the assumptions mentioned above. The reader is referred to the discussion in Art. 13.4 regarding significance tests.

13.5.5 Point Estimates and Confidence Intervals for the Linear Model

The point estimates and interval estimates for the various parameters are given in Table 13.40.

13.5.6 Predicting an Interval Within Which a Future Observation y^* Corresponding to x^* Will Lie

Table 13.40 presents a confidence interval for the parameter, the mean value of y^* corresponding to x^*. It is often the case that a confidence statement about the mean value is unimportant, whereas a probability statement about a future observation is relevant. For example, tensile strength of cement is related to the curing time (t) by: strength $= \alpha e^{-\beta/t}$. This function is transformed by taking logarithms:

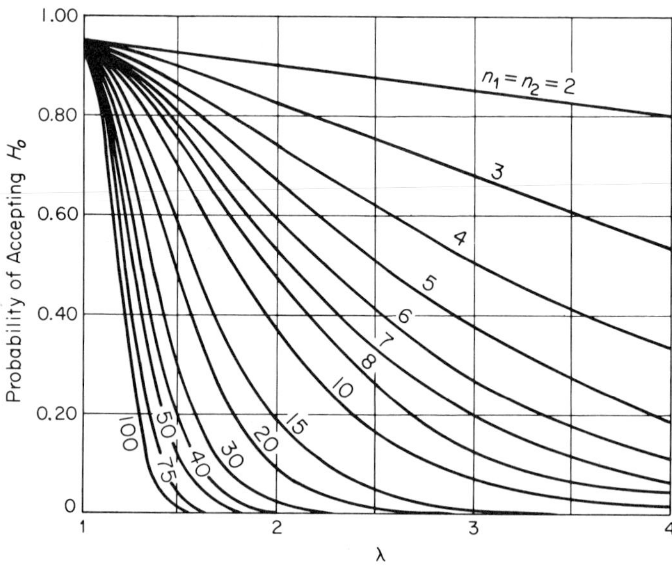

Fig. 13.43 *Operating characteristics of the one-sided* F *test for a level of significance equal to 0.05. Reproduced by permission from "Operating Characteristics for the Common Statistical Tests of Significance" by Charles D. Ferris, Frank E. Grubbs, Chalmers L. Weaver,* Annals of Mathematical Statistics, *June, 1946.*

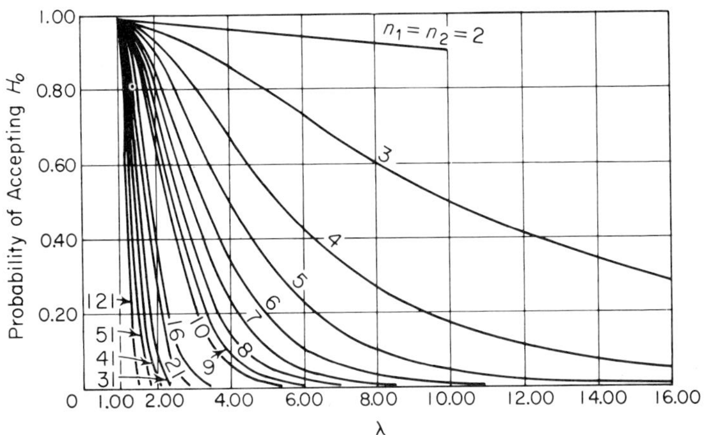

Fig. 13.44 *Operating characteristics of the one-sided* F *test for a level of significance equal to 0.01.*

$$\ln \text{strength} = \ln \alpha - \frac{\beta}{t} = \ln \alpha - \beta z$$

where $z = 1/t$, which is now linear in z and can be estimated by the method of least squares. The cement manufacturer is interested in the average tensile strength of his cement after a particular period of time, i.e., a confidence statement, whereas a builder is interested in the tensile strength of his particular batch of cement to determine whether it will carry the required load. After a specified period of time, say 28 days, the builder would like to have such a statement as the probability is $1 - \alpha$ that the tensile strength of his batch of cement will lie in a specified interval.

The following statement can be made. The probability is $1 - \alpha$ that the value of a future observation y^* corresponding to x^* will lie in the interval

$$b_0 + b_1 x^* \pm t_{(\alpha/2; n-2)} s_{y|x} \times \sqrt{1 + \frac{1}{n} + \frac{(x^* - \bar{x})^2}{\sum_{i=1}^{n} (x_i - \bar{x})^2}}$$

13.5.7 Correlation

A measure of the degree of association between two variables is the correlation coefficient ρ. An estimate of ρ is given by the sample correlation coefficient r, which is defined

$$r = \frac{\sum_{i=1}^{n} (x_i - \bar{x})(y_i - \bar{y})}{\sqrt{\sum (x_i - \bar{x})^2 \sum (y_i - \bar{y})^2}}$$

In engineering applications, the correlation coefficient does not play a very important role. The correlation coefficient can be derived from the slope of the fitted least squares line, e.g.,

$$r = b_1 \sqrt{\frac{\sum (x_i - \bar{x})^2}{\sum (y_i - \bar{y})^2}}$$

Consequently it is clear that the correlation coefficient does not contain any additional information. However, a significance test for $\rho = 0$ is equivalent to testing whether $B_1 = 0$, i.e., whether or not a linear relation is present. The test for $B_1 = 0$ is given in Table 13.39.

To test the hypothesis that $\rho = 0$, reject when

$$|t| = \left| \frac{r}{\sqrt{1 - r^2}} \right| \sqrt{n - 2} \geq t_{\alpha/2}$$

where values of t_α can be found in Table 13.25.

13.5.7.1 Example. A screw manufacturer is interested in giving out data to his customers on the relation between nominal and actual lengths. The following results were observed:

Nominal x	Actual y		
$\frac{1}{4}$ in.	0.262	0.262	0.245
$\frac{1}{2}$	0.496	0.512	0.490
$\frac{3}{4}$	0.743	0.744	0.751
1	0.976	1.010	1.004
$1\frac{1}{4}$	1.265	1.254	1.252

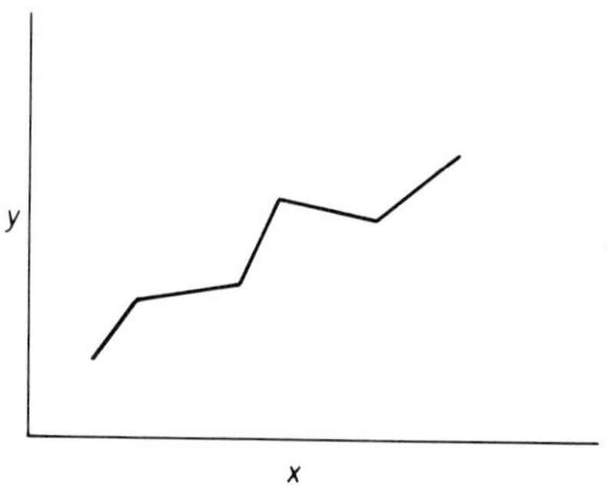

Fig. 13.45 *Relation between* x *and* y *obtained by connecting points.*

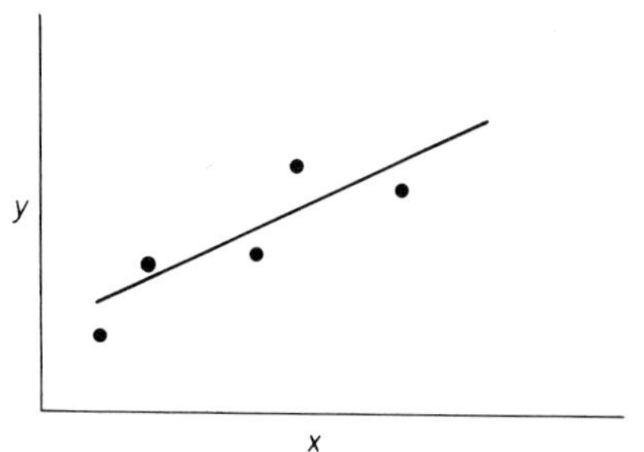

Fig. 13.46 *Relation between* x *and* y *obtained by drawing in best line by eye.*

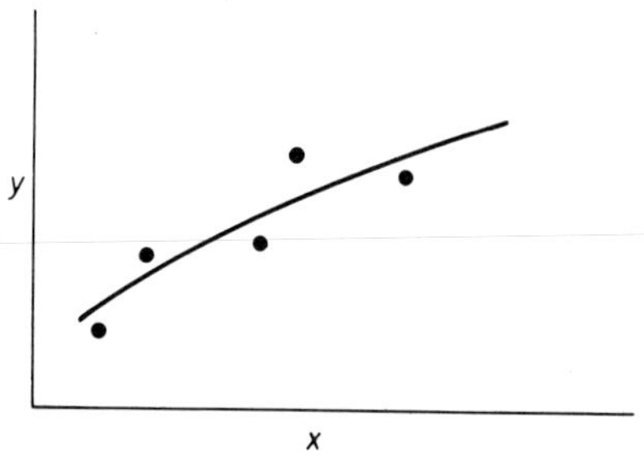

Fig. 13.47 *Relation between* x *and* y *obtained by drawing in best curve by eye.*

$1\frac{1}{2}$	1.498	1.518	1.504
$1\frac{3}{4}$	1.738	1.759	1.750
2	2.005	1.992	1.992

In this problem, there is a definite physical relation

$$\text{actual value} = B_0 + B_1 \text{ (nominal value)}$$

If the manufacturing process were perfect, it would be expected that $B_0 = 0$ and $B_1 = 1$. It should be noted that without using regression, one could make a confidence statement about the average of length of "1-inch" screws on the basis of the three numbers in the table; but a much more precise statement can be made by using all the data, together with the fact that linear regression is applicable. (1) Estimate the above relation. (2) For nominal 1-in. screws, find a confidence interval for the average actual value of screw length. (3) For a 1-in. screw find an interval such that the actual value will lie in this interval with probability 0.95. From Table 13.41a.

1. Estimated line:

$$\tilde{y} = 0.000672882 + 0.998603x$$

2. Confidence interval for the average actual value of 1–in. screws:

$$b_0 + b_1x^* \pm t_{(\alpha/2;n-2)}s_{y/x}\sqrt{\frac{1}{n} + \frac{(x^* - \bar{x})^2}{\sum(x_i - \bar{x})^2}}$$

$$0.000672882 + 0.998603 \pm 2.074 \times 0.002136$$

$$0.99485 \leq \text{average actual value} \leq 1.00371$$

3. Interval such that the probability is 0.95 that the true length of a 1-in. screw will be in this interval:

$$0.999275 \pm 0.01045 \times 2.074$$
$$0.97760 \leq \text{true length} \leq 1.02095$$

13.5.7.2 Example. The following results on the tensile strength of specimens of cold drawn copper have been recorded in a laboratory.

These variables fall into the second category, namely, a degree of association between x and y. Furthermore, it is reasonable to assume that x and y have a bivariate normal distribution, so that the population mean value of $y = B_0 + B_1x$. Hence, it remains to find the estimated line $\tilde{y} = b_0 + b_1x$ by the method of least squares. (1) Test the hypothesis that $B_0 = -100{,}000$. (2) Test the hypothesis that $B_1 = 1900$. (3) Get 95 per cent confidence intervals for B_1; B_0; the mean value of tensile strength corresponding to a Brinell hardness of 105.0; and $\sigma_{y/x}$. (4) Find an interval such that the probability is 0.95 that the value of tensile strength for a Brinell hardness of 105.0 will lie in this interval.

y Tensile Strength, psi	x Brinell Hardness Number
38,871	104.2
40,407	106.1

39,887	105.6
40,821	106.3
33,701	101.7
39,481	104.4
33,003	102.0
36,999	103.8
37,632	104.0
33,213	101.5
33,911	101.9
29,861	100.6
39,451	104.9
40,647	106.2
35,131	103.1

From Table 13.41b

1. Hypothesis: $B_0 = -100{,}000$; reject if

$$|t| = \left|\frac{b_0 + 100{,}000}{\sqrt{s^2_{y|x}[1/n + \bar{x}^2/\sum(x_i - \bar{x})^2]}}\right|$$
$$\geq t_{.025,13} = 2.160$$
$$|t| = \left|\frac{-149{,}786.5 + 100{,}000}{11{,}879.8}\right|$$
$$= 4.19 \geq 2.160$$

Therefore reject the hypothesis that $B_0 = -100{,}000$.

2. Hypothesis: $B_1 = 1900$; reject if

$$|t| = \left|\frac{b_1 - 1900}{\sqrt{s^2_{y|x}(1/\sum(x_i - \bar{x})^2)}}\right| \geq t_{.025,13}$$
$$= 2.160$$
$$|t| = \left|\frac{1799.025 - 1900}{114.5}\right| = 0.881$$
$$< 2.160$$

Therefore accept the hypothesis that $B_1 = 1900$.

3. Confidence intervals:
 a. for B_1

$$b_1 \pm \frac{t_{\alpha/2;n-2}s_{y|x}}{\sqrt{\sum(x_i - \bar{x})^2}}$$
$$1799.025 \pm 2.160(114.5)$$
$$1551.7 \leq B_1 \leq 2046.3$$

 b. for B_0

$$b_0 \pm t_{\alpha/2;n-2}s_{y|x}\sqrt{\frac{1}{n} + \frac{\bar{x}^2}{\sum(x_i - \bar{x})^2}}$$
$$-149{,}786.5 \pm 2.160(11{,}879.8)$$
$$-175{,}446.9 \leq B_0 \leq -124{,}126.1$$

 c. for mean value of tensile strength corresponding to a Brinell hardness of 105.0

$$b_0 + b_1x^* \pm t_{\alpha/2;n-2}s_{y|x}\sqrt{\frac{1}{n} + \frac{(x^* - \bar{x})^2}{\sum(x_i - \bar{x})^2}}$$
$$-149{,}786.5 + 1{,}799.025(105.0)$$
$$\pm 2.160(252.56) = 39{,}111.1 \pm 545.53$$

38,565.6 $\leq$ mean value of tensile strength $\leq$ 39,656.6

 d. for $\sigma^2_{y|x}$

$$\frac{(n-2)s^2_{y|x}}{\chi^2_{(\alpha/2;n-2)}}; \qquad \frac{(n-2)s^2_{y|x}}{\chi^2_{(1-\alpha/2;n-2)}}$$
$$\frac{13(649{,}546)}{5.01}; \qquad \frac{13(649{,}546)}{24.7}$$
$$1{,}685{,}449 \geq \sigma^2_{y|x} \geq 341{,}866$$

4. Interval such that the probability is 0.95 that the value of tensile strength for a Brinell hardness of 105.0 will be in this interval.

$$b_0 + b_1x^* \pm t_{(\alpha/2;n-2)}s_{y|x} \times \sqrt{1 + \frac{1}{n} + \frac{(x^* - \bar{x})^2}{\sum(x_i - \bar{x})^2}}$$
$$39{,}111.1 \pm 2.160(844.59)$$

37,286.8 $\leq$ tensile strength $\leq$ 40,935.4

If we wished to estimate hardness from tensile strength, we would have simply reversed the roles of x and y, since we are working in Case 2.

13.5.8 Relations Between Several Variables

In Pars. 13.5.1 to 13.5.7 the problem considered was that of finding a relation which would enable one to predict a variable from a knowledge of some one other variable. However, in many cases several factors are relevant to the prediction of just one variable. The relations between height and weight of individuals are different at different ages and different for men and women; an estimate of weight based on height, age, and sex will be better than one based on height alone. The relation between tensile strength and Rockwell hardness may depend on the density of the specimens. For each hardness-density combination the tensile strength will vary about a certain mean; we are interested in a formula giving the mean for each combination. As in the case of a single variable, it is assumed that the variance will be the same for each combination.

TABLE 13.38. *SUMMARY OF ESTIMATES BY CONFIDENCE INTERVALS AND BY SINGLE POINTS*

Parameters	Notation	Qualifying Conditions	Formula	Tables to Find Percentage Points	Point Estimates
Mean of a normal distribution	μ	known σ	$\bar{x} - K_{\alpha/2}(\sigma/\sqrt{n}) \leq \mu \leq \bar{x} + K_{\alpha/2}(\sigma/\sqrt{n})$	Normal table, Table 13.3	$\bar{x}$
Standard deviation of a normal distribution	σ		$s\sqrt{\frac{n-1}{\chi^2_{\alpha/2;n-1}}} \leq \sigma \leq s\sqrt{\frac{n-1}{\chi^2_{1-\alpha/2;n-1}}}$	Chi-square table, Table 13.27	s
Mean of a normal distribution	μ	unknown σ	$\bar{x} - (t_{\alpha/2;n-1})\frac{s}{\sqrt{n}} \leq \mu \leq \bar{x} + (t_{\alpha/2;n-1})\frac{s}{\sqrt{n}}$	t tables, Table 13.25	$\bar{x}$
Difference between the means of two normal distributions	$\Delta = \mu_x - \mu_y$	$x_1, \ldots, x_{n_x}$ are observations in first sample; $y_1, \ldots, y_{n_y}$ are observations in second sample.	$\bar{x} - \bar{y} - K_{\alpha/2}\sqrt{\left(\frac{\sigma_x^2}{n_x} + \frac{\sigma_y^2}{n_y}\right)} \leq \Delta \leq \bar{x} - \bar{y} + K_{\alpha/2}\sqrt{\left(\frac{\sigma_x^2}{n_x} + \frac{\sigma_y^2}{n_y}\right)}$		$\bar{x} - \bar{y}$
Difference between the means of two normal distributions	$\Delta = \mu_x - \mu_y$	$\sigma_x = \sigma_y = \sigma$ unknown σ	$(\bar{x} - \bar{y}) - t_{\alpha/2;n_x+n_y-2}\sqrt{\frac{1}{n_x} + \frac{1}{n_y}}\sqrt{\frac{\sum(x_i - \bar{x})^2 + \sum(y_i - \bar{y})^2}{n_x + n_y - 2}}$ $\leq \Delta \leq (\bar{x} - \bar{y}) + [t_{\alpha/2;n_x+n_y-2}]$ $\left[\sqrt{\frac{1}{n_x} + \frac{1}{n_y}}\sqrt{\frac{\sum(x_i - \bar{x})^2 + \sum(y_i - \bar{y})^2}{n_x + n_y - 2}}\right]$		$\bar{x} - \bar{y}$

TABLE 13.39. *SIGNIFICANCE TESTS FOR COEFFICIENTS OF STRAIGHT LINE*

Hypothesis	Test Statistic†	Rule of Rejection‡
$B_0 = B'_0$	$t = \dfrac{b_0 - B'_0}{\sqrt{s^2_{y\mid x}\left(\dfrac{1}{n} + \dfrac{\bar{x}^2}{\sum (x_i - \bar{x})^2}\right)}}$	$\lvert t \rvert \geq t_{\alpha/2;n-2}$
$B_1 = B'_1$	$t = \dfrac{b_1 - B'_1}{\sqrt{s^2_{y\mid x}(1/\sum (x_i - \bar{x})^2)}}$	$\lvert t \rvert \geq t_{\alpha/2;n-2}$
$B_0^{xy} = B_0^{uv}$ where mean $y = B_0^{xy} + B_1^{xy}x$; mean $v = B_0^{uv} + B_1^{uv}u$	$t = \dfrac{b_0^{xy} - b_0^{uv}}{\sqrt{\dfrac{(n_{xy} - 2)s^2_{y\mid x} + (n_{uv} - 2)s^2_{v\mid u}}{n_{xy} + n_{uv} - 4}}\sqrt{\dfrac{1}{n_{xy}} + \dfrac{1}{n_{uv}} + \dfrac{\bar{x}^2}{\sum (x_i - \bar{x})^2} + \dfrac{\bar{u}^2}{\sum (u_i - \bar{u})^2}}}$	$\lvert t \rvert \geq t_{\alpha/2;n_{xy}+n_{uv}-4}$
$B_1^{xy} = B_1^{uv}$ where mean $y = B_0^{xy} + B_1^{xy}x$; mean $v = B_0^{uv} + B_1^{uv}u$	$t = \dfrac{b_1^{xy} - b_1^{uv}}{\sqrt{\dfrac{(n_{xy} - 2)s^2_{y\mid x} + (n_{uv} - 2)s^2_{v\mid u}}{n_{xy} + n_{uv} - 4}}\sqrt{\dfrac{1}{\sum (x_i - \bar{x})^2} + \dfrac{1}{\sum (u_i - \bar{u})^2}}}$	$\lvert t \rvert \geq t_{\alpha/2;n_{xy}+n_{uv}-4}$

OC Curve	Notation
Compute $d = \dfrac{\lvert B'_0 - B^*_0 \rvert}{\sigma\sqrt{[1/n + \bar{x}^2/\sum (x_i - \bar{x})^2](n-1)}}$ and refer to Fig. 13.31 or 13.32, using the curve for $n - 1$.	Here σ is the standard deviation of y about the mean $B_0 + B_1x$, $s^2_{y\mid x}$ is the estimate of σ^2, and is equal to $s^2_{y\mid x} = \dfrac{\sum_{i=1}^{n} [y_i - (b_0 + b_1x_i)]^2}{n-2} = \dfrac{\sum (y_i - \bar{y})^2 - \dfrac{[\sum (x_i - \bar{x})(y_i - \bar{y})]^2}{\sum (x_i - \bar{x})^2}}{n-2}$ where B_0^* is the value of the true intercept for use in referring to the OC curve.
Compute $d = \dfrac{\lvert B'_1 - B^*_1 \rvert}{\sigma\sqrt{(n-1)/\sum (x_i - \bar{x})^2}}$ and refer to Fig. 13.31 or 13.32, using the curve for $n - 1$.	Here B^*_1 is the value of the true slope for use in referring to the OC curve.

TABLE 13.39. *Cont.*

OC Curve	Notation
Compute $d = \dfrac{\lvert\delta_0\rvert}{\sqrt{n_{xy} + n_{uv} - 3}\,\sigma_{y\mid x}\sqrt{\dfrac{1}{n_{xy}} + \dfrac{1}{n_{uv}} + \dfrac{\bar{x}^2}{\sum (x_i - \bar{x})^2} + \dfrac{\bar{u}^2}{\sum (u_i - \bar{u})^2}}}$ and refer to Fig. 13.31 or 13.32, using the curve for $n_{xy} + n_{uv} - 3$.	This is the problem of comparing coefficients from two straight lines. The subscripts x and y refer to one line, whereas u and v refer to the second line. Then $\sigma_{y\mid x}$ is the unknown standard deviation of y about the mean $B_0^{xy} + B_1^{xy}x$, and $\sigma_{v\mid u}$ is the unknown standard deviation of v about the mean $B_0^{uv} + B_1^{uv}u$. It is assumed that $\sigma_{y\mid x} = \sigma_{v\mid u}$; δ_0 is the difference in intercept of the two lines for use in referring to the OC curve. Here δ_1 is the difference in slopes of the two lines for use in referring to the OC curve.
Compute $d = \dfrac{\lvert\delta_1\rvert}{\sigma_{y\mid x}\sqrt{\left(\dfrac{1}{\sum (x_i - \bar{x})^2} + \dfrac{1}{\sum (u_i - \bar{u})^2}\right)(n_{xy} + n_{uv} - 3)}}$ and refer to Fig. 13.31 or 13.32, using the curve for $n_{xy} + n_{uv} - 3$.	

† If the population standard deviation is known, these values should be used in place of the sample estimates, and K_α should be used instead of t_α.
‡ These values are for two-sided tests, and the values of t_α are given in Table 13.25.

TABLE 13.40. *ESTIMATION OF PARAMETERS FOR LINEAR MODEL*

Parameter	Symbol for Estimate	Estimation Computation Formula	100 $(1 - \alpha)$ Per Cent Confidence Interval	Notes
B_1	b_1	$\frac{\sum (x_i - \bar{x})(y_i - \bar{y})}{\sum (x_i - \bar{x})^2}$	$b_1 \pm \frac{t_{(\alpha/2;n-2)}s_{y\|x}}{\sqrt{\sum (x_i - \bar{x})^2}}$	
B_0	b_0	$\bar{y} - b_1\bar{x}$	$b_0 \pm t_{(\alpha/2;n-2)}s_{y\|x}\sqrt{\frac{1}{n} + \frac{\bar{x}^2}{\sum (x_i - \bar{x})^2}}$	
Average value of y corresponding to x^*, i.e., $B_0 + B_1x^*$	$\bar{y}^*$	$b_0 + b_1x^*$	$b_0 + b_1x^* \pm t_{(\alpha/2;n-2)}s_{y\|x}\sqrt{\frac{1}{n} + \frac{(x^* - \bar{x})^2}{\sum (x_i - \bar{x})^2}}$	
Value of x corresponding to observed value of y' for the case when there is an underlying physical relationship.	x'	$\frac{y' - b_0}{b_1} = x'$	$x' + b_1\frac{(y' - \bar{y})}{c}$ $\pm t_{(\alpha/2;n-2)}\frac{s_{y\|x}}{c}\sqrt{\left(1 + \frac{1}{n}\right)c + A(y' - \bar{y})^2}$	$c = b_1^2 - t^2_{\alpha/2}(s^2_{y\|x})(A)$ $A = 1/\sum (x_i - \bar{x})^2$ If $(x' - \bar{x})^2/\sum (x_i - \bar{x})^2$ is small, then we have for the confidence interval, $(y' - b_0)/b_1 \pm t_{\alpha/2}s_{y\|x}\sqrt{(1 + 1/n)}/b_1$
$\sigma^2_{y\|x}$	$s^2_{y\|x}$	$\frac{\sum [y_i - (b_0 + b_1x_i)]^2}{n - 2}$ $= \sum (y_i - \bar{y})^2 - \frac{[\sum (x_i - \bar{x})(y_i - \bar{y})]^2/\sum (x_i - \bar{x})^2}{n - 2}$	$\frac{(n - 2)s^2_{y\|x}}{\chi^2_{\alpha/2;n-2}}; \frac{(n - 2)s^2_{y\|x}}{\chi^2_{1-\alpha/2;n-2}}$	

TABLE 13.41. *WORKSHEET FOR THE FITTING OF STRAIGHT LINES BY THE METHOD OF LEAST SQUARES*

General Data	Data for Confidence Intervals and Predictions for $x = x^*$
x represents________	Data for the equation of line
$\sum_{i=1}^{n} x_i = \underline{\quad\quad}; \bar{x} = \sum_{i=1}^{n} \frac{x_i}{n} = \underline{\quad\quad}; \left(\sum_{i=1}^{n} x_i\right)^2 / n = \left(\sum_{i=1}^{n} x_i\right)(\bar{x}) = \underline{\quad\quad}$	$b_1 = \dfrac{\sum_{i=1}^{n} (x_i - \bar{x})(y_i - \bar{y})}{\sum_{i=1}^{n} (x_i - \bar{x})^2} = \underline{\quad\quad}$
$\sum_{i=1}^{n} x_i^2 = \underline{\quad\quad}; \sum_{i=1}^{n} x_i^2 - \left(\sum_{i=1}^{n} x_i\right)^2 / n = \sum_{i=1}^{n} (x_i - \bar{x})^2 = \underline{\quad\quad}$	
y represents________	$b_0 = \bar{y} - b_1\bar{x} = \underline{\quad\quad}$
$\sum_{i=1}^{n} y_i = \underline{\quad\quad}; \bar{y} = \sum_{i=1}^{n} \frac{y_i}{n} = \underline{\quad\quad}; \left(\sum_{i=1}^{n} y_i\right)^2 / n = \left(\sum_{i=1}^{n} y_i\right)(\bar{y}) = \underline{\quad\quad}$	equation of line $= \tilde{y} = b_0 + b_1 x$ $= \underline{\quad\quad} + \underline{\quad\quad}\, x$ Estimate of standard deviation
$\sum_{i=1}^{n} y_i^2 = \underline{\quad\quad}; \sum_{i=1}^{n} y_i^2 - \left(\sum_{i=1}^{n} y_i\right)^2 / n = \sum_{i=1}^{n} (y_i - \bar{y})^2 = \underline{\quad\quad}$	$(n - 2)(s^2_{y\mid x}) = \sum_{i=1}^{n} (y_i - \bar{y})^2$
$n = \underline{\quad\quad}; \frac{1}{n} = \underline{\quad\quad}; \left(\sum_{i=1}^{n} x_i\right)\left(\sum_{i=1}^{n} y_i\right) / n = (\bar{x}) \sum_{i=1}^{n} y_i = \underline{\quad\quad}$	$- \dfrac{\left[\sum_{i=1}^{n} (x_i - \bar{x})(y_i - \bar{y})\right]^2}{\sum_{i=1}^{n} (x_i - \bar{x})^2} = \underline{\quad\quad}$
$\sum_{i=1}^{n} x_i y_i = \underline{\quad\quad}; \sum_{i=1}^{n} x_i y_i - (\sum x_i)(\sum y_i)/n = \sum (x_i - \bar{x})(y_i - \bar{y}) = \underline{\quad\quad}$	$s^2_{y\mid x} = \underline{\quad\quad}; s_{y\mid x} = \sqrt{s^2_{y\mid x}} = \underline{\quad\quad}$
$\dfrac{\left[\sum_{i=1}^{n} (x_i - \bar{x})(y_i - \bar{y})\right]^2}{\sum (x_i - \bar{x})^2} = \underline{\quad\quad}$	

Data for Significance Tests	Data for Confidence Intervals and Predictions for $x = x^*$
$\dfrac{s^2_{y\mid x}}{\sum_{i=1}^{n}(x_i - \bar{x})^2} = \underline{\qquad}$; $\sqrt{\dfrac{s^2_{y\mid x}}{\sum_{i=1}^{n}(x_i - \bar{x})^2}} = \underline{\qquad}$; $\bar{x}^2 = \underline{\qquad}$	$(x^* - \bar{x})^2 = \underline{\qquad}$; $\dfrac{(x^* - \bar{x})^2}{\sum_{i=1}^{n}(x_i - \bar{x})^2} = \underline{\qquad}$
$\dfrac{\bar{x}_2^2}{\sum_{i=1}^{n}(x_i - \bar{x})^2} = \underline{\qquad}$; $\dfrac{1}{n} + \dfrac{\bar{x}^2}{\sum_{i=1}^{n}(x_i - \bar{x})^2} = \underline{\qquad}$	$\dfrac{1}{n} + \dfrac{(x^* - \bar{x})^2}{\sum_{i=1}^{n}(x_i - \bar{x})^2} = \underline{\qquad}$
$(s^2_{y\mid x})\left[\dfrac{1}{n} + \dfrac{\bar{x}^2}{\sum_{i=1}^{n}(x_i - \bar{x})^2}\right] = \underline{\qquad}$	$(s^2_{y\mid x})\left[\dfrac{1}{n} + \dfrac{(x^* - \bar{x})^2}{\sum_{i=1}^{n}(x_i - \bar{x})^2}\right] = \underline{\qquad}$
$\sqrt{(s^2_{y\mid x})\left[\dfrac{1}{n} + \dfrac{\bar{x}^2}{\sum_{i=1}^{n}(x_i - \bar{x})^2}\right]} = \underline{\qquad}$	$\sqrt{(s^2_{y\mid x})\left[\dfrac{1}{n} + \dfrac{(x^* - \bar{x})^2}{\sum_{i=1}^{n}(x_i - \bar{x})^2}\right]} = \underline{\qquad}$
	$1 + \dfrac{1}{n} + \dfrac{(x^* - \bar{x})^2}{\sum_{i=1}^{n}(x_i - \bar{x})^2} = \underline{\qquad}$
	$(s^2_{y\mid x})\left[1 + \dfrac{1}{n} + \dfrac{(x^* - \bar{x})^2}{\sum_{i=1}^{n}(x_i - \bar{x})^2}\right] = \underline{\qquad}$
	$\sqrt{(s^2_{y\mid x})\left[1 + \dfrac{1}{n} + \dfrac{(x^* - \bar{x})^2}{\sum_{i=1}^{n}(x_i - \bar{x})^2}\right]} = \underline{\qquad}$

TABLE 13.41a.

General Data	Data for Confidence Intervals and Predictions for $x = x^*$
x represents nominal value $\sum_{i=1}^{n} x_i = 27.0;\ \bar{x} = \sum_{i=1}^{n} \frac{x_i}{n} = 1.125;\ \left(\sum_{i=1}^{n} x_i\right)^2 / n = \left(\sum_{i=1}^{n} x_i\right)(\bar{x}) = 30.375$ $\sum_{i=1}^{n} x_i^2 = 38.2500;\ \sum_{i=1}^{n} x_i^2 - \left(\sum_{i=1}^{n} x_i\right)^2 / n = \sum_{i=1}^{n} (x_i - \bar{x})^2 = 7.875$ y represents actual value $\sum_{i=1}^{n} y_i = 27.022;\ \bar{y} = \sum_{i=1}^{n} \frac{y_i}{n} = 1.1259;\ \left(\sum_{i=1}^{n} y_1\right)^2 / n = \left(\sum_{i=1}^{n} y_i\right)(\bar{y}) = 30.424552$ $\sum_{i=1}^{n} y_i^2 = 38.279818;\ \sum_{i=1}^{n} y_i^2 - \left(\sum_{i=1}^{n} y_i\right)^2 / n = \sum_{i=1}^{n} (y_i - \bar{y})^2 = 7.855266$ $n = 24;\ \frac{1}{n} = 0.041667;\ \left(\sum_{i=1}^{n} x_i\right)\left(\sum_{i=1}^{n} y_i\right) / n = (\bar{x}) \sum_{i=1}^{n} y_i = 30.39975$ $\sum_{i=1}^{n} x_i y_i = 38.26375;\ \sum_{i=1}^{n} x_i y_i - (\sum x_i)(\sum y_i)/n = \sum (x_i - \bar{x})(y_i - \bar{y}) = 7.86400$ $\frac{\left[\sum_{i=1}^{n} (x_i - \bar{x})(y_i - \bar{y})\right]^2}{\sum (x_i - \bar{x})^2} = $ ———	Data for the equation of line $b_1 = \frac{\sum_{i=1}^{n} (x_i - \bar{x})(y_i - \bar{y})}{\sum_{i=1}^{n} (x_i - \bar{x})^2} = 0.998603$ $b_0 = \bar{y} - b_1\bar{x} = -\ 0.000672882$ equation of line $= \tilde{y} = b_0 + b_1 x$ $= 0.000672882 + 0.998603x$ Estimate of standard deviation $(n - 2)(s^2_{y \mid x}) = \sum_{i=1}^{n} (y_i - \bar{y})^2 - \frac{\left(\sum_{i=1}^{n} (x_i - \bar{x})(y_i - \bar{y})\right)^2}{\sum_{i=1}^{n} (x_i - \bar{x})^2} = 0.0023$ $s^2_{y \mid x} = 0.00010455;\quad s_{y \mid x} = \sqrt{s^2_{y \mid x}} = 0.010225$

Data for Significance Tests	Data for Confidence Intervals and Predictions for $x = x^*$
$\dfrac{s^2_{y\mid x}}{\sum_{i=1}^{n}(x_i - \bar{x})^2} = \underline{\qquad}$; $\sqrt{\dfrac{s^2_{y\mid x}}{\sum_{i=1}^{n}(x_i - \bar{x})^2}} = \underline{\qquad}$; $\bar{x}^2 = \underline{\qquad}$	$(x^* - \bar{x})^2 = 0.015625$; $\dfrac{(x^* - \bar{x})^2}{\sum_{i=1}^{n}(x_i - \bar{x})^2} = 0.001984$
$\dfrac{\bar{x}^2}{\sum_{i=1}^{n}(x_i - \bar{x})^2} = \underline{\qquad}$; $\dfrac{1}{n} + \dfrac{\bar{x}^2}{\sum_{i=1}^{n}(x_i - \bar{x})^2} = \underline{\qquad}$	$\dfrac{1}{n} + \dfrac{(x^* - \bar{x})^2}{\sum_{i=1}^{n}(x_i - \bar{x})^2} = 0.043651$
$(s^2_{y\mid x})\left[\dfrac{1}{n} + \dfrac{\bar{x}^2}{\sum_{i=1}^{n}(x_i - \bar{x})^2}\right] = \underline{\qquad}$	$(s^2_{y\mid x})\left[\dfrac{1}{n} + \dfrac{(x^* - \bar{x})^2}{\sum_{i=1}^{n}(x_i - \bar{x})^2}\right] = 0.000004564$
$\sqrt{(s^2_{y\mid x})\left[\dfrac{1}{n} + \dfrac{\bar{x}^2}{\sum_{i=1}^{n}(x_i - \bar{x})^2}\right]} = \underline{\qquad}$	$\sqrt{(s^2_{y\mid x})\left[\dfrac{1}{n} + \dfrac{(x^* - \bar{x})^2}{\sum_{i=1}^{n}(x_i - \bar{x})^2}\right]} = 0.002136$
	$1 + \dfrac{1}{n} + \dfrac{(x^* - \bar{x})^2}{\sum_{i=1}^{n}(x_i - \bar{x})^2} = 1.043651$
	$(s^2_{y\mid x})\left[1 + \dfrac{1}{n} + \dfrac{(x^* - \bar{x})^2}{\sum_{i=1}^{n}(x_i - \bar{x})^2}\right] = 0.0001091$
	$\sqrt{(s^2_{y\mid x})\left[1 + \dfrac{1}{n} + \dfrac{(x^* - \bar{x})^2}{\sum_{i=1}^{n}(x_i - \bar{x})^2}\right]} = 0.01045$

TABLE 13.41b.

General Data	Data for Confidence Intervals and Predictions for $x = x^*$
x represents Brinell hardness number	Data for the equation of line
$\sum_{i=1}^{n} x_i = 1556.3;\ \bar{x} = \sum_{i=1}^{n} \frac{x_i}{n} = 103.753;\ \left(\sum_{i=1}^{n} x_i\right)^2 / n = \left(\sum_{i=1}^{n} x_i\right)(\bar{x}) = 161{,}471.3126$	$b_1 = \dfrac{\sum_{i=1}^{n} (x_i - \bar{x})(y_i - \bar{y})}{\sum_{i=1}^{n} (x_i - \bar{x})^2} = 1{,}799.025$
$\sum_{i=1}^{n} x_i^2 = 161{,}520.87;\ \sum_{i=1}^{n} x_i^2 - \left(\sum_{i=1}^{n} x_i\right)^2 / n = \sum_{i=1}^{n} (x_i - \bar{x})^2 = 49.56$	$b_0 = \bar{y} - b_1\bar{x} = -149{,}786.5$
y represents tensile strength	equation of line is $\tilde{y} = b_0 + b_1 x$
$\sum_{i=1}^{n} y_i = 553{,}016;\ \bar{y} = \sum_{i=1}^{n} \frac{y_i}{n} = 36{,}867.7;\ \left(\sum_{i=1}^{n} y_i\right)^2 / n = \left(\sum_{i=1}^{n} y_i\right)(\bar{y}) = 20{,}388{,}446{,}420$	$= -149{,}786.5 + 1{,}799.025\,x$
$\sum_{i=1}^{n} y_i^2 = 20{,}557{,}291{,}078;\ \sum_{i=1}^{n} y_i^2 - \left(\sum_{i=1}^{n} y_i\right)^2 / n = \sum_{i=1}^{n} (y_i - \bar{y})^2 = 168{,}844{,}660$	Estimate of standard deviation
$n = 15;\ \frac{1}{n} = 0.0666667;\ \left(\sum_{i=1}^{n} x_i\right)\left(\sum_{i=1}^{n} y_i\right)/n = (\bar{x}) \sum_{i=1}^{n} y_i = 57{,}377{,}253.38$	$(n - 2)(s^2_{y\mid x}) = \sum_{i=1}^{n} (y_i - \bar{y})^2 - \dfrac{\left[\sum_{i=1}^{n} (x_i - \bar{x})(y_i - \bar{y})\right]^2}{\sum_{i=1}^{n} (x_i - \bar{x})^2} = 8{,}444{,}093$
$\sum_{i=1}^{n} x_i y_i = 57{,}466{,}413.1;\ \sum_{i=1}^{n} x_i y_i - (\sum x_i)(\sum y_i)/n = \sum (x_i - \bar{x})(y_i - \bar{y}) = 89{,}159.7$	$s^2_{y\mid x} = 649{,}546;\ s_{y\mid x} = \sqrt{s^2_{y\mid x}} = 805.945$
$\dfrac{\left[\sum_{i=1}^{n} (x_i - \bar{x})(y_i - \bar{y})\right]^2}{\sum (x_i - \bar{x})^2} = 160{,}400{,}567$	

Data for Significance Tests	Data for Confidence Intervals and Predictions for $x = x^*$
$\frac{s^2_{y\mid x}}{\sum_{i=1}^{n}(x_i-\bar{x})^2} = 13{,}106; \quad \sqrt{\frac{s^2_{y\mid x}}{\sum_{i=1}^{n}(x_i-\bar{x})^2}} = 114.5; \; \bar{x}^2 = 10{,}764.7$	$(x^*-\bar{x})^2 = 1.555; \quad \frac{(x^*-\bar{x})^2}{\sum_{i=1}^{n}(x_i-\bar{x})^2} = 0.03135$
$\frac{\bar{x}^2}{\sum_{i=1}^{n}(x_i-\bar{x})^2} = 217.2051; \quad \frac{1}{n} + \frac{\bar{x}^2}{\sum_{i=1}^{n}(x_i-\bar{x})^2} = 217.2718$	$\frac{1}{n} + \frac{(x^*-\bar{x})^2}{\sum_{i=1}^{n}(x_i-\bar{x})^2} = 0.0982$
$(s^2_{y\mid x})\left[\frac{1}{n} + \frac{\bar{x}^2}{\sum_{i=1}^{n}(x_i-\bar{x})^2}\right] = 141{,}128{,}029$	$(s^2_{y\mid x})\left[\frac{1}{n} + \frac{(x^*-\bar{x})^2}{\sum_{i=1}^{n}(x_i-\bar{x})^2}\right] = 63{,}785.4$
$\sqrt{(s^2_{y\mid x})\left[\frac{1}{n} + \frac{\bar{x}^2}{\sum_{i=1}^{n}(x_i-\bar{x})^2}\right]} = 11{,}879.8$	$\sqrt{(s^2_{y\mid x})\left[\frac{1}{n} + \frac{(x^*-\bar{x})^2}{\sum_{i=1}^{n}(x_i-\bar{x})^2}\right]} = 252.56$
	$1 + \frac{1}{n} + \frac{(x^*-\bar{x})^2}{\sum_{i=1}^{n}(x_i-\bar{x})^2} = 1.0982$
	$(s^2_{y\mid x})\left[1 + \frac{1}{n} + \frac{(x^*-\bar{x})^2}{\sum_{i=1}^{n}(x_i-\bar{x})^2}\right] = 713{,}331.4$
	$\sqrt{(s^2_{y\mid x})\left[1 + \frac{1}{n} + \frac{(x^*-\bar{x})^2}{\sum_{i=1}^{n}(x_i-\bar{x})^2}\right]} = 844.59$

13.5.8.1 Least Squares Equations. Suppose, for example, we are interested in predicting y, and let x_1 and x_2 be the independent variables. Assume that the true underlying relationship is of the form

$$\text{average value of } y = B_0 + B_1 x_1 + B_2 x_2$$

If we denote the estimates of the B_i by b_i, the least squares estimate is $\tilde{y} = b_0 + b_1 x_1 + b_2 x_2$, where b_1 and b_2 are found from the following set of equations:

$$\sum_{i=1}^{n} (y_i - \bar{y})(x_{1i} - \bar{x}_1) = b_1 \sum_{i=1}^{n} (x_{1i} - \bar{x}_1)^2 + b_2 \sum_{i=1}^{n} (x_{1i} - \bar{x}_1)(x_{2i} - \bar{x}_2)$$

$$\sum_{i=1}^{n} (y_i - \bar{y})(x_{2i} - \bar{x}_2) = b_1 \sum_{i=1}^{n} (x_{1i} - \bar{x}_1)(x_{2i} - \bar{x}_2) + b_2 \sum_{i=1}^{n} (x_{2i} - \bar{x}_2)^2$$

$$b_0 = \bar{y} - b_1 \bar{x}_1 - b_2 \bar{x}_2$$

These equations are known as normal equations and can be solved by solving the first for b_1 in terms of b_2 and substituting this expression for b_1 in the second, or by any other method. In general, we might consider a regression equation of the sort

$$\text{average value of } y = B_0 + B_1 x_1 + B_2 x_2 + \cdots + B_k x_k$$

with estimates

$$\tilde{y} = b_0 + b_1 x_1 + b_2 x_2 + \cdots + b_k x_k$$

where $b_1, b_2, \ldots, b_k$ are found by the solution of a set of k simultaneous equations in k unknowns. For $k = 2$ and 3, detailed instructions are given in Tables 13.44 and 13.45 later in the text.

13.5.9 Test of Significance

There are many hypotheses which we might be interested in testing, but the most important question is whether the independent variables taken as a whole lead to an improvement in our ability to forecast y.

For $k = 2$. This is equivalent to testing the hypothesis that $(B_1, B_2) = (0, 0)$, i.e., mean value of $y = B_0$. The results may be arranged into Table 13.42, where c_1 and c_2 are defined in the computational Table 13.44 and $\sum_{i=1}^{n} (y_i - \tilde{y}_i)^2$ is obtained by subtraction.* Reject the hypothesis that $(B_1, B_2) = (0, 0)$ if

$$F = \frac{(b_1 c_1 + b_2 c_2)/2}{\sum_{i=1}^{n} (y_i - \tilde{y}_i)^2/(n-3)} \geq F_{\alpha;2,n-3}$$

where $F_{\alpha;2,n-3}$ is the α percentage point of the F distribution found in Table 13.32.

For $k = 3$. To test $H_0: (B_1, B_2, B_3) = (0,0,0)$, arrange the results into Table 13.43, where c_1, c_2, and c_3 are defined in the computational Table 13.45 and $\sum_{i=1}^{n} (y_i - \tilde{y}_i)^2$ is obtained by subtraction.†

Reject the hypothesis that $(B_1, B_2, B_3) = (0,0,0)$ if

$$F = \frac{(b_1 c_1 + b_2 c_2 + b_3 c_3)/3}{\sum_{i=1}^{n} (y_i - \tilde{y}_i)^2/(n-4)} \geq F_{\alpha;3,n-4}$$

where $F_{\alpha;3,n-4}$ is the α percentage point of the F distribution found in Table 13.32.

---

* The following is an identity:

$$\sum_{i=1}^{n} (y_i - \bar{y})^2 = b_1 c_1 + b_2 c_2 + \sum_{i=1}^{n} (y_i - \tilde{y}_i)^2$$

† The following is an identity:

$$\sum_{i=1}^{n} (y_i - \bar{y})^2 = b_1 c_1 + b_2 c_2 + b_3 c_3 + \sum_{i=1}^{n} (y_i - \tilde{y}_i)^2$$

TABLE 13.42. *REGRESSION TABLE FOR* $k = 2$

Variation	Sum of Squares	Degrees of Freedom	Mean Square
Due to regression	$b_1c_1 + b_2c_2$	2	$(b_1c_1 + b_2c_2)/2$
About regression	$\sum_{i=1}^{n} (y_i - \tilde{y}_i)^2$	$n - 3$	$\sum_{i=1}^{n} (y_i - \tilde{y}_i)^2/(n - 3)$
Total	$\sum_{i=1}^{n} (y_i - \bar{y})^2$	$n - 1$	

TABLE 13.43. *REGRESSION TABLE FOR* $k = 3$

Variation	Sum of Squares	Degrees of Freedom	Mean Square
Due to regression	$b_1c_1 + b_2c_2 + b_3c_3$	3	$(b_1c_1 + b_2c_2 + b_3c_3)/3$
About regression	$\sum_{i=1}^{n} (y_i - \tilde{y}_i)^2$	$n - 4$	$\sum_{i=1}^{n} (y_i - \tilde{y}_i)^2/(n - 4)$
Total	$\sum_{i=1}^{n} (y_i - \bar{y})^2$	$n - 1$	

13.5.10 Nonlinear Regression

Nonlinear regression falls into the above model. For example,

$$\text{average value of } y = B_0 + B_1Z + B_2Z^2$$

Since the Z's are fixed variables, we can write the quadratic equation in the form

$$\text{average value of } y = B_0 + B_1x_1 + B_2x_2$$

where $x_1 = Z$ and $x_2 = Z^2$. Hence the least squares estimate of the B's can be obtained using the procedure presented above. (See Tables 13.44 and 13.45.)

13.5.11 Example

During the years 1951 and 1952, the County of Santa Clara, in California, contracted for a cloud-seeding operation.* In order to evaluate the effectiveness of this operation, the county and the surrounding area were divided geographically into regions. The rainfall in the Los Gatos, Palo Alto, and Santa Clara region (y) was to be predicted from the rainfall in the Berkeley, Crockett, Oakland, and San Francisco area (x_1); the Merced, Modesto, and Stockton area (x_2); and the Big Sur, Coalinga, King City, and Paso Robles area (x_3). These areas were determined in such a manner that they contained long-record stations, and all the stations within a group were geographically located close to one another. In making this rainfall study, storm totals, summed over all rainfall stations within an area, were chosen as the index to be considered. From preliminary analysis it was determined that the cube root of the total rainfall over all the stations within a region in the county, per storm, can be predicted from a linear combination of the cube roots of the total rainfall over all the stations within a region in the control area, per storm, i.e.,

$$\text{average value of } y = B_0 + B_1x_1 + B_2x_2 + B_3x_3$$

From past data, these coefficients were estimated so that

$$\tilde{y} = b_0 + b_1x_1 + b_2x_2 + b_3x_3$$

Assuming no seeding effect in the areas outside the county, the rainfall within the county can be

* G. J. Lieberman, *An Evaluation of the 1951–1952 Cloud Seeding Experiment in Santa Clara County*, Stanford University, Stanford, California.

TABLE 13.44. *COMPUTATIONAL SCHEME FOR SOLVING THE NORMAL EQUATIONS*

$$\tilde{y} = b_0 + b_1x_1 + b_2x_2$$

The normal equations presented in Par. 13.5.8 are of the form

$$a_{11}b_1 + a_{12}b_2 = c_1 \qquad a_{21}b_1 + a_{22}b_2 = c_2$$

where

$$a_{11} = \sum_{i=1}^{n} (x_{1_i} - \bar{x}_1)^2 = \sum_{i=1}^{n} x_{1_i}{}^2 - n\bar{x}_1{}^2 = \text{——}$$

$$a_{21} = a_{12} = \sum_{i=1}^{n} (x_{1_i} - \bar{x}_1)(x_{2_i} - \bar{x}_2) = \sum_{i=1}^{n} x_{1_i}x_{2_i} - n\bar{x}_1\bar{x}_2 = \text{——}$$

$$a_{22} = \sum_{i=1}^{n} (x_{2_i} - \bar{x}_2)^2 = \sum_{i=1}^{n} x_{2_i}{}^2 - n\bar{x}_2{}^2 = \text{——}$$

$$c_1 = \sum_{i=1}^{n} (x_{1_i} - \bar{x}_1)(y_i - \bar{y}) = \sum_{i=1}^{n} x_{1_i}y_i - n\bar{x}_1\bar{y} = \text{——}$$

$$c_2 = \sum_{i=1}^{n} (x_{2_i} - \bar{x}_2)(y_i - \bar{y}) = \sum_{i=1}^{n} x_{2_i}y_i - n\bar{x}_2\bar{y} = \text{——}$$

operation	b_1	b_2	c	check
(1)	a_{11}	a_{12}	c_1	$a_{11} + a_{12} + c_1$
(2)	a_{21}	a_{22}	c_2	$a_{21} + a_{22} + c_2$
(3) (1) repeated	μ_{11}	μ_{12}	μ_{13}	
(4) (3) divided by a_{11}	1	V_{12}	V_{13}	
(5) (2) $-$ $V_{12} \times$ (3)	0	μ_{22}	μ_{23}	
(6) (5) divided by μ_{22}	0	1	V_{23}	

solution

(7) $b_2 = V_{23}$

(8) $b_1 = V_{13} - V_{12}b_2$

(9) $b_0 = \bar{y} - b_1\bar{x}_1 - b_2\bar{x}_2$

The check is started with line 3. All the operations from line 3 on are performed on the values in the check column. For each line, then, the sum of the entries in all the columns should check with the entry in the check column.

predicted from the above relationship, during the seeding period, and compared with the actual values. If seeding is effective, we should expect the actual rainfall for seeded storms to exceed the predicted values rather consistently.

The data are given in the list which is presented on the page at right.

13.6 ANALYSIS OF VARIANCE

13.6.1 Introduction

In Art. 13.4, the problem of testing whether two means are equal was considered. The t test was used for making such a comparison. This problem is actually a special case of the more general problem of comparing several means. A possible solution to this general problem is to enumerate every possible pair, and test whether each pair differs by means of the t test. The disadvantage of this procedure is clear upon a little reflection. Even if the population mean is the same for every sample mean, we must expect that if there are enough sample means some will be extremely large and some extremely small. This will lead to one or more fictitious significant differences—with probability much higher than α. This distortion in the significance level introduces another difficulty. Suppose that the 45 possible comparisons

$n = 168$ $\quad \sum_{i=1}^{168} x_{1_i} y_i = 332.176315$ $\quad \sum_{i=1}^{168} x_{3_i}{}^2 = 349.784479$

$\bar{x}_1 = 1.37761$ $\quad \sum_{i=1}^{168} x_{2_i} y_i = 251.280625$ $\quad \sum_{i=1}^{168} x_{1_i} x_{2_i} = 284.67716$

$\bar{x}_2 = 1.05103$ $\quad \sum_{i=1}^{168} x_{3_i} y_i = 315.567472$ $\quad \sum_{i=1}^{168} x_{1_i} x_{3_i} = 352.654387$

$\bar{x}_3 = 1.27261$ $\quad \sum_{i=1}^{168} x_{1_i}{}^2 = 384.327981$ $\quad \sum_{i=1}^{168} x_{2_i} x_{3_i} = 270.377485$

$\bar{y} = 1.20063$ $\quad \sum_{i=1}^{168} x_{2_i}{}^2 = 221.509607$ $\quad \sum_{i=1}^{168} y_i^2 = 296.885778$

$\sum_{i=1}^{168} x_{1_i}{}^2 - n\bar{x}_1{}^2 = 65.494584$ $\quad \sum_{i=1}^{168} x_{1_i} x_{2_i} - n\bar{x}_1\bar{x}_2 = 41.428439$

$\sum_{i=1}^{168} x_{1_i} x_{3_i} - n\bar{x}_1\bar{x}_3 = 58.122085$ $\quad \sum_{i=1}^{168} x_{2_i}{}^2 - n\bar{x}_2{}^2 = 35.926129$

$\sum_{i=1}^{168} x_{2_i} x_{3_i} - n\bar{x}_2\bar{x}_3 = 45.668373$ $\quad \sum_{i=1}^{168} x_{3_i}{}^2 - n\bar{x}_3{}^2 = 77.701072$

$\sum_{i=1}^{168} x_{1_i} y_i - n\bar{x}_1\bar{y} = 54.303488$ $\quad \sum_{i=1}^{168} x_{2_i} y_i - n\bar{x}_2\bar{y} = 39.281616$

$\sum_{i=1}^{168} x_{3_i} y_i - n\bar{x}_3\bar{y} = 58.873775$

operation	b_1	b_2	b_3	c	check
(1)	65.494584	41.428439	58.122085	54.303488	219.348596
(2)	41.428439	35.926129	45.668373	39.281616	162.304551
(3)	58.122085	45.668373	77.701072	58.873775	240.365305
(4) (1) repeated	65.494584	41.428439	58.122085	54.303488	219.348596
(5) (4) divided by 65.494584	1	0.63254756	0.88743346	0.82912944	3.34911046
(6) (2) −0.63254756 × (4)	0	9.720671	8.903390	4.932078	23.556139
(7) (6) divided by 9.720671	0	1	0.9159234	0.5073804	2.4233038
(8) (3) −0.88743346 (4) −0.9159234 × (6)	0	0	17.966766	6.165637	24.132403
(9) (8) divided by 17.966766	0	0	1	0.3431690	1.3431690

(10) $b_3 = 0.3431690$

(11) $b_2 = 0.5073804 - (0.9159234)(0.3431690)$
$= 0.1930639$

(12) $b_1 = 0.82912944 - (0.88743346)(0.3431690)$
$- (0.63254756)(0.1930639) = 0.4024677$

(13) $b_0 = 1.20063 - (0.4024677)(1.37761) - (0.1930639)((1.05103)$
$- (0.3431690)(1.37761) = 0.60655$

Equation:

$$\tilde{y} = 0.0066 + 0.4025x_1 + 0.1931x_2 + 0.3432x_3$$

TABLE 13.45. *COMPUTATIONAL SCHEME FOR SOLVING THE NORMAL EQUATIONS*

$$\tilde{y} = b_0 + b_1x_1 + b_2x_2 + b_3x_3$$

The normal equations are of the form:

$$\begin{aligned} a_{11}b_1 + a_{12}b_2 + a_{13}b_3 &= c_1 \\ a_{21}b_1 + a_{22}b_2 + a_{23}b_3 &= c_2 \\ a_{31}b_1 + a_{32}b_2 + a_{33}b_3 &= c_3 \end{aligned}$$

where

$$a_{11} = \sum_{i=1}^{n} (x_{1_i} - \bar{x}_1)^2 = \sum_{i=1}^{n} x_{1_i}{}^2 - n\bar{x}_1{}^2 = \text{——}$$

$$a_{21} = a_{12} = \sum_{i=1}^{n} (x_{1_i} - \bar{x}_1)(x_{2_i} - \bar{x}_2) = \sum_{i=1}^{n} x_{1_i}x_{2_i} - n\bar{x}_1\bar{x}_2 = \text{——}$$

$$a_{22} = \sum_{i=1}^{n} (x_{2_i} - \bar{x}_2)^2 = \sum_{i=1}^{n} x_{2_i}{}^2 - n\bar{x}_2{}^2 = \text{——}$$

$$a_{13} = a_{31} = \sum_{i=1}^{n} (x_{1_i} - \bar{x}_1)(x_{3_i} - \bar{x}_3) = \sum_{i=1}^{n} x_{1_i}x_{3_i} - n\bar{x}_1\bar{x}_3 = \text{——}$$

$$a_{23} = \sum_{i=1}^{n} (x_{2_i} - \bar{x}_2)(x_{3_i} - \bar{x}_3) = \sum_{i=1}^{n} x_{2_i}x_{3_i} - n\bar{x}_2\bar{x}_3 = \text{——}$$

$$a_{33} = \sum_{i=1}^{n} (x_{3_i} - \bar{x}_3)^2 = \sum_{i=1}^{n} x_{3_i}{}^2 - n\bar{x}_3{}^2 = \text{——}$$

$$c_1 = \sum_{i=1}^{n} (x_{1_i} - \bar{x}_1)(y_i - \bar{y}) = \sum_{i=1}^{n} x_{1_i}y_i - n\bar{x}_1\bar{y} = \text{——}$$

$$c_2 = \sum_{i=1}^{n} (x_{2_i} - \bar{x}_2)(y_i - \bar{y}) = \sum_{i=1}^{n} x_{2_i}y_i - n\bar{x}_2\bar{y} = \text{——}$$

$$c_3 = \sum_{i=1}^{n} (x_{3_i} - \bar{x}_3)(y_i - \bar{y}) = \sum_{i=1}^{n} x_{3_i}y_i - n\bar{x}_3\bar{y} = \text{——}$$

operation	b_1	b_2	b_3	c	check
(1)	a_{11}	a_{12}	a_{13}	c_1	$a_{11} + a_{12} + a_{13} + c_1$
(2)	a_{21}	a_{22}	a_{23}	c_2	$a_{21} + a_{22} + a_{23} + c_2$
(3)	a_{31}	a_{32}	a_{33}	c_3	$a_{31} + a_{32} + a_{33} + c_3$
(4) (1) repeated	μ_{11}	μ_{12}	μ_{13}	μ_{14}	
(5) (4) divided by a_{11}	1	V_{12}	V_{13}	V_{14}	
(6) (2) $-$ $V_{12} \times$ (4)	0	μ_{22}	μ_{23}	μ_{24}	
(7) (6) divided by μ_{22}	0	1	V_{23}	V_{24}	
(8) (3) $-$ $V_{13} \times$ (4) $-$ $V_{23} \times$ (6)	0	0	μ_{33}	μ_{34}	
(9) (8) divided by μ_{33}	0	0	1	V_{34}	

solution

(10) $b_3 = V_{34}$

(11) $b_2 = V_{24} - V_{23}b_3$

(12) $b_1 = V_{14} - b_3V_{13} - b_2V_{12}$

(13) $b_0 = \bar{y} - b_1\bar{x}_1 - b_2\bar{x}_2 - b_3\bar{x}_3$

The check is started with line 4. All the operations from line 4 on are performed on the values in the check column. For each line, then, the sum of the entries in all the columns should check with the entry in the check column.

among ten means have all been made, and four are significant by an 0.05 level t test; the experimenter must now wonder *which* (if any) of these four significant differences are really indicative of true differences.

The appropriate solution to the problem is the analysis of variance. The analysis of variance, however, has a much wider application than this simple generalization, and is probably the most powerful procedure in the field of experimental statistics. In this article, three problems will be treated: (1) A test for homogeneity of means will be given. (2) If the hypothesis of homogeneity is rejected, a procedure for comparing the means will be given. (3) Estimates of the variability due to a particular effect will be found.

13.6.2 One-Way Classification

Suppose four different quantities of carbon are to be added to a batch of raw material in the manufacture of steel. Six specimens are taken for each of the four quantities, and it is desired to determine the effect of these percentages on the tensile stress of the resultant steel. The data are given in Table 13.46.

It is reasonable to assume that the tensile strength of a specimen consists of the sum of two components; namely, the population mean effect attributed to the addition of the particular percentage of carbon, and a random effect. The random effects are assumed to be independently normally distributed with mean 0 and variance σ^2. Thus, the third specimen having 0.20 per cent of carbon has a tensile strength:

46700 = population mean effect due to 0.20% carbon + random effect

All the specimens having 0.20 per cent carbon have the same mean effect and differ only because of the random effect. More generally, these assumptions can be summarized as follows:

$$X_{ij} = \zeta_i + \varepsilon_{ij}$$

where X_{ij} is the outcome of the jth specimen corresponding to the ith treatment; $i = 1, 2, \ldots, r$; $j = 1, 2, \ldots, s$; ζ_i is the population mean effect attributed to the ith treatment; and ε_{ij} is the random effect assumed to be independently normally distributed with mean 0 and variance σ^2.

A few words about the selection of the specimens of carbon steel to be used in the experiment are in order. The tensile strength of steel depends on many factors besides the percentage of carbon—the temperature, presence of other alloys, and so forth. It is important in running experiments to determine the effect of carbon on a particular kind of steel to make sure that differences which are asserted to be due to carbon do not arise from some other factors, say all the 0.10 per cent carbon bars were treated at a higher temperature than the others. In some cases other factors which affect the steel may be isolated as factors in the experiment and their effect removed by the techniques of two-way, three-way, and other analysis of variance techniques discussed in the subsequent articles. If this is not done, the specimens to be used with each percentage of carbon should be chosen at random with respect to the other factors which are not specified; this random choice will ensure that the conclusions are not vitiated by a coincidence of effects. A table of random numbers is useful in assigning specimens to the specific treatments.

13.6.2.1 Test for Homogeneity. A major problem arising in the analysis of variance is to test the hypothesis that r means are equal, i.e., to test whether all the carbon effects are the same in the example.

This hypothesis can be written as H_0: $\zeta_1 = \zeta_2 = \zeta_3 = \cdots = \zeta_r$, and may be answered by making complete the analysis of variance table, Table 13.47.

TABLE 13.46. *TENSILE STRENGTH OF HOT ROLLED STEEL FOR DIFFERENT PERCENTAGE OF CARBON*

								$\bar{x}$
Percentage 1	0.10%	23050	36000	31100	32650	30900	31400	30850
Percentage 2	0.20%	41850	25650	46700	34500	36650	31450	36133
Percentage 3	0.40%	47050	43450	43000	38650	41850	35450	41575
Percentage 4	0.60%	49650	73900	66450	74550	62400	63750	65117

TABLE 13.47. *ANALYSIS OF VARIANCE TABLES, ONE-WAY CLASSIFICATION*

Source	Sum of Squares	Degrees of Freedom	Mean Square	Average Mean Square	Test
Between treatments	$SS_3 = s\sum_{i=1}^{r}(\bar{X}_{i.} - \bar{X}_{..})^2$	$r - 1$	$SS^*_3 = SS_3/(r-1)$	$\sigma^2 + \dfrac{s\sum_{i=1}^{r}(\zeta_i - \bar{\zeta})^2}{r-1}$	$F = \dfrac{SS^*_3}{SS^*_2}$
Within treatments	$SS_2 = \sum_{i=1}^{r}\sum_{j=1}^{s}(X_{ij} - \bar{X}_{i.})^2$	$r(s - 1)$	$SS^*_2 = SS_2/r(s-1)$	σ^2	
Total	$SS = \sum_{i=1}^{r}\sum_{j=1}^{s}(X_{ij} - \bar{X}_{..})^2$	$rs - 1$	$SS^* = SS/(rs-1)$	$\sigma^2 + \dfrac{s\sum_{i=1}^{r}(\zeta_i - \bar{\zeta})^2}{rs-1}$	

In this table $\bar{X}_{i.} = \sum_{j=1}^{s}\frac{X_{ij}}{s}$; $\bar{X}_{..} = \sum_{i=1}^{r}\sum_{j=1}^{s}\frac{X_{ij}}{rs}$; $\bar{\zeta} = \sum_{i=1}^{r}\frac{\zeta_i}{r}$.

Computational procedure

1. Calculate totals for each treatment:

$$R_1, R_2, \ldots, R_r$$

2. Calculate over-all total:

$$T = R_1 + R_2 + \cdots + R_r$$

3. Compute crude total sum of squares:

$$\sum_{i=1}^{r}\sum_{j=1}^{s} X_{ij}^2 = X_{11}^2 + X_{12}^2 + \cdots + X_{rs}^2$$

4. Calculate crude sum of squares between treatments:

$$\frac{\sum_{i=1}^{r} R_i^2}{s} = (R_1^2 + R_2^2 + \cdots + R_r^2)/s$$

5. Calculate correction factor due to mean $= T^2/rs$.

From the above quantities compute

$$6.\ SS_3 = (4) - (5) = \sum_{i=1}^{r} R_i^2/s - \frac{T^2}{rs}$$

$$7.\ SS = (3) - (5) = \sum_{i=1}^{r}\sum_{j=1}^{s} X_{ij}^2 - \frac{T^2}{rs}$$

$$8.\ SS_2 = (7) - (6) = SS - SS_3$$

If $$F = \frac{SS_3/(r-1)}{SS_2/r(s-1)} = \frac{SS_3^*}{SS_2^*} \geq F_{\alpha;r-1,r(s-1)}$$

we reject the hypothesis

$$\zeta_1 = \zeta_2 = \zeta_3 = \cdots = \zeta_r$$

where $F_{\alpha;r-1,r(s-1)}$ is the upper α percentage point of the F distribution given in Table 13.32.

* J. Tukey, *Allowances for Various Types of Error Rates*, unpublished invited address presented before a joint meeting of the Institute of Mathematical Statistics and the Eastern North American Region of the Biometric Society on March 19, 1952, at Blacksburg, Va.

13.6.2.2 Test for Comparisons of Mean. If the hypothesis

$$H_0 : \zeta_1 = \zeta_2 = \zeta_3 = \cdots = \zeta_r$$

is rejected, it is possible to make statements about which means differ. For a long time there was no satisfactory solution to this problem. However, Tukey and Scheffé† have presented solutions whereby it becomes posslble to make contrasts of the form $\zeta_k - \zeta_l$. Tukey's procedure leads to probability statements in the form of confidence intervals for the contrasts $\zeta_k - \zeta_l$; i.e., the probability is $1 - \alpha$ that the values $\zeta_k - \zeta_l$ for *all* such contrasts ($k = 1, 2, \ldots, r$; $l = 1, 2, \ldots, r$) simultaneously satisfy

$$\bar{X}_{k\cdot} - \bar{X}_{l\cdot} - c \leq \zeta_k - \zeta_l \leq \bar{X}_{k\cdot} - \bar{X}_{l\cdot} + c$$

where c is a factor obtained from Table 13.48 depending on whether $\alpha = 0.05$ or 0.01 and from the analysis of variance table. The means are said to differ significantly when the appropriate confidence interval fails to include 0. Furthermore, if the interval includes only positive values, the difference is said to differ significantly from zero and to be positive. Similarly, if the confidence interval includes only negative values, the difference is said to differ significantly from zero and to be negative. If the interval includes 0, the means are said not to differ significantly. Summarizing then, the F test is made for the homogeneity of the means. If this hypothesis is rejected, Tukey's procedure is applied to determine which of the means differ.

A possible short-cut method for dctermining the homogeneity of all the treatment effects is to check whether $\bar{X}_{\max} - \bar{X}_{\min} \pm c$ includes 0. If this contrast includes 0, every other contrast of this type includes 0, and hence all the treatment effects are called equal. If the above contrast fails to include zero, the corresponding treatment effects, and perhaps others, differ significantly. This short-cut method has a level of significance exactly equal to α. However, the data in the analysis of variance table are necessary for determining the factor c and for other results, so that this table should be completed regardless.

† H. Scheffé, "A Method for Judging All Contrasts in the Analysis of Variance," *Biometrika*, Vol. 40, June 1953.

TABLE 13.48. *TABLE OF FACTORS c**

5 Per Cent Significance Level‡

ν \ n‡	2	3	4	5	6	7	8	9	10	11	12	13	14	15	16	17	18	19	20
1	18.0	26.7	32.8	37.2	40.5	43.1	45.4	47.3	49.1	50.6	51.9	53.2	54.3	55.4	56.3	57.2	58.0	58.8	59.6
2	6.09	8.28	9.80	10.89	11.73	12.43	13.03	13.54	13.99	14.39	14.75	15.08	15.38	15.65	15.91	16.14	16.36	16.57	16.77
3	4.50	5.88	6.83	7.51	8.04	8.47	8.85	9.18	9.46	9.72	9.95	10.16	10.35	10.52	10.69	10.84	10.98	11.12	11.24
4	3.93	5.00	5.76	6.31	6.73	7.06	7.35	7.60	7.83	8.03	8.21	8.37	8.52	8.67	8.80	8.92	9.03	9.14	9.24
5	3.64	4.60	5.22	5.67	6.03	6.33	6.58	6.80	6.99	7.17	7.32	7.47	7.60	7.72	7.83	7.93	8.03	8.12	8.21
6	3.46	4.34	4.90	5.31	5.63	5.89	6.12	6.32	6.49	6.65	6.79	6.92	7.04	7.14	7.24	7.34	7.43	7.51	7.59
7	3.34	4.16	4.68	5.06	5.36	5.61	5.82	6.00	6.16	6.30	6.43	6.55	6.66	6.76	6.85	6.94	7.02	7.09	7.17
8	3.26	4.04	4.53	4.89	5.17	5.40	5.60	5.77	5.92	6.05	6.18	6.29	6.39	6.48	6.57	6.65	6.73	6.80	6.87
9	3.20	3.95	4.42	4.76	5.02	5.24	5.43	5.60	5.74	5.87	5.98	6.09	6.19	6.28	6.36	6.44	6.51	6.58	6.65
10	3.15	3.88	4.33	4.66	4.91	5.12	5.30	5.46	5.60	5.72	5.83	5.93	6.03	6.12	6.20	6.27	6.34	6.41	6.47
11	3.11	3.82	4.26	4.58	4.82	5.03	5.20	5.35	5.49	5.61	5.71	5.81	5.90	5.98	6.06	6.14	6.20	6.27	6.33
12	3.08	3.77	4.20	4.51	4.75	4.95	5.12	5.27	5.40	5.51	5.61	5.71	5.80	5.88	5.95	6.02	6.09	6.15	6.21
13	3.06	3.73	4.15	4.46	4.69	4.88	5.05	5.19	5.32	5.43	5.53	5.63	5.71	5.79	5.86	5.93	6.00	6.06	6.11
14	3.03	3.70	4.11	4.41	4.64	4.83	4.99	5.13	5.25	5.36	5.46	5.56	5.64	5.72	5.79	5.86	5.92	5.98	6.03
15	3.01	3.67	4.08	4.37	4.59	4.78	4.94	5.08	5.20	5.31	5.40	5.49	5.57	5.65	5.72	5.79	5.85	5.91	5.96
16	3.00	3.65	4.05	4.34	4.56	4.74	4.90	5.03	5.15	5.26	5.35	5.44	5.52	5.59	5.66	5.73	5.79	5.84	5.90
17	2.98	3.62	4.02	4.31	4.52	4.70	4.86	4.99	5.11	5.21	5.31	5.39	5.47	5.55	5.61	5.68	5.74	5.79	5.84
18	2.97	3.61	4.00	4.28	4.49	4.67	4.83	4.96	5.07	5.17	5.27	5.35	5.43	5.50	5.57	5.63	5.69	5.74	5.79
19	2.96	3.59	3.98	4.26	4.47	4.64	4.79	4.92	5.04	5.14	5.23	5.32	5.39	5.46	5.53	5.59	5.65	5.70	5.75
20	2.95	3.58	3.96	4.24	4.45	4.62	4.77	4.90	5.01	5.11	5.20	5.28	5.36	5.43	5.50	5.56	5.61	5.66	5.71
24	2.92	3.53	3.90	4.17	4.37	4.54	4.68	4.81	4.92	5.01	5.10	5.18	5.25	5.32	5.38	5.44	5.50	5.55	5.59
30	2.89	3.48	3.84	4.11	4.30	4.46	4.60	4.72	4.83	4.92	5.00	5.08	5.15	5.21	5.27	5.33	5.38	5.43	5.48
40	2.86	3.44	3.79	4.04	4.23	4.39	4.52	4.63	4.74	4.82	4.90	4.98	5.05	5.11	5.17	5.22	5.27	5.32	5.36
60	2.83	3.40	3.74	3.98	4.16	4.31	4.44	4.55	4.65	4.73	4.81	4.88	4.94	5.00	5.06	5.11	5.15	5.20	5.24
120	2.80	3.36	3.69	3.92	4.10	4.24	4.36	4.47	4.56	4.64	4.71	4.78	4.84	4.90	4.95	5.00	5.04	5.09	5.13
∞	2.77	3.32	3.63	3.86	4.03	4.17	4.29	4.39	4.47	4.55	4.62	4.68	4.74	4.80	4.84	4.89	4.93	4.97	5.01

† This table is taken by permission from "Extended and Corrected Tables of the Upper Percentage Points of the Studentized Range," by Joyce M. May, *Biometrika*, Vol. 39, Part 122, May 1952, p. 192; and from "Corrigenda to Tables of Percentage Points of the Studentized Range," by H. O. Hartley. *Biometrika*, Vol. 40, Part 122, June 1953, p. 236.

‡ Here n is the number of effects being studied.

1 Per Cent Significance Level†

ν \ n‡	2	3	4	5	6	7	8	9	10	11	12	13	14	15	16	17	18	19	20
1	90.0	134	164	186	202	216	227	237	246	253	260	266	272	277	282	286	291	295	298
2	14.0	18.9	22.3	24.7	26.6	28.2	29.5	30.7	31.7	32.6	33.4	34.2	34.8	35.5	36.0	36.5	37.0	37.5	38.0
3	8.26	10.56	12.17	13.34	14.25	15.00	15.65	16.20	16.69	17.13	17.53	17.89	18.23	18.54	18.83	19.09	19.33	19.56	19.79
4	6.51	8.08	9.17	9.97	10.58	11.10	11.55	11.93	12.26	12.56	12.84	13.09	13.32	13.53	13.73	13.92	14.09	14.25	14.40
5	5.70	6.97	7.80	8.42	8.91	9.32	9.67	9.97	10.24	10.48	10.70	10.89	11.08	11.24	11.40	11.55	11.68	11.81	11.93
6	5.24	6.32	7.03	7.56	7.97	8.31	8.61	8.87	9.10	9.30	9.49	9.65	9.81	9.95	10.08	10.21	10.32	10.43	10.54
7	4.95	5.92	6.54	7.01	7.37	7.68	7.94	8.17	8.37	8.55	8.71	8.86	9.00	9.12	9.24	9.35	9.46	9.55	9.65
8	4.74	5.63	6.20	6.63	6.96	7.24	7.47	7.68	7.86	8.03	8.18	8.31	8.44	8.55	8.66	8.76	8.86	8.95	9.03
9	4.60	5.42	5.96	6.35	6.66	6.91	7.13	7.33	7.50	7.65	7.79	7.91	8.02	8.13	8.23	8.33	8.41	8.50	8.58
10	4.48	5.26	5.77	6.14	6.43	6.67	6.88	7.06	7.22	7.36	7.49	7.60	7.71	7.81	7.91	7.99	8.08	8.15	8.23
11	4.39	5.14	5.62	5.98	6.25	6.47	6.67	6.84	6.99	7.13	7.25	7.36	7.46	7.56	7.65	7.73	7.81	7.88	7.95
12	4.32	5.04	5.50	5.84	6.10	6.32	6.51	6.67	6.81	6.94	7.06	7.17	7.26	7.36	7.44	7.52	7.60	7.67	7.73
13	4.26	4.96	5.40	5.73	5.98	6.19	6.37	6.53	6.67	6.79	6.90	7.01	7.10	7.19	7.27	7.35	7.42	7.49	7.55
14	4.21	4.89	5.32	5.64	5.88	6.08	6.26	6.41	6.54	6.66	6.77	6.87	6.96	7.05	7.13	7.20	7.27	7.34	7.40
15	4.17	4.83	5.25	5.56	5.80	5.99	6.16	6.31	6.44	6.55	6.66	6.76	6.85	6.93	7.00	7.07	7.14	7.20	7.26
16	4.13	4.78	5.19	5.49	5.72	5.91	6.08	6.22	6.35	6.46	5.56	6.66	6.74	6.82	6.90	6.97	7.03	7.09	7.15
17	4.10	4.73	5.14	5.43	5.66	5.85	6.01	6.15	6.27	6.38	6.48	6.57	6.66	6.73	6.81	6.87	6.94	7.00	7.05
18	4.07	4.70	5.09	5.38	5.60	5.79	5.95	6.08	6.20	6.31	6.41	6.50	6.58	6.65	6.73	6.79	6.85	6.91	6.97
19	4.05	4.66	5.05	5.34	5.55	5.73	5.89	6.02	6.14	6.25	6.34	6.43	6.51	6.58	6.65	6.72	6.78	6.84	6.89
20	4.02	4.63	5.02	5.30	5.51	5.69	5.84	5.97	6.09	6.19	6.28	6.37	6.45	6.52	6.59	6.66	6.71	6.77	6.82
24	3.96	4.54	4.91	5.17	5.37	5.54	5.69	5.81	5.92	6.02	6.11	6.19	6.26	6.33	6.39	6.45	6.51	6.57	6.61
30	3.89	4.45	4.80	5.05	5.24	5.40	5.53	5.65	5.76	5.85	5.93	6.01	6.08	6.14	6.20	6.26	6.31	6.36	6.41
40	3.82	4.36	4.70	4.93	5.11	5.26	5.39	5.50	5.60	5.69	5.77	5.84	5.90	5.96	6.02	6.07	6.12	6.17	6.21
60	3.76	4.28	4.60	4.82	4.99	5.13	5.25	5.36	5.45	5.53	5.60	5.67	5.73	5.78	5.83	5.88	5.93	5.98	6.01
120	3.70	4.20	4.50	4.71	4.87	5.00	5.12	5.21	5.30	5.38	5.44	5.50	5.56	5.61	5.66	5.71	5.75	5.79	5.83
∞	3.64	4.12	4.40	4.60	4.76	4.88	4.99	5.08	5.16	5.23	5.29	5.35	5.40	5.45	5.49	5.53	5.57	5.61	5.64

† This table is taken by permission from "Extended and Corrected Tables of the Upper Percentage Points of the Studentized Range", May; and from Hartley, "Corrigenda to Tables of Percentage Points of the Studentized Range."

‡ Here n is the number of effects being studied.

The factor c can be obtained from the data in the analysis of variance table and Table 13.48 depending on the level of significance. Table 13.48 is entered with the indices degrees of freedom and the number of treatments studied (r). The proper degrees of freedom are those corresponding to the degrees of freedom of the within treatments source in the analysis of variance table, i.e., $r(s-1)$. The factor c^* is read from the table; c is obtained from $c = c^* \sqrt{SS_2^*/s}$.

Scheffé's procedure also leads to confidence statements and if we apply his test when the F test of homogeneity is rejected, the level of significance is preserved. On the other hand, if we apply Tukey's test when the F test of homogeneity is rejected, we increase the significance level *slightly*. However, if the only type of comparison of interest has the form $\zeta_k - \zeta_l$, Tukey's procedure will result in shorter confidence intervals for $\zeta_k - \zeta_l$ than would Scheffé's procedure.† Consequently, we are going to use the Tukey method, and proceed as if the significance level suffered a negligible increase.

13.6.2.3 Estimating the Variability Due to the Sources. In the analysis of variance table there is a column headed "average mean square." Estimates of these quantities are given in the column headed "mean square." If all the treatment means are equal, the entries in the "average mean square" column would be σ^2. Consequently, it is expected, if all the means are equal, that the estimates given in the "mean square" column will be close together. On the other hand, if the treatments are in fact different, then the average value of the mean square for treatments will be increased by $[s/(r-1)] \sum(\zeta_i - \bar{\zeta})^2$, which will tend to increase the value of F, leading to a significant result and the detection of the differences.

† If other types of contrasts are also desired, e.g., $(\zeta_1+\zeta_2+\zeta_3)/3-(\zeta_4+\zeta_5+\zeta_6)/3$, Scheffé's procedure should be used.

13.6.2.4 Example. Returning to the data on the tensile strength of carbon steel, Table 13.49 is the complete analysis of variance table.

The completed computational procedure is given below:

(1) $R_1, R_2, \ldots, R_r = 185{,}100;\ 216{,}800;\ 249{,}450;\ 390{,}700$

(2) $T = \sum_{i=1}^{4} R_i = 185{,}100 + 216{,}800 + 249{,}450 + 390{,}700 = 1{,}042{,}050$

(3) $\sum_{i=1}^{4} \sum_{j=1}^{6} x_{ij}^2 = (23{,}050)^2 + (36{,}000)^2 + \cdots + (63{,}750)^2 = 50{,}224{,}067{,}500$

(4) $\dfrac{\sum_{i=1}^{4} R_i^2}{6} = \dfrac{(185{,}100)^2 + (216{,}800)^2 + (249{,}450)^2 + (390{,}700)^2}{6} = 49{,}356.007{,}000$

(5) $\dfrac{T^2}{24} = \dfrac{1{,}085{,}868{,}202{,}500}{24} = 45{,}244{,}508{,}400$

(6) $SS_3 = 49{,}356{,}007{,}000 - 45{,}244{,}508{,}400 = 4{,}111{,}498{,}600$

(7) $SS = 50{,}224{,}067{,}500 - 45{,}244{,}508{,}400 = 4{,}979{,}559{,}100$

(8) $SS_2 = 4{,}979{,}559{,}100 - 4{,}111{,}498{,}600 = 868{,}060{,}500$

Since $F_{.05;3,20} = 3.10$ and $F = 31.576 > 3.10$, the hypothesis that $\zeta_1 = \zeta_2 = \zeta_3 = \zeta_4$ is rejected and we conclude that the tensile strengths differ. In order to determine the magnitude of these differences, we compute c for $r = 4$, $r(s-1) = 20$. Here $c^* = 3.96$, where the factor 3.96 is obtained from Table 13.48.

Therefore, $c = 3.96(2689) = 10{,}648$, and

TABLE 13.49. *ANALYSIS OF VARIANCE TABLE FOR TENSILE STRENGTH*

Source	Sum of Squares	Degrees of Freedom	Mean Square	Average Mean Square	Test
Between treatments	4,111,498,600	3	1,370,499,533		$F = 31.576$
Within treatments	868,060,500	20	43,403,025		
Total	4,979,559,100	23	216,502,670		

any two means differing by this much differ significantly at $\alpha = 0.05$. Thus the following differences are significant.

0.60 from 0.40
0.60 from 0.20
0.60 from 0.10
0.40 from 0.10

Confidence intervals of size 0.95 for the differences can be given, e.g., the confidence interval for difference $\zeta_{.60} - \zeta_{.10}$ is:

$$\bar{X}_{.60} - \bar{X}_{.10} \pm c = 34{,}267 \pm 10{,}648$$

so that

$$23{,}619 \leq \zeta_{.60} - \zeta_{.10} \leq 44{,}915$$

Similarly,

$$\bar{X}_{.60} - \bar{X}_{.40} \pm c = 23{,}542 \pm 10{,}648$$

so that

$$12{,}894 \leq \zeta_{.60} - \zeta_{.40} \leq 34{,}190$$

13.6.2.5 Randomization Tests in the Analysis of Variance. It has been emphasized in the discussion of the t test in the analysis of variance that the random effects are assumed to be normally and independently distributed. These tests actually have a wider justification if they are considered special cases of what are called permutation tests. Suppose, for example, we are interested in testing the difference between two effects, say percentage of carbon, and suppose that the experimenter takes eight samples (let them be named a, b, c, etc.) and chooses four with a table of random numbers to receive treatment A and the other four treatment B. The results of this experiment are:

	A		B
a	2	f	17
h	15	e	19
b	10	c	18
d	9	g	12

He could test the hypothesis that there is no difference between treatments by calculating a t test or an F test, or consider it a one-way analysis of variance, calculating an F with 1 and 6 degrees of freedom; the value of F is 6 in this problem. Suppose he is interested in a level of significance of 0.05. Even if he is unwilling to assume normality, he can obtain a test of the hypothesis as follows: Consider all possible pairs of samples that can be formed with the eight numbers above; there are $\binom{8}{4} = 70$ different ways for the experiment to turn out (one for each way of selecting four of the eight numbers for treatment A). If there is no dlfference between treatments, each of these outcomes is equally likely. Suppose we adopt the procedure that we will reject the hypothesis of equality of treatments if we have one of the samples with the four largest differences in the mean; the probability of rejection when the hypothesis is true is $4/70 = 0.057$, since all 70 possibilities are equally likely. If the hypothesis is false, large differences are more likely.

A partial list of the seventy outcomes (starting with most extreme ones) is:

A	B	Difference in Means $\bar{x}_A - \bar{x}_B$	F
2 9 10 12	15 17 18 19	−9	14.84
15 17 18 19	2 9 10 12	+9	
2 9 10 15	12 17 18 19	−7.5	6.00
12 17 18 19	2 9 10 15	+7.5	

A	B	Difference in Means $\bar{x}_A - \bar{x}_B$	F
2 9 10 17	12 15 18 19	−6.5	3.61
12 15 18 19	2 9 10 17	+6.5	
2 9 12 15	10 17 18 19	−6.5	
10 17 18 19	2 9 12 15	+6.5	

Observe that only four of the seventy possibilities give values of F as large as that in the particular sample obtained.

The significance level of the test is 0.057. If $F = 6$ is looked up in the table it is seen to be significant at the 0.05 level (using the assumption of normality).

This kind of test is called a randomization test. It has the drawback that for medium or large-sized samples the computational burden is enormous. For example, if there are twenty objects divided at random into two groups of ten, then the number of possibilities is $\binom{20}{10} = 184{,}756$, and listing the 5 per cent most extreme cases would be prohibitive.

Both numerical and analytical investigations* have shown, however, that for medium and large samples, the distribution of F obtained through enumeration of the possibilities is closely approximated by the tabulated F distribution. So without assuming normality we *can* test the hypothesis of no treatment effects by making a randomization test, but instead of exactly evaluating the significance

---

* The agreement between the two values 0.05 and 0.057 above is not merely by coincidence.

TABLE 13.50. *NEW COUNTING RATES LESS* 25.00

Specimen Number	Experiment Number 1	2	3	4	Specimen Average
1	1.46	2.00	1.48	1.51	1.61
2	2.58	3.03	2.82	2.90	2.83
3	4.15	4.61	4.31	4.13	4.30
4	4.54	4.52	4.53	4.15	4.44
Experiment average	3.18	3.54	3.28	3.17	

TABLE 13.51. *TABLE OF POPULATION MEANS FOR COUNTING RATES*

Specimen Number	Experiment Number 1	2	3	4	Specimen Mean	Differences
1	ζ_{11}	ζ_{12}	ζ_{13}	ζ_{14}	$\zeta_{1.}$	$\varphi_1 = \zeta_{1.} - \zeta_{..}$
2	ζ_{21}	ζ_{22}	ζ_{23}	ζ_{24}	$\zeta_{2.}$	$\varphi_2 = \zeta_{2.} - \zeta_{..}$
3	ζ_{31}	ζ_{32}	ζ_{33}	ζ_{34}	$\zeta_{3.}$	$\varphi_3 = \zeta_{3.} - \zeta_{..}$
4	ζ_{41}	ζ_{42}	ζ_{43}	ζ_{44}	$\zeta_{4.}$	$\varphi_4 = \zeta_{4.} - \zeta_{..}$
Experiment mean	$\zeta_{.1}$	$\zeta_{.2}$	$\zeta_{.3}$	$\zeta_{.4}$	$\zeta_{..}$	
Differences	$\gamma_1 = \zeta_{.1} - \zeta_{..}$	$\gamma_2 = \zeta_{.2} - \zeta_{..}$	$\gamma_3 = \zeta_{.3} - \zeta_{..}$	$\gamma_4 = \zeta_{.4} - \zeta_{..}$		

level by actual counting we content ourselves with an approximate evaluation, as yielded by the F tables.

13.6.3 Two-Way Classification

The reader is advised to read the section on the one-way classification before going on to this section. Experiments can be conducted in such a way as to study the effects of several variables in the same experiment. For each variable a number of levels may be chosen for study. When observations are made for all possible combinations of levels the experiment is called a factorial experiment. This section is concerned with a factorial experiment where the effects of two variables at several levels are to be studied. For example, W. J. Youden† reports the following experiment. Counts are made on four samples of radium. If the four specimens are run consecutively, the four counts may be called experiment 1. The experiment may be repeated as often as desired, varying the order of the specimens at random in each experiment. The results of this investigation are given in Table 13.50 where the data are given as the net counting rates less 25.00. The experimenter is interested in knowing whether these samples of radium are different. However, it is possible that the line voltage or other conditions that influence the counting rate vary from experiment to experiment. Thus the model can be written

$$X_{ij} = \zeta_{ij} + \varepsilon_{ij}; \qquad i = 1,2,3,4; j = 1,2,3,4$$

where

X_{ij} is the counting rate of the ith specimen in the jth experiment;

ζ_{ij} is the population mean counting rate of the ith specimen in the jth experiment;

ε_{ij} is the random effect assumed to be normally independently distributed with mean 0 and variance σ^2.

13.6.3.1 Interaction. Before proceeding to make inferences about the structure of the ζ_{ij}, it is useful to refer to Table 13.51, which is a table of population means.

Here $\zeta_{i.} = \frac{1}{4}\sum_{j=1}^{4} \zeta_{ij}$; $\zeta_{.j} = \frac{1}{4}\sum_{i=1}^{4} \zeta_{ij}$; $\zeta_{..} = \frac{1}{16}\sum_{i=1}^{4}\sum_{j=1}^{4} \zeta_{ij}$. Thus a mean ζ_{ij} can be written

† W. J. Youden, "The Interpretation of Chemical Data," *Industrial Quality Control*, May 1952.

$$\begin{aligned}\zeta_{ij} &= \zeta_{..} + (\zeta_{i.} - \zeta_{..}) + (\zeta_{.j} - \zeta_{..}) \\ &\quad + (\zeta_{ij} - \zeta_{i.} - \zeta_{.j} + \zeta_{..}) \\ &= \zeta_{..} + \varphi_i + \gamma_j + \eta_{ij}\end{aligned}$$

where $\eta_{ij} = \zeta_{ij} - \zeta_{i.} - \zeta_{.j} + \zeta_{..}$ and is known as the interaction term. We say that an interaction between two effects, φ_i and γ_j, exists if the joint effect of the two taken simultaneously is different from the sum of their separate effects. Thus either lye or muriatic acid may be an effective cleanser, but taken together they are ineffective, and would be said to interact. Or again, suppose there are five machines together with four workmen and it is desired to test whether machines differ in the number of units produced per day. It is quite possible that the second man may work better on the third machine than any other machine because he has worked for 20 years on such a machine. The resultant increased production cannot be assumed to be a characteristic of this man or of this particular machine, but a combination of only this man with only this machine, and we would say that interaction was present. In the counting rate problem we expect any cause raising the reading for one specimen in an experiment to raise all readings similarly, and so we may confidently assume the interactions between experiments and specimens to be zero, i.e., $\eta_{ij} = 0$ for all i and j.

Returning to the table of means, we find the following linear relationships:

$$\sum_{i=1}^{4} \varphi_i = 0; \quad \sum_{j=1}^{4} \gamma_j = 0;$$

$$\sum_{i=1}^{4} \eta_{ij} = 0; \quad \sum_{j=1}^{4} \eta_{ij} = 0$$

Furthermore, instead of the original 16 means, we have introduced new parameters which will supply the results. By writing $\zeta_{ij} = \zeta_{..} + \varphi_i + \gamma_j + \eta_{ij}$ we can write the mean for the ith specimen and jth experiment as a constant, $\zeta_{..}$, plus an effect due to the ith specimen which is constant over all columns, φ_i, plus an effect due to the jth experiment which is constant over all rows, γ_j, plus the interaction between the ith specimen and jth experiment. Testing whether (1) the specimens are homogeneous, (2) the experiments are homogeneous, and (3) whether there is any interaction is equivalent to testing:

(1) $\varphi_1 = \varphi_2 = \varphi_3 = \varphi_4 = 0$

(2) $\gamma_1 = \gamma_2 = \gamma_3 = \gamma_4 = 0$

(3) $\eta_{ij} = 0$

for all i and j.

13.6.4 Two-Way Classification: One Observation Per Cell and No Interaction Assumed

In order to make an analysis with only one observation per cell it is necessary to assume that the interaction is 0. The model is as follows:

$$X_{ij} = \zeta_{..} + \varphi_i + \gamma_j + \varepsilon_{ij}$$

where

X_{ij} is the outcome of the experiment using the ith row effect and jth column effect, $i = 1, 2, \ldots, r$; $j = 1, 2, \ldots, s$;

$\zeta_{..}$ is a general mean;

φ_i is the effect of adding the ith row treatment; $\sum_{i=1}^{r} \varphi_i = 0$;

γ_j is the effect of adding the jth column treatment; $\sum_{j=1}^{s} \gamma_j = 0$;

ε_{ij} is the random effect, which is independently normally distributed with mean 0 and variance σ^2.

13.6.4.1 Test for Homogeneity. Complete the following analysis of variance table (Table 13.52):

Computational procedure for analysis of variance table is:

1. Calculate row totals:

   $R_1, R_2, \ldots, R_r$

2. Calculate column totals:

   $C_1, C_2, \ldots, C_s$

3. Calculate over-all total:

   $T = R_1 + R_2 + \cdots + R_r$

4. Calculate crude total sum of squares:

   $$\sum_{i=1}^{r} \sum_{j=1}^{s} X_{ij}^2 = X_{11}^2 + X_{12}^2 + \cdots + X_{rs}^2$$

5. Calculate crude sum of squares between rows:

TABLE 13.52. *ANALYSIS OF VARIANCE TABLE, TWO-WAY CLASSIFICATION, ONE OBSERVATION PER CELL*

Source	Sum of Squares	Degrees of Freedom	Mean Square	Average Mean Square	Test
Between columns	$SS_4 = r \sum_{j=1}^{s} (\bar{X}._j - \bar{X}..)^2$	$s - 1$	$SS^*_4 = SS_4/(s - 1)$	$\sigma^2 + \frac{r}{s-1} \sum_{j=1}^{s} \gamma_i^2$	$F_4 = SS^*_4/SS^*_2$
Between rows	$SS_3 = s \sum_{i=1}^{r} (\bar{X}_i. - \bar{X}..)^2$	$r - 1$	$SS^*_3 = SS_3/(r - 1)$	$\sigma^2 + \frac{s}{r-1} \sum_{i=1}^{r} \varphi_i^2$	$F_3 = SS^*_3/SS^*_2$
Residue	$SS_2 = \sum_{i=1}^{r} \sum_{j=1}^{s} (X_{ij} - \bar{X}_i. - \bar{X}_j. + \bar{X}..)^2$	$(r - 1)(s - 1)$	$SS^*_2 = SS_2/(r - 1)(s - 1)$	σ^2	
Total	$SS = \sum_{i=1}^{r} \sum_{j=1}^{s} (X_{ij} - \bar{X}..)^2$	$rs - 1$	$SS^* = SS/(rs - 1)$	$\sigma^2 + \frac{r}{rs-1} \sum_{j=1}^{s} \gamma_j^2 + \frac{s}{rs-1} \sum_{i=1}^{r} \varphi_i^2$	

In this table, $\bar{X}._j = \sum_{i=1}^{r} X_{ij}/r$; $\bar{X}_i. = \sum_{j=1}^{s} X_{ij}/s$; $\bar{X}.. = \sum_{i=1}^{r} \sum_{j=1}^{s} X_{ij}/rs$

$$\sum_{i=1}^{r} \frac{R_i^2}{s} = (R_1^2 + R_2^2 + \cdots + R_r^2)/s$$

6. Calculate crude sum of squares between columns:

$$\sum_{j=1}^{s} \frac{C_j^2}{r} = (C_1^2 + C_2^2 + \cdots + C_s^2)/r$$

7. Calculate correction factor due to mean $= T^2/rs$. From the above quantities compute

8. $SS_4 = (6) - (7) = \sum_{j=1}^{s} \frac{C_j^2}{r} - T^2/rs$

9. $SS_3 = (5) - (7) = \sum_{i=1}^{r} \frac{R_i^2}{s} - T^2/rs$

10. $SS = (4) - (7) = \sum_{i=1}^{r} \sum_{j=1}^{s} X_{ij}^2 - T^2/rs$

11. $SS_2 = (10) - (9) - (8) = SS - SS_3 - SS_4$

If $$F_4 = \frac{SS_4/(s-1)}{SS_2/(r-1)(s-1)} = \frac{SS_4^*}{SS_2^*} \geq F_{\alpha;s-1,(r-1)(s-1)}$$

we reject the hypothesis that there is no column effect, i.e.,

$$\gamma_1 = \gamma_2 = \gamma_3 = \cdots = \gamma_s = 0$$

where $F_{\alpha;s-1,(r-1)(s-1)}$ is the upper α percentage point of the F distribution given in Table 13.32.

If $$F_3 = \frac{SS_3/(r-1)}{SS_2/(r-1)(s-1)} = \frac{SS_3^*}{SS_2^*} \geq F_{\alpha;r-1,(r-1)(s-1)}$$

we reject the hypothesis that there is no row effect, i.e.,

$$\varphi_1 = \varphi_2 = \varphi_3 = \cdots = \varphi_r = 0$$

where $F_{\alpha;r-1,(r-1)(s-1)}$ is the upper α percentage point of the F distribution given in Table 13.32.

13.6.4.2 Tests for Comparison of Mean Effect. If the hypothesis that there is no row effects or no column effects, or both, is rejected, Tukey's method can again be applied to get confidence intervals on the difference between two row effects or two column effects. The probability is $1 - \alpha$ that the value $(\varphi_k - \varphi_l)$ for *all* such contrasts (for $k = 1, 2, \ldots, r$; $l = 1, 2, \ldots, r$) simultaneously satisfy

$$\bar{X}_{k\cdot} - \bar{X}_{l\cdot} - c \leq \varphi_k - \varphi_l \leq \bar{X}_{k\cdot} - \bar{X}_{l\cdot} + c$$

and the probability is $1 - \alpha$ that the values $(\gamma_m - \gamma_n)$ for *all* such contrasts (for $m = 1, 2, \ldots, s$; $n = 1, 2, \ldots, s$) simultaneously satisfy

$$\bar{X}_{\cdot m} - \bar{X}_{\cdot n} - c \leq \gamma_m - \gamma_n \leq \bar{X}_{\cdot m} - \bar{X}_{\cdot n} + c$$

Again two mean effects are judged significantly different when 0 is not included in the confidence interval.

The factor c for row effects and for column effects is obtained from the analysis of variance table and from Table 13.48 depending on the level of significance. Table 13.48 is entered with the indices degrees of freedom, and the number of row effects being studied (r) if row effects are of interest. The proper degrees of freedom are those corresponding to the degrees of freedom of the residual source in the analysis of variance table, i.e., $(r-1)(s-1)$. The factor c^* is read from the table, and

$$c(\text{for rows}) = c^* \sqrt{SS_2^*/s}$$

Similarly, if column effects are of interest, the table is entered with indices degrees of freedom and the number of column effects being studied (s). The factor c^* is read from the table and

$$c(\text{for columns}) = c^* \sqrt{SS_2^*/r}$$

13.6.4.3 Estimating the Variability Due to the Sources. If there were no row effects and no column effects, the column headed "average mean square" in the analysis of variance table would contain σ^2 for the row source entry and σ^2 for the column source entry. The residual entry is always σ^2. The entry for the residual source, SS_2^*, in the "mean square" column estimates σ^2. The values in the "mean square" column for row effects, SS_3^*, and column effects, SS_4^*, estimate the corresponding value in the "average mean square" column. The value of SS^* is an estimate of the total variability of the process (corresponding to the "average mean square" entry). Since $[(r-1)/$

$s](SS_3^* - SS_2^*)$ is an estimate of $\sum_{i=1}^{r} \varphi_i^2$, and

$[(s-1)/r](SS_4^* - SS_2^*)$ is an estimate of $\sum_{j=1}^{s} \gamma_j^2$,

the contribution of these main effects to the total variability of the process can be estimated.

In some investigations, concern with reducing the total variability around the general mean is paramount. In such a case, this kind of analysis can be of value in determining the relative sizes of various contributions to the variability. In many other experiments, differences around the general mean are more to be identified than reduced, and this kind of analysis is irrelevant.

Many experiments which can be handled as a two-way analysis of variance might alternatively be attacked as a simpler one-way analysis of variance. For example, in the experiment on radium counts, rows correspond to specimens (difference among which we wish to estimate) and columns to experiments, where each specimen is handled in each experiment; this is a two-way plan. Alternatively, we *could* simply take 16 readings in random order, with each specimen being measured four times. The analysis of this second plan is arithmetically easier but it would seem to be a less precise experiment. Using the data and information given in the "average mean square" column we can estimate the increase in precision resulting from using the two-way plan. This is done in the following way.

Calculate $SS_2^* + [r/(rs-1)] \sum \gamma_j^2$. This is an estimate of what the residual mean square would be had the experiment been done as a one-way analysis of variance. Clearly, if $\sum \gamma_j^2$ is large this means that the residual mean square for the one-way plan would be inflated from the value obtained by using the two-way plan. The ratio

$$e = \frac{SS_2^* + [r/(rs-1)] \sum \gamma_j^2}{SS_2^*}$$

is called the efficiency of the two-way plan relative to the one-way plan. And we can say approximately that for the same precision as that given by n observations in the two-way plan, $e \times n$ would be needed for a one-way plan.

13.6.4.4 Example. Let us return to the example of the counting rates for different specimens of radium using different experiments. The completed analysis of variance table (Table 13.53) is obtained by following the computational procedure.

1. $R_1, R_2, R_3, R_4 = 6.45, 11.33, 17.20, 17.74$
2. $C_1, C_2, C_3, C_4 = 12.73, 14.16, 13.14, 12.69$
3. $T = 6.45 + 11.33 + 17.20 + 17.74 = 52.72$
4. $\sum_{i=1}^{4} \sum_{j=1}^{4} X_{ij}^2 = (1.46)^2 + (2.00)^2 + \cdots + (4.15)^2 = 195.6948$
5. $\sum_{i=1}^{4} \frac{R_i^2}{4} = \frac{(6.45)^2 + (11.33)^2 + (17.20)^2 + (17.74)^2}{4} = 195.1298$
6. $\sum_{i=1}^{4} \frac{C_j^2}{4} = \frac{(12.73)^2 + (14.16)^2 + (13.14)^2 + (12.69)^2}{4} = 174.06355$
7. $T^2/16 = (52.72)^2/16 = 173.7124$
8. $SS_4 = 174.06355 - 173.7124 = 0.3512$
9. $SS_3 = 195.1298 - 173.7124 = 21.4174$
10. $SS = 195.6948 - 173.7124 = 21.9824$
11. $SS_2 = 21.9824 - 21.4174 - 0.3512 = 0.2138.$

TABLE 13.53. *COMPLETED ANALYSIS OF VARIANCE TABLE FOR COUNTING RATE PROBLEM*

Source	Sum of Squares	Degrees of Freedom	Mean Square	Test
Between experiments	0.3512	3	0.1171	$F_4 = 4.93$
Between specimens	21.4174	3	7.1391	$F_3 = 300.5$
Residual	0.2138	9	0.0238	
Total	21.9824	15	1.4655	

$F_{.05;3,9} = 3.86$

Consequently, we conclude that the specimen effects differ significantly, and the experiment effects also differ significantly. Hence, determining the largest effects is in order.

For experiments and from Table 13.48 for $\alpha = 0.05$ and for $r = 4, s = 4$,

$$c^* = 4.42$$

so that

$$c = 4.42\sqrt{\frac{SS_2^*}{4}} = 0.3408$$

Making the comparison between the second and fourth experiment effect,

$$\begin{aligned}(3.54 - 3.17) - 0.34 \le \gamma_2 - \gamma_4 \\ \le (3.54 - 3.17) + 0.34 \\ 0.03 \le \gamma_2 - \gamma_4 \le 0.71\end{aligned}$$

it is evident that they differ significantly. Furthermore, the difference $\gamma_2 - \gamma_4$ can be said to lie between 0.03 and 0.71.

It is evident by inspection that the only other significant difference is between γ_2 and γ_1. Hence we can conclude that experiment 2 produced an effect on the counting rate which was significantly different from the others. For specimens, c^* also equals 4.42, so that c is again 0.3408.

Making the comparison between the third and fourth specimen effects,

$$\begin{aligned}(4.44 - 4.30) - 0.34 \le \varphi_4 - \varphi_3 \\ \le (4.44 - 4.30) + 0.34 \\ -0.20 \le \varphi_4 - \varphi_3 \le 0.48\end{aligned}$$

It appears that these two specimens do *not* differ. From inspection of the data, it is evident that these are the only two specimens that do not differ significantly. The difference between specimen effect 4 and specimen effect 1 can be said to lie between 2.49 and 3.17, i.e.,

$$2.49 \le \varphi_4 - \varphi_1 \le 3.17$$

13.6.4.5 Estimates of Variability. The estimate of

$$\sigma^2 + \frac{r}{s-1}\sum_{j=1}^{s} \gamma_j^2 = \sigma^2 + \frac{4}{3}\sum_{j=1}^{4} \gamma_j^2$$

is 0.1171. The estimate of σ^2 is 0.0238. Hence, the estimate of $\sum_{j=1}^{4} \gamma_j^2$ is

$$\tfrac{3}{4}(0.1171 - 0.0238) = 0.0699$$

Similarly, the estimate of

$$\sigma^2 + \frac{s}{r-1}\sum_{i=1}^{r} \varphi_i^2 = \sigma^2 + \frac{4}{3}\sum_{i=1}^{4} \varphi_i^2$$

is 7.1391. Hence the estimate of $\sum_{i=1}^{4} \varphi_i^2$ is

$$\tfrac{3}{4}(7.1391 - 0.0238) = 5.3364$$

We may now compute the value of

$$\begin{aligned}SS_2^* + \frac{r}{rs-1}\sum \gamma_j^2 = 0.0238 \\ + \frac{4}{(4)(4) - 1}(0.0699) = 0.0424\end{aligned}$$

and the efficiency of the two-way plan used, relative to a one-way plan (where the 16 readings would not be taken in sets of four) is: $0.0424/0.0238 = 1.78$, which means that about $16 \times 1.78 = 29$ observations taken in random order would be needed to give as precise results as those obtained by using the two-way plan.

13.6.5 Two-Way Classification: More Than One Observation Per Cell—Interaction

The model is as follows:

$$X_{ijk} = \zeta_{..} + \varphi_i + \gamma_j + \eta_{ij} + \varepsilon_{ijk}$$

where

X_{ijk} is the outcome of the kth experiment using the ith row effect and jth column effect, $i = 1, 2, \ldots, r$; $j = 1, 2, \ldots, s$; $k = 1, 2, \ldots, v$;

$\zeta_{..}$ is a general mean;

φ_i is the effect of adding the ith row treatment, $\sum_{i=1}^{r} \varphi_i = 0$;

γ_j is the effect of adding the jth column treatment, $\sum_{j=1}^{s} \gamma_j = 0$;

η_{ij} is the interaction of the ith row treatment with the jth column treatment,

$$\sum_{i=1}^{r} \eta_{ij} = 0, \sum_{j=1}^{s} \eta_{ij} = 0;$$

ε_{ijk} is the random effect which is independently normally distributed with mean 0 and variance σ^2.

13.6.5.1 Tests for Homogeneity. Complete Table 13.54. Computational procedure for analysis of variance table:

1. Calculate row totals: $R_1, R_2, \ldots, R_r$
2. Calculate column totals: $C_1, C_2, \ldots, C_s$
3. Calculate within cell totals: $w_{11}, w_{12}, \ldots, w_{rs}$
4. Calculate over-all total: $T = R_1 + R_2 + \cdots + R_r$
5. Calculate crude sum of squares:
$$\sum_{i=1}^{r}\sum_{j=1}^{s}\sum_{k=1}^{v} X_{ijk}^2 = X_{111}^2 + X_{112}^2 + \cdots + X_{rsv}^2$$
6. Calculate crude sum of squares between columns:
$$\sum_{j=1}^{s} \frac{C_j^2}{rv} = \frac{(C_1^2 + C_2^2 + \cdots + C_s^2)}{rv}$$
7. Calculate crude sum of squares between rows:
$$\sum_{i=1}^{r} \frac{R_i^2}{sv} = \frac{(R_1^2 + R_2^2 + \cdots + R_r^2)}{sv}$$
8. Calculate crude sum of squares between cells:
$$\sum_{i=1}^{r}\sum_{j=1}^{s} \frac{w_{ij}^2}{v} = \frac{(w_{11}^2 + w_{12}^2 + \cdots + w_{rs}^2)}{v}$$
9. Calculate correction factor due to mean T^2/rsv.

From the above quantities compute:

10. $SS_4 = (6) - (9) = \sum_{j=1}^{r} \frac{C_j^2}{rv} - \frac{T^2}{rsv}$
11. $SS_3 = (7) - (9) = \sum_{i=1}^{r} \frac{R_i^2}{sv} - \frac{T^2}{rsv}$
12. $SS_1 = (5) - (8)$
$$= \sum_{i=1}^{r}\sum_{j=1}^{s}\sum_{k=1}^{v} X_{ijk}^2 - \sum_{i=1}^{r}\sum_{j=1}^{s} w_{ij}^2/v$$
13. $SS = (5) - (9)$
$$= \sum_{i=1}^{r}\sum_{j=1}^{s}\sum_{k=1}^{v} X_{ijk}^2 - \frac{T^2}{rsv}$$
14. $SS_2 = (13) - (12) - (11) - (10)$
$= SS - SS_1 - SS_3 - SS_4$

If there is more than one observation per cell and it is assumed *before* the *experiment is performed* that the interaction is zero, the interaction and within cells sums of squares are

Source	Sum of Squares	Degrees of Freedom	Mean Square	Average Mean Square
Residual	$SS'_1 = SS_1 + SS_2$	$(r-1)(s-1) + rs(v-1) = rsv - r - s + 1$	$SS'^*_1 = SS'_1/(rsv - r - s + 1)$	σ^2

combined into a new row, where $SS'_1 = SS_1 + SS_2$. In all formulas, SS_1 and SS_1^* are replaced by SS'_1 and SS'^*_1, respectively. In making tests of significance, when interaction may be present, it is important to first test for interaction.

If $F_2 = SS_2^*/SS_1^* \geq F_{\alpha;(r-1)(s-1),rs(v-1)}$ we reject the hypothesis that there is no interaction present, i.e., that $\eta_{ij} = 0$ for all i, j, where $F_{\alpha;(r-1)(s-1),rs(v-1)}$ is the upper α percentage point of the F distribution given in Table 13.32.

If this hypothesis is rejected, the tests for significance of row effects and column effects are meaningless under the present formulation of the problem. For example, suppose there is interaction present between workmen and machines, and suppose that there are also significant machine effects. If a particular machine effect is positive, and interaction is present, there is no guarantee that this machine will produce more units consistently. Yet, this is what is implied by saying that this machine has a positive effect. Consequently, if interaction is present, tests for row effects and column effects are irrelevant.

If the hypothesis that there is no interaction is accepted, we reject the hypothesis that there are no column effects, i.e.,

$$\gamma_1 = \gamma_2 = \gamma_3 = \cdots = \gamma_s = 0$$

if $F_4 = SS_4^*/SS_1^* \geq F_{\alpha;s-1,rs(v-1)}$.

TABLE 13.54. *ANALYSIS OF VARIANCE TABLE: TWO-WAY CLASSIFICATION — INTERACTION*

Source	Sum of Squares	Degrees of Freedom	Mean Square	Average Mean Square	Test
Between columns	$SS_4 = rv \sum_{j=1}^{s} (\bar{X}_{\cdot j \cdot} - \bar{X}_{\cdot\cdot\cdot})^2$	$s - 1$	$SS^*_4 = SS_4/(s-1)$	$\sigma^2 + \frac{rv}{s-1} \sum_{j=1}^{s} \gamma_j^2$	$F_4 = SS^*_4/SS^*_1$
Between rows	$SS_3 = sv \sum_{i=1}^{r} (\bar{X}_{i\cdot\cdot} - \bar{X}_{\cdot\cdot\cdot})^2$	$r - 1$	$SS^*_3 = SS_3/(r-1)$	$\sigma^2 + \frac{sv}{r-1} \sum_{i=1}^{r} \varphi_i^2$	$F_3 = SS^*_3/SS^*_1$
Interaction	$SS_2 = v \sum_{i=1}^{r} \sum_{j=1}^{s} (\bar{X}_{ij\cdot} - \bar{X}_{i\cdot\cdot} - \bar{X}_{\cdot j\cdot} + \bar{X}_{\cdot\cdot\cdot})^2$	$(r-1)(s-1)$	$SS^*_2 = SS_2/(r-1)(s-1)$	$\sigma^2 + \frac{v}{(r-1)(s-1)} \sum_{i=1}^{r} \sum_{j=1}^{s} \eta_{ij}^2$	$F_2 = SS^*_2/SS^*_1$
Within cells	$SS_1 = \sum_{i=1}^{r} \sum_{j=1}^{s} \sum_{k=1}^{v} (X_{ijk} - \bar{X}_{ij\cdot})^2$	$rs(v-1)$	$SS^*_1 = SS_1/rs(v-1)$	σ^2	
Total	$SS = \sum_{i=1}^{r} \sum_{j=1}^{s} \sum_{k=1}^{v} (X_{ijk} - \bar{X}_{\cdot\cdot\cdot})^2$	$rsv - 1$	$SS^* = SS/(rsv-1)$	$\sigma^2 + \frac{rv}{rsv-1} \sum_{j=1}^{s} \gamma_j^2 + \frac{sv}{rsv-1} \sum_{i=1}^{r} \varphi_i^2 + \frac{v}{rsv-1} \sum_{i=1}^{r} \sum_{j=1}^{s} \eta_{ij}^2$	

In this table

$$\bar{X}_{\cdot j\cdot} = \sum_{i=1}^{r} \sum_{k=1}^{v} X_{ijk}/rv \qquad \bar{X}_{i\cdot\cdot} = \sum_{j=1}^{s} \sum_{k=1}^{v} X_{ijk}/sv$$

$$\bar{X}_{ij\cdot} = \sum_{k=1}^{v} X_{ijk}/v \qquad \bar{X}_{\cdot\cdot\cdot} = \sum_{i=1}^{r} \sum_{j=1}^{s} \sum_{k=1}^{v} X_{ijk}/rsv$$

We reject the hypothesis that there are no row effects, i.e.,

$$\varphi_1 = \varphi_2 = \varphi_3 = \cdots = \varphi_r = 0$$

if $F_3 = SS_3^*/SS_1^* \geq F_{\alpha;r-1,rs(v-1)}$.

13.6.5.2 Tests for Comparison of Mean Effects. If the interactions are not significant, the procedure for comparing row or column effects is exactly the same as for the two-way classification with one observation per cell except for the following changes.

The factor c for row effects and for column effects is obtained from the analysis of variance table and from Table 13.48, depending on the level of significance. Table 13.48 is entered with the indices degrees of freedom, and the number of row effects being studied (r) if row effects are of interest. The proper degrees of freedom are those corresponding to the degrees of freedom of the within cell source in the analysis of variance table, i.e., $rs(v-1)$. The factor c^* is read from the table and

$$c(\text{for rows}) = c^*\sqrt{SS_1^*/vs}$$

Similarly, if column effects are of interest, tne table is entered with indices degrees of freedom, and the number of column effects being studied (s). The factor c^* is read from the table and

$$c(\text{for columns}) = c^*\sqrt{SS_1^*/vr}$$

Of course, for finding confidence intervals, the calculated terms that are of interest are $\bar{X}_{i\cdot\cdot}$ and $\bar{X}_{\cdot j\cdot}$.

13.6.5.3 Estimating the Variability Due to the Sources. In many problems to which this experimental plan is applicable the main concern is with estimating the size of row effect differences, column effect differences, and magnitudes of interactions. However, if the investigation is principally concerned with estimating the relative sizes of contributions to the total variance around the general mean, these can be estimated.

The column of the analysis of variance table headed "mean square" is an estimate of the corresponding entry in the column headed "average mean square." Consequently, we can calculate estimates of $\sum_{i=1}^{r} \varphi_i^2$, $\sum_{j=1}^{s} \gamma_j^2$, and $\sum_{i=1}^{r}\sum_{j=1}^{s} \eta_{ij}^2$.

$$\text{Estimate of} \sum_{j=1}^{s} \gamma_j^2 = (SS_4^* - SS_1^*)\frac{s-1}{rv}$$

$$\text{Estimate of} \sum_{i=1}^{r} \varphi_i^2 = (SS_3^* - SS_1^*)\frac{r-1}{sv}$$

$$\text{Estimate of} \sum_{i=1}^{r}\sum_{j=1}^{s} \eta_{ij}^2 = (SS_2^* - SS_1^*)\frac{(r-1)(s-1)}{v}$$

We cannot appropriately calculate the efficiency of this design to a one-way plan, since the two plans are not really comparable; the information on interactions cannot be obtained from the one-way plan.

13.6.5.4 Example. Youden† reports the following experiment to determine the effect of time of aging in the strength of cement (see Table 13.55). Two mixes of cement were prepared and six specimens were made from each mix. Three specimens from each mix were tested after three days and after seven days. Interaction cannot be ignored since it is quite possible that the two mixes might not differ

† W. J. Youden, *Statistical Methods for Chemists* (New York: John Wiley & Sons, Inc., 1951), pp. 64–65.

TABLE 13.55. *YIELD LOADS FOR CEMENT SPECIMENS*

	Three-Day Test	Seven-Day Test	$\bar{X}_i$
Mix 1	660 674 648	979 1038 1051	841.7
Mix 2	661 624 652	1070 1066 1053	854.3
$\bar{X}_{\cdot j}$	653.2	1042.8	

after a short period of time, but would after a longer period. This is equivalent to saying that the effect of an additional period of time is different for the two mixes, i.e., interaction is present. The test specimens are 2-in. cubes that yielded under the indicated loads, which are in units of 10 lb.

The following computations lead to the analysis of variance table (Table 13.56).

1. $R_1, R_2 = 5050, 5126$
2. $C_1, C_2 = 3919, 6257$
3. $w_{11}, w_{21}, w_{12}, w_{22} = 1982, 1937, 3068, 3189$
4. $T = 5050 + 5126 = 10{,}176$
5. $\sum_{i=1}^{2}\sum_{j=1}^{2}\sum_{k=1}^{3} X_{ijk}^2 = (660)^2 + (674)^2 + \cdots + (1053)^2 = 9{,}091{,}732$
6. $\sum_{j=1}^{2} \frac{C_j^2}{6} = \frac{(3919)^2 + (6257)^2}{6} = 9{,}084{,}768$
7. $\sum_{i=1}^{2} \frac{R_i^2}{6} = \frac{(5050)^2 + (5126)^2}{6} = 8{,}629{,}729$
8. $\sum_{i=1}^{2}\sum_{j=1}^{2} \frac{w_{ij}^2}{3} = \frac{(1982)^2 + (1937)^2 + (3068)^2 + (3189)^2}{3} = 9{,}087{,}546$
9. $\frac{T^2}{12} = \frac{(10{,}176)^2}{12} = 8{,}629{,}248$
10. $SS_4 = 9{,}084{,}768 - 8{,}629{,}248 = 455{,}520$
11. $SS_3 = 8{,}629{,}729 - 8{,}629{,}248 = 481$
12. $SS_1 = 9{,}091{,}732 - 9{,}087{,}546 = 4{,}186$
13. $SS \;= 9{,}091{,}732 - 8{,}629{,}248 = 462{,}484$
14. $SS_2 = 462{,}484 - 4{,}186 - 481 - 455{,}520 = 2{,}297$

The F values in the table for interaction and mixes are well below the critical 5 per cent value for 1 and 8 degrees of freedom, i.e., $F_{.05;1,8} = 5.32$. Hence it is concluded that the data do not give sufficient evidence of the presence of interaction or any mix effect. On the other hand, the F test for times is significant, so that it is concluded that there is an effect from the additional four days of aging. Furthermore, a comparison of the column means indicates that

$$\bar{X}_{.2} - \bar{X}_{.1} - c \leq \gamma_2 - \gamma_1 \leq \bar{X}_{.2} - \bar{X}_{.1} + c$$

or $(1042.8 - 653.2) - 30.4 = 359.2 \leq$ mean effect for 7 days $-$ mean effect for 3 days $\leq (1042.8 - 653.2) + 30.4 - 420.0$ where c is obtained from Table 13.48, using as indices: $s = 2$; degrees of freedom $= rs(v-1) = 8$; $c^* = 3.26$; so that $c = 30.4$. In this problem we are not interested in finding the contributions of the main effects to the total variability about the mean. However, for illustrative purposes, we calculate the estimates of the magnitude of the main effects and interactions.

Estimate of time effects

$$\sum \gamma_j^2 = (SS_4^* - SS_1^*)\frac{s-1}{rv} = (455{,}520 - 523.25)\frac{1}{6} = \frac{454{,}996.75}{6} = 75{,}832.79$$

Estimate of mix effects

$$\sum \varphi_i^2 = (SS_4^* - SS_1^*)\frac{r-1}{sv} = (481 - 523.25)\frac{1}{6} = 0 \text{ since } \sum \varphi_i^2 \geq 0$$

Estimate of interaction effects

$$\sum\sum \eta_{ij}^2 = (SS_2^* - SS_1^*)\frac{1}{3} = \frac{(2297 - 523.25)}{3} = \frac{1773.75}{3} = 591.25$$

TABLE 13.56. *ANALYSIS OF VARIANCE TABLE FOR YIELD LOADS ON CEMENT SPECIMENS*

Source	Sum of Squares	Degrees of Freedom	Mean Square	Test
Between times	455,520	1	455,520	$F = 870.54$
Between mixes	481	1	481	$F = 0.919$
Interaction	2,297	1	2,297	$F = 4.39$
Within cells	4,186	8	523.25	
Total	462,484	11	42,044	

TABLE 13.57. *ANALYSIS OF VARIANCE TABLE: THREE-WAY CLASSIFICATION—INTERACTION*

Source	Sum of Squares	Degrees of Freedom
Between treatments of factor C	$SS_5 = rsv \sum_{k=1}^{t} (\bar{X}_{..k.} - \bar{X}_{....})^2$	$t - 1$
Between treatments of factor B	$SS_4 = rtv \sum_{j=1}^{s} (\bar{X}_{.j..} - \bar{X}_{....})^2$	$s - 1$
Between treatments of factor A	$SS_3 = stv \sum_{i=1}^{r} (\bar{X}_{i...} - \bar{X}_{....})^2$	$r - 1$
Interaction AB	$SS_2^{AB} = vt \sum_{i=1}^{r} \sum_{j=1}^{s} (\bar{X}_{ij..} - \bar{X}_{i...} - \bar{X}_{.j..} + \bar{X}_{....})^2$	$(r - 1)(s - 1)$
Interaction AC	$SS_2^{AC} = vs \sum_{i=1}^{r} \sum_{k=1}^{t} (\bar{X}_{i.k.} - \bar{X}_{i...} - \bar{X}_{..k.} + \bar{X}_{....})^2$	$(r - 1)(t - 1)$
Interaction BC	$SS_2^{BC} = vr \sum_{j=1}^{s} \sum_{k=1}^{t} (\bar{X}_{.jk.} - \bar{X}_{.j..} - \bar{X}_{..k.} + \bar{X}_{....})^2$	$(s - 1)(t - 1)$
Interaction ABC	$SS_2^{ABC} = v \sum_{i=1}^{r} \sum_{j=1}^{s} \sum_{k=1}^{t} (\bar{X}_{ijk.} - \bar{X}_{ij..} - \bar{X}_{i.k.} - \bar{X}_{.jk.} + \bar{X}_{i...} + \bar{X}_{.j..} + \bar{X}_{..k.} - \bar{X}_{....})^2$	$(r - 1)(s - 1)(t - 1)$
Within cells	$SS_1 = \sum_{i=1}^{r} \sum_{j=1}^{s} \sum_{k=1}^{t} \sum_{l=1}^{v} (X_{ijkl} - \bar{X}_{ijk.})^2$	$rst(v - 1)$
Total	$SS = \sum_{i=1}^{r} \sum_{j=1}^{s} \sum_{k=1}^{t} \sum_{l=1}^{v} (X_{ijkl} - \bar{X}_{....})^2$	$rstv - 1$

In this table

$$\bar{X}_{..k.} = \sum_{i=1}^{r} \sum_{j=1}^{s} \sum_{l=1}^{v} X_{ijkl}/rsv \qquad \bar{X}_{.j..} = \sum_{i=1}^{r} \sum_{k=1}^{t} \sum_{l=1}^{v} X_{ijkl}/rtv$$

$$\bar{X}_{i...} = \sum_{j=1}^{s} \sum_{k=1}^{t} \sum_{l=1}^{v} X_{ijkl}/stv \qquad \bar{X}_{ij..} = \sum_{k=1}^{t} \sum_{l=1}^{v} X_{ijkl}/tv$$

$$\bar{X}_{i.k.} = \sum_{j=1}^{s} \sum_{l=1}^{v} X_{ijkl}/sv \qquad \bar{X}_{.jk.} = \sum_{i=1}^{r} \sum_{l=1}^{v} X_{ijkl}/rv$$

$$\bar{X}_{ijk.} = \sum_{l=1}^{v} X_{ijkl}/v \qquad \bar{X}_{....} = \sum_{i=1}^{r} \sum_{j=1}^{s} \sum_{k=1}^{t} \sum_{l=1}^{v} X_{ijkl}/rstv$$

TABLE 13.57. *Cont.*

Mean Square	Average Mean Square	Test
$SS_5^* = \dfrac{SS_5}{t-1}$	$\sigma^2 + \dfrac{rsv}{t-1}\sum_{k=1}^{t}\lambda_k^2$	$F = \dfrac{SS_5^*}{SS_1^*}$
$SS_4^* = \dfrac{SS_4}{s-1}$	$\sigma^2 + \dfrac{rtv}{s-1}\sum_{j=1}^{s}\gamma_j^2$	$F = \dfrac{SS_4^*}{SS_1^*}$
$SS_3^* = \dfrac{SS_3}{r-1}$	$\sigma^2 + \dfrac{stv}{r-1}\sum_{i=1}^{r}\varphi_i^2$	$F = \dfrac{SS_3^*}{SS_1^*}$
$SS_2^{AB*} = \dfrac{SS_2^{AB}}{(r-1)(s-1)}$	$\sigma^2 + \dfrac{vt}{(r-1)(s-1)}\sum_{i=1}^{r}\sum_{j=1}^{s}\eta_{ij}^2$	$F = \dfrac{SS_2^{AB*}}{SS_1^*}$
$SS_2^{AC*} = \dfrac{SS_2^{AC}}{(r-1)(t-1)}$	$\sigma^2 = \dfrac{vs}{(r-1)(t-1)}\sum_{i=1}^{r}\sum_{k=1}^{t}\psi_{ik}^2$	$F = \dfrac{SS_2^{AC*}}{SS_1^*}$
$SS_2^{BC*} = \dfrac{SS_2^{BC}}{(s-1)(t-1)}$	$\sigma^2 + \dfrac{vr}{(s-1)(t-1)}\sum_{j=1}^{s}\sum_{k=1}^{t}\mu_{jk}^2$	$F = \dfrac{SS_2^{BC*}}{SS_1^*}$
$SS^{ABC*} = \dfrac{SS_2^{ABC}}{(r-1)(s-1)(t-1)}$	$\sigma^2 + \dfrac{v}{(r-1)(s-1)(t-1)}\sum_{i=1}^{r}\sum_{j=1}^{s}\sum_{k=1}^{t}\rho_{ijk}^2$	$F = \dfrac{SS_2^{ABC*}}{SS_1^*}$
$SS_1^* = \dfrac{SS_1}{rst(v-1)}$	σ^2	
$SS^* = \dfrac{SS}{rstv-1}$	$\sigma^2 + \dfrac{rsv}{rstv-1}\sum_{k=1}^{t}\lambda_k^2 + \dfrac{rtv}{rstv-1}\sum_{j=1}^{s}\gamma_j^2 + \dfrac{stv}{rstv-1}\sum_{i=1}^{r}\varphi_i^2$ $+ \dfrac{vt}{rstv-1}\sum_{i=1}^{r}\sum_{j=1}^{s}\eta_{ij}^2 + \dfrac{vs}{rstv-1}\sum_{i=1}^{r}\sum_{k=1}^{t}\psi_{ik}^2$ $+ \dfrac{vr}{rstv-1}\sum_{j=1}^{s}\sum_{k=1}^{t}\mu_{jk}^2 + \dfrac{v}{rstv-1}\sum_{i=1}^{r}\sum_{j=1}^{s}\sum_{k=1}^{t}\rho_{ijk}^2$	

13.6.6 Three-Way Classification

The reader is cautioned to read the section on two-way classification, since the present article is merely a simple extension. The model is as follows:

$$X_{ijkl} = \zeta\ldots + \varphi_i + \gamma_j + \lambda_k + \eta_{ij} + \psi_{ik} + \mu_{jk} + \rho_{ijk} + \varepsilon_{ijkl}$$

where

X_{ijkl} is the outcome of the lth experiment using the ith treatment of factor A, the jth treatment of factor B, and the kth treatment of factor C; $i = 1, 2, \ldots, r$; $j = 1, 2, \ldots, s$; $k = 1, 2, \ldots, t$; $l = 1, 2, \ldots, v$;

$\zeta\ldots$ is a general mean;

φ_i is the effect of adding the ith treatment of factor A, $\sum_{i=1}^{r} \varphi_i = 0$;

γ_j is the effect of adding the jth treatment of factor B, $\sum_{i=1}^{s} \gamma_i = 0$;

λ_k is the effect of adding the kth treatment of factor C, $\sum_{k=1}^{t} \lambda_k = 0$;

η_{ij} is the interaction of the ith treatment of factor A with the jth treatment of factor B,

$$\sum_{i=1}^{r} \eta_{ij} = \sum_{j=1}^{s} \eta_{ij} = 0;$$

ψ_{ik} is the interaction of the ith treatment of factor A with the kth treatment of factor C,

$$\sum_{i=1}^{r} \psi_{ik} = \sum_{k=1}^{t} \psi_{ik} = 0;$$

μ_{jk} is the interaction of the jth treatment of factor B with the kth treatment of factor C,

$$\sum_{j=1}^{s} \mu_{jk} = \sum_{k=1}^{t} \mu_{jk} = 0;$$

ρ_{ijkl} is the interaction of the ith treatment of factor A, the jth treatment of factor B, and the kth treatment of factor C, $\sum_{i=1}^{r} \rho_{ijk} =$

$$\sum_{j=1}^{s} \rho_{ijk} = \sum_{k=1}^{t} \rho_{ijk} = 0;$$

ε_{ijkl} is the random effect, which is independently normally distributed with mean 0 and variance σ^2.

13.6.6.1 Tests for Homogeneity. Complete the analysis of variance table (Table 13.57).

Computational Procedure for the Analysis of Variance Table: For simplicity let $r = 2$, $s = 4$, $t = 3$, $v = 2$, and the data will appear in a table such as Table 13.58.

Form Tables 13.59, 13.60, and 13.61.

For Table 13.59:

1. Calculate total $= X_{1+++} + X_{2+++}$

$$= \sum_{i=1}^{2} \sum_{j=1}^{4} \sum_{k=1}^{3} \sum_{l=1}^{2} X_{ijkl}$$

2. Calculate crude sum of squares and divide by the number of original individuals that have been summed to give the individuals in this table:

$$\sum_{i=1}^{2} \sum_{j=1}^{4} \frac{X_{ij++}^2}{6} = \frac{X_{11++}^2 + X_{12++}^2 + \cdots + X_{24++}^2}{6}$$

3. Calculate crude sum of squares between rows and divide by the number of original individuals that have been summed to give the row totals:

$$\sum_{i=1}^{2} \frac{X_{i+++}^2}{24} = \frac{X_{1+++}^2 + X_{2+++}^2}{24}$$

4. Calculate crude sum of squares between columns and divide by number of original individuals that have been summed to give the column total:

$$\sum_{j=1}^{4} \frac{X_{+j++}^2}{12} = \frac{X_{+1++}^2 + X_{+2++}^2 + X_{+3++}^2 + X_{+4++}^2}{12}$$

5. Calculate correction factor due to mean which is total squared divided by total number of observations:

$$\frac{(1)^2}{48} = \frac{(\sum\sum\sum\sum X_{ijkl})^2}{48}$$

From the above factors compute:

6. (2) − (5)
7. $SS_3 = (3) - (5)$

$$= \sum_{i=1}^{2} \frac{X_{i+++}^2}{24} - \frac{(\sum\sum\sum\sum X_{ijkl})^2}{48}$$

8. $SS_4 = (4) - (5)$

$$= \sum_{j=1}^{4} \frac{X_{+i++}^2}{12} - \frac{(\sum\sum\sum\sum X_{ijkl})^2}{48}$$

TABLE 13.58. *EXAMPLE OF DATA FOR THREE-WAY CLASSIFICATION*

		B_1	B_2	B_3	B_4
A_1	C_1	X_{1111}	X_{1211}	X_{1311}	X_{1411}
		X_{1112}	X_{1212}	X_{1312}	X_{1412}
	C_2	X_{1121}	X_{1221}	X_{1321}	X_{1421}
		X_{1122}	X_{1222}	X_{1322}	X_{1422}
	C_3	X_{1131}	X_{1231}	X_{1331}	X_{1431}
		X_{1132}	X_{1232}	X_{1332}	X_{1432}
A_2	C_1	X_{2111}	X_{2211}	X_{2311}	X_{2411}
		X_{2112}	X_{2212}	X_{2312}	X_{2412}
	C_2	X_{2121}	X_{2221}	X_{2321}	X_{2421}
		X_{2122}	X_{2222}	X_{2322}	X_{2422}
	C_3	X_{2131}	X_{2231}	X_{2331}	X_{2431}
		X_{2132}	X_{2232}	X_{2332}	X_{2432}

9. $SS_2^{AB} = (6) - (7) - (8)$

In a similar manner, using the other two tables, the following can be computed:

10. SS_5
11. SS_2^{BC}
12. SS_2^{AC}

To get the remaining factors:

13. Calculate within cell totals from original data:

$$w_{111}, w_{112}, \ldots, w_{423}$$

14. Calculate crude sum of squares between cells and divide by number of observations per cell:

$$\sum_{i=1}^{2}\sum_{j=1}^{4}\sum_{k=1}^{3} \frac{w_{ijk}^2}{2} = \frac{w_{111}^2 + w_{112}^2 + \cdots + w_{243}^2}{2}$$

15. Calculate crude sum of squares:

$$\sum_{i=1}^{2}\sum_{j=1}^{4}\sum_{k=1}^{3}\sum_{l=1}^{2} X_{ijkl}^2 = X_{1111}^2 + X_{1112}^2 + \cdots + X_{2432}^2$$

TABLE 13.59. *TABLE FOR FACTORS A AND B*

	B_1	B_2	B_3	B_4	Totals
A_1	X_{11++} = total of all X's whose first two subscripts are 1,1	X_{12++} = total of all X's whose first two subscripts are 1,2	X_{13++} = total of all X's whose first two subscripts are 1,3	X_{14++} = total of all X's whose first two subscripts are 1,4	X_{1+++}
A_2	X_{21++} = total of all X's whose first two subscripts are 2,1	X_{22++} = total of all X's whose first two subscripts are 2,2	X_{23++} = total of all X's whose first two subscripts are 2,3	X_{24++} = total of all X's whose first two subscripts are 2,4	X_{2+++}
Totals:	X_{+1++}	X_{+2++}	X_{+3++}	X_{+4++}	

TABLE 13.60. *TABLE FOR FACTORS B AND C*

	B_1	B_2	B_3	B_4	Totals
C_1	X_{+11+} = total of all X's whose second subscript is 1 and third is 1	X_{+21+} = total of all X's whose second subscript is 2 and third is 1	X_{+31+} = total of all X's whose second subscript is 3 and third is 1	X_{+41+} = total of all X's whose second subscript is 4 and third is 1	X_{++1+}
C_2	X_{+12+} = total of all X's whose second subscript is 1 and third is 2	X_{+22+} = total of all X's whose second subscript is 2 and third is 2	X_{+32+} = total of all X's whose second subscript is 3 and third is 2	X_{+42+} = total of all X's whose second subscript is 4 and third is 2	X_{++2+}
C_3	X_{+13+} = total of all X's whose second subscript is 1 and third is 3	X_{+23+} = total of all X's whose second subscript is 2 and third is 3	X_{+33+} = total of all X's whose second subscript is 3 and third is 3	X_{+43+} = total of all X's whose second subscript is 4 and third is 3	X_{++3+}
Totals:	X_{+1++}	X_{+2++}	X_{+3++}	X_{+4++}	

TABLE 13.61. *TABLE FOR FACTORS A AND C*

	C_1	C_2	C_3	Totals
A_1	X_{1+1+} = total of all X's whose first subscript is 1 and third is 1	X_{1+2+} = total of all X's whose first subscript is 1 and third is 2	X_{1+3+} = total of all X's whose first subscript is 1 and third is 3	X_{1+++}
A_2	X_{2+1+} = total of all X's whose first subscript is 2 and third is 1	X_{2+2+} = total of all X's whose first subscript is 2 and third is 2	X_{2+3+} = total of all X's whose first subscript is 2 and third is 3	X_{2+++}
Totals:	X_{++1+}	X_{++2+}	X_{++3+}	

From the above quantities compute:

16. $SS_1 = (15) - (14)$

$$= \sum\sum\sum\sum X^2_{ijkl} - \frac{\sum\sum\sum w^2_{ijk}}{2}$$

17. $SS = (15) - (5)$

$$= \sum\sum\sum\sum X^2_{ijkl} - \frac{(\sum\sum\sum\sum X_{ijkl})^2}{48}$$

18. $SS_2^{ABC} = SS - SS_5 - SS_4 - SS_3 - SS_2^{AB} - SS_2^{AC} - SS_2^{BC} - SS_1$
$= (17) - (10) - (8) - (7) - (9) - (11) - (12) - (16)$

In making tests of significance when interaction may be present, it is important to test first for interaction between ABC.

If $F_2^{ABC} = SS_2^{ABC*}/SS_1^* \geq F_{\alpha;(r-1)(s-1)(t-1),rst(v-1)}$ we reject the hypothesis that there is no interaction between ABC present, i.e., that $\rho_{ijk} = 0$ for all ijk.

If this hypothesis is rejected, the tests for significance of effects A, effects B, and effects C are again meaningless under the present formulation of the problem (see Par. 13.6.5.1 for a discussion).

If F_2^{ABC} is not rejected, we make tests for interaction between AB, BC, and AC.

If $F_2^{AB} = SS_2^{AB*}/SS_1^* \geq F_{\alpha;(r-1)(s-1),rst(v-1)}$ we reject the hypothesis that there is no interaction between AB present, i.e., that $\eta_{ji} = 0$ for all i, j.

If $F_2^{AC} = SS_2^{AC*}/SS_1^* \geq F_{\alpha;(r-1)(t-1),rst(v-1)}$ we reject the hypothesis that there is no interaction between AC present, i.e., that $\psi_{ik} = 0$ for all i, k.

If $F_2^{BC} = SS_2^{BC*}/SS_1^* \geq F_{\alpha;(s-1)(t-1),rst(v-1)}$ we reject the hypothesis that there is no interaction between BC present, i.e., that $\mu_{jk} = 0$ for all j, k.

The presence of interaction between any two effects eliminates the need for tests of significance for these effects, since such a test is meaningless under the present formulation.

If all four types of interaction may be present, it is necessary that there be more than one observation per cell to make the necessary tests. However, if it is assumed that the interaction between ABC is 0, i.e., $\rho_{ijk} = 0$ for all ijk, a test can be formulated when there is only one observation in each cell, i.e., $v = 1$. The sources "within cells" and "interaction ABC" are eliminated in the analysis of variance table. They are replaced by the source "residual." All the tests are then made with the new source residual ($SS_1'^*$). For ease of computation SS_1' is obtained by subtraction. Returning to tests of significance, if the interaction terms are nonsignificant, the tests for effects A, B, and C are made in the usual manner.

Source	Sum of Squares	Degrees of Freedom	Mean Square	Average Mean Square
Residual	$SS'_1 = \sum_{i=1}^{r}\sum_{j=1}^{s}\sum_{k=1}^{t} (\bar{X}_{ijk.} - \bar{X}_{ij..} - \bar{X}_{i.k.} - \bar{X}_{.jk.} + \bar{X}_{i...} + \bar{X}_{.j..} + \bar{X}_{..k.} - \bar{X}_{....})^2$	$(r-1)(s-1)(t-1)$	$SS'^*_1 = \dfrac{SS'_1}{(r-1)(s-1)(t-1)}$	σ^2

13.6.6.2 Tests for Comparison of Mean Effects. If the interactions are nonsignificant, the procedure for comparing effects for factor A, effects for factor B, and effects for factor C is exactly the same as that given for the two-way classification except for the following changes.

The value of c for factor A, for factor B, and for factor C is obtained from the analysis of variance table and from Table 13.48 depending on the level of significance. Table 13.48 is entered, using as indices the degrees of freedom and the number of A effects being studied (r) if A effects are of interest. The proper degrees of freedom are those corresponding to the degrees of freedom of the within cells source in the analysis of variance table, i.e., $rst(v - 1)$. The factor c^* is read from the table, and

$$c \text{ (for } A \text{ effects)} = c^* \sqrt{SS_1^*/vst}$$

Similarly, if B effects are of interest, the table is entered with indices degrees of freedom and the number of B effects being studied (s). The factor c^* is read from the table, and

$$c \text{ (for } B \text{ effects)} = c^* \sqrt{SS_1^*/vrt}$$

If C effects are of interest, the table is entered with indices degrees of freedom and the number of C effects being studied (t). The factor c^* is read from the table, and

$$c \text{ (for } C \text{ effects)} = c^* \sqrt{SS_1^*/vrs}$$

Of course, in considering confidence intervals, the calculated terms that are of interest are $\bar{X}_{i\ldots}$, $\bar{X}_{.j..}$, and $\bar{X}_{..k..}$

13.6.6.3 Estimating the Variability Due to the Sources. Estimates of the entries in the "average mean square" column are obtained from the "mean square" column. Hence estimates of

$$\sum_{i=1}^{r} \varphi_i^2, \quad \sum_{j=1}^{s} \gamma_j^2, \quad \sum_{k=1}^{t} \lambda_k^2, \quad \sum_{i=1}^{r}\sum_{j=1}^{s} \eta_{ij}^2,$$

$$\sum_{i=1}^{r}\sum_{k=1}^{t} \psi_{ik}^2, \quad \sum_{j=1}^{r}\sum_{k=1}^{t} \mu_{jk}^2,$$

and $\sum_{i=1}^{r}\sum_{j=1}^{s}\sum_{k=1}^{t} \rho_{ijk}^2$ can be calculated:

$$\text{Estimate of} \sum_{i=1}^{r} \varphi_i^2 = (SS_3^* - SS_1^*)\frac{r-1}{stv}$$

$$\text{Estimate of} \sum_{j=1}^{s} \gamma_j^2 = (SS_4^* - SS_1^*)\frac{s-1}{rtv}$$

$$\text{Estimate of} \sum_{k=1}^{t} \lambda_k^2 = (SS_5^* - SS_1^*)\frac{t-1}{rsv}$$

$$\text{Estimate of} \sum_{i=1}^{r}\sum_{j=1}^{s} \eta_{ij}^2 = (SS_2^{AB*} - SS_1^*)\frac{(r-1)(s-1)}{tv}$$

$$\text{Estimate of} \sum_{i=1}^{r}\sum_{k=1}^{t} \psi_{ik}^2 = (SS_2^{AC*} - SS_1^*)\frac{(r-1)(t-1)}{sv}$$

$$\text{Estimate of} \sum_{k=1}^{s}\sum_{k=1}^{t} \mu_{jk}^2 = (SS_2^{BC*} - SS_1^*)\frac{(s-1)(t-1)}{rv}$$

$$\text{Estimate of} \sum_{i=1}^{r}\sum_{j=1}^{s}\sum_{k=1}^{t} \rho_{ijk}^2 = (SS_2^{ABC*} - SS_1^*)\frac{(r-1)(s-1)(t-1)}{v}$$

13.6.6.4 Example. An electronic manufacturer was having difficulty with a particular type of tube where the mutual conductance was running below the bogie value. Actually this tube is interchangeable in radio sets with an alternate. An experiment was performed to determine the effect of the exhaust variables, plate temperature, and filament lighting on the electrical characteristics of the tubes. Two levels of plate temperature and four levels of filament lighting current were used. The data are given in Table 13.62.

The analysis of variance table is completed by following the prescribed procedure.

1. $X_{1+++} + X_{2+++} = 103{,}767 + 100{,}816 = 204{,}583$

2. $$\sum_{i=1}^{2}\sum_{j=1}^{4} \frac{X_{ij++}^2}{6} = \frac{(25{,}505)^2 + (26{,}092)^2 + \cdots + (26{,}270)^2}{6} = 872{,}531{,}808$$

3. $$\sum_{i=1}^{2} \frac{X_{i+++}^2}{24} = \frac{(103{,}767)^2 + (100{,}816)^2}{24} = 872{,}144{,}006$$

4. $\sum_{j=1}^{4} \frac{X^2_{+j++}}{12} = \frac{(50{,}835)^2 + (50{,}562)^2 + (50{,}538)^2 + (52{,}648)^2}{12}$

$= 872{,}217{,}868$

5. $\frac{(1)^2}{48} = \frac{(204{,}583)^2}{48} = 871{,}962{,}581$

6. $(2) - (5) = 872{,}531{,}808 - 871{,}962{,}581$
$= 569{,}227$

7. $SS_3 = 872{,}144{,}006 - 871{,}962{,}581$
$= 181{,}425$

8. $SS_4 = 872{,}217{,}868 - 871{,}962{,}581$
$= 255{,}287$

9. $SS_2^{AB} = 569{,}227 - 255{,}287 - 181{,}425$
$= 132{,}515$

1. $X_{++1+} + X_{++2+} = 102{,}552 + 102{,}031$
$= 204{,}583$

2. $\sum_{k=1}^{2} \sum_{j=1}^{4} \frac{X^2_{+jk+}}{6} = \frac{(25{,}388)^2 + (25{,}582)^2 + \cdots + (26{,}408)^2}{6}$

$= 872{,}252{,}486$

3. $\sum_{k=1}^{2} \frac{X^2_{++k+}}{24} = \frac{(102{,}552)^2 + (102{,}031)^2}{24}$

$= 871{,}968{,}236$

4. $\sum_{j=1}^{4} \frac{X^2_{+j++}}{12} = \frac{(50{,}835)^2 + (50{,}562)^2 + (50{,}538)^2 + (52{,}648)^2}{12}$

$= 872{,}217{,}868$

5. $\frac{(1)^2}{48} = \frac{(204{,}853)^2}{48} = 871{,}962{,}581$

6. $(2) - (5) = 872{,}252{,}486 - 871{,}962{,}581$
$= 289{,}905$

TABLE 13.62. *TRANSCONDUCTANCE*

		Filament Lighting Conditions L_1	L_2	L_3	L_4
Plate temperature T_1	Tube	3774	4710	4176	4540
		4364	4180	4140	4530
		4374	4514	4398	3964
	Alt.	4138	4276	4228	4292
		4503	4168	4570	4386
		4352	4244	4280	4660
Plate temperature T_2	Tube	4216	3828	4122	4484
		4524	4170	4280	4332
		4136	4180	4226	4390
	Alt.	4074	4224	4108	4326
		4030	3922	4070	4312
		4350	4146	3940	4426

TABLE 13.63. *TABLE FOR FACTORS: TEMPERATURE AND LIGHTING CONDITIONS*

	L_1	L_2	L_3	L_4	X_{i+++}
T_1	25,505	26,092	25,792	26,378	103,767
T_2	25,330	24,470	24,746	26,270	100,816
X_{+j++}	50,835	50,562	50,538	52,648	

TABLE 13.64. *TABLE FOR FACTORS: TUBE AND LIGHTING CONDITIONS*

	L_1	L_2	L_3	L_4	X_{++k+}
Tube	25,388	25,582	25,342	26,240	102,552
Alternate	25,447	24,980	25,796	26,408	102,031
X_{+j++}	50,835	50,562	50,538	52,648	

TABLE 13.65. *TABLE FOR FACTORS, TUBES, AND TEMPERATURES*

	Tube	Alternate	X_{i++}
T_1	51,664	52,103	103,767
T_2	50,888	49,928	100,816
X_{++k+}	102,552	102,031	

7. $SS_5 = 871{,}968{,}236 - 871{,}962{,}581 = 5655$†
8. $SS_4 = 872{,}217{,}868 - 871{,}962{,}581 = 255{,}287$
9. $SS_2^{BC} = 289{,}905 - 5655 - 255{,}287 = 28{,}963$‡

1. $X_{1+++} + X_{2+++} = 103{,}767 + 100{,}816 = 204{,}583$

2. $$\sum_{i=1}^{2}\sum_{k=1}^{2} \frac{X_{i+k+}^2}{12} = \frac{(51{,}664)^2 + (52{,}103)^2 + (50{,}888)^2 + (49{,}928)^2}{12} = 872{,}190{,}435$$

3. $$\sum_{i=1}^{2} \frac{X_{i+++}^2}{24} = \frac{(103{,}767)^2 + (100{,}816)^2}{24} = 872{,}144{,}006$$

4. $$\sum_{k=1}^{2} \frac{X_{++k+}^2}{24} = \frac{(102{,}552)^2 + (102{,}031)^2}{24} = 871{,}968{,}236$$

5. $$\frac{(1)^2}{48} = \frac{(204{,}583)^2}{48} = 871{,}962{,}581$$

6. $(2) - (5) = 872{,}190{,}435 - 871{,}962{,}581 = 227{,}854$
7. $SS_3 = 872{,}144{,}006 - 871{,}962{,}581 = 181{,}425$
8. $SS_5 = 871{,}968{,}236 - 871{,}962{,}581 = 5655$
9. $SS_2^{AC} = 227{,}854 - 181{,}425 - 5655 = 40{,}774$§

13. $w_{111}, w_{112}, \ldots, w_{242} = 12{,}512; 13{,}404; \ldots; 13{,}064$

14. $$\sum_{i=1}^{2}\sum_{j=1}^{4}\sum_{k=1}^{2} \frac{w_{ijk}^2}{3} = \frac{(12{,}512)^2 + (13{,}404)^2 + \cdots + (13{,}064)^2}{3} = 872{,}772{,}468$$

15. $$\sum_{i=1}^{2}\sum_{j=1}^{4}\sum_{k=1}^{2}\sum_{l=1}^{3} X_{ijkl}^2 = (3774)^2 + (4364)^2 + \ldots + (4426)^2 = 873{,}947{,}177$$

16. $SS_1 = 873{,}947{,}177 - 872{,}772{,}468 = 1{,}179{,}709$
17. $SS = 873{,}947{,}177 - 871{,}962{,}581 = 1{,}984{,}596$
18. $SS_2^{ABC} = 1{,}984{,}596 - 5655 - 255{,}287 - 181{,}425 - 132{,}515 - 40{,}774 - 28{,}963 - 1{,}174{,}709 = 165{,}268$

The analysis of variance table (Table 13.66) reveals that only plate conditions give significant results. For plate temperatures,

$$\bar{X}_{1\ldots} = 4323.6 \quad \text{and} \quad \bar{X}_{2\ldots} = 4200.7$$

From Table 13.48, for 32 degrees of freedom and $r = 2$, $c^* = 2.88$, so that $c = 112.8$. Hence,

$$(4323.6 - 4200.7) - 112.8 = 10.1 \le T_1 \text{ effect} - T_2 \text{ effect} \le (4323.6 - 4200.7) + 112.8 = 235.7$$

Thus to increase the mutual conductance, the tubes can be manufactured under plate temperature T_1.

13.6.7 Latin Square

In a three-way classification with four levels for each factor, there must be $4^3 = 64$ observations taken in order to analyze the experiment. Furthermore, it may be impossible to obtain observations for all combinations of all levels of the factors. For example, suppose that in the counting rate problem discussed in Pars. 13.6.3 and 13.6.4.4 there was an additional factor, order, besides the factors, specimens and experiments; i.e., a short-time effect may exist. It is impossible to obtain observations for all combinations of all levels, since only one specimen can be measured first in the first experiment. (It must be remembered that experiments were actually long-time differences.) The difficulty is met by setting up the procedure so that all factors appear at the same number of levels. This procedure is an arrangement of the levels into a Latin square (see Table 13.67), which is simply a square array of letters such that every letter appears once and only once in every row and column.

† This result is denoted by (10) in the computational procedure.

‡ This result is denoted by (11) in the computational procedure.

§ This result is denoted by (12) in the computational procedure.

TABLE 13.66. *ANALYSIS OF VARIANCE TABLE FOR ELECTRONIC TUBE EXPERIMENT*

Source	Sum of Squares	Degrees of Freedom	Mean Square	Average Mean Square	Test
Between tube or alternate	5,655	1	5,655		$F = 0.154 \leq F_{.05;1,32}$
Between plate temperatures	181,425	1	181,425		$F = 4.94 \geq F_{.05;1,32}$
Between lighting conditions	255,287	3	85,095.7		$F = 2.32 \leq F_{.05;3,32}$
Interaction lighting conditions and plate temperatures	132,515	3	44,171.7		$F = 1.20 \leq F_{.05;3,32}$
Interaction lighting conditions and tube or alternate	28,963	3	9,654.3		$F = 0.260 \leq F_{.05;3,32}$
Interaction tube or alternate and plate temperatures	40,774	1	40,774		$F = 1.11 \leq F_{.05;1,32}$
Interaction plate temperatures, lighting conditions and tube or alternate	165,268	3	55,089.3		$F = 1.50 \leq F_{.05;3,32}$
Within cells	1,174,709	32	36,710		
Total	1,984,596	47			

TABLE 13.67. *LATIN SQUARE TABLE*

Row Factor	Column Factor 1	2	3	4
1	*A*	*C*	*B*	*D*
2	*B*	*D*	*A*	*C*
3	*D*	*A*	*C*	*B*
4	*C*	*B*	*D*	*A*

We can identify the letters with the four specimens. The column factor is the experiment and the row factor is the order. Thus the third specimen (*C*) is used second during the fourth experiment. Note that only sixteen observations are required compared with the 64 if the factorial design were used (if it could be used). This saving is at the expense of a loss in degrees of freedom for estimating σ^2. Furthermore, the analysis of such a Latin square design requires that there be no interactions present. In spite of these two drawbacks, the Latin square, when applicable, is of great importance from the point of view of industrial experimentation.

The general model is

$$X_{ij(k)} = \zeta\ldots + \varphi_i + \gamma_j + \lambda_{(k)} + \varepsilon_{ij(k)}$$

where the subscript i refers to rows, j to columns, and k to letters in the square. The k is enclosed in parentheses to indicate that it is not independent of i and j. Then

$X_{ij(k)}$ is the outcome of the experiment using the ith row effect, jth column effect, and (k)th letter effect, $i = 1, 2, \ldots, n$; $j = 1, 2, \ldots, n$; $k = 1, 2, \ldots, n$;

$\zeta\ldots$ is a general mean;

φ_i is the effect of adding the ith row treatment,

$$\sum_{i=1}^{n} \varphi_i = 0;$$

γ_j is the effect of adding the jth column treatment, $\sum_{j=1}^{n} \gamma_j = 0$;

TABLE 13.68. *ANALYSIS OF VARIANCE TABLE, LATIN SQUARE*

Source	Sum of Squares	Degrees of Freedom	Mean Square	Average Mean Square	Test
Between letters	$SS_5 = n \sum_{(k)=1}^{n} (\bar{X}_{(k)} - \bar{X}_{..})^2$	$n - 1$	$SS_5^* = \frac{SS_5}{n-1}$	$\sigma^2 + \frac{n}{n-1} \sum_{(k)=1}^{n} \lambda_{(k)}^2$	$F_5 = \frac{SS_5^*}{SS_2^*}$
Between columns	$SS_4 = n \sum_{j=1}^{n} (\bar{X}_{.j} - \bar{X}_{..})^2$	$n - 1$	$SS_4^* = \frac{SS_4}{n-1}$	$\sigma^2 + \frac{n}{n-1} \sum_{j=1}^{n} \gamma_j^2$	$F_4 = \frac{SS_4^*}{SS_2^*}$
Between rows	$SS_3 = n \sum_{i=1}^{n} (\bar{X}_{i.} - \bar{X}_{..})^2$	$n - 1$	$SS_3^* = \frac{SS_3}{n-1}$	$\sigma^2 + \frac{n}{n-1} \sum_{i=1}^{n} \varphi_i^2$	$F_3 = \frac{SS_3^*}{SS_2^*}$
Residual	$SS_2 = \sum_{i=1}^{n} \sum_{j=1}^{n} (X_{ij(k)} - \bar{X}_{i.} - \bar{X}_{.j} - \bar{X}_{(k)} + 2\bar{X}_{..})^2$	$(n-1)(n-2)$	$SS_2^* = \frac{SS_2}{(n-1)(n-2)}$	σ^2	
Totals	$SS = \sum_{i=1}^{n} \sum_{j=1}^{n} (X_{ij(k)} - \bar{X}_{..})^2$	$n^2 - 1$	$SS^* = \frac{SS}{n^2-1}$	$\sigma^2 + \frac{n}{n^2-1} \left(\sum_{(k)=1}^{n} \lambda_{(k)}^2 + \sum_{j=1}^{n} \gamma_j^2 + \sum_{i=1}^{n} \varphi_i^2 \right)$	

In this table $\bar{X}_{(k)}$ = average over-all observations with the kth letter;

$$\bar{X}_{.j} = \sum_{i=1}^{n} X_{ij(k)}/n; \quad \bar{X}_{i.} = \sum_{j=1}^{n} X_{ij(k)}/n; \quad \bar{X}_{..} = \sum_{i=1}^{n} \sum_{j=1}^{n} X_{ij(k)}/n^2.$$

$\lambda_{(k)}$ is the effect of adding the (k)th letter treatment, $\sum_{(k)=1}^{n} \lambda_{(k)} = 0$;

$\varepsilon_{ij(k)}$ is the random effect which is independently normally distributed with mean 0 and variance σ^2.

13.6.7.1 Tests for Homogeneity. Compute the following analysis of variance table (Table 13.68). Computational procedure is as follows:

1. Calculate row totals: $R_1, R_2, \ldots, R_n$
2. Calculate column totals: $C_1, C_2, \ldots, C_n$
3. Calculate letter totals: $L_1, L_2, \ldots, L_n$
4. Calculate over-all total: $T = R_1 + R_2 + \cdots + R_n$
5. Calculate crude sum of squares:

$$\sum_{i=1}^{n}\sum_{j=1}^{n} X_{ij(k)}^2 = X_{11}^2 + X_{12}^2 + \cdots + X_{nn}^2$$

6. Calculate crude sum of squares between rows:

$$\sum_{i=1}^{n} \frac{R_i^2}{n} = \frac{R_1^2 + R_2^2 + \cdots + R_n^2}{n}$$

7. Calculate crude sum of squares between columns:

$$\sum_{j=1}^{n} \frac{C_j^2}{n} = \frac{C_1^2 + C_2^2 + \cdots + C_n^2}{n}$$

8. Calculate crude sum of squares between letters:

$$\sum_{(k)=1}^{n} \frac{L_{(k)}^2}{n} = \frac{L_1^2 + L_2^2 + \cdots + L_n^2}{n}$$

9. Calculate correction factor due to mean $= T^2/n^2$.

From the quantities above compute:

10. $SS_5 = (8) - (9) = \sum_{(k)=1}^{n} \frac{L_{(k)}^2}{n} - \frac{T^2}{n^2}$

11. $SS_4 = (7) - (9) = \sum_{j=1}^{n} \frac{C_j^2}{n} - \frac{T^2}{n^2}$

12. $SS_3 = (6) - (9) = \sum_{i=1}^{n} \frac{R_i^2}{n} - \frac{T^2}{n^2}$

13. $SS = (5) - (9) = \sum_{i=1}^{n}\sum_{j=1}^{n} X_{ij(k)}^2 - \frac{T^2}{n^2}$

14. $SS_2 = (13) - (12) - (11) - (10) = SS - SS_3 - SS_4 - SS_5$

If $F_5 = SS_5^*/SS_2^* \geq F_{\alpha;n-1,(n-1)(n-2)}$ we reject the hypothesis that there are no letter effects, i.e., that

$$\lambda_{(1)} = \lambda_{(2)} = \cdots = \lambda_{(n)} = 0$$

If $F_4 = SS_4^*/SS_2^* \geq F_{\alpha;n-1,(n-1)(n-2)}$ we reject the hypothesis that there are no column effects, i.e., that

$$\gamma_1 = \gamma_2 = \cdots = \gamma_n = 0$$

If $F_3 = SS_3^*/SS_2^* \geq F_{\alpha;n-1,(n-1)(n-2)}$ we reject the hypothesis that there are no row effects, i.e., that

$$\varphi_1 = \varphi_2 = \cdots = \varphi_n = 0$$

13.6.7.2 Tests for Comparison of Mean Effects. The procedure for comparing letter effects, row effects, and column effects is exactly the same as for the two-way classification except for the following changes.

The factor c for all three effects is obtained from the analysis of variance table and from Table 13.48, depending on the level of significance. Table 13.48 is entered, using as indices the degrees of freedom, and the number of levels of each factor (n). The proper degrees of freedom are those corresponding to the degrees of freedom of the residual source in the analysis of variance table, i.e., $(n - 1)(n - 2)$.

The factor c^* is read from the table for all three effects, and c is obtained from

$$c = c^*\sqrt{SS_2^*/n}$$

Of course, for finding confidence intervals, the calculated terms that are of interest are $\bar{X}_{i\cdot}$, $\bar{X}_{\cdot j}$, and $\bar{X}_{(k)}$.

13.6.7.3 Estimating the Variability Due to the Sources. Estimates of the entries in the "average mean square" column are obtained from the "mean square" column. Hence estimates of $\sum_{i=1}^{n} \varphi_i^2$, $\sum_{j=1}^{n} \gamma_j^2$, and $\sum_{(k)=1}^{n} \lambda_{(k)}^2$ can be calculated:

$$\text{Estimate of} \sum_{i=1}^{n} \varphi_i^2 = (SS_3^* - SS_2^*)\frac{n-1}{n}$$

$$\text{Estimate of} \sum_{j=1}^{n} \gamma_j^2 = (SS_4^* - SS_2^*)\frac{n-1}{n}$$

$$\text{Estimate of} \sum_{(k)=1}^{n} \lambda_{(k)}^2 = (SS_5^* - SS_2^*)\frac{n-1}{n}$$

Often an experiment which can be treated as a Latin square might alternatively be attacked as a two-way plan, neglecting either the row or column grouping. For example, the counting rate problem has already been presented where the specimens were measured in random order in each experiment. We might hope to increase the precision by requiring that each specimen be measured first in one experiment, second in another, third in another, and fourth in another. This experimental plan is a more complicated one to carry out, and it will be worthwhile to assess how much the precision is increased as a guide in deciding how future investigations of a similar sort should be conducted.

The efficiency is calculated as:

$$e = \frac{SS_2^* + \dfrac{1}{n-1}\sum_{i=1}^{n} \varphi_i^2}{SS_2^*}$$

when the row grouping is neglected and

$$e = \frac{SS_2^* + \dfrac{1}{n-1}\sum_{j=1}^{n} \gamma_j^2}{SS_2^*}$$

when the column grouping is neglected.

We can say approximately that for the same precision as that given by the n^2 observations in the Latin square, $e \times n^2$ would be needed for a two-way plan.

13.6.7.4 Example. Consider the problem of the net counting rates described at the beginning of the article, where the factor "order" has been included (Table 13.69). The letters represent specimens. The analysis of variance table (Table 13.70) is obtained by following the computational procedure.

1. $R_1, R_2, R_3, R_4 = 13.04, 12.71, 13.75, 13.22$
2. $C_1, C_2, C_3, C_4 = 12.73, 14.16, 13.14, 12.69$

TABLE 13.69. *NET COUNTING RATES LESS* 25.00

	Experiment Number				
Order	1	2	3	4	Order Average
1	$A_{1.46}$	$C_{4.61}$	$B_{2.82}$	$D_{4.15}$	3.26
2	$B_{2.58}$	$D_{4.52}$	$A_{1.48}$	$C_{4.13}$	3.18
3	$D_{4.54}$	$A_{2.00}$	$C_{4.31}$	$B_{2.90}$	3.44
4	$C_{4.15}$	$B_{3.03}$	$D_{4.53}$	$A_{1.51}$	3.30
Experiment average	3.18	3.54	3.28	3.17	
Specimen average	$A_{1.61}$	$B_{2.83}$	$C_{4.30}$	$D_{4.44}$	

TABLE 13.70. *ANALYSIS OF VARIANCE TABLE FOR LATIN SQUARE DESIGN OF THE COUNTING RATE EXPERIMENT*

	Sum of Squares	Degrees of Freedom	Mean Square	Average Mean Square	Test
Between letters (specimens)	21.4174	3	7.1391		$F = 594.9$
Between columns (experiments)	0.3512	3	0.11707		$F = 9.76$
Between rows (order)	0.1418	3	0.0473		$F = 3.94$
Residual	0.0720	6	0.0120		
Totals	21.9824	15	1.4655		

3. $L_{(1)}, L_{(2)}, L_{(3)}, L_{(4)} = 6.45, 11.33, 17.20, 17.74$

4. $T = 13.04 + 12.71 + 13.75 + 13.22 = 52.72$

5. $\sum_{i=1}^{4}\sum_{j=1}^{4} X^2_{ij(k)} = (1.46)^2 + (4.61)^2 + \cdots + (1.51)^2 = 195.6948$

6. $\sum_{i=1}^{4} \frac{R_i^2}{4} = \frac{(13.04)^2 + (12.71)^2 + (13.75)^2 + (13.22)^2}{4} = 173.8542$

7. $\sum_{j=1}^{n} \frac{C_j^2}{4} = \frac{(12.73)^2 + (14.16)^2 + (13.14)^2 + (12.69)^2}{4} = 174.06355$

8. $\sum_{(k)=1}^{4} \frac{L_{(k)}^2}{4} = \frac{(6.45)^2 + (11.33)^2 + (17.20)^2 + (17.74)^2}{4} = 195.1298$

9. $T^2/n^2 = \frac{(52.72)^2}{16} = 173.7124$

10. $SS_5 = 195.1298 - 173.7124 = 21.4174$

11. $SS_4 = 174.06355 - 173.7124 = 0.3512$

12. $SS_3 = 173.8542 - 173.7124 = 0.1418$

13. $SS = 195.6948 - 173.7124 = 21.9824$

14. $SS_1 = 21.9824 - 21.4174 - 0.3512 - 0.1418 = 0.0720$

For a critical value $F_{.05;3,6} = 4.76$, we see that the data indicate that both the specimens and experiments reveal significant differences. To judge which specimens differ, we enter Table 13.48 with $n = 4$ and degrees of freedom $= 6$, and $c^* = 4.90$, so that

$$c = 4.90 \times \tfrac{1}{2}\sqrt{0.0120} = 0.2695$$

and we can make such statements as: specimen 1 differs from specimens 2, 3, and 4; specimen 2 also differs from specimens 3 and 4; and furthermore with a confidence level of 0.95 we can make such statements as
$(4.44 - 1.61) - 0.27 = 2.56 \leq$ mean effect of specimen 4 − mean effect of specimen 1 $\leq (4.44 - 1.61) + 0.27 = 3.10$
Similarly experiment 2 differs from experiment 1 and from experiment 4. To estimate the magnitude of the main effects we find

For specimens, $\sum \lambda^2_{(k)} = (7.1391 - 0.0120)\frac{3}{4} = 5.3453$

For experiments, $\sum \gamma_j^2 = (0.11707 - 0.0120)\frac{3}{4} = 0.08690$

For order, $\sum \varphi_i^2 = (0.0473 - 0.0120)\frac{3}{4} = 0.0265$

We can compute now the efficiency of the Latin square relative to the two-way plan for these data where order is ignored:

$$e = \frac{0.0120 + \frac{1}{4-1}(0.0265)}{0.0120} = 1.74$$

This indicates that we have increased our information per observation by more than half, by taking account of order (in addition to experiments) in the design. In the earlier treatment of these data, using only a two-way analysis, the value of c for getting confidence intervals was 0.34; in the present analysis the value of c is 0.27, which exhibits tangibly the gain resulting from the use of the Latin square design.

13.6.8 Components of Variance Model

Throughout this article we have treated the problem where each observation is composed of two components, namely, the unknown mean of the population from which the observation is drawn and which is common to all observations within a set, and the deviation from this mean. Depending on the number of factors involved, each set mean could be further subdivided, and tests made on the component parts.

For example, in a small laboratory four thermometers are used interchangeably to make all temperature measurements. A natural question to ask is whether these thermometers differ. The model can be described as an observation consisting of a mean value common to a particular thermometer, and a deviation from this mean. Observations on a thermometer at the same temperature differ only because of these random deviations. A difference in thermometers is equivalent to speaking of a difference in the thermometer means. The techniques described above handle such problems.

It is possible, however, that the problem at hand does not fit this model. For example, suppose that there are hundreds of thermometers used interchangeably to make temperature measurements. To perform an experiment with all the thermometers is too costly, and a random sample of 4 thermometers is drawn to determine whether thermometers differ. Naturally, the experimental results will depend heavily on which thermometers are chosen. Yet inferences are to be made about all the thermometers in the laboratory. The measurement may be considered as consisting of the two random components: (1) the deviation of the thermometer measurement from the mean value of the particular thermometer used, and (2) the deviation of the thermometer mean from the over-all mean of all the thermometers in the laboratory.

We are interested in determining not whether these particular thermometers chosen at random differ, but whether all the thermometers in the laboratory, of which these particular four thermometers are samples, differ. Another way of expressing this is to determine whether the deviation of the population means of the different thermometers from the over-all mean is zero; i.e., the thermometer variance is zero. This type of problem can be characterized in that each observation is regarded as the sum of three components, namely, an unknown constant, the mean, which is common to all observations, and two random components which give rise to the total variability; one component producing the variation within the sets of observations, and the other producing the variation between the sets of observations.

Although it is important to distinguish between the two models formally, the analysis of variance table, computations, and procedures for significance tests are identical in both cases, with the following exceptions: (1) The column in the analysis of variance table headed "average mean square" is replaced by a column headed "components of variance." For example, Table 13.52 for the two-way classification would have the following entry:

Sources	Components of variance
Between rows:	$\sigma^2 + s\sigma^2$ (rows)
Between columns:	$\sigma^2 + r\sigma^2$ (columns)
Residual:	σ^2

The entries in the "mean square" column are now estimates of the entries in the column headed "components of variance," so that $\sigma^2_{(\text{rows})}$ and $\sigma^2_{(\text{columns})}$ can be estimated. (2) If the possibility of interactions is allowed in the model, the sum of squares for main effects are usually compared with the proper interaction sum of squares in tests of significance. For example, for the two-way classification, with interaction, the test for interaction is still $F_2 = SS_2^*/SS_1^*$. However, the test for column effects involves SS_4^*/SS_2^*, and the test for row effects involves SS_3^*/SS_2.

13.6.9 Hartley's Test for Homogeneity of Variances

Since a primary assumption in the analysis of variance is that the within-group variability be constant for all groups, we may wish to test the hypothesis that $\sigma_1^2 = \sigma_2^2 = \cdots = \sigma_k^2$.

The following test, known as Hartley's $F_{\text{max.}}$ test, will be suitable for the purpose. Reject the hypothesis that the variances are equal when

$$H = \frac{\text{largest } s_i^2}{\text{smallest } s_i^2} \geq H_\alpha$$

where

$$s_i^2 = \sum_{j=1}^{n} \frac{(X_{ij} - \bar{X}_i)^2}{n-1}$$

$$= \sum_{j=1}^{n} \frac{X_{ij}^2}{n-1} - \frac{n\bar{X}_i^2}{n-1}$$

and n = number of observations within each group (assumed the same for each group). Values of H_α are found in Table 13.71 by entering with n and k (the number of variances being considered).

13.7 ANALYSIS OF ENUMERATION DATA: CHI-SQUARE TESTS

13.7.1 Introduction

The problems considered in Art. 13.4 dealt with the measurements on certain variables; there are many experiments in which we count the number of cases which fall into specified categories. For example, we may record the number of defective items produced by various machines, the number of errors made by several operators, the number of hits and misses made by several firing control devices, the number of accidents as a function of the shift.

We will assume that each sample observation must fall into one and only one of k categories; let $0_1, 0_2, \ldots, 0_k$ be the observed frequencies for each category. We are interested in testing hypotheses about the true relative frequencies.

Category	Observed Frequency	Theoretical Frequency
1	0_1	E_1
2	0_2	E_2
—	—	—
—	—	—
—	—	—
k	0_k	E_k

The test statistic is

$$\chi^2 = \sum_{i=1}^{k} \frac{(0_i - E_i)^2}{E_i}$$

which for large enough samples has an approximate χ^2 distribution with a number of degrees of freedom which depends on how the data are used in computing the E_i.* Several problems of this sort are described in the following articles.

13.7.2 The Hypothesis Completely Specifies the Relative Frequencies of the Categories

Consider $p_1, p_2, \ldots, p_k$, where p_i is the true relative frequency of the ith category. In this type of problem, the theoretical frequencies are calculated from the formula $E_i = Np_i$ and the χ^2 statistic has $k - 1$ degrees of freedom.

13.7.2.1 Example. The total number of defective units in a day's production was tabulated by shifts.

	Observed Frequency	Theoretical Frequency
Shift 1:	20	26.67
Shift 2:	36	26.67
Shift 3:	24	26.67

It is of interest to see whether the variation from shift is due to chance or whether there is a real difference in the occurrence of defectives. That is, we test the hypothesis that the true relative frequencies of defectives are the same on all shifts, i.e., $p_1 = p_2 = p_3 = \frac{1}{3}$. Our test statistic is

$$\begin{aligned}\chi^2 &= \frac{(20 - 26.67)^2}{26.67} + \frac{(36 - 26.67)^2}{26.67} + \frac{(24 - 26.67)^2}{26.67} \\ &= \frac{44.4889}{26.67} + \frac{87.0489}{26.67} + \frac{7.1289}{26.67} \\ &= 1.6681 + 3.2639 + 0.26730 \\ &= 5.1993 < \chi^2_{.05;2}\end{aligned}$$

Therefore we accept the hypothesis and conclude that the data do not indicate that the frequencies differ.

13.7.3 Test of Independence in a Two-Way Classification

Often frequency data are tabulated according to two criteria, with a view toward testing whether the criteria are associated. Consider the following analysis of the 157 machine breakdowns during a given quarter. We are interested in whether the same percentage of breakdowns occurs on each machine during each shift or whether there is some difference due perhaps to untrained operators or other factors peculiar to a given shift.

If the number of breakdowns is independent of shifts and machines, the probability of a breakdown occurring in the first shift and in the first machine can be estimated as

$$p_{11} = \frac{41}{157} \times \frac{33}{157} = 0.05489$$

If there are 157 breakdowns, the expected number of breakdowns on this shift and machines is estimated as

$$E_{11} = 157 \times p_{11} = 8.6177$$

Similarly for the third shift and second machine

$$p_{32} = \frac{54}{157} \times \frac{28}{157} = 0.06134$$

$$E_{32} = p_{32} \times 157 = 9.6306$$

This is done for all categories, and

$$\begin{aligned}\chi^2 &= \sum_{i=1}^{3} \sum_{j=1}^{4} \frac{(0_{ij} - E_{ij})^2}{E_{ij}} \\ &= \frac{(10 - 8.6177)^2}{8.6177} + \frac{(6 - 7.3120)^2}{7.3120} \\ &\quad + \cdots + \frac{(18 - 17.885)^2}{17.885} = 2.02\end{aligned}$$

This is to be compared with the χ^2 statistic given in Table 13.27 for $(3 - 1)(4 - 1) = 6$ degrees of freedom, i.e., $\chi^2_{.05;6} = 12.6$. Hence we accept the hypothesis and conclude that the data do not indicate different percentages of breakdowns on each machine during each shift. In general, for an r by s table, we have

$$\chi^2 = \sum_{i=1}^{r} \sum_{j=1}^{s} \frac{(0_{ij} - E_{ij})^2}{E_{ij}}$$

---

* The following rules of thumb can be used to assess adequacy of sample size:

1. If χ^2 has from 2 to 15 degrees of freedom, all E_i should be at least 2.5.

2. If χ^2 is computed from a 2×2 table (which will be discussed below), all E_i should be at least 5; however, if all but one of them is at least 5, the remaining one may be as small as 1 with little distortion in significance level.

which has an approximate χ^2 distribution with $(r-1)(s-1)$ degrees of freedom.

13.7.4 Computing Form for Test of Independence in a 2-by-2 Table

An important special case of the test of independence arises when both criteria of classification have two categories; the data may be represented in a 2–by–2 table. The data may be represented as in Table 13.73. In this case the chi-square statistic which has one degree of freedom may be written

$$\chi^2 = \frac{(ad - bc)^2(a + b + c + d)}{(a + b)(a + c)(b + d)(c + d)}$$

TABLE 13.71. *UPPER PERCENTAGE POINTS OF THE RATIO S^2_{max}/S^2_{min} IN A SET OF k MEAN SQUARES, EACH BASED ON n OBSERVATIONS (NORMAL VARIATION ASSUMED)*†

n \ k	5 Per Cent Points										
	2	3	4	5	6	7	8	9	10	11	12
3	39.0	87.5	142	202	266	333	403	475	550	626	704
4	15.4	27.8	39.2	50.7	62.0	72.9	83.5	93.9	104	114	124
5	9.60	15.5	20.6	25.2	29.5	33.6	37.5	41.1	44.6	48.0	51.4
6	7.15	10.8	13.7	16.3	18.7	20.8	22.9	24.7	26.5	28.2	29.9
7	5.82	8.38	10.4	12.1	13.7	15.0	16.3	17.5	18.6	19.7	20.7
8	4.99	6.94	8.44	9.70	10.8	11.8	12.7	13.5	14.3	15.1	15.8
9	4.43	6.00	7.18	8.12	9.03	9.78	10.5	11.1	11.7	12.2	12.7
10	4.03	5.34	6.31	7.11	7.80	8.41	8.95	9.45	9.91	10.3	10.7
11	3.72	4.85	5.67	6.34	6.92	7.42	7.87	8.28	8.66	9.01	9.34
13	3.28	4.16	4.79	5.30	5.72	6.09	6.42	6.72	7.00	7.25	7.48
16	2.86	3.54	4.01	4.37	4.68	4.95	5.19	5.40	5.59	5.77	5.93
21	2.46	2.95	3.29	3.54	3.76	3.94	4.10	4.24	4.37	4.49	4.59
31	2.07	2.40	2.61	2.78	2.91	3.02	3.12	3.21	3.29	3.36	3.39
61	1.67	1.85	1.96	2.04	2.11	2.17	2.22	2.26	2.30	2.33	2.36
∞	1.00	1.00	1.00	1.00	1.00	1.00	1.00	1.00	1.00	1.00	1.00

n \ k	1 Per Cent Points										
	2	3	4	5	6	7	8	9	10	11	12
3	199	448	729	1036	1362	1705	2063	2432	2813	3204	3605
4	47.5	85	120	151	184	216	249	281	310	337	361
5	23.2	37	49	59	69	79	89	97	106	113	120
6	14.9	22	28	33	38	42	46	50	54	57	60
7	11.1	15.5	19.1	22	25	27	30	32	34	36	37
8	8.89	12.1	14.5	16.5	18.4	20	22	23	24	26	27
9	7.50	9.9	11.7	13.2	14.5	15.8	16.9	17.9	18.9	19.8	21
10	6.54	8.5	9.9	11.1	12.1	13.1	13.9	14.7	15.3	16.0	16.6
11	5.85	7.4	8.6	9.6	10.4	11.1	11.8	12.4	12.9	13.4	13.9
13	4.91	6.1	6.9	7.6	8.2	8.7	9.1	9.5	9.9	10.2	10.6
16	4.07	4.9	5.5	6.0	6.4	6.7	7.1	7.3	7.5	7.8	8.0
21	3.32	3.8	4.3	4.6	4.9	5.1	5.3	5.5	5.6	5.8	5.9
31	2.63	3.0	3.3	3.4	3.6	3.7	3.8	3.9	4.0	4.1	4.2
61	1.96	2.2	2.3	2.4	2.4	2.5	2.5	2.6	2.6	2.7	2.7
∞	1.00	1.0	1.0	1.0	1.0	1.0	1.0	1.0	1.0	1.0	1.0

† This table is taken by permission from "Upper 5 and 1% Points of the Maximum F–Ratio," by H. A. David, *Biometrika*, Vol. 39, Part 324, December 1952, p. 424.

TABLE 13.72. *NUMBER OF BREAKDOWNS*

	Machine				Total per Shift
	A	B	C	D	
Shift 1	10	6	12	13	41
Shift 2	10	12	19	21	62
Shift 3	13	10	13	18	54
Total per machine	33	28	44	52	157

TABLE 13.73. *2-BY-2 TABLE*

	First Criteria		
Second Criteria	a	b	$a+b$
	c	d	$c+d$
	$a+c$	$b+d$	

TABLE 13.74. *NUMBER OF MACHINE BREAKDOWNS*

Machine	Supplier *A*	Supplier *B*	
I	4	9	13
II	15	3	18
	19	12	31

13.7.4.1 Example. From the data in Table 13.74, determine whether the same percentage of breakdowns occurs on each machine using material from each supplier.

$$\chi^2 = \frac{(12 - 135)^2(31)}{(13)(19)(12)(18)} = 8.8 \geq \chi^2_{.05;1} = 3.84$$

Therefore, we reject the hypothesis that the percentage of breakdowns is the same for each machine using material from each supplier.

13.7.5 Comparison of Two Percentages

A problem which in many cases is equivalent to the test of independence in a 2–by–2 table is the problem of testing equality of two percentages. Suppose, following Wallis,* we have the following data on two fire control devices.

	Hits	Misses	
Old	3	197	200
New	4	196	200
			400

Suppose we want to test the hypothesis that the percentage of hits is the same and want to reject when the new is superior, i.e., has a larger percentage of hits. If we let P_E be the observed proportion for the new experimental method, P_S be the observed proportion for the standard, and N be the number of trials with each method, then

$$U = \frac{\sqrt{N}\,(P_E - P_S)}{\sqrt{(P_E + P_S)[1 - (P_E + P_S)/2]}}$$

has approximately the normal distribution, and we would reject when $U \geq K_\alpha$. For the above data $P_S = 0.015$, $P_E = 0.020$, and

$$U = \frac{\sqrt{200}(0.005)}{\sqrt{0.020 + 0.015(1 - 0.035/2)}}$$
$$= 0.3795 < U_{.05} = 1.645$$

and we accept the hypothesis and conclude that the data do not reveal that the new device is better than the old one.

In terms of the notation of the last section,

$$U = \frac{\sqrt{a+b+c+d}(ad - bc)}{\sqrt{(a+b)(a+c)(b+d)(c+d)}}$$

If we are interested in rejecting whenever the proportions differ, we could reject when $|U| \geq K_{\alpha/2}$ or when $U^2 \geq K^2_{\alpha/2}$. This is exactly the same as analyzing the data by the method in the last article.

Figure 13.48 is a nomogram for determining sample size in problems of this sort. It requires a guess at the true values of the percentage of hits with the standard and experimental method. Here are detailed instructions for the nomogram.

1. Select values of α, β, P_S, and P_E (see explanation).
2. Locate the value of α on the scale marked "α."
3. Locate the value of β on the scale marked "β."
4. Locate the point at which a straight line from α to β cuts the scale marked "*A*."
5. Locate the value of P_S on the scale marked "P_S." If P_S exceeds 0.50 use $1-P_S$ and make a corresponding change of P_E to $1-P_E$.
6. Locate the value of P_E on the scale marked "P_E."

* Eisenhart, Hastay, and Wallis, *Techniques of Statistical Analysis*.

7. Locate the point at which a straight line from P_S to P_E cuts the scale marked "*B*."
8. Locate the point at which the scale marked "*N*" is cut by a straight line from the point on "*B*" to the point on "*A*." The "*N*" scale is calibrated so that the required sample size *N* is given opposite this point.

Explanation: Call the two percentages being compared the "standard" and the "experimental." α is defined as follows: We will conclude that the experimental method is better only if it shows in the experiment a superiority so great that the probability of this much superiority would be only α if the true long-run percentages were equal. In other words, α is the probability that we will erroneously judge the experimental method better when, in fact, the two methods are equally good.

β is the probability that we will fail to judge the experimental method better when, in fact, it is better. In other words, it is the probability that the experimental method, although actually better in the long run, will show a smaller superiority in the experiment than is required by the value of α for concluding that it is better.

P_S is an estimate of the true long-run probability for the standard method. This estimate can be based on prior experience, scientific judgment, etc.

P_E is an estimate of the true long-run probability for the experimental method. This estimate may be determined like that for P_S; or it may be the smallest figure above P_S for which it would be important to detect the superiority. For example, if $P_S = 0.10$, and if it would be important to detect any superiority of more than 50 per cent, we set $P_E = 0.15$.

N is the number of trials with *each* method required in the experiment, as given by the formula

$$N = \frac{1}{2}\left(\frac{K_\alpha + K_\beta}{\text{arc sin}\sqrt{P_S} - \text{arc sin}\sqrt{P_E}}\right)^2$$

where K_ε is given by

$$\frac{1}{\sqrt{2\pi}}\int_{k_\varepsilon}^{\infty} e^{-\frac{x^2}{2}}\, dx = \varepsilon$$

and angles are measured in radians.

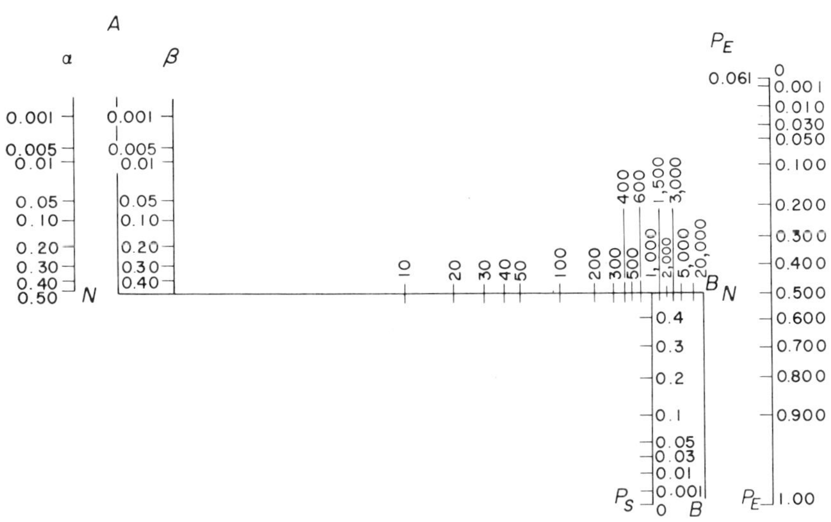

Fig. 13.48 *Number of cases required for comparing two percentages. Reprinted by permission from* Techniques of Statistical Analysis *by C. Eisenhart, M. W. Hastay, and W. A. Wallis. Copyright 1947. McGraw-Hill Book Company, Inc.*

THOMAS C. MCDERMOTT *is manager of operations for the Navigation and Computers Division, Autonetics, North American Rockwell Corporation. In that capacity, he is responsible for purchasing, manufacturing and quality assurance programs for all division products.*

He received his B.S. in industrial engineering from the Pennsylvania State University, and his M. S. in industrial administration from the Carnegie Institute of Technology. He has engaged in additional graduate studies at the University of Southern California. He is a member of Sigma Tau and Alpha Pi Mu honorary fraternities. He is a member of the American Management Association and a fellow of the American Society for Quality Control. He has actively participated in many of the ASQC conferences and was national president for 1969–70.

Prior to his current assignment, Mr. McDermott held the position of vice president of quality and reliability assurance at the Space Division. Prior to joining the Space Division in 1966, Mr. McDermott was division director of quality assurance and logistics of the Autonetics Division of North American Aviation, Inc. Prior to joining North American Aviation, he was an Air Force officer for ten years, serving as chief of the plans branch, Quality Control Office in AMC headquarters and chief of plans and engineering in the Quality Assurance Office of the Ballistic Missile Center.

Mr. McDermott has been directly responsible for many Air Force documents relating to quality and reliability and has authored many papers and reports.

DANA M. COUND *is the director of quality assurance for the Autonetics Division of North American Rockwell Corporation, where he is responsible to the president for directing all quality assurance activities associated with the products of the division.*

Since joining North American Aviation in 1953, Mr. Cound has held a number of key positions, including director of quality engineering and manager of quality assurance administration in the Space Division, chief of quality assurance, head of Quality Control Administrative Services, and senior quality control project administrator for Navy programs in the Autonetics Division.

He is a senior member of the American Society for Quality Control and a regular participant in its national conferences. He is active in several areas of quality training and education and is a member of the Education and Training Institute Board of the ASQC. He is a regular faculty member of the institute and teaches quality management subjects. He has also conducted after-hours training courses for North American Rockwell employees under the sponsorship of the manpower development program of the company.

Mr. Cound is a graduate of California State College, from which he received the bachelor of science degree in industrial relations.

SECTION
14

INSPECTION AND QUALITY CONTROL

T. C. MCDERMOTT
AND
D. M. COUND

The purpose of this section is to provide a general overview of the quality control function within an industrial organization. Throughout the section reference is made to sources that provide more detailed information on quality control concepts and methodology. A bibliography is provided for those interested in more extensive study of quality control.

14.1 NATURE AND SCOPE

14.1.1 Introduction

During the past several decades, quality control has emerged as a primary function in modern industrial enterprise. The importance of quality control has been strongly influenced by several basic industrial trends; viz., the high degree of specialization of labor, the increase in precision and complexity of products, improved communication, and, hence, greater discrimination on the part of consumers.

There is also a growing realization of the economic implications of marginal quality. The result has been continual management emphasis on the quality control function. The survival of an industrial enterprise today is strongly dependent on two factors: how well the nature of product quality is understood and effectiveness of the methods employed in achieving the company's quality objectives.

14.1.2 Definitions

14.1.2.1 Quality. *The sum of the attributes or properties that describe the product.* These are generally expressed in terms of specific product characteristics such as length, width, color, specific gravity, etc. To be meaningful in an industrial sense, these characteristics must be quantitatively expressed in terms that can be objectively measured or observed (see Par. 14.9.4.3). Cases do arise where subjective measures are required and are appropriate, but these are generally held to an absolute minimum.

14.1.2.2 Quality Creation. *Those activities involved in the selection of the specific characteristics required to achieve the desired quality and the processing or fabrication of materials to conform to the specific characteristics selected.* Quality creation involves almost all organizational elements of the enterprise and is the basic objective toward which most activity is directed.

14.1.2.3 Quality Control. *Those activities which assure that quality creation is performed in such a manner that the resulting product will in fact perform its intended function.* When used in this sense, quality control can be divided into two fundamental endeavors: assurance that the product characteristic selected will achieve the intended result and assurance that items produced contain the specified characteristics.

In a more limited sense, quality control is frequently used to refer to a specific organization within the industrial enterprise which is assigned responsibility for many of the work activities necessary to achieve quality objectives. In actual practice there is a wide variance between industries and individual companies within industries on the range of activities that are specifically assigned to the quality control organization. A discussion of organization considerations is contained in Art. 14.2.

14.1.3 Quality Control Functions

Quality objectives are best achieved through developing a systematic approach to the control of quality, which generally includes the performance of the following subfunctions. (This list is not all-inclusive but provides a basic outline of quality control functions.)

14.1.3.1 Control of Engineering Quality

A. Assist in the evaluation of customer requirements to assure a clear understanding of the product quality objectives

B. Review design documentation for conformance to design standards and practices and for identification of potential quality problems

C. Validate the accuracy and completeness of design proof tests and qualification tests

D. Audit the release and distribution of design documentation to assure that all drawings and specifications in use are current and correct

E. Provide information on previous quality problems encountered for consideration in new product designs or current product improvement

14.1.3.2 Control of Purchased Material Quality

A. Assist in the evaluation and selection of potential suppliers or subcontractors

B. Review purchase orders and subcontracts for correctness and completeness of quality requirements

C. Assure that purchased material conforms to the requirements of purchase orders and specifications

D. Initiate corrective action with suppliers and subcontractors when purchased material is not of an acceptable quality level

14.1.3.3 Control of Manufacturing Quality

A. Evaluate and approve manufacturing equipment, processes, testing, and test equipment

B. Assure that measuring and test equipment is properly calibrated and maintained

C. Establish points of inspection and develop inspection methods and instructions

D. Perform inspections and tests at selected points in the production processes

E. Collect and analyze inspection and test data and provide information on process and product quality levels

F. Initiate corrective action on out-of-control conditions and related quality problems

G. Conduct follow-up to assure that corrective action is accomplished in a timely manner

H. Control the handling, preservation, and packaging of material and equipment from receipt through shipment of the final product

14.1.3.4 Actions Supporting the Product After Delivery

A. Assure that product service publications are clear and correct

B. Assure that spare parts conform to quality requirements

C. Assure that company-performed repair and modification are performed in accordance with company quality requirements

D. Gather and analyze complaint data from the field to measure the degree of customer satisfaction and initiate appropriate corrective action

14.1.4 Relationship to Reliability

In recent years there has been considerable emphasis on the reliability of products, particularly on those purchased by the military services. The relationship between quality control and reliability, therefore, deserves some consideration.

In its basic sense, reliability is an attribute or property of the product. It is the probability of equipment operating for a specified period of time under established environmental parameters. Although it generally involves a more complex set of considerations than other product attributes, it utilizes the same basic approach. That is, reliability must be quantitatively expressed, translated into specific design characteristics, and these characteristics controlled throughout manufacture and use of the product. Achieved reliability can be measured through such activity as life tests, stress tests, and actual performance in use. Quality control activities, in assuring that this requirement is attained, involve the basic quality control functions previously stated, although the degree of application may be significantly different as reliability levels increase.

14.2 ORGANIZATION

14.2.1 Purpose of Organizing

The purpose of organizing is to establish a relationship between the numerous industrial functions so they will aid one another in reaching common goals and eliminate friction and cross-purposes. The organization establishes lines of authority, responsibility, and communication within the enterprise. The form of the quality control organization may vary considerably between companies. An organization is only "right" in the sense that it properly satisfies the basic objectives of the enterprise. It is important, therefore, that the development of the quality control organization consider some basic organizational criteria.

14.2.2 Location Within the Total Enterprise

14.2.2.1 Prime Considerations. Quality control can be located at various levels within the enterprise. The determination of the proper level should consider the following:

A. Level at which management wants such trade-off decisions as quality versus cost and quality versus schedule to be made

B. Extent to which quality control represents the company to the customer

C. Impact of a quality system failure on the company

D. Number of technologies or specialties included in the total quality function

E. Customer policy

14.2.2.2 Organizational Trade-off Level. The separate interests of quality, cost, and schedule will frequently come into conflict. Usually dependence is placed upon the managers of the individual functions to jointly arrive at the best decision in terms of the total enterprise. Occasionally such resolution cannot be reached at this level, and some executive must make the decision regarding which is to be satisfied and which subordinated. This possibility should be considered in placing the quality function within the organization. For example, an organization engaged in the manufacture of clay flower pots might not consider the quality trade-off to be nearly so sensitive as the cost and schedule and may assign quality control as a part of the production department. However, an aerospace manufacturer or pharmaceutical corporation may consider quality of paramount concern, never subordinate to cost and schedule, and may establish quality control as a major functional department. In this manner any compromise decisions will be made by the highest executive of the enterprise.

14.2.2.3 Extent to Which Quality Control Represents the Company to the Customer. Normally, the greater the extent to which the quality control function represents the company to the customer, the higher will be the level of the quality control function within the enterprise. In some businesses the quality control function is involved only in the economics of manufacture and, once manufactured, the product is not differentiated on the basis of quality. On the other hand, if one of the functions of quality control is customer service or warranty administration, the quality control department will normally be established higher in the organization in order to have more ready access to top management.

14.2.2.4 Impact of Quality System Failure. The more serious the impact of failures of the quality system, the higher the quality control department will be located within the organization. In the case of the flower-pot manufacturer, a breakdown in the quality system might lead to such things as mislocation of the drain hole in the bottom, variation in wall thickness, or variations in the scrap rate of the finished product. Since none of these could reasonably be expected to have serious impact on the existence of the business, one would expect to see the quality function subordinated within the organization. However, a breakdown in the quality system of the pharmaceutical manufacturer, which could lead to the production and delivery of toxic substances represented to be medicine, could easily lead to enormous financial loss and eventually to destruction of the enterprise. As such, the quality control function must be considered a major function.

14.2.2.5 Number of Technologies. If the quality control department consists of a product inspection function only, it will be possible and may be appropriate to establish it as a department under the production branch. However, if the quality control function includes product inspection, receiving inspection, quality engineering, customer service, and quality laboratory functions, it will require a manager of sufficient scope to effectively harmonize and employ all these technologies. Thus it would probably be established as a major functional department reporting to a senior executive.

14.2.2.6 Customer Policy. In a monopsony situation, the policies of the enterprise are strongly influenced by the policies of the customer. These policies may govern the location of the quality control function within the contractor's organization.

14.2.3 Internal Organization of Quality Control

14.2.3.1 Prime Considerations. Once the level of the quality control function is established, the internal organization of quality control must be considered. The factors bearing on the internal organization include the number of people employed, geographic dispersion of the activity, technological dispersion, and organizations with which quality control interfaces.

14.2.3.2 Size. The quality control department may be undifferentiated as to type and caliber of employees but still may employ large numbers of them. If so, the sole factor governing internal organization would be the ideal size of the work unit. This unit may be an inspection department under a chief inspector who in turn has three identical general foremen who each in turn supervise three foremen, etc., in order to provide proper supervision to a large number of people performing identical work.

14.2.3.3 Geographical Dispersion. An organization with a relatively large number of people performing identical work may require subdivision based on geographical dispersion. Perhaps one foreman would supervise the work of three crews of five men each if they are working in the same area. However, each of these crews may be assigned to a different building or product line, making it impossible for him to supervise all three groups. In this case, the organization may be more appropriately subdivided geographically.

14.2.3.4 Technological Dispersion. This is perhaps the most common form of organization within a complex quality control department. Frequently a quality control department will be comprised of an inspection function, a quality engineering function, and a laboratory function. The inspection function in turn may include the inspection of machine parts, precision assemblies, and a functional test group. The laboratory may include chemical and metallurgical analysis sections, an environmental test section, and a measurement standards laboratory. These functions are formed by grouping common technological skills.

14.2.3.5 The Organizations with Which Quality Control Interfaces. Frequently this becomes a major consideration in the establishment of an organization. One of the principles of a sound organization is that the formal organization (the one printed on paper) should be as similar as possible to the informal organization (the way people naturally tend to work). It may be well to consider the separation of functions in the interest of improving their interfaces with the organizations with which they work.

14.2.3.6 Other Considerations. The final step remaining after developing the internal organizational structure is the assignment of activities to the internal organizational blocks. Given an internal organization, the assignment of the major functions is very clear. Inspection activities belong in the inspection department, statistical analysis activities belong in the quality engineering department, and chemical analysis activities belong in the laboratory department. But what of the many activities that are not clearly assigned to one of these segments? They will include such things as the disposition of nonconforming material, supplier survey activities, inspection stamp control, tool proofing, test equipment calibration, etc. What guidelines exist to help determine the proper assignment of these activities within quality control?

A. Assignment by similarity. Perhaps the most common basis for the assignment of activities is the grouping of activities based upon their similarity. This might indicate that shipping inspection, receiving inspection, and packaging and crating inspection be assigned to the product inspection department simply because the inspection aspect is a common significant characteristic.

B. Intimate association. This guideline suggests that an activity for which a manager has the most use should normally be assigned to him. For example, the function concerned with the review and disposition of nonconforming material is frequently assigned to the product inspection manager simply because he has the ultimate responsibility for the acceptance or rejection of material.

C. Competition. There are many situations in which a particularly desirable activity does not flourish because such a development is feared by certain executives, because its possibilities are not recognized, or because it fails to receive vigorous direction. Frequently, establishing such a function as a separate entity allows the activity to flourish.

D. Suppressed competition. This guideline, the exact reverse of the one that encourages competition, is applicable under a different set of conditions. It dictates combination of functions under certain circumstances. As the advantages reaped from competition become more fully realized, the executive of the enterprise may feel the growing need for coordination between these functions, that is, more cooperation and less competition.

E. Coordination at a point. The coordination of activities frequently involves two factors: the things to be done must be timed properly, and the effort to be placed behind each activity must be integrated. An example of this would be an audit inspection activity. Many manufacturers of complex equipment maintain within the quality control department an inspection audit activity which duplicates inspections on a sample basis to assure that the line inspectors are themselves maintaining a satisfactory level of performance, neither rejecting good material nor accepting bad to an excessive degree. Quite a case can be made for separating this activity from product inspection since it is a check on that function. However, the "point coordination" necessary to perform an audit inspection must be recognized. The audit inspection needs to be timed with the original inspection so that the item to be audited is intercepted after it is inspected but before it is shipped or reworked. It also must be intercepted at such a place and time that any defective performance is still traceable to the inspector. Also, inspection tooling and inspection instructions must be available to both functions. The recognition for the necessity of this point coordination might cause these functions to be combined under the leadership of the same manager even though such combination would apparently violate the principle of separation.

F. Separation. The successful operation of businesses of all types requires that certain activities be undertaken purely as a check upon the effectiveness of other functions. It is not enough to assign them solely on the basis of interest lest the very purpose of their creation be jeopardized. They must usually be carried out by a manager independent of the executive whose work is being evaluated. It is this principle that most frequently leads to establishing the quality control function separately from the production functions being controlled. Similar issues are involved in the employment of outside auditors and the separation of accounting and finance functions. The application of this principle would frequently lead to the assignment of corrective action responsibilities to the quality engineering department, particularly if the corrective action frequently resulted in discipline of the inspection process.

14.3 PERSONNEL

14.3.1 Introduction

The selection of personnel is an important element in the management of any business function. This is certainly true for the quality control function because of the broad scope of quality control work activity. The initial steps in any program of organization staffing involve the identification of the work to be performed by an individual and the definition of the specific skills and abilities that the individual must possess to perform the work properly. The latter activity is usually facilitated by the development of job specifications or job descriptions. These documents provide a systematic method of analyzing the jobs to be done and the education, experience, and special physical attributes required.

Quality control work activity involves a broad spectrum of job requirements ranging from the unskilled to the scientific classifications. A discussion of some of the more important considerations in the selection of individuals in these various classifications is provided below.

14.3.2 Labor

14.3.2.1 Unskilled Labor. Unskilled laborers are normally used in relatively small numbers for general utility jobs in order to increase the effectiveness of higher-paid employees. Examples of such jobs would be drawing samples of paints, chemicals, and other supplies upon receipt and delivering them to the test or inspection functions and transporting products from their place of manufacture to their place of inspection and from their place of inspection to storage or the next step in the production process. The significant point is that, normally, unskilled labor is used in the quality control organization as facilitating labor to provide services to higher-paid quality control personnel in order to increase their effectiveness upon the job.

Selection of unskilled labor is usually based upon the most rudimentary criteria. Some jobs obviously require men while others might require women. Some jobs require physical strength, others manual dexterity. In any event, the requirements of the job are usually quite obvious and the qualifications of the individual easily discerned.

14.3.2.2 Semiskilled Labor. Semiskilled jobs would include simple, highly repetitive tests and certain quickly learned physical inspections. This is an area where many organizations can economize on labor costs. For example, many organizations will have chemists or skilled technicians performing highly routine and repetitive tests on chemicals or materials being purchased or fabricated when, in fact, sufficient work exists to justify the employment of semiskilled labor to perform these operations at much less cost. Some excellent screening tests have been designed which are easily administered and highly sensitive to aptitude characteristics.*

14.3.2.3 Skilled Labor. Skilled labor is the backbone of the quality control organization in most enterprises producing complex hardware. This classification would include such personnel as machine parts inspectors, precision assembly inspectors, systems test inspectors, and functional test inspectors for complex products. Also, laboratory personnel performing repetitive tests of a complex nature or where the accept/reject decision requires experienced judgment would fall into this category. Notice that all of the jobs in the skilled category involve the use of tooling or equipment for which special training or experience is required (gage blocks, height gages, electronic test equipment, etc.). These skilled jobs usually will require the ability to make special setups, require a greater degree of judgment in making acceptance or rejection decisions, or require the ability to identify certain obscure modes of nonconformance.

The selection of skilled personnel almost always is based upon the dual criteria of training and experience, with experience normally bearing the heavier weight. This explains why the line of progression for these personnel is normally from semiskilled jobs, where they have the opportunity to gain experience, into the skilled jobs, based upon their accumulated experience. Training may come from different sources, depending upon the jobs in question. For example, in the past 10 to 15 years a major source of skilled electronics personnel has been personnel trained in electronics specialties by the military, whereas most skilled mechanical labor has been developed through an informal apprenticeship system. These jobs generally are not so concerned with an individual's ability to perform a simple inspection rapidly or continuously without fatigue but, rather, with the ability of an individual to perform a relatively complex and diverse range of inspection with skill and with minimum supervision.

14.3.3 Engineering and Scientific Personnel

When quality control employs scientific and engineering personnel, it is normally for the performance of quality laboratory or quality engineering fuuctions. Engineers in quality control will frequently be found performing such functions as the certification of the performance of complex special test equipment, solution of complex quality problems, design review and specification review, statistical analysis, and similar functions of applied engineering. Scientists from the fields of chemistry, physics, metallurgy, and metrology will be found mostly in laboratory analysis and process control functions. In these areas the term "quality control engineering" is frequently used. Is the quality control engineer a person trained in quality control engineering or is he a person trained in one of the basic engineering disciplines who has specialized and supplemented his training with quality control tools? It is the opinion of the authors that most applications require the latter. That is, they require a mechanical engineer with additional background in the area of statistical analysis, measurement, etc., or a graduate electronic engineer with additional training and experience in statistical analysis, testing, etc. In any event, an engineer or scientist should be selected primarily on the dual criteria of training and experience, with the heavier weight on the former.

* For an example of screening tests, see "Development and Validation of an Aptitude Test for Inspectors of Electronic Equipment" by Dr. Douglas H. Harris, *Journal of Industrial Psychology*, 1964, Vol. 2, pp. 29–35.

Note that the selection of both skilled labor and scientific and engineering personnel involves evaluation of training and experience, with one weighted heavier in favor of training and the other in favor of experience.

14.3.4 Supervisory and Management Personnel

14.3.4.1 Supervisory Personnel. Supervision, as it is discussed here, involves the elements of foremanship; that is, the ability of an individual to select, motivate, discipline, train, and correct employees in a manner that satisfies the objectives of the enterprise with maximum efficiency. Most of his job involves dealing with people. The decisions that he makes are short-range decisions, but ones which have long-range significance. His decisions on when to support his personnel and when to reverse their decisions; the quality standards that he holds; and the actions he takes in training, disciplining, hiring, firing, and promoting his people are mostly decisions made with a minimum of formal long-range planning. However, all these decisions have serious long-range consequences to the quality organization. A series of unwise decisions to "cooperate" with manufacturing results in the establishment of a poor quality standard that may not be perceived by management until great damage has been done. Similarly, a series of overly simplified decisions to "back his people up" may result in incurring unnecessary cost to the enterprise. Both these characteristics—cooperation and the courage to support his subordinates—are admirable, but a series of decisions excessively biased one way or the other can have serious long-range consequences to the company. Therefore, the selection of supervisory personnel requires the careful attention of quality control management.

Selection of supervisory personnel has almost always been based upon experience, and the authors are not prepared to cite a better basis than this. It is often pointed out that the tendency to promote the best employee to first-level supervision is not sound, recognizing that the best employee is not always the best supervisor. While this is true, it is also true that we rarely have any indication at all of his foremanship ability. Certainly the best employee is apt to be the most respected by his fellow employees for his skill. He is apt to be the one best qualified to forward the training and skill of the men in the group. The decision to promote the best employee to the first level of supervision, given no other evidence, is not so amateurish as it might appear and perhaps it will suffice until a better selection device comes along. Certainly this assumes that there are no serious defects in the man's personality or character. At the first level of supervision a man begins to learn the skills of foremanship. As supervisors are promoted up the ladder of supervision, the importance of the skill they personally possess in the area being supervised diminishes. A first-line supervisor over a machine shop inspection group should be an experienced machine shop inspector; however, a third- or fourth-level supervisor who is perhaps responsible for machine shop, sheet metal shop, assembly, and test departments is not required to be so highly skilled in all these areas.

14.3.4.2 Management Personnel. The functions of management involve making decisions regarding policy, product, markets, and the administration of the business that are essential to its survival and expansion. Notice that while every manager is a supervisor, his supervisory duties per se receive a minor portion of his attention. Most of his work is concerned with policies, quality standards, organization, long-range plans, and budgets. As a member of management, the decisions and judgments he makes are normally oriented to the needs of the total organization and tend to transcend strict functional boundaries.

A manager, as in many other positions, is selected on the basis of training and experience. Traditionally, managers have been promoted from the ranks of supervision. The trend in recent years, however, has modified this type of progression. Today management tends to be a professional specialty of its own, requiring less detailed knowledge of the work being performed and a more detailed knowledge of the problems, goals, and administration of the enterprise. As a result, an increasing number of managers are being taken from the administrative functions where these skills and understandings are more fully developed than in the straight line functions. This observation is made not to suggest that the reader select personnel for management positions from administrative staff assignments, but to encourage managers to rotate key line supervisors into staff positions, and vice versa. In this way each can broaden his experience and training so that personnel of sufficient depth and scope will be available to meet all the demands incumbent upon future management.

14.4 TRAINING

14.4.1 Training and Labor Positions

14.4.1.1 Importance of Job Training. Job training is a very important aspect of labor management in any firm. As the size of the firm or complexity of its product increases, the nature of the training may change, but it exists in all firms and its importance cannot be overstated. Job training will usually take one of three forms or a combination of these, depending upon need and availability of training resources.

14.4.1.2 On-the-Job Training. On-the-job training (OJT) is the most common form of job training and is present to some degree in all jobs, regardless of skill level. As the name implies, it is conducted on the job, usually by a supervisor or senior employee. OJT will vary from familiarization with company products and procedures to detailed "how-to" instructions. OJT is most applicable to unskilled labor since it is usually the only form of training given this level of labor. Moreover, since the unskilled laborer, by definition, brings no skill to the job, any that is required must be learned on the job.

14.4.1.3 Instruction Manuals. Instruction manuals are one form of training and are particularly important if there is a wide range of employees, usually unskilled or semiskilled performing jobs or if the labor force has a relatively large turnover rate. Instruction (or "how-to") manuals usually represent a significant investment to develop and publish but, once published, are usually less expensive than 100 per cent supervisory conducted OJT. For this reason OJT is preferred if few workers are involved or labor turnover (and hence replacement training requirements) is low. Examples of instruction manuals would include equipment operation manuals, "how to" read a micrometer, "how to" conduct specific lab tests, and similar highly detailed, step-by-step technical procedures. It should be remembered that equipment operation manuals can usually be secured from the equipment manufacturer at a nominal cost.

14.4.1.4 Industrial Training. This form of training is particularly applicable to the employee who wishes to qualify himself on his own time for advancement or to the firm which finds it desirable to provide such training on company time. Off-the-job industrial training can usually be secured through local junior college classes, evening adult education programs, or home training courses. Such training is generally available today in a full range of skills from vocational to academic, including, for example, machine tool operation, electronics, computer programing, statistics, reliability, and supervisory techniques.

Many companies, by virtue of their own personnel policies or rapidly changing employee skill requirements, find it expedient to provide such training on company time and premises, using company-paid instructors. There are a number of reasons to recommend such a program in spite of its obvious expense. For instance, there may not be sufficient numbers of trained people in the labor market to satisfy the need, and training may be less expensive than recruiting. It may be desirable to provide specific product familiarization which obviously cannot be secured from public sources. Company-sponsored training increases the opportunity to promote from within the firm and enhances employee morale and company prestige.

14.4.2 Training for Engineering and Scientific Employees

Formal supplementary training for engineering and scientific personnel almost always takes the form of graduate work or academic study in a particular specialty. Such training is secured through local colleges and universities. If company training facilities exist, a significant portion will usually be devoted to informal supplementary training, particularly for engineers. Such training may include product familiarization on the engineering level and seminars on areas of particular interest. Another source of supplementary training not to be overlooked is the wide range of professional societies. Their seminars, conferences, and journals provide a wealth of up-to-date "state-of-the-art" information.

14.4.3 Training for Supervisory and Management Personnel

14.4.3.1 Supervisory Training. There is an obscure but definite dividing line between the vocation of supervision and the profession of management. This dividing line becomes more tangible when the sources and nature of specialized training are considered. Most supervisory training is presented as vocational training and as such is available through vocational training sources; i.e., high school and junior college adult education divisions. This training usually includes such courses as elements of supervision, employee counseling and motivation, applied employee relations, etc. Company courses in the same general areas are common and are frequently taught by a qualified management staff member if a formal training department does not exist.

14.4.3.2 Management Training. Management training is offered as a formal college or university major, and this is the source of most such education. Areas of specialty such as finance, accounting, industrial management, or employee relations are usually offered within the major. A broad range of courses is offered in the areas of personnel, accounting, marketing, production control, statistics, and industrial engineering which can be taken with a degree objective or as supplementary training for personnel who are emerging from the ranks of practicing engineers or foremen and moving into the field of management.

14.5 QUALITY SYSTEMS AND PROCEDURES

14.5.1 Requirement for Systems and Procedures

One of the most important prerequisites for operating an effective quality control program is establishing a formal procedural structure. Today, industrial firms of even moderate size are complex organizations with a high degree of specialization of labor and function. In such a situation, the success of the firm depends not only on the competence of the individual employees but also upon the skill with which their individual activities are linked and integrated to form an efficient, well-disciplined program. It must be recognized that quality control is ultimately a management system, not a production system. It is the control device that management depends upon to accurately sense the quality of the product, measure it against a standard, report variations, initiate corrective action, and, if appropriate, withhold acceptance of the product. Any process performing such a crucial function must be well disciplined in its internal processes if its decisions are to be taken at face value (Sec. 1).

14.5.2 Systems and Procedures Defined

Procedures are the means by which all repetitive business action is initiated, carried forward, controlled, and stopped. Accepting this, a system is a network of procedures developed according to an integrated scheme for performing a major activity of the business.* Each of these terms exists in a business sense as well as a technical sense. In this section we are concerned with the quality control business systems and procedures.

14.5.3 System and Procedure Analysis

The most formidable job facing most managements with respect to procedures is their maintenance and use. Large organizations tend to be exceptionally dynamic. Departments and functions are created, separated, combined, assigned, and eliminated. Such organizational motion has a profound effect on procedures because procedures are an intrinsic part of organization. Procedures occupy a key role in the organization in that they define authorities, responsibilities, and communication and, as such, convert a potential organization into a kinetic organization. Maintaining procedurally defined relationships in consonance with a constantly changing organization chart is a formidable task at best, requiring constant analysis to assure that the necessary and sufficient relationships are maintained.

The process of system or procedures analysis requires that each system or procedure be flow charted in order to strip it of its language and reduce it to a diagram of series, parallel, and series-parallel activities in a process flow format. Any procedure or system can be viewed as a processor with input(s), output(s), control(s), and feedback, as shown in Fig. 14.1.

Space will not permit an exhaustive treatment of the analysis function.† Suffice it to say that

---

* For a detailed discussion of these definitions, see Richard F. Neuschel, *Management by System* (New York: McGraw-Hill Book Company, 1960).

† For further information on the subject of system analysis, see *System Analysis for Business Management* by Stanford L. Optner, published by Prentice-Hall, 1960.

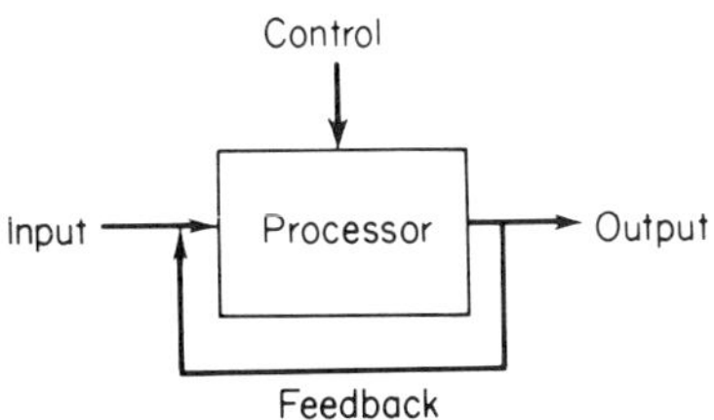

Fig. 14.1 *System or procedure generalized.*

this technique is valuable in detecting a number of very common system or procedural defects which can easily contribute to increased costs and lessened control. Examples of these defects would be:

A. Procedures may be found with an input and a complex series of activites, but no output.

B. Procedures may be found that are circular in nature with no input and no output.

C. Procedures may be found which require an input, but no corollary procedures exist to provide this input.

D. Procedures may exist which have an output, but no procedures or system exists to operate on this output.

E. Procedures may exist consisting of a series of events wherein the chain is broken by virtue of no linkage between the events which must act in series.

14.5.4 The Quality Control Manual

14.5.4.1 Size and Complexity of the Quality Control Manual. The procedures making up the quality control system are normally organized into a manual separate from other company instructional material. Size of the quality manual is immaterial in and of itself. One misconception regarding quality manuals is "the larger the better" on the basis that a larger manual is more detailed and, as a result, superior.

Another misconception is the one which presumes that the complexity of the manual is proportional to the complexity of the product. This fallacy is more difficult to detect; but one must realize that some very complex products have been manufactured by small "garage shop" operations with no procedures at all. In reality, manual complexity is largely dictated by organization complexity. If your organization is simple and relatively straightforward, your manual is likely to be the same. If your organization is complex, your manual will have to be the same if it is to execute its basic function of defining authorities and responsibilities and establishing necessary coordinating links.

14.5.4.2 Criteria for the Manual Structure. The organization of the manual should reflect the quality system and be easy to comprehend, easy to manage, and easy to communicate to others (supervision, employees, customers, and management). It should harmonize with the other company business systems. The manual structure should provide discipline at the procedure level since procedures sometimes tend to grow like Topsy without regard to the needs of the total system. The manual structure must provide for the future development of the system. Occasionally systems are so rigidly structured that necessity for expansion results in complete redesign and reissuance of the manual. Finally, the manual structure must recognize that few, if any, of the holders are concerned with the entire manual and should be designed so as to avoid the necessity for each holder to be intimately familiar with all sections.

14.6 QUALITY COSTS

14.6.1 Introduction

It is an axiom of business that in the long run all costs must be recovered in the price if a firm is to succeed. Certainly a firm can operate at a loss on an item or for a time, but eventually it must arrive at a position of charging a price that covers all costs. Simultaneously, another force is applied to the firm and this is the force of competition. The firm must provide a product of comparable quality at an equal or lower price than its competitors. Within this narrow band, bounded on the one side by the cost of operations and on the other by the price dictated by competition, the firm must operate and, if possible, attain an advantage over its competitors. To gain a competitive advantage, the firm must provide either a more desirable product at the same price or an equally desirable product at a lower price. However, either of these alternatives usually resolves itself into a problem of lowering production costs, either to allow a lowering of price or to provide the opportunity to increase the quality of the product through more expensive materials, more elaborate design, or increased performance.

14.6.2 Classes of the Firm's Cost

14.6.2.1 Overhead Costs. A tight rein is maintained by most firms on overhead costs. These include the relatively fixed costs of supervision, clerical, financial, janitorial, maintenance, and facilities (light, heat, rent, electricity, etc.). Costs can frequently be reduced in this area, but the probability of reducing costs sufficiently to provide a competitive advantage to an otherwise competitive firm is not too high.

14.6.2.2 Design Costs. These are critical costs for they establish the base cost of the product. The design defines the materials to be used, number of parts, and tolerances to be met which establish the theoretical minimum labor and material costs of the product. However, once the design is fixed, the theoretical costs remain relatively stable and do not normally require continuous monitoring. Quality control can make a significant contribution to establishing minimum design cost by participating in design review and value engineering programs during the design phase (see Par. 14.9.5.2).

14.6.2.3 Production Costs. These are the variable costs of factory operations, essentially direct labor and material. The differential between the theoretical design costs and the actual production costs is largely caused by inefficiencies in the utilization of labor, and many of these can be identified as quality costs and losses. Certainly the "learning curve" and "normal production scrap" explain a part of this differential, but this sum can be easily overshadowed by quality costs and losses in the factory.

14.6.3 Quality Costs and Losses

Quality costs are the total of all costs incurred to assure the production of an acceptable product. Quality costs include costs which are sometimes classed as "prevention, detection, and corrective action." They are the costs incurred in "striving for" perfect production.

Quality losses are simply the costs incurred because of failure to achieve perfection in production. This class of costs includes such items as scrap and rework due to defective material or labor, customer returns, and warranty costs.

Some managers are preoccupied with scrap and rework costs while others work to reduce the cost of quality control. The significant point is that neither of these costs has meaning without reference to the other. Obviously the cheapest quality control organization is no quality control organization at all. Similarly, the lowest scrap and rework costs are none at all. But no reasonable manager would pursue either of these costs to the ultimate. It has been demonstrated that these costs tend to reciprocate: to minimize one is to maximize the other. Clearly, the objective is to minimize the total of the two (see Fig. 14.2).

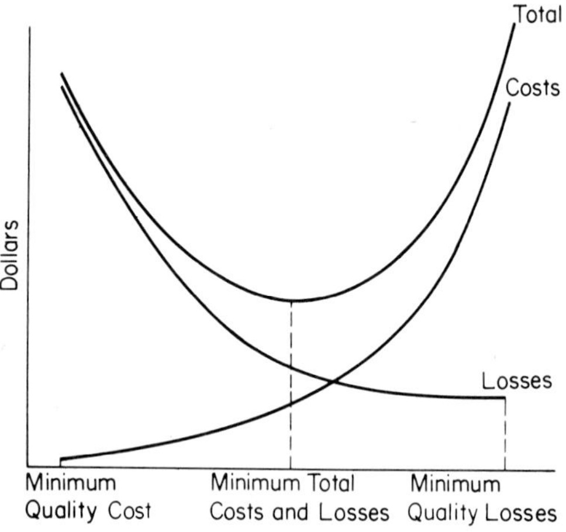

Fig. 14.2 *Quality cost model.*

14.6.4 Implementation

Many firms fail to implement formal quality cost systems because of the difficulty of defining the classical cost curve shown in Fig. 14.2. It must be recognized that failure to define the classical supply and demand curve did not prevent the firm from going into business. The firm operates in many areas on an empirical basis and can also operate on an empirical basis in the area of quality costs.

A very effective empirical approach to quality costs can be made by a series of successive approximations with management in relentless pursuit of minimum quality costs. As an example, assume a department produces functional subassemblies and operates with an inspection force of X per cent of manufacturing, costing $1,200 per unit time. However, malfunctions in the next assembly result in quality losses of $2,500 per unit time in replacement and rework costs. A decision could be made to increase inspection cost to $1,500 and "see what happens." If losses drop to $1,800, it was clearly a good decision. Perhaps it was not the best since another move might have reduced losses more, but clearly it was a good decision since it resulted in a net quality cost/loss reduction.

The manager might further assume that since a $300 increase in cost produced a $700 decrease in losses, more gold may be in the mine, and he will apply further iterations. Once interdepartmental costs are minimized, he might attack intradepartmental cost, substituting process control for some of the physical inspection and testing the result. In such a manner he might continually test for the minimum of the total cost curve much as a man might test for the bottom of a snowbank with a broom handle, never seeing the lay of the land but nevertheless effectively mapping it.

14.6.5 Accounting for Quality Costs and Losses

Even the most rudimentary quality cost management systems require knowledge of actual costs and losses if decisions are to be made. Many accounting systems fail to provide necessary data for management of quality costs and losses because they fail to recognize certain management requirements. The specific cost elements being managed must be clearly and separately identified, including, for example, the costs of receiving inspection, product inspection, functional test, failure analysis, corrective action, scrap, rework, and warranty actions.

Each element must be identifiable to a specific manager if responsive cost management is to be expected. Acute management of a cost element cannot reasonably be expected if several functions contribute to it and their contributions cannot be identified. Costs that are everyone's business tend to become no one's business.

Each quality cost element must be identifiable as a cost or a loss. These can be viewed separately or summed at every level of the organization at which quality cost decisions are to be made. The separate classes of "quality costs" and "quality losses" can each be subdivided into its component functional parts; for example, quality control, manufacturing, purchasing, and engineering. The next level could, for example, subdivide quality control into product inspection, quality engineering, procurement quality, and quality program management, if each of these is headed by a separate manager. The next lower level could divide functional elements into activity elements. In this way quality costs can be made to flow together from activities into subfunctional increments, into related functional increments, into quality cost classes, and finally into total quality costs. This results in accounts at each level being identifiable as to activity and responsible manager, allowing the functionally responsible member of management to trace his costs up and down for analysis purposes.

14.7 QUALITY MOTIVATION

14.7.1 Introduction

Prior to industrialization, skilled crafts existed in an environment that made special attention to the subject of motivation unnecessary. Workers went through a careful period of apprenticeship, became journeymen, and the way was clear to become master craftsmen and eventually own their own enterprises. Workers were responsible for complete units of production and, if they were so inclined, had something tangible in which to take pride.

With the coming of the Industrial Revolution, labor became highly specialized. Work that was previously done by individuals was subdivided and spread among many individuals. With mass production came the need for standardized parts and, in turn, products engineered in minute detail, further constraining the contribution of people. Eventually many industries arrived at a situation where labor was no longer considered to be the producer but simply another means of production. Workers were used to feed and tend the machines and in some ways were assumed to be inferior to the machines.

But with all the industrialization that has taken place, the role of the individual has never been more important than it is today. While much of our production is automated, the bulk of it still has a high proportion of labor content. Competitive pressures are enormous, and labor can contribute greatly to the success of the business. In order to secure the finest contribution from labor, each person must be made to realize and appreciate the importance of his own role in the production system and must be given something tangible in which he can take pride. Recognizing these needs, many companies have established formal motivation programs.

14.7.2 Elements of a Motivation Program

There are three essential elements in any formal employee motivation program if it is to succeed, and all must be present: inspiration, measurement, and communication.

14.7.2.1 Inspiration. The most frequent mistake made in well-meaning employee motivation programs is that of centering the entire program about the element of inspiration. People can be inspired to great heights by the sheer power of emotion, but the benefits of such are short-lived. To merely attain the same peak the second time requires an even more inspiring appeal. The law of diminishing returns sets in very quickly, and programs based upon this appeal alone are doomed to be flash-in-the-pan successes. Inspiration has the greatest value early in the program to get the ball rolling. From that point forward, primary dependence for program success should be placed on the tools of measurement and communication.

Occasional inspirational programs are very valuable to indoctrinate new employees and revive the enthusiasm of old employees. It is important for the employee to realize that the appeal is directed to him as an individual, a valuable and essential member of the production team. It should give him status and recognition in the eyes of his family, his community, and above all, himself. Signs, posters, and pass-outs at the plant gate all have their place in a total program, but too frequently they do not give the employee status and do not recognize him as an individual. Some techniques that have been found to be extremely valuable include:

A. Employee gatherings where management and distinguished members of industry and/or community speak face-to-face with the employees.

B. Closed-circuit TV broadcasts into the factory have been used successfully by some large companies. If these can be made relatively spontaneous and be used by management to announce successes and achievements in which the entire production team can take pride, so much the better.

C. Community and civic activities in which management and labor alike share in a common face to the community raise the status of the employee and give him recognition in the eyes of the community.

D. Executive plant tours are extremely valuable, particularly in large organizations where the employee rarely has the opportunity to meet executives on a face-to-face basis. Such tours should involve one or two executives touring a production area, talking with individuals, and getting to know them and their jobs. The key point in these tours is that the executives must talk *with* the employee, not *to* the employee. Care should be taken to avoid the insulating effect created by the presence of department heads and local supervision in the touring group. This is sometimes done by having the executives meet first and separately with supervision, then allowing the executive to tour the factory areas alone.

E. Mail-outs to the home are valuable to an employee motivation program, provided they are done well. A high-quality Christmas card with a thoughtful and appropriate message, key chains, calendars, and similar items have great value when mailed to the employee or his family. These must be sent for a reason and must be accompanied by a well-prepared letter explaining that reason to assure that they are received as a thoughtful gesture and not as a promotional gimmick.

The importance of the inspirational aspects of the program cannot be overstated, but again the program cannot depend solely on their use.

14.7.2.2 Measurement. Measurement is perhaps the most difficult and important aspect of any employee motivation program. Remember, the employee must be given something tangible in which he can take pride. The silversmith had completed units of production that he could appreciate and that would bear his hallmark. He could further sense and enjoy the appreciation of customers who sought out his work. The stonecutter had the advantage of being able to see the amount and the quality of his production at the end of each day and compare it to that of his neighbors. The employee similarly must be given something tangible in which he can take pride.

The unit of measure that is used should be designed to give weight to that which is important in the job. The index should be designed to be as fair as possible so that an employee will not feel that he is victimized by an index that leaves him at a disadvantage. If necessary, several indexes should be established so that an appropriate one will be available for each group of employees. Quality measures that have been used to good advantage include number of shifts of defect-free performance, mean production hours between defects, and consecutive number of defect-free units.

People are natural competitors. Given the opportunity they will compete with each other and, if no one else is available, they will compete with themselves, provided they are given an equitable measure of success with which to gage the competition. Informal competitions are the best, and such competitions have been spontaneously generated between departments, between shifts of the same department, between gangs or crews on the same shift, and between members of the same gang or crew. Sometimes all these competitions take place at the same time.

14.7.2.3 Communication. This is the final link in the motivation cycle. The employee has been inspired, he has been measured, and now the results of that measurement must be communicated to him in a significant manner. This is an extremely crucial aspect of the program. At this point a mediocre program can be made good or an excellently conceived program can result in destruction of morale, the loss of superior employees, and long-lasting labor problems. The key point is the spirit of the communication. The employee must be told how he is doing, but he must be told in a positive, friendly, and encouraging manner. Motivation must be separated from compensation. Most successful programs have a common set of characteristics regarding communication:

A. They tell the employee (or the gang, shift, or department) how well he is doing, not how badly he is doing.

B. They concentrate on successes, improvements, and people.

C. The motivation program is kept clearly separate from the disciplining program. Granted, there is a fine line of difference in that both functions tend to use the same data or information; however, it must be recognized that if an employee is disciplined for failure to achieve a standard, you cannot expect him to achieve that standard out of pride or craftsmanship. He is pursuing that standard out of fear of discipline. The successful programs scrupulously avoid disciplining the employee with the same index with which they motivate him. For example, if his defect rate is high because of his propensity to produce a particular defect, he should be counseled or trained with respect to that type of defect. In this manner, while both programs are using the same information, there is a clear and legitimate separation of the two programs in the minds of management and employee alike.

To secure the maximum benefit, communication of performance should take a public form. This can be done under certain circumstances by posting performance figures on a department bulletin board. In other cases the maintenance of performance charts at a leadman's bench or departmental wall charts may be more appropriate. The important thing is to keep a score and publish that score to the employee and his peer group in a friendly and constructive way. Someone once said regarding the game of baseball that he was not really sure whether batting averages are kept publicly because they are important or important because they are kept publicly.

14.7.3 Motivation and Quality Control

Why is the subject of motivation programs mentioned in a quality control section? By and large, the performance being measured is that of production people, not that of quality people, although they, too, can and should be motivated. Such motivation programs are and should be of key concern to the quality executive. Remember that people make or allow the mistakes that degrade quality that escape to the customer that cost money to prevent, detect, and correct. Machines cannot be blamed for mistakes. A strongly motivated employee feels his errors more personally and works harder to eliminate them. The highest level of operation for a quality program is the one at which active measures are taken to prevent quality problems. One of the strongest tools in such a prevention program is an intelligently conceived and professionally executed employee motivation program.

14.8 QUALITY AUDIT

Most medium to large quality organizations contain an audit function. The purpose of audit is to provide additional discipline over the internal processes of the quality function and validate the findings of product inspection upon which many management decisions are made. A complete audit function will consist of three separate parts: systems and procedures conformance audit, systems and procedures effectiveness audit, and product audit. Although the three audits together comprise a total audit activity, the first two are distinctly different from the third in regard to skills required and processes used.

14.8.1 Systems and Procedures Conformance Audit

The purpose of this audit is to assure that the systems and procedures governing the activities of quality functions are, in fact, being followed. Frequently, new procedures are developed to meet operating needs when existing procedures would suffice if they were being followed. The conformance audit is appropriate to organizations of any size; however, the formality with which the audits are conducted will increase with the size of the organization. In a small organization audits may be performed by a member of management on a basis so informal that no one, including the manager, realizes that they are in fact an audit. In a medium-sized organization the function will usually be performed by a full-time auditor reporting to a member of management. In a very large organization there will frequently be a full-time audit staff complete with its own supervision, management procedures, and management reporting systems.

14.8.2 Systems and Procedures Effectiveness Audit

This form of audit is normally found in larger companies. Assuming that systems and procedures exist and that conformance audits have provided reasonable assurance that they are being followed, the question will remain as to how effective the systems are in accomplishing their main purpose. For example, a conformance audit would determine whether personnel are completing appropriate corrective action forms when required by procedure and whether they are filling them out correctly and forwarding them to the proper parties. An effectiveness audit then would determine the effectiveness of the system in identifying conditions requiring corrective action and securing such action in a timely and effective manner. To maintain absolute objectivity, conformance and effectiveness audits are frequently conducted by completely separate and distinct groups.

14.8.3 Product Audit

Product audit can be one of the most powerful tools in the quality manager's program. It can provide him with a valid measure of the effectiveness of his inspectors, the true quality of the outgoing product, a true estimate of the quality level being submitted by manufacturing, and a tool for identification of troublesome product designs. A product audit should be conducted by senior inspectors, carefully selected for their ability and objectivity. Product audits should be conducted after the inspector has completed his inspection and before the item has been further processed or reworked by manufacturing. The item should be reinspected, using the same or equivalent tooling, blueprints, and other aids. If sampling was used on the production line, the auditor must reinspect the actual parts that were inspected by the original inspector. The identity of the original inspector must be maintained until after the audit so that any defects found by the auditor, and not by the original inspector, can be traced to that inspector.

A word on the statistical treatment of product audit findings is in order. Like any other manager, the quality manager is constantly manipulating a combination of resources, each of which is relatively fixed. He attempts to manipulate these resources in such a manner as to maximize their total effectiveness. This is a "game" that is impossible to play effectively on a basis of pure hunch, suspicion, or judgment. It requires information. The quality of the product leaving the plant is a result of two factors: first, the quality level being submitted to inspection by manufacturing; and second, the effectiveness of inspection in detecting defects. For example, if the submitted quality level to inspection is excellent, say 99 per cent acceptable, the outgoing quality level is excellent even if inspection never detects a defect. Similarly, if the submitted quality level is very poor, say 25 per cent acceptable, a significant number of escapes will occur even if the inspection effectiveness is excellent. For the middle ranges of effectiveness and submitted quality level, each has the ability to offset the effect of the other to a very large degree.

14.8.3.1 Submitted Quality Level (SQL). Where no audit is conducted, SQL can be assumed to be represented by the statistic:

$$\frac{D}{N}$$

where D = defects detected by inspection
N = submittals inspected by inspection

This presumes that the inspector is 100 per cent effective in determining defects, which is categorically incorrect. If an audit is conducted, we can represent the SQL as:

$$\frac{D\left(\frac{n}{N}\right)}{n} + d$$

where d = defects detected by the auditor
n = submittals inspected by the auditor

This formula contains the assumption that the auditor is 100 per cent effective. Although this is also incorrect, it can be assumed that the auditor, being of the highest caliber available and working without production pressure, will approach perfection sufficiently closely to allow this treatment of the statistics.

14.8.3.2 Effectiveness. The term "effectiveness" as used here is the ability of the inspector to detect defects. It can be computed only if audit results are available, using the following formula:

$$\frac{D\left(\frac{n}{N}\right)}{D\left(\frac{n}{N}\right) + d}$$

This, in essence, is the percentage of total defects detected by the inspector.

14.8.3.3 Estimated Outgoing Quality Level (EOQL). This statistic, measuring the escape of defects from the inspection process, can be represented as d/n. This assumes that the auditor is 100 per cent effective, as do the other calculations, a technically incorrect but usually practical assumption.

14.8.3.4 Confidence. Statements regarding inspection effectiveness must be "taken with a grain of salt." That grain of salt is the confidence that we have in the statement by virtue of its statistical origin. If we make measurements for each inspector in the department and sum these to arrive at departmental inspection effectiveness, we probably have a relatively high confidence in the department statement by virtue of the merging of sample sizes. Our confidence will obviously be less for any individual inspector by virtue of the relatively small sample size. Confidence may be represented as $D(n/N) + d$. The confidence afforded by a factor of a particular size can be noted and this knowledge used to temper action taken on the basis of the effectiveness statistic.

14.8.3.5 Use of Product Audit Data. Use of the above statistics will remove much of the guesswork from quality resource management. SQL's can be measured and used to identify highly defective processes requiring greater attention by manufacturing, engineering, or quality control. Extremely good processes can be identified, and the inspection controls can be lessened appropriately. Inspection effectiveness can be measured, and if it is low for a particular operation, quality engineering can be applied to the problem to provide improved instructions, tools, etc. Individual inspectors, if ineffective, can be identified and provided with additional training. The task of compiling all these statistics and making the calculations, no matter how simple, is a formidable task and one ideally suited for the computer if one is available. If not available, there is a significant amount of clerical work involved, but the work will be of extreme value in the long run.

14.8.4 Organizing for Audit

Ideally, the audit functions should be organizationally separated from the groups being audited in recognition of the principle of separation (see Art. 14.2).

Effectiveness audits are frequently conducted by a staff reporting to general management. Such audits are concerned not just with quality systems and procedures, but with all company systems and procedures. This audit recognizes that conformance is primarily an *intrafunctional* discipline while the measurement of effectiveness requires *interfunctional* measurements and judgments to be performed. If no interfunctional effectiveness audit operation exists, another good approach is to form individual ad hoc teams of specialists from line functions to perform the audits. This approach avoids the necessity of maintaining a staff of specialists and allows the organization to draw upon the most talented personnel available, regardless of their normal line function. Caution must be taken to assure the objectivity of such teams, however, since their normal line assignment may bias their judgment.

The product audit function is usually a part of the quality engineering function, although it may report to the chief inspector. It should be staffed by inspectors carefully selected for their objectivity, experience, and ability. Regardless of the organization to which they are assigned, they should not report to the supervisor of the group being audited.

14.8.5 Conducting the Audit

A formal routine must be maintained for conducting any audit if its results are to be taken at face value. Audits should be conducted on a random, unannounced basis, or at least sufficiently unannounced to prevent local personnel from "scrubbing up" operations for the benefit of the auditor. Audits should be conducted using written check lists traceable to documented requirements. Check-list items based upon interpretation may be included for the purpose of gathering information, but they cannot be used for determining conformance. It must be assumed that any procedure, blueprint, or specification means "what it says, all of what it says, and nothing but what it says."

Effectiveness audits frequently involve two or more people and often cover weeks of time. Prior to beginning these audits there should be a definite, written, and agreed-upon objective. Effectiveness audit groups can spend a great amount of time pursuing what they think is the objective of a survey when, in fact, they misinterpreted it, or perhaps even management themselves did not have a clear view of what they wanted to learn. After the agreed-upon statement of objective is documented, the audit team should develop a plan of action, specific personnel assignments, and a time schedule. This plan of action should have the concurrence of appropriate management and be published to the functional managers who are expected to work with the audit team members. The final report should be reviewed with the concerned managers and, if at all possible, agreement should be reached prior to publication.

The important thing to remember regarding product audit is that the audit is concerned with the quality of a product emanating from a production operation and the effectiveness of inspection in controlling that quality. Product audit is not a qualification test, design proof test, or any other form of engineering evaluation; therefore, the parts being audited must be randomly selected after inspection and reinspected by production-experienced auditors, using the same or identical tools and blueprints that the original inspector used. If the auditor is measuring a part with a height gage that the inspector's instructions tell him to measure with a 6-inch scale, the auditor's results cannot be considered as valid criticism of the inspector's effectiveness.

14.8.6 Reporting Audit Results

If management is to take action, audit results must be timely, understandable, objective, and constructive.

Reporting of systems and procedures conformance audits should begin at the time the audit is conducted. Upon entering the department, the auditor should announce his presence to a member of local supervision. This is a must. Hopefully, the local supervisor will be free to accompany the auditor, and, if so, this will actually be the first level of reporting. If sufficiently definitive check sheets are used, the auditor will be able to leave a copy of the audit deficiencies with the supervisor as he leaves the department.

Procedures should exist requiring corrective action by local supervision and response to the auditor within a defined period of time. Upon receipt of notice of corrective action, the auditor should schedule a reaudit in order to measure the effectiveness of the corrective action. Corrective action reporting and follow-up reaudit are an absolute must if audit is to operate as a closed system and yield tangible results.

In conducting effectiveness audits, the audit team must formally review a preliminary draft of the final report with the management being audited. If at all possible, agreement should be reached prior to publication of the final report. If agreement cannot be reached, line management must be given their "day in court" but cannot be allowed to impose their will upon the audit team. This "day in court" may involve a meeting with upper management at which time line management may present their case or a minority report may be included in the final issue. The final report may or may not include recommendations; however, procedures must require that, at some specified time in the reporting and follow-up process, recommendations for corrective action are made and scheduled completion dates are assigned. The audit staff should be responsible for follow-up on the resulting actions to assure that they occur as scheduled and that the actions are effective in solving the problem.

Product audit is an on-going process and as such does not lend itself to "final reports" as described in the previous paragraphs. Rather, these statistics are accumulated, computations made, and reports published periodically—weekly, monthly, or quarterly. They must be available to those members of management in manufacturing and quality control who are responsible for the training of inspectors, the management of submitted quality level, and the allocation of quality resources.

14.8.7 Staffing for Audit

An effective quality audit system is one of the most powerful tools available to the quality manager. Paradoxically, most companies which do not operate such functions state that they cannot afford the cost, while most companies which do operate such systems state that they cannot afford the costs that would result from giving them up. Audit staffs need not take the form of layering on of personnel. For example, many companies that desire a product audit staff of 10 per cent of the line inspection force do not hire 10 per cent more inspectors, but instead segregate a fraction of the inspectors already available. In this manner the staff is not increased, but rather personnel resources have been reallocated.

14.9 CONTROL OF ENGINEERING QUALITY

14.9.1 Engineering's Role in Quality Creation

In most industrial enterprises the engineering organization has the basic responsibility for the initial steps in creating quality; viz., establishing quality objectives, selection of specific characteristics to achieve these objectives, and the development and publication of documentation which defines the characteristics that have been selected. These functions provide the foundation upon which most of the subsequent actions are taken by other elements of the organization. Assurance that these functions have been performed effectively is essential. An error in this stage of quality creation usually results in subsequent error during purchasing, fabrication, and test of the product, thereby amplifying the consequence of the error.

14.9.2 Establishing Quality Objectives

The design and development of a product usually start with the identification of a customer need and a general concept of how this need might be satisfied. Government organizations and large industrial customers frequently define their own needs and outline in a general specification the product performance that is required. It is then the responsibility of the manufacturer to create the detailed design which will result in the achievement of the performance that has been specified. The customer may also include with the requirements detail specifications which restrict the designer to specific design techniques or processing methods.

The quality control organization frequently is assigned the task of comparing customer detail specification requirements with internal process specifications and practices in order to determine differences. These differences may be eliminated through negotiation with the customer if the company can demonstrate that its internal specifications and standards are equal to or better than those requested by the customer. If the customer requirement must be introduced into the engineering system, care must be exercised to assure that appropriate information is disseminated throughout the organization. The quality control organization is assigned responsibility in this area for two reasons: an intimate familiarity with detailed company standards and practices is inherent within quality control; and secondly, quality control must conduct an analysis prior to first item acceptance so that check lists for first article inspection can be developed.

In producing consumer goods, the manufacturer is frequently faced with the problem of determining consumer needs and establishing basic concepts of how these needs may be fulfilled. Market research techniques provide a valuable aid in measuring the customer's need and his probable reaction to new products. In like manner, evaluation of competitive products will also provide information upon which quality objectives can be based. Quality control engineers and inspectors frequently are utilized in conducting a detailed evaluation of a competitor's product.*

Regardless of the methods utilized, a clear understanding of the customer's need is a prerequisite to the development of design concepts and subsequent translation of these concepts to specific quality characteristics. There are many examples in industry of an outstanding design that satisfied the wrong need. There are also examples of designs that satisfied the basic need but failed to conform to some specifically stated customer requirement. It is extremely important that both the basic requirements and the detailed requirements be clearly defined and that this information be disseminated to those responsible for quality creation and control.

14.9.3 Selection of Quality Characteristics

The process of selecting quality characteris-

* For additional information regarding evaluation of competitors' products to establish quality objectives, see J. M. Juran, *Quality Control Handbook*, 2nd ed., McGraw-Hill, 1962, pp. 1–16 to 1–19.

tics involves the entire spectrum of engineering activity and is beyond the scope of this section; however, a few comments with regard to controls that influence the selection process are appropriate.

Most companies develop design standards and associated documents which provide guidelines to the individual performing the detailed design. These documents serve two basic purposes: first, to accumulate information that helps the designer to avoid previously encountered pitfalls and, second, to promote standardization in the materials, components, and processes that the designer may select. The quality control organization can play a major role in providing information on quality problems that have been encountered in the past. Other information can help to improve the design standards and, hopefully, reduce the incidence of quality problems attributed to poor design. The quality control organization can also provide detailed information on machine and process capability, thereby providing the designer with improved criteria for selection of quality characteristics.

14.9.4 Specification of Quality

14.9.4.1 Method of Specification. Quality is specified by enumerating in detail the specific characteristics of the material to be used, the processes to be employed, and the physical and functional characteristics that must be present in the completed item. The term "design documentation" is generally used to refer to the various documents which contain this information. These documents fall into three basic categories; viz., specifications, drawings, and standards.

Specifications exist at various levels of product definition and are used as primary documents for defining many of the detailed product and process quality requirements. Within most industrial enterprises there are a number of different types of specifications that are utilized. These include end item product specifications, process specifications, material specifications, part specifications, test specifications, and packaging specifications.

Drawings are the most frequently used documents to describe the physical characteristics of the product. In addition to dimensional requirements, drawings usually reference material and process specifications, parts specifications, and other information necessary to build the product.

Standards are used primarily to outline specific criteria for product definition or standard methods and techniques that are used to evaluate conformance to requirements. Examples of standards include standard tests for material requirements, workmanship standards, standard sampling procedures, and related documents.

The above types of documents represent the basic foundation to which most future quality control actions relate. They also form the primary element for determining product acceptability.

14.9.4.2 Responsibility for Specification. In a complex industrial organization, many functions may contribute to the development of drawings, specifications, and standards. It is advisable, however, to have a single organization responsible for issuing and revising these documents, usually the engineering department. To assure that all segments of the organization have a voice in formulating design documentation, most companies have a standards committee and also provide for formal coordination of these documents prior to release.

When the design documentation has been agreed upon by all functions involved, it is officially issued as the "law" of the industrial community. If chaos and confusion are to be prevented, adherence to design documentation is essential. It is the responsibility of the quality control department to assure that the requirements of the drawings, specifications, and standards are enforced. If it becomes necessary to deviate from the documentation for any reason, the engineering organization should either change the documentation or authorize a temporary deviation by means of a controlled change mechanism. There must be only one "law." Serious consequences will result if the shop decides to follow its own set of rules instead of the authorized documents.

Since most organizations have a large number of drawings, specifications, and standards, and since the documents change with time, it is extremely important that the dissemination of this information be carefully controlled. Document control stations or "cribs" are usually established to provide this control mechanism. These stations are assigned the responsibility for maintaining accountability and control of all documentation. Quality control is often assigned the responsibility to make periodic audits of the "cribs" as well as documents in use in the shop to assure that the system is operating as specified.

14.9.4.3 Specification of Characteristics. Characteristics are usually specified in terms of a standard that can be objectively measured or verified. A linear dimension cannot simply be long. It can only be proportionally as long as something else, namely, a standard. Such terms as "pure," "smooth," "good," and similar terms have little meaning in the specification of quality requirements. It is surprising, however, how many times such phrases do actually occur through carelessness on the part of the designer. The quality control organization is perhaps more sensitive to this problem since it is faced with the decision to accept or reject the product and must squarely face the problem of acceptance criteria. Quality control participation in design review (see Par. 14.9.5.2) will assist in overcoming these problems at an early stage.

At the source of most measurement standards is the National Bureau of Standards (NBS), charged by Congress to provide the central base for a uniform system of physical measurement throughout the United States. NBS also serves as the focal point within the federal government for activities involving the relationship of the United States measurement system to those used in other countries.

Other standards may be established on an industry or company basis. For example, many workmanship standards, such as soldering criteria, may be established by individual companies based upon laboratory study and previous experience. These standards may be communicated through the use of visual aids or standard samples.

There are some product attributes that do not lend themselves to objective measurement; for example, there is no objective standard for taste in foods. In such cases, a standard may be determined on the basis of subjective measurement by a group of individuals, but subjective standards should be held to an absolute minimum.

14.9.4.4 Specification of Tolerances. It is practically impossible to manufacture one article exactly like another, or one batch like another. Variability is one of the fundamental factors influencing quality control methods. Tolerances on a dimension of a part tell the shop how small or large the part may be and still fit into an assembly so that it can function properly when it is assembled with the other parts. Tolerances are established not only for dimensions, but also for other quality characteristics, such as temperature, pressure, and volume.

The selection of tolerances is very important. They should not be set needlessly tight, since the cost of manufacture usually rises rapidly as the tolerance is decreased. On the other hand, the tolerances must be sufficient to assure interchangeable manufacture, where possible, and to assure proper functioning of the product. Engineers are able to set tolerances more accurately if they are assured that the factory processes will operate within statistical control and if process or machine capability studies are available. The quality control organization is usually the source of this information.

14.9.4.5 Tolerance Buildups. Unless dimensions of all component parts are near the upper limit of tolerance or the lower limit of tolerance, the assembly will have a variation that is less than the sum of the tolerances of the components. Use of this fact can be made either to increase the tolerances of the component parts, thereby reducing manufacturing cost, or to decrease the tolerance of the assembly to facilitate further assembly or improve quality.

The ideal case results where the dimensions of each individual part are normally distributed about its respective mean and where each distribution lies between its tolerance limits in such a way that the three sigma limits of the distribution correspond to the tolerance limits. In such an ideal case, and where the parts are randomly selected, the standard deviation of the sum of a number of independent parts is equal to the square root of the sum of the squares of the independent standard deviations. Expressed mathematically:

$$\sigma_A = \sqrt{\sigma_1^2 + \sigma_2^2 + \cdots + \sigma_n^2}$$

where σ_A = the standard deviation of the assembly

σ_1 = the standard deviation of the first component part

σ_n = the standard deviation of the nth component part.

Since the ideal condition described above is not consistently met, actual practice lies somewhere between that theoretical sum and the arithmetic sum of the deviations or tolerances.*

14.9.4.6 Classification of Characteristics. Since variations in product characteristics will affect end item performance to differing degrees, it is frequently desirable to classify characteristics with regard to their relative importance. Through such action, quality control planning and operations can be adjusted to provide a higher confidence level on attainment of the characteristics which are more critical to end item performance.

Classification should be performed by personnel most knowledgeable of the effect that product variation will have on end item performance. This is usually the organization which established the original requirements. It may be desirable to have classifications established by two functions, one familiar with the engineering relationships and the other familiar with the methods and techniques of classification. Classification programs vary widely as to general usage, classification methods, and documentation techniques.

14.9.4.7 Revision of Design Documentation. Changing conditions brought about by a variety of causes make it necessary to revise design documentation from time to time. Although requests for changes may be initiated by various functions, only one organization should be responsible for formal release of the changes, usually the engineering organization.

It is essential that rigid control be exercised in the change control system so that changes are introduced at the proper time and that everyone is working to the same documentation. This involves accurate identification of the changes and a system for assuring that obsolete documentation is withdrawn or that change information is attached to the documentation in *all* cases. This is frequently achieved through document cribs that maintain document accountability and recall or "track down" documents in use to assure that the changes are made. The importance of change control cannot be overstressed in assuring that the quality of the product is maintained.

* For a more detailed review of tolerance considerations, see J. M. Juran, *Quality Control Handbook*, 2nd ed., McGraw Hill, 1962, pp. 3–19 to 3–38.

14.9.5 Evaluation of Engineering Quality

14.9.5.1 Responsibility for Evaluation. Once the specific quality characteristics have been selected and specified, they are usually evaluated from an analytical point of view and from actual performance demonstration by various types of tests. This is usually an iterative process involving reviews and tests at various stages of product development. Although responsibility for this evaluation varies widely, it is frequently assigned to the engineering organization with assistance from other interested organizations. In some companies, the function is assigned to the quality control organization or to another function independent of engineering. When performed within the engineering organization, it is frequently assigned to a group which is independent of the one responsible for design. This is done to assure objectivity in the evaluation and to remove possible bias that may occur because of close personal involvement in the design.

14.9.5.2 Design Review. One of the techniques used to evaluate the adequacy of design is the design review. A team consisting of specialists who are knowledgeable in specialized technical fields or related activities upon which the design may have an important bearing is used to perform this function. The nature of the design will influence the team composition; however, the following is typical of the types of individuals utilized:

A. Stress analysts
B. Circuit analysts
C. Reliability specialists
D. Value engineers
E. Quality engineers
F. Production engineers
G. Maintenance engineers

The purpose of the review is to provide the designer with a constructive evaluation and suggestions for improving the design. Because of the complexity of modern science and technology, a single individual cannot possess all the knowledge and skills necessary to select the optimum characteristics for a product and the processes by which it is to be fabricated. If properly utilized, design review can assist the designer in moving closer toward the optimum. (See also Sec. 15.)

Quality control participation in the design review activity will include review of the following:

A. Have the requirements of detail customer specifications been considered and included?
B. Have the requirements of detailed company standards and practices been considered and included?
C. Is the design consistent with existing machine and process capabilities?
D. Does the design include potential quality problems that have been encountered previously or might reasonably be expected?

The design review provides an opportunity to detect and prevent potential problems early in the quality creation cycle when the impact of a design change will be minimal. It also provides the quality control organization with information at a time when effective planning can be initiated.*

14.9.5.3 Documentation Review. Documentation review may be performed as part of the design review activity or as a subsequent review prior to final release of the documentation. Since this documentation will be utilized by many individuals, it is important that it communicate the intent of the designer. Most companies establish design standards and practices that provide a common "language" for communicating engineering information. These documents establish standard methods for identifying characteristics, tolerances, and additional information required. The purpose of documentation review is to assure that the design documentation conforms to these standards and will, in fact, communicate the intent of the designer.

14.9.5.4 Design and Qualification Tests. Various types of design and qualification tests are normally performed to assure that the quality characteristics that have been selected will achieve the quality objectives. These tests usually include functional and environmental tests (vibration, shock, temperature, humidity, etc.) and may include various reliability life tests and demonstrations.

In many companies the engineering organization will fabricate breadboards or engineering models to be utilized in these tests. Inspection of these breadboards or models is desirable in order to clearly establish the characteristics; it is important that the "as tested" configuration be clearly defined and documented. Inspection is also desirable to assure that factors not directly related to concepts that are being evaluated are not, in fact, influencing the final evaluation results. For example, a designer may desire to evaluate a circuit design; however, standard parts used in the circuit may influence the results if they, in fact, do not fall within their prescribed parameters.

It is important to assure that the test equipment has the required characteristics and that it has been calibrated. Equally important is the assurance that the tests are performed in the specified manner and that the test results are accurately recorded. The quality control organization is frequently assigned responsibility for these actions.

14.10 CONTROL OF PURCHASED MATERIAL QUALITY

14.10.1 Purchasing's Role in Quality Creation

After the engineering organization has established the specific characteristics required for a product and has published these requirements in appropriate design documentation, it is necessary to acquire raw materials, parts, and components for further processing and assembly of the product. With few exceptions, industrial enterprises are dependent upon suppliers for these items. This purchasing function is usually assigned to an organization which is charged with the responsibility to acquire items that conform to the requirements specified, to obtain these items in time to support manufacturing schedules, and to procure the items at a minimum cost to the enterprise.

Since material, parts, and components will directly influence the characteristics of the final product, it is necessary to establish controls which provide reasonable assurance that the required characteristics are present in the purchased items. One of the major roles of the quality control organization is to provide this assurance.

14.10.2 Make-or-Buy Committees

Since many organizations have the ability to make, as well as buy, a particular item, it is

* For a more detailed discussion of design review methods, see R. M. Jacobs and H. D. Hulme, "Commercial Design Review and Data Analysis Program," *Nineteenth Annual Technical Conference Transactions*, American Society for Quality Control.

necessary to first establish the items that will be purchased. To assist in making this determination, many companies establish a committee composed of various interested functional groups. Quality control participation on this committee is primarily concerned with an evaluation of quality factors related to the make-or-buy decision. For example, what is the previous quality history of suppliers of this type of item? What quality problems might be anticipated for a supplier that might not be experienced within the buyer's organization because of specialized skills or equipment? Certainly these points must be evaluated as basic factors in the decision to make or buy a particular item, and there are others that will be revealed as committee action progresses.

14.10.3 Source Selection

The second step in acquiring raw material, parts, and components is to select suppliers who have the capability of furnishing the required items at the proper time and at a cost that is fair and equitable to the buyer and the seller. The participation of the quality control organization in source selection is principally directed toward the determination of the supplier's continuing ability to furnish items which meet the requirements of the design documentation; i.e., material, part, or component specifications. This determination can be made in several ways, which are discussed in the following paragraphs.

14.10.3.1 Supplier's Performance History. If the supplier has previously furnished items to the purchasing organization, a review of the supplier's previous quality history may provide the basis for determining the adequacy of the supplier's quality system. This method is good since the previous history is a record of demonstrated capability rather than potential capability. On the other hand, if the history is not current or if the product to be purchased is significantly different from products previously manufactured by that supplier, its previous history may not be valid.

14.10.3.2 Supplier Evaluation Questionnaire. A supplier evaluation questionnaire can provide information to assist in arriving at a decision on the acceptability of a supplier's quality system. This type of questionnaire may be used to supplement historical information about the supplier. It can also assist in planning or making a decision to conduct an on-site quality evaluation survey. Finally, it can be used to make a decision regarding the adequacy of a supplier's quality system when the articles involved can be evaluated by receiving inspection without extensive disassembly or destructive testing.

14.10.3.3 Supplier Quality Evaluation Survey. An on-site evaluation survey of the supplier's quality system is another method for determining the ability of the supplier to continually furnish acceptable items. The survey is generally used when the purchase involves the following:

A. Major items of equipment

B. Items that cannot be evaluated by receiving inspection without disassembly or destructive testing

C. A supplier whose performance history is incomplete, inconclusive, or below acceptable quality standards

D. Articles sensitive to quality degradation because of storage and handling conditions

Supplier surveys provide valuable information for determining the adequacy of the supplier's quality system, but surveys are also expensive and time-consuming. If a survey is determined to be the most effective method, it should be conducted in a straightforward, businesslike manner. Personnel conducting the surveys should be mature, experienced, and capable specialists in this field. Standard evaluation criteria and associated check lists should be utilized to make the survey as objective as possible. Findings of different surveyors should be periodically evaluated to detect overemphasis, underemphasis, or bias regarding particular criteria. Maintaining good vendor-vendee relationships depends very strongly upon the fairness and objectivity of the survey.

14.10.4 Procurement Document Review

Once a supplier has been selected, the normal method of establishing the responsibilities of the supplier and the purchaser is by means of a contract or purchase order. Specific quality requirements, schedule requirements, and prices for the items to be acquired are established in these documents or by reference to other documents. They represent the formal and legal communication channel between the two organizations. Correctness and completeness of information furnished to the supplier are fundamental in assuring a satisfactory relationship between the two organizations and in avoiding quality problems.

In many companies the quality control organization reviews each subcontract or purchase order, or those of a selected level of item complexity or cost, to assure that information pertaining to quality requirements in procurement documents is adequate. This review will include the following considerations:

A. Are the specifications accurate and complete?

B. Are the specifications and related documents sufficient? (In this regard, errors can occur when items previously manufactured internally are subcontracted without consideration of related information, procedures, standards, or other specifications which are utilized internally in conjunction with the referenced specification.)

C. Is source inspection planned and, if so, has provision been made for this requirement?

D. Does the procurement document reference special requirements for test and inspection procedures and equipment, qualification testing, deliverable data, first article inspection, packaging, identification, and similar needs?

In a complex industrial organization the above factors may be overlooked without a systematic method of checking the procurement documents. By checking early in the procurement cycle, many potential problems may be avoided to benefit both the supplier and the purchaser.

14.10.5 Post-Award Communication

14.10.5.1 Post-Award Conferences. Once a contract or purchase order has been awarded to a supplier, it is necessary to assure that continued understanding is maintained between the buyer and seller. If the purchase involves complex items, it is frequently desirable to hold periodic conferences with the supplier to maintain a clear understanding of all requirements and to resolve problems of engineering interpretation. As the supplier begins to perform the work required, a review of his drawings and his first fabricated items may point up areas of misunderstanding. It is much better to resolve these problems early than to be "surprised" with unusable parts when the first items are received and the production line is waiting.

14.10.5.2 Resident Representative. Representatives of the company quality control organization may be assigned at the supplier's facility. This is done to assist in improving communication, to provide guidance and help to the supplier, and to evaluate the progress of the supplier in meeting his quality requirements. Resident representatives are usually assigned when the purchase involves highly complex equipment or processes or if the magnitude of the work effort requires the need for close coordination.

14.10.5.3 Supplier Information Request. Some organizations encourage suppliers to use special forms provided by the company to trigger action if questions or problems arise. The important point in using these documents is to actively encourage communication and to provide a positive response to the supplier's request for information. Careless attention to these requests or improper responses will only result in closing the communication channel, thereby limiting the understanding.

14.10.6 Source Inspection

Inspection of purchased items is normally performed at the time the items are received at the purchaser's facility although, in some cases, source inspectors may be used on a resident or itinerant basis to inspect the products at the supplier's plant. Normally source inspection is used for the following reasons:

A. The level of assembly prevents adequate inspection at the purchaser's plant.

B. The quality of manufacturing processes cannot be determined solely by inspection or test of completed products.

C. The required environment or test equipment is not available at the purchaser's plant.

D. The items are to be shipped directly from the supplier to the purchaser's customer or to an outside facility.

E. For other reasons it is determined to be more economical to inspect at the supplier's plant.

Since source inspectors are frequently located at facilities that are some distance from the purchaser's plant, it is important that their work requirements be clearly and completely stated. Some organizations prepare detailed instruction sheets to guide these individuals for each of the major contracts or purchase orders

TYPE OF INSPECTION	SOURCE INSPECTION INSTRUCTIONS	
MECHANICAL ______ ☐ ELECTRICAL ______ ☐ ELECTRO MECH ______ ☐		PART NUMBER ______ PART NAME ______ PROGRAM ______ MFG. MODEL ______
CONTROLLING DOCUMENTS PURCHASE ORDER AND: NA-5 ______ DWG. ______ PROCESS SPEC. ______ IID ______ MIL. SPEC ______ OTHER ______	SUPPLIER ______ ADDRESS ______ CITY ______ TELEPHONE ______ CONTACT ______ MSFC # ______	**IDENTIFICATION** TAG ______ STAMPED ______ A/N NUMBER ______ MFG. NUMBER ______ OTHER ______

INSPECTION PARAMETERS	SOURCE INSPECTION POINTS
DIMENSIONS PER ______	DETAIL PARTS ______
FINISH PROCESS PER ______	INPROCESS ______
HARDWARE INSTALLATION ______	FINAL PHYSICAL ______
MARKING ______	FINISH ______
FUNCTIONAL TEST ______	FUNCTIONAL ______
WORKMANSHIP ______	DOCUMENTATION ______
SERIALIZATION ______	VARIABLE DATA ______
CONTINUITY ______	PACKAGING ______
PACKAGING ______	TRACEABILITY DATA ______
TRACEABILITY ______	OTHER ______

REMARKS AND SPECIAL INSTRUCTIONS

______

______

______

______

______

______

______

______

______

RESPONSIBLE PERSONNEL	
	USED ON ______
PROCUREMENT Q.C. ______	PREPARED BY ______
PURCHASING ______	APPROVED BY ______
ENGINEERING ______	DATE ______
PURCHASED LABOR ______	REVISED ______

Fig. 14.3 *Typical source inspection instruction sheet.*

which include source inspection. Figure 14.3 is an example of such an instruction sheet.

Although source inspection presents many advantages, it also results in technical as well as administrative problems. These include obtaining individuals with the necessary diversified skills since many of the in-plant specialists are not available for assistance, difficulties in effective supervision because of geographical dispersion, and other administrative functions which become more difficult when communication channels are lengthened. Source inspection, therefore, requires greater attention in personnel selection and supervision than usually is involved for in-plant operations.

14.10.7 Receiving Inspection

Receiving inspection is a major point of control for assuring that purchased items are in conformance with the established design documentation. Even when items are source inspected, they usually flow through the receiving inspection operation to detect damage which may have occurred during shipment from ths supplier's plant. Generally, receiving inspection is involved with a large number of widely differing items; therefore, a carefully planned and organized approach is required to assure that flow time is not excessive and the required product assurance is obtained in the most economical manner.

14.10.7.1 Inspection Planning. In addition to the general procedures governing the performance of the receiving inspection activity, many companies use detailed inspection instructions which guide the efforts of the inspector for specific products. These instructions usually include the following information:

A. Specific outline of the characteristics to be checked

B. Detailed description of special techniques or methods which are required

C. Instructions on sampling tables to be used

D. List of special tools required to perform the inspection

E. List of all reference material required by the inspector

F. Special visual aids appropriate for the specific item being inspected

G. Instructions regarding frequency and type of samples to be forwarded to the laboratory for chemical or metallurgical analysis

As with other work instructions, it is necessary to assure that the instructions are controlled and that changes in product specifications are reflected in corresponding changes to the inspection instruction.

14.10.7.2 First Article Inspection. The initial item produced by the supplier using his standard production tooling is usually given an exhaustive evaluation, often called first article inspection. This inspection is performed on the initial production item on a contract or the initial item following a major revision to an existing configuration. Since many of the sample items furnished by the supplier for evaluation and qualification may be fabricated in a research laboratory, the first articlc inspection provides a basis for determining the supplier's ability to duplicate the product under normal manufacturing conditions. First article inspections are usually more exhaustive than the normal receiving inspection and frequently include environmental and destructive type tests. In a sense, it is a major checkpoint to assure that the production process of the supplier is capable of producing the item in accordance with the requirements contained in the contract or purchase order.

14.10.7.3 Sampling Inspection. Sampling inspection is used extensively in receiving inspection operations because of the large number of items processed and the ability of sampling techniques to provide a high level of assurance at a minimum cost. For a discussion of types of sampling plans and criteria for their application to specific situations, see Sec. 13.

Sampling inspection during receiving operations has an advantage over acceptance sampling of finished product in that during subsequent manufacturing operations, the purchased items will receive additional checks, thereby reducing the risk of a defective item being included in the finished article. The cost and losses associated with the detection of this defective item later in the production process must be balanced with other considerations in determining the acceptable risk level and associated acceptable quality levels (AQL's).

14.10.7.4 Inspection Records. The results of the inspections made during the receiving cycle are usually recorded in a file in such a manner that history on a supplier's performance on a specific part can be readily determined. If the volume of incoming material transactions is significant, a mechanized data-processing system may be appropriate for accumulating inspection data.

These data can then be used in several ways:

A. A history of the supplier's actual performance can be provided and may be used to select the appropriate level of inspection; i.e., reduced, normal, or tightened.

B. Information for rating a supplier's performance can be provided for use in determining the most effective suppliers for award of future contracts and purchase orders.

C. Unfavorable trends and problems that require corrective action can be determined.

14.10.8 Corrective Action

Effective communication and coordination with suppliers early in the procurement program should materially reduce the number of instances of unsatisfactory supplier quality performance. These actions, however, cannot be expected to preclude all problems, and instances will arise where corrective action is necessary.

Problems are frequently identified during source inspection or receiving inspection which require immediate notification to the supplier and necessary action on his part. An analysis of the supplier's history records may also reflect the existence of a more subtle quality problem or an unfavorable trend in performance.

When corrective action is required, the supplier should be notified in writing with sufficient detailed information to clearly identify the nature and magnitude of the problem. Seriousness or urgency may dictate telephonic communication, but this should be followed with written confirmation for record and follow-up purposes. The supplier should be requested to investigate the problem and to advise, by a specific date, the results of his investigation and the specific action he will take to correct the basic cause of the problem. A suspense file should be maintained to assure that the supplier responds by the date specified. If the supplier fails to respond, follow-up action should be taken, which may include the decision to withhold future acceptance of the product until effective corrective action is instituted by the supplier.

14.10.9 Surveillance of Warehouse and Storage Facilities

After material has been accepted, it is usually stored in warehouses or other storage facilities until such time as it is required by the manufacturing organization. Periodic inspection of material in these facilities is necessary to assure that material is not degraded during the storage period. These inspections usually include a check of the following factors:

A. Proper identification and storage to prevent comingling of items and possible loss of identification

B. Proper palleting, shelving, and packaging to prevent damage from atmospheric exposure, excessive weight, or related factors

C. Provision for stock rotation to prevent excessive aging and deterioration of parts

D. Provision for identification and control for items that have shelf life requirements; e.g., certain rubber goods, plastics, chemicals, etc.

Storage inspection is usually accomplished on a periodic basis by a roving inspector. The type of inspection should be based on the requirements for the material stored in the warehouse and may involve the use of check lists as well as selection of random samples for detailed verification that all requirements are being satisfied.

14.11 CONTROL OF MANUFACTURING QUALITY

14.11.1 Manufacturing's Role in Quality Creation

The manufacturing function is basically concerned with the conversion of raw material, parts, and components into finished products and preparation of the products for delivery. This involves processing, machining, fabricating, assembly, packaging, packing, and shipping operations. The design documentation establishes the specific product and process characteristics required and provides the framework within which the manufacturing operations are performed. The quality control function plays a major role in assuring that the manufacturing operations are properly performed and that the intended characteristics exist in the finished product.

14.11.2 Evaluation of the Production Process*

14.11.2.1 Purpose of Evaluation. The production of a specific product containing specific

* The term "process" is used here in its broadest sense to mean a summation of all equipments, methods, and actions required in the conversion of material to finished product.

characteristics usually involves the application of various types of processing equipment, tools, machines, and measuring equipment as well as actual physical labor of individuals. It is essential that these elements possess the inherent capability of performing functions in a manner that will provide the desired characteristics in the end item. For example, a machine that produces parts with a variability of 0.001 inch cannot be expected to produce parts to an accuracy of 0.0001 inch except by chance occurrence. The purpose of process evaluation, therefore, is to determine the capability of the process to consistently produce a product within specified limits of variation.

14.11.2.2 Nature and Type of Evaluation. Ideally, the production process should be evaluated fully prior to the initiation of production; however, this is seldom possible in a dynamic, complex production operation. In actual practice the nature and type of process evaluation develop with time. Certainly, knowledge of past process performance provides a basis for process decisions even though formal studies have never been performed. Some process elements are extremely critical whereas others have little influence on the end product. Sometimes the sequence in which the steps of the process are performed is extremely important, but in other cases it is immaterial. Therefore, the nature and degree of process evaluation techniques that are used will vary considerably in type and complexity. In general, two basic types of process evaluation techniques are utilized; viz., process capability studies and mandatory process evaluation points.

14.11.2.3 Process Capability Studies. Process capability studies constitute a tool for determining the inherent capability of any element or group of elements in the production process. For example, they may be used to evaluate a machine, a chemical processing operation, a measuring device, a console of interconnected items of electronic test equipment, or even the capability of individuals to perform a specific operation.

A simple example of a machine capability study may help to clarify the application of this technique. Every machine has an inherent variability that can be evaluated by determining its standard deviation (σ') on the basis of a series of individual measurements for the quality characteristic under consideration. A controlled process can be expected to produce individual articles with measurements spread over a band $6\sigma'$ units wide. For example, if a milling operation had a σ' of 0.0008 inch, the total spread for the machine would be 6(0.0008) or 0.0048 inch. It would be expected that the thickest piece would be approximately 0.005 inch thicker than the thinnest piece.

If the process is to be capable of meeting the specification tolerance, the specified tolerance must be at least as great as ± 0.0025 inch (a total spread of 0.005 inch), but for practical purposes should be ± 0.0035 inch (a total spread of 0.007 inch) to allow for variations in set-ups and tool wear. A good rule of thumb is to have $6\sigma'$ of the process equal to two-thirds the total spread of the specified tolerance.

Information derived from studies of this type may be utilized by the engineering organization to establish realistic tolerances based upon current process capabilities, to determine the need for different processes or new equipment, or to use alternate design approaches for obtaining a specific characteristic. In manufacturing, the information may be used for selecting specific equipment or people for a particular operation, for establishing maintenance cycles, or for methods or equipment improvement studies.*

14.11.2.4 Mandatory Process Evaluation Points. Since it is difficult, if not impossible, to evaluate every variable in the production process, most companies establish mandatory process evaluation points for those variables that are critical in achieving quality objectives. These process elements are usually evaluated and approved by the quality control organization prior to or concurrent with initial production. The following process elements are examples of some of these mandatory process evaluation points:

A. Operator certification. When specialized critical skills are involved, operator certification programs are recommended. Certification must include training of the individual as well as actual demonstration by the individual that he possesses the necessary capability (knowledge and skill) to perform the particular operation in the correct manner. Welding, soldering, and the operation of certain critical

* For a more detailed discussion of methods and techniques used in performing process capability studies, see J. M. Juran, *Quality Control Handbook* 2nd ed., McGraw-Hill, 1962, pp. 11–15 to 11–34.

equipments are typical of activities requiring certification. Operators should not be permitted to perform these operations until the quality control organization has certified that they possess the basic capability required. Periodic recertification may be required to assure continued ability of the individual to perform the operation.

B. Tool proofing. Many companies require that tools and fixtures used in the production process be evaluated and approved prior to their use in production operations. This may be done by measuring the actual tool or fixture or, alternatively, the first few items produced using the tool or fixture.

C. Critical processes. Critical processing operations, such as heat treating or plating, may require evaluation and approval prior to use. This is particularly true of processes for which the presence or absence of a characteristic in the product is difficult to determine after the fact without destructive testing.

D. Test operations. Complex testing operations frequently are evaluated and approved prior to their use to demonstrate that when properly used the equipment will test the product to the requirements of the functional test specification.

The nature and type of mandatory evaluation points will vary, depending on the nature of the product, criticality of the element, and other factors. Properly used, they are effective tools in the prevention of deficiencies in the final product.

14.11.3 Measurement and Measuring Equipment

14.11.3.1 Nature of Measurement. The design documentation specifies the quality characteristics that a product is to have. The presence or absence of a characteristic or the degree to which it is present or absent is determined by one or more of the human senses; i.e., sight, sound, feel, taste, or smell. When the characteristic cannot be detected or measured directly by one of the senses, it is necessary to *transform* the characteristic into some phenomenon that can be sensed by means of suitable instruments, apparatus, or test equipment. For example, a wattmeter transforms watts into length (on a scale) that can be evaluated by sight.

Sometimes it is possible to detect a characteristic but impossible to evaluate it to a sufficiently fine degree with the unaided human senses. In such cases, an instrument, apparatus, or gage may be used to multiply the phenomenon. The micrometer employs a screw to multiply length so that 0.001 inch between the anvils is approximately 1/16 inch on the thimble. Many instruments both transform and multiply the characteristic to be evaluated.

14.11.3.2 Characteristics of Measuring Equipment. Measuring equipment has three basic characteristics: sensitivity, precision, and accuracy. These three factors directly influence the selection and control of measuring equipment in industrial use.

A. Sensitivity involves the degree to which measurement variation can be observed; for example, a micrometer is more sensitive than a rule. The rule may be extremely accurate; the ability to observe the accuracy that may be present is hampered by its sensitivity.

B. Precision involves the degree to which an instrument will reproduce a given measurement under constant conditions. If an instrument is used to make a series of measurements on the same item, the readings will vary slightly due to the inherent construction of the instrument. The degree to which the instrument varies is a measure of its precision. The less the variance, the greater the precision.

C. Accuracy involves the degree to which an instrument measures the true value as represented by an accepted standard such as those provided by the National Bureau of Standards. An instrument may be very precise (i.e., yielding consistent values repeatedly) and yet be inaccurate in that it is misaligned in relation to an accepted standard. On the other hand, the precision of an instrument has a direct bearing on its accuracy.

Sensitivity and precision of instruments have an important bearing on the selection of specific instruments for a particular measurement requirement. Accuracy is generally maintained by periodic comparison of the instrument against standards of known accuracy. This is usually referred to as instrument calibration (see Par. 14.11.3.4).

14.11.3.3 Selection of Measuring Equipment. The selection of measuring instruments involves a series of considerations and is a complex subject in its own right. The following are some of the basic factors which must be considered.

A. Sensitivity. As a general rule of thumb,

the instrument should be sensitive enough to permit dividing the total allowable range of the characteristic to be measured into tenths. For example, a 1,000-watt electrical heating unit with a tolerance of $\pm$ 25 watts (975 to 1,025) would have a total allowable range of 50 watts. The wattmeter used to measure the heating unit should be capable of measuring down to 5 watts (10 per cent of 50 watts).

B. Precision. The degree of instrument precision may be determined by making a number of measurements under the same conditions on the same unit of the product. The values observed can be used to calculate the standard deviation of the instrument. The standard deviation of the instrument can be stated in terms of instrument tolerance. As a general rule, the measuring instrument tolerance should be one-tenth of the tolerance of the product which is to be measured. When the tolerance ratio is greater than one-tenth (i.e., one-fifth or one-half), the acceptance tolerances on the product should be adjusted to consider the error introduced by the measuring instrument. The techniques used for adjusting tolerances are the same as those outlined in Par. 14.9.4.5.

C. Fixed versus indicating instruments. Fixed instruments (such as plug gages, ring gages, adjustable snap gages) merely show that the part is undersize, within limits, or oversize. On the other hand, indicating instruments (such as micrometers and dial indicators) provide more useful information (actual readings), which shows where the process is set and how much variation is occurring. Indicating instruments are necessary for process control when small periodic samples are measured and charted, but fixed instruments are suitable for little more than sorting. Since modern quality control concepts emphasize control of the process and avoidance of defects and sorting, the trend is away from the use of fixed gages and toward indicating gages. Future advances will lead in the direction of recording and process-regulating types of instruments.

14.11.3.4 Measurement Echelons and Traceability. The accuracy of measuring equipment is established by calibration of the instrument, utilizing a standard of known accuracy. The results of this comparison are used to adjust the instrument or to record the differences for the preparation of correction charts which may then be used to adjust the data derived from the instrument. The source of our measurement standards in the United States is the National Bureau of Standards. At one end of the chain are the National Prototype Standards, NBS Derived Standards, National Reference Standards, and National Working Standards that are maintained by the NBS. Farther down the chain are Interlaboratory Standards, which are sent to the NBS by industrial laboratories for calibration and certification, and which are used to calibrate "reference" standards in the industrial laboratories. The reference standards are used to calibrate Transfer Standards and/or Working Standards. And finally, at the bottom of the traceability chain, are the instruments that make the measurements in the shop, usually called "working" instruments. Thus from the worker's bench there is, ideally, an unbroken chain of calibrations to the NBS.

14.11.3.5 Industrial Standards Laboratories. Many industrial firms establish their own "standards laboratory" where they maintain the company's "reference" standards. Stability and reliability are characteristics of major importance in reference standards; therefore, they are usually of fixed and permanent construction and seldom are appropriate for use in the shop by virtue of their design. The reference standards are the highest-accuracy devices in the company, and their calibration requires special techniques and usually a considerable time period (e.g., to measure drift). The nature of reference standards dictates that they not be used any more often than absolutely necessary. In the case of dimensional standards, wear will result in loss of accuracy and excessive replacement cost. Also, progressive drifts are easily introduced to values of electrical standards (e.g., saturated cells, which are used as voltage references).

Reference standards must be maintained under highly controlled environments which, in turn, dictate special facility requirements. For further information on the development of facility requirements and suggested environments for standards laboratories, the reader is referred to the bibliography.

The requirement that reference standards should not be used any more often than absolutely necessary and the requirement for special environmental controls establish a need for a tier of standards between the reference standards and the working instruments; viz., working standards. Working standards require less stringent environmental controls than reference standards; therefore, they are usually maintained by secondary-level laboratories which are located near the shops or laboratories using working instruments.

The tremendous increase in industrial activity experienced recently, particularly in the aerospace industry, has led to an unprecedented demand for precision measurement which, in turn, is bringing about the establishment of hundreds of new standards laboratories. To aid these new laboratories in transmitting the accuracies of the national standards to the industrial shops, the NBS has prepared Handbook 77.* This compilation of publications by the Bureau staff is recommended for technical personnel who have specific interest in precision measurements.

14.11.3.6 Elements of Calibration Control. Calibration control systems are established to assure that measuring and test equipment is adequate and properly calibrated to assure continuous conformance of supplies and services to customer requirements. The types of controls vary, depending upon both the product and the customer; however, the following are basic functional concepts which apply to any calibration control system:

A. The highest order of standards should be traceable to a national standard, derived from accepted values of natural physical constants or derived by the ratio type of self-calibration techniques.

B. Measurement standards should be carefully selected with consideration given to accuracy, stability, and range requirements. Adequacy of devices proposed for use as standards should be proved through appropriate testing and use of statistical techniques.†

C. Accuracy ratios between measurement tiers should be based on trade-offs among instrument cost, test cost, availability, and accuracy. Arbitrary accuracy ratio requirements between measurement tiers should be avoided. Many authorities recommend that instruments be calibrated using standards having ten times the accuracy of the instruments; however, state-of-the-art accuracy limitations, cost of standards, and testing costs often make the 10 to 1 ratio an unrealistic requirement.

D. The measurement system environment should be controlled to the extent necessary to continuously assure measurements of required accuracy.

E. An instrument recall system should be established to assure that calibration is scheduled for measuring and test equipment consistent with accuracy requirements and usage conditions. In larger companies, the recall system is often mechanized to provide automatic issuance of calibration-due notices to equipment users.

14.11.4 Process Control

14.11.4.1 Objectives of Process Control. A primary function of the quality control activity is the prevention of defects. This function is best served by effective process control. If the process is statistically controlled at the level specified for a certain characteristic, no defectives will be produced for that characteristic. Such an ideal situation would mean no scrap, no rework, no detailed inspection or sorting, and a product conforming to specification. These are the objectives of process control.

14.11.4.2 Fundamental Concepts of Process Control. No process can produce one unit exactly like another unit time after time. Although a process may appear to be precise, a sufficiently sensitive measurement of a quality characteristic will reveal an inherent variation in the product. This variation in the product is caused by a variation in the process. Even a simple process involves matter and energy by which some physical or chemical change is made on the article or substance being processed. Since there is an inherent nonuniformity in matter, and since the release or conversion of energy is influenced by heat, pressure, and other variable factors, it is impossible to divorce the process from the ever-present *cause system* that brings about variability. Although statistical quality control recognizes this inherent variability, it also recognizes that such variability follows a characteristic pattern for a given process and that the variation falls within certain limits. The process pattern is determined by measurements of the quality characteristics of the product. It is generally necessary to measure one hundred or more items to realize a sample of sufficient size to give a truly representative pattern. In the majority of cases encountered in industry, this pattern approximates the normal frequency distribution (see Sec. 13).

---

* Available from the Superintendent of Documents, U.S. Government Printing Office, Washington D. C. 20025.

† See U.S. Department of Commerce, NBS Handbook 91, "Experimental Statistics."

There are two features of the frequency distribution (pattern) that are of primary interest:

1. The spread of measurements (also called "scatter" or "dispersion").
2. The placement of the mean or average with respect to some scale of measurement.

The first feature is measured by a statistic known as the standard deviation and designated by the Greek lower-case letter sigma (σ). This measure of spread tells us whether the normal process variations will permit the manufacture of the product within specification limits. The second feature is measured by the arithmetic average and is *set* at the proper level or is *directed* at the desired value.

Besides knowing the pattern for the process, quality personnel are interested in the stability of the pattern and hence the stability of the process. A periodic increase or decrease in spread or a shift in mean (setting) of the process shows that it is *unstable*. If a process is controlled, it must be a stable process. (In some cases where a change in spread or mean can be predicted and adjusted for, the process can be held in control. An example is a shift in turned diameter due to tool wear. A periodic adjustment or sharpening of the tool will prevent too great a departure from the desired mean.)

When the process is stable and produces a consistent, nonchanging pattern of variation, the process is said to be operating under a *constant cause system*.* Under these conditions, the process is said to be in a state of statistical control and its future performance can be predicted. If, on the other hand, a shift in the normal pattern of variation occurs, the process is said to be operating under an *assignable cause system*. In other words, one or more extraneous causes for variation exerted their influence on the process. These causes can usually be identified with the help of quality control charts that tell when the extraneous cause entered the system. If these charts show a lack of statistical control, the future performance of the process cannot be predicted and the process is said to be erratic or unstable.

14.11.4.3 Quality Control Charts for Process Control. The construction of $\bar{X},R$ charts and p charts has been covered in Sec. 13 on industrial statistics. It will be noted that the $\bar{X}$ charts for variables measurements use averages of subgroups of individual measurements. This practice gives greater sensitivity to the chart for detecting shifts in mean than does the plotting of individual measurements. One disadvantage is that the points on the chart show average measurements instead of individual measurements. Shop personnel are accustomed to thinking in terms of individual measurements, and the specification is always referred to individual measurements. Some quality control personnel believe that the $\bar{X},R$ chart should be used only as a "laboratory tool" for quality control analysts; others feel that it is a very valuable tool for the operator in the shop.

Certainly if the process is going to be effectively controlled, the operator must know frequently and promptly what the process is doing so that he can make necessary adjustments at the proper time. The control chart should be at the location of the process and under the constant observation of the operator. If the operator is trained to take proper action when a point falls outside control limits, the control chart has served its purpose. It does not matter if the control limits are based on an average of subgroups so long as correct action is obtained at the correct time to maintain control of the process. The R chart gives a measure of spread for the process. It is related to the standard deviation of the process by the formula

$$\sigma' = \frac{\bar{R}}{d_2}$$

where σ' = standard deviation of the process
$\bar{R}$ = average range
d_2 = factor dependent on subgroup size

What has been stated for $\bar{X},R$ charts as far as location of the charts is concerned also applies for per cent defective charts (p charts), number of defectives charts (pn charts), or number of defects per sample charts (c charts).† Unfortunately, the latter charts require a larger sample than the $\bar{X},R$ chart to provide sufficiently accurate information. Because of this, points are usually not plotted on p charts as frequently as in the case of $\bar{X},R$ charts. Often points on p charts are based on the results of the total inspection for the day. Time for processing the data may cause delays, and if an investigation is

* Some authorities also call this a *chance cause system*, on the basis that all variation under a constant cause system is due to chance variation.

† p, pn, and c charts are presented in Sec. 13 on industrial statistics.

necessary before action can be taken, a delay of several days may result before corrective action is realized. Obviously, the lines of communication are too long and take too much time.

14.11.4.4 Size of Sample. For $\bar{X},R$ charts, samples should preferably consist of a subgroup of four, five, or six items. A subgroup size of two or three results in lack of sensitivity for the $\bar{X}$ chart. Subgroup size of over ten results in loss of efficiency for R in giving a measure of the standard deviation. A series of twenty-five subgroups should be used for calculation of trial control limits and central lines (see Sec. 13). Limits should be reviewed at the end of each run of twenty-five subgroups and adjusted until standard values for average and range can be established. Even after standard values are established, periodic review of the process against the standard should be made.

For p, pn, and c charts, the samples should be large enough to provide 3σ control limits that have a reasonable degree of sensitivity.

14.11.4.5 Method of Taking Samples. Samples should be taken to obtain rational subgroups. A rational subgroup is defined as a group *within* which variations may, for engineering reasons, be considered to be caused by nonassignable chance causes only, but *between* which there may be differences because of assignable causes whose presence is considered possible. All members of a subgroup should be produced under essentially the same conditions; hence, items comprising the subgroup should be taken from a very short interval of time. For example, it would be preferable to take a sample from the product made during a five-minute interval rather than taking it at random from the product accumulated over an hour's time. (If sampling is done for acceptance purposes, rather than for control-chart purposes, a random sample taken from several hours' production is desirable in order to represent all periods of production. (See Acceptance Sampling, Par. 14.11.6).

14.11.4.6 Frequency of Sample. The frequency of sample depends on the period of time one would be willing to operate the process "out of control" without being aware that it was out of control. Economic considerations usually influence the decision. For example, if a sample showed the process to be out of control, it might be desirable to screen all the products made after the preceding sample that indicated that the process was in control.

When very close control is desired, samples may be taken as frequently as every fifteen minutes. Common practice is to sample approximately once each hour. For processes that have remained in good or excellent control over long periods of time, a sample taken once every four hours or once a shift should be adequate.

14.11.4.7 Maintenance of the Control Chart. When a control chart is started, it is usual practice for a quality control engineer or quality control analyst to obtain the initial measurements, calculate control limits, and set up the chart. Also, any later revision of limits is the responsibility of such individuals.

Once the chart is established, either the operator or inspector on the job may obtain the measurements, make the sample calculations involved, and plot the points. If the operation permits, it is preferable to have the operator run his own chart. He is more likely to feel that the chart is a tool to help him control the process rather than a device that gives the inspector a way of checking on him. There may be some objection to this on the basis that the operator would be tempted to bias the results to make a good showing on the chart. An occasional audit point identified by the foreman or inspector would reveal such bias if present. Another objection is that the operator is hired to manufacture, not "to keep books." But a little "paper work" might add to the interest of the job.

14.11.4.8 Number of Control Charts. Maintaining a chart on every characteristic of every part and product is usually unnecessary. Charts should be maintained only on important characteristics and then only if they will "pay their way." It costs money to provide and maintain test equipment, obtain measurements, calculate results, and plot and interpret charts. If the chart saves more than these costs, then it should be kept. If not, it should be eliminated. If a chart is not quite paying its way, it is often advisable to decrease the frequency of sampling to reduce costs rather than lose entirely the values the chart provides.

14.11.5 Inspection

14.11.5.1 Nature of Inspection. In previous sections we have discussed specific quality control activities which are directed toward the prevention of quality problems. Even with the best planning and prevention activity, it is obvious that errors will occur during the production process. The detection of these errors and the feedback of the information for correction of the basic cause of the errors are vital links in assuring an acceptable product. The act of determining conformance or nonconformance of the process or product characteristics to the design documentation is the function of inspection.

Even in the manufacture of a simple item, the product usually progresses through a series of production operations to produce the desired characteristics in the end item. It is necessary to determine the specific inspection activities that are required to assure that the end items conform to the requirements of the design documentation. This involves the determination of what to inspect, when to inspect, where to inspect, and how to inspect. A prime consideration in inspection planning is the selection of specific action from the possible alternatives such that the total quality costs are a minimum.

14.11.5.2 What to Inspect. In specifying quality, we have enumerated in detail the specific characteristics of the product and the process by which it will be fabricated, tested, and accepted. These characteristics are imparted to the product through various manufacturing operations which may include chemical processing, machining fabrication, and assembly. Generally speaking, there are two basic elements that may be inspected; viz., the operation producing the product or the product itself. For example, in heat-treating a part the decision may be made to inspect and verify the temperature of the oven and the time of processing rather than to inspect the presence of hardness in the completed unit. If the appropriate process characteristics are controlled, the product characteristics will in fact be present in the completed product. Process control is usually the only method of inspection that can be employed on a routine basis when the presence of a characteristic in the end product can only be determined by destructive testing.

When special tools or fixtures are used, the first article produced on the fixture should be inspected to verify the adequacy of the tool or fixture. Subsequent inspections may then be performed at some predetermined time to detect tool or fixture wear. Care must be exercised in utilizing the approach since some variables influencing the characteristic may not be controlled by the specific tool or fixture and, therefore, introduce error in individual units.

14.11.5.3 When to Inspect. When an item involves a series of operations, it usually is not economical to inspect after each operation. In general, an inspection is made after a series of manufacturing operations has been performed. A logical time to inspect is when a part leaves one department or section and enters another section for further processing. The determination of when to inspect usually involves an economic trade-off of the cost of inspection compared to the cost of subsequent processing of the item if the inspection is not performed until some later point in the process. In this regard it is generally advisable to inspect:

A. Prior to a very expensive processing operation,

B. Prior to assembling several parts if the nature of the assembly precludes adequate inspection, or

C. Prior to assembling several parts if the detection of an error would involve expensive disassembly and reassembly for correction

The final assembly or finished product is usually given a final physical inspection and functional test at the completion of all operations. If progressive inspection has taken place, the final inspection will include a verification (usually by check of records or stamps) that the required inspections have been completed and all inspection "squawks" cleared.

The establishment of points of inspection is usually done in conjunction with the manufacturing planning function and is frequently reflected on the job tickets or operation routing sheets. The sheets, for example, may call for several manufacturing operations and then an inspection operation. The operator is thereby advised that inspection is required at that point in the manufacturing operation and prior to performing the next operation. Inspection personnel normally "stamp off" the inspection operation after the inspection is completed, not only for a permanent record, but also to authorize the manufacturing operator to proceed with subsequent steps.

14.11.5.4 Where to Inspect. Inspection may be performed in one of several basic locations: inspection cribs, at inspection stations in the production line, or a series of production locations serviced on a roving basis. Generally speaking, crib inspection results in the most efficient utilization of inspection equipment and personnel. Lot quantities of parts are delivered to a specific location set aside for inspection purposes. These lots are inspected and returned to the next production location. Having the inspectors and equipment in a central location permits the supervisor to more effectively assign work and provide closer surveillance of the work performed by the inspector.

If production flows from operation to operation on a production line, it is frequently desirable to include the inspection stations at selected points along the line. In this regard the inspection effort must be consistent with the time allowed in the line flow to permit an adequate inspection. If the inspection operation is not balanced with the line, it may either become a bottleneck on one hand, or result in idle inspection time on the other hand.

In some cases, the item to be inspected may be large and difficult to move, or for some other reason it may be more desirable for the inspector to go to the location of the work. In such cases, an inspection "call system" is used to notify the inspector when his services are required. Other inspections may require that the inspector periodically visit a specific location on a random or time-sequenced basis.

14.11.5.5 How to Inspect. Some inspection operations involve a series of generally repetitive tasks while others may involve a series of unique tasks which vary considerably from item to item. In the case of repetitive-type tasks, the inspector usually receives training in the inspection equipment he must use and the general procedures that apply. The job then is to perform the inspection to the specifications or drawing, using this basic knowledge or experience. This is typical of machine part inspection.

In the case of unique inspection requirements, it is desirable to prepare a specific inspection instruction which outlines the equipment to be used, the specific steps in the inspection, and the specific techniques that must be applied. These instructions may also include visual aids, templates, and other aids to the inspector in performing the operation. When inspection instructions are used, it is important that they be controlled and kept current with changes that occur in the basic design documentation.

14.11.6 Acceptance Sampling

14.11.6.1 Purpose of Acceptance Sampling. Acceptance sampling is used to make a decision regarding acceptance or rejection of a lot without having to examine the entire lot, thereby providing economy of inspection.

14.11.6.2 Types of Acceptance Sampling.

A. Attribute: When acceptance criteria are based on the number of defectives or defects found in the sample

B. Variable: When acceptance criteria are based on "mean" and "spread" of a number of individual measurements on articles making up the sample

14.11.6.3 Acceptance Sampling Plans. Various sampling plans and their characteristics have been discussed in Sec. 13, Industrial Statistics. A plan specifies a sample size for a given lot size and the acceptance criteria; i.e., the number of permissible defects in the sample. Each sampling plan has stated risks. It is important for the quality control engineer to be thoroughly familiar with the operating characteristics of the plan being used so that the producer and consumer risks will be known. It should be emphasized that the average outgoing quality (AOQ) associated with a sampling plan is a long-term average. Individual lots having a per cent defective considerably above this average may be accepted, depending upon the quality submitted. The operating characteristics (OC) curve provides this information.

Single, double, or multiple acceptance sampling plans may be used. These are selected to provide either lot quality protection or average outgoing quality protection. The former, LTPD or p_t (lot tolerance per cent defective), corresponds to $p_a = 0.10$ on the OC curve for most commercial applications. A considerable amount of inspection (large samples) is necessary to ensure a lot p_t.

Average quality or average outgoing quality limit (AOQL) is generally the basis upon which sampling plans are selected, for it is the average that is most generally dealt with from a cost standpoint.

14.11.6.4 Selection of Plan. After it has been decided whether inspection is to be on an attribute or variables basis, and after a decision has been made between lot quality protection and average quality protection, a decision still must be made regarding single, double, or multiple sampling if attribute inspection is used.

Single sampling is the most easily administered and the least complicated as far as procurement of samples is concerned; however, when inspection cost per unit is high, the smaller amount of sampling afforded by double and multiple sampling will be an offsetting advantage. Single sampling gives the most information on the quality level of the lot.

When submitted quality is better than the AQL for the plan (as it usually should be), double and multiple sampling require a lower average amount of inspection than single sampling. When AQL values are so low that two or three multiple samples are required prior to acceptance, the advantage for multiple sampling decreases.

Double and multiple sampling have a psychological advantage in that they afford the producer a "second chance." When submitted quality is considerably worse than the AQL, double and multiple sampling require less average inspection because of rejection on the first sample. When quality is intermediate, double and multiple sampling may require more inspection than single sampling. When double sampling is used, personnel who use the tables should fully understand that the *second acceptance number is the number of permissible defects in the combined first and second sample* and does not apply to the second sample alone.

14.11.6.5 Importance of Process Average. In selecting sampling plans, it should be recognized that single sampling gives the most information concerning the process average, which provides a basis for decision between normal, tightened, and reduced inspection.*

Because of the importance of the process average, the specified first sample should be completed regardless of what plan was chosen and regardless of the fact that the rejection number may be reached prior to completion of inspection for the first sample. The percentage of defectives found in the first sample provides a measure or estimate of the process average. In double-sampling plans, inspection may be terminated as soon as the rejection number is reached *during the inspection of the second sample*. In multiple sampling, inspection may be terminated during or beyond the second sample. A suitable record card should be provided which will tabulate inspection results on first samples.

14.11.6.6 Tightened Inspection. Tightened inspection provides protection against acceptance, over an extended period of time, of lots that have quality slightly worse than the acceptable quality level (AQL). The tightened inspection reduces the probability of accepting a lot of a given per cent defective and is used when it becomes apparent that below-standard quality is being submitted.

14.11.6.7 Reduced Inspection. Unless reduced inspection will actually permit a reduction in inspection force or permit the "fixed" inspection force to devote more time to closer inspection of critical characteristics, it should not be used. When a larger number of items qualify for reduced inspection, real economy may be realized and the inspection force may be reduced. It must be realized, however, that protection against an occasional bad lot will be greatly reduced through the use of reduced inspection.

14.11.6.8 Formation of Inspection Lots. The more homogeneous the lot being sampled, the better chance that the sample will correctly evaluate the lot. For this reason, it is advantageous to have the producer separate those articles that are produced under essentially the same conditions. If items produced from one machine could be considered as a lot and sampled as a lot, it would be better than making up a lot from two or more machines.

Subject to the foregoing limitations, inspection lots should be as large as possible. A large lot requires larger samples, thereby improving discrimination between good and bad lots. Also, the sample specified for large lots is a smaller percentage of the total items submitted than is the case with smaller lots.

14.11.6.9 Randomness in Sampling. All sampling plans are based upon the sample being selected in a random manner; i.e., all items in the lot have an equal chance of being selected in the sample. The calculated risks, quality limits, etc., are likely to be in error unless the sample is randomly selected. This is one of the more important requirements in acceptance sampling, yet it is the one most often

---

* For a more complete discussion on selection of sampling plans, see E. L. Grant, *Statistical Quality Control*, 3rd ed., McGraw-Hill, 1964, pp. 397–409.

neglected. It takes discipline and effort to get a random sample. The inspection supervisor must provide means for shifting pallets and opening crates to ensure a random sample. A good disciplinary device is the use of a table of random numbers for selecting prenumbered articles from the lot. A "sampling thief," such as that used in the milling industry for sampling grain, is a useful device for sampling kegs of small rivets, nuts, bolts, screws, and similar items.

14.11.6.10 Resubmitted Lots. Lots that have been rejected, returned to the producer for 100 per cent inspection, and resubmitted for the consumer's acceptance sampling should be properly identified so that:

A. Results of inspection on 100 per cent detailed lots will not be included in process average calculations.

B. More severe acceptance criteria may be used to assure that practically all the defectives have been removed. (This is a requirement if the AOQL of the plan is to be relied upon.)

A problem will arise if rejected lots are resubmitted unchanged, without detailed inspection, on the chance that they will be accepted during the next sampling inspection. The only protection the consumer has is to insist upon proper identification of resubmitted lots. This is a requirement in MIL-STD-105D, used for government procurement.

14.11.6.11 Acceptance Sampling by Variables. Because of its simplicity, acceptance sampling by attributes is more widely used than acceptance sampling by variables; however, the latter will fill a specific need when it is important to measure certain specific quality characteristics. As cited by Grant,† acceptance sampling by variables has the following advantages over attribute sampling:

1. Better quality protection with a small sample size
2. Measures degree or extent of conformance or nonconformance
3. Better basis for guidance toward quality improvement
4. Better basis for giving weight to quality history in acceptance decisions
5. Errors of measurement more likely to be disclosed by variables information

Although the calculations based on sample measurements are not particularly complicated, they are generally difficult for the average inspector or inspection supervisor. A quality control engineer or statistician will usually process the inspection results and reach the decision to accept or reject.

The lot-plot plan developed by Shainin makes use of a frequency distribution of a sample of fifty measurements. This plan can be operated by the inspector and is being successfully used by many companies. A set of concise rules are given, involving calculation of 3σ from R based on ten subgroups of five measurements each and $\bar{X}$.‡

14.11.7 Quality Information

14.11.7.1 Nature of Quality Information. As with most business functions, quality control is strongly dependent upon the acquisition and utilization of information to effectively perform its function. Information requirements of quality control generally fall into three basic categories: status information, historical information, and action information. Status information consists of records and reports which are used to provide visibility of the status of actions that occur during the flow of the product through the plant. Historical information consists of records and reports which provide objective evidence of actions and can be used in future problem investigation. Action information includes records and reports used as primary tools in operational and management decisions. Although few records and reports fall exclusively into one of these three categories, the categories form a good basis for discussion of quality information requirements.

14.11.7.2 Quality Status Information. In most industrial operations, the product proceeds through a series of steps that require various actions by both manufacturing and quality control personnel. To assure that ap-

* The problem of how many containers should be opened is discussed in J. M. Juran, *Quality Control Handbook*, 2nd ed., McGraw-Hill, 1962, pp. 421 and 422, footnote.

† E. L. Grant, *Statistical Quality Control*, 3rd ed., McGraw-Hill, 1964, p. 422.

‡ For a complete presentation of this plan, see Dorian Shainin, "The Hamilton Standard Lot Plot Method of Acceptance Sampling by Variables," *Industrial Quality Control*, Vol. 7, No. 1, July 1950, pp. 15–34.

propriate actions are taken when required, certain status information is essential. For example, most organizations assign work through the use of "job tickets," "shop travelers," or similar documents. These documents travel through the various operations with the product and provide a vehicle for maintaining current status of the product. When a manufacturing worker has completed his operation, he signs or initials the ticket. Similarly, when an inspection operation has been completed, the inspector usually "stamps off" the inspection operation on the ticket. If the inspector discovers discrepancies during the inspection, a notation is made on the back of the ticket as to the exact nature of the discrepancy. The inspection operation will usually not be "stamped off" until the discrepancies listed have been corrected, at which time the inspector will place his acceptance stamp next to the inspection operation. This procedure provides for a continuous record of the exact status of the items being manufactured. The "job tickets" will normally include other information such as the drawing change letter, latest engineering order, and other information necessary to establish the exact configuration of the item produced. Figure 14.4 is an example of such a record.

In most companies, a series of documents, tags, and related status devices will be used, depending upon whether the product is serialized or unserialized, the size of the item, the physical nature of the item, and the type of processing activity involved. Physical stamping of the part is frequently used to provide status information, particularly when it is difficult or impossible to identify the specific parts with the corresponding records.

Items that do not conform to the design documentation are frequently segregated from acceptable material through the application of a deficiency or reject stamp or tag, or through physical segregation in a controlled stockroom. Scrap items are usually mutilated or positively identified (e.g., painted red) to assure that they are not returned to the production process.

When certification of processes or equipment is required, a decal is usually applied to the equipment to indicate the status of certification and provide for the acceptance stamp of the individual who approved the certification. Many examples may be cited on types of status records and techniques which are used throughout industry; however, in the final analysis, the types required must be based on the specific products and business systems in the organization. It is sufficient to state that effective status control is fundamental to any quality control program.

14.11.7.3 Historical Information. Many of the records and reports used to provide status information also form the basis for historical records; however, additional information may be useful for many reasons. For example, variables test data may be recorded during the calibration of an instrument, not for the purpose of providing status information but for future use in making decisions regarding calibration intervals. Frequently, it is desirable to maintain certain information for use if problems occur subsequently and the analyst desires to review information from previous operations. Figure 14.5 is an example of a record of historical test data collected during functional test of an electronic device.

In some cases, a customer may require that selected historical information be maintained for the purpose of auditing the quality system. In other cases, a customer may require that historical information be delivered with the product.

The amount of historical information that is retained will vary considerably, based on the needs of the particular company or on the expressed needs of customers. It should be recognized, however, that the collection and retention of information are expensive and that appropriate trade-offs must be made between the cost of the information and its value.

14.11.7.4 Quality Action Information. The third type of quality information is information that is necessary to make specific decisions in the operation and management of the quality program. Generally, there are two levels of decisions which must be made using quality action information.

The first level involves the specific action to accept or reject a product or to take a specific action regarding the process. This information is usually obtained by the inspector through direct observation of a measuring instrument or visual comparison with a standard. Control charts are a good example of first-level quality action data. The data recorded on these charts are viewed in terms of some pre-set decision criteria, such as control limits or action limits. When the limits are approached or exceeded, certain actions are called for on the part of the operator or inspector.

The second level of decisions is usually man-

SUPPLIER | REQ. NO. | P. O. NUMBER

PART NUMBER

NAR-ASG MANUFACTURING ORDER

1. | RC | DATE ISSUED | CONTROL NO. | SPLIT | IW

DIV. | PLAN DEPT. | PAGE OF | C

PART NO. | 2. | SC-SCN | SCHEDULE REF. | INITIAL STORES | FINAL STORES

PART NAME | 3. | ITEM/UNIT | ACCOUNT | ORDER NO. | SUB.-ACCT. | NO. UNITS | LH/N | (PCS) | RH

TYPE ORDER

PLANNED BY | TC | 4. | WORK FOR

IDENT. SERIAL NO. | AUTHORIZATION | TOT. THIS ORD.

NOTED A.

B. | USED ON PART NO. / NEXT ASSY. | SPLIT BY:

C.

D. | QTY. PER UNIT/NA LH/N RH | MODEL | INDEX NO.

E.

F. | THIS COPY USED ON | THIS PART EFF. ON

G.

LAST MECH. INSP. CVR ☐ MOCK-UP ☐ NO. CHG. ☐ CHG LTR NO E.O. ☐ E.O. NO. DATE:

LINE	UNIT QTY. DEPT.	OPERATION / MATERIAL 1	2	DATE	MECH/LDM.	QTY. ACCEPTED LH/N	RH	DATE	INSP.	TOOL/CAT NO. E.O. DWG. REV.	DEPTH OF CUT TOOL TYPE	SET-UP SPEED SEQ. NO.	RUN FEED HOURS	SCN MTC LINE	FLOW COOL'T SCHED.
1.															
2.															
3.															
4.															
5.															
6.															
7.															
8.															
9.															
10.															
11.															
12.															
13.															
14.															
15															
16															
17															
18.															
19															
20															
21.															
22.															
23.															
24.															
25															
26.															
27.															
28.															
29.															
30.															
31.															
32.															
33.															
34.															
35.															
36.															
37.															
38.															
39.															
40.															
41.															
42.															
43.															

FORM 135-C-4 (NEW 2-67) | NO. OF PIECES RECEIVED | LH/N | RH | BY STOCKROOM NO. | CLERK'S SERIAL NO. | DATE

Fig. 14.4 *Typical shop traveler.*

agement oriented and is not concerned with the acceptance or rejection of individual units of product. For example, we know we have accepted most of the units submitted to inspection. We have returned some for rework, scrapped some, and downgraded some others to be sold at marginal profit. But what is the "quality" of the total production flow? Is it

SECTION "A"

E.O. NUMBER (AND DWG. IF APPLICABLE)	CHG. CTR.	TYPE	CHG. PT.	DESCRIPTION OF CHANGE/REMARKS	ORIG. DEPT.	ENTRY DATE	MECH. DATE	INSP. SQ REQ	INSP. STAMP

MATERIAL REVIEW/INSPECTION RECORD

ACTION NO	O.K. TO USE LH/N	O.K. TO USE RH	SCRAP LH/N	SCRAP RH	ACCOUNT NO.	DATE	M/R STAMP	M/R ENG. SIGN	INSPECTION SUPERVISION SIGN	MRD NO
1										
2										

SQUAWKS/DEFECTS/REWORK INSTRUCTIONS

SQ/ DEF. NO.	RWK. NO	DEPT.			DATE	MECH. LDM.	QTY. ACCEPTED LH/N	QTY. ACCEPTED RH	DATE	INSP.	DEPTH OF CUT	SET-UP SPEED SEQ. NO.	RUN FEED HOURS	SCN MTC LINE	FLOW COOL.T SCHED.

SQUAWK/DEFECT VERIFICATION				REWORK AUTHORIZATION				
SQ/ DEF. NO.	INSPECTOR SIGNATURE	DATE	PROPORTION WITHHELD	RWK. NO.	MR INSP. SIGNATURE	DATE	INSPECTOR SIGNATURE	DATE
	INSP. SUPVR. SIGNATURE	DATE			MR ENG. SIGNATURE	DATE	INSP. SUPVR. SIGNATURE	DATE
SQ/ DEF. NO.	INSPECTOR SIGNATURE	DATE	PROPORTION WITHHELD	RWK. NO.	MR INSP. SIGNATURE	DATE	INSPECTOR SIGNATURE	DATE
	INSP. SUPVR. SIGNATURE	DATE			MR ENG. SIGNATURE	DATE	INSP. SUPVR. SIGNATURE	DATE

Fig. 14.4 (*Cont.*)

good enough? Are there specific equipments, operations, or operators that are causing quality problems? These second-order questions are usually much more difficult to answer and frequently much more important in the long run.

To answer these questions, it is necessary to collect, group, and analyze certain quality data. Most companies establish a formal reporting

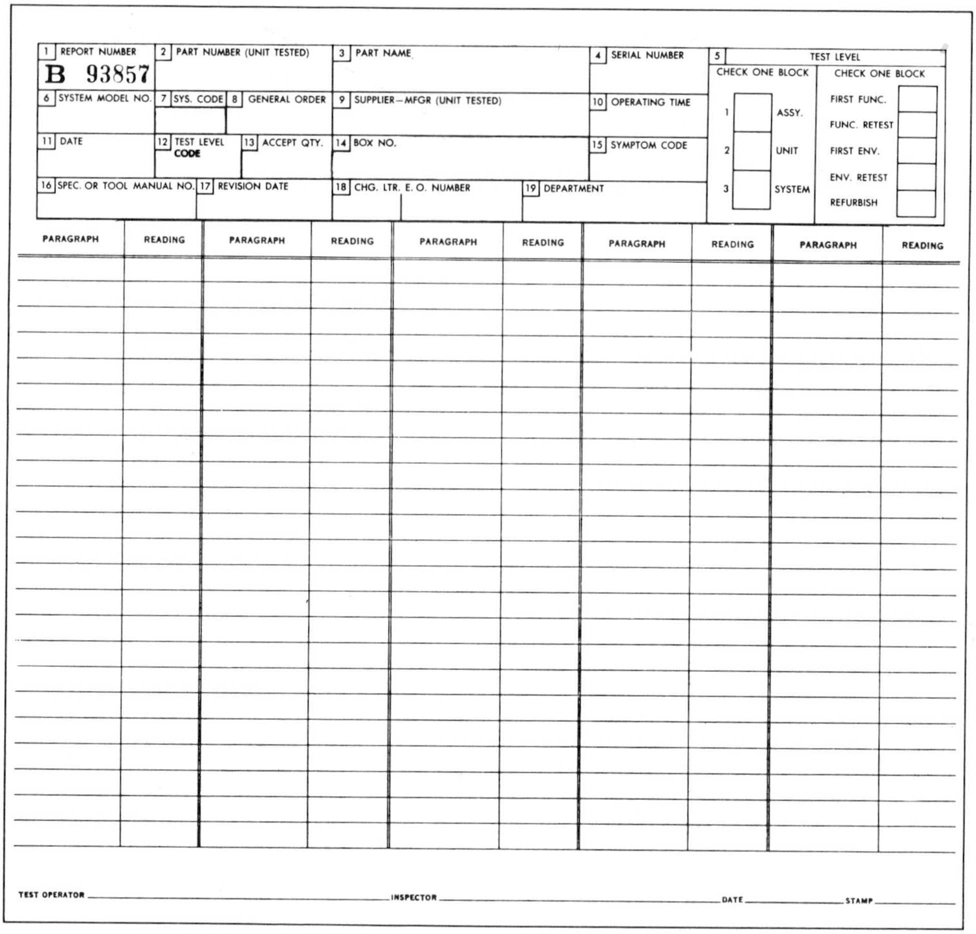

1 REPORT NUMBER	2 PART NUMBER (UNIT TESTED)		3 PART NAME		4 SERIAL NUMBER	5 TEST LEVEL			
B 93857						CHECK ONE BLOCK		CHECK ONE BLOCK	
6 SYSTEM MODEL NO.	7 SYS. CODE	8 GENERAL ORDER	9 SUPPLIER—MFGR (UNIT TESTED)		10 OPERATING TIME	1	ASSY.	FIRST FUNC.	
								FUNC. RETEST	
11 DATE	12 TEST LEVEL CODE	13 ACCEPT QTY.	14 BOX NO.		15 SYMPTOM CODE	2	UNIT	FIRST ENV.	
								ENV. RETEST	
16 SPEC. OR TOOL MANUAL NO.	17 REVISION DATE		18 CHG. LTR. E. O. NUMBER	19 DEPARTMENT		3	SYSTEM	REFURBISH	

PARAGRAPH	READING	PARAGRAPH	READING	PARAGRAPH	READING	PARAGRAPH	READING	PARAGRAPH	READING

TEST OPERATOR ________ INSPECTOR ________ DATE ________ STAMP ________

Fig. 14.5 *Typical functional test data sheet.*

system to provide the data necessary to answer these more subtle questions and to form the basis for decisions. Figure 14.6 is an example of a transmittal which provides data on the result of product inspections and is exemplary of the type of data needed.

14.11.7.5 Quality Information Processing. In developing and implementing a quality information system, there are a number of factors to consider. The following are some of the more important elements:

A. Data versus information. Contrary to popular opinion, data are not necessarily information. Many information systems have been designed by determining the data that should be collected and then attempting to manipulate the data to provide certain information. Frequently, the information provided in this manner is of little value in decision-making. As difficult as it may be, the design of a quality information system must be done in the reverse order from which it is operated. The design process should follow this sequence: what questions need to be answered, what information is required to answer these questions, and what data are necessary to develop this information? Adherence to this approach will usually result in obtaining the required information at a minimum of cost and effort.

B. Manual versus mechanized. Recent advances in electronic data-processing equipment used in many industrial organization pose the question of whether quality information should be processed by manual or mechanized means. Since quality information usually involves a large number of transactions which must be grouped in several ways to provide meaningful information, it is ideally suited for the application of mechanized processing systems. Generally speaking, the decision whether to use a manual or mechanized system is a question of economic analysis of the alternatives. This analysis should consider the cost implications of having greater data flexibility when using the various systems.

C. Timeliness of data. Timely data are not necessarily instant data. Instant data usually cost a lot of money—wasted money if they are not required. Product acceptance information is needed immediately, but the inspector generates that. Process control information is not required immediately but can be accumulated on a longer time base. Management data can usually afford an even longer time base. For this reason, most data systems will have multiple reporting time bases, usually weekly and monthly, and sometimes daily. If necessary, "instant upon demand" capability can also be included.

D. Regular versus exception reporting. A classic argument in quality information circles concerns the question of regular versus exception reporting.

Regular reporting implies periodic reports generated in an identical fashion, regardless of the state of the process being controlled. Proponents of this approach state that those responsible for acting on the data must be kept in constant touch with the process. They must watch for trends as well as symptoms of a potential out-of-control condition, many of which could not be programed into an exception-reporting scheme.

Opponents argue that no member of management should have to review data from a properly operating process. These people prefer exception reporting; that is, a system that reports when, and only when, predetermined control limits have been exceeded. A combina-

QUALITY CONTROL INFORMATION TRANSMITTAL LOG SHEET

DEPT. | WK CTR

ITEM	PART NUMBER	SERIAL NO	TYPE INSP			GEN ORDER NO	DATE	INSPEC STAMP NO	WORK AUTH			LOT		INSPEC OPER NO	MANUF		DEFECTS			QUANT DEFECTIVE	ACTION C.A.R. NO
			N	R	A				P	E	M	SIZE	DEFECTIVES		L/M NO	MECH NO	CODE	CL	QTY		
1																					
2																					
3																					
4																					
5																					
6																					
7																					
8																					
9																					
10																					
11																					
12																					
13																					
14																					
15																					
16																					
17																					
18																					
19																					
20																					
21																					
22																					
23																					
24																					
25																					
26																					
27																					

Fig. 14.6 *Typical physical inspection data transmittal.*

tion of these two approaches is usually appropriate, providing reporting at a summary level of in-control processes and a "spill-out" of detailed data by exception.

14.11.7.6 Quality Information Analysis. Analysis of quality information will provide visibility of the overall quality status of products and processes and provide a basis for the allocation of quality resources to correct or improve the existing situation. Analytical methods will vary considerably and will depend upon the nature of the product and the processes employed. Space does not permit a detailed review of analytical approaches; however, there are certain primary considerations in the analysis of most quality information, and these will be briefly discussed.

Generally speaking, the majority of quality problems will be associated with a fairly small number of products or processing operations. The analytical techniques used should, therefore, be directed toward isolating the principal problems in relation to the total deficiencies that exist. This will provide a means for determining which items are the most important in allocating resources for further analysis.

Information systems seldom provide the information to determine the specific cause of a problem but can provide a "road map" to guide further investigation. If the data, for example, reveal that there are excessive solder defects on a part, floor investigation will usually be necessary to determine if the solder being used is the proper type, if the operator has been properly trained, or if the construction of the part makes a particular soldering operation extremely difficult. None of these questions can be answered merely by reviewing the data in the quality information system.

Normally data are most meaningful if related to a specified base. Functional failures, for example, have little meaning unless related to the number of items tested or the number of hours of test time involved. Bases used in quality control are defects per unit, per cent defective, defects per production hour, and similar criteria which provide a measure of perspective. In some cases, it may be desirable to establish expected product or process discrepancy rates based upon previous history or analysis of the operation. These may then be used as criteria for exception reporting and analysis when the rates are exceeded.

14.11.7.7 Corrective Action. The primary purpose of obtaining quality information is to provide a basis for correcting unsatisfactory conditions. Corrective action is usually one of the most difficult but most meaningful activities of the quality control function.

Corrective action includes the identification of a problem, the determination of the basic cause of the problem, and the correction of the conditions or factors which caused the problem. Problems must be investigated in sufficient depth to identify the basic cause. For example, superficial investigation might conclude that "operator error" was the cause of the problem. However, a detailed review might show that the instructions to the operator were incorrect or ambiguous, that the tools provided to him were in error, or a number of other factors were involved such that the operator himself was not at fault.

It is normally advisable to document corrective action so that subsequent follow-up can be made to assure that the recommended action was taken and was effective in reducing or eliminating the problem. Figure 14.7 is an example of a form used to obtain and document corrective action. The form includes a section for defining the condition requiring action, a section to document the results of further investigation and related information, and a section outlining the specific action taken to correct the problem. The importance of establishing and maintaining an effective corrective action system cannot be overstressed.

14.11.8 Packaging

14.11.8.1 Delivered Quality. Unless the product is delivered to the customer in a satisfactory condition, the quality job has not been carried to completion. The product may be perfect when it leaves the factory, but if it is received by the customer in a damaged condition, much customer dissatisfaction and monetary loss may result.

14.11.8.2 Special Protection Requirements. The nature of the product may require special protection from the following: elevated temperatures, freezing or sub-zero temperatures, sudden changes in temperature, humidity (high or low), sunlight, dust, microorganisms, air (oxygen), vibration, shock, abrasion (from packing materials), water damage, absorption

CORRECTIVE ACTION REQUEST NO. A 27451

PROBLEM AREA	DATE	G.O.	REASON CODE	SUPPLIER CODE	TO: DEPT: DATE:

PART NO./PROCESS	SERIAL NO.	NOMENCLATURE	FROM: DEPT:

NEXT ASSY NO.	DEPT.	WORK CENTER	"P" STAMP NO.	MECH. NO.	INSP. NO.	INSP. OPERATION	MFG. OPER.

ITEM	CONDITION REQUIRING ACTION	DEFECT CODE	CLASS	QTY	REF. - SYM. NO.

ORIGINATOR	DEPT.	EXT.	DISCREPANCY CONCURRENCE FOR CERTIFIED OPERATIONS

ITEM	CAUSE OF CONDITION AND/OR HISTORICAL DATA	MAT DISP	CAUSE CODE

QUALITY CONTROL CONCURRENCE	DEPT: DATE:	ANALYSIS PERFORMED BY:	DEPT: DATE:

RETURN COMPLETED REPORT TO ______________ DEPT ______ WITHIN ______ DAYS AFTER RECEIPT.

ITEM	CORRECTIVE ACTION	CA CODE	CA DOC NO.

RESPONSIBLE INDIVIDUAL	DEPT: DATE:	QUALITY CONTROL APPROVAL	DEPT: DATE:

Fig. 14.7 *Typical corrective action form.*

of off-odors or off-flavors, and many other factors. It is possible to protect the product from any or all of these factors by means of special packaging and handling. Such protection may involve hermetically sealed packages, refrigeration, desiccation, grease or plastic coatings, special wrappings, and floating packages. The job of the packaging engineer is to specify the proper protection and package for a particular product, but it is often the responsibility of the quality control engineer to assure the adequacy of such protection.

14.11.8.3 Packaging Tests. Packaging tests may be classified as follows:

A. Tests on the material of containers, such as the Mullen or Cady tests.*

B. Tests on the packaged product to measure adequacy of the package to protect the product against handling or in-transit damage caused by impact. Such tests would include the drop test, compression test, vibration test, revolving drum tests, and the Conbur test.†

C. Actual trial shipping test with examination of product at destination.

All three types of tests have their purpose and should be used. Tests should be conducted for quality characteristics of packaging material as for other purchased material.

Carefully controlled tests, such as the second type, offer an advantage of comparing different packages under controlled conditions. This comparison overcomes a serious drawback to the third type, where it is difficult to assure the same handling for various trial shipments and difficult to assure that representative severity of handling was present during the respective trial shipments.

Since it is not always possible to duplicate actual conditions in laboratory tests, it is advisable to have the added protection of actual shipping tests. These tests should duplicate the mode of transportation and distances that will be encountered in getting the finished product to the customer.

14.12 ACTIONS SUPPORTING THE PRODUCT AFTER DELIVERY

14.12.1 Product Support

After the product has been designed, fabricated, and delivered, it is often assumed that the quality control activity is completed. However, continued successful functioning of the enterprise is largely dependent on maintaining satisfied customers who will be interested in subsequent purchases from the firm. For many products, the degree of customer satisfaction is strongly influenced by the effectiveness of the company in supporting the product after delivery. This includes adequate instructions for operation or maintenance of the item, satisfactory spare parts, and other factors. When the amount of activity subsequent to delivery is large, many companies assign these functions to a product support organization. Quality control frequently plays an active role in assuring that the work performed by this organization is proper insofar as quality factors are concerned.

14.12.2 Control of Service Publications

Many products require the use of instructions for operation or service of the articles sold. Since the first impression of product quality is often influenced by the quality of these instructions, their adequacy and completeness may be equally as important as the item itself. Most people have used assembly instructions for Christmas toys and have experienced the frustration that occurs when the instructions are incomplete, incorrect, or ambiguous. It is important, therefore, that adequate controls be established to assure the "quality" of these documents. In this regard, several techniques are quite useful.

The first technique is review of the document by an independent group or individual not intimately familiar with the product. Frequently,

* Standard procedure for making these tests can be procured from the American Society for Testing Materials.

† The Conbur Incline Testing Device is described in *Bulletin No. 511* of the Freight Loading and Container Bureau of the Association of American Railroads, 59 East Van Buren Street, Chicago, Illinois 60605. The Freight Loading and Container Bureau also provides many pamphlets covering the packaging of a variety of articles of commerce. They also maintain a Container Research and Development Laboratory in Chicago to aid industry in eliminating the causes for loss and damage due to faulty shipping practices. The American Management Association, 330 West 42nd Street, New York, New York 10018, has also published a packaging series of pamphlets which are helpful in analyzing and solving packaging problems.

the individual preparing the instruction is so familiar with the product that he overlooks the inclusion of information that he takes for granted but that may not be generally understood by the user. This type of review can point up areas of incomplete or ambiguous information.

The second technique is validation of the document through a demonstration of its use; i.e., actually following the instructions on a step-by-step basis to detect errors or omissions. This technique is particularly useful with complex products because of the difficulty the writer may have in visualizing the many interrelationships that exist. Problems encountered during the demonstration should be documented and reviewed against the final draft to assure that corrections have been made.

14.12.3 Control of Spare Parts

Many companies manufacture and distribute spare parts for use in the maintenance and repair of their product. These items, of course, are controlled by the normal techniques which apply to regular production items processed through the shop. However, several additional factors must be considered for spare parts which may not apply to the manufacture of the regular product.

The amount and type of receiving inspection, for example, may be based on subsequent inspections and tests that the item receives during fabrication and assembly. If this item is delivered directly as a spare part, it will not receive the benefits of these inspections and tests. A good example is a bolt that is used in large quantities in a production line. It may be much cheaper to throw away an occasional bad bolt in the production line and reach in the bin for another than to establish extensive receiving inspection operations for bolts. However, if this bolt is sold as a spare part and the customer buys only one, he will probably be extremely dissatisfied if it is not acceptable and cannot be used. Thus it can be seen that the quality requirements for spare parts need to be carefully reviewed separately from the requirements governing the use of the part in production and additional inspections and tests included.

14.12.4 Modification and Repair

It is not uncommon for a company to conduct modification and repair of the product. This work may be performed in the same manufacturing areas as new production or in "service centers" specifically set up for such work. An important difference exists in these operations as contrasted with new production; viz., the quality criteria or standards which are used. Many standards that are appropriate to new production are not consistent with modification and repair standards. For example, deficiencies in appearance which would not be tolerated in new products may be acceptable in a product that has been used if the cost of correcting the condition is greater than the value received. As a result, it is important that criteria used for determining acceptability be clearly established. In this regard, many companies prepare specific quality standards for use in overhaul and repair operations.

14.12.5 Field Results

14.12.5.1 Field Results as a Measure of Quality Performance. Since customer satisfaction is an important goal of the quality control program, it is desirable to obtain some measure of customer satisfaction. Customer complaints provide a measure of dissatisfaction, an inverse measure of satisfaction. Caution must be exercised in using the complaint rate as a quality index since other factors are involved, such as market conditions, inventories, styles, and general attitude of the buying public.

14.12.5.2 Measuring the Complaint Rate. Several different bases may be used for measuring the complaint rate. The following are examples:

A. Percentage of total units produced that resulted in complaints. Usually a definite warranty period is established on a product. The percentage should be calculated from the ratio of units on which in-warranty complaints were received compared to the number of units in the field on which the warranty is valid. The latter figure can be obtained from shipments, allowing suitable time lag for the product to get into the hands of the customer.

B. Ratio of in-warranty expense to manufacturing cost of units exposed in the field. Unfortunately, this index is affected by factors related to product service costs, such as increased labor rates for repairmen, improvement in service through establishment of more and better service repair shops, and increased handling costs due to changes in shipping rates. This is also a quality cost and should be available from the quality cost system (see Art. 14.6).

C. Service calls per thousand units sold and on which warranty is still valid.

The chief value in these indices is in showing trends; however, caution must be used in evaluating extraneous causes for trends, as cited above.

14.12.5.3 Determination of Causes for Complaints. This is usually a difficult task. In fact, it is often impossible to get a statement on the specific complaint of the customer for small appliances, since the dealer sends the customer's item in with a notation "repair-in-warranty." This leaves the complaint a "guess" on the part of the repairman. Even after the complaint has been stated, there may be several possible causes, none of which is clearly defined. These problems are not as great for major industrial products or military hardware since the company can maintain skilled representatives in the field to accumulate this information.

The manager of quality control endeavors to obtain causes for complaints from the repairman or field service organization. The specific information may be collected using check lists or reports that the repairman or field representative prepares for each article returned for repair or replacement. Since a specific defect might be described in several different ways, it is advisable to issue a glossary of terms so that all personnel will standardize on the description of a defect or cause for complaint.

Sometimes it is too expensive to process all reports from the field. In such cases, a sample may be taken or only certain districts may be requested to report; however, care should be taken to avoid a biased sample caused by differences between districts, such as climate and conditions of use. Replacement-part consumption data also provide information on causes for failure of the product in the field.

Complete and sole reliance should not be placed on written reports or check lists from the field. An occasional visit to repair stations by quality control personnel can be of tremendous value. There is no substitute for first-hand investigation and analysis of items returned because of complaints. In some instances, the quality control engineer should request that certain failed products be sent to him at the factory for his personal investigation if the size of the item permits.

14.12.5.4 Use of Information Obtained from the Field. *The only real value of field data lies in the corrective action they initiate.* Such action may be improved design, improved test and inspection methods, better packaging, or even a better instruction book. The quality control organization has a responsibility to see that the proper individuals in the organization take the corrective action that is needed.

BIBLIOGRAPHY

Feigenbaum, A. V., *Total Quality Control.* New York: McGraw-Hill Book Co., 1961.

Grant, Eugene L., *Statistical Quality Control*, 3rd ed. New York: McGraw-Hill Book Co., 1964.

Harris, Douglas H., "Development and Validation of an Aptitude Test for Inspectors of Electronic Equipment," *Journal of Industrial Psychology*, Vol. 2, 1964.

Harris, Forest K., *Electrical Measurements.* New York: John Wiley and Sons, Inc.

ISA F-6 Environmental Committee Report, "Recommended Environments for Standards Laboratories." *ISA Transactions*, Vol. 3, Issue 4, October 1964.

Javitz, A. E., "Editor's Commentary on Fundamentals of Measurement." *Electro-Technology*, May 1963. C-M Technical Publication Corp., New York, N. Y.

Juran, J. M. *Quality Control Handbook*, 2nd ed. New York: McGraw-Hill Book Co., 1962.

Koontz, Harold and Cyril O'Donnell, *Principles of Management*, 4th ed. New York: McGraw-Hill Book Co., 1968.

Optner, Stanford L., *System Analysis for Business Management*, 2nd ed. Englewood Cliffs, N. J.: Prentice-Hall, Inc., 1968.

U. S. Department of Commerce, *National Bureau of Standards Handbook* 77, a Compilation of Previously Published Technical Papers by the Staff of the NBS. Vols. I, II, and III.

U. S. Government Printing Office, MIL-C-45662A, "Calibration System Requirements," February 9, 1962.

U. S. Government Printing Office, MIL–I–45208A, "Inspection System Requirements," December 17, 1963.

U. S. Government Printing Office, MIL–Q–9858A, "Quality Program Requirements," December 17, 1963.

MYRON LIPOW *is manager of product assurance, Science and Technology Division, of TRW Systems Group, TRW, Inc. He joined TRW Systems in 1958 and has participated in planning, management, and evaluation of reliability programs for the Thor, Atlas, Titan and Minuteman rocket engines. He now has responsibility for managing the reliability and quality assurance programs for ballistic missile and spacecraft mechanical and propulsion subsystems and for studies related to military and space operational systems.*

Before joining TRW, Mr. Lipow's experience was in the application of mathematics and statistics to reliability analysis, quality engineering, and performance analysis of solid and liquid-fueled rocket engines at the Aerojet-General Corporation. He did both theoretical and experimental work on combustion stability and thermal stress analysis of solid fuel rocket motors as well as design and development testing of various rocket engines.

Mr. Lipow is a member of the American Institute of Aeronautics and Astronautics and the American Society for Quality Control. He is coauthor with D. K. Lloyd of the book Reliability: Management, Methods, and Mathematics. *He also has authored many papers for national and regional meetings of AIAA and ASQC. He received his B.S. degree in mathematics from the California Institute of Technology.*

DAVID K. LLOYD, *reliability manager, Science and Technology Division, TRW Systems, received his Bachelor's and Master's degrees in mathematics and statistics from Oxford University in England. For the last seventeen years he has devoted himself to the application of statistical techniques to reliability engineering and evaluation. He has made significant contributions to the development of management and statistical techniques used by NASA and the Air Force in reliability evaluation and demonstration programs for solid-and liquid-propellant rocket engines and reentry systems.*

As reliability manager, he has been responsible for directing the reliability programs for both DOD and NASA contracts. These contracts have included propulsion, mechanical, and ordnance subsystems on ballistic missiles and spacecraft.

Prior to joining TRW, Mr. Lloyd was a member of the corporate staff, reliability control activities, of Aerojet-General Corporation. While there he developed experimental programs for qualification testing of both solid- and liquid-fueled rocket motors.

He is coauthor with M. Lipow of the book Reliability: Management, Methods, and Mathematics, *and has authored many other reports and papers on the various aspects of reliability.*

SECTION 15

RELIABILITY METHODS

MYRON LIPOW
AND
DAVID K. LLOYD

The subject of reliability has developed considerably since the mid 1950's but because of space limitations, we can present only a small portion of the topics concerning the management and statistical techniques used in reliability programs here. We have therefore attempted to give only an outline of the subject in the section on reliability methodology and to include only the most useful analytical techniques and tables in the section on reliability mathematics.

Over two dozen books and hundreds of papers have been published since 1948, when the earliest articles on reliability were written. Thus for a treatment in depth the reader should consult the references listed at the end of this section.

15.1 RELIABILITY METHODOLOGY

15.1.1 A Definition of Reliability

Reliability is the probability of the successful operation of a device in the manner and under the conditions of intended customer use. Examining this definition closely, we see that reliability is a probability statement which contains qualifying conditions open to interpretation. Thus we must understand the subject of probability and the associated framework of definitions and practical considerations before making a reliability statement which is free from ambiguity.

15.1.2 Reliability as a Probability Concept

To evaluate reliability one must utilize statistical estimation theory in all except trivial cases, such as when the whole population has been used; e.g., "In 1957 the reliability of rockets used for aircraft-assisted take-off was 99.9 per cent." This figure is simply the ratio of all successes to all attempted operations. No attempt is made in this statement to predict what might have happened in 1959 or, if in 1957 only half the rockets had been used, what the expected reliability of the remainder would have been. Probability is introduced when statements are to be made about a population based on observations obtained from only a sample or portion of that population or when we are trying to predict in advance the outcome of events.

In predictions, statistical theory allows us to associate levels of assurance or confidence with any reliability estimate we make. Thus, we can be 50 per cent sure that the true reliability will be above or between certain specified values. On the other hand, we might wish to be more confident about our predictions. We might want to be 90 per cent sure, 95 per cent sure, or perhaps 99 per cent sure, or even more. We can never be 100 per cent sure except where (as in the exceptional case cited above) we have complete experience of our item, and of course, here we are no longer predicting. However, the greater the assurance we wish to have in any predictions, the more conservative our predictions must be if the observed sample size is not increased. Thus, the reliability limits of our predictions must be wider for increased assurance. Alternatively, if we wish to remain above or within specified limits, such as in contractual reliability demonstration, we must be prepared to take a larger sample or to reduce the confidence level. To illustrate these ideas with numbers consider the following example.*

Suppose we had tested fifty items and experienced two failures. What could we say about the reliability? One estimate could be 96 per cent, another estimate 90 per cent, and still another 84 per cent. To any reader not acquainted with statistics, having three different values of reliability (in fact, an infinity of values) estimated from our observations is confusing. However, there is no contradiction here, just a necessity for understanding the ground rules that these estimates represent. The first value, 96 per cent, is simply the ratio of successes to the number of tests, the point estimate. The second and third reliability numbers are both confidence limits; i.e., lower bounds for the true reliability. In the second case, we are 90 per cent confident that the true reliability is greater than 90 per cent. In the third case, we are 99 per cent confident that the true reliability is greater than 84 per cent. As we pointed out earlier, with higher confidence, our statements become more conservative. The first value, 96 per cent, the point estimate of reliability, sometimes called the best estimate, does not have any specific confidence level associated with it. It would also represent the true reliability if, for instance, fifty items were the total number manufactured, and only two failed in all the experience with them. However, if we suppose this observation is only a sample of perhaps a thousand such items, we might ask, "What is wrong with our point estimate?" Why should we not always use this estimate? The answer is that nothing is wrong. However, the point estimate is not very discriminating. The actual value of the point estimate is the same whether there is one failure in twenty-five tests or twenty failures in 500 tests; i.e., both sets of numbers give estimates of 96 per cent. It is obvious, though, that we should have more confidence in any statement made from 500 observations than in that made from twenty-five or fifty. Statistical theory provides us with this confidence in a quantitative form. Thus, with twenty failures in 500 tests, we can state with 90 per cent confidence that the true reliability is greater than 95 per cent; whereas, with one failure in twenty-five tests and still with 90 per cent confidence, we can state that the true reliability is greater than only 85 per cent.

Reliability estimates are open to a great amount of statistical manipulation. None of the above figures is inconsistent; but when we are specifying, estimating, or otherwise quoting reliability numbers, we must be quite clear as to their statistical meaning as well as their practical implications in terms of numbers and cost of tests.

15.1.3 Reliability as a Function of Success Criteria

The definition of reliability contains the phrase *successful operation*. Consider the defini-

* The numbers quoted in this discussion are obtained from Fig. 15.1.

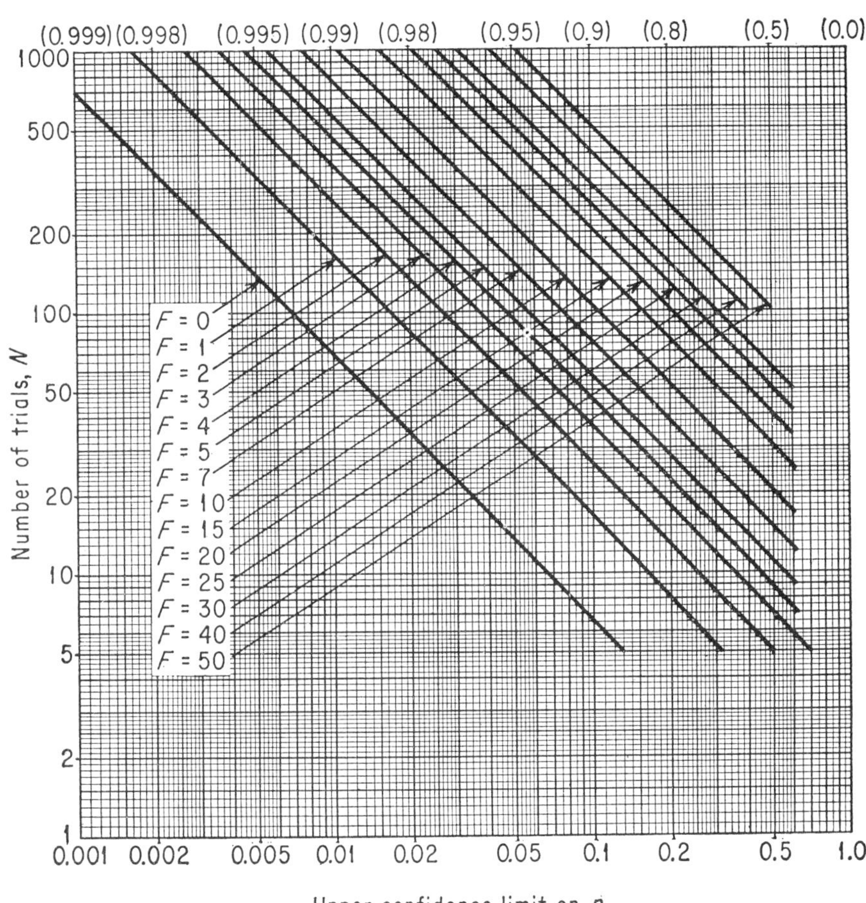

Fig. 15.1(a) *Lower confidence limit on* R. *Upper confidence limit on unreliability (one minus lower confidence limit on reliability). Number of trials* N. *Observed failures* F. *(a) Confidence coefficient* $\gamma = 0.50$. *Graphs a-e are from David K. Lloyd & Myron Lipow,* Reliability: Management, Methods, and Mathematics. *© 1962. Reprinted by permission of Prentice-Hall, Inc. Englewood Cliffs, N.J.*

tion of success and some of the questions it raises. If we have a piece of equipment which operates without failing for a given time and if this time is not equal to the customer's expected use time, is this test a success? To begin, it should be stated that it is not a failure. If it is a success, is it a complete success? If not, should it have some weighting factor associated with it before being placed into a reliability equation? For example, what is the degree of success of 1 hour of operation of a computer required to operate for only 2 minutes? Again, what is the degree of success of a rocket engine test fired without failure for 20 seconds when it has to fire for 2 minutes in operational use? Further, what is the degree of success when under test conditions we cannot simulate all the environmental conditions which will be experienced in operational use?

To see how the test criteria should relate to the end usage we can consider the following example. A rocket engine giving only 95 per cent of the total impulse necessary to get its satellite into orbit is a complete failure; on the other hand, a rocket engine giving 95 per cent of the total impulse necessary to send its warhead to a maximum-range destination would destroy 100 per cent of a close-in target. Therefore in choosing the ground rules for success and failure of any test, we should carefully

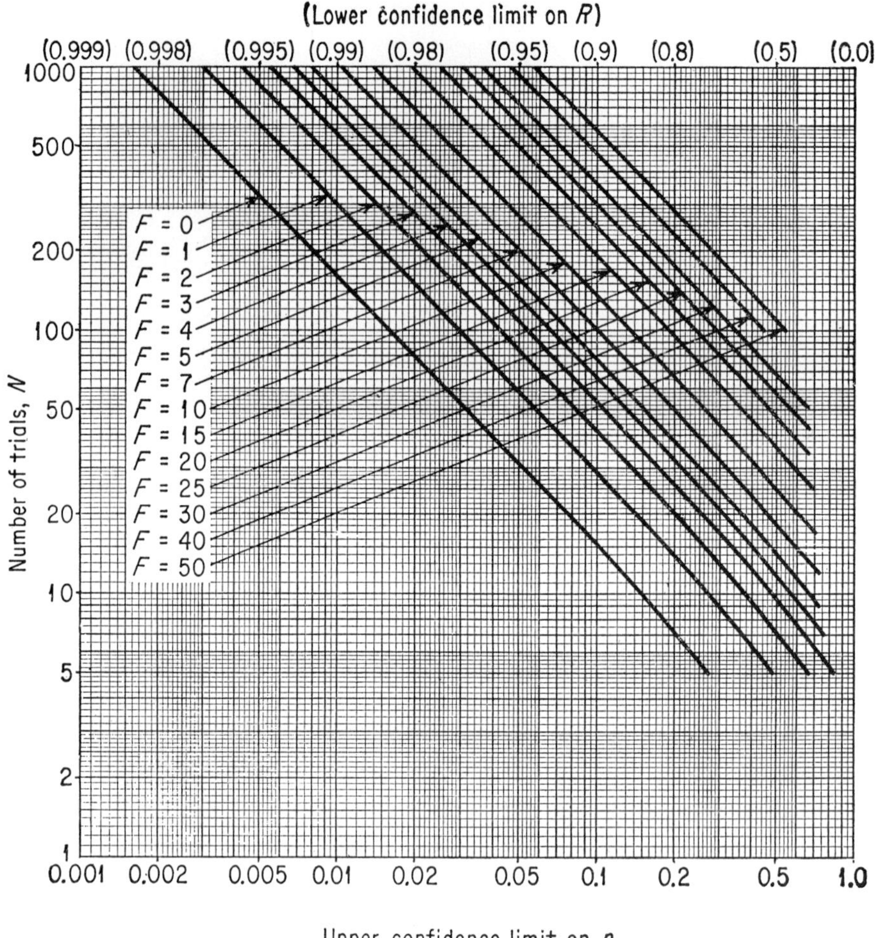

Fig. 15.1(b) *Lower confidence limit on R. Upper confidence limit on unreliability (one minus lower confidence limit on reliability). Number of trials* N. *Observed failures* F. *Confidence coefficient* $\gamma = 0.80$.

relate them to the various modes of operation. There is, however, no reason why we cannot establish several sets of criteria to evaluate the data against different end requirements.

15.1.4 Nonrepresentative Configurations and Tests

Another feature of reliability evaluation with which we must contend is that of interpreting test results of nonrepresentative equipment. During any development program we have a continuing modification of items or configurations or both, but we must constantly know how good our equipment is and how rapidly it is improving toward the end product. Therefore, we must consider how to utilize the results of tests of systems or subassemblies which have components that are nonrepresentative of the final configuration. Similarly, we must determine the use we can make of separate component tests and how subscale test results are related to the full-scale unit. We must also consider how to use data from tests which have specifically been made to fail or which have failed because of human error or test-equipment malfunction.

These, then, are some general problems of reliability estimation. Obviously, we must take great care to use the pertinent data, to make the correct interpretation of the results, and to apply the most appropriate statistical methods for reliability estimation. We can solve many of the problems by use of appropriate statistical

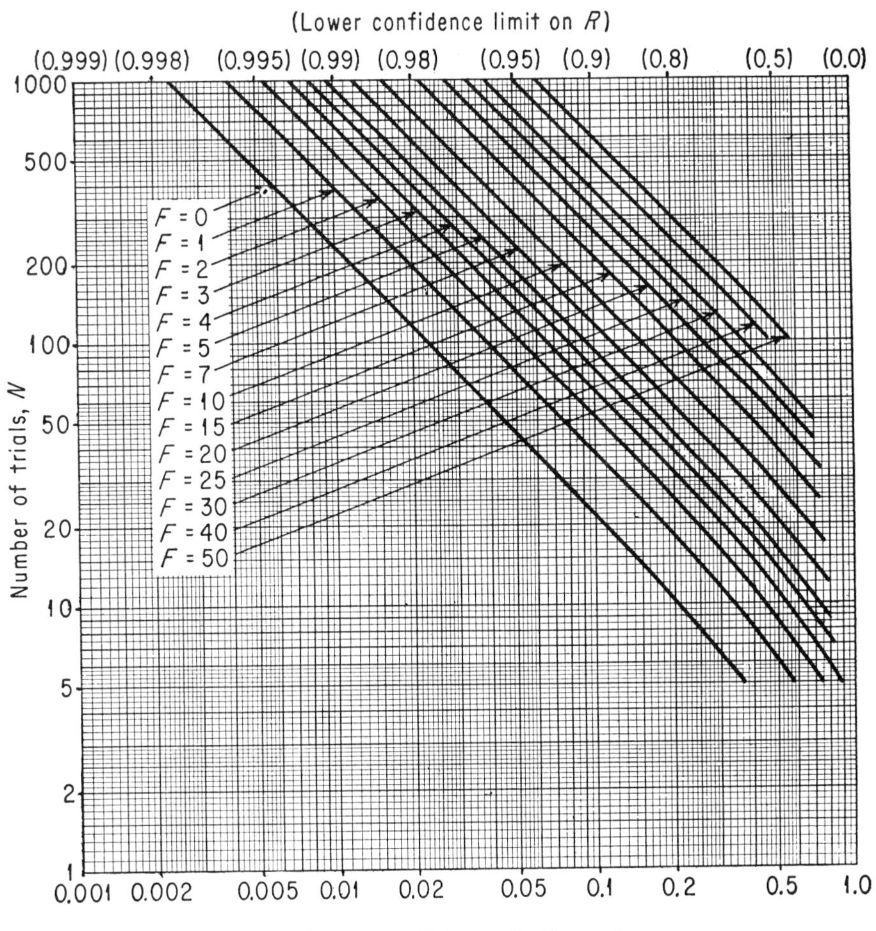

Fig. 15.1(c) *Lower confidence limit on R. Upper confidence limit on unreliability (one minus lower confidence limit on reliability). Number of trials* N. *Observed failures* F. *Confidence coefficient* $\gamma = 0.90$.

techniques, others by the introduction of careful controls of the reliability information, by common sense, or by arbitrary decision. Section III of Lloyd and Lipow, *Reliability: Management, Methods, and Mathematics** discusses these problems at length in terms of specific examples.

15.1.5 Reliability Program Requirements

A specific numerical reliability requirement is the most effective way of directing attention toward reliability whether or not it is a contractual requirement. The most important reliability number is that for the highest level of assembly or system for which the organization concerned has responsibility. Thus, if the organization were one of the armed services, the reliability of prime concern would be that of the complete weapon system. If the organization were a small company, it might be the reliability of, for example, a miniaturized computer.

The first step is to see what the system reliability requirement demands from its subsystems and components. We must then relate these demands to what can be achieved. In order to do this, we must simultaneously review the reliability interrelationships among the components of the system, their actual and

* Hereafter, reference to this text will be denoted by *R:MMM*.

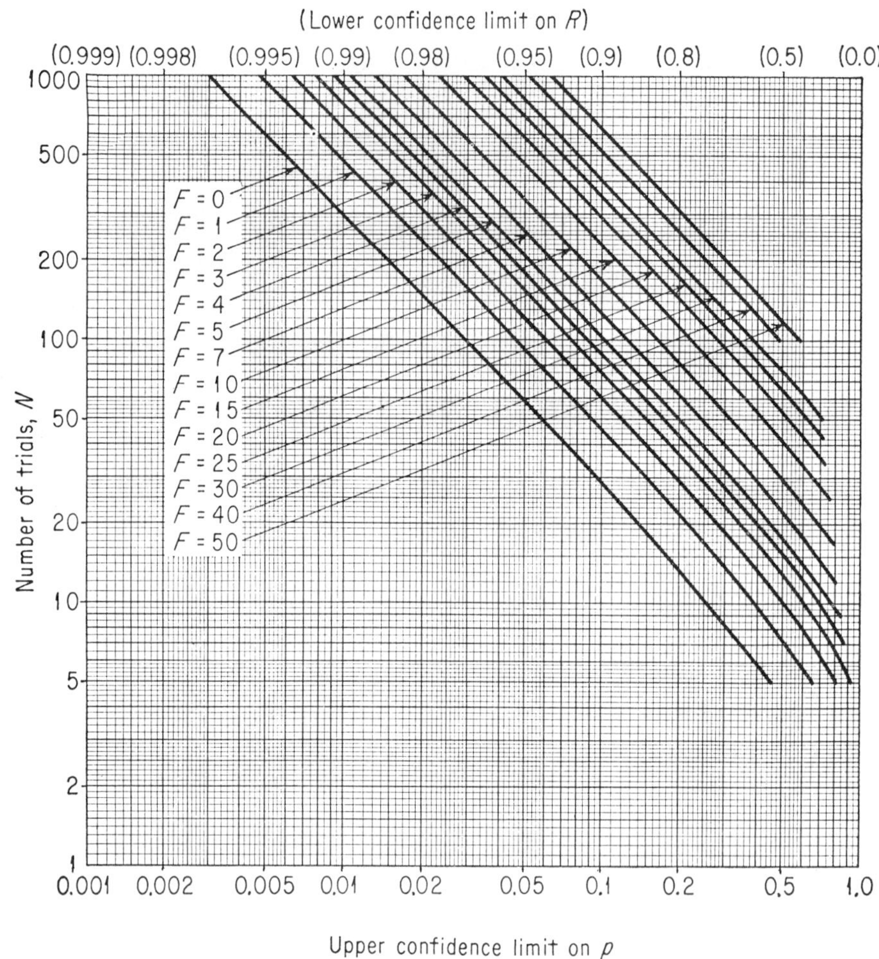

Fig. 15.1(d) *Lower confidence limit on R. Upper confidence limit on unreliability (one minus lower confidence limit on reliability). Number of trials* N. *Observed failures* F. *Confidence coefficient* $\gamma = 0.95$.

potential or expected reliabilities, and the size and adequacy of the planned development program. The techniques for carrying out this study are reliability apportionment, reliability design analysis, including failure mode, effects, and criticality analysis, and test and experimentation review (as discussed in Chaps. 12 and 13 of *R:MMM*). As we follow the development program we continuously reappraise these reliability objectives, and in addition, we introduce those activities necessary for the efficient handling of the program. These activities are design control, vendor control, reliability data reporting and follow up, and reliability activity and information coordination. We shall now discuss these topics in detail.

15.1.6 Reliability Apportionment

Reliability apportionment is the process of subdividing the system reliability requirement into subsystem and component requirements. In apportionment the reliability requirements established for each of the subsystems are made compatible with their current state of development, expected improvement, and amount of testing and money budgeted for their development. When the component requirements are recombined, their reliability interrelationship should be such that they satisfy the system requirement. Apportionment has its greatest value at the first level of breakdown of a system

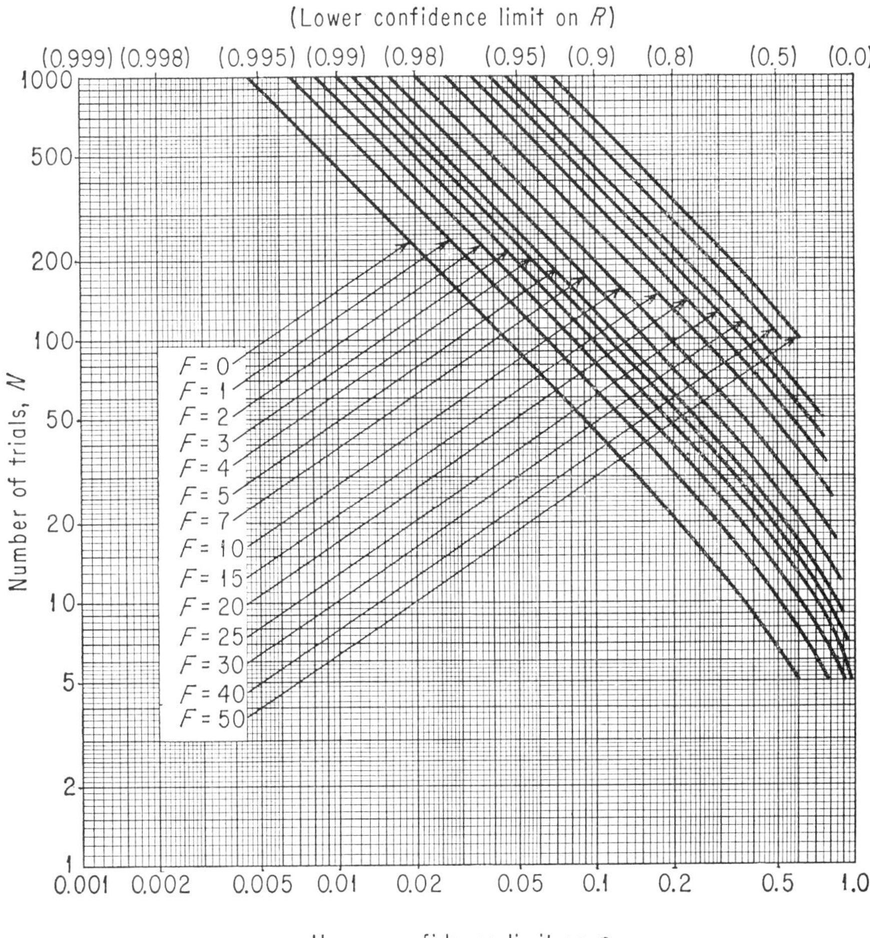

Fig. 15.1(e) *Lower confidence limit on R. Upper confidence limit on unreliability (one minus lower confidence limit on reliability). Number of trials* N. *Observed failures* F. *Confidence coefficient* $\gamma = 0.99$.

into its major subsystems. It is also often necessary at this level, for frequently each of the major subsystems is produced by a separate contractor, and, as mentioned above, it is advantageous to put a numerical reliability requirement into each contract. This situation holds for example in the case of an ICBM weapon system. The entire system is planned to have a certain reliability which is then apportioned into requirements for the propulsion contractor, requirements for the guidance contractor, etc. A still further breakdown of these subsystems into major components may be made and component reliability requirements may be established; for example, if the propulsion system consists of several rocket engines, requirements for each engine might be established.

Reliability apportionment becomes very critical if the apportioned requirements are to be demonstrated. Obviously the more apportionments which require statistical demonstration, the greater the chance of failing to demonstrate, particularly at the component or part level where requirements approach 100 per cent reliability. Apportionment also has its limitations. As it is frequently applied, its mathematics are based on the assumptions of statistical independence and on the form of the reliability structure model which describes the relationship between the components and the system. At the higher levels of assembly, the independence and the model may be obvious; at lesser levels the model may not be discernible, and statistical independence may not necessarily apply. It is not necessary for independence to exist, but the mathematics become increasingly difficult if it does not hold. Of course, these arguments apply only when we are referring to reliability numbers estimated from separate component or subsystem tests. When we are referring to subsystems tested within the complete system, then correct functional, environmental, and statistical interrelationships are automatically present and the objection does not hold. As an example, it is perfectly valid (though it may not be feasible) to require an igniter to demonstrate its apportioned 99.9 per cent reliability requirement on rocket-engine tests. On the other hand, it is not valid (in this example) to state that we have demonstrated our apportioned reliability by igniter tests conducted independently of the engine because we do not know the extent of the interaction between the igniter and the propellant. Generally, we attempt to establish the limits of interaction by means of the specification, but these cannot always sufficiently define the system, as in the above case. This comment always applies to those systems in which the components exhibit strong mutual interdependence, such as solid-propellant rocket engines. Generally, electronic equipment has weak mutual interdependence, and liquid rocket engines are somewhere in between.

The physical makeup of the system also plays an important role in determining how far down into the assembly we should apportion our system requirement. Electronic equipment, for example, frequently consists of many repeated components, such as tubes and printed circuits, which are identical. In addition, functional and environmental effects are generally mild or controllable. Under these circumstances, then, the evaluation or demonstration of component reliability is worthwhile since information on one component frequently relates to all the other components of that type. On the other end of the scale is a liquid-propellant rocket engine. Here we have a large number of completely different components which, in addition, experience severe system-generated environments of vibration and temperature. Even if we were able to establish the various component reliability requirements, the difficulty and cost of separately demonstrating the necessarily high number for every component and of simulating the environments makes this task impractical.

We have seen that under certain circumstances we can establish a subsystem or component reliability requirement which can be demonstrated by valid test results. When reliability requirements cannot be validly demonstrated, because of either the unrealistic test circumstances or the lack of a sufficient number of units for testing, the apportioned figures are still useful in indicating the relative stage of development of the various components. We may indicate this development by comparing the relative magnitude of the percentage defective or failure rate with the apportioned percentage defective or failure rate (where we have statistically translated reliability into the appropriate failure measure). In addition, we should compare the failure rates for any single component in its different environments. Reliability apportionment forces us to look closely at our test schedules at all levels of equipment testing to see how much information and assurance we can derive from the test program. As a consequence, it will be evident whether we have organized or budgeted our test plan as wisely as possible; and, if not, how we should change the test plan in light of the above considerations.

15.1.7 Design Review Analysis for Reliability

In the initial stages of conception of a system, the reliability group should review the proposed configuration with respect to the reliability of its major assemblies and components. In order to be able to make this review they must make surveys of different components and assemblies to determine their relative merits with regard to performance, weight, reliability, cost, etc. For instance, they might compare pump-fed systems with pressure-fed systems, turbogenerator power supplies with chemical-electrical power supplies, liquid rocket engines with solid rocket engines, one type black box with a second type, one computer with another. With this comparison they conduct a paper study to obtain prediction of the expected reliability of the system.

They then review in detail the system configuration decided upon. They make analytical studies to determine whether good reliability practices have been used. These practices include the use of items at derated outputs, the establishment of safety factors in marginal cases, and the use of redundancy where possible. They also ensure that the design is sufficiently flexible to accommodate later modifications and that modified designs or alternate items are available as back-up in the event of difficulties in the program. When new designs are used, the reliability group ascertains that they are significantly better than standard or off-the-shelf items by comparing them with approximately similar designs on which they have failure data. They review the new design to see which modes of failure have been eliminated by modification or by elimination of components or parts. They must also estimate by design analysis and judgment the potential failure modes introduced by any new design. In addition, the reliability group ascertains that the design takes into account human factors and ease of maintenance.

An activity sometimes called out separately from reliability analysis is failure mode, effects, and criticality analysis (FMECA). This is the process of reviewing the design from the negative viewpoint by asking the questions: What will make it fail? How will it fail? What will the effect of the failure be in terms of criticality with respect to its end effect on the system? Frequently the FMECA also includes an estimate of the probability of failure for each failure mode listed.

Reliability design review is essential; it is not a duplication of the effort of the design engineer. The designer's mind tends to be oriented toward new designs and configurations, and he is not probability conscious. A feasible design is not necessarily a reliable one. The designer's drive for maximum performance is natural; however, if there is too great a tendency to change and an insufficiency of experience or of opportunity to learn by analysis and test, we experience initial failures which we cannot always afford in money or time. Maximum performance is frequently associated with low reliability. We must, therefore, evaluate our performance objectives with techniques of design review performed independently of the original designer. In this way we arrive at an effective system.

Formal design reviews are in addition to the continuing reliability design analysis cited above. Usually there are three reviews, but sometimes only two are performed. The first is held at the conceptual stage, the second at the developmental stage, and the third at the preproduction stage of the program. This last review is sometimes omitted. Formal design reviews are presentations to a committee consisting of project management and technical specialists outside the project organization. Reliability presentations are given by both design and reliability specialists showing how reliability considerations have been accounted for and what the predicted value of reliability is for the design.

15.1.7.1 An Example of a Reliability Design-Review Check List

A. Is the item an off-the-shelf device or is it especially developed for a particular function?

B. Does the item perform more than one function?

C. How many critical and noncritical parts does it have?

D. If an off-the-shelf item, is it being used as in previously experienced environments and at normal operating levels? What are the limits of satisfactory performance and environmental usage?

E. Is the item to be used at the limits of its strength or performance capabilities?

F. Could an existing off-the-shelf item be used? If not, why?

G. Could an existing off-the-shelf item be suitably modified? How might the modification affect its performance and behavior?

H. What have the problems been on similar or subscale designs? Are these likely to recur on the present design? What can be done to eliminate them?

I. What are the previous hypothesized modes of failure? What are the new expected modes of failure? Why do these differ?

J. What is the failure history of this item? Is this a critical item; i.e., would its failure create system failure?

K. What steps have been taken in the application of the item or in the system design to eliminate similar failures?

L. What steps have been taken to prevent new failures from being introduced?

M. What is the expected or estimated numerical value of the reliability of the item?

N. Is this a sufficient value for the apportioned or allocated reliability requirements of the item?

O. Is it possible to introduce the concepts of redundancy or to use the item at derated performance levels or both?

P. Is the part completely interchangeable with that of another manufacturer? If not, how does it differ in its failure experience?

Q. What is known about its storage life, operating time or cycles; i.e., how much time or how many cycles operating and nonoperating may be accumulated without significantly, if at all, degrading the reliability?

R. State methods of inspecting and testing for the modes of failure in I.

S. If the item is newly developed, what are its critical weaknesses? What provision has been made in the design so that modifications can be made at the earliest possible time if these or other weaknesses show up in testing?

T. Has the item been designed as simply as possible? Have potential human errors, such as reversed wiring and misassembly been designed out of the item?

U. Is it physically and functionally compatible with its neighboring components; i.e., is its performance likely to drift outside the range of its neighbors? Are tolerances and clearances adequate? Will the physical location affect the performance or reliability?

V. Have physical features which might adversely affect performance or reliability been taken into account; i.e., are there any sharp corners which might damage parts or short-circuit wiring? Is the design sufficiently stressed against vibration, temperature, humidity, dust, etc.?

W. How might the design be modified to improve reliability? Would this compromise such factors as performance, cost, weight, availability, schedules, maintainability, etc.?

X. Has the item or system been designed for ease of production and assembly, maintenance, and inspection?

Y. Are unusual quality-control or vendor problems expected?

Z. Have factors of handling, transportation, packaging, and environments, other than the specified operational environments, been taken into account?

15.1.8 Design Control and Specification, Materials, and Processing Review

These functions are the controls by which the physical hardware is made to conform to the design. Specifications are reviewed to assure that reliability requirements are correctly and adequately expressed and appropriate test requirements are specified. Component and system specifications are reviewed to determine their compatibility—e.g., that the output limits of one component are not greater than the input limits of its functional neighbors. Specifications are reviewed for tightness to ensure that the statistical buildup of component tolerance limits will not result in system inadequacy. Dimensions, weights, pressures, currents, times, and tolerances which are critical with respect to modes of failures are determined.

Material, inspection, and manufacturing procedures are examined to assure that the quality of the material and the sampling plans are compatible with reliability requirements. Processing input levels are investigated to determine whether they are operating at an optimum combination. Processing outputs are reviewed for reproducibility and adequacy.

Design drawings are checked and copies retained in a central file. A master list of system components is drawn up when the prototype configuration is established. This list contains those components which in the judgment of the component engineer and reliability group are the most reliable at any given point in time. Only one component design should be listed for each component of the system, and all components must appear on the list. The purpose of the list is to completely define the most reliable system at any given time. It also allows a check to be kept on any test configuration and helps ascertain whether any experimental or obsolete components are part of the configuration and whether failure may be attributed to them. A record of all engineering change orders is maintained, and the consequences on the reliability are analyzed and noted. The master list is reviewed periodically and brought up to date. With these controls we can establish whether it is the basic design or the hardware or the procedures which are deficient.

15.1.9 Subcontractor and Vendor Control

The area of subcontractor and vendor control is frequently the most critical in a reliability program. This is particularly true with small vendors, who often do not have or cannot afford reliability organizations. Vendors are assessed for their engineering and manufacturing capability on meeting program requirements. Under certain circumstances, when cost and feasibility permit, they may be required to demonstrate apportioned or specified reliability requirements or to provide reliability input data on their products. The (prime) contractor's reliability group should be represented on the vendor survey team; the other members are usually from the engineering, quality-control, and purchasing departments. The survey team evaluates the vendor's program to determine whether sufficient emphasis is placed on reliability. If necessary, as it might be in the case of small companies, the prime contractor should help organize and must monitor the vendor's reliability program. The reliability group should review all purchase orders to ensure that reliability is not being compromised on account of cost.

For design and process control of both contractor and vendor components, reliability tests and quality tests should be performed under specified environments and criteria to assure that all designs and manufacturing techniques will satisfy the system performance requirements.

15.1.10 Failure and Reliability Reporting

The establishment, operation, and adequacy of the failure and reliability reporting system are the responsibility of the reliability department. The customer's requirements must be satisfied and so must the contractor's internal information needs. Sometimes the contract specifies the reporting system; on other occasions only the minimum data needs without regard to format are asked for. Normally for managerial and technical reasons the contractor's reporting requirements are more extensive than the customer's.

We do not discuss failure and reliability reporting formats here. For that aspect of reporting we refer the reader to Chap. 3 of *R:MMM* and Chap. 4 of Von Alven, *Reliability Engineering*, with the comment that the format can be very critical in encouraging the proper use of the system.

The failure reporting system has the following requirements.

1. There must be a clear, unambiguous definition of when a failure report is to be initiated. Sometimes confusion results because the definitions used for failure in initiating a failure report are often (and legitimately) different from the definition of failure used in a reliability estimate. This is one of the reasons why a failure reporting system is often given other names such as trouble, discrepancy, or unplanned-event reporting. Thus the failures used in computing reliability are only a portion of those for which a failure report is initiated. Failure reporting is for the control and prevention of problems as well as for the provision of the basis for reliability estimates.

2. Points in the assembly cycle must be specified as to where and when failure report forms are to be initiated. This procedure prevents failures from being reported on, for example, strictly internal manufacturing problems, where such failed parts would never be delivered to the customer or even reach final acceptance inspection. Thus, the level of assembly as well as the stage of development of the equipment on which failure reports may be initiated should be defined. For example, failure reports may be made out only on parts, components, assemblies, etc., which have passed an acceptance test. Failures prior to this point are frequently reported as manufacturing discrepancies and are normally the responsibility of quality assurance although the same report form may be used.

3. It is mandatory that failure reporting, analyzing, recording, corrective action, and verification of effectivity be closed-loop so that reporting and correction activities and results are as effective as possible. Also time limits are specified as to when these activities must be completed.

4. It is important to assure that the failure reporting system is compatible with the reporting systems of other organizations which use the equipment after delivery. This requirement includes compatibility with maintainability reporting systems as well as, for example, with the field failure reporting system.

The reliability reporting system has similar considerations with regard to what data are applicable. Procedures can be set up to protect the integrity of the data. One such method is to use a *declaration form*, in which the test is declared applicable or not according to rules agreed to prior to the tests being undertaken. The use of such forms (described in Sec. III of *R:MMM*) helps to prevent a great amount of subjectivity in interpretation of test results.

Often the need for reliability data is so great that test results are considered on configurations and generated under conditions deviating from those of the end product and its end use. Provisions must therefore be made to collect time/cycle, functional, environmental, and configurational information so that any reliability estimates computed can be well understood by the framework in which they were generated.

15.1.11 Internal Coordination

As reliability covers so many interrelated functions, close surveillance must be maintained over the reliability-oriented activities of all the separate departments. Coordination, therefore, is important in keeping management, engineering, testing, quality assurance, and manufacturing abreast of the reliability status of the various components and the overall system. More important, it assures that all possible means of increasing the reliability are being used. Contract proposals, work statements, schedules, program plans, and specifications should all be be reviewed to ascertain that reliability is adequately covered. Methods for ensuring coordination vary depending on the activity; however, positive controls must be established which guarantee that coordination is obtained.

A signature is the most effective method of ensuring that the reliability group has reviewed the document or activity. Committee membership is another formal method of providing the reliability-group representative with an opportunity to coordinate all the reliability activities. Other methods, by personal contact or other informal relationship, are generally not satisfactory since personnel change and lack of time frequently prevent informal contacts from being made and maintained. It is therefore mandatory to establish a formal and cooperative working relationship between reliability and the other departments.

15.1.12 Conclusion and Summary: The Reliability Program Plan

Experience has shown that it is valuable at the beginning of the development program to formalize the reliability activities in the form of a reliability-program plan. This plan serves at least four purposes. First, a detailed document helps organize the approach of the reliability group to the problem. Second, during the formulation of the document, it provides the reliability personnel with an indication of the amount of cooperation or enthusiasm they may expect from the other departments and of the backing they will receive from management as well as with an opportunity to sell their contribution to the development program. Third, a detailed reliability decument can be regarded as a charter which delineates the areas of responsibilities and describes the activities of the reliability group as well as the associated reliability activities of other departments. Fourth, it is an important contractual document which should explain to the procuring agency or customer the specific details of the contractor's internal reliability program as it relates to his specific equipment or system, and it should permit the recipient to determine its adequacy.

The reliability-program plan should perform these functions.

1. State the system reliability requirements.
2. Specify the tests, ground rules, and statistical methods of reliability demonstration.
3. Describe the size, technical capabilities, organization, and responsibilities of the reliability group.
4. Define the position of the reliability group, its effectiveness, and its formal relationships within the company or organization.
5. List the test plans and schedules for all levels of equipment.
6. Describe objectives, sample sizes, and state of development of equipment in these various tests.
7. Define which apportionment goals will be demonstrated and the circumstances of demonstration.
8. Describe in detail the reliability activities for evaluation, analysis, and product improvement as discussed in general under the subheadings of this section.

15.2 RELIABILITY MATHEMATICS

15.2.1 Reliability Distribution Models

This section presents some of the statistical reliability models found useful for reliability measurement and evaluation. For further details see *R:MMM*, Chap. 6. These models, or statistical distributions, can be conveniently put into two categories, discrete and continuous.

15.2.1.1 Discrete Distribution Models. The discrete distributions are used to evaluate data when there are two or more outcomes of a test. The binomial distribution is the most important discrete probability distribution used in reliability calculations. It is most frequently used when a statement (or estimate) is to be made about an infinite (or large) population from sample observations and when the observations are limited to two exclusive outcomes. Examples of such outcomes are *success* or *failure* and *exceeding* or *not exceeding a specified time*. A further example is ignition versus nonignition, or more precisely ignition within certain defined ignition transient characteristics versus not igniting or not performing within the limits. In other words, the binomial distribution may be used in estimating probabilities not only from observed discrete events but also from events measured by variables categorized into one of two classes by applying limits to the variables. Related distributions are (1) the hypergeometric distribution and (2) the multinomial distribution.

The hypergeometric distribution is used to make probability statements about finite populations from an observed sample when the outcome of an event is measured as one of two discrete classifications. This distribution has not been used extensively in reliability work for several reasons. First, most probability estimates are considered representative of an infinite population. Second, the hypergeometric distribution has an additional parameter N, the population size, to be considered, with the result that tables of the hypergeometric distribution tend to be very extensive and rather cumbersome to use. Third, as $N \rightarrow \infty$, the form approaches the binomial; consequently, binomial tables can be used as reasonably accurate approximations for most situations except when N is small (see Owen, *Handbook of Statistical Tables*).

The multinomial distribution is a direct extension of the binomial distribution; however, here the outcome of the experiment can be categorized into more than two mutually exclusive categories. For example, we could imagine the outcome of a test to be one of the following: success, critical failure, major failure, minor failure, or exclusion—i.e., five mutually exclusive events.

Another discrete distribution is the Poisson distribution, which can be used to measure the number of failures occurring in a given time interval or observation period.

The negative binomial distribution describes the situation when testing is terminated after a given number of failures (or successes).

Table 15.1 presents a mathematical description of the binomial, Poisson, and negative binomial distribution models.

15.2.1.2 Continuous Distribution Models. The continuous distribution models are most often used to describe time to failure, although they also have been used to describe the stress value, or level at which failure occurs. Table 15.2 presents a description of the continuous distribution models found most useful in reliability evaluation.

15.2.2 Reliability Evaluation and Demonstration

15.2.2.1 General Considerations. The considerations regarding evaluation and demonstration fall into two groups. One is statistical, the other is concerned with practical problems. The statistical questions relate to the techniques by which the data will be assessed and the assumptions that can or need be made about the underlying failure distributions. The practical concerns apply to the manner in which the data are generated—i.e., whether they are valid for use in reliability evaluation and demonstration programs. The questions raised in Par. 15.1.4 must be addressed—i.e., whether equipment which is not completely representative of the end item or which is tested under conditions different from those of its final use or both can provide data sufficiently valid so that reliability estimates can be obtained. When reliability demonstration is being contemplated, the emphasis is not only on a highly reliable product but also on whether an adequate sample size from one source or another is available. If the requirements for sample size and reliability are high, the producer may have to use for reliability demonstration data from experimental development tests, which are undertaken for reasons other than reliability demonstration. Unless other provisions are made this situation might force the producer into taking an unnecessarily high risk—i.e., having a high probability of failing the demonstration program. One such provision is to allow the producer to declare the applicability of any test for reliability-demonstration purposes prior to the test's taking place. This allows the producer to exclude from reliability counting the use of those tests in which the outcome is so uncertain (or so certain of failure) because of the configuration or the extreme environment that a failure would not reflect a real condition and therefore should not count against reliability. Of course, a test of the operational configuration under end-use conditions could not normally be ex-

TABLE 15.1. *DISCRETE DISTRIBUTION MODELS*

Name	Probability Function	Parameter Range	Moment Generating Function	Mean	Variance	Notes and Table References
Binomial	$P(v = j) = \binom{n}{j} p^j q^{n-j}$	$0 < p < 1$ $q \equiv 1 - p$ $j = 0, 1, 2, \ldots, n$	$(q + pe^t)^n$	np	npq	a. $\sum_{j=r}^{n} \binom{n}{j} p^j q^{n-j} \equiv E(n, r, p)$ b. $E(n, r, p) = I_p(r, n - r + 1)$, the Incomplete Beta Function $I_x(a, b)$, tabulated in [T-5]*. $E(n, r, p)$ tabulated in [T-3], [T-4], [T-6], [T-7].
Poisson	$P(v = j) = (\lambda^j e^{-\lambda})/j!$	$\lambda > 0$ $j = 0, 1, 2, \ldots$	$e^{-\lambda + \lambda e^t}$	λ	λ	a. Tables of $\sum_{j=r}^{\infty} \lambda^j e^{-\lambda}/j! \equiv P(r, \lambda)$. $P(r, \lambda)$ tabulated in [T-8], [T-9]. b. $P(r, \lambda) = \frac{1}{\Gamma(r)} \int_0^{\lambda} x^r e^{-r}\, dx = I\left(\frac{\lambda}{\sqrt{r+1}}, r\right)$. $I(u, p)$ tabulated in [T-2].
Negative binomial	$P(v = n) = \binom{n-1}{c-1} p^c q^{n-c}$	$n = c, c + 1, \ldots$ $0 < p < 1$ $q \equiv 1 - p$	$\left(\frac{pe^t}{1 - qe^t}\right)^c$	c/q	cq/p^2	a. $\sum_{j=c}^{n} \binom{j-1}{c-1} p^c q^{j-c} = E(n, c, p)$

* [T–] Refers to the table references given at the end of this section.

cluded. However, whether completely representative of the end item or only partially, once a test is declared applicable its result, success or failure, stands. Another advantage of pretest declaration, by use of a *Pre-Test Reliability Declaration Form*, is the discipline it encourages. The purpose of the test is thoroughly reviewed; the configuration is completely identified; and the risks are evaluated meaningfully. It provides a measure of the producer's knowledge of his product. In reliability evaluation, generally there is no emphasis on sample size (nor perhaps on the contractual implications often associated with reliability demonstration); thus, the producer can often preselect only completely representative tests.

The choice of sampling technique, or the method of obtaining the data, is also influenced by many factors. Generally, the more information used in making a decision or taking an action, the less chance of making an error. To increase the information and thereby decrease this risk, we can increase the sample size or the number of tests used to reach a decision. However, suppose the risk can be expressed in terms of dollar loss. If testing is also costly in terms of dollars, the attempt to decrease the risk to a negligible value and thereby to prevent incurring a large dollar loss is nullified by the expense of testing.

Cost is not the only nonstatistical factor that influences the choice of a sampling plan. Before any one plan can be formulated, the practical circumstances of the situation must be thoroughly examined. In fact, consideration of these circumstances invariably results in a narrowing down of the variety of plans from which to choose. For instance, the type of equipment being tested and the method of testing determine whether we observe attribute or variables data. Of course, there may be some freedom of choice in the type of data obtained. For example, if the test items are continuously operating, perhaps it would be desirable to measure times to failure. However, since it might be costly or impractical to install automatic equipment or to have observers working full-time taking these measurements, we may have to be satisfied with counting numbers of failures at the end of discrete periods (e.g., each morning). This procedure would therefore change the sampling model. Again, because the failure characteristics of the equipment itself or its complexity of operation or both might not indicate any amenable time-to-failure distribution, an attribute test result of failure or success would probably be the most realistic observation to make.

Further considerations which would strongly influence the choice of a sampling plan result from the manner in which the items were made available for testing as well as from the urgency of generating the test results. For example, it might be efficient to use a sequential plan whether all items were available at one time or were being produced sequentially. On the other hand, perhaps the schedules and the high reliability goals would require more data than could be generated with the testing of single items, one at a time.

The method of truncating the testing might depend on such factors as the necessity to stop testing by a given date (as for a scheduled milepost requirement) or the fact of having a limited number of items available for a reliability demonstration program or that of having a limited amount of test-facility time available.

The restricting considerations given above, although typical of those found in industry, are by no means inclusive, and the reader can probably add many of his own from the particular field with which he is most familiar.

Therefore, the purpose of establishing a reliability-demonstration sampling plan is to plan the testing and sampling so that, within the framework of the above practical considerations, the risks of making wrong decisions are acceptably small in some sense.

15.2.2.2 Estimates of Reliability Functions. The various statistical reliability models described in Par. 15.2.1 are useful because they provide methods of extracting as much information as possible from tests in a rigorous and consistent manner. Obviously, any model, such as the exponential distribution of time to failure, must also describe the reliability characteristics of a device more informatively and more accurately than, for example, the simple binomial model can, before the exponential distribution assumption can be used. In general, the exponential distribution appears to have been empirically justified and can be theoretically justified under general conditions. (See Drenick, "The Failure Law of Complex Equipment"; *R:MMM*; and Klimov, "Limiting Law for Reliability of Sequential Systems.") Further refinements have been made, such as replacing the exponential with the Weibull distribution (see Table 15.2).

TABLE 15.2 *CONTINUOUS DISTRIBUTION MODELS FOR TIME TO FAILURE*

Name	Reliability Function $R(t),\ t \geq l$ $(R(t) = 1,\ t < l)$	Parameter Ranges	Mean Time to Failure	Variance of Time to Failure	Notes and Table References
Exponential	$e^{-\lambda(t-l)}$	$\lambda > 0$ $l \geq 0$	$l + 1/\lambda$	$1/\lambda^2$	Tables [T-1]. Special case of gamma and Weibull distributions. When $l = 0$, we have the ordinary exponential distribution.
Gamma	$1 - I\left(\frac{\lambda}{\sqrt{\alpha}} \bullet (t - l),\ \alpha - 1\right)$	$\lambda > 0$ $\alpha > 0$ $l \geq 0$	$l + \alpha/\lambda$	α/λ^2	$I(u, p) = \frac{1}{\Gamma(p+1)} \bullet \int_0^{u\sqrt{p+1}} \nu^p e^{-\nu} d\nu$ tabulated in [T-2]. Special case of exponential gamma distribution.
Exponential-gamma	$e^{-\mu(t-l)}\left\{1 - I\left(\frac{\lambda}{\sqrt{N}}(t - l),\ N - 1\right)\right\}$	$\mu > 0$ $\lambda > 0$ $N > 0$ $l \geq 0$	$m = l + \frac{1}{\mu} \bullet \left[1 - \left(\frac{\lambda}{\lambda + \mu}\right)^N\right]$	$\frac{2m}{\mu} - \frac{2N}{\mu(\lambda + \mu)} \bullet (1 - \mu m) - m^2$	Tables $I(u, p)$ in [T-2]. Derivation in R: MMM.
Weibull	$e^{-\left[\frac{(t-l)^\beta}{\alpha}\right]}$	$\alpha > 0$ $\beta > 0$ $l \geq 0$	$l + \alpha^{1/\beta}\Gamma\left(\frac{1}{\beta} + 1\right)$	$\alpha^{2/\beta}\left\{\Gamma\left(\frac{2}{\beta} + 1\right) - \left[\Gamma\left(\frac{1}{\beta} + 1\right)\right]^2\right\}$	Tables of $\Gamma(x)$ in [T-10].
Extreme value	$e^{-\alpha[e^{\beta(t-l)} - 1]}$	$\alpha < 0$ $\beta < 0$ $l \geq 0$	$l + 1/\beta\alpha \bullet [1 - e^\alpha E_2(\alpha)]$	$e^\alpha/\beta^2\,[2E_1^{(2)}(\alpha) - e^\alpha E_1(\alpha)]$	Tables of $E_2(x)$ in [T-10] and [T-11c]. Tables of $E_1^{(2)}(x)$ in [T-11]. $E_1(x) = (e^{-x} - E_2(x))/x$.
Normal	$\simeq \Phi\left(\frac{\mu - t}{\sigma}\right) \equiv (2\pi)^{-1/2} \int_{-\infty}^{\frac{\mu - t}{\sigma}} e^{-x^2/2} dx$	$\mu >> 0$ $\mu/\sigma >> 0$ $\sigma > 0$	μ	σ^2	$\simeq$ indicates that probability for $t < 0$ is assumed to be negligible. $2\Phi(x) - 1$ tabulated in [T-12].

Two kinds of reliability estimators are given in Tables 15.3 and 15.4 for some of the models described in Par. 15.2.1; they are *point estimates* and *lower γ confidence limits.* The former are sometimes referred to as best estimates of the actual reliability. The latter estimators are called *interval estimators* in that a degree of assurance (confidence) γ ($\gamma = 90$ per cent, for example) is associated with the statement that the actual reliability value lies in an interval $(\hat{R}_L, 1)$.

Table 15.4 gives a fairly complete set of formulas for finding lower γ confidence limits on reliability for the exponential distribution. This table may be directly applied to the Weibull distribution when the shape parameter β is known by replacing all real-time data (x_j, t_j, T, T^*) by x_j^β, t_j^β, T^β, $T^{*\beta}$). A more general

TABLE 15.3. *RELIABILITY ESTIMATES FOR OBSERVED ATTRIBUTES (SUCCESS-FAILURE) DATA*

Observed Data: Distribution Model	Reliability Estimate $\hat{R}$	Lower Confidence Limit $\hat{R}_L$	Upper Confidence Limit $\hat{R}_U$	Notes and References
f failures in N trials; f has binomial distribution: $P(\nu = f) = \binom{N}{f} p^f R^{N-f}$	$\frac{N-f}{N}$	$X(f, N, \gamma_L)$	$1 - X(N - f, N, \gamma_L)$	1. $\hat{R}_L < R$ with *at least* confidence γ_L; $R < \hat{R}_U$ with *at least* confidence γ_U; $\hat{R}_L < R < \hat{R}_U$ with *at least* confidence $\gamma_L + \gamma_U - 1$. $p \equiv 1 - R$ 2. $X(f, N, \gamma)$ satisfies $\sum_{j=0}^{f} \binom{N}{j} X^{N-j}(1-X)^j = 1 - \gamma$
n trials to obtain c failures; n has negative binomial distribution: $P(\nu = n) = \binom{n-1}{c-1} p^c R^{n-c}$	$\frac{n-c}{n-1}$	$X(c - 1, n - 1, \gamma_L)$	$1 - X(n - c, n - 1, \gamma_U)$	3. Fig. 15.1 gives values of $X(f, N, \gamma)$. 4. Alternately $X(f, N, \gamma_L) = \frac{N-f}{N - f + (f+1)F_{1-\gamma_L;2(f+1),2(N-f)}}$ where $F_{\varepsilon;n_1,n_2}$ is the deviate of the F distribution with n_1, n_2 degrees of freedom, exceeded with probability ε (Reference ([T-10]).
K observations of n_j failures; $j = 1, 2, \ldots, K$; each observation for a fixed time interval t; $\sum n_j \equiv f$ has Poisson distribution: $P(\nu = f) = \frac{(K\lambda t)^f e^{-K\lambda t}}{f!}$ Reliability $R \equiv e^{-\lambda T}$, where T is required life, λ is failure rate.	$e^{-Tf/Kt}$	$\exp\left(-\frac{T\chi^2_{2(f+1);1-\gamma_L}}{2Kt}\right)$	$\exp\left(-\frac{T\chi^2_{2f;\gamma_U}}{2Kt}\right)$	5. Reference [T-2], pp. 138-161, gives values of $\chi^2_{n;\alpha}$

TABLE 15. 4. *LOWER CONFIDENCE LIMITS AND RELIABILITY DEMONSTRATION*

Description of Test Data and Definitions	Termination Rule	Type
1. n items on test at time zero. $x_1, x_2, \ldots, x_r$ are times to failure in order, r given in advance. T_r = total lives.	Stop testing when rth failure occurs.	Replacement $T_r = nx_r$
2. n items on test at time zero. $x_1, x_2, \ldots, x_r$ are times to failure in order, r given in advance. T_r = total lives.	Stop testing when rth failure occurs.	Nonreplacement $T_r = \sum_{i=1}^{r} x_i + (n-r)\,x_r$
3. n items on test at time zero. r = number of failures. T^* = truncation time (given in advance).	Stop testing at time T^*.	Replacement
4. n items on test at time zero. x_{r_0} = time at r_0th failure (r_0 given in advance). T^* = truncation time (given in advance).	Stop testing at time min (x_{r_0}, T^*).	Replacement
5. n items on test at time zero. r = number failures in $(0, T^*)$. T^* = truncation time(given in advance).	Stop testing at time T^*.	Nonreplacement
6. n items on test at time zero. $x_1, x_2, \ldots, x_r$ are times to failure in order. r = number of failures in $(0, T^*)$. T^* = truncation time (given in advance). $\tau(t) \equiv$ total lives $\begin{cases} \tau(T^*) = \sum_{i=1}^{r} x_i + (n-r)T^* \\ \tau(x_n) = \sum_{i=1}^{n} x_i \end{cases}$	Stop testing at time T^*.	Nonreplacement

CRITERIA BASED ON VARIOUS LIFE TESTS FOR EXPONENTIAL DISTRIBUTION

Lower Confidence Limit $\hat{R}_L$	Demonstration Criteria $\hat{R}_L$ at Confidence γ	Remarks
$\hat{R}_L = \exp\left[-\dfrac{T\chi^2_{2r;1-\gamma}}{2T_r}\right]$	$T_r \geq \dfrac{T\chi^2_{2r;1-\gamma}}{2\log(1/\hat{R}_L)}$	$T \equiv$ required life $\chi^2_{f;\alpha} \equiv$ chi-square deviate with degrees of freedom, exceeded with probability α. [Example: $\chi^2_{4;0.10} = 7.77944$]
$\hat{R}_L = \exp\left[-\dfrac{T\chi^2_{2r;1-\gamma}}{2T_r}\right]$	$T_r \geq \dfrac{T\chi^2_{2r;1-\gamma}}{2\log(1/\hat{R}_L)}$	Replacement $\equiv$ items replaced as soon as they fail, so that n items are kept on test continuously.
$\hat{R}_L = \exp\left[-\dfrac{T\chi^2_{2r+2;1-\gamma}}{2nT^*}\right]$	$r \leq r_d$ $r_d = \max j$ such that $\chi^2_{2j+2;1-\gamma} \leq (2nT^*/T)\log(1/\hat{R}_L)$	Nonreplacement $\equiv$ items which fail are not replaced.
$\hat{R}_L = \begin{cases} \exp\left[-\dfrac{T\chi^2_{2k+2;1-\gamma}}{2nT^*}\right] & \text{if number of failures } k \text{ prior to terminating test at time } T^* \text{ is less than } r_0. \\ \exp\left[-\dfrac{T\chi^2_{2r_0;1-\gamma}}{2nx_{r_0}}\right] & \text{if number of failures equals } r_0. \end{cases}$	a. Termination at x_{r_0} $T^* \geq \dfrac{T\chi^2_{2r_0;1-\gamma}}{2n\log(1/\hat{R}_L)}$ b. Termination at T^* $k \leq k_d < r_0$ $k_d = \max j$ such that $\chi^2_{2j+2;1-\gamma} \leq (2nT^*/T)\bullet$ $\bullet\log(1/\hat{R}_L)$	
$\hat{R}_L = \left[1 + \dfrac{r+1}{n-r}\bullet \bullet F_{1-\gamma;2r+2,2n-2r}\right]^{-T/T^*}$ or $\hat{R}_L = [X(r, n, \gamma)]^{T/T^*}$	$r \leq r_d$ $r_d = \max j$ such that $\sum_{i=0}^{j} \binom{n}{i} \hat{R}_L^{(T^*/T)(n-i)} (1 - \hat{R}_L^{T^*/T})^i \leq 1 - \gamma$	$F_{\alpha;\nu_1,\nu_2} \equiv F$ distribution deviate with ν_1, ν_2 degrees of freedom exceeded with probability α. [Example: $F_{0.10;4,10} = 2.61$] $X(r, n, \gamma)$ defined by $\sum_{j=0}^{r} \binom{n}{j} X^{n-j}(1-X)^j = 1 - \gamma$
$\hat{R}_L \simeq \begin{cases} \exp\left[-\dfrac{T\chi^2_{2r+2;1-\gamma}}{2\tau(T^*)}\right], & r = 0, 1, \ldots, n-1. \\ \exp\left[-\dfrac{T\chi^2_{2n;1-\gamma}}{2\tau(x_n)}\right], & r = n \end{cases}$	a. Termination with $r = n$ failures $\tau(x_n) \geq \dfrac{T\chi^2_{2n;1-\gamma}}{2\log(1/\hat{R}_L)}$ b. Termination with $r < n$ failures $\log(1/\hat{R}_L) \geq \dfrac{T\chi^2_{2r+2;1-\gamma}}{2\tau(T^*)}$	$\simeq$ means approximately.

TABLE 15.4 *Cont.*

Description of Test Data and Definitions	Termination Rule	Type
7. n items on test at time zero. $x_1, x_2, \ldots, x_r$ are times to failure in order. x_{r_0} = time of r_0th failure (r_0 given in advance). T^* = truncation time (given in advance). $\tau(t)$ = total lives $\begin{cases} \tau(T^*) = \sum_{i=1}^{k} x_i + (n-k)T^* \\ \tau(x_{r_0}) = \sum_{i=1}^{r_0} x_i + (n-r_0)x_{r_0} \end{cases}$	Stop testing at time min (x_{r_0}, T^*).	Nonreplacement
8. One item tested at a time (see remarks, however). $t_1, t_2, \ldots, t_r$ are times to failure. r = number of items failing prior to time T^* (r given in advance). T^* = truncation time (given in advance). L_r = Total lives $= \sum_{i=1}^{r} t_i + (n-r)T^*$.	Test each item for at most time T^*. Stop testing when rth item fails. n = number of items tested (n unknown in advance).	Nonreplacement

situation may be dealt with as described in *R:MMM*, p. 307. Other cases are discussed in Resinikoff and Lieberman, *Tables of the Non-Central* t–*Distribution*; Pugh, "The Best Estimate of Reliability in the Exponential case"; Kaufman and Lipow, "Reliability-Life Test Analysis Using the Weibull Distribution"; Johns and Lieberman, "An Exact Asymptotically Efficient Confidence Bound for Reliability in the Case of the Weibull Distribution"; and Owen, *Factors for One-Sided Tolerance Limits and for Variables Sampling Plans*. The reader should note, particularly in Table 15.4, the effect of the type of test being performed and of the type of data used upon the evaluation of the lower γ confidence limit on reliability.

15.2.2.3 Reliability-Demonstration Methods. The methods for reliability demonstration given in this section are some of those most commonly used. The first method is the simplest binomial success-failure sampling plan. It has broad usefulness since reliability test data can generally be transformed into binomial trials, although some information is lost in the process. The second method presented here is a somewhat more generalized binomial success-failure sampling plan in that it does not require a fixed sample size.

Other methods of reliability demonstration where the underlying model is the exponential distribution of time to failure can be found in Table 15.4, as discussed under C below. Characteristics of such life tests, such as expected time to termination of the test, operating-characteristic functions, etc., may be found in *Statistical Techniques in Life Testing*.

A. Binomial sampling: fixed sample size. Reliability R is to be demonstrated with confidence γ by making N binomial trials in which either success or failure is observed. The criterion for meeting the demonstration requirement is that the observed number of failures should not exceed a number F. F is called the *acceptance number*, and $\beta \equiv 1 - \gamma$ is called the *consumer's* (or *customer's*) *risk* in

Lower Confidence Limit $\hat{R}_L$	Demonstration Criteria $\hat{R}_L$ at Confidence γ	Remarks
$\hat{R}_L \simeq \begin{cases} \exp\left[-\dfrac{T\chi^2_{2k+2;1-\gamma}}{2\tau(T^*)}\right] & \text{if number of failures } k \text{ prior to terminating test at time } T^* \text{ is less than } r_0. \\ \exp\left[-\dfrac{T\chi^2_{2r_0;1-\gamma}}{2\tau(x_{r_0})}\right] & \text{if number of failures equals } r_0. \end{cases}$	a. Termination at x_{r_0} $\tau(x_{r_0}) \geq \dfrac{T\chi^2_{2r_0;1-\gamma}}{2\log(1/\hat{R}_L)}$ b. Termination at T^* $\log(1/\hat{R}_L) \geq \dfrac{T\chi^2_{2k+2;1-\gamma}}{2\tau(T^*)}$	$\simeq$ means approximately.
$\hat{R}_L = \exp\left[-\dfrac{T\chi^2_{2r;1-\gamma}}{2L_r}\right]$	$L_r \geq \dfrac{T\chi^2_{2r;1-\gamma}}{2\log(1/\hat{R}_L)}$	Alternately, testing could have proceeded in stages: (1) test r items simultaneously and stop testing if all $t_j \leq T^*$, $j = 1, \ldots, r$. Otherwise (2) test the number of items equal to the number needed to make the total accumulated number of failures equal to r. Continue, until r items have failed, each prior to time T^*.

statistical quality-control parlance. For given N, R, γ, the number F should be chosen so that

$$E(N, F+1, 1-R) \equiv \sum_{j=F+1}^{N} \binom{N}{j} (1-R)^j R^{N-j} \geq \gamma \tag{15.1}$$

Values of F may be found in [T–3] or [T–4].

Example: $N = 50$, $R = 0.90$, $\gamma = 0.95$. Reference [T–4] shows on p. 69, for $p \equiv 1 - R = 0.10$, $n \equiv N = 50$, that when $r \equiv F + 1 = 2$, the tabled probability just exceeds $\gamma = 0.95$. Hence $F = r - 1 = 1$. When R and γ, but not N, are specified by the customer, the producer, who must demonstrate the reliability R with confidence γ, can determine both N and F by selecting the maximum risk α he wishes to take that the sampling plan will not demonstrate the requirements when the actual reliability is some high value $R_1 > R$. This is done by simultaneously solving

$$E(N, F+1, 1-R_1) \leq \alpha \tag{15.2}$$

$$E(N, F+1, 1-R) \geq \gamma \tag{15.3}$$

Example: $R = 0.90$, $\gamma = 0.95$, $R_1 = 0.99$, $\alpha = 0.05$. We find from [T–4] that for $F = 0$ or 1, no value of N satisfies both inequalities. However, for $F = 2$, any N in the range $61 \leq N \leq 82$ allows both inequalities to be satisfied. Consequently $N = 61$, $F = 2$.

Table 15.5 gives the minimum value of N, the number of tests required to demonstrate a given reliability R at confidence level γ, for a given number of failures F. Table 15.5 may also be used for solving the previous example. The value of N corresponding to the requirement (R, γ) is found for $F = 0$. This is compared to the value N_1 corresponding to (R_1, α) also for $F = 0$. If the number N_1 does not exceed N, then the process is repeated for $F = 1$, and so on. Thus for $F = 0$, $N_1 = 6$, $N = 29$; for $F = 1$, $N_1 = 36$, $N = 46$; for $F = 2$, $N_1 = 83$, $N = 61$; hence the smallest value of

TABLE 15.5. *MINIMUM NUMBER OF TESTS NEEDED TO DEMONSTRATE RELIABILITY AND CONFIDENCE LEVELS*

Confidence Level γ / R	Number of Failures = 0 0.01	0.05	0.10	0.20	0.50 N	0.80	0.90	0.95	0.99
0.9999	101	513	1054	2232	6931	16094	23025	29956	46049
0.9998	51	257	527	1116	3466	8047	11512	14977	23024
0.9997	34	171	352	744	2311	5364	7675	9985	15349
0.9996	26	129	264	558	1733	4023	5756	7488	11511
0.9995	21	103	211	447	1386	3219	4605	5990	9209
0.9994	17	86	176	372	1155	2682	3837	4992	7673
0.9993	15	74	151	319	990	2299	3289	4279	6577
0.9992	13	65	132	279	867	2011	2878	3744	5755
0.9991	12	57	118	248	770	1788	2558	3328	5115
0.9990	11	52	106	224	693	1609	2302	2995	4603
0.9980	6	26	53	112	347	804	1151	1497	2301
0.9970	4	18	36	75	231	536	767	998	1533
0.9960	3	13	27	56	173	402	575	748	1149
0.9950	3	11	22	45	139	322	460	598	919
0.9940	2	9	18	38	116	268	383	498	766
0.9930	2	8	15	32	99	230	328	427	656
0.9920	2	7	14	28	87	201	287	373	574
0.9910	2	6	12	25	77	179	255	332	510
0.9900	1	6	11	23	69	161	230	299	459
0.9800	1	3	6	12	35	80	114	149	228
0.9700	1	2	4	8	23	53	76	99	152
0.9600	1	2	3	6	17	40	57	74	113
0.9500	1	1	3	5	14	32	45	59	90
0.9400	1	1	2	4	12	27	38	49	75
0.9300	1	1	2	4	10	23	32	42	64
0.9200	1	1	2	3	9	20	28	36	56
0.9100	1	1	2	3	8	18	25	32	49
0.9000	1	1	1	3	7	16	22	29	44
0.8900	1	1	1	2	6	14	20	26	40
0.8800	1	1	1	2	6	13	19	24	37
0.8700	1	1	1	2	5	12	17	22	34
0.8600	1	1	1	2	5	11	16	20	31
0.8500	1	1	1	2	5	10	15	19	29
0.8000	1	1	1	1	4	8	11	14	21
0.7500	1	1	1	1	3	6	9	11	17
0.7000	1	1	1	1	2	5	7	9	13
0.6500	1	1	1	1	2	4	6	7	11
0.6000	1	1	1	1	2	4	5	6	10
0.5500	1	1	1	1	2	3	4	6	8
0.5000	1	1	1	1	1	3	4	5	7

F for which $N_1 > N$ is $F = 2$. $N = 61$ is the smallest N meeting the requirement; thus the plan is to select $(N, F) = (61, 2)$, as before.

B. Binomial sampling: multiple demonstration criteria. The method given in this section allows a demonstration requirement to be met with zero failures in N_0 trials, or one failure in N_1 trials, or two failures in N_2 trials, etc.

Although, for example, 0.90 reliability with 0.90 confidence is demonstrated with zero failures in twenty-two trials, one failure in thirty-eight trials, two failures in fifty-two

TABLE 15.5 *Cont.*

Confidence Level γ / R	0.01	0.05	0.10	0.20	Number of Failures = 1 0.50 N	0.80	0.90	0.95	0.99
0.9999	1486	3554	5319	8244	16783	29942	38896	47437	66381
0.9998	743	1777	2660	4122	8392	14971	19447	23717	33189
0.9997	496	1185	1773	2749	5595	9981	12965	15812	22126
0.9996	372	889	1330	2062	4196	7485	9723	11858	16594
0.9995	298	712	1064	1649	3357	5988	7778	9486	13274
0.9994	248	593	887	1375	2797	4990	6482	7905	11062
0.9993	213	508	760	1178	2398	4277	5556	6776	9481
0.9992	187	445	665	1031	2098	3742	4861	5928	8296
0.9991	166	396	592	917	1865	3327	4321	5270	7374
0.9990	149	356	533	825	1679	2994	3889	4742	6636
0.9980	75	179	267	413	839	1497	1944	2371	3317
0.9970	50	119	178	275	560	998	1296	1580	2210
0.9960	38	90	134	207	420	748	971	1185	1657
0.9950	31	72	107	165	336	598	777	947	1325
0.9940	26	60	89	138	280	499	647	789	1104
0.9930	22	52	77	118	240	427	555	676	946
0.9920	20	45	67	104	210	374	485	592	827
0.9910	17	40	60	92	187	332	431	526	735
0.9900	16	36	54	83	168	299	388	473	662
0.9800	8	19	27	42	84	149	194	236	330
0.9700	6	13	18	28	56	99	129	157	219
0.9600	5	10	14	21	42	74	96	117	164
0.9500	4	8	11	17	34	59	77	93	130
0.9400	3	7	10	14	28	49	64	78	108
0.9300	3	6	8	12	24	42	55	66	92
0.9200	3	5	7	11	21	37	48	58	81
0.9100	3	5	7	10	19	33	42	51	71
0.9000	2	4	6	9	17	29	38	46	64
0.8900	2	4	6	8	15	27	34	42	58
0.8800	2	4	5	7	14	24	31	38	53
0.8700	2	4	5	7	13	23	29	35	49
0.8600	2	3	5	6	12	21	27	32	45
0.8500	2	3	4	6	11	19	25	30	42
0.8000	2	3	3	5	9	14	18	22	31
0.7500	2	2	3	4	7	11	15	18	24
0.7000	2	2	3	3	6	9	12	14	20
0.6500	2	2	2	3	5	8	10	12	16
0.6000	2	2	2	3	4	7	9	10	14
0.5500	2	2	2	2	4	6	8	9	12
0.5000	2	2	2	2	3	5	7	8	11

trials, etc., it is not correct to apply all these criteria simultaneously in order to meet the 0.90 reliability, 90 per cent confidence requirement. Rather, a single one of these criteria is chosen in advance, and the reliability requirement is demonstrated if the number of failures is equal to or less than the number allowed by the criterion. The choice of one criterion is dictated by consideration of the producer's risk and the cost of testing as discussed previously.

To allow demonstration of the reliability

TABLE 15.5 *Cont.*

Confidence Level γ	Number of Failures = 2								
	0.01	0.05	0.10	0.20	0.50	0.80	0.90	0.95	0.99
R					N				
0.9999	4361	8178	11021	15351	26740	42789	53222	62956	84056
0.9998	2181	4089	5511	7676	13370	21394	26610	31477	42027
0.9997	1455	2727	3675	5118	8914	14263	17740	20984	28017
0.9996	1091	2045	2756	3838	6685	10697	13305	15738	21012
0.9995	873	1636	2205	3071	5348	8557	10643	12590	16809
0.9994	728	1364	1838	2559	4457	7131	8869	10491	14007
0.9993	624	1169	1575	2194	3820	6112	7602	8992	12006
0.9992	546	1023	1379	1920	3343	5348	6652	7868	10505
0.9991	486	910	1225	1706	2971	4754	5913	6994	9337
0.9990	437	819	1103	1536	2674	4278	5321	6294	8403
0.9980	219	410	552	768	1337	2139	2660	3146	4200
0.9970	147	274	368	512	892	1426	1773	2097	2799
0.9960	110	206	276	384	669	1069	1329	1572	2099
0.9950	88	165	221	308	535	855	1063	1258	1678
0.9940	74	137	185	257	446	713	886	1048	1398
0.9930	64	118	158	220	382	611	759	898	1198
0.9920	56	103	139	193	334	534	664	785	1048
0.9910	50	92	123	171	297	475	590	698	931
0.9900	45	83	111	154	268	427	531	628	838
0.9800	23	42	56	77	134	213	265	313	418
0.9700	16	28	38	52	89	142	176	208	277
0.9600	12	22	29	39	67	106	132	156	207
0.9500	10	17	23	31	54	85	105	124	165
0.9400	9	15	19	26	45	71	88	103	137
0.9300	8	13	17	23	38	60	75	88	117
0.9200	7	11	15	20	34	53	65	77	102
0.9100	6	10	13	18	30	47	58	68	91
0.9000	6	9	12	16	27	42	52	61	81
0.8900	5	9	11	15	24	38	47	56	74
0.8800	5	8	10	14	22	35	43	51	67
0.8700	5	7	9	13	21	32	40	47	62
0.8600	5	7	9	12	19	30	37	43	57
0.8500	4	7	8	11	18	28	34	40	53
0.8000	4	5	7	8	14	21	25	30	39
0.7500	3	4	5	7	11	16	20	23	31
0.7000	3	4	5	6	9	14	16	19	25
0.6500	3	4	4	5	8	12	14	16	21
0.6000	3	3	4	5	7	10	12	14	18
0.5500	3	3	4	4	6	9	10	12	16
0.5000	3	3	3	4	5	8	9	11	14

requirement $R = 1 - p$ with $1 - \beta$ confidence with multiple criteria, the maximum number of failures f to be allowed is decided upon. Then the desired customer risk β (= 1 − confidence) is split into a sum $\beta = \beta_0 + \beta_1 + \ldots + \beta_f$. (The effect of choosing different β's is indicated in an example below). Then the methods of Chap. 10 of *R:MMM* can be used to calculate the exact probabilities of reaching the demonstration points as a function of p; $N_0, N_1, \ldots, N_f$; $\beta_1, \beta_2, \ldots, \beta_f$, where $N_0 < N_1 < \ldots < N_f$.

The equations for $f = 2$ failures maximum are given here. Calculate N_0 from

TABLE 15.5 *Cont.*

Confidence Level γ / R	0.01	0.05	0.10	0.20	Number of Failures = 3 0.50 N	0.80	0.90	0.95	0.99
0.9999	8234	13664	17448	22968	36720	55149	66806	77534	100448
0.9998	4117	6832	8725	11485	18360	27574	33402	38766	50222
0.9997	2746	4556	5817	7657	12240	18383	22268	25844	33481
0.9996	2060	3417	4363	5743	9180	13787	16701	19382	25110
0.9995	1648	2734	3491	4594	7344	11029	13360	15505	20087
0.9994	1374	2279	2909	3829	6120	9191	11133	12921	16739
0.9993	1178	1953	2494	3282	5246	7878	9543	11075	14347
0.9992	1031	1709	2182	2872	4590	6893	8350	9690	12553
0.9991	916	1519	1940	2553	4080	6127	7422	8613	11158
0.9990	825	1368	1746	2298	3672	5514	6679	7752	10042
0.9980	413	684	873	1149	1836	2757	3339	3875	5020
0.9970	276	457	583	766	1224	1838	2226	2583	3345
0.9960	207	343	437	575	918	1378	1669	1936	2508
0.9950	166	275	350	460	735	1102	1335	1549	2006
0.9940	139	229	292	384	612	918	1112	1290	1671
0.9930	119	197	250	329	525	787	953	1106	1432
0.9920	105	172	219	288	459	689	834	967	1253
0.9910	93	153	195	256	408	612	741	860	1113
0.9900	84	138	176	231	367	551	667	773	1001
0.9800	43	70	88	116	184	275	333	386	499
0.9700	29	47	59	77	123	183	221	257	332
0.9600	22	35	45	58	92	137	166	192	248
0.9500	18	29	36	47	74	110	132	153	198
0.9400	15	24	30	39	61	91	110	127	164
0.9300	13	21	26	34	53	78	94	109	140
0.9200	12	18	23	30	46	68	82	95	122
0.9100	11	17	21	26	41	60	73	84	109
0.9000	10	15	19	24	37	54	65	76	97
0.8900	9	14	17	22	34	49	59	69	88
0.8800	9	13	16	20	31	45	54	63	81
0.8700	8	12	15	19	28	42	50	58	74
0.8600	8	11	14	17	26	39	46	53	69
0.8500	7	11	13	16	25	36	43	50	64
0.8000	6	8	10	12	19	27	32	37	47
0.7500	5	7	8	10	15	21	25	29	37
0.7000	5	6	7	9	12	18	21	24	30
0.6500	4	5	6	8	11	15	18	20	25
0.6000	4	5	6	7	9	13	15	17	22
0.5500	4	5	5	6	8	11	13	15	19
0.5000	4	4	5	6	7	10	12	13	17

$$(1 - p)^{N_0} = \beta_0 \tag{15.4*}$$

Calculate N_1 from

$$N_0 p(1 - p)^{N_1 - 1} = \beta_1 \tag{15.5}$$

(Use N_0 from Eq. (15.1).) Calculate N_2 from

$$(N_0/2)(2N_1 - N_0 - 1)\, p^2(1 - p)^{N_2 - 2} = \beta_2 \equiv \beta - \beta_0 - \beta_1 \tag{15.6}$$

* Use the next larger integer values for N_0, N_1, N_2 in this and the next two equations since the indicated calculation does not generally give an exact integer value.

TABLE 15.5 *Cont.*

Confidence Level γ / R	0.01	0.05	0.10	0.20	Number of Failures = 4 0.50 N	0.80	0.90	0.95	0.99
0.9999	12792	19703	24327	30896	46709	67208	79934	91533	116042
0.9998	6397	9852	12164	15448	23354	33604	39966	45765	58019
0.9997	4266	6569	8110	10299	15570	22402	26644	30510	38679
0.9996	3200	4927	6083	7725	11677	16802	19982	22882	29008
0.9995	2560	3942	4866	6180	9342	13441	15986	18305	23206
0.9994	2134	3285	4056	5150	7785	11201	13321	15254	19338
0.9993	1829	2816	3476	4415	6673	9601	11418	13074	16575
0.9992	1601	2464	3042	3863	5839	8400	9990	11440	14502
0.9991	1423	2191	2704	3434	5190	7467	8880	10168	12890
0.9990	1281	1972	2434	3090	4671	6720	7992	9151	11601
0.9980	641	987	1218	1546	2336	3360	3995	4575	5799
0.9970	428	658	812	1031	1557	2239	2663	3049	3865
0.9960	322	494	609	773	1168	1679	1997	2286	2898
0.9950	258	396	488	619	934	1343	1597	1829	2318
0.9940	215	330	407	516	779	1119	1331	1523	1931
0.9930	185	283	349	442	667	959	1140	1306	1654
0.9920	162	248	305	387	584	839	998	1142	1447
0.9910	144	220	272	344	519	746	887	1015	1286
0.9900	130	199	245	310	467	671	798	913	1157
0.9800	66	100	123	155	234	335	398	456	577
0.9700	45	67	82	104	156	223	265	303	383
0.9600	34	51	62	78	117	167	198	227	287
0.9500	27	41	50	63	94	134	158	181	229
0.9400	23	34	42	52	78	111	132	150	190
0.9300	20	30	36	45	67	95	113	129	162
0.9200	18	26	32	40	59	83	98	112	142
0.9100	16	23	28	35	52	74	87	100	126
0.9000	15	21	26	32	47	66	78	89	113
0.8900	14	19	23	29	43	60	71	81	102
0.8800	13	18	22	27	39	55	65	74	93
0.8700	12	17	20	25	36	51	60	68	86
0.8600	11	16	19	23	34	47	56	63	79
0.8500	11	15	18	22	31	44	52	59	74
0.8000	8	11	14	16	24	33	38	44	55
0.7500	7	10	11	13	19	26	30	34	43
0.7000	6	8	10	11	16	21	25	28	35
0.6500	6	7	8	10	14	18	21	24	30
0.6000	5	7	8	9	12	16	18	21	25
0.5500	5	6	7	8	11	14	16	18	22
0.5000	5	6	6	7	9	12	14	16	19

Examples:

a. $\beta = 0.10$, $p = 0.10$. Let $\beta_0 = 0.03$, $\beta_1 = 0.03$, $\beta_2 = 0.04$. Then from Eqs. (15.4), (15.5), (15.6)

$N_0 = 34 \qquad N_1 = 46 \qquad N_2 = 55$

Note that the price of allowing multiple demonstration criteria is the possibility of needing more trials (compare $N_2 = 55$ trials (two failures) with fifty-two trials (two failures) if there were a single criterion).

b. Let $\beta_0 = 0.08$, $\beta_1 = 0.01$, $\beta_2 = 0.01$, where again $p = 0.10$. Here the situation is more nearly like that of having a single criterion where all the risk $\beta = 0.10$ is concentrated on demonstrating the require-

TABLE 15.5 *Cont.*

Confidence Level γ / R	Number of Failures = 5								
	0.01	0.05	0.10	0.20	0.50	0.80	0.90	0.95	0.99
					N				
0.9999	17854	26131	31520	39037	56701	79058	92745	105128	131081
0.9998	8928	13066	15760	19519	28351	39529	46371	52562	65538
0.9997	5953	8712	10508	13013	18901	26352	30914	35041	43691
0.9996	4465	6534	7881	9760	14176	19764	23185	26280	32768
0.9995	3573	5228	6305	7808	11340	15811	18548	21024	26213
0.9994	2978	4357	5255	6507	9450	13176	15456	17519	21844
0.9993	2553	3735	4504	5578	8100	11293	13248	15016	18723
0.9992	2234	3268	3941	4881	7088	9882	11592	13139	16382
0.9991	1986	2905	3504	4338	6300	8783	10304	11679	14561
0.9990	1787	2615	3153	3905	5670	7905	9273	10511	13105
0.9980	895	1308	1577	1953	2835	3952	4636	5254	6551
0.9970	597	873	1052	1302	1890	2634	3090	3502	4366
0.9960	448	655	789	977	1418	1976	2317	2626	3274
0.9950	359	524	632	782	1134	1580	1853	2100	2618
0.9940	300	437	527	652	945	1317	1544	1750	2181
0.9930	257	375	452	559	810	1128	1323	1500	1869
0.9920	225	328	395	489	709	987	1158	1312	1635
0.9910	200	292	352	435	630	877	1029	1166	1453
0.9900	181	263	317	391	567	790	926	1049	1307
0.9800	91	132	159	196	284	394	462	523	652
0.9700	62	89	106	131	189	263	308	348	433
0.9600	47	67	80	99	142	197	230	261	324
0.9500	38	54	64	79	114	157	184	208	259
0.9400	32	45	54	66	95	131	153	173	215
0.9300	28	39	46	57	81	112	131	148	184
0.9200	24	34	41	50	71	98	114	129	160
0.9100	22	31	36	44	63	87	101	115	142
0.9000	20	28	33	40	57	78	91	103	127
0.8900	18	26	30	37	52	71	83	93	116
0.8800	17	24	28	34	47	65	76	85	106
0.8700	16	22	26	31	44	60	70	79	97
0.8600	15	20	24	29	41	55	65	73	90
0.8500	14	19	23	27	38	52	60	68	84
0.8000	11	15	17	21	29	39	45	50	62
0.7500	10	12	14	17	23	31	35	40	49
0.7000	8	11	12	14	19	25	29	33	40
0.6500	8	9	11	12	16	22	25	28	34
0.6000	7	9	10	11	14	19	21	24	29
0.5500	7	8	9	10	13	16	19	21	25
0.5000	6	7	8	9	11	15	17	18	22

ments with $f = 0$ failures. The results are

$N_0 = 24 \qquad N_1 = 53 \qquad N_2 = 67$

Again, the ability to meet demonstration requirements by allowing zero or one failure is offset by the larger number of trials required for two failures ($N_2 = 67$, compared with 52).

C. Life tests: exponential distribution. Table 15.4 presents reliability-demonstration criteria for the life tests described, all based on the exponential distribution. The user should note for each life test which parameters are known in advance and which data are observed outcomes of the tests. The same remark as that given at the end of Par. 15.2.2.2 regarding the Weibull distribution with known shape para-

TABLE 15.5 *Cont.*

Confidence Level γ / R	0.01	0.05	0.10	0.20	0.50	0.80	0.90	0.95	0.99
	Number of Failures = 6								
	N								
0.9999	23304	32855	38949	47337	66696	90752	105318	118421	145702
0.9998	11653	16428	19475	23669	33348	45375	52658	59209	72849
0.9997	7770	10953	12984	15780	22232	30250	35105	39472	48565
0.9996	5828	8215	9738	11835	16674	22687	26328	29604	36423
0.9995	4663	6572	7791	9468	13339	18150	21062	23682	29137
0.9994	3886	5477	6493	7891	11116	15125	17552	19735	24281
0.9993	3331	4695	5566	6764	9528	12964	15044	16915	20811
0.9992	2915	4109	4870	5918	8337	11343	13163	14801	18209
0.9991	2591	3652	4329	5261	7411	10083	11701	13156	16186
0.9990	2333	3287	3896	4735	6670	9074	10530	11840	14567
0.9980	1167	1645	1949	2368	3335	4537	5264	5919	7282
0.9970	779	1097	1300	1579	2223	3024	3509	3945	4853
0.9960	585	823	975	1185	1668	2268	2631	2958	3639
0.9950	468	659	781	948	1334	1814	2105	2366	2910
0.9940	391	549	651	790	1112	1512	1754	1971	2425
0.9930	335	471	558	677	953	1295	1503	1689	2078
0.9920	294	413	488	593	834	1133	1315	1478	1818
0.9910	261	367	434	527	741	1007	1168	1313	1615
0.9900	235	330	391	475	667	906	1051	1182	1453
0.9800	119	166	196	238	334	453	525	590	725
0.9700	80	111	131	159	222	301	349	392	482
0.9600	61	84	99	119	167	226	262	294	360
0.9500	49	68	79	96	134	180	209	234	288
0.9400	41	57	66	80	111	150	174	195	239
0.9300	36	49	57	69	95	129	149	167	204
0.9200	32	43	50	60	84	112	130	146	178
0.9100	28	38	45	54	74	100	115	129	158
0.9000	26	35	41	48	67	90	104	116	142
0.8900	24	32	37	44	61	81	94	105	129
0.8800	22	29	34	41	56	75	86	96	118
0.8700	20	27	32	38	51	69	79	89	108
0.8600	19	25	29	35	48	64	73	82	100
0.8500	18	24	28	33	45	59	68	76	93
0.8000	14	18	21	25	34	44	51	57	69
0.7500	12	15	17	20	27	35	40	45	54
0.7000	10	13	15	17	22	29	33	37	44
0.6500	9	11	13	15	19	25	28	31	37
0.6000	9	10	12	13	17	22	24	27	32
0.5500	8	10	10	12	15	19	21	24	28
0.5000	8	9	10	11	13	17	19	21	25

meter β, applies to the reliability-demonstration criteria of Table 15.4.

15.2.3 Reliability Structure Models

15.2.3.1 General Considerations. Reliability structure models are concerned with the relationship between the components of a system and their effect on the performance of that system. This relationship consists of two parts: the output or performance of an individual component and the functional relationships among all the components which together form the reliability structure of the system.

The output of a component may be observed

TABLE 15.5 *Cont.*

Confidence Level γ / R	0.01	0.05	0.10	0.20	Number of Failures = 7 0.50 N	0.80	0.90	0.95	0.99
0.9999	29063	39810	46562	55761	76692	102324	117707	131478	159995
0.9998	14533	19906	23282	27881	38346	51161	58852	65737	79995
0.9997	9690	13271	15522	18588	25564	34107	39234	43824	53329
0.9996	7268	9954	11642	13941	19173	25580	29425	32868	39996
0.9995	5815	7964	9314	11153	15339	20464	23540	26294	31996
0.9994	4846	6637	7762	9295	12782	17053	19616	21911	26663
0.9993	4154	5689	6653	7967	10956	14617	16814	18780	22853
0.9992	3635	4978	5822	6971	9587	12790	14712	16433	19996
0.9991	3232	4425	5175	6197	8522	11368	13077	14606	17774
0.9990	2909	3983	4658	5577	7669	10231	11769	13146	15996
0.9980	1456	1992	2330	2789	3835	5115	5884	6571	7996
0.9970	971	1329	1554	1860	2557	3410	3922	4380	5329
0.9960	729	997	1166	1395	1917	2557	2941	3284	3996
0.9950	584	798	933	1116	1534	2045	2352	2627	3196
0.9940	487	665	778	931	1278	1704	1960	2189	2663
0.9930	418	571	667	798	1096	1461	1680	1876	2282
0.9920	366	500	584	698	959	1278	1469	1641	1996
0.9910	325	444	519	621	852	1136	1306	1458	1774
0.9900	293	400	467	559	767	1022	1175	1312	1596
0.9800	148	201	234	280	384	510	587	655	796
0.9700	99	135	157	187	256	340	390	436	529
0.9600	75	102	118	141	192	255	292	326	396
0.9500	61	82	95	113	154	204	234	260	316
0.9400	51	68	79	94	128	169	194	217	263
0.9300	44	59	68	81	110	145	166	185	225
0.9200	39	52	60	71	96	127	145	162	196
0.9100	35	46	53	63	85	113	129	143	174
0.9000	32	42	48	57	77	101	116	129	156
0.8900	29	38	44	52	70	92	105	117	141
0.8800	27	35	41	48	64	84	96	107	129
0.8700	25	33	38	44	59	78	89	98	119
0.8600	23	31	35	41	55	72	82	91	110
0.8500	22	29	33	38	51	67	77	85	103
0.8000	17	22	25	29	39	50	57	63	76
0.7500	14	18	20	24	31	40	45	50	60
0.7000	13	15	17	20	26	33	37	41	49
0.6500	11	14	15	17	22	28	32	35	41
0.6000	10	12	14	15	19	24	27	30	36
0.5500	10	11	12	14	17	21	24	26	31
0.5000	9	10	11	13	15	19	21	23	27

as attribute data or variables data. In the case of attribute data we are concerned only with the result of the operation as success (S) or as its complement, failure (F). However, a variable output can be used in several ways. First, we can observe whether the output falls inside or outside given limits, such as specification limits, and thus classify the test as a success or failure accordingly; i.e., we can use the output simply as attribute data. Here the output can be a single parameter indicative of reliability or it can consist of several parameters, some of which might be functionally related but no one of which could be completely identified with reliability. In this case the event, success, can be said to have occurred when each and all of the

TABLE 15.5 *Cont.*

Confidence Level γ / R	0.01	0.05	0.10	0.20	0.50	0.80	0.90	0.95	0.99
				Number of Failures = 8	N				
0.9999	35077	46954	54326	64286	86689	113796	129945	144343	174022
0.9998	17540	23478	27164	32143	43345	56897	64971	72170	87009
0.9997	11694	15653	18110	21430	28897	37931	43314	48113	58005
0.9996	8771	11740	13583	16072	21673	28448	32485	36084	43502
0.9995	7018	9393	10867	12858	17338	22758	25987	28867	34801
0.9994	5849	7828	9056	10715	14448	18965	21656	24055	29000
0.9993	5013	6710	7762	9185	12384	16256	18562	20618	24857
0.9992	4387	5871	6792	8037	10836	14224	16241	18041	21749
0.9991	3900	5219	6038	7144	9632	12643	14437	16036	19332
0.9990	3510	4697	5434	6430	8669	11379	12993	14432	17398
0.9980	1756	2350	2718	3216	4335	5689	6495	7215	8697
0.9970	1172	1567	1813	2144	2890	3792	4330	4809	5797
0.9960	880	1176	1360	1608	2167	2844	3247	3606	4346
0.9950	704	941	1088	1287	1734	2275	2597	2884	3476
0.9940	587	785	907	1073	1445	1895	2164	2403	2896
0.9930	504	673	778	920	1239	1624	1854	2059	2482
0.9920	441	589	681	805	1084	1421	1622	1802	2171
0.9910	392	524	605	716	963	1263	1442	1601	1929
0.9900	354	472	545	644	867	1137	1297	1441	1736
0.9800	178	237	273	323	434	568	648	719	866
0.9700	120	159	183	216	289	378	431	478	576
0.9600	90	120	138	162	217	283	323	358	431
0.9500	73	96	110	130	174	226	258	286	344
0.9400	61	80	92	108	145	188	215	238	286
0.9300	53	69	79	93	124	161	184	203	244
0.9200	47	61	70	82	109	141	160	178	213
0.9100	42	54	62	73	96	125	142	158	189
0.9000	38	49	56	66	87	113	128	142	170
0.8900	35	45	51	60	79	102	116	128	154
0.8800	32	41	47	55	72	94	106	117	141
0.8700	30	38	44	51	67	86	98	108	130
0.8600	28	36	41	47	62	80	91	100	120
0.8500	26	34	38	44	58	75	85	93	112
0.8000	21	26	29	33	44	56	63	69	83
0.7500	17	21	24	27	35	44	50	55	65
0.7000	15	18	20	23	29	37	41	45	53
0.6500	13	16	18	20	25	31	35	38	45
0.6000	12	14	16	18	22	27	30	33	39
0.5500	11	13	14	16	19	24	27	29	34
0.5000	10	12	13	14	17	21	24	26	30

parameters has been evaluated and together have been found to satisfy the requirements for success. Second, we can use several observations of the output and, without classifying the individual observations as success or failure, can use variables analysis to determine the probability of the output falling within certain limits (again, these limits might be specifications). Third, although each individual component might not have any output limits *per se*, the composite contribution of the outputs of all the components, resulting in the system output, has limits of its own within which to perform. In turn the system output can be attribute or variables data and can be analyzed accordingly.

TABLE 15.5 *Cont.*

Confidence Level γ / R	0.01	0.05	0.10	0.20	0.50	0.80	0.90	0.95	0.99
	Number of Failures = 9				N				
0.9999	41304	54256	62214	72893	96687	125186	142057	157049	187826
0.9998	20653	27129	31108	36447	48343	62592	71027	78523	93911
0.9997	13770	18087	20740	24299	32229	41728	47351	52348	62606
0.9996	10328	13566	15555	18224	24172	31296	35513	39260	46953
0.9995	8263	10853	12444	14580	19338	25036	28410	31408	37562
0.9994	6887	9045	10371	12150	16115	20863	23675	26173	31301
0.9993	5903	7753	8889	10415	13813	17883	20292	22433	26829
0.9992	5166	6784	7779	9113	12086	15647	17755	19629	23475
0.9991	4592	6031	6914	8100	10743	13908	15782	17447	20866
0.9990	4133	5428	6223	7291	9669	12517	14204	15702	18779
0.9980	2068	2715	3113	3646	4835	6258	7101	7850	9387
0.9970	1380	1811	2076	2431	3223	4172	4733	5232	6257
0.9960	1035	1359	1557	1824	2417	3128	3549	3923	4691
0.9950	829	1087	1246	1459	1934	2502	2839	3138	3752
0.9940	691	906	1039	1216	1612	2085	2366	2615	3126
0.9930	593	777	891	1043	1381	1787	2027	2241	2679
0.9920	519	680	780	913	1209	1564	1774	1960	2343
0.9910	462	605	693	811	1074	1390	1576	1742	2083
0.9900	416	545	624	730	967	1251	1418	1568	1874
0.9800	209	274	313	366	484	625	708	782	935
0.9700	141	183	209	244	322	416	471	521	622
0.9600	106	138	157	184	242	312	353	390	465
0.9500	86	111	126	147	194	249	282	311	371
0.9400	72	93	106	123	161	207	235	259	309
0.9300	62	80	91	106	138	178	201	221	264
0.9200	55	70	80	92	121	155	175	193	230
0.9100	49	63	71	82	108	138	156	172	204
0.9000	44	57	64	74	97	124	140	154	183
0.8900	41	52	59	68	88	113	127	140	166
0.8800	37	48	54	62	81	103	116	128	152
0.8700	35	44	50	57	75	95	107	118	140
0.8600	33	41	46	53	69	88	99	109	130
0.8500	31	39	43	50	65	82	93	102	121
0.8000	24	30	33	38	49	61	69	76	89
0.7500	20	24	27	31	39	49	55	60	70
0.7000	17	21	23	26	32	40	45	49	58
0.6500	15	18	20	22	28	34	38	42	49
0.6000	14	16	18	20	24	30	33	36	42
0.5500	13	15	16	18	22	26	29	32	37
0.5000	12	14	15	16	19	24	26	28	33

The functional relationship of the system organizes the outputs of the components so that they perform in a certain specified manner and order. The system might be designed so that for satisfactory operation all the components must themselves necessarily function satisfactorily. A system of this type is called a *serial system*. On the other hand, the system might be designed so that in the event of failure of any one of the components another is already available within the system to take over the operation. Systems which contain this provision are called *parallel systems*. Systems frequently include both serial and parallel relationships. The following paragraphs explain the operation of systems composed of

TABLE 15.5 *Cont.*

Confidence Level γ / R	0.01	0.05	0.10	0.20	Number of Failures = 10 0.50 N	0.80	0.90	0.95	0.99
0.9999	47715	61692	70209	81571	106685	136505	154064	169619	201442
0.9998	23859	30847	35105	40786	53342	68252	77030	84808	100718
0.9997	15907	20566	23404	27191	35562	45501	51353	56538	67144
0.9996	11931	15425	17554	20394	26671	34125	38514	42403	50357
0.9995	9546	12340	14043	16315	21337	27300	30811	33921	40285
0.9994	7955	10284	11703	13596	17781	22750	25676	28267	33570
0.9993	6819	8815	10032	11654	15241	19500	22007	24229	28774
0.9992	5967	7714	8778	10198	13336	17062	19256	21200	25176
0.9991	5305	6857	7803	9065	11854	15166	17116	18844	22378
0.9990	4774	6171	7023	8158	10669	13649	15404	16959	20140
0.9980	2389	3087	3512	4080	5334	6824	7701	8478	10068
0.9970	1594	2059	2342	2720	3556	4549	5133	5651	6710
0.9960	1196	1545	1757	2041	2667	3411	3849	4238	5032
0.9950	957	1236	1406	1633	2134	2729	3079	3389	4024
0.9940	798	1031	1172	1361	1778	2274	2566	2824	3353
0.9930	685	884	1005	1167	1524	1949	2199	2420	2873
0.9920	600	774	880	1021	1334	1705	1924	2117	2514
0.9910	533	688	782	908	1186	1515	1710	1882	2234
0.9900	480	619	704	817	1067	1364	1538	1693	2010
0.9800	242	311	353	409	534	681	768	845	1003
0.9700	162	208	236	273	356	454	511	562	667
0.9600	122	157	178	205	267	340	383	421	499
0.9500	99	126	142	165	214	272	306	336	398
0.9400	83	105	119	137	178	226	255	280	331
0.9300	71	91	102	118	153	194	218	239	283
0.9200	63	80	90	103	134	169	190	209	247
0.9100	56	71	80	92	119	150	169	185	219
0.9000	51	64	72	83	107	135	152	167	197
0.8900	47	59	66	76	97	123	138	151	178
0.8800	43	54	61	69	89	112	126	138	163
0.8700	40	50	56	64	82	104	116	127	150
0.8600	37	47	52	60	76	96	108	118	139
0.8500	35	44	49	56	71	90	100	110	130
0.8000	27	33	37	42	54	67	75	82	96
0.7500	23	27	30	34	43	53	59	65	76
0.7000	19	23	26	29	36	44	49	53	62
0.6500	17	20	22	25	31	38	42	45	53
0.6000	16	18	20	22	27	33	36	39	45
0.5500	14	17	18	20	24	29	32	34	40
0.5000	13	15	16	18	21	26	28	30	35

such functional structures in terms of probability relationships, when the performance of the components is measured either as attribute or as variables data.

Understanding the reliability structure permits us to recognize the weaknesses or potential weaknesses of the system from a reliability viewpoint. More important, it provides us with the knowledge and techniques for eliminating many unreliable areas. For example, we can compute how many redundant components we might need as back-ups to provide a system with a required level of reliability. Or, we can determine the probable reliability of a system composed of components with known or estimated reliabilities. Or, inversely, we can establish how the system reliability requirements can be apportioned into component reliability requirements and thereby determine the allocation of development time and funds to attain these goals. The techniques presented also illustrate how we can determine reliability safety margins when there are fixed limits for successful operation (e.g., specification limits) or variable limits consisting of the variable input capability of the neighboring component. This analysis is frequently referred to as reliability *stress-vs.-strength analysis*.

15.2.3.2 Serial Systems. A serial system is one in which failure of any of its components causes system failure. Let R_S be the reliability of the system, and $R_1, R_2, \ldots, R_n$ be the reliabilities of its n components. By making appropriate assumptions, we can derive the relationships shown below.

A. Independent serial systems. If the failure (or success) of any one or more components is uninfluenced by the failure (or success) or any other one or more components, the system is said to be an *independent* serial system. In this case the product rule holds:

$$R_S = R_1 R_2 \ldots R_n \tag{15.7}$$

For example, in a system with 100 components and $R_1 = R_2 = \ldots = R_n = 0.999$, $R_S = (0.999)^{100} = 0.905$. Table 15.6 gives values of $R_S = R^k$ for R in the range 0.90 to 0.9999999 for $k = 2, \ldots, 10$.

B. Weakest-link serial systems. We can show that the product rule is the limiting case when failure is produced by an environmental stress. If the components of the system have large variabilities in strength to resist the environment and in turn the environment has very low variability, then the product rule is approximately correct. However in the same situation, if the environment has relatively a much larger variability than have the strengths of the individual components, then the system tends to exhibit the weakest-link reliability, namely

$$R_S = \min(R_1, R_2, \ldots, R_n) \tag{15.8}$$

Thus, when an environment induces failure of a serial system, the reliability of the system tends to lie between a lower limit given by the product rule Eq. (15.7) and an upper limit given by Eq. (15.8). The weakest-link system is one in which failure of one distinguished component is always associated with system failure (e.g., if one component has the lowest melting point and heat is the environment and if the system fails, this one component must have failed, although others may also have failed). This system is a *dependent* serial system.

C. General serial systems. In any case, i.e., whatever the cause of failure or the assumptions of independence or dependence, as long as the system is serial, then it can be shown that

$$R_1 + R_2 + \ldots + R_n - (n - 1) \leq R_S \leq \min(R_1, R_2, \ldots, R_n) \tag{15.9}$$

Interestingly enough the left-hand side of this inequality yields, for example, $R_S \geq 0.900$ when $R_1 = R_2 = \ldots = R_{100} = 0.999$ (compare to $R_S = 0.905$ for an independent serial system).

15.2.3.3 Parallel Systems. A parallel system is the dual of a serial system. Thus, failure of a parallel system can occur if and only if all its components fail. Let $p_P = 1 - R_P$ be the unreliability of the system and $p_1, p_2, \ldots, p_n$ be the unreliabilities of its components, where $p_k \equiv 1 - R_k$, $k = 1, 2, \ldots, n$. By making appropriate assumptions, we can derive the relationships which follow.

A. Independent parallel systems. As in the case of the serial system, the independence of failures (or successes) yields the (dual) product rule:

$$p_P = p_1 p_2 \cdots p_n \tag{15.10}$$

or

$$R_P = 1 - (1 - R_1)(1 - R_2)(\ldots)(1 - R_n) \tag{15.11}$$

Table 15.7 gives values of $1 - (1 - R)^k$ for R in the range 0.50 to 0.9999 for $k = 2, \ldots, 10$.

B. General parallel systems. In general, no matter what assumptions are made, as long as the system is parallel, the dual of inequalities (15.9) is true; namely

$$p_1 + p_2 + \ldots + p_n - (n - 1) \leq p_P \leq \min(p_1, p_2, \ldots, p_n) \tag{15.12}$$

Or, in terms of reliabilities, the equivalent of (15.9) is

$$\max(R_1, R_2, \ldots, R_n) \leq R_P \leq R_1 + R_2 + \ldots + R_n \tag{15.13}$$

for a parallel system. Of course, the right-hand inequality of (15.13) is of little use.

TABLE 15.6. *VALUES OF R^k FOR AN INDEPENDENT SERIAL SYSTEM WITH k COMPONENTS*

R	R^2	R^3	R^4	R^5	R^6	R^7	R^8	R^9	R^{10}
0.9000000	0.8100000	0.7290000	0.6561000	0.5904900	0.5314410	0.4782969	0.4304672	0.3874205	0.3486784
0.9100000	0.8281000	0.7535710	0.6857496	0.6240321	0.5678693	0.5167610	0.4702525	0.4279298	0.3894161
0.9200000	0.8464000	0.7786880	0.7163930	0.6590815	0.6063550	0.5578466	0.5132189	0.4721614	0.4343885
0.9300000	0.8649000	0.8043570	0.7480520	0.6956884	0.6469902	0.6017009	0.5595818	0.5204111	0.4839823
0.9400000	0.8836000	0.8305840	0.7807490	0.7339040	0.6898698	0.6484776	0.6095689	0.5729948	0.5386151
0.9500000	0.9025000	0.8573750	0.8145062	0.7737809	0.7350919	0.6983373	0.6634204	0.6302494	0.5987369
0.9600000	0.9216000	0.8847360	0.8493466	0.8153727	0.7827578	0.7514475	0.7213896	0.6925340	0.6648326
0.9700000	0.9409000	0.9126730	0.8852928	0.8587340	0.8329720	0.8079828	0.7837434	0.7602311	0.7374241
0.9800000	0.9604000	0.9411920	0.9223682	0.9039208	0.8858424	0.8681255	0.8507630	0.8337478	0.8170728
0.9900000	0.9801000	0.9702990	0.9605960	0.9509900	0.9414801	0.9320653	0.9227447	0.9135172	0.9043821
0.9910000	0.9820810	0.9732423	0.9644831	0.9558027	0.9472005	0.9386757	0.9302276	0.9218556	0.9135589
0.9920000	0.9840640	0.9761915	0.9683820	0.9606349	0.9529498	0.9453262	0.9377636	0.9302615	0.9228194
0.9930000	0.9860490	0.9791467	0.9722926	0.9654866	0.9587282	0.9520171	0.9453530	0.9387355	0.9321643
0.9940000	0.9880360	0.9821078	0.9762151	0.9703578	0.9645357	0.9587485	0.9529960	0.9472780	0.9415944
0.9950000	0.9900250	0.9850749	0.9801495	0.9752488	0.9703725	0.9655206	0.9606930	0.9558896	0.9511101
0.9960000	0.9920160	0.9880479	0.9840957	0.9801594	0.9762387	0.9723338	0.9684444	0.9645707	0.9607124
0.9970000	0.9940090	0.9910270	0.9880539	0.9850897	0.9821345	0.9791881	0.9762505	0.9733217	0.9704018
0.9980000	0.9960040	0.9940120	0.9920240	0.9900399	0.9880598	0.9860837	0.9841116	0.9821433	0.9801790
0.9990000	0.9980010	0.9970030	0.9960060	0.9950100	0.9940150	0.9930210	0.9920279	0.9910359	0.9900449
0.9991000	0.9982008	0.9973024	0.9964049	0.9955081	0.9946121	0.9937170	0.9928226	0.9919291	0.9910364
0.9992000	0.9984006	0.9976019	0.9968038	0.9960064	0.9952096	0.9944134	0.9936179	0.9928230	0.9920287
0.9993000	0.9986005	0.9979015	0.9972029	0.9965049	0.9958073	0.9951103	0.9944137	0.9937176	0.9930220
0.9994000	0.9988004	0.9982011	0.9976022	0.9970036	0.9964054	0.9958076	0.9952101	0.9946129	0.9940162
0.9995000	0.9990002	0.9985007	0.9980015	0.9975025	0.9970037	0.9965052	0.9960070	0.9955090	0.9950112
0.9996000	0.9992002	0.9988005	0.9984010	0.9980016	0.9976024	0.9972034	0.9968045	0.9964058	0.9960072
0.9997000	0.9994001	0.9991003	0.9988005	0.9985009	0.9982013	0.9979019	0.9976025	0.9973032	0.9970040
0.9998000	0.9996000	0.9994001	0.9992002	0.9990004	0.9988006	0.9986008	0.9984011	0.9982014	0.9980018
0.9999000	0.9998000	0.9997000	0.9996001	0.9995001	0.9994001	0.9993002	0.9992003	0.9991004	0.9990004
0.9999100	0.9998200	0.9997300	0.9996400	0.9995501	0.9994601	0.9993702	0.9992802	0.9991903	0.9991004
0.9999200	0.9998400	0.9997600	0.9996800	0.9996001	0.9995201	0.9994401	0.9993602	0.9992802	0.9992003
0.9999300	0.9998600	0.9997900	0.9997200	0.9996500	0.9995801	0.9995101	0.9994401	0.9993702	0.9993002
0.9999400	0.9998800	0.9998200	0.9997600	0.9997000	0.9996401	0.9995801	0.9995201	0.9994601	0.9994002
0.9999500	0.9999000	0.9998500	0.9998000	0.9997500	0.9997000	0.9996501	0.9996001	0.9995501	0.9995001
0.9999600	0.9999200	0.9998800	0.9998400	0.9998000	0.9997600	0.9997200	0.9996800	0.9996401	0.9996001
0.9999700	0.9999400	0.9999100	0.9998800	0.9998500	0.9998200	0.9997900	0.9997600	0.9997300	0.9997000
0.9999800	0.9999600	0.9999400	0.9999200	0.9999000	0.9998800	0.9998600	0.9998400	0.9998200	0.9998000
0.9999900	0.9999800	0.9999700	0.9999600	0.9999500	0.9999400	0.9999300	0.9999200	0.9999100	0.9999000

0.9999910	0.9999820	0.9999730	0.9999640	0.9999550	0.9999460	0.9999370	0.9999280	0.9999190	0.9999100
0.9999920	0.9999840	0.9999760	0.9999680	0.9999600	0.9999520	0.9999440	0.9999360	0.9999280	0.9999200
0.9999930	0.9999860	0.9999790	0.9999720	0.9999650	0.9999580	0.9999510	0.9999440	0.9999370	0.9999300
0.9999940	0.9999880	0.9999820	0.9999760	0.9999700	0.9999640	0.9999580	0.9999520	0.9999460	0.9999400
0.9999950	0.9999900	0.9999850	0.9999800	0.9999750	0.9999700	0.9999650	0.9999600	0.9999550	0.9999500
0.9999960	0.9999920	0.9999880	0.9999840	0.9999800	0.9999760	0.9999720	0.9999680	0.9999640	0.9999600
0.9999970	0.9999940	0.9999910	0.9999880	0.9999850	0.9999820	0.9999790	0.9999760	0.9999730	0.9999700
0.9999980	0.9999960	0.9999940	0.9999920	0.9999900	0.9999880	0.9999860	0.9999840	0.9999820	0.9999800
0.9999990	0.9999980	0.9999970	0.9999960	0.9999950	0.9999940	0.9999930	0.9999920	0.9999910	0.9999900
0.9999991	0.9999982	0.9999973	0.9999964	0.9999955	0.9999946	0.9999937	0.9999928	0.9999919	0.9999910
0.9999992	0.9999984	0.9999976	0.9999968	0.9999960	0.9999952	0.9999944	0.9999936	0.9999928	0.9999920
0.9999993	0.9999986	0.9999979	0.9999972	0.9999965	0.9999958	0.9999951	0.9999944	0.9999937	0.9999930
0.9999994	0.9999988	0.9999982	0.9999976	0.9999970	0.9999964	0.9999958	0.9999952	0.9999946	0.9999940
0.9999995	0.9999990	0.9999985	0.9999980	0.9999975	0.9999970	0.9999965	0.9999960	0.9999955	0.9999950
0.9999996	0.9999992	0.9999988	0.9999984	0.9999980	0.9999976	0.9999972	0.9999968	0.9999964	0.9999960
0.9999997	0.9999994	0.9999991	0.9999988	0.9999985	0.9999982	0.9999979	0.9999976	0.9999973	0.9999970
0.9999998	0.9999996	0.9999994	0.9999992	0.9999990	0.9999988	0.9999986	0.9999984	0.9999982	0.9999980
0.9999999	0.9999998	0.9999997	0.9999996	0.9999995	0.9999994	0.9999993	0.9999992	0.9999991	0.9999990

15.2.3.4 Partially Parallel Systems. A partially parallel system is one of n components, not all of which need to operate successfully for the system to operate successfully. In the simplest case each component performs the same function, and all components have the same reliability, independent of the number of operating components.

A. General case. Let R = component reliability. Then the probability that exactly k out of n components operate successfully is

$$Q_k = \binom{n}{k} R^k p^{n-k}$$

where $p \equiv 1 - R$. Let P_k be the probability that k components operating will result in the system operating successfully. Then system reliability is given by

$$R_S = \sum_{k=1}^{n} P_k Q_k$$

B. Special cases. If $P_k = 0$ when $k < j$, and $P_k = 1$ when $k \geq j$, then

$$R_S = \sum_{k=j}^{n} \binom{n}{k} R^k p^{n-k} \qquad (15.14)$$

When $j = n$, $R_S = R^n$; i.e., the system is equivalent to an independent serial system. When $j = 1$, $R_S = 1 - p^n$; i.e., the system is equivalent to an independent (completely) parallel system. Figure 15.2 gives values of R_S as a function of R for $n = 2, \ldots, 10$, for $j \leq n$, based on Eq. (15.14).

15.2.3.5 Stand-by Systems. A stand-by system is two or more components arranged so that, if one fails, the next one may then operate to perform the function. It is desirable to accomplish the switchover passively—i.e., without using active sensing and switching devices—to keep system reliability as high as possible. When a sensing and switching device is incorporated, the possibility of erroneous switchover may also need to be considered. In addition, reliability may be a function of time, not only for the operating components, but also for the sensing and switching device.

A. Two components in stand-by with one sensing-switching device.

Table 15.7. *VALUES OF* $1-(1-R)^k$ *FOR AN INDEPENDENT PARALLEL SYSTEM WITH k COMPONENTS*

R	$1-(1-R)^2$	$1-(1-R)^3$	$1-(1-R)^4$	$1-(1-R)^5$	$1-(1-R)^6$	$1-(1-R)^7$	$1-(1-R)^8$	$1-(1-R)^9$	$1-(1-R)^{10}$
0.5000000	0.7500000	0.8750000	0.9375000	0.9687500	0.9843750	0.9921875	0.9960938	0.9980469	0.9990234
0.5100000	0.7599000	0.8823510	0.9423520	0.9717525	0.9861587	0.9932178	0.9966767	0.9983716	0.9992021
0.5200000	0.7696000	0.8894080	0.9469158	0.9745196	0.9877694	0.9941293	0.9971821	0.9986474	0.9993507
0.5300000	0.7791000	0.8961770	0.9512032	0.9770655	0.9892208	0.9949338	0.9976189	0.9988809	0.9994740
0.5400000	0.7884000	0.9026640	0.9552254	0.9794037	0.9905257	0.9956418	0.9979952	0.9990778	0.9995758
0.5500000	0.7975000	0.9088750	0.9589938	0.9815472	0.9916962	0.9962633	0.9983185	0.9992433	0.9996595
0.5600000	0.8064000	0.9148160	0.9625190	0.9835084	0.9927437	0.9968072	0.9985952	0.9993819	0.9997280
0.5700000	0.8151000	0.9204930	0.9658120	0.9852992	0.9936786	0.9972818	0.9988312	0.9994974	0.9997839
0.5800000	0.8236000	0.9259120	0.9688830	0.9869309	0.9945110	0.9976946	0.9990317	0.9995933	0.9998292
0.5900000	0.8319000	0.9310790	0.9717424	0.9884144	0.9952499	0.9980525	0.9992015	0.9996726	0.9998658
0.6000000	0.8400000	0.9360000	0.9744000	0.9897600	0.9959040	0.9983616	0.9993446	0.9997379	0.9998951
0.6100000	0.8479000	0.9406810	0.9768656	0.9909776	0.9964813	0.9986277	0.9994648	0.9997913	0.9999186
0.6200000	0.8556000	0.9451280	0.9791486	0.9920765	0.9969891	0.9988558	0.9995652	0.9998348	0.9999372
0.6300000	0.8631000	0.9493470	0.9812584	0.9930656	0.9974343	0.9990507	0.9996488	0.9998700	0.9999519
0.6400000	0.8704000	0.9533440	0.9832038	0.9939534	0.9978232	0.9992164	0.9997179	0.9998984	0.9999634
0.6500000	0.8775000	0.9571250	0.9849938	0.9947478	0.9981617	0.9993566	0.9997748	0.9999212	0.9999724
0.6600000	0.8844000	0.9606960	0.9866366	0.9954565	0.9984552	0.9994748	0.9998214	0.9999393	0.9999794
0.6700000	0.8911000	0.9640630	0.9881408	0.9960865	0.9987085	0.9995738	0.9998594	0.9999536	0.9999847
0.6800000	0.8976000	0.9672320	0.9895142	0.9966446	0.9989263	0.9996564	0.9998900	0.9999648	0.9999887
0.6900000	0.9039000	0.9702090	0.9907648	0.9971371	0.9991125	0.9997249	0.9999147	0.9999736	0.9999918
0.7000000	0.9100000	0.9730000	0.9919000	0.9975700	0.9992710	0.9997813	0.9999344	0.9999803	0.9999941
0.7100000	0.9159000	0.9756110	0.9929272	0.9979489	0.9994052	0.9998275	0.9999500	0.9999855	0.9999958
0.7200000	0.9216000	0.9780480	0.9938534	0.9982790	0.9995181	0.9998651	0.9999622	0.9999894	0.9999970
0.7300000	0.9271000	0.9803170	0.9946856	0.9985651	0.9996126	0.9998954	0.9999718	0.9999924	0.9999979
0.7400000	0.9324000	0.9824240	0.9954302	0.9988119	0.9996911	0.9999197	0.9999791	0.9999946	0.9999986
0.7500000	0.9375000	0.9843750	0.9960938	0.9990234	0.9997559	0.9999390	0.9999847	0.9999962	0.9999990
0.7600000	0.9424000	0.9861760	0.9966822	0.9992037	0.9998089	0.9999541	0.9999890	0.9999974	0.9999994
0.7700000	0.9471000	0.9878330	0.9972016	0.9993564	0.9998520	0.9999660	0.9999922	0.9999982	0.9999996
0.7800000	0.9516000	0.9893520	0.9976574	0.9994846	0.9998866	0.9999751	0.9999945	0.9999988	0.9999997
0.7900000	0.9559000	0.9907390	0.9980552	0.9995916	0.9999142	0.9999820	0.9999962	0.9999992	0.9999998
0.8000000	0.9600000	0.9920000	0.9984000	0.9996800	0.9999360	0.9999872	0.9999974	0.9999995	0.9999999
0.8100000	0.9639000	0.9931410	0.9986968	0.9997524	0.9999530	0.9999911	0.9999983	0.9999997	0.9999999
0.8200000	0.9676000	0.9941680	0.9989502	0.9998110	0.9999660	0.9999939	0.9999989	0.9999998	1.0000000
0.8300000	0.9711000	0.9950870	0.9991648	0.9998580	0.9999759	0.9999959	0.9999993	0.9999999	
0.8400000	0.9744000	0.9959040	0.9993446	0.9998951	0.9999832	0.9999973	0.9999996	0.9999999	
0.8500000	0.9775000	0.9966250	0.9994938	0.9999241	0.9999886	0.9999983	0.9999997	1.0000000	
0.8600000	0.9804000	0.9972560	0.9996158	0.9999462	0.9999925	0.9999989	0.9999999		

0.8700000	0.9831000	0.9978030	0.9997144	0.9999629	0.9999952	0.9999994	0.9999999
0.8800000	0.9856000	0.9982720	0.9997926	0.9999751	0.9999970	0.9999996	1.0000000
0.8900000	0.9879000	0.9986690	0.9998536	0.9999839	0.9999982	0.9999998	
0.9000000	0.9900000	0.9990000	0.9999000	0.9999900	0.9999990	0.9999999	
0.9100000	0.9919000	0.9992710	0.9999344	0.9999941	0.9999995	1.0000000	
0.9200000	0.9936000	0.9994880	0.9999590	0.9999967	0.9999997		
0.9300000	0.9951000	0.9996570	0.9999760	0.9999983	0.9999999		
0.9400000	0.9964000	0.9997840	0.9999870	0.9999992	1.0000000		
0.9500000	0.9975000	0.9998750	0.9999938	0.9999997			
0.9600000	0.9984000	0.9999360	0.9999974	0.9999999			
0.9700000	0.9991000	0.9999730	0.9999992	1.0000000			
0.9800000	0.9996000	0.9999920	0.9999998				
0.9900000	0.9999000	0.9999990	1.0000000				
0.9910000	0.9999190	0.9999993					
0.9920000	0.9999360	0.9999995					
0.9930000	0.9999510	0.9999997					
0.9940000	0.9999640	0.9999998					
0.9950000	0.9999750	0.9999999					
0.9960000	0.9999840	0.9999999					
0.9970000	0.9999910	1.0000000					
0.9980000	0.9999960						
0.9990000	0.9999990						
0.9991000	0.9999992						
0.9992000	0.9999994						
0.9993000	0.9999995						
0.9994000	0.9999996						
0.9995000	0.9999998						
0.9996000	0.9999998						
0.9997000	0.9999999						
0.9998000	1.0000000						
0.9999000							

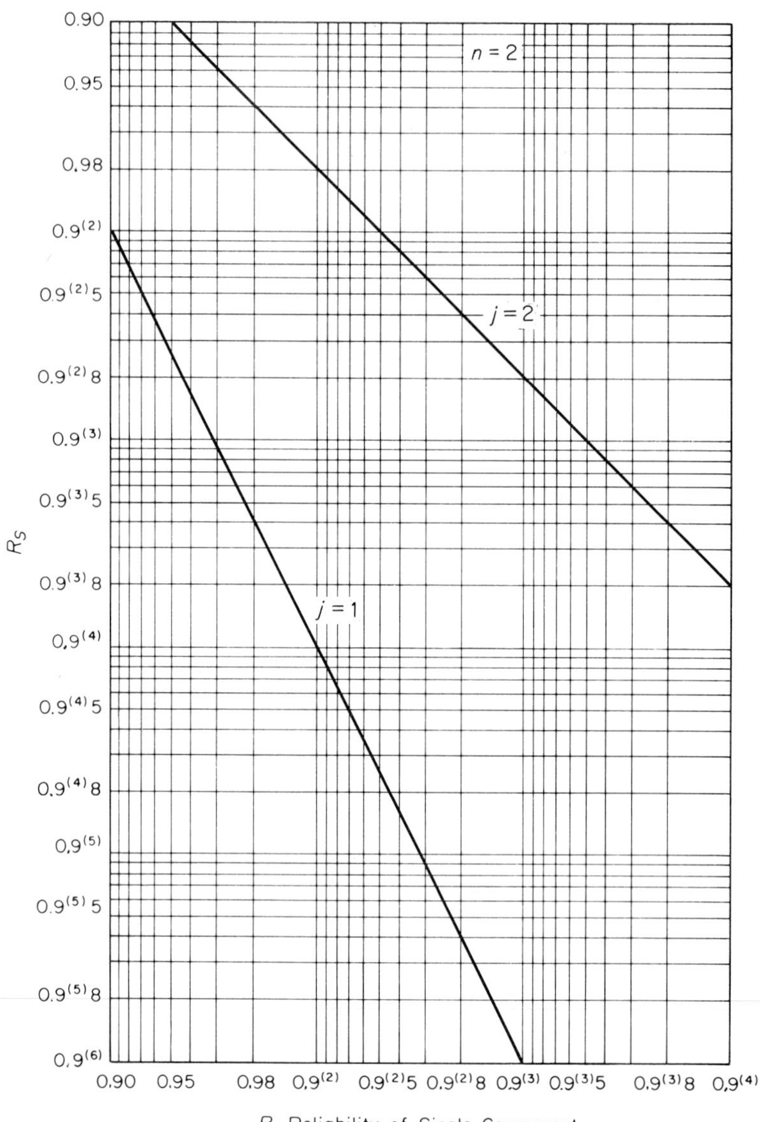

Fig. 15.2 R_S *is probability of success of a system of* n *components, each of reliability* R, *when* j *or more successful components are necessary for system success.*

$$R_S = R_1 + (1 - R_1)\, R_{ss}\, R_2 \tag{15.15}$$

where R_{ss} is the reliability of the sensing-switching device.

B. Three components in stand-by with two sensing-switching devices. The second sensing-switching device can operate only when the second component fails.

$$R_S = R_1 + (1 - R_1)R_{ss1}R_2 + (1 - R_1)R_{ss1}(1 - R_2)R_{ss2}\, R_3 \tag{15.16}$$

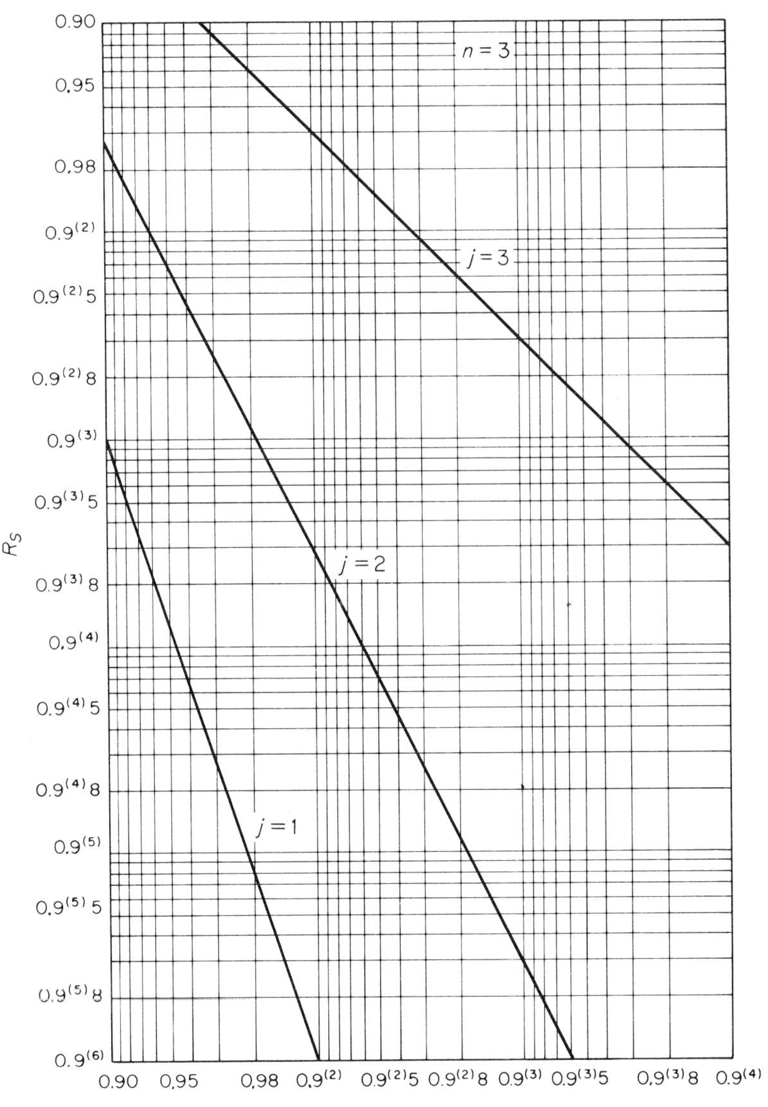

Fig. 15.2 *Cont.*

C. n components in stand-by with $n - 1$ sensing-switching devices. The ith sensing-switching device can operate only when the ith component fails. All components have equal reliability: $R_1 = R_2 = \ldots = R_n = r$. All sensing-switching devices have equal reliability.

$$R_{SS1} = R_{SS2} = \ldots = R_{SS(n-1)} = R_{SS}$$

$$R_S = r\left[\frac{1 - (1 - r)^n R_{SS}^n}{1 - (1 - r)R_{SS}}\right] \tag{15.17}$$

D. Two components in stand-by with one sensing-switching device. Components 1 and 2

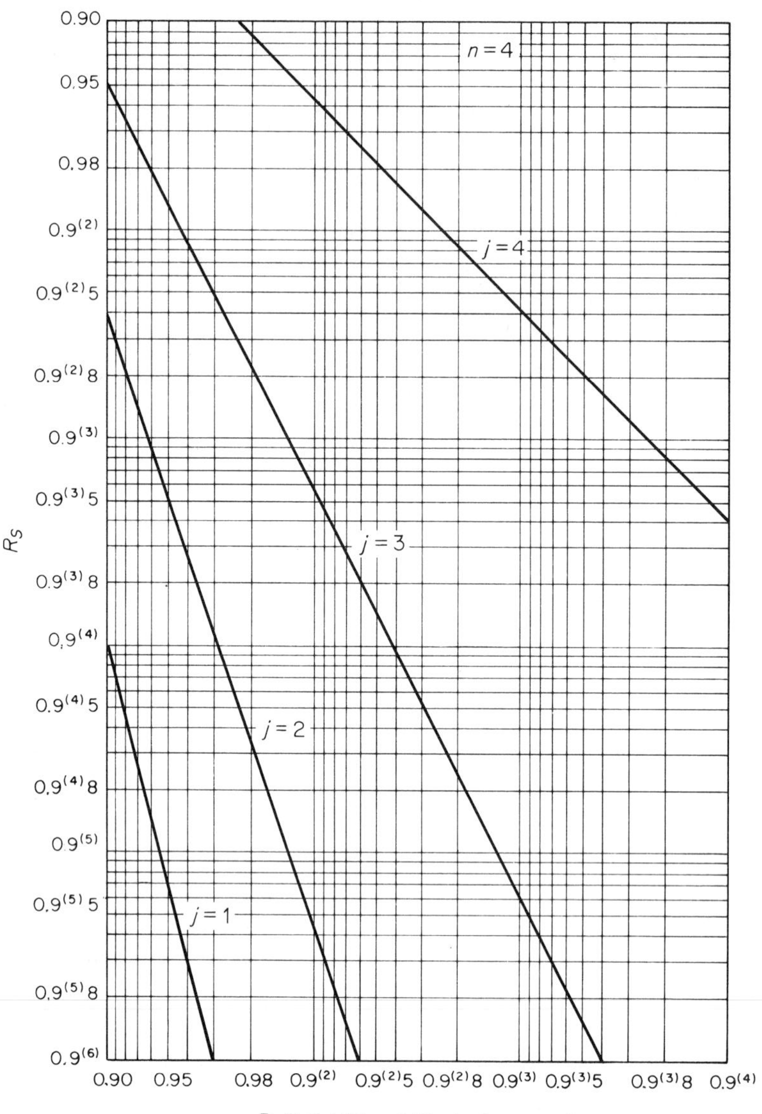

Fig. 15.2 *Cont.*

have exponential time-to-failure distributions while operating: λ_{10} and λ_{20}, respectively; and Component 2 has nonoperating failure rate λ_{2n}. The sensing-switching device has probability R_{ss} of successfully operating should Component 1 fail.

$$R = e^{-\lambda_{10}t} + \frac{\lambda_{10} R_{ss}}{\lambda_{10} + \lambda_{2n} - \lambda_{20}} [e^{-\lambda_{20}t} - e^{-(\lambda_{10}+\lambda_{2n})t}] \tag{15.18}$$

E. Same as D; however the sensing-switching device has exponential time-to-failure distribution with failure rate λ_{ss}.

$$R = e^{-\lambda_{10}t} + \frac{\lambda_{10}}{\lambda_{10} + \lambda_{ss} + \lambda_{2n} - \lambda_{20}} [e^{-\lambda_{20}t} - e^{-(\lambda_{10}+\lambda_{ss}+\lambda_{2n})t}] \tag{15.19}$$

F. Reliability of stand-by system E exceeds

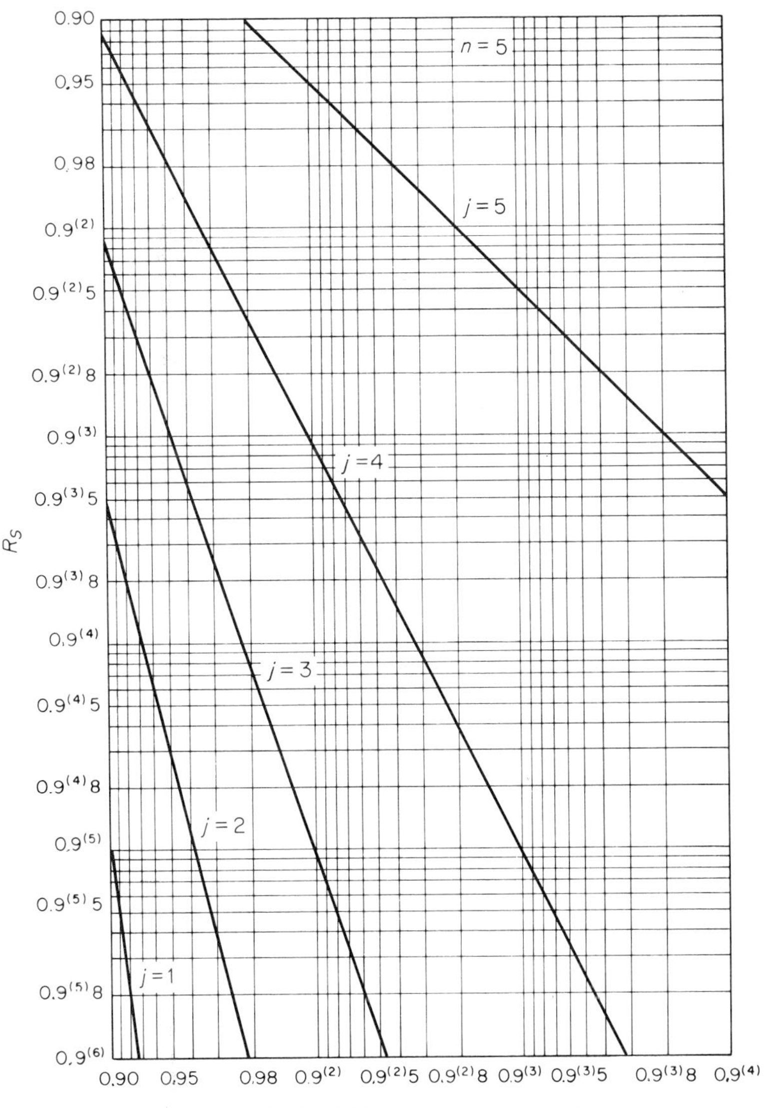

Fig. 15.2 *Cont.*

reliability of simple parallel system (no sensing-switching device required, both components operate starting at $t = 0$) when and only when

$$\lambda_{ss} < \lambda_{20} - \lambda_{2n} \tag{15.20}$$

That is, the sensing-switching device failure rate must be less than the difference between the operating and nonoperating failure rates of Component 2.

15.2.3.6 Analysis of Multifailure Mode Systems. This paragraph discusses the reliability analysis of systems whose components may have several failure modes, any one or more of which may occur simultaneously. In addition, failure of at least two of the components in a certain mode may be required in order to cause

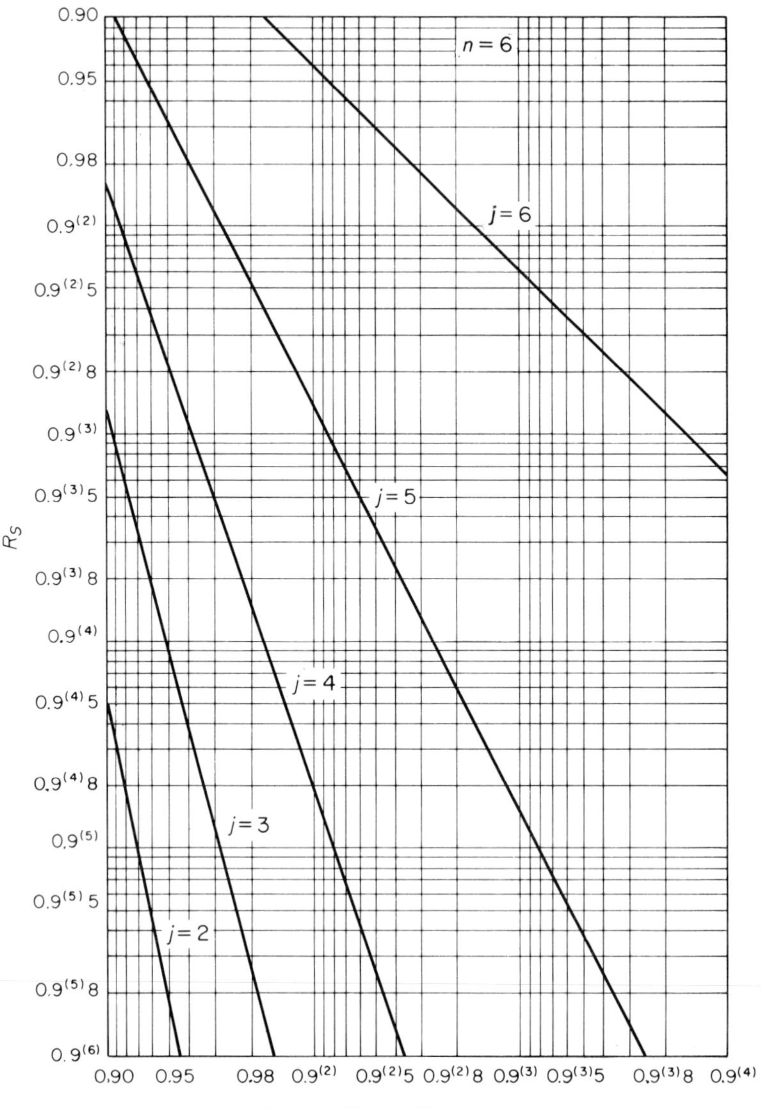

Fig. 15.2 *Cont.*

system failure. In other words, the system may be in parallel with respect to one or more modes of failure, but serial with respect to other modes of failure.

A simple example will be used to illustrate the methods employed in the analysis. It will be seen that the methods can be easily extended to a general multifailure mode system.

Consider the use of a propellant valve of a rocket propulsion system in a spacecraft, which takes nearly 2 years to move into the vicinity of Jupiter after being launched from Earth. At that time the propellant valve, consisting of two solenoid valves, either of which feeds sufficient propellant to the rocket engines for required total impulse, is actuated. However, should either of the two valves leak during the long coast period, insufficient propellants will

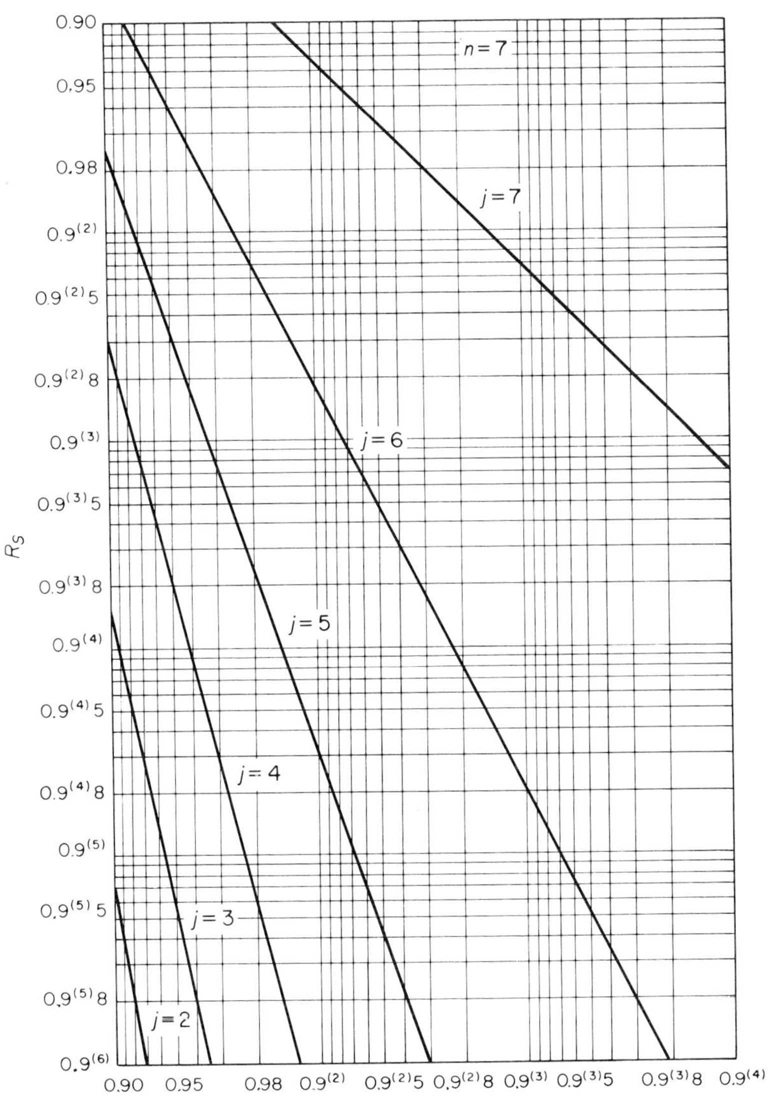

Fig. 15.2 *Cont.*

be remaining to allow the propulsion system to put the spacecraft into the desired orbit of Jupiter. Thus if *both* of the solenoid valves fail to actuate or if *either* leaks, the propellant valve fails; otherwise it operates successfully.

Let F_{ij} denote the event: Failure mode j in Component i occurs. Here, $i, j = 1, 2$, where Failure Mode 1 for both solenoid valves is the fail-to-actuate mode, and Failure Mode 2 is the leakage mode. The symbol $\bar{F}_{ij}$ denotes the complementary event: Failure Mode j in Component i does not occur, and if $P(F_{ij})$ denotes the probability of event F_{ij}, then $P(\bar{F}_{ij}) = 1 - P(F_{ij})$. For simplicity we denote $P_{ij} \equiv P(F_{ij})$ and $Q_{ij} \equiv 1 - P_{ij}$.

The first step is to assign all combinations of occurrence and nonoccurrence of failure modes to each component, and whether the result of

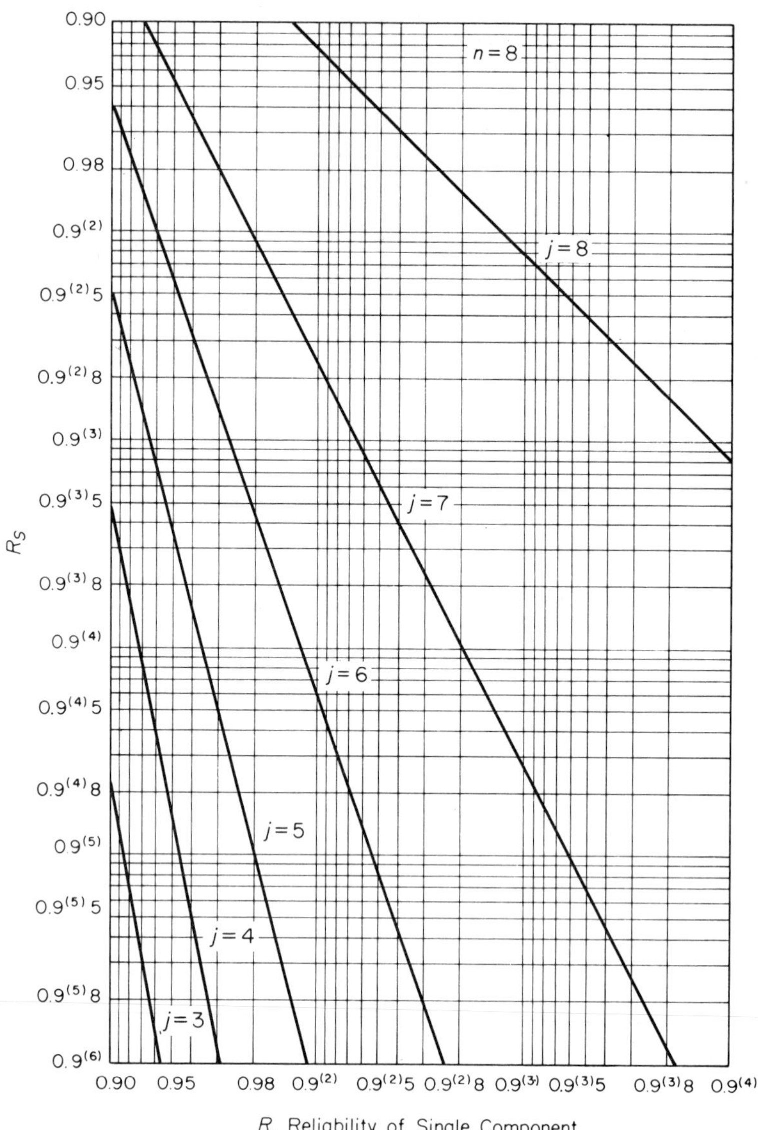

Fig. 15.2 *Cont.*

the combination is success or failure of the system. Usually, one can tell in advance which are the fewer number of combinations, those resulting in system failure or those resulting in system success, and list only the smaller number. The system probability of failure or success is then the sum of the probabilities of all the possible combinations which lead to the corresponding outcome. The foregoing analysis can be shown as in Table 15.8.

For simplicity we will assume that all events resulting in a given outcome are *independent;* e.g., the probability that actuation of Valve 1 occurs or does not occur is unchanged whether or not it is leaking or for that matter whether or not Valve 2 is leaking or failing to actuate, etc.

More generally, assumptions that the events are *dependent* could be made. For example,

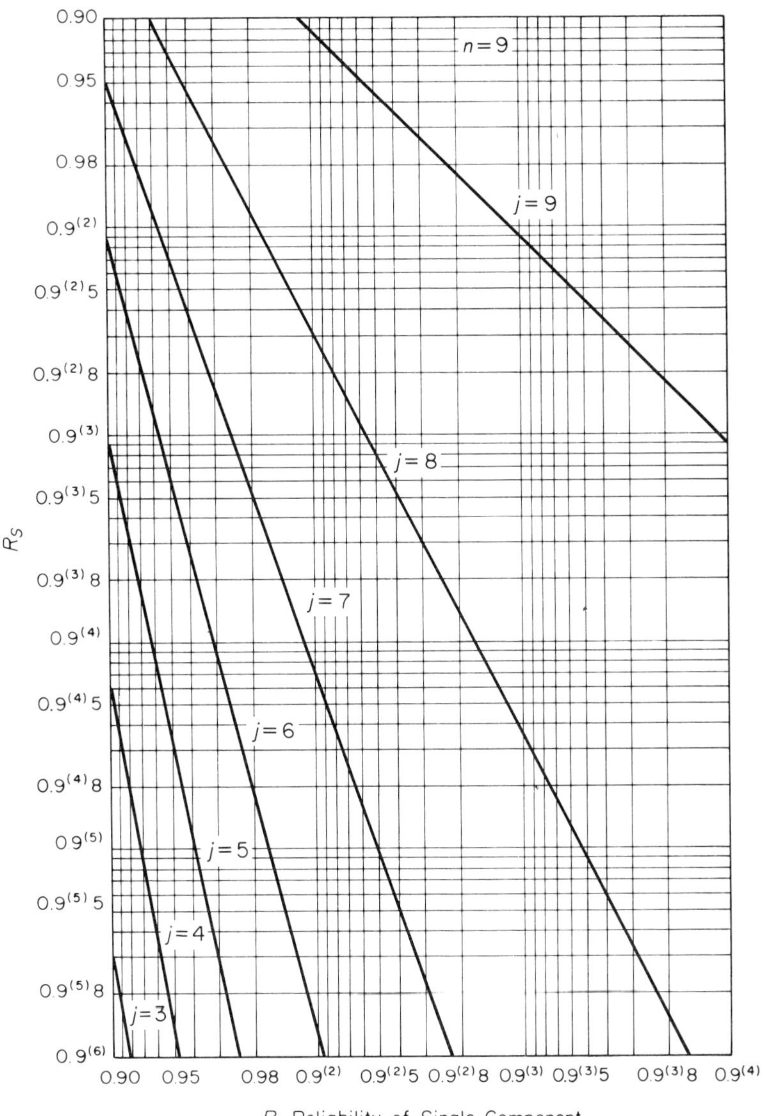

Fig. 15.2 *Cont.*

leakage, if caused by contamination, would tend to occur in both valves together or not at all. Also, for example, if failure to actuate is nearly always caused by an electrical failure elsewhere in the system, then both valves would tend to fail simultaneously if at all. One type of dependence occurs when the occurrence of one failure mode precludes the occurrence of the other; for example open vs. short circuits. In this case the failure modes are said to be *mutually exclusive*. Consequently, probability of system success (from the first three rows of Table 15.8) is

$$R_S = P(\bar{F}_{11} \cap \bar{F}_{12} \cap \bar{F}_{21} \cap \bar{F}_{22}) + P(F_{11} \cap \bar{F}_{12} \cap \bar{F}_{21} \cap \bar{F}_{22}) + P(\bar{F}_{11} \cap \bar{F}_{12} \cap F_{21} \cap \bar{F}_{22})$$

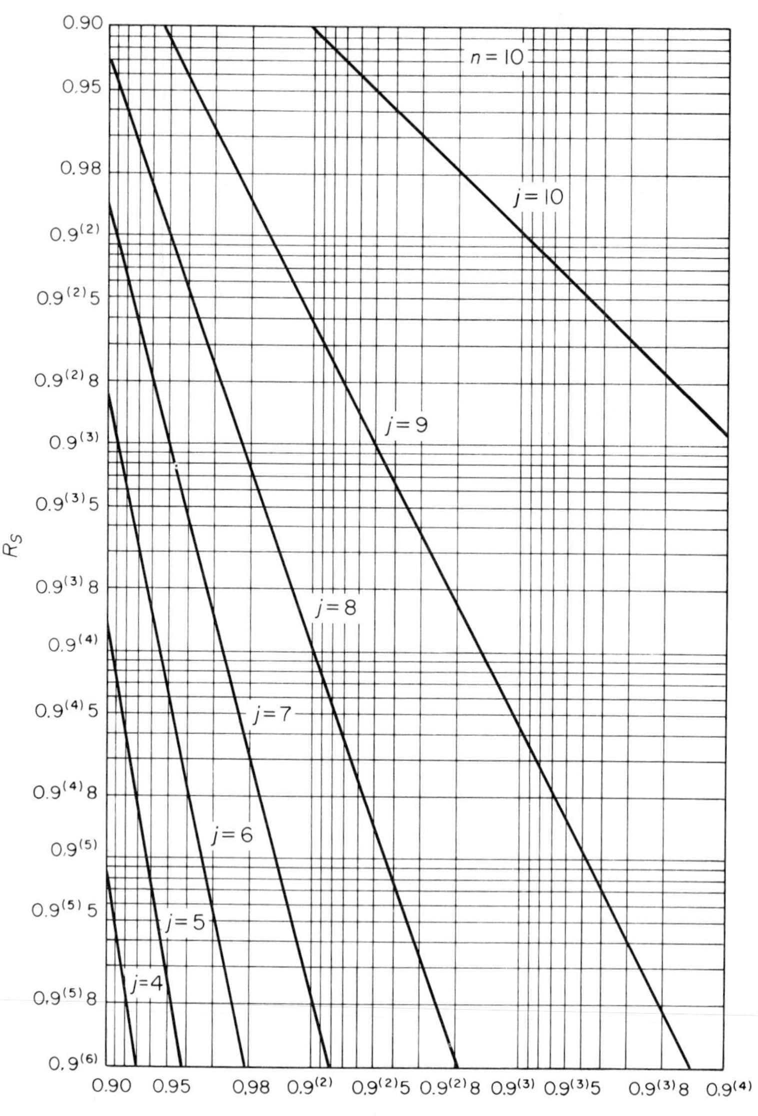

Fig. 15.2 *Cont.*

where $\cap$ denotes *and* (also see *R:MMM*, Chap. 5). Since independence implies that each of the above probabilities is the product of the probabilities of the events in the parentheses, we have

$$R_S = Q_{11}\,Q_{12}\,Q_{21}\,Q_{22} + P_{11}\,Q_{12}\,Q_{21}\,Q_{22} + Q_{11}\,Q_{12}\,P_{21}\,Q_{22}$$

Now, if the reliability of each solenoid valve is defined as the probability that neither failure mode occurs, then

$$R_1 = Q_{11}\,Q_{12}$$
$$R_2 = Q_{21}\,Q_{22}$$

are these reliabilities respectively for solenoid Valves 1 and 2. Thus,

TABLE 15.8. *EXAMPLE OF ANALYSIS OF MULTIFAILURE MODE SYSTEM*

Component 1		Component 2		System	
(Actuation)	(Leakage)	(Actuation)	(Leakage)	Outcome	Remarks
$\bar{F}_{11}$	$\bar{F}_{12}$	$\bar{F}_{21}$	$\bar{F}_{22}$	Success	No failures
F_{11}	$\bar{F}_{12}$	$\bar{F}_{21}$	$\bar{F}_{22}$	Success	Valve 1 fails to actuate
$\bar{F}_{11}$	$\bar{F}_{12}$	F_{21}	$\bar{F}_{22}$	Success	Valve 2 fails to actuate
$\bar{F}_{11}$	F_{12}	$\bar{F}_{21}$	$\bar{F}_{22}$	Failure	Valve 1 leaks
$\bar{F}_{11}$	$\bar{F}_{12}$	$\bar{F}_{21}$	F_{22}	Failure	Valve 2 leaks
$\bar{F}_{11}$	F_{12}	$\bar{F}_{21}$	F_{22}	Failure	Both valves leak
F_{11}	$\bar{F}_{12}$	F_{21}	$\bar{F}_{22}$	Failure	Both valves fail to actuate
—	—	—	—		
(Nine more combinations all resulting in system failure)					

$$
\begin{aligned}
R_S &= R_1R_2 + P_{11}Q_{12}R_2 + R_1P_{21}Q_{22} \\
&= R_1R_2\left(1 + \frac{P_{11}}{Q_{11}} + \frac{P_{21}}{Q_{21}}\right)
\end{aligned}
$$

which shows that the system reliability exceeds that of a simple serial system. Actually, the system is serial with respect to the leakage failure mode and parallel with respect to the actuation failure mode, and for this simple example can be written directly as

$$
\begin{aligned}
R_S &= [Q_{12}Q_{22}]\,[Q_{11}Q_{21} + (1 - Q_{11})Q_{21} \\
&\quad + Q_{11}(1 - Q_{21})] \\
&= [Q_{12}Q_{22}]\,[1 - (1 - Q_{11})(1 - Q_{21})]
\end{aligned}
$$

The first bracketed expression is the product of the probabilities of no leakage failure, showing a serial combination, and the second shows that the probabilities of no actuation failure combine in parallel.

15.2.3.7 System Performance as a Function of Component Performance

A. Introduction. A problem of frequent importance is to determine the variation in performance of a system in terms of the values and variations of the parameters describing inputs to the system and the transfer functions of the system. For example, in a vacuum and for a gravitation-free trajectory, the incremental velocity of an upper-stage rocket-propelled vehicle is

$$V = g_c I_{sp} \log\left(1 + \frac{W_p}{W_M}\right) \tag{15.21}$$

where $g_c = 32.174$ ft-sec^{-2}-lbm-lbf^{-1}, I_{sp} is the specific impulse of the rocket motor (lbf-sec-lbm^{-1}), W_p is the propellant weight, and W_M is the weight of the vehicle after the propellant has been expended. One problem would be: What is the probability that a specified velocity increment will not be attained? If the result were (for example) 10 per cent or more, it would be necessary to modify the performance of other parts of the overall vehicle (e.g., increase firing duration for one or more booster stages, etc.). The following methods and formulas are useful in answering this and similar questions.

B. General formulas. Let Y represent an output performance parameter, and X_1, $X_2, \ldots, X_n$ represent input variables; assume that Y is some known function of the X's.

$$Y = H(X_1, X_2, \ldots, X_n) \tag{15.22}$$

Then the following formulas are approximately true.

$$\bar{Y} \simeq H(\bar{X}_1, \bar{X}_2, \ldots, \bar{X}_n) \tag{15.23}$$

$$
\begin{aligned}
\sigma_Y^2 \simeq &\sum_{j=1}^{n} \left(\frac{\partial H}{\partial X_j}\right)^2 \sigma_j^2 \\
&+ 2 \sum_{i<j} \left(\frac{\partial H}{\partial X_i}\right)\left(\frac{\partial H}{\partial X_j}\right) \rho_{ij}\sigma_i\sigma_j
\end{aligned} \tag{15.24}
$$

where $\bar{X}_1$, $\bar{X}_2, \ldots, \bar{X}_n$, $\bar{Y}$ are the respective means or nominal values of the X's and Y; the partial derivatives are evaluated at the point $(\bar{X}_1, \bar{X}_2, \ldots, \bar{X}_n)$; and σ_j and ρ_{ij} are the standard deviation of the jth input variable and the correlation coefficient for the ith and jth input variables, respectively.

C. Special cases

i. $Y = \sum_{j=1}^{n} a_j X_j$

$$\bar{Y} = \sum_{j=1}^{n} a_j \bar{X}_j \qquad \text{(exact)} \tag{15.25}$$

$$\sigma_Y^2 = \sum_{j=1}^{n} a_j^2 \sigma_j^2 + 2 \sum_{i<j} a_i a_j \rho_{ij} \sigma_i \sigma_j \qquad \text{(exact)} \tag{15.26}$$

ii. $Y = \prod_{j=1}^{n} X_j^{b_j}$

$$\bar{Y} \simeq \prod_{j=1}^{n} \bar{X}_j^{b_j} \tag{15.27}$$

$$\frac{\sigma_{\bar{Y}}^2}{\bar{Y}^2} \simeq \sum_{j=1}^{n} b_j^2 \frac{\sigma_j^2}{\bar{X}_j^2} + 2 \sum_{i<j} b_i b_j \rho_{ij} \frac{\sigma_i}{\bar{X}_i} \frac{\sigma_j}{\bar{X}_j} \tag{15.28}$$

iii. A special case of ii is the quotient X_1/X_2 where

$$n = 2 \qquad b_1 = 1 \qquad b_2 = -1$$

iv. When the variables $X_1, X_2, \ldots, X_n$ are independent or merely uncorrelated, $\rho_{ij} = 0$.

D. Tolerances. Define $100(1 - \alpha)$ per cent tolerances on the variable X as the two numbers $\bar{X} \pm a$ so that

$$P[\bar{X} - a \leq X \leq \bar{X} + a] = 1 - \alpha$$

Assuming that the variable X has a normal probability distribution, then

$$\sigma = a/K_{\alpha/2}$$

where σ is the standard deviation of X, and K_ε is the standard normal deviate exceeded with probability ε (e.g., $K_{0.005} = 2.576$).

E. Example. Based on the example in Par. 15.2.3.7A, assuming $100(1 - \alpha)$ per cent tolerances on all variables, compute the maximum allowable tolerance on W_M given that

$$I_{sp} = 290 \pm 2.9$$
$$W_p = 200 \pm 1.0$$
$$V = 3140 \pm 41.2$$

Answer: $W_M = 500 \pm 4.6$ lbm
$= 500 \pm 0.92$ per cent

For $1 - \alpha = 0.99$, σ_{W_M} should be at most $4.6/2.576 \simeq 1.8$ lbm.

15.2.3.8 Confidence Limits for Reliability of Systems. In this section we give methods and formulas for finding confidence limits on system reliability as a function of the observed results on subsystem tests.

A. General results. Let $\hat{R}_{L1}, \hat{R}_{L2}, \ldots, \hat{R}_{Lk}$ be lower confidence limits on the reliability of subsystems, each with confidence coefficients at least $\gamma_1, \gamma_2, \ldots, \gamma_k$. The following results do not depend on the manner in which the subsystem confidence limits are obtained—i.e., by binomial trials or any other means.

i. Serial system, independent subsystems, independent tests of subsystems:

$$\hat{R}_L = \hat{R}_{L1} \hat{R}_{L2} \ldots \hat{R}_{Lk}$$

is a lower confidence limit on system reliability with confidence γ, where

$$\gamma \geq \gamma_1 \gamma_2 \ldots \gamma_k$$

ii. Serial system. No assumptions of independence of subsystems, but tests of subsystems performed independently.

$$\hat{R}_L = \hat{R}_{L1} + \hat{R}_{L2} + \cdots + \hat{R}_{Lk} - (k - 1)$$

and $\gamma \geq \gamma_1 \gamma_2 \ldots \gamma_k$

iii. Serial system. No assumption of independence of either subsystems or tests of subsystems.

$$\hat{R}_L = \hat{R}_{L1} + \hat{R}_{L2} + \cdots + \hat{R}_{Lk} - (k - 1)$$

and $\gamma \geq \gamma_1 + \gamma_2 + \ldots + \gamma_k - (k - 1)$.

Examples: Let $k = 3, \hat{R}_{L1} = \hat{R}_{L2} = \hat{R}_{L3} = 0.90$; $\gamma_1 = \gamma_2 = \gamma_3 = 0.90$. Then under Condition i, $R_L = 0.729$, $\gamma \geq 0.729$; Condition ii, $R_L = 0.700$, $\gamma \geq 0.729$; Condition iii, $R_L = 0.700$, $\gamma \geq 0.700$.

B. Independent serial system, independent tests of subsystems, binomial trials. Let $N_1, \ldots, N_k$ and $F_1, \ldots, F_k$ be the numbers of trials and failures of the k subsystems, respectively.

Define $N_m = \min(N_1, \ldots, N_k)$, and using the point estimate for system reliability

$$\hat{R} = \prod_{j=1}^{k} \left(1 - \frac{F_j}{N_j}\right)$$

define $\bar{F} = N_m(1 - \hat{R})$.
Then use Fig. 15.1 with $F \equiv \bar{F}$ and $N \equiv N_m$ to obtain $\hat{R}_L$. This method is believed to be conservative (see *R:MMM*, Chap. 9; and Zelen, *Statistical Theory of Reliability*).

Example. $N_1 = N_2 = 10$; $F_1 = 0$, $F_2 = 1$. Hence $N_m = 10$, $\bar{F} = 1$, and from Fig. 15.1, $\hat{R}_L = 0.66$ with confidence at least 0.90. Methods Ai, ii, and iii yield $\hat{R}_L \equiv 0.45, 0.35, 0.35$, respectively, each with confidence at least 0.90.

15.2.3.9 Stress-vs.-Strength Models

A. General considerations. The reliability models discussed in this section take into account the environment (stress) that may cause the failure. The ability of a part to resist the applied stress is called the strength of the part. Strength and stress are measured in the same physical units, which may be other than the common stress units such as lbf-in.$^{-2}$ (psi). For example, temperature in degrees Fahrenheit may be the more significant variable representing stress or strength. Time to failure can always be interpreted as a strength; therefore the operating time is interpreted as a stress. In general, strength and stress are highly variable or uncertain quantities. Consequently, it is useful to assume they are random variables with some known or partially known probability distribution.

Let $F(x)$ be the distribution function of part strength X and $G(y)$ be the distribution function of stress Y applied to the part. The strength of the part is defined as the stress at which failure occurs. The reliability of the part is defined as the probability that the strength of the part is greater than the stress applied to it. Thus

$$R = \int_{-\infty}^{\infty} f(x)\, G(x)\, dx = \int_{-\infty}^{\infty} g(y)\,[1 - F(y)]\, dy \qquad (15.29)$$

where $f(x) = d/dx\, F(x)$ is the probability density function of strength and $g(x) = d/dx\, G(x)$ is the probability density function of applied stress.

TABLE 15. 9. *RELIABILITY IN TERMS OF STRENGTH AND STRESS*

	Strength Distribution Function $F(x)$	Stress Distribution Function $G(y)$	Probability R That Strength Exceeds Stress
Normal	$\Phi\left(\frac{x - \mu_f}{\sigma_f}\right)$	Normal $\Phi\left(\frac{y - \mu_g}{\sigma_g}\right)$	$\Phi\,(k)$
Truncated normal	$0, \quad x < y_0$; $\dfrac{\Phi\left(\frac{x - \mu_f}{\sigma_f}\right)}{\Phi(h)}, \ x \geq y_0$	Normal $\Phi\left(\frac{y - \mu_g}{\sigma_g}\right)$	$1 - [1 - \Phi\,(k) - L\,(h, k, r)]\,[\Phi\,(h)]^{-1}$
	$h \equiv (\mu_g - y_0)/\sigma_g$	$\Phi(z) = \frac{1}{(2\pi)^{1/2}} \int_{-\infty}^{z} \exp\,[\,-\,t^2/2]\, dt$	
	$k \equiv (\mu_f - \mu_g)/\sqrt{\sigma^2{}_f + \sigma^2{}_g}$	$L\,(h, k, r) = \frac{1}{2\pi\,(1 - r^2)^{1/2}} \int_h^{\infty} dx \int_k^{\infty} dy \exp\left\{-\frac{1}{2\,(1 - r^2)}\,[x^2 - 2\,xyr + y^2]\right\}$	
	$r \equiv \sigma_g/\sqrt{\sigma^2{}_f + \sigma^2{}_g}$	$L\,(h, k, r)$ tabulated in [T-13].	

B. Tables. Table 15.9 gives formulas for R calculated from Eq. (15.29) for normal stress distributions and both normal and truncated-normal strength distributions. An example of the latter situation is given in Lipow and Eidemiller "Application of the Bivariate Normal Distribution to a Stress vs. Strength Problem in Reliability Analysis."

C. Applications. Many components fail because of a structural failure, described by such terms as rupture, burst, crack, yield, etc. The symptoms of failure are fluid leakage of varying degrees, breakage or distortion, binding, etc. The cause of structural failure may be sudden or prolonged application of one of several more or less independent stresses. Often one stress is predominant, however, and when failure occurs it is almost always due to this stress. The choice of size, weight, material, etc. is generally based on considerations of the predominant stress. Because of independent considerations, other kinds of stresses or loading conditions may require that a part be designed to a higher margin of safety than necessary for the predominant stress alone. In these cases the failure probability calculated by the method shown here is correspondingly conservatively high. When prolonged or repetitive application of stresses is more significant than sudden application, the methods of evaluating fatigue failure must be used. (See Weibull, *Fatigue Testing and the Analysis of Results;* Sarhan and Greenberg, *Contributions to Order Statistics;* Crandall and Mark, *Random Vibration in Mechanical Systems;* and Bendat and Piersol, *Measurement and Analysis of Random Data.*)

The design of a structural member or a part whose predominant mode of failure can be termed a structural failure caused by a suddenly applied load is a function of the allowable stress, the applied load, the safety factor, and the margin of safety MS. These quantities are related by the equation

$$MS = \frac{\text{allowable stress}}{(\text{applied load})\, k} - 1 \qquad (15.30)$$

where k is the safety factor.* The applied load

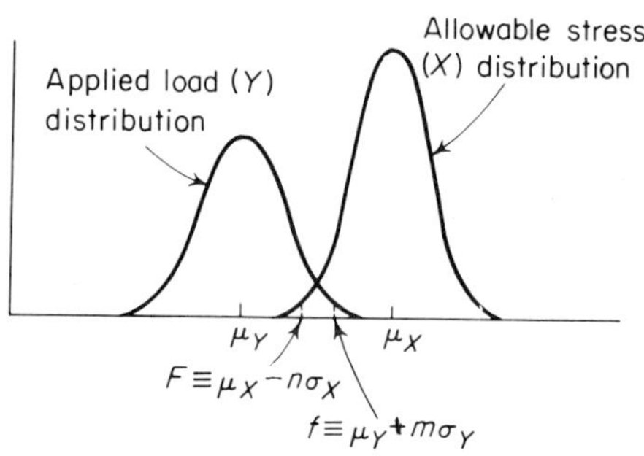

Fig. 15.3

is usually taken as some conservative value above expected load. The allowable stresses are determined from data on material properties, such as from MIL–HDBK–5, and other characteristics of the part. If possible a 99 per cent or a 90 per cent guaranteed value is used (see for example Chap. 3 of MIL–HDBK–5).

Fig. 15.3 shows the relationships of stress and applied load in terms of probability density distributions which, for convenience, are assumed to be normal.

In terms of the normal distribution model and the choice of allowable stresses and applied loads conservatively below and above, respectively, the expected or nominal values, the above equation may be rewritten as

$$MS = \frac{\mu_X - n\sigma_X}{(\mu_Y + m\sigma_Y)k} - 1 = \frac{F}{fk} - 1 \qquad (15.31)$$

where μ_X, μ_Y, and σ_X, σ_Y are the respective means and standard deviations of allowable stress and applied load. The quantities $F \equiv \mu_X - n\sigma_X$, $f \equiv \mu_Y + m\sigma_Y$, k (and consequently MS) are known or assumed known. However, one must know the quantities m, n, σ_X, and σ_Y in order to calculate R.

* Many definitions for safety factor and margin of safety in the literature differ from those used here. Consequently, the reader is urged to make sure that definitions of the factors frequently called out in contractual requirements are consistent with those he is accustomed to use. (See Kecicioglu and Haugen, "A Unified Look at Design Safety Factors, Safety Margins and Measures of Reliability.") The safety factor is used to derate the allowable stress because of uncertainties in the theory, actual application, usage environment, quality control of the material, manufacturing of the part, and also risk to human beings in case of failure. Thus the allowable stress has a statistical variability under a controlled process, as indicated in the next paragraph, but also is derated because of uncertainty in ability to control.

Thus, from Table 15.9 and the definitions of F and f

$$R = \Phi\left(\frac{F - f - m\sigma_Y + n\sigma_X}{\sqrt{\sigma_X^2 + \sigma_Y^2}}\right) \tag{15.32}$$

Since, however, it is convenient to use the quantities $v_X \equiv \sigma_X/\mu_X$ and $v_Y \equiv \sigma_Y/\mu_Y$ (the coefficients of variation of, respectively, allowable stress and applied load) we may rewrite the last equation as

$$R = \Phi\left[\frac{1 - \left(\dfrac{1 - nv_X}{1 + mv_Y}\right)\dfrac{f}{F}}{\sqrt{v_X^2 + v_Y^2\left(\dfrac{1 - nv_X}{1 - mv_Y}\right)^2\left(\dfrac{f}{F}\right)^2}}\right] \tag{15.33}$$

Consequently, given the quantities m, n, v_X, and v_Y, we may calculate R from the last expression.

We may often assume $m = n = 3$; that is, 3-sigma conservative values are assigned to applied load and allowable stress. Also, values of $v_X = v_Y = 0.1$ seem to be conservative (values of v_X may be calculated in some cases from MIL–HDBK–5; v_Y seems to be larger in general than v_X and is generally more difficult to estimate). With these assumptions we have approximately

$$R = \Phi\left[\frac{10 - 5(f/F)}{\sqrt{1 + 0.3(f/F)^2}}\right] \tag{15.34}$$

15.2.4 Growth Models

15.2.4.1 General Considerations. A complex system undergoes development testing so that the final design, manufacturing process, maintenance procedures, etc., may be eventually established for the system to meet its required performance under the specified conditions. During such testing, deficiencies in the design, manufacturing procedure, operating procedures, etc., are revealed and changes are made to correct these problems. Consequently the system may be said to have reliability growth if it is reasonable to assume that the changes are for the better or at least that they do not have an adverse effect. Evaluation of this growth process is important for making decisions relative to the status of the development program. Is further development testing necessary? Can we go into full-scale production?

One reliability growth model is presented in the following paragraphs based on Barlow and Scheuer, "Reliability Growth During a Development Testing Program." It is relatively simple, with a minimum of assumptions, and estimates of reliability can be easily obtained. Other models utilizing further assumptions may be found in Chap. 11 of *R:MMM* and in Corcoran, Weingarten, and Zehna, "Estimating Reliability After Corrective Action." These models assume that only successes or failures are recorded; however, the one presented here assumes that the failures can be correctly classified into one of two categories: *inherent* and *correctable*.

15.2.4.2 Description of a Reliability Growth Model. Development tests are conducted in K stages or groups. The probability of inherent failure q_0 is constant throughout development. It is assumed that the probability of a correctable failure q_i does not increase from stage to stage and is constant for all tests in the ith stage. The data for the ith stage are the numbers a_i of inherent failures, b_i of correctable failures, and c_i of successes.

15.2.4.3 Estimates of Parameters. The maximum likelihood estimate of q_0 is

$$\hat{q}_0 = \sum_{i=1}^{K} a_i \Big/ \sum_{i=1}^{K} (a_i + b_i + c_i) \tag{15.35}$$

The maximum likelihood estimates $\bar{q}_i$, $i = 1, 2, \ldots, K$, are obtained in this way: Inspect the values of $b_i/(b_i + c_i)$ in order. Ordinarily these will be nonincreasing as i goes from 1 to K. Suppose you find that $b_j/(b_j + c_j) < b_{j+1}/(b_{j+1} + c_{j+1})$, indicating a reversal. Combine the data of the jth and $(j + 1)$st stages into a new jth stage with new b_j and c_j equal to the sum, respectively, of the bs and cs from the old jth and $(j + 1)$st stages. Should the new result indicate a reversal with the old $(j - 1)$st or new $(j + 1)$st stage, combine these results into one stage and so on until all reversals are eliminated. Then

$$\bar{q}_i = (1 - \hat{q}_0)\,\bar{b}_i/(\bar{b}_i + \bar{c}_i) \tag{15.36}$$

the overbars indicating that the new ith stage may be the result of combining data from stages adjacent to the original ith stage. All these (original) stages are assigned the value

TABLE 15.10

Stage i	Inherent Failures a_i	Correctable Failures b_i	Successes c_i	Trials $a_i + b_i + c_i$	$b_i/(b_i + c_i)$
1	0	2	3	5	0.400
2	1	1	8	10	0.111
3	0	0	10	10	0.000
4	0	1	7	8	0.125
5	0	1	8	9	0.111
6	0	0	7	7	0.000
Totals	1	5	43	49	

TABLE 15.11

i	b_i	c_i	$b_i/(b_i + c_i)$	First Combination	Second Combination
1	2	3	0.400	0.400	0.400
2	1	8	0.111	0.111	0.111
3	0	10	0.000}	0.059}	
4	1	7	0.125}		0.080
5	1	8	0.111	0.111}	
6	0	7	0.000	0.000	0.000

q_i. The example below clarifies the above description.

15.2.4.4 Example. Development tests yield the results of Table 15.10. From which, $\hat{q}_0 = 1/49 = 0.0204$.

Table 15.11, which retabulates some of the previous results for convenience, indicates a reversal of stages 3 and 4; these data are combined; however, the new stages 3 and 4 still show increasing failure probability. The latter data are then combined, resulting in the maximum likelihood estimates of failure probability in final decreasing order.

Hence, $\bar{q}_1 = 0.392$, $\bar{q}_2 = 0.1087$, $\bar{q}_3 = \bar{q}_4 = \bar{q}_5 = 0.0784$, $q_6 = 0.000$. The maximum likelihood estimate for r_6, the reliability of the item in the configuration tested in the sixth stage, is

$$\hat{r}_6 = 1 - \hat{q}_0 - \bar{q}_6$$
$$= 0.9796 \qquad (15.37)$$

If no assumption of reliability growth were made, then the maximum likelihood estimate of r_6 would be

$$\hat{r}_6 = \sum c_i / \sum (a_i + b_i + c_i)$$
$$= 43/49 = 0.878$$

BIBLIOGRAPHY

Barlow, R. E., and E. M. Scheuer, "Reliability Growth During a Development Testing Program," *Technometrics*, **8,** No. 1 (Feb. 1966), 53–60.

Bendat, J. S., and A. G. Piersol, *Measurement and Analysis of Random Data.* New York: John Wiley & Sons, Inc., 1966.

Corcoran, W. J., H. Weingarten, and P. W. Zehna, "Estimating Reliability After Corrective Action," *Management Science*, **10,** No. 4 (July 1964), 786–95.

Crandall, S. H., and W. D. Mark, *Random Vibration in Mechanical Systems.* New York: Academic Press Inc., 1963.

Drenick, R. F., "The Failure Law of Complex Equipment," *Journal of the Society for Industrial and Applied Mathematics*, **8,** No. 4 (Dec. 1960), 680–90.

Johns, M. V., Jr., and G. J. Lieberman, "An Exact Asymptotically Efficient Confidence Bound for Reliability in the Case of the Weibull Distribution," Technometrics, **8,** No. 1 (Feb. 1966), 135–75. This reference gives very complete tables of exact lower confidence limits on reliability for the two-parameter Weibull distribution (unknown shape and scale parameters), based on censored random samples.

Kaufman, N., and M. Lipow, "Reliability-Life Test Analysis Using the Weibull Distribution,"

Western Conference Proceedings, *American Society for Quality Control* (April 9–11, 1964), pp. 175–87. This reference gives a conservative lower confidence limit on reliability for the Weibull distribution, for unknown shape, scale, and location parameters, based on complete random samples.

Kecicioglu, D., and E. B. Haugen, "A Unified Look at Design Safety Factors, Safety Margins and Measures of Reliability," Seventh Reliability and Maintainability Conference, American Society of Mechanical Engineers, San Francisco, Calif., July 14–17, 1968, pp. 520–601.

Klimov, B. G., "Limiting Law for Reliability of Sequential Systems" (1962), Foreign Technology Division Translation FTD–TT–63–470, Aug. 1963.

Lipow, M., and R. L. Eidemiller, "Application of the Bivariate Normal Distribution to a Stress vs. Strength Problem in Reliability Analysis," *Technometrics*, **6**, No. 3 (Aug. 1964), 325–28.

Lloyd, D. K., and M. Lipow, *Reliability: Management, Methods, and Mathematics*. Englewood Cliffs, N. J.: Prentice-Hall, Inc., 1962. Appendix 9B.

Madansky, A., *Approximate Confidence Limits for the Reliability of Series and Parallel Systems*, RAND Research Memorandum 2552. Santa Monica, Calif.: RAND Corporation, 1960.

Owen, D. B., *Factors for One-Sided Tolerance Limits and for Variables Sampling Plans*, Sandia Corporation Monograph SCR–607. 1963. (available from the Office of Technical Services, Department of Commerce, Washington, D. C. 20025.

———, *Handbook of Statistical Tables*. Reading, Mass.: Addison-Wesley Publishing Co., Inc., 1962. Sec. 18. This reference gives (indirectly) exact lower confidence limits on reliability for two-parameter normal (Gaussian) distribution (mean and standard deviation unknown).

Pugh, E. L., "The Best Estimate of Reliability in the Exponential Case," *Operations Research*, **11** (1963), 57–61. This reference gives a minimum-variance unbiased estimate of reliability for the exponential distribution of time to failure.

Resnikoff, G. J., and G. J. Lieberman, *Tables of the Non-Central* t-*Distribution*. Stanford, Calif.: Stanford University Press, 1957. This reference gives a minimum-variance unbiased estimator and lower confidence limit (pp. 17–18 and 22–23, respectively) for reliability for the two-parameter normal distribution of time to failure. See also Owen, *Factors for One-Sided Tolerance Limits and for Variables Sampling Plans*.

Sarhan, A. E., and B. G. Greenberg, eds., *Contributions to Order Statistics*. New York: John Wiley & Sons, Inc., 1962. Chap. 12C.

Statistical Techniques in Life Testing, PB 171580. Washington, D. C.: U. S. Department of Commerce, Office of Technical Services, July 1961.

Weibull, W., *Fatigue Testing and the Analysis of Results*. New York: Pergamon Press, 1961.

Zelen, M., ed., *Statistical Theory of Reliability*. Madison, Wis.: The University of Wisconsin, 1963, pp. 115–48.

Reliability Texts

Barlow, R. E., and F. Proschan, *Mathematical Theory of Reliability*, New York: John Wiley & Sons, Inc., 1965.

Bazovsky, I., *Reliability Theory and Practice*, Englewood Cliffs, N. J.: Prentice-Hall, Inc., 1961.

Calabro, S. R., Reliability Principles and Practices. New York: McGraw-Hill Book Company, 1962.

Chorafas, D. N., *Statistical Processes and Reliability Engineering*, Princeton, N. J.: D. Van Nostrand Company, Inc., 1960.

Gnedenko, B. V., K. Belyayev, and A. D. Solovyev, *Mathematical Methods of Reliability Theory*. New York: Academic Press Inc., 1969.

Haviland, R. P., *Engineering Reliability and Long Life Design*, Princeton, N. J.: D. Van Nostrand Company, Inc., 1964.

Ireson, W. G., ed., *Reliability Handbook*, New York: McGraw-Hill Book Company, 1966.

Landers, R. R., *Reliability and Product Assurance*, Englewood Cliffs, N. J.: Prentice-Hall, Inc., 1963.

Lloyd, D. K., and M. Lipow, *Reliability: Management, Methods, and Mathematics*, Englewood Cliffs, N. J.: Prentice-Hall, Inc., 1962.

Myers, R. H., K. L. Wong, and H. M. Gordy, eds., *Reliability Engineering for Electronic Systems*. New York: John Wiley & Sons, Inc., 1964.

Pieruschka, E., *Principles of Reliability*, Englewood Cliffs, N. J.: Prentice-Hall, Inc., 1963.

Polovko, A. M., *Reliability Theory*. New York: Academic Press Inc., 1968.

Roberts, N. H., *Mathematical Methods in Reliability Engineering*. New York: McGraw-Hill Book Company, 1964.

Sandler, G. H., *System Reliability Engineering*. Englewood Cliffs, N. J: Prentice-Hall, Inc., 1963.

Shooman, M. L., *Probabilistic Reliability: An Engineering Approach*. New York: McGraw-Hill Book Company, 1968.

Von Alven, William H., ed., *Reliability Engineering*. Englewood Cliffs, N. J.: Prentice-Hall, Inc., 1964.

Reliability and Life Testing Bibliographies

Buckland, W. R., *Statistical Methods and the Life Characteristic*. London: Griffin, 1962.

Govindarajulu, Z., "A Supplement to Mendenhall's Bibliography on Life Testing and Related Topics," *Journal of the American Statistical Association*, **59** (1964), 1231–91.

Mendenhall, William, "A Bibliography on Life Testing and Related Topics," *Biometrika*, **45**, Part 4 (1958), 521–43.

Motes, J. H., "KWIC Index to Reliability and Quality Control Literature," *Proceedings Ninth National Symposium on Reliability and Quality Control*, (Jan. 22–24, 1963), pp. 556–81.

Yurkowsky, W., R. E. Schafer, and J. M. Finkelstein, "Accelerated Testing Technology," Technical Report No. RADC–TR–67–420, Vol. 1, Rome Air Development Center, Grifiss Air Force Base, N. Y., 1967, pp. 6–0 to 6–43.

Reliability Abstracts

"Quality Control and Applied Statistics Abstracts," monthly report published by Executive Sciences Institute, Inc., Whippany, N. J.

"Reliability Abstracts and Technical Reviews," prepared by Research Triangle Institute, of Durham, N. C. for NASA. Address inquiries to NASA, Scientific and Technical Information Division, CODE US, Washington D.C. 20546.

Reliability Information Retrieval System, P.O. Box 215, Goleta, Calif.

Government Publications and Standards

MIL-HDBK-5A(DOD), "Materials and Elements for Aerospace Vehicle Structures," February 1966.

MIL-HDBK-217A (DOD), "Reliability Stress and Failure Rate Data for Electronic Equipment," 1965.

MIL-STD-721B (DOD), "Definitions of Effectiveness Terms for Reliability, Maintainability, Human Factors, and Safety," August 1966.

MIL-STD-756A (DOD), "Reliability Prediction," May 1963.

MIL-STD-757 (DOD), "Reliability Evaluation from Demonstration Data," June 1964.

MIL-STD-781B (DOD), "Reliability Tests: Exponential Distribution," November 1967.

MIL-STD-785, "Requirements for Reliability Program (for Systems and Equipments)," June 1965.

MIL-STD-810A (USAF), "Environmental Test Methods for Aerospace and Ground Equipment," June 1964.

NHB-5300.4 (1A), "Reliability Program Provisions for Aeronautical and Space System Contractors," April 1970.

Table References

T–1. National Bureau of Standards, *Tables of the Exponential Function e^x*, Applied Mathematics Series AMS14. Washington, D. C., June 1951.

T–2. Harter, H. Leon, *New Tables of the Incomplete Gamma-Function Ratio and of Percentage Points of the Chi-square and Beta Distributions*, Washington, D. C.: Aerospace Research Laboratories, Office of Aerospace Research, U.S. Air Force. October 1963.

T–3. Weintraub, Sol, *Tables of the Cumulative Binomial Probability Distribution for Small Values of p*. New York: The Free Press, 1963.

T–4. *Tables of the Cumulative Binomial Probability Distribution*. Cambridge, Mass.: Harvard University Press, 1953.

T–5. Pearson, Karl. *Tables of the Incomplete Beta Function*. Cambridge, Mass.: Cambridge University Press for the Biometrika Trustees, 1934.

T–6. *Tables of the Cumulative Binomial Probabilities*, Ordnance Corps Pamphlet ORDP 20–1. Washington, D. C.: Ordnance Corps, September 1952; Change 2, April 1956.

T–7. National Bureau of Standards, *Tables of the Binomial Probability Distribution*, Applied Mathematics Series AMS 6, Washington, D. C., 1950.

T–8. Defense Systems Department, General Electric Company, *Tables of the Individual and Cumulative Terms of Poisson Distribution*, Princeton, N. J.: D. Van Nostrand Company, Inc., 1962.

T–9. Molina, E. C., *Poisson's Exponential Binomial Limit*, Princeton, N. J.: D. Van Nostrand Company, Inc., 1942.

T–10. National Bureau of Standards, *Handbook of Mathematical Functions with Formulas, Graphs and Mathematical Tables*, Applied Mathematics Series AMS 55. Washington, D. C., June 1964.

T–11. a. Van de Hulst, H. C., "Scattering in a Planetary Atmosphere," *Astrophysical Journal* **107**, (1948). 245.

b. Chandrasekhar, S., and F. H. Breen, "On the Radiative Equilibrium of a Stellar Atmosphere. XXIV," *Astrophysical Journal* **108** (1948), 110.

c. Kourganoff, V., and I. W. Busbridge, *Basic Methods in Transfer Problems*. Oxford: Clarendon Press, 1952, p. 267.

T–12. National Bureau of Standards, *Tables of*

Normal Probability Functions, Applied Mathematics Series AMS 23. Washington, D. C., 1953.

T–13. National Bureau of Standards, *Tables of the Bivariate Normal Distribution Function and Related Functions*, Applied Mathematics Series AMS 50. Washington, D. C., 1959.

Merkle Press

BERTRAM GOTTLIEB *received his B.S. and M.S. degrees from the Illinois Institute of Technology. Since then he has served as either instructor or professor at the Illinois Institute of Technology, the University of Wisconsin, the the University of Connecticut, and the University of Iowa. He has also been a guest lecturer at many other universities throughout the United States.*

He has been a factory worker, consultant, and a member of the UAW, the AFT, and the American Newspaper Guild. Presently he is a member of the International Association of Machinists. For several years Mr. Gottlieb served as staff adviser on industrial engineering, automation, manpower, and related subjects for the AFL-CIO headquarters in Washington, D.C. In that capacity he investigated complaints, handled grievances, and arbitrated cases involving disputes over industrial-engineering installations.

Mr. Gottlieb is a member of the American Institute of Industrial Engineers (senior), Editorial Board of the Journal of Industrial Engineering, *Department of Labor advisory committees on productivity and technological development and on wages and industrial relations, Industrial Relations Research Association, and Sigma Iota Epsilon. He is the author of many articles on industrial engineering and industrial relations in several professional journals and union publications.*

Mr. Gottlieb is currently director of research for the Transportation Institute, Washington, D.C. He is also National Director of the Division of Industrial and Labor Relations of the AIIE, registered professional engineers.

SECTION 16

THE ATTITUDES OF ORGANIZED LABOR TOWARD INDUSTRIAL-ENGINEERING METHODS

BERTRAM GOTTLIEB

16.1 INTRODUCTION

Of all engineering disciplines, industrial engineering is unique in one important aspect—its concern with integrated systems which must be directed and manned by human beings. Unfortunately, and with all too few exceptions, theorists and practitioners have done little more than pay lip service to the special problems caused by the fact that human behavior cannot be predicted with the ease with which physical phenomena can.

The founders of scientific management in the United States were engineers, as Aitken pointed out in his excellent study of *Taylorism at the Watertown Arsenal.*

It is no coincidence that Taylor and most of his immediate disciples were engineers, for it was by

> way of engineering that scientific analysis had made its most powerful and continuing impact upon industrial production. They accepted without question the engineering approach that had already proved itself in the design of physical objects, and they extended it to the analysis and control of the activities of people. The essential core of scientific management, regarded as a philosophy, was the idea that human activity could be measured, analyzed, and controlled by techniques analogous to those that had proved successful when applied to physical objects.*

With few exceptions this philosophy of scientific management, which dominated the movement for most of its early history, was responsible for much of the conflict between scientific management on one hand and workers and their unions on the other.

Many writers have described the early years of scientific management thoroughly in a variety of forms. In addition to Taylor's own speeches, articles, and books, his views probably were best summarized in Frank Copley's two-volume biography.† Much insight into the ideas of Taylor and his colleagues and of the conflict they generated can be found in the three volumes of the *Hearings Before the Special Committee of the House of Representatives to Investigate the Taylor and Other Systems of Shop Management* (Washington, D.C.: Government Printing Office, 1912). Excellent studies by McKelvey and Nadworny‡ have described relations between unions and early industrial engineers, while Aitken's account of the work of Taylor, Barth, and Merrick at the Watertown Arsenel is a thorough exposition of the events which led to the first strike by workers against a scientific management application. William Gomberg in the first edition of this *Handbook* not only reviewed much of this material but also analyzed the significant years of World War II and the decade which followed.§ These publications are only a small sample of the wealth of available information on the history of industrial engineering and its relations with unions.

To avoid redundancy, therefore, this section will explore only current industrial-engineering policies and practices of unions. The history of these policies and practices will be referred to only when it is necessary for an understanding of present activities.

16.2 HUMAN INFLUENCES ON INDUSTRIAL ENGINEERING

In order for the industrial engineer to treat the human component in his efforts to integrate systems of men, materials, and equipment, training in mathematics, statistics, physics, chemistry, and the engineering sciences is not enough. The industrial engineer must bring to his work a basic knowledge of people, their motivations and capabilities, and their strengths and limitations. Obtaining this knowledge requires more than a casual study of the social sciences—of economics, sociology, psychology, and anthropology—and it requires a thorough grounding in physiology. Working at what Gomberg describes as "the bridgehead where technological problems merge into social questions," the industrial engineer is daily faced with the need to evaluate, predict, and develop controls for human performance.‖ This need is present regardless of the degree of sophistication of the industrial-engineering activity. Considering the influence of human beings is just as essential in operations research as in work measurement. And, of course, taking into consideration the influence that systems have on man is of as much importance as considering man's influence on systems.

These influences—the system on man and man on the system—plunge the industrial engineer into a nonscientific, frequently irrational, opinionated, and emotional arena—an arena with a variety of names such as human relations, labor relations, industrial relations, and collective bargaining.

---

* Reprinted by permission of the publishers from Hugh G. Aitken, *Taylorism at the Watertown Arsenal* (Cambridge, Mass.: Harvard University Press, 1960), pp. 15–16.

† *Frederick W. Taylor* (New York: Harper & Row, Publishers, 1923).

‡ Jean Trepp McKelvey, *AFL-CIO Attitudes Toward Production 1900–1932* (Ithaca, N.Y.: Cornell University Press), 1952. Milton J. Nadworny, *Scientific Management and the Unions* (Cambridge, Mass.: Harvard University Press), 1955.

§ "Trade Unions and Industrial Engineering," in *Handbook of Industrial Engineering and Management* (Englewood Cliffs, N.J.: Prentice-Hall, Inc., 1955), pp. 1121–83.

‖ "Trade Unions and Industrial Engineering," p. 1121.

Undoubtedly, industrial engineers have become increasingly aware of the influence of their work on workers. Witness, for example, the following statement by American Manufacturers Association Vice-President Keith J. Louden.

> The work of the industrial engineer has greater impact on the individual worker than perhaps any other function in the business. The degree of success the industrial engineer is able to achieve in the area of human relations will determine the degree of final success he will achieve in the performance of his duties.*

Louden was also a member of the committee which developed the American Institute of Industrial Engineers' (AIIE) "Code of Work Measurement Principles," which states in part these principles.

> Industrial engineers recognize that the part of their work which deals with work measurement, goal-setting, and wage payment has an important effect on management, labor, and the success of the enterprise as a whole. . . . The articles of the Code represent standards by which industrial engineers, management, and labor can judge the soundness and propriety of the courses of action they are following or contemplating.†

That workers and their unions influence the work of the industrial engineer is less apparent, and recognition of this fact seems to come with much reluctance. The Code, for example, also states that "these articles are derived from an analysis of the practices which over the years have resulted not only in the economic success of the enterprises using them but in enduring, satisfactory labor-management relations."‡ In spite of this declaration, an objective observer, discovering that the committee which drafted the Code was made up almost entirely of management consultants, might well be pessimistic of the chances that the twenty articles of the Code will contribute to the achievement of satisfactory labor-management relations.

The committee did not include and did not seek the advice of any of AIIE members employed as industrial engineers by unions. Yet, these members have great influence in determining whether the articles of the Code lessen labor-management conflict over industrial-engineering activities. In fact the very disinclination of many managements and industrial engineers to recognize the influence that workers and their unions have on industrial engineering contributes substantially to assuring the continuance of the present conflict. But inclination or disinclination is not of basic importance. Most industrial-engineering systems are specifically designed to affect the wages and working conditions of the people involved in the systems. Since industrial-engineering activities have this effect, they become mandatory subjects of collective bargaining. A union has not only the legal right to bargain on these matters but also a moral responsibility to do so. In addition, the National Labor Relations Board (NLRB) and the federal courts have consistently interpreted the Labor-Management Relations Act of 1947 (Taft-Hartley) as requiring employers to make available to unions all manuals, instructions, records, and studies which form the basis of industrial-engineering applications.

Union staff representatives are being trained in increasing numbers in the skills necessary to analyze industrial-engineering data intelligently. At the same time the number of graduate industrial engineers employed by unions is growing slowly. Obviously, then, workers through their unions will have an increasingly important and fundamental influence on the work of the industrial engineer. Therefore, industrial engineers should be thoroughly familiar with union attitudes, policies, and practices in the industrial-engineering field.

16.3 LEGAL CONSIDERATIONS§

Before analyzing current union policies and practices, we shall look in greater detail at some of the legal aspects with which industrial engineers must be familiar in order to practice their profession in an ethical manner. The assump-

---

* "Management's Use of Industrial Engineering." Chapter 5 from *Industrial Engineering Handbook* by H. B. Maynard. Copyright © 1963. McGraw-Hill Book Company, New York. Used by permission.

† Reprinted from the "Code of Work Measurement Principles," published by the American Institute of Industrial Engineers, Inc., 345 East 47th Street, New York, New York 10017.

‡ "Code of Work Measurement Principles."

§ The author wishes to acknowledge with gratitude the assistance of Professor Theodore J. St. Antoine of the Law School of the University of Michigan in preparing this section. Errors of fact or interpretation are the author's.

tion here is that knowingly violating the law, advising others to do so, or even passively observing an employer doing so is in fact a violation of the "Canons of Ethics for Engineers"—canons to which all professional industrial engineers are supposed to subscribe.*

Of course, simple adherence to minimum legal requirements may not be enough to assure that industrial engineering will be practiced in a fruitful as well as in a peaceful manner. But the industrial engineer who not only practices within these requirements but also insists on adherence by operating management, personnel and industrial-relations specialists, and others associated with industrial-engineering applications is at least making it possible to minimize conflict and is providing the basis of ethical, professional industrial-engineering practice.

16.3.1 Collective Bargaining and Industrial-Engineering Applications

Section 8(a)(5) of the National Labor Relations Act makes it an unfair labor practice for an employer to refuse to bargain with a union representing his employees. Section 8(d) defines this obligation as follows.

> For the purpose of this section, to bargain collectively is the performance of the mutual obligation of the employer and the representative of the employees to meet at reasonable times and confer in good faith with respect to wages, hours, and other terms and conditions of employment, or the negotiation of an agreement, or any question arising thereunder. . . .

The National Labor Relations Board and the federal courts have, since the passage of the Act in 1947, been frequently called upon to apply Sec. 8 to many practical situations involving industrial-engineering practices. By 1965 the decisions and awards of the NLRB and the courts had thoroughly defined the obligation of an employer to bargain with the representative of his employees on almost any matter affecting "wages, hours, and other terms and conditions of employment." This obligation as it applies to classic industrial-engineering techniques is so well established that no attempt will be made here to fully trace its development.

The following quote from a case involving a job-evaluation plan should be sufficient to illustrate the underlying reasons for the development of the collective-bargaining obligation.

> We think it is hardly open to dispute but that the total point evaluation assigned by petitioner to the eleven involved factors was not only directly related to the pay determined for each job, but that it constitutes an important consideration in petitioner's entire wage structure.†

Job-evaluation proponents have always claimed that job evaluation results in a rational wage structure in which individual jobs are properly related. However, the 7th Circuit judges accepted the concept that job evaluation does in fact have a subjective effect on wages as a result of the relating of jobs and the rationalizing of the structure.

Similar reasoning is behind the requirement that piece rates, work loads, incentive plans, and production standards, as well as the techniques used to establish them, are subjects for collective bargaining under the National Labor Relations Act.

16.3.2 Scope of Employer's Duty to Furnish Information

The provisions of the Act which set the bargaining obligation also form the statutory basis for the employer's duty to provide a union with information needed in the negotiation and administration of a collective-bargaining contract. If an employer violates his statutory obligations, the NLRB is empowered to order him to cease and desist from refusing to bargain with the union and to take the affirmative action of furnishing the union with the job information or related data involved.

An employer has the duty to furnish a union, upon request, with all the relevant information (pertaining to jobs in the bargaining unit) needed by the union in negotiating and administering a collective contract. An employer cannot refuse to supply information on the ground that it is confidential and solely for internal management use or on the ground that furnishing it would be burdensome, unless he can show compelling need to keep it confidential or can

* The "Canons of Ethics for Engineers" have been included with the "Code of Work Measurement Principles."

† *Taylor Forge and Pipe Works*, 113 NLRB 65, 36 LRRM 1358 (1955), enforced 234 F. 2d 227 (7th Cir. 1956).

show that furnishing it would be unduly burdensome.*

Certain data, such as current wage information, are presumed to be relevant to the negotiating and administering of a labor contract. Thus, a union asking for these data does not have to make an initial showing of how the information fits into its bargaining needs, unless the particular data sought clearly appear to be irrelevant.†

The right of a union to job data is not limited to data for pending contract negotiations. A union is entitled to information needed for the intelligent discharge of three distinct aspects of its representational function: (1) the conduct of bargaining over the terms and conditions of a new contract: (2) the administration of a current contract, including the handling of disputes through the grievance machinery and the resolution of new problems not covered by the existing agreement; and (3) the preparation for coming negotiations.‡

An employer must comply with all reasonable union requests for relevant job information in such a manner as to make the data furnished meaningful and understandable without undue difficulty. He does not, however, have to supply information in the exact form specified by a union.§ In this area as in others the facts of the individual case are far more important than general rules. In effect, the Board tries to balance how burdensome, costly, and time consuming it will be for the employer to have to supply the information in the particular form requested against how necessary and appropriate it is for the union to have the information in that form. For example, sometimes an employer can convey information orally, at other times, when complicated data are involved, a written presentation of the information is required.‖

The duty of an employer to furnish these data is not an obligation which the union forces upon him but is a statutory obligation. However, a union can waive its right to information —for example, through an express clause in the labor contract spelling out the right of the union to receive only certain limited data. But any such waiver must be clear and unmistakable.# The inclusion in a contract of a provision requiring the employer to supply certain specified data does not mean that the union has waived its statutory right to other information. Similarly, the inclusion of a general grievance procedure does not relieve the employer of his statutory obligation to provide necessary data.**

16.3.3 Data on Job Descriptions, Classifications, Evaluations, and Wage Rates

A union is entitled to the existing job descriptions for the bargaining unit it represents, even though the descriptions contain mistakes which the employer is correcting.†† The unrevised, faulty job descriptions might become quite important to everyone concerned in the processing of a grievance or in contract negotiations if the union claims that inequities exist by reason of the errors in question.

An employer may have the right to impose certain restrictions on the use of some data in order to protect its competitive position. For example, where job descriptions are so precise that if they fall into competitors' hands vital information on certain chemical processes is revealed, the NLRB has held that a company can restrict the union from removing the des-

* *Aluminum Ore Co.*, 39 NLRB 1286, 10 LRRM 49 (1942), enforced 131 F. 2d 485 (7th Cir. 1942); *Utica Observer-Dispatch* v. *NLRB*, 229 F. 2d 575 (2nd Cir. 1956); *J. I. Case Co.* v. *NLRB*, 253 F. 2d 149 (7th Cir. 1958).

† *NLRB* v. *F. W. Woolworth Co.*, 352 U.S. 938 (1956); *NLRB* v. *Yawman & Erbe Mfg. Co.*, 187 F. 2d 947 (2nd Cir. 1951).

‡ *NLRB* v. *Whitin Machine Works*, 217 F. 2d 593 (4th Cir. 1954); *J. I. Case Co.* v. *NLRB*, 253 F. 2d 149 (7th Cir. 1958); *NLRB* v. *Otis Elevator Co.*, 208 F. 2d 176 (2nd Cir. 1953).

§ *Cincinnati Steel Casting Co.*, 86 NLRB 592, 24 LRRM 1657 (1949); *Westinghouse Electric Corp.*, 129 NLRB 98, 47 LRRM 1072 (1960).

‖ *Cincinnati Steel Casting Co.*, 86 NLRB 592, 24 LRRM 1657 (1949); *J. I. Case Co.* v. *NLRB*, 253 F. 2d 149 (7th Cir. 1958).

J. I. Case Co. v. *NLRB*, 253 F. 2d 149 (7th Cir. 1958); *Westinghouse Air Brake Co.*, 119 NLRB 1118, 41 LRRM 1252 (1957).

** *California Portland Cement Co.*, 101 NLRB 1436, 31 LRRM 1220 (1953); *Leland Gifford Co.*, 95 NLRB 1306, 28 LRRM 1443 (1951), enforced 200 F. 2d 620 (1st Cir. 1952); *Timken Roller Bearing Co.*, 138 NLRB 1, 50 LRRM 1508 (1962).

†† *American Sugar Refining Co.*, 130 NLRB 81, 47 LRRM 1361 (1961).

criptions from the premises.* A company seeking to impose such restrictions must be prepared to prove the need for them. Note, however, that this restriction does not remove the obligation to make the data available. In fact, in the above case the company was willing to make the data available to officials of the national union but only on company premises.

The NLRB and the courts have broadly sustained the obligation to supply data on job classifications, wage rates, and rate ranges, including the number of employees in each classification and the number of employees at each rate in each rate range.† This obligation was determined to include the furnishing of the point breakdowns on each of eleven factors for all evaluated jobs in one plant.‡

Of course, wage data may not have to be supplied in the precise manner specified by the union. For example, in one case, a company kept figures on the average earnings of all its employees on a quarterly basis. The union was not upheld in its demand that the employer provide monthly average earnings of employees in each of two dozen separate bargaining units, going back 2 years. The NLRB said the information was not readily available in the form requested and would be unduly burdensome to assemble.§

When an employer uses industry or area wage surveys or makes comparisons with other plants of his company, he should be prepared to furnish the data from these sources.‖

16.3.4 Time-Study and Wage-Incentive Data

The need to supply wage data may vary depending on the wage system in effect or under consideration. Thus, an employer has to supply a union with all the wage information used as a basis for establishing and maintaining an incentive wage plan and needed by the union to understand and deal with the plan.# Information on bonus payments must also be supplied, and where a merit system is in operation, the employer must make available data on the standards used in determining the merit ratings, as well as information on the performance ratings of each employee covered by the system.**

In the case of piece rates, the employer has to provide breakdowns of the itemized rates on all piece-rate operations, depending on the particular product being made. This requirement also covers methods and rates of payment for time lost because of machine breakdowns and for other reasons.††

In one case the NLRB broadly upheld the right of the union to the following information in connection with handling pending grievances.

1. Original time-study sheets and other documents relative to both the prior rates and the new rates
2. All data, studies, and other information used to determine the rate of pay for each job
3. All documents, studies, and other information used to evaluate such jobs, both prior to the change and thereafter

Furthermore, the NLRB upheld the right of the union to the following information in connection with both the handling of the grievances and the general administering of the labor contract.

1. Time-study manuals, instructions, and procedures used in making time studies of jobs in the plants of the employer, including full information on the weights given to each factor used to arrive at a final decision on the established rate and on the factors considered in making such decisions
2. Manuals, instructions, and procedures used in the development of standard data and the application thereof in the development of job rates in the plants

In the same case the NLRB expressly held that the fact that the contract contained a grievance procedure providing for the adjust-

---

* *American Cyanamid Co.*, 129 NLRB 77, 47 LRRM 1039 (1960).

† *Dixie Corp.*, 105 NLRB 49, 32 LRRM 1259 (1953); *Lock Joint Pipe Co.*, 141 NLRB 82, 52 LRRM 1410 (1963).

‡ *Taylor Forge and Pipe Works*, 113 NLRB 65, 36 LRRM 1358 (1955), enforced 234 F. 2d 227 (7th Cir. 1956).

§ *Westinghouse Electric Corp.*, 129 NLRB 98, 47 LRRM 1072 (1960).

‖ *Sherwin-Williams Co.*, 34 NLRB 651, 9 LRRM 27 (1941), enforced 130 F. 2d 255 (3rd Cir. 1942); *Hollywood Brands*, 142 NLRB 25, 53 LRRM 1012 (1963).

City Packing Co., 98 NLRB 203, 30 LRRM 1004 (1952).

** *Montgomery Ward & Co.*, 90 NLRB 180, 26 LRRM 1333 (1950); *B. F. Goodrich Co.*, 89 NLRB 139, 26 LRRM 1090 (1950).

†† *Skyland Hosiery Mills*, 108 NLRB 190, 34 LRRM 1254 (1954).

ment of any complaint did not prevent a union from going to the NLRB in a dispute over the extent of the statutory right of the union to bargaining information.*

16.3.5 Independent Union Time Study

In contrast to the many cases involving the data-furnishing obligation, very few involve a request by a union to make its own observation, time study, or other study of disputed standards or rates.

The lack of NLRB and court cases is probably due to a combination of the following factors.

1. In the interest of expeditiously and equitably handling time-study grievances, many managements have been willing to allow the union to make its own study. In fact, it is not unusual for management to suggest that the union make such a study.

2. Many unions have negotiated this right into labor-management agreements. Here is a sample of such a clause.

> At any step in the grievance procedure, whether on disputes over the standards or over changes in standards, the union shall have the right to call in a representative of the international union or any outside time-study man or both. Such representative or outside time-study man or both shall be allowed to time study the job or changes in a job to help the union determine whether the union's position is valid. Such representative of the union or outside time-study man or both shall then have the right to participate in all steps of the grievance procedure, including arbitration.

3. In most instances, before requesting a union time study, the union requests that the company supply all necessary data. A refusal results in a charge to the NLRB, based on the refusal to furnish the requested data. When such a charge is finally resolved and the data are furnished, companies have usually allowed unions to make any studies or checks necessary to evaluate the furnished data.

Three cases illustrate the current union time-study situation. In 1951 the NLRB ruled that (1) a union is entitled to be furnished with the original time-study data developed by the employer's experts in determining job standards and (2) a union is entitled to have its own time-study expert examine the job in dispute to check the employer's figures and procedures. In 1953 the reviewing court, however, agreed only that the union had a right to the employer's data and refused to allow the union expert to go on the employer's premises to conduct an independent study.† The court reasoned in this way.

> Whether the standards set by the respondent do in fact give a reasonable leeway for its employees affected by them to earn premium wages by the exertion of extra effort depends, of course, upon whether the amount of effort required for a unit of production is more than it reasonably would be. That is knowledge which the union already has or should have acquired through the trial by its members of the wage-incentive system as implemented by the respondent's standards. No invasion of the respondent's plant is needed to get that. Obviously, by changing the standards to decrease the amount of production per unit the opportunity to earn premium wages will be *pro tanto* increased and no invasion of the respondent's plant is needed to show that. Given a fair opportunity to study the data used by the respondent in fixing the standards, the union will have been advised as to the manner in which they were established and will have an adequate basis on which to determine, in the light of the actual experience of its members, what position it should take in respect to the processing of the grievance concerning them pursuant to the terms of the contract.‡

In spite of the clarity of this statement, the issue itself was not so clear to the three-man court. In his dissent, Justice Charles E. Clark made this statement.

> As for the independent time study sought by the union, this differs at most in degree and not in kind from the other sources of information thus allowed it. The data in respondent's possession do, or at least may, present an incomplete picture. If the union is really to attempt to appraise or correct this, it must supply its own substitute.
>
> Indeed the Trial Examiner actually found the union entitled to such a study, although he did not carry through to hold the respondent's refusal a violation of the Act since he found that an independent study would be available to an arbitrator at the next step of the grievance procedure and that such studies had been allowed in the past. And respondent, in supporting the Trial Examiner's ruling as to the availability of an independent study conducted by an arbitrator, in effect concedes the ultimate need of such a

---

* *Timken Roller Bearing Co.*, 138 NLRB 1, 50 LRRM 1508 (1962).

† *Otis Elevator Co.*, 102 NLRB 72, 31 LRRM 1334 (1953).

‡ *NLRB* v. *Otis Elevator Co.*, 208 F. 2d 176 (2nd Cir. 1953).

study if adequate information for settlement of the grievance is to be had.

But if this is so, if the union may go behind a standard to examine its foundation and if on occasion or ultimately such a study is needed in the examination thus permitted, no good reason is perceived for its denial at an early and effective stage in the procedure. If the union is to be allowed to acquire some information, it should not be stopped short of the most useful data it can develop; nor should it be forced to grope somewhat blindly through the very stages of grievance procedure where adequate information is most likely to lead the parties to amicable agreement, to await an arbitrator-conducted study to the same end. In the writer's view the Board at least acted within its powers in overruling the Trial Examiner and ordering the respondent to permit the union to conduct its own time study. As the Board found, the possibility of a time study at the arbitration stage was inadequate to satisfy the legitimate union request. The contract was silent as to any such right, and the fact that respondent had allowed arbitrators in the past to obtain independent time studies and indicated its intention to do so in the future did not eradicate respondent's consistent denial of any union right thereto. The Board was therefore settling the doubt by a definite ruling.*

In spite of the majority ruling in the Otis case, most employers did not close their shop-floor doors to the union representative and engineer. Thousands of job-standard disputes were settled and still are settled by having union and company engineers exchange information developed individually and collectively from studies of jobs in operation.

Finally, however, in 1964, the NLRB came out with a clear ruling which may finally determine this issue. For some time the UAW representative responsible for providing service and assistance to the local union at Fafnir Bearing Company was concerned with the number and quality of production-standard cases going through the grievance procedure and arbitration. When, in early 1963, grievances were filed by the local on piecework rates involving four jobs, the international union representative determined to have expert assistance in evaluating these grievances prior to making a decision on the advisability of proceeding to the final steps of the grievance procedure and to arbitration. Accordingly, he requested such assistance from the UAW headquarters.

* *NLRB* v. *Otis Elevator Co.*, 208 F. 2d 176 (2nd Cir. 1953).

Kermit Mead, the director of the Time Study and Industrial Engineering Department of the UAW, personally attempted to investigate and to make an intelligent engineering evaluation of these grievances. Mead, after discussing the problem with the workers involved, union representatives, and company engineers, analyzed the time-study data which had been used by the company in establishing piecework rates. At the completion of this procedure, Mead explained to both union and company representatives that he was unable to determine from the data set forth in the company studies (1) whether the piecework rates were established according to the contract and (2) whether proper industrial-engineering procedure had been used in establishing them. In fact, company engineers informed Mead that they were unable to account for some of the items which appeared in one of the time studies. Mr. Mead then asked permission to perform a time study of the disputed operations in order to obtain sufficient information to enable him to advise the union whether the grievance should be arbitrated. The company refused, and the UAW filed an unfair-labor-practice charge with the NLRB.

The NLRB decision stated that merely studying the information supplied by the company was not sufficient to permit the union to judge the accuracy of company times studies and to make an intelligent decision whether to proceed to arbitration. Therefore, the NLRB ordered the company to cease and desist from "refusing to permit the union to perform independent time studies, through its own experts, on jobs involved in grievances arising under the parties' collective-bargaining agreement."†

The company appealed the NLRB order, and the case was heard by the United States Court of Appeals in 1966. Its decision upheld the NLRB.

A time study is conducted by an industrial engineer to determine how many units per hour an average employee working at a normal pace can produce. This figure is called the "standard." . . .

There is, however, no demonstrably accurate method for setting "standards"; and, as can be readily predicted, this area is a fertile source of disputes requiring invocation of the company-union grievance machinery. . . .

The governing agreement established only

† *Fafnir Bearing Co.*, 146 NLRB 179, 56 LRRM 1108 (1964).

general guidelines as to how "standards" were to be formulated . . . .

While it is true that in *Otis* . . . we declined to enforce that portion of the Board's order directing a company to permit a union to conduct an independent time study which the union sought for the purpose of assessing the accuracy of proposed piecework prices, we believe *Otis* is distinguishable in at least two vital respects. First, the company in *Otis* not only refused to permit the union to make a "live" study but rejected outright the union's request for whatever data the company had in its possession. Thus, *Otis* reached this Court on a sparse record. Neither the trial examiner nor the Board had been afforded an opportunity to analyze the company's time-study material. Consequently, no determination had been made as to whether this data alone would have been sufficient for the union to make an informed judgment to accent the company's proposed rates or seek arbitration. Second, and even more significant, this Court in *Otis*, on the relatively skimpy record before it, concluded that whatever additional information the union sought could have been obtained merely by interviewing its members. In view of this circumstance, the Court determined that no "invasion of respondent's plant" was necessary. 208 F. 2d at 177.

Seizing upon this language, the company urges that in the instant case, the union had alternative sources for the information it sought and therefore did not really require independent time studies. But, there was ample evidence before the trial examiner and the Board to justify the rejection of this contention out of hand. Mead testified in detail as to the absolute need for a "live" time study, stating that without it the union could not make an informed judgment on the company's proposed rates. The testimony of Mead and Bertram Gottlieb, staff industrial engineer of the AFL CIO, also gave the trial examiner a full view of how time studies were generally made: Piecework prices were established by a ratesetter (usually an industrial engineer) on the basis of a combination of subjective and objective factors. Preliminarily, the ratesetter examined the place where the work was performed and noted environmental factors such as heat, light, and other conditions which might have a bearing on productivity. He then reduced the particular job under study into its component parts, designated "cyclic" and "noncyclic" elements. Cyclic elements were those performed in the production of every unit; noncyclic elements comprised factors such as going to the lavatory, consulting with the foreman, or getting stock from a storage area. Next, the ratesetter observed several operators performing their jobs and made a record of the "actual performance of the job in terms of the motion pattern used." If the ratesetter believed that a particular employee under examination was above or below average pace, he would "normalize" the performance by adjusting the actual timing of the job upward or downward. He then added an allowance, based upon his personal observation and judgment, for "personal fatigue and delay," and other elements of the job.

Mead testified, as we have indicated, that he could not assess the accuracy of the company's studies without making his own investigations. As illustrative, he noted that with respect to one of the piece rates which was the subject of a grievance the company had employed a normalizing factor of 90 per cent. But, without conducting his own study, Mead had no way of determining whether this adjustment was appropriate. The record is replete with numerous similar examples. Based upon all the testimony with respect to the need for "live" studies, we believe that unlike *Otis*, there was substantial evidence in this record to support the Board's finding that union-conducted "live" studies provided the only means for full assessment and verification of the company's proposed rates.

. . . While it is true that the material the union sought, because of its subjective nature, was not in the company's filing cabinets, it was nevertheless within the company's exclusive control . . . . Consequently, on the evidence and findings presented in this proceeding, we believe the Board, exercising its expertise and special competence, could properly determine that the union's need to conduct these tests outweighed the company's interests in closing its doors to outsiders. . . .

The reason the union desired to make "live" tests was not to usurp the function of the arbitrator and thereby interfere with standard grievance procedures; it was needed, instead, to determine whether to take the grievances to arbitration in the first place. By preventing the union from conducting these studies, the company was, in essence, requiring it to play a game of blind man's bluff. Only by guesswork could the union determine which piece rates to protest and which to accept. Of course, the union always had the option of demanding that *all* grievances on proposed piecework prices be reviewed by the arbitrator; but that alternative obviously would be impractical, expensive, and indeed strife-breeding, considering that there were over 10,000 new proposed rates each year. Thus, if the company-union grievance machinery was to work smoothly, it was essential that the union obtain the information necessary for a considered judgment.

Equally without merit is the company's argument that the failure of the governing agreement to specifically provide for independent time studies is dispositive of this proceeding. While it is true that the union had not requested permission to make these tests in the past, we do not find that determinative. Our main concern here is with statutory rights. Thus, the agreement did not expressly prohibit union-conducted studies and its silence in these circumstances cannot be construed as a "clear and unmistakable" waiver.

16.3.6 Legal Considerations and Ethical Implications

Since most industrial-engineering activity is conducted by industrial engineers as employees of or as consultants to management, the industrial engineer has special problems which other professionals seldom face. For example, a patient desiring a second medical opinion is free to seek it. A physician attempting to deny this right is guilty of unethical conduct. A professional industrial engineer should have no objection to having his work critically scrutinized by others, assuming, of course, that he accepts the responsibility for doing a thorough and complete job which will stand scrutiny. When a company refuses to allow such scrutiny, the profession of industrial engineering suffers. When such a company is involved with a union, its refusal may be based on many considerations, none of which may take account of the ethics of an individual industrial engineer or of his profession. The refusal to show data or allow independent study may be based on concepts of property rights, on management prerogatives, or on antiunion policies. The refusal may simply involve an interpretation of an existing collective-bargaining agreement or may even involve collective-bargaining strategy or tactics. Whatever the reason for the refusal, it is looked upon as an indication that the company has something to hide. Since in this case the work of the industrial engineer is involved, naturally the industrial engineer's work becomes the hidden something open to suspicion.

Industrial engineers cannot hope to gain recognition as practitioners of a recognized profession if they continue to acquiesce in the veiling of their activities. The articles of the "Code of Work Measurement Principles" are supposed to "represent standards by which industrial engineers, management, *and labor* can judge the soundness and propriety of the courses of action they are following or contemplating."* (Emphasis supplied.) It is hardly ethical practice to allow industrial engineers and management to make this judgment by study, observation, and analysis of a job as it is performed under existing working conditions, while labor is denied the same opportunity. When labor is denied on-the-job study of disputed operations, the denial is frequently taken as proof that the standards involved were not "soundly conceived and applied with the competence and integrity" that the "Code of Work Measurement Principles" prescribes.

* Preamble.

16.4 AFL-CIO POLICY

Industrial engineers have long recognized worker and union objections to many of their techniques and applications. The host of articles on how-to-sell this or that tool or technique which have appeared in professional and trade journals attest to this recognition. Modern writers tend to agree that past union attitudes were justified by the abuses of the old-style efficiency experts—the industrial patent-medicine men who sold get-rich schemes to employers. That these schemes resulted in speed-ups, rate-cutting, and unfair work loads is not denied. Employers guilty of these activities are today labeled "shortsighted and unscrupulous."†

Today, however, the claim is that these activities, if they exist at all, occur with such infrequency as to be insignificant. That this attitude is not shared by U.S. unions should be of concern to industrial engineers. In 1961, the AFL–CIO adopted, at its Fourth Constitutional Convention, the following policy resolution highly critical of industrial engineers and industrial engineering.

> The last decade has seen tremendous changes in work and in the work environment. Many of these changes are associated with accelerated mechanization and rapidly changing technology and have had profound effects on workers and their unions, on working conditions, on employment and unemployment, on the general character of the work force, and on collective bargaining.
>
> Concurrently with these "automation" changes, other techniques have been introduced to the workplace and the collective-bargaining arena which, while not generally accompanied by as much fanfare and publicity, have had equally disturbing effects. These are the innovations generally "credited" to the industrial engineer

† Benjamin Niebel, *Motion and Time Study* (3rd ed.) (Homewood, Ill.: Richard D. Irwin, Inc., 1962), p. 12.

and include predetermined motion-time systems, standard data, work sampling, and other "statistical" techniques for setting production standards, as well as the development of electronic timing and control devices. Labor, after several years of trial, is forced to conclude that these newer techniques are based on the same or similar false assumptions which have historically characterized the work of the management engineer. When applied to the workplace these techniques yield results which are neither accurate, reliable, nor valid.

Unions have historically opposed the use of stop-watch time-study, piecework and wage-incentive systems and job evaluation. This opposition was originally based on the arbitrary and abusive use of these methods by management. But unions learned that even when attempts were made to use the systems in an objective manner the results were inequitable. The systems themselves, as well as their applications, were found lacking.

Labor has found that the new techniques are nothing more than subtle forms of the old. "Improvements" represent little more than techniques of confusion, making it more and more difficult for workers to understand and cope with problems raised by the "new methods." All too often workers find that the "scientific" method merely is a device to circumvent established collective-bargaining arrangements.

In spite of proven shortcomings, work-measurement techniques, new and old, are still being used to set unfair production standards and work loads. Incentive systems of questionable validity are used to induce excessive work pace, and job-evaluation techniques are being used to downgrade workers and reduce legitimate wage gains.

Research into basic problems of human work; of the measurement and effects of physical and mental fatigue; of the effects of the automated workplace; of job design and job enlargement; of fitting the job to the worker, all are largely ignored by the industrial engineer.

When these management schemes exist unions must be prepared to negotiate maximum contractual protection in order to minimize harmful effects. But contract language is not enough. Unionists must be as fully informed as possible to cope with the "traditional" as well as the "modern" industrial-engineering techniques.

A few unions have effectively educated local union members and officers and union staff not only to understand management engineering techniques, but to thoroughly recognize the limitation of these devices and their harmful effects.

In order to answer the need for trained full-time union staff, the AFL-CIO, for the past 4 years, has conducted an annual AFL-CIO Union Industrial Engineering Institute. Held for 2 weeks each summer in Madison, Wis., this program is staffed by leading industrial-engineering authorities from AFL-CIO and its affiliated unions and from the University of Wisconsin's School for Workers. The AFL-CIO Union Industrial Engineering Institutes' comprehensive, concentrated training has been of tremendous value to the workers represented by the approximately 150 full-time union staff from thirty national unions who have participated. But these efforts have, at best, scratched the surface. Much remains to be done if unions are to provide workers with needed protection. Exploitation is still being carried out under the halo of "science." Now therefore, be it

RESOLVED: That this Convention call upon all AFL-CIO affiliates:

1. To use every available means to make all union members aware of the threat of current industrial-engineering practices to established working conditions and collective-bargaining procedures,
2. To join in the fight to stop the spread of abusive and arbitrary applications of questionable industrial-engineering techniques,
3. To attempt to influence the practitioners of industrial engineering to channel their efforts to basic human problems of work; to research in job design and the adjusting of the workplace to the worker,
4. To take full advantage of schools such as the annual AFL-CIO Union Industrial Engineering Institutes to train as many union representatives as possible to effectively represent workers faced with management's misuse of industrial-engineering techniques.*

Note that the resolution is critical of industrial-engineering systems as well as their application. Note also that new techniques do not fare better than the old. As far as the AFL-CIO is concerned the past of industrial engineering is still with us. But the AFL-CIO is normally not faced with industrial-engineering applications. These problems are the province of the AFL-CIO-affiliated national unions, whose elected officers constitute the majority of delegates to an AFL-CIO convention.

The AFL-CIO is a confederation of 128 national unions. These national unions are federations of local unions with membership employed in almost every occupation in almost every industry in the United States. Today over 60,000 local unions of AFL-CIO affiliates have a combined membership of over 16 million in the United States and Canada. Over 115,000

---

* *Proceedings of the AFL-CIO 4th Constitutional Convention* (Washington, D. C., 1961), pp. 358–360.

collective-bargaining agreements cover these members.*

Most resolutions coming before an AFL-CIO convention are submitted by AFL-CIO-affiliated national unions; by state, county, or city AFL-CIO councils; and by the various AFL-CIO departments. These resolutions usually have been passed by conventions of these bodies with the proviso that they also be submitted to the AFL-CIO convention for action. Often resolutions all dealing with the same general topic are received from many affiliates. Wherever possible, these resolutions are combined by the Resolutions Committee and its staff into one resolution.

Since each resolution passed by the AFL-CIO convention becomes labor-movement policy, they are given careful consideration by the members of the Resolutions Committee. Frequently, the committee holds hearings to help it understand the problems which led to a resolution submission. The Resolutions Committee, which recommended adoption of the Resolution on Industrial Engineering, was composed of the elected officials of forty-four national unions and three state councils. The resolution was carried by the 1961 convention without opposition. This, of course, does not indicate that each affiliate was 100 per cent in accord with every word of the resolution since every resolution reflects the various experiences, practices, and opinions of the extremely heterogeneous makeup of the AFL-CIO membership. The resolution does represent, however, the posture of an overwhelming majority of the 128 AFL-CIO national unions and their members.

16.5 NATIONAL UNION POLICY

Policy statements and convention resolutions highly critical of many industrial-engineering practices have been adopted in recent years by such unions as the Allied Industrial Workers; International Brotherhood of Boiler Makers; Iron Ship Builders; Blacksmiths, Forgers and Helpers; International Brotherhood of Electrical Workers (IBEW); International Union of Electrical Radio and Machine Workers (IUE); International Association of Machinists and Aerospace Workers; International Molders and Allied Workers; United Federation of Postal Clerks; International Association of Sheet Metal Workers; and the Upholsterers' International Union. The March 1964 convention of the United Automobile, Aerospace and Agricultural Workers (UAW), claiming increasing dehumanization of the automobile industry workplace, called for an end to unfair work loads and speed-up. The return of human dignity to the workplace was a major goal of the 1964 negotiations, and as a result personal time was increased by 50 per cent. This increase was secured at the bargaining table without the benefit of engineering analysis.†

Generally unions have been concerned with only those tools of industrial engineering that directly affect the worker and the workplace. Historically this meant stop-watch time study, piecework, piece rates, wage incentives, and job evaluation. More recently standard data, work sampling, and predetermined motion-time systems have aroused union concern and attention. In addition, the extension of certain industrial-engineering applications to indirect service workers, particularly highly skilled craftsmen, has invoked considerable anxiety.

The use of cameras, closed-circuit-television metering devices, and other forms of surveillance by industrial engineers has raised serious questions which cannot be ignored. While the industrial engineer may look at these devices simply as a means of gathering accurate data for methods improvement, work measurement, and allowance determination, the union must analyze them from other angles. The right of workers to perform their assigned tasks under working conditions which do not invade their privacy or degrade their dignity must be maintained.

National union attitudes toward industrial engineering have been developing since the first strike at the Watertown Arsenal in 1911 by members of the Molders Union. This strike led to an AFL resolution urging all affiliates to resist the extension of Taylor's systems. Many people credit AFL pressure for the 1915 federal prohibition against time study and wage incentives in government installations. This pro-

* *Directory of National and International Unions in the United States*, Bulletin No. 1596 (Washington, D. C.: Department of Labor, 1967), pp. 45–46.

† Many union members would probably feel this worker gain was secured without the hindrance of engineering analysis.

hibition was reenacted as a rider on appropriation bills each year until 1949. The famous sit-down strikes of the 1930s had as a major goal the elimination of incentive programs in the automobile industry.

The war, sometimes cold, sometimes hot, which generally characterized union–industrial engineering relations was temporarily set aside during the years of the defense build-up and of World War II. The great production demands of this period led to the establishment of labor-management committees to attack production problems. Cost-plus contracts and the production-at-any-price philosophy did little to encourage efficient production methods. Tremendous production records were achieved during these years primarily through the dedication of highly motivated workers, managers, technicians, and scientists. Frequently these production increases were not matched by productivity increases. Many industries maintained this same high-cost, high-production philosophy during the decade following the war in the rush to satisfy pent-up consumer demand.

More recently, however, cost-conscious management has used industrial engineers with greater frequency and intensity. This increased exposure has resulted in the corresponding heating up of relations between unions and industrial engineers. It has already been pointed out that the AFL–CIO policy as expressed by the 1961 convention resolution represents a consensus of its affiliates—a consensus which embraces various policies, attitudes, and practices. It should be of interest, however, to see how one AFL–CIO affiliate (IUE) expresses its view of industrial engineering.

> The last two decades have witnessed a tremendous growth in the influence of industrial engineers, not only in planning methods and machinery but also in the development of systems to evaluate jobs and establish work and incentive standards. The great changes taking place in technology, with new machinery, including automation, are creating increasingly greater problems in this field.
>
> While we recognize that industrial engineering improves methods, increases productivity, and develops more orderly methods of handling problems, these improvements often come at the expense of the worker. Systems of job evaluation and work standards in the hands of unscrupulous consulting firms or greedy employers tend to undermine the standards of those we represent.
>
> Most impartial engineering experts agree that these systems are subject to considerable error and the variations of human judgment, and even experienced engineers will differ considerably in the measurement of given jobs.
>
> The attempt by some engineering groups to give a specific mathematical definition to the concept of a "fair day's work" has completely failed.
>
> Our primary interest is to assure that those we represent are fairly compensated for a fair day's work and that when tasks are set they are consistent with health, well-being, and the maintenance of maximum employment.
>
> As a result of our experience with all kinds of methods for evaluating jobs and determining work standards and incentive payments WE HEREBY RESOLVE, that:
>
> Where incentive methods of payment exist, a transfer to day work should be made if it can be negotiated on a satisfactory basis.
>
> 1. Any change from incentive to day work may be made only by the mutual consent of the union and the employer. We oppose unilateral action on the part of employers. The governing principle must be that the earnings of incentive workers be maintained. The contract should make doubly sure that no aspect can be modified or eliminated except by mutual consent.
>
> 2. Where change to day work under proper conditions is inadvisable or unattainable, full protection should be secured in the agreement against abuses of incentive systems by the employer and to assure at least a 1 per cent increase in total pay for each 1 per cent increase in production.
>
> 3. Where day-work methods of payment exist, we oppose establishment of fixed work standards obligatory for individuals or groups.
>
> If the general level of plant production is satisfactory, nothing more should be demanded.
>
> 4. In any case, whether day work or incentive, where a form of work standards exists, there must be adequate allowances for fatigue, personal needs, and delays.
>
> Adequate rest periods should be negotiated.
>
> 5. We oppose the introduction of so-called "metering devices" attached to the machines, such as Productograph, until satisfactory limitations on their operations are negotiated.
>
> 6. We should examine with great care any job-evaluation plan for establishment of labor grades. None of the job-evaluation plans is scientific. Many are specifically devised in favor of the employer and are full of "booby traps." The bases of such systems and the job descriptions should be examined with great care to make sure that they conform to the type of operation and qualities required for the jobs in the plant.
>
> Where high speed, labor savings, or automated machines are introduced, the current job-evaluation systems must be revised to reflect the emphasis on new responsibilities.

7. There is no better method to determine what is fair than by collective bargaining to resolve a dispute.

8. Employers must supply union representatives with material that completely describes these systems, whether they be job evaluation, predetermined time systems, incentive, or day work, and must supply a copy of the standards used for setting the rate on any particular job.*

16.6 UNION INDUSTRIAL ENGINEERS

Now let us turn from the area of union attitudes to that of union industrial-engineering practices and personnel. With few exceptions, a union industrial engineer does not engage in the practice of industrial engineering in the same manner as does the employed or consulting industrial engineer. In fact, only ten of the 128 AFL-CIO national unions employ anyone full-time in the capacity of an industrial engineer or to do work normally done by industrial engineers. Of the forty-six people employed in this capacity thirty-eight are employed by four unions—sixteen by the Steelworkers, nine by Rubber, seven by the UAW, and six by the International Ladies' Garment Workers' Union (ILGWU). The Meat Cutters employ two industrial engineers, while five others—Brick and Clay; Textile; IBEW; Pulp, Sulphite; and the Upholsterers'—employ one industrial engineer each. In addition, AFL-CIO employs one staff industrial engineer at its Washington, D. C., headquarters.

The variation in the education of union industrial engineers (Table 16.1) is probably similar to the variation found in industry among those predominantly performing work-measurement, wage-incentive, and job-evaluation duties.

TABLE 16.1. *UNIVERSITY DEGREES OF FULL-TIME INDUSTRIAL ENGINEERS*

No degrees	28
Bachelor of Science	8
Bachelor of Science in Industrial Engineering	7
Master of Science	2
Master of Science in Industrial Engineering	1
Total	46

The majority of industrial engineers work out of the national headquarters of their unions; however, some are assigned to work in a specific region, usually on the staff of a vice-president or regional director.

Added to the full-time union industrial engineers are approximately twenty members of union research and education departments assigned some industrial-engineering responsibilities. The members of this group typically spend less than 25 per cent of their time on such activities. Educationally, this group is composed primarily of labor economists with bachelor's, master's, and doctorates.

While most U.S. unions do not employ, either full- or part-time, industrial-engineering or research personnel to handle industrial-engineering activities, these unions are not without resources when problems arise as a result of management's industrial-engineering applications.

16.6.1 Training Union Representatives

Many U.S. unions, including several of those employing full- or part-time industrial engineers, feel that most, if not all, union industrial-engineering problems should be handled in the same manner as other collective-bargaining matters and by the same personnel. Individual national unions sometimes set up special schools for training their national and business representatives, although since 1957 most union representatives have been trained in the annual AFL-CIO Union Industrial Engineering Institutes. These institutes, jointly sponsored by AFL-CIO and the University of Wisconsin, provide concentrated but comprehensive training in stop-watch time study, standard data, predetermined motion-time systems, work sampling, wage incentives, and job evaluation. While much time is spent on understanding the theory and application of these techniques, emphasis is placed on their shortcomings and collective-bargaining implications. The training and experience of these union representatives may be limited to a rather specific area of industrial engineering; however, their effectiveness can be attested to by numerous graduate industrial engineers who have been confronted by these unionists during grievance and arbitration hearings and at the negotiating table.

---

* *Resolution on Industrial Engineering*, Convention of International Union of Electrical, Radio and Machine Workers (Washington, D.C., 1964).

Elected local union officers and stewards handle most industrial-engineering problems along with all their other duties. Special time-study stewards or job-evaluation committeemen also handle a large number of these problems. Most local unionists handling industrial-engineering problems are not university graduates and have little if any training in industrial-engineering techniques. These shortcomings are frequently more than overcome by their intimate knowledge of jobs and of the workers manning them.

16.6.2 Union Industrial-Engineering Activities

In most instances there is a great overlapping in the activities of the various union personnel handling industrial-engineering activities. The local union official may negotiate or assist in negotiating contract clauses involving the application and administration of work-measurement, production-standards, wage-incentive, and job-evaluation plans. He also handles, at least initially, grievances which arise from the application of these systems.

The national union representative is also involved in the negotiation phase. He not only brings to the negotiation table his wide experience but also may be responsible for coordinating activities of many locals as well as negotiating contract clauses which agree with national union policy. The union representative normally becomes involved in those grievances which local union officers have been unable to settle. He is responsible for planning and presenting arbitration cases as well as for running short training programs in time study and job evaluation for the local unions in his territory. A few of these representatives have become so proficient in handling industrial-engineering disputes and negotiations that they tend to specialize in these activities, especially in large unions, where a regional vice-president or district director may have a large staff under his direction.

The full-time union industrial engineer also engages in negotiation, grievance, and arbitration activities, either on his own or as an assistant to local or national representatives or union attorneys. In these activities, because of his background and experience, the average union industrial engineer is probably superior to his company-employed or consulting colleague.

In addition to these collective-bargaining activities, the union industrial engineer performs a variety of other functions. His prime responsibility in some unions is as staff advisor to the elected national officers and to the appropriate committees of the union. In this regard he may be vitally concerned with broad problems and policies—for example, in the fields of automation, manpower, and unemployment—as well as with narrower industrial-engineering matters.

The union industrial engineer is normally responsible for the preparation of brochures, pamphlets, educational materials, and sample contract clauses in the industrial-engineering field. Often he is responsible for the education of union members in industrial-engineering subjects. He may alone or in cooperation with the union education staff plan and conduct time-study and job-evaluation courses. He also analyzes company policies and procedures for their effects on workers and unions.

In a very few unions—the ILGWU is a notable example—the union industrial engineer assists employers with general engineering surveys and suggests improvements. These consulting activities occur infrequently and usually only in the small plants characteristic of a large part of the garment industry.

For many unions, the only industrial-engineering service available and acceptable is from the AFL-CIO. While the AFL-CIO industrial engineer's prime responsibilities are of a staff rather than of a line variety, all the activities described here as being performed by industrial engineers and representatives employed by national unions are performed on a consulting (no-fee) basis by the AFL-CIO industrial engineer for affiliated unions.

The full-time graduate union industrial engineer is in a unique position. More than his colleagues who are employed by other organizations, he is free to practice his profession with few restraints. While his use of techniques may be limited, his opportunities for exercising his own initiative and ingenuity in a variety of situations are great. He influences policy and contributes directly to the welfare of his fellow man.

In many cases, the union industrial engineer is responsible for convincing a company to adopt policies and procedures which enable the company industrial engineer to operate in an environment conducive to ethical, professional practice.

16.7 INDUSTRIAL-ENGINEERING TECHNIQUES

Historically, as mentioned earlier, unions have been concerned primarily with only those tools of the industrial engineer that directly affect the worker and the workplace.

Probably with time unions will also become concerned with job design, plan layout, work physiology, information systems, queueing theory, systems simulation, linear programing, operations research, and other "sophisticated" industrial-engineering techniques. So far, and at least at the local union level, the direct or even the indirect effects of these techniques have been too remote to be of concern. Or if there has been a direct effect—for example, a reduction in work force—the connection between the lay-off and the work of the industrial engineer has not been conveyed to workers or their unions.

16.7.1 Stop-Watch Time Study

Unions are now faced with a dilemma. The use of some form of standard-setting procedure has become almost universal in all production jobs. In many cases this procedure has been introduced in spite of militant union opposition. Historically, of course, this opposition centered on stop-watch time study. To overcome union opposition and the growing ability of unions to handle time-study problems, many managements were sold on and adopted alternative standard-setting procedures, such as standard data and predetermined motion-time systems. In some cases the conversion was even suggested by a union representative. The current dilemma has developed as unions have come to recognize that these newer techniques are inferior to stop-watch time study from both a collective-bargaining and a technical standpoint. It is not unusual, today, to find a union with a long anti–time-study history bargaining to retain the stop watch rather than to allow the use of a newer technique.

Current union attitudes toward stop-watch time study (as well as the current dilemma) probably are best illustrated in the article "Time Study and Union Safeguards" from the AFL–CIO "Collective Bargaining Report." The remainder of Par. 16.7.1 is that report.*

---

* Bertram Gottlieb, "Collective Bargaining Report," *AFL-CIO American Federationist* (Nov. 1965), pp. 15–19. This article is nontechnical because it was prepared for laymen, primarily local union members and officers, and not for union professional and technical staff.

16.7.1.1 Time Study and Union Safeguards. Time study, which is widely used for determining work loads and wage-incentive standards, is an imprecise tool and lends itself to easy abuse. Unions confronted with it must be continually on guard against the use of arbitrary, unreasonable, and unrealistic time-study results.

Of the whole field of so-called "scientific management," time study is the area in which most of labor's distrust and suspicions are centered. Ever since its introduction in the 1880s most unions have opposed the use of stop-watch time study.

Labor's distrust stems from its practical experiences with time study. Problems arise because of the inherent shortcomings of the time-study process itself as well as the application of the technique in industry.

Time study is usually represented to unions as "scientific." But time study produces results which are simply judgments. They are not, and cannot be, scientific or accurate; at best they represent approximations, and at worst they are no better than wild guesses.

16.7.1.2 What Is Time Study? Time study is supposed to be a method of determining the time which should be allowed for a worker to perform a defined job according to a specific method and under prescribed conditions.

Time studies usually are made by timing workers with a stop watch while they do a certain job. This time is then adjusted for such factors as unavoidable delays, rest, personal needs, and incentives. The result is frequently called a standard. The job or time standard may be in terms of units per hour, standard hours per 100 units, or time per unit. If an incentive plan exists, the standard also may be expressed in monetary terms, such as 1¢ per piece produced.

Since most time studies are taken of an individual worker or a small group of workers, disagreements over time studies usually are expressed as grievances on specific job standards at the local union level.

While international unions can and do provide expert assistance and information to their locals, the investigation and processing of time-study disputes remain primarily a local union problem.

Based on experience, local unions have devised approaches to time study which fall into these categories:

1. Some locals prohibit time study altogether.
2. Some locals allow management to use any method of setting job standards it desires, but the locals reserve the right to bargain on the results.
3. Some locals participate directly with management in making time studies and in setting standards.
4. A majority of locals allow management to

make time studies but insist on bargaining on both the methods used and their application.

16.7.1.3 The Right to Time-Study Data. Unions faced with time study must be certain they have complete information on how it is used by their companies. To provide essential protection to members, this information must not be limited in any respect. It must include not only the results of time study, that is, the individual job standards, but also the plan in use by the company and the exact procedures followed.

That this information is essential to collective bargaining is apparent. Without it, a union would be unable to sensibly process grievances or discuss pertinent contract clauses.

The union's legal right to such information has been clearly established by arbitrators, the National Labor Relations Board, and the federal courts.

In spite of this, some unions still find it difficult to secure all the necessary data. Some managements still claim that such information is confidential. When union pressure forces compliance, some managements attempt to restrict the information they provide or the means of providing it.

To ensure immediate availability of time-study data, without question or limitation, most unions insist on a contract clause like the following:

> "The company shall furnish to the union a copy of the time-study plan presently being used. It shall make available for inspection by the union any and all records pertaining to time study and the setting of production standards, including original time-study observation sheets. Upon request by any shop steward or union officer, the company will furnish copies of any of the above information, including copies of time studies."

Some companies have been required by contract to furnish the union with photostatic copies of the time-study observation sheet each time a study is made.

16.7.1.4 On the Accuracy of Time Studies. "We do not bargain standards!" is a frequent assertion by management when faced with union time-study grievances.

Management claims that time studies produce facts and "of course, facts are not subject to bargaining or compromise."

If time study did result in "facts," then this management position might be sound and the union might be able to bargain only on whether or not time study should be used in the plant.

But such is not the case with time study. At best, standards developed from time study are only approximations. They involve the use of considerable judgment by the time-study man at every step of the time-study procedure.

The actual recording of the time involves the least judgment of all, but even here a 10 per cent error is expected and recognized by time-study experts.

Among some of the variable factors which can significantly affect a time-study result are:

1. The selection of the worker to be studied
2. The conditions under which the work is performed during the time study
3. The manner in which the operation is broken down into parts or elements
4. The method of reading the stop watch
5. The duration of the time study
6. The rating of the worker's performance
7. The amount of allowances for personal needs, fatigue, and delays
8. The method of applying the allowances
9. The method of computing the job standard from the timing data

16.7.1.5 Weaknesses in the Rating Process. After a time-study man has completed the stopwatch timing, he has obtained a figure which represents the time, on the average, that it took the particular worker he observed to perform the job.

This time could be used to set standards *only* if the worker who has been time studied could be considered a "qualified," "average" employee, working at a "normal pace" and exhibiting a "normal" amount of skill.

If the worker observed does not meet these specifications, then the observed time must be adjusted to make it conform to the "normal" time. This adjusting procedure most frequently is called "rating" but is sometimes called "leveling" or "normalizing."

Rating is the step of the time-study procedure where the most judgment or guesswork is involved.

Rating has been defined as the process whereby the time study man compares the actual performance which he observes with a concept or idea of what normal work performance would be on the job being studied. This concept of "normal" must by its nature be carried around in the time-study man's head.

During the rating process, the time-study man must make two distinctly personal judgments. First, he must formulate in his mind the concept of what a normal performance should be on the job being studied; second, he must numerically compare the observed performance with his mentally conceived normal performance.

The very nature of rating opens it to abuse. By manipulating this rating factor, it is easy for a time-study man to end up with practically any result he chooses.

In fact, as many unionists know, the rating factor is often used to enable a time-study man to end up with a standard determined before the time study was taken. In other words, time study is often used to "prove" to the workers that a work load or standard set by the company is "fair."

For example, a company may decide, without measurement of any kind, that a certain item should be produced at the rate of sixty pieces an hour or one piece per minute. Its time-study man then studies the job and finds that the average time it took the operator to make one piece was $1\frac{1}{4}$ min. This means that only forty-eight pieces would be produced in an hour. Obviously, this falls far short of management's set goal of sixty pieces per hour.

So the time-study man simply decides that the operator was working below "normal" during the time study and that a "normal" operator would have worked faster. The time-study man then adjusts his finding; he cuts the $1\frac{1}{4}$ min that it took the "slow" operator by 20 per cent—to come up with the standard that the company wanted of one piece per minute, or sixty pieces per hour.

It is practically impossible for a union to prove this kind of deliberate deceit since the "normal" operator is only a figment of the time-study man's imagination.

So it is easy to see how unscrupulous time-study men can readily manipulate time studies.

But the situation, unfortunately, is not much better even when management and its time-study men are trying to be completely honest and objective.

16.7.1.6 The Fallibility of Time-Study Men. Most time-study men claim they can judge worker pace within 5 per cent, and some even have claimed that, with experience, this could be reduced to an average error of 2 per cent. But these claims of precision have been proved false by university researchers and management groups who studied the ability of time-study men to rate workers' performance.

The studies show that time-study men will, in more than half their ratings, misjudge by more than 10 per cent the variations in work pace. "Errors" of as much as 40 per cent are not at all uncommon among "qualified" and experienced time-study men.

Specifically, the Society for the Advancement of Management [SAM], the largest national organization of industrial engineers and supervisors, conducted one of the most extensive rating studies ever undertaken.

The study used films of workers performing a variety of industrial operations. Time-study men rated different paces of these operations. (From the calibration of the film, it was possible to get an accurate measurement of variations in pace.)

Trade unionists were not surprised at the published results of the SAM study.* They showed, for example, that for 599 time-study men:

1. The average error in estimating variations in work pace was 10.57 per cent.
2. Of all the time-study men, 50 per cent had average errors larger than 10 per cent, 20 per cent averaged less than 10 per cent error, and less than 12 per cent had errors averaging below 5 per cent.

One man's errors averaged as high as 22 per cent. These, of course, are average errors. Some individual ratings were lower and some were higher than the averages. The SAM did not publish the range of errors, but similar studies have shown errors over 40 per cent.

Some managements have agreed that wide variations and inconsistencies like these can be expected—but not from full-time, experienced time-study men.

The SAM study answered this question, too. It proved that length of experience or the amount of time spent on time study had no relation to accuracy. In fact, the average error of those with over 15 years' experience was greater than for those with less than 6 months' experience. And men who spent less than half their time taking time studies rated slightly more accurately than those spending all their time in time-study work.

The SAM study, as well as many others, proves conclusively that the results of time studies are simply approximations and cannot be considered factual. They reflect the individual judgment of the time-study man.

As far as rating is concerned, the opinion of workers is just as valid as that of time-study men. In fact, the results of experiments at Purdue University indicate that people actually working on a job can judge variations in work pace more accurately than time-study men observing the same job.

Incidentally, in checking hundreds of time-study men for its study, SAM decided not to use data of some 700 of them—about 40 per cent of all the time-study men it examined from companies in all sections of the country. Even though these men were actually working as company time-study men and were daily setting standards for workers, SAM in effect decided they were not qualified enough to take time studies and set job standards.

But even if these "errors" in rating operator performance were reduced and time-study men could recognize differences in work-pace variations with greater reliability and validity, the basic problem of determining a "normal" work pace would still be with us.

That workers cannot leave this determination to management was dramatically demonstrated

* *A Fair Day's Work* (New York: Society for the Advancement of Management, 1954), p. 45.

by Dr. C. L. M. Kerkhoven of Syracuse University. Dr. Kerkhoven subjected two of the operations in the SAM films to physiological study. He concluded that these operations, when performed at a work pace equivalent to SAM's "normal," would require workers to work at a pace approximately $2\frac{1}{2}$ times faster than a physiologically permissible work speed.*

It is not surprising that Kerkhoven found that SAM's concept of normal is "physiologically inadmissable" and certain to overstrain the heart. In one of these operations, one man working at SAM's "normal" work pace is required to lift and move 11,560 cartons weighing 40 lb each during an 8-hr work day. This adds up to 440,000 lb, or 220 tons.

One obvious conclusion can be drawn. Time study cannot be accepted without question by unionists. Every aspect of the time-study procedure must be subject to union review through collective bargaining and grievance procedure.

16.7.1.7 The Burden on Grievance Procedures. Time study has become an important issue in union-management relations. Unions have found that grievance-handling problems are greatly increased in plants where time study exists. Not only are there more grievances, but a greater amount of time must be spent in investigating and processing time-study grievances. And, in spite of a general decline in the number of plants using work measurement, there has been no corresponding decline in the percentage of grievances going to arbitration. This figure has remained at about 20 per cent for the last 10 years, according to published reports of both the American Arbitration Association and the Federal Mediation and Conciliation Service.

While local unions may seek expert help in special cases, the majority of time-study grievances not only should but can be successfully handled by shop stewards.

The most important factor to remember is that no time study is accurate. Union judgment is on a par with management's. The worker on the job is as accurate a judge as anyone of the propriety of any job standard.

Because management does not like to bargain on standards and because there is so little that is factual about time study, a large proportion of time-study grievances are taken to arbitration.

Unfortunately, while many arbitrators may not be biased in favor of the company or union, they all too often accept time study as being scientific and mathematically precise. They do not recognize the shortcomings of time study and therefore are more likely to accept its results.

16.7.1.8 Unions and Time-Study Grievances. Despite these difficulties and shortcomings, local unions can do a good job of protecting workers against unfair management time studies. To do an effective job, however, locals need informed and alert officers and stewards. Above all, union representatives should not be "snowed under" by the so-called scientific procedures and arguments of management time-study men.

In locals where trouble with time study persists, the international union should be requested to give advice and assistance. Various international unions have staff members with specialized experience in handling time-study difficulties.

Time-study grievances should be handled the same as any other grievances. The most important step is that of getting the facts.

While some knowledge of time study is helpful, it is not necessary for the shop steward or other union representative to be a time-study man in order to process a time-study grievance.

The union representative should:

1. Secure a copy of the company's record of the operation in dispute.

2. Make certain that the record of job conditions and the job description are complete. If either is incomplete, it will be impossible to reproduce the job as it was when the time study was made and therefore the company's time study cannot be checked. This alone is grounds for rejecting the study.

3. If the time-study sheet does contain sufficient information as to how and under what conditions the job was being performed when the time study was made, then it is necessary to determine whether the job is still being performed in exactly the same way.

4. Usually the total operation or job cycle is broken down for timing purposes into parts, which are called elements.

Check the elemental breakdown of the job on the time sheet. See that the beginning and ending point of each element is clearly defined. If they are not, any attempt to measure elements and give them a time value was pure guesswork.

5. Check the descriptions of each element to see if they describe what the operator is presently required to do. Any change affecting time invalidates the original study.

6. Make sure that everything the operator is required to do as part of the job has been recorded and timed on the time-study sheet. Watch for tasks which are not part of every cycle.

Such things as getting stock, adjusting machines, changing tools, reading blueprints, and waiting for materials are examples of items most often missed.

---

* Kerkhoven, C. L. M., "The Rating of Performance Levels of the S.A.M. Films," *The Journal of Industrial Engineering* (July–August 1963), pp. 170–74.

7. Check for "strike-outs." A time-study finding is based on a number of different timings of the same job. The time-study man may discard some of his timings as "abnormal."

If the time-study man has discarded any of his recorded times, he must record his reason for doing so. This enables the union to intelligently determine if the "strike-out" is valid. The fact that a particular time is larger or smaller than other times for the same element is not sufficient reason for discarding it.

8. Determine if the time study was long enough to accurately reflect all of the variations and conditions which the operator can be expected to face. Was it a proper sample of the whole job? If not, the time study should be rejected, as its results are meaningless.

9. See that only a simple average, and not the median, mode, or other arithmetic device was used in calculating the elemental times. The simple average is the only proper method for time-study purposes.

10. Check the rating factor on the time-study sheet. Try to find out if the time-study man recorded his rating factor before leaving the job or after he computed the observed times. Ask the operator who was time studied if he feels the rating factor is a proper one.

Watch the worker who was timed work at the pace he considers proper and then at the pace required to produce the company's work load. The judgment of the worker and steward is as valid as that of the time-study man.

11. Make sure that allowances for personal time, rest, and delays have been provided for in proper amounts.

12. And, finally, check all the arithmetic.

Most union representatives have found it unwise to take additional time studies themselves as a check, except as a last resort. It is more effective to show the errors in management's study than to try to prove a new union time study is a proper one.

A union time study is still only the result of judgment. Even when proper methods are used, they tend merely to reduce the inconsistencies of time study, not eliminate them.

Historically, the words "time study" have referred to the process explained in this report which involves the use of a stop watch or other timing device. Recently, some companies have introduced other methods of setting production standards such as standard data and predetermined motion-time systems. When unions object, companies have claimed that these systems are simply forms of time study and may rightfully be used even if a contract has a clause such as: "Standards shall be set by time study."

To guard against this type of subterfuge where stop-watch time study is presently used, unions should make sure that future contracts contain this type of clause: "All production standards are to be set by *stop-watch* time study."

16.7.2 Standard Data

As with time study, union policies and practices relating to the application of standard data have also been dealt with in a "Collective Bargaining Report," which follows.*

"You will never again see a time-study man with a stop watch in this department."

This is a statement heard with greater frequency each year by union members. To the unionist who has experienced the frustrations and disappointments of handling time-study grievances the demise of the stop watch must seem like a blessing. Unfortunately, experience has shown the trend away from the stop watch may bring effects even worse than those experienced when production standards are set by stop-watch time study.

When management eliminates the use of stop-watch time study it usually is not eliminating work measurement or the setting of production or work standards. In place of the "older" methods, "newer" techniques may be substituted. Frequently these newer methods are introduced to the plant by company or consulting engineers who "trumpet" the superiority of their new approach as a "cure-all" for the ills of time study.

While in some cases there may be legitimate "engineering" or "economic" justifications for the use of the newer techniques, possibly the chief reason is the growing ability of unions to cope with the use of stop-watch time study. Informed union officers and members have taken the "sting" out of time-study standards by negotiating protective contract clauses and by handling grievances effectively.

It is not the method used by management in setting production standards that is the prime concern of unions but rather the effect of a method on work loads and earnings. Unions must protect workers against unreasonable work loads and ensure that workers receive equitable compensation for their efforts. To accomplish this, union representatives must be equipped to meet these "newer" techniques just as they have learned to handle stop-watch time-study problems.

The purpose here is to describe the development, use, and limitations of "standard data"—one of the new work-measurement techniques.

16.7.2.1 Development of Standard Data. While the organized development and use of

* *AFL-CIO American Federationist* (June 1962), pp. 16–20.

standard data are rather new, the principles upon which it is based were recognized over 50 years ago.

It was not uncommon in early piecework shops for the price of a new job to be determined by comparing the new job to other jobs in the shop on which "acceptable" prices were in effect. If an old job could be found which was "similar" to the new job, the old job price was established for the new job.

In modern applications of standard data the comparisons are based on time rather than money, and parts of jobs (called work elements) are usually the basis for comparison rather than the whole job. In its simplest form, the use of standard data involves applying the times for work elements of prior time studies to similar work elements on jobs set up in the future.

A company which has been doing stop-watch time study for several years normally would have a rather large file of time-study sheets of many different jobs. An inspection of these time studies might reveal that similar elements appear on many of these sheets. Common work elements might be "press button to start machine" and "press button to stop machine." It also might be found that there is very little variation from job to job of the average time allowed to perform these elements. The allowed time from each study then might be added together and the average computed. The average time then would become the allowed time for these elements whenever they occur in future jobs.

This type of element would be called a constant element. This does not mean the element time would be constant for all kinds of work in the plant but rather the time for this element will remain approximately the same for a given range and class of work. For example, "press button" might be given a constant allowed time in the drill-press department, while a different constant time would be allowed for the "press button" element in the lathe department.

Inspecting the time-study sheets probably would reveal other constant elements as well as many variable elements.

Variable elements are those whose times will vary for a given type of work due to different job requirements. For example, in the drill-press department practically every time-study sheet would have the element "drill hole." It would not be proper to average out the "drill hole" times in the same manner as was done for the "press button" times. Differences in hole diameters, hole depths, tolerances allowed, kind of material drilled, speed of machine and type of drill, individually and in combination, would result in wide variations in drilling time.

It would be necessary in this case to set up tables to take into account these variables. This would be done by separating the time studies according to the variables and averaging out the times for each set of circumstances. One table might be developed showing the times for drilling tool steel with tungsten-tipped drills. The times for various combinations of hole diameter and hole depths would be shown. Similar tables might be prepared which would show times for the same depth and diameter combinations for other types of materials and drill bits.

A third type of element is the machine-controlled element, where the performance time is mechanically controlled by the speed, feed, length of cut, etc., that the machine may be set for.

16.7.2.2 Claimed Advantages of Standard Data. Management engineers claim that, when compared to times developed by stop-watch time study, standard data:

1. Eliminates or reduces the need for stop-watch time study
2. Is more accurate and constant
3. Is more economical to apply
4. Helps to predict costs and schedules
5. Helps balance production lines and improve methods before production begins

Before analyzing these claims it should be pointed out that an additional factor must be considered when evaluating any work-measurement technique: the effect of the technique on workers and unions. Even if all claimed advantages of standard data were true, unions would still have to put standard data to this collective-bargaining test.

Let's see how the advantages claimed for the use of standard data square with the facts.

16.7.2.3 Claim No. 1: Standard Data Eliminates or Reduces the Need for Stop-Watch Time Study. There is no question that once standard-data values are established for certain elements, stopwatch timing of those elements no longer will be performed. On many jobs data will be available for all its elements, and thus the time standard can be synthesized completely.

In practice, however, stop-watch time studies will have to be made on classes of work not previously studied and should be made to check and maintain the standard data. Management frequently points out that unions should be pleased when standard data reduces the use of stop-watch time study since this results in a corresponding reduction in the use of performance rating.

Performance rating is an essential step in stop-watch time study. It is necessary for the time-study man to adjust the times he has recorded during a time study according to his estimate of the work pace of the operator studied. If, for example, the time-study man thinks the worker he studied was working 10 per cent faster than "normal" during the time study, then the times would have to be increased by 10 per cent. If in this case the faster-than-"normal" worker took, on the average, 1.00 min to perform a given task, the time would have to be increased to 1.10 min. The 1.10 min would be the time for a "normal" operator to perform the same task.

Unions have always been critical of the ability of time-study men to judge the speed or pace of workers. Studies by university researchers as well as by management organizations have proved the nonscientific nature of performance rating. These studies uniformly show the inability of time-study men to rate worker speed within accepted limits of accuracy and consistency.

It is true the person developing the individual, standard-data-determined production standard does not rate operator performance, but most standard-data values are based on times developed from studies which included pace judgments. Speed ratings are therefore not eliminated but rather are built right into the standard data.

Some management engineers assert that, since standard-data values are determined from many studies, the rating errors of time-study men are "averaged out" and therefore the standards are better. This assumes there were an equal number of "tight" and "loose" ratings and the degree of "tightness" and "looseness" was about equal. In practice, since management attempts to control the rating process, it is more likely that ratings show a definite bias, and this bias becomes part of the standard data.

16.7.2.4 Claim No. 2: Standard-Data Times Are More Accurate and Consistent than Stop-Watch Times. This claim is also based on an "averaging out" concept. It is felt that if standard-data values are based on enough time studies of a particular element, errors will cancel out. As pointed out in the discussion of pace rating, this would only be true if the time values had a "normal" distribution—that the high and low times were of equal number and size. There is no evidence to prove this assumption. It should be obvious that standard data can be no better than the time studies upon which it is based. While standard data may be more accurate than individual stopwatch studies, it is not necessarily so and in many applications may be less accurate.

Consistency is another matter. Obviously applying times from tables will result in more consistent time standards as far as the standards themselves are concerned. The real question is how well these "consistent" values, when added together to form a production standard, fit the actual work that must be performed. In some cases the consistency of the time-value–work-requirement relationship might be quite good. In other cases there may be little relationship between the two.

It already has been pointed out that accuracy and consistency of standard data will depend to a large measure on the caliber of the time studies used to build the data. Inaccurate, inconsistent, sloppy time-study work can only result in equally poor standard data. But even the best constituted standard data is of no value until applied, and it is in the application of the data that further problems arise.

Standard data is valid only for work of the same type and under the same working conditions as existed during the studies upon which the data are based. It should be obvious, for example, that data collected from a new, well-lit, air-conditioned factory with new machines, tools, and equipment should not be used in an old plant where these conditions do not exist.

Moreover skill of the work force, labor-relations policy and history, type of wage-payment plan, wage levels, and many other factors other than physical conditions alone also influence production. In spite of these influences many companies attempt to use standard data developed in other plants of which they have little if any knowledge. There is no justification for this practice.

16.7.2.5 Claim No. 3: Standard Data Is More Economical. The management engineer usually can develop a time standard from a series of tables and charts in a fraction of the time that would have to be spent on a proper stop-watch time study. Whether this economy of individual standard-setting results in overall savings will depend on other factors. Developing the data for even relatively simple operations can easily take thousands of hours of work by qualified industrial engineers. The cost of developing the standard data will have to be spread over many applications to realize any economies.

To development costs will have to be added the cost of maintaining the standards and keeping them current, as well as the costs of processing grievances and checking disputed standards.

Savings anticipated from the use of standard data may never materialize. Even if standard data were always cheaper to use, this would hardly be justification for using it to set production standards unless it is superior to other techniques.

When standard data is used to set production standards which workers must meet in order to remain employed or when it is used as the basis of an incentive system, then the tests must be the accuracy and consistency of the results and the collective-bargaining implications rather than the cheapness of the standard-setting procedure.

16.7.2.6 Claims No. 4 and 5: Standard Data Helps to Predict Costs and Schedules and Helps to Balance Production Lines and Improve Methods Before Production Begins. These claims are justified. Standard data does provide a planning and predictive tool for management. Unions have no quarrel with this use as long as management recognizes the limitations of the data and expects to make adjustments as they are needed after production begins.

16.7.2.7 Collective-Bargaining Problems. The use of standard data poses special problems for unions. When management uses stop-watch time studies of individual operations as the means of setting production standards, unions may question through the grievance procedure any individual standard without questioning or threatening the total system. Adjustments may be made by management in any individual time study without fear that the adjustment will cause any chain reaction.

But when unions question a production standard set from standard data, a different situation exists. If an investigation shows the times for one or more elements taken from the tables of standard data are wrong, then every job which contains the same elements also may have an incorrect standard.

Much of the "flexibility" which makes it possible for both unions and managements to work with production standards is lost when standard data is used. Agreements on adjustments to production standards are harder to arrive at.

16.7.2.8 The Need for Contract Clauses. Because standard data may provide a useful tool for management for predicting, scheduling, and planning purposes, frequently management may develop and use it for these purposes and then later attempt to extend the use of the data to the setting of production standards. Unions must be extremely cautious of this practice.

Management will claim the standard data has "proved" itself in these other uses and therefore the system should be okay for setting production standards. No one doubts that time standards can be set by using standard data. It is the quality of the standard that is questioned. The degree of accuracy and consistency may be acceptable to management for budgetary purposes but not be "acceptable" to the workers having to meet the required production.

Too often management has attempted to install a standard-data system without fully understanding the limitations of standard data and the effects its use might have on labor relations. Frequently management is "sold" a standard-data system by a management consultant who "proclaims" the benefits of his "product" but disclaims any problems. Some of these "sellers" even falsely claim union approval of their systems.

Experience has led many unions to oppose the use of standard data in any form. Some unions in plants with production standards have effectively barred the use of standard data by negotiating a specific clause such as: "Production standards shall be established only through the use of stop-watch time studies of each operation."

There are no "union" methods of setting time standards, and some unions prefer to let management make the decision on which method it will use. When proper safeguards are negotiated, when management fully recognizes the limitations of standard data, and when the data are properly and carefully developed and applied, then standard data may provide an economical and at the same time "acceptable" means for setting time standards.

If management is using standard data for setting time standards either because the union cannot or does not choose to prevent its use, then contract clauses should be negotiated to provide that:

a. Production standards be established only through the use of stop-watch time studies or by the use of standard data developed from stop-watch time studies on jobs and in the plant covered by the agreement.

b. Upon request by the union the company make a stop-watch time-study check of any production standard developed from standard data and questioned by the union.

c. The check time study is made by a different time-study man than the one who set the original standard.

d. All standard data tables, charts, and formulas-furnished are to the union, as well as copies of all the time studies or other materials upon which the standard data are based.

e. Copies of all standard-data application sheets showing how individual standards are developed are furnished to the union.

f. Specified allowance time for personal needs, rest periods, and contingencies or delays are be added to all standards.

g. The union is allowed to bring a representative of its own choosing to check on any standard in any manner which the union desires.

h. All production standards are subject to questioning through grievance procedures.

i. Either unsettled production-standard grievances are handled through normal arbitration procedures or the union retains the right to strike.

One final point should be made about the use of standard data. Historically the setting of time standards usually has been limited to repetitive-type production work. While management always wanted to extend the use of work measurement to nonrepetitive, skilled, maintenance, and craft jobs, stop-watch time study was not suited for this purpose. The expense involved in making time studies of these types of skilled work to meet even the loosest standards of accuracy could seldom be justified. Where attempts were made the results were, most frequently, unusable.

With the advent of standard data, once more management has turned its "work-measurement" eye on its skilled workers. Many companies have developed tables of work times for the various tasks these workers perform. Some companies are attempting to use these time data as the bases of an incentive wage system.

One of the leading time-and-motion-study texts relates how time standards were developed for tool and die work, which in turn was the basis of a wage-incentive plan for the tool and die makers. The text points out how, using the tables, one person determines the time standard for 125 toolmakers, taking about 3 to 5 min per standard.*

Another company has recently instituted an incentive plan for electricians. Task times are determined from tables by a clerk after recording the size and length of wires and cables needed, the number of connections, switch boxes, terminals, etc.

It should be obvious that these procedures pose serious dangers to the status and skill of these workers, putting the premium on quantity rather than quality of work. The extension of work measurement and wage incentives to skilled workmen should be strongly resisted by all unions.

16.7.3 Predetermined Motion-Time Systems

As of the date of publication of this *Handbook*, AFL-CIO did not have a collective-bargaining report devoted to predetermined motion-time systems (PMTS). However, in 1961 two articles in *The Journal of Industrial Engineering* contained claims of union acceptance of PMTS. The following section is from the publication *Union Attitudes Toward Predetermined Motion Time Systems*, which was published by the AFL-CIO in response to these false claims.†

16.7.3.1 Union Attitudes Toward Predetermined Motion-Time Systems. The two articles by H. O. Davidson in recent issues of *The Journal* came as a breath of professional fresh air after the Bailey-Sellie-Taggart "Comments on 'An Experimental Evaluation of the Validity of Predetermined Elemental Time Systems,' " in previous issues of *The Journal*.‡

The use of the technique of "proof by proclamation" so aptly described by Professor Abruzzi§ uniformly characterized the three articles by these purveyors of predetermined motion-time systems. Such articles might best be left to the "paid-advertisement" category.

Davidson points out quite clearly that the Schmidtke-Stier research report which led to all of these articles simply confirmed "the only tenable position" reached by all objective researchers, namely, that no known predetermined motion-time system has scientific validity.‖ The articles by predetermined motion-time system "sellers" did not question this conclusion nor did these sellers shed any light on the claimed "operational validity" of their "sales products."

Had the first Davidson article been the last of the series there probably would have been no need for further comments from this or any other source. However, Honeycutt's "Comments" as well as the second Davidson article, both making reference to union attitudes toward predeter-

---

* Ralph H. Barnes, *Motion and Time Study, Design and Measurement of Work* (5th ed.) (New York: John Wiley & Sons, Inc., 1963), Chap. 30, pp. 462–76.

† (Washington, D.C., 1962). A resume of this publication appeared in *The Journal of Industrial Engineering* (May-June 1963), pp. 137, 138.

‡ H. O. Davidson, "On Balance—The Validity of Predetermined Elemental Time Systems," *The Journal of Industrial Engineering* (May–June 1962), pp. 162–65. H. O. Davidson, "Concerning the Right of Open Forum," *The Journal of Industrial Engineering* (July–Aug. 1962), pp. 242–44. G. B. Bailey, "Comments on 'An Experimental Evaluation of the Validity of Predetermined Elemental Time Systems, ' " *The Journal of Industrial Engineering* (Sept.–Oct. 1961), pp. 328–30. C. Sellie, "Comments on 'An Experimental Evaluation of the Validity of Predetermined Elemental Time Systems,' " *The Journal of Industrial Engineering* (Sept.–Oct. 1961), pp. 330–33. J. B. Taggart, "Comments on 'An Experimental Evaluation of the Validity of Predetermined Elemental Time Systems,' " *The Journal of Industrial Engineering* (Nov.–Dec. 1961), pp. 422–27.

§ Adam Abruzzi, *Work, Workers, and Work Measurement* (New York: Columbia University Press, 1956), pp. 3–4.

‖ H. Schmidtke and F. Stier, "An Experimental Evaluation of the Validity of Predetermined Elemental Time Systems," *The Journal of Industrial Engineering* (May–June 1961), pp. 182–204. Davidson, "On Balance," p. 162.

mined motion-time systems, call for additional discussion.*

16.7.3.2 Claims of Union Acceptance of Work Factor. Davidson† referred to the fact that Taggart based his defense of the work pace required by work factor on the assertion that:

> "We consider it obvious that, were this not the case, Work Factor would have been discarded long since through the efforts of dedicated labor unions. On the contrary, many unions (along with the managements of their respective companies) praise the fairness and equity of Work-Factor time standards."‡

That Taggart is forced to rely on alleged union acceptance to support the "fairness and equity" of Work Factor probably indicates that Work Factor has no other proof available.

Surely even the most naive observer of labor-management relations knows that the existence of a particular technique does not necessarily indicate approval of the technique by either party to a collective-bargaining agreement. To assume that the installation of Work Factor in an organized plant signifies union approval would be as ridiculous as to assume that management approves of the union shop, strict seniority, payment of average earnings to incentive workers on down-time, or any other contract clause.

Practicing industrial engineers know that many industrial-engineering activities are engaged in without the approval of and frequently in direct opposition to the desires of local unions and their members.

Davidson's answer to Taggart's assertion relies on statements attributed to union representatives at a 1953 meeting of the American Standards Association. This meeting resulted in the refusal of the ASA to sponsor a project, initiated by H. B. Maynard, to adopt Methods-Time Measurement or any other predetermined motion-time system as an American standard for manual work times.

16.7.3.3 Survey "Proof" of Union Acceptance. Before bringing the record up to date on union experiences and present attitudes it might be well to turn to Honeycutt's claim of union acceptance of predetermined motion-time systems:

> "An interesting insight into union attitudes was obtained through the Stewart, Dougall and Associates survey. Many people have been influenced to believe that unions in general are very much opposed to predetermined time systems. The facts do not justify this conclusion. Of the ninety-six plants in the survey that made full-fledged use of predetermined time systems, sixty-two were unionized. It was found that at the time the system was first introduced, 44 per cent of the local unions involved were in favor of using a predetermined time system. At the time of the survey in Sept. 1959, the sixty-two unionized plants reported that their unions were in favor of the system in a ratio of 6 to 1. It was also found that fully half of the unionized plants felt that the introduction of a predetermined time system actually improved labor relations."§

As Honeycutt points out, interviews were reportedly conducted with 497 *companies.* Questions were asked over the telephone, supposedly of people having knowledge of and responsibility in the industrial-engineering field.

Comments on this survey seem appropriate:

1. The survey was made by persons so unfamiliar with predetermined motion-time systems that the interviewers had to be instructed in the meanings of MTM, WF, BMT, etc.
2. The interviews were conducted by phone. The interviewers had no means of knowing if the respondent was the proper company official to supply the information requested.
3. The objectivity of the respondent could not be checked. It would not be illogical to assume that in many cases the respondent was the very company official responsible for the installation of a predetermined motion-time system.
4. The competence of the respondent also could not be checked. A careful analysis of survey responses clearly indicates that many respondents were not qualified to answer questions on predetermined motion-time systems. Many respond-

---

* John M. Honeycutt, Jr., "Comments on 'An Experimental Evaluation of the Validity of Predetermined Elemental Time Systems,' "*The Journal of Industrial Engineering* (May–June 1962), pp. 171–179.

† Davidson, "Concerning the Right of Open Forum," p. 244.

‡ Taggart, "Comments," p. 425. Reprinted from the November–December, 1961, issue of *The Journal of Industrial Engineering*, Official Publication of the American Institute of Industrial Engineers, Inc., 345 East 47th Street, New York, New York 10017.

§ Honeycutt, "Comments," p. 174. Reprinted from the May–June, 1962, issue of *The Journal of Industrial Engineering*, Official Publication of the American Institute of Industrial Engineers, Inc., 345 East 47th Street, New York, New York 10017.

ents were unable to identify the particular predetermined motion-time systems used in their plants.*

Not only would the responses to all questions of respondents who could not identify their own predetermined motion-time system be open to suspicion, but one might legitimately question the value of all aspects of such a survey.

5. The interviews covered 497 plants with 100 or more employees in five Standard Industrial Classification Groups in twenty-two city areas.†

a. The 1958 Census of Manufacturers‡ lists a total of 26,990 plants with 100 or more employees; 497 is less than 2 per cent of this total.

b. The Census lists a total of sixty-eight city areas. Many important city areas were therefore not included in the twenty-two surveyed.

c. Eighteen of the twenty-two city areas surveyed, or about 80 per cent, were large city areas with 1000 or more establishments, whereas less than 50 per cent are actually in this category.

d. The survey is heavily biased with cities in the Atlantic Coast area states (twelve of the eighteen states represented). Only three city areas (Dallas, Los Angeles, and Minneapolis-St. Paul) were west of the Mississippi and only four were southern.

e. The five industrial classifications surveyed and described in the report as "typical of U. S. industry as a whole" § were:

(1) Textile mill products
(2) Apparel manufacturing
(3) Fabricated metal products
(4) Machinery (except electrical)
(5) Electrical machinery, equipment, and supplies

Not surveyed was food and kindred products, the industry with the largest number of plants of 100 or more workers. Nor were any of the following important industries surveyed:

(1) Tobacco products
(2) Lumber and wood products
(3) Furniture and fixtures
(4) Paper and allied products
(5) Printing and publishing
(6) Chemicals and allied products
(7) Petroleum and coal products
(8) Rubber and plastic products
(9) Leather and leather products
(10) Stone, clay, and glass products
(11) Primary metals
(12) Transportation equipment
(13) Instruments and related products
(14) Miscellaneous manufacturing

These industries which were excluded from the survey include over 60 per cent of the plants with 100 or more workers.

6. The survey reports that ninety-six of the 497 plants make "full-fledged" use of predetermined motion-time systems and that 62 per cent of these were union. No definition of "full-fledged" was applied to respondents. Predetermined motion-time systems have been put to many uses which have no immediate effect on workers. For example, according to the survey, 89 per cent of the plants using predetermined motion-time systems for work-measurement purposes applied the results for cost-estimating purposes, 75 per cent for production planning, 60 per cent for budgetary controls, and 46 per cent for prebalancing assembly lines.

Gotterer has reported: ‖

> "Unions normally do not object if PMTS (predetermined motion-time study) is used in a manner which has no apparent effect on workers. If a company uses a system for estimating future manpower needs, plant layout, etc., it can expect little interest on the part of the union."

These and other aspects of the 1959 Stewart, Dougall survey lead this observer to question Honeycutt's conclusion that the survey was "scientifically constructed" or that the findings of the survey "arc projectable." #

It is interesting to note that one of the conclusions of the survey was that "7 per cent of the plants in the United States larger than 100 employees use Methods-Time Measurement as a primary engineering or work-measurement tool." ** Seven per cent of 26,984 (the number of plants of this size) is 1889. This figure is in conflict with published statements by H. B. Maynard and by the MTM Association for Standards and Research. In 1959 Maynard sent to this writer a copy of a speech titled "History and Development of MTM." In this speech Maynard said: "Six years ago, it was estimated that over 3000 companies in the United States alone were using MTM in one way or another. Today,

* Stewart, Dougall and Associates, *A Use Survey of MTM Within the Predetermined Time System Field—A Study Prepared for the MTM Association* (New York, Jan. 1960), Question 3.

† Stewart, Dougall and Associates, *A Use Survey . . .*, p. 5.

‡ U.S. Bureau of the Census, *United States Census of Manufacturers 1958*, Vol. 1, "Summary Statistics" (Washington, D.C.: Government Printing Office, 1961), pp. 2–3.

§ Stewart, Dougall and Associates, *A Use Survey . . .*, p. 5.

‖ Malcolm H. Gotterer, "Union Reactions to Unilateral Changes in Work Measurement Procedures," *Personnel Psychology*, **14**, No. 4 (Winter 1961), 436.

Honeycutt, "Comments," p. 174.

** Stewart, Dougall and Associates, *A Use Survey . . .*, p. 4.

although no accurate estimates are available, it is safe to say that the number of MTM users is much greater."

The MTM Association recently published a pamphlet describing the Association and claiming: "More than 3500 companies now use MTM for various purposes."

It is strange, to say the least, that Honeycutt is willing to base conclusions of union acceptance of MTM on a survey which shows 1611 fewer users (3500–1889) than claimed by Maynard and the MTM Association. Of course, it may be that the 1611 installations "lost" in the survey were in plants of fewer than 100 employees. If this were the case, however, why limit the survey to plants of 100 or more workers?

Before moving on, it may be well to point out that Honeycutt does not even accurately report the results of his own survey. Honeycutt states: "The sixty-two unionized plants reported that their unions were in favor of the system in a ratio of 6 to 1."*

This statement is not true. The Stewart, Dougall survey gives the following figures:†

Total unionized plants	62
At the present time:	
Union favors	65 per cent
Union does not favor	11 per cent
Union not involved	5 per cent
Don't know, no answer	19 per cent

It is obvious that Honeycutt's 6-to-1 citation refers to the 65 per cent to 11 per cent relationship and therefore only includes 47 plants [(0.65+0.11)62] and not 62 plants.

However, even if all of the above points had no weight and the survey were, in fact, an unbiased, scientifically constituted survey, the conclusions of the surveyors, as well as of Honeycutt, do not follow from the survey.

Based on this survey, Honeycutt and Stewart, Dougall claim union acceptance of predetermined motion-time systems, *yet not one union member, steward, local officer, technician, industrial engineer, or national union officer was contacted.*

Union acceptance is "deduced" from answers by company representatives to the following questions:‡

10. Does your company have a union?
    Yes ( )
    No ( )
    If "no," skip to top of page 4.
11. What is the name of the union?
12. When the predetermined time system was first introduced in your plant, did the union generally favor the system?
    Union favored ( )
    Union did not favor ( )
13. At the present time, does the union generally favor the system?
    Union favors ( )
    Union does not favor ( )

14a. Are any union personnel trained in the system?
    Yes ( )
    No ( )
    If "yes"

14b. How many union people are trained in the system?

15. Is the predetermined time system part of your union contract?
    Yes ( )
    No ( )

16a. Would you say that the use of the system has generally improved your labor relations?
    Yes ( )
    No ( )
    If "yes"

16b. In what ways has the system improved labor relations?

The answers to the above questions may shed some light on how management *thinks* unions feel about predetermined motion-time systems, but they certainly cannot be relied upon to mirror true union reactions. Any attempt to define union attitudes by researching management attitudes is without foundation.

16.7.3.4 National Unions Against PMTS. Now let us look at union experiences with and attitudes toward predetermined motion-time systems.

The AFL-CIO resolution on industrial engineering [see Art. 16.4] was highly critical of predetermined motion-time systems, calling them "techniques of confusion." When the resolution was presented to the convention delegates, E. C. Hallbeck, president of the United Federation of Postal Clerks, strongly urged its adoption. Hallbeck called the introduction of a predetermined motion-time system in the post offices "a speed-up system designed to wring the last ounce of effort out of an employee." The Postal Clerks, in convention, have since made the elimination of BMT a "paramount issue."§

The International Union of Electrical Workers adopted a resolution, introduced at its 1962 national convention, which among other items stated:

* Honeycutt, "Comments," p. 174.

† Stewart, Dougall and Associates, *A Use Survey . . .*, Table 12.

‡ Stewart, Dougall and Associates, *A Use Survey . . .*, p. 8.

§ "Postal Clerks to Fight Work Measurement," *AFL-CIO News* (Washington, D.C.), Sept. 1, 1962, p. 11.

> "We should oppose all so-called 'predetermined time systems' such as MTM, MTS, Work Factor, etc., as being unfair, unworkable, and not applicable to the establishment of work standards. All impartial industrial-engineering opinion opposes these systems for the establishment of work standards. Our experience with them has been very unsatisfactory.
>
> "Where the system cannot be eliminated or the introduction successfully opposed, we should insist that standards set should be subject to time-study check and adequate protection be provided against abuses."

The United Automobile Workers in a 1961 convention resolution on "The Problem of Production Standards" used the following quote from an arbitrator's award to back the UAW's anti-predetermined-motion-time-system stand:

> "There is no purely scientific system of setting production standards. All who are competent and conversant with the field recognize this fact. All impartial specialists do recognize this fact. Such industrial-engineering procedures as MTM are not pure science; they are at best 'scientific art.' The exercise of human judgment in all methods of standard-setting is inevitable and inherent. Nor is it likely ever to be otherwise because of the nature of the procedures themselves. In addition to the continuing presence of human judgment, there is always the further problem of the degree of variation in the exercise of human judgment. This degree of variation depends fundamentally upon one's philosophy of life, on an individual's general attitudes toward the goals of human existence, on one's conscious or unconscious evaluation of the goals of workers vs. the goals of management, on an individual's level of impartiality, and upon numerous other similar conditioning factors."

One international union leader stated privately that the policy to be followed would be to resist the introduction or use of this "unscientific management tool" in union rate bargaining situations. He advised that under no circumstances should local unions agree to be bound by, or to accept, any of the findings of management that were based on such a system. If management, despite all efforts at discouraging use of such motion-time study systems, insisted on unilaterally using such a system to set standards of piece rates, he counselled that local unions should serve notice that they were reserving the right to protest and bargain "on all cases" without regard to the MTM or any other system.

16.7.3.5 Local Unions Oppose PMTS. The examples above hardly indicate acceptance or approval of predetermined motion-time systems by national unions. Local union policy is not different. Some examples should prove illustrative.

A. The Upholsterers' International Union *Journal* of June 1960 contains a news item which lists "elimination or drastic revision of MTM, a speed-up plan masquerading as a standard-setting incentive-pay system," as a high-priority demand of four Upholsterers' International Union Locals. The article defines MTM as "More Toil, Man!"

B. A local of the Molders and Allied Workers Union in Louisville, Ky., returned to work after an unsuccessful 7-week strike. The main issue was the local's opposition to the introduction of MTM. MTM is now in use in this plant. Taggart and Honeycutt would probably call this "union acceptance."

C. A few years ago a major manufacturer of air-conditioning equipment concluded a first agreement with a local of the Sheet Metal Workers International Association. The local was successful in negotiating the following clause:

> "All time standards are to be set by stop-watch time study or by standard-data or prorated data based on stop-watch study of jobs in the plants of the company covered by this agreement."

While this clause prevents the use of any predetermined motion-time system to set standards on direct workers, the company was able to negotiate the following clause affecting indirect workers:

> "The time standards for indirect labor shall be set or changed according to the same procedures as specified in the direct-labor wage-incentive portion of this agreement, with the following additions: 'Since methods-time measurement (MTM) has been in use prior to negotiation of this agreement, the company may continue to use MTM in conjunction with stop-watch time study for setting time standards for indirect labor groups."

The union reluctantly agreed to the above clause only after an additional clause was negotiated providing for "stop-watch time study of any operation when requested by the union within 60 days after the installation of the standard regardless of whether a written grievance has been filed."

D. In Hartford, Conn., a manufacturer of office equipment attempted to introduce a predetermined motion-time system. The local union of the International Association of Machinists resisted and in the latest agreement negotiated the following clauses:

1. Incentive rates shall be established from either detailed elemental time studies or standard data developed from detailed elemental time studies of operations performed in the Hartford plant unless the parties mutually agree on new systems or methods of setting standards.

2. *No predetermined motion-time system of establishing time standards, such as MTM, Work Factor, BMT, etc., or any other method of setting standards shall be used.*

In the same city another Machinists' local was engaged in protracted negotiations. The main item preventing agreement was the insistence by the local that the company eliminate MTM, which had been installed about 2 years before. Part of a local union release stated:

> "MTM must go.
>
> "Our most serious problem and our greatest dispute are over MTM. Once we rid ourselves, both company and union, of this albatross around our necks, we should be able to conclude a contract that is fair to both parties.
>
> "We realize the company has expended large sums of money to install MTM in the plant. (We believe this money was wasted.) If we as a union, however, can help the company salvage its investment and at the same time get rid of MTM, we will be happy to cooperate for this purpose.
>
> "But MTM must go!
>
> "We open negotiations, therefore, with the express purpose of searching for solutions, but under no conditions will we continue the humbug of setting standards under MTM.
>
> "To the management, we express the wish that they listen no longer to outside hucksters, whose only purpose is to sell a service and reap a profit, at the expense of both company and workers."*

E. In Chicago a national union recently won, by a substantial majority, bargaining rights at the plants of a division of a major electrical-appliance manufacturer. This division had successfully prevented many previous organizing attempts. In fact the same union had been soundly beaten about 3 years earlier.

One of the main issues in the successful organizing drive was the introduction of the corporation's private predetermined motion-time system. The union representative in charge of the campaign claimed this was the deciding factor and elimination of the predetermined motion-time system was a major demand in negotiations for the first agreement.

16.7.3.6 Why PMTS Exists in Union Plants. It cannot be denied that, in spite of the above illustrations, predetermined motion-time systems do exist in organized plants.

It might be well, therefore, to examine how this apparent ambivalence comes about.

A. In many cases the predetermined motion-time system existed before the plant was organized and the union has not been able to eliminate it. In contrast to Taggart's statement, even a "dedicated labor union" may not be able to force a company to discard a predetermined motion-time system any more than a "dedicated" police force can eliminate crime.

B. Frequently, management will anticipate union opposition and devise means to nullify it. These devices usually involve some sort of "package deal" based on "sweeteners." These "sweeteners" often consist of granting to the workers some benefit that the union has demanded and the company has strongly resisted in the past, providing, of course, that the union accept the predetermined motion-time system. Sometimes these "bribes" really amount to the company substituting proper industrial-engineering techniques for historic and improper methods.

In one major U.S. company, for example, two "gimmicks" were used to gain acceptance of its own system of predetermined motion times. In some cases the introduction of a predetermined motion-time system was coupled with changing the method of applying incentive earnings. In several plants of this company, workers received less than a 1 per cent increase in earnings for a 1 per cent increase in production. The unions involved had never been successful in gaining a one-for-one incentive plan.

But when the company wanted to install a predetermined motion-time system, it offered to change at the same time to a one-for-one incentive system.

In some plants of the same company a sizeable portion of workers' earnings was "added on" after incentive earnings were calculated on a low "incentive base rate." In these plants the company offered to include the "add-ons" into the incentive base rate before incentive earnings were calculated, providing, of course, that change in incentive computation be coupled with the installation of a predetermined motion-time system.

It is not surprising that, when inducements such as these are offered, involving large increases in earnings, contingent upon the installation of a predetermined motion-time system, workers in local unions allow installation.

C. Many local unions have become completely

* The intensity of worker feeling about MTM is probably best illustrated by the fact that the workers called MTM "Maynard's Terrible Mess" and characterized those working under MTM standards as "Maynard's Trained Monkeys."

fed up with stop-watch time study and the usual arguments over rating factors. Some of these unions have been victimized by the "predetermined motion-time system eliminates rating" claims of system purveyors.

D. Inconsistent production standards are as much a problem for local unions as for management. The false claim of consistency for predetermined motion-time system standards has led some unions to agree to the installation of a system.

E. Some locals have been induced to "try out" a predetermined motion-time system for a stated time period. Frequently management will attempt to gain acceptance during the trial period by first applying predetermined-motion-time-system analysis to known "tight-rate" jobs. The resulting work standards may allow higher incentive earnings, and thus some union members become "sold" on the "equity" of the new method.

F. Claims of acceptance by other unions have led some locals to allow a predetermined motion-time system to be introduced.

G. The policy of some unions may be stated as follows: "We don't care how management sets standards. Standards may be set with a stop watch, crystal ball, ouija board, or tea leaves. All we are concerned with is the result."

Where this attitude prevails there is no bar to management's installing predetermined motion-time systems, but this hardly signifies acceptance.

H. Unions have, in some instances, been successful in negotiating sufficient safeguards and limitations so that the effect on workers of the application of predetermined motion-time systems is so minimized as to no longer be of major concern to the locals.

I. In a few cases unions have fallen prey to the claims that a predetermined motion-time system is necessary in order for the company to remain "competitive" and maintain employment.

J. Finally, many contracts are so written that management has the right, for the duration of the agreement, to use any system it may desire.

For these and probably other reasons predetermined motion-time systems will probably continue to be installed in organized shops, and the predetermined motion-time system sellers will continue to claim union approval.

16.7.3.7 Union Opposition Documented. But one does not have to go to union sources to document union attitudes toward predetermined motion-time systems.

A recent study investigated the reactions of unions to the introduction of predetermined motion-time systems. The researcher, Dr. Malcolm H. Gotterer, was at the time of the study a vice-president of the MTM Association for Standards and Research. Dr. Gotterer described his study of thirteen predetermined motion-time system installations thus:

> "Intensive interviews were conducted with factory managers, foremen, industrial-relations personnel, industrial-engineering personnel, and in most but not all cases with officers and representatives of the local and international union representing the workers in the company. These interviews were conducted over approximately a 3-month period of time. The companies were selected from lists of clients provided by management-consulting companies that specialize in the installation of predetermined motion-time systems. The basis of selection would, therefore, tend to weight the study toward the experiences of firms which had indicated satisfaction with the systems and the consultant's services. These companies varied in size from 50 to 60,000 employees. Their products were both consumer and industrial goods.*
>
> "With different objectives on the part of the company and the union concerning these newer management techniques, it is often difficult for the company to get the agreement necessary for their use. Frequently, therefore, company officials proceed without such agreement. In some instances the wording of the labor-management agreement may permit the company to use whatever procedure it might desire. The contract may be mute on the point, and the managers feel they have the right to make such changes. Or, finally, the company may be of the opinion that the contract does not specifically prohibit the use of the newer procedures. Therefore, with any one of these conditions present the executives of the company may proceed to use one of the newer techniques without regard to the union attitude or sentiment."†

Gotterer points out that the attempt by management to install a predetermined motion-time system "would almost always result in massive union resistance to the new procedure."‡

The Taggart and Honeycutt claims of union approval seem quite surprising in light of the following Gotterer statements:

> "Labor's distrust of these new synthetic time systems has been widely publicized.§
>
> "Unions have not been secretive about their attitudes toward the new predetermined time systems. Officials of both local and

---

* Gotterer, "Union Reactions . . . ," pp. 435–36.
† Gotterer, "Union Reactions . . . ," p. 434.
‡ "Union Reactions . . . ," p. 434.
§ "Union Reactions . . . ," p. 435.

international unions have written articles and have given talks before both union and management groups in which they have expressed their opposition to these systems. Therefore, when management decides to adopt such a system for establishing production standards, it need not be, and generally is not, ignorant of labor's position. This knowledge, in most of the companies studied, led to actions either to obtain the support of the union or an attempt to neutralize their opposition."*

Some of Gotterer's conclusions follow:

"The use of predetermined times for setting production standards is almost certain to lead to vigorous resistance by organized labor. This reaction is prompted by the union's fear that it will lose the ability to negotiate with management over individual standards, by its disbelief of the claims made regarding the accuracy and scientific premises of these systems, by its doubts concerning the training and ability of the engineers applying these systems, and frequently by the union's general distrust of any new management-sponsored work-measurement program.†

"Management is generally not ignorant of the union's attitudes. But most of the company's actions to gain acceptance of the new techniques are founded on an erroneous assumption that the union resistance is the result of a lack of knowledge by the union leaders and members about the technical aspects of the new procedure. As a result, management's efforts to educate workers and union officials do not result in a decrease in union resistance.

"A brief reflection on the nature and outcome of this labor-management struggle raises the question of whether or not the gains that might be derived from the new procedure are worth the efforts required. The gains being discussed are not the potential advantages of the system, but those benefits that were actually realized after labor and management had resolved their differences and found a mutually satisfactory relationship.

"The gains actually realized in the companies studied were surprisingly few. In no instance was it found that the new work-measurement procedure was-actually being used exclusively for determining production standards. The union resistance resulted in an enormous expenditure of management's time and efforts. Numerous meetings to discuss the system itself or its application took place in almost every company. While no records had been maintained, a significant number of executives commented on the large amount of time expended in this way. But this investment of effort resulted in only a slight difference in procedures. These changes would then appear to offer little advantage to the company."‡

There has been no attempt here to evaluate predetermined motion-time systems nor has there been an attempt to justify or rationalize union attitudes toward predetermined motion-time systems. This article is simply designed to set the record straight on the claims of union "acceptance" or "approval" of these systems.

Stated plainly, unions on the national and local levels have not accepted or approved of predetermined motion-time systems as a work-measurement technique, nor have they accepted predetermined motion time systems as superior to other work-measurement techniques.

Any claim to general union acceptance or approval of predetermined motion-time systems is without foundation.

16.7.3.8 Technical Evaluation of PMTS. In addition to the unfounded claim that predetermined motion-time systems improve industrial relations and are accepted by workers and their unions, PMTS purveyors base their sales efforts on two other general advantages:

1. PMTS are scientific and their use yields universally applicable, accurate, and consistent time standards.
2. The systems work.

A major problem in analyzing these claims is the lack of the kind of data needed to make objective judgments. Too often the claims of predetermined motion-time systems are characterized by what Abruzzi has called "proof by proclamation."

> To be fully effective, this technique of proof by proclamation requires that the proclamation be trumpeted at frequent intervals so that acceptance may be achieved by noise level rather than by verifiable content. Proof by proclamation must decree that results can be verified but only "if properly applied" by "properly trained observers" using "proper methods." It must decree that the results can be stated in multidecimal terms arrived at by arbitrary, whimsical, or shallow methods. It must decree that assumption and hypotheses be mentioned either not at all or phrased in terms of pontifical gibberish. Also, of

* "Union Reactions . . . ," p. 442.

† "Union Reactions . . . ," p. 448.

‡ "Union Reactions . . . ," p. 449.

> course, it must decree that objectives be phrased in terms that can neither be contested nor have any concrete meaning.
>
> The technique of proof by proclamation has been found highly useful in popularizing certain political and other doctrines.*

Unfortunately, the trumpeters of predetermined motion-time systems have found the doctrine equally useful in providing economic and other advantages for themselves. As a result the nontrumpeting segment of industrial-engineering society has suffered.

16.7.3.9 Claim: PMTS Are Scientific. In spite of the lack of objective data, the truth and merit of PMTS claims must be analyzed. Let's begin with Claim 1: Predetermined time systems are "scientific," yielding universally applicable, accurate, and consistent time standards.

While it may possibly be more fashionable today to describe predetermined motion-time systems as systematic rather than as scientific, the word *science* is still used, as by one PMTS seller in describing the development of predetermined motion-time systems.

> And only in the last half century have these techniques developed to the point of being a science, where two engineers given the same data arrive at comparable results.

Predetermined motion-time systems are then described as "this science of time study without a stop watch."†

To enhance the salability of their product, we can expect predetermined-motion-time-systems peddlers to continue to use words like systematic or scientific in spite of the conclusions of such objective researchers as Davidson that "the only tenable position" is that no known predetermined motion-time study has scientific validity.‡

None of the systems are based on scientific investigation by objective researchers. The motion times have been developed from inadequate samples of biased populations. Witness the development of MTM walk times on an oval course laid out on a gymnasium floor§ or the use of University of Michigan athletes in the "research" conducted by the MTM Association.‖

The MTM developers also proclaim, with great pride, that the reach times were developed from "painstaking analysis" of 1350. ft of motion-picture film of thirty-six different drill-press operations.# This sounds like a lot of film, but 1350 ft. represents only 56.4 min, or 1.6 min per operation. In this time, 242 values were obtained for five cases of reach.

16.7.3.10 Basic Assumptions. Even if the systems were based on studies of representative samples of unbiased populations, claims of accuracy and consistency would still be unjustified. The development and use of predetermined motion-time systems are based on a series of assumptions, all of which are open to serious question if not outright rejection. Here are some of these assumptions.

1. It is possible to define with scientific accuracy the motions people use in performing work.
2. The defined motions can be isolated by motion analysis.
3. Times derived from one type of work can be applied with scientific accuracy to all other types of work.
4. The average performance time for one group can be applied with scientific accuracy to an individual or group anywhere.
5. Motions should be confined to the lowest possible classifications in order to reduce fatigue.
6. The time required for a series of consecutive motions is the sum of the time for the individual motions in the series. Each motion is independent of other motions in a group, and the predetermined time for each motion is additive. Positions of motions are of no importance.

---

* Reprinted from Adam Abruzzi, *Work, Workers, and Work Measurement*, p. 3 (copyright © 1956, Columbia University Press) by permission of the publisher.

† Clifford Sellie, *Time Study Without a Stop Watch!* (Chicago: Standards Engineering Co., 1958).

‡ Davidson, "On Balance," p. 162.

§ Harold B. Maynard, G. J. Stegemerten, and John Schwab, *Methods Time Measurement*, (New York: McGraw-Hill Book Company, 1948), p. 105.

‖ Barbara Goodman, "MTM Research on Apply Pressure," *Journal of Methods-Time Measurement*, V, No. 4 (Nov.–Dec., 1958), 8.

Maynard *et al.*, *Methods Time Measurement*, p. 29.

16.7.3.11 "Scientific" Definitions of Motions. When analyzing Assumption 1, one might question the use of the word *scientific*. But the definition of motions in these systems is in terms not only of words but also of time. MTM uses tenths of a TMU to define and differentiate movements. A TMU is defined as one hundred-thousandth of an hour (0.00001), and a Work-Factor time unit is equal to one ten-thousandth of a minute (0.0001).

Whether one uses the word *scientific* to describe these claimed accuracies is unimportant. What is important is that the basic data in these systems were developed by the use of either stop-watch or fixed-speed motion-picture cameras. Stop watches are normally calibrated in hundredths of a minute (0.01). Even with complete accuracy of stop-watch readings (an untenable assumption) no averaging technique could yield accurate results to more than hundredths of a minute. Time values stated in ten- or hundred-thousandths of a minute are ridiculous. Similarly, camera speeds of sixteen frames per second can yield results under optimum conditions to the second decimal but not to four or five. Laymen might be excused for accepting times developed to insignificant decimal places by mathematical manipulation, but the graduate industrial engineer cannot.

16.7.3.12 Isolation of Motions. The second assumption involves the problem of isolating and recognizing the defined motions. Bailey and Presgrave kill this assumption.

> To start with, judgment is required to decide what motions an operator being observed appears to be using. On the surface this question would seem to be strictly a matter of observation rather than judgment. Actually, basic motions occur in such rapid succession in many cases that it is virtually impossible for the observer to see clearly each individual motion, particularly when the work is performed mainly by the fingers and hands. He usually sees groups of motions rather than individual motions as he observes an operator at work. He makes the detailed motion analysis in part from what he observes and in part from his knowledge of the job and what, in his judgment, the operator must do to complete the work cycle properly.
>
> An experienced observer will take into account what might be referred to as "cause and effect." If an operator is considered to be using a particular set of motions at one point in a work cycle, he may have to use another equally definite set of motions later in order to complete the cycle properly.*

16.7.3.13 Universality of Data. The third and fourth assumptions can be combined for our purposes since they both claim the universality of predetermined motion times. Assumption 3 concerns the application of times derived from one type of work to other types of work under a wide variety of environmental conditions. Assumption 4 requires from one population a level of performance developed from another and possibly very different population.

In 1950 Sylvester recognized the falseness of the claims of universality.

> Tables of standard time values can have neither significance nor accuracy unless they represent a combination of the mechanical, human, and conditional components of a situation, which time values are the result of a constant-chance cause system. In other words, the situation to which basic data apply must be an area in which similar machines and equipment, workers of similar skills, strength, abilities, and sex (if relevant), and standardized working conditions are used to perform work within a known range of size, complexity, and accuracy. The simple significance of the foregoing is that basic data cannot be transferred from plant to plant, nor to dissimilar classes of products, nor to different groups of workers unless checked and corrected for changed values of the human and conditional components. Even the mechanical component, as reflected in the summation of modal times, may be altered if the basic data are transferred from one situation to another. An important change in posture or other physical conditions can raise or lower the modal values. And, also, the modal values can be affected by different habitual rhythms.†

These different rhythms may be caused by variations in lot sizes, in quality and tolerance requirements, in equipment, in layout, and in environmental and many other factors. So much for "universality."

16.7.3.14 Low-Order Motions and Fatigue. The fifth assumption is that motions should be

---

* From *Basic Motion Time Study* by Gerald B. Bailey and Ralph Presgrave. Copyright © 1958, McGraw-Hill Book Company, New York, pp.16–17. Used by permission.

† From *Advanced Time-Motion Study*, 1950, by Arthur Sylvester (New York: Funk & Wagnalls Company, Inc.), p. 221.

confined to the lowest possible classifications in order to reduce fatigue and thus to increase productivity. You all know this claim. It appears, for example, in Karger and Bayha's *Engineered Work Measurement* as the second law of motion economy: "Motions should be confined to the lowest order possible." The lowest order is equated by these authors with that motion which carries the least time value.*

In 1954, Jennings commented on this problem in *The Journal of Industrial Engineering*. Another aspect of the assumption that there is an elemental motion with a universal time standard is the statement that "motion should be confined to the lowest possible classifications in order to reduce fatigue." In effect, of the five classifications of motions, finger motions are most preferable. One study investigating the motions involved in the simple task of tapping found wide difference in ability to tap ranging from 1.5 to 5 taps per second. However, the highest tapping rate is obtained when most of the arm is used either completely freely or at least with the whole arm from the elbow. In effect, people cannot usually tap as fast when they use only wrist motions, and tapping is worse when only the fingers are used. This study also sheds light on the existence of an elemental motion in that it reveals that the tapping rate is not particularly stable. It can be changed by noise and attitude influence. All of these conditions are present in variable amounts in the actual work situation.†

The now-famous Schmidtke-Stier evaluation of predetermined motion-time systems appeared in the *Journal* in 1961. The conclusion of these independent researchers was that both Work Factor and MTM yielded working times so low as to make excessive physiological demands on operators—demands which could be injurious to health.‡ It is, of course, not surprising that the effects of fatigue on body members do not appear in time values resulting from studies of an almost instantaneous nature.

16.7.3.15 Independence vs. Interdependence. The sixth assumption is probably the one which has had the most attention from predetermined motion-time systems critics: The time required by a series of consecutive motions is the sum of the time for the individual motions in the series. This assumption is based on the concepts that each motion is independent of other motions in a group of motions and that the predetermined time for each motion is additive. The sum of the times for each motion is the "proper" time for the group of motions constituting an industrial operation.

The assumption here is that the position of a motion in a series is of no importance. But this is not the case. Reported research overwhelmingly proves that motions are interdependent rather than independent. The time required to perform any given motions or therbligs is influenced by preceding and succeeding motions or therbligs. In the *Journal* article by Jennings and in research reported by Barnes and Mundel, by Karl Smith and Associates, by Nadler and Associates, by Abruzzi, and by Schmidtke and Stier, the interdependence of motion times is proven.§

In the *Journal of Methods-Time Measurement*, the official MTM Association publication, J. Varkevisser, a member of the MTM Association of The Netherlands, reported on his research in an article titled "The Relation Between Motion Pattern and Performance Time of Therbligs." Varkevisser first reviewed the various interdependence criticisms of MTM

---

* Delmar Karger and Franklin Bayha, *Engineered Work Measurement* (New York: Industrial Press, 1957), p. 511.

† Eugene Jennings, "The Validity of Basic Assumptions in Motion and Time Study," pp. 16–18. Reprinted from Vol. V, No. 2, March 1954. *The Journal of Industrial Engineering* official publication of the American Institute of Industrial Engineers, Inc., 345 East 47th Street, New York, New York 10017.

‡ Schmidtke and Stier, "An Experimental Evaluation . . . ," p. 204.

§ Jennings, "The Validity of Basic Assumptions . . . ." Ralph Barnes and Marvin Mundel, "Studies of Hand Motions and Rhythm Appearing in Factory Work," Bulletin No. 12, *Studies in Engineering*, University of Iowa, Iowa City (1938). (See also Bulletin Nos. 17 and 22.) Donald Hecker, Donovan Green, and Karl U. Smith, "Dimensional Analysis of Motion: X. Experimental Evaluation of a Time Study Problem," *The Journal of Applied Psychology*. **40**, No. 4 (1956), 220–27. (Nos. I through IX of the series "Dimensional Analysis of Motion" appear in *The Journal of Applied Psychology* from 1951 through 1956.) Gerald Nadler and J. W. Wilkes, "Studies in Relationships of Therbligs," *Advanced Management*, **18**, No. 2 (February 1953), 20–22. Abruzzi, *Work, Workers, and Work Measurement*, p. 203. Schmidtke and Stier, "An Experimental Evaluation . . . ," pp. 182–204.

and then reported on experiments he performed "to check whether the critical statements were right when a reasonable work pace was maintained." He concluded that "there is without any doubt a relation between the time needed to perform the reaches and the places that these reaches occupy in the cycle." In Varkevisser's study also, the time for performing the simple task of pushing a button varied on the average by more than 10 per cent, and MTM analysis of the operation produced a result significantly less than the measured times.* In other words, the MTM concept of normal is faster than MTM practitioner Varkevisser's reasonable work pace.

But predetermined-motion-time-systems peddlers die hard. Bailey, in his comments in 1961 on the Schmidtke-Stier research report, still claimed that proper research would prove the independence of motion time.† And Honeycutt, research director of the Maynard Foundation, insulted every industrial engineer with his *Journal* "Comments" when he "proved" that MTM motion times are additive because one gets "five pennies change for a nickel. Twelve eggs still make a dozen, and 5,280 ft still make a mile."‡

The claims that predetermined motion-time systems are scientific and that their use yields universally applicable, accurate, and consistent time standards were probably best debunked by Gotterer, who stated: "Any management considering the installation of a predetermined motion-time system should realize that any claims made concerning their scientific foundations are unverified."§ When he wrote this, Gotterer was a vice-president of the MTM Association.

16.7.3.16 Do Predetermined Motion-Time Systems Work? Does all this evidence worry the predetermined-motion-time-system purveyors? Definitely not! After all, why worry about these criticisms when the systems "work"? If some naive, prospective client should be so bold as to raise any objections, simply reassure him, as does one leading PMTS company, that objections to the assumptions of PMTS are not provable since they are based on reasoning theory and laboratory research. If such an answer is unsatisfactory to the client, the next step in the hard sell is to claim that since few of those conducting the research have had training approved by the PMTS seller, the results of their research must be suspected.‖ This claim is made even though the researchers referred to are in most cases leading professors of industrial engineering from major universities, many of whom are heads of industrial-engineering departments. After all, the systems do work.

Is there merit to this claim? Do the systems, in fact, work? The answer is yes! Of course, they work! If one is willing to accept rather loose approximations, the systems may have utility when used with discretion for estimating times of future task performance, for comparing alternative methods, for examining equipment-purchase proposals, for estimating manpower and space requirements, for developing tentative layouts, and for other general uses. This is the way Nadler explained the working of the systems.

> The systems *can* determine a "standard" time. The accuracy of the resultant standard is what is in question. They are not any more accurate or consistent, and frequently less accurate and consistent, at the present time than any other work-load determination procedure.

It is precisely in the area of work-load determination that the conflict appears. The systems should be used neither for setting standards which workers must meet as a condition of employment nor as the basis for an incentive system. For these uses, the systems do not work.

Abruzzi had this to say about the systems.

> The "It Works" Criterion. The most persistent claim made for each kind of system is that "it works." This claim is neither surprising nor convincing. The same claim is made with respect to standards derived from ordinary time studies; it is invalid in both cases, primarily because "it works" means only that there is a certain measure of acceptance. For example, admittedly com-

---

* "The Relation Between Motion Patterns on Performance Time of Therbligs," *Journal of Methods-Time Measurement*, V, No. 6 (March–April 1958), 10–11.

† Bailey, "Comments," p. 329.

‡ Honeycutt, "Comments," p. 176.

§ Malcolm H. Gotterer, "Predetermined Time Systems: Their Role and Function in Manufacturing Companies" (unpublished D.B.A. thesis, Harvard University Graduate School of Business Administration, 1959), p. 12.

‖ From *Motion and Time Study* by Gerald Nadler. Copyright © 1955, McGraw-Hill Book Company, New York, p. 507. Used by Permission.

petitive and even contradictory motion systems can coexist only in a framework where acceptance may be granted or withheld on the basis of the results obtained in local application. This is certainly not the same thing as having validity from an estimation standpoint.

In some cases the "it works" criterion takes the form of "test" verification. The "tests" consist of comparing standards developed from some standard-data system to standards developed from direct time studies. This is a completely meaningless activity.*

16.7.3.17 Future Acceptance of PMTS. Before unions begin to seriously evaluate predetermined motion-time systems from a collective-bargaining standpoint, they will have to be convinced that technically the systems have been proven by properly designed research under objective sponsorship. The likelihood of obtaining such proof is remote. While some excellent work has been accomplished at individual universities, the disjointed, fragmented, uncommunicated nature of university industrial-engineering research to date has greatly distracted from its usefulness.

Research by organizations committed to widening acceptance of a particular brand of predetermined motion-time system (such as the MTM Association for Standards and Research) can hardly be considered objective by unions. The economic opportunism of the peddlers of predetermined motion-time systems if continued will guarantee continued union objection.

MTM purveyors describe their system as "being the preeminent predetermined time system for work study and measurement" and as being "preeminent among predetermined time systems."† Work-Factor salesmen obviously do not agree with MTM preeminence since they claim that "Work Factor is now acknowledged as the most widely used and accepted system . . . ."‡ Work Factor is also proclaimed as being technically superior to other methods. In a brochure advertising a programed Work Factor self-instruction course, the Work Factor technique is described as being a simple, easy-to-learn method of motion analysis which eliminates the judgment inaccuracies of stopwatch time study methods, as well as the "complicated, often inconsistent" techniques of MTM, BMT, and other such predetermined motion-time study methods. The General Electric brand of predetermined motion-time system, Motion-Time Survey, is described as "having more desirable features" than other systems have. § Thus, PMTS purveyors should not be shocked when they read the following statement.

> The jungle warfare between peddlers of various predetermined time systems continues unabated. If anything, the fracas is getting noisier. Each of the partisans seems to be saying, in effect, that his system is very similar to all others—only better. How much of this is professional conviction and how much is opportunism is not always clear.‖

These predetermined-motion-time-systems backers should not be surprised therefore if workers and unions prefer to stay out of the predetermined-motion-time-systems jungle.

16.7.4 Wage-Incentive Plans#

Probably no technique has been the subject of so much recent debate in the public press and in management, personnel, psychology, sociology, and industrial-engineering journals as have wage-incentive plans. This public scrutiny is only natural since rapid technological change and the advent of the automated workplace obviously reduce or eliminate the ability of workers to influence rates of production.

Unions have long recognized the disadvantages of wage incentives, and with few exceptions, they have been opposed to wage-incentive systems, both because of past experience with the abuses under such plans and because of the difficulties and ill effects inherent in them.

---

* Reprinted from Adam Abruzzi, *Work, Workers, and Work Measurement*, p. 199 (copyright © 1956, Columbia University Press), by permission of the publisher.

† Karger and Bayer, *Engineered Work Measurement*, pp. 45–68.

‡ Joseph H. Quick, James H. Duncan, and James A. Malcolm, Jr., *Work-Factor Time Standards* (New York: McGraw-Hill Book Company, 1962), p. vii.

§ *Motion Time Survey* (Schenectedy, N.Y.: Manufacturing Services Division, General Electric, 1954), p. 1.

‖ Roger Christian, "Work Measurement Today," pp. 123–28. Reprinted by special permission of *Factory*, September 1963. Copyright by McGraw-Hill, Inc.

Much of the material in this section is from "Collective Bargaining Report," *AFL-CIO Federationist* (Dec. 1957 and Nov. 1960), pp. 73–78 and 65–72, respectively.

Unions which have accepted them or permitted them to continue have usually done so only with reluctance and misgivings. It simply has not always been practical or expedient to actively oppose or to eliminate such plans. A few unions, primarily in the rubber and needle-trade industries, where wage incentives are most firmly entrenched, have at least temporarily accepted incentives as part of their collective-bargaining programs.

Many industrial engineers agree that wage incentives have been abused in the past but maintain that this is no longer true. In fact, although some of the worst abuses have been toned down, primarily through union action, a variety of ill effects and strains on workers are still very much in evidence.

An incentive system invariably brings special problems in the education, representation, and protection of workers. It puts a strain on the entire collective-bargaining process, making it more difficult, more complex, and more costly.

The value of any such plan is also highly questionable because, even though it may initially yield increased earnings, it inevitably requires a speed-up of work efforts, creates friction among workers, and produces continual wrangling over production standards.

16.7.4.1 Nature of Incentive Plans. Employers install incentive plans because they expect them to lead to higher profits through reduced production costs. Essentially, these plans all attempt to induce workers to produce more than a fair day's work by promising a monetary reward. They are also based on the notion that workers do not perform an honest day's work unless they are bribed by the promise of extra money.

Management attempts to sell these plans to workers by claiming and emphasizing that an incentive system will result in earnings higher than straight hourly wages do.

While it may be true that, in theory, management and labor can both gain via the incentive route, in practice, over any extended period of time, they seldom do. And labor has found that the disadvantages normally outweigh the advantages.

16.7.4.2 Value to Management. Management may not gain enough from incentive plans, when all things are considered, to make them worthwhile.

Some costs, usually called *factory overhead* or *factory burden*, remain relatively constant over a fairly large range of production. They include such items as executive and supervisory salaries, certain taxes, and machinery and building costs. Naturally, as workers increase production, these fixed costs are allocated to more units. The resulting decrease in fixed cost per unit produced can result in substantial savings to management. These savings appeal to managements and cause them to install incentive plans.

But once incentives are installed, management discovers many offsetting increases in cost. Research indicates that, in many cases, the costs of setting up time-study and wage-incentive departments, coupled with such typical by-products of incentive plans as inferior quality, customer discontent, low employee morale, grievances, and increased accident-insurance rates, more than offset any savings.

Studies of incentive plans at 416 companies indicate that some managements recognize these incentive shortcomings. Of 100 companies surveyed in one study, 40 per cent acknowledged that their incentive plans were not satisfactory. Another 17 per cent said they were only partially satisfied; their comment was that incentive benefits fell short of justifying the bonuses paid to workers.* In another study of the experience of 316 companies, covering a 15-year period, 78 per cent of their wage-incentive plans either had failed or had developed such major weaknesses that the companies were completely dissatisfied.†

Some companies have expressly stated that efficient supervision and reasonable labor-management relations can result in more efficient and less expensive production than can reliance on an incentive system.

16.7.4.3 Shortcomings of Incentive Plans. The many different plans vary in detail and have different descriptions but are fundamentally the same. The reasonableness or practicality of any plan cannot be determined merely by examining a written description of the way it is supposed to work. Some of the basic short-

* "Is Your Incentive Plan Headed for Success—Or Failure?" *Factory Management and Maintenance* (May 1955), pp. 128–30.

† Bruce Payne, "Incentives That Work," in *Proceedings, Annual Fall Conference* (New York: Society for Advancement of Management, 1951), pp. 23–28.

comings of virtually all plans are discussed below.

Incentive plans require the setting of production standards, usually on the basis of time study, with all its inconsistencies, inaccuracies, and unreliability. The union may negotiate certain basic wage guarantees, but earnings under an incentive plan depend on time standards or rates set by time-study men. Because time-study men cannot set standards accurately or consistently, earnings bear little if any relation to workers' efforts. Some workers are assigned to jobs having loose rates, where exceptionally high earnings can be made. On other jobs rates are tight, so that equal or greater effort does not yield so much in incentive earnings as it does on jobs with looser rates. Frequently workers on jobs with tight, unrealistic production standards must work at a killing pace in order to just make standard and earn base rates.

The supervisor who allocates work has a tremendous weapon in his hands when workers are on incentive. He can easily give his friends and informers the gravy jobs while giving the trade unionist the tight ones. On the other hand, some managements may hope to influence shop stewards or union officers by allotting them jobs with loose rates. In any case, wage incentives lend themselves to abuse. Workers with similar jobs and pay grades frequently have large differences in earnings. The opportunities for discrimination are obvious.

16.7.4.4 Rate-Cutting and Speed-Up. There is also considerable room for jockeying with standards as jobs gradually change. Piece rates or job standards which may be fair when they are set often become loose as workers gain experience, improve skills, and develop new short-cut work methods, and workers' earnings rise accordingly. Although management usually assures workers that standards will not change simply because earnings go up, workers often find these promises to be empty. Management frequently seeks some excuse to tighten the rate on jobs where earnings increase. Usually these excuses take the form of changes in methods, materials, etc. Frequently the very improvement made by a worker to increase his own earnings becomes management's excuse for cutting the rate!

Speed-up is easily accomplished throughout a plant in the same manner. A periodic tightening up of rates, under the guise of methods improvements, forces workers to work harder and faster than previously in order to maintain earnings.

Another shortcoming is a different type of steady pressure on workers to work harder. According to the theory of incentives, workers determine their own work pace. They are supposed to be free to work at a faster than normal pace or not as they see fit. But again theory and practice are different. Workers who produce less than a predetermined amount, usually an average, are accused of loafing and goaded to greater efforts. Management tries continually to raise production by eliminating below-average workers. Thus workers are forced to work faster in order to keep up with the always rising speed of the average worker. Older workers, particularly, find such competition difficult.

All workers are also faced with the fact that the risk of accidents is increased when work pace is stepped up.

16.7.4.5 Divisive Effect on Union. Approximately 25 per cent of industrial-plant workers are covered by some form of wage-incentive plan. The largest proportion of these workers are on an individual rather than on a group incentive basis; their earnings are influenced primarily by their individual production. But, when workers are paid according to their individual efforts, the union's function of securing high guaranteed wages for all workers becomes difficult. The ability of the local union to present a unified position for base-rate increases is weakened.

Incentive systems often account for serious disputes among members of the same union. Workers in one job or department may have higher earnings than workers in another job or department even though they all have the same base rates. The differences in earnings may have little if any relation to differences in work pace or production but may be due simply to the inconsistencies of time study or to other causes unrelated to work effort.

16.7.4.6 Group-Plan Pressure on Workers. Some incentive plans tie wages to the production not of the individual worker but of a group of workers. Incentive earnings are given to all in the group on the basis of the group's total output. The group may be small or large, involving as few as two or as many as several hundred or even thousands of workers.

Group incentive systems can be especially

vicious since one worker's earnings are made to depend (at least partially) on another's work. No matter how hard someone may work, factors beyond his control or beyond the control of the group may limit total output and limit his incentive earnings.

These systems easily degenerate, causing workers to assume management's functions. Management plays workers against each other by trying to get them to police each other. Management may expect one union member to pressure another who may have caused group earnings to decrease because of absenteeism, lateness, etc. In addition, young workers become impatient with older workers who cannot keep up.

16.7.4.7 Incentive vs. Nonincentive Workers. Incentive systems may provide higher earnings to a semiskilled group than to skilled workers or may create disproportionate differences between incentive and nonincentive workers. Such situations can contribute to a serious breakdown of worker morale, with the wage-incentive systems pitting one group of union members against another. The problem can be met somewhat by providing for additional payments to nonincentive workers. Often such payments are determined by the level of productivity of the incentive workers.

16.7.4.8 Weakening Effect on Union. Wage incentives deemphasize the role of the union in securing higher wages. Combined with the creation of friction among workers and the development of conflicting factions, this lessened role may threaten the existence of the union. Its ability to secure decent contractual protection in other areas, such as grievance procedures, seniority, and vacations, may be placed in jeopardy.

Serious difficulties also arise in properly and adequately protecting and representing workers under these systems. Their complicated nature as well as their dependence on time study and other work-measurement techniques make it difficult for local union officers to handle day-to-day problems.

International and business representatives usually have to be called in to spend a disproportionate amount of time on incentive shops, as compared to day-work shops, with a resulting drain on time needed for other grievance problems, for preparation for negotiations, for organizing, and for other union activities.

16.7.4.9 Elimination of Incentive Plans. The dangers and inequities of wage-incentive systems and their possible antiunion effects have led many unions to actively attempt to eliminate them at every opportunity. Many unions have been successful in bargaining incentive plans out of contracts where they previously existed, and others have succeeded in persuading management not to install new plans.

Workers often feel so strongly about the evils of incentive systems that they are willing to strike rather than to permit incentives to be installed or continued unreasonably. Thus, members of Local 375 of the International Brotherhood of Pulp, Sulphite and Paper Mill Workers struck at the National Container Corp. of Bristol, Pa., when the company attempted unilaterally to install an incentive system. The company had refused to sign an agreement unless the union agreed to its incentive plan. It took an 11-week strike to convince the company that these members would not work under an incentive scheme.

There is one bright spot on the incentive horizon. Most experts agree that incentive systems may be on the way out because of technological improvements and automation. As production becomes more automatic, workers have less control over output. The supposed gains to management of incentive systems decrease, and thus presumably many of these plans will be eliminated.

Some companies lacking new automatic equipment may try to cut costs through incentive systems in order to compete with their more mechanized or automated competitors. Workers in such companies will be forced to work faster and harder, but they will not be able to win a race with automatic equipment. Workers cannot produce enough to save incompetent or inefficient management.

Unions desiring to eliminate incentives must be careful that they do not trade one set of inequities for another. Some companies have agreed to do away with incentives providing they can substitute measured day work. In other words, management sets standards of production (usually by time study) which workers must meet. The workers receive straight hourly earnings with no incentive pay. Frequently the standards set require workers to work at an incentive pace for nonincentive pay. Management can thus secure the equivalent of incentive production without paying for it. When incentives are eliminated, unions should insist that, if any production standards are to be required on the nonincentive operations, such standards be developed through direct bargaining with the union. The union must also concentrate on the level of the all-important new hourly pay scales. With or without incentives, guaranteed hourly wages must be negotiated high enough to yield a fair day's pay to workers.

16.7.4.10 Protection of Workers Under Incentive Plans. It is impossible to build an incentive system that is fair to all workers or one that does not create serious problems for unions. If a union must accept a wage-incentive plan, it should insist as strongly as it can on certain principles and practices to minimize arbitrary and abusive applications of such a plan. The refusal of a company to accept and observe these principles must be recognized as a sign that it is seeking to manipulate the plan unfairly to its own advantage at the expense of its workers.

1. *There must be adequate, guaranteed, minimum hourly payments to all incentive workers.* Workers should be assured of certain reasonable minimum earnings regardless of the amount they may produce in any particular period. Such an hourly guarantee is normally referred to as the *guaranteed base rate* in an incentive program.

2. *The guaranteed base rates on which incentives are built must be realistic.* Frequently, workers and unions are lulled into a false sense of security by periods of high incentive earnings. They neglect to insist that the guaranteed base rates be kept up to date. Management may deliberately allow a loosening up of incentive systems in order to keep from raising guaranteed base rates. Workers become more dependent on incentive earnings as the guaranteed base rate lags farther behind.

3. *Guaranteed base rates should be the same as nonincentive hourly rates.* A realistic incentive base rate is one which is at least equal to the hourly rate for nonincentive workers on similar jobs in the industry or area.

4. *To apply general wage increases properly to incentive workers, they should be added to the base rates.* Unions sometimes agree on smaller general wage increases to be added to the base rates of incentive workers than those provided for day workers. This differential increase is designed to give the same increase in gross pay to both groups. This plan is unfair to the incentive workers, however, since they must work at an incentive pace in order to earn the same increases as the hourly worker does, working at a nonincentive pace. When increases are negotiated, they should appy equally to incentive base rates and nonincentive hourly rates to ensure that workers get equal pay for equal work.

General increases should not be applied as "add-ons" or "side payments," under which negotiated increases are added to wages after incentive earnings are figured at old base rates. For example, assume that in 1960 a worker was on an incentive job that had a base rate of $1 per hour and a production standard of 100 pieces per hour—he received 1¢ per piece. If the worker produced 150 pieces in an hour, he received $1.50 for that hour.

Assume that by 1967 negotiated increases totaled 50¢. Applied properly—that is, put on the base rate—this increase would raise the base rate to $1.50 per hour, or to 1½¢ per piece. For production of 150 pieces, the worker would now earn $2.25 per hour, receiving the same amount of money for each piece he produced.

Many companies take advantage of workers, however, by applying the negotiated increases, not to base rates, but as payments on the side, as add-ons. Under the add-on method, the more a worker produces, the less he receives per piece. Using the figures in the example above and assuming that general increases are not added to the base rate but are paid as an add-on, we find that a worker who produces only the standard or normal requirement of the job, 100 pieces, receives $1 per hour base rate plus the 50¢ general increase add-on for a total of $1.50 or 1½¢ per piece. If he produces 150 pieces, he is paid the $1 base rate, plus 50¢ incentive, plus the 50¢ "add-on," or $2. These earnings of $2 for the 150 pieces mean only 1⅓¢ per piece (as against the 1½¢ per piece if the 50¢ add-on were in the base rate). As he increases production, his pay per piece decreases, not only for the pieces above the standard, or for the increased production, but also for every piece he produces.

5. *Incentive plans must provide a realistic opportunity for earnings above hourly rates.* Any incentive system which does not provide an opportunity to all workers to earn reasonable extra earnings is unfair. There must be no artificial barriers to incentive earnings and no ceiling on earnings.

6. *Incentive payments should increase at least in direct proportion to production.* Workers' earnings should never vary proportionately less

than output. In fact, since it is always more difficult to produce extra units and since cost reductions are greater as production is increased, payments to workers should really be increased more than output.

7. *Incentive earnings should not be averaged for the purpose of reducing total earnings.* Incentive earnings should never be calculated for periods longer than 1 day. Days of high earnings should not be lowered by poor days. In no case should below-standard production be used to equalize above-standard production. When workers normally work on short-run jobs, their earnings should be calculated by the job. A job with a fair rate should not be used to cover up or equalize a job with a tight rate. This can result only in reduced earnings for the worker.

For example, assume a worker during one day is assigned to two different jobs, both having a production standard of 100 pieces per hour. He works on Job *A* for 4 hr. and produces 600 pieces, or 150 pieces an hour. Working on Job *B* for the remaining 4 hr. and at the same pace as on Job *A*, he is only able to produce 320 pieces, or eighty pieces an hour. In other words, on Job *A* he has produced at the rate of 150 per cent of standard and on Job *B* at the rate of 80 per cent, even though he worked at the same pace on both jobs. Obviously the fault is with the tight or unfair standard on Job *B*. It would be unfair to reduce this worker's incentive earnings on Job *A* by tying them to his low Job *B* results, as some companies do. This worker should be paid at least a reasonably adequate base rate for the 4 hr. he worked on Job *B*.

Assuming his base rate is $1.60 per hour, his earnings should be calculated as follows:

4 hr. × $1.60 base rate × 150 per cent = $9.60
4 hr. × $1.60 base rate × 100 per cent = 6.40

Pay = $16.00

The inequitable method used by some companies would be as follows:

4 hr. × $1.60 base rate × 150 per cent = $9.60
4 hr. × $1.60 base rate × 80 per cent = 5.12

Pay = $14.72

8. *Changed standards should provide the same earning opportunity as original standards did.* One way of limiting unreasonable management tampering with production standards is to provide that, when a legitimate change in methods or equipment occurs, only the time for the changed portion of the job should be adjusted and only in the amount of the change. Management should not be permitted to use a minor change in a job as an excuse for revising the production standard for the entire job.

The new standard must allow the operator to earn at least as much as before. Changes in earning opportunities should be made, if at all, by mutual agreement through collective bargaining. They should never be determined by time study.

9. *Fair payments must be made to incentive operators assigned to nonincentive work.* In most such cases, the base rate is not enough. There may be some justification for paying the base rate when no incentive work is available for short periods for causes beyond the control of the company. Base rates may also be appropriate for limited down-time or unavoidable delays, but only when base rates are relatively high and reasonable.

Workers on an incentive system should be assured that their usual incentive earnings will not be cut if they are temporarily shifted to nonincentive work. The most equitable arrangement is to pay prior average hourly earnings, including incentive pay, or the hourly rate on the nonincentive job, whichever is higher.

10. *The whole incentive system must be easily understood by workers and their representatives.* No system should be applied which can be understood by only engineers and mathematicians. Workers must be able (1) to understand how the standards of production set by the company were arrived at and (2) to calculate easily and quickly their own pay.

Many managements have attempted to prove the fairness of an incentive system by the absence of incentive grievances. Unfortunately their absence is often due to the complicated nature of the incentive plan. Workers cannot intelligently process a grievance about something they cannot understand.

11. The final point is most important. *Management must permit every phase of its incentive system to be reviewed by the union through collective-bargaining procedures.* Unions, of course, have the legal right to bargain with management on anything which affects the earnings of union members. Unions must be furnished with all wage-incentive information that the company has, as well as with information on the methods, such as time study, upon which incentives are based.

A refusal by management to give a union requested wage-incentive or time-study data, a refusal to discuss individual and general problems associated with incentives, or a refusal to resolve incentive complaints through the grievance procedure (including arbitration, if necessary) indicates that management has something to hide, that it is not dealing in good faith with the union, and that the plan involved is hardly likely to be beneficial to the union members.

16.7.4.11 Extent of Incentive Pay. The majority of American workers have always been paid on a time basis rather than on a piece-rate or other wage-incentive system. Various management groups have sought to promote the use of incentive systems, but wage incentives are not spreading. The proportion of workers on an incentive-pay basis is smaller now than it was in the post-World War II years. Signs also indicate the trend away from piecework or other incentive methods toward paying wholly on an hourly, weekly, or other time basis.

Worker distaste for incentive schemes and resulting union opposition account for the failure of incentive plans to spread and for their gradual decline. But many managements, too, have had reservations about or have turned against incentive methods, either because they are reluctant to take on the problems associated with use of incentives or because they have had direct experience which has persuaded them to eliminate or to avoid incentive plans.

How many workers are paid on an incentive basis—that is, have their wages tied directly to their output? The proportion varies by industry, but in manufacturing as a whole, three-quarters of production workers are now paid on a time basis, and only about one-quarter are paid on an incentive basis. In nonmanufacturing, the proportion on incentives is smaller. Although precise data are not available, it is probable that no more than 10 per cent are under incentive-pay systems (even if commission systems for sales employees are considered incentive systems). Such industries as transportation, communications, electric and gas, construction, warehousing, and most services make little or virtually no use of incentive-pay arrangements.

Thus, incentive systems are a minority practice, even though many management-consulting firms actively sell incentive plans and have distributed a steady flood of literature on the need for wage incentives and their proper administration. Moreover, it appears that the proportion of workers on incentive systems is gradually decreasing for a combination of reasons discussed later.

The principal growth period of incentives was during World War II. With war-time wage control, unions frequently allowed or cooperated in the installation of incentive programs as a means of obtaining increased wages. Government agencies also encouraged incentive programs in the belief that they would contribute to needed increases in war-time production.

The U.S. Department of Labor had, from time to time, compiled some information on the extent of incentive pay in individual industries or areas, but its studies did not provide an overall picture. In 1960, however, the Bureau of Labor Statistics concluded the first comprehensive, authoritative survey of incentive coverage among production workers in all manufacturing. It reported that, "of the 11¼ million production and related workers employed in the nation's manufacturing industries in May 1958, 27 per cent were paid on an incentive basis."*

The study, based on a survey of all factory workers' earnings in May 1958 plus separate, more detailed surveys of individual industries, provides separate data for each of seventy-three industries. The industries with the largest proportion of workers on incentives are the men's and boys' apparel industries and the leather footwear industry, in each of which approximately 70 per cent of production workers are covered by incentive-pay systems. On the low side, fewer than 10 per cent are under incentives in such large manufacturing industries as aircraft and aircraft parts, bakery products, beverages, printing, industrial chemicals, sawmilling, and millwork. Here is a tabulation of the seventy-three industries.

Percentage of Production Workers on Incentive Pay	Number of Industries
Under 10 per cent	10
10 to 24 per cent	29
25 to 49 per cent	23
50 per cent or more	11

In several respects, the Department of Labor

---

* U.S. Department of Labor, "Extent of Incentive Pay in Manufacturing," *Monthly Labor Review* (May 1960), pp. 460–63.

figures tend to overstate the number of workers on an incentive basis. For some industries, the surveys used an employment cut-off of twenty or more workers. Since incentive-pay plans tend to be more common among the larger establishments, the prevalence figures are in some cases slightly larger than they would have been if all establishments were surveyed. Also, the figures include as incentive workers some who are paid incentive bonuses even though their personal output has little relation to their wages. This figure includes indirect service or maintenance workers paid a bonus to correspond to that paid production workers. It also includes workers paid a bonus based on overall plant production, such as total tonnage produced. While workers on a plant bonus system are paid additional incentive wages if total production exceeds certain levels, many of them have little if any influence on total production, for that is usually determined in such plants by management decisions ("We will produce 44 tons today") rather than by worker decision on pace or effort.

16.7.4.12 Geographic Prevalence. The Department of Labor studies show some notable geographic differences in the prevalence of incentive systems. Incentive pay ordinarily is considerably less common in the Pacific region and somewhat less frequent in the South than in other regions, with some Midwestern cities having the highest frequency of incentive workers. In the footwear industry, for example, 70 per cent or more of the workers were on incentives in every region except the Pacific, where the percentage was 38 per cent. In the leather-tanning and -finishing industry, over 40 per cent of the workers in each region were on incentives except for the Southeast, where the proportion was 30 per cent, and the Pacific, where the proportion was 13 per cent.

In its surveys of seventeen large cities in 1957–58, the Department of Labor checked the percentages of plant (nonoffice) workers paid on an incentive basis. Its findings are listed in Table 16.2.

TABLE 16.2. *PERCENTAGE OF PLANT WORKERS ON INCENTIVE PAY**

City	Manu-facturing	Nonmanu-facturing
Northeast		
Boston	36 per cent	14 per cent
Newark-Jersey City	29	13
New York City	24	7
Philadelphia	34	14
South		
Atlanta	23	17
Baltimore	30	10
Memphis	24	20
New Orleans	15	13
North-Central		
Chicago	33	11
Cleveland	30	6
Milwaukee	43	29
Minneapolis-St. Paul	20	11
St. Louis	29	9
West		
Denver	20	12
Los Angeles	13	5
Portland, Ore.	10	12
San Francisco-Oakland	7	11
Weighted average, all cities	27 per cent	11 per cent

* These percentages are overstated somewhat because (1) the surveys covered only plants over a specified size, usually 50 or 100 employees, and incentives are less common in the smaller firms not included in the surveys, and (2) the surveys did not include construction, railroads, and government operations, in each of which incentives are rather rare.

16.7.4.13 Trend of Incentive Coverage. What is the trend in the prevalence of incentives? The Department of Labor compared its finding of 27 per cent of manufacturing production workers on incentives in May 1958 with an estimate it made in 1945–46 of 30 per cent. However the 1945–46 data covered only part of manufacturing. Narrowing its comparison to the same industries, the Department stated that the proportion on incentive in 1958 was fairly unchanged from that in 1945–46.

Several surveys, however, point out a definite decline in the number of workers covered by incentive systems since a peak point sometime after 1945–46.

In 1964 the Bureau of Labor Statistics published an analysis of wage-payment plans based on its occupation wage surveys for the period July 1961 through June 1963.* This study showed that only 20 per cent of plant workers were paid by incentive methods. In manufactur-

* John H. Cox, "Wage Payment Plans in Metropolitan Areas," in *Monthly Labor Review* (Washington, D. C.: U.S. Department of Labor, July 1964), pp. 794–96.

ing alone the percentage on incentives was 26, or one percentage point less than that in the 1960 report. A decline of one point represents an approximate 4 per cent decline in a 4-year period.

A nongovernmental source indicated a similar trend. In 1965 *Factory* magazine published a survey of 751 manufacturing plants in twenty-one industries and compared the results with those of a previous (1959) survey. The decline over this period in the number of plants having wage-incentive plans was about 20 per cent.* The comparisons made in this article were not wholly justified since the samples differed in the 1959 and 1965 surveys. The magnitude of the claimed decline, however, probably at least justifies the conclusion that there was a decrease.

For several reasons, which are gaining widening recognition, incentives are likely to dwindle further in the years ahead.

1. Managements and unions are becoming increasingly aware of the special problems that accompany an incentive system and are therefore less likely than in the past to embark on a wage-incentive program.

2. Many existing incentive programs are losing whatever initial attraction they had, as experience falls short of expectations and changes in production operations require revision of the program. In a growing number of situations, the decision is to scrap the program and to return to a time-payment system.

3. Increasingly technological advances are making production speed a function of machine pace rather than of worker pace. Where output depends on machines rather than on worker effort, there is little room for wage incentives for the workers.

16.7.4.14 Difficulties Under Incentive Plans. In theory, wage-incentive plans are, from the standpoint of management, supposed to increase production and reduce unit costs, thereby contributing to increased profits. For workers, the plans are supposed to provide an opportunity for increased earnings.

But experience has taught many managements that such plans breed difficulties and disadvantages which quickly outweigh the hoped-for advantages and that savings from reductions in direct labor costs tend to be canceled out by increases in other costs.†

16.7.4.15 Earnings Under Incentive Plans. Proponents of wage incentives often believe that workers paid by the hour or day do not perform an honest day's work but that if they are promised extra money they can be induced to produce a fair day's work or more.

Undoubtedly when a new wage-incentive plan is installed, production and worker earnings increase. However, since most incentive-plan installations are accompanied by other changes or improvements in methods, tooling, layout, work flow, supervision, and shop routine, it is virtually impossible to determine the effect of the incentive plan itself.

Initially, too, workers are frequently led by a new plan to step up their work pace and effort, often with a substantial increase in earnings (which may correspond to but is frequently less than the increase in their effort). But whatever the immediate benefits of a new plan, experience shows that average incentive earnings slide back over time or else hold up only at the expense of excessive strain on the workers involved.

Through tightening of production standards as well as through natural human let-down in the initial unsustainable work pace, incentive earnings rarely stay so high as workers have been led to expect. This experience holds particularly if general wage increases are applied as add-ons to pay rather than as part of the base rates.

But, even after initial increased earnings ease down, is it nonetheless true that incentive workers make more than time workers? There is no persuasive proof that incentive workers earn more over a period of time. Many comparisons offered by incentive-plan proponents show that the average earnings of a group of incentive workers are higher than those of another group doing similar work on an hourly pay basis. But the wage differential in these comparisons is typically or often due to factors other than the incentive plan—factors such as union strength, the size and location of the establishment, the work pace, the type of product, the economic status and wage policies of the company, and the degree of mechanization.

That incentives do not assure a higher level

---

* Gregory Schultz, "Plant Incentives Slump Badly over Last Six Years," *Factory* (June 1965), pp. 68–79.

† See Par. 16.7.4.2.

of earnings is abundantly clear from the fact that, in every wage survey covering both incentive and nonincentive workers on similar jobs, many of the time workers frequently earn more than the incentive workers. On this point, the region of the country in which worker earnings tend to be the highest, the Pacific Coast area, is the same region in which wage incentives are the least frequent.

16.7.4.16 Technological Improvements. Wage-incentive programs are based on the idea that the incentive of extra money will induce workers to produce extra output. The assumption is that workers are able to influence production by applying more physical or mental effort or both to the tasks specified by the company. Further, both the tasks and the workers' role in their accomplishment can be accurately defined and measured.

Historically, workers have obviously had some direct control of their production. But with the technological advances of recent years, with increased mechanization and automation, such direct control has been steadily declining. Many management representatives agree that the need for incentives is being eliminated and their effectiveness is being reduced because of such reduction of worker control of production pace.

Unfortunately, some managements have not been so perceptive and have tended to blame workers and unions for the declining effectiveness of incentives. Others have tried unsuccessfully to doctor an archaic piecework system to fit the modern industrial scene, and they too have cast about for some reason for their failure. Sometimes they blame lack of cooperation by unions when the fact is that their machine-oriented operations are simply no longer so suitable as before for incentive methods.

Modern management is coming to recognize that wage-incentive systems are not compatible with new production methods and equipment. It is also recognizing that wage incentives can be eliminated with no loss of production or profits. Intelligent supervision and reasonable labor-relations policies can ensure a high level of worker cooperation in the productive process without a system of wage incentives.

16.7.4.17 Elimination of Incentive Plans. Unions generally have been more successful in preventing installation of new incentive systems than in eliminating existing systems. Most managements develop an attachment to their incentive schemes and strongly resist their removal, in part because they believe that removal might be interpreted as acknowledgment that their plan was improper to begin with. Workers, too, are often reluctant to drop plans because they come to consider incentive pay as something extra they would lose if the plan were eliminated.

But unions have tended to underestimate their ability to eliminate incentive plans. Management attachment to existing plans is now often more psychological than practical, particularly if the union has effectively policed the existing plan. When management can cut incentive rates and institute speed-ups, when it can pay workers less per unit as production increases, and when its time-study and incentive applications are not subject to grievance and arbitration procedures, it probably can make the incentive system pay off. But when unions negotiate proper protection to minimize abuses, then management's zeal for retaining incentive systems is greatly reduced. Thus, unions are increasingly finding responsive attitudes from management when they propose elimination of incentive plans. And where they have successfully bargained incentives out of existence, they have done it without loss of earnings for the workers involved.

16.7.5 Work Sampling

Work sampling was developed as a technique for determining delay ratios. When used for this purpose, it is undoubtedly superior, in many instances, to the techniques that preceded it. Unfortunately, the very simplicity of work sampling has led to its use for purposes for which it was not intended (e.g., setting work standards) and by people who understood neither the technique nor its limitations.

With few exceptions, even relatively sophisticated work-study practitioners have been guilty of erroneous use of work sampling. A considerable amount of published material on work sampling has appeared in recent years. Much of it shows a serious lack of understanding of the theoretical foundations of work sampling, and thus many wild and unfounded claims are made. Work sampling, carefully used with full understanding of its limitations, can provide useful work and delay data. But unless the limitations are understood, serious errors may result.

16.7.5.1 Limitations of Work Sampling. Let us look at some of the limitations and problems connected with work sampling. Many practitioners have completely ignored the effects of work pace on work-sampling results, and they usually apply these results without pace corrections. When the question of pace was first raised, it was felt that since samples are usually taken of more than one operator on a given type of activity, pace should have little if any effect.* However, the effects of pace have now been recognized, and the evidence is conclusive that, for accuracy, it is necessary to rate operator performance. For example, discussing the results of a series of studies of factory operations, Nadler concludes, "These results indicated that when accuracy is required the operator's pace definitely affects work-sampling results."† The same authority, in another article, points out that "a person who performs more rapidly than another does *not* take the same proportion of time of the total day for many of the activities."‡ Even under the best circumstances, pace ratings result in errors of ± 10 per cent more than 50 per cent of the time. Performance rating errors of instantaneous observations must be infinitely greater.

Method variations and variations in the skill of operators also influence the amount of time spent on irregular elements.§ Obviously the mere presence of an observer has an influence on the kind of activity engaged in by workers. This problem is illustrated by a rather amusing incident which occurred in a West Coast plant. A group of workers reacted unfavorably to a management decision to make work-sampling studies in their department. Management had taken great pains to explain to the workers how work samplings were made, what the company expected to find, and how the results would be used. The explanations gave the workers enough knowledge of the technique so that they knew that, if they refrained from work whenever the observer was around, he would be able to gather no useful data. And this they did. While the plan was not 100 per cent successful, the observer realized that each time he entered the department for a series of random observations of various operations and workers, the workers simply stopped all productive work. Personal needs were satisfied, water was drunk, machines were oiled, tools were changed, etc., but no parts were produced. After several days of this activity, the company threatened to discipline the workers if they did not cooperate.

Once more knowledge of the technique came in handy. The workers simply reversed their strategy. Each time a worker noted the approach of an observer (or had it called to his attention), he would work furiously on his normal job cycle until the observer left his vicinity. While it took the company longer to discover this ruse, after a while it became obvious that these data were also useless. But how do you discipline employees for working?

Davidson‖, as well as Martin and Thomson, recognizes that even in the absence of an extreme situation like the one above, the observer does influence worker activity. Martin and Thomson found a significant difference between mean work levels when the sampling was done by university engineering students and when it was done by supervisors. The students' data revealed a mean work level of 80.5 per cent as compared to the supervisors' sampling results of 88.5 per cent. Martin and Thomson concluded that this 10 per cent increase in work activity "was due to the closer supervision generated by the supervisors' sampling activities."#

Frequently, one sees the statement that "work sampling works because the law of probability works." The fact is that probability is a theory, not a law. Mathematical models may be useful for solving practical problems, but when abstract mathematical theory is applied to any real phenomena, such as the workplace, applicability is fortuitous. The best results can yield only tendencies, not facts.

Even if the mathematical model were appropriate to the workplace, the binomial equa-

---

* Nadler, *Motion and Time Study*, p. 478.

† Gerald Nadler, "Pace as a Factor in Work Sampling," *Factory Management and Maintenance* (July 1961), pp. 172–73.

‡ Gerald Nadler, "New Applications for Work Sampling," *Proceedings, 15th Annual S.A.M.-A.S.M.E. Management Engineering Conference* (April 1960).

§ A.B. Drui, "Industrial Research on the Effects of Pace on Work Sampling Results," *Journal of Industrial Engineering*, **XII**, No. 1 (Jan.–Feb. 1961), 28–31.

‖ H. O. Davidson, "Work Sampling—Eleven Fallacies," *Journal of Industrial Engineering*, **XI**, No. 5 (Sept.–Oct. 1960), 367–71.

H. W. Martin and R. J. Thomson, "Work Management and Productivity in Engineering Design," *IRE Transactions on Engineering Management* (Sept. 1960), pp. 83–95.

tion or the normal-distribution curve may not be the correct model to use for work-sampling applications. Data collected from random observations, when plotted, frequently reveal a skewness, making it impossible to fit the data to a normal-distribution curve.

A limitation of all forms of work measurement, that of predicting future performance from present data, is also a limitation of the work-sampling technique. Delay allowances are set after completion of the sampling procedure and applied to future delays. But the estimates and confidence levels obtained from statistical theory apply only to the period from which the sample data were obtained. This fallacy seems not to be recognized by most writers and practitioners. Forecasting future delays based on past performance assumes that future conditions will be exactly the same as those which existed during the period of observation. This assumption is hardly tenable in the modern industrial scene.

Other fallacies associated with the use of work sampling require a discussion of mathematical theory and application not appropriate in a chapter of this nature. For those familiar with the mathematics of probability theory, Davidson's "Work Sampling—Eleven Fallacies" is probably the best critique.

16.7.5.2 Work Sampling and Unions. In spite of its shortcomings, work sampling can supply data of use to employers and worker representatives. Union industrial-engineering training programs are giving the technique more attention, and as a result an increasing number of union representatives are becoming familiar with it. Union representatives often suggest the use of work sampling when disputes occur over the adequacy of delay allowances.

When a company uses work sampling for arriving at delay allowances, either as a regular practice or when grievances arise, the union can be expected to insist on its right to bargain collectively over the sampling methods as well as over the confidence level and accuracy criteria. The union should insist on receiving full information from the company on all aspects of the work-sampling study and results.

16.7.6 Job Evaluation

The AFL–CIO "Collective Bargaining Report" on job evaluation,* from which most of the material in this section was taken, was published in 1957. Since that time there have been too few changes either in job-evaluation practices or in union attitudes to warrant a revision. Undoubtedly, however, the rapid increase of formal job-evaluation plans in the last two or three decades has brought a variety of special problems to unions.

A major danger is that these plans, which use technical and often inflexible methods to decide how much each job is worth as compared to other jobs, may supplant or unduly cramp or limit collective bargaining on wage rates for individual jobs.

The difficulties in working with job-evaluation plans are compounded by the fact that there are no generally accepted, uniform practices. Wide variations in types of plans and in their administration exist. Also, since the plans tend to be plantwide, rather than industry- or even companywide, the bargaining normally has to be carried out on the local union level. The one major exception is the United Steelworkers, who have negotiated a plan extending throughout the basic steel industry.

The following material discusses briefly several of the questions and problems involved in job-evaluation plans.

16.7.6.1 Union Attitudes. Union experience with job-evaluation plans has varied widely. As a result, there is no single or overall union attitude or policy toward job evaluation. Some unions have resisted its introduction. Others have joined in or even requested its installation.

For unions which already have such plans in effect, the immediate question is not whether to accept or reject job evaluation, but rather, "How can we represent our members most effectively under the plan? Should we participate in it and if so, how?" The degree to which unions desire to participate varies considerably. In some cases unions have decided on a job-evaluation plan jointly with management; they have analyzed jobs, have written job descriptions, have evaluated jobs, and have determined labor grades and rate ranges, all on a joint basis. Other unions have allowed management to perform these functions but have reserved the right to question the results of each step through the grievance procedure. In still other cases unions have simply stated that they are interested only in the wages paid to their members and that management may use whatever means it desires to arrive at what it thinks wages should be. The union simply reserves the right

* *AFL-CIO Federationist* (June 1957), pp. 33–44.

to bargain on the results of any management wage determination, regardless of how it is made.

Whether unions are for or against job evaluation on principle, all agree that where a plan exists it is necessary to provide protection against any arbitrary and abusive application by management. This protection must be gained primarily through collective bargaining on applications of the plan and on the makeup of the plan itself.

16.7.6.2 Job Evaluation and Collective Bargaining. No job-evaluation plan should be considered sacred; no job-evaluation finding should be considered a scientific fact not subject to challenge. Job-evaluation plans involve considerable judgment, with room for wide variation. Also, the very nature of such plans ordinarily does not enable them to take into account adequately special human and economic factors which may arise occasionally.

If the plan is regarded as a rigid formula from which the parties cannot depart, it becomes the master rather than the servant, a hindrance rather than a tool of sound labor relations. A plan should be considered a guide for collective bargaining, not the final authority. If findings are unsatisfactory, the plan or the findings in question should be adjusted.

The following case illustrates how a job-evaluation plan when considered as a governing authority rather than as a guide soon becomes a barrier to satisfactory solution of wage-determination problems. A Midwest company, producing machined forgings, decided to set up its own forge shop to supply its existing machine shops. When the time came to man the new forge shop, the company asked some of its best machine-shop operators to transfer there. As an inducement, the company promised that they would be paid their prior average hourly earnings (under the incentive plan) while the forge shop was getting under way and then would earn more when full production was reached. When the forge shop was in full production the company evaluated and time-studied the operations and set job rates and production standards—and resulting earnings were lower than those the men had received in the machine shop. The company refused to adjust the earnings upward, and the men struck. The union called in a consultant to determine the reasons for the lower earnings.

The consultant was taken into the plant superintendent's office by the union president and was greeted with these words by the superintendent, "I don't know why the union called you in. I know we promised these workers more money and we want to deliver on our promise, but our bible won't let us." The bible he pointed to was a copy of the company job-evaluation manual, an application of the National Metal Trades Association (NMTA) plan.

A check of the time studies of the jobs in question revealed that the allowances for fatigue and personal time were inadequate, and the company agreed to liberalize them, but earnings still fell below machine-shop earnings.

Review of the job-evaluation plan by the consultant led to the suggestion that the plan be revised to provide a higher value for the factors physical demand, working conditions, and hazards. The company balked. Although it had readily agreed to changes in time studies, it strongly resisted any tampering with its $50,000 investment, the job-evaluation plan.

It took several more days of strike and pressure from customers before the company agreed to revise the plan, with the adjustments resulting in about 15¢ more per hour on the base rates. The men were then satisfied and returned to work.

At no time did the company take the position that it did not want to increase the pay for these men. It wanted to live up to its promise. But its attitude toward the job-evaluation plan was such that it did not even consider adjusting it to meet the practical situation, and it then resisted strongly efforts to adapt the plan to a special need.

Job evaluation must be subordinate to collective bargaining or it should not be used at all. Where wage adjustments for particular jobs are warranted for special reasons, a job-evaluation plan should not be permitted to stand in the way of such adjustments. Unions accepting job-evaluation plans not only must retain the right to question through the grievance procedure and to gain adjustment of the results of job evaluation, but also must be able to bargain on and change the plan itself. Job evaluation must be flexible enough to meet the practical demands of any specific situation.

16.7.6.3 Accuracy of Job Evaluation. Many companies have taken the position that job evaluation is a scientific process yielding factual results. Since facts are not bargainable, these companies say the results of job evaluation are not proper matters for discussion at the bargaining table. Union pressure has resulted in some companies' modifying this position. Many companies, however, still try to develop the attitude among union officers and members that job evaluation yields results superior to those gained by collective bargaining.

Job-evaluation plans are not scientific in any sense. They represent the subjective judgments of fallible human beings. These judgments, whether made by an industrial engineer or a supervisor, should be subject to collective bargaining between union and management.

16.7.6.4 Need for Bargaining on Job-Evaluation Plans. The following list of the steps involved in setting up a job-evaluation plan includes some of the considerations which require judgment and which should therefore be reviewed and decided through collective bargaining.

1. *Studying jobs.* Before a plan is developed, information must be gathered on the nature of the jobs in the plant, usually by observing jobs, by interviewing workers and supervisors, and sometimes by questionnaires. Notes are made of the various tasks which make up the job. Judgment is used to determine what information is recorded for each job. The local union should not accept as gospel truth the job studies made by an engineer or management official.

2. *Analyzing results of study.* The notes made are then analyzed to determine what should be included in the next step—the writing of job descriptions. The analyst usually decides what to include on the basis of what he believes the company is willing to pay for. He may consider certain tasks unimportant and delete them. Although tasks do differ in importance, he should record all tasks which contribute to the total job. This analysis step is not scientific either; it too should be subject to bargaining.

3. *Describing jobs.* This is probably the most important step in the whole process since the evaluations are based on the descriptions. The entire wage scale as well as the pay for individual jobs is influenced by the descriptions. Much judgment is used in this step not only as to what appears in the description but also as to how it appears. Union review is necessary.

4. *Deciding on factors.* After the jobs are described it becomes necessary to choose factors common to some extent in all jobs. Again this step relies heavily on judgment, and the variation possible can readily be seen in the multitude of job-evaluation plans with different factors in plants doing similar work.

5. *Defining factors.* Once the factors are decided upon, it is necessary to define exactly what each means. For example, is the factor of hazards to be measured by past accident experience as to frequency and severity or in terms of probable severity if an accident should occur, regardless of frequency? A definition is necessary to ensure, as much as possible, uniform evaluation. Union viewpoints may differ from those of management; they should be reflected in bargaining.

6. *Deciding on degrees of each factor.* Recognizing that the factors apply in varying degrees to each job, the analyst must break down the factors into smaller units and assign values to each. For example, it is not a question of whether a job requires education but rather how much education. Judgment is necessary to determine the number of degrees in each factor. Often the number of degrees decided upon provides a nice, neat plan with an equal number of degrees for each factor, but this assignment may be unrealistic and have little value. Again, this is properly a bargainable matter.

7. *Defining degrees.* Exactly what is meant by each degree must be spelled out as closely as possible so that the evaluators may judge what distinctions to consider in deciding whether, for example, a job requires the second degree of experience (44 points under the NMTA plan) or the third degree (66 points). If the words used in these definitions are too general, the evaluators may have inadequate guidance.

8. *Weighting factors.* This is probably the most difficult and most abused step in the setting up of a job-evaluation plan because it requires the most significant exercise of judgment. The total number of points of a job-evaluation plan is not important. What is important is the percentage of the total applied to each factor.

The NMTA plan is so weighted that education is 2.8 times as important as job hazards. Experience is 4.4 times more important than hazards. Thus, more points are given to a job requiring experience of as little as 3 months than for one which subjects the worker to accidents causing loss of an eye or limb. Such weighting may be appropriate in some plants or occupations where the chance of such an accident is highly improbable, but certainly it is not appropriate for hundreds of plants where the NMTA plan is in use.

It is difficult to determine proper weighting,

but it should be related to the jobs in the particular plant. Because of the variation possible, the union has a stake in seeing that weighting takes realistic account of the particular situation.

9. *Weighting degrees.* Most plans apply the points to degrees within each factor equally, indicating a lack of judgment rather than the use of judgment. The weightings of degrees should correspond to the degree definitions. For example, the NMTA plan allows 10 points for a job in which probable damage is seldom over \$25. But where probable damage is likely to be forty times as great (\$1000) it allows only twice as many points (20). Thus, in one case a point is given for each \$2.50 of probable damage, and in the other for \$50 of probable damage.

10. *Evaluating jobs.* Now the plan is ready to be applied. It should be noted first, however, that many companies do not actually go through all of the above nine steps to set up a plan, but rather install a canned plan, devised by management consulting engineers and manufacturers' associations, which usually has little if any relation to job requirements and working conditions in a particular plant.

Whether the plan is tailor made or store bought, however, the evaluation itself, that is, the application to the jobs in the plant or, more specifically, to the job descriptions, is the same. The factor and degree definitions must be compared to the job descriptions and determination made as to which degree of each factor the job falls in. The points allowed for each factor are then simply added to arrive at a point total for the job.

That considerable judgment is involved in this evaluation process is evident. The individual evaluator, regardless of the care used in developing the plan, still has wide latitude. For example, he must decide whether a job requires some training or broad training. He must judge whether it needs 12 or 13 months of experience. Light or moderate effort? Continuous or concentrated visual attention? Two people applying the same definition on such distinctions as these can easily differ by many points in their final totals. Thus, evaluation results cannot be accepted automatically and must be subject to bargaining.

11. *Turning points into money.* The points must of course be translated into specific wage rates. Most people uninitiated in job evaluation would expect management to simply say a point is worth a certain amount. Unions would expect to decide this through collective bargaining. This figure would then be multiplied by the number of points for a job to determine the hourly rate for that job. For example, if a point were to be worth 1¢ and a job were evaluated at 150 points, then it would carry a rate of \$1.50 per hour. A job having 300 points would pay \$3 per hour. But this is not the procedure.

The points assigned are only one of the determinants of the hourly rate for the job. Most companies first compare the point values with present job rates and rates for similar jobs in the industry and the area. This procedure usually results in the points and money having little, if any, direct relationship. If there is a relationship, it usually is one in which jobs with the highest number of points get the least amount of money per point.

Many companies set up a wage structure based on labor grades and not jobs, with the result that jobs with different point values pay the same wages. NMTA recommends a score range of 21 points for each grade. For example, a job with 206 points is in the same labor grade and pays the same hourly rate as a job with 227 points. Yet a job with 205 points pays less money than the 206-point job, and a 228-point job pays more money than a 227-point job.

Development of a wage structure and determination of wage rates for each job are fundamental areas in which unions must protect members' interests through collective bargaining. This responsibility is not eased or changed when formal job evaluation is used. As long as unions morally and legally have the right to bargain collectively on wages, they must subject every aspect of job evaluation to close scrutiny and question. If they are working with job evaluation, voluntarily or otherwise, they must recognize that no part of the job-evaluation process, either in its development or in its application, can be removed from the bargaining table.

16.7.6.5 Right to All Information on Job-Evaluation Plan. When a job-evaluation plan is used, unions must have information on all aspects of it, including all information on job descriptions, the nature of the plan, labor grades and corresponding wages, and the number of workers in each job and labor grade. Such information is necessary if the union is to bargain on a reasonably informed and intelligent basis and is to adequately represent its members.

The right of unions to such information is

apparent and has been upheld by the NLRB, various court actions, and in arbitration cases. To help avoid wrangling in individual cases, some unions have negotiated contract provisions spelling out their right to complete information.

The right to information should not be crippled by unreasonable limitations. Thus, unions should not be limited merely to seeing the information but should rather be entitled to receive copies for use as their own time and resources require. A union should be able to study, check, and analyze at its own convenience data submitted by the company, so that it may fully understand all phases of the plan and any problems of its application.

16.7.6.6 Red-Circle Rates. Since they are interested in gaining employee acceptance of a new job-evaluation plan, most companies guarantee that no wage reductions will take place upon installation of job evaluation. When an employee's wage rate is higher than the rate determined by job evaluation, the company agrees to his retaining his higher rate as a personal rate. Such personal rates, higher than evaluated rates, have become known as *red-circle rates.*

Selling the plan to the union and its members may also include the promise of raises to those below the evaluated rates and may even include an across-the-board wage increase. These promises, coupled with the red-circle guarantee, may lead unions to accept the adoption of a job-evaluation plan as a way of getting wage increases for most workers while at the same time protecting those members carrying high personal rates.

The presence of red-circle rates should be inspected closely for many reasons. Account must be taken in advance of the problems they are likely to present in the future. When a job-evaluation rate is significantly lower than a personal rate, it may be an indication that, instead of the personal rate's being too high, the job was evaluated incorrectly and should carry a higher rate. The job may have been evaluated properly according to the job-evaluation plan, but the error may be in the plan itself. The plan may not take into consideration certain aspects peculiar to the job, or it may not consider them properly in their relation to other job factors.

If there are many red-circle situations, the plan may not fit the plant at all. After all, it is foolish to say a man is being paid more than his job is worth if the company would have to pay as much to replace him. In other words, if results of job evaluation are completely unrealistic in terms of wages for similar jobs in other companies, then the job evaluation is wrong and not the present rate.

Even when a union believes the job is properly evaluated, red-circle rates raise special questions. How long can the individual retain his higher red-circle rate? "As long as he remains on that particular job," the company says. But will the company allow him to stay there or will it find some means of displacing him? How will the red-circle employee fare with future wage increases? What about red-circle rates that will come up in the future as jobs change or new jobs are created? How will they affect future wage levels? And how will they affect the value of future negotiated wage increases?

Every red-circle rate presents a challenge to management. Unions have often found that red circlers become marked men and that management finds some excuse for eliminating them or their wage advantage. Some managements have transferred such men to higher-rated jobs in the same wage bracket as their red-circle job. Often because of lack of training the man may not be able to perform the higher-skill job and may have to accept a lower-paying job in order to remain employed. Some managements have been willing to give a red circler an opportunity to learn the skills required on the higher-ranked job, and some have even provided this training. These job transfers are fine until they clash with established seniority practices. Is a less senior man to be trained and transferred to a higher-rated job simply because he carries a red-circle rate?

Then there is the problem of future wage increases. Management will desire to eliminate the difference between evaluated job rates and red-circle rates as quickly as possible. It often proposes that when wage increases are negotiated they should apply to everyone except those whose personal rates exceed the evaluated rates. That unions cannot accept this procedure is obvious. Confronted with union refusal, management may offer to compromise by seeking union approval to a plan whereby some part of future wage increases would not be applied to these higher personal rates. Thus, some union members would receive substantially less in wage increases over a period of time, dissatisfied groups may develop within the union, and possible serious frictions could lead to a decrease in collective-bargaining strength.

16.7.6.7 Changes in Job Content. Red-circle rates occasioned by the introduction of a job-evaluation plan disappear in time because of normal turnover, transfers, promotions, job eliminations, or other procedures. However, their disappearance does not eliminate red-circle problems. As management improves production methods, many jobs have lower requirements. The new job evaluations result in job rates lower than those currently being paid and thus create new red-circle rates.

While many unions have been able to negotiate red-circle protection when a job-evaluation plan is installed, they have found it more difficult to do so after the content of a job has been diminished. Many unions have discovered that a large part if not all of negotiated wage increases may be eaten away by downgrading of jobs and of workers' wage rates as job content changes. The factors in plans such as that of the NMTA are weighted in such a way that the usual changes in jobs brought about by methods improvement and new equipment normally reduce the total number of points. For this reason the job-evaluation plan itself, as well as the application of the plan, should be subject to negotiation.

Changing of job content by management after the installation of a job-evaluation plan and after agreement has been reached on the wage structure poses other problems. Does the changed job description portray changes in the job itself? If the job has been changed, does the change require a reevaluation? For example, in the NMTA point-rating plan, a job receives the third degree of the education factor, or 42 points, if it requires the use of a *variety* of precision-measuring instruments. If the job requires the use of a *wide variety* of precision-measuring instruments, it receives the fourth degree, or 56 points. The plan does not define a variety vs. a wide variety. Would the elimination of one instrument from a wide variety make a variety? Would the company add 14 points if one instrument were added to a job in the variety category? In all job-evaluation plans factors are defined with words subject to wide interpretation. In other words, jobs can be changed without the necessity of changing descriptions or evaluations. What was considered a wide variety before may remain a wide variety even after a minor change in the job.

Unions should check any changed descriptions or evaluations carefully to make certain that they reflect changes and are not used as a handy excuse for watering down the job and its rating. They should check, too, to see whether any added duties did not appear in the changed description and whether they may balance any deletions.

Frequently a change in a job results in a particular task's being performed less frequently but still being required as part of the job. The position of some companies has been to decrease point allocations for almost all factors. Is this a legitimate position? Consider one example in detail. The job of an engine-lathe operator requires doing a variety of operations on different parts. One operation, threading, is done on a few jobs put primarily on a heavy steel shaft weighing about 75 lb. Now suppose that the company discontinues the manufacture of this shaft. The elimination of this part reduces the time spent threading, and so the company may seek to reevaluate the job and apply fewer points for education, experience, and physical demand. When questioned by the union the company position may be that threading will now take much less of the total job time and therefore instead of broad training (56 points under the education factor) only some training (42 points) is required. Where previously more than 1 year of experience (66 points) was required, now less than 1 year of experience is needed (44 points). Also, since the job will now rarely require lifting heavy material, it may be given 30 points for less physical demand instead of the previous 40 points for occasional lifting of heavy material. When the physical demand of a job is less, it is understandable that it will receive fewer points, but one must be careful to recognize that frequent lifting of light material can be equally if not more demanding than occasional heavy lifting.

Complex considerations are involved in weighing education or experience factors. The frequency with which a job requires certain specific knowledge, skills, or abilities should not be the criterion. If the job requirements are such that a worker must have the knowledge, skill, and ability to cut close-tolerance threads on an engine lathe, then the company must be willing to pay for these attributes regardless of how often they are exhibited.

16.8 INDUSTRIAL ENGINEERING, UNIONS, AND COLLECTIVE BARGAINING

Undoubtedly industrial engineers and industrial-engineering techniques have not been accepted by the majority of American workmen and their unions. The lack of acceptance of techniques is easily understood. As with most engineering disciplines, the work of the industrial engineer is not finite. At best, it can be stated only in statistical terms of validity and reliability. Unfortunately for the industrial engineer, the luxury of the safety factor is not available to him. This margin of ignorance would enable the industrial engineer to state his production standard as 0.675 ± 0.068 standard hours per hundred units. But a tolerance such as this would create havoc in the minds of those concerned with scheduling, wage administration, and cost estimating.

It was pointed out in Art. 16.7 that unions have been primarily concerned with those industrial-engineering tools which directly affect workers. In the foreseeable future, enough progress will probably not be made in the application of these techniques to greatly reduce inaccuracies and to increase confidence levels. In fact, the use of the word *inaccuracies* is probably improper. In order to know that something is inaccurate one must know what is accurate. Where human work is concerned, acceptability replaces accuracy.

Acceptance by unions of the work of the industrial engineer will depend not on the status of his tools but on the industrial engineer himself. And here the future is bleak, for the industrial engineer seems unable to analyze and deal objectively with unions—a major factor of the industrial-engineering system.

16.8.1 Industrial Engineers Are Antiunion

Workers and their unions are industrial facts of life. The industrial engineer's general antiunion bias—his feeling that unions are an obstruction to his system rather than a part of it—can lead only to continued chaotic relations.

> For whatever reason or combination of reasons, perhaps because engineering faculties are predominantly antiunion, perhaps because he feels that he must be loyal to his employer, perhaps because of his family background, the industrial engineer's sentiments are practically always antiunion. This to him means that he must oppose unions and anything a union stands for. When these feelings translate themselves into his daily activities, they evidence themselves in his attitude toward the problems which are a by-product of union-management relations. Unlike his attitude toward production problems, materials problems, equipment problems, his attitude becomes belligerent. He refuses to analyze and solve; he fights and attempts to force.*

The industrial engineer can overcome his antiunion bias. Indeed, if he is to engage in the profession of industrial engineering, he must overcome it. After all, this is not an either-or situation. One does not have to be for management and against unions or for unions and against management. Certainly, the industrial engineer should not have to be for or against either, at least as far as his engineering work is concerned. All that unions expect from industrial engineers is neutrality in all aspects of collective bargaining.

16.8.2 Organizing Campaigns

Usually during organizing campaigns company officials call together the management team. Frequently the industrial engineer is given specific assignments as part of the keep-the-union-out drive, but he is in special danger because of his general lack of understanding of the law and of labor-management relations. More important, when the organizing drive is over, the industrial engineer will be working with a system which contains workers, union or nonunion. If during the union campaign the industrial engineer becomes a protagonist, he is forever stopped from claiming his work is unbiased, objective, and professional. He has become a management tool in the eyes of the workers. His work is now suspect. In situations such as these industrial engineering meets the fork in the road—one branch leading to industrial engineering as a profession, the other to the trade of industrial engineering.

Some writers and practitioners have claimed that union organization leads to more efficiency and greater profit. Perhaps if the industrial engineer were to put his analytical skills to work—if he were to research and simulate for his computer the game of collective bargaining—he might conclude that unionization is the point of optimization of the company position on the union-management matrix.

---

* Paul N. Lehoczky, "Industrial Engineering and Collective Bargaining." Paper presented at 17th Annual Institute Conference and Convention, American Institute of Industrial Engineers, May 27, 1966, San Francisco, Calif. Published in *The Labor Law Journal*, **17**, No. 7 (July 1966), 393–400.

16.8.3 Contract Negotiations

After a plant is organized and the workers become union members, a negotiation leads (hopefully) to an agreement. This document is called the contract, the agreement, the labor-management agreement, or the collective-bargaining agreement. It sets out the rules of the game and delineates to a great extent the roles of union and management in the months ahead. This document may be quite specific in spelling out what an industrial engineer may do and how and when he may do it.

Code 4 of the "Code of Work Measurement Principles" states: "We will encourage during contract negotiations the use of industrial-engineering counsel on matters pertaining to work measurement and wage payment."* In spite of this recommendation, Arbitrator Paul Lehoczky's description probably best defines the role of industrial engineers at this stage of collective bargaining.

> He [the industrial engineer] invariably loses his fight in the long run and he does his employer a disservice; he is excluded from the very meetings which dictate his future. This self-imposed attitude breeds an even more serious consequence: He is absent from collective-bargaining sessions. It is at these sessions that a representation of the industrial-engineering function is most important. This does not imply that the industrial-relations personnel would fail to represent him, but when the time arrives that something has to give in order to consummate the final move, it is the absentee's interests which usually are most likely to suffer. Thus is created the indeterminate language of the "reasonable production" variety, expressions which are seemingly definitive yet can be interpreted differently by different people. Nondefinitive language often finds its way into the text of the agreement for this very reason: It permits either party to "assume" that what it wanted has been agreed to. This despite the fact that their positions are diametrically opposite, as are their interpretations of what had been agreed to.
>
> The parties agree to this type of solution only because they hope that the question it allegedly answers will never arise. When it does appear, as it frequently does in the industrial-engineering area, it usually ends up in arbitration. The outcome will necessarily depend upon the credibility and persuasiveness of the presentations and, to a degree, upon the background and/or philosophy of the arbitrator. The aloofness expressed by the industrial engineer simply underscores the union's claims that the union position in all probability went unchallenged, hence was accepted during the negotiations. Thus are created decisions which tend to upset the entire industrial-engineering function. Whose fault is it? I suggest that the fault rests with the indifference, the hostility, and possibly the competence of the industrial-engineering function as it relates to union-management relations.
>
> "I should add at this point that a simple presence at the negotiations, at grievance meetings, and/or arbitration hearings guarantees nothing. To comprehend what is going on, one must have had experience with negotiations, one must have some understanding of the fundamentals of labor-management relations, and one must have a sympathetic understanding of the specific problems of the union's leadership. A union is a political organization, a democracy which is dependent upon votes and which is subject to partisan pressures. There are concessions which it can never make save via arbitration, others it can make only for heavy returns, while still other concessions are of a "small change" nature. The industrial engineer must be able to understand and evaluate these characteristics when he makes his request for changes in the established procedure.†

16.8.4 Grievance Procedures

Almost all labor-management agreements contain some procedure by which workers, through the union, can question management action. Frequently these disputes center on production standards, piece rates, and job evaluation. Among other causes, they arise when a production requirement is raised, when a piece rate is cut, or when a job is reevaluated downward.

Once more the industrial engineer is in a difficult spot. His difficulty becomes serious if he dogmatically states: "My standard is correct! There will be no restudy and no change." Workers know he has been wrong in the past and may be this time. If on the other hand the industrial engineer allows worker pressure to influence his actions and he raises a rate without doing the necessary restudy, he is in equally hot water. Workers soon learn that the squeaking wheel gets the grease.

---

* Reprinted from the "Code of Work Measurement Principles," published by the American Institute of Industrial Engineers, Inc., 345 East 47th Street, New York, New York 10017.

† Lehoczky, "Industrial Engineering and Collective Bargaining," p. 9.

At all steps of the grievance procedure the industrial engineer has the opportunity to influence worker and union attitudes toward industrial engineers. His job depends on the confidence of both sides. He can be discharged by the employer. But, equally, any group of workers has the means of getting rid of any industrial engineer engaged in standards-setting work.

When a grievance is filed claiming an erroneous production standard, piece rate, or job evaluation, the industrial engineer frequently looks upon it as a personal attack. He feels that the worker does not trust him or that the union is out to get him. He then attempts to prove that he is not antiworker or antiunion. While this may be a laudable objective, it has little to do with the problem at hand. A prounion industrial engineer may still use tools of analysis which lead to results having relatively low accuracy and confidence levels.

The industrial engineer's role in the grievance procedure depends basically on how thorough a job he does as a matter of course. If, for example, in standards-setting the industrial engineer follows the prescription of Code 9 of the "Code of Work Measurement Principles,"* he is on sound ground. He can say the same thing to the worker, the union representative, the foreman, and the labor-relations man: "I have done my job well. I have followed ethical procedures. I believe the standard on this job is correct, but the nature of work measurement is such that even when I do my best the result may be poor a certain percentage of the time. I will recheck this job to see if I can find the reasons for your belief that the standard is incorrect."

If the industrial engineer then approaches the review procedure with an open mind and evaluates the standard rather than trying to prove the standard, he is most likely to achieve a satisfactory operating climate. If the standard needs changing, it should be changed. If, on the other hand, an objective review of the standard leads the industrial engineer to conclude that the original standard was correct, he can say so with confidence.

* "We will accept the responsibility for doing a thorough and complete job on every study we make, including recording detailed methods descriptions, obtaining sufficient data, recording reasons for rejecting any data, identifying value judgments, and making sure that the time for all work the operator is expected to do is included in the standard." Reprinted from the "Code of Work Measurement Principles," published by the American Institute of Industrial Engineers, Inc., 345 East 47th Street, New York, New York 10017.

16.8.5 Arbitration

Provisions for arbitration of some or all grievance disputes and incorporated in over 94 per cent of labor-management agreements according to a Bureau of Labor Statistics survey.† Seventy per cent of these agreements provide that all disputes not resolved at the last step of the grievance procedure be referred to arbitration. Among the items not normally excluded are job classification and evaluation, incentive or piece rates, and production standards.

16.8.5.1 Arbitrators of Industrial-Engineering Disputes. The Federal Mediation and Conciliation Service (FMCS) during the fiscal year 1965 received over 5000 requests for panels of arbitrators. The normal panel submitted by the FMCS consists of seven names from which the parties attempt to choose one. During this period the FMCS roster contained 838 arbitrators, but only sixty were considered technically qualified to handle industrial-engineering cases.

The American Arbitration Association, another major source of arbitrators, lists only fifty-seven industrial-engineering graduates or educators in the field. The National Academy of Arbitrators, in a 1962 survey, found only six of 193 of its members had engineering degrees.

This paucity of industrial-engineering arbitrators was emphasized when Wiggins surveyed 306 cases reported in the *Labor Arbitration Reporter* of the Bureau of National Affairs.‡ Wiggins selected these cases as being the most representative of various industrial-engineering problems. Only twenty-eight of the 139 arbitrators who heard the 306 cases had engineering or business training.§

There are no reliable statistics on the number

† *Arbitration Procedures*, BLS Bulletin 1425–6 (Washington, D.C.: Superintendent of Documents, June 1966), p. 5.

‡ Ronald L. Wiggins, *Arbitration of Industrial Engineering Cases* (unpublished doctoral dissertation, University of California, Los Angeles, 1965).

§ Wiggins did not provide data on the type of engineering degree, and he did not give separate figures for engineering and business categories.

of cases of an industrial-engineering nature arbitrated each year. These cases however are on the increase. In the sources cited above also, the overwhelming number of arbitration cases involving industrial-engineering applications are not handled by industrial engineers. These nonindustrial–engineer arbitrators continually hear cases and render decisions which determine what industrial engineers may do and how they may do it.

16.9 ETHICAL CONSIDERATIONS AND CONCLUSIONS

The industrial engineer, called as an expert witness during the arbitration procedure, is faced with special problems. He is called upon to prove that which is unprovable. He is expected to discuss technical problems in nontechnical language. He is torn between the pressures put upon him as an employee and the ethical requirements of his profession.

Faced with this conflict between management goals and ethical practice, the industrial engineer is in a real dilemma. This dilemma is caused by the absence of an operative ethical-practices code which specifically applies to him. The "Canons of Ethics for Engineers" provide some general guidelines but are of little help in the important human aspect of industrial engineering, which distinguishes industrial engineering from other engineering disciplines. The "Code of Work Measurement Principles," in spite of its inadequacies, is an important initial step. But the simple adoption of a code is useless unless effective sanctions are developed which protect the employed industrial engineer from being subjected to pressures to act in an unethical manner.

The current effort to develop standard industrial-engineering terminology is another example of a step in the right direction. For too long the standards sellers have operated with weasel-word definitions of tools and techniques. What is time study? What is standard data? How is normal defined? For example, how can a worker who for years has used and has heard management representatives and industrial engineers use the words *time study* in connection with on-the-job observation and stopwatch timing be expected to have respect for the industrial engineer when suddenly management declares that time study means predetermined motion-time systems, standard data, work sampling, or some other system? Manipulation of industrial-engineering terminology to satisfy nonindustrial–engineering goals must end.

Of equal importance is the need for the AIIE to begin to take some responsibility for the techniques and tools used by its members and for the character of the claims made about them. In order to do this the AIIE must plan, develop, sponsor, supervise, coordinate, and control research involving industrial-engineering methods. The evaluation of industrial-engineering tools must not be left to their purveyors or to individual universities, where much excellent work has not been systematically planned, analyzed, or communicated. AIIE must therefore set up an independent research facility. It could do worse than to use the Max-Planck Institute in Dortmund, Germany, as its model.

In addition, AIIE must concern itself with broadening the practice of industrial engineering to match its definition. It must coordinate the industrial engineer's work with that of doctors, psychologists, and sociologists. If it does not, large portions of what is claimed as the jurisdiction of the industrial engineer will pass by default to practitioners of other disciplines.

Finally, industrial-engineering education must be improved and expanded. The recent emphasis on mathematics and statistics, while necessary, must not bury other needs. Options in industrial psychology and sociology and in physiology must be encouraged. Industrial-engineering schools must provide a basic understanding of workers and their unions and of their legitimate concern with the work of industrial engineers. Industrial-engineering students should not be taught to be pro- or antiunion, but workers and unions will have a profound influence on the work of the industrial engineer for many years to come, and industrial engineering must recognize this influence.

Unquestionably the industrial engineer of tomorrow will be technically capable. But will he be wholly capable? Will industrial engineering of the future be in the hands of the university-trained technician and salesman? If so, it will be in the same relatively unsatisfactory master-mechanic position that it is in today and might best have no future at all.

On the other hand, if industrial engineering stops looking for scapegoats for its own shortcomings, if it takes a good, long, hard look at its own conscience, if it assumes the responsibility for its own mistakes, if it desires to objectively and open-mindedly analyze and understand all of the factors of production, and, finally, if it puts the concept of service before economic gain, then industrial engineering will have an excellent chance to achieve true professional status.

W. J. FABRYCKY, *associate dean of engineering at Virginia Polytechnic Institute and State University, earned his B. S. degree at Wichita State University, his M.S. degree at the University of Arkansas, and his Ph.D. at Oklahoma State University. He has held teaching and research positions at the University of Arkansas and at Oklahoma State University.*

In addition to teaching a number of undergraduate and graduate courses, Dr. Fabrycky has maintained an active research program. He was director of the Operations Research Group at Oklahoma State University and is chairman of the Systems Engineering Group in the College of Engineering at Virginia Tech. His teaching and research has resulted in a number of publications which include the texts Engineering Economy, *with G. J. Thuesen;* Operations Economy: Industrial Applications of Operations Research *with P. E. Torgersen; and* Procurement and Inventory Systems: Theory and Analysis *with Jerry Banks.*

Dr. Fabrycky has augmented his academic experience with regular or consulting appointments with Cessna Aircraft Company, Chance Vought Corporation, the United States Air Force, Ethyl Corporation, Brown Engineering, and other companies. His professional assignments include those of design engineer, operations research analyst, and principal engineer.

Dr. Fabrycky is a member of A.I.I.E., A.S.E.E., O.R.S.A., S.O.L.E., and the Virginia Academy of Science. He is also a member of Alpha Pi Mu, Sigma Tau, and Sigma Xi and is a registered professional engineer.

PAUL E. TORGERSEN, *dean of engineering at Virginia Polytechnic Institute and State University, earned his B. S. degree at Lehigh University and his M. S. and Ph. D. degrees at Ohio State University. He has held teaching and research positions at Ohio State University, the Battelle Memorial Institute, and Oklahoma State University. He received the Outstanding Teacher Award from the College of Engineering at Oklahoma State in 1963.*

Dr. Torgersen's research has covered many aspects of industrial engineering with special emphasis on management gaming and quality control. He has organized and directed a number of short courses, conferences, and seminars. His publications are numerous and include the texts A Concept of Organization, *and* Operations Economy: Industrial Applications of Operations Research *with W. J. Fabrycky. Dr. Torgersen is a member of the publications policy board of* A.I.I.E.

Dr. Torgersen has had industrial experience with New Moon Homes, Guy James Industries, Magee-Hale Park-O-Meter, Aluminum Company of America, and Continental Can Company. He is a member of A.I.I.E., A.S.E.E., A.S.Q.C., American Arbitration Association, the Human Factors Society, Alpha Pi Mu, and Sigma Tau and has held a number of regional and national offices in these societies. He is a registered professional engineer.

SECTION 17

LINEAR PROGRAMING AND ITS APPLICATIONS*

W. J. FABRYCKY AND PAUL E. TORGERSEN

17.1 INTRODUCTION

Linear programing is a mathematical means for providing management with a basis for resolving complex operational alternatives. It is applicable to a general category of optimization problems involving the interaction of many variables which are subject to certain constraints. These constraints usually exist because the activities under consideration compete for scarce resources. Linear programing is thus a quantitative procedure for determining the best allocation of limited resources to achieve a given objective.

A basic supposition in linear programing is the existence of linearity. Formally stated, the objective sought is the optimization of a linear effectiveness function subject to linear constraints. This may require the minimization of time, distance, cost, or the maximization of

* Much of the material in this section was taken from W. J. Fabrycky and Paul E. Torgersen, *Operations Economy: Industrial Applications of Operations Research* (Englewood Cliffs, N. J.: Prentice-Hall, Inc., 1966).

profit, depending upon the problem under consideration. The result of the computational procedure is a program outlining a particular plan of action.

The initial mathematical formulation of the general linear programing model, together with the simplex method for its solution, was developed by Dantzig and his associates in 1947. This group was engaged by the U.S. Department of the Air Force to investigate the applicability of quantitative techniques to military programing and planning problems. Like many other results of operations research effort, linear programing found many successful applications in government, business, and industry. The development of the high-speed digital computer, together with the availability of prepared computer programs, has contributed to its widespread acceptance as an aid in decision making.

Prior to the development of the general linear programing model, other problems were recognized as being of the type requiring the optimization of a linear effectiveness function subject to linear constraints. In 1941 Hitchcock and independently in 1947 Koopmans formulated the transportation model, a special case of the general linear programing model. This model deals with the common problem that arises when a number of origins possess units needed by a number of destinations. The quantity of units from each origin to be allocated to each destination is to be determined. A second special case of the general linear programing model is the assignment model. The problem of assigning each of a number of means to an equal number of requirements on a one-for-one basis is explained by this model. A theorem derived by the Hungarian mathematician König in 1916 provides the basis for Kuhn's development of the Hungarian method of assignment published in 1955.

This section will open with the development and application of the assignment model. Next, the transportation model will be developed and illustrated by simple example applications. These distribution models provide a convenient introduction to the class of problems that can be solved by linear programing. Development of the general linear programing model together with examples of its application is deferred until the last half of the section.

Each example application presented in this section was chosen to illustrate the theoretical and computational aspects of linear programing. Specifically, these include problems in procurement, shipping, personnel assignment, distribution, production scheduling, and feed mixing. These are only a few of the areas in which linear programing may be successfully applied. Identification of areas for the application of linear programing is facilitated by a thorough understanding of the available models. Also, understanding of the models is beneficial in that it contributes to a better insight regarding the complex operations which the models explain. The objective of this section is to help the reader gain an understanding of the computational management applications of linear programing.

17.2 THE ASSIGNMENT MODEL

The assignment model of linear programing may be described as follows. There exist n requirements together with n means for satisfying them. Associated with the assignment of the ith means to be the jth requirement, x_{ij}, is a certain effectiveness coefficient, e_{ij}. It is required that $x_{ij} = 1$ if the ith means is used to satisfy the jth requirement, and that $x_{ij} = 0$ if the ith means is not used to satisfy the jth requirement. Since one means can be associated with only one requirement, the assignment problem can be stated mathematically as that of optimizing the effectiveness function

$$E = \sum_{i=1}^{n} \sum_{j=1}^{n} e_{ij} x_{ij}$$

subject to

$$\sum_{i=1}^{n} x_{ij} = 1 \qquad j = 1, 2, \ldots, n$$

$$\sum_{j=1}^{n} x_{ij} = 1 \qquad i = 1, 2, \ldots, n$$

Optimization will require either minimization or maximization, depending upon the measure of effectiveness involved. The decision maker has control of a matrix of assignments, x_{ij}. Not directly under his control is the matrix of effectiveness coefficients, e_{ij}.

17.2.1 The Assignment Matrix

The assignment matrix will contain $n \times n$ elements with $n!$ possible arrangements. Solution by direct enumeration is ordinarily not possible. For example, if eight requirements are to be satisfied by eight means there will be 40,320 possible assignments. In the following paragraphs the Hungarian method for finding the optimal assignment will be presented.

Consider the following application of the assignment model. Four different assemblies are to be produced by four contractors. Each contractor is to receive only one assembly to produce. The cost of each assembly is determined by bids submitted by each contractor. This information, when arranged in tabular form, gives an effectiveness matrix shown in Table 17.1.

The differences in the bid prices are due to differences in the work to be done and preferences for certain assemblies. The objective is to determine the assignment of assemblies to contractors that will result in a minimum cost.

The first step in the Hungarian method is to alter the effectiveness matrix to obtain a reduced matrix. This is accomplished by subtracting the minimum element in each row from all elements in the row, then subtracting the minimum element in each column from all elements in the column. The first reduced matrix for the procurement problem of Table 17.1 is shown in Table 17.2. An assignment that minimizes the total for a matrix reduced in this manner also minimizes the total for the original effectiveness matrix.

The elements of the first reduced matrix will always be zero or positive. As a result, the total cannot be negative for any assignment. Therefore, all assignments that might be made will have a total that is zero or positive. If an assignment can be made that has a zero total, there cannot be an assignment with a lower total.

Reference to Table 17.2 indicates that an assignment with a zero total can be made. This requires assigning assembly 1 to contractor *B*, assembly 2 to contractor *A*, assembly 3 to contractor *D*, and assembly 4 to contractor *C*. The resulting assignment matrix is shown in Table 17.3. Its total cost may be found from Table 17.1 and Table 17.3 as $14,000 + $12,000 + $12,000 + $15,000 = $53,000. No other assignment will result in a lower total cost.

The problem described above conforms exactly to the assignment model. Sometimes, however, a problem can be made to conform to the assignment model by adding dummy rows or dummy columns. Consider the following example. At the end of a cycle of schedules, a trucking firm has a surplus of one vehicle in

TABLE 17.1. *COSTS FOR PROCURING ASSEMBLIES, THOUSANDS OF DOLLARS*

		Contractor			
		A	*B*	*C*	*D*
Assembly	1	16	14	15	18
	2	12	13	16	14
	3	14	13	11	12
	4	16	18	15	17

TABLE 17.2. *FIRST REDUCED COST MATRIX FOR PROCUREMENT*

		Contractor			
		A	*B*	*C*	*D*
Assembly	1	2	0	1	3
	2	0	1	4	1
	3	3	2	0	0
	4	1	3	0	1

TABLE 17.3. *ASSIGNMENT MATRIX FOR PROCUREMENT*

		Contractor			
		A	*B*	*C*	*D*
Assembly	1	0	1	0	0
	2	1	0	0	0
	3	0	0	0	1
	4	0	0	1	0

TABLE 17.4. *DISTANCES FOR MOVING VEHICLES, MILES*

		To City					
		A	*B*	*C*	*D*	*E*	*F*
From City	1	13	11	16	23	19	9
	2	11	19	26	16	17	13
	3	12	11	4	9	6	10
	4	7	15	9	14	14	13
	5	9	13	12	8	14	11

cities 1, 2, 3, 4, and 5, and a deficit of one vehicle in cities *A*, *B*, *C*, *D*, *E*, and *F*. The distances between cities with a surplus and cities with a deficit are shown in Table 17.4. The objective is to find the assignment of surplus vehicles to deficit cities that will result in a minimum total distance.

Inspection of Table 17.4 indicates that there exist more requirements than means. When this is the case, an $n \times n$ effectiveness matrix will result by adding a dummy row, or dummy rows if required. Elements in the dummy row are zeros since no cost is associated with not moving a vehicle. The adjusted effectiveness matrix is shown in Table 17.5. If there had been more

TABLE 17.5. *ADJUSTED COST MATRIX FOR MOVING VEHICLES*

	To City					
From City	*A*	*B*	*C*	*D*	*E*	*F*
1	13	11	16	23	19	9
2	11	19	26	16	17	13
3	12	11	4	9	6	10
4	7	15	9	14	14	13
5	9	13	12	8	14	11
Dummy	0	0	0	0	0	0

TABLE 17.6. *FIRST REDUCED COST MATRIX FOR MOVING VEHICLES*

	To City					
From City	*A*	*B*	*C*	*D*	*E*	*F*
1	4	2	7	14	10	0
2	0	8	15	5	6	2
3	8	7	0	5	2	6
4	0	8	2	7	7	6
5	1	5	4	0	6	3
Dummy	0	0	0	0	0	0

TABLE 17.7. *STEP ONE: THE MINIMUM SET OF LINES*

	To City					
From City	*A*	*B*	*C*	*D*	*E*	*F*
1	4	2	7	14	10	0
2	0	8	15	5	6	2
3	8	7	0	5	2	6
4	0	8	2	7	7	6
5	1	5	4	0	6	3
Dummy	0	0	0	0	0	0

TABLE 17.8. *STEP TWO: MODIFIED REDUCED MATRIX*

	To City					
From City	*A*	*B*	*C*	*D*	*E*	*F*
1	4	0	7	14	8	0
2	0	6	15	5	4	2
3	8	5	0	5	0	6
4	0	6	2	7	5	6
5	1	3	4	0	4	3
Dummy	2	0	2	2	0	2

means than requirements, dummy columns would have been added with zeros as elements. Dummy rows mean that some requirements are not met, and dummy columns mean that some means are not used. Adjustment of the effectiveness matrix in this manner makes the problem conform to the assignment model.

The first reduced matrix for this problem is shown in Table 17.6. Inspection indicates that an assignment with a zero total cannot be made. When this is the case, additional computations are necessary to produce more zeros.

17.2.2 Iteration Toward an Optimal Assignment

The following iterative scheme may be employed when the first reduced matrix does not yield an assignment with a zero total:

1. Draw the minimum number of lines that will pass through all zeros in the reduced matrix. There may be several minimum sets. A set is known to be minimum when the number equals the number of independent zeros used in constructing an assignment. Any minimum set may be chosen.
2. Select the smallest element in the reduced matrix that does not have a line through it. Add this element to all elements that occur at the intersection of two lines and subtract it from all elements that do not have a line through them. The other elements of the matrix remain unchanged.

Application of step one to the first reduced matrix of the vehicle assignment example results in Table 17.7. The minimum number of lines that can be drawn is five. An assignment with a zero total cannot be made when the minimum number of lines is less than the matrix size. Application of step two gives the matrix shown in Table 17.8. It may be called a modified reduced matrix. An assignment with a zero total still cannot be made as is indicated by the application of step one shown in Table 17.9. Step two now yields the final reduced matrix shown in Table 17.10.

An optimal assignment can now be made as is shown in Table 17.11. This will require assigning vehicle 1 to city *B*, vehicle 2 to city *F*, vehicle 3 to city *C*, vehicle 4 to city *A*, vehicle 5 to city *D*, and no vehicle to city *E*. The total distance for this assignment may be found from Table 17.5 and Table 17.11 as $11 + 13 + 4 + 7 + 8 + 0 = 43$ miles. No other assignment will result in less total distance for moving the surplus vehicles. However, an alternate assignment with the same total distance is revealed by Table 17.10. It involves assigning vehicle 1 to city *F*, vehicle 2 to city *A*, vehicle 3 to city *E*, vehicle 4 to city *C*, vehicle 5 to city *D*, and no vehicle to city *B*. There may be some reason why this alternate optimal solution would be preferred.

It can be shown that iteration as demonstrat-

TABLE 17.9. *STEP ONE: A MINIMUM SET OF LINES*

	To City					
	A	*B*	*C*	*D*	*E*	*F*
From City 1	4	0	7	14	8	0
2	0	6	15	5	4	2
3	8	5	0	5	0	6
4	0	6	2	7	5	6
5	1	3	4	0	4	3
Dummy	2	0	2	2	0	2

TABLE 17.10. *STEP TWO: FINAL MODIFIED REDUCED MATRIX*

	To City					
	A	*B*	*C*	*D*	*E*	*F*
From City 1	6	0	7	14	8	0
2	0	4	13	3	2	0
3	10	5	0	5	0	6
4	0	4	0	5	3	4
5	3	3	4	0	4	3
Dummy	4	0	2	2	0	2

TABLE 17.11. *ASSIGNMENT MATRIX FOR MOVING VEHICLES*

	To City					
	A	*B*	*C*	*D*	*E*	*F*
From City 1	0	1	0	0	0	0
2	0	0	0	0	0	1
3	0	0	1	0	0	0
4	1	0	0	0	0	0
5	0	0	0	1	0	0
Dummy	0	0	0	0	1	0

TABLE 17.12. *PROFITS FOR SALESMAN ASSIGNMENT, DOLLARS*

	District			
	A	*B*	*C*	*D*
Salesman 1	14	8	12	9
2	12	9	13	13
3	13	13	11	10
4	11	10	12	13

TABLE 17.13. *FIRST REDUCED MATRIX FOR SALESMAN ASSIGNMENT*

	District			
	A	*B*	*C*	*D*
Salesman 1	0	6	2	5
2	1	4	0	0
3	0	0	2	3
4	2	3	1	0

ed with this example will always lead to an optimal assignment in a finite number of cycles. The optimum assignment will occur when the minimum number of lines that can be drawn is equal to the matrix size. This optimum assignment is guaranteed by the existence of one or more assignments with a zero total in the reduced effectiveness matrix.

17.2.3 Finding A Maximum Profit Assignment

Effectiveness is often expressed in terms of profit instead of cost. This will require maximizing the effectiveness function. Maximization may be accomplished by replacing each element of the effectiveness matrix by its negative and proceeding as for minimization. This procedure is valid, since minimizing the negative of a function is equivalent to maximizing the function.

Consider the following maximization problem. A sales manager has four salesmen and four sales districts. After considering the capabilities of the salesmen and the nature of the districts, he estimates that the profit per day for each salesman in each district would be as shown in Table 17.12. The objective is to find the assignment of salesmen to districts that will result in maximum profit.

The first step is to replace each element in Table 17.12 by its negative. Next, the most negative element in each row is subtracted from all elements in the row, then each column is subtracted from all elements in the column. The resulting first reduced matrix is shown in Table 17.13.

Inspection indicates that an assignment with a zero total can be made, as is shown in Table 17.14. This requires assigning salesman 1 to district *A*, salesman 2 to district *C*, salesman 3 to district *B*, and salesman 4 to district *D*. The total profit from this assignment may be found from Table 17.12 and Table 17.14 as $14 + $13 + $13 + $13 = $53. No other assignment will result in a greater profit. If an assignment with a zero total was not produced at this

TABLE 17.14. *ASSIGNMENT MATRIX FOR SALESMAN ASSIGNMENT*

	District			
	A	*B*	*C*	*D*
Salesman 1	1	0	0	0
2	0	0	1	0
3	0	1	0	0
4	0	0	0	1

point, it would have been necessary to use the iterative scheme described previously.

When the effectiveness matrix is not square, dummy rows or dummy columns can be added as before. Elements in the dummy rows or dummy columns are zeros since no profit is associated with them. The maximization process then proceeds as outlined above.

17.3 THE TRANSPORTATION MODEL

The transportation model of linear programing is a generalization of the assignment model. It may be described as follows. There exist m origins, with the ith origin possessing a_i units, and n destinations, with the jth destination requiring b_j units. It is not required that m be equal to n, but the sum of the units available at the origins must equal the sum of the requirements at the destinations. Associated with the allocation of one unit from the ith origin to the jth destination is a certain effectiveness coefficient, e_{ij}. If x_{ij} is the number of units allocated from the ith origin to the jth destination, the transportation problem may be stated mathematically as that of optimizing the effectiveness function

$$E = \sum_{i=1}^{m} \sum_{j=1}^{n} e_{ij} x_{ij}$$

subject to

$$\sum_{i=1}^{m} x_{ij} = b_j \qquad j = 1, 2, \ldots, n$$

$$\sum_{j=1}^{n} x_{ij} = a_i \qquad i = 1, 2, \ldots, m$$

where

$$\sum_{i=1}^{m} a_i = \sum_{j=1}^{n} b_j$$

Optimization will require either minimization or maximization, depending on the measure of effectiveness involved. The decision maker has under his control the allocation matrix, x_{ij}. Not directly under his control is the matrix of effectiveness coefficients, e_{ij}.

17.3.1 The Transportation Matrix

The transportation matrix will contain $m \times n$ nonnegative allocations. Selection of that set of allocations that optimizes the effectiveness function is sought. The paragraphs that follow will present a solution improvement algorithm. It starts with any initial allocation and terminates with an optimal allocation.

Consider the following application of the transportation model. Four dealers place orders for new automobiles that are to be shipped from three plants. Dealer A requires six automobiles, dealer B requires five, dealer C requires four, and dealer D requires four. Plant 1 has seven automobiles in stock, plant 2 has thirteen, and plant 3 has three. The cost of shipping one automobile from the ith plant to the jth dealer is shown in Table 17.15. For this situation to be classified as a linear programing problem, the assumption is made that the cost of shipping more than one automobile is proportional to the number shipped.

The number of automobiles available at the three plants is twenty-three, whereas the number required by the four dealers is only nineteen. Therefore, to make this problem conform to the transportation model, it will be necessary to add a dummy destination. This means that some of the available automobiles will not be shipped. The costs associated with the dummy destination are zero.

A transportation matrix exhibiting all information relevant to the situation described is shown as Table 17.16. The number of automobiles available at the respective plants are given in the last column, and the number required at each dealer are given in the last row. The dummy destination is. indicated by the next-to-last column and receives four units. As a result, the number of automobiles available is made equal to the number required. The cost per unit shipped for each possible route is entered in the small squares from Table 17.15. The objective is to find the allocation of auto-

TABLE 17.15. *COSTS FOR SHIPPING AUTOMOBILES, DOLLARS*

		Dealer			
		A	*B*	*C*	*D*
	1	50	80	60	70
Plant	2	80	50	60	60
	3	70	60	80	60

TABLE 17.16. *TRANSPORTATION MATRIX FOR SHIPPING AUTOMOBILES*

Plant	Dealer					Available
	A	*B*	*C*	*D*	Dummy	
1	50	80	60	70	0	7
2	80	50	60	60	0	13
3	70	60	80	60	0	3
Required	6	5	4	4	4	23

TABLE 17.17. *INITIAL ALLOCATION BY THE NORTHWEST CORNER RULE*

Plant	Dealer					Available
	A	*B*	*C*	*D*	Dummy	
1	50 6	80 1	60	70	0	7
2	80	50 4	60 4	60 4	0 1	13
3	70	60	80	60	0 3	3
Required	6	5	4	4	4	23

mobiles to dealers that will satisfy the requirements, and that will result in a minimum total cost.

17.3.2 Finding an Initial Allocation

One formal means for finding an initial allocation is to employ the northwest corner rule. This requires allocating units to the northwest cell in the transportation matrix in such a magnitude that either the origin capacity is exhausted, or the destination requirement is satisfied, or both. If the origin capacity is exhausted first, an allocation is made to column one from the second origin. This will either exhaust the origin capacity of row two or satisfy the remaining requirement of column one. If the first allocation satisfies the requirement of column one, an allocation is made in column two. This will either exhaust the capacity of row one or will satisfy the requirement of column two or both. Continuing in this manner, satisfying the destination requirements and exhausting the origin capacities, one at a time, results in a movement toward the southeast corner with all row and column requirements being satisfied in the process.

Application of the northwest corner rule to the transportation matrix of Table 17.16 results in the allocations shown in Table 17.17. Since seven automobiles were available at plant 1, dealer *A* has his requirements completely satisfied from this source. The remaining unit at plant 1 is allocated to dealer *B*. Four automobiles from plant 2 are used to finish satisfying the requirement of dealer *B*. Of the remaining nine units at plant 2, four are allocated to dealer *C*, four to dealer *D*, and one to the dummy destination. The three automobiles available at plant 3 are allocated to the dummy destination. The resulting solution contains seven allocations. The total cost of this solution may be found from Table 17.17 as 6($50) + 1($80) + 4($50) + 4($60) + 4($60) + 1($0) + 3($0) = $1060.

The northwest corner rule does not take into consideration the cost of each allocation. Therefore, it is likely that a scheme considering

TABLE 17.18. *INITIAL ALLOCATION BY INSPECTION*

Plant	Dealer *A*	Dealer *B*	Dealer *C*	Dealer *D*	Dealer Dummy	Available
1	[50] 6	[80]	[60] 1	[70]	[0]	7
2	[80]	[50] 5	[60] 3	[60] 1	[0] 4	13
3	[70]	[60]	[80]	[60] 3	[0]	3
Required	6	5	4	4	4	23

these costs will result in a lower total cost. One method is to make an allocation to the cell with the lowest cost, up to the maximum allowed by the origin and destination involved. Then an allocation is made to the next lowest cost cell in view of the remaining capacities and requirements. This process is continued until all origins are emptied and all destinations are filled. If a tie occurs between lowest cost cells, judgment may be used in making the allocation. For small problems, this procedure may result in an optimum solution.

Applying this procedure to the transportation matrix of Table 17.16 results in the allocations shown in Table 17.18. Six units were allocated to dealer *A* and five units were allocated to dealer *B* since each involves a minimum shipping cost. The requirements of dealer *D* are then satisfied from plant 3 at a cost of \$60 per unit. Four units from plant 2 are allocated to the dummy destination at no cost. Finally, the remaining requirements are satisfied by allocating one unit from plant 1 and three units from plant 2 to dealer *C*, and one unit from plant 2 to dealer *D*. The total cost of this inspection allocation is 6(\$50) + 1(\$60) + 5(\$50) + 3(\$60) + 1 (\$60) + 4(\$0) + 3(\$60) = \$1030. This scheme resulted in a \$30 improvement over the northwest corner rule.

17.3.3 Testing for Optimality

A basic feasible solution to the transportation problem is one that contains $m + n - 1$ positive allocations. When a basic feasible solution has been obtained, it may be tested to see if a lower cost allocation can be made. This can be accomplished as follows:

1. Set up a matrix containing the costs associated with the cells for which allocations have been made.
2. Enter a set of numbers, v_j, across the top of the matrix and a set of numbers, u_i, down the left side so that their sums equal the costs entered in step one.
3. Fill the vacant cells in step one with the sums of the u_i and v_j.
4. Subtract the u_i plus v_j values from the original unit cost matrix.
5. If any of the cell evaluations are negative, the basic feasible solution is not optimal.

Table 17.19 illustrates the application of this optimality test to the initial solution of the automobile shipment problem in Table 17.18. The u_i plus v_j matrix was constructed by entering costs in those cells for which allocations were made. These are the boxed values shown. Next, v_1 was arbitrarily set equal to zero to

TABLE 17.19. *TESTING TABLE 17.18 FOR OPTIMALITY*

u_i Plus v_j Matrix

u_i \ v_j	0	0	10	10	−50
50	[50]	50	[60]	60	0
50	50	[50]	[60]	[60]	[0]
50	50	50	60	[60]	0

Cell Evaluation Matrix

•	30	•	10	0
30	•	•	•	•
20	10	20	•	0

begin the determination of u_i and v_j numbers. This action forces u_1 to be 50 which, in turn, forces v_3 to be 10, u_2 to be 50, v_2 to be 0, v_4 to be 10, v_5 to be -50, and u_3 to be 50. The vacant cells were then filled with the sum of the u_i and v_j giving the u_i plus v_j matrix shown. By subtracting the elements of the u_i plus v_j matrix from those in the unit cost matrix of Table 17.15, the cell evaluation matrix results. Since no elements in the cell evaluation matrix are negative, the initial solution by the inspection allocation is optimal.

If the solution by inspection is optimal, the solution by the northwest corner rule cannot be, since it has a higher total cost. This fact is demonstrated by the optimality test illustrated in Table 17.20. Here three cells in the cell evaluation matrix are negative. When this is the case, an iterative scheme may be used to find an optimal allocation.

17.3.4 Iteration Toward an Optimal Allocation

Since the northwest corner rule did not yield an optimal allocation, the following iterative scheme can be employed:

1. From the cell evaluation matrix, identify the cell with the most negative entry. Make a choice when a tie is involved.
2. Trace a path in the transportation matrix consisting of a series of segments which are alternately horizontal and vertical. The path begins and terminates in the cell identified in step one. All corners of the path occur in cells for which allocations have been made.
3. Mark the cell identified in step one plus and each cell at a corner of the path alternately minus, plus, minus, etc.
4. Make a new allocation in the cell identified in step one by entering the smallest allocation on the path which has been given a minus sign.
5. Add and subtract the quantity of the new allocation in step four to all cells at the corners of the path maintaining all row and column requirements.

TABLE 17.20. *TESTING TABLE 17.17 FOR OPTIMALITY*

u_i Plus v_j matrix

u_i \ v_j	0	30	40	40	−20
50	50	80	90	90	30
20	20	50	60	60	0
20	20	50	60	60	0

Cell Evaluation Matrix

•	•	−30	−20	−30
60	•	•	•	•
50	10	20	0	•

Table 17.21 illustrates the application of step one and step two to the initial solution of Table 17.17. The "plus-minus" path begins and terminates with the cell that was evaluated to be minus 30. Table 17.22 shows the allocation of one unit into this cell. This is the maximum amount that could be reallocated among the members of the loop and still maintain the row and column requirements. The new solution has a total cost of 6($50) + 1($0) + 5($50) + 4 ($60) + 4($60) + 3($0) = $1,030. This is equal to the optimal cost found by the inspection allocation and is also optimal. Whenever the cell evaluation matrix contains zero elements, alternate optimal solutions exist.

17.3.5 Treating Degeneracy

The allocation given in Table 17.22 is a degenerate basic feasible solution, since it contains less than $m + n - 1$ allocations. When degeneracy exists, a "plus-minus" path cannot begin in all cells for which allocations have not been made. If further iteration had been necessary, this degeneracy would have had to be resolved before applying the optimality test.

Consider the degenerate solution of Table 17.23 obtained by the northwest corner rule. It is only possible to trace a "plus-minus" path for cells 1–*B* and 4–*C*. In Table 17.22 degeneracy occurred because the reallocation of one unit caused two previous allocations to become zero.

Degeneracy can be resolved at any stage in the solution by placing an infinitesimally small allocation ε in an appropriate cell. This small allocation is made by inspection and assumed not to affect the row and column totals. In Table 17.23, "plus-minus" paths can be traced for all open cells if ε is placed in any cell other than 1–*B* or 4–*C*. The ε allocation is then manipulated in accordance with the rules eatablished previously. It is treated no differently than any

TABLE 17.21. *TRACING THE "PLUS-MINUS" PATH*

Plant	Dealer A	Dealer B	Dealer C	Dealer D	Dealer Dummy	Available
1	[50] 6	[80] −1	[60]	[70]	[0] +	7
2	[80]	[50] +4	[60] 4	[60] 4	[0] −1	13
3	[70]	[60]	[80]	[60]	[0] 3	3
Required	6	5	4	4	4	23

TABLE 17.22. *REALLOCATION BY ADDING AND SUBTRACTING*

Plant	Dealer A	Dealer B	Dealer C	Dealer D	Dealer Dummy	Available
1	[50] 6	[80]	[60]	[70]	[0] 1	7
2	[80]	[50] 5	[60] 4	[60] 4	[0]	13
3	[70]	[60]	[80]	[60]	[0] 3	3
Required	6	5	4	4	4	23

TABLE 17.23. *A DEGENERATE BASIC FEASIBLE SOLUTION*

From	To A	To B	To C	To D	Available
1	5				5
2	3	6			9
3			5	1	6
4				6	6
Required	8	6	5	7	26

other allocation. When an optimal solution is found, ε is set equal to zero, thus regaining the original problem.

It can be shown that degeneracy can occur only if the sum of some subset of the row requirements equals the sum of some subset of the column requirements. Since degeneracy can exist only when this condition occurs, any adjustment that makes this equality impossible also makes degeneracy impossible. This can be accomplished by making a small perturbation of the row requirements and one of the column requirements before any calculations are made. Application of the perturbation technique to

TABLE 17.24. *PERTURBATION TO ELIMINATE DEGENERACY*

	To				
From	*A*	*B*	*C*	*D*	Available
1	$5+\varepsilon$				$5+\varepsilon$
2	$3-\varepsilon$	$6+2\varepsilon$			$9+\varepsilon$
3		-2ε	5	$1+3\varepsilon$	$6+\varepsilon$
4				$6+\varepsilon$	$6+\varepsilon$
Required	8	6	5	$7+4\varepsilon$	$26+4\varepsilon$

the problem given by Table 17.23 results in the nondegenerate basic feasible solution of Table 17.24. The optimality test and the iterative process can now be applied. Again, the ε and the multiples of ε are treated no differently than other allocations. When an optimal solution is reached, ε is set equal to zero.

17.3.6 Finding a Maximum Profit Allocation

Consider the following maximization problem. A carpet manufacturer has two plants and produces three styles of carpet. For the coming week, plant *A* has a capacity of 8400 yards and plant *B* has a capacity of 5800 yards. This total capacity is to be used to produce 4200 yards of style 1, 6100 yards of style 2, and 3900 yards of style 3 for the week. The estimated profit per yard for each style at each plant is given in Table 17.25. The profit per yard depends upon the style and upon the plant at which it is manufactured.

Minimizing the negative of a function is equivalent to maximizing the function. Therefore, the first step in finding a maximum profit allocation of styles to plants is to make all elements in the profit matrix of Table 17.25 negative. The transportation method may then be applied as for minimization.

TABLE 17.25. *PROFITS FOR CARPET MANUFACTURING, DOLLARS*

		Plant	
		A	*B*
Style	1	0.82	0.76
	2	0.34	0.41
	3	0.66	0.60

TABLE 17.26. *TRANSPORTATION MATRIX FOR CARPET MANUFACTURING*

	Plant		
Style	*A*	*B*	Production
1	4200 [−0.82]	[−0.76]	4200
2	300 [−0.34]	5800 [−0.41]	6100
3	3900 [−0.66]	[−0.60]	3900
Capacity	8400	5800	14,200

Table 17.26 exhibits the transportation matrix applicable to the carpet manufacturing problem. Since the required total production equals the total capacity available, no dummy origins or destinations are needed. Each profit element carries a negative sign since profit is to be maximized. The basic feasible solution shown was determined by judgment. It represents a total profit of 4200($0.82) + 300($0.34) + 5800($0.41) + 3900($0.66) = $8498. Application of the optimality test is shown in Table 17.27. Since all cells in the cell evaluation matrix are positive, this solution is optimal. No other allocation will result in a higher profit for the coming week.

17.3.7 Vogel's Approximation Method

Vogel's approximation method is offered as an alternative to the method presented previously. The procedure is simple and fast by comparison. Its advocates claim that it will give an optimal allocation for a majority of problems, and that the approximation is very good for the

TABLE 17.27. *TESTING TABLE 17.26 FOR OPTIMALITY*

	u_i Plus v_j Matrix	
	v_j	
u_i	0	−0.07
−0.82	−0.82	−0.89
−0.34	−0.34	−0.41
−0.66	−0.66	−0.73

Cell Evaluation Matrix

•	0.13
•	•
•	0.13

remainder. In some applications, the final result of this approximation method may be accepted as is. In others it might be desirable to apply the optimality test and then iterate toward an optimal solution if needed. Normally, use of this scheme to find an initial solution will reduce the number of iterations required by the conventional method.

17.3.7.1 An Example Application. Suppose that there are three machine centers in a job shop that can process any one of four orders. The capacities of each machine center as well as the time needed by each order are expressed in standard machine hours. The capacity for machine center *A* is 95 standard hours, for machine center *B* is 115 standard hours, and for machine center *C* is 50 standard hours. Order 1 needs 66 standard machine hours, order 2 needs 45 standard hours, order 3 needs 82 standard hours, and order 4 needs 44 standard hours. The costs per standard machine hour for each order at each machine center is given in Table 17.28. The costs are a function of the order and the nature of the machine center relative to the order.

TABLE 17.28. *COSTS PER STANDARD MACHINE HOUR, DOLLARS*

		Machine Center		
		A	*B*	*C*
Order	1	8.40	7.90	7.60
	2	9.20	6.10	8.70
	3	6.00	7.50	9.10
	4	7.80	8.00	6.90

Table 17.29 exhibits the transportation matrix for this production problem. Since there are 260 standard machine hours available and only 237 required by the four orders, it is necessary to add a dummy order. This dummy order will absorb the extra capacity, thus making the problem conform to the transportation model. Elements in the dummy row are set equal to zero since no cost is associated with a dummy order that will not be processed.

17.3.7.2 The Solution Procedure. The solution procedure for Vogel's approximation method may be pursued with the aid of the original transportation matrix. It is not necessary to develop a new matrix for each cycle. Each cycle involves a series of steps that are repeated in exactly the same manner until the final allocation is made.

Since a minimum cost assignment is required, the first step in the cycle is to inspect the costs in each row and take the difference between the two smallest of these. If the problem is one of maximization, the difference between the two highest values would be taken. These differences form the first column of values to the right of the matrix in Table 17.29. This difference process is now applied to each column in the matrix and the results form the first row of values across the bottom.

The second step involves identification of the largest difference in the column and row just developed. This occurs in column 3 for the problem under consideration. Thus, column 3 is the first candidate for an allocation.

Step 3 involves the allocation of needed standard machine hours to the column with the largest difference, up to the maximum allowed by the row and column totals. This allocation is made in the row with the lowest cost or highest profit. Since this row is the dummy with zero cost, twenty-three hours are allocated. Since all needed machine hours have been allocated, all cells in the row are marked out. If the largest difference had occurred in a row instead of a column, the assignment would have been made to the cell in that row with the lowest cost or highest profit. The procedure is the same in either case. This completes the first cycle.

Each cycle follows the steps outlined above and is applied to the reduced matrix; that is, the

TABLE 17.29. *TRANSPORTATION MATRIX FOR VOGEL'S APPROXIMATION METHOD*

Order	Machine A	Machine B	Machine C	Needed					
1	X (8.40)	66 (7.90)	X (7.60)	66	0.30	0.30	0.30	0.30	0.50
2	X (9.20)	45 (6.10)	X (8.70)	45	2.60	2.60	—	—	—
3	82 (6.00)	X (7.50)	X (9.10)	82	1.50	1.50	1.50	—	—
4	13 (7.80)	4 (8.00)	27 (6.90)	44	0.90	0.90	0.90	0.90	0.20
Dummy	X (0)	X (0)	23 (0)	23	0.00	—	—	—	—
Available	95	115	50	260					
	6.00	6.10	6.90						
	1.80	1.40	0.70						
	1.80	0.40	0.70						
	0.60	0.10	0.70						
	0.60	0.10	—						

matrix remaining after some rows or columns or both are marked out. The difference column and difference row from the previous cycle are disregarded. A new difference column and row are developed for the reduced matrix at the beginning of each cycle. These are shown for each cycle in Table 17.29. Allocations are made in accordance with the largest differences until all allocations are completed. The resulting solution may now be tested for optimality if desired.

17.4 THE GENERAL LINEAR PROGRAMING MODEL

The general linear programing problem may be stated symbolically as that of optimizing the effectiveness function

$$E = \sum_{j=1}^{n} e_j x_j$$

subject to the constraints

$$\sum_{j=1}^{n} a_{ij} x_j = b_i \qquad i = 1, 2, \ldots, m$$

$$x_j \geq 0 \qquad j = 1, 2, \ldots, n$$

Optimization requires either maximization or minimization depending on the measure of effectiveness involved. The decision maker has control of the vector of variables, x_j. Not directly under his control is the vector of effectiveness coefficients, e_j, the matrix of constants, a_{ij}, and the vector of constants, b_i.

The symbolic statement of the general linear programing model may be explained in graphical terms. There exist n variables that define an n dimensional space. Each restriction corresponds to a hyperplane in this space. These restrictions surround the region of feasible solution by hypersurfaces so that the region is the interior of a convex polyhedron. Since the effectiveness function is linear in the n variables, the requirement that this function have some constant value gives a hyperplane that may or may not cut through the polyhedron. If it does, one or more feasible solutions exist. By changing the value of this constant, a family of hyperplanes parallel to each other is generated. The distance from the origin to a member of this family is proportional to the value of the effectiveness function.

Two limiting hyperplanes may be identified. One corresponds to the largest value of the effectiveness function for which the hyperplane just touches the polyhedron, and the other corresponds to the smallest value which just touches. In most cases, the limiting members just touch a vertex of the limiting polyhedron. This outermost limiting point is the solution that optimizes the effectiveness function. Although this is a graphical description, a graphical solution is not convenient when more than three variables are involved. The paragraphs that follow will illustrate solution of the general linear programing problem by graphical means for problems having two and three activities.

17.4.1 Graphical Maximization for Two Activities

When two activities compete for scarce resources, a two-dimensional space is defined. Each restriction corresponds to a line on this surface. These restrictions identify a region of feasible solutions. The effectiveness function is also a line, its distance from the origin being proportional to its value. The optimum value for the effectiveness function occurs when it is located so that it just touches an extreme point of the region.

Consider the following production example. Two products are to be manufactured. A single unit of product A will require 2.4 minutes of punch press time and 5.0 minutes of assembly time. The profit for product A is $0.60 per unit. A single unit of product B will require 3.0 minutes of punch press time and 2.5 minutes of welding time. The profit for product B is $0.70 per unit. The capacity of the punch press department available for these products is 1200 minutes per week. The welding department has idle capacity of 600 minutes per week, and the assembly department can supply 1500 minutes of capacity per week. The manufacturing and marketing data for this production situation are summarized in Table 17.30.

TABLE 17.30. *MANUFACTURING AND MARKETING DATA FOR TWO PRODUCTS*

Department	Product A	Product B	Capacity
Punch press	2.4	3.0	1200
Welding	0.0	2.5	600
Assembly	5.0	0.0	1500
Profit	$ 0.60	$ 0.70	

In this example, two products will compete for the available production time. The objective is to determine the quantity of product A and the quantity of product B to produce so that total profit will be maximized. This will require maximizing

$$TP = \$0.60A + \$0.70B$$

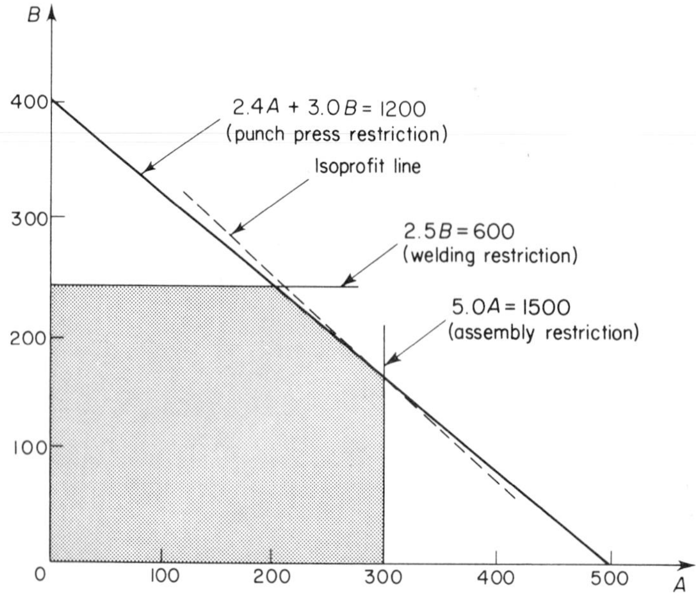

Fig. 17.1 *Maximizing profit for two products.*

subject to

$$\begin{aligned} 2.4A + 3.0B &\le 1200 \\ 0.0A + 2.5B &\le 600 \\ 5.0A + 0.0B &\le 1500 \\ A \ge 0 \text{ and } B &\ge 0 \end{aligned}$$

The graphical equivalent of the algebraic statement of this two-product problem is shown in Fig. 17.1. The set of linear restrictions define a region of feasible solutions. This region lies below $2.4A + 3.0B = 1200$ and is restricted further by the requirement that $B \le 240$, $A \le 300$, and that both A and B be nonnegative. Thus, the scarce resources determine which combinations of the activities are feasible and which are not feasible.

The production quantity combinations of A and B that fall within the region of feasible solutions constitute feasible production programs. That combination or combinations of A and B which maximize profit is sought. The relationship between A and B is $A = 1.167B$. This relationship is based on the relative profit of each product. The total profit realized will depend upon the production quantity combination chosen. Thus, there is a family of isoprofit lines, one of which will have at least one point in the region of feasible production quantity combinations and be a maximum distance from the origin. The member that satisfies this condition intersects the region of feasible solutions at the extreme point $A = 300$, $B = 160$. This is shown as a broken line in Fig. 17.1, and represents a total profit of \$0.60 (300) + \$0.70 (160) = \$292. No other production quantity combination would result in a higher profit.

Alternate production programs with the same profit might exist in some cases. This occurs when the isoprofit line lies parallel to one of the limiting restrictions. For example, if the relative profits of product A and product B were $A = 1.25B$, the isoprofit line in Fig. 17.1 would coincide with the restriction $2.4A + 3.0B = 1200$. In this case, the isoprofit line would touch the region of feasible solutions along a line instead of at a point. All production quantity combinations along the line would maximize profit.

17.4.2 Graphical Maximization for Three Activities

When three activities compete for scarce resources, a three-dimensional space is defined. Each restriction is a plane in this space, and all restrictions taken together identify a volume of feasible solutions. The effectiveness function is also a plane, its distance from the origin being proportional to its value. The optimum value for the effectiveness function occurs when this plane is located so that it is at the extreme point of the volume of feasible solutions.

As an example, suppose that the production operations for the previous example are to be expanded to include a third product designated product C. A single unit of product C will require 2.0 minutes of punch press time, 1.5 minutes of welding time, and 2.5 minutes of assembly time. The profit associated with product C is \$0.50 per unit. Manufacturing and marketing data for this revised production situation are summarized in Table 17.31.

In this example, three products will compete for the available production time. The objective is to determine the quantity of product A, the quantity of product B, and the quantity of product C to produce so that total profit will be maximized. This will require maximizing

$$TP = \$0.60A + \$0.70B + \$0.50C$$

subject to

$$\begin{aligned} 2.4A + 3.0B + 2.0C &\le 1200 \\ 0.0A + 2.5B + 1.5C &\le 600 \\ 5.0A + 0.0B + 2.5C &\le 1500 \\ A \ge 0,\ B \ge 0,\ \text{and } C &\ge 0 \end{aligned}$$

The graphical equivalent of the algebraic statement of this three-product production situation is shown in Fig. 17.2. The set of restricting planes defines a volume of feasible solutions. This region lies below $2.4A + 3.0B + 2.0C =$

TABLE 17.31. *MANUFACTURING AND MARKETING DATA FOR THREE PRODUCTS*

Department	Product A	Product B	Product C	Capacity
Punch press	2.4	3.0	2.0	1200
Welding	0.0	2.5	1.5	600
Assembly	5.0	0.0	2.5	1500
Profit	\$0.60	\$0.70	\$0.50	

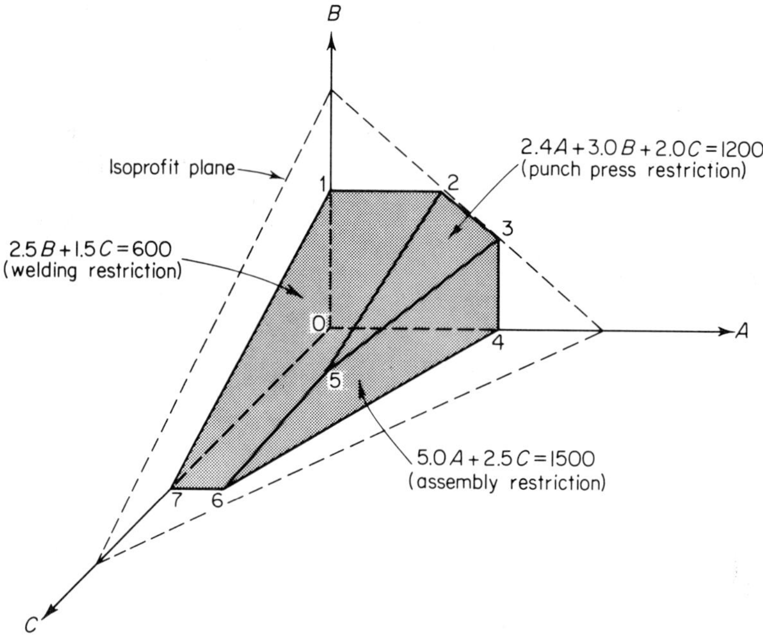

Fig. 17.2 *Maximizing profit for three products.*

1200 and is restricted further by the requirement that $2.5B + 1.5C \leq 600$, $5.0A + 2.5C \leq 1500$, and that A, B, and C be nonnegative. Thus, the scarce resources determine which combinations of the activities are feasible and which are not feasible.

The production quantity combinations of A, B, and C that fall within the volume of feasible solutions constitute feasible production programs. That combination or combinations of A, B, and C which maximizes total profit is sought. The expression $0.60A + 0.70B + 0.50C$ gives the relationship among A, B, and C based on the relative profit of each product. The total profit realized will depend upon the production quantity combination chosen. Thus, there exists a family of isoprofit planes, one for each value of total profit. One of these planes will have at least one point in the volume of feasible solutions and will be a maximum distance from the origin. The plane that maximizes profit will intersect the volume at an extreme point. This calls for the computation of total profit at each extreme point as given in Table 17.32. The coordinates of each extreme point were found from the restricting planes, and the associated profit was calculated from the total profit equation.

TABLE 17.32. *TOTAL PROFIT COMPUTATIONS AT EXTREME POINTS OF FIG. 17. 2*

Point	Coordinate			Profit
	A	*B*	*C*	
0	0	0	0	$ 0
1	0	240	0	$ 168.00
2	200	240	0	$ 288.00
3	300	160	0	$ 292.00
4	300	0	0	$ 180.00
5	180	96	240	$ 295.20
6	100	0	400	$ 260.00
7	0	0	400	$ 200.00

Inspection of the total profit values in Table 17.32 indicates that profit is maximized at point five which has the coordinates 180, 96, and 240. This means that if 180 units of product A, 96 units of product B, and 240 units of product C are produced, profit will be maximized. No other production quantity combination will result in a higher profit. Also, no alternate optimum solutions exist since the total profit plane intersects the volume of feasible solutions at only a single point.

Addition of a third product to the two-product production system increased total profit from \$292.00 per week to \$295.20 per week. This increase in profit results from reallocation of the idle capacity of the welding department. The number of units of product *A* was reduced from 200 to 180 and the number of units of *B* was reduced from 240 to 96. This made it possible to add 240 units of product *C* to the production program with the resulting increase in profit.

17.4.3 Graphical Minimization for Two Activities

As an example of the minimization problem in linear programing, consider a feed mixing operation that can be described in terms of two activities. The required mixture must contain four kinds of nutrient ingredients designated w, x, y, and z. Two basic feeds, designated *A* and *B*, containing the required ingredients are available on the market. One pound of feed *A* contains 0.1 pound of w, 0.1 pound of y, and 0.2 pound of z. Likewise, one pound of feed *B* contains 0.1 pound of x, 0.2 pound of y, and 0.1 pound of z.

The feed mixture is to be used to fatten steers which have a daily per-head requirement of at least 0.4 pound of ingredient w, 0.6 pound of ingredient x, 2 pounds of ingredient y, and 1.8 pounds of ingredient z. Feed *A* can be purchased for \$0.07 per pound and feed *B* can be purchased for \$0.05 per pound. The availabilities, requirements, and costs are summarized in Table 17.33.

TABLE 17.33. *AVAILABILITIES, REQUIREMENTS, AND COSTS FOR FEED MIXING*

Nutrient	Feed *A*	Feed *B*	Requirement
w	0.1	0.0	0.4
x	0.0	0.1	0.6
y	0.1	0.2	2.0
z	0.2	0.1	1.8
Cost	\$ 0.07	\$ 0.05	

The objective is to determine the quantity of feed *A* and the quantity of feed *B* to include in the mixture so that the total cost will be a minimum. This will require minimizing

$$TC = \$0.07A + \$0.05B$$

subject to

$$0.1A + 0.0B \geq 0.4$$
$$0.0A + 0.1B \geq 0.6$$
$$0.1A + 0.2B \geq 2.0$$
$$0.2A + 0.1B \geq 1.8$$
$$A \geq 0 \text{ and } B \geq 0$$

The graphical equivalent of the algebraic

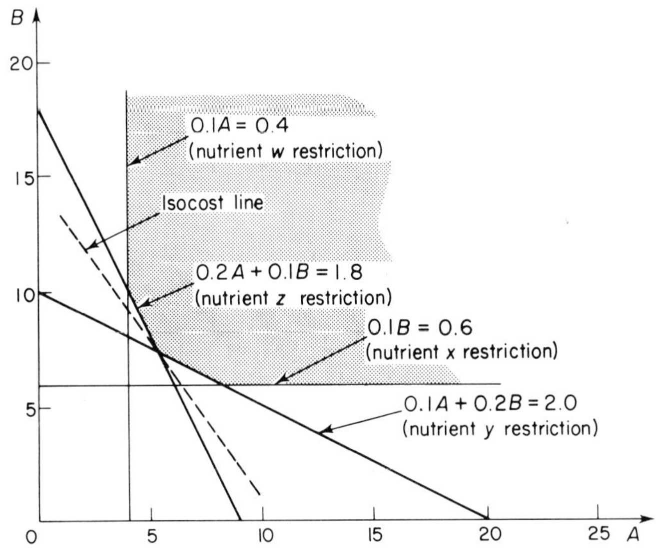

Fig. 17.3 *Minimizing cost for feed mixing.*

statement of this problem is shown in Fig. 17.3. The set of linear restrictions defines a region of feasible solutions. This region is of a greater than or equal to type, due to the fact that the problem requires minimization. The region lies above $0.1A + 0.2B = 2.0$ and $0.2A + 0.1B = 1.8$, and is further restricted by $A \geq 4$, $B \geq 6$, and the requirements that A and B be nonnegative. Thus, the minimum nutrient requirements specify which combinations of feed A and feed B are feasible and which are not feasible.

The quantity combinations of feed A and feed B that fall within the region of feasible solutions constitute feasible feed mixes. That combination or combinations of A and B which minimize total cost is sought. There is a family of isocost lines, one of which will have at least one point in the region of feasible solutions and will be a minimum distance from the origin. The member that satisfies this condition intersects the region of feasible solutions at the extreme point $A = 5.33$, $B = 7.34$. This is shown as a broken line in Fig. 17.3 and represents a total cost of $0.07 (A) + $0.05 (B) = $0.74. No other feed mixture will yield a lower total cost. Also, no alternate optimum solutions exist.

17.5 THE SIMPLEX METHOD

The simplex method is an algorithm that makes possible the numerical solution of the general linear programing problem. It is not restricted to problems involving three or less activities as are the graphical solution methods. The simplex method is an iterative process that begins with a feasible solution, tests for optimality, and proceeds toward an improved solution. It can be shown that the algorithm will finally lead to an optimal solution if such a solution exists. The simplex method will be applied to the three-product maximization problem and to the minimization problem presented graphically. Reference to the graphical solution will explain certain facets of the computational procedure.

17.5.1 The Simplex Method for Maximization

The three-product production problem under consideration requires the maximization of a total profit equation subject to certain restrictions. These restrictions must be converted to equalities of the form specified by the general linear programing model. This requires the addition of three "slack" variables to remove the inequalities. Thus the restrictions become

$$\begin{aligned} 2.4A + 3.0B + 2.0C + S_1 &= 1200 \\ 0.0A + 2.5B + 1.5C + S_2 &= 600 \\ 5.0A + 0.0B + 2.5C + S_3 &= 1500 \end{aligned}$$

The amount of departmental time not used in the production program is represented by a slack variable. Each slack variable takes on whatever value is necessary for the equality to exist. If nothing is produced, the slack variables assume values equal to the total production time available in each department. This gives an initial feasible solution expressed as $S_1 = 1200$, $S_2 = 600$, and $S_3 = 1500$. Each slack variable is one of the x_j variables in the model. However, since no profit can be derived from idle capacity, total profit may be expressed as

$$TP = \$0.60A + \$0.70B + \$0.50C + \$0S_1 + \$0S_2 + \$0S_3$$

The initial matrix required by the simplex algorithm may now be set up as shown in Table 17.34. The first column is designated e_i and gives the profit coefficients applicable to the

TABLE 17.34. *INITIAL MATRIX FOR A THREE-PRODUCT PRODUCTION PROBLEM*

e_j			0	0	0	0.60	0.70	0.50		
e_i	Sol	b	S_1	S_2	S_3	A	B	C	θ	
0	S_1	1200	1	0	0	2.4	3.0	2.0	400	
0	S_2	600	0	1	0	0	2.5	1.5	240	r
0	S_3	1500	0	0	1	5.0	0	2.5	∞	
E_j		0	0	0	0	0	0	0		
$e_j - E_j$			0	0	0	0.60	0.70	0.50		
							k			

initial feasible solution. These are all zero since the initial solution involves the allocation of all production time to the slack variables. The second column is designated Sol and gives the variables in the initial solution. These are the slack variables that were introduced. The third column is designated b and gives the number of minutes of production time associated with the solution variables of the previous column. These reflect the total production capacity in the initial solution. The next three columns are headed by slack variables with elements of zero or unity, depending upon which equation is served by which slack variable. The e_j heading for these columns carries an entry of zero, corresponding to a zero profit. The last three columns are headed by the activity variables with elements entered from the restricting equations. The e_j heading for these columns is the profit associated with each activity variable. The last column, designated θ, is utilized during the computational process.

17.5.1.1 Testing for Optimality. After an initial feasible solution has been obtained, it must be tested to see if a program with a higher profit can be found. The optimality test is accomplished with the aid of the last two rows in Table 17.34. The required steps are:

1. Enter values in the row designated E_j from the expression $E_j = \sum e_i a_{ij}$, where a_{ij} are the matrix elements in the ith row and the jth column.
2. Calculate $e_j - E_j$ for all positions in the row designated $e_j - E_j$.
3. If $e_j - E_j$ is positive for at least one j, a better program is possible.

Application of the optimality test to the initial feasible solution is shown in the last two rows of Table 17.34. The first element in the E_j row is calculated as $0(1200) + 0(600) + 0(1500) = 0$. The second is $0(1) + 0(0) + 0(0) = 0$. All values in this row will be zero since all e_i values are zero in the initial feasible solution. The first element in the $e_j - E_j$ row is $0 - 0 = 0$, the second is $0 - 0 = 0$, the third is $0 - 0 = 0$, the fourth is $0.60 - 0 = 0.60$, and so forth. Since $e_j - E_j$ is positive for at least one j, this initial solution is not optimal. This will always be the case when the initial feasible solution is obtained by allocating all capacity to the slack variables.

17.5.1.2 Iteration Toward an Optimal Program. If the optimality test indicates that an optimal program has not been found, the following iterative procedure may be employed:

1. Find the maximum value of $e_j - E_j$ and designate this column k. The variable at the head of this column will be the incoming variable.
2. Calculate entries for the column designated θ from $\theta_i = b_i/a_{ik}$.
3. Find the minimum positive value of θ_i and designate this row r. The variable to the left of this row will be the outgoing variable.
4. Set up a new matrix with the incoming variable substituted for the outgoing variable. Calculate new elements, a'_{ij}, as $a'_{rj} = a_{rj}/a_{rk}$ for $i = r$ and $a'_{ij} = a_{ij} - a_{ik}a'_{rj}$ for $i \neq r$.
5. Apply the optimality test.

Steps one, two, and three of the above procedure are applied to the initial matrix. In Table 17.34 step one designates B as the incoming variable. Values for θ_i are calculated from step two. Step three designates S_2 as the outgoing variable. The affected column and row are marked with a k and an r respectively in Table 17.34.

Steps four and five require a new matrix as is shown in Table 17.35. The incoming variable B, together with its associated profit, replaces the outgoing variable S_2 with its profit. All other elements in this row are calculated from the

TABLE 17.35. *FIRST ITERATION FOR A THREE-PRODUCT PRODUCTION PROBLEM*

e_j			0	0	0	0.60	0.70	0.50		
e_i	Sol	b	S_1	S_2	S_3	A	B	C	θ	
0	S_1	480	1	−1.20	0	2.40	0	0.20	200	r
0.70	B	240	0	0.40	0	0	1	0.60	∞	
0	S_3	1500	0	0	1	5.00	0	2.50	300	
E_j		168	0	0.28	0	0	0.70	0.42		
$e_j - E_j$			0	−0.28	0	0.60	0	0.08		
						k				

TABLE 17.36. *SECOND ITERATION FOR A THREE-PRODUCT PRODUCTION PROBLEM*

e_j			0	0	0	0.60	0.70	0.50		
e_i	Sol	b	S_1	S_2	S_3	A	B	C	θ	
0.60	A	200	0.416	−0.50	0	1	0	0.084	2381	
0.70	B	240	0	0.40	0	0	1	0.60	400	
0	S_3	500	−2.08	2.50	1	0	0	2.08	240	r
E_j		288	0.25	−0.02	0	0.60	0.70	0.47		
$e_j - E_j$			−0.25	0.02	0	0	0	0.03		
								k		

TABLE 17.37. *THIRD ITERATION FOR A THREE-PRODUCT PRODUCTION PROBLEM*

e_j			0	0	0	0.60	0.70	0.50	
e_i	Sol	b	S_1	S_2	S_3	A	B	C	θ
0.60	A	180	0.50	−0.60	−0.04	1	0	0	
0.70	B	96	0.60	−0.32	−0.29	0	1	0	
0.50	C	240	−1	1.20	0.48	0	0	1	
E_j		295.20	0.22	0.016	0.016	0.60	0.70	0.50	
$e_j - E_j$			−0.22	−0.016	−0.016	0	0	0	

first formula of step four. Elements in the remaining two rows are calculated from the second formula of step four. The optimality test indicates that an optimal solution has not yet been reached. Note that after this first iteration, the profit at point one of Fig. 17.2 appears. Comparison of the results in Table 17.34 and Table 17.35 with the total profit computations in Table 17.32 indicates that the isoprofit plane which began at the origin has now moved away from this initial position to point one. The gain from this iteration was \$168 − \$0 = \$168.

Since the first iteration did not yield an optimal solution, it is necessary to repeat steps one through five. Steps one, two, and three are applied to Table 17.35 designating A as the incoming variable and S_1 as the outgoing variable. The incoming variable, together with its associated profit, replaces the outgoing variable as shown in Table 17.36. All other elements in this new matrix are calculated from the formulas of step four. Application of the optimality test indicates that the solution indicated is still not optimal. Table 17.32 and Fig. 17.2 show that the isoprofit plane is now at point two. The gain from this iteration was \$288 − \$168 = \$120.

Table 17.36 did not yield an optimal solution, requiring the application of steps one through five again. Steps one, two, and three designate C as the incoming variable and S_3 as the outgoing variable. This incoming variable, together with its associated profit, replaces the outgoing variable as shown in Table 17.37. All other elements in the matrix are calculated from the formulas of step four. Application of the optimality test indicates that the solution exhibited by Table 17.37 is optimal. Table 17.32 and Fig. 17.2 indicate that the isoprofit plane is now at point five. The gain from this iteration was \$295.20 − \$288 = \$7.20.

17.5.2 The Simplex Method for Minimization

The computational scheme of the previous section may be used without modification for problems requiring minimization if the signs of the cost coefficients are changed from positive to negative. The principle that maximizing the negative of a function is the same as minimizing the function then applies. If these coefficients are entered in the simplex matrix with their negative signs, the value of the solution will decrease as the computations proceed. In the following paragraphs the feed-mixing problem presented graphically will be solved by the simplex method.

The problem under consideration requires the minimization of a total cost equation subject to certain restrictions. Changing the sign of each cost coefficient in the total cost equation gives

$$TC = -\ \$0.07A - \$0.05B$$

As before, the restrictions must be converted to equalities so the problem will conform to the general linear programing model. This is accomplished by subtracting a slack variable from the left-hand side of each. The restrictions then become

$$\begin{aligned}
0.1A + 0.0B - S_1 \qquad\qquad\qquad &= 0.4\\
0.0A + 0.1B \qquad - S_2 \qquad\qquad &= 0.6\\
0.1A + 0.2B \qquad\qquad - S_3 \qquad &= 2.0\\
0.2A + 0.1B \qquad\qquad\qquad - S_4 &= 1.8
\end{aligned}$$

Although the restrictions are now equalities, a feasible solution does not exist. If A and B are set equal to zero, $S_1 = -0.4$, $S_2 = -0.6$, $S_3 = -2.0$, and $S_4 = -1.8$. An initial feasible solution of the correct algebraic form can be forced by adding artificial variables resulting in

$$\begin{aligned}
0.1A + 0.0B - S_1 + A_1 \qquad\qquad\qquad &= 0.4\\
0.0A + 0.1B \qquad - S_2 + A_2 \qquad\qquad &= 0.6\\
0.1A + 0.2B \qquad\qquad - S_3 + A_3 \qquad &= 2.0\\
0.2A + 0.1B \qquad\qquad\qquad - S_4 + A_4 &= 1.8
\end{aligned}$$

Artificial variables with a value greater than zero destroy the equality required by the general linear programing model. Therefore, they must not appear in the final solution. To assure that they are forced out of the solution, a large penalty will be associated with each. This penalty will be larger than any other effectiveness coefficient and is designated $-$M. With these changes, the total cost equation becomes

$$\begin{aligned}
TC = &-\$0.07A - \$0.05B\\
&-\$0S_1 - \$0S_2 - \$0S_3 - \$0S_4\\
&-\$\mathrm{M}A_1 - \$\mathrm{M}A_2 - \$\mathrm{M}A_3 - \$\mathrm{M}A_4
\end{aligned}$$

The initial matrix required by the simplex algorithm may now be set up as shown in Table 17.38. The initial solution variables are the artificial variables. Effectiveness coefficients for each artificial variable, slack variable, and activity variable are entered in the second row. All other matrix elements are entered from the restrictions. The simplex method may be applied to this matrix in the manner outlined for the maximization problem.

17.5.2.1 Performing the Calculations for Minimization. The last two rows of Table 17.38 are used in the optimality test. Since at least one $e_j - E_j$ is positive, the initial feasible solution is not optimal. Therefore, B becomes the incoming variable. Computation of θ_1 identifies A_2 as the outgoing variable. The incoming variable and the outgoing variable are designated k and r, respectively, in Table 17.38. They provide the basis for the first iteration.

The incoming variable B is entered in place of the outgoing variable A_2 in Table 17.39 together with its associated effectiveness coefficient. Each element in this new matrix is calculated from step four of the iterative procedure outlined in the previous section. Application of the optimality test indicates that the solution is still not optimal. This makes the second iteration given in Table 17.40 necessary. Proceeding in this manner results in the third iteration given in Table 17.41, the fourth iteration given in Table 17.42, and the fifth iteration given in Table 17.43. The program indicated by the fifth iteration is optimal. This final table indicates that 5.33 pounds of feed A and 7.34 pounds of feed B should be used in the mixture. The total cost of the mixture will be \$0.74, neglecting the negative sign. These results agree with the solution found graphically.

17.5.3 Duality in Linear Programing

The simplex examples in the previous paragraphs may be called primal problems. These same examples may be solved by transposing the rows and columns of the algebraic statement of the problem. Inverting the problem in this way results in a dual problem. A solution to the dual problem may be found in a manner similar to that used for the primal. Once a solution is obtained, it may be converted to a solution of the primal. Solution of the dual problem normally requires less computation if the primal problem contains a large number of rows and a small number of columns. For this reason, it offers an advantage in many applications.

Consider the following algebraic statement of the general linear programing problem, called the primal. Minimize

$$e_1x_1 + e_2x_2 + \cdots + e_nx_n$$

subject to the constraints

$$\begin{aligned}
a_{11}x_1 + a_{12}x_2 + \cdots + a_{1n}x_n &\geq b_1\\
a_{21}x_1 + a_{22}x_2 + \cdots + a_{2n}x_n &\geq b_2\\
\cdot \quad \cdot \quad \cdot \quad \cdot \quad \cdot \quad \cdot \quad \cdot \quad &\cdot \quad \cdot\\
a_{m1}x_1 + a_{m2}x_2 + \cdots + a_{mn}x_n &\geq b_m
\end{aligned}$$

TABLE 17.38. *INITIAL MATRIX FOR A MINIMIZATION PROBLEM*

e_j			0	−M	0	−M	0	−M	0	−M	−0.07	−0.05		
e_i	Sol	b	S_1	A_1	S_2	A_2	S_3	A_3	S_4	A_4	A	B	θ	
−M	A_1	0.4	−1	1	0	0	0	0	0	0	0.1	0	∞	
−M	A_2	0.6	0	0	−1	1	0	0	0	0	0	0.1	6	r
−M	A_3	2.0	0	0	0	0	−1	1	0	0	0.1	0.2	10	
−M	A_4	1.8	0	0	0	0	0	0	−1	1	0.2	0.1	18	
E_j		−4.8M	M	−M	M	−M	M	−M	M	−M	−0.4M	−0.4M		
$e_j - E_j$			−M	0	−M	0	−M	0	−M	0	--0.07 + 0.4M	−0.05 + 0.4M		
												k		

TABLE 17.39. *FIRST ITERATION FOR A MINIMIZATION PROBLEM*

e_j			0	−M	0	−M	0	−M	0	−M	−0.07	−0.05		
e_i	Sol	b	S_1	A_1	S_2	A_2	S_3	A_3	S_4	A_4	A	B	θ	
−M	A_1	0.4	−1	1	0	0	0	0	0	0	0.1	0	∞	
−0.05	B	6	0	0	−10	10	0	0	0	0	0	1	−1.6	
−M	A_3	0.8	0	0	2	−2	−1	1	0	0	0.1	0	0.4	r
−M	A_4	1.2	0	0	1	−1	0	0	−1	1	0.2	0	1.2	
E_j		−0.3−2.4M	M	−M	0.5 − 3M	−0.5 + 3M	M	−M	M	−M	−0.4M	−0.05		
$e_j - E_j$			−M	0	−0.5 + 3M	0.5 − 4M	−M	0	−M	0	−0.07 + 0.4M	0		
					k									

TABLE 17.40. *SECOND ITERATION FOR A MINIMIZATION PROBLEM*

e_j			0	−M	0	−M	0	−M	0	−M	−0.07	−0.05		
e_i	Sol	b	S_1	A_1	S_2	A_2	S_3	A_3	S_4	A_4	A	B	θ	
−M	A_1	0.4	−1	1	0	0	0	0	0	0	0.1	0	∞	
−0.05	B	10	0	0	0	0	−5	5	0	0	0.5	1	−2	
0	S_2	0.4	0	0	1	−1	−0.5	0. 5	0	0	0.05	0	−0.8	
−M	A_4	0.8	0	0	0	0	0.5	−0. 5	−1	1	0.15	0	1.6	r
E_j		−0.5 − 1.2M	M	−M	0	0	0.25 − 0.5M	−0.25 + 0.5M	M	−M	−0.025 − 0.25M	−0.05		
$e_j - E_j$			−M	0	0	−M	−0.25 + 0.5M	0.25 − 1.5M	−M	0	−0.045 + 0.25M	0		
							k							

TABLE 17.41. *THIRD ITERATION FOR A MINIMIZATION PROBLEM*

e_j			0	−M	0	−M	0	−M	0	−M	−0.07	−0.05		
e_i	Sol	b	S_1	A_1	S_2	A_2	S_3	A_3	S_4	A_4	A	B	θ	
−M	A_1	0.4	−1	1	0	0	0	0	0	0	0.1	0	4	r
−0.05	B	18	0	0	0	0	0	0	−10	10	2.0	1	9	
0	S_2	1.2	0	0	1	−1	0	0	−1	1	0.2	0	6	
0	S_3	1.6	0	0	0	0	1	−1	−2	2	0.3	0	5.33	
E_j		−0.9 − 0.4M	M	−M	0	0	0	0	0.5	−0.5	−0.1M − 0.1	−0.05		
$e_j - E_j$			−M	0	0	−M	0	−M	−0.5	−M + 0.5	+0.03 + 0.1M	0		
											k			

TABLE 17.42. *FOURTH ITERATION FOR A MINIMIZATION PROBLEM*

e_j			0	−M	0	−M	0	−M	0	−M	−0.07	−0.05		
e_i	Sol	b	S_1	A_1	S_2	A_2	S_3	A_3	S_4	A_4	A	B	θ	
−0.07	A	4	−10	10	0	0	0	0	0	0	1	0	−0.4	
−0.05	B	10	20	−20	0	0	0	0	−10	10	0	1	0.5	
0	S_2	0.4	2	−2	1	−1	0	0	−1	1	0	0	0.2	
0	S_3	0.4	3	−3	0	0	1	−1	−2	2	0	0	0.133	r
E_j		−0.78	−0.3	0.3	0	0	0	0	0.5	−0.5	−0.07	−0.05		
$e_j - E_j$			0.3	−0.3 − M	0	−M	0	−M	−0.5	0.5 − M	0	0		
			k											

TABLE 17.43. *FIFTH ITERATION FOR A MINIMIZATION PROBLEM*

e_j			0	−M	0	−M	0	−M	0	−M	−0.07	−0.05	
e_i	Sol	b	S_1	A_1	S_2	A_2	S_3	A_3	S_4	A_4	A	B	θ
−0.07	A	5.33	0	0	0	0	3.30	−3.30	−6.70	6.70	1	0	
−0.05	B	7.34	0	0	0	0	−6.60	6.60	3.40	−3.40	0	1	
0	S_2	0.134	0	0	1	−1	−0.66	0.66	0.34	−0.34	0	0	
0	S_1	0.133	1	−1	0	0	0.33	−0.33	−0.67	0.67	0	0	
E_j		−0.74	0	0	0	0	0.099	−0.099	0.299	−0.299	−0.07	−0.05	
$e_j - E_j$			0	−M	0	−M	−0.099	0.099 − M	−0.299	0.299 − M	0	0	

The algebraic statement of the dual problem may be presented by transposing rows and columns, the vector of constants b_i and the effectiveness function, reversing the inequalities, and maximizing instead of minimizing. The use of dual variables $y_1, y_2, \ldots, y_m$ with $y_j \geq 0$ results in the dual. Maximize

$$b_1y_1 + b_2y_2 + \cdots + b_my_m$$

subject to the constraints

$$\begin{aligned} a_{11}y_1 + a_{21}y_2 + \cdots + a_{m1}y_m &\leq e_1 \\ a_{12}y_1 + a_{22}y_2 + \cdots + a_{m2}y_m &\leq e_2 \\ \cdot \quad \cdot \quad \cdot \quad \cdot \quad \cdot \quad & \cdot \quad \cdot \\ a_{1n}y_1 + a_{2n}y_2 + \cdots + a_{mn}y_m &\leq e_n \end{aligned}$$

If the inequalities of the primal were not of the greater than or equal to type, algebraic manipulation would have to be used to put them in this form.

As an example of the application of the dual, consider the feed mixing problem presented previously. The original algebraic statement of this problem can be transformed to the dual. Maximize

$$0.4y_1 + 0.6y_2 + 2.0y_3 + 1.8y_4$$

subject to the constraints

$$\begin{aligned} 0.1y_1 + 0.0y_2 + 0.1y_3 + 0.2y_4 &\leq 0.07 \\ 0.0y_1 + 0.1y_2 + 0.2y_3 + 0.1y_4 &\leq 0.05 \end{aligned}$$

Adding slack variables results in the initial matrix of Table 17.44. The simplex method may now be applied as was outlined previously. This results in Table 17.45 and Table 17.46.

Values for the activity variables of the primal appear in the columns of slack variables in the dual with negative signs. These are indicated in Table 17.46. Finally, the value for the effectiveness function is in the same location in both the primal and dual, but with opposite sign. Table 17.43 should be compared with Table 17.46.

17.5.4 Degeneracy in the Simplex Method

Degeneracy becomes evident in the simplex method at the time that the outgoing variable is being selected. In step three of the iterative process, the minimum positive value of θ_i determines the outgoing variable. If two or more values of θ_i are minimal, the problem is degenerate. An arbitrary choice of one of the tied variables may result in more iterations than some other choice. More serious, however, is the condition of cycling which might occur due to a poor choice from among the tied variables.

Although cycling seldom occurs in practical problems, it can be prevented by applying the following procedure:

1. Divide each element in the tied rows by the positive coefficients of the kth column progressing from left to right.

TABLE 17.44. *DUAL MATRIX FOR PRIMAL OF TABLE 17.38*

e_j			0	0	0.4	0.6	2.0	1.8		
e_i	Sol	b	S_1	S_2	y_1	y_2	y_3	y_4	θ	
0	S_1	0.07	1	0	0.1	0.0	0.1	0.2	0.70	
0	S_2	0.05	0	1	0.0	0.1	0.2	0.1	0.25	r
E_j		0	0	0	0	0	0	0		
$e_j - E_j$			0	0	0.4	0.6	2.0	1.8		
							k			

TABLE 17.45. *FIRST ITERATION FOR DUAL PROBLEM*

e_j			0	0	0.4	0.6	2.0	1.8		
e_i	Sol	b	S_1	S_2	y_1	y_2	y_3	y_4	θ	
0	S_1	0.045	1	−0.5	0.1	−0.05	0	0.15	0.30	r
2.0	y_3	0.250	0	5	0	0.5	1	0.50	0.50	
E_j		0.50	0	10	0	1	2.0	1.0		
$e_j - E_j$			0	−10	0.4	−0.4	0	0.81		
								k		

2. Compare the resulting ratios from left to right.

3. The row which first contains the smallest algebraic ratio is designated r.

As an example of the above procedure, consider the degenerate problem of Table 17.47. A dilemma exists regarding the choice of row one or row two as the outgoing variable. The first ratio for row one is $^1/_4$. Likewise, the first ratio for row two is $^0/_6$. Since the second row yields the smallest ratio, it is designated r and the simplex procedure is continued.

17.5.5 Treating Equality Constraints

Some problems that are solvable by the simplex method of linear programing require that a given resource be fully utilized. For example, a production manager might specify that the entire capacity of the assembly department be utilized in the interest of stabilizing manpower. In such cases, the algebraic statement of the problem will contain an equality constraint. This is unlike the problems of the previous pages which utilized only inequalities.

As an example of a situation in which an equality constraint is needed, consider the case of a manufacturer who has two manufacturing resources, machine hours and labor hours. In each working day, the manager has 180 machine hours and 240 labor hours to spend on three products, A, B, and C. Product A requires 4 machine hours and 3 labor hours per unit, product B requires 4 machine hours and 5 labor hours per unit, and product C requires 2 machine hours and 6 labor hours per unit. The profit per unit of product A is \$12, per unit of product B is \$8, and per unit of product C is \$10.

The production manager wishes to maximize profit subject to the limitation on machine time and labor hours. However, he requires that all labor hours be fully utilized so that his manpower level will remain constant. Therefore, the algebraic statement of this problem requires the maximization of

$$TP = \$12A + \$8B + \$10C$$

subject to

$$3A + 5B + 6C = 240$$
$$4A + 4B + 2C \leq 180$$
$$A \geq 0,\ B \geq 0,\ \text{and } C \geq 0$$

The initial simplex matrix for this problem is given in Table 17.48. Note that an artificial variable is employed to yield an initial feasible solution. Although this is a maximization problem, the effectiveness coefficient associated with the artificial variable is $-M$. This is required since artificial variables are only used to obtain an initial feasible solution and must be forced out of solution as the computations proceed.

TABLE 17.46. *FINAL MATRIX FOR DUAL PROBLEM*

e_j			0	0	0.4	0.6	2.0	1.8	
e_i	Sol	b	S_1	S_2	y_1	y_2	y_3	y_4	θ
1.8	y_4	0.3	6.6667	−3.3334	0.6667	−0.3334	0	1	
2.0	y_2	0.1	−3.3350	6.6667	−0.3334	0.6667	1	0	
E_j		0.74	5.333	7.333	0.5468	0.7333	2.0	1.8	
$e_j - E_j$			−5.333	−7.333	−0.1468	−0.1333	0	0	

TABLE 17.47. *A DEGENERATE SIMPLEX PROBLEM*

e_j			0	0	0	40	60	80	
e_i	Sol	b	S_1	S_2	S_3	A	B	C	θ
0	S_1	0	1	0	0	−2	0	4	0
0	S_2	0	0	1	0	3	0	6	0
60	B	60	0	0	1	2	1	1	60
E_j		360	0	0	0	120	60	60	
$e_j - E_j$			0	0	0	−80	0	20	
								k	

TABLE 17.48. *INITIAL SIMPLEX MATRIX FOR EQUALITY CONSTRAINT*

e_j			$-M$	0	12	8	10	
e_i	Sol	b	A_1	S_1	A	B	C	θ
$-M$	A_1	240	1	0	3	5	6	
0	S_1	180	0	1	4	4	2	
E_j								
$e_j - E_j$								

The simplex method may be applied to Table 17.48 to find a maximum profit program. However, total profit may be less with the equality constraint than it would be without such a constraint. The production manager may incur a penalty for his desire to fully utilize the available labor hours.

BIBLIOGRAPHY

Charnes, A., and W. W. Cooper, *Management Models and Industrial Applications of Linear Programming*, Volumes I and II. New York: John Wiley and Sons, Inc., 1961.

Chung, A., *Linear Programming*. Columbus, Ohio: Charles E. Merrill Books, Inc., 1963.

Churchman, C. W., R. L. Ackoff, and E. L. Arnoff, *Introduction to Operations Research*. New York: John Wiley and Sons, Inc., 1957.

Ferguson, R. O., and L. F. Sargent, *Linear Programming*. New York: McGraw-Hill Book Company, Inc., 1958.

Garvin, W. W., *Introduction to Linear Programming*. New York: McGraw-Hill Book Company, Inc., 1960.

Gass, S. I., *Linear Programming*, 2nd Ed. New York: McGraw-Hill Book Company, Inc., 1964.

Greenwald, D. U., *Linear Programming*. New York: The Ronald Press Company, 1957.

Hadley, G., *Linear Programming*. Reading, Mass.: Addison-Wesley Publishing Company, Inc., 1962.

Koopmans, T. C. (ed.), *Activity Analysis of Production and Allocation*, Cowles Commission Monograph No. 13. New York: John Wiley and Sons, Inc., 1951.

Kuhn, H. W., "The Hungarian Method for the Assignment Problem," *Nav. Res. Log. Quart. 2*, Nos. 1 and 2, Mar.-June, 1955.

Llewellyn, R. W., *Linear Programming*. New York: Holt, Rinehart, and Winston, Inc., 1964.

Loomba, N. P., *Linear Programming*. New York: McGraw-Hill Book Company, Inc., 1964.

Metzger, R. W., *Elementary Mathematical Programming*. New York: John Wiley and Sons, Inc., 1958.

Reinfeld, N. V., and W. R. Vogel, *Mathematical Programming*. Englewood Cliffs, New Jersey: Prentice-Hall, Inc., 1958.

Sasieni, M., A. Yaspan, and L. Friedman, *Operations Research—Methods and Problems*. New York: John Wiley and Sons, Inc., 1958.

Vaszonyi, A., *Scientific Programming in Business and Industry*. New York: John Wiley and Sons, Inc., 1958.

INDEX

B

C

M

Q

T

Y